						2 **He** Helium 4.00
		III	IV	V	VI	VII	
		5 **B** Boron 10.82	6 **C** Carbon 12.01	7 **N** Nitrogen 14.01	8 **O** Oxygen 16.00	9 **F** Fluorine 19.00	10 **Ne** Neon 20.17
		13 **Al** Aluminum 26.98	14 **Si** Silicon 28.09	15 **P** Phosphorus 30.97	16 **S** Sulfur 32.06	17 **Cl** Chlorine 35.45	18 **Ar** Argon 39.95
28 **Ni** Nickel 58.71	29 **Cu** Copper 63.54	30 **Zn** Zinc 65.38	31 **Ga** Gallium 69.72	32 **Ge** Germanium 72.60	33 **As** Arsenic 74.91	34 **Se** Selenium 78.96	35 **Br** Bromine 79.90	36 **Kr** Krypton 83.80
46 **Pd** Palladium 106.4	47 **Ag** Silver 107.880	48 **Cd** Cadmium 112.41	49 **In** Indium 114.82	50 **Sn** Tin 118.70	51 **Sb** Antimony 121.87	52 **Te** Tellurium 127.61	53 **I** Iodine 126.91	54 **Xe** Xenon 131.30
78 **Pt** Platinum 195.09	79 **Au** Gold 197.0	80 **Hg** Mercury 200.61	81 **Tl** Thallium 204.39	82 **Pb** Lead 207.21	83 **Bi** Bismuth 209.00	84 **Po** Polonium (210)	85 **At** Astatine (210)	86 **Rn** Radon (222)

63 **Eu** Europium 152.0	64 **Gd** Gadolinium 157.26	65 **Tb** Terbium 158.93	66 **Dy** Dysprosium 162.51	67 **Ho** Holmium 164.94	68 **Er** Erbium 167.27	69 **Tm** Thulium 168.94	70 **Yb** Ytterbium 173.04	71 **Lu** Lutetium 174.99
95 **Am** Americium (243)	96 **Cm** Curium (247)	97 **Bk** Berkelium (249)	98 **Cf** Californium (251)	99 **Es** Einsteinium (254)	100 **Fm** Fermium (253)	101 **Md** Mendele-vium (256)	102 **No** Nobelium (253)	103 **Lr** Lawrencium (257)

General, Organic, and Biochemistry
Second Edition

William H. Brown
Beloit College

Elizabeth P. Rogers
University of Illinois

Willard Grant Press, Boston

PWS PUBLISHERS

Prindle. Weber & Schmidt · ⚚ · Willard Grant Press · **wᴳ** · Duxbury Press · ♠
Statler Office Building · 20 Providence Street · Boston. Massachusetts 02116

Library of Congress Cataloging in Publication Data

Brown, William Henry
    General, organic, and biochemistry.

    Includes index.
    1. Chemistry.   I. Rogers, Elizabeth P.   II. Title.
QD31.2.B79   1983   540   82-18385
ISBN 0-87150-762-5

ISBN 0-87150-762-5

Cover: photomicrograph of wing scales of a large moth, *Hyalophora cecropia,* enlarged to
150 diameters. Cover photo by Fritz Goro. Previously published in *Scientific American,*
November 1981, p. 141.

Text designed by Trisha Hanlon. Cover designed by David Foss. Art for this edition drawn
by J & R Services. Composed in Times Roman and Avant Garde by Syntax International.
Cover printed by Lehigh Press, Inc. Text printed and bound by Halliday Lithograph.
Printed in the United States of America.

87 86 85 84 83 — 10 9 8 7 6 5 4 3 2

Preface

to the Second Edition

This is the second edition of our text written for students of the life sciences, particularly those planning careers in one of the many health professions. While we recognize that these students are not planning to become professional chemists, we also know that they need to understand how living systems depend on chemistry. In this survey of general, organic, and biochemistry, we have tried to reveal this relationship and to provide the necessary chemical foundations for further study.

The category of life and health sciences includes a large variety of academic majors. Although all students in these majors need a knowledge of chemistry, some need to know more about certain areas than do others. An inhalation therapist, for example, needs to know more about the properties of gases than does a physical education major; a radiation technologist requires a greater background in radiochemistry than does a dietician. These diverse needs impose special requirements on a text.

First, in addition to a few basic concepts, the text must contain a wide choice of material from which instructors can tailor courses to fit the needs of their particular students. For this reason, we have included more material than most courses will cover. We leave the decision on topic selection to each individual instructor.

Second, the organization of the text must be flexible enough to allow the selected topics to be presented in an order that is logical for each particular course. The text is organized in a way that seems logical to us, but we realize that other approaches will work equally well. For this reason, we have divided chapters into many freestanding sections and subsections to give each instructor as much latitude as possible in choosing and arranging the material to be covered. For example, nuclear chemistry is presented in Chapter 2, but the chapter is so designed that this material can, at the discretion of the user, be omitted or included later in the course.

Third, the presentation must be even, neither slighting one topic nor going overboard on another. To this end, we have attempted throughout to bring a high degree of pedagogical, organizational, and stylistic unity to this book.

iii

This text is divided into three major sections: general, organic, and biochemistry.

General

Chapters 1–8 present the fundamental concepts of chemistry. Because quantitative thinking is often a major stumbling block for beginning students, we have given special attention in this section to calculations and problem solving. Problem solving by unit analysis is introduced in the first chapter with a detailed algorithm that enables students to analyze a problem, identify the data, and arrange the problem in solvable form. The same algorithm is used throughout the book. Chapters 2 and 3 concentrate on atomic and electron structure, including radioactivity and the use of radioisotopes in medicine. Chapter 4 discusses compounds, their composition and their properties, with particular emphasis on bonding and the geometry of molecules and ions. Chemical reactions are presented in Chapter 5, the kinetic molecular theory of matter in Chapter 6, and the properties of solutions and colloids in Chapter 7. This first section of the text concludes with Chapter 8 on chemical equilibrium, with emphasis on the dissociation of weak acids.

Organic

Chapter 9 considers covalent bonding in organic compounds and introduces the concepts of structural and functional group isomerism. In this chapter we introduce the hybridization of atomic orbitals and covalent bond formation by the overlap of atomic orbitals. Chapters 10–13 present the chemistry of specific types of compounds: hydrocarbons, unsaturated hydrocarbons, alcohols, aldehydes, and ketones. These chapters are similarly organized to include structure, nomenclature, physical properties, preparation, and reactions. Chapter 14 on optical isomerism properly precedes the chemistry of carbohydrates in Chapter 15.

Biochemistry

The major classes of biomolecules are introduced in Chapters 15–22. Chapter 15 discusses the chemistry of carbohydrates. Chapters 16 and 17 on carboxylic acids and the functional derivatives of acids lay the foundation for discussion of the structure and function of lipids in Chapter 18. The introduction of amines in Chapter 19 leads into the chemistry of amino acids and proteins, and of enzymes in Chapters 20 and 21. Chapter 22 discusses the structure and function of nucleic acids. Chapter 23 introduces metabolism and clearly delineates the several stages in the oxidation of foodstuffs and the generation of ATP. This chapter introduces bioenergetics. The metabolism of carbohydrates, fatty acids, and amino acids is presented in Chapters 24–26. These chapters constantly point out that the metabolism of these foodstuff molecules is interrelated and precisely regulated.

Several features of this edition make it an especially effective teaching tool.

Example Problems with step-by-step solutions appear in each chapter, and each example is followed by a similar problem for the student to solve. Answers to all in-chapter problems are found in the back of the book.

End-of-Chapter Problems are grouped according to chapter section. This feature ensures a balanced and representative group of problems for each section of the text.

Key Terms and Concepts are listed at the end of each chapter. The section reference following each term/concept directs the student to the appropriate place in the text for its definition and use.

Fourteen Mini-Essays, five of them new to this edition, are included. They have several purposes: they bridge the gap between the study of chemistry and the projected vocational areas of life science students; they demonstrate some of the creative excitement inherent in chemistry; and finally, they offer a glimpse of the human involvement in research and development.

Extensive Use of Graphics, much of it new and designed specifically for this edition, serves to enhance the visual appeal and pedagogical effectiveness of the text.

Consultation with users of the first edition have identified the strengths of that edition. We have retained those strengths and built on them in the following ways.

· The first eight chapters, which worked well in the first edition, are essentially unchanged. Users of that edition will notice some changes introduced to improve the clarity of presentation. They will also notice somewhat more emphasis on the role of energy in chemical processes.

· Reorganization in the organic section. Carbohydrates (Chapter 15) now follows Aldehydes and Ketones (Chapter 13) and Optical Isomerism (Chapter 14); Lipids (Chapter 17) now follows Functional Derivatives of Carboxylic Acids (Chapter 16); and Amino Acids and Proteins (Chapter 20) now follows Amines (Chapter 19).

· Expanded discussion in Chapter 20 of the acid-base properties of amino acids and of isoelectric points and isoelectric precipitation.

· A new section in Chapter 20 on serum proteins, their separation and function, and the role of antibodies.

· Expanded discussion in Chapter 21 of coenzymes and vitamins, the mechanism of enzyme catalysis, and the regulation of enzyme activity.

· Considerably expanded discussion in Chapter 22 of DNA replication, the biosynthesis of proteins, inhibition of protein synthesis, and point mutations.

· Fuller treatment in Chapter 24 of digestion and absorption of carbohydrates, including the glucose tolerance test and the pentose phosphate pathway. Also added to this chapter is material on the regulation of glycolysis and of the tricarboxylic acid cycle.

In addition to the text, we have prepared a study guide, a laboratory manual, and an instructor's guide.

We wish to acknowledge all those who provided advice and assistance in developing this edition, including: Mabel Armstrong, Lane Community College; P. Wayne Ayers, East Carolina University; Toby Chapman, University of Pittsburgh; Alan Cunningham, Monterey Peninsula College; Sylvia Horowitz, California State University at Los Angeles; William G. Movius, Kent State University; David J. Rislove, Winona State University; Marvin A. Smith, Brigham Young University; Harris O. VanOrden, Utah State University; and Brent Wurfel, Beloit College, Wisconsin.

We are also deeply grateful to the editors of Willard Grant Press who have been so supportive of our efforts and so tolerant of our foibles. In this regard, we are especially grateful to David Foss who has guided this book with such skill, insight, and imagination from a rough manuscript to a finished book. And it goes without saying that the whole thing would have been impossible without the support and encouragement of our families and of each other.

William H. Brown   Elizabeth P. Rogers
Beloit, Wisconsin   *Champaign, Illinois*

Contents

4 Compounds and Chemical Bonding   98

5 Chemical Reactions   142

9

Organic Chemistry: The Study of the Compounds of Carbon   300

10

Saturated Hydrocarbons: Alkanes and Cycloalkanes   320

14

Optical Isomerism 467

15

Carbohydrates 487

16

Carboxylic Acids 514

17

Functional Derivatives of Carboxylic Acids   542

18

Lipids   580

19

Amines   602

20

Amino Acids and Proteins   622

21

Enzymes   663

22

Nucleic Acids and the Synthesis of Proteins   689

23

The Flow of Energy in the Biological World   720

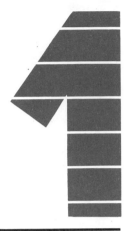

Matter
and Its Properties

Chemistry is defined as *the study of matter and its properties*. **Matter** is defined as *everything that has mass and occupies space*. Although these are proper definitions, they do not explain why a person interested in a career in the health professions needs to know chemistry. To answer this question, think for a moment about the fact that your body is a complex chemical factory that uses chemical processes to change the food you eat and the air you breathe into bones, muscle, blood, and tissue. When you consider that illnesses prevent parts of these processes from functioning correctly, you will realize how important a knowledge of chemistry is to health professionals and to everyone else as well.

The chemistry of life is complex. No sensible person would try to understand such a complex process without first understanding some of the simpler processes that make up the complex one. For this reason, our study will begin with those fundamentals of chemistry that you must first understand before you can move on to organic chemistry and biochemistry, complex areas more closely allied with the health professions.

In carrying out our study, we will build on and expand observations that you have been making all your life. For example, you have learned to identify objects by their physical properties: some are heavy, others are light, and all have characteristic colors and textures. You have discovered that changes occur in matter when its temperature changes: lakes freeze, snow melts, and water boils. You have seen toy balloons inflated with helium and watched them float up into the sky and out of sight. You have used words like alkaline, acid, and octane number and talked about substances such as baking soda, sugar, lime, DDT, TNT, alcohol, and petroleum. These are the kinds of knowledge we shall make use of in our study of chemistry.

1.1 Matter: Simple Substances and Mixtures

We have already defined chemistry as the study of matter and its properties. In this introductory section of the text we will not study *all* types of matter. Rather, we will concentrate on *simple substances*, the properties that identify simple substances, the changes these substances undergo, and the energy associated with these changes.

A. Simple Substances

A **simple** or **pure substance** consists of a single kind of matter. It always has the same composition and the same set of properties. For example, baking soda is a single kind of matter known chemically as sodium hydrogen carbonate. A sample of pure baking soda, regardless of its source or size, will be a white solid containing 57.1% sodium, 1.2% hydrogen, 14.3% carbon, and 27.4% oxygen. The sample will dissolve in water. When heated to 270°C, the sample will decompose, giving off carbon dioxide and water vapor and leaving a residue of sodium carbonate. Thus, by definition, baking soda is a pure substance, since it has a constant composition and a unique set of properties, some of which we have listed. Notice that the properties we have described hold true for *any* sample of baking soda, not just the one mentioned above. These are the kinds of properties in which we are interested.

A note about the term *pure:* in this text, the word *pure* means "chemically pure." As used by the U.S. Food and Drug Administration, the term *pure* means "fit for human consumption." Milk, whether whole, 2% fat, or skim, may be "pure" (fit for human consumption) by public health standards, but it is hardly "pure" in the chemical sense. Milk is a mixture of a great many substances, including water, butterfat, proteins, and sugars. Each of these substances has a unique set of properties and is present in different amounts in each of the different kinds of milk (Figure 1.1).

Matter, whether pure or not, can exist in three different **physical states:** solid, liquid, or gas. Under normal conditions, for example, baking soda is a solid, ethyl alcohol is a liquid, and oxygen is a gas. Matter can also be changed from one state to another. A familiar example is water, which can be changed from a liquid to a solid by freezing and from a liquid to a gas by boiling (Figure

Figure 1.1 The labels on a carton of milk and a box of baking soda show that milk is a mixture and baking soda is a pure substance.

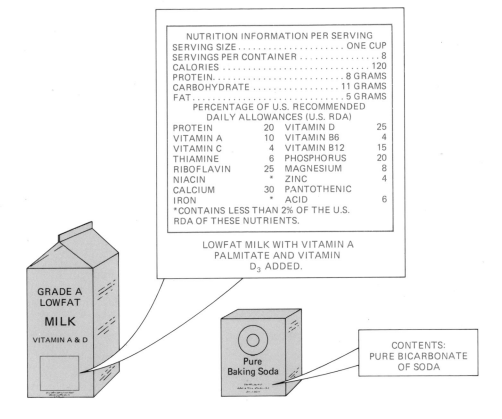

1.2). For each pure substance, the temperatures at which these changes occur are characteristic properties, unique to that substance. A freezing point of 0°C and a boiling point of 100°C are characteristic properties of the simple substance water.

Some of the characteristics that we have mentioned, such as color, texture, and freezing and boiling points, can be observed without altering the

Figure 1.2 The three physical states of water: ice, liquid water, and steam.

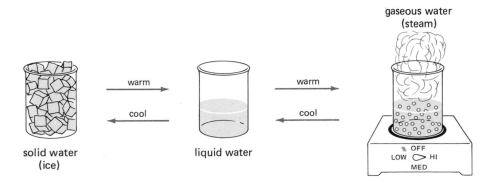

Figure 1.3 Chopping wood physically changes its size but not its composition. Burning wood changes it chemically, turning it into other substances.

identity of the sample: these are **physical properties.** Other properties, such as decomposition, cannot be observed without destroying the sample. Properties whose observation transforms the sample into other substances of different composition are **chemical properties** (see Figure 1.3). You can observe the decomposition of baking soda, but after you make this observation, you no longer have baking soda. Instead you have carbon dioxide, water, and sodium carbonate. A **physical change** alters only physical properties: size, shape, and so on. A **chemical change** alters chemical properties, such as composition.

B. Mixtures

A **mixture** consists of two or more pure substances. Most of the matter we see around us is composed of mixtures. Seawater contains dissolved salts; river water contains suspended mud; hard water contains salts of calcium, magnesium, and iron. Both seawater and river water also contain dissolved oxygen, without which fish and other aquatic life could not survive.

The composition of a mixture can be changed, and the properties of the mixture depend on the percentage of each pure substance in it. Steel is an example of a mixture. All steel starts with the pure substance iron. Refiners then add varying percentages of carbon, nickel, chromium, vanadium, or other substances to obtain steels of a desired hardness, tensile strength, corrosion resistance, and so on. The properties of a particular type of steel depend not only on which substances are mixed with the iron, but on the relative percentage of each. One type of chromium-nickel steel contains 0.6% chromium and 1.25% nickel. Its surface is easily hardened, a property that makes it valuable in the manufacture of automobile gears, pistons, and transmissions. The stainless steel used in the manufacture of surgical instruments, food-processing equipment, and kitchenware is also a mixture of iron, chromium, and nickel; it contains 18% chromium and 8% nickel. Steel with this composition can be polished to a very smooth surface and is very resistant to rusting.

Like pure substances, mixtures can exist in three different physical states: solid, liquid, or gas. Air is a gaseous mixture of approximately 78% nitrogen, 21% oxygen, and varying percentages of several other gases. Rubbing alcohol is a liquid mixture of approximately 70% isopropyl alcohol and 30% water. Steel is a solid mixture of iron with other pure substances.

You can often tell, from the appearance of a sample, whether or not it is a mixture. For example, if river water is clouded with mud or silt particles, you know that it is a mixture. If there is a layer of brown haze over a city, you know that the air is mixed with atmospheric pollutants. However, the appearance of a sample is not always sufficient evidence on which to judge its purity. A sample of matter may look like a pure substance without being one. For instance, rubbing alcohol is a mixture of isopropyl alcohol and water, but pure water, pure isopropyl alcohol, and rubbing alcohol are all clear, colorless liquids. However, two of them are pure substances and the third is a mixture. As another example, you cannot look at a surgical instrument and know whether it is pure iron or a mixture of iron, chromium, and nickel.

1.2 Scientific Measurement

A. The SI System

Chemists study the properties of matter; in order to do this, they must measure these properties. Measurements in the scientific world and, increasingly, in the nonscientific world are made in **SI** (*Système Internationale*) **units.** The system was established in order to allow comparison of measurements made in one country with those made in another. SI units and their relative values were adopted by an international association of scientists meeting in Paris in 1960. Table 1.1 lists the basic SI units and derived units. Notice that metric units are part of this system.

Table 1.1 SI units, derived units, and the relationships of these units to those in the English system.

Property Being Measured	SI Unit	Derived Units	Relationship to English Unit
length	meter (m)	—	1 m = 39.37 inches
		kilometer (km) (1000 m = 1 km)	1.61 km = 1 mile
		centimeter (cm) (100 cm = 1 m)	2.54 cm = 1 inch
mass	kilogram (kg)	—	1 kg = 2.20 pounds
		gram (g) (1000 g = 1 kg)	453.6 g = 1 pound
volume	cubic meter (m^3)	liter (L) (1 L = 0.001 m^3)	1 liter = 1.057 quarts
		cubic centimeter (cm^3) (1000 cm^3 = 1 L)	
		milliliter (mL) (1 mL = 1 cm^3)	
temperature	Kelvin (K)	Celsius (°C) (K = °C + 273.2)	Fahrenheit (°F) °C = $\frac{5}{9}$(°F − 32)
energy	Joule (J)	calorie (cal) (1 cal = 4.184 J) kilocalorie (kcal) (1000 cal = 1 kcal)	

Prefix	Symbol	Multiply Base Unit by
kilo-	k	10^3 or 1000
deci-	d	10^{-1} or 0.1
centi-	c	10^{-2} or 0.01
milli-	m	10^{-3} or 0.001
micro-	μ	10^{-6} or 0.000001
nano-	n	10^{-9} or 0.000000001

The system still in common, nonscientific use in the United States is called the **English system,** even though England, like most other developed countries, now uses metric units. Anyone using units from both the English and SI systems needs to be aware of a few simple relationships between the two systems. These relationships are given in Table 1.1.

Two features of the SI system make it easy to use. First, it is a base-10 system; that is, the various units of a particular dimension vary by multiples of ten. Once a base unit is defined, units larger and smaller than the base unit are indicated by prefixes added to the name of the base unit. Table 1.2 summarizes these prefixes, along with the abbreviation for each and the numerical factor relating it to the base unit. The second feature that increases the usefulness of the SI system is the direct relationship between base units of different dimensions. For example, the unit of volume (cubic meter) is directly related to the unit of length (meter). We shall see later how the unit of mass is related to the unit of volume.

The base unit of length in the SI system is the **meter (m).** The meter, approximately 10% longer than a yard, is equivalent to 39.37 inches, or 1.094 yards. The metric units of length most commonly used in chemistry are listed in Table 1.3 and are illustrated in Figure 1.4.

Unit of Length	Relationship to Base Unit
kilometer (km)	1 kilometer = 1000 meters
meter (m)	—
decimeter (dm)	10 decimeters = 1 meter
centimeter (cm)	100 centimeters = 1 meter
millimeter (mm)	1000 millimeters = 1 meter
micrometer (μm)	10^6 micrometers = 1 meter
nanometer (nm)	10^9 nanometers = 1 meter

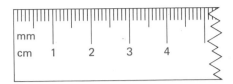

Figure 1.4 Each centimeter contains 10 millimeters (shown actual size).

The base unit of volume in the SI system is the **cubic meter (m^3).** Other commonly used units of volume are the liter (L), the cubic centimeter (cm^3), and the milliliter (mL), the most common one being the liter.

One **liter** has a volume equal to 0.001 m^3. The nearest unit of comparable volume in the English system is the quart: 1.000 liter equals 1.057 quarts. The SI volume units are summarized in Table 1.4 and illustrated in Figure 1.5.

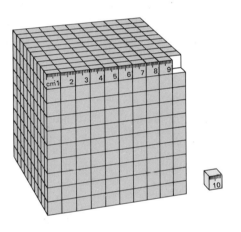

Figure 1.5 The large cube measures 10 centimeters on a side and has a volume of 1000 cm^3, or 1 liter. The small cube on top has a volume of 1 cm^3, or 1 mL.

Table 1.4 Units of volume in the SI system.

Unit of Volume	Relationship to Liter
liter (L)	—
milliliter (mL)	1000 milliliters = 1 liter
cubic centimeter (cm^3, cc)	1000 cubic centimeters = 1 liter
microliter (μL)	10^6 microliters = 1 liter

The base unit of mass in the SI system is the **kilogram.** In a safe in Sevres, France, is a metal cylinder with a mass of exactly one kilogram. The mass of that cylinder is the same as the mass of 1000 cm^3 (1 L) of water at 4°C, thereby relating mass to volume. The most commonly used SI units of mass are listed in Table 1.5.

Table 1.5 Units of
mass in the SI system.

Unit of Mass	Relationship to Base Unit
kilogram (kg)	—
gram (g)	1000 g = 1 kilogram
milligram (mg)	1000 milligrams = 1 gram
microgram (μg)	10^6 micrograms = 1 gram

B. Mass and Weight

In discussing SI units we have used the term *mass* instead of the more familiar term *weight*. **Mass** is a *measure of the amount of matter in a particular sample*. The mass of a sample does not depend on its location; it is the same whether measured on Earth, on the moon, or anywhere in space. **Weight** is a *measure of the pull of gravity on a sample* and depends on where the sample is weighed.

    Astronauts travelling in space and landing on the moon have experienced the difference between mass and weight. Consider the following example. In Earth's gravitational field at sea level, a particular astronaut weighs 198 pounds. On the surface of the moon, the astronaut still has the same mass, but his weight (33 pounds) is only one-sixth of what it is on Earth, because the gravitational field of the moon is much weaker than that of the Earth. In outer space he is weightless, but his mass remains unchanged.

Figure 1.6 Balances of several types and a scale. (a) A classical balance, where the weighing pans are suspended from a straight beam. (b) A common laboratory balance, which weighs to 0.01 g, hence called a *centigram balance*. The three beams give rise to the balance's other, less precise name, *triple beam balance*. (c) An electric balance, which weighs rapidly to 0.0001 g. (d) A common bathroom scale, which measures weight by the distortion of a spring.

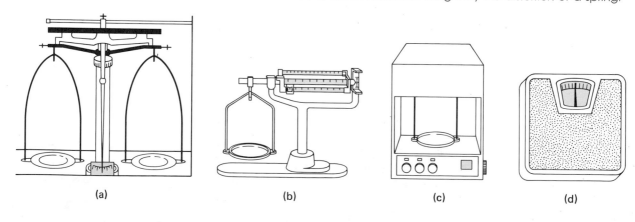

(a)   (b)   (c)   (d)

Weight and mass are measured on different instruments. Mass is measured on a **balance** (Figure 1.6). An object of unknown mass is put at one end of a straight beam and objects of known mass are added to the other end until their mass exactly balances that of the object whose mass is being measured. Because both ends of the beam, at the moment of balancing, are the same distance from the center of the Earth, this measurement is independent of gravity. Weight, on the other hand, is measured on a **scale.** A scale determines weight by measuring the distortion of a spring. Such a measurement depends on the pull of gravity. You could lose weight by weighing yourself at the top of a mountain, but you would not lose any mass.

In spite of the clear difference in meaning between the terms *mass* and *weight*, a measurement of the mass of an object is often called "weighing," and the terms *mass* and *weight* are frequently and incorrectly used interchangeably.

1.3 Recording Measurements

A. Accuracy and Precision

Chemistry is an exact science; its development has been based on careful measurements of properties of matter and careful observations of changes in these properties. Measurements in chemistry must be both *accurate* and *precise* (Fig. 1.7). An **accurate** measurement is one that is close to the actual value being measured. The accuracy of a measurement depends on the calibration of the tool used to make the measurement. For example, if you are measuring the distance between two cities by driving between them, an accurate measurement requires that the odometer in your car reads one kilometer for each kilometer driven. If it only reads 0.95 km for each kilometer driven, the accuracy of your measurement will be reduced.

A **precise** measurement is one that can be reproduced. For example, the distance between Detroit and Chicago is 493 km. An accurate odometer, one that reads 1.00 km for every kilometer traveled, measures the distance between

Figure 1.7 Accurate measurements are close to a correct value. Precise measurements are close to each other.

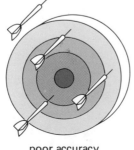
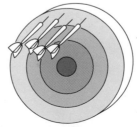
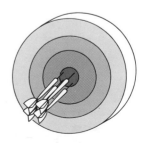

poor accuracy
poor precision

poor accuracy
good precision

good accuracy
good precision

these two cities as 493 km. However, an odometer that reads 1.00 km for every 0.95 km traveled will measure the Detroit–Chicago distance as 519 km. Each time it is used on the trip, the inaccurate odometer records the same value, 519 km. The odometer reading is *precise* because it can be reproduced time after time. However, it is not *accurate* because the odometer itself is not properly calibrated.

The importance of obtaining measurements that are both accurate and precise is rarely greater than it is in a medical laboratory. Patients and doctors alike want to be certain that instruments give readings that are not only precise but accurate. For this reason, instruments used in medical laboratories are calibrated each day (and often at the beginning of each shift) against samples of known concentrations. Periodically, accrediting agencies send the laboratories samples to be analyzed. The results obtained on these samples must be accurate to within the range allowed by the accrediting agencies. Precision is not enough; the determinations must also be accurate.

B. Exponential Notation

Measurements in chemistry often involve very large or very small numbers. An example of a very small number is the mass of a small particle called a *proton* (discussed in Section 2.6):

$$\text{mass of proton} = 0.000000000000000000000000167 \text{ gram}$$

At the opposite extreme is the mass of the Earth:

$$\text{mass of the Earth} = 5,993,000,000,000,000,000,000,000 \text{ grams}$$

Numbers written like this are hard to deal with, difficult to copy without making mistakes, and almost impossible to say. To simplify writing and tabulating such numbers, we use **exponential notation.** When exponential notation is used to express a number *greater* than one, the original number is expressed as a number between one and ten, multiplied by ten to the nth power, where n is the number of places the decimal point was moved to the left.

Example 1.1

Express the following numbers in exponential notation.

a. 436207   **b.** 106000000

Solution

a. Express 436207 as a number between one and ten, multiplied by 10^n:

$$4.36207 \times 10^n$$

Next, determine the value of n. In going from 436207 to 4.36207, the decimal point was moved five places to the left. Therefore, the value of n is 5 and the number becomes

$$4.36207 \times 10^5$$

b. Express 106000000 as a number between one and ten, times 10^n:

$$1.06000000 \times 10^n$$

Determine the value of n. In going from 106000000 to 1.06000000, the decimal point was moved eight places to the left. Therefore, the value of n is 8. The number becomes

$$1.06000000 \times 10^8$$

Problem 1.1

a. Express the mass of the Earth in exponential notation.

b. The planet Pluto is 5908215300 km from the sun. Express this distance in exponential notation.

If the number to be expressed in exponential notation is *less* than one, the original number is expressed as a number between one and ten, multiplied by ten to the minus n power, where n equals the number of places that the decimal point was moved to the right.

Example 1.2

Express the following numbers in exponential notation:

a. 0.00639   **b.** 0.0000104

Solution

a. 1. Express 0.00639 as a number between 1 and 10, times 10^{-n}:

$$6.39 \times 10^{-n}$$

2. Determine the value of n. In going from 0.00639 to 6.39, the decimal point was moved three places to the right. Therefore, the value of n is 3.

$$0.00639 = 6.39 \times 10^{-3}$$

b. 1. Express 0.0000104 as a number between 1 and 10, times 10^{-n}:

$$1.04 \times 10^{-n}$$

2. Determine the value of *n*. In going from 0.0000104 to 1.04, the decimal point was moved five places to the right. Therefore, the value of *n* is 5.

$$0.0000104 = 1.04 \times 10^{-5}$$

Problem 1.2

a. Express the mass of a proton in exponential notation.

b. The radius of a proton is 0.0000000154 m. Express this number in exponential notation.

C. Uncertainty in Chemical Measurement

Each time we make a measurement of length, volume, mass, or any other physical quantity, there is some degree of uncertainty in the measurement. The following example illustrates this uncertainty. Suppose you have a quantity of liquid and wish to measure its volume. You are given three different containers in which you might make the measurement: a 50-mL beaker, a 50-mL graduated cylinder, and a 50-mL buret. Figure 1.8 shows these containers, each holding an identical volume of liquid.

Figure 1.8 Experimental uncertainty in measuring volume.

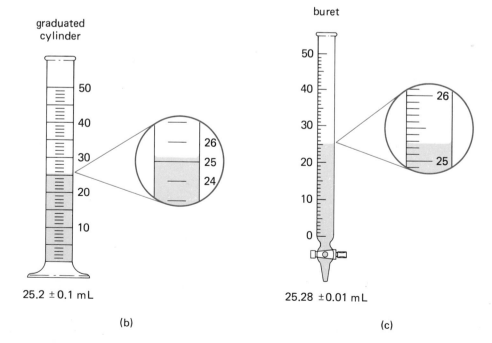

Look first at the 50-mL beaker (Figure 1.8(a)). It has divisions or cali-brations every 10 mL. You can see that the level of the liquid in the beaker is between the 20-mL and 30-mL marks. If you look more closely, you can see that the level of liquid is approximately midway between the two marks; you estimate that the volume is 25 mL. However, there is some uncertainty. The volume could be as little as 24 mL or it could be as large as 26 mL. If you record this volume as

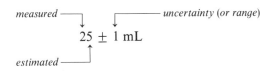

you can show the number you are certain of (the 20 mL), the number you think is the best estimate (5 mL), and the range within which you are certain the number falls (± 1 mL), which is known as the **uncertainty,** or **range,** of the reading.

In Figure 1.8(b), the same volume of sample has been placed in a 50-mL graduated cylinder. Divisions on the cylinder are marked every 1 mL. You can read that the volume is between 25 mL and 26 mL, and estimate that it is about 0.2 mL above the 25-mL mark. However, it could be as small as 25.1 mL or as large as 25.3 mL. Therefore, you should record the volume of the liquid as

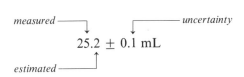

Finally, you measure the same volume of liquid in the 50-mL buret (Figure 1.8(c)). Calibration marks on the buret are 0.1 mL apart. You can read that the volume is between 25.2 mL and 25.3 mL, and estimate that it is 0.08 mL above the 25.2-mL mark. Therefore, you should report the volume of the liquid as

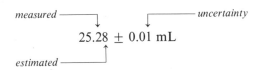

In conclusion, for any measurement, the uncertainty is assumed to be ± 1 in the last recorded digit. This uncertainty is rarely shown but is under-stood to be present. For example, if we write a measurement as 372, we under-stand that the uncertainty is ± 1; if we write 0.017, we understand that the uncertainty is ± 0.001.

D. Significant Figures

Uncertainty in measurements is indicated by the number of significant figures used. **Significant figures** are all those that are measured plus one that is estimated. Using our volume measurements taken from Figure 1.8:

25 mL   contains 2 significant figures
25.2 mL   contains 3 significant figures
25.28 mL   contains 4 significant figures

Often, after calculations, your result has more significant figures than were in any of the original measurements. This is incorrect, since no calculation may produce a result that is more accurate than the least accurate of the factors involved. In these cases, it is necessary to **round off** the result to the correct number of significant figures. The rules for rounding off are:

1. Digits are dropped one-by-one, starting from the right, until the answer contains the correct number of significant figures.

2. If the digit following the last to be kept is less than 5, simply discard all unwanted digits. For example, to report the quantity 36.723 mL using four significant figures, write 36.72 mL; to report it using three significant figures, write 36.7 mL.

3. If the digit following the last one to be kept is 5 or greater, the digit to be kept is increased by one. Thus, to report 36.723 mL using only two significant figures, write 37 mL.

The number of significant figures in a measurement is best shown using exponential notation. For example, consider the number 56,075. As written, it contains five significant figures. To report it with only three significant figures, express the number in exponential notation with only three digits:

$$5.61 \times 10^4$$

1. When is a zero a significant figure? When we work with significant figures, zeroes require special consideration. A zero can either report a measurement or locate the decimal point, thereby determining the magnitude of the number. We have said that a digit is significant if it reports a measurement; similarly, a zero is significant if it reports a measurement. In the number 4.034, the zero is significant, for it shows that the measurement in the tenths place was zero. Any time a zero occurs between two non-zero digits, it is significant. If a zero only locates the decimal point, it is not significant. In the number 0.00456, the zeroes show that measurement started in the thousandths place. They report only the magnitude of the number, so they are not significant. To determine whether a zero is significant, express the number in exponential notation. Any zeroes that disappear in the exponential form of the number

Table 1.6 Significant figures and exponential notation.

Number	Significant Figures	Expressed Exponentially
560,000	two	5.6×10^5 (The zeroes show the location of the decimal point.)
30,290	four	3.029×10^4 (The first 0 is between two digits and is significant; the last 0 shows only the location of the decimal point.)
0.0160	three	1.60×10^{-2} (The first two zeroes show the location of the decimal point and are not significant; the last zero can have no reason for being present other than to show a measurement, and therefore must be significant.)

are not significant. For example, the mass of a proton has been given as

$$0.00000000000000000000000167 g$$

In exponential notation this number becomes

$$1.67 \times 10^{-24} \text{ g}$$

Since the zeroes in the number have disappeared, we know that they merely showed the location of the decimal point and the magnitude of the number; they were not significant. Similarly, the mass of the Earth expressed exponentially is

$$5.993 \times 10^{24} \text{ g}$$

The zeroes shown in the original expression of the measurement (Section 1.3B) have disappeared; they were not significant. Table 1.6 gives further examples.

Notice how simply exponential notation solves the problem of significant zeroes. If we measured a distance and found it to be *exactly* 480,000 miles, we would report the distance as 4.80000×10^5 miles. Because the zeroes are not needed to locate the decimal point, it is clear that they are reporting a measurement and must therefore be significant.

Example 1.3

Express the following numbers in exponential notation to three significant digits:

a. 506,001   **b.** 0.005278   **c.** 50192   **d.** 0.08263

Solution

a. 5.06×10^5. The decimal point has been moved five places to the left, giving 10^5. The numbers in the last three places are dropped because only three significant figures can remain.

b. 5.28×10^{-3}. The decimal point has been moved three places to the right, giving 10^{-3}. The zeroes are not significant because they show only the location of the decimal point. The 8 is dropped because it would be a fourth significant figure. The 7 is raised to 8 because the dropped 8 is greater than 4.

c. 5.02×10^4. The decimal point has been moved four places to the left, giving 10^4. The 2 and 9 are dropped and the 1 is raised to 2, because the 9 is greater than 4.

d. 8.26×10^{-2}. The decimal point has been moved two places to the right, giving 10^{-2}. The 3 is dropped because it would be the fourth significant figure. The 6 is not increased because 3 is less than 5.

Problem 1.3

Express the following numbers in exponential notation to three significant figures:

a. 109810   **b.** 90360   **c.** 0.0000006101   **d.** 0.7008

2. The use of significant figures in calculations. Most often, the measurements that we make are not final answers in themselves. Rather, they are used in further calculations involving addition, subtraction, multiplication, or division. In these calculations, it is necessary to remember that the accuracy of a measurement cannot be improved by calculation.

In calculations involving multiplication and division, the answer can contain no more significant figures than that measurement in the calculation having fewest significant figures.

Example 1.4

A sample of straight carbon steel (the kind commonly used to make railroad-track bolts and automobile axles) has a mass of 0.795 g and contains 3.6×10^{-3} g carbon. What is the percentage of carbon in this steel alloy?

Solution

The arithmetic solution to the problem is:

$$\frac{3.6 \times 10^{-3}\text{g}}{0.795 \text{ g}} \times 100\% = 0.45283\%$$

According to the rule for determining the number of significant figures in calculations involving division or multiplication, the answer to this problem must be rounded off to two significant figures, since there are only two in the number 3.6×10^{-3}.

0.45% or $4.5 \times 10^{-1}\%$

In calculations involving addition or subtraction, the answer can show only as many places to the right of the decimal point as are common to all of the measurements used in the calculations. In other words, the answer can be no more accurate than the least accurate number in the calculation. In these cases, the number of significant figures in the different factors is unimportant.

Example 1.5

A 2.65-mL sample is withdrawn from a bottle containing 375 mL of alcohol. What is the volume of liquid remaining in the bottle?

Solution

$375 - 2.65 = 372.35$ mL. One of the numbers in the calculation (375 mL) is reported only to the units place. The units place is the smallest place common to both numbers; therefore, the answer can be reported only to the units place. Thus, the correct answer is 372 mL. In addition and subtraction problems, it is not the number of significant figures in the numbers being added or subtracted (there are three in 375 and three in 2.65) but the location of the decimal point that guides you in determining how many places to use in reporting the answer.

Problem 1.4

Three samples of blood are drawn from a patient. One sample has a volume of 0.51 cm^3, the second a volume of 0.01 cm^3, and the third a volume of 15.0 cm^3. What is the total volume drawn from the patient?

Problem 1.5

Research has shown that trace amounts of the metal zinc are necessary for good health. The body of a person weighing 78 kg should contain 1.61×10^{-4} g zinc. Calculate the percentage of zinc in a healthy body.

3. Significant figures and pocket calculators. The use of pocket electronic calculators has increased enormously the importance of understanding significant figures and of observing the rules governing their determination. A calculator with an eight-digit display capability may display eight digits in the answer to a calculation, regardless of the number of digits punched in. For example, dividing 5.0 by 1.67 on a calculator may give the following answer:

$$\frac{5.0}{1.67} = 2.9940119$$

The correct answer, 3.0, has only two significant figures, the same as in the least accurate number (5.0) in the problem. All other digits displayed by the calculator are insignificant.

1.4 Conversion Factors and Problem Solving by Unit Analysis

A. Conversion Factors

Measurements made during a chemical experiment are often used to calculate another property. Frequently, it is necessary to change measurements from one unit to another: inches to feet, meters to centimeters, or hours to seconds. Calculations of this type can be done very simply by using conversion factors. A **conversion factor** measures the same quantity using two different units, for example:

$$1 \text{ yard} = 3 \text{ feet}$$

This equality can also be expressed as a fraction:

$$\frac{1 \text{ yard}}{3 \text{ feet}} \text{ is identical to } \frac{3 \text{ feet}}{3 \text{ feet}} \text{ which equals } 1$$

Conversion factors that define relationships, such as 3 feet = 1 yard or 1 liter = 1000 mL, are said to be **infinitely significant.** This means that the number of figures in these factors does not affect the number of significant figures in the answer to the problem.

B. Solving Problems by Unit Analysis

The use of conversion factors in chemical problem solving requires careful attention to units and to the conversion factors relating those units. We call it **problem solving by unit analysis.** The name may be new to you but the method is familiar.

For example, if you were asked how many inches there are in 6 feet, you would reply without hesitation that 6 feet equals 72 inches. In doing this calculation, you would be using a familiar relationship between inches and feet, namely, 12 inches = 1 foot. There are two ways to write this as a conversion factor:

$$\frac{12 \text{ inches}}{1 \text{ foot}} \text{ and } \frac{1 \text{ foot}}{12 \text{ inches}}$$

To convert 6 feet to inches, you would use the conversion factor on the left because it allows you to cancel the units or dimensions you do not want (feet) and arrive at an answer with the units you do want (inches).

$$6 \text{ feet} \times \frac{12 \text{ inches}}{1 \text{ foot}} = 72 \text{ inches}$$

Note that although the units cancel, the numerical values remain.

Problem solving by unit analysis can be divided into the following five steps.

1. Determine what quantity is wanted and in what units.
2. Determine what quantity is given and in what units.
3. Determine what conversion factor or factors can be used to convert from the units given to the units wanted.
4. Determine how the quantity and units given and the appropriate conversion factors can be combined into an equation, in such a way that the unwanted units cancel and only the wanted units remain.
5. Perform the mathematical calculations and express the answer using the proper number of significant figures. After you have done the calculation, it is always wise to look at the equation again and estimate the answer. If your estimate is close to the calculated answer, all is probably well. If it is quite different, check your calculations. Be sure that you performed all operations correctly and that you have not misplaced the decimal point.

Example 1.6

How many centimeters are there in 1.63 meters?

Solution

Wanted:
Length in centimeters (? cm).

Given:
1.63 meters. We can write the partial equation:

? cm = 1.63 meters × conversion factor

Conversion factors:
1 meter = 100 cm (from Table 1.1). This conversion factor can be written as either:

$$\frac{1 \text{ meter}}{100 \text{ centimeters}} \quad \text{or} \quad \frac{100 \text{ centimeters}}{1 \text{ meter}}$$

Equation:
By combining these quantities and conversion factors in the following way, meters cancel and centimeters remain, which are the units we want in the answer.

$$? \text{ cm} = 1.63 \text{ m} \times \frac{100 \text{ cm}}{1 \text{ m}}$$

Arithmetic:

Performing the calculation gives 163 cm. The result should then be expressed in exponential notation to the proper number of significant figures, which gives the final answer:

$$1.63 \times 10^2 \text{ cm}$$

Many problems require two or more conversion factors for conversion from the units given to the units wanted.

Example 1.7

A typical birth weight in the United States is 8.45 pounds. How many kilograms is this?

Solution

Wanted:

Weight in kilograms (? kg).

Given:

8.45 pounds. Therefore, we can start the equation:

$$? \text{ kg} = 8.45 \text{ lb} \times \text{conversion factor(s)}$$

Conversion factors:

We know from Table 1.1 that 1 lb = 453.6 g. However, the problem asks for the number of kilograms, not the number of grams. Therefore, we also need the conversion factor relating grams and kilograms, 1000 g = 1 kg. Thus, the conversion factors we need in solving this problem can be stated as:

$$\frac{1 \text{ lb}}{453.6 \text{ g}} \quad \text{or} \quad \frac{453.6 \text{ g}}{1 \text{ lb}} ; \quad \frac{1000 \text{ g}}{1 \text{ kg}} \quad \text{or} \quad \frac{1 \text{ kg}}{1000 \text{ g}}$$

Equation:

The quantity given and the appropriate conversion factors are combined in the following equation so that the unwanted units (lb, g) cancel and only the wanted units (kg) remain:

$$? \text{ kg} = 8.45 \text{ lb} \times \frac{453.6 \text{ g}}{1 \text{ lb}} \times \frac{1 \text{ kg}}{1000 \text{ g}}$$

Arithmetic:

Performing the mathematical calculation gives 3.8329 kg. Expressing this result using only three significant figures (the number of significant figures in 8.45 lb), we arrive at the final answer:

$$3.83 \text{ kg}$$

Example 1.8

Blood donors typically give 1.00 pint of blood during each visit to the blood bank. Calculate the volume in liters.

Solution

Wanted:
Volume in liters (? L blood).

Given:
1.00 pint blood.

$$? \text{ L blood} = 1.00 \text{ pint} \times \text{conversion factor(s)}$$

Conversion factors:
1 quart = 2 pints;   1 liter = 1.057 quart

Equation:

$$? \text{ L blood} = 1.00 \text{ pint} \times \frac{1 \text{ qt}}{2 \text{ pints}} \times \frac{1 \text{ liter}}{1.057 \text{ qt}}$$

Answer:
0.473 L blood

Problem 1.6

A football player runs the opening kickoff back 45 yards. If football were to convert to the metric system, how many meters would this run be?

Problem 1.7

Aspirin tablets weigh 5.00 grains. If one grain is 2.29×10^{-3} ounce and there are 16 ounces in a pound, what is the mass in mg of one aspirin tablet?

Problem 1.8

The gas tank of a car holds 19.5 gallons. How many liters will it hold?

As you can see from these examples, problem solving by unit analysis is a straightforward method to help you organize your thinking and attack problems in a systematic way. We will use this method and these same five steps for all the numerical problems that follow.

1.5 Physical Properties

A. Density and Specific Gravity

In Section 1.2 we discussed measurements of three properties: length, mass, and volume. The mass and volume of a sample can be combined to express

Table 1.7 Mass, volume, and density of iron samples.

Sample	Volume (cm³)	Mass (g)	Density (g/cm³)
A	1.05	8.25	7.86
B	25.63	201.5	7.86
C	90.7	713	7.86
D	0.02471	0.1942	7.86

another property of matter: *density*. By definition, **density** is the *ratio of the mass of a sample to its volume*:

$$\text{density} = \frac{\text{mass}}{\text{volume}}$$

Table 1.7 shows the mass, volume, and density of four different samples of iron. Although the samples differ in mass and volume, all have the same ratio of mass to volume, or density. All samples of the same kind of matter under the same conditions have the same density. Density is a physical property by which a particular kind of matter can be characterized and identified (Figure 1.9). Table 1.8 lists the densities of some common solids, liquids, and gases under normal conditions for each.

The densities of solids and liquids are usually given in grams per cubic centimeter, as shown in the table. Using these units, the density of water is 1.000 g/cm³ at 4°C. Based on the information in the table, some basic observations can be made. The densities of most metals are greater than that of water. The densities of liquids vary; some are less dense than water, while others are more dense. For example, the density of gasoline is about 30% less than that of water, and the density of chloroform is 50% greater. Gases are so much lighter than solids and liquids that their densities are commonly given in

Figure 1.9 Mercury is one of the heaviest liquids known (d = 13.6 g/cm³); 1 cm³ of mercury balances 13.6 cm³ of water (d = 1.0 g/cm³).

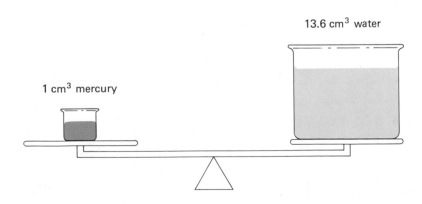

13.6 cm³ water

1 cm³ mercury

Table 1.8 Densities of some common solids, liquids, and gases under normal conditions.

Metals (g/cm³)		Other Solids (g/cm³)		Liquids (g/cm³)		Gases (g/L)	
aluminum	2.70	bone	1.85	chloroform	1.49	air	1.20
gold	19.32	butter	0.86	ethyl alcohol	0.791	carbon dioxide	1.83
lead	11.34	cork	0.24	gasoline	0.67	carbon monoxide	1.16
magnesium	1.74	diamond	3.51	olive oil	0.918	chlorine	2.99
mercury	13.59	marble	2.7	water	1.000	helium	0.17
platinum	21.45	sugar	1.59	(at 4°C)		hydrogen	0.08
sodium	0.97	table salt	2.16			oxygen	1.33
zinc	7.14					ozone	1.99

units of grams per liter. The lightest gas of all, hydrogen, has a density of 0.08 g/L; the density of dry air is 1.20 g/L.

Density relates mass to volume and can be used to convert between mass and volume.

Example 1.9

A sample of ethyl alcohol has a mass of 2.02 g. What is the volume of this sample in cubic centimeters?

Solution

Wanted:
Volume in cubic centimeters (? cm³).

Given:
2.02 grams ethyl alcohol.

$$? \text{ cm}^3 = 2.02 \text{ g} \times \text{conversion factor(s)}$$

Conversion factors:
Density of ethyl alcohol $= 0.791$ g/cm³. This can also be stated as: 1.00 cm³ ethyl alcohol weighs 0.791 g.

Equation:

$$? \text{ cm}^3 = 2.02 \text{ g} \times \frac{1.00 \text{ cm}^3}{0.791 \text{ g}}$$

Answer:
2.55 cm³.

whole blood	1.052–1.064
plasma	1.024–1.030
urine	1.002–1.040
cerebro-spinal fluid	1.006–1.008
amniotic fluid	1.006–1.008
saliva	1.010–1.020

Table 1.9 Specific gravities of various body fluids.

Problem 1.9

A lead sinker for fishing weighs 16.5 g. What is the volume of this piece of lead?

Often, particularly in discussing biological fluids, *specific gravity* is reported rather than density. The **specific gravity** of a substance is the *ratio of its density to that of a reference substance:*

$$\text{specific gravity} = \frac{\text{density of substance}}{\text{density of reference substance}}$$

Generally, water is the reference substance for comparing solids and liquids. Air is the usual reference substance for comparing gases.

In reporting specific gravity, it is essential to indicate the temperature at which the measurements were made, because densities vary with temperature. For example, the ratio between the density of benzene measured at 20°C and the density of water at 4°C is 0.8794. This fact is indicated in the following way:

$$\text{sp gr}_4^{20} = 0.8794$$

Table 1.9 lists the specific gravities of several body fluids. That of urine is particularly important because it indicates the condition of the kidneys. Diseased kidneys are unable to handle as much fluid as healthy ones. A higher than normal specific gravity for urine indicates trouble, usually either acute nephritis, fever, or diabetes mellitus.

B. Temperature

1. Temperature scales. In addition to measuring mass, length, and volume, chemists are often concerned with *temperature* and changes in temperature. **Temperature** *measures how hot or cold a sample is in comparison to some arbitrary standard.* Temperature is measured with a thermometer and is most commonly reported using one of three different scales: **Fahrenheit (F), Celsius (C)** (sometimes called *Centigrade*), and **Kelvin (K)** (sometimes called *Absolute*).

The relationship among these three scales is straightforward if you understand how a thermometer is constructed and calibrated. Two essential features of a thermometer are: (1) a substance that expands as it is heated and contracts as it is cooled; and (2) some means to measure the expansion and contraction. In the thermometer with which you may be most familiar, the substance that expands and contracts is mercury. In order to measure its expansion or contraction, the mercury is confined within a small, thin-walled glass bulb connected to a narrow capillary tube. When the temperature increases, the mercury expands and its level in the capillary tube rises. This increase in height is proportional to the increase in temperature.

A thermometer is calibrated in the following manner. First, the mercury bulb of a new thermometer is immersed in a mixture of ice and water. When the height of the mercury in the column becomes constant, a mark is made. This mark is one reference point. The ice-water mixture is then heated to boiling and kept at that temperature until the height of the mercury in the column rises to a new constant level. Another mark is made on the column at this level; this mark is a second reference point. Now the manufacturer must decide whether this thermometer will measure temperature on the Celsius, Fahrenheit, or Kelvin scale. If it is to be on the Celsius scale, the reference point for the ice-water mixture is labeled 0°C and that for boiling water is labeled 100°C. The distance between these two reference points is divided into 100 equal divisions. If the thermometer is to measure temperature on the Fahrenheit scale, the reference point for the ice-water mixture is labeled 32°F and that for boiling water is labeled 212°F. The distance between 32°F and 212°F is divided into 180 equal divisions. If the thermometer is to measure temperature on the Kelvin scale, the ice-water reference point is labeled 273.2, the boiling water reference point is labeled 373.2, and the distance between these two marks is divided equally into 100 divisions. As you can see, there is no difference in the temperatures measured by any of these thermometers; the difference is in the units in which each temperature is reported. The relationships between the three temperature scales are illustrated in Figure 1.10.

2. Conversions between the temperature scales. A temperature reading on any one of the three scales can be converted to a reading on any other. First, consider a conversion from degrees Celsius to degrees Fahrenheit. Figure 1.10 shows that between the temperature readings of the ice-water and boiling water marks, there are 180 Fahrenheit degrees but only 100 Celsius degrees. This relationship can be written as a conversion factor:

$$180°F = 100°C \quad \text{or} \quad \frac{180°F}{100°C} = \frac{9°F}{5°C} = \frac{1.8°F}{1°C}$$

In other words, a temperature increase of 9 Fahrenheit degrees is equivalent to an increase of 5 Celsius degrees. Figure 1.10 also shows that the numerical values assigned to the two ice-water reference points differ by 32 degrees; a

Figure 1.10 Fahrenheit, Celsius, and Kelvin thermometers.

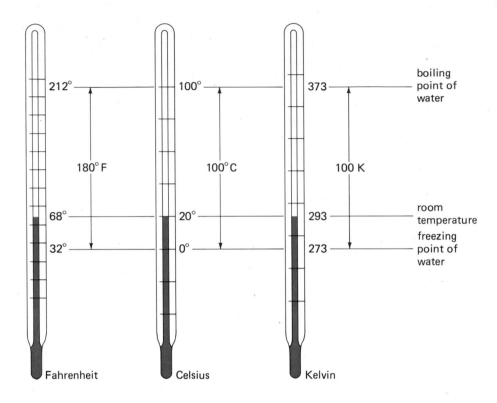

reading of 0° on the Celsius scale corresponds to a reading of 32° on the Fahrenheit scale. Putting these facts together in an equation gives:

$$°F = \tfrac{9}{5}(°C) + 32 \quad \text{or} \quad °F = 1.8(°C) + 32$$

This equation can be rearranged to give the reverse conversion equation:

$$\tfrac{9}{5}(°C) = °F - 32 \quad \text{or} \quad °C = \tfrac{5}{9}(°F - 32)$$

Example 1.10

a. A recommended temperature setting for household hot water heaters is 140°F. What is this temperature on the Celsius scale?

b. The boiling point of pure ethyl alcohol is 78.5°C. What is its boiling point on the Fahrenheit scale?

Solution

a. $°C = \tfrac{5}{9}(140 - 32) = 60°C$ **b.** $°F = \tfrac{9}{5}(78.5) + 32 = 173°F$

What is the relationship between the Celsius and Kelvin scales? As each scale has exactly 100 divisions or degrees between the ice-water temperature and the boiling water temperature, the size of a degree on the Celsius scale is the same as a degree on the Kelvin scale. The difference between the scales lies in the readings at the ice-water reference points: the reading is 0° on the Celsius scale and 273.2 on the Kelvin scale. Therefore, to convert a reading on the Celsius scale to one on the Kelvin scale, simply add 273.2.

$$K = °C + 273.2$$

You should note that the symbol K stands for "degrees Kelvin"; it is not preceded by the degree symbol, °. The symbols for Fahrenheit and Celsius do require the degree symbol; for example, 212°F, 100°C, but 373 K.

Problem 1.10

Perform the following temperature conversions:

a. 110°F to °C   **b.** 43°C to K   **c.** 25°C to °F

C. Specific Heat

We have seen that each substance has a number of physical properties that characterize it. One such property is *specific heat*. **Specific heat** is the *amount of energy needed to raise the temperature of a given mass of a substance by 1°C.* Energy can be measured in **joules** or in **calories,** quantities that are related by the equation:

$$1 \text{ calorie (cal)} = 4.184 \text{ joules (J)}$$

Specific heat has the units:

$$\frac{\text{calories}}{(\text{g})(°\text{C})} \quad \text{or} \quad \frac{\text{joules}}{(\text{g})(°\text{C})}$$

The specific heat of water is 1 cal/(g)(°C) or 4.184 J/(g)(°C).

Table 1.10 lists the specific heat of water and a variety of other substances. Notice that water has the highest specific heat. This means that if the same

Table 1.10 Specific heats of some common substances.

Metals	cal/(g)(°C)	Liquids	cal/(g)(°C)
aluminum	0.215	benzene	0.406
copper	0.092	ether	0.529
gold	0.031	ethyl alcohol	0.581
iron	0.108	glycerine	0.540
lead	0.038	ice (solid)	0.492
silver	0.057	olive oil	0.471
		water	1.00

amount of heat energy is added to equal masses of water and almost any other substance, the temperature of the other substance will increase considerably more than that of the water. For example, the average specific heat of air is 0.24 cal/(g)(°C), much lower than that of water (1.00 cal/(g)(°C)). Because of this great difference, the temperature of a large body of water, such as a lake or an ocean, increases much more slowly in the spring than does the temperature of the surrounding air. Conversely, the temperature of a large body of water falls much more slowly in autumn than does that of the surrounding air. Due to these conditions, islands and coastal areas experience less temperature variation than inland locations.

The following examples show how specific heats are calculated and used. As you solve these problems, notice that the mass of sample is the same in each example, but the energies required for the same temperature change are very different.

Example 1.11

a. A sample of aluminum weighs 56 g. 712 calories are required to raise the temperature from 20°C to 80°C. Calculate the specific heat of aluminum.

b. How many calories are required to raise the temperature of 56 grams of water from 20°C to 80°C?

c. Calculate the final temperature of a 56-g sample of water, originally at 25°C, to which is added 712 cal.

Solution

a. Specific heat has the units

$$\frac{\text{calories}}{\text{grams} \times \text{temperature change in } °C}$$

We are given a value for each of these quantities in the problem. Substituting them into the definition for specific heat gives:

$$\text{sp ht of aluminum} = \frac{712 \text{ cal}}{56 \text{ g} \times 60°C} = 0.21 \frac{\text{cal}}{\text{g}°C}$$

b. Wanted:
The amount of heat in calories (? cal).

Given:
A 56-g sample and a temperature increase of 60°C. We also know that the number of calories depends on both the mass of the sample and the size of the temperature change. Therefore, we start the equation:

$$? \text{ cal} = 56 \text{ g} \times 60°C \times \text{conversion factor(s)}$$

Conversion factors:
The specific heat of water (from Table 1.10):

$$\frac{1.00 \text{ cal}}{1.0 \text{ g} \times 1.0°C}$$

Equation:

$$? \text{ cal} = 56 \text{ g} \times 60°C \times \frac{1.00 \text{ cal}}{1.0 \text{ g} \times 1.0°C}$$

Answer:
3.4×10^3 cal   or   3.4 kcal

Notice that the mass of the water sample and the temperature change in this part of the example are the same as the mass of aluminum and the temperature change in part (a). Because of the difference in specific heat between aluminum and water, almost five times as many calories were required in part (b) as were required in part (a).

c. Wanted:
The final temperature in °C. Use two steps to solve this problem. First, calculate the temperature change; second, add this value to the initial temperature of 25°C.

Given:
A 56-g sample of water at 25°C to which is added 712 cal. Remember that the number of calories required depends on the mass of the sample, the temperature change, and the specific heat.

Conversion factors:

$$\text{specific heat} = \frac{\text{calories}}{\text{grams} \times \text{temperature change}}$$

Rearranging this relationship gives:

$$\text{temperature change} = \frac{\text{calories}}{\text{grams} \times \text{specific heat}}$$

We also need the specific heat of water:

$$\frac{1.00 \text{ cal}}{1.0 \text{ g} \times 1.0°C}$$

Equation:

$$? \, °C = \frac{712 \text{ cal}}{56 \text{ g} \times \dfrac{1.00 \text{ cal}}{1.0 \text{ g} \times 1.0°C}} = \frac{712 \text{ cal}}{56 \text{ g}} \times \frac{1.0 \text{ g} \times 1.0°C}{1.00 \text{ cal}}$$

Answer:
13°C. Adding this to the initial temperature of the sample (25°C) gives 38°C as the final temperature of the water.

Problem 1.11

The specific heat of ether is 0.529 cal/g°C. How many calories are required to raise the temperature of 25 g of ether from room temperature (23°C) to its boiling point (35°C)?

Key Terms and Concepts

accuracy (1.3A)

calorie (1.5C)

Celsius (1.5B)

chemical property (1.1A)

constant composition (1.1A)

conversion factor (1.4A)

density (1.5A)

English system (1.2A)

exponential notation (1.3B)

Fahrenheit (1.5B)

Joule (1.5C)

Kelvin (1.5B)

mass (1.2B)

matter (1.1A)

meter (1.2A)

metric units (1.2A)

mixtures (1.1B)

physical property (1.1A)

physical states (1.1A)

precision (1.3A)

pure substance (1.1A)

rounding off (1.3D)

SI prefixes (1.2A)

SI system (1.2A)

significant figures (1.3D)

specific gravity (1.5A)

specific heat (1.5C)

temperature (1.5B)

unit analysis (1.4B)

Problems

Significant Figures and Exponential Notation (Section 1.3)

1.12 Report the following numbers in exponential notation using three significant figures.

a. 160502	b. 0.006059	c. 678000
d. 132.419	e. 1.00605	f. 0.43689
g. 0.0006004	h. 0.000650	i. 0.1000234
j. 800.923	k. 4263529	l. 1007855

1.13 Carry out the following calculations and report your answers using the correct number of significant figures.
   a. $396 + 1.05 + 16203.526 + 2900 =$ b. $14705 + 0.0001 + 20.35 =$
   c. $14.70 \times 0.0025 \times 9.2 =$ d. $65 \div 3256 =$
   e. $29.62 - 1.009 =$ f. $0.002395 \times 12.625 \times 6.02 =$
   g. $0.00159 + 0.01956 =$ h. $13.49(1.23 + 44.6) =$
   i. $35.78 + 32.0 + 5.765 =$ j. $8.98 \div 0.68 =$
   k. $45000 + 987 + 54.8 + 0.786 =$

The Use of Conversion Factors (Section 1.4)

1.14 Carry out the following conversions. Be certain that your answers have the correct number of significant figures.
   a. 7.6 g to milligrams   b. 72 kg to grams
   c. 17.8 cm^3 to liters   d. 16.8 cm to meters
   e. 20.6 m to millimeters   f. 34.3 L to milliliters
   g. 163 g to kilograms   h. 28.4 cm^3 to milliliters
   i. 3.0 L to cubic centimeters   j. 14 mg to grams
   k. 0.034 mg to nanograms   l. 0.95 mL to microliters

1.15 Carry out the following conversions. Be certain that your answers have the correct number of significant figures.
   a. 39 cm to inches   b. 145 g to pounds
   c. 13.0 yd to meters   d. 1.6 qt to liters
   e. 166 km to miles   f. 565 lb to kilograms
   g. 12.6 in. to centimeters   h. 250 cm^3 to quarts
   i. 3.65 sq ft to square meters   j. 2.63 m to feet

1.16 Carry out the following conversions. Be certain that your answers have the correct number of significant figures.
   a. 6246 m to kilometers   b. 1963 g to milligrams
   c. 15960 g to kilograms   d. 235616 cm to meters
   e. 22412 mL to liters   f. 1963529 mg to kilograms

1.17 What is the metric mass of 2.5 lb of peaches?

1.18 Are you exceeding the speed limit if you are driving 97 km/hr in a 55 mi/hr zone?

1.19 A person weighs 96.4 kilograms and is 1.90 meters tall. What are this person's weight and height as measured in the English system?

1.20 According to the Food and Nutrition Board of the National Research Council, the recommended weight for a woman five feet, four inches tall is 122 ± 10 pounds. Calculate the woman's height in centimeters and her recommended weight in kilograms.

1.21 Calculate your weight and height as measured in the metric system.

1.22 Driving in Europe, you need to buy 6.3 gallons of gasoline. The pump registers in the metric system. How many liters will you buy?

1.23 The Earth is 9.29×10^7 miles from the sun.
   a. What is this distance in meters?
   b. If light travels 3.0×10^8 m/sec, how long does it take light to travel from the sun to the Earth?

1.24 When "burned" as a metabolic fuel, one gram of body fat is equivalent to 7.7 kilo-calories of energy. How many kilocalories of energy must an adult use to lose the equivalent of one pound of body fat?

1.25 The table below lists the energy expended by various activities, given in terms of kilocalories per kilogram of body weight per hour. The rates are applicable to either sex.

Activity	$\dfrac{\text{kcal}}{\text{kg} \times \text{hr}}$
bicycling (moderate speed)	7.6
carpentry (heavy)	11.4
lying still (awake)	0.1
playing Ping-Pong	4.4
running	7.0
sawing wood	5.7
sitting quietly	0.4
skating	3.5

    a. Calculate the number of kilocalories expended per hour by a 160-pound male in each of these activities.

    b. Calculate the number of kilocalories expended per hour by a 130-pound female in each of these activities.

    c. How many hours must each run to burn off the equivalent of one pound of body fat?

1.26 The city of Chicago declares an ozone alert when the concentration of ozone in the air reaches 137 micrograms per cubic meter.

    a. Express this ozone concentration in micrograms per liter and in nanograms per liter.

    b. The lung capacity per breath of an adult is about 2 quarts. What is the mass of ozone taken into the lungs per breath, by an adult breathing air containing 137 micrograms ozone/m^3 air?

1.27 For an adult in good health and with an adequate diet, the concentration of ascorbic acid (vitamin C) in the blood is about 0.2 milligram per 100 milliliters of blood.

    a. What is this concentration in grams per liter? In milligrams per milliliter?

    b. The average person has about 5 liters of blood. Calculate the total number of milligrams of ascorbic acid in the blood of an adult in good health eating an adequate diet.

    c. According to the National Research Council, the recommended daily allowance for ascorbic acid is 45 milligrams. How does this compare with the number of milligrams of ascorbic acid in the blood of a healthy adult with an adequate diet?

1.28 The average values for red cell volume, plasma volume, and total blood volume, expressed in milliliters per kilogram of body weight for a normal adult, are:

	Male	Female
red cell volume	28	24
plasma volume	44	42
total blood volume	72	66

Calculate your red cell volume, plasma volume, and total blood volume, based on these figures and your body weight. Express your answer in liters using the proper number of significant figures.

Density (Section 1.5A)

1.29 Calculate the missing quantities:

Mass	Volume	Density
a. 13.6 g	21.9 cm^3	———
b. 4.6 g	1.2 L	———
c. 155.1 g	13.2 cm^3	———
d. 5.23 g	6.9 cm^3	———
e. ———	23 cm^3	1.45 g/cm^3
f. 5.6 g	———	0.831 g/cm^3
g. ———	11.4 cm^3	5.4 g/cm^3
h. ———	0.45 L	1.3 g/cm^3

1.30 A piece of iron weighs 6.53 g and has a density of 7.86 g/cm^3. What is the volume of the piece of iron?

1.31 At 20°C, the volume of a colorless liquid is 9.43 cm^3 and its density is 0.789 g/cm^3. What is the mass of the sample? Could this liquid be water?

1.32 A sample of magnesium has a volume of 7.43 cm^3 and a density of 1.74 g/cm^3. What is the mass of the sample?

1.33 A person has a mass of 113 pounds. What volume (in liters) of mercury has the same mass?

1.34 A given sample of gas may be pure oxygen or it may be oxygen mixed with another gas. The sample weighs 2.75 g and has a volume of 2.40 L. Is the sample pure oxygen? If not, does the other gas have a density greater or less than that of oxygen?

1.35 19.8 g of a liquid has a volume of 25 cm^3. Identify the liquid using the data in Table 1.8.

1.36 Using the principles of density, explain why a balloon filled with helium rises while one filled with air does not.

Temperature Conversions (Section 1.5B)

1.37 Complete the following temperature chart.

°F	°C	K
55	13	286
——	165	——
——	——	450
——	−40	——
−25	——	——

1.38 Normal body temperature is 98.6°F. What is normal body temperature on the Celsius scale? On the Kelvin scale?

1.39 If body temperature is elevated by 6° on the Fahrenheit scale (from 98.6°F to 104.6°), by how many degrees is it elevated on the Celsius scale?

1.40 The boiling point of benzene is 80.1°C. At what temperature on the Fahrenheit scale does benzene boil?

Specific Heat (Section 1.5C)

1.41 The specific heat of magnesium is 0.235 cal/g°C. How many calories must be added to 7.15 g of magnesium to raise its temperature by 15°C?

1.42 a. How many calories are required to raise the temperature of 15.6 g of water from 0°C to 10°C?

 b. How many calories are required to raise the temperature of 15.6 g of ice from −10°C to 0°C? (Use 0.49 cal/(g)(°C) for the specific heat of ice.)

1.43 How many calories must be removed from 555 g of water to cool the sample from 25°C to 0°C?

1.44 The specific heat of ethyl alcohol is 0.581 cal/g°C. How many calories are needed to raise the temperature of 75 g of ethyl alcohol from 21°C to 37°C (body temperature)?

1.45 The average specific heat of dry air is 0.24 cal/g°C.

 a. If 125 cal of heat are added to 75 g of dry air at 20°C, what is the final temperature of the air?

 b. The average specific heat of water is 1.00 cal/g°C. If 125 calories are added to 75 g of water at 20°C, what is the final temperature of the water sample?

 c. Why are temperature variations between day and night more extreme in a desert than near an ocean?

Elements

From a chemist's point of view, matter in its simplest form exists as *elements*. Elements are found in different concentrations, in different physical states, and in different combinations, throughout the universe. In this chapter, we will characterize elements.

Elements exist as *atoms*. We will describe the properties and composition of atoms and how atoms differ from one another. We will also talk about *radioisotopes* and their uses in medicine and industry.

The following discussion contains many examples of the *scientific method*. The scientific method has, over the past 200 years, provided a pattern for the development of chemical knowledge, changing it from a collection of chance observations and random experiments to an organized body of knowledge from which further insights can be developed.

2.1 The Scientific Method

The **scientific method** requires that the hypotheses, theories, and laws of science be based on careful deductions made from carefully planned and executed experiments. As natural as it is for us today to accept the importance of experiments in chemistry, their use dates only from about the end of the eighteenth century. Prior to that date, philosophers contemplated the nature of matter but did not perform experiments as we know them today. Instead, it was often the theory of the most respected teacher that was most widely accepted. Alchemists, of course, did use experiments in their efforts to isolate the mystical "Philosopher's Stone," which they believed could be used to turn other substances into gold. However, their experiments were not designed to form new theories on the nature of matter. During the same period, artisans sought to discover better methods of refining metals, preparing dyes, and so on, but this work was applied science and, like that of the alchemists, was not aimed at discovering fundamental knowledge or formulating new laws about the properties of matter.

During the eighteenth century, when science became a popular hobby of the rich, it was common for noblemen to have laboratories in their homes. There they did experiments, considered the implications of the experimental findings, and formulated theories that could be tested by new experiments. These experimentalists met with one another to discuss their work and formulate theories on the nature of matter. This approach to science formed the basis for the pattern of experimentation that we call the scientific method.

According to the scientific method, new knowledge and understanding of the world around us are most easily gained if the observer organizes his work around the following steps:

1. Careful observations are collected about a given event in nature. These may be direct observations of nature or ones that others have made.

2. A **hypothesis** or model is constructed that explains or consolidates these observations.

3. New experiments to test the hypothesis are planned and carried out.

4. The original hypothesis is modified so that it is consistent with both the new and original observations.

A hypothesis that survives extensive testing becomes accepted as a **theory.** Although our present hypotheses and theories are the best we have been able to devise so far, there is no guarantee that they are final. See Figure 2.1, which diagrams the course of the scientific method. Regardless of how many experiments have been done to test a given theory and how much data have been accumulated to support it, a single experiment that can be repeated by other scientists and whose results contradict that theory forces its modification or rejection. Some of our currently accepted theories on the nature of matter may in the future have to be modified or even rejected on the basis of data from new experiments. It is essential to keep an open mind and be ready to accept new data and new theories.

Science is a dynamic process, moving by fits and starts. It may leap forward or it may be bogged down for lack of a workable hypothesis. Science

Figure 2.1 The steps of the scientific method. The steps of hypothesizing and data collecting continue to alternate for some time before the hypothesis earns the right to be called a theory.

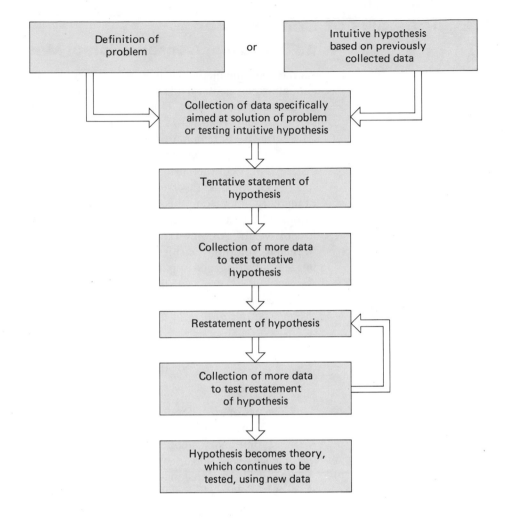

is less like a wide, smoothly flowing river than it is like a mountain stream, sometimes rushing ahead, at other times scarcely moving, and even wandering off into dead-end ravines. Yet its overall movement is unquestionably forward, toward greater understanding of the world.

An inherent part of the scientific method is the element of creativity. The scientist assembles all of the observations that have been made in a particular area and combines this knowledge in a new way. Out of it comes something original and unique. For some scientists, it is a new concept; for others, it is the refinement and clarification of an existing concept. We shall see many examples of creativity in science as we move through this text, including several in this chapter on the structure of the atom.

2.2 The Laws of Conservation

A. The Law of Conservation of Mass

As the scientific method gained acceptance during the late eighteenth century, enough experimental evidence was collected to prove the long-suspected **Law of Conservation of Mass.** This law states: *In a chemical reaction, the total mass of the products equals, within detectable limits, the total mass of the reactants.* An alternative statement of the law is: *The total mass of the universe is constant.* One of the earliest experiments to verify this law was a careful study of the decomposition of mercuric oxide. Pure mercuric oxide is a red powder; when it is heated, mercury (a silvery liquid) is obtained. It was known that the weight of mercury obtained in this experiment was always less than that of the original sample of mercuric oxide. There were many imaginative explanations for this loss of weight. Antoine Lavoisier, a French nobleman later guillotined in the French Revolution, considered the problem and then carried out the experiment in a specially designed apparatus using a very accurate balance. His experiment showed that oxygen as well as mercury is formed by the reaction and that the combined weights of mercury and oxygen always equalled the weight of the original sample of mercuric oxide:

$$\text{mercuric oxide} \xrightarrow{\text{heat}} \text{mercury} + \text{oxygen}$$
$$\text{(red powder)} \qquad \text{(silvery} \qquad \text{(colorless}$$
$$\text{liquid)} \qquad \text{gas)}$$

The use of an accurate balance by Lavoisier and others to make *quantitative* measurements of the starting materials and products of chemical reactions was a revolutionary step. From that time on, chemistry became a *quantitative* science. This step forward, like so many others in science, depended on technology; in this instance, on the development of an accurate and precise balance. For his pioneering work in the establishment of the Law of Conservation of Mass and the organization of chemical data, Lavoisier has been called the father of modern chemistry.

B. The Law of Conservation of Energy

The **Law of Conservation of Energy** is also basic to the study of chemistry. This law states: *Energy can neither be created nor destroyed.* An alternative statement of the law is: *The total amount of energy in the universe remains constant.* Just as mass can be changed from one kind of matter to another without any loss (as in Lavoisier's experiment, where mercuric oxide was changed to mercury and oxygen), energy can be changed from one kind to another without any loss. In a light bulb, electric energy is changed into light and heat energy. An electric motor changes electric energy into mechanical energy. In all such changes, however, the total amount of energy remains constant.

C. The Law of Conservation of Mass/Energy

It remained for Albert Einstein (1879–1955) to show that the Law of Conservation of Mass and the Law of Conservation of Energy are really one law. *Energy can be transformed into matter and matter into energy;* in any reaction, the sum of matter and energy is conserved. Therefore, the fundamental law concerning mass and energy is the **Law of Conservation of Mass/Energy.** The equation relating these two quantities is:

$$\Delta E = c^2 \, \Delta m \qquad \text{where } \Delta E = \text{energy change}$$

$$c = \text{speed of light } (3.00 \times 10^8 \text{ m/sec})$$

$$\Delta m = \text{mass change}$$

The transformation of matter into energy is not encountered in ordinary circumstances, but it does occur in nuclear reactions, during the course of which a tiny amount of mass is converted to an enormous amount of energy.

2.3 The Atomic Theory of Matter

By the end of the eighteenth century, it was well established that each pure substance had its own characteristic set of properties, such as density, specific heat, melting point, and boiling point. It was also established that certain quantitative relationships, such as the Law of Conservation of Mass, governed all chemical changes. But there was still no understanding of the nature of matter itself. Was matter continuous, like a ribbon from which varying amounts could be snipped, or was it granular, like a string of beads from which only whole units or groups of units could be removed? Some scientists believed strongly in the continuity of matter, while others believed equally strongly in granular, or corpuscular, matter, but both reasonings were based solely on speculation and philosophy.

In 1803, an English schoolmaster named John Dalton summarized and extended the current theory of matter. The postulates of his theory, changed only slightly from their original statement, form the basis of modern atomic theory. Today, we express these four postulates as:

1. Matter is made up of tiny particles called **atoms.** A typical atom has a mass of approximately 10^{-23} gram and a radius of approximately 10^{-10} meter. Atoms are made up of even smaller particles (see Section 2.6) and some decompose by a process called *radioactive disintegration* (see Section 2.9).

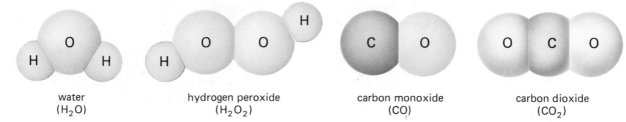

water
(H_2O)

hydrogen peroxide
(H_2O_2)

carbon monoxide
(CO)

carbon dioxide
(CO_2)

Figure 2.2 Atoms of the same elements combine in different ratios to form different compounds.

2. There are over 100 different kinds of atoms; each kind is called an **element.** A list of the elements is on the inside back cover of this text. All the atoms of a particular element are alike chemically but can vary slightly in mass and other physical properties. Atoms of the same element that differ in mass are called **isotopes.** For example, there are some chlorine atoms with a relative mass of 35 and others with a relative mass of 37. These are two isotopes of chlorine. (Isotopes are discussed further in Section 2.6D).

3. Atoms of different elements combine in small, whole-number ratios to form **compounds.** Hydrogen and oxygen atoms combine in the ratio of 2:1 to form the compound water. Carbon and oxygen atoms combine in the ratio of 1:2 to form the compound carbon dioxide. Iron and oxygen atoms combine in a ratio of 2:3 to form the familiar substance rust.

4. Atoms combine in different whole-number ratios to form different compounds. As we just observed, hydrogen and oxygen atoms combine in a ratio of 2:1 to form water; they also combine in a 1:1 ratio to form hydrogen peroxide (Figure 2.2). Carbon and oxygen atoms combine in a 1:2 ratio to form carbon dioxide; they also combine in a 1:1 ratio to form carbon monoxide.

2.4 Elements

Elements are pure substances. They cannot be decomposed by ordinary chemical means, such as heating, dissolving in acid, or combining with other elements. The atoms of each element are chemically distinct and differ from those of any other element.

As this book goes to press, 105 elements have been discovered and characterized, and their existence has been accepted by the International Union of Pure and Applied Chemistry (IUPAC). There is considerable evidence that element 106 has also been identified, but it has not yet been accepted and named by IUPAC.

A. Names and Symbols of the Elements

Each element has a **name.** Many of these are already familiar to you, including gold, silver, copper, chlorine, platinum, carbon, oxygen, and nitrogen. The names of elements provide an insight into the history of chemistry. Consider several elements whose names begin with the letter *c*. Cerium was discovered in 1803 and was named for the asteroid Ceres (discovered in 1801). Cesium was discovered in 1860 by the German chemist Bunsen (the inventor of the Bunsen burner). Because this element imparts a blue color to a flame, Bunsen named it cesium, from the Latin word *caesius*, meaning "sky blue." Californium is an example of an element named for its place of discovery. This element does not occur in nature. It was first produced in 1950 in the Radiation Laboratory at the University of California, Berkeley, by a team of scientists headed by Glenn Seaborg. Seaborg was also the first to identify curium, which was prepared in 1944 at the metalurgical laboratory of the University of Chicago, now Argonne National Laboratory. Curium is named for Pierre and Marie Curie, pioneers in the study of radioactivity. Marie Curie, a French scientist of Polish birth, was awarded the Nobel Prize in Physics in 1903 for her studies of radioactivity. She was also awarded the Nobel Prize in Chemistry in 1911 for her discovery of the elements polonium (named after Poland) and radium (Latin *radius*, "ray"). Seaborg himself became a Nobel laureate in 1951 in honor of his pioneering work in the preparation of hitherto unknown elements.

B	boron
C	carbon
F	fluorine
H	hydrogen
I	iodine
N	nitrogen
O	oxygen
P	phosphorus
K	potassium
S	sulfur
W	tungsten
U	uranium
V	vanadium
Y	yttrium

Each element has a **symbol,** one or two letters that represent the element much as your initials represent you. The symbol of an element stands for one atom of that element. For fourteen of the elements, the symbol is one letter. With the possible exceptions of yttrium (Y) and vanadium (V), you are probably familiar with the names of all elements having one-letter symbols listed at the top of the margin. For twelve of these elements, the symbol is the first letter of the name. In addition, there is potassium, discovered in 1807 and named for potash, the substance from which potassium was first isolated. Its symbol, K, comes from *kalium*, the Latin word for potash. Tungsten, discovered in 1783, has symbol W, from wolframite, the mineral from which tungsten was first isolated.

Cd	cadmium
Ca	calcium
Cf	californium
Ce	cerium
Cs	cesium
Cl	chlorine
Cr	chromium
Co	cobalt
Cu	copper
Cm	curium

All other elements have two-letter symbols. In these two-letter symbols, the first letter is always capitalized and the second is always lower case. The need to use two-letter symbols becomes obvious when you consider the number of elements whose names begin with the same letter. Eleven elements have names beginning with the letter *c*. One of these, carbon, has a one-letter symbol, C. The other ten have two-letter symbols (see margin).

B. Distribution of the Elements

The 105 known elements are not equally distributed throughout the world. Only 91 are found in either the Earth's crust, oceans, or atmosphere; the remaining 14 have been produced in laboratories. It is quite possible that these

Table 2.1 Distribution of elements in the Earth's crust, oceans, and atmosphere.

Element	Percent of Total Mass	Element	Percent of Total Mass
oxygen	49.2	titanium	0.58
silicon	25.7	chlorine	0.19
aluminum	7.50	phosphorus	0.11
iron	4.71	manganese	0.09
calcium	3.49	carbon	0.08
sodium	2.63	sulfur	0.06
potassium	2.40	barium	0.04
magnesium	1.93	nitrogen	0.03
hydrogen	0.87	fluorine	0.03
		all others	0.49

14 did or do exist in nature, but efforts to find them have been unsuccessful thus far. However, the search continues, and you might read of its success or of the isolation of new elements as you are taking this course.

Table 2.1 lists the 18 elements that are most abundant in the Earth's crust, oceans, and atmosphere, along with their relative percentages. One of the most striking points about this list is the remarkably uneven distribution of the elements (see Figure 2.3). Oxygen is by far the most abundant element. It forms 21% of the volume of the atmosphere and 89% of the mass of water. Oxygen in air, water, and elsewhere makes up 49.2% of the mass of the Earth's crust, oceans, and atmosphere. Silicon is the Earth's second most abundant element (25.7% by mass). Silicon is not found free in nature but occurs combined with oxygen as silicon dioxide, SiO_2, in sand, quartz, rock crystal, amethyst, agate, flint, jasper, and opal, as well as in various silicate minerals such as granite, asbestos, clay, and mica. Aluminum is the most abundant metal in the Earth's crust (7.50%), but it is always found combined in nature. Most of the aluminum used today is obtained by processing bauxite, an ore rich in aluminum oxide. These three elements, plus iron, calcium, sodium, potassium, and magnesium, make up more than 97% of the Earth's crust, oceans, and atmosphere. Another surprising feature of the distribution of elements is that several of the metals most important to our civilization are among the rarest; these include lead, tin, copper, gold, mercury, silver, and zinc.

The distribution of elements in the cosmos is quite different from that on Earth. According to present knowledge, hydrogen is by far the most abundant element in the universe, accounting for as much as 75% of its mass. Helium and hydrogen together make up almost 100% of the mass of the universe.

Table 2.2 lists the biologically important elements, those found in a normal, healthy body. The first four of these elements—carbon, oxygen, hydrogen, and nitrogen—make up about 96% of total body weight (see

Figure 2.3 Relative percentages by mass of elements in the Earth's crust, oceans, and atmosphere.

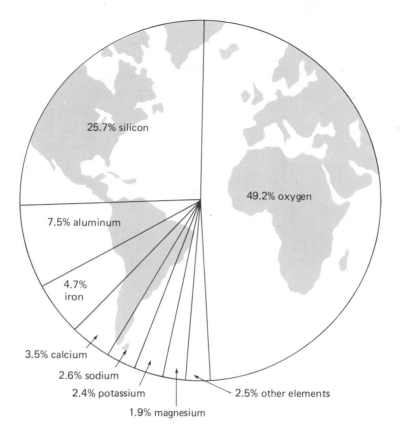

25.7% silicon

49.2% oxygen

7.5% aluminum

4.7% iron

3.5% calcium

2.6% sodium

2.4% potassium

1.9% magnesium

2.5% other elements

Table 2.2 Biologically important elements.

Major Elements	Approximate Amount (kg) in 70 kg Male	Elements Present in Less Than 1 mg Amounts in 70 kg Male (in Alphabetical Order)
oxygen	45.5	arsenic
carbon	12.6	chromium
hydrogen	7.0	cobalt
nitrogen	2.1	copper
calcium	1.0	fluorine
phosphorus	0.70	iodine
magnesium	0.35	manganese
potassium	0.24	molybdenum
sulfur	0.18	nickel
sodium	0.10	selenium
chlorine	0.10	silicon
iron	0.003	vanadium
zinc	0.002	

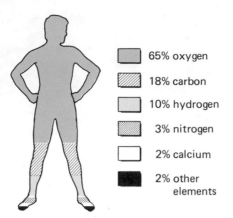

Figure 2.4 Distribution of biologically important elements in the human body.

65% oxygen

18% carbon

10% hydrogen

3% nitrogen

2% calcium

2% other elements

Figure 2.4). The other elements listed in Table 2.2, although present in only trace (tiny) amounts, are nonetheless necessary for good health.

2.5 The Occurrence of Elements in Nature

A. Uncombined Atoms

Although the smallest unit of an element is an atom, only a few elements are found in nature as single, uncombined atoms. Table 2.3 lists these elements. Under normal conditions all of these elements are gases; collectively, they are known as the **noble gases.** They are also known as **monatomic** gases, meaning that they exist uncombined as single atoms (*mono* means "one").

B. Molecules

Seven elements are found in nature as pairs of atoms bonded together in molecules. A **molecule** is the smallest unit of matter that can exist independently and still retain the characteristic properties of that kind of matter. Strictly speaking, the freely existing atoms of the noble gases could be called molecules. In practice, however, the term *molecule* implies more than one atom. A molecule containing two atoms is a **diatomic** molecule; molecules containing more than two atoms are **polyatomic.** The seven elements found in nature as diatomic molecules are listed in Table 2.4.

Some elements exist in nature as molecules containing more than two atoms. For example, elemental sulfur is found as a molecule containing eight sulfur atoms, S_8; phosphorus is found as P_4. Other elements, particularly metals, are encountered in nature in an uncombined state but neither as single atoms nor as clearly defined polyatomic molecules. This group includes copper, gold, silver, and others. In referring to these elements, the symbol, meaning one atom, is used.

Table 2.3 The noble gases.

Symbol	Name
He	helium
Ne	neon
Ar	argon
Kr	krypton
Xe	xenon
Rn	radon

Table 2.4 Diatomic elements.

Formula	Name	Normal State
H_2	hydrogen	colorless gas
N_2	nitrogen	colorless gas
O_2	oxygen	colorless gas
F_2	fluorine	pale yellow gas
Cl_2	chlorine	greenish-yellow gas
Br_2	bromine	dark red liquid
I_2	iodine	violet-black solid

C. Combined in Compounds

The majority of elements, including those listed in Table 2.2, are found in compounds. **Compounds** are pure substances containing two or more different elements. The composition of a compound is expressed by a **formula** containing the symbols of the elements combined in the compound. Each symbol is followed by a subscript indicating the number of atoms of the element in the simplest unit of the compound. The subscript 1 is not shown.

Water is a compound. It has the formula H_2O, meaning that one molecule of water contains two hydrogen atoms and one oxygen atom. The compound diethyl ether has the formula $C_4H_{10}O$; a molecule of diethyl ether contains 4 atoms of carbon, 10 atoms of hydrogen, and 1 atom of oxygen.

Example 2.1

A molecule of hydrogen sulfide contains two atoms of hydrogen and one atom of sulfur. Write the formula for hydrogen sulfide.

Solution

1. Write the symbols of each element: H   S

2. Follow each symbol with a subscript indicating the number of atoms per molecule: H_2S

Example 2.2

The formula of glucose is $C_6H_{12}O_6$. What is the composition of a molecule of glucose?

Solution

The symbols tell us which elements are contained in glucose:

carbon   hydrogen   oxygen

The subscripts tell how many atoms of each element are in each molecule of glucose:

6 atoms carbon, 12 atoms hydrogen, 6 atoms oxygen

Problem 2.1

Ethanol is composed of two atoms of carbon, six atoms of hydrogen, and one atom of oxygen. Write the formula for ethanol.

Problem 2.2

The formula of vinyl chloride is C_2H_3Cl. What is the composition of a molecule of vinyl chloride?

The properties of a compound are quite unlike those of the elements from which it is formed. This is apparent if we compare the properties of carbon dioxide (a colorless gas used in fire extinguishers) with those of carbon (a black, combustible solid) and oxygen (a colorless gas necessary for burning). We will discuss the properties of compounds in Chapter 4.

2.6 The Composition of Atoms

A. Subatomic Particles

In Section 2.3 we said that all atoms are composed of smaller, subatomic particles. Evidence for the existence of such particles had begun to accumulate even as Dalton was developing his atomic theory of matter. By the end of the nineteenth century, two of these particles, the *electron* and the *proton*, had been identified, and there was considerable evidence for the existence of a third particle, the *neutron*. Positive proof of the neutron's existence came in the 1930s. Other subatomic particles have since been discovered, but we will confine our discussion to electrons, protons and neutrons, the subatomic particles most important in chemistry.

The **electron,** the first of the three to be discovered, is an extremely tiny particle with a mass of 9.110×10^{-28} gram and a negative charge. All atoms contain electrons. All atoms also contain **protons.** The mass of a proton is 1.6726×10^{-24} gram, about 1836 times that of an electron. The proton carries a positive charge equal in magnitude to the charge on the electron. These charges are referred to in relative terms; hence the charge on the electron is -1 and the charge on the proton is $+1$.

The third subatomic particle of interest to us is the **neutron.** Its mass of 1.6749×10^{-24} g is very close to that of the proton. A neutron carries no charge. With the exception of the lightest isotope of hydrogen, all atoms contain one or more neutrons. The properties of these three particles are summarized in Table 2.5.

Table 2.5 Properties of the three major subatomic particles.

Particle	Mass (g)	Relative Mass	Relative Charge
proton	1.6726×10^{-24}	1.007	$+1$
neutron	1.6749×10^{-24}	1.008	0
electron	9.110×10^{-28}	5.45×10^{-4}	-1

B. Atomic Numbers

Each element has an **atomic number.** The names, symbols, and atomic numbers of the known elements are listed on the inside back cover of this text. The atomic number of an element equals the number of protons in one atom of the element. Because atoms are electrically neutral, the atomic number of an element also equals the number of electrons in its neutral atoms.

Different elements have different atomic numbers; therefore, atoms of different elements contain different numbers of protons (and electrons). Oxygen has the atomic number 8; its atoms contain 8 protons and 8 electrons. Uranium has the atomic number 92; its atoms contain 92 protons and 92 electrons.

The relationship between atomic number and protons and electrons can be stated as follows:

$$\text{atomic number} = \text{number of protons per atom}$$
$$= \text{number of electrons per atom}$$

C. Mass Numbers

Each atom also has a **mass number.** The mass number of an atom is equal to the number of protons plus the number of neutrons that it contains. In other words, the number of neutrons in any atom is its mass number minus its atomic number:

$$\text{number of neutrons} = \text{mass number} - \text{atomic number}$$

We indicate the atomic number and the mass number of an atom of an element by writing the mass number as a superscript and the atomic number as a subscript, placing them both before the symbol of the element:

$$^{\text{mass number}}_{\text{atomic number}}\text{symbol of the element}$$

For example, an atom of gold (symbol Au), atomic number 79, has a mass number of 196:

$$^{196}_{79}\text{Au}$$

Example 2.3

What is the composition of a silver atom, $^{107}_{47}\text{Ag}$?

Solution

The atomic number of this atom is given by the subscript 47; it therefore contains 47 protons and 47 electrons. The mass number of this atom is given by the superscript 107; it therefore contains $(107 - 47)$, or 60, neutrons.

Problem 2.3

What is the composition of an atom of phosphorus, $^{31}_{15}\text{P}$?

D. Isotopes

Although all atoms of a given element must have the same atomic number, they need not all have the same mass number. For example, some atoms of carbon (atomic number 6) have a mass number of 12, others have a mass number of 13, and still others have a mass number of 14. We refer to these different kinds of atoms as **isotopes.** Isotopes are atoms that have the same atomic number (and are therefore of the same element) but different mass numbers. The compositions of atoms of the naturally occurring isotopes of carbon are shown in Table 2.6.

Table 2.6 The naturally occurring isotopes of carbon.

Isotope	Protons	Electrons	Neutrons
$^{12}_{6}\text{C}$	6	6	6
$^{13}_{6}\text{C}$	6	6	7
$^{14}_{6}\text{C}$	6	6	8

The various isotopes of an element can be designated by using superscripts and subscripts to show mass number and atomic number, as in Table 2.6. They can also be identified by following the name of the element with the mass number of the particular isotope. For example, instead of writing $^{12}_{6}\text{C}$, $^{13}_{6}\text{C}$, and $^{14}_{6}\text{C}$, we can write carbon-12, carbon-13, and carbon-14. We will generally designate isotopes in this second way.

About 350 isotopes occur naturally on Earth and another 1500 have been produced artificially. The isotopes of a given element are by no means equally abundant. For example, 98.89% of all carbon in nature is carbon-12, 1.11% is carbon-13, and only a trace is carbon-14. Some elements occur as only one isotope. Table 2.7 lists the naturally occurring isotopes of several common elements, along with their relative abundance.

Table 2.7 Relative abundance of naturally occurring isotopes of several elements.

Isotope	Abundance (%)	Isotope	Abundance (%)
hydrogen-1	99.985	silicon-28	92.21
hydrogen-2	0.015	silicon-29	4.70
hydrogen-3	trace	silicon-30	3.09
carbon-12	98.89	phosphorus-31	100
carbon-13	1.11	chlorine-35	75.53
carbon-14	trace	chlorine-37	24.47
nitrogen-14	99.63	iron-54	5.82
nitrogen-15	0.37	iron-56	91.66
oxygen-16	99.76	iron-57	2.19
oxygen-17	0.037	iron-58	0.33
oxygen-18	0.204	cobalt-59	100
aluminum-27	100		

Example 2.4

There are three isotopes of hydrogen:

> hydrogen-1 (protium)
> hydrogen-2 (deuterium)
> hydrogen-3 (tritium)

Give the symbol and atomic composition of each of these isotopes.

Solution

^1_1H; 1 proton, 0 neutrons, 1 electron
^2_1H; 1 proton, 1 neutron, 1 electron
^3_1H; 1 proton, 2 neutrons, 1 electron

Problem 2.4

Naturally occurring uranium is 99.3% uranium-238 and 0.7% uranium-235. Give the composition of an atom of each isotope.

2.7 Atomic Weights and the Mole

A. Atomic Weights

The **atomic weight** of an element is the average mass of the naturally occurring atoms of that element relative to the mass of carbon-12, an atom whose mass has been set at 12.00. From our discussion of isotopes, you can see why these atomic weights are averages. A collection of naturally occurring carbon atoms contains 98.89% carbon-12 atoms and 1.11% carbon-13 atoms, along with a trace percentage of carbon-14 atoms. The atomic weight of carbon (12.01) reflects the relative abundance of these three isotopes. The atomic weight of chlorine (35.45) reflects the fact that 75.53% of naturally occurring chlorine is chlorine-35 and 24.47% is chlorine-37.

The table of elements on the inside back cover lists the atomic weight for each element. Atomic weights range from 1.008 for hydrogen to 257 for lawrencium. For some very rare elements, the atomic weight is enclosed in parentheses. These elements have not been found in nature. The atomic weight listed for these elements is an estimate based on the extremely small amounts of these elements that scientists have been able to detect and study.

B. The Atomic Mass Unit

Atomic weights are given in **atomic mass units (amu).** An atomic mass unit is defined as 1/12 of the mass of an atom of carbon-12, which has a mass of 12.00 amu. In other words, the atomic weight of an element is always measured relative to·the mass of an atom of carbon-12.

C. The Mole

In doing chemical calculations, it is often necessary to know how many atoms or molecules a sample contains. For this purpose, we use a counting unit called **Avogadro's number:**

$$\text{Avogadro's number} = 6.02 \times 10^{23}$$

Just as the number 12 is described by the term *dozen*, Avogadro's number is described by the term **mole.** A mole of electrons is 6.02×10^{23} electrons; and a mole of Ping-Pong balls is 6.02×10^{23} Ping-Pong balls. This is a huge number—a mole of Ping-Pong balls would cover the surface of the Earth in a layer approximately 60 miles thick! Fortunately, atoms are not as large as Ping-Pong balls.

Thus, one mole of a substance contains 6.02×10^{23} units of that substance. Equally important is the fact that one mole of a pure substance has a weight in grams numerically equal to the formula weight of that substance. One mole of an element has a weight in grams equal to the atomic weight of the element and contains 6.02×10^{23} atoms of the element.

These definitions allow a new definition of atomic weight: The atomic weight of an element is the weight in grams of one mole of naturally occurring atoms of that element.

Example 2.5

The atomic weight of sulfur is 32.06. How many moles of sulfur are contained in 5.05 g of sulfur?

Solution

Wanted:
Moles of sulfur

Given:
5.05 g S

Conversion factor:
1 mole sulfur = 32.06 g sulfur

Equation:

$$? \text{ moles sulfur} = 5.05 \text{ g S} \times \frac{1 \text{ mole S}}{32.06 \text{ g S}}$$

Answer:
0.158 mole S

Problem 2.5

The atomic weight of carbon is 12.01. What is the mass of 1.62 moles of carbon?

Example 2.6

How many copper atoms are present in 5.43 g copper?

Solution

Wanted:
Atoms of copper

Given:
5.43 g copper

Conversion factors:
1 mole copper weighs 63.54 g
1 mole copper contains 6.02×10^{23} atoms

Equation:

$$? \text{ atoms copper} = 5.43 \text{ g copper} \times \frac{1 \text{ mole copper}}{63.54 \text{ g}} \times \frac{6.02 \times 10^{23} \text{ atoms}}{1 \text{ mole copper}}$$

Answer:
5.14×10^{22} atoms copper

Problem 2.6

A sample of uranium contains 5.15×10^{20} atoms. What is the mass of the sample?

These definitions give us some insight into the size of atoms. The mass of one mole of aluminum is 26.98 grams. Knowing that the density of aluminum is 2.70 g/cm³ (Table 1.8), you can calculate that the volume of one mole of aluminum is only 10.0 cm³. A sample of aluminum weighing 27.0 g has a volume of only 10.0 cm³, yet it contains 6.02×10^{23} atoms of aluminum.

Example 2.7

Calculate the volume of 2.51 moles of lead, density 11.34 g/cm³.

Solution

Wanted:
? cm³ of lead

Given:
2.51 moles of lead

Conversion factors:
11.34 g lead occupies 1 cm³
1 mole of lead has a mass of 207.2 g

Equation:

$$? \text{ cm}^3 \text{ lead} = 2.51 \text{ moles lead} \times \frac{207.2 \text{ g}}{1 \text{ mole lead}} \times \frac{1 \text{ cm}^3}{11.34 \text{ g}}$$

Answer:
45.9 cm³ lead

Problem 2.7

Calculate the volume of 5.34 moles of sodium, density 0.97 g/cm³.

2.8 The Inner Structure of the Atom

So far, we have discussed electrons, protons, and neutrons and ways to determine how many of each a particular atom contains. The question remains: Are these particles randomly distributed inside the atom like blueberries in a muffin, or does an atom have some organized inner structure? Scientists at the beginning of the twentieth century were trying to answer this question. Various theories had been proposed but none had been verified by experiment. In our discussion of the history of science (Section 2.1), we suggested that at various points in its development science has marked time until someone performs a key experiment that provides new insights. In the history of the study of atoms, a key experiment was performed in 1911 by Ernest Rutherford and his colleagues.

In Rutherford's experiment, a beam of small, positively charged particles, called *alpha particles* (discussed in Section 2.9B), was directed at a piece of gold foil only a few atoms thick, so thin as to be translucent (Figure 2.5A).

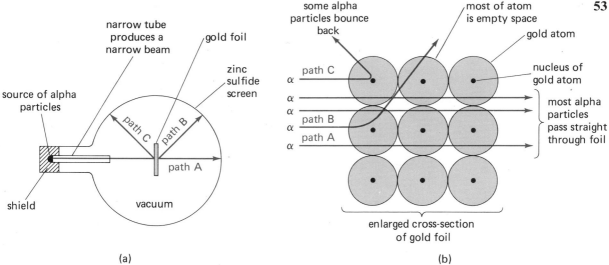

Figure 2.5 Rutherford's experiment. (a) Cross-section of the apparatus used. (b) Enlarged cross-section of the gold foil in the apparatus, showing the deflection of α particles by the nuclei of the gold atoms.

The foil was surrounded by a zinc sulfide screen which flashed every time it was struck by an alpha particle. From the location of these flashes, the effect of the gold foil on the path of the alpha particles could be determined.

The three paths shown in Figure 2.5 (paths A, B, and C) are representative of those observed. Most of the alpha particles followed path A; they passed directly through the foil as if it were not there. Some were deflected slightly from their original path, as in path B, and an even smaller number bounced back from the foil as if they had hit a solid wall (path C).

You may be surprised that any alpha particles passed through the gold foil. Rutherford was not—he had expected that almost all would pass straight through (path A). He had also expected that, due to the presence in the atom of positively charged protons, some alpha particles would follow a slightly deflected path (path B). The fact that some alpha particles bounced back (path C) astounded Rutherford. It suggested that the particles had smashed into a dense region of mass and bounced back. To use Rutherford's analogy, the possibility of such an event was as unlikely as a cannonball bouncing off a piece of tissue paper.

Careful consideration of these results convinced Rutherford (and the scientific community) that an atom contains a very small, dense **nucleus** and a large amount of extranuclear space. According to Rutherford's theory, the nucleus contains all the neutrons and protons. The protons give the nucleus a positive charge. Because like charges repel each other, positively charged alpha particles passing close to the nucleus are deflected (path B). The nucleus, containing all the protons and neutrons, has a greater density than an alpha particle. Therefore, an alpha particle striking the much denser nucleus of the gold atoms bounces back from the collision, as did those following path C. If the nucleus contains virtually all the mass of the atom, it must be extremely dense. Its diameter is about 10^{-12} cm, about 1/10,000 of that of the whole atom. Given this model, if the nucleus were the size of a marble, the atom with its extranuclear electrons would be 300 meters in diameter. If a marble had the same density as the nucleus of an atom, it would weigh 3.3×10^{10} kg!

Outside the nucleus, in the relatively enormous extranuclear space of the atom, are the tiny electrons. Because electrons are so small relative to the space they occupy, the extranuclear space of the atom is essentially empty. In Rutherford's experiment, alpha particles encountering this part of the atoms in the gold foil passed through the foil undeflected (path A).

The model of the atom based on Rutherford's work is, of course, no more than a model, for we cannot see these subatomic particles nor their arrangement within the atom. However, it does give us a way of thinking about the atom that coincides with observations made about its properties. We can now determine not only what subatomic particles a particular atom contains, but also where they are found in the atom. For example, an atom of carbon-12 ($^{12}_{6}C$) contains 6 protons and 6 neutrons in its nucleus and 6 electrons outside the nucleus.

Example 2.8

Iodine-131 is used in thyroid therapy. What is the composition of an atom of this iodine isotope?

Solution

The atomic number of iodine is 53 (from the table of the elements). Therefore, an atom of iodine contains 53 protons in the nucleus and 53 electrons outside the nucleus. The mass number of this isotope is 131. Hence, in addition to 53 protons, the nucleus of an atom of iodine-131 contains $(131 - 53)$, or 78, neutrons.

Problem 2.8

Cobalt-60 is used in cancer therapy. What is the composition of an atom of this isotope?

2.9 Radioactivity

A. General Characteristics

From the discussions in the previous section, we know that the atoms of any element have two distinct parts: the nucleus that contains the protons and neutrons and the extranuclear space that contains the electrons. The electrons in the atom, particularly those furthest from the nucleus, determine the chemical properties of the element. We will discuss electrons and the chemical properties of elements in detail in Chapter 3.

In the remainder of this chapter, we will describe properties of the nucleus and, in particular, the characteristics of **radioactivity** or **radioactive decay** of the nucleus. Nuclei of radioactive atoms decay spontaneously to form other nuclei, a process that always results in a loss of energy and often involves the release of one or more small particles. Some atoms are naturally radioactive. Others that are normally stable can be made radioactive by bombarding them with subatomic particles. Often, one isotope of an element is radioactive and others of the same element are stable.

Radioactivity is a common phenomenon. Of the 350 isotopes known to occur in nature, 67 are radioactive. Over a thousand radioactive isotopes have been produced in the laboratory. Every element, from atomic number 1 to 106, has at least one natural or artificially produced radioactive isotope. Of the three known isotopes of hydrogen, one is radioactive: hydrogen-3, more commonly known as tritium. Oxygen, the Earth's most abundant element, has 8 known isotopes, 5 of which are radioactive (oxygen-13, -14, -15, -19, and -20). Iodine, an element widely used in nuclear medicine, has 22 known isotopes ranging in mass from 117 to 139; of these, only iodine-127 is stable. This isotope is the only naturally occurring one. Uranium has 14 known isotopes, all of which are radioactive.

B. Radioactive Emissions

Nuclei undergoing radioactive decay release various kinds of emissions. We will discuss three of these: *alpha particles*, *beta particles*, and *gamma rays*. All three are forms of **ionizing radiation**, so called because their passage through matter leaves a trail of *ions* (Section 3.6B) and molecular debris.

1. Alpha particles. An **alpha particle** is a helium atom stripped of its two extranuclear electrons; that is, it contains 2 protons and 2 neutrons. Because there are no electrons to balance the positive charges of the two protons, an alpha particle has a charge of $+2$ and can be represented as He^{2+}. Another symbol for this particle is $_2^4He$. When ejected from a decaying nucleus, alpha particles interact with all matter in their path, whether it be photographic film, lead shielding, or body tissue. Alpha particles strip electrons from other atoms as they move through matter. In their wake, they leave a trail of positive ions (atoms from which electrons have been removed) and free electrons. A single alpha particle, ejected at high speed from a nucleus, can create up to 100,000 ions along its path before it takes on 2 electrons to become a neutral helium atom.

In air, an alpha particle travels about 4 cm before gaining two electrons to become a neutral helium atom. Within body tissue, its average path is only a few thousandths of a centimeter long. An alpha particle is unable to penetrate the outer layer of human skin. Because of this limited penetrating power, external exposure to alpha particles is not nearly so serious as internal exposure. If a source of alpha emissions is taken internally, the alpha radiation can do massive damage to the surrounding tissue. For this reason, alpha emitters are never used in nuclear medicine. Notice in Table 2.11 that none of the radioisotopes widely used in nuclear medicine are alpha emitters.

2. *Beta particles*. A **beta particle** is a high-speed electron ejected from a decaying nucleus; it carries a charge of -1. (Section 2.10A discusses how a nucleus can eject an electron even though it does not contain electrons.) It can be represented as $_{-1}^{0}e$ or $_{-1}^{0}\beta$. Like alpha particles, beta particles cause ionization by interacting with whatever matter is in their path. However, because they are far less massive than alpha particles and have only one-half the charge, beta particles produce less ionization and travel farther through matter before coming to rest in combination with a positive ion. The penetrating power of a beta particle is about 100 times that of an alpha particle. About 25 centimeters of wood, 1 centimeter of aluminum, or 0.5 centimeter of body tissue will stop a beta particle. The lower ionization levels of beta particles make them more suitable for use in radiation therapy, since the likelihood of damage to healthy tissue is greatly reduced. Beta emitters such as calcium-46, iron-59, cobalt-60, and iodine-131 are widely used in nuclear medicine.

3. *Gamma rays*. The release of either alpha or beta particles from a decaying nucleus is generally accompanied by the release of nuclear energy in the form of **gamma rays,** symbol $_{0}^{0}\gamma$. Gamma rays have no charge or mass and are equivalent to X-rays, except that they have higher energy. Even though they bear no charge, gamma rays are able to produce ionization as they pass through matter. The degree of penetration of gamma rays through matter is very much greater than that of either alpha or beta particles. Because of their penetrating power, gamma rays are especially easy to detect. Virtually all radioactive isotopes used in diagnostic nuclear medicine are gamma emitters. Each of the beta emitters listed in the previous paragraph is also a gamma emitter. Additional gamma emitters commonly used in nuclear medicine include chromium-51, arsenic-74, technetium-99, and gold-198.

The characteristics of alpha particles, beta particles, and gamma rays are summarized in Table 2.8.

Table 2.8 Characteristics of radioactive emissions.

Name	Symbol	Charge	Mass (amu)	Penetration Through Matter
alpha particle	$_{2}^{4}He$	$+2$	4	4.0 cm air 0.005 cm tissue no penetration in lead
beta particle	$_{-1}^{0}e$ or $_{-1}^{0}\beta$	-1	5.5×10^{-4}	6–300 cm air 0.006–0.5 cm tissue 0.0005–0.03 cm lead
gamma ray	$_{0}^{0}\gamma$	0	0	400 m air 50 cm tissue 3 cm lead

2.10 Characteristics of Nuclear Reactions

A. Equations for Nuclear Reactions

We have shown that radioactivity is the decay or disintegration of the nucleus of an atom. During the process, either alpha or beta particles may be emitted. Energy, in the form of gamma rays, may also be released by this process. All of these characteristics and more can be shown by using an equation to describe the radioactive process.

Because this is the first time that chemical equations have been used in this text, we will define the term and describe its parts. A **chemical equation** uses chemical formulas instead of the names of substances. In an equation, the reacting substance(s) come(s) first, followed by an arrow that means "reacts to form." The arrow is followed by the formula(s) of the product(s) of the reaction. An equation must be **balanced** with respect to mass, charge, and energy. First, the *total mass of the products must equal the total mass of the reactants.* Second, the *total charge of the reactants (the sum of their atomic numbers) must equal the total charge of the products (the sum of their atomic numbers).* Third, *any energy lost by the reactants must be accounted for in the products.*

Consider the equation for the decay of radium-226 to radon-222, with the simultaneous loss of an alpha particle and energy in the form of a gamma ray. Radium-226 is the reactant; radon, an alpha particle, and a gamma ray are the products. The equation is:

$$^{226}_{88}\text{Ra} \longrightarrow ^{222}_{86}\text{Rn} + ^{4}_{2}\text{He} + ^{0}_{0}\gamma$$

radium-226   radon-222   alpha   energy
                                       particle

The superscripts show the mass of the particles; the subscripts show the charges. The charge on each of these particles is its atomic number, the number of protons each contains. The equation is balanced with respect to mass because the sum of the masses of the reactants (226) equals the sum of the masses of the products (4 + 222 + 0). The equation is balanced with respect to charge because the sum of the atomic numbers of the reactants (88) equals the sum of the atomic numbers of the products (86 + 2 + 0). The energy change accompanying the reaction is accounted for by the release of gamma rays.

A similar equation can be written for nuclear decay by beta emission. Iodine-131 is a beta emitter commonly used in nuclear medicine. The equation for its decay is:

$$^{131}_{53}\text{I} \longrightarrow ^{131}_{54}\text{Xe} + ^{0}_{-1}\beta + ^{0}_{0}\gamma$$

Note that both charge and mass are balanced and that iodine-131 emits a gamma ray at the same time as it emits the beta particle. For this reason, iodine-131 is known as a beta-gamma emitter. Carbon-14, the isotope widely used in radiodating of archaeological artifacts containing carbon, is also a beta emitter:

$$^{14}_{6}C \longrightarrow {}^{14}_{7}N + {}_{-1}^{0}\beta$$

In the decay of carbon-14, only a beta particle is emitted. The energy released by the reaction is not of sufficient intensity to be called a gamma ray. Carbon-14 is therefore known as a pure beta emitter. Phosphorus-32 is also a pure beta-emitter.

How can nuclei give off beta particles, which are actually electrons, if there are no electrons in the nucleus? The process is not yet understood, but it may occur through the disintegration of a neutron to form a proton and the emitted electron:

$$neutron \longrightarrow proton + electron$$

The electron is ejected and the proton remains in the nucleus. In beta emission, the atomic number of the product nucleus is one greater than that of the reactant nucleus because the nucleus now contains one more proton. The mass of the product nucleus is approximately the same as that of the reactant nucleus because an electron's mass is negligible with respect to that of a proton.

Emission of a gamma ray changes neither the mass nor the charge of the nucleus. It accompanies the rearrangement of a nucleus from a less stable, more energetic nuclear configuration to a more stable, less energetic form. The identity and mass of the nucleus stay the same. The changes caused by the emission of the three types of radiation are summarized in Table 2.9.

Given the atomic number and mass number of a radioactive isotope and the type of radiation emitted during its decay, it is an easy matter to predict the mass, atomic number, and identity of the new element formed.

Radiation Emitted	Change in	
	Atomic Number	Mass Number
alpha particle	-2	-4
beta particle	$+1$	0
gamma ray	0	0

Table 2.9 Changes in atomic number and atomic mass resulting from the emission of an α particle, β particle, or γ ray.

Example 2.9

Cobalt-60 decays by emission of a beta particle. Predict the atomic number and mass number of the new isotope formed. Which element has this atomic number?

Solution

First, write an equation showing the radioactive decay.

$$^{60}_{27}\text{Co} \longrightarrow ^{\text{mass number}}_{\text{atomic number}}\text{element} + ^{0}_{-1}\beta$$

Second, determine the mass number and atomic number of the new isotope. Since mass number is unchanged in beta emission, the isotope formed must also have a mass number of 60. Because the atomic number increases by one in beta emission, the isotope formed must have an atomic number of 28. Therefore, we can write:

$$^{60}_{27}\text{Co} \longrightarrow ^{60}_{28}\text{element} + ^{0}_{-1}\beta$$

Third, consult the table of the elements on the inside back cover and find the element whose atomic number is 28: nickel. Thus, the complete equation for the radioactive decay of cobalt-60 is:

$$^{60}_{27}\text{Co} \longrightarrow ^{60}_{28}\text{Ni} + ^{0}_{-1}\beta$$

Problem 2.9

Radium-226 decomposes with the loss of an alpha particle. Write the equation for this reaction and identify the second product.

B. Half-Life

The rate of decay of a radioactive isotope (also called a *radioisotope*) is measured in terms of its **half-life.** Half-life is defined as the length of time required for one-half of the sample to decay. Half-lives vary from fractions of a second for some isotopes to billions of years for others. Table 2.10 lists half-lives and modes of decay for several isotopes.

Iodine-131 has a half-life of 8.1 days. If you start today with a sample containing 25 mg of iodine-131, after 8.1 days that sample will contain only

Table 2.10 Half-lives of some radioisotopes

Isotope	Emissions	Half-Life
hydrogen-3	$^{0}_{-1}\beta$	12.3 years
carbon-14	$^{0}_{-1}\beta$	5730 years
calcium-47	$^{0}_{-1}\beta, ^{0}_{0}\gamma$	4.5 days
cobalt-60	$^{0}_{-1}\beta, ^{0}_{0}\gamma$	5.26 years
gold-198	$^{0}_{-1}\beta, ^{0}_{0}\gamma$	2.7 days
iodine-131	$^{0}_{-1}\beta, ^{0}_{0}\gamma$	8.1 days
iron-59	$^{0}_{-1}\beta, ^{0}_{0}\gamma$	45.1 days
molybdenum-99	$^{0}_{-1}\beta, ^{0}_{0}\gamma$	67.0 hours
phosphorus-32	$^{0}_{-1}\beta$	14.3 days
sodium-24	$^{0}_{-1}\beta, ^{0}_{0}\gamma$	15.0 hours

Figure 2.6 Rate of
decay of iodine-131 as
a function of time.

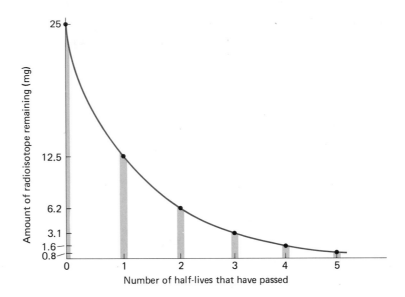

12.5 mg of iodine-131. At the end of 16.2 days, the sample will contain only 6.2 mg of iodine-131. Of course, the matter in the sample does not gradually disappear; it simply changes to another element, the product of the radioactive decay of iodine-131. Figure 2.6 shows the amounts of iodine-131 remaining after the passage of several half-lives, given an initial sample containing 25 mg of the isotope.

Knowing the identity of the radioisotope, its half-life, and the type of radiation it emits, you can calculate the identity of the product and the amount formed in a given period of time.

Example 2.10

Phosphorus-32 is a beta emitter with a half-life of 14.3 days.

a. Write the equation for the radioactive disintegration of phosphorus-32.

$$^{32}_{15}P \longrightarrow \, _{-1}^{0}\beta + \, ?$$

b. A package containing 5 mg of phosphorus-32 was shipped at 5:00 A.M. on February 1, 1980, but did not arrive at its destination until 7:30 P.M. on February 29, 1980. How much phosphorus-32 did the package contain on arrival?

Solution

a. In beta emission, the mass number remains the same and the atomic number increases by 1. The element with atomic number 16 is sulfur. Thus, the equation is:

$$^{32}_{15}P \longrightarrow \, _{-1}^{0}\beta + \, ^{32}_{16}S$$

b. How many half-lives have elapsed?

$$28 \text{ days} + 14.5 \text{ hours} = 28.6 \text{ days}$$

$$28.6 \text{ days} \times \frac{1 \text{ half-life}}{14.3 \text{ days}} = 2 \text{ half-lives}$$

5 mg of phosphorus-32 would decrease by half during one half-life (to 2.50 mg) and by another half in the second half-life. Therefore, only one-fourth (1.25 mg) of the original sample would be phosphorus-32; the rest would be sulfur-32.

Problem 2.10

Strontium-90, a major product of the disintegration of uranium, decays by beta emission and has a half-life of 28 years. The equation for its decay is:

$$^{90}_{38}\text{Sr} \longrightarrow \ ^{90}_{39}\text{Y} + \ ^{0}_{-1}\beta$$

Once absorbed into the bones, strontium-90 emits beta rays, damaging the bone marrow. The maximum permissible dose of strontium-90 for an adult is 6.9×10^{-9} g. If an adult received this dose at age 20, and it was all incorporated into his bone tissue, what mass of strontium-90 would be found in his body after death at the age of 104?

2.11 Applications of Radioactivity in the Health and Biological Sciences

The use of radioisotopes is widespread in chemistry, biology, medicine, and many other areas of science and industry. All of these uses balance the good and bad aspects of the characteristics of radioisotopes and nuclear decomposition listed below:

1. The chemical properties and reactions of a radioisotope are exactly the same as those of a nonradioactive isotope of the same element.
2. Radiation can be detected some distance from its source.
3. Each radioisotope has a characteristic half-life.
4. Radioactive emissions interfere with normal cell growth.

Several uses of radioisotopes in the health and biological sciences illustrate ways in which they can provide information that would be difficult or impossible to obtain by any other means.

A. The Use of Tracers

Because the chemical properties and reactions of a radioisotope are exactly the same as those of a nonradioactive isotope of the same element, a radioisotope can be substituted for a stable isotope of the same element in a molecule or compound, without changing the chemical properties of the compound. Such a compound is said to be **tagged** or **labeled.** Since the radiation emitted by the radioisotope can be detected some distance from the radiating atom, the progress of those atoms through the body can be followed or traced. Hence, the term for such labeled compounds is **tracer.**

Among the first scientists to use labeled molecules in the biological sciences were Melvin Calvin and his colleagues at the University of California at Berkeley. In the mid-1940s, they were studying photosynthesis, a biological process that has intrigued scientists for centuries, but about which very little was known at the time. (We will discuss photosynthesis in Section 5.8A.) Carbon-14, a radioisotope discovered in 1940, had become available in quantity in 1945 as a result of nuclear research conducted under the Atomic Energy Commission. In that same year, Calvin began his study of the incorporation of carbon dioxide into glucose. He allowed the alga *Chlorella* to photosynthesize in a stream of carbon dioxide containing carbon-14. After exposure to the labeled $^{14}CO_2$, the algae were killed and analyzed to determine which molecules and compounds contained carbon-14. Calvin discovered that after an exposure of only ten seconds, at least six different compounds contained carbon-14. He determined what these molecules were and how *Chlorella* used carbon dioxide to build them. For this pioneering work in the chemistry of photosynthesis, Calvin was awarded the Nobel prize in Chemistry in 1961.

Radioactive tracers are used in medicine for diagnosis. For example, both chromium-51, in the form of sodium chromate (Na_2CrO_4), and iron-59, in the form of iron citrate ($FeC_6H_6O_7$), are routinely used in the determination of red blood-cell volume, the rates of red blood-cell production and destruction, and iron metabolism. Sodium chromate is used because it is soluble in blood, easily penetrates the red cell membrane, and then becomes firmly attached to hemoglobin molecules within the red blood cells. Iron may also be used because it is a natural component of hemoglobin molecules. To estimate the survival time of red blood cells, a blood sample is taken from a patient, the red blood cells are tagged with chromium-51, and the sample is reinjected. Samples of blood are withdrawn after 24 hours and then every 3–4 days in order to determine the time it takes the blood radioactivity to fall to one-half of its initial value. This time, corrected for the amount of decay of the chromium-51, represents the half-life of red blood cells. These studies show that in a normal adult in good health, the half-life of a red blood cell averages 29 days. A knowledge of red blood cell survival time can be very valuable to the physician in diagnosing and treating cases of unexplained anemia.

Iodine-131 is the most commonly used radioisotope for the study of iodine metabolism in humans. The thyroid gland has a remarkable ability to ex-

Figure 2.7 The uptake of iodine by the thyroid gland can be measured by tracing atoms of the radioisotope iodine-131. The plot at the right of the figure shows the location in the thyroid of the source of each radioactive emission detected by the counter. Note that the radioactivity counts show more iodine in one lobe of the thyroid gland than in the other.

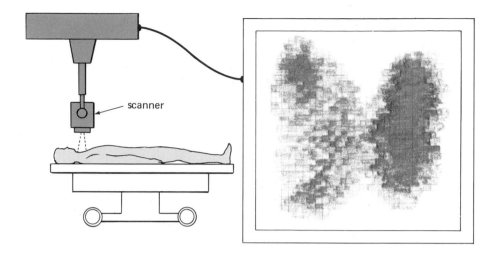

scanner

tract iodine from the blood stream and use it to produce the thyroid hormones, thyroxine (T_4), and triiodothyronine (T_3). These two hormones have a direct effect on the body's metabolism. Very small amounts of iodine-131, in the form of sodium iodide, can be injected into the blood stream and within minutes, begin to concentrate in the thyroid gland. By monitoring the accumulation of radioactivity, it is possible to estimate the size and shape of the thyroid gland and to determine whether any part of it is functioning abnormally (Figure 2.7).

Several other radioisotopes commonly used in nuclear medicine are listed in Table 2.11. Notice that all have comparatively short half-lives and

Table 2.11 Radioisotopes used in nuclear medicine for clinical diagnosis.

Radioisotope	Radiation Emitted	Half-Life	Used to Study
chromium-51	$_0^0\gamma$	27.8 days	red blood cells
cobalt-57	$_{-1}^0\beta$, $_0^0\gamma$	270 days	absorption, storage and metabolism of vitamin B-12
iodine-123	$_0^0\gamma$	13.3 hours	thyroid activity and iodine metabolism
iodine-131	$_{-1}^0\beta$, $_0^0\gamma$	8.1 days	
iron-59	$_{-1}^0\beta$, $_0^0\gamma$	45.1 days	red blood cells
molybdenum-99	$_{-1}^0\beta$, $_0^0\gamma$	67 hours	general metabolism
phosphorus-32	$_{-1}^0\beta$, $_0^0\gamma$	14.3 days	liver function
strontium-87	$_0^0\gamma$	2.8 hours	bone metabolism
technetium-99	$_0^0\gamma$	6.0 hours	bones, liver, lungs, and thyroid
xenon-133	$_{-1}^0\beta$, $_0^0\gamma$	5.27 days	lungs

that all are pure gamma or beta-gamma emitters, a requirement for radio-isotopes to be used in medicine (see Section 2.9).

B. Biological Effects of Radiation

While the exact manner in which radiation causes damage to tissues and cells is not fully understood, it is clear that cellular damage can occur any time alpha, beta, or other ionizing radiation (Section 2.9B) passes through the cell. The effects of ionizing radiation range from minor damage to specific molecules to the death of the cell. Ionizing radiation is especially damaging to the cell nucleus, particularly nuclei undergoing rapid division and nuclei of younger, less mature cells. Many types of cancer cells are especially sensitive to gamma radiation because they are growing rapidly and are therefore less mature than cells of surrounding noncancerous tissue. This sensitivity is the reason behind the use of radiation to destroy cancer cells. The apparatus used for such treatments is designed so that the radiation can be sharply focused on the cancer cells, thus minimizing the damage to nearby healthy cells (Figure 2.8).

While no cell is completely resistant to irradiation, the susceptibility of particular cells depends on their type. White blood cells (leukocytes) and red blood cells (erythrocytes), along with the tissues that produce blood cells (bone marrow, spleen, and lymph nodes), are especially sensitive to irradiation. White blood cells have a life span of only 2 days, so damage to leukocyte-producing tissue shows up very quickly. White blood cells are the first to disappear from the circulation after irradiation, and the decrease in concentration of these cells in the blood stream can be used as an indication of the extent of radiation exposure. Red blood cells have a life span of about 120 days, so damage to tissues producing red cells does not show up as quickly. The ovaries in the

Figure 2.8 Cancer treatment with cobalt-60. The source can move along a circular track, thus rotating the radioactive beam around the patient so that only the tumor receives continuous radiation.

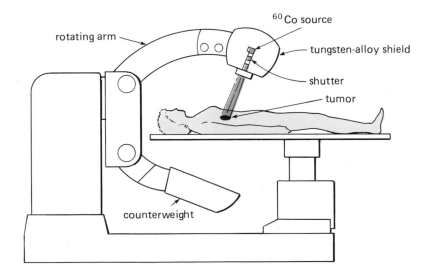

female and the testes in the male are also highly sensitive to irradiation. In these tissues, radiation produces chromosome mutations and aberrations from which there is no recovery. In addition, the skin, gastrointestinal tract, and eyes are particularly sensitive to irradiation. Other organs, such as the heart, liver, kidney, pancreas, and central nervous system, are much more resistant to the effects of ionizing radiation.

One quantitative measure of the danger inherent in exposure to a particular form of radiation is its **LET,** or **linear energy transfer,** which measures the amount of energy transferred by that radiation to the matter through which it passes. The LET depends on the type of matter affected (the **target**) and how deeply the radiation penetrates. Alpha particles have a high LET; gamma rays and beta particles have lower LET's. Because low LET radiation is more diffuse, tissues are better able to repair damage from low LET radiation than from high LET radiation. In conclusion, the discussions in this section have shown that it is the combination of the nature of the radiation, its intensity, its target, and its depth of penetration that determines its potential for tissue damage.

2.12 The Use of Nuclear Reactions to Produce Energy

A. Fission

In Section 2.11 we listed four characteristics of radioactivity and nuclear decay that form the basis for the use of radioisotopes in the health and biological sciences. A fifth characteristic of nuclear reactions is that they release enormous amounts of energy. The first nuclear reactor to achieve controlled nuclear disintegration was built in the early 1940s by Enrico Fermi and colleagues at the University of Chicago. Since that time, a great deal of effort and expense has gone into developing nuclear reactors as sources of energy. The nuclear reactions presently used or under study by the nuclear-power industry fall into two categories: *fission* reactions and *fusion* reactions. In a **fission reaction,** a large nucleus is split into two medium-sized nuclei. Only a few nuclei are known to undergo fission. Nuclear power plants currently in use depend primarily on the fission of uranium-235 and plutonium-239.

When a nucleus of U-235 undergoes fission, it splits into two smaller atoms and, at the same time, releases neutrons (designated as $_0^1n$) and energy. Some of these neutrons are absorbed by other atoms of U-235. In turn, these atoms split apart, releasing more energy and more neutrons. A typical reaction is:

$$_{92}^{235}U + _0^1n \longrightarrow [_{92}^{236}U] \longrightarrow _{56}^{139}Ba + _{36}^{94}Kr + 3\,_0^1n + energy$$

The brackets around $_{92}^{236}U$ indicate that it has a highly unstable nucleus. Under proper conditions, the fission of a few nuclei of uranium-235 sets in motion a

Figure 2.9 Diagram of a nuclear fission chain reaction.

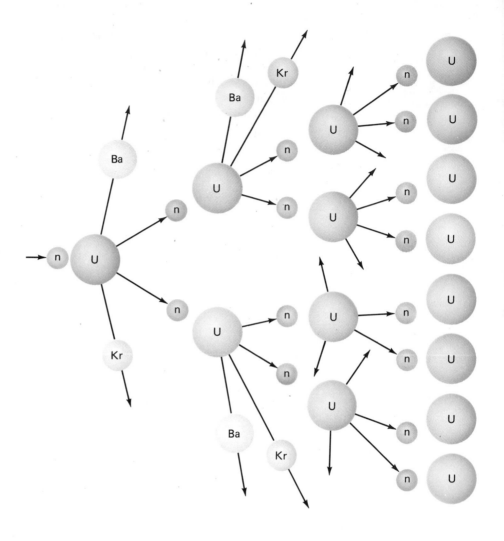

chain reaction (Figure 2.9) which can proceed with explosive violence if not controlled. In fact, this reaction is the source of energy in the atomic bomb.

In nuclear power plants, the energy released by the controlled fission of U-235 is collected in the reactor and used to produce steam in a heat exchanger. The steam then drives a turbine to produce electricity. Energy generation can be regulated by inserting control rods between the fuel rods in the reactor to absorb excess neutrons, thereby controlling the rate of the chain reaction. Figure 2.10 shows a schematic diagram of a typical nuclear power plant.

The first nuclear power plant in this country to produce electricity for commercial use began operation in 1957 in Shippingport, Pennsylvania. A typical nuclear power plant in operation today uses about 2 kilograms of uranium-235 to generate 1000 megawatts of electricity. About 5600 tons of coal are required to produce the same amount of electricity in a conventional power plant.

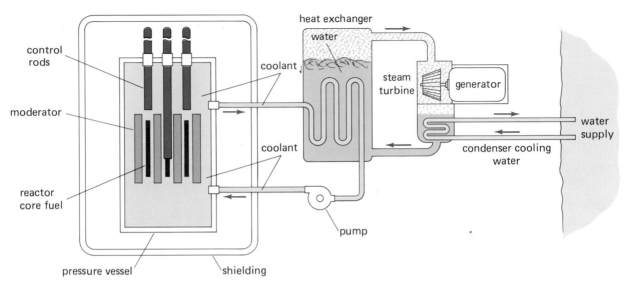

Figure 2.10 Diagram of a nuclear power plant.

Uranium-235 (natural abundance 0.71%) is very scarce and difficult to separate from uranium-238 (natural abundance 99.28%). The much more abundant U-238 does not undergo fission and therefore cannot be used as a fuel for nuclear reactors. However, if U-238 is bombarded with neutrons (from U-235, for example), it absorbs a neutron and is transformed into U-239. This isotope undergoes beta emission to generate neptunium-239, which, in turn, undergoes another beta emission to produce plutonium-239:

$$^{238}_{92}\text{U} + ^{1}_{0}\text{n} \longrightarrow ^{239}_{92}\text{U}$$

$$^{239}_{92}\text{U} \longrightarrow ^{239}_{93}\text{Np} + ^{0}_{-1}\beta$$

$$^{239}_{93}\text{Np} \longrightarrow ^{239}_{94}\text{Pu} + ^{0}_{-1}\beta$$

Plutonium-239 also undergoes fission, with the production of more energy and more neutrons. These neutrons can then be used to breed more plutonium-239 from uranium-238. Thus, a so-called **breeder reactor** can produce its own supply of fissionable material. Several breeder reactors are now functioning in Europe.

Nuclear reactors using fissionable materials pose several serious dangers to the environment. First, there is always the chance that leaks, accidents, or acts of sabotage will release radioactive materials from the reactor into the environment. This problem has become a continuing concern since the Three Mile Island nuclear plant accident in March, 1979. Second, many of the products of nuclear fission are themselves radioactive. The radioactivity from spent nuclear fuel and from the products of nuclear fission will remain lethal for thousands of years, and the safe disposal of these materials is a problem that has not yet been solved. Third, obsolete generating plants will also present a problem to future generations, for they will contain much radioactive material. One suggestion has been to encase such plants in concrete for 100 or more years. This may become necessary, but it is hardly a simple or permanent solution.

B. Fusion

Nuclear fusion, the other process currently under study for the generation of atomic energy, depends on the putting together or fusing of two nuclei to form a single nucleus. One of the most promising fusion reactions generates energy by the fusion of two deuterium atoms to form an atom of helium-3:

$$\mathrm{^{2}_{1}H + {}^{2}_{1}H \longrightarrow {}^{3}_{2}He + {}^{1}_{0}n + energy}$$

Reactions such as this require enormously high energy to force the two positively charged nuclei close enough together to fuse. Once the nuclei fuse, however, much more energy is released than is required for the reaction. Nuclear fusion occurs in the core of the sun, where the temperature is approximately 40 million degrees Celsius. Unfortunately, man has not yet found a way to produce and control nuclear fusion on Earth. Controlled nuclear fusion produces almost no radioactive wastes and would therefore be a nonpolluting source of energy. Hence, there is a massive effort going on in this country and abroad to find ways to harness this energy source.

Key Terms and Concepts

alpha particle (2.9B1)

atom (2.3)

atomic mass unit (2.7B)

atomic number (2.6B)

atomic theory (2.3)

atomic weight (2.7A)

Avogadro's number (2.7C)

beta particle (2.9B2)

compounds (2.5C)

diatomic (2.5B)

electron (2.6A)

element (2.3)

equations (2.10A)

formula of a compound (2.5C)

gamma rays (2.9B3)

half-life (2.10B)

ionizing radiation (2.9B)

isotope (2.6D)

Law of Conservation of Energy (2.2B)

Law of Conservation of Mass (2.2A)

Law of Conservation of Mass/Energy (2.2C)

LET (2.11B)

mass number (2.6C)

mole (2.7C)

molecule (2.5B)

neutron (2.6A)

noble gases (2.5A)

nuclear fission (2.12A)

nuclear fusion (2.12B)

nucleus, atomic (2.8)

polyatomic (2.5B)

proton (2.6A)

radioactivity (2.9A)

Rutherford (2.8)

scientific method (2.1)

symbols of the elements (2.4A)

tracers (2.11A)

Problems

Scientific Method (Section 2.1)

2.11 What are the steps of the scientific method? Do scientists always follow these steps? Explain.

Names and Symbols of the Elements (Section 2.4A)

2.12 Give the symbol for each of the following elements. Try to do this without looking in your text.
a. phosphorus   b. oxygen   c. cobalt
d. calcium   e. chlorine   f. bromine
g. potassium   h. copper   i. magnesium
j. iron

2.13 Give the names of the elements that have the following symbols. Try to do this without looking them up.
a. Na   b. S   c. I   d. Ce   e. Al
f. N   g. Cl   h. Mn   i. Zn   j. F

2.14 Make a list of the names and symbols of those elements found in Table 2.1 but not in Table 2.2. Do the same for those in Table 2.2 that are not in Table 2.1.

Distribution of the Elements (Section 2.4B)

2.15 Have all the known elements been found in nature? Is it possible that new elements will be discovered in the future?

Composition of Atoms (Section 2.6)

2.16 Make a table showing the relative mass and charge of the following particles:
a. proton   b. alpha particle
c. neutron   d. electron

2.17 Complete the following table:

| | | | Number of | | |
Element	Atomic Number	Mass Number	Electrons	Protons	Neutrons
___	11	23	___	___	___
sulfur	___	34	___	___	___
barium	___	___	56	___	81
___	20	40	___	___	___
___	___	___	___	8	8

2.18 Describe the difference in composition between:

antimony-121   and   antimony-123
gallium-69   and   gallium-71

2.19 Write symbols showing the mass number and the atomic number for the isotopes in Problem 2.18.

2.20 Why is the atomic weight of an element rarely a whole number?

2.21 Give the atomic composition of the atoms having the following properties. Name the isotope.

	a.	b.	c.
Atomic number	17	19	28
Mass number	37	39	60

2.22 Describe the apparatus, the process studied, and the results of Rutherford's gold foil experiment.

Moles (Section 2.7C)

2.23 Complete the following table:

Mass of Sample	Moles of Sample	Atoms in Sample
____	0.20 mole sodium	____
5.0 g barium	____	
____	____	1.0×10^{23} atoms calcium
____	0.42 mole potassium	____
9.3 g lithium	____	____

2.24 Calculate the mass of 0.15 mole of the following elements:
   a. carbon   b. helium   c. zinc   d. lead
   e. phosphorus   f. calcium   g. boron   h. iron

2.25 Calculate the number of atoms in 2.5 g of each of the elements in Problem 2.24.

2.26 A normal, healthy body contains 34 parts per million by weight of zinc. How many zinc atoms would be found in the body of a healthy person weighing 52.3 kg?

2.27 According to the Food and Drug Administration, fish sold for human consumption can contain no more than 5 parts per million of mercury by weight. How many atoms of mercury would there be in one pound of fish that contained this amount of mercury?

2.28 At a busy street intersection, the lead concentration in the air may be 9 micrograms per cubic meter. If 40% of the lead passing through the lungs is absorbed and a typical adult breathes 20 m³ of air per day, how many moles of lead are absorbed during an eight-hour working day by a newspaper vendor at this corner?

2.29 It is estimated that 1200 tons of uranium will be needed to operate the nuclear plants that might be in operation in this country in the year 2000. How many moles of uranium is this?

2.30 Regulations permit no more than 2 micrograms of vanadium per cubic meter of air occupied by humans. How many atoms of vanadium would be contained in a room 3.0 m × 2.5 m × 4.0 m that contained the maximum allowable amount of vanadium?

Radioactivity (Section 2.9A and 2.10)

2.31 Complete the following equations.

$$^{103}_{46}\text{Pd} + \,^{0}_{-1}\beta \longrightarrow \underline{\hspace{1cm}}$$

$$^{210}_{84}\text{Po} \longrightarrow \,^{4}_{2}\text{He} + \underline{\hspace{1cm}}$$

$$^{9}_{4}\text{Be} + \,^{4}_{2}\text{He} \longrightarrow \,^{1}_{0}\text{n} + \underline{\hspace{1cm}}$$

2.32 What is the half-life of an isotope if 6.00 g of the isotope decays to 0.75 g in 27 days?

2.33 Lead-210 has a half-life of 22 years. If a 1.0-g sample of lead-210 is buried in 1988, how much of it will remain as lead in 2076? Lead-210 decays with the loss of an alpha particle. Write the equation for this decay.

2.34 What happens to the atomic number and mass number of an atom if it loses: (a) an alpha particle? (b) a gamma ray? (c) a beta particle? In which cases will the atom change its chemical nature?

2.35 Why is radioactivity measured in half-lives rather than in the time required for total decay?

Atoms
and the Properties
of Electrons

An atom is not a tiny sphere of uniform density. As we saw in the previous chapter, it consists of a very small, dense nucleus and a relatively large extranuclear space containing negatively charged electrons. By studying the light given off by heated atoms, scientists have learned about the behavior of electrons in the extranuclear space. In this chapter, we will study the behavior and configuration of electrons in the atom and learn how this configuration determines the properties of an element.

3.1 Radiant Energy

Light is one form of energy. It moves at a constant speed through space in **waves,** much as waves move through a body of water. Energy, like light, that travels in waves is called **radiant energy.** Not all waves have the same **wave-**

Figure 3.1 Different waves have different wavelengths.

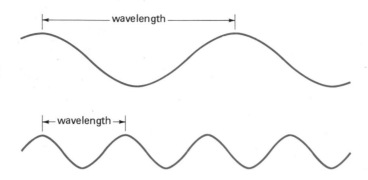

length (Figure 3.1). Around the edges of major oceans, long "combers" roll onto the shore, but on a small lake or pond, short, choppy waves are more common. Different kinds of light also have different wavelengths. Red light, for example, has a longer wavelength than blue light. Some radiant energy is invisible. Infrared light, microwaves, television, and radio waves are forms of radiant energy invisible to the eye, with wavelengths longer than that of visible light. Ultraviolet light, cosmic rays, X-rays, and gamma rays, all of which have wavelengths shorter than that of visible light (see Figure 3.2), are also invisible forms of radiant energy. The entire range of radiant energy is called the **electromagnetic spectrum.**

When an object is heated, it radiates energy, often in the form of visible light. Our sun is probably the most familiar example of a heated body giving off light. The "white" light from the sun is really a collection of light of all wavelengths and is called **continuous light.** When continuous light passes through a prism, the various wavelengths are separated. Sunlight passed through a prism separates into the colors of the rainbow. If, however, a sample of an element is heated until it gives off light, and this light is passed through

Figure 3.2 The electromagnetic spectrum.

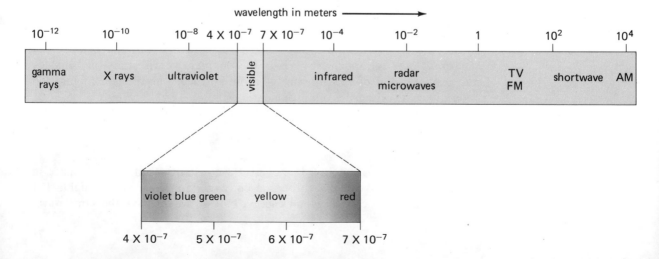

Figure 3.3 Atomic spectra of hydrogen, neon, and sodium in the visible range. The numbers below the figure are wavelengths (in meters).

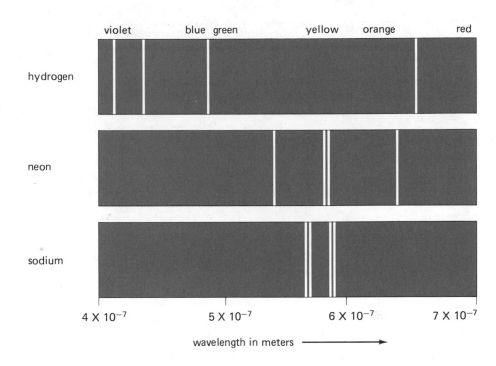

a prism, instead of a rainbow we see a series of brightly colored lines. The light given off by the heated element is not continuous, but of only a few wavelengths. The pattern of wavelengths (or lines) is unique for each element and is called its **atomic spectrum** (plural, *spectra*). The pattern is so characteristic that it can be used to show the presence of that element in even the tiniest amounts. Figure 3.3 shows the atomic spectra of hydrogen, sodium, and neon. Note that the figure includes only the visible part of these spectra. The spectra of these elements, as well as those of all the other elements, show another pattern of lines in the invisible parts of the electromagnetic spectrum. The light from sodium-vapor street lamps is that of the yellow-orange lines of the sodium spectrum. Red neon signs use the red lines of the neon spectrum to produce their glow.

3.2 The Energy of the Electron

Light is a form of energy, and different colors (wavelengths) of light have different amounts of energy. When an atom absorbs light, its electrons gain energy. When an atom gives off light, its electrons lose energy. The energy (or wavelength) of the light given off measures the amount of energy lost.

As we saw, the atomic spectra of the elements are not continuous (Figure 3.3). The lines of these and other spectra show that electrons lose or gain only certain amounts of energy. We can express this fact by saying that the energy changes of an electron are **quantized.** What does it mean to say that energy is quantized?

If you are pouring a soft drink from a can, you can pour out as much or as little as you like. However, if you are buying a soft drink from a machine, you can only buy a certain amount. You cannot buy half a can of soda or a third of a can; you can only buy a whole can or several cans. Soft drinks dispensed by a machine are only available in multiples of a set volume, or **quantum.** Thus, the dispensing of soft drinks by machine has been quantized.

Energy can also be quantized. If you are climbing a ladder, you can only stop on the rungs; you cannot stop between them. The energy needed to climb the ladder is used in finite amounts that lift your body from one rung to the next. To move upward, you must use enough energy to move your feet to the next higher rung. If the available energy is only enough to move partway up to the next rung, you cannot move higher because you cannot stop between rungs. In going up the ladder, your expenditure of energy is quantized. If you are going up a hill instead of a ladder, your energy expenditure is not quantized. You can go straight up the hill or you can zigzag back and forth, going up gradually. You can take big steps or little steps; there are no limitations on where you can stop or on how much energy you use.

Let us apply the analogy of the ladder and its rungs to an atom and its electrons. In climbing the ladder, there are only certain places (rungs) where you can put your feet. In an atom, there are only certain distances from the nucleus, called **energy levels,** where electrons may be found. Unlike a ladder, which has a limited length, the energy levels of an atom extend infinitely far from the nucleus. The energy levels are not evenly spaced. Moving outward from the nucleus, the levels get closer together and contain more energy (Figure 3.4). The energy of an electron in one of the levels at a considerable distance from the nucleus is greater than that of an electron in a closer-in level.

For an electron to move from one energy level to the next higher level, it must gain exactly the right amount of energy. If less than that amount is available, the electron must stay where it is. Electrons always move from level

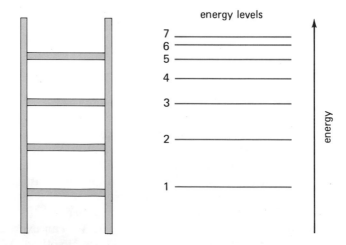

Figure 3.4 The energy levels in an atom are similar to the rungs of a ladder.

to level; they cannot stop in between. This means that there are regions of space within an atom where an electron can be, and there are also regions where an electron cannot be.

3.3 An Atomic Model

Our present model of the atom is based on the concept of energy levels for electrons within an atom and on the mathematical interpretation of detailed atomic spectra. The requirements for our model are:

1. Each electron in a particular atom has a unique energy that depends on the negatively charged electron's relationship to the positively charged nucleus and to the other negatively charged electrons in the atom.
2. The energy of an electron in an atom can increase or decrease, but only by specific amounts.

A. Energy Levels

We can picture an atom as a small nucleus surrounded by a much larger volume of extranuclear space containing the electrons. The electrons must move rapidly; they are so small that only by being in constant, high-speed motion can they fill the comparatively large extranuclear space of the atom. This space is divided into regions called **principal energy levels,** numbered 1, 2, 3, 4, . . . , ∞ outward from the nucleus. These energy levels are also called **shells** and are sometimes designated by letters. The first energy level is sometimes called the K shell, the second level the L shell, and so on, through the remainder of the alphabet. The lower the number of the principal energy level, the closer the negatively charged electron in it is to the positively charged nucleus, and the more difficult it is to remove this electron from the atom.

Each principal energy level can contain up to $2n^2$ electrons, where n is the number of the level. Thus, the first level can contain up to 2 electrons, the second up to 8 electrons, the third up to 18, and so on. Only seven energy levels are needed to contain all of the electrons in the atoms of those elements now known.

B. Orbitals

When an electron is in a particular energy level, it is more likely to be found in some parts of that level than in others. These parts are called **orbitals.** Each orbital can contain a maximum of two electrons. Each principal energy level has one orbital, called an s orbital, which can contain one or two electrons. Electrons in this orbital are called s electrons and have the lowest energy of any electrons in that principal energy level. The first principal energy level contains only an s orbital; therefore, it can hold a maximum of two electrons.

Each principal energy level above the first contains one *s* orbital and three *p* orbitals. A set of three *p* orbitals can hold a maximum of 6 electrons. Therefore, the second level can contain a maximum of 8 electrons, that is, 2 in the *s* orbital and 6 in the three *p* orbitals.

Each principal energy level above the second contains, in addition to an *s* orbital and three *p* orbitals, five *d* orbitals. The five *d* orbitals can hold up to 10 electrons. Thus, the third level can hold up to 18 electrons: 2 in the *s* orbital, 6 in the three *p* orbitals, and 10 in the five *d* orbitals.

The fourth and higher levels also contain seven *f* orbitals, which can hold a maximum of 14 electrons. Thus, the fourth level can hold up to 32 electrons: 2 in the *s* orbital, 6 in the three *p* orbitals, 10 in the five *d* orbitals, and 14 in the seven *f* orbitals. The orbitals of the first four principal energy levels and the maximum number of electrons that each type of orbital can contain are summarized in Table 3.1.

To distinguish which of the *s*, *p*, *d*, and *f* orbitals we are talking about, we use the number of the principal energy level the particular orbital is in. For example, the *s* orbital of the second principal energy level is designated 2*s*; the *s* orbital of the third principal energy level is designated 3*s*; and so on. The number of electrons in a particular type of orbital is shown by a superscript after the symbol of the orbital. The notation

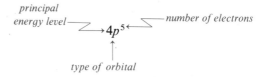

means that there are 5 electrons in the *p* orbitals of the fourth energy level.

Table 3.1 Orbitals of the first four energy levels.

| Level | Orbitals Present | | Total Possible Occupying Electrons |
	Type	Number	
1	*s*	1	2
2	*s*	1	2 ⎫ 8
	p	3	6 ⎭
3	*s*	1	2 ⎫
	p	3	6 ⎬ 18
	d	5	10 ⎭
4	*s*	1	2 ⎫
	p	3	6 ⎪ 32
	d	5	10 ⎬
	f	7	14 ⎭

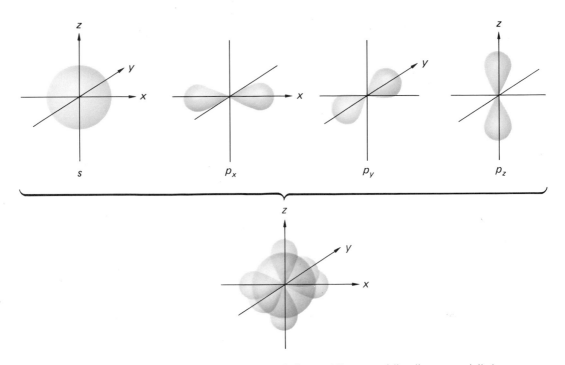

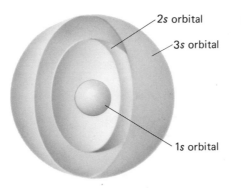

Figure 3.5 Perspective representations of the *s* and the three *p* orbitals.

1. Orbital shapes and sizes. The shapes of the *s* and *p* orbitals are shown in Figure 3.5. In these diagrams, the nucleus is at the origin of the axes. The *s* orbitals are spherically symmetrical about the nucleus. The 2*s* orbital is a larger sphere than the 1*s* orbital, the 3*s* orbital is larger than the 2*s* orbital, and so on (see Figure 3.6). The three *p* orbitals (p_x, p_y, and p_z) are more or less dumbbell-shaped with the center of the dumbbell at the nucleus. They are oriented at right angles to each other along the *x*, *y*, and *z* axes. Like the *s* orbitals, the *p* orbitals increase in size as the number of the principal energy level they occupy increases. A 4*p* orbital is larger than a 3*p* orbital.

Figure 3.6 Cross-sectional view representing the relative sizes of the 1*s*, 2*s*, and 3*s* orbitals.

2. Orbital energies. Within a given principal energy level, *p* orbitals are always at a higher energy than *s* orbitals, *d* orbitals are always at a higher energy than *p* orbitals, and *f* orbitals are always at a higher energy than *d* orbitals. For example, within the fourth principal energy level we have:

lowest energy ⎤
⎣ *highest energy*

$$4s < 4p < 4d < 4f$$

In addition, the energy associated with an orbital increases as the number of the principal energy level of the orbital increases. For instance, the energy of a 3*p* orbital is always higher than that of a 2*p* orbital, and the energy of a 4*d* orbital is always higher than that of a 3*d* orbital. The same is true of *s* orbitals:

lowest energy ⎤
⎣ *highest energy*

$$1s < 2s < 3s < 4s < 5s$$

There is some overlap among the energies of certain orbitals in different principal energy levels. For example, the energy of a 3*d* orbital is slightly higher than that of a 4*s* orbital, and that of a 4*d* orbital is a little higher than that of a 5*s* orbital. Figure 3.7 shows the relationships among the energies of different orbitals. Note especially the overlap of orbitals in the higher principal energy levels.

The two electrons in a particular orbital differ very slightly in energy. This is believed to be due to a property called **electron spin.** The theory of electron spin states that the two electrons in a single orbital spin in opposite directions on their axes, causing an energy difference between them. Like many models, this is an oversimplification, but for the purpose of this course it is a useful description.

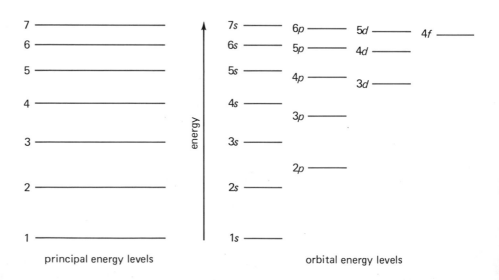

Figure 3.7 Principal energy levels and orbital energy levels.

principal energy levels   orbital energy levels

C. Our Model and the Differences among Atoms

Our model of the atom says that electrons are distributed among the energy levels and orbitals of the atom according to certain rules, and that each electron has a unique energy determined by the position of its orbital within the atom. When an atom absorbs the right amount of energy, an electron moves from its original orbital to a higher-energy orbital in which there is a vacancy. Similarly, when an atom gives off energy, the electron drops back to a lower energy orbital that has a vacancy. For example, an electron in a $3s$ orbital can drop to the $2p$ orbital, the $2s$ orbital, or the $1s$ orbital. The energy given up by an electron in dropping to a lower energy orbital is released in the form of light and determines the lines in the spectrum of that element.

When all the electrons of an atom are in the lowest possible energy states (meaning that the energy levels are filled in order of increasing energy), the atom and its electrons are in the **ground state.** If one of these electrons moves to a higher energy level, the atom is in an **excited state.**

We know that each element has a unique spectrum. These spectra show that the energy differences among the electrons in an atom vary from one element to another. What causes this variation?

Recall that the nucleus of an atom is positively charged, that electrons carry a negative charge, and that oppositely charged bodies attract one another. The atoms of one element differ from those of another element in the number of protons in the nucleus and, consequently, in the charge on the nucleus. The attraction for an electron, and therefore its energy, will differ from one element to the next according to differences in nuclear charge. In addition, the atoms of one element contain a different number of electrons than do atoms of any other element. The energy of each electron within the atom depends not only on its interaction with the positively charged nucleus, but also on its interaction with the other electrons in the atom. Therefore, the energies of the electrons of one element will differ from those of the electrons of another element. Considering these two variables, nuclear charge and the number of electrons, it is apparent that each element must have a unique spectrum derived from its unique set of allowed electron energy levels.

3.4 Electron Configurations of Atoms

The **electron configuration** of an atom lists the number of electrons in each kind of orbital in each energy level of the ground state atom. To obtain the electron configuration of a particular atom, start at the nucleus and add electrons one by one until the number added equals the number of protons in the nucleus. Each added electron is assigned to the available orbital of lowest energy. The first orbital filled will be the $1s$ orbital, then the $2s$ orbital, the $2p$ orbital, the $3s$, $3p$, $4s$, $3d$, and so on. This order is difficult to remember and often hard to determine from energy level diagrams such as Figure 3.7. A more convenient

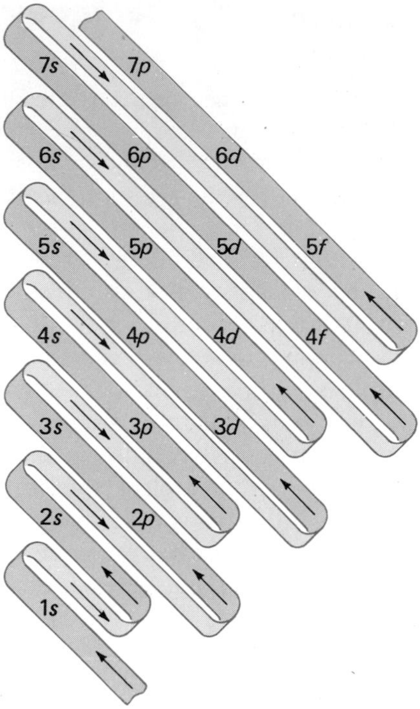

Figure 3.8 Order of filling of orbitals.

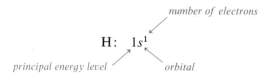

Order of filling: 1s, 2s, 2p, 3s, 3p, 4s, 3d,
                   4p, 5s, 4d, 5p, 6s, 4f, 5d,
                   6p, 7s, 5f, 6d, 7p . . .

way to remember the order is to use Figure 3.8. Here the energy levels are listed in rows, starting with the 1s level. To use this figure, read along the diagonal lines in the direction of the arrows. The order is summarized under the diagram.

An atom of hydrogen (atomic number 1) has one proton and one electron. The single electron is assigned to the 1s orbital, the lowest energy orbital in the lowest energy level. Therefore, the electron configuration of hydrogen is written:

number of electrons

H: $1s^1$

principal energy level   *orbital*

For helium (atomic number 2), which has two electrons, the electron configuration is:

He: $1s^2$

Two electrons completely fill the first energy level. Because the helium nucleus is different from the hydrogen nucleus, neither of the helium electrons will have exactly the same energy as the single hydrogen electron.

The element lithium (atomic number 3) has three electrons. In order to write its electron configuration, we must first determine from Figure 3.8 that the $2s$ orbital is next higher in energy after the $1s$ orbital. Therefore, the electron configuration of lithium is:

$$\text{Li:} \quad 1s^2 2s^1$$

Boron (atomic number 5) has five electrons. Four electrons fill both the $1s$ and $2s$ orbitals. The fifth electron is added to a $2p$ orbital, the orbital next higher in energy (Figure 3.8). The electron configuration of boron is:

$$\text{B:} \quad 1s^2 2s^2 2p^1$$

Table 3.2 shows the electron configurations of the elements with atomic numbers $1-18$.

The electron configuration of elements with higher atomic numbers can be written by following the orbital-filling chart in Figure 3.8.

Table 3.2 Electron configurations of the first 18 elements.

Element	Atomic Number	Electron Configuration
hydrogen	1	$1s^1$
helium	2	$1s^2$
lithium	3	$1s^2 2s^1$
beryllium	4	$1s^2 2s^2$
boron	5	$1s^2 2s^2 2p^1$
carbon	6	$1s^2 2s^2 2p^2$
nitrogen	7	$1s^2 2s^2 2p^3$
oxygen	8	$1s^2 2s^2 2p^4$
fluorine	9	$1s^2 2s^2 2p^5$
neon	10	$1s^2 2s^2 2p^6$
sodium	11	$1s^2 2s^2 2p^6 3s^1$
magnesium	12	$1s^2 2s^2 2p^6 3s^2$
aluminum	13	$1s^2 2s^2 2p^6 3s^2 3p^1$
silicon	14	$1s^2 2s^2 2p^6 3s^2 3p^2$
phosphorus	15	$1s^2 2s^2 2p^6 3s^2 3p^3$
sulfur	16	$1s^2 2s^2 2p^6 3s^2 3p^4$
chlorine	17	$1s^2 2s^2 2p^6 3s^2 3p^5$
argon	18	$1s^2 2s^2 2p^6 3s^2 3p^6$

Example 3.1

Write the electron configuration of vanadium.

Solution

The atomic number of vanadium is 23; we must therefore account for 23 electrons. The first orbitals to fill are $1s$, $2s$, $2p$, $3s$, $3p$, and $4s$, which can hold 20 electrons. The remaining three electrons must be in the $3d$ orbital, giving the configuration:

$$V: \quad 1s^2 2s^2 2p^6 3s^2 3p^6 4s^2 3d^3$$

Problem 3.1

Write the electron configuration of arsenic (atomic number 33).

3.5 The Periodic Table

You have seen alphabetical lists of the elements, as for example the one on the inside back cover of this text. Now that you understand more about the structure and electron configuration of atoms, the elements can be grouped together in an entirely different and more meaningful way called the **periodic table** (Figure 3.9).

Figure 3.9 Periodic table of the elements.

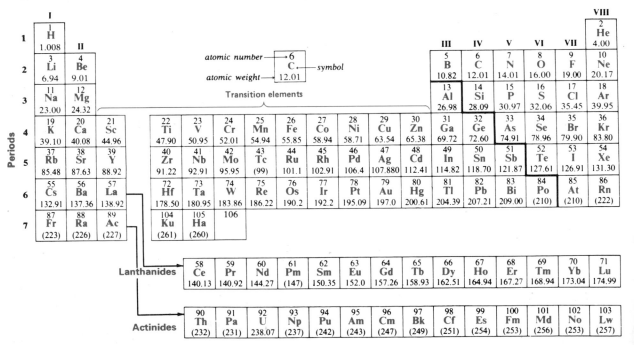

In the periodic table, the elements are arranged in horizontal rows, in order of increasing atomic number. Each row is called a **period.** The periods are numbered 1 through 7 and differ in the number of elements they contain. Period 1 contains 2 elements, periods 2 and 3 contain 8 elements apiece, periods 4 and 5 contain 18 elements each, and there is room for 32 elements in both periods 6 and 7, although period 7 only contains 20 elements at present. Looking at Figure 3.9, you will notice that in order to fit the periodic table onto a normal book page, certain elements have been placed below the table: those elements of period 6 with atomic numbers 58 through 71 and those elements of period 7 with atomic numbers 90 through 103.

A. Electron Configuration and the Periodic Table

What does the arrangement of the elements within the periodic table have to do with their electron configuration? Compare the number of elements in each period of the table with the number of electrons added to an atom in building its electron configuration. The first energy level is filled when it contains two electrons. Similarly, the first period of the table contains only two elements, hydrogen (atomic number 1) and helium (atomic number 2). The second energy level is filled when it contains eight electrons, and the second period contains eight elements, moving from lithium (atomic number 3) to neon (atomic number 10). The third energy level can add only eight electrons before electrons enter the fourth energy level. Likewise, the third period contains eight elements: sodium (atomic number 11) through argon (atomic number 18).

The fourth energy level, like the third, can add only eight electrons before electrons are added to the fifth energy level. However, ten $3d$ electrons are added between the two $4s$ and the six $4p$ electrons. This gives eighteen electrons in all, and there are eighteen elements in the fourth period of the table. Eight of these are in the same columns used by elements in the first, second, and third periods; the remaining ten elements start new columns. The addition of electrons to the fifth energy level follows the same pattern as the fourth level: eighteen electrons are added to the fifth level before electrons enter the sixth level. The fifth period has eighteen elements, placed below those in the fourth period.

Two $6s$ electrons are added in the sixth energy level, followed by fourteen $4f$ electrons and ten $5d$ electrons and then the six $6p$ electrons. This is an addition of 32 electrons, and there are 32 elements in the sixth period, 18 in the main body of the table and 14 below it. The fourteen elements outside the main body of the table (cerium through lutetium) are known as the **lanthanides** or **rare earths.**

The seventh period is incomplete, although it, too, could contain 32 elements. Elements francium and radium correspond to the addition of the two $7s$ electrons. The fourteen $5f$ electrons account for the formation of elements 90 (thorium) through 103 (lawrencium), collectively known as the

actinides. All of them are radioactive and only the first three, thorium, protactinium, and uranium, have been found in nature. The others are the products of experiments of nuclear scientists. Of the elements that would represent the filling of the $6d$ orbitals, only two, kurchatovium (atomic number 104) and hahnium (atomic number 105), have been accepted by the International Union of Pure and Applied Chemistry. Element 106 would continue this group. For a discussion of theoretical elements of atomic number greater than 106, see the mini-essay on the periodic table.

Before closing this discussion of the relationship between the periodic table and electron configuration, we should take a closer look at the electron configuration of the **noble gases,** the elements in column VIII. The electron configurations of these elements are:

Element	Atomic Number	Electron Configuration
He	2	$1s^2$
Ne	10	$1s^2\|2s^22p^6$
Ar	18	$1s^22s^22p^6\|3s^23p^6$
Kr	36	$1s^22s^22p^63s^23p^6\|3d^{10}4s^24p^6$
Xe	54	$1s^22s^22p^63s^23p^63d^{10}4s^24p^6\|4d^{10}\quad 5s^25p^6$
Rn	86	$1s^22s^22p^63s^23p^63d^{10}4s^24p^64d^{10}\|4f^{14}\|5s^25p^6\|5d^{10}6s^26p^6$

The brackets in the display illustrate the fact that the core of each noble gas atom has the same configuration as the noble gas immediately above it in the periodic table. Note that the listing of electrons in these configurations does not follow the filling order shown in Figure 3.8; the electrons have been rearranged by level. The two arrangements are equally correct.

A comparison of these electron configurations leads to two observations:

1. All of the noble gases have the same outer-shell electron configuration: filled s and p orbitals.

2. None have any partially filled orbitals.

3. Each of these elements ends a row of the periodic table.

B. Core Notation

The electron configurations of the noble gases are often used in describing the electron configurations of other elements. In doing so, the symbol of the noble gas, enclosed in brackets, represents its electron configuration. Thus, [He] stands for a $1s^2$ configuration and [Ne] symbolizes a $1s^22s^22p^6$ configuration. Similarly, [Kr] represents a configuration identical to that of its 36 electrons.

Let us use this method to show the electron configuration of iodine. The complete electron configuration of iodine is:

$$I: \quad 1s^2 2s^2 2p^6 3s^2 3p^6 4s^2 3d^{10} 4p^6 5s^2 4d^{10} 5p^5$$

Using the symbol of krypton (atomic number 36) to represent the first thirty-six electrons, we can write the electron configuration of iodine as:

$$I: \quad [Kr] 5s^2 4d^{10} 5p^5$$

This way of writing electron configuration is called **core notation.** Core notation emphasizes the outer-shell electrons, which are involved in chemical activity. In writing electron configurations using this method, remember that even though the inner configuration of an element may be written the same as that of a noble gas, the energies of these inner electrons are slightly different.

Example 3.2

Using core notation, write the electron configuration of lead, atomic number 82.

Solution

Lead has 82 electrons. The nearest noble gas of fewer electrons is xenon, atomic number 54. Let $[Xe]$ represent the first 54 electrons of lead. Xenon is at the end of period 5 of the table, therefore its last electrons must be $5p$ electrons. The configuration of the next 28 electrons can be determined by picking up the arrow in Figure 3.9, starting after $5p$. Filling the orbitals in proper order, we obtain:

$$Pb: \quad [Xe] 6s^2 4f^{14} 5d^{10} 6p^2$$

Problem 3.2

Using core notation, write the electron configuration of:

a. strontium **b.** bromine

C. Categories of Elements within the Periodic Table

Within the periodic table, the elements can be grouped into three large categories: *representative* elements, *transition* elements, and *inner transition* elements. The **representative elements** are those found in the columns that start in the first and second periods of the table: I, II, III, IV, V, VI, VII, and VIII. For these elements, the last added electron is a s or p electron, and the number of the column equals the sum of the number of s and p electrons in the element's outermost shell. Helium, with 2 electrons, is the only exception and is in

Table 3.3 Electron configurations of elements in columns I and VI using core notation.

I		VI	
H	$1s^1$	O	$[He]2s^22p^4$
Li	$[He]2s^1$	S	$[Ne]3s^23p^4$
Na	$[Ne]3s^1$	Se	$[Ar]4s^23d^{10}4p^4$
K	$[Ar]4s^1$	Te	$[Kr]5s^24d^{10}5p^4$
Rb	$[Kr]5s^1$	Po	$[Xe]6s^24f^{14}5d^{10}6p^4$
Cs	$[Xe]6s^1$		
Fr	$[Rn]7s^1$		

column VIII. Table 3.3 uses core notation to show the electron configuration of the elements in columns I and VI of the periodic table.

The **transition elements** are those in the part of the table between columns II and III. For these elements, the last added electron is usually a d electron, and their structure is characterized by partially filled d orbitals. There are one or two s electrons in the outermost shells of the transition elements. Many familiar metals are in this block of elements, including the coinage metals: gold, silver, copper, and platinum. Iron, the principal ingredient of steel, is here, as well as those elements (chromium, nickel, manganese, and others) that are added to iron to make particular kinds of steel. Many of the chemical properties of these elements are due to the fact that their two outermost electron shells are only partially filled.

The lanthanides and the actinides, the elements in the two rows below the main body of the table, are known collectively as the **inner transition elements,** because their last added electron, an f electron, has been added to a shell two levels in from the outside of the atom.

Example 3.3

Carbon is in column IV, period 2 of the periodic table. What is its electron configuration?

Solution

Because carbon is in period 2, the carbon atoms must have a filled first energy level with outer-shell electrons in the second energy level. Because carbon is in column IV, it must have 4 outer-shell electrons. Therefore, its electron configuration is:

$$1s^22s^22p^2$$

Problem 3.3

From their positions in the periodic table, predict the electron configurations of:

a. gallium

b. sulfur

D. Valence Electrons

In discussing the chemical properties of an element, we often focus on electrons in the outermost energy levels. These outer-shell electrons are called **valence electrons,** and the energy level they are in is called the **valence shell.** Valence electrons participate in chemical bonding and chemical reactions. The valence electrons of an element are shown by using a representation of the element called an **electron-dot structure** or **Lewis structure,** named after G. N. Lewis, the twentieth-century American chemist who first pointed out the importance of outer-shell electrons. A Lewis structure shows the symbol of the element surrounded by a number of dots equal to the number of electrons in the outer energy level of the element.

Looking back at Table 3.3, we see that the core notation for sodium is $[Ne]3s^1$. This tells us that a sodium atom has one electron in its outer shell, so its Lewis structure is Na$\cdot$. The core notation for selenium is $[Ar]3d^{10}4s^24p^4$. Its Lewis structure is $\cdot\ddot{Se}\!:$. The $3d$ electrons of selenium are not shown because they are not in the outer shell. Lewis structures for the elements in the first three periods and group II of the periodic table are shown in Table 3.4.

Table 3.4 Lewis structures for the elements of the first three periods and group II.

Period	I	II	III	IV	V	VI	VII	VIII
1	H$\cdot$							He$:$
2	Li$\cdot$	Be$:$	$\overset{\cdot}{B}:$	$\cdot\overset{\cdot}{C}:$	$\cdot\overset{\cdot}{N}:$	$\cdot\overset{\cdot\cdot}{O}:$	$:\overset{\cdot\cdot}{F}:$	$:\overset{\cdot\cdot}{Ne}:$
3	Na$\cdot$	Mg$:$	$\overset{\cdot}{Al}:$	$\cdot\overset{\cdot}{Si}:$	$\cdot\overset{\cdot}{P}:$	$\cdot\overset{\cdot\cdot}{S}:$	$:\overset{\cdot\cdot}{Cl}:$	$:\overset{\cdot\cdot}{Ar}:$
4		Ca$:$						
5		Sr$:$						
6		Ba$:$						
7		Ra$:$						

Example 3.4

Give the Lewis structure of **(a)** gallium and **(b)** bromine.

Solution

a. Gallium is in column III. Its atoms have 3 electrons in the highest energy level. The Lewis structure is $\overset{\cdot}{Ga}:$.

b. Bromine is in column VII. Its atoms have 7 electrons in the highest energy level. Its Lewis structure is $:\overset{\cdot\cdot}{Br}:$.

Problem 3.4

Give the Lewis structure of **(a)** potassium and of **(b)** arsenic.

E. Development of the Periodic Table: Families of Elements

The periodic table did not originally have the form that we have shown in Figure 3.9. In 1868, when the Russian chemist Dmitri Mendeleev (1834–1907) first organized the periodic table, atomic numbers and electron configurations were unknown. For the elements then recognized, only approximate atomic weights and some physical and chemical properties were known. In constructing his table, Mendeleev arranged the known elements roughly in order of increasing atomic weight, placing elements of similar physical and chemical properties in the same column. Similarity of properties took precedence over exact order of increasing atomic weight. Mendeleev called each of these columns of like elements a **family.**

Mendeleev also assumed, quite rightly, that there were elements yet to be discovered and left spaces in his table for them. Arsenic was known; in physical and chemical properties, arsenic resembles phosphorus and was placed under it in column V. This left empty spaces under aluminum and silicon. The subsequent discoveries of gallium in 1875 and germanium in 1886 were due, in part, to Mendeleev's strikingly accurate prediction of their properties, based on the known properties of aluminum and silicon, respectively.

In calling his chart a *periodic* table, Mendeleev was referring to the fact that a periodic property is one that recurs at regular intervals. Spring is periodic; it recurs regularly at the end of winter. Many properties of the elements are also periodic, recurring at regular intervals of atomic number. We have already seen that outer-shell electron configurations are periodic, recurring after certain increments in atomic number. The periodic table was so named because its original construction related physical or chemical properties periodically to atomic weight.

The periodic table shown in Figure 3.9 is the modern equivalent of Mendeleev's table. Note that the elements 27 (cobalt) and 28 (nickel) and 52 (tellurium) and 53 (iodine) illustrate the point that atomic weights do not always increase regularly through the table. For a more complete study of the history of the periodic table, see the mini-essay "The Past and Future of the Periodic Table."

F. Groups of Elements in the Table

1. Metals and nonmetals. Figure 3.9 has a heavy stepwise line crossing the table from B (boron) to At (astatine). This line separates metallic elements (those below and to the left of the line) from nonmetals (those above and to the right of the line). Elements bordering on this line have some metallic and some nonmetal properties; they are called **semimetals** (sometimes **metalloids**).

Characteristically, a metal is lustrous and a good conductor of heat and electricity. It can be hammered into a thin sheet or drawn into a fine wire. Except for mercury, all metals are solid at room temperature.

The physical properties of the nonmetals are more varied than those of metals. Some nonmetals (nitrogen and oxygen, for example), are gases, bromine is a liquid, and other nonmetals are solids. Some solid nonmetals are soft (sulfur) and some hard (carbon, in the form of diamond). Nonmetals are usually poor conductors of either heat or electric currents. The solid nonmetals are frequently dull in appearance and quite brittle.

*2. **Groups of elements.*** As we have seen, the elements in a single column of the periodic table are often much alike in their physical and chemical characteristics, and can be thought of as a **group,** or family, of elements. Table 3.5 summarizes the characteristics of the different groups of the periodic table.

The elements of group I are known as the **alkali metals:** lithium, sodium, potassium, rubidium, cesium, and francium. They are soft, silvery-gray solids which are all very active chemically. These elements do not occur free in nature, but are found combined with other elements, often with chlorine. The most common of these compounds is sodium chloride (table salt). Sodium chloride is the substance that makes seawater salty; it is also found in huge underground deposits (salt mines), from which it is easily mined.

Table 3.5 Electrons and the periodic table.

Group	Last Electron Added	Type of Element or Name of Group	Valence Electrons	Typical Lewis Structure
I	s electron	alkali metal	1	Na·
II	s electron	alkaline earth metal	2	Ca:
		s orbital has been filled		
short columns	d electron	transition metals		
		d orbitals have been filled		
III	p electron	metal	3	Ȧl:
IV	p electron	nonmetal (C) to metal (Pb)	4	·Ċ:
V	p electron	nonmetal (N) to metal (Bi)	5	·N̈:
VI	p electron	nonmetal	6	·S̈:
VII	p electron	halogen	7	:C̈l:
VIII	p electron	noble gas	8 or 0	:Är:
		p orbitals have been filled		

Table 3.6 Properties of halogens.

	Symbol	Atomic Number	Lewis Structure	Atomic Weight	Radius × 10⁻⁸ cm	Melting Point, °C	Boiling Point, °C	Color and State
fluorine	F	9	:F̈:	19.0	0.72	−219.6	−188	pale yellow gas
chlorine	Cl	17	:C̈l:	35.5	0.99	−101.0	−34.7	greenish-yellow gas
bromine	Br	35	:B̈r:	79.9	1.14	−7.2	58.0	dark red liquid
iodine	I	53	:Ï:	126.9	1.33	113.7	183	shiny black solid
astatine	At	85	:Ät:	(210)*	1.45	302	—	—

* The longest-lived isotope of this element, astatine-210, has a half-life of only 8.3 hours. Certain radioisotopes of astatine do exist in nature, together with uranium-233. However, the total amount of astatine in the Earth's crust is probably less than one gram.

The elements of group II are the **alkaline earth metals,** which are harder and stronger than the alkali metals. They, too, are found in combination with other elements. While all alkaline earth metals have similar chemical reactivity, they have a wider range of reactivity than do the alkali metals. Beryllium and magnesium are unaffected by water, calcium reacts slowly with boiling water, and barium reacts violently with even cold water.

The elements of group VII are known as the **halogens.** Chemically, these elements are very similar; physically, they are less alike. The lightest halogen, fluorine, is a pale yellow gas; iodine, the heaviest, is a shiny, black solid. Several characteristic properties of halogens are shown in Table 3.6. Notice how these properties change as the atomic number increases.

The elements of group VIII are the noble gases. These elements are all monatomic gases and occur free and uncombined. They are all singularly unreactive, so much so that they were known earlier as the *inert gases.* Only in 1960 were any of them shown to take part in chemical reactions. Even now, only krypton and xenon are known to form chemical compounds with other elements.

The elements in other columns of the table do not show striking similarities in property. These columns are crossed by the line that separates metals from nonmetals so they have nonmetals at the top and metals at the bottom. For example, group IV is headed by the nonmetal carbon and has lead, a typical metal, as its heaviest member. Group V has the nonmetal nitrogen at the top, and the metal bismuth at the bottom.

3.6 Some Periodic Properties

Many properties of the elements change in fairly regular ways as location in the table shifts downward or to the right.

A. Atomic Radius

Figure 3.10 shows the relative sizes of the atoms of the representative elements. Notice that the size of the atoms increases from top to bottom in a column and decreases from left to right across a row.

This trend is related to electron configuration. Going down column I, for example, the single valence electron is in a successively higher principal energy level, farther and farther away from the positively charged nucleus; hence, the atomic radius increases going from top to bottom of the column. This same regular increase in size can be observed in each column of the periodic table.

Going across a period from left to right, the atoms decrease in size. For elements within a period, electrons are being added one by one to the same principal energy level. At the same time, protons are also being added one by

Figure 3.10 Relative atomic sizes of the representative elements.

$(2 \times 10^{-10} \text{m})$

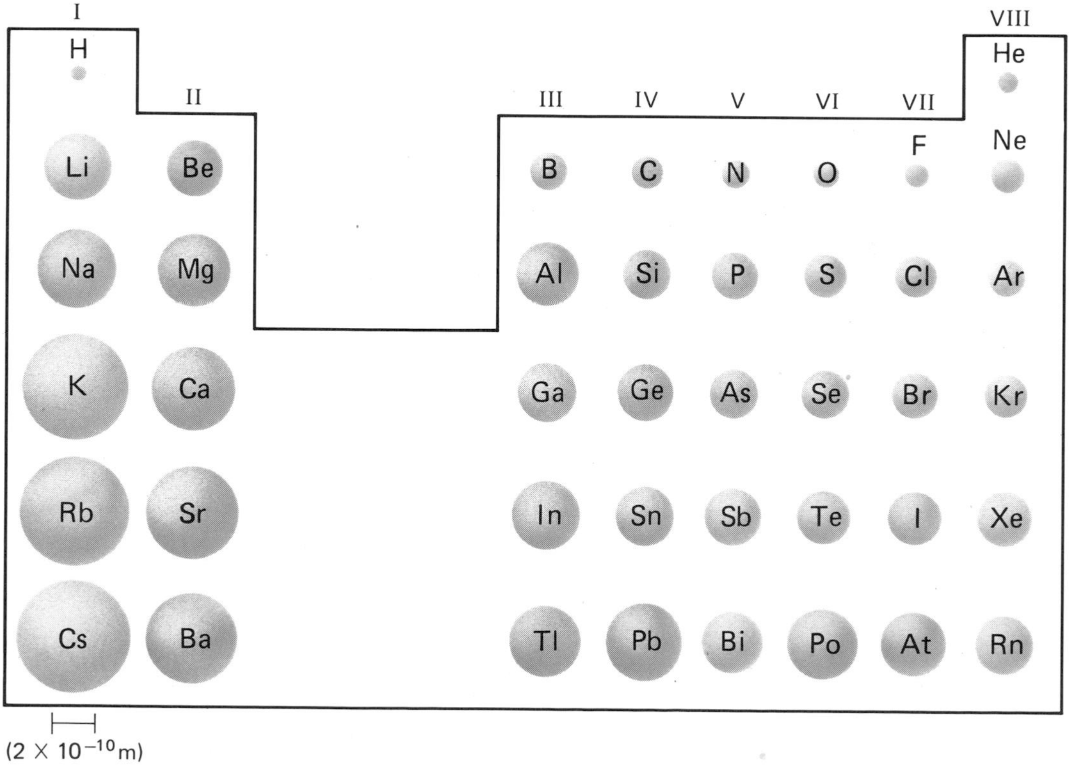

one to the nucleus, increasing its positive charge. This increasing positive charge pulls each electron shell closer to the nucleus, decreasing the atom's radius. Thus, atomic size is a periodic property which increases from top to bottom within a column and decreases from left to right across a period.

B. The Formation of Ions

Atoms are electrically neutral because the number of positively charged protons in the nucleus equals the number of negatively charged electrons outside the nucleus. If, when an atom reacts, electrons are added or lost, the atom acquires a charge and becomes an **ion.** Since group VIII elements *never* react by the loss or gain of electrons, they are not discussed in the rest of this section.

1. Positive ions, or cations. When a neutral atom loses an electron, it forms a positively charged ion called a **cation.** In general, metals lose electrons to form cations. The atom thereby attains the electron configuration of the noble gas next below it in atomic number.

For example, an alkali metal loses one electron to form a cation with a single positive charge. Sodium loses its single $3s$ valence electron to form the ion Na^+, which has the electron configuration of neon:

$$Na\cdot \longrightarrow Na^+ + e^-$$

An alkaline earth metal loses two electrons to form a cation with a charge of $+2$. In forming the magnesium ion, Mg^{2+}, a magnesium atom loses two electrons:

$$Mg: \longrightarrow Mg^{2+} + 2\,e^-$$

Aluminum loses three electrons to form a cation with charge $+3$:

$$\dot{Al}: \longrightarrow Al^{3+} + 3\,e^-$$

The names of these cations are the same as the metals from which they are formed:

Alkali Metal Cations		Alkaline Earth Metal Cations		Other Metal Cations	
Li^+	lithium ion	Mg^{2+}	magnesium ion	Al^{3+}	aluminum ion
Na^+	sodium ion	Ca^{2+}	calcium ion		
K^+	potassium ion	Sr^{2+}	strontium ion		
Rb^+	rubidium ion	Ba^{2+}	barium ion		
Cs^+	cesium ion				

Table 3.7 Naming of cations.

Symbol	IUPAC	Old	Symbol	IUPAC	Old
Co^{2+}	cobalt(II)	cobaltous	Cr^{2+}	chromium(II)	chromous
Co^{3+}	cobalt(III)	cobaltic	Cr^{3+}	chromium(III)	chromic
Cu^{+}	copper(I)	cuprous	Fe^{2+}	iron(II)	ferrous
Cu^{2+}	copper(II)	cupric	Fe^{3+}	iron(III)	ferric

Transition elements and metals to their right frequently form more than one cation. For example, iron forms Fe^{2+} and Fe^{3+}; cobalt forms Co^{2+} and Co^{3+}. The names of these ions must indicate the charge they carry. The preferred system of nomenclature is that recommended by the International Union of Pure and Applied Chemistry (IUPAC). In this system, the name of the metal is followed by a Roman numeral showing the charge on the ion. No extra space is left between the name and the number. Thus, Fe^{2+} is iron(II) (say: "iron two"), and Fe^{3+} is iron(III). In the old system, the name of the cation of lower charge ends in -*ous*. The name of the cation of higher charge ends in -*ic*. Examples of both systems of naming are given in Table 3.7.

None of the cations listed in Table 3.7 have a charge greater than $+3$. In forming ions, the electrons are pulled off one by one from the atom. Thus, the first electron is removed from a neutral atom, the second electron from an ion of charge $+1$, the third electron from an ion of charge $+2$, and so on. The amount of energy necessary to remove an electron increases dramatically as the positive charge of the ion increases. To remove a fourth electron and form an ion of charge $+4$ is energetically unlikely.

2. Negative ions, or anions. When a neutral atom gains an electron, it forms a negatively charged ion called an **anion.** Typically, nonmetals form anions, gaining enough electrons to acquire the electron configuration of the noble gas of next higher atomic number. Elements of group VI form anions by gaining two electrons, and the halogens form anions by gaining one electron. The names of these anions include the root name of the element and the ending -*ide*. Below are several anions and their names; in each case, the root of the name is *italicized*.

F^{-}	*fluor*ide ion	O^{2-}	*ox*ide
Cl^{-}	*chlor*ide ion	S^{2-}	*sulf*ide
Br^{-}	*brom*ide ion		
I^{-}	*iod*ide ion		

3. Polyatomic ions. The ions described in the preceding paragraphs are monatomic ions, that is, they each contain only one atom. Many polyatomic ions are also known. These are groups of atoms bonded together that carry a charge due to an excess or deficiency of electrons. The formulas and names of several common polyatomic ions follow. The symbols in the formula show which elements are present. The subscripts ("1" is understood) tell how many atoms of each element are present in the ion.

charge $+1$ NH_4^+ ammonium ion   *charge* -2 CO_3^{2-} carbonate ion
                                                      SO_4^{2-} sulfate ion

charge -1 OH^- hydroxide ion
           NO_3^- nitrate ion
           HCO_3^- bicarbonate ion   *charge* -3 PO_4^{3-} phosphate ion

C. Ionization Energy

The **ionization energy** of an element is the minimum energy required to remove an electron from a gaseous atom of that element. Electrons are held in the atom by the attractive force of the positively charged nucleus. The farther the outermost electrons are from the nucleus, the less tightly they are held. Thus, the ionization energy within a group of elements decreases as the elements increase in atomic number. Among the alkali metal atoms, the single valence electron of cesium is farthest from the nucleus (in the sixth energy level), and we can correctly predict that the ionization energy of cesium is the lowest of all the alkali metals.

Moving across a period, the ionization energy increases. The number of protons in the nucleus (the nuclear charge) increases, yet the valence electrons of all the elements are in the same energy level. It becomes increasingly more difficult to remove an electron from the atom. The ionization energy of chlorine is much greater than that of sodium, an element in the same period.

The ionization energies of elements 1 through 36 are plotted against their atomic numbers in Figure 3.11. The peaks of the graph are the high ionization energies of the noble gases. The height of the peaks decreases as the number of the highest occupied energy level increases. The low points of the graph are the ionization energies of the alkali metals, which have only one electron in their valence shells. These points, too, decrease slightly as the number of the highest occupied energy level increases. The graph shows that ionization energy is periodically related to atomic number. Even within a row of the periodic table, the variations in ionization energy are closely related to electron configuration.

The ionization energy of an element measures its metallic nature. From Figure 3.11, we see that the alkali metals have the lowest ionization energies of all the elements. Therefore, they are the most metallic elements. Moving up the periodic table or moving from left to right across it, the metallic nature of the elements decreases.

Nonmetals, located in the upper righthand section of the periodic table, have high ionization energies. Except for the noble gases, fluorine has the highest ionization energy; therefore, fluorine is the least metallic or most nonmetallic element. Moving down the column or to the left of fluorine, elements become more metallic.

Figure 3.11 Ionization energies of the first 36 elements versus their atomic numbers.

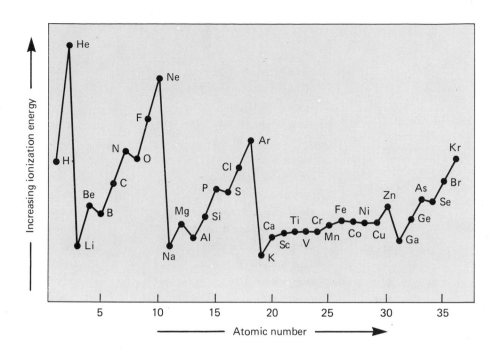

3.7 The Octet Rule

The **octet rule** states: In general, an atom reacts in ways that give it a complete octet of electrons in its outer shell. An atom with one, two, or three valence electrons loses these electrons to acquire the electron configuration of the noble gas next below it in atomic number. An atom with six or seven valence electrons will add enough electrons to acquire a complete octet of valence electrons. Other atoms may attain a complete octet by *sharing* electrons with a neighboring atom (discussed in Section 4.3A).

Example 3.5

Show how calcium follows the octet rule in forming its ion, Ca^{2+}.

Solution

The electron configuration of calcium is: $1s^2 2s^2 2p^6 3s^2 3p^6 4s^2$

The electron configuration of Ca^{2+} is: $1s^2 2s^2 2p^6 3s^2 3p^6$

Ca^{2+} has a complete octet in its outermost shell; its configuration is identical to that of argon.

Problem 3.5

Show how the sulfide ion and the aluminum ion illustrate the octet rule.

Key Terms and Concepts

actinides (3.5)

alkali metals (3.5F2)

alkaline earth metals (3.5F2)

anions (3.6B2)

atomic radius (3.6A)

atomic spectrum (3.1)

cations (3.6B1)

continuous light (3.1)

core notation (3.5B)

electromagnetic spectrum (3.1)

electron configuration (3.4)

electron-dot structure (3.5D)

electron spin (3.3B)

energy level (3.2)

excited state (3.3C)

ground state (3.3C)

groups of elements (3.5F2)

halogens (3.5F2)

inner transition elements (3.5C)

ionization energy (3.6C)

ions (3.6B)

Lewis structure (3.5D)

Mendeleev (3.5E)

noble gases (3.5A, E2)

octet rule (3.7)

orbitals (3.3B)

period (3.5)

periodic table (3.5)

polyatomic ions (3.6B3)

principal energy levels (3.3A)

quantization (3.2)

radiant energy (3.1)

representative elements (3.5C)

s, p, d, f orbitals (3.3B)

semimetals (3.5F1)

transition elements (3.5C)

wavelength (3.1)

valence electrons (3.5D)

Problems

Electron Configuration (Section 3.4)

3.6 Write the complete electron configuration of the following elements:

    a. magnesium   b. phosphorus   c. argon

    d. oxygen   e. chlorine   f. sodium

3.7 Write the complete electron configuration of the following elements:

    a. silicon   b. tin   c. vanadium   d. lead

    e. aluminum   f. sulfur   g. iodine   h. barium

3.8 What characterizes the electron configuration of:

    a. elements in the same column of the periodic table?

    b. elements in the same period of the table?

    c. the noble gases?

3.9 By noting their positions in the periodic table, write the complete electron configuration of the following elements:

    a. calcium   b. antimony   c. boron   d. radium

    e. potassium   f. gallium   g. germanium

3.10 Write the complete electron configuration of each of the alkaline earth metals.

3.11 For the elements in period 4 of the periodic table, (a) give the complete electron configuration and (b) give the electron configuration using core notation.

3.12 Draw the Lewis structures of the alkali metals.

3.13 Draw the Lewis structures of atoms of the following elements:

    a. sulfur   b. chlorine   c. magnesium   d. tellurium

    e. carbon   f. boron   g. lithium   h. barium

Periodic Properties (Section 3.5C)

3.14 a. How is each of the following properties related to position in the periodic table?

metallic nature   nonmetallic properties
ionization energy   atomic radius

b. How is each property in part (a) related to electron configuration?

3.15 Classify the following elements as metals or nonmetals:
a. vanadium   b. palladium   c. selenium
d. sulfur   e. zinc   f. fluorine

3.16 Element number 117 has not yet been discovered. When (and if) it is discovered, we expect it will be in Column VII of the table; explain why. Predict the following properties for this element, comparing them to those of a known element.

metallic nature   atomic radius   ionization energy   Lewis structure

3.17 Which element in each of the following pairs is more metallic?

cesium/cerium   arsenic/bismuth
aluminum/silicon   iodine/bromine

3.18 Which element in each of the following pairs has a larger atomic radius?

potassium/rubidium   nitrogen/arsenic
aluminum/sulfur   hydrogen/oxygen

3.19 Arrange the following elements in order of increasing ionization energy:
a. Be, Mg, Sr   b. Na, Al, S   c. Bi, Cs, Ba

Naming Ions (Section 3.6B)

3.20 Name the following ions, using both naming systems.
Fe^{2+}, Fe^{3+} Cr^{2+}, Cr^{3+} Cu^{2+}, Cu^{+} Ni^{2+}, Ni^{3+}

3.21 Write the formulas of the following cations:
a. iron(III)   b. silver(I)   c. platinum(II)
d. osmium(III)   e. lead(II)   f. mercury(I)

3.22 Write the formula and give the name of the monatomic anion formed by:
a. iodine   b. oxygen   c. bromine
d. sulfur   e. fluorine

3.23 Write the formula of the monatomic cation formed by the following elements:
a. aluminum   b. strontium   c. cesium
d. lithium   e. magnesium   f. potassium

3.24 Complete the following chart:

Name of Ion	Formula	Name of Ion	Formula
_____	NH_4^{+}	Sulfate	_____
Nitrate	_____	_____	CO_3^{2-}
_____	HCO_3^{-}	Phosphate	_____

Octet Rule (Section 3.7)

3.25 Write the complete electron configuration of the following ions:
a. Se^{2-} b. Br^{-} c. Rb^{+} d. Sr^{2+}
Which noble gas has the same electron configuration?

3.26 The following ions are known to exist:
a. Cs^{+} b. Ga^{+} c. Te^{2+} d. Bi^{3+} e. Pb^{2+}
Which are exceptions to the octet rule?

Elements in the Body

Each of us is aware of the importance of diet in maintaining good health. A good diet contains vitamins and minerals as well as fats, carbohydrates, and proteins. The elements present in the largest amounts in our bodies are carbon, hydrogen, oxygen, sulfur, and nitrogen, the same elements that are found in fats, carbohydrates, and proteins. Lesser quantities of other elements, called **minerals,** are also essential for good health. Table 1 lists those elements, along with the daily amounts required. Group A of the table lists minerals needed in substantial quantities for the formation of bones, cartilage, and teeth. Group B contains the elements whose ions maintain electrochemical balance across cell walls. Sodium and potassium cations and chloride, sulfate, carbonate and phosphate anions are small ions; hence, they can migrate easily through the cell membranes to maintain charge balance on both sides of the cell membranes.

The third group of elements in Table 1 are called **trace elements.** These elements are required for good health but only in very small amounts.

Figure 1 shows the groups of elements discussed above, arranged in the periodic table (the lanthanides and actinides are not shown). The five elements shaded in light gray are those present in the largest quantities in our bodies. The elements from group A of

Table 1 Daily mineral requirements for 70 kg male aged 23–50.

From Recommended Dietary Allowances, 9th ed. National Research Council, National Academy of Sciences, Washington, D.C., 1980.

Element	Amount required daily (mg)	Element	Amount required daily (mg)
Group A		**Group C**	
calcium	800	zinc	15
phosphorus	800	iron	10
magnesium	350	manganese	2.5
		copper	2.0
Group B		fluorine	1.5
potassium	1875–5625	iodine	0.15
chlorine	1700–5100	molybdenum	0.15
sodium	1100–3300	chromium	0.05
		selenium	0.05
		vanadium	
		cobalt	minimum daily
		nickel	requirement not
		arsenic	yet established
		silicon	

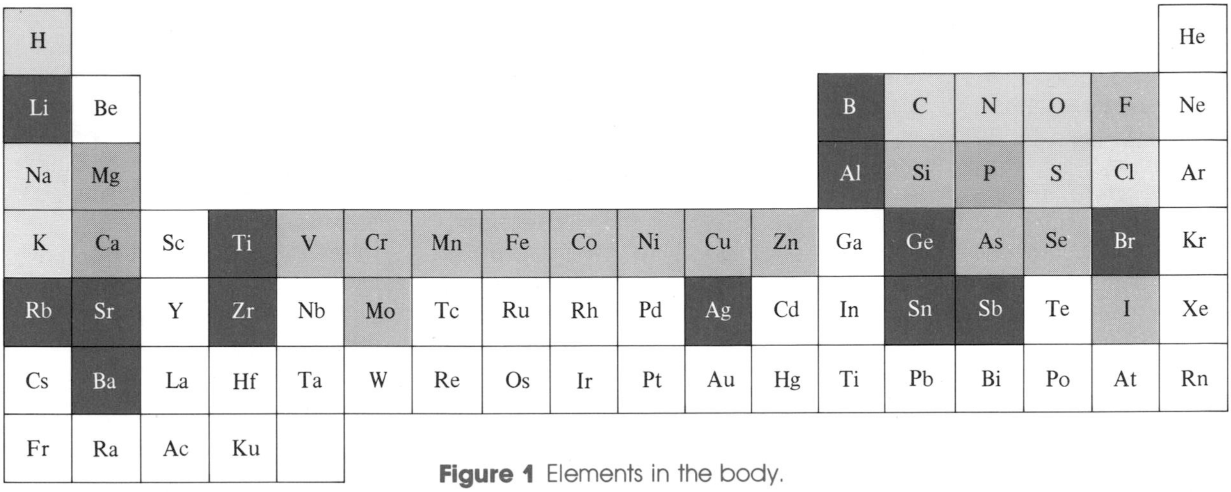

Figure 1 Elements in the body.

Table 1 are shaded in dark gray. The elements shaded in light color are the ones in group B. The trace elements (group C of Table 1) are shaded in darker color. The elements shown in darkest color, with white letters, are consistently found in human tissue in trace amounts. The role of these elements has not yet been defined. It has not yet been established whether they are essential or just so prevalent in the environment that they become part of every living thing. Notice that, of the elements in the first four rows of the table, all but the noble gases, beryllium, scandium, and gallium are regularly found in human tissue.

The metabolic roles played by the elements in the body are varied. Sometimes, these elements function structurally. Many are incorporated into the structure of large molecules such as enzymes and hormones, compounds involved in metabolism. Sulfur occurs in the amino acids cystine and cysteine. These amino acids occur in many proteins but in especially large concentrations in keratin, the protein in hair. Iodine is incorporated by the thyroid gland into thyroxin, the hormone important in growth and metabolism. Absence of the proper concentration of this hormone, resulting from a dietary deficiency of iodine, causes goiter, which is the chronic enlargement of the thyroid gland. Iron is in the heme fraction of the hemoglobin molecule, the blood protein that carries oxygen throughout the body. Cobalt is in vitamin B_{12}, which plays a key role in the formation of blood. A deficiency of vitamin B_{12} causes severe anemia. Fluorine is found in tooth enamel, particularly in enamel that has proven resistant to decay. Throughout the U.S. during the last quarter century, fluorine has been added to drinking water and applied to teeth in an effort to decrease the incidence of tooth decay in the general population.

All of the elements in Table 1 act as **cofactors** (activators) of enzymes, the catalysts of biological reactions. The exact function of an enzyme cofactor is quite variable, ranging from promoting the correct spatial orientation of the reactants to forming an intermediate with one of the molecules. Either way they aid the reaction. Zinc, iron, chromium, molybdenum, and copper are particularly important as cofactors in the body. For example, zinc activates the enzyme that catalyzes the decomposition of carbonic acid to carbon dioxide and water. Zinc is also a cofactor for other enzymes, such as kidney phosphatase, necessary for phosphorylation reactions in the kidneys. Calcium is a cofactor in the reactions of heart muscle.

For each of the elements discussed above, there is an optimum concentration in the body (Table 1). Below that amount, symptoms of the element's deficiency appear; above that amount, the element is toxic. Certainly, the potential toxicity of the trace element arsenic has been well established.

Selenium has recently received a great deal of publicity as a trace element required for good health. For years, only its toxic properties were recognized. It has now been established that selenium plays several beneficial metabolic roles. As the selenate ion (SeO_4^{2-}), it activates the enzyme glutathione peroxidase; it is involved in vitamin E activity; and it is required for the efficient oxidation of the sulfhydryl ($-SH$) group of proteins. Selenium is found in heart muscle, and increased selenium intake is associated with lower incidence of heart disease.

Natural sources of selenium include whole-wheat grain cereals and products derived from such cereals, certain yeasts, liver, kidney, and seafoods. The amount of selenium in the diet depends on the selenium content of these natural sources, which depends, in turn, on the selenium content of the soil. This interrelationship has been explored in studies showing the antagonism of selenium to cancer. In one such study, statistics from twenty-seven countries showed that the death rate attributable to cancer is inversely proportional to the dietary intake of selenium and to the selenium content of the soils in the area. Another study tabulated the birth months of 180,000 cancer patients and found that an unexpectedly large number of the patients were born in the winter, while surprisingly few were born in the summer. This difference has been attributed to seasonal variations in the amount of selenium in the diet. In a different study, statistics from seventeen countries showed the selenium content of female blood to be inversely proportional to the incidence of breast cancer.

Some elements interfere with the toxic effects of other metals. Selenium has been shown to be an antidote in cases of cadmium poisoning. Increased amounts of dietary selenium seem to protect against tumors caused by high levels of dietary zinc. Increased intake of arsenic will offset a toxic dose of selenium. The mechanisms causing these antagonistic effects are unknown.

In the future, perhaps, other light elements will be identified as essential. Aluminum *in vitro* activates succinic dehydrogenase but does not seem an essential cofactor *in vivo*. Is aluminum a cofactor of another enzyme? Lithium has been used in the treatment of mental disorders, such as schizophrenia. Is a deficiency of lithium the cause of this disease? Why is boron found so frequently in mammalian tissue? Is it essential?

Which heavy elements might be identified in the future as essential, albeit in trace amounts? For generations, gold has been used to treat certain forms of arthritis. Gold has been found in mammalian liver tissue in concentrations of 2–3 parts per billion (ppb). In some Asian countries, mercury has been used to treat the common cold. Rubidium has been found in human tissue in concentrations of 20–40 ppm. Are these, too, essential elements? Are lead and cadmium toxic in all amounts, or are they beneficial in trace amounts? As our methods of analysis become more sophisticated and our studies of biochemical processes become more refined, we can look forward to answers to these questions.

References

Bowen, H.M.J. 1966. *Trace elements in biochemistry.* New York: Academic Press.

Mertz, W. 1981. Essential trace elements. *Science* 273: 1332.

National Research Council. 1980. *Recommended dietary allowances.* 9th ed. Washington, D.C.: National Academy of Sciences.

Schrauzer, G.N. 1977. *Nutritional aspects of cancer: a conference report.* New York: Plenum Press.

Sevin, M.J., ed. 1960. *Metal binding in medicine.* Philadelphia: J.B. Lippincott.

Underwood, E.J. 1977. *Trace elements in human and animal nutrition.* 4th ed. New York: Academic Press.

Mini-Essay 2

The Past and Future of the Periodic Table

As we discussed in Chapter 3, the elements in the modern periodic table are arranged in order of increasing atomic number. The configuration of rows and columns in the table is such that elements with the same outer-shell electron configurations fall in the same column. The electron configuration of an element can be predicted from its position in the table. Chemically similar elements are in the same column. This table is the product of the efforts of many scientists to arrange the known elements in a meaningful way. The history of the periodic table gives some interesting insights into scientific progress.

One of the first to recognize that there were regular similarities in the properties of groups of elements was the German chemist Johann Döbereiner (1780–1849). In 1829, Döbereiner suggested that many elements belong to three-member groups, or triads, and that within each triad there are noticeable similarities and trends in properties. The halogens chlorine, bromine, and iodine form such a triad. They have similar chemical properties. The physical and chemical properties of bromine are midway between those of chlorine and those of iodine.

John Newlands (1838–1898), a British chemist, was the first to attempt a meaningful arrangement of all the elements. In 1866, Newlands organized all known elements in groups of seven, in order of increasing atomic weight (Figure 1). The first members of each group are, indeed, simi-

Figure 1 Newland's table of the elements (1866). Note the following: In some cases, two elements are in the same box; some rows contain elements with dissimilar properties (chlorine and platinum, for example); and the order of elements is vertical rather than horizontal.

1 H	8 F	15 Cl	22 Co, Ni	29 Br	36 Pd	42 I	50 Pt, Ir
2 Li	9 Na	16 K	23 Cu	30 Rb	37 Ag	44 Ca	51 Os
3 Be	10 Mg	17 Ca	24 Zn	31 Sr	38 Cd	45 Ba, V	52 Hg
4 B	11 Al	19 Cr	25 Y	33 Ce La	40 U	46 Ta	53 Tl
5 C	12 Si	18 Ti	26 In	32 Zr	39 Sn	47 W	54 Pb
6 N	13 P	20 Mn	27 As	34 Di, Mo	41 Sb	48 Nb	55 Bi
7 O	14 S	21 Fe	28 Se	35 Ro, Ru	43 Te	49 Au	56 Th

97d

Table 1 Comparison of Mendeleev's predictions for the properties of the undiscovered element eka-silicon and the actual properties of the element germanium.

Properties	Predicted by Mendeleev for Eka-silicon (1871)	Actual Properties of Germanium (1886)
atomic weight	72	72.6
density	5.5 g/cm³	5.36 g/cm³
appearance	gray metal	gray metal
melting point	very high	960°C
specific heat	0.073 cal/g°C	0.076 cal/g°C
formula of oxide	EsO_2	GeO_2
density of oxide	4.7 g/cm³	4.70 g/cm³
formula of chloride	$EsCl_4$	$GeCl_4$
density of chloride	1.9 g/cm³	1.88 g/cm³
boiling point of chloride	100°C	83°C

lar. However, Newlands was an avid musician and used the term *Law of Octaves* in describing the ordering. The term was ridiculed by other scientists, causing a general rejection of his entire theory. One chemist even suggested that an alphabetical arrangement might be more useful. Britain's Royal Chemical Society refused to publish Newland's paper. Much later, in 1887, after Newland's work had been shown to be basically correct, the Royal Chemical Society awarded him the Davy Medal, their highest honor.

A much more successful arrangement of the elements was conceived independently in 1869 by both Lothar Meyer (1830–95) in Germany and Dmitri Mendeleev (1834–1907) in Russia. Mendeleev is usually given most of the credit for the construction of the periodic table because he not only described the or-

dering of known elements, but also used this arrangement to predict the existence of elements not yet discovered. In fact, he even predicted what the properties of the unknown elements would be. Figure 2 shows Mendeleev's periodic table as it appeared in 1871. His insistence that the elements be ordered by similar chemical properties led him to leave gaps in the table for undiscovered elements, and he used the general trends established in the table to predict the properties of these missing elements. Table 1 lists the properties predicted by Mendeleev in 1871 for the element he called *eka-silicon,* meaning "next after silicon," (predicted atomic weight, 72). The table also lists the properties of the element germanium, atomic weight 72.6, discovered in 1886. In the modern periodic table, germanium is below silicon. The remarkable agreement between Mende-

leev's predictions for eka-silicon and the actual values for germanium show that his theories work. The publication of Mendeleev's table and his predictions of new elements and their properties spurred chemists to search for these elements, using the predicted properties as a guide for isolating and identifying the elements. Each additional element discovered further confirmed the accuracy of Mendeleev's table.

In constructing his table, Mendeleev found some apparent disagreements among properties of known elements. For example, the values of the atomic weights for gold and platinum accepted at this time indicated that platinum was heavier than gold. However, in grouping these elements in families according to similar properties, Mendeleev observed that platinum should come before gold, which caused him to question the

Series	Group I — R_2O	Group II — RO	Group III — R_2O_3	Group IV RH_4 RO_2	Group V RH_3 R_2O_5	Group VI RH_2 RO$_3$	Group VII RH R_2O_7	Group VIII — RO_4
1	H = 1							
2	Li = 7	Be = 9.1	B = 11	C = 12	N = 14	O = 16	F = 19	
3	Na = 23	Mg = 24.4	Al = 27	Si = 28	P = 31	S = 32	Cl = 35.5	
4	K = 39.1	Ca = 40	— = 44	Ti = 48.1	V = 51.2	Cr = 52.3	Mn = 55	{ Fe = 56, Ni = 58.5, Co 59.1, Cu 63.3.
5	(Cu) = 63.3	Zn = 65.4	— = 68	— = 72	As = 75	Se = 79	Br = 80	
6	Rb = 85.4	Sr = 87.5	Y = 89	Zr = 90.7	Nb = 94.2	Mo = 95.9	— = 100	{ Rh = 103, Ru = 103.8, Pd = 108, Ag = 107.9.
7	(Ag) = 107.9	Cd = 112	In = 113.7	Sn = 118	Sb = 120.3	Te = 125.2	I = 126.9	
8	Cs = 132.9	Ba = 137	La = 138.5	Ce = 141.5	Di = 145	—	—	— — —
9	(—)	—	—	—	—	—	—	
10	—	—	Yb = 173.2	—	Ta = 182.8	W = 184	—	{ Ir = 193.1, Pt = 194.8, Os = 200, Au = 196.7.
11	(Au) = 196.7	Hg = 200.4	Tl = 204.1	Pb = 206.9	Bi = 208	—	—	
12	—	—	—	Tb = 233.4	—	U = 239	—	— — —

Figure 2 Mendeleev's periodic table (1871). Note the spaces left for undiscovered elements. Columns are arranged according to the formulas of the hydrides and oxides of the elements. Note that several atomic weights are incorrect, according to present knowledge.

accepted atomic weights for these elements. Other chemists redid their measurements, thinking to prove Mendeleev wrong; instead, they found that gold has a higher atomic weight than platinum.

By 1892, Mendeleev's periodic table was generally accepted by chemists. In that year, however, a new challenge to Mendeleev's theory arose: the discovery of an element that did not fit anywhere in the table. This was the element argon, discovered by the British physicist Baron Rayleigh (1842–1919). Baron Rayleigh had observed a discrepancy of one part per thousand between the density of nitrogen isolated from the atmos-

phere and the density of nitrogen obtained by the decomposition of ammonia. After repeating his experiments, in an effort to find his error, he became convinced that he had detected a new gas in the atmosphere. Eventually, he was able to isolate this gas and determine its properties. The physical properties he observed were not unusual, but the chemical properties were quite singular: the new gas did not react with anything!

During the next few years, several attempts were made to account for argon within Mendeleev's periodic table. Some scientists proposed that it was really a diatomic or

even a triatomic gas, since no monatomic gases were known at that time, but all attempts to break down molecules of the gas failed. Argon's atomic weight (39.9) meant it should fit between potassium and calcium in the periodic table, but that would mean a new column and several undiscovered elements. Investigation of this possibility led Rayleigh to the discovery of a second new element, also nonreactive, which he called helium. Its atomic weight of 4 compounded the problem created by argon. The riddle was still not solved. By this point, most scientists had such belief in the periodic table that they hoped the new elements would prove to be compounds, since this would avoid any change in the table. Not until 1898, when Rayleigh and his coworker, Sir William Ramsay (1852–1916), announced the discovery of three more nonreactive, monatomic gases (neon, krypton, and xenon), did chemists finally accept the idea of a new family of elements. Even so, the importance of these elements in understanding electron configuration was not appreciated for many years. They were considered an unsolved puzzle that confused rather than simplified the periodic table. The problem caused by the atomic weight of argon falling between the atomic weights of potassium and calcium was not cleared up until 1914, when the English physicist Henry Moseley (1887–1915) showed that arrangement of the elements by atomic number rather than atomic weight gave a truer alignment of properties within the table columns. His new ordering of the elements based on their atomic numbers left more spaces for undiscovered elements and led to the search for, and eventual discovery of, technetium, promethium, rhenium, and hafnium. These discoveries confirmed Moseley's theory. Unfortunately, Moseley did not live long enough to know this; he was killed in the First World War at age 27.

Various new formats for the periodic table were tried during the first half of the twentieth century, including spirals, 3-D tubes, and other shapes. The form most commonly used at present, with the central block of transition elements, was not developed until the 1930s, when the significance of electron configurations began to be recognized. The row for the lanthanides was added during this period. The concept of the actinides row was not proposed until 1944, when the American chemist Glenn Seaborg (b. 1912) developed the idea that the new elements he was trying to synthesize might have properties that were more similar to the lanthanides than to the transition elements. This conjecture led him to devise new experiments which isolated the elements americium and curium. The announcement of his discoveries was first made on a radio quiz show in 1945, on which Seaborg was appearing as a guest.

By 1982, chemists were certain of the identities of 105 elements, including 92 found in nature and 13 manmade ones. The synthetic elements are difficult to study because they are highly radioactive and quickly break up into smaller atoms. The half-life of element 104, kurchatovium, is about 1 minute; that of hahnium, number 105, is about 40 seconds; and the half-life of a proposed element 106, whose name and exact nature are not yet agreed on, is about 1 second. The brevity of these half-lives has led to predictions that all elements heavier than 106, the so-called superheavy elements, have even shorter half-lives, breaking apart into smaller atoms so quickly that their existence as single atoms cannot be detected.

In June, 1976, data published by a team of nuclear scientists suggested the existence of element 126 in a crystal sample of the mineral monazite. The data were in the form of a peak in an X-ray spectrum that corresponded to the predicted energy of a nucleus of this element, tentatively named *bicentenium* in honor of the U.S. bicentennial. The find was exciting to physicists because the rock sample had been dated as being about 1 billion years old, suggesting a much higher stability for this element than theorists had predicted. However, additional experiments by the same researchers indicated that a peak of the exact same energy as the X-ray peak for element 126 could be generated by gamma rays emitted from an excited nucleus of praseodymium. They also found that praseodymium could be created from cerium (a principal constituent of monazite) during the proton-bombardment process used to generate the original X-ray spectra. Experiments performed on the same mineral samples using different identification techniques failed to reproduce the original data. Although it is not a closed question, it is now commonly believed that element 126 has not yet been found in nature.

Will elements heavier than atomic number 106 ever

1 H																	2 He
3 Li	4 Be											5 B	6 C	7 N	8 O	9 F	10 Ne
11 Na	12 Mg											13 Al	14 Si	15 P	16 S	17 Cl	18 Ar
19 K	20 Ca	21 Sc	22 Ti	23 V	24 Cr	25 Mn	26 Fe	27 Co	28 Ni	29 Cu	30 Zn	31 Ga	32 Ge	33 As	34 Se	35 Br	36 Kr
37 Rb	38 Sr	39 Y	40 Zr	41 Nb	42 Mo	43 Tc	44 Ru	45 Rh	46 Pd	47 Ag	48 Cd	49 In	50 Sn	51 Sb	52 Te	53 I	54 Xe
55 Cs	56 Ba	57 La	72 Hf	73 Ta	74 W	75 Re	76 Os	77 Ir	78 Pt	79 Au	80 Hg	81 Tl	82 Pb	83 Bi	84 Po	85 At	86 Rn
87 Fr	88 Ra	89 Ac	104 Rf	105 Ha	106	(107)	(108)	(109)	(110)	(111)	(112)	(113)	(114)	(115)	(116)	(117)	(118)
(119)	(120)	(121)	(154)	(155)	(156)	(157)	(158)	(159)	(160)	(161)	(162)	(163)	(164)	(165)	(166)	(167)	(168)

Lanthanides

58 Ce	59 Pr	60 Nd	61 Pm	62 Sm	63 Eu	64 Gd	65 Tb	66 Dy	67 Ho	**68** Er	69 Tm	70 Yb	71 Lu

Actinides

90 Th	91 Pa	92 U	93 Np	94 Pu	95 Am	96 Cm	97 Bk	98 Cf	99 Es	100 Fm	101 Md	102 No	103 Lr

Superactinides (122) (123) (124) (125) (126) ... (153)

Figure 3 Seaborg's suggested periodic table of the future.

be found? While most scientists believe that no superheavy elements will be found in nature, there is no agreement on the possibility of synthesizing superheavy elements. The problem is partly one of finding a technique for fusing two nuclei into one without using so much energy that the resulting nucleus undergoes spontaneous fission. Even so, there are suggestions in the literature of methods for preparing elements with atomic numbers up to 119, along with predicted properties for these elements.

What about the future of the periodic table? Again, without the existence of superheavy elements as evidence, there is no agreement as to how the table might continue. The energy levels in these elements become so close together that no clear-cut pattern can be predicted as to when one electron shell will be filled and another begin. Glenn Seaborg has suggested the theoretical periodic table shown in Figure 3, but whether this will be the periodic table used by students taking this course 100 years from now is for the future to decide.

References

Robinson, A.L. 1977. Superheavy elements: confirmation fails to materialize. *Science* 195: 473.

Seaborg, G.T. 1979. The periodic table, tortuous path to manmade elements. *Chem. & Eng. News*, 16 April 1979, p. 46.

Seaborg, G.T.; Loveland, W.; and Morrissey, D.J. 1979. Superheavy elements — a crossroads. *Science* 203: 711.

Wolfenden, J.H. 1969. The noble gases and the periodic table. *J. Chem. Ed.* 46: 569.

Compounds and Chemical Bonding

Thus far, our discussion has centered on elements and their properties. In this chapter we consider compounds, which are formed by the chemical combination of elements. Although there are only 105 elements currently accepted, there are millions of compounds. Our discussion will focus on: how compounds result from the formation of bonds between atoms; the nature of these chemical bonds; and some of the relationships between the structure of a compound and its properties.

4.1 What Is a Compound?

A **compound** is formed by the chemical combination of two or more elements in a definite and constant proportion by weight. The properties of each compound are unique, different from both those of other compounds and those of the elements of which it is composed. For example, sodium is a silvery-gray, soft metal, toxic to humans; chlorine is a pale green gas, also toxic to humans.

Table 4.1 Names and formulas of some familiar compounds.

Common Name	Chemical Name	Formula
ammonia	ammonia	NH_3
baking soda	sodium bicarbonate or sodium hydrogen carbonate	$NaHCO_3$
bleach	sodium hypochlorite	$NaOCl$
chalk	calcium carbonate	$CaCO_3$
dry ice	carbon dioxide	CO_2
lime	calcium oxide	CaO
LPG	propane	C_3H_8
milk of magnesia	magnesium hydroxide	$Mg(OH)_2$
vinegar	acetic acid	$C_2H_4O_2$
water	water	H_2O

Yet, sodium combines with chlorine to produce sodium chloride, which we know as table salt, a part of our daily diet. Another familiar compound is water, formed by chemical combination of the flammable gas hydrogen and another gas, oxygen. It is obvious that the properties of each of these compounds are quite different from those of the elements the compound contains.

Many other compounds are familiar to you. Some of these are listed in Table 4.1, along with the chemical name and formula of each. As you look over the table, think about the differences between the properties of each compound and those of the elements from which it is formed.

The same elements may combine in different ratios to form different compounds. Table 4.2 lists several of the many thousands of compounds composed of carbon, hydrogen, and oxygen in different atomic ratios. You are undoubtedly familiar with many of these compounds and aware of their different properties.

Table 4.2 Common compounds composed of carbon, hydrogen, and oxygen.

Common Name	Chemical Name	Formula	mp (°C)
sugar	sucrose	$C_{12}H_{22}O_{11}$	185
formalin	formaldehyde	CH_2O	-92
ethyl alcohol	ethanol	C_2H_6O	-117
wood alcohol	methanol	CH_4O	-98
acetone	propanone	C_3H_6O	-95
blood sugar	glucose	$C_6H_{12}O_6$	146

Mixtures are also formed by the combination of two or more pure substances. However, there are three important differences between a compound and a mixture:

1. The composition of a compound is constant; the composition of a mixture can vary over a wide range.
2. The properties of a compound are always the same; those of a mixture vary, depending on the exact composition of the mixture.
3. The separation of a compound into its elements requires chemical decomposition.

A mixture can be separated into its components by physical means, because each has a different set of physical properties. For example, a sample of the compound sodium chloride always contains 39.4% sodium and 60.6% chlorine by mass. The sample will be white, crystalline, and melt at 801°C. However, if sodium chloride is part of a mixture, as in salt water, the properties of the mixture (composition, density, melting point, taste, and so on) will vary, depending on how much salt was added to how much water. The mixture can be easily separated into its components by heating. If we heat the salt water to 100°C, the water will evaporate, leaving the salt (bp 1413°C) behind. In contrast, heating the compound sodium chloride will not release chlorine gas.

4.2 Formulas of Compounds

The definite and constant proportion of component elements in a compound is expressed by the compound's formula. A chemical **formula** shows which elements combine to form the compound and how many atoms of each element are present in the simplest unit of the compound. The formula of ammonia, NH_3, shows that the simplest unit of this compound contains one atom of nitrogen and three atoms of hydrogen. Calcium carbonate, the principal component of marble, has the formula $CaCO_3$, showing that this compound's simplest unit contains atoms of calcium, carbon, and oxygen in the ratio 1:1:3.

The formula of a compound starts with the symbol of the most metallic element, followed by the other elements in the formula in order of decreasing metallic character. As we saw in Section 3.6C, the relative metallic character of an element can be predicted from its position in the periodic table; metallic character increases from top to bottom within a column and decreases from left to right within a period.

Example 4.1

Write the formula of the compound that contains:

a. 1 atom of carbon, 2 atoms of potassium, 3 atoms of oxygen

b. 3 atoms of oxygen, 1 atom of sulfur

Solution

a. Among these three elements, potassium is the most metallic and oxygen the least metallic. Therefore, the elements are written in the order: potassium, carbon, and oxygen. Each symbol is followed by a subscript that shows how many atoms of each element are in the compound. Hence, the formula for this compound is:

$$K_2CO_3$$

b. Sulfur is more metallic than oxygen (it is below oxygen in column VI of the periodic table), so it should be listed first. The formula is:

$$SO_3$$

Problem 4.1

Write the formula of the compound that contains:

a. 4 atoms of oxygen, 1 atom of magnesium, 1 atom of sulfur

b. 3 atoms of oxygen, 2 atoms of nitrogen

Some compounds contain polyatomic ions (Section 3.6B) that occur more than once in the formula. The formula of such a compound shows the formula of the polyatomic ion in parentheses, followed by a subscript that states how many of these ions are in the compound. For example, the formula of calcium hydroxide, $Ca(OH)_2$, shows that for each calcium ion there are two hydroxide ions. The formula of ammonium sulfate, $(NH_4)_2SO_4$, shows that there are two ammonium ions for each sulfate ion.

Example 4.2

How many hydrogen atoms are in a unit of ammonium phosphate, $(NH_4)_3PO_4$?

Solution

Each ammonium phosphate unit contains 3 ammonium ions. Each ammonium ion contains 4 hydrogen atoms. Therefore, each ammonium phosphate unit contains (3×4) or 12 hydrogen atoms.

Problem 4.2

How many oxygen atoms are in each unit of calcium nitrate, $Ca(NO_3)_2$?

A. Empirical and Molecular Formulas

The **molecular formula** of a compound states the number of atoms of each element found in the simplest freely existing unit of the compound. The molecular formula is not always the simplest ratio of atoms. For example, the molecular formula of butene, C_4H_8, shows that each freely existing molecule of butene contains four atoms of carbon and eight atoms of hydrogen. One molecule of ethylene (molecular formula, C_2H_4) contains two atoms of carbon and four atoms of hydrogen. Both butene and ethylene contain two hydrogen atoms for each carbon atom. We can express this ratio by saying that butene and ethylene have the same empirical formula, CH_2. The **empirical formula** of a compound shows the smallest whole-number ratio among atoms of different elements in that compound; the molecular formula of a compound shows the *actual* number of atoms in the simplest freely existing unit of that compound. The molecular formula of a compound is always a multiple of its empirical formula, except when the two formulas are identical. For example, the molecular formula of hydrogen peroxide (H_2O_2) is twice its empirical formula (HO). For some molecules, such as H_2O, the molecular formula is the same as the empirical formula.

As we shall see later, the formulas of compounds containing ions are always empirical, showing the simplest ratio between atoms. Compounds that do not contain ions exist as molecules. The molecular formula is frequently a multiple of the empirical formula.

Example 4.3

Acetic acid has the molecular formula $C_2H_4O_2$.

a. What is its empirical formula?

b. Does acetic acid have the same empirical formula as ethylene glycol, whose molecular formula is $C_2H_6O_2$?

Solution

a. The subscripts of the acetic acid formula are all multiples of 2. Therefore, the empirical formula of acetic acid is CH_2O.

b. Dividing the subscripts of the molecular formula of ethylene glycol by 2 gives the empirical formula CH_3O, which is different from the empirical formula of acetic acid.

Problem 4.3

List those compounds in Table 4.2 that have the same empirical formula, and give that common formula. For each compound, determine the multiple that its molecular formula is of the common empirical formula.

B. Formula Weights

The **formula weight** of a compound or ion is the sum of the atomic weights of all elements in the compound or ion, with each element's atomic weight being multiplied by the number of atoms of that element appearing in the compound or ion's formula.

Example 4.4

a. The molecular formula of sulfuric acid is H_2SO_4. Calculate its formula weight.

b. What is the formula weight of the nitrate ion, NO_3^-?

Solution

a.

	Atomic Weight		Number of Atoms in the Formula		Amu Contributed by Element
hydrogen	1.008	×	2	=	2.016
sulfur	32.06	×	1	=	32.06
oxygen	16.00	×	4	=	64.00
			Answer:		98.076 or 98.08 amu
b. nitrogen	14.01	×	1	=	14.01
oxygen	16.00	×	3	=	48.00
			Answer:		62.01 amu

Problem 4.4

a. Aspirin has the molecular formula $C_9H_8O_4$. Calculate its formula weight.

b. What is the formula weight of the ammonium ion, NH_4^+?

In Section 2.7C, the mole was defined as 6.02×10^{23} units. At that time, we talked about moles of atoms and you learned that one mole of atoms of a particular element has a weight in grams equal to its atomic weight in atomic mass units. In Example 4.4, we calculated that the formula weight of sulfuric acid is 98.08 amu. We then know that one mole of sulfuric acid weighs 98.08 g. We also know that one mole of sulfuric acid contains:

> 2 moles hydrogen atoms weighing (2 × 1.008)g or 2.016 g
> 1 mole sulfur atoms weighing (1 × 32.06)g or 32.06 g
> 4 moles oxygen atoms weighing (4 × 16.00)g or 64.00 g

As another example, one mole of carbon tetrachloride, CCl_4, weighs 153.81 g and contains:

> 1 mole carbon atoms weighing (1 × 12.01)g or 12.01 g
> 4 moles chlorine atoms weighing (4 × 35.45)g or 141.80 g

Some problems require samples with a mass that is greater or less than one mole. In these cases, the mass of one mole of the compound can be used as a conversion factor, as in the following example.

Example 4.5

An experimental procedure requires 1.76 moles of glucose, $C_6H_{12}O_6$. What mass of glucose is required?

Solution

The mass of 1 mole glucose is:

carbon   6 × 12.01 = 72.06 g
hydrogen 12 × 1.008 = 12.096 g
oxygen   6 × 16.00 = 96.00 g
                            180.156 g or 180.16 g

Using the steps developed in Section 1.4B:

Wanted:
? g glucose

Given:
1.76 moles glucose

Conversion factor:
1 mole glucose = 180.16 g

Equation:

$$? \text{ g glucose} = 1.76 \text{ moles glucose} \times \frac{180.16 \text{ g glucose}}{1 \text{ mole glucose}}$$

Answer:
317 g glucose (the answer is given to three significant figures because that is the number of significant figures in 1.76 moles glucose)

Example 4.6

How many atoms of oxygen are in 0.262 g carbon dioxide, CO_2?

Solution

The weight of 1 mole carbon dioxide is:
carbon 1 × 12.01 = 12.01 g
oxygen 2 × 16.00 = 32.00 g
                        44.01 g

Wanted:
? atoms oxygen

Given:
0.262 g carbon dioxide

Conversion factors:
1 mole CO_2 = 44.01 g
1 mole CO_2 contains 2 moles of oxygen atoms
1 mole of atoms is 6.02×10^{23} atoms

Equation:

$$? \text{ atoms oxygen} = 0.262 \text{ g CO}_2 \times \frac{1 \text{ mole CO}_2}{44.01 \text{ g CO}_2}$$

$$\times \frac{2 \text{ moles oxygen atoms}}{1 \text{ mole CO}_2} \times \frac{6.02 \times 10^{23} \text{ atoms}}{1 \text{ mole atoms}}$$

Answer:

7.17×10^{21} atoms oxygen

In Example 4.6, notice that each factor using the mole states the chemical composition of the mole: "1 mole CO_2" and "2 moles oxygen atoms." As problems become increasingly complex, this bookkeeping habit becomes especially important. Notice also that the example deals with *atoms* of oxygen; it is not concerned with the fact that oxygen exists in nature as a diatomic molecule, O_2.

Example 4.7

A solution of glucose contains 9.00 g glucose per 100 cm^3 of solution. How many moles glucose are contained in a liter of this solution?

Solution

Wanted:
? moles glucose/liter of solution

Given:
9.00 g glucose/100 cm^3 solution

Conversion factors:
From Example 4.5, 1 mole glucose = 180.16 g
1 liter = 1000 cm^3

Equation:

$$\frac{? \text{ moles glucose}}{1 \text{ liter solution}} = \frac{9.00 \text{ g glucose}}{100 \text{ cm}^3} \times \frac{1 \text{ mole glucose}}{180.16 \text{ g}} \times \frac{1000 \text{ cm}^3}{1 \text{ liter}}$$

Answer:
0.500 mole glucose/liter of solution

Problem 4.5

Calculate the mass of 0.875 mole of carbon dioxide.

Problem 4.6

Calculate the number of hydrogen atoms in 5.32×10^{-3} g of ammonia, NH_3.

Problem 4.7

Calculate the number of moles of sulfuric acid, H_2SO_4, in 1 liter of solution, if 200 cm^3 of solution contain 6.23 g of the acid.

C. Percent Composition

Percent means parts per hundred. The **percent composition** of a compound is the number of grams of each element or group of elements in one hundred grams of the compound, expressed as a percent. For example, the percent composition of sodium chloride (NaCl) can be calculated from the atomic weights of sodium and chlorine and the formula weight of NaCl.

Formula weight of NaCl: 23.00 g + 35.45 g = 58.45 g

$$\text{Percent sodium:} \frac{23.00 \text{ g Na}}{58.45 \text{ g NaCl}} \times 100\% = 39.35\% \text{ sodium}$$

$$\text{Percent chlorine:} \frac{35.45 \text{ g Cl}}{58.45 \text{ g NaCl}} \times 100\% = 60.65\% \text{ chloride}$$

Example 4.8

Calculate the percent nitrogen in the fertilizer ammonium sulfate $(NH_4)_2SO_4$.

Solution

1. Calculate the formula weight of $(NH_4)_2SO_4$:

The formula weight of NH_4^+ is:

$$
\begin{array}{lll}
\text{N:} & 1 \times 14.01 = & 14.01 \\
\text{4H:} & 4 \times 1.008 = & 4.032 \\
\hline
 & & 18.04 \\
\end{array}
$$

The formula weight of SO_4^{2-} is:

$$
\begin{array}{lll}
\text{S:} & 1 \times 32.06 = & 32.06 \\
\text{O:} & 4 \times 16.00 = & 64.00 \\
\hline
 & & 96.06 \\
\end{array}
$$

The formula weight of $(NH_4)_2SO_4$ is:

$$
\begin{array}{lll}
(NH_4)_2^+ : & 2 \times 18.04 = & 36.08 \\
SO_4^{2-} : & 1 \times 96.06 = & 96.06 \\
\hline
 & & 132.14 \text{ amu} \\
\end{array}
$$

2. Calculate the percent nitrogen:

Each formula unit contains 2 nitrogen atoms.
Therefore, there are 28.02 g (2 × 14.01) of nitrogen in 132.14 g ammonium sulfate.

$$\text{Percent nitrogen} = \frac{28.02 \text{ g N}}{132.14 \text{ g } (NH_4)_2SO_4} \times 100\% = 21.20\% \text{ N}$$

increasing electronegativity

	I	II											III	IV	V	VI	VII
1	H 2.1																
2	Li 1.0	Be 1.5											B 2.0	C 2.5	N 3.0	O 3.5	F 4.0
3	Na 0.9	Mg 1.2			Transition elements								Al 1.5	Si 1.8	P 2.1	S 2.5	Cl 3.0
4	K 0.8	Ca 1.0	Sc 1.4	Ti 1.5	V 1.6	Cr 1.7	Mn 1.6	Fe 1.8	Co 1.9	Ni 1.9	Cu 2.0	Zn 1.6	Ga 1.8	Ge 2.0	As 2.2	Se 2.6	Br 2.8
5	Rb 0.8	Sr 1.0	Y 1.2	Zr 1.3	Nb 1.6	Mo 2.2	Tc —	Ru 2.2	Rh 2.3	Pd 2.2	Ag 1.9	Cd 1.7	In 1.8	Sn 1.8	Sb 2.0	Te 2.1	I 2.5
6	Cs 0.79	Ba 0.9						Pt 2.3	Au 2.5	Hg 2.0	Tl 2.0	Pb 2.3	Bi 2.0	Po —			

increasing electronegativity

Figure 4.1 Electronegativities of common elements (Pauling scale).

It is more accurate to say that these two types of bonds represent the two extremes between which most bonds fall. The reason for this is that elements differ in the degree to which they attract electrons. This attraction is measured by the electronegativity of an element. **Electronegativity** is defined as a measure of the tendency of an atom to attract the pair of electrons it shares with another atom in a chemical bond.

The scale of electronegativities was developed by the American chemist Linus Pauling (b. 1901). On this scale, fluorine, the most electronegative element, has an electronegativity of 4.0. Carbon has an electronegativity of 2.5, hydrogen 2.1, and sodium 0.9. Figure 4.1 shows the electronegativities of the elements with which we deal most often. Notice that the electronegativity of most metals is close to 1.0 and that the electronegativity of a nonmetal, although dependent on its location in the table, is always greater than 1.0. In general, the electronegativity of an element increases going up a column and from left to right across a period.

When two atoms combine, the nature of the bond between them is determined by the difference between their electronegativities.

Example 4.12

Calculate the difference in electronegativity between the following pairs of atoms.

a. Na and Cl **b.** C and H **c.** Mg and Cl

Solution

a. $3.0 - 0.9 = 2.1$

b. $2.5 - 2.1 = 0.4$

c. $3.0 - 1.2 = 1.8$

Problem 4.12

Calculate the difference in electronegativity between the following pairs of atoms.

a. O and H   **b.** N and H   **c.** K and F

When two bonding atoms differ by more than about two units in electronegativity, the bond formed is ionic. The more electronegative atom will have completely removed one (or more) electron(s) from the less electronegative atom. When the difference in electronegativity is less than two, the electrons will be shared more or less evenly, depending on the exact difference in electronegativity. The bond formed is covalent. For example, in sodium chloride, the difference in electronegativity between sodium and chlorine is 2.1 units, so the bond between sodium and chlorine is ionic. In methane, CH_4, the difference in electronegativity between carbon and hydrogen is 0.4 units, only a small difference. Thus, the carbon-hydrogen bond in methane is covalent.

4.4 Ionic Compounds

Ionic compounds are neutral combinations of positive and negative ions. These ions may be single atoms that carry a charge because they have lost or gained electrons, or they may be **polyatomic ions,** groups of atoms bonded together covalently and carrying a positive or negative charge.

A. Names and Formulas of Ionic Compounds

The formula of an ionic compound is an empirical one, because these compounds exist not as discrete molecules with definite molecular formulas but as neutral collections of ions.

In naming an ionic compound, the cation is named first, then the anion. (For the names of the individual ions, review Section 3.6B.) Thus:

$(NH_4)_2CO_3$ is named ammonium carbonate
$FeCl_2$ is named iron(II) chloride or ferrous chloride

The number of ions in a single formula unit of a compound is not part of the name. However, it is part of the formula, because there must be as many units of positive charge contributed by the cations as there are units of negative charge contributed by the anions.

Example 4.13

Name the following ionic compounds:

a. $CaCl_2$ b. Na_2SO_4 c. $FeCl_3$ d. $Hg(NO_3)_2$ e. BaO

Solution

a. calcium chloride

b. sodium sulfate

c. iron(III) chloride or ferric chloride

d. mercury(II) nitrate or mercuric nitrate

e. barium oxide

Example 4.14

Write the formulas of the following ionic compounds:

a. potassium nitrate   b. magnesium sulfate

c. tin(II) fluoride   d. iron(III) nitrate

e. copper(II) oxide   f. copper(I) oxide

Solution

		Sum of ionic charges
a.	KNO_3	$1 + (-1) = 0$
b.	$MgSO_4$	$2 + (-2) = 0$
c.	SnF_2	$2 + 2(-1) = 0$
d.	$Fe(NO_3)_3$	$3 + 3(-1) = 0$
e.	CuO	$2 + (-2) = 0$
f.	Cu_2O	$2(1) + (-2) = 0$

Notice that the sum of the ionic charges is zero in all cases.

Problem 4.13

Name the following ionic compounds:

a. $BaSO_4$ b. $Fe(NO_3)_2$ c. Cr_2O_3 d. LiF

Problem 4.14

Write the formulas of the following ionic compounds:

a. lead(II) chloride   b. ammonium nitrate

c. iron(II) sulfide   d. potassium bicarbonate

e. magnesium oxide   f. cobalt(III) nitrate

B. Solutions of Ionic Compounds

When an ionic compound dissolves in water, the solution contains ions, rather than neutral particles. For example, when sodium chloride dissolves in water, the solution contains sodium ions (Na^+) and chloride ions (Cl^-), rather than neutral units of NaCl. In a solution of sodium nitrate, there are sodium ions and nitrate ions; there are no nitrogen ions or oxygen ions. In a solution of ammonium sulfate, there are ammonium ions and sulfate ions. A polyatomic ion does not break up into separate atoms in solution.

The presence of ions in a solution of an ionic compound can be demonstrated with an apparatus like that shown in Figure 4.2. The apparatus consists of two electrodes, one connected to a power source, the other connected to a light bulb which is, in turn, connected to the other pole of the power source. If the two electrodes touch, an electric current flows through the completed circuit and the bulb lights (Figure 4.2b). If the two electrodes are separated and immersed in water, no current flows (Figure 4.2c). Pure water cannot carry an electric current. However, if the pure water is replaced by an aqueous (water) solution of an ionic compound, the bulb lights, indicating that there is a flow of electricity through the solution (Figure 4.2d). The electric current is carried through the solution by the dissolved ions. Ionic compounds are **electrolytes,** compounds whose aqueous solutions conduct electricity. Compounds whose solutions in water do not conduct electricity are called **non-electrolytes.**

Ionic compounds exhibit enormous differences in their solubility in water. Table 4.4 lists the solubilities of several ionic solids in cold water. Some ionic solids, such as barium iodide and silver nitrate, are very soluble in water, while others, such as lead chloride, are only slightly soluble. As a rule of thumb, if

Figure 4.2 The conduction of electricity by an aqueous solution of an ionic compound.

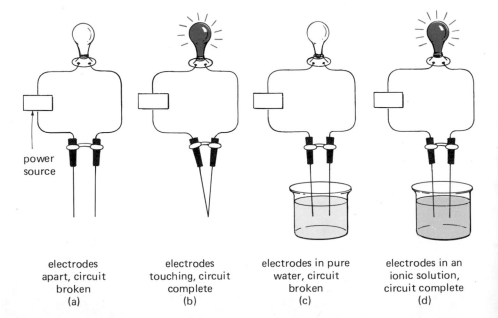

power source

electrodes apart, circuit broken
(a)

electrodes touching, circuit complete
(b)

electrodes in pure water, circuit broken
(c)

electrodes in an ionic solution, circuit complete
(d)

Table 4.4 The solubility of ionic solids in water.

Compound	Solubility g/100 mL H_2O	Compound	Solubility g/100 mL H_2O
BaI_2	170	$NaHCO_3$	6.9
$AgNO_3$	122	$PbCl_2$	0.99
NH_4NO_3	118	$CaCO_3$	1.4×10^{-3}
NaOH	42	$BaSO_4$	2.2×10^{-4}
NaCl	35.7	AgCl	8.9×10^{-5}

less than 0.1 g of an ionic solid will dissolve in 100 cm^3 of water, the compound is said to be insoluble. Thus, calcium carbonate, barium sulfate, and silver chloride are classified as insoluble compounds.

C. Other Characteristic Properties of Ionic Compounds

Ionic compounds are usually solids at room temperature and have very high melting points. The melting point of sodium chloride is 801°C, that of calcium nitrate is 561°C, and that of barium oxide is 1923°C. Ionic compounds are crystalline solids. The ions of the compound are arranged in a **crystal lattice,** in which the nearest neighbors of the cations are anions and the nearest neighbors of the anions are cations. The photograph in Figure 4.3 (left) shows the cubic crystals of sodium chloride. Notice the sharp angles and smooth planes of the crystals. Figure 4.3 (right) is a drawing of the ionic lattice of sodium chloride. Notice the alternating anions and cations. It is the strong electrostatic bonds between the oppositely charged ions that makes ionic compounds strongly resistant to melting and, at times, only minimally soluble in water.

Figure 4.3 (Left) Crystals of rock salt (sodium chloride). (Photo courtesy of Morton Salt, division of MortonNorwich.) (Right) The crystal lattice of sodium chloride.

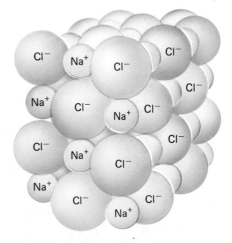

4.5 Structures Containing Covalent Bonds

In contrast to ionic compounds, which exist as neutral collections of anions and cations, wholly covalent compounds exist as molecules. The atoms they contain are joined by the sharing of electron pairs. Covalent bonds may also occur in polyatomic ions, such as sulfate ion (SO_4^{2-}), nitrate ion (NO_3^-), and ammonium ion (NH_4^+). Molecular compounds have no charge; the polyatomic ions have an excess or deficiency of electrons, giving them either a negative or a positive charge, respectively. Whereas ionic compounds are usually solid under normal conditions, wholly covalent compounds may be solid (sugar), liquid (water), or gaseous (hydrogen).

A. Characteristics of a Covalent Molecule

The simplest example of a covalent bond is that found in the hydrogen molecule. When two hydrogen atoms combine, the single valence electron from each combines to form an electron pair that is shared between the two atoms. The shared pair serves to complete the valence shell of each hydrogen, so that each atom has the electron configuration of helium.

$$H\cdot + \cdot H \longrightarrow H\!:\!H \qquad \text{a shared pair of electrons}$$

Hydrogen chloride (HCl), water (H_2O), ammonia (NH_3), and methane (CH_4) are compounds containing chemical bonds formed by the sharing of electron pairs. The structures of these molecules are shown in Table 4.5, using Lewis structures, which we discussed in Chapter 3. In these compounds, each nucleus has an electron configuration resembling that of a noble gas. In the structure for water, each hydrogen atom shares one electron pair with the oxygen atom, so each has a filled outer shell resembling that of helium. The

$$H\cdot + :\ddot{\underset{\cdot\cdot}{C}}l: \longrightarrow H\!:\!\ddot{\underset{\cdot\cdot}{C}}l: \qquad \text{hydrogen chloride}$$

$$2H\cdot + \cdot\ddot{\underset{\cdot\cdot}{O}}: \longrightarrow H\!:\!\underset{\overset{\displaystyle|}{H}}{\ddot{O}}: \qquad \text{water}$$

$$3H\cdot + \cdot\ddot{N}: \longrightarrow H\!:\!\underset{\overset{\displaystyle|}{H}}{\ddot{N}}\!:\!H \qquad \text{ammonia}$$

$$4H\cdot + \cdot\ddot{C}: \longrightarrow H\!:\!\overset{\overset{\displaystyle H}{|}}{\underset{\overset{\displaystyle|}{H}}{C}}\!:\!H \qquad \text{methane}$$

Table 4.5 Lewis structures of some simple molecules with covalent bonds.

oxygen atom in water is surrounded by eight electrons (four shared with the hydrogens and four unshared) and has a filled outer shell like that of neon. Of the eight electrons surrounding oxygen, two pair are part of covalent bonds and are called **bonding electrons.** The other two pair of electrons are not involved in bonding and are called **nonbonding electrons,** or **unshared electron pairs.** Notice that in these compounds the electronegativity difference among the bonding atoms is less than 2 and that each atom is showing its characteristic *valence* (Section 4.3A).

B. Lewis Structures of Molecules

Many properties of molecular compounds and polyatomic ions can be predicted from the arrangement of atoms and electrons in the compound or ion. The most convenient way to show this arrangement is by means of a Lewis structure. The structures shown in Table 4.5 are Lewis structures; in this section we will describe how to draw these structures.

The following guidelines are used:

1. The atomic skeleton of the structure must be established. To do this, determine the **central atom.** This atom will be the most electropositive (least electronegative) atom and the one with the highest valence. In some structures there will be two or more similar atoms, all roughly equal in electropositivity and valence; these will form the backbone of the molecule.

2. Arrange the other atoms symmetrically around the central atom(s). Hydrogen and the halogens, both having valences of one, must be at the edge of the structure.

3. Determine the number of valence electrons in the molecule. This will be the sum of the valence electrons in the individual atoms. For ions, add one electron for each unit of negative charge on the ion, or subtract one electron for each unit of positive charge on the ion.

4. Arrange the electrons in pairs around the atoms so that each atom in the molecule or ion has a complete outer shell of electrons. Each hydrogen atom must be surrounded by two electrons; most other atoms must be surrounded by eight valence electrons, although some, like sulfur and phosphorus in the third row of the periodic table, can have ten or twelve valence electrons.

5. Show a pair of electrons involved in a covalent bond (bonding electrons) as a dash; show an unshared pair of electrons (nonbonding electrons) as a pair of dots. In Table 4.5, all electrons are shown as dots to emphasize the filled shells. In the future, we will follow this rule.

6. Two atoms may be bonded together with single, double, or triple bonds. In a single bond, they share one pair of electrons; in a double bond, they share two pairs of electrons; and in a triple bond, they share three pairs of electrons.

Table 4.6 Lewis structures of several small molecules. The number of valence electrons is given in parentheses.

H—B̈r:	H—C—C̈l: with :C̈l: above and :C̈l: below	:C̈l—P̈—C̈l: with :C̈l: below	H—C≡N:
HBr (8) hydrogen bromide	CHCl$_3$ (26) chloroform	PCl$_3$ (26) phosphorus trichloride	HCN (10) hydrogen cyanide

H, H on C=C, H, H	H—C≡C—H	H, H on C=Ö:	Ö above, H—C—H with Ö below left and Ö below right
C$_2$H$_4$ (12) ethylene	C$_2$H$_2$ (10) acetylene	CH$_2$O (12) formaldehyde	H$_2$CO$_3$ (24) carbonic acid

Shown in Table 4.6 are the names, molecular formulas, and Lewis structures for several small molecules. After the name of each is shown the number of valence electrons it contains. Notice that in these neutral molecules, each hydrogen atom is surrounded by two valence electrons and each atom of carbon, nitrogen, and chlorine is surrounded by eight valence electrons. Furthermore, each carbon atom has four bonds; each nitrogen atom has three bonds and one unshared pair of electrons; and each oxygen atom has two bonds and two unshared pair of electrons. Each chlorine has one bond and three unshared pair of electrons. Notice also that the most electropositive atom is central in each molecule and that hydrogen is always at the edge of the structure.

Example 4.15

Draw Lewis structures, showing all valence electrons, for the following molecules.

a. H$_2$O$_2$　　　　　　　　**b.** CO$_2$　　　　　　　　**c.** CH$_3$OH

Solution

a. Since oxygen is more electronegative than hydrogen and the two oxygens are alike, they must be bonded together to form the backbone of the molecule. The most symmetric structure has one hydrogen bonded to each oxygen, giving the skeletal arrangement:

H　O　O　H

The Lewis structure for hydrogen peroxide, H$_2$O$_2$, must show 14 valence electrons: 12 from the two oxygen atoms and 2 from the two hydrogen atoms:

H—Ö—Ö—H

With three shared pairs of electrons and four unshared pairs of electrons, this structure has the correct number of valence electrons. Each hydrogen has 2 valence electrons and an electron configuration like that of a helium atom. Each oxygen has 8 valence electrons and an electron configuration like that of a neon atom.

b. Carbon is the most electropositive atom and therefore central. A symmetric structure has both oxygens bonded to the central carbon, giving the skeleton:

O   C   O

There are 16 valence electrons to show. Sixteen electrons in single bonds will not give each atom a complete outer shell. Therefore, the molecule must have either two carbon-oxygen double bonds or one carbon-oxygen single bond and one carbon-oxygen triple bond. Symmetry suggests that both oxygens are bonded in the same way to the central carbon and, therefore, favors the two carbon-oxygen double bonds. This gives the Lewis structure:

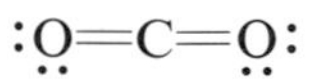

$$:\ddot{O}{=}C{=}\ddot{O}:$$

With 4 shared pairs and 4 unshared pairs of electrons, this structure has the required 16 electrons. Furthermore, each atom of carbon and oxygen has a complete octet. Notice that in this Lewis structure, carbon has four bonds and oxygen has two bonds and two unshared pairs of electrons.

c. The formula CH_3OH implies three hydrogens bonded to the carbon and one hydrogen bonded to oxygen. Carbon will be the central atom, giving the skeletal arrangement:

```
        H

   H    C    O    H

        H
```

The structure must show 14 valence electrons: 4 from the single carbon atom, 4 from the four hydrogen atoms, and 6 from the oxygen atom. Distributing the electrons, we get:

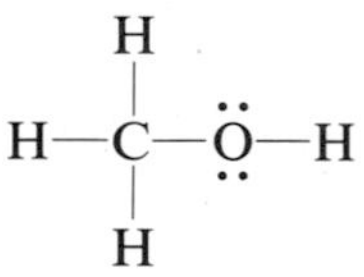

```
        H
        |
   H — C — Ö — H
        |
        H
```

This structure shows five single bonds and two unshared pairs of electrons. It has the correct number of valence electrons. In addition, each hydrogen has 2 valence electrons, and carbon and oxygen each have 8 valence electrons. Notice that oxygen has two covalent bonds and two unshared pairs of electrons.

Problem 4.15

Draw Lewis structures, showing all valence electrons, for the following molecules:

a. CH_3Cl **b.** HCN **c.** CS_2

C. Lewis Structures of Ions; Formal Charges

To draw a Lewis structure of an ion, follow the same steps you have just used for molecules. First, calculate the number of valence electrons contributed by the individual atoms in the ion. Then, add one additional electron for each negative charge on the ion and subtract one electron for each positive charge on the ion.

In discussing the properties of these covalently bonded ions and molecules, it is frequently useful to know which atom bears the positive or the negative charge. This so-called **formal charge** can be easily calculated from the Lewis structure.

To determine formal positive or formal negative charges, first assign to each particular atom all its unshared (nonbonding) electrons and half of all its shared (bonding) electrons. Second, subtract this number from the number of valence electrons in the neutral, unbonded atom. The difference is the formal charge.

Example 4.16

Draw Lewis structures for the following ions. Show which atom in the ion bears the formal charge.

a. H_3O^+ **b.** NH_4^+ **c.** HCO_3^-

Solution

a. The hydronium ion, H_3O^+, contains 8 valence electrons: 6 from oxygen, 3 from the three hydrogens, less 1 for the single positive charge. Each hydrogen in H_3O^+ is assigned one valence electron, just as in an isolated hydrogen atom. Therefore, each hydrogen has $1 - 1 = 0$ formal charge. Oxygen is assigned 5 valence electrons, one less than in an isolated oxygen atom. Thus, the oxygen has a formal charge of $+1$.

$$H-\overset{\displaystyle ..+}{\underset{\displaystyle |}{O}}-H \qquad \textit{assigned 5 valence electrons;}$$
$$\underset{\displaystyle H}{} \qquad \textit{formal charge of } +1$$

b. There are 8 valence electrons in the ammonium ion. Nitrogen is assigned 4 valence electrons, one less than in an isolated nitrogen atom, so the nitrogen has a formal charge of $+1$.

$$H-\overset{\overset{\displaystyle H}{|}}{\underset{\underset{\displaystyle H}{|}}{N^+}}-H \qquad \textit{assigned 4 valence electrons;}\\ \textit{formal charge of } +1$$

c. The bicarbonate ion contains 24 valence electrons: 18 from the three oxygens, 4 from carbon, 1 from hydrogen, plus an additional electron for the single negative charge. Because carbon is assigned four valence electrons, the same number in an isolated carbon atom, it has a formal charge of 0. Two of the oxygens are assigned 6 valence electrons each and also have a formal charge of 0. The third oxygen is assigned 7 valence electrons, one more than a neutral, unbonded oxygen atom, so it has a formal charge of −1.

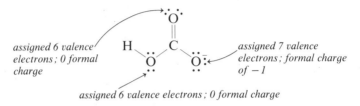

assigned 6 valence electrons; 0 formal charge

assigned 7 valence electrons; formal charge of −1

assigned 6 valence electrons; 0 formal charge

Problem 4.16

Draw Lewis structures for the following ions and show which atom in the ion bears the formal charge.

a. CO_3^{2-} **b.** OH^- **c.** NO_3^-

D. Resonance

As chemists began to work with the Lewis structures, it became more and more obvious that, for a great many molecules and ions, no single Lewis structure provided a truly accurate representation. For example, a Lewis structure for the carbonate ion, CO_3^{2-}, shows carbon bonded to three oxygen atoms by a combination of one double bond and two single bonds. The three possible Lewis structures for CO_3^{2-} are shown in Figure 4.4. Each implies that one carbon-oxygen bond is different from the other two. However, this is not the case; rather, it has been shown that all three bonds are identical.

Figure 4.4 Three possible Lewis structures for the carbonate ion.

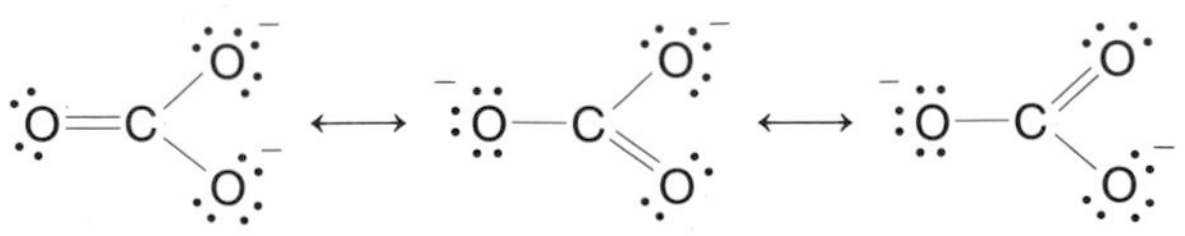

Figure 4.5 The carbonate ion represented as a resonance hybrid of three contributing structures.

(a)   (b)   (c)

To describe molecules and ions, like the carbonate ion, for which no single Lewis structure is adequate, the theory of **resonance** was developed by Linus Pauling in the 1930s. According to this theory, many molecules and ions are best described by writing two or more Lewis structures and considering the real molecule or ion as a hybrid of these structures. The individual Lewis structures are called **contributing structures.** We show that the real structure is a hybrid of the various contributing structures by connecting them with double-headed arrows. The resonance hybrid for the carbonate ion is shown in Figure 4.5.

It is important to remember that the carbonate ion or any other compound we describe in this way has one, and only one, real structure. The problem is that our systems of representation are not adequate to describe the real structures of molecules and ions. The resonance method is a particularly useful way to describe the structure of these compounds, for it retains the use of Lewis structures with electron-pair bonds. While we fully realize that the carbonate ion is not accurately represented by contributing structure (a), (b), or (c), nevertheless, we will continue to represent this ion using one of these contributing structures for convenience, understanding that what is intended is the resonance hybrid.

Example 4.17

Show that sulfur trioxide can be represented by a resonance hybrid of three contributing structures.

Solution

The possible Lewis structures of sulfur trioxide are:

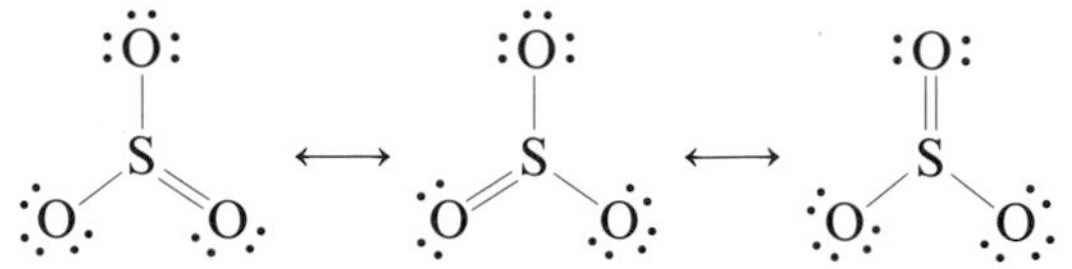

All are equivalent; therefore, the molecule exhibits resonance and is a hybrid of these structures.

Problem 4.17

Show that ozone, O_3, is a resonance hybrid.

4.6 Bond Angles and Shapes of Molecules

In the preceding section, we used a shared pair of electrons as the fundamental unit of the covalent bond. We then drew Lewis structures for several small molecules and ions containing various combinations of single, double, and triple bonds. In this section we will learn how to predict bond angles in these and other covalent molecules and ions, using the **valence-shell electron-pair repulsion (VSEPR) model.**

The VSEPR model can be explained in the following way. We know that an atom has an outer shell of valence electrons. These valence electrons may be involved in the formation of single, double, or triple bonds, or they may be unshared. Each combination of electrons in a bond creates a negatively charged region of space. We have already learned that like charges repel each other. Therefore, the VSEPR model states that the various regions of high electron density around an atom spread out so that each is as far apart from the others as possible.

A. Structures with a Central Atom Surrounded by Four Regions of High Electron Density

Let us use the VSEPR model to predict the shape of methane, CH_4. The Lewis structure of methane (Figure 4.6) shows a central carbon atom surrounded by four separate regions of high electron density. Each region consists of a pair of electrons forming a bond from the carbon to a hydrogen atom. According to the VSEPR model, these regions of high electron density spread out from the central carbon atom in such a way that they are as far from each other as possible.

Figure 4.6 The shape of the methane molecule, CH_4. (a) Lewis structure; (b) the tetrahedral shape; (c) a ball-and-stick model; (d) a space-filling model.

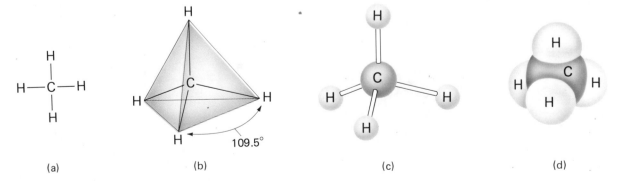

(a)   (b)   (c)   (d)

You can approximate the shape of a methane molecule with a Styrofoam ball and four toothpicks. Poke the toothpicks into the ball, making sure that the free ends of the toothpicks are as far from one another as possible. If you have done this correctly, the angle between any two toothpicks will be 109.5°. If you now cover this model with four triangular pieces of paper, you will have built a four-sided figure called a *regular tetrahedron*. Figure 4.6 shows the Lewis structure for methane (a), the tetrahedral arrangement of the four regions of high electron density around the central carbon atom (b), a so-called **ball-and-stick** model of methane (c), and a so-called **space-filling model** of this molecule (d).

According to the VSEPR model, the H—C—H bond angle in methane should be 109.5°. This angle has been measured experimentally and found to be 109.5°. Thus, the bond angle predicted by the VSEPR model is identical to that observed.

We can predict the shape of the ammonia molecule in exactly the same manner. The Lewis structure of NH_3 (Figure 4.7) shows a central nitrogen atom surrounded by four separate regions of high electron density. Three of these regions contain single pairs of electrons forming covalent bonds with hydrogen atoms; the fourth region contains an unshared pair of electrons. According to the VSEPR model, the four regions of high electron density around the nitrogen are arranged in a tetrahedral manner (Figure 4.7), so we predict that each H—N—H bond angle should be 109.5°. The observed bond angle is 107.3°. This small difference between the predicted angle and the observed angle can be explained by proposing that the unshared pair of electrons on nitrogen repels the adjacent bonding pairs more strongly than the bonding pairs repel each other.

Figure 4.7 The shape of the ammonia molecule, NH_3. (a) Lewis structure; (b) the tetrahedral shape; (c) a ball-and-stick model; (d) a space-filling model. Unshared pairs of electrons are not shown in parts (c) and (d).

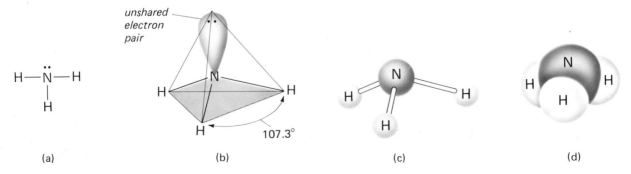

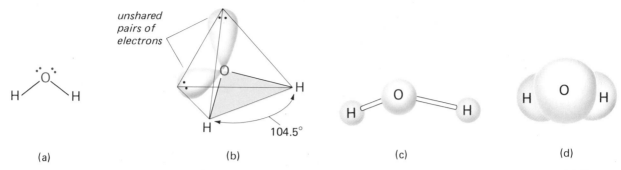

Figure 4.8 The shape of the water molecule H_2O. (a) Lewis structure; (b) the tetrahedral shape; (c) a ball-and-stick model; (d) a space-filling model. Unshared pairs of electrons are not shown in parts (c) and (d).

Figure 4.8 shows the Lewis structure of the water molecule. In H_2O, a central oxygen atom is surrounded by four separate regions of high electron density. Two of these regions contain pairs of electrons used to form covalent bonds with hydrogens; the other two regions contain unshared electron pairs. The four regions of high electron density in water are arranged in a tetrahedral manner around oxygen. Based on the VSEPR model, we predict an H—O—H bond angle of 109.5°. Experimental measurements show that the actual bond angle is 104.5°. The difference between the predicted and observed bond angles can be explained by proposing, as we did for NH_3, that unshared pairs of electrons repel adjacent bonding pairs more strongly than the bonding pairs repel each other. Note that the variation from 109.5° is greatest in H_2O, which has two unshared pairs of electrons; it is smaller in NH_3, which has one unshared pair; and there is no variation in CH_4.

A general prediction emerges from our discussions of the shapes of methane, ammonia, and water: Any time there are four separate regions of high electron density around a central atom, we can accurately predict a tetrahedral distribution of electron density and bond angles of approximately 109.5°.

B. Structures with a Central Atom Surrounded by Three Regions of High Electron Density

In many of the molecules we encounter, a central atom is surrounded by three regions of high electron density. According to the VSEPR model, a *double* bond is treated as a *single* region of electron density. For this reason, the central carbon atom in formaldehyde is considered to be surrounded by three regions of

Figure 4.9 Shapes of the formaldehyde and ethylene molecules. (a) Lewis structures; (b) planar arrangement of the three regions of high electron density around the carbon atom; (c) ball-and-stick models; (d) space-filling models. In these figures, ➤ represents a bond projecting in front of the plane of the paper and ⋯⋯ represents a bond projecting behind the plane of the paper.

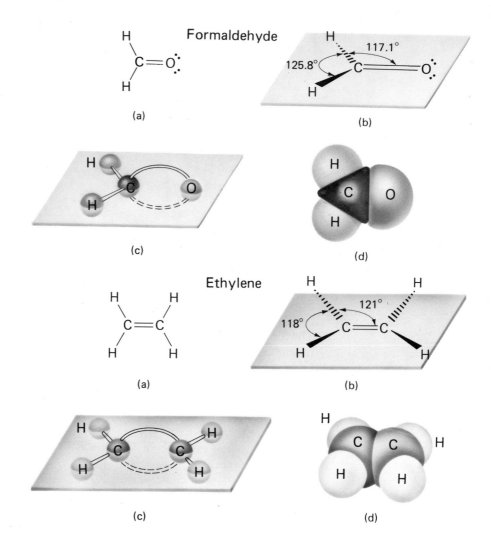

high electron density (Figure 4.9). Two regions contain single pairs of electrons forming bonds to hydrogen atoms; the third region contains two pair of electrons forming a double bond to oxygen. In ethylene, each carbon atom is also surrounded by three regions of high electron density (Figure 4.9): two contain single pairs of electrons and one contains two pair of electrons. If you again experiment with the Styrofoam ball and toothpicks, this time using only three toothpicks, you will find that the free ends of the toothpicks are farthest apart if they are all in the same plane and make angles of 120° with each other in that plane. Thus, the H—C—H and H—C—O bond angles in formaldehyde are predicted to be 120°; the H—C—H and H—C—C bond angles in ethylene are also predicted to be 120°. We describe such an arrangement of three atoms as *trigonal planar*.

C. Structures with a Central Atom Surrounded by Two Regions of High Electron Density

In another type of molecule, a central atom is surrounded by only two regions of high electron density. Shown in Figure 4.10 are Lewis structures of carbon dioxide, CO_2, and acetylene, C_2H_2. In carbon dioxide, the central carbon atom is surrounded by two regions of high electron density. Each region contains a double pair of electrons and forms a double bond to an oxygen atom. I.. acetylene, each carbon is also surrounded by two regions of high electron density. One region contains a single pair of electrons and forms a single bond to a hydrogen atom; the other region contains a triple pair of electrons and forms a triple bond to a carbon atom. In each case, the two regions of high electron density are farthest apart if they form a straight line through the central carbon atom, creating a bond angle of 180°. Because of their bond angles, carbon dioxide and acetylene are called *linear molecules*.

Predictions of bond angles and molecular shapes based on the valence-shell electron-pair repulsion model are summarized in Table 4.7 (next page).

Example 4.18

Predict all bond angles in the following molecules and ions.

a. CH_3Cl **b.** HCN **c.** CH_3COOH

Solution

a. In the Lewis structure of CH_3Cl, carbon is surrounded by four regions of high electron density, each of which forms a single bond. Based on the VSEPR

Figure 4.10 Shapes of (a) carbon dioxide and (b) acetylene molecules.

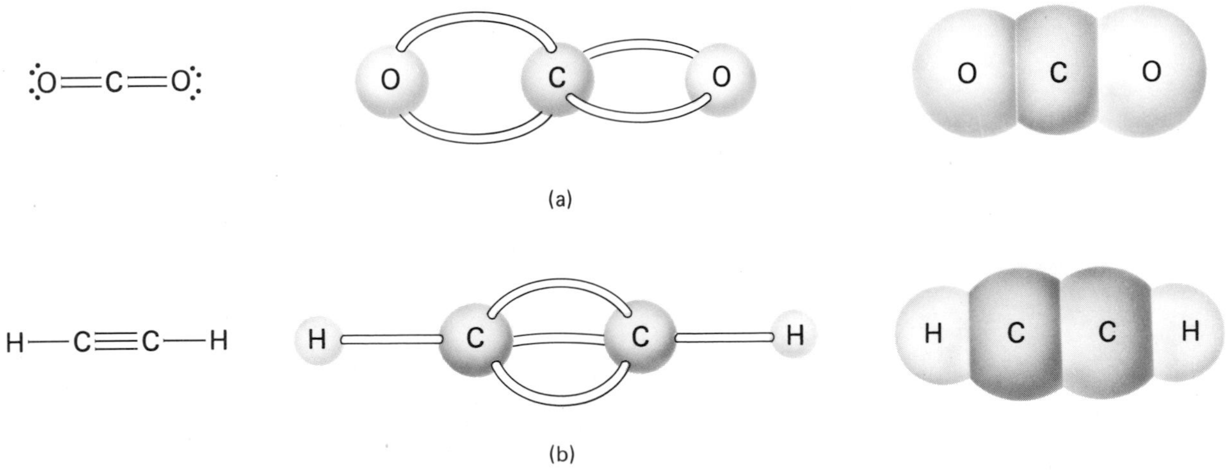

(a)

(b)

Table 4.7 Molecular shapes and bond angles.

Number of Regions of High Electron Density Around Central Atom	Arrangement of Regions of High Electron Density in Space	Predicted Bond Angles	Example
4	tetrahedral	109.5°	CH_4 methane NH_3 ammonia H_2O water
3	trigonal planar	120°	H_2CO formaldehyde H_2CCH_2 ethylene
2	linear	180°	CO_2 carbon dioxide HCCH acetylene

model, we predict a tetrahedral distribution of electron pairs around carbon, H—C—H and H—C—Cl bond angles of 109.5°, and a tetrahedral shape for the molecule.

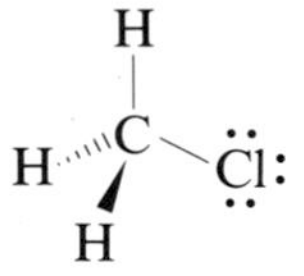

all bond angles
109.5°
(predicted)

b. In the Lewis structure of HCN, carbon is surrounded by two regions of high electron density. Therefore, we predict 180° for the H—C—N bond angle. HCN is a linear molecule.

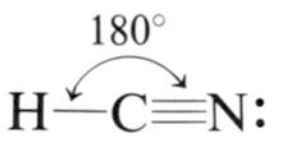

c. The Lewis structure of acetic acid is:

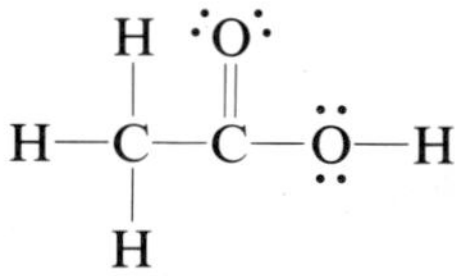

Both the carbon bonded to three hydrogens and the oxygen bonded to carbon and hydrogen are part of tetrahedral structures. The central carbon will have 120° bond angles:

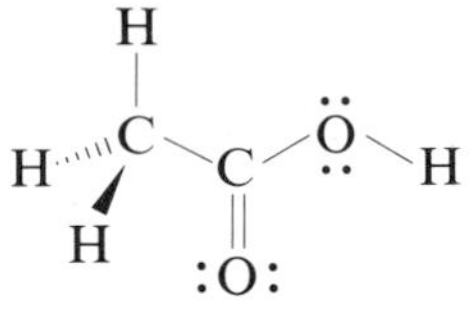

Problem 4.18

Predict all bond angles for the following molecules and ions.

a. CH_3CHO **b.** CO_3^{2-} **c.** CH_2Cl_2 **d.** NH_4^+

4.7 Polarity

A. The Polarity of Covalent Bonds

While all covalent bonds involve the sharing of electrons between atoms, they differ widely in just how equally the electrons are shared. For example, consider the covalent bond between carbon and sulfur. Because the electronegativities of carbon and sulfur are identical, electrons in a C—S covalent bond are shared equally. Covalent bonds in which the sharing of electrons is approximately equal are called **nonpolar covalent bonds.**

A different situation arises when two atoms joined by a covalent bond have different electronegativities. For example, the difference in electronegativity between carbon and fluorine is 1.5 units. The C—F bond is covalent, but the sharing of electrons is quite unequal; the electrons of the bond are attracted to fluorine much more strongly than to carbon. An important consequence of this unequal sharing of electrons is that one atom has a higher concentration of electrons around it; hence, it has a *partial negative charge.* This charge is indicated by the symbol $\delta-$. The other atom in such a bond has a *partial positive charge*, indicated by the symbol $\delta+$. Covalent bonds in which the sharing of electrons is unequal are called **polar covalent bonds.**

Chemical bonds can be classified as nonpolar covalent, polar covalent, or ionic, according to the following guidelines:

nonpolar covalent bond: The electronegativity difference between the atoms bonded together is between 0.0 and 0.4 units.

polar covalent bond: The electronegativity difference between the atoms bonded together is between 0.5 and 1.9 units. In a polar covalent bond, the more electronegative atom has a partial negative charge; the less electronegative atom has a partial positive charge.

ionic bond: The electronegativity difference between the atoms bonded together is 2.0 units or greater.

Figure 4.11 Sharing of electrons in (a) non-polar and (b) polar covalent bonds; (c) transfer of electrons in an ionic bond. Electrons are shared equally in (a) and unequally in (b).

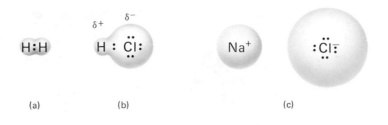

(a)   (b)   (c)

The distribution of electrons in a nonpolar covalent bond, a polar covalent bond, and an ionic bond are illustrated in Figure 4.11.

Example 4.19

Classify the bonds between the following pairs of atoms as nonpolar covalent, polar covalent, or ionic. For each polar covalent bond, show which atom bears the partial positive charge and which the partial negative charge.

a. S—O   **b.** C—O   **c.** Al—F

Solution

a. The electronegativity of oxygen is 3.5 and that of sulfur is 2.5. The difference is 1.0 unit; we predict the S—O bond to be polar covalent. The oxygen is partially negative and the sulfur is partially positive.

$$\overset{\delta^+}{S}—\overset{\delta^-}{O}$$

b. The electronegativity difference between oxygen and carbon is 1.0 unit (3.5 − 2.5). Therefore, we predict the C—O bond to be polar covalent. Since oxygen is the more electronegative of the two, it carries the negative charge.

$$\overset{\delta^+}{C}—\overset{\delta^-}{O}$$

c. The electronegativity difference between fluorine and aluminum is 2.5 units (4.0 − 1.5). Therefore, we predict the Al—F bond to be ionic.

Problem 4.19

Is the bond between carbon and chlorine in a C—Cl pair nonpolar covalent, polar covalent, or ionic? In this pair of atoms, which bears the partial positive charge and which bears the partial negative charge?

B. The Polarity of Covalent Molecules

We have seen that a chemical bond may be nonpolar covalent, polar covalent, or ionic. The polarity of a diatomic molecule is the same as the polarity of the single bond it contains. In order to describe the polarity of a molecule containing more than two atoms, we need to consider not only the polarity of each bond in the molecule but also the bond angles and the shape of the molecule. Shown

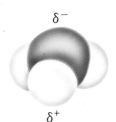

δ^-

δ^+

Figure 4.12 The three-dimensional structure of ammonia, showing polarity.

in Figure 4.12 is the three-dimensional representation of ammonia, NH_3. The difference in electronegativity between nitrogen and hydrogen is 0.9 units. Hence, there will be a partial positive charge on the hydrogens and a partial negative charge on the nitrogen.

The water molecule also contains polar covalent bonds. Oxygen is more electronegative than hydrogen, so both H—O bonds are polar covalent, with hydrogen bearing a partial positive charge and oxygen a partial negative charge. The centers of positive and negative charge within the water molecule do not coincide, so H_2O is a polar molecule (Figure 4.13).

Figure 4.13 The polarity of the water molecule.

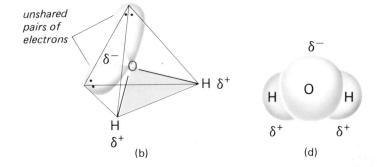

unshared pairs of electrons

δ^-

O

H δ^+

H

δ^+

(b)

δ^-

H   O   H

δ^+ δ^+

(d)

Example 4.20

Predict whether the following molecules are polar or nonpolar.

a. CH_2O **b.** CO_2 **c.** SO_2

Solution

a. The Lewis structure of CH_2O is:

$$\begin{array}{c} H \\ \\ H \end{array} \!\! C \overset{\delta^+}{=} \overset{\delta^-}{\ddot{O}}:$$

The electronegativity difference between carbon and oxygen is 1.0. Thus, the C—O bond will be polar with a partial positive charge on the carbon and a partial negative charge on the oxygen. The C—H bond is not polar because there is only a slight electronegativity difference between carbon and hydrogen. Therefore, the molecule is polar.

b. The Lewis structure of carbon dioxide is:

$$:\overset{\delta^-}{\ddot{O}} = \overset{\delta^+}{C} = \overset{\delta^-}{\ddot{O}}:$$

Both carbon-oxygen bonds are polar but, because they are at opposite ends of the linear molecule, their net effect makes the molecule nonpolar.

c. The Lewis structure of sulfur dioxide is:

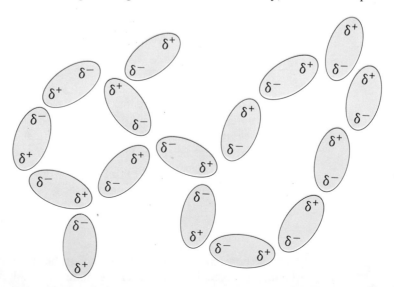

The sulfur-oxygen bond is polar, since their electronegativity difference is 1.0. The molecule is not linear, so the charges cannot cancel each other as in carbon dioxide. There will be a partial positive charge on sulfur and a partial negative charge on the oxygens.

Problem 4.20

Predict the polarity of the following molecules.

a. HBr　　　　　　　　　　**b.** CCl_4　　　　　　　　　　**c.** PCl_3

    Polar molecules interact with one another. This interaction is governed by the fact that like charges repel and opposite charges react. Figure 4.14 shows a collection of polar molecules in a liquid, aligned so that the partially positive end of one molecule is near the partially negative end of another.

    Polar molecules aligned by charge need not be all of the same compound. We've seen that water is a polar compound. When another polar compound such as ammonia is added to water, it dissolves because of the interaction between the two types of molecules. Figure 4.15 shows the polarity of water and ammonia molecules and the interactions between them.

    The same kind of interaction occurs when an ionic solid dissolves in water. As shown in Figure 4.16, each ion of the solid crystal becomes surrounded by water molecules, with the negative end of the water molecules approaching closest to the positive cations, and the positive end of the water molecules surrounding the negative anions. In this way, the ions are pulled away from

Figure 4.14 The orientation of polar molecules in a liquid.

Figure 4.15 Inter-action between water and ammonia mole-cules.

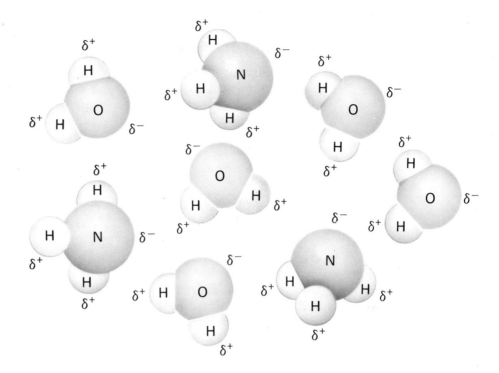

the rest of the crystal one by one. Those ionic compounds that are virtually insoluble in water have such strong interactions between their ions that the pull of the polar water molecules is not strong enough to break the ions apart.

Figure 4.16 Sodium chloride dissolving in water, showing the interaction of polar water molecules with ions.

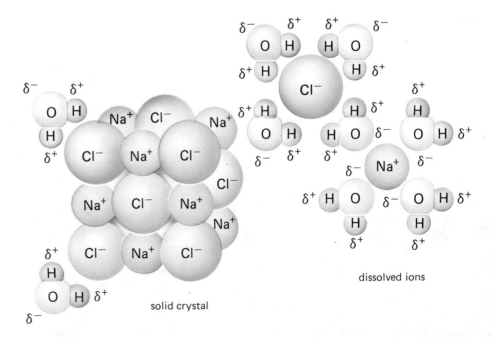

dissolved ions

solid crystal

4.8 Acids and Bases

A. Traditional Definitions

Our discussion of compounds has thus far divided them into two classes: ionic and covalent. Before ending our discussion, we should introduce two special categories of compounds: *acids* and *bases*.

Acids are covalent in the pure state; in aqueous solution, they ionize to produce hydrogen ions. An acid is *a compound whose aqueous solution contains more hydrogen ions than hydroxide ions*. The compound HNO_3 is classified as an acid because when dissolved in water, it dissociates into hydrogen ions and nitrate ions:

$$HNO_3 \xrightarrow{\text{H}_2\text{O}} H^+ + NO_3^-$$

The sour taste of vinegar is due to the presence of hydrogen ions from acetic acid; the sour taste of lemon juice is due to the presence of hydrogen ions from citric acid. Some dyes have characteristic colors in acid solution and are called *indicators*, for by their color they indicate whether or not a solution contains an acid. The names and molecular formulas of several common acids are listed in Table 4.8, along with the anions they form in solution.

Each acid listed in Table 4.8 contains at least one ionizable hydrogen. Some, like sulfuric acid, contain two ionizable hydrogens; phosphoric acid

Table 4.8 Common acids.

Acid	Molecular Formula	Anion Formed in Solution	Anion Name
hydrochloric acid	HCl	Cl^-	chloride
sulfuric acid	H_2SO_4	HSO_4^-	bisulfate (hydrogen sulfate)
	HSO_4^-	SO_4^{2-}	sulfate
sulfurous acid	H_2SO_3	HSO_3^-	bisulfite (hydrogen sulfite)
	HSO_3^-	SO_3^{2-}	sulfite
nitric acid	HNO_3	NO_3^-	nitrate
nitrous acid	HNO_2	NO_2^-	nitrite
carbonic acid	H_2CO_3	HCO_3^-	bicarbonate (hydrogen carbonate)
	HCO_3^-	CO_3^{2-}	carbonate
phosphoric acid	H_3PO_4	$H_2PO_4^-$	dihydrogen phosphate
	$H_2PO_4^-$	HPO_4^{2-}	hydrogen phosphate
	HPO_4^{2-}	PO_4^{3-}	phosphate

contains three ionizable hydrogens. The terms monoprotic, diprotic, and triprotic are used to show the number of ionizable hydrogens in an acid. A **monoprotic acid** has one ionizable hydrogen, a **diprotic acid** has two ionizable hydrogens, and a **triprotic acid** has three ionizable hydrogens. Diprotic and triprotic acids ionize in steps:

$$H_3PO_4 \xrightarrow{H_2O} H^+ + H_2PO_4^- \qquad H_2SO_4 \xrightarrow{H_2O} H^+ + HSO_4^-$$

$$H_2PO_4^- \xrightarrow{H_2O} H^+ + HPO_4^{2-} \qquad HSO_4^- \xrightarrow{H_2O} H^+ + SO_4^{2-}$$

$$HPO_4^{2-} \xrightarrow{H_2O} H^+ + PO_4^{3-}$$

Bases are *compounds whose aqueous solution contains more hydroxide ions* (OH^-) *than hydrogen ions*. Such a solution turns litmus blue and feels slippery. Typical ionic bases are sodium hydroxide (NaOH) and potassium hydroxide (KOH). By our definition, ammonia (NH_3) is also a base. Although ammonia does not have a hydroxide group in its formula, hydroxide ions are formed when ammonia dissolves in water:

$$NH_3 + H_2O \longrightarrow NH_4^+ + OH^-$$

B. The Brønsted-Lowry Definitions

The definitions of acids and bases that we have used so far describe what acids and bases do when dissolved in water. In 1923, T. M. Lowry in England and J. M. Brønsted in Denmark proposed a system that defines acids and bases in terms of *structure*. According to the Brønsted-Lowry definition:

> an acid is a proton (H^+) donor
> a base is a proton (H^+) acceptor

Because a hydrogen ion consists of a nucleus containing a single proton, the terms *hydrogen ion* and *proton* are synonymous.

By this definition, compounds such as hydrochloric acid are acids:

$$HCl \longrightarrow H^+ + Cl^-$$
(*acid*)

but so are the ammonium ion,

$$NH_4^+ \longrightarrow NH_3 + H^+$$
(*acid*)

the water molecule,

$$H_2O \longrightarrow H^+ + OH^-$$
(*acid*)

and the bicarbonate ion.

$$\underset{(acid)}{HCO_3^-} \longrightarrow H^+ + CO_3^{2-}$$

According to the Brønsted-Lowry definition, a base is any substance that accepts a proton. By this definition, hydroxide ion is a base,

$$\underset{(base)}{OH^-} + H^+ \longrightarrow H_2O$$

and so is ammonia,

$$\underset{(base)}{NH_3} + H^+ \longrightarrow NH_4^+$$

and the carbonate ion.

$$\underset{(base)}{CO_3^{2-}} + H^+ \longrightarrow HCO_3^-$$

Water may also be a base. For example, when an acid is dissolved in water, H_2O acts as a base: it accepts a proton to form the hydronium ion, H_3O^+.

$$H^+ + \underset{(base)}{H_2O} \longrightarrow H_3O^+$$

Following are Lewis structures for the bases we have just discussed.

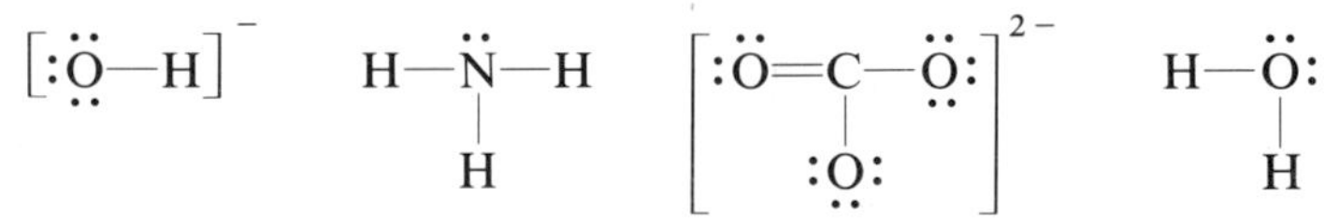

OH^- NH_3 CO_3^{2-} H_2O

In each of these bases, there is at least one unshared pair of electrons on a very electronegative atom. The hydrogen ion forms a covalent bond with one of these unshared electron pairs, as shown below:

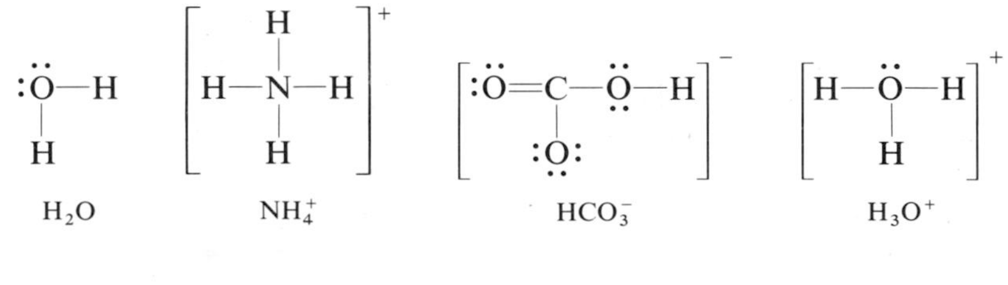

H_2O NH_4^+ HCO_3^- H_3O^+

Example 4.21

Use the Lewis structure for HCO_3^- to show how it acts as a Brønsted-Lowry base.

Solution

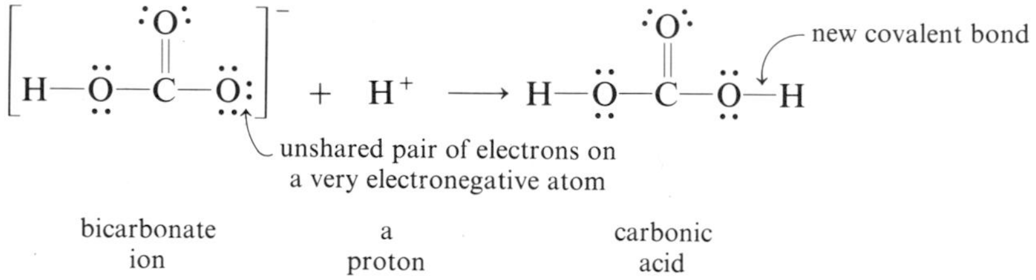

bicarbonate ion + a proton → carbonic acid

Problem 4.21

Draw the Lewis structure of the bisulfide ion, HS^-, and show how it acts as a Brønsted-Lowry base.

Acids, then, are compounds that contain hydrogen bonded covalently to a very electronegative atom. When the acid is dissolved in water, the hydrogen ion is released. A base is either: a compound containing a metallic cation and the hydroxide anion; or a compound, such as ammonia, which can remove hydrogen from a water molecule to release hydroxide ions in solution.

Key Terms and Concepts

acids (4.8)

bases (4.8)

bond angles (4.6)

bonding electrons (4.5A)

Brønsted-Lowry definition (4.8B)

chemical bond (4.3)

compound (4.1)

contributing structures (4.5D)

covalent bond (4.3A)

covalent compounds (4.5)

crystal lattice (4.4C)

diprotic acids (4.8A)

electrolytes (4.4B)

electronegativity (4.3B)

empirical formula (4.2A)

formal charge (4.5C)

formula (4.2)

formula weight (4.2B)

ionic bond (4.3)

ionic compounds (4.4)

linear structure (4.6C)

mixtures (4.1)

molecular formula (4.2A)

monoprotic acids (4.8A)

names of ionic compounds (4.4A)

nonbonding electrons (4.5A)

nonelectrolyte (4.4B)

nonpolar covalent bond (4.7)

octet rule (4.3A)

percent composition (4.2C)

planar structure (4.6B)

polar covalent bonds (4.7A)

polar molecules (4.7B)

polyatomic ions (4.5)

resonance (4.5D)

resonance hybrid (4.5D)

tetrahedral structure (4.6A)

trigonal planar structure (4.6B)

triprotic acids (4.8A)

unshared electron pair (4.5B)

valence (4.3A)

VSEPR (4.6)

Problems

Empirical and Molecular Formulas (Section 4.2A)

4.22 A compound contains 53.3% carbon, 11.1% hydrogen, and 35.6% oxygen by weight. Its formula weight is 90.1. What is its empirical formula? Its molecular formula?

4.23 A sulfide of arsenic contains 60.9% by weight of arsenic and 39.1% sulfur. What is the empirical formula of this compound?

4.24 A compound has the empirical formula C_4H_6. Its formula weight is 162. What is its molecular formula?

4.25 Which compounds in the following groups have the same percent composition?
a. C_2H_4, C_6H_{12}, C_5H_8, $C_{10}H_{22}$
b. N_2O_4, NO_2, N_2O_5
c. $C_2H_4O_2$, $C_6H_8O_6$, $C_3H_6O_3$

Formulas and Formula Weights (Section 4.2B)

4.26 State which elements and how many atoms of each are in a formula unit of:
a. naphthalene, $C_{10}H_8$
b. potassium sulfate, K_2SO_4
c. lysergic acid diethylamide (LSD), $C_{20}H_{25}N_3O$

4.27 Write the formula of the compound whose formula unit contains:
a. 4 atoms phosphorus, 10 atoms oxygen
b. 1 atom calcium, 2 atoms carbon, 4 atoms oxygen
c. 1 atom bromine, 6 atoms carbon, 5 atoms hydrogen

4.28 Calculate the formula weight of:
a. nitric acid (HNO_3)
b. methane (CH_4)
c. sulfur dioxide (SO_2)
d. citric acid ($C_6H_8O_7$)
e. ascorbic acid ($C_6H_8O_6$) (vitamin C)
f. cholesterol ($C_{27}H_{46}O$)

4.29 Calculate the formula weight of:
a. chloroform ($CHCl_3$)
b. ammonium sulfate ($(NH_4)_2SO_4$)
c. magnesium oxide (MgO)
d. camphor ($C_{10}H_{16}O$)
e. aspirin ($C_9H_8O_4$)
f. vitamin $B_1(C_{12}H_{18}Cl_2N_4OS)$

Moles of Compounds (Section 4.2B)

4.30 Calculate the mass of 0.155 mole of each of the compounds in Problem 4.29.

4.31 Calculate the mass of 2.13 moles of each of the compounds in Problem 4.29.

4.32 Calculate the moles present in each of the following samples.
   a. 6.85 g sodium iodide (NaI)
   b. 437.1 g sulfur dioxide (SO_2)
   c. 0.442 g nitrogen dioxide (NO_2)
   d. 7.41 g lithium oxide (Li_2O)
   e. 32.5 g glucose ($C_6H_{12}O_6$)
   f. 0.27 g oxygen gas
   g. 32.5 g sucrose ($C_{12}H_{22}O_{11}$)

4.33 Calculate the moles of sodium present in the following samples.
   a. 0.155 g sodium chloride (NaCl)
   b. 0.155 g sodium sulfate (Na_2SO_4)
   c. 0.155 g sodium phosphate (Na_3PO_4)
   d. 0.155 g sodium bicarbonate ($NaHCO_3$)
   e. 0.155 g sodium benzoate ($NaC_7H_5O_2$)
   f. 0.155 g monosodium glutamate ($NaC_5H_8NO_4$)

Percent Composition (Section 4.2C)

4.34 What is the percent composition of:
   a. octane, C_8H_{18}
   b. sodium bicarbonate, $NaHCO_3$
   c. phosphorus trichloride, PCl_3
   d. lactic acid, $C_3H_6O_3$

4.35 Calculate the percent composition of:
   a. ethanol, C_2H_6O
   b. citric acid, $C_6H_8O_7$
   c. magnesium chloride, $MgCl_2$

Ionic and Covalent Bonds and Valence (Section 4.3A)

4.36 Chlorine bonds with each of the elements of period 3 to form the compounds:

$$NaCl \quad MgCl_2 \quad AlCl_3 \quad SiCl_4 \quad PCl_3 \quad SCl_2 \quad Cl_2$$

Which of these bonds would you predict to be ionic, which polar covalent, and which nonpolar covalent?

4.37 What is the valence of:
   a. bromine   b. silicon   c. carbon
   d. sulfur   e. nitrogen   f. hydrogen

4.38 In each of the following polar bonds, indicate which atom has a slightly negative charge $(\delta-)$ and which atom has a slightly positive charge $(\delta+)$.

$$Br—Cl \quad S—F \quad N—O \quad Se—O \quad I—Cl$$

Ionic Formulas and Names (Section 4.4A)

4.39 Name the following compounds:
   a. $Fe(NO_3)_2$ b. $KHCO_3$ c. Fe_2O_3 d. Na_2S
   e. $MgSO_4$ f. LiF   g. $Ca_3(PO_4)_2$ h. NH_4F
   i. $NaNO_2$ j. $Mn(NO_3)_2$ k. $CaHPO_4$ l. $AlBr_3$

4.40　Write the formula for each of the following compounds:
　　a. copper(II) chloride　　　　　　b. iron(III) nitrate
　　c. ammonium sulfate　　　　　　　d. cesium chloride
　　e. lithium carbonate　　　　　　　f. aluminum iodide
　　g. potassium sulfite　　　　　　　h. calcium sulfide
　　i. tin(II) nitrate

4.41　Complete the following chart as done in line 1.

	SO_4^{2-}	HCO_3^-	SO_3^{2-}	Br^-	PO_4^{3-}
Na^+	Na_2SO_4	$NaHCO_3$	Na_2SO_3	$NaBr$	Na_3PO_4
NH_4^+	——	——	——	——	——
Fe^{2+}	——	——	——	——	——
Mg^{2+}	——	——	——	——	——

4.42　Name the compounds in Problem 4.41.

Lewis Structures and Molecular Geometry (Section 4.5B)

4.43　Draw the Lewis structure of $SiCl_4$, SCl_2, $CCl_3—CH_3$, and $Cl_2C═CHCl$.

4.44　Predict the geometry of each compound in Problem 4.43.

4.45　Draw the Lewis structure of:
　　a. hydrogen iodide, HI　　　　　　b. bromine, Br_2
　　c. chloromethane, CH_3Cl　　　　d. sulfur dioxide, SO_2
　　e. nitrate ion, NO_3^-

4.46　Predict the geometry of the structures in Problem 4.45.

4.47　Draw the Lewis structure and predict the geometry of:
　　a. acetylene, C_2H_2　　　　　　b. carbon disulfide, CS_2

4.48　The following substances contain both ionic and covalent bonds. Draw a Lewis structure for each. Show each ion as a separate structure.
　　a. NaOH　　　　　　b. CH_3ONa　　　　　　c. $NaHCO_3$
　　d. Na_2CO_3　　　　e. NH_4Cl　　　　　　f. $Ca(NO_3)_2$

Lewis Structures and Formal Charge (Section 4.5C)

4.49　Not all electrons are shown in the following structures. Add enough pairs of electrons to complete the valence shells of each atom. Assign formal positive or negative charges as appropriate.

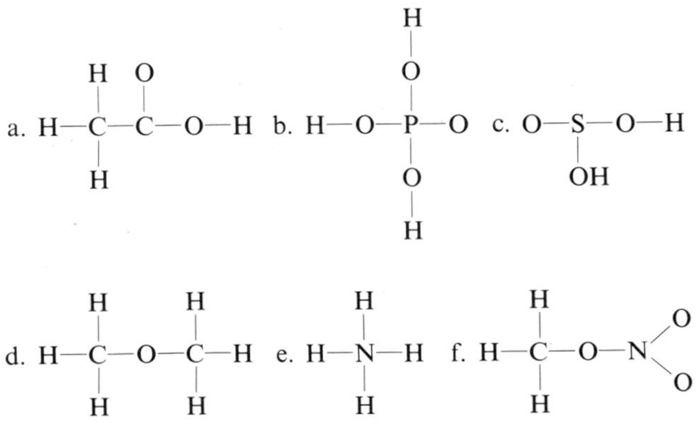

**Polarity of Molecules
(Section 4.7)**

For the next three questions, you will probably want to draw the Lewis structure of the molecules under consideration.

4.50 Why is ammonia polar but the ammonium ion (NH_4^+) nonpolar?

4.51 Silicon and chlorine have different electronegativities. Why, then, is silicon tetrachloride not a polar molecule?

4.52 Is methyl alcohol, CH_3OH, a polar molecule? Why or why not?

**Acids and Bases
(Section 4.8)**

4.53 Draw Lewis structures for the following acids and bases. Show how each acts as a Brønsted-Lowry acid or base.

    a. CH_3COOH b. methyl amine, CH_3NH_2

    c. CN^- d. formate ion, $HCOO^-$

4.54 Each of the following can act as either a Brønsted-Lowry acid or base. Show this by appropriate equations.

    a. HSO_4^- b. $H_2PO_4^-$

Chemical Reactions

Many people picture a chemist as a person dressed in a white coat standing in front of a laboratory bench crowded with mysteriously shaped glass containers. The contents of these containers bubble, change color, and give off strange vapors. One feels certain that the bubbling and fuming substances are about to combine and form a new miracle drug or fiber.

In spite of this picture, our study of chemistry has included little thus far about what happens when substances combine. In this chapter we will discuss *chemical reactions*, the ways in which elements and compounds interact chemically to form other substances. We will lay the groundwork for our discussions in later chapters of how the food we eat and the air we breathe become flesh and bone and provide the energy we need to survive.

5.1 Chemical Changes

In Section 1.1A we divided the variety of observable properties of matter into two classes: physical and chemical. Physical properties are those properties whose observation does not involve a change in the composition of the sample.

Figure 5.1 The crushing of limestone is a physical change; it does not alter the composition of the limestone. The heating of limestone is a chemical change; the limestone changes into two other substances, lime and carbon dioxide.

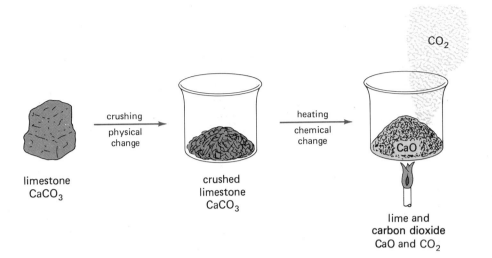

crushing
physical
change

heating
chemical
change

CO_2

limestone
$CaCO_3$

crushed
limestone
$CaCO_3$

CaO

lime and
carbon dioxide
CaO and CO_2

A sample of water does not change its composition if it is poured from a tall pitcher into a flat bowl. It can be frozen to a solid or vaporized to steam. Yet, it remains water, with the formula H_2O. Through all of these changes, the composition of water is unchanged. Another physical change, the crushing of limestone, is illustrated in Figure 5.1. Even though the limestone is crushed to smaller particles, the composition of the limestone is unchanged.

The observation of a chemical property involves a change in the composition of the sample and a **chemical reaction.** When an electric current is passed through water containing a few drops of sulfuric acid, the water decomposes to hydrogen and oxygen. Water molecules are no longer present; instead, we have two new substances, hydrogen and oxygen. Methane (natural gas) is used as a fuel for home heating and cooking throughout large areas of the United States. When methane burns (combines chemically with oxygen), carbon dioxide and water are formed. Both methane and oxygen undergo chemical changes and two new substances are formed. Although it is theoretically possible to reverse this change, there is no easy way to prepare methane and oxygen from carbon dioxide and water. Another example of a chemical change is illustrated in Figure 5.1. When it is heated, limestone is converted into two new substances, lime and carbon dioxide.

5.2 Chemical Equations

A. Characteristics

Arithmetical equations describe mathematical operations; for example, $2 + 3 = 5$. **Chemical equations** can be used to describe both physical and chemical changes in atoms, molecules, and ions. For example, the following chemical equation describes the vaporization of water, a physical change:

$$H_2O(l) \longrightarrow H_2O(g)$$

The letters in parentheses show the physical state of the substances in the equation; **(l)** means that the substance is a liquid, **(g)** means that it is a gas. If the substance in an equation is a solid, its formula is followed by **(s)**; if it is in aqueous (water) solution, its formula is followed by **(aq).**

The equations in Section 2.10A showed nuclear changes. The radioactive decay of radium is described by the equation:

$$^{226}_{88}Ra \longrightarrow {}^{222}_{86}Rn + {}^{4}_{2}He + {}^{0}_{0}\gamma$$

In Section 3.6B(1), equations were used to illustrate the formation of ions. The formation of the Mg^{2+} ion is shown by the equation:

$$Mg: \longrightarrow Mg^{2+} + 2\,e^-$$

In this chapter, we use equations to describe chemical reactions. For example, the burning of propane can be described by the equation:

$$C_3H_8(g) + 5\,O_2(g) \longrightarrow 3\,CO_2(g) + 4\,H_2O(l) \qquad \Delta H = -2,220 \text{ kJ}$$
$$(-530.6 \text{ kcal})$$

Those substances whose formulas precede the arrow are called **reactants;** the arrow means "yield"; and those substances whose formulas follow the arrow are the **products** of the reaction. The numbers preceding each molecular formula are called **coefficients** and show the numerical ratio among the kinds of molecules taking part in the reaction. The ΔH term following the equation gives the **enthalpy change,** the energy change that accompanies the reaction. We would read this equation as:

One molecule of gaseous propane reacts with five molecules of gaseous oxygen to yield three molecules of gaseous carbon dioxide and four molecules of liquid water. For each mole of propane reacting, 2,220 kilojoules (or 530.6 kilocalories) of energy are released.

In the SI system, the joule is the preferred unit of energy. For many people, however, the calorie (1 calorie = 4.184 joule) is still the unit of choice. In this chapter we will give energy values in both units.

The equation given for the burning of propane is a complete equation. When the discussion centers on the reactants and products of a reaction rather than on the associated energy change, the physical states and the energy change are often omitted.

B. Writing Chemical Equations

A correctly written equation obeys certain rules.

Rule 1: The formulas of all reactants and products must be correct.

Correct formulas must be used because the wrong formulas represent different substances and completely change the meaning of the equation. For example, the equation

$$2\,H_2O_2 \longrightarrow 2\,H_2O + O_2$$

describes the decomposition of hydrogen peroxide. This is quite a different reaction from the electrolysis of water, described by the equation

$$2\,H_2O \longrightarrow 2\,H_2 + O_2$$

When an element occurs uncombined in an equation, the symbol is used to represent one atom unless the element is diatomic. The diatomic elements are:

hydrogen, H_2 fluorine, F_2 bromine, Br_2
nitrogen, N_2 chlorine, Cl_2 iodine, I_2
oxygen, O_2

When these elements appear uncombined in equations, the diatomic formulas must be used, as they have been in the above equations.

Rule 2: An equation must be **balanced by mass.**

An equation is balanced by mass when the number of atoms of each element in the reactants equals the number of atoms of that element in the products. For example, the equation shown above for the electrolysis of water has four atoms of hydrogen in the two reacting molecules of water and four atoms of hydrogen in the two molecules of hydrogen gas produced; therefore, hydrogen is balanced. It has two atoms of oxygen in the two reacting molecules of water and two atoms of oxygen in the single molecule of oxygen produced; therefore, oxygen is also balanced.

$$2\,H_2O \longrightarrow 2\,H_2 + O_2$$

four H on the left = four H on the right

two O on the left = two O on the right

When the atoms are balanced, the mass is balanced and the equation obeys the Law of Conservation of Mass.

Rule 3: An equation must be **balanced by charge,** if it includes charged particles (ions or electrons).

The charges in an equation are balanced when the total charge carried by the reactants equals the total charge carried by the products. Consider the equation for the ionization of magnesium:

$$Mg \longrightarrow Mg^{2+} + 2\,e^-$$

The charge on the reactant magnesium atom is zero. The sum of the $+2$ charge on the magnesium ion and the total negative charge of -2 furnished by the two electrons is also zero.

You can write and balance equations in four steps:

1. Write the correct formulas of the reactants. Put an arrow after them. After the arrow, write the correct formulas of the products.

2. Count the number of atoms of each element on each side of the equation.

3. Change the coefficients as necessary so that the number of atoms of each element on the left side of the equation is the same as that on the right side. *Only the coefficients may be changed in balancing an equation; the subscripts in a formula must never be changed.*

4. Balance charges on the two sides of the equation by adding electrons to the more positive side.

Example 5.1

The brilliant white light in some fireworks displays is produced by burning magnesium. The reactants are Mg and O_2 (remember that uncombined oxygen is diatomic); the product is magnesium oxide, MgO. Write the balanced equation for this reaction.

Solution

1. Write the correct formulas for all reactants, followed by an arrow; then write the correct formulas of the products:

$$Mg + O_2 \nrightarrow MgO$$

The line through the arrow indicates that the equation may not yet be balanced.

2. Count the atoms on each side:

reactants		products	
magnesium	1	magnesium	1
oxygen	2	oxygen	1

3. Change the coefficients in the equation so that the atoms of each element are balanced. This can be done by placing a coefficient of 2 in front of Mg and MgO.

4. Check to see if there is a charge imbalance. After step 3, the charge on the left is zero and the charge on the right is also zero. Thus, charges are balanced. Therefore, the balanced equation for the combustion of magnesium is:

$$2\,Mg + O_2 \longrightarrow 2\,MgO$$

Example 5.2

Write a balanced equation for the combustion of ethanol, C_2H_6O, to carbon dioxide and water.

Solution

1. Write the correct formulas:

$$C_2H_6O + O_2 \not\longrightarrow CO_2 + H_2O$$

2. Count the atoms:

reactants		*products*	
carbon	2	carbon	1
hydrogen	6	hydrogen	2
oxygen	3	oxygen	3

3. Balance the equation by mass. Because it is often easiest to start with the element present in the fewest compounds, we will start with carbon. There are two carbons on the left, so we need two carbons on the right, and we write $2\,CO_2$. Now, hydrogen: because there are six hydrogens on the left, we need to write $3\,H_2O$ on the right.

$$C_2H_6O + O_2 \not\longrightarrow 2\,CO_2 + 3\,H_2O$$

Finally, we balance the oxygen. There are now seven oxygens on the right and only three on the left; we need four more oxygens on the left. We can achieve this by changing the coefficient in front of O_2 from 1 to 3. The balanced equation is:

$$C_2H_6O + 3\,O_2 \longrightarrow 2\,CO_2 + 3\,H_2O$$

Example 5.3

Metallic magnesium reacts with hydrogen ion to form magnesium ion, Mg^{2+}, and hydrogen gas. Write the balanced equation for this reaction. (Remember that hydrogen gas is diatomic.)

Solution

1. Write the correct formulas:

$$Mg + H^+ \not\longrightarrow Mg^{2+} + H_2$$

2. Count the atoms:

reactants		*products*	
magnesium	1	magnesium	1
hydrogen	1	hydrogen	2

3. The element that is not balanced is hydrogen, therefore put a coefficient of 2 in front of H^+:

$$Mg + 2\,H^+ \not\longrightarrow Mg^{2+} + H_2$$

4. Balance the charges. The charge on the reactants is $+2$, the same as the charge on the products. Thus, the mass and charge are now balanced:

$$Mg + 2\,H^+ \longrightarrow Mg^{2+} + H_2$$

Problem 5.1

Lightning causes nitrogen and oxygen in the atmosphere to react to form nitrogen oxide (NO). Write the balanced equation for this reaction.

Problem 5.2

Isooctane (C_8H_{18}) is a component of gasoline. Isooctane reacts with oxygen in an automobile engine to form carbon dioxide and water. Write the balanced equation for this reaction.

Problem 5.3

Chlorine gas reacts with bromide ion to yield free bromine and chloride ion. Write the balanced equation for this reaction.

5.3 Classification of Chemical Reactions

One common way of classifying chemical reactions separates them into four categories: *combination, decomposition, displacement,* and *metathesis.* Several examples of each type follow, partly so that you will understand the scope of each category and partly so that you can gain experience in balancing equations.

Figure 5.2 The extremely hot flame of an acetylene torch is created by the reaction of acetylene and oxygen. This combination reaction produces carbon dioxide, water, and intense heat. The flame of an acetylene torch may be used to cut through steel.

A. Combination Reactions

In a combination reaction, two substances combine to form a single compound. Two examples are the reaction of magnesium with oxygen to form magnesium oxide

$$2\,Mg + O_2 \longrightarrow 2\,MgO$$

and the reaction of hydrogen with chlorine to form hydrogen chloride:

$$H_2 + Cl_2 \longrightarrow 2\,HCl$$

Figure 5.2 illustrates an example of a combination reaction.

Other combination reactions involve compounds, such as the reaction of carbon dioxide with calcium oxide to form calcium carbonate:

$$CaO + CO_2 \longrightarrow CaCO_3$$

Example 5.4

Write balanced equations for the following combination reactions:

a. The reaction of phosphorus with chlorine to form phosphorus trichloride.

b. In spaceships, exhaled carbon dioxide is absorbed by lithium hydroxide to form lithium carbonate and water.

Solution

a. The reactants are phosphorus, P, and chlorine, a diatomic gas, Cl_2. The formula of the product, PCl_3, is apparent from the name. The unbalanced equation is:

$$P + Cl_2 \nrightarrow PCl_3$$

The number of chlorine atoms in the equation must be divisible by 2 to give a whole number of chlorine molecules and also by 3 to give a whole number of PCl_3 molecules. Six atoms of chlorine meet this requirement, giving the equation:

$$P + 3\,Cl_2 \not\longrightarrow 2\,PCl_3$$

To complete the balancing of this equation, a total of 2 phosphorus atoms is needed on the left side. The balanced equation is:

$$2\,P + 3\,Cl_2 \longrightarrow 2\,PCl_3$$

b. The reactants are lithium hydroxide, LiOH, and carbon dioxide, CO_2. The products are lithium carbonate, Li_2CO_3, and water, H_2O. The unbalanced equation is:

$$LiOH + CO_2 \not\longrightarrow Li_2CO_3 + H_2O$$

To balance the lithium atoms, place a coefficient of 2 in front of LiOH. This will also balance the oxygen and hydrogen atoms. Thus, the balanced equation is:

$$2\,LiOH + CO_2 \longrightarrow Li_2CO_3 + H_2O$$

Problem 5.4

Write balanced equations for the following combination reactions:

a. When carbon burns in a limited supply of oxygen, the gas carbon monoxide (CO) is formed. This gas is deadly to humans because it combines with hemo-globin in the blood, making it impossible for the blood to transport oxygen. Write the equation for the formation of carbon monoxide.

b. Barium oxide reacts with water to form barium hydroxide.

B. Decomposition Reactions

In a **decomposition reaction,** a compound is decomposed to its component elements or to other compounds. In each of the following reactions, one of the products is a gas. We show this with an upward-pointing arrow after the formula of the gas.

The antiseptic hydrogen peroxide decomposes in light to form water and oxygen (Figure 5.3):

$$2\,H_2O_2 \xrightarrow{\text{light}} 2\,H_2O + O_2\uparrow$$

Oxygen can be prepared by heating potassium chlorate:

$$2\,KClO_3 \longrightarrow 2\,KCl + 3\,O_2\uparrow$$

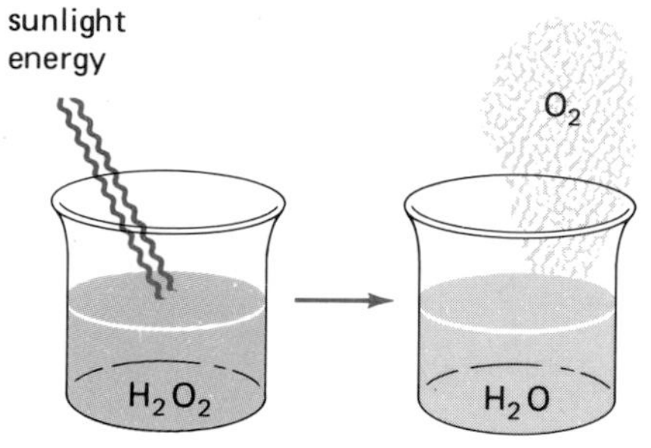

sunlight
energy

O_2

H_2O_2

H_2O

Figure 5.3 The exposure of hydrogen peroxide to light causes it to decompose into water and oxygen. That is why bottles containing hydrogen peroxide are always made of dark-colored glass.

When slaked lime, $Ca(OH)_2$, is heated, quicklime (CaO) and water vapor are produced:

$$Ca(OH)_2 \longrightarrow CaO + H_2O\uparrow$$

Example 5.5

Write balanced equations for the following decomposition reactions:

a. Ammonium carbonate decomposes at room temperature to ammonia, carbon dioxide, and water. (Because of the ease of decomposition and the penetrating odor of ammonia, ammonium carbonate can be used as smelling salts.)

b. On heating, lead(II) nitrate decomposes to lead(II) oxide, oxygen, and nitrogen dioxide.

Solution

a. The unbalanced equation for the decomposition of ammonium carbonate is

$$(NH_4)_2CO_3 \nrightarrow NH_3\uparrow + CO_2\uparrow + H_2O$$

Inspection of this equation indicates that 2 nitrogen atoms, therefore 2 ammonia molecules, are needed on the right. With this change, all other atoms are balanced:

$$(NH_4)_2CO_3 \longrightarrow 2 NH_3\uparrow + CO_2\uparrow + H_2O$$

b. Writing the formulas for the reactant and the products in the form of an equation gives:

$$Pb(NO_3)_2 \nrightarrow PbO + O_2 + NO_2$$

By putting a coefficient of 2 in front of PbO and a coefficient of 4 in front of NO_2, we obtain a balanced equation:

$$2 Pb(NO_3)_2 \longrightarrow 2 PbO + O_2 + 4 NO_2$$

Problem 5.5

Write balanced equations for the following decomposition reactions:

a. Silver(I) oxide decomposes on heating to yield silver and gaseous oxygen.

b. In the chemical test for arsenic, the compound arsine, AsH_3, is prepared. When the arsine is decomposed by heating, arsenic deposits as a mirror-like coating on the surface of the glass container and hydrogen comes off as a gas. Write the balanced equation for the decomposition of arsine.

C. Displacement Reactions

In **displacement reactions,** an uncombined element reacts with a compound, displacing an element from the compound. For example, bromine is found in seawater as sodium bromide. When chlorine is bubbled through seawater, bromine is released and sodium chloride is formed:

$$2\,NaBr + Cl_2 \longrightarrow 2\,NaCl + Br_2$$

When an iron nail is dropped into a solution of copper(II) sulfate, iron(II) sulfate is formed and metallic copper is deposited:

$$CuSO_4 + Fe \longrightarrow FeSO_4 + Cu$$

Example 5.6

Write balanced equations for the following displacement reactions:

a. Aluminum displaces hydrogen from hydrochloric acid, forming aluminum chloride and hydrogen gas.

b. Bromine reacts with benzene, C_6H_6, to form dibromobenzene, $C_6H_4Br_2$, and hydrogen bromide.

Solution

a. The unbalanced equation is:

$$Al + HCl \xrightarrow{} AlCl_3 + H_2$$

Hydrogen and chlorine are in a 1:1 ratio in hydrochloric acid. In the products we need a chlorine multiple of three and a hydrogen multiple of two. The lowest common multiple is six. Use 6 HCl and balance the other substances to give:

$$2\,Al + 6\,HCl \longrightarrow 2\,AlCl_3 + 3\,H_2$$

b. The unbalanced equation is:

$$C_6H_6 + Br_2 \not\longrightarrow C_6H_4Br_2 + HBr$$

Because bromine is diatomic, there must be an even number of bromine atoms in the equation. For each molecule of dibromobenzene, two molecules of hydrogen bromide must be formed. The balanced equation is:

$$C_6H_6 + 2\,Br_2 \longrightarrow C_6H_4Br_2 + 2\,HBr$$

Problem 5.6

Write balanced equations for the following displacement reactions:

a. When a piece of zinc is dropped into sulfuric acid, bubbles of hydrogen appear. Zinc sulfate, $ZnSO_4$, is also formed.

b. If a cloth bag containing mercury is suspended in a solution of silver(I) nitrate, silver crystals form on the surface of the bag. The second product, mercury(I) nitrate, is found in the surrounding solution.

D. Metathesis or Double Replacement Reactions

In **metathesis reactions,** two ionic compounds react to form two different compounds. The reactions fall into two categories: (1) those in which an acid reacts with a base to form a salt and water; (2) those in which one of the products is insoluble.

1. Reaction of an acid with a base: a neutralization reaction. In **neutralization reactions,** an acid reacts with a base to form a salt and water. A **salt** is defined as *an ionic compound in which the cation is not hydrogen and the anion is not hydroxide.* These reactions are called neutralization reactions because the base neutralizes the acid. Some examples are:

1. The reaction of sodium hydroxide with hydrochloric acid to form sodium chloride and water:

$$NaOH + HCl \longrightarrow NaCl + H_2O$$

2. The reaction of magnesium hydroxide with phosphoric acid to form magnesium phosphate and water:

$$3\,Mg(OH)_2 + 2\,H_3PO_4 \longrightarrow Mg_3(PO_4)_2 + 6\,H_2O$$

When a polyprotic acid is one of the reactants, neutralization may be incomplete and an **acid salt** formed. An example of this is:

$$NaOH + H_2SO_4 \longrightarrow NaHSO_4 + H_2O$$

In this reaction, only one of the hydrogens of the diprotic acid reacts and the product is an acid salt, sodium hydrogen sulfate. The addition of more sodium hydroxide neutralizes the second hydrogen of this diprotic acid:

$$NaHSO_4 + NaOH \longrightarrow Na_2SO_4 + H_2O$$

In the Brønsted-Lowry system (Section 4.8B), a base is a proton acceptor (for example, OH^-, NH_3, and CO_3^{2-}) and an acid is a proton donor. Acid-base reactions in this system need not have water as one of the products. For example, the reaction of a proton with ammonia, a base, is an acid-base reaction in which the ammonium ion is formed.

$$\underset{(proton)}{H^+} + \underset{(base)}{NH_3} \longrightarrow NH_4^+$$

The reaction of a proton from an acid with the carbonate ion, a base, is an acid-base reaction in which the bicarbonate ion is formed.

$$\underset{(proton)}{H^+} + \underset{(base)}{CO_3^{2-}} \longrightarrow HCO_3^-$$

Example 5.7

Write balanced equations for the following neutralization reactions:

a. The complete reaction of sulfuric acid with calcium hydroxide.

b. The reaction of hydroxide ion with bicarbonate ion.

Solution

a. The formulas of the reactants are H_2SO_4 and $Ca(OH)_2$. The products are a salt and water. The salt formed by the combination of calcium ion with sulfate ion is calcium sulfate, $CaSO_4$. The unbalanced equation is:

$$H_2SO_4 + Ca(OH)_2 \xrightarrow{\quad} CaSO_4 + H_2O$$

In balancing the equation, note that the two hydrogen ions and two hydroxide ions combine to form two molecules of water. The balanced equation is:

$$H_2SO_4 + Ca(OH)_2 \longrightarrow CaSO_4 + 2\,H_2O$$

b. The reactants are OH^- and HCO_3^-. This is an acid-base reaction in which the hydroxide ion is the base and the bicarbonate ion is the acid, or proton donor. Because one of the reactants carries a charge, we must balance the equation for both mass and charge.

$$OH^- + HCO_3^- \longrightarrow H_2O + CO_3^{2-}$$

Problem 5.7

Write balanced equations for the following acid-base reactions:

a. Aluminum hydroxide and hydrochloric acid.

b. Acetate ion and hydrochloric acid.

2. Metathesis reactions that form insoluble ionic products. The second group of metathesis reactions results in the formation of insoluble ionic compounds. To understand these equations, we require some knowledge of the solubility of ionic compounds (Section 4.4B). Table 5.1 lists some rules by which the solubility of an ionic compound in water can be predicted.

Example 5.8

Write the formulas of the following salts and predict whether each is soluble in water.

a. lead(II) nitrate   **b.** iron(II) chloride

c. ammonium sulfide   **d.** barium sulfate

Table 5.1 The solubility of ionic compounds in water.

NH_4^+	All common salts of ammonium ion are soluble.
Na^+ K^+	All common salts of sodium and potassium are soluble.
NO_3^-	All nitrates are soluble.
$CH_3CO_2^-$	All acetates are soluble except iron(III) acetate, $Fe(C_2H_3O_2)_3$.
Cl^- Br^- I^-	All chlorides, bromides, and iodides are soluble except those of Ag^+, Hg^+, and Pb^{2+}. $PbCl_2$ and $PbBr_2$ are slightly soluble in hot water.
SO_4^{2-}	All sulfates are soluble except $CaSO_4$, $BaSO_4$, $PbSO_4$, and Ag_2SO_4.
PO_4^{3-} CO_3^{2-}	Only alkali metal and NH_4^+ carbonates and phosphates are soluble.
S^{2-}	Only alkali metal and NH_4^+ sulfides are soluble.
OH^-	Only alkali metal and NH_4^+ hydroxides are soluble. Ca^{2+}, Ba^{2+}, and Sr^{2+} hydroxides are slightly soluble.

Solution

	Formula	Solubility	Reason
a. lead(II) nitrate	$Pb(NO_3)_2$	soluble	it is a nitrate
b. iron(II) chloride	$FeCl_2$	soluble	it is a chloride but not one of the listed exceptions
c. ammonium sulfide	$(NH_4)_2S$	soluble	it is an ammonium salt
d. barium sulfate	$BaSO_4$	insoluble	listed as an insoluble sulfate

Problem 5.8

Write formulas for the following salts and predict whether each is soluble in water.

a. barium acetate   **b.** silver(I) sulfide

c. ammonium phosphate   **d.** calcium carbonate

e. chromium(III) nitrate   **f.** sodium sulfate

g. aluminum hydroxide

Examples of metathesis reactions that form insoluble products are:

1. The reaction between two salts. For instance, the reaction between barium chloride and sodium sulfate:

$$BaCl_2(aq) + Na_2SO_4(aq) \longrightarrow 2\,NaCl(aq) + BaSO_4\downarrow$$

From Table 5.1 we know that sodium salts are soluble. The insoluble product is barium sulfate.

2. The reaction between a salt and a soluble hydroxide. For example, the reaction of potassium hydroxide with iron(II) chloride:

$$2\,KOH(aq) + FeCl_2(aq) \longrightarrow Fe(OH)_2\downarrow + 2\,KCl(aq)$$

From Table 5.1 we know that potassium salts are soluble. The insoluble product is iron(II) hydroxide.

In the equations for these reactions, the formula of the insoluble product, or **precipitate,** is followed by a downward-pointing arrow.

Example 5.9

Write balanced equations for the following metathesis reactions. Indicate with a downward-pointing arrow any precipitate formed; name the precipitate.

a. Solutions of lead(II) nitrate and sodium iodide react to form a yellow precipitate.

b. The reaction between a solution of copper(II) nitrate and one of potassium sulfide yields a heavy, black precipitate.

Solution

a. The formulas for the reactants are $Pb(NO_3)_2$ and NaI. The formulas of the products of a metathesis reaction between these two compounds show an interchange of anions, yielding PbI_2 and $NaNO_3$. Arranging these in an unbalanced equation gives:

$$Pb(NO_3)_2(aq) + NaI(aq) \nrightarrow PbI_2 + NaNO_3$$

Balancing this equation requires 2 iodide ions, therefore 2 sodium iodide. Two sodium nitrate are formed:

$$Pb(NO_3)_2(aq) + 2\,NaI(aq) \longrightarrow PbI_2\!\downarrow + 2\,NaNO_3(aq)$$

Because all sodium salts are soluble, the precipitate must be lead(II) iodide; an arrow is put after that formula.

b. The formulas of the reactants are $Cu(NO_3)_2$ and K_2S. The formulas of the products are CuS and KNO_3. Because potassium nitrate is soluble, the precipitate must be CuS, copper(II) sulfide. The unbalanced equation is:

$$Cu(NO_3)_2(aq) + K_2S(aq) \nrightarrow CuS\!\downarrow + KNO_3(aq)$$

Balancing this equation requires one potassium sulfide and two potassium nitrate. The balanced equation is:

$$Cu(NO_3)_2(aq) + K_2S(aq) \longrightarrow CuS\!\downarrow + 2\,KNO_3$$

Problem 5.9

Write balanced equations for the following metathesis reactions and indicate any precipitate by a downward-pointing arrow. Name the products.

a. When chromium(III) chloride is added to a solution of sodium hydroxide, a green precipitate forms.

b. When sulfuric acid is added to a solution of barium chloride, a white precipitate forms.

5.4 Oxidation-Reduction Reactions

Oxidation-reduction reactions involve a transfer of electrons. Many of the reactions we have discussed are oxidation-reduction reactions. For example, in the reaction of sodium with chlorine

$$2\,Na + Cl_2 \longrightarrow 2\,NaCl$$

electrons are transferred from sodium to chlorine. Each sodium atom loses an electron to form a sodium ion:

$$Na\cdot \longrightarrow Na^+ + e^-$$

Each chlorine atom gains an electron to form a chloride ion:

$$2\,e^- + \mathbf{:}\ddot{\underset{\cdot\cdot}{C}}l\!-\!\ddot{\underset{\cdot\cdot}{C}}l\mathbf{:} \longrightarrow 2\,\mathbf{:}\ddot{\underset{\cdot\cdot}{C}}l\mathbf{:}^-$$

The element that loses electrons is **oxidized.** In the reaction of sodium with chlorine, sodium is oxidized. The element that gains electrons is **reduced.** In this reaction, chlorine is reduced.

Displacement reactions are usually oxidation-reduction reactions. A typical displacement reaction is that of copper with silver nitrate:

$$Cu + 2\,AgNO_3 \longrightarrow 2\,Ag + Cu(NO_3)_2$$

In this reaction, the copper loses electrons (is oxidized):

$$Cu \longrightarrow Cu^{2+} + 2\,e^-$$

and the silver ion gains electrons (is reduced):

$$Ag^+ + e^- \longrightarrow Ag$$

Oxidation is always accompanied by reduction; one cannot happen without the other.

A. Oxidation Numbers

The **oxidation number** of an element represents the positive or negative character of an atom of that element in a particular bonding situation. Oxidation numbers are assigned according to the following rules:

1. The oxidation number of an uncombined element is 0. In the equation

$$Zn + 2\,HCl \longrightarrow H_2 + ZnCl_2$$

the oxidation number of zinc is 0. The oxidation number of hydrogen in H_2 is 0.

2. The oxidation number of a monatomic ion is the charge on that ion. In $ZnCl_2$, the oxidation number of chlorine is -1 and that of zinc is $+2$. In Ag_2S, the oxidation number of silver is $+1$ and that of sulfur is -2.

3. Hydrogen usually has the oxidation number $+1$.

4. Oxygen usually has the oxidation number -2. Peroxides are an exception to this rule; in hydrogen peroxide, for example, the oxidation number of oxygen is -1.

5. The sum of the oxidation numbers of the atoms in a compound is 0. For example, in the compound $ZnCl_2$, the oxidation number of zinc is $+2$ and that of each chlorine is -1. The sum of these oxidation numbers ($+2$ for zinc and -2 for the two chloride ions) is 0.

6. In a polyatomic ion, the charge on the ion is the sum of the oxidation numbers of the atoms in the ion. For example, in the NO_3^- ion, the sum of the oxidation number of nitrogen plus three times the oxidation number of oxygen (-2) must equal -1. We can use this rule to calculate the oxidation number of nitrogen by setting up the following equation:

$$\text{(oxidation number of nitrogen)} + 3(-2) = -1$$

By rearranging, this becomes:

$$\text{(oxidation number of nitrogen)} = -1 - 3(-2) = +5$$

Example 5.10

Assign an oxidation number to each atom in the following compounds and polyatomic ions:

a. CO_2 b. SO_4^{2-} c. NH_4^+

Solution

a. CO_2: oxygen -2 (rule 4), carbon $+4$ (rule 5)

b. SO_4^{2-}: oxygen -2 (rule 4), sulfur $+6$ (rule 6)

c. NH_4^+: hydrogen $+1$ (rule 3), nitrogen -3 (rule 6)

Problem 5.10

Assign an oxidation number to each atom in the following compounds:

a. Fe_2O_3 b. $NaMnO_4$ c. NO_2

Oxidation numbers have many uses, but the one that concerns us here is their role in determining whether or not a particular reaction involves oxidation-reduction. In an oxidation-reduction reaction, at least two elements change oxidation numbers. The element that is oxidized increases its oxidation number and the element that is reduced decreases its oxidation number. In the reaction of sodium with chlorine, sodium atoms are oxidized to sodium ions; the oxidation number of sodium increases from 0 to +1. In this reaction, chlorine is reduced to chloride ions; the oxidation number of chlorine decreases from 0 to −1.

B. Identifying Oxidation-Reduction Reactions

By assigning oxidation numbers to all the elements in the reactants and the products of a reaction, we can determine whether the reaction results in a change in oxidation number. If a change does occur, the reaction is an oxidation-reduction reaction. For example, consider the reaction between magnesium and oxygen:

$$2 \, Mg + O_2 \longrightarrow 2 \, MgO$$
$$\quad 0 \qquad 0 \qquad\qquad +2, -2$$

Under each element in each substance in the equation is written its oxidation number. The oxidation number of magnesium increases from 0 to +2; magnesium is oxidized. The oxidation number of oxygen decreases from 0 to −2; oxygen is reduced. We conclude that the reaction of magnesium with oxygen is an oxidation-reduction reaction.

A reaction that is not an oxidation-reduction reaction will cause no changes in oxidation numbers. Consider the reaction of sodium hydroxide with hydrochloric acid:

$$NaOH + HCl \longrightarrow NaCl + H_2O$$
$$+1, -2, +1 \quad +1, -1 \qquad +1, -1 \quad +1, -2$$

Under the equation are written the oxidation numbers of the elements. Because none have changed, we know that this neutralization reaction is not an oxidation-reduction reaction.

Example 5.11

For the following reactions decide:

1. Is it an oxidation-reduction reaction?
2. If so, which element is oxidized and which element is reduced?

a. Bromine can be prepared by bubbling chlorine gas through a solution of sodium bromide. The equation for this reaction is:

$$2\ NaBr + Cl_2 \longrightarrow 2\ NaCl + Br_2$$

b. If you blow through a straw into limewater, the solution becomes milky. In chemical terms, if carbon dioxide is bubbled through a solution of calcium hydroxide in water, a milky-white precipitate of calcium carbonate forms:

$$CO_2 + Ca(OH)_2 \longrightarrow CaCO_3 \downarrow + H_2O$$

Solution

a. 1. Write the oxidation number under each element in the equation.

$$2\ NaBr + Cl_2 \longrightarrow 2\ NaCl + Br_2$$
$${+1,\ -1}0{+1,\ -1}0$$

2. Do any elements change oxidation number? Yes, both chlorine and bromine do. Therefore, this is an oxidation-reduction reaction.

3. The oxidation number of chlorine changes from 0 to -1; chlorine is reduced. The oxidation number of bromine changes from -1 to 0; bromine is oxidized.

b. $CO_2 + Ca(OH)_2 \longrightarrow CaCO_3 + H_2O$
${+4,\ -2}{+2,\ -2,\ +1}{+2,\ +4,\ -2}{+1,\ -2}$

Under each element is written its oxidation number. None of these numbers changed during the reaction; the reaction is not an oxidation-reduction reaction.

Problem 5.11

Determine which of the following are oxidation-reduction reactions. For those reactions that are oxidation-reduction reactions, identify the element oxidized and the element reduced.

a. $NH_3(g) + HCl(g) \longrightarrow NH_4Cl(s)$
b. $Zn(s) + 2\ HCl(aq) \longrightarrow ZnCl_2(aq) + H_2(g)$

In an oxidation-reduction reaction, the substance that gains electrons is the **oxidizing agent.** The substance that loses electrons is the **reducing agent.** In the reaction of magnesium with oxygen,

$$2\ Mg\ +\ O_2\ \longrightarrow\ 2\ MgO$$

oxidation numbers:	0	0	+2, −2
	loses electrons	gains electrons	
	is oxidized	is reduced	
	is the reducing	is the oxidizing	
	agent	agent	

Example 5.12

In the reaction of sodium with chlorine to form sodium chloride, which substance is the oxidizing agent? Which is the reducing agent?

Solution

Write the equation for the reaction and assign oxidation numbers.

$$\underset{0}{2\,Na} + \underset{0}{Cl_2} \longrightarrow \underset{+1,\,-1}{2\,NaCl}$$

Because chlorine changes oxidation number from 0 to -1, it is reduced; it is the oxidizing agent. Because sodium changes oxidation number from 0 to $+1$, it is oxidized; it is the reducing agent.

Problem 5.12

The reaction of metallic copper with mercury(II) nitrate yields metallic mercury and copper(II) nitrate. Show that this is an oxidation-reduction reaction. Identify the oxidizing and reducing agents.

The characteristics of oxidation and reduction are summarized in Table 5.2.

Table 5.2 Characteristics of oxidation and reduction.

Substance Oxidized	Substance Reduced
loses electrons	gains electrons
attains a more positive oxidation number	attains a more negative oxidation number
is the reducing agent	is the oxidizing agent

C. Half-Reactions and Their Use

Although oxidation and reduction proceed simultaneously and an oxidation-reduction reaction can be shown in a single equation, the separate processes of oxidation and reduction are often shown as separate equations known as **half-reactions.** We have already encountered several examples of half-reactions. The half-reactions for the oxidation of sodium and magnesium are:

$$Na \longrightarrow Na^+ + e^-$$
$$Mg \longrightarrow Mg^{2+} + 2\,e^-$$

In these oxidation half-reactions, electrons are found as products. Similarly, we have already encountered the reduction half-reactions for chlorine and oxygen:

$$Cl_2 + 2\,e^- \longrightarrow 2\,Cl^-$$

$$O_2 + 4\,e^- \longrightarrow 2\,O^{2-}$$

In a reduction half-reaction, electrons are reactants.

Simple oxidation-reduction equations can be balanced by inspection. Consider the reaction of copper with silver nitrate. The balanced equation for this reaction is:

$$Cu + 2\,Ag(NO_3) \longrightarrow 2\,Ag + Cu(NO_3)_2$$
$$\;\;0 \qquad +1, +5, -2 \qquad\quad 0 \qquad +2, +5, -2$$

Under the equation, we have written the oxidation numbers of the elements present. By inspection we see that copper is oxidized. The half-reaction for this oxidation is:

$$Cu \longrightarrow Cu^{2+} + 2\,e^-$$

Silver is reduced, by the following half-reaction:

$$Ag^+ + e^- \longrightarrow Ag$$

In the complete balanced equation, the number of electrons lost must equal the number of electrons gained. Therefore, in this reaction, two silver ions must be reduced for each copper atom oxidized:

$$2(Ag^+ + e^- \longrightarrow Ag)$$

The oxidation and reduction half-reactions can be added together in the following manner:

$$
\begin{aligned}
2\,Ag^+ + \qquad\quad 2\,e^- &\longrightarrow 2\,Ag \\
Cu \qquad\qquad &\longrightarrow \qquad\qquad Cu^{2+} + 2\,e^- \\
\hline
2\,Ag^+ + Cu + 2\,e^- &\longrightarrow 2\,Ag + Cu^{2+} + 2\,e^-
\end{aligned}
$$

Two electrons appear on each side of this equation. We can cancel them, giving an ionic equation for the oxidation of copper by silver ion:

$$2\,Ag^+ + Cu \longrightarrow 2\,Ag + Cu^{2+}$$

Notice that this ionic equation omits the nitrate ion, which does not change during the reaction. Adding the necessary nitrate ions to both the reactants and products gives us the original balanced equation:

$$Cu + 2\,AgNO_3 \longrightarrow 2\,Ag + Cu(NO_3)_2$$

Example 5.13

Write the equation for the reaction of zinc with hydrochloric acid. Isolate the half-reactions involved. Show that the balanced equation can be obtained from these half-reactions.

Solution

The equation is:

$$Zn + 2\,HCl \longrightarrow H_2 + ZnCl_2$$

Zinc's oxidation number changes from 0 to $+2$; it is oxidized.

$$Zn \longrightarrow Zn^{2+} + 2\,e^-$$

Hydrogen's oxidation number changes from $+1$ to 0; it is reduced.

$$2\,H^+ + 2\,e^- \longrightarrow H_2$$

Both half-reactions use two electrons; therefore, they can be added as they stand to give:

$$Zn + 2\,H^+ + 2\,e^- \longrightarrow H_2 + Zn^{2+} + 2\,e^-$$

Cancelling the electrons and adding 2 chloride ions to each side of the equation gives

$$Zn + 2\,HCl \longrightarrow H_2 + ZnCl_2$$

which is identical to the original equation.

Problem 5.13

Write the equation for the reaction of bromine with sodium iodide to form sodium bromide and free iodine. Isolate the half-reactions involved. Show that the balanced equation can be obtained by the addition of these half-reactions.

D. Balancing More Complicated Oxidation-Reduction Equations

Many oxidation-reduction equations cannot be balanced by inspection. Such equations are balanced according to the following series of steps:

1. Write the unbalanced equation for the reaction. For example, in aqueous solution, sulfuric acid reacts with potassium iodide to form free iodine

and hydrogen sulfide (a gas with the smell of rotten eggs):

$$H_2SO_4 + KI \nrightarrow H_2S + I_2$$

2. Assign oxidation numbers to each element:

$$H_2SO_4 \quad + \quad KI \quad \nrightarrow \quad H_2S \quad + \; I_2$$
$$\underset{+1, +6, -2}{} \qquad \underset{+1, -1}{} \qquad \quad \underset{+1, -2}{} \quad \underset{0}{}$$

3. Identify the elements that have changed oxidation number and write an unbalanced half-reaction for each, using the formula of the ion or molecule in which it occurs:

   a. $SO_4^{2-} \nrightarrow H_2S$ b. $I^- \nrightarrow I_2$

4. Balance these half-reactions.

   By mass. In balancing these half-reactions, certain options may be employed. If the reaction takes place in aqueous solution, water molecules may participate in the reaction as either reactant or product. For a reaction occurring in acid solution, hydrogen ions may participate in the reaction, so H^+ and H_2O may be used as needed to balance the equation. In alkaline solution, OH^- and H_2O may be used as needed to balance the equation. Remember that because hydrogen ions and hydroxide ions react immediately to form water, hydroxide ions cannot be used to balance the equation for a reaction that takes place in acid solution, nor can hydrogen ions be used to balance the equation for a reaction that takes place in alkaline solution. The equations for the half-reactions in this example can be balanced by mass in the following way:

   a. $10\,H^+ + SO_4^{2-} \nrightarrow H_2S + 4\,H_2O$ (acid solution)

   b. $2\,I^- \nrightarrow I_2$

   By charge. Add electrons to the less negative side of each equation so that the total charge on each side of each equation is the same.

   a. $8\,e^- + 10\,H^+ + SO_4^{2-} \longrightarrow H_2S + 4\,H_2O$ (electrons are gained; therefore, reduction occurs)

   b. $2\,I^- \longrightarrow I_2 + 2\,e^-$ (electrons are lost; therefore, oxidation occurs)

5. Equate the electrons in the two equations by multiplying by appropriate whole numbers. In our example, the reduction half-reaction requires 8

electrons. The equation for the oxidation of I^- yields only 2 electrons. To equate the loss and gain of electrons, the iodide oxidation reaction must occur 4 times for each sulfate ion reduced.

a. $8 e^- + 10 H^+ + SO_4^{2-} \longrightarrow H_2S + 4 H_2O$

b. $8 I^- \longrightarrow 4 I_2 + 8 e^-$

6. Add the two balanced half-reactions.

$$8 e^- + 10 H^+ + SO_4^{2-} + 8 I^- \longrightarrow H_2S + 4 H_2O + 4 I_2 + 8 e^-$$

7. If all the steps have been carried out correctly, the number of reactant electrons equals the number of product electrons, and all elements are balanced. In our equation, the number of product and reactant electrons are equal and the sulfur, iodine, hydrogen, and oxygen atoms balance. The total charge on the right (-8) equals the total charge on the left (-8). Thus, the equation is balanced by mass and charge. Cancelling the electrons, we have:

$$10 H^+ + SO_4^{2-} + 8 I^- \longrightarrow H_2S + 4 H_2O + 4 I_2$$

The potassium ions, which do not take part in the reaction, are not shown.

Example 5.14

The reaction of permanganate ion (MnO_4^-) in aqueous acid (H^+) with oxalate ion $(C_2O_4^{2-})$ forms manganese(II) ion (Mn^{2+}) and carbon dioxide (CO_2).

Solution

1. **Unbalanced equation:** $MnO_4^- + H^+ + C_2O_4^{2-} \not\longrightarrow CO_2 + Mn^{2+}$

2. **Oxidation numbers:** $+7, -2$ $+1$ $+3, -2$ $+4, -2$ $+2$

3. **Isolate unbalanced half-reactions:**
    a. $MnO_4^- \not\longrightarrow Mn^{2+}$
    b. $C_2O_4^{2-} \not\longrightarrow CO_2$

4. **Balance half-reactions by mass:** Because the reaction takes place in acid solution, H^+ and H_2O can be used.
    a. $MnO_4^- + 8 H^+ \not\longrightarrow Mn^{2+} + 4 H_2O$
    b. $C_2O_4^{2-} \not\longrightarrow 2 CO_2$
    Balance half-reactions by charge: Add electrons to the less negative side.
    a. $5 e^- + MnO_4^- + 8 H^+ \longrightarrow Mn^{2+} + 4 H_2O$
    b. $C_2O_4^{2-} \longrightarrow 2 CO_2 + 2 e^-$

5. **Equate electrons in the two half-reactions:**
    a. $2(5 e^- + MnO_4^- + 8 H^+ \longrightarrow Mn^{2+} + 4 H_2O)$
    b. $5(C_2O_4^{2-} \longrightarrow 2 CO_2 + 2 e^-)$

6. **Add half-reactions:**
$$10 \, e^- + 2 \, MnO_4^- + 16 \, H^+ + 5 \, C_2O_4^{2-} \longrightarrow$$
$$2 \, Mn^{2+} + 8 \, H_2O + 10 \, CO_2 + 10 \, e^-$$

7. **Cancel electrons and check balancing:**
$$2 \, MnO_4^- + 16 \, H^+ + 5 \, C_2O_4^{2-} \longrightarrow 2 \, Mn^{2+} + 8 \, H_2O + 10 \, CO_2$$

Problem 5.14

When bromide ion reacts with sulfuric acid, the products are sulfur dioxide, water, and free bromine. Using half-reactions, write the balanced equation for this oxidation-reduction reaction.

5.5 Mass Relationships in an Equation

A balanced equation is a quantitative statement of a reaction. It relates the mass of the reactants to the mass of the products. Let us see what this statement means in terms of a particular reaction. Pentane, C_5H_{12}, burns in oxygen to form carbon dioxide and water. Burning of pentane also releases energy. The balanced equation for the combustion of pentane is:

$$C_5H_{12} + 8 \, O_2 \longrightarrow 5 \, CO_2 + 6 \, H_2O$$

In qualitative terms, this equation shows that pentane reacts with oxygen to form carbon dioxide and water. In quantitative terms, the equation states that one molecule of pentane reacts with 8 molecules of oxygen to form 5 molecules of carbon dioxide and 6 molecules of water. If we had 15 molecules of pentane, we would need (8×15) or 120 molecules of oxygen for complete reaction; (5×15) molecules of carbon dioxide and (6×15) molecules of water would be formed. We could start with any number (n) molecules of pentane and form $5n$ molecules of carbon dioxide and $6n$ molecules of water.

If 6.02×10^{23} molecules (one mole) of pentane is burned, $[8 \times (6.02 \times 10^{23})]$ molecules (8 moles) of oxygen are needed. The reaction would form $[5 \times (6.02 \times 10^{23})]$ molecules (5 moles) of carbon dioxide and $[6 \times (6.02 \times 10^{23})]$ molecules (6 moles) of water. These quantitative relationships are summarized in Table 5.3.

Table 5.3 Quantitative relationships in a chemical equation.

72 g	+ 256 g	⟶ 220 g	+ 108 g
C_5H_{12}	+ 8 O_2	⟶ 5 CO_2	+ 6 H_2O
1 molecule	+ 8 molecules	⟶ 5 molecules	+ 6 molecules
1 mole	+ 8 moles	⟶ 5 moles	+ 6 moles

A balanced equation gives the ratio between moles of reactants and moles of products. Given the number of moles of one component in a reaction, the number of moles or grams of any other component can be calculated. Such calculations are called **stoichiometry.**

A stoichiometric problem can be stated in many ways, but it always contains the following parts:

1. The reaction involved.
2. A stated amount of one component of the reaction.
3. A question asking "how much" of another substance is needed or formed in the reaction.

The problems in previous chapters were solved by answering a series of questions:

1. What is asked for?
2. What is given?
3. What conversion factors are needed to go from "given" to "wanted"?
4. How should the arithmetic equation be set up so that the units of the "given" are converted to the units of the "wanted"?

Stoichiometric problems can be solved by answering the same set of questions, the only difference being that the conversion factors are derived from the balanced chemical equation for the reaction involved. The steps to follow in solving a stoichiometric problem are:

1. Write the balanced equation for the reaction.
2. Decide which substance is asked for and in what units.
3. Decide which substance is given and in what units and what amount.
4. Determine the conversion factors required to convert:
   a. the amount of given substance into moles.
   b. moles of the given substance into moles of the asked-for substance.
   c. moles of asked-for substance into the units wanted in the problem.
5. Combine the amount of given substance and its units along with the appropriate conversion factors into an equation in such a way that only the asked-for substance in the proper units remains.

Example 5.15

How many grams of carbon dioxide are formed when 61.5 g of pentane are burned in oxygen?

Solution

Equation:
$$C_5H_{12} + 8 O_2 \longrightarrow 5 CO_2 + 6 H_2O$$

Wanted:
? g CO_2

Given:
61.5 g C_5H_{12}

Conversion factors:
(a) 1 mole C_5H_{12} = 72.2 g C_5H_{12}
(b) 1 mole C_5H_{12} yields 5 moles CO_2
(c) 1 mole CO_2 = 44.0 g CO_2

Arithmetic equation:

$$? \text{ g } CO_2 = 61.5 \text{ g } C_5H_{12} \times \frac{1 \text{ mole } C_5H_{12}}{72.2 \text{ g } C_5H_{12}} \times \frac{5 \text{ moles } CO_2}{1 \text{ mole } C_5H_{12}} \times \frac{44.0 \text{ g } CO_2}{1 \text{ mole } CO_2}$$

Answer:
187 g CO_2

Several points should be emphasized. First, the name and units of each item in the equation are always shown. This prevents confusion and errors. Second, no arithmetic is done until the whole equation is written out. Third, all units in the final equation must cancel except for those asked for in the answer.

Example 5.16

How many molecules of hydrogen will be formed by the reaction of 2.65×10^{-3} g of zinc with hydrochloric acid?

Solution

Equation:
$$Zn + 2 HCl \longrightarrow ZnCl_2 + H_2$$

Wanted:
? molecules H_2

Given:
2.65×10^{-3} g Zn

Conversion factors:
(a) 1 mole zinc = 65.4 g zinc
(b) 1 mole zinc yields 1 mole hydrogen (from equation)
(c) 1 mole hydrogen contains 6.02×10^{23} molecules H_2

Arithmetic equation:

$$? \text{ molecules } H_2 = 2.65 \times 10^{-3} \text{ g } \overline{Zn} \times \frac{1 \text{ mole } \overline{Zn}}{65.4 \text{ g } \overline{Zn}} \times \frac{1 \text{ mole } H_2}{1 \text{ mole } \overline{Zn}}$$

$$\times \frac{6.02 \times 10^{23} \text{ molecules } H_2}{1 \text{ mole } H_2}$$

Answer:

2.44×10^{19} molecules H_2

Problem 5.15

Hydrogen burns in oxygen to form water. What mass of oxygen is necessary for the complete combustion of 1.74 g of hydrogen?

Problem 5.16

Bromine is prepared by the reaction of chlorine with sodium bromide. How many grams of chlorine are necessary for the preparation of 2.12 g of bromine?

5.6 Percent Yield

In the solutions of the problems in Section 5.5, we have assumed two things. First, that all reactants are completely converted to products. In Example 5.15 we assumed that all of the pentane available (61.5 g) was burned and completely converted to carbon dioxide and water; in Example 5.16 we assumed that all of the zinc reacted to form zinc chloride and hydrogen. Second, we assumed there was enough of the second reactant present to complete the reaction. In Example 5.15 we assumed there was enough oxygen present to react with all of the pentane; in Example 5.16 we assumed there was enough hydrochloric acid present to react with all of the zinc.

These two assumptions are not always justified. In some instances, a chemical reaction produces small amounts of substances other than those shown in the equation. In other instances, there is not enough of one reactant present to allow complete reaction of the other. In both instances, the amount of product obtained is less than that predicted by calculation.

The **theoretical yield** is the maximum amount of product that can be formed by complete conversion of reactants to the desired products. The **percent yield** compares the actual yield to the theoretical yield. In Example 5.15, we calculated that the complete combustion of 61.5 g of pentane would have a theoretical yield of 187 g carbon dioxide. Suppose you actually burned 61.5 g of pentane and obtained only 161 g of carbon dioxide. What would be the percent yield?

$$\% \text{ yield} = \frac{161 \text{ g}}{187 \text{ g}} \times 100\% = 86.1\%$$

If a problem asks for percent yield, first calculate the theoretical yield and then determine what percent of that amount was actually obtained.

Example 5.17

The reaction of ethane with chlorine yields ethyl chloride and hydrogen chloride:

$$C_2H_6 + Cl_2 \longrightarrow C_2H_5Cl + HCl$$

ethane   ethyl
chloride

When 5.6 g of ethane is reacted with chlorine, 8.2 g of ethyl chloride is obtained. Calculate the percent yield of ethyl chloride.

Solution

Because this is a percent yield problem, the first step is to calculate the theoretical yield. This can be done using the steps outlined in Section 5.5.

Equation:
$$C_2H_6 + Cl_2 \longrightarrow C_2H_5Cl + HCl$$

Wanted:
? g C_2H_5Cl if the yield is 100%

Given:
5.6 g C_2H_6

Conversion factors:
(a) 1 mole ethane = 30.1 g C_2H_6
(b) 1 mole C_2H_6 yields 1 mole C_2H_5Cl
(c) 1 mole C_2H_5Cl = 64.5 g C_2H_5Cl

Arithmetic equation:

$$? \text{ g } C_2H_5Cl = 5.6 \text{ g } C_2H_6 \times \frac{1 \text{ mole } C_2H_6}{30.1 \text{ g } C_2H_6} \times \frac{1 \text{ mole } C_2H_5Cl}{1 \text{ mole } C_2H_6}$$

$$\times \frac{64.5 \text{ g } C_2H_5Cl}{1 \text{ mole } C_2H_5Cl} = 12.0 \text{ g } C_2H_5Cl$$

The theoretical yield is 12.0 g C_2H_5Cl. The actual yield is 8.2 g ethyl chloride.

Answer:

$$\text{percent yield} = \frac{8.2 \text{ g } C_2H_5Cl}{12.0 \text{ g } C_2H_5Cl} \times 100\% = 68\%$$

Example 5.18

When a sample of impure zinc weighing 7.45 g reacts with an excess of hydrochloric acid, 0.214 g of hydrogen is released. Calculate the percent zinc in the sample. Note that whenever a reactant is said to be present **in excess,** it means that more than enough is present to cause complete reaction of all other reactants.

Analysis of the problem:

1. The word percent is used; somewhere in the solution of the problem there must be a calculation of percent.

2. The zinc sample is impure. The part that is zinc reacts to form hydrogen, the rest does not.

3. We can calculate a theoretical yield, or how much hydrogen would be obtained if the sample were all zinc. This theoretical yield will be greater than 0.214 g of hydrogen.

4. The percent yield will equal the percent of zinc in the sample.

Solution

1. Calculate the theoretical yield using the steps from Section 5.5.

Equation:
$$Zn + 2\,HCl \longrightarrow ZnCl_2 + H_2$$

Wanted:
? g hydrogen

Given:
7.45 g zinc

Conversion factors:
1 mole zinc = 65.4 g zinc
1 mole zinc yields 1 mole hydrogen gas
1 mole hydrogen gas = 2.02 g hydrogen gas

Arithmetic equation:

$$? \text{ g } H_2 = 7.45 \text{ g Zn} \times \frac{1 \text{ mole Zn}}{65.4 \text{ g Zn}} \times \frac{1 \text{ mole } H_2}{1 \text{ mole Zn}} \times \frac{2.02 \text{ g } H_2}{1 \text{ mole } H_2}$$

$$= 0.230 \text{ g } H_2$$

2. Calculate the percent yield, which will equal the percentage of zinc in the sample.

Answer:

$$\% \text{ zinc} = \frac{0.214 \text{ g}}{0.230 \text{ g}} \times 100\% = 93.0\%$$

Problem 5.17

When 1.6 g of oxygen reacts with an excess of nitrogen, 1.3 g of nitrogen oxide is formed. Calculate the percent yield of nitrogen oxide. The balanced equation is:

$$N_2 + O_2 \longrightarrow 2\,NO$$

Problem 5.18

When a piece of impure copper weighing 0.54 g is added to a solution of silver nitrate, 1.5 g of silver precipitates. What is the percent purity of the copper sample?

Each of the previous examples dealt with a percent calculation. Example 5.17 dealt with an incomplete reaction and Example 5.18 involved a reactant that was less than 100% pure. All problems dealing with percent calculations are variations on one or the other of these two types. After calculating the theoretical yield, you must use the actual yield to calculate either the percent yield of a product or the percent purity of a reactant.

5.7 Problems Involving a Limiting Reactant

In all of the stoichiometry problems encountered thus far, there have been two reactants, but the amount was given for only one of them. In calculating the solutions, we assumed that enough of the second reactant was present to allow complete reaction of the first. This is not always the case. We will encounter many problems in which we know the amounts of all reactants and must calculate which is present in excess before we can calculate a theoretical yield.

A problem of this type might be encountered if you owned a bicycle shop. Among other things, each bicycle requires two wheels and one set of handlebars. Suppose that, after a busy season, your stockroom contains only 14 wheels and 6 sets of handlebars, although there are huge quantities of all the other necessary parts. How many bicycles can you build before a new shipment of wheels and handlebars arrives? You have enough wheels for seven bicycles (14/2). You have enough handlebars for six bicycles (6/1). Clearly, you can put together only six bicycles. There will be two unused wheels. No matter how you juggle the parts, there are enough handlebars for only six bicycles. The number of handlebars **limits** the number of bicycles you can make.

Similarly, in a chemical reaction where the amount of each reactant is known, one is usually present in excess; the amount of the other limits the amount of product that can be formed. The problem becomes how to decide which is the **limiting reactant.**

Figure 5.4 The amount of product formed by a chemical reaction is limited by the amounts of reactants. In (a) and (b), the reactants are present in the ratio of the balanced equation, $H_2 + Cl_2 \rightarrow 2\ HCl$, so they react completely. In (c) and (d), one of the reactants is present in excess; some of this reactant is left unreacted after the reaction is complete.

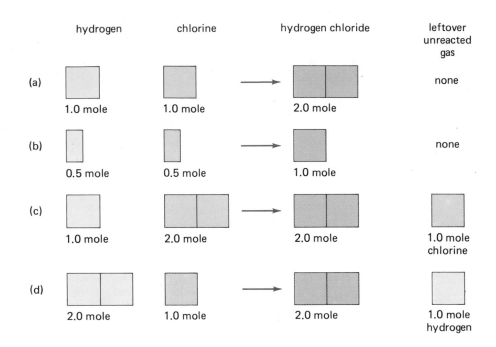

Consider the reaction of hydrogen with chlorine to form hydrogen chloride. The balanced equation for this reaction is:

$$H_2 + Cl_2 \longrightarrow 2\ HCl$$

According to the equation, each mole of hydrogen reacting requires one mole of chlorine to form 2 moles of hydrogen chloride (Figure 5.4a). If only 0.5 mole of hydrogen is present, only 0.5 mole of chlorine is needed and 1.0 mole of hydrogen chloride is formed (Figure 5.4b). If 2.0 moles of hydrogen are to react, then 2.0 moles of chlorine are necessary and 4.0 moles of hydrogen chloride are formed. Suppose you have 2.0 moles of hydrogen but only 1.0 mole of chlorine (Figure 5.4d). Here, only 1.0 mole of hydrogen can react because there is only 1.0 mole of chlorine. In this instance, chlorine is the limiting reactant. Only 2.0 moles of hydrogen chloride are formed and the second mole of hydrogen remains unreacted.

A limiting reactant problem, then, consists of two simpler problems. Suppose we have the reaction

$$A + B \longrightarrow C$$

and we know how much A and B are available. To determine how much C will be formed, we must calculate (1) how much C can be prepared from the given amount of A, and (2) how much C can be prepared from the given amount of B. The smaller of these two amounts, as in the case of the bicycles, is the theoretical yield of C.

Example 5.19

What mass of lithium chloride can be formed by the reaction of 5.00 g of lithium with 5.00 g of chlorine?

Analysis of the problem:
We know how much of each reactant is present. An estimated yield of product can be calculated from either. The theoretical yield is the smaller of the two.

Solution

Equation:
$$2 \text{ Li} + \text{Cl}_2 \longrightarrow 2 \text{ LiCl}$$

Wanted:
? g LiCl

Given:
Amounts of the two reactants: 5.00 g lithium and 5.00 g chlorine

Conversion factors:
1 mole lithium = 6.94 g Li
1 mole chlorine = 70.9 g Cl_2
1 mole lithium yields 1 mole lithium chloride
1 mole chlorine yields 2 moles lithium chloride
1 mole lithium chloride = 42.4 g LiCl

Arithmetic equation:
There are two arithmetic equations:
A. Based on the amount of lithium present.
B. Based on the amount of chlorine present.

Equation A is:

$$? \text{ g LiCl} = 5.00 \text{ g Li} \times \frac{1 \text{ mole Li}}{6.94 \text{ g Li}} \times \frac{1 \text{ mole LiCl}}{1 \text{ mole Li}} \times \frac{42.4 \text{ g LiCl}}{1 \text{ mole LiCl}}$$

$$= 30.6 \text{ g LiCl}$$

Equation B is:

$$? \text{ g LiCl} = 5.00 \text{ g Cl}_2 \times \frac{1 \text{ mole Cl}_2}{70.9 \text{ g Cl}_2} \times \frac{2 \text{ moles LiCl}}{1 \text{ mole Cl}_2} \times \frac{42.4 \text{ g LiCl}}{1 \text{ mole LiCl}}$$

$$= 5.98 \text{ g LiCl}$$

We now ask ourselves, "Will this reaction yield 30.6 g or 5.98 g of lithium chloride?" Remembering the bicycles, we choose the smaller amount, 5.98 g. There is only enough chlorine to prepare 5.98 g of lithium chloride; there is no way to get 30.6 g.

Problem 5.19   Chlorine reacts with ethane to form ethyl chloride:

$$Cl_2 + C_2H_6 \longrightarrow C_2H_5Cl + HCl$$

ethane   ethyl chloride

How many grams of C_2H_5Cl will be formed by the reaction of 2.56 g of chlorine with 8.51 g of ethane?

5.8 Energy Changes Accompanying Chemical Reactions

All changes, whether chemical or physical, are accompanied by a change in **energy.** There are several ways of measuring energy changes. The two kinds of energy of most interest to us are: (1) **free energy (G),** the energy available to do work (discussed in Chapter 8); (2) **enthalpy (H),** heat energy measured at constant pressure, which we will discuss here. Most chemical reactions, especially those in living organisms, take place under the constant pressure of the atmosphere. The energy released or absorbed by such reactions is a change in enthalpy, ΔH (say "delta H"), which can be shown as

$$\Delta H_{reaction} = H_{products} - H_{reactants}$$

In reporting values of ΔH, a superscript is used to show the temperature at which the measurements were made. For example, the symbol $\Delta H^{25°C}$ shows that the change in enthalpy is for a reaction carried out at 25°C. Each reacting molecule possesses a certain amount of energy due to the nature of its chemical bonds. So does each product molecule. As the bonds of the reacting molecules break and the new bonds of the products are formed, energy is released or absorbed depending on whether the reactants have higher or lower enthalpy than the products.

A. Exothermic and Endothermic Reactions

If the enthalpy of the products of a reaction is greater than that of the reactants, the change in enthalpy is positive. Energy is *absorbed* from the surroundings and the reaction is **endothermic.** The following reactions are endothermic.

1. The melting of ice:

$$H_2O(s) \longrightarrow H_2O(l) \qquad \Delta H^{0°C} = +6.00 \text{ kJ/mole } H_2O$$
$$\text{or } +1.44 \text{ kcal/mole } H_2O$$

2. The formation of hydrogen iodide:

$$\tfrac{1}{2} H_2(g) + \tfrac{1}{2} I_2(s) \longrightarrow HI(g) \qquad \Delta H^{25°C} = +25.9 \text{ kJ/mole HI}$$
$$(+6.19 \text{ kcal/mole HI})$$

3. The decomposition of water:

$$H_2O(l) \longrightarrow H_2(g) + \tfrac{1}{2} O_2(g) \qquad \Delta H^{25°C} = +285.8 \text{ kJ/mole } H_2O$$
$$(+68.3 \text{ kcal/mole } H_2O)$$

If the enthalpy of the products is less than that of the reactants, the change in enthalpy is negative. Energy is *released* to the surroundings and the reaction is **exothermic.** The following reactions are exothermic.

1. The combustion of methane:

$$CH_4(g) + 2 O_2(g) \longrightarrow CO_2(g) + 2 H_2O(g)$$
$$\Delta H^{25°C} = -891 \text{ kJ/mole } CH_4$$
$$(-213 \text{ kcal/mole } CH_4)$$

2. The formation of water:

$$H_2(g) + \tfrac{1}{2} O_2(g) \longrightarrow \tfrac{1}{2} H_2O(l) \qquad \Delta H^{25°C} = -285.8 \text{ kJ/mole } H_2O$$
$$(-68.3 \text{ kcal/mole } H_2O)$$

3. The metabolism of sucrose:

$$C_{12}H_{22}O_{11}(s) + 12 O_2(g) \longrightarrow 12 CO_2(g) + 11 H_2O(l)$$
$$\Delta H^{25°C} = -5.64 \times 10^3 \text{ kJ/mole sucrose}$$
$$(-1.35 \times 10^3 \text{ kcal/mole sucrose})$$

Notice that the decomposition of water is endothermic and requires the *input* of 285.8 kilojoules of energy per mole of water decomposed. The formation of one mole of water from hydrogen and oxygen is exothermic and *releases* 285.8 kJ of energy. The amount of energy is the same, but the sign of the energy term is different.

Another example of the relationship between energy change and the direction of a reaction that is much more important to us is the formation and decomposition of glucose. Glucose ($C_6H_{12}O_6$) is formed from carbon dioxide and oxygen in the cells of green plants in the process called **photosynthesis.** Photosynthesis is an endothermic reaction. The source of the energy for the formation of glucose is light (radiant energy), usually from the sun.

$$6 \ CO_2(g) + 6 \ H_2O(l) \xrightarrow{\text{photosynthesis}} C_6H_{12}O_6(s) + 6 \ O_2(g)$$
$$\Delta H^{25°C} = +2.80 \times 10^3 \ \text{kJ/mole glucose}$$
$$(+670 \ \text{kcal/mole glucose})$$

In the reverse of this reaction, the glucose formed is metabolized (broken down) in plant and animal cells to form carbon dioxide, water, and energy:

$$C_6H_{12}O_6(s) + 6 \ O_2(g) \xrightarrow{\text{metabolism}} 6 \ CO_2(g) + 6 \ H_2O(l)$$
$$\Delta H^{25°C} = -2.80 \times 10^3 \ \text{kJ/mole glucose}$$
$$(-670 \ \text{kcal/mole glucose})$$

Thus, green plants have the remarkable ability to trap the energy of sunlight and use that energy to produce glucose from carbon dioxide and water. The energy is stored in the glucose. Animal and plant cells have the equally remarkable ability to metabolize glucose and use the energy released to maintain body temperature or do biological work, such as contracting muscles or thinking. We will discuss the metabolism of glucose and this flow of energy in the biological world in some detail in Chapters 23 and 24.

Example 5.20

For each of the following: (1) decide whether the reaction is exothermic or endothermic; (2) write the equation for the reverse reaction and state the accompanying enthalpy change.

a. $N_2(g) + O_2(g) \longrightarrow 2 \ NO(g)$ $\Delta H = +181 \ \text{kJ} \ (+43.3 \ \text{kcal})$

b. $2 \ NO_2(g) \longrightarrow N_2O_4(g)$ $\Delta H = -92.0 \ \text{kJ} \ (-22.0 \ \text{kcal})$

c. $PCl_3(g) + Cl_2(g) \longrightarrow PCl_5(g)$ $\Delta H = -126 \ \text{kJ} \ (-30.1 \ \text{kcal})$

Solution

a. The enthalpy change in positive; the reaction is endothermic. The reverse reaction is:

$$2 \ NO(g) \longrightarrow N_2(g) + O_2(g) \qquad \Delta H = -181 \ \text{kJ} \ (-43.3 \ \text{kcal})$$

b. The enthalpy change is negative; the reaction is exothermic. The reverse reaction is:

$$N_2O_4(g) \longrightarrow 2 \ NO_2(g) \qquad \Delta H = +92.0 \ \text{kJ} \ (+22.0 \ \text{kcal})$$

c. The enthalpy change is negative; the reaction is exothermic. The reverse reaction is:

$$PCl_5(g) \longrightarrow PCl_3(g) + Cl_2(g) \qquad \Delta H = +126 \ \text{kJ} \ (+30.1 \ \text{kcal})$$

Problem 5.20

For each of the following: (1) state whether the reaction is exothermic or endothermic; (2) write the equation for the reverse reaction and state its enthalpy change.

a. $N_2(g) + 3 H_2(g) \longrightarrow 2 NH_3(g)$ $\Delta H = -46.0$ kJ $(-11.0$ kcal$)$

b. $2 H_2O(g) + 2 Cl_2(g) \longrightarrow 4 HCl(g) + O_2(g)$

$$\Delta H = -120 \text{ kJ } (-28.7 \text{ kcal})$$

c. $C_2H_5OH(l) + 3 O_2(g) \longrightarrow 2 CO_2(g) + 3 H_2O(l)$

$$\Delta H = -1.37 \times 10^3 \text{ kJ } (-327 \text{ kcal})$$

B. The Stoichiometry of Energy Changes

The energy change associated with a reaction is a stoichiometric quantity and can be treated arithmetically, as mass changes were in Section 5.5. For many reactions, enthalpy changes have been determined and tabulated in the chemical literature. The changes listed in such sources apply only to the form of the equation they accompany. Consider the formation of water vapor from hydrogen and oxygen gas. When the equation is written

$$H_2(g) + \tfrac{1}{2}O_2(g) \longrightarrow H_2O(g)$$

the enthalpy change is:

$$\Delta H = -241 \text{ kJ } (-57.6 \text{ kcal})$$

If the equation is written

$$2 H_2(g) + O_2(g) \longrightarrow 2 H_2O(g)$$

the enthalpy term is doubled

$$\Delta H = -482 \text{ kJ } (-115 \text{ kcal})$$

because the amount of water formed is doubled.

In these equations, the physical state of the components must be given. When *liquid* water is formed from its elements, the enthalpy change is different from that given above for the formation of water vapor, because the enthalpy change for the reaction includes the enthalpy change when water vapor condenses to a liquid.

$$H_2(g) + \tfrac{1}{2}O_2(g) \longrightarrow H_2O(l) \qquad \Delta H = -286 \text{ kJ } (-68.4 \text{ kcal})$$

Example 5.21

Calculate the enthalpy change for the combustion of 35.5 g gaseous propane (C_3H_8).

$$C_3H_8(g) + 5 O_2(g) \longrightarrow 3 CO_2(g) + 4 H_2O(l)$$

$$\Delta H = -2.22 \times 10^3 \text{ kJ } (-531 \text{ kcal})$$

Solution

Equation:
Given above

Wanted:
? kJ released

Given:
35.5 g C_3H_8

Conversion factors:
one mole of propane = 44.1 g C_3H_8
one mole of propane on combustion yields 2.22×10^3 kJ

Arithmetic equation:

$$? \text{ kJ} = 35.5 \text{ g } C_3H_8 \times \frac{1 \text{ mole } C_3H_8}{44.1 \text{ g } C_3H_8} \times \frac{-2.22 \times 10^3 \text{ kJ}}{1 \text{ mole } C_3H_8}$$

$$= -1.79 \times 10^3 \text{ kJ}$$

Problem 5.21

Calculate the enthalpy change when 45.6 g liquid water is formed by the reaction of gaseous hydrogen with gaseous oxygen according to the equation:

$$H_2(g) + \tfrac{1}{2} O_2(g) \longrightarrow H_2O(l) \qquad \Delta H = -286 \text{ kJ } (-68.4 \text{ kcal})$$

Key Terms and Concepts

acid salts (5.3D1)

coefficients in an equation (5.2A)

combination reactions (5.3A)

decomposition reactions (5.3B)

displacement reactions (5.3C)

endothermic reactions (5.8A)

enthalpy (5.8)

exothermic reactions (5.8A)

half-reactions (5.4C)

insoluble salts (5.3D2)

limiting reactant (5.7)

metathesis reactions (5.3D)

neutralization reactions (5.3D1)

oxidation (5.4)

oxidation numbers (5.4A)

oxidizing agent (5.4B)

percent yield (5.6)

photosynthesis (5.8A)

precipitate (5.3D2)

products (5.2A)

reactants (5.2A)

reducing agent (5.4B)

reduction (5.4)

soluble salts (5.3D2)

stoichiometry (5.5)

theoretical yield (5.6)

Problems

**Chemical Equations
(Section 5.2)**

5.22 Why should an equation be balanced?

5.23 Why can't you change the subscripts in a formula to balance an equation? For example, if you are trying to balance

$$Mg + O_2 \nrightarrow MgO$$

why is it incorrect to write:

$$Mg + O_2 \longrightarrow MgO_2$$

5.24 Write a balanced equation for the following reaction and explain what the letters in parentheses mean:

$$Na(s) + H_2O(l) \nrightarrow NaOH(aq) + H_2(g)$$

5.25 What is a neutralization reaction?

**Classifying and
Balancing Chemical
Equations (Section 5.3)**

5.26 Write balanced equations for the following displacement reactions:

a. $Hg + AgNO_3 \nrightarrow Hg(NO_3) + Ag$

b. $Cl_2 + KI \nrightarrow KCl + I_2$

c. $Ca + H_2O \nrightarrow Ca(OH)_2 + H_2$

d. $Ni + HCl \nrightarrow NiCl_2 + H_2$

5.27 Write balanced equations for the following combination reactions:

a. $K + O_2 \nrightarrow K_2O$

b. $Se + O_2 \nrightarrow SeO_2$

c. $Cr + O_2 \nrightarrow Cr_2O_3$

d. $S + O_2 \nrightarrow SO_3$

5.28 Write balanced equations for the following combination reactions:

a. $Cu + S \nrightarrow Cu_2S$

b. $Al + N_2 \nrightarrow Al_2N_3$

c. $N_2 + I_2 \nrightarrow NI_3$

d. $Ag + S \nrightarrow Ag_2S$

5.29 Write balanced equations for the following displacement reactions:

a. $Fe + Cu(NO_3)_2 \nrightarrow Fe(NO_3)_2 + Cu$

b. $Mg + HCl \nrightarrow MgCl_2 + H_2$

c. $Li + H_2O \nrightarrow LiOH + H_2$

d. $Al + H_2SO_4 \nrightarrow Al_2(SO_4)_3 + H_2$

5.30 An important industrial chemical process is the electrolysis of a water solution of sodium chloride to give chlorine gas, sodium hydroxide (known in the chemical industry as caustic soda), and hydrogen gas. Write a balanced equation for this important process.

Oxidation-Reduction Reactions (Section 5.4)

5.31 What characterizes an oxidation-reduction reaction?

5.32 Write balanced equations for the following reactions and identify those that are oxidation-reduction.
a. $Ca(OH)_2 + HCl \longrightarrow CaCl_2 + H_2O$
b. $SO_3 + BaO \longrightarrow BaSO_4$
c. $AgNO_3 + Fe \longrightarrow Fe(NO_3)_2 + Ag$
d. $Na_2SO_4 + BaCl_2 \longrightarrow BaSO_4 + NaCl$
e. $NaI + Cl_2 \longrightarrow NaCl + I_2$

5.33 In the oxidation-reduction reactions in Problem 5.32, what is oxidized and what is reduced? What changes in oxidation number have occurred?

5.34 Determine the oxidation number of nitrogen in each of the following compounds:
a. N_2O b. NH_4Cl c. $NaNO_2$
d. HNO_3 e. N_2 f. N_2O_4

5.35 Determine the oxidation number of chromium in each of the following compounds:
a. K_2CrO_4 b. Cr_2O_3 c. $Cr(NO_3)_2$
d. $CrCl_3$ e. $Na_2Cr_2O_7$ f. $(NH_4)_2Cr_2O_7$

5.36 Explain why an oxidation number is not always the charge on an ion. Illustrate this using compounds from Problem 5.35.

5.37 Complete and balance the following half-reactions by using H_2O, H^+, and electrons as needed. State whether each reaction is oxidation or reduction.
a. $Fe^{2+} \longrightarrow Fe^{3+}$ b. $NO_3^- \longrightarrow NO_2$
c. $Cl_2 \longrightarrow Cl^-$ d. $MnO_4^- \longrightarrow MnO_2$
e. $Cr \longrightarrow Cr^{3+}$ f. $S \longrightarrow S^{2-}$

Stoichiometry (Section 5.5)

5.38 Chlorine dioxide is used for bleaching paper. It is prepared by the reaction:

$$2\,NaClO_2 + Cl_2 \longrightarrow 2\,ClO_2 + 2\,NaCl$$

a. Name the oxidizing and reducing agents in this reaction.
b. Calculate the weight of chlorine dioxide that would be prepared by the reaction of 5.50 kg of sodium chlorite ($NaClO_2$).

5.39 Write the balanced equation for the reaction of chlorine with sodium to form sodium chloride. Calculate the weight of sodium that will react completely with 5.00 g of chlorine.

5.40 Glucose, $C_6H_{12}O_6$, burns to form carbon dioxide and water.
a. Write the balanced equation for this reaction.
b. Calculate the moles of glucose that would be needed to form 1.55 moles of carbon dioxide.
c. Calculate the mass of water formed by the reaction in (b).

5.41 Pure aluminum is prepared by electrolysis of aluminum oxide according to the balanced equation:

$$2\,Al_2O_3 \longrightarrow 4\,Al + 3\,O_2$$

What mass of aluminum would be prepared from 6.06 g of aluminum oxide?

5.42 a. Write the balanced equation for the reaction of sodium hydroxide with sulfuric acid to form sodium sulfate.

b. Calculate how many moles of sulfuric acid would be neutralized by 8.00 g of sodium hydroxide.

5.43 Each of the reactions in Problem 5.29 produces a free element. For each equation, calculate the number of grams of this element that would be formed by the reaction of 5.15 g of the first reactant with an excess of the second reactant.

5.44 For Problem 5.29, calculate the mass of free element formed by the reaction of 23.6 g of the first reactant with an excess of the second.

5.45 a. Write the balanced equation for the reaction of octane (C_8H_{18}) with oxygen to form carbon dioxide and water.

b. Calculate the mass of water formed by the combustion of 1.8 L of octane (density = 0.775 g/cm^3).

5.46 When silver carbonate is heated, it decomposes to silver, oxygen, and carbon dioxide according to the unbalanced equation:

$$Ag_2CO_3 \nrightarrow Ag + O_2 + CO_2$$

a. Balance the equation.

b. Calculate the weight of silver that would be isolated by heating 0.565 g of silver carbonate.

5.47 Aspirin is formed by the reaction of salicylic acid with acetyl chloride according to the balanced equation:

$$C_7H_6O_3 \quad + \quad C_2H_3OCl \quad \longrightarrow \quad C_9H_8O_4 + HCl$$

salicylic acid   acetyl chloride   aspirin

What weight of aspirin is formed by the reaction of 5.00 g of salicylic acid with excess acetyl chloride?

Percent Yield (Section 5.6)

5.48 Chlorine can be prepared by the reaction of manganese(IV) oxide with excess hydrogen chloride according to the balanced equation:

$$MnO_2 + 4\,HCl \longrightarrow MnCl_2 + Cl_2 + 2\,H_2O$$

If 0.85 g of chlorine is obtained by the reaction of 1.35 g of manganese(IV) oxide, what is the percent yield of the reaction?

5.49 A 1.116-g sample of a mixture of potassium chloride and potassium chlorate ($KClO_3$) was heated, yielding 0.38 g of oxygen. The balanced equation for the reaction is:

$$2\,KClO_3 \longrightarrow 2\,KCl + 3\,O_2$$

What percent of the sample is potassium chlorate?

5.50 When bromine reacts with benzene, bromobenzene is the product:

$$C_6H_6 + Br_2 \longrightarrow C_6H_5Br + HBr$$

If 5.05 g of benzene yields 7.81 g of bromobenzene, what is the percent yield of the reaction?

**Limiting Reactants
(Section 5.7)**

5.51 a. What is a limiting reactant?
　　 b. What weight of which reactant remains after 6.34 g of hydrogen are reacted with 6.24 g of oxygen to form water?

5.52 Carbon dioxide will react with calcium oxide and water to form calcium bicarbonate.
　　 a. Write the balanced equation for this reaction.
　　 b. Calculate the yield of calcium bicarbonate if 4.65 g of carbon dioxide react with 5.10 g of calcium oxide.

5.53 Calculate the weight of magnesium oxide formed by the reaction of 1.56 g of magnesium with 2.63 g of oxygen.

5.54 Calculate the mass of aspirin formed by the reaction of 4.67 g salicylic acid with 7.64 g of acetyl chloride. See Problem 5.47 for the balanced equation for this reaction.

Enthalpy (Section 5.8)

5.55 For the following reactions, calculate the energy change when 5.0 g of the underlined reactant is used. Express your answer in both kilojoules and kilocalories. State whether the energy is absorbed or released.

$$H_2(g) + \underline{CO_2(g)} \longrightarrow CO(g) + H_2O(l) \qquad \Delta H = -1883 \text{ kJ/mole}$$
$$(-450 \text{ kcal/mole})$$

$$C_{12}H_{22}O_{11}(s) + 12\ O_2(g) \longrightarrow 12\ CO_2(g) + 11\ H_2O(l) \qquad \Delta H = -5641 \text{ kJ/mole}$$
$$(-1348.2 \text{ kcal/mole})$$
　　 sucrose

$$C_3H_6O_3(l) + 3\ O_2(g) \longrightarrow 3\ CO_2(g) + 3\ H_2O(l) \qquad \Delta H = -1367 \text{ kJ/mole}$$
$$(-326.8 \text{ kcal/mole})$$
　　 lactic acid

5.56 a. Calculate the energy required for the formation of 0.5 mole glucose from carbon dioxide and water. (The equation is shown in Section 5.8A.) Express your answer in both kilojoules and kilocalories.
　　 b. Calculate the energy change when 0.50 mole of glucose is metabolized to carbon dioxide and water. Express your answer in both kilojoules and kilocalories.

5.57 Calculate the energy released by the metabolism of 6.5 g ethyl alcohol according to the equation

$$C_2H_5OH(l) + 3\ O_2(g) \longrightarrow 2\ CO_2(g) + 3\ H_2O(l) \qquad \Delta H = -326.7 \text{ kcal/mole}$$
$$(-1367 \text{ kJ/mole})$$

Express your answer in both kilojoules and kilocalories.

5.58 A plant requires 4178.8 kcal for the formation of 1.00 kg starch from carbon dioxide and water. Calculate the energy required for the formation of 6.32 g starch. Express your answer in both kilojoules and kilocalories. Is this reaction endothermic or exothermic?

Energy and States of Matter

Every object possesses two kinds of energy: *kinetic energy* because of its motion and *potential energy* because of its composition, structure, and position. At the molecular level, it is largely the interplay of these two kinds of energy that determines whether molecules are in the solid, liquid, or gaseous state. In this chapter we will consider the energy of molecules and how their physical state reflects the relationship between their kinetic and potential energy.

6.1 Kinetic Energy

Kinetic energy is the energy of motion. The kinetic energy (KE) of an object is determined by the equation

$$\text{KE} = \tfrac{1}{2}mv^2 \qquad \text{where } m = \text{mass}$$
$$v = \text{velocity}$$

A. The Distribution of Kinetic Energy

In a sample of a pure molecular substance, each molecule has a kinetic energy that can be calculated by the above equation. Although the molecules have a constant mass, they differ in velocity. Therefore, in such a collection of molecules there will be a wide range of kinetic energies, from very low to very high. Figure 6.1 shows a typical distribution of kinetic energies in a collection of molecules. In the graph, kinetic energy is plotted along the horizontal axis, and the percent of molecules having a particular kinetic energy is shown by the height of the graph at that point. There are several observations that can be made by studying the graph.

1. The area under the curve represents the total number of molecules in the sample. Between two points on the horizontal axis, the area under the curve represents the number of molecules that have kinetic energies in that range. For example, the shaded area between A and B represents the number of molecules that have kinetic energies between A and B.

2. The peak of the curve shows the **most probable kinetic energy.** More molecules have this energy than any other.

3. Some molecules have a kinetic energy much higher than the most probable value.

4. Some molecules have a kinetic energy much lower than the most probable value.

Figure 6.1 Distribution of kinetic energy in a collection of molecules.

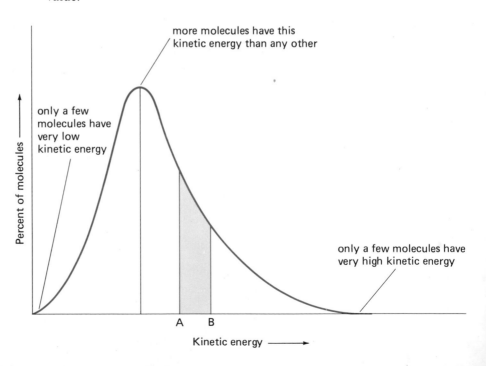

Figure 6.2 Graph of test grades. Each bar represents the number of students who received a particular grade.

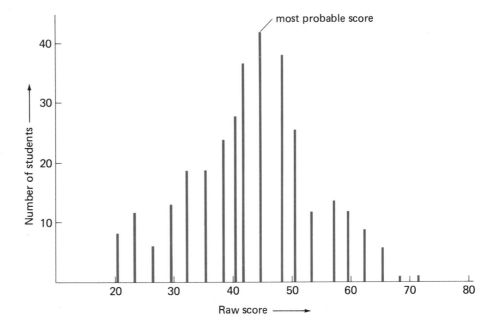

Notice that the distribution of energies is much like the distribution of grades on a test. Figure 6.2 shows the distribution of grades on a chemistry test. Note that there is a most probable grade. Most of the grades fall close to the most probable score. Some grades are much higher than the most probable, while others are much lower.

B. Kinetic Energy and Temperature

The kinetic energy of a collection of molecules is proportional to its temperature. At **absolute zero** ($-273°C$), the molecules have a minimum kinetic energy. As the temperature of the sample increases, so does its kinetic energy. As the temperature rises, there is also a change in the distribution of kinetic energies among the molecules in the sample. Figure 6.3 (next page) shows the distribution of kinetic energies in the sample from Figure 6.1, but at two different temperatures. Curve A is at the lower temperature, curve B at the higher temperature. Notice the following differences between the two curves:

1. The peak of curve B is lower and broader than the peak of curve A. This means that, at the higher temperature, fewer molecules have the most probable kinetic energy and the distribution of energies is more spread out.
2. The peak of curve B is at a higher kinetic energy than the peak of curve A.

Figure 6.3 Distribution of kinetic energy in the same collection of molecules at two different temperatures.

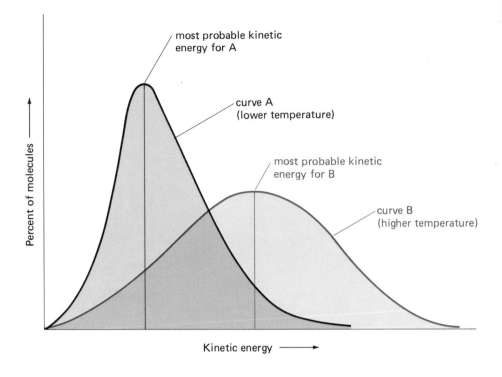

We can conclude that as the temperature of a sample increases, not only does the most probable kinetic energy increase but also the distribution of energies is spread out.

6.2 Intermolecular Forces and Potential Energy

In addition to kinetic energy, a sample of matter has **potential energy,** stored largely in interatomic and intermolecular bonds. The addition of energy to a sample does not always cause a change in temperature; sometimes it may cause a change of state. For example, when 6.02 kJ (1.44 kcal) of energy is added to 18 g of ice at 0°C, the ice melts to liquid water, also at 0°C. There is no change in temperature, only a change of state from solid to liquid. The added energy counteracted the forces that held the water molecules in the rigid ice structure. It did not break the bonds *within* molecules, only those *between* molecules.

The bonds between atoms, ions, or molecules in a substance range from very strong to comparatively weak. As a rule of thumb, the melting and boiling points of a substance are a measure of the strength of its intermolecular, interatomic, or interionic bonds. A high-melting substance contains strong intermolecular bonds; a low-melting substance has comparatively weak intermolecular bonds.

The bonding in substances with melting points well above room temperature falls into three categories, discussed below. These bonds are all interatomic or interionic.

1. ***Network covalent bonds.*** In Chapter 4 we discussed covalent bonds in small molecules. Most of these are low-melting. In network covalent solids (Figure 6.4a) every atom is covalently bonded to other atoms so that the entire structure can be thought of as a huge covalent molecule of enormous molecular weight. Such compounds have very high melting points. Diamonds (mp 3550°C) and quartz (SiO_2; mp 1610°C) are typical covalent solids.

2. ***Ionic bonds.*** In Chapter 4 we discussed ionic compounds, pointing out that the particles in an ionic compound are not atoms or molecules but ions. In the solid state, the ions are held in a regular rigid structure (Figure 6.4b) by the electrostatic forces of attraction between oppositely charged particles, as noted in Section 4.4C. Ionic solids usually have high melting points, indicating that electrostatic forces between ions are very strong.

3. ***Metallic bonds.*** The bonding in metals differs from that in ionic or covalent solids. We know that metals typically have one, two, or three valence electrons. One model of a metallic solid pictures these valence electrons as a fluid within which the nuclei and inner electron shells of the metal atoms float (Figure 6.4c). The conductivity of metals is due to

Figure 6.4 The bonding in (a) a network covalent solid, (b) an ionic solid, and (c) a metal.

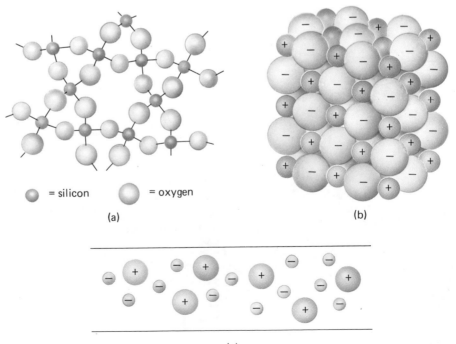

= silicon   = oxygen

(a)

(b)

(c)

the movement of this "sea" of electrons. Hammering a metal into a thin sheet spreads out the fluid; similarly, drawing a metal into a thin wire is a rearrangement of the sea of electrons and metal nuclei into a thin stream. Metal fatigue, a problem in the aircraft industry and in nuclear power plants, is associated with metals losing their "fluid" nature and assuming a rigid structure. The melting points of metals vary over a wide range. Mercury is the lowest-melting ($-39°C$); tungsten is one of the highest-melting ($3410°C$).

The bonding in substances with low melting points or substances that are gases or liquids at room temperature is due to dipole-dipole interaction. These bonds are weaker than ionic, covalent, or metallic bonds. Dipole-dipole interaction can be subdivided into three categories, discussed below.

1. *Interaction between polar molecules.* In Section 4.7B, we discussed the polarity of molecules and the interaction between polar molecules. Although a polar compound is apt to have a higher melting point than a nonpolar compound of the same molecular weight, the strength of the interaction between the molecules is much less than that of a covalent, ionic, or metallic bond. Figure 6.5 shows the orientation of polar molecules in a crystal lattice.

2. *Dispersion forces.* Dispersion forces, also called London or Van der Waal's forces, are weak forces of attraction between nonpolar molecules. To understand the origin of these forces, we need to remember that even though a molecule may have no permanent dipole, it does have a cloud

Figure 6.5 The orientation of polar molecules in a crystal lattice.

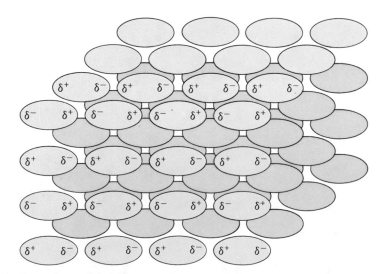

Molecules

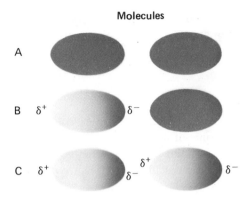

Figure 6.6 The development of temporary dipoles in molecules: (a) neutral electron clouds; (b) temporary distortion of left cloud, causing a temporary dipole; (c) induced distortion of right cloud caused by presence of dipole in left cloud, also resulting in a temporary dipole.

of rapidly moving electrons. If this cloud is distorted, no matter how briefly, the molecule will then have a temporary negative charge at one end and a temporary positive charge at the other end (Figure 6.6). In other words, the molecule will have a **temporary dipole.** This temporary dipole can distort the electron clouds of nearby molecules so that they, too, have temporary dipoles. The forces of attraction between the temporary partial positive charges of some molecules and the temporary partial negative charges of neighboring molecules are called **dispersion forces.** The number and therefore the strength of dispersion forces between molecules increases as the number of electrons within them increases. For this reason, we observe an increase in boiling point as the molecular weight of nonpolar molecules increases (Table 6.1).

3. *Hydrogen bonding.* Hydrogen bonding is the attraction between a partially negative atom in one molecule and a partially positive hydrogen atom in another molecule. The attractive force between water molecules is an example of hydrogen bonding. Because of the difference in electronegativity between hydrogen and oxygen, the O—H bond is polar

Table 6.1 The boiling points of nonpolar gases. Note the increase in boiling point with increased molecular weight.

Name	Molecular Weight	Boiling Point (C°)
helium (He)	4.00	− 268.9
hydrogen (H$_2$)	2.02	− 252.5
oxygen (O$_2$)	32.0	− 183.0
acetylene (C$_2$H$_2$)	26.0	− 84.0
Freon-12 (CCl$_2$F$_2$)	120.9	− 29.8

covalent. The oxygen atom bears a partial negative charge and the hydrogen atom bears a partial positive charge.

$$\overset{\delta^-}{O}$$
$$\delta^+ H \qquad H \delta^+$$

When two water molecules come close to each other, there is an interaction between one of the partially positive hydrogen atoms of one water molecule and the partially negative oxygen atom of the other. We say that the two molecules of water are *held together* by a hydrogen bond.

Figure 6.7 shows several water molecules as they might be found in ice. The hydrogen bonds between neighboring molecules are shown as dashed lines. The overall effect is a network of bonds with no separate molecules.

Hydrogen bonds are quite strong. We have already pointed out that the boiling point of a substance is a measure of the strength of its intermolecular forces. Thus, we expect hydrogen-bonded compounds to have higher boiling points than compounds that do not involve hydrogen bonding. Boiling points

Figure 6.7 Hydrogen bonding in ice. Each water molecule is bonded to four others. Its two hydrogen atoms are attracted to the oxygen atoms in two other water molecules, and its oxygen atom attracts hydrogens in two more water molecules. Note how the structure resembles that of a network covalent solid (Figure 6.4a).

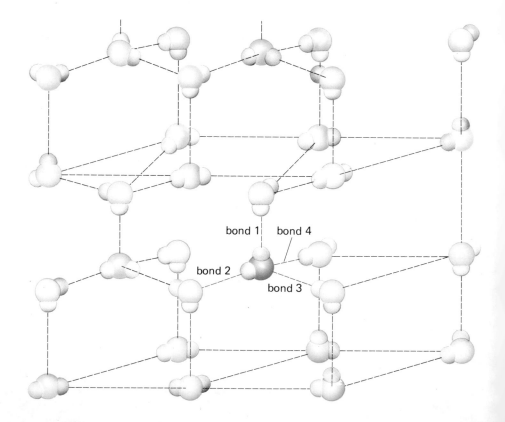

Figure 6.8 The boiling points of hydrides of the elements in groups IV, V, VI, and VII. The dotted lines show the projected boiling points for H_2O, HF, and NH_3. These compounds have abnormally high boiling points.

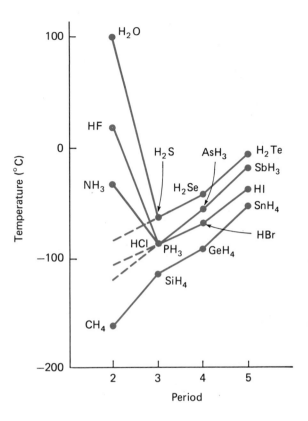

are also related to molecular weight. Within a group of molecules of similar structure, increased molecular weight usually means a higher boiling point. Figure 6.8 plots the boiling points of the hydrides of the elements in groups IV, V, VI, and VII of the periodic table. Those of group IV show the regular increase in boiling point we would expect with increased molecular weight. What stands out in Figure 6.8 is that the first members of groups V, VI, and VII have abnormally high boiling points. The dotted lines in this figure show what the predicted boiling points of H_2O, HF, and NH_3 would be if molecular weight were the only factor determining boiling point. Notice that the actual boiling point of H_2O is approximately 200° higher than the predicted value. In each of these molecules, hydrogen is bonded to a very electronegative element and the O—H, N—H, and F—H bonds are polar covalent. Within each pure substance, each molecule interacts with its neighbors by hydrogen bonding to produce a network of molecules all interconnected by hydrogen bonds. It is the strength of this network of hydrogen bonds that causes the abnormally high boiling points of H_2O, HF, and NH_3.

Although a hydrogen bond has only about ten percent of the strength of a covalent bond within a molecule, it is strong enough to have important effects on the physical and chemical properties of compounds whose structures contain them. We have seen one effect of hydrogen bonds in low-molecular-weight hydrides, where they cause abnormally high boiling points. We will see other effects later in the structures of proteins and other biologically important molecules.

Of all these attractive forces, dipole-dipole interaction, dispersion forces, and hydrogen bonds are of most interest to us, for they are the intermolecular forces most often encountered in the world of organic and biochemistry. Metallic bonding applies only to metals. Network covalent compounds will be rarely encountered, and only a basic understanding of the nature and strength of ionic bonds is necessary.

6.3 Characteristics of the Solid, Liquid, and Gaseous States

We have seen that the strength of the intermolecular attractions in a compound partially determines the temperature at which the compound melts, changing from solid to liquid. These attractive forces also affect the temperature at which the compound boils, changing from liquid to gas. Different compounds have different melting and boiling points. Nevertheless, all solids have certain properties in common as do all liquids and all gases (Figure 6.9). It is these common properties which we now wish to consider.

A. Shape and Volume

A solid has a fixed shape and volume which do not change with the shape of its container. Consider a rock and how its size and shape stay the same, regard-

Figure 6.9 Volume, shape, and mass in the three states of matter.

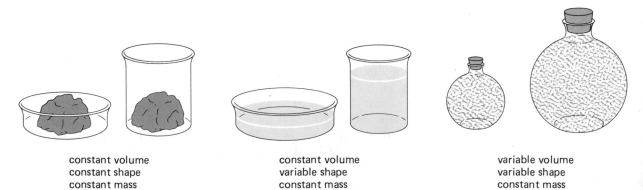

constant volume
constant shape
constant mass

constant volume
variable shape
constant mass

variable volume
variable shape
constant mass

less of where you put it. A liquid has a constant volume but its shape conforms to the shape of its container. Consider a sample of milk. Its volume stays the same, whether you put it in a saucer for the cat to drink or in a glass for yourself. Its shape changes to match the shape of the saucer or glass. A gas changes both its shape and its volume to conform to the shape and volume of its container. Consider a sample of air. It will fill a balloon, a tire, or a rubber raft. Its shape and volume conform to the shape and volume of the container you put it in. Note that in all three states of matter, the mass is constant.

B. Density

The densities of solids and liquids are measured in grams/milliliter and change very little as the temperature of the sample changes. Gases have much lower densities, so much lower that gas densities are measured in grams/liter instead of grams/milliliter. The density of a gas varies considerably as the temperature of the gas changes.

C. Compressibility

The volume of a solid or a liquid does not change with pressure. You cannot change the volume of a brick by squeezing it, nor can you squeeze one liter of liquid into a 0.5-L bottle. The volume of a gas does change with pressure; you can squeeze a 1.0-L balloon into a 0.5-L space.

From these commonly observed properties, it is possible to suggest the following:

1. From observing how the shape and volume of samples do or do not change to match the shape and volume of their containers, we can postulate that the molecules of a solid are held together by rigid bonds; the molecules of a liquid are strongly attracted to one another but this attraction does not take the form of rigid bonds; and the molecules of a gas are quite independent of each other with very little attraction between them.

2. From observing the differences in density and in compressibility among different samples, we can postulate that the molecules of solids and liquids are quite close together. The molecules of a gas are, on the other hand, quite far apart, with a great deal of empty space between them. This empty space decreases when pressure is applied to the gas. The empty space also causes gases to have very low densities, while the absence of empty space in solids and liquids makes their densities higher.

Let us now consider each of the physical states in detail.

6.4 The Properties of Gases

Under ordinary conditions, such as at 0°C and at sea level, all gases behave more or less alike. This behavior is explained by the **kinetic molecular theory.** The postulates of this theory describe the properties of gas molecules in the following ways. Molecules of a gas are very far apart with no attraction between them. Only a negligible percent of the volume of a gas sample is due to the volume of the molecules themselves; most of the volume is empty space. Molecules of a gas are in constant, rapid motion, i.e., they have kinetic energy, and the speed of this motion increases as the temperature increases. The molecules collide frequently with each other and with the walls of the container, and no kinetic energy is lost in these collisions.

Clearly, the actual properties of individual gases vary somewhat from these postulates, for their molecules do have a real volume and there is some attraction between the molecules. However, our discussion will ignore these variations and concentrate on an "ideal" gas, one that does not vary from this model.

A. Measuring Gas Samples

A gas sample obeys a number of laws that relate its volume to its pressure, temperature, and mass. These properties can be measured.

Mass and volume can be measured by the same methods used for samples of liquids and solids. Temperature can be measured on any scale, Celsius, Fahrenheit, or Kelvin. However, if the temperature is to be used in a calculation, the Kelvin scale must be used. **Standard temperature** for gases, the temperature at which the properties of different gases are usually compared, is 273 K.

Pressure is defined as force per unit area and measured in units that have dimensions of force per unit area. For example, the air pressure in tires is measured in pounds per square inch (psi). The pressure of the atmosphere is frequently measured with a mercury **barometer.** Pressure can be more easily understood if we consider how a barometer measures pressure. The basic features of a mercury barometer are shown in Figure 6.10. A glass tube at least 760 mm long and closed at one end is filled with mercury, then carefully inverted into a pool of mercury. The level of the mercury in the column will fall slightly and then become steady.

The height of the column of mercury measures the pressure of the atmosphere. To understand why this is so, consider the pressure on the surface of the mercury pool at the base of the column. Above this surface rises the "sea" of air (the atmosphere) which surrounds the Earth. On each square centimeter of the surface we can visualize a 20 km-column of air pressing down. On the square centimeter of surface under the mercury column the mercury is pressing down. The two pressures must be equal—if they were not, mercury would be flowing into or out of the column and the height of the column would not be steady. Therefore, the atmosphere must be exerting a pressure equal to that exerted by the mercury column.

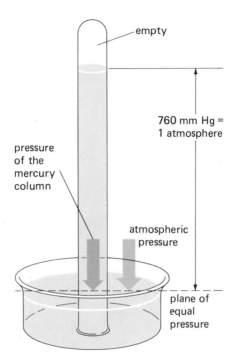

Figure 6.10 A mercury barometer. The height of the mercury column in the tube measures the atmospheric pressure on the surface of the mercury in the dish.

empty

760 mm Hg = 1 atmosphere

pressure of the mercury column

atmospheric pressure

plane of equal pressure

When this experiment is performed in dry air at sea level and at 0°C, the column of mercury is 760 mm high; therefore, we say that the atmosphere is exerting a pressure equal to that of 760 mm of mercury. This amount of pressure has been defined as **one atmosphere** of pressure and designated as **standard pressure.**

Pressure can be measured in other units that are related to standard pressure:

1 atmosphere (atm) = 76 cm or 760 mm mercury

= 760 torr (one torr = the pressure exerted by 1 mm mercury)

= 29.92 inches mercury (used to report atmospheric pressure in weather reports)

= 1.013 bar (used in meteorology) (1 cm mercury = 13.3 millibars)

= 101,325 Pascals (Pa) (a Pascal is the SI unit of pressure)

Each of these relationships can be used as a conversion factor, as shown in the following problems.

Example 6.1

a. How many atmospheres pressure is exerted by a column of mercury 654 mm high?

b. What is this pressure in Pascals?

Solution

a. Wanted:
pressure in atmospheres

Given:
a column of mercury 654 mm high

Conversion factor:
1 atmosphere = 760 mm Hg

Equation:

$$? \text{ atm} = 654 \text{ mm Hg} \times \frac{1 \text{ atm}}{760 \text{ mm Hg}} = \frac{654}{760} \text{ atm}$$

Answer:
0.861 atm

b. Wanted:
pressure in Pascals

Given:
a pressure of 0.861 atm

Conversion factor:
1 atm = 1.01325×10^5 Pa

Equation:

$$? \text{ Pa} = 0.861 \text{ atm} \times \frac{1.01325 \times 10^5 \text{ Pa}}{1 \text{ atm}}$$

Answer:
8.72×10^4 Pa

Problem 6.1

What pressure in torr is equal to a pressure of 1.65 atm?

In dry air at sea level, the average air pressure is one atmosphere. Atmospheric pressure decreases as altitude increases because the sea of air above becomes less deep. Our bodies become adjusted to the normal pressure of the altitude at which we live. Minor problems of adjustment can occur when we move from sea level to the mountains, and vice versa. Major problems develop

Figure 6.11 A manometer. The height difference between the mercury levels in the two sides of the tube measures the pressure difference between the gas sample and the atmosphere.

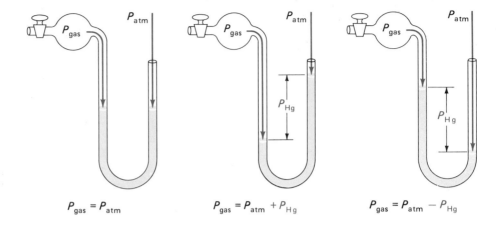

$P_{gas} = P_{atm}$ $P_{gas} = P_{atm} + P_{Hg}$ $P_{gas} = P_{atm} - P_{Hg}$

at higher altitudes. Commercial jet aircraft routinely fly at 35,000 feet; the air pressure at this altitude is about 0.3 atmosphere. Cabins of jet aircraft must therefore be pressurized, because passengers could not survive such low pressure. Travellers in space must wear pressurized suits. The mini-essay entitled "Skiing, Diving, and the Gas Laws" discusses similar situations.

Barometers measure the pressure of the atmosphere. **Manometers** measure the pressure of isolated gas samples. Some manometers measure pressure with a column of mercury, like a mercury barometer. This type of manometer has a U-shaped tube partially filled with mercury (Figure 6.11). One end of the tube is open to a chamber holding a gas sample and the other end is open to the atmosphere. If the mercury level on the side of the tube open to the gas sample is lower than that on the side open to the atmosphere, the pressure of the gas is greater than that of the atmosphere by an amount equal to the difference in height between the two mercury columns. If the mercury level on the side of the gas sample is higher than that on the side open to the atmosphere, the pressure of the gas is less than atmospheric pressure by the difference in the heights of the two columns.

B. The Gas Laws

1. Boyle's Law. **Boyle's Law** states: If the temperature of a gas sample is kept constant, the volume of the sample will vary inversely as the pressure varies. If the pressure increases, the volume will decrease. If the pressure decreases, the volume will increase. This law can be expressed as an equation which relates the initial volume (V_1) and pressure (P_1) to the final volume (V_2) and pressure (P_2).

At constant temperature: $V_2 = V_1 \times \dfrac{P_1}{P_2}$ or $P_2 V_2 = P_1 V_1$

Figure 6.12 Boyle's Law: At constant temperature, the volume of a gas sample is inversely proportional to the pressure. The curve is a graph based on the data listed in the figure.

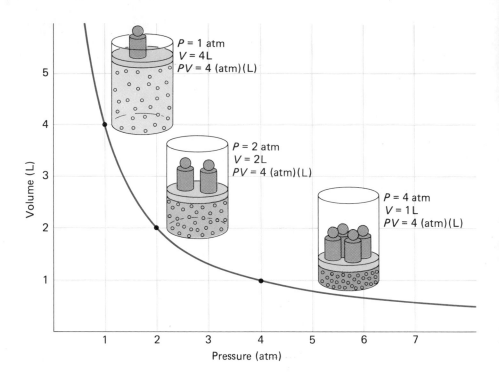

Boyle's Law is illustrated in Figure 6.12, which shows a sample of gas enclosed in a container with a movable piston. The container is kept at a constant temperature. When the piston is stationary, the pressure it exerts on the gas sample is equal to the pressure the gas exerts on it. When the pressure on the piston is doubled, it moves downward until the pressure exerted by the gas equals the pressure exerted by the piston. At this point the volume of the gas is halved. If the pressure on the piston is again doubled, the volume of gas decreases to one-fourth its original volume. At the molecular level, the pressure of a gas depends on the number of collisions its molecules have with the walls of the container. If the pressure on the piston is doubled, the volume of the gas decreases by one-half, at which point gas molecules collide with the walls of the container twice as often and their pressure once again equals that of the piston.

Example 6.2

A sample of gas has a volume of 6.20 L at 20°C and 0.980 atm pressure. What is its volume at the same temperature and at a pressure of 1.11 atm?

Solution

1. Tabulate the data.

	Initial conditions	**Final conditions**
volume	$V_1 = 6.20$ L	$V_2 = ?$
pressure	$P_1 = 0.980$ atm	$P_2 = 1.11$ atm

2. Check the pressure units. If they are different, use a conversion factor to make them the same. (Pressure conversion factors are found in Section 6.4A.)

3. Substitute in the Boyle's Law equation:

$$V_2 = V_1 \times \frac{P_1}{P_2} = 6.20\text{ L} \times \frac{0.980\text{ atm}}{1.11\text{ atm}} = 5.47\text{ liters}$$

4. Check that your answer is reasonable. The pressure has increased, the volume should decrease. The calculated final volume is less than the initial volume, as predicted.

Problem 6.2

A sample of gas has a volume of 253 cm^3 at 0.50 atm. What is its volume at the same temperature and at a pressure of 1.0 atm?

2. Charles' Law. Charles' Law states: If the pressure of a gas sample is kept constant, the volume of the sample will vary directly as the temperature in degrees Kelvin varies (Figure 6.13). As the temperature increases, so will the volume; if the temperature decreases, the volume will decrease. This can be expressed by an equation relating the initial volume (V_1) and temperature (T_1 measured in K) to the final volume (V_2) and temperature (T_2 measured in K).

$$\text{At constant pressure:} \quad V_2 = V_1 \times \frac{T_2}{T_1} \quad \text{or} \quad \frac{V_2}{T_2} = \frac{V_1}{T_1}$$

Figure 6.13 Charles' Law: At constant pressure, the volume of a gas sample is directly proportional to the temperature in degrees K.

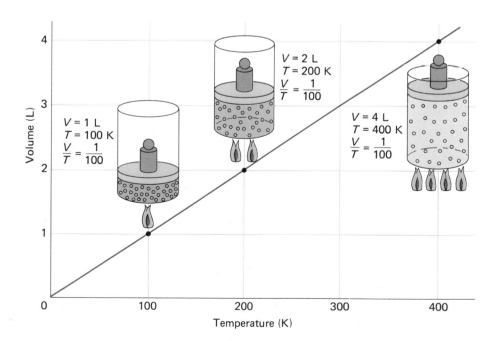

In terms of molecular behavior, as the temperature increases, the kinetic energy of the molecules increases. Although the number of molecules remains the same, their increased speed allows them to occupy a larger volume.

The next example shows how Charles' Law can be used in calculations.

Example 6.3

The volume of a gas sample is 746 cm^3 at 20°C. What is its volume at body temperature (37°C)? Assume the pressure remains constant.

Solution

1. Tabulate the data.

	Initial conditions	Final conditions
volume	$V_1 = 746$ cm^3	$V_2 = ?$
temperature	$T_1 = 20°$C	$T_2 = 37°$C

2. Charles' Law requires that the temperature be measured in degrees Kelvin in order to give the correct numerical ratio. Therefore, change the given temperature to Kelvin:

$$T_1 = 20 + 273 = 293 \text{ K} \qquad T_2 = 37 + 273 = 310 \text{ K}$$

3. Calculate the new volume:

$$V_2 = 746 \text{ cm}^3 \times \frac{310 \text{ K}}{293 \text{ K}} = 789 \text{ cm}^3$$

This volume is larger than the original volume, as was predicted from the increase in temperature.

Problem 6.3

A balloon has a volume of 1.56 L at 25°C. If the balloon is cooled to −10°C, what will be its new volume? Assume the pressure remains constant.

3. *The combined gas law.* Frequently, a gas sample is subjected to changes in both temperature and pressure. In such cases, the Boyle's Law and Charles' Law equations can be combined into a single equation, representing the **Combined Gas Law:**

$$V_2 = V_1 \times \frac{T_2}{T_1} \times \frac{P_1}{P_2}$$

Here, V_1, P_1, and T_1 are the initial conditions and V_2, P_2, and T_2 are the final conditions. This equation can be rearranged into a form that is much easier to remember:

$$\frac{P_1 V_1}{T_1} = \frac{P_2 V_2}{T_2}$$

Example 6.4

A gas sample occupies a volume of 2.5 L at 10°C and 0.95 atm. What is its volume at 25°C and 0.75 atm?

Solution

Initial conditions	Final conditions
$V_1 = 2.5$ L	$V_2 = ?$
$P_1 = 0.95$ atm	$P_2 = 0.75$ atm
$T_1 = 10°C = 283$ K	$T_2 = 25°C = 298$ K

Check that P_1 and P_2 are measured in the same units and that both temperatures have been changed to Kelvin. Substitute in the equation:

$$\frac{0.95 \text{ atm} \times 2.5 \text{ L}}{283 \text{ K}} = \frac{0.75 \text{ atm} \times V_2}{298 \text{ K}}$$

Solving this equation, we get:

$$V_2 = 3.3 \text{ L}$$

This is a reasonable answer. Both the pressure change and the temperature change would cause an increased volume.

Problem 6.4

A sample of gas occupies a volume of 5.7 L at 37°C and 732 mm pressure. What is its volume at standard temperature and pressure?

4. Avogadro's hypothesis and molar volumes. At constant temperature and pressure, the volume of a gas is directly proportional to the number of molecules (or the number of moles) in the sample. Like Charles' and Boyle's Laws, this relationship between volume and number of molecules does not depend on the nature of the gas in the sample—it is true for all gases. In other words, if one liter of nitrogen at a particular temperature and pressure contains 1.0×10^{20} molecules, then one liter of any other gas at the same temperature and pressure also contains 1.0×10^{20} molecules. **Avogadro's Hypothesis** states this in a slightly different fashion: At the same temperature and pressure,

Figure 6.14 Avogadro's Hypothesis: At the same temperature and pressure, equal volumes of different gases contain the same number of molecules.

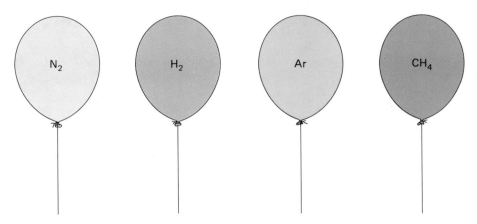

Each balloon holds 1.0 L of gas at 20°C and 1 atm pressure.
Each contains 0.045 mole or 2.69 X 10^{22} molecules of gas.

equal volumes of gases contain equal numbers of molecules (Figure 6.14). It has been determined that one mole (6.02×10^{23} molecules) of any gas at 273 K and 1 atmosphere pressure (standard conditions) occupies a volume of 22.4 liters. This volume is called the **molar volume.**

Since the concept of molar volume relates the mass and volume of a sample, molar volume can be used to calculate gas densities under standard conditions. The equation for this is:

$$\text{density} = \frac{\text{mole weight in grams}}{22.4 \text{ liters}}$$

Example 6.5

Calculate the density of nitrogen under standard conditions (STP).

Solution

The mole weight of nitrogen is (2×14.0) or 28.0 g. The molar volume is 22.4 L. Substituting in the equation, we get:

$$\text{density of nitrogen at STP} = \frac{28.0 \text{ g}}{22.4 \text{ L}} = 1.25 \text{ g/L}$$

Problem 6.5

Calculate the density of helium under standard conditions.

5. *The ideal gas equation.* The various equations relating to pressure, volume, temperature, and number of moles of a gas sample can be combined into one statement:

The volume (V) occupied by a gas is directly proportional to its Kelvin temperature (T) and the number of moles (n) it contains, and is inversely proportional to its pressure (P).

In mathematical form this becomes:

$$V = \frac{nRT}{P}, \text{ where } V = \text{volume}$$

$$n = \text{moles of sample}$$
$$P = \text{pressure}$$
$$T = \text{temperature in K}$$
$$R = \text{a proportionality constant known as the } \textbf{gas constant}$$

This equation is known as the **ideal gas equation** and is seen most often in the form:

$$PV = nRT$$

We can determine the value of the gas constant R by substituting in the equation the known values for 1 mole of gas at standard conditions.

$$R = \frac{PV}{nT} = \frac{1 \text{ atm} \times 22.4 \text{ L}}{1 \text{ mole} \times 273 \text{ K}} = 0.0821 \frac{\text{L atm}}{\text{mole K}}$$

Example 6.6

What volume is occupied by 5.50 g of carbon dioxide at 25°C and 742 torr?

Solution

1. Identify the variables in the equation and convert units to match those of the gas constant.

$$P = 742 \text{ torr} \times \frac{1 \text{ atm}}{760 \text{ torr}} = 0.976 \text{ atm}$$

$$V = ? \text{ liter}$$

$$n = 5.50 \text{ g} \times \frac{1 \text{ mole}}{44.0 \text{ g}} = 0.125 \text{ mole}$$

$$T = 25°C + 273 = 298 \text{ K}$$

2. Substituting these values in the ideal gas equation:

$$V = \frac{nRT}{P}$$

$$= 0.125 \text{ mole} \times \frac{0.0821 \text{ liter atm}}{\text{mole K}} \times 298 \text{ K} \times \frac{1}{0.976 \text{ atm}}$$

$$= \frac{0.125 \times 0.0821 \times 298}{0.976} \text{ liters} = 3.13 \text{ liters}$$

Notice how nicely the units cancel when you use the ideal gas equation. However, you have now done enough problems of this type to know that if the units hadn't cancelled, the arithmetic equation had been set up incorrectly.

Problem 6.6

What mass of oxygen will occupy 1.23 L at 37°C and 0.752 atm?

The ideal gas equation can also be used in stoichiometric problems involving gases.

Example 6.7

What volume of carbon dioxide at 37°C (body temperature) and 740 torr is produced by the metabolism of 1.0 g of ethyl alcohol (C_2H_6O)? The balanced equation is:

$$C_2H_6O + 3 O_2 \longrightarrow 2 CO_2 + 3 H_2O$$

Solution

To solve this problem, follow the steps listed in Section 5.5 to find the number of moles of CO_2 formed. Then, use the ideal gas equation to convert the number of moles of CO_2 to a volume of CO_2.

Wanted:
liters of CO_2

Given:
1.0 gram of C_2H_6O

Conversion factors:
1 mole C_2H_6O = 46.1 grams C_2H_6O
1 mole C_2H_6O yields 2 moles CO_2
37°C = 310 K

Arithmetic equation:

$$? \text{ moles } CO_2 = 1.0 \text{ g } C_2H_6O \times \frac{1 \text{ mole } C_2H_6O}{46.1 \text{ g } C_2H_6O} \times \frac{2 \text{ moles } CO_2}{1 \text{ mole } C_2H_6O}$$

$$= 0.043 \text{ mole } CO_2$$

Substituting this value in the ideal gas equation gives:

$$V = \frac{nRT}{P} = 0.043 \text{ mole} \times \frac{0.0821 \text{ liter atm}}{\text{mole K}} \times 310 \text{ K} \times \frac{1}{740 \text{ torr}} \times \frac{760 \text{ torr}}{1 \text{ atm}}$$

Answer:
1.1 liter CO_2

Problem 6.7

What volume of oxygen measured at 40°C and 1.1 atm reacts completely with 5.0 g of propane? The balanced equation for the reaction is:

$$C_3H_8 + 5 O_2 \longrightarrow 3 CO_2 + 4 H_2O$$

C. Mixtures of Gases

1. Partial pressures. According to the kinetic molecular theory, the volume occupied by a gas is largely empty space and the volume of the molecules themselves is negligible. Because of this, each gas in a mixture of two or more nonreacting gases behaves as if it alone occupied the entire volume of the container. The pressure exerted by a mixture of gases and the pressure of each component gas obey **Dalton's Law of Partial Pressures.** (This is the same Dalton who proposed the atomic theory described in Section 2.3.) The Law of Partial Pressures states that each gas in a mixture of gases exerts a pressure, known as its **partial pressure,** equal to the pressure that gas would exert if it were the only gas present; and the total pressure of the mixture is the sum of the partial pressures of all the gases present. A mathematical expression of this statement is:

$$P_T = P_1 + P_2 + P_3 + \cdots$$

where P_T equals the total pressure of the mixture and P_1, P_2, and P_3 are the partial pressures of the gases present in the mixture.

Suppose we have one liter of oxygen at 1 atm pressure in one container, one liter of nitrogen at 0.5 atm pressure in a second container, and one liter of hydrogen at 3 atm pressure in a third container (Figure 6.15). If we combine the samples in a single 1-L container, the total pressure is 4.5 atm (1 atm + 0.5 atm + 3 atm). The partial pressure of oxygen, P_{O_2}, is 1 atm (the pressure it alone exerted in its container). Similarly, the partial pressure of nitrogen, P_{N_2}, is 0.5 atm, and that of hydrogen, P_{H_2}, is 3.0 atm.

Figure 6.15 The total pressure of a mixture of gases equals the sum of the individual gas pressures.

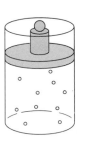
$P_{O_2} = 1$ atm

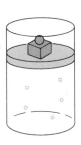
$P_{N_2} = 0.5$ atm

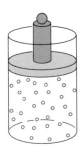
$P_{H_2} = 3$ atm

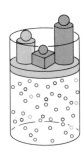
$P_T = P_{O_2} + P_{N_2} + P_{H_2} = 4.5$ atm

From the total pressure of a mixture of gases and its percent composition, we can calculate the partial pressure of the individual gases.

Example 6.8

The three major components of dry air along with the percent of each are 78.08% nitrogen, 20.95% oxygen, and 0.93% argon. Calculate the partial pressure of each gas in a sample of dry air at 760 torr. Also calculate the total pressure exerted by the three gases combined.

Solution

1. Equation: $P_T = P_{N_2} + P_{O_2} + P_{Ar}$
2. Calculate the partial pressure of each gas:

$$P_{N_2} = \frac{78.08}{100} \times 760 = 593.4 \text{ torr} \qquad P_{O_2} = \frac{20.95}{100} \times 760 = 159.2 \text{ torr}$$

$$P_{Ar} = \frac{0.93}{100} \times 760 = 7.1 \text{ torr}$$

3. Total pressure (P_T) = 159.2 + 593.4 + 7.1 = 759.7 torr. The difference between the total pressure of the three gases and the total pressure of the air sample is due to the partial pressures of other gases present in dry air, including carbon dioxide.

Problem 6.8

What is the percent composition by volume of air in the trachea (windpipe) if it contains oxygen at a partial pressure of 149.0 torr, carbon dioxide at a partial pressure of 0.3 torr, water vapor at a partial pressure of 47.0 torr, and nitrogen at a partial pressure of 563.7 torr?

2. Gaseous diffusion. **Gaseous diffusion** is the process by which a gas spreads through a space and mixes with any other gases present. The rate of diffusion depends on the mass of the molecules of the gas. **Graham's Law of Gaseous Diffusion** states that the relative rates of diffusion of two gases are inversely proportional to the square roots of their molecular weights. Using this law, we can compare the rates of diffusion of two different gases. The law can be expressed mathematically as:

$$\frac{r_1}{r_2} = \frac{\sqrt{MW_2}}{\sqrt{MW_1}}, \text{ where } r_1 = \text{diffusion rate of gas 1}$$

$$r_2 = \text{diffusion rate of gas 2}$$
$$MW_1 = \text{molecular weight of gas 1}$$
$$MW_2 = \text{molecular weight of gas 2}$$

Suppose we have two vials of gas, one containing ammonia (NH_3, mol. wt. 17) and the other containing hydrogen sulfide (H_2S, mol. wt. 34). Each gas has a characteristic odor by which it can be identified. If you are standing at one end of a long hall and the vials of H_2S and NH_3 are opened simultaneously at the other end, you will detect the lighter gas (NH_3) first. The rate at which ammonia diffuses is approximately 1.4 times the rate at which hydrogen sulfide diffuses.

$$\frac{\text{rate } NH_3}{\text{rate } H_2S} = \frac{\sqrt{\text{mol. wt. } H_2S}}{\sqrt{\text{mol. wt. } NH_3}} = \frac{\sqrt{34}}{\sqrt{17}} = \frac{5.8}{4.1} = 1.4$$

Example 6.9

Calculate the ratio between the rates of diffusion of natural gas (methane, CH_4) and LPG (propane, C_3H_8).

Solution

The molecular weight of methane is 16 and that of propane is 44. The ratio between their rates of diffusion will be the inverse of the ratio between the square roots of their molecular weights.

$$\frac{\text{rate } CH_4}{\text{rate } C_3H_8} = \frac{\sqrt{\text{mol. wt. } C_3H_8}}{\sqrt{\text{mol. wt. } CH_4}}$$

Substituting:

$$\frac{\text{rate } CH_4}{\text{rate } C_3H_8} = \frac{\sqrt{44}}{\sqrt{16}} = \frac{6.6}{4.0} = 1.7$$

Methane will diffuse approximately 1.7 times faster than propane.

Problem 6.9

Calculate the ratio between the rate of diffusion of hydrogen chloride and that of hydrogen bromide.

D. Real Gases

Thus far in this chapter, we have assumed that all gases are ideal and behave in accordance with the postulates of the kinetic molecular theory and the ideal gas equation. Under standard conditions of temperature and pressure, and also at higher temperatures and lower pressures, the behavior of most real gases such as oxygen, nitrogen, and carbon dioxide is that predicted by the gas laws and the kinetic molecular theory. It is for this reason that we study ideal gases. However, as the temperature of a gas is decreased, the kinetic energy of the molecules decreases, their movement becomes more sluggish,

Figure 6.16 The interaction of polar molecules: (a) gas molecules moving freely at STP, no interaction; (b) interaction occurs when the gas is at low temperature or high pressure, causing temporary dipoles.

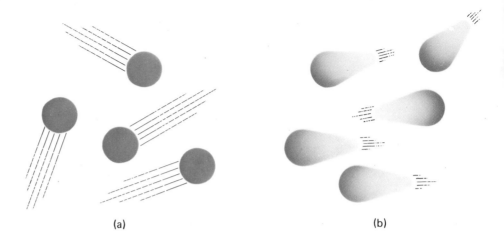

(a) (b)

and the attractive forces that exist between real molecules play a larger role in determining the behavior of the sample. Likewise, if the pressure is increased and the volume decreased until the volume of space between the molecules equals the volume of the molecules themselves, the molecules can no longer act as the wholly independent particles postulated by the kinetic molecular theory.

The intermolecular forces that come into play in these circumstances are dipole-dipole interactions and dispersion forces, discussed in Section 6.2. Under standard conditions of temperature and pressure (STP), molecules move freely without intermolecular attraction, as illustrated in Figure 6.16(a). In Figure 6.16(b), the molecules are moving more slowly, are closer together, and they interact. A temporary dipole in one molecule affects the electron cloud of its neighbor. The lower the temperature and the closer the molecules are together (a result of higher pressure), the more effective are these forces in preventing the free movement of molecules required by the kinetic molecular theory. At some temperature and pressure, the molecules of these substances come close enough together to become a liquid. The closer this temperature is to absolute zero, the weaker are the intermolecular forces between the molecules and the more closely the behavior of the gas resembles that of an ideal gas.

In summary, we expect the behavior of real gases to deviate more and more from the ideal as the polarity (either real or induced) and the molecular weight of the molecules increase.

6.5 The Properties of Liquids

A gas becomes a liquid when the average kinetic energy of the molecules is no longer great enough to overcome the attractive forces operating between the molecules. The sample then has the properties common to liquids that were outlined in Section 6.3. The molecules are close enough together that most of the volume of a liquid is that of the molecules themselves. There are strong attractive forces between the molecules and the molecules have an average kinetic energy which is proportional to the temperature of the sample. In the remainder of this section we will relate these properties to the observed properties of liquids.

A. Attractive Forces in Liquids

In liquids such as water, oil, and gasoline the intermolecular forces operating are dispersion forces, dipole-dipole interactions, and hydrogen bonding. Evidence of the strength of these forces in liquids comes from a liquid's **viscosity,** or resistance to flow (as in heavy oils and thick syrups). Further evidence is in the tendency of a liquid to form drops when poured. Think of the large drops formed by a heavy oil, which has strong intermolecular forces, as opposed to the tiny drops formed by gasoline, in which the intermolecular forces are relatively weak.

B. Vapor Pressure

In spite of the strong intermolecular forces operating in a liquid, some molecules on its surface have enough energy to escape. It is these escaped molecules that reach your nose when you smell a liquid. The partial pressure exerted by the escaped molecules is the **vapor pressure** of the liquid.

Just as there is a distribution of kinetic energies among the molecules of a gas sample, there is a distribution of energies among the molecules of a liquid sample. Figure 6.17 shows this distribution for the same sample at two different temperatures. Point A on the graph represents the kinetic energy a molecule must have to overcome the intermolecular forces that hold it in the body of the liquid. The shaded area under the curves to the right of point A shows the percent of molecules that have at least that much energy. The lower,

Figure 6.17 Distribution of kinetic energies among the molecules of a liquid at two different temperatures.

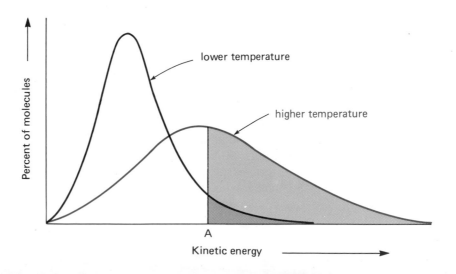

broader curve is the distribution of energies at a higher temperature. The shaded area under this curve is larger than the shaded area under the lower temperature curve. This means that, at the higher temperature, more molecules have enough energy to escape from the liquid.

When a liquid absorbs heat from its surroundings, its molecules absorb this energy. Some molecules gain enough kinetic energy to escape from the surface. Topical anesthetics function by this mechanism. These substances are low-boiling liquids. When sprayed on the skin, the molecules of the anesthetic absorb heat from the body, gaining enough kinetic energy to vaporize. They remove heat from treated areas and the temperature of the skin drops to the point where the skin becomes numb.

We know that low-boiling liquids evaporate when their containers are left open to the atmosphere. They do not evaporate if kept in a closed container. Even ethyl ether (bp 34.5°C) can be stored at room temperature in a closed container. Yet, if the container is open, it evaporates. How can we explain this difference? When a liquid is put in a closed container, some molecules will have enough energy to become gaseous and leave the liquid. The vaporized molecules in the space above the liquid also have a typical kinetic energy distribution. Those at the lower end of the distribution have comparatively low kinetic energy. Some of these low-energy molecules condense, returning to the liquid state. At all times, some molecules are escaping from the liquid to the gaseous state, while others are returning from the gaseous to the liquid state. After a time, the number of molecules making the transition from liquid to gas equals the number changing from gas to liquid. This balance can be stated as:

$$\text{rate of evaporation} = \text{rate of condensation}$$

The sample is now in a state of **dynamic equilibrium,** in which two opposing processes proceed at the same rate so that there is no net change occurring. Equilibrium cannot be attained when the sample is in an open container because the vapor does not accumulate. Figure 6.18(a) shows equilibrium in a closed container, and Figure 6.18(b) shows nonequilibrium—the evaporation process.

At equilibrium, the partial pressure of the molecules in the vapor phase is the vapor pressure of the liquid at that temperature. The vapor pressure of a substance increases as the temperature increases. Figure 6.19 shows this relationship for water and carbon tetrachloride. The boiling point of a substance is that temperature at which its vapor pressure equals the external pressure. Notice in the figure that when the vapor pressure of water is 760 mm Hg, the water is at 100°C, its normal boiling point.

For normal boiling points, the external pressure is 1 atmosphere. In a pressure cooker, the vapor is confined and builds up a pressure greater than that of the atmosphere. The boiling point of water in a pressure cooker is above 100°C because of the greater pressure. This means that the water can

Figure 6.18 (a) Equilibrium between vapor and liquid in a closed container; (b) nonequilibrium (evaporation) in an open container; equilibrium cannot be established because the vapor does not collect.

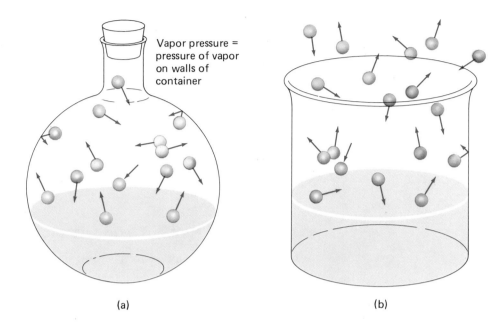

Vapor pressure = pressure of vapor on walls of container

(a)   (b)

be hotter than 100°C and still remain a liquid. Foods cook in a shorter time because they are cooking in water that is at a higher temperature. At high altitudes, such as in the mountains, liquids boil at lower temperatures than normal because the atmospheric pressure is lower. Foods take longer to cook because the water boils at a lower temperature.

Figure 6.19 The relationship between vapor pressure and temperature for water and carbon tetrachloride (CCl_4). A liquid boils when its vapor pressure equals the external pressure. At 760 mm (1 atm) pressure, the boiling points of H_2O and CCl_4 are 100°C and 78°C, respectively.

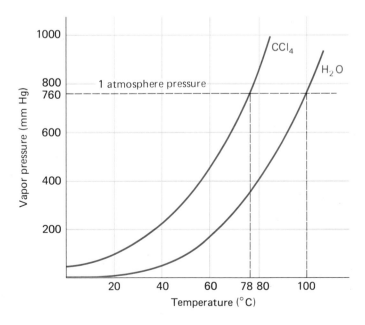

C. The Specific Heat of a Liquid

Recall from Section 1.5C that specific heat relates the heat energy and mass of a sample to changes in its temperature:

$$\text{specific heat} = \frac{\text{calories}}{(g)(^{\circ}C)}$$

At the molecular level, the temperature of a liquid is proportional to the average kinetic energy of the molecules within the liquid, and any change in temperature corresponds to a change in the average kinetic energy of liquid molecules.

D. The Molar Heat of Vaporization

When heat energy is added to a liquid, its temperature rises to the boiling point at a rate that is dependent on the specific heat of the liquid. Energy added to a liquid at its boiling point does not cause a change in temperature. Instead, this added energy counteracts the intermolecular forces in the liquid and the liquid vaporizes, becoming a gas. The amount of energy required to vaporize one mole of a liquid at its boiling point is its **molar heat of vaporization** (ΔH_{vap}).

Table 6.2 shows the molecular weight, boiling point, specific heat, and molar heat of vaporization for several liquids.

The molar heat of vaporization of a liquid can be calculated from experimental data, as the following example shows.

Table 6.2 Molar heats of vaporization of several substances.

Name	Molecular Weight	bp (°C)	$\Delta H_{vap}\left(\dfrac{kcal}{mole}\right)$	Specific Heat (cal/g °C)
carbon disulfide (CS_2)	76.1	46	6.79	0.239
chloroform ($CHCl_3$)	119.4	61.7	7.50	0.231
ethyl alcohol (C_2H_6O)	46.1	78.5	9.67	0.586
octane (C_8H_{18})	114.2	126	9.22	0.532
water (H_2O)	18.0	100	9.72	1.000

Example 6.10

128 calories are required to vaporize 136 grams of benzene at its boiling point. Calculate the molar heat of vaporization of benzene.

Solution

Wanted:
the molar heat of vaporization in calories per mole

Given:
128 calories are required to vaporize 136 grams

Conversion factors:
1 mole of benzene = 78.1 g benzene

Equation:

$$? \text{ cal/mole} = \frac{128 \text{ cal}}{136 \text{ g}} \times \frac{78.1 \text{ g benzene}}{1 \text{ mole benzene}} = \frac{128 \times 78.1}{136} \frac{\text{cal}}{\text{mole}}$$

Answer:
73.5 cal/mole

Problem 6.10   Calculate the heat required to vaporize 95 g of water at 100°C.

6.6 The Properties of Solids

In Section 6.3 we discussed the properties of solids. These properties suggest that the particles in a solid occupy fixed positions from which they cannot easily move. This orderly arrangement is called the **crystal structure** or **crystal lattice** of the solid. Strong attractive forces between the particles of a solid keep them in this arrangement. Even so, each particle in a solid has some kinetic energy (unless the solid is at a temperature of absolute zero). The ions, atoms, or molecules of which the solid is composed vibrate, rotate, and even move about within their assigned space in the crystal structure (Figure 6.20).

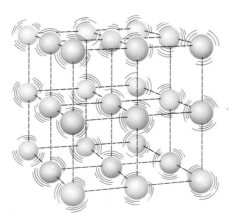

Figure 6.20 The crystal structure of a solid, showing the movement of the component ions, atoms, or molecules within their assigned space.

As in gas and liquid samples, particles in a solid have a distribution of kinetic energies that depends on the temperature of the sample. At every temperature, some particles have enough energy to overcome the forces that hold them in place and escape from the solid as a vapor. The process by which atoms or molecules go directly from the solid state to the gaseous state is called **sublimation.** A solid has a vapor pressure just as a liquid does. Solids that sublime have unusually high vapor pressures.

If a solid is heated, there is a temperature at which it melts and becomes a liquid. At this temperature, the average kinetic energy of the particles is greater than the energy of the forces holding them in a rigid lattice. The melting point of a substance is the temperature at which the liquid and solid states are in equilibrium at a pressure of one atmosphere. We show this equilibrium in equation form by writing *solid* as a reactant and *liquid* as a product and connecting them by a double (equilibrium) arrow:

$$\text{solid} \; \underset{\text{freezing}}{\overset{\text{melting}}{\rightleftharpoons}} \; \text{liquid}$$

During melting, both solid and liquid are present until all of the solid is converted to a liquid. Conversely, during freezing, solid and liquid are also present until all of the liquid is converted to a solid. These processes are indicated by the word *melting* over the forward equilibrium arrow and the word *freezing* under the reverse equilibrium arrow. Because freezing is just the reverse of melting, the freezing point of a substance is the same as its melting point.

The **molar heat of fusion** (ΔH_{fus}) of a substance is the amount of heat that must be supplied to convert one mole of that substance from a solid to a liquid at its melting point. The melting point and molar heat of fusion of several substances are given in Table 6.3. Calculations involving heats of fusion are similar to those involving heats of vaporization.

Table 6.3 Physical properties of several nonionic solids.

Compound	Mol. Wt.	mp (°C)	ΔH_{fus} (kcal/mole)
decane ($C_{10}H_{22}$)	142	−29.7	6.8
iodine (I_2)	252	114	4.0
naphthalene ($C_{10}H_8$)	128	80.2	4.5
phenol (C_6H_6O)	94	40.9	2.7
stearic acid ($C_{18}H_{36}O_2$)	284	71	13.5
water (H_2O)	18.0	0	1.44

Example 6.11

The molar heat of fusion of carbon tetrachloride (CCl_4) is 0.78 kcal/mole. How many calories must be supplied to 34 g of solid carbon tetrachloride at its melting point ($-23°C$) to change the sample to a liquid?

Solution

Wanted:
? cal

Given:
34 g carbon tetrachloride

Conversion factors:

$$\Delta H_{fus} = \frac{0.78 \text{ kcal}}{\text{mole}}$$

mol. wt. = 153.8
1 mole CCl_4 = 153.8 g CCl_4

Equation:

$$? \text{ cal} = 34 \text{ g} \times \frac{1 \text{ mole}}{153.8 \text{ g}} \times \frac{0.78 \text{ kcal}}{\text{mole}} \times \frac{1000 \text{ cal}}{1 \text{ kcal}}$$

$$= \frac{34 \times 1 \times 0.78 \times 1000}{153.8} \text{ cal}$$

Answer:
1.7×10^2 cal

Problem 6.11

How many calories are required to melt 15 g of ice at $0°C$? The molar heat of fusion of water at $0°C$ is 1.44 kcal/mole.

6.7 Transitions from the Solid through the Liquid to the Gaseous State

We have now discussed the properties of matter in its three states: solid, liquid, and gas. In each case, we have talked about the average kinetic energy of molecules in that state and about changes in average kinetic energy as the temperature is changed. We have also talked about changes from one state to another.

Figure 6.21 The change in temperature of a sample as heat energy is added and the sample changes from a solid to a liquid to a gas.

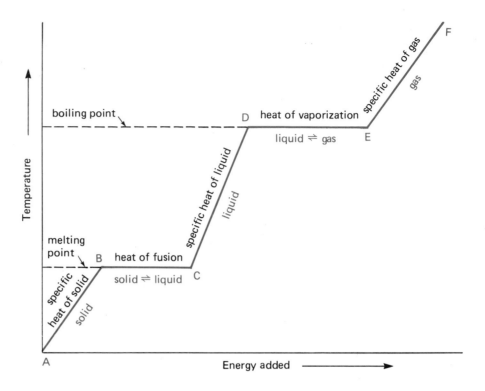

Now let us put all of this together and discuss what happens to a sample of matter when it changes from a solid to a liquid to a gas. Figure 6.21 shows the change in temperature of a sample as heat energy is added at a constant rate. At the left end of the graph, the sample is a solid; at the right end, it is a gas. Pressure remains constant at 1 atmosphere throughout the change.

The three segments AB, CD, and EF of the graph represent changes in temperature as energy is added to the solid, liquid, or gas. The rate of temperature change in each of these segments is determined by the specific heat (cal/g°C) for each state. These segments represent not only increases in temperature, but also increases in the average kinetic energy of the molecules in the sample.

The two other segments of the graph, BC and DE, indicate no change in temperature, but a considerable change in energy. Segment BC is at the melting point of the solid and represents the heat of fusion, the energy required to overcome the attractive forces holding the particles in the rigid lattice of the solid. Notice that while this change of state is going on, there is no temperature change; the solid and the liquid states are both present. Segment DE occurs at the boiling point and represents the heat of vaporization, the energy necessary to overcome the attractive forces between the particles and allow the liquid to become a gas.

Remember that the intermolecular forces in a solid or liquid depend on the composition and structure of the molecules. The slopes of segments AB, CD, and EF depend on specific heat and therefore vary from substance to substance. Similarly, the heats of vaporization and of fusion are unique for each substance.

The amount of energy required to change a sample from one state and temperature to another state and temperature can be calculated if the various physical constants are known.

Example 6.12

How many calories are needed to change 55.0 g of ice at $-5°C$ to steam at 110°C?

specific heat of ice = 0.49 cal/g°C
specific heat of liquid water = 1.00 cal/g°C
specific heat of steam = 0.48 cal/g°C
molar heat of fusion of water = 1.44 kcal/mole
molar heat of vaporization of water = 9.72 kcal/mole

Solution

We can calculate the answer to this problem by calculating the energy change for each segment of the graph in Figure 6.21.

1. To raise the temperature of ice from $-5°C$ to 0°C, a 5-degree change:

$$? \text{ cal} = 55.0 \text{ g} \times \frac{0.49 \text{ cal}}{\text{g} \cdot °C} \times 5°C = 135 \text{ cal}$$

2. To melt the ice at 0°C:

$$? \text{ cal} = 55.0 \text{ g} \times \frac{1 \text{ mole}}{18.0 \text{ g}} \times \frac{1.44 \text{ kcal}}{\text{mole}} \times \frac{1000 \text{ cal}}{1 \text{ kcal}} = 4400 \text{ cal}$$

3. To raise the temperature of liquid water from 0°C to 100°C:

$$? \text{ cal} = 55.0 \text{ g} \times \frac{1.0 \text{ cal}}{\text{g} \cdot °C} \times 100°C = 5500 \text{ cal}$$

4. To vaporize liquid water at 100°C:

$$? \text{ cal} = 55.0 \text{ g} \times \frac{1 \text{ mole}}{18.0 \text{ g}} \times \frac{9.72 \text{ kcal}}{\text{mole}} \times \frac{1000 \text{ cal}}{1 \text{ kcal}}$$

$$= 29,700 \text{ cal}$$

5. To heat water vapor (steam) from 100°C to 110°C:

$$? \text{ cal } = 55.0 \text{ g } \times \frac{0.48 \text{ cal}}{\text{g } °C} \times 10°C = 264 \text{ cal}$$

The sum of these energy changes is the total amount of energy required by the change:

$$? \text{ cal } = 135 \text{ cal } + 4400 \text{ cal } + 5500 \text{ cal } + 29,700 \text{ cal } + 264 \text{ cal } = 39,999 \text{ cal}$$

Correcting to three significant figures:

$$? \text{ cal } = 4.00 \times 10^4 \text{ calories.}$$

Problem 6.12

Calculate the number of calories released when 60 g of ice melt at 0°C and come up to room temperature (25°C).

Looking at the previous examples, notice how many more calories are involved in a change of state than in a change of temperature. In Example 6.12, 4400 calories were needed to melt the sample and 29,700 calories to vaporize it; only 5900 calories were needed for all the temperature changes. Does this suggest why steam burns are worse than those caused by hot water? Or why an ice pack provides more cooling than cold water?

Key Terms and Concepts

atmospheric pressure (6.4A)

Avogadro's Hypothesis (6.4B4)

barometer (6.4A)

boiling point (6.5B)

Boyle's Law (6.4B1)

Charles' Law (6.4B2)

Combined Gas Law (6.4B3)

dipole-dipole interaction (6.2)

dispersion forces (6.2)

dynamic equilibrium (6.5B)

gas constant (6.4B5)

gaseous diffusion (6.4C2)

hydrogen bonding (6.2)

ideal gas (6.4)

ideal gas equation (6.4B5)

intermolecular forces (6.2)

ionic bonds (6.2)

kinetic energy (6.1)

kinetic molecular theory (6.4)

manometer (6.4A)

melting point (6.6)

metallic bonds (6.2)

molar heat of fusion (6.6)

molar heat of vaporization (6.5D)

molar volume (6.4B4)

most probable kinetic energy (6.1A)

network covalent solids (6.2)

partial pressures (6.4C1)

potential energy (6.2)

pressure (6.4A)

real gas (6.4)

sublimation (6.6)

temporary dipole (6.2)

vapor pressure (6.5B)

Key Equations

Boyle's Law: $V_2 = V_1 \times P_1/P_2$ (at constant temperature)   or $P_2 V_2 = P_1 V_1$

Charles' Law: $V_2 = V_1 \times T_2/T_1$ (at constant pressure)   or $V_2/T_2 = V_1/T_1$

Graham's Law of Gaseous Diffusion: $\dfrac{r_1}{r_2} = \dfrac{\sqrt{MW_2}}{\sqrt{MW_1}}$

Ideal gas equation: $PV = nRT$

Kinetic energy: $KE = \frac{1}{2}mv^2$

Partial pressures: $P_T = P_1 + P_2 + P_3 + \cdots$

Problems

Intermolecular Forces (Section 6.2)

6.13 The melting point of several substances at 1 atm is given below. Suggest what type of intermolecular force may be in effect in each case.

	mp (°C)
propane, C_3H_8	−190
gallium, Ga	29.8
potassium nitrate, KNO_3	334
sulfur dioxide, SO_2	−73
nitrogen oxide, NO	−164
chlorine, Cl_2	−101
lanthanum oxide, La_2O_3	2307

6.14 Molecular weights and boiling points are given below for three pairs of gases. For each pair, explain why there is a difference between the boiling points and why the direction of that difference is as reported.

	Formula	Mol Wt.	bp (°C)
fluorine	F_2	38.0	−188
bromine	Br_2	159.8	58.8
hydrogen chloride	HCl	36.5	−84.9
chlorine	Cl_2	70.9	−34.6
ammonia	NH_3	17.0	−33.3
methane	CH_4	16.0	−164

6.15 Why do low-formula-weight ionic compounds usually have higher melting points than low-formula-weight covalent compounds?

Compressibility (Section 6.3C)

6.16 Why is it difficult to compress a liquid or a solid?

Kinetic Molecular Theory (Section 6.4)

6.17 State the postulates of the kinetic molecular theory of gases. Explain how they predict Charles' and Boyle's Laws.

Pressure (Section 6.4A)

6.18 The pressure exerted by a column of water 10 cm high with a cross section of 0.5 cm^2 is twice that of a column 5 cm high with a 1.0 cm^2 cross section. Explain why there is a difference between the two pressures.

6.19 In Section 6.4A we described how a barometer is made. What would happen to the height of the mercury column in Figure 6.10 if the closed end of the glass tube were cut off? Why?

The Gas Laws (Section 6.4B)

6.20 Complete the following table:

V_1	T_1	P_1	V_2	T_2	P_2
546 L	43°C	6.5 atm	_____	65°C	1.9 atm
43 cm^3.	−56°C	865 torr	_____	43°C	1.5 atm
4.2 L	234 K	0.87 atm	3.2 L	29°C	_____
1.3 L	25°C	1.89×10^5 Pa	_____	0°C	1.0 atm

6.21 Complete the following table:

Sample	Initial Conditions			Final Conditions		
	V_1	T_1	P_1	V_2	T_2	P_2
A	6.35 L	10°C	0.75 atm	_____	0°C	1 atm
B	75.6 L	0°C	1 atm	_____	35°C	735 torr
C	1.06 L	75°C	0.55 atm	0.76 L	0°C	_____

6.22 Why do aerosol cans carry the warning "do not incinerate"?

6.23 Does the density of a gas increase or decrease as the pressure increases at constant temperature? As the temperature increases at constant pressure?

6.24 A sample of gas has a volume of 1.0 L at 1.0 atm. The temperature is kept constant and the pressure is raised first to 2.0 atm, then to 4.0 atm, to 8.0 atm, and finally to 16.0 atm. Calculate the volume of the sample at each pressure. If the pressure continues to increase, will the volume ever reach zero?

6.25 Convert 1.65 atm to torr and to Pascals.

6.26 What is the pressure in atmospheres on a scuba diver at a depth of 50 ft? (Remember, 33 feet of water = one atm.)

6.27 An average pair of lungs has a volume of 6.5 L. If the air they contain is 21% oxygen, how many molecules of oxygen do they contain at 1 atm and 37°C?

6.28 What is the density of hydrogen at standard conditions? At 25°C and 0.925 atm?

6.29 A balloon filled with a 1.2 L sample of air at 25°C and 0.98 atm pressure is submerged in liquid nitrogen at −196°C. Calculate the final volume of the gas in the balloon.

6.30 A sample of gas collected at 38°C and 740 torr weighs 0.0630 g and occupies 26.2 cm^3. What is its volume at STP? What is its molecular weight?

6.31 If a 156-g block of dry ice, $CO_2(s)$, is sublimed at 25°C and 740 torr, what volume will the gas occupy?

6.32 1.65 L of a gas is collected at 25°C and 450 mm pressure. What will its volume be at 40°C and 550 mm pressure?

6.33 How many moles of gas are contained in the sample of gas in Problem 6.32?

6.34 Calculate the volume 1.1 g oxygen occupies at 2.0 atm and 5.0°C.

6.35 Calculate the molecular weight of a gas if 3.03 g of the gas occupies 660 cm^3 at 735 mm Hg and 27°C.

6.36 The density of liquid octane, C_8H_{18}, is 0.7025 g/cm^3. If 1.00 cm^3 of liquid octane is vaporized at 100°C and 725 torr, what volume will the vapor occupy?

6.37 Calculate the final pressure if 2.63 L of gas at 25°C and 1.00 atm pressure is allowed to expand to 8.45 L at the same temperature.

6.38 Calculate the density of ethane (C_2H_6) at STP.

6.39 Arrange the following gases in order of increasing rate of diffusion:
a. sulfur dioxide   b. hydrogen sulfide   c. ammonia
d. sulfur trioxide   e. argon   f. chlorine

Stoichiometry of Gases (Section 6.4B5)

6.40 What volume of hydrogen at 55°C and 0.95 atm can be prepared by the reaction of 1.5 g of magnesium with hydrochloric acid? The balanced equation is:

$$Mg + 2\,HCl \longrightarrow H_2 + MgCl_2$$

6.41 Oxygen can be prepared in the laboratory by heating potassium chlorate. The balanced equation is:

$$2\,KClO_3 \longrightarrow 2\,KCl + 3\,O_2$$

What volume of oxygen is prepared at 40°C and 737 torr by the decomposition of 63.5 g of potassium chlorate?

Vapor Pressure (Section 6.5B)

6.42 The equilibrium vapor pressure of water is 31.8 torr at 30°C and 9.21 torr at 10°C. Explain why, on a hot humid day, moisture forms on the outside of a glass containing a cold drink.

Properties of Solids (Section 6.6)

6.43 Woolens are often packed in moth balls (naphthalene) over the summer. In the fall, the moth balls are no longer there. Explain.

Changes of State (Section 6.7)

6.44 How many calories are needed to melt 25 g of ice at 0°C and to raise its temperature to 85°C?

6.45 How many calories are released when 10 g of steam at 100°C are condensed and cooled to body temperature (37°C)? How many calories are released when 10 g of water at 100°C are cooled to 37°C? Why are steam burns more painful than hot-water burns?

6.46 Compare the number of calories absorbed when 100 g of ice at 0°C are changed to water at 37°C with the number absorbed when 100 g of water are warmed from 0°C to 37°C.

6.47 Calculate the final temperature if 11,055 calories are added to 74 g of ice at −10°C. (The specific heat of ice is 0.49 cal/g°C.)

6.48 The density of liquid water at 0°C is 0.9999 g/cm³. The density of solid water (ice) at 0°C is 0.9168 g/cm³. How does the volume of a sample of water change when it freezes? Why do water pipes burst when the water in them freezes?

6.49 Temperature is a measure of kinetic energy. At the melting point of a substance:
   a. Is there a change in its kinetic energy? Explain.
   b. Is there a change in temperature? Explain.

6.50 Example 6.12 lists some thermal properties of water in calories. Convert those values to joules; to kilojoules.

6.51 Calculate the molar heat of vaporization and the specific heat in joules of the compounds whose values are given in Table 6.2.

6.52 Calculate the molar heat of fusion in joules for the compounds whose data are listed in Table 6.3.

6.53 Calculate the final temperature if 5783 joules of energy are added to 54 g ice at 0°C.

Skiing, Diving, and the Gas Laws

We live on the surface of the Earth under a blanket of air that extends miles above us. This blanket of air is approximately one-fifth (21%) oxygen; the rest is primarily nitrogen with small amounts of other gases, mostly carbon dioxide, argon, and water vapor. At sea level, air exerts a pressure of one atmosphere, the same pressure exerted by a column of water 10.3 m (33.9 ft) high. The pressure of the oxygen alone is 0.21 atmosphere. For those of us who live close to sea level, our bodies have become adjusted to this oxygen concentration. In normal quiet breathing, we inhale approximately half a liter of air per breath, containing approximately 0.02 mole of oxygen. Under these conditions, the hemoglobin in our blood is 97% saturated with oxygen and the pressure of oxygen in arterial blood is 0.13 atm. For the hemoglobin to become 100% saturated, we must breathe about twice as much air, so that the pressure of oxygen in arterial blood becomes approximately 0.26 atm. However, when the hemoglobin is 97% saturated, enough oxygen is transported to our cells for metabolism to proceed normally.

Although the percent composition of air remains constant at all altitudes, its density decreases with increasing altitude. This decreasing density results in a decreasing atmospheric pressure and a consequent decrease in the partial pressure due to oxygen. At 1500 m (5000 ft), atmospheric pressure is 0.83 atm and the partial pressure of oxygen is 0.17 atm. At four times that height (6000 m or 20,000 ft), atmospheric pressure is only one-third of 0.83 atm, or 0.26 atm, and the partial pressure of oxygen is 0.051 atm. What effect would these changes have on someone flying in from sea level for a few days of skiing or mountain climbing at this altitude?

Suppose you flew in to 3600 m (12,000 ft). At this altitude, the mountains are really spectacular; both skiing and climbing are more challenging than on the lower slopes. However, the partial pressure of oxygen at this altitude is only 0.14 atm, or two-thirds the partial pressure of oxygen at sea level. Each breath you take delivers to your lungs only two-thirds the amount of oxygen they are accustomed to receiving. In order to get the amount of oxygen your body needs, you breathe deeper and more rapidly to use the reserve capacity of your lungs, approximately 5 L. What then happens in your body?

By breathing more deeply and more often, you exhale more carbon dioxide than normally. This decreases the acidity of the blood, causing **alkalosis.** The respiratory center of the brain reacts to alkalosis by sending a message to the lungs to stop getting rid of so much carbon dioxide – in other words, to stop breathing so fast. You then breathe more slowly and even less oxygen reaches the lungs.

Because there is less oxygen in the lungs, the blood leaving the lungs is less than normally saturated with the gas and has less to deliver to the tissues. The heart pumps harder because the blood must circulate more rapidly to supply the normal amount of oxygen to the tissues; arterial blood pressure increases, and the capillaries dilate. In an additional effort to get more oxygen to the lungs, you start to breathe through your mouth. The air delivered to the lungs by this route is drier and colder than that which comes through the longer nasal passages. The epithelial (passage-lining) tissues quickly become dry and crusted. This situation is aggravated by the normally low moisture content of mountain air.

The severity and duration of these symptoms vary from individual to individual and are unrelated to sex or previous high-altitude experience. They are most severe when the change in altitude is rapid. Considerable data on these effects have been collected at the base camps at Mount Everest in the Himalayas. Climbers flying in to base camps at 2800 m are more apt to have difficulty adjusting to these changes than those who hike in from lower altitudes. The difficulties are frequently more severe in younger rather than older climbers. This may be related to the rate of ascent, younger climbers trying to cover more ground in less time than older ones. When most severe, these problems result in what is known as **high-altitude sickness.** The symptoms of this illness are a headache unrelieved by aspirin; nausea; insomnia; dizziness; shortness of breath that does not diminish when resting; and severe and unexplained lassitude. Severe cases of high-altitude sickness are sufficiently like a heart attack to indicate the need for prompt removal to a lower altitude. The need is not imaginary in cases where cerebral edema (swelling) occurs.

Mountaineers climbing without the aid of supplementary oxygen supplies at high altitudes give vivid pictures of the severe lassitude of high-altitude sickness. Good readers are unable to comprehend even simple sentences at 8200 m altitude. Temporary insanity and amnesia have been observed. In one account, after months of planning and weeks of effort, a climber found himself at 8500 m, about 300 m from the summit of Mount Everest, and was too weak to continue. He remembers no particular sense of frustration or disappointment at being unable to go on to the summit. Starting down alone after a brief rest, he became engulfed in a blinding blizzard which he found neither unpleasant nor exciting, but mildly interesting.

Does all this indicate that you should never ski or climb in the mountains if you live at sea level? Not at all. If the change in altitude is made slowly, allowing the body to adjust gradually to the decreasing oxygen pressure, difficulties can be minimized. If this is not possible, you can prepare for the change by doing exercises that increase lung capacity. After arriving at a high altitude, you can anticipate the problems and try not to do too much too soon. Knowing the problems caused by the decreased oxygen supply will help you to minimize their effects. The symptoms disappear as the body adjusts.

The problems discussed so far have been related to low partial pressures of oxygen. Diving, particularly scuba diving, subjects the body to greater than normal pressures, as each 10 m of water exerts a pressure of 1 atmosphere on the body. Thus, at a depth of 10 meters, a diver is under 2 atm pressure, one from the atmosphere and one from the water. At 20 m depth, he is under 3 atm pressure, and at 30.5 m (100 ft), the total pressure is 4.1 atm. The increased pressure can cause many problems.

What are the effects of these pressure changes? When our bodies are subjected to greater than normal air pressures, three kinds of effects are encountered: those due to overall changes in pressure; those due to changes in the partial pressures of oxygen and nitrogen; and those due to the increased solubility of gases in liquids at high pressures. There are many spaces in the body that contain gases rather than liquids. The lungs are the most obvious; other air spaces are the stomach and intestines, the sinuses, and the ear canals. Tiny air pockets may even be present under tooth fillings. As a diver descends, increasing pressure acts on all of these volumes as well as on the outside of the body, and it becomes difficult to keep the internal pressure equalized using the narrow passages that interconnect these spaces. We encounter this problem when going up (or down) in an express elevator. The swallowing we do instinctively in an elevator serves to equalize the effect of changing pressure on the gases in our body cavities. A descending diver may experience a squeeze in the ear canals, particularly if they are blocked because of allergies or a head cold. If he shuts off the air passages by holding his breath, he may experience a squeeze on the lungs. A tooth filling may become excruciatingly painful. Normal breathing during descent prevents this.

The decrease in pressure coming up from a dive reverses this process. On the average, the total lung volume is 5.5 L. In ascending to the surface from a depth of 10 m, the air in the lungs will double in volume. It is obviously essential to exhale this excess 5.5 L during ascent from a dive, to prevent the lungs from bursting.

We have already seen that too little oxygen, a situation encountered at high altitudes, is inimical to life. Too much oxygen is also toxic. The toxicity becomes severe at oxygen pressures of 2 atm. This oxygen pressure would be encountered breathing pure oxygen at a depth of 10 m or breathing compressed air at a depth of 90 m. The symptoms of oxygen poisoning are twitching, nausea, difficulties with vision and hearing, anxiety, confusion, and fatigue. They combine to make the diver less aware of his actions

and their consequences. The onset of these symptoms varies from one individual to another. Luckily, the symptoms disappear when the oxygen pressure decreases to the normal partial pressure of oxygen (0.21 atm).

Nitrogen is also a narcotic at high pressures. The term **"rapture of the deep"** refers to an overdose of nitrogen. The symptoms of nitrogen narcosis begin to occur at a depth of 31 m. Here the partial pressure of nitrogen has increased from 0.79 atm at the surface to 3.2 atm. Below this depth, the symptoms become more intense. The symptoms vary from one diver to another, as does the precise depth at which they are noticeable. Euphoria and a general inability to perform or to recognize the need to perform simple motor tasks are the most obvious symptoms. Dizziness, inability to communicate, and a false sense of well-being may also be present. The effect at 100 feet is similar to the effect of one dry martini; this relationship is nicknamed "Martini's Law." Divers subject to nitrogen intoxication have been known to be wholly irresponsible for themselves and their companions. Decreasing the nitrogen pressure by ascending is the best treatment for this condition.

The dangers associated with the increased solubility of gases at increased pressures are encountered in ascending after a dive. Everyone is familiar with the appearance of bubbles in a

carbonated beverage when the cap is removed. The sudden decrease in pressure allows the dissolved carbon dioxide to come out of solution throughout the liquid. A sudden ascent from a dive has the same effect on the gases that have dissolved in the blood in ever-increasing amounts as the depth of the dive and the consequent pressure on the diver have increased. The amount of dissolved gas also depends on the duration of the dive. The diver must ascend slowly so that the gases are released in the lungs rather than as bubbles in the blood stream. Oxygen, being continually burned by the body, does not cause any problems during ascent. Nitrogen, on the other hand, makes up about 80% of the dissolved gases and is the chief cause of decompression problems. A bubble in the blood stream stops the passage of blood; it acts as an embolism, or clot, and can be fatal. Once the bubbles occur, their size and location determine their effect on the diver. The most frequent sign of decompression sickness is pain, which is usually localized in joints, muscles, tendons, or ligaments. The term **bends** is derived from the temporary deformaties caused by the stricken diver's inability to straighten his or her joints. The diver is literally "bent out of shape." This condition is treated in recompression chambers. The affected diver is subjected again to the same high pressure encountered in a deep dive, and the pressure is then decreased slowly to sea-level pressure. This slow decrease of pressure permits the dissolved gases to be transferred from the blood to the lungs in the normal way. The practice of breathing a mixture of helium and oxygen is one way of lessening the dangers of the bends, for helium is much less soluble in the blood than is nitrogen. The only sure way of avoiding the bends is a slow ascent; scuba-diving manuals recommend no more than 60 ft/minute. After a dive to 100 ft, the ascent should take 4 1/2 minutes, with a 3-minute stop at 10 feet to allow for complete equilibration between the blood and the lungs.

Female divers should be aware that fetuses are much more sensitive than adults to the increased pressures encountered in diving. The following data come from a study done with pregnant ewes, animals chosen because the dynamics of fetal and maternal blood flow in ewes are very similar to those in humans. On a dive to 30 m, humans can accept no more than 25 minutes at this depth without needing decompression stops during the ascent. Ewes in the final trimester of pregnancy were subjected to dives of that depth and duration. In all cases, examination of the fetuses after these dives showed many bubbles in the fetal blood stream, probably enough to kill the fetus unless it received decompression treatment. Fetuses developed bubbles in nearly all dives deeper than 10 m. Since sheep are more resistant to decompression sickness than humans, there is little doubt that scuba diving when pregnant carries with it potential danger to the fetus, particularly in deep dives.

Respiratory therapy involves treatment of medical conditions where too little oxygen is available to the blood. Diseases involving the lungs and the diaphragm or circulatory difficulties often require such treatment. The therapist, working under a doctor's supervision, determines what partial pressures of oxygen should be administered to the patient in order to increase the amount of oxygen carried in the blood. Such treatment must be carefully monitored, for, as we have seen, excessive oxygen pressure is just as dangerous as insufficient pressure. For example, oxygen is often administered to newborn babies, particularly to those born prematurely or those who have been subjected to a particularly traumatic birth. But these babies, when exposed to high oxygen concentrations, are subject to retrolental fibroplasia, a disease that results in irreversible blindness. Hence, great care must be taken to ascertain that the baby gets enough, but not too much, oxygen.

Reference

Chem. and Eng. News 6 November 1978, p. 52.

The Importance of Water

Water is so much a part of our daily life that we rarely consider how unusual are its properties or their importance to our survival. In our bodies, water provides a medium for the chemical reactions that occur. Water transports nutrients to the cells through the various circulating systems of the body and carries away the waste products. It acts to regulate the temperature of the body. Outside the body, water provides a means for cooking, laundry, and bathing. In the larger view, the enormous bodies of water on our planet act to regulate its temperature. Water transportation is one of the oldest and still one of the cheapest methods of transportation. A large supply of water is essential to agriculture, to manufacturing, and to other activities too numerous to list.

What are the properties of water that make it so important to life? Recall that the water molecule is tiny, containing only three atoms, two hydrogens and one oxygen. Because it is such a small molecule, water can pass through the membranes of the body by osmosis, thereby maintaining fluid balance in the cells and tissues. Water is a polar molecule, with the two hydrogen-oxygen bonds forming an angle of 105° and giving the molecule a slightly positive end and a slightly negative end. Because of this polarity, water can dissolve the many polar organic and biochemical molecules involved in metabolism. In addition, because water is not a very reactive molecule, it can serve as an inert medium for the various metabolic reactions.

In addition to its usefulness as a solvent in the body, water has a unique set of physical properties which allow it to regulate the temperature of our bodies. These properties are related to the intermolecular structure of water. The hydrogen atoms of the water molecule, although covalently bonded to the oxygen atom of the molecule they are part of, are attracted to the electronegative oxygen atoms of other water molecules, forming hydrogen bonds between molecules. These intermolecular bonds give water the characteristics of a loosely bonded but enormous molecule. The physical properties of water reflect this structure. The boiling point is considerably higher than would be expected; compounds of similar molecular weight, such as methane, are gases at room temperature. The specific heat (1 cal/g°C) and the heat of vaporization (540 cal/g at 40°C) are similarly higher than would be expected for a molecule of molecular weight 18. These last two properties are the ones that help regulate our body temperature, as explained below.

Heat is generated in the body by reactions occurring in the cells. Blood flowing through the body absorbs this heat. The blood then flows through tiny capillaries near the surface of the skin or lungs. As it does so, heat is transferred from the blood to the layer of moisture normally on these surfaces. This moisture evaporates into the air, taking with it the calories it has absorbed. When the body is overheated, the blood flows more rapidly and the capillaries dilate. The flushed complexion associated with exertion or fever is due to an increased amount of blood in the capillaries just below the surface of the skin. When the body is overheated, exhaled breath becomes moister and more water molecules move through the cell walls of the blood vessels to the surface of the skin as perspiration. Each gram of perspiration that evaporates and each gram of water in the exhaled breath represent the loss of about 540 calories from the body. By increasing or decreasing the amount of perspiration and the moisture content of the exhaled breath, the body regulates its temperature.

Following this line of reasoning, you would expect a person with a fever to sweat profusely in order to reduce the temperature of the body. Why, then, does a person with a fever usually have dry skin? The reason is that the regulatory mechanism is not working efficiently and the feverish person is losing water from the skin faster than it can be replenished. The fluid intake and fluid balance of a person with a high fever should therefore be a matter of concern. In connection with this, note how the treatment of patients with high fevers has changed over the years. Earlier, patients were wrapped in heavy blankets — now they are covered lightly, if at all. The more easily heat can be released to the surroundings, the better. Frequently, the patient is placed on a water bed through which cold water is circulated. Sponge baths may be given. These are all ways of reducing the temperature of the body based on the principles of specific heat and heat of vaporization.

You yourself may have observed how the rate of evaporation from the skin affects body temperature. In the absence of moving air, the moisture layer on your skin tends to reach equilibrium with the water vapor in the air around it. If a breeze is constantly changing that air, there is no possibility for equilibrium and evaporation continues. This effect is particularly noticeable when you come out of a swimming pool on a windy day. The moving air, regardless of its temperature, blows away the moisture-saturated layer of air surrounding your skin. More evaporation takes place and you may be cold enough to shiver.

The discomfort associated with a hot, humid, windless day is also related to this evaporation process. **Humidity** measures the amount of water vapor in the air. At 100% humidity, the air is saturated with water vapor. At high humidity, water cannot evaporate from the skin because the air above the skin cannot absorb any more vapor. We are miserably uncomfortable. Why do we instinctively wish for a breeze? Because the moving air will remove the water-saturated air surrounding the skin and allow evaporation to occur.

The home use of humidifiers during winter months involves the same process. If the air is dry, evaporation from the skin takes place rapidly and we feel cold. By putting moisture into the air, we lower the rate of evaporation and feel comfortable at a lower temperature.

The importance of water in our lives is even greater than suggested by the examples discussed in the previous paragraphs. The average person uses about one ton of water per year for basic needs. Industries require enormous amounts of water for cooling (as in power plants) and removal of wastes (as in paper factories). The demands of agriculture for more and more water are always with us (witness the constant battle for the water of the Colorado River basin). Growing one bushel of corn requires between 10 and 20 tons of water; producing one pound of steer beef requires between 15 and 20 tons of water. All of this water is part of the **water cycle,** which provides for the recycling of all the water on Earth. Water evaporates into the atmosphere from the oceans and other bodies of water and falls as rain or snow. One-quarter of the rain falls over land and makes its way through lakes, rivers, and underground water supplies back to the ocean and reevaporation. As it moves toward the ocean, we divert it for our uses.

Much of the water used by industry and almost all of that used by agriculture requires no treatment. The water used for public water supplies usually does require treatment. The treatment for a particular water supply depends on what has happened to the water since it precipitated from the atmosphere, including any opportunities it has had to dissolve other substances. Because of its contact with different rocks and soils, it will probably contain some dissolved minerals, such as calcium, magnesium, and iron, as well as bicarbonate and sulfate ions. Because it has been in contact with air, it will contain some dissolved gases: carbon dioxide, oxygen, and nitrogen. Because of vegetation, aquatic life, or previous use as a water supply, it may contain waste chemicals and bacteria, as well as decaying vegetable and animal matter.

In a treatment plant, water may travel first through settling basins in which the suspended solids are deposited. Chlorine or ozone is added to kill any bacteria present. Recent discoveries that chlorine will react with various organic molecules present in the water to form carcinogenic compounds has favored the use of ozone for this bacterial control. Activated charcoal is sometimes added to absorb undesired organic material. The water is then filtered and pumped out into the system.

Dissolved minerals present in appreciable quantity make the water **hard.** The taste characteristic of various water supplies may also be due to these dissolved minerals. The minerals that make the water hard are an expensive nuisance. When hard

water is heated, carbonates precipitate, forming boiler scale in hot water pipes and teakettles. This scale prevents efficient transfer of heat and requires frequent replacement of the pipes. Soaps will not lather in hard water. Soap is a sodium or potassium salt of a large organic acid. When soap is added to hard water, the calcium or magnesium salt of this acid is formed; such salts are insoluble, curdy substances which form a gray scum but no lather. Chapter 16 contains a thorough discussion of soaps and how they work.

The current widespread use of detergents in hard water is a result of this problem. Detergents are also salts of organic acids, but the calcium and magnesium salts of these acids are soluble in water. Detergents lather easily in both hard and soft water. In fact, one of the ecological problems first associated with the widespread use of detergents was the enormous amount of foam produced. This foam persisted through sewage treatment plants and even into the streams where treated sewage was released. Detergent manufacturers quickly responded to this situation by producing detergents that were more biodegradable, generally eliminating the problem.

Hard water can be softened by treatment with lime at the water treatment plant or by passing it over an ion exchange resin. These resins, whether synthetic or naturally

occurring zeolites, are essentially complex sodium salts. As the water flows over their surface, the sodium ions dissolve and the calcium, magnesium, and other cations precipitate onto the surface of the resin. Treated water in which the only cation is sodium is "soft" water. When chemically pure or **deionized** water is required, the resin used replaces unwanted cations with hydrogen and unwanted anions with hydroxide.

The task of providing a safe, adequate water supply to cities and towns becomes more and more difficult as these areas continue to grow. The problem becomes particularly acute as towns grow in areas far from a natural supply of fresh water (as in the desert) or take the water for their supply from a river into which upstream municipalities have dumped their wastes (as in New Orleans, at the mouth of the Mississippi River).

Earlier, we talked about the usefulness of water as a solvent and as a temperature regulator. These properties depend on the polarity of the water molecule and its high specific heat. Hydrogen bonds play a role in causing the high specific heat of water, and they are even more important in determining the density of water in the liquid and solid states. In the liquid state, the hydrogen bonds are rather randomly oriented. As freezing takes place, they become arranged in a regular crystal lattice, and the molecules cannot be squashed together as they are in the liquid. Hence, the volume of the sample increases on becoming solid. The density of liquid water at 4°C is 1.000 g/cm³ and at 0°C it is 0.9998 g/cm³. The density of solid water (ice) at 0°C is less, 0.9168 g/cm³. Although this difference is important to humans, it is even more important to fish. When a lake freezes, the lighter ice rises to the surface. Thus, a lake freezes from the top down. The denser liquid water sinks to the bottom of the lake, providing a winter habitat for fish and other aquatic life. The surface layer of ice provides an insulating layer above. Lakes rarely freeze all the way to the bottom. In the spring, as the air temperature rises, the ice melts to form a liquid more dense than the ice. This liquid moves to the bottom of the lake, forcing the warmer and lighter water to the top. The currents thus produced mix not only the water but also those nutrients previously concentrated on the bottom that are essential to aquatic life. If ice were more dense than liquid water, this mixing would not occur in large bodies of water. Ice would remain on the bottom all summer, insulated from the warm air by the upper layers of liquid water. The consequences of such a reversal are hard to imagine but there is little doubt that they would extend farther than just making the swimming hole a little colder.

Highway engineers and plumbers will tell you that the expansion of water caused by freezing is not all good. Water freezing in pipes causes them to burst. Water freezing in tiny cracks in street pavements expands, breaking up the pavement and leading to the potholes that plague the highway engineers and drivers in northern climates.

Solutions and Colloids

Our discussions thus far have concentrated on the properties of pure substances: elements, compounds formed by the combination of elements, and the reactions of these pure substances. In actual experience, we do not often encounter pure substances. More often, the matter we see and use is a mixture. In this chapter we will discuss two particular kinds of mixtures: *solutions* and *colloids*.

7.1 Solutions, Colloids, and Suspensions

Consider the following mixtures, all of which contain water: tap water, milk, and a mixture of sand and water (Figure 7.1). Tap water is a **solution.** A solution is a mixture that is clear and uniform throughout; one substance (or more) has been dissolved in another. Tap water contains some dissolved substances: sometimes calcium and magnesium ions, which make it hard; fluoride ion added to prevent tooth decay; and probably chlorine molecules added to destroy bacteria and make the water safe for consumption. Milk is a **colloid.** A colloid is uniform throughout and sometimes clear. Milk is a mixture of fats, water, and many other things. The third mixture, sand and water, is a heterogeneous

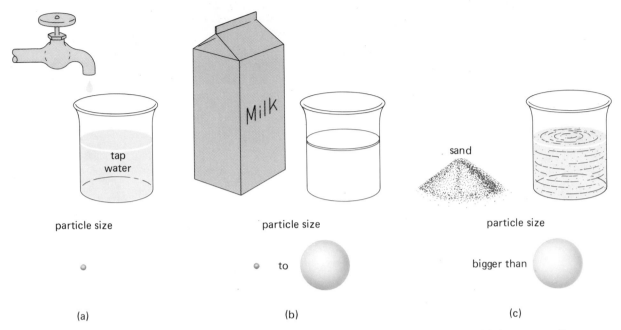

Figure 7.1 Examples of (a) a solution, (b) a colloid, and (c) a suspension, along with the relative sizes of the particles in each.

suspension. In spite of vigorous stirring, the grains of sand settle to the bottom of the container. It is a mixture that is not uniform throughout and in which it is easy to see the two **phases** or components.

A. The Components of Solutions and Colloids

There are two components in either a solution or a colloid. A solution consists of the **solute,** or the substance dissolved, and the **solvent,** the substance the solute is dissolved in. In a colloid, one substance, the **dispersed phase,** is uniformly distributed throughout the **dispersing medium.** In mud, for example, the dispersed phase is soil and the dispersing medium is water. In fog, the dispersed phase is water and the dispersing medium is air.

You are familiar with solutions in which the solute is a gas (carbonated beverages), a liquid (rubbing alcohol), or a solid (sugar solutions). The solvent need not be a liquid—it may be a gas or a solid. Similarly, in colloids the dispersed phase may be a solid (smoke), a liquid (mayonnaise), or a gas (whipped cream). The dispersing medium can also be either solid, liquid, or gas. Many physiological fluids are colloidal dispersions of large molecules (for example, proteins) in water.

B. Differences among Solutions, Colloids, and Suspensions

The most fundamental difference among these mixtures is the size of the particles that are mixed with the solvent. In solutions, the dissolved particles are ions or small molecules whose diameter is usually less than 10^{-7} cm (see Figure 7.1). The suspended particles of a colloid range in diameter from about 10^{-7} cm to about 10^{-4} cm; they are large molecules or clusters of smaller molecules. The grains of sand that settle out of a suspension usually have diameters greater than 10^{-4} cm.

The particles in solutions and in colloids are small enough to be kept in constant motion by the buffeting of the solvent molecules that surround them. This constant motion keeps both solutions and colloids uniform throughout. The sand particles in the mixture are too big to be bounced around by the relatively small water molecules and are also big enough to be affected by the pull of gravity. Hence, they settle to the bottom.

Colloidally suspended particles and dissolved particles are sufficiently different in size to differ in their appearance in visible light. The dissolved particles of a solution are too small to reflect visible light. Those suspended in a colloid are larger and do reflect light. This difference gives rise to two unique properties of colloids that can be used to differentiate a solution and a colloid. The first one, the **Tyndall effect,** is illustrated in Figure 7.2, in which a flashlight beam is shining through a solution and a colloid. The path of light through the solution is invisible, but it is clearly visible through the colloid as the light is reflected by the tiny colloidally suspended particles. When you see the beam of a searchlight sweeping the sky on a foggy night, you are observing the Tyndall effect, for fog is a colloid formed by the suspension of very tiny water drops in air. The Tyndall effect can be observed even in colloids so clear as to look like solutions to the naked eye.

You have also probably seen **Brownian movement,** the second property unique to colloids. When sunlight streams into a slightly dusty, slightly darkened room, dancing specks of light are noticeable in the sunlight. These are reflections of light from colloidally dispersed dust particles. The particles are in constant

Figure 7.2 The Tyndall effect. The path of light is invisible in the solution but visible in the colloid.

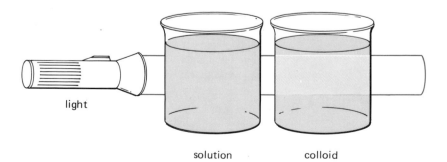

light   solution   colloid

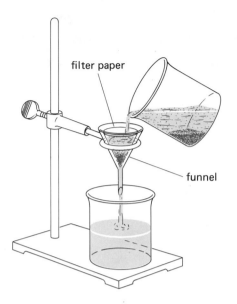

Figure 7.3 Filtration. By pouring a suspension through a filter, the solid suspended material can be collected in the filter.

filter paper

funnel

Table 7.1 Properties of solutions, colloids, and suspensions.

	Solutions	Colloids	Suspensions
particle size:	less than 10^{-7} cm	10^{-7}–10^{-4} cm	greater than 10^{-4} cm
appearance:	uniform, clear	uniform, cloudy	not uniform, two separate components visible
path of light beam (see Figure 7.2):	invisible	visible	varies
method of separation:	distillation	distillation	filtration

Examples of Solutions and Colloids:

States of matter Combined	Examples of Solutions	Examples of Colloids
gas in liquid	capped soft drink	whipped cream
gas in solid	—	foam
liquid in gas	humid air	fog
liquid in liquid	rubbing alcohol	mayonnaise, egg white
liquid in solid	dental amalgam (silver fillings)	fruit pulp
solid in gas	—	smoke
solid in liquid	salt water	gelatin
solid in solid	brass (zinc dissolved in copper)	ruby glass (a colloid of gold in glass)

motion, which gives the dancing effect. Brownian movement is characteristic of colloidally suspended particles. The constant, irregular movement of the dust particles as they reflect the light is much like that postulated for gas molecules.

Most suspensions of solids in liquids can be separated by **filtration** (Figure 7.3). Our mixture of sand and water can be separated into its two components using this method, by pouring it through a filter paper or a cotton plug. A solution or colloid cannot be separated by filtration. The dissolved particles of a solution and the suspended particles of a colloid are small enough to go through the filter. The difference in property is illustrated by the difference between jam and jelly. If you have ever watched anyone make jelly, you have observed this difference. Fruit jelly is a colloidal suspension of pectin (a large molecule) in a water solution of sugar and fruit juices. Fruit jams contain solid bits of fruit pulp and seeds as well as sugar and fruit juices. When you make jam, the solids are not removed. When you make jelly, the mixture is strained through a fine cloth to remove the solids. The colloidal jelly passes through the cloth.

Distillation (boiling off the more volatile component) is one way to separate the components of either a solution or a colloid. Table 7.1 lists several properties of solutions, colloids, and suspensions, along with examples of each.

7.2 Solubility

Although there are many types of solutions (Table 7.1), the word *solution* usually implies that the solvent is a liquid. The **solubility** of a substance is the amount of the substance that will dissolve in a given amount of solvent. A substance is said to be **soluble** if more than 0.1 g of that substance dissolves in 100 mL of solvent. A substance is **insoluble** when less than 0.1 g dissolves in 100 mL of solvent. **Miscible** has the same meaning as *soluble* but is usually used when both solute and solvent are liquids. **Immiscible** has the same meaning as insoluble. For example, vinegar is miscible with (soluble in) water; oil is immiscible with (insoluble in) water. Italian salad dressing is a good illustration of the basic principles involved.

Figure 7.4 (a) Alcohol is miscible (soluble) in water; when added to water, it forms a clear solution. (b) Oil is immiscible (insoluble) in water; when added to water, it remains separate, forming a layer on top of the water.

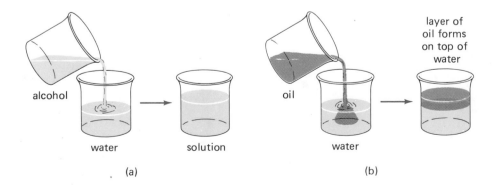

alcohol

water

solution

(a)

oil

water

layer of oil forms on top of water

(b)

A. Determining Solubility

How is the solubility of a substance determined? A known amount of the solvent, for example, 100 cm³, is put into a container. Then the substance whose solubility is to be determined is added until, even after vigorous and prolonged stirring, some of that substance does not dissolve but remains at the bottom of the container. The amount of substance then in solution is the solubility of the substance at that temperature in that solvent. Doing this experiment with water as the solvent and sodium chloride as the solute, we find that at 20°C, 35.7 g of the salt dissolves in 100 cm³ of water. The solubility of sodium chloride is, then, 35.7 g/100 cm³ water. Sodium chloride is a moderately soluble salt. The solubility of sodium nitrate is 92.1 g/100 cm³ water; sodium nitrate is a very soluble salt. At the opposite end of the scale is barium sulfate, which has a solubility of 2.2×10^{-4} g/100 cm³ water. Barium sulfate is an "insoluble" salt. See Table 4.4 for the solubility of other compounds and Table 5.1 for rules predicting the solubility of ionic compounds.

B. Equilibrium in Saturated Solutions

In the container in Figure 7.5 is a saturated solution of sodium chloride, and at the bottom of the container is some undissolved sodium chloride. If we could see the individual ions of sodium chloride, we would see that some ions are going into solution and an equal number are coming out of solution. There is a dynamic **equilibrium** between these two processes, which can be shown by the equation:

$$NaCl(s) \rightleftharpoons NaCl(aq)$$

Because solution and precipitation occur at equal rates, the total amount of dissolved salt does not change. The dissolved solute is in equilibrium with the undissolved portion. A solution in which this equilibrium exists is a **saturated solution.** The saturated solution of sodium chloride contains ions. A saturated solution of sucrose (sugar), which is not ionic, contains molecules. The equilibrium existing in a saturated sugar solution is between dissolving and precipi-

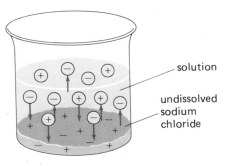

solution

undissolved sodium chloride

Figure 7.5 When solid sodium chloride is added to a saturated sodium chloride solution, an equilibrium arises between ions going into solution and ions coming out of solution.

tating molecules. See Sections 6.5B and 6.6 for other examples of equilibrium.

An unsaturated solution contains less solute than can dissolve in a given amount of solvent at a given temperature. An unsaturated solution of sodium chloride contains less than 35.7 g of sodium chloride in each 100 cm³ of water.

C. Factors Affecting Solubility

Many factors affect the solubility of one substance in another:

1. The nature of the intermolecular or interionic forces in both the solute and the solvent. In Section 6.2, we listed the kinds of forces operating in solids and liquids. When one substance dissolves in another, the attractive forces in both must be overcome to allow the solute to move into the solvent molecule by molecule or ion by ion. If the solute has a separation of charges, due to its ionic or polar nature, it is most apt to dissolve in a polar solvent such as water, whose molecules also have a separation of charges. There must be an interaction between the two sets of charges to allow the dissolving of the solute. Dispersion forces represent only a weak interaction between molecules. Such forces are not strong enough to break up the network of molecules in a polar solvent. Therefore, nonpolar substances will not usually dissolve in a polar solvent. We have, then, a rule of thumb: like dissolves like. Ionic and polar compounds are most apt to be soluble in polar solvents like water or liquid ammonia. Nonpolar compounds are most apt to be soluble in nonpolar solvents such as carbon tetrachloride and hydrocarbon solvents like gasoline.

The solubility of gases in water depends a great deal on the polarity of the gas molecules. Those gases with polar molecules are much more soluble in water than are nonpolar gases. Ammonia, a strongly polar molecule, is very soluble in water (89.9 g/100 g H_2O); so is hydrogen chloride (82.3 g/100 g H_2O). Helium and nitrogen are nonpolar molecules. Helium is only slightly soluble (0.94 g/100 g H_2O); so is nitrogen (2.33 g/100 g H_2O).

Table 7.2 shows specific examples of different kinds of compounds and their relative solubilities in water, a polar solvent, in alcohol, a less polar solvent, and in benzene, a nonpolar solvent.

Table 7.2 Different compounds and their relative solubilities.

Kinds of Bonds	Example	Solubility in:		
		Water	Alcohol	Benzene
ionic	sodium chloride	very soluble	slightly soluble	insoluble
polar covalent	vitamin C (ascorbic acid)	very soluble	soluble	insoluble
nonpolar covalent	naphthalene	insoluble	soluble	very soluble

Table 7.3 The effect of temperature on solubility.

Compound	Type of Bonding	Solubility (g/100 cm³ water) at 25°C	at 100°C
sodium chloride	ionic	35.7	39.1
potassium chloride	ionic	34.7	56.7
barium sulfate	ionic	2.3×10^{-4}	4.1×10^{-4}
sucrose	polar covalent	179	487
ammonia	polar covalent	89.9	7.4
hydrogen chloride	polar covalent	82.3	56.1
oxygen	nonpolar covalent	3.2	2.3

2. Temperature. The solubility of solids and liquids usually increases as the temperature increases. Table 7.3 shows the solubility of several substances in water at 20°C and 100°C. Several gases are included in this table; notice that their solubilities *decrease* with increasing temperature. This is typical of gases and causes our concern for the fish population of lakes, oceans, and rivers threatened with thermal (heat) pollution. Fish require dissolved oxygen to survive. If the temperature of the water increases, the concentration of dissolved oxygen decreases, and survival of the fish becomes questionable.

3. Pressure. The pressure on the surface of a solution has very little effect on the solubilities of solids and liquids. However, it does have an enormous effect on the solubility of gases. As the partial pressure of a gas in the atmosphere above the surface of a solution increases, the solubility of that gas increases. A carbonated beverage is bottled and capped under a high partial pressure of carbon dioxide so that a great deal of carbon dioxide dissolves. When the bottle is uncapped, the partial pressure of carbon dioxide above the liquid drops to that in the atmosphere, the solubility of carbon dioxide decreases, and the gas, CO_2, comes out of solution as fine bubbles.

This same phenomenon is of concern in diving. As a diver descends, he is under ever-increasing pressure due to the increasing mass of water above him. Each thirty-three feet of water exerts a pressure of one atmosphere. As the diver descends, the pressure on the air in the lungs increases rapidly and the blood dissolves more than normal amounts of both nitrogen and oxygen. As the diver comes up, the pressure decreases and the gases come out of solution. If the change in pressure is too swift, the gases are released into the blood stream and tissues, instead of into the lungs. The bubbles cause sharp pain wherever they occur, usually in the joints and ligaments. This dangerous and sometimes fatal condition is called **bends,** the name deriving from the temporary deformities suffered by the affected diver being unable to straighten his or her joints. Because helium, at all pressures, is less soluble than nitrogen, divers often breathe a mixture of helium and oxygen instead of air, a nitrogen-oxygen mixture. For a more complete discussion of this phenomenon, see the mini-essay on the gas laws immediately following Chapter 6.

The term **hyperbaric** means greater than normal atmospheric pressure. In hyperbaric medicine, patients are subjected to a pressure greater than atmospheric. This increases the amount of oxygen dissolved in the blood. Treatment in hyperbaric, or high-pressure, chambers is of particular value for patients who are suffering from severe anemia, hemoglobin abnormalities, or carbon monoxide exposure, for it increases the amount of oxygen transported by the blood to the tissues.

7.3 Properties of Solutions Containing Nonvolatile Solutes: Colligative Properties

The physical properties of a solution are different from those of the pure solvent. Many differences in physical properties are predictable if the solute in the pure state is nonvolatile—that is, it has very low vapor pressure. Sugar, sodium chloride, and potassium nitrate are nonvolatile solutes. Those physical properties of solutions of nonvolatile solutes that depend only on the number of particles present in a given amount of solution are called **colligative properties.** We will consider four colligative properties: vapor pressure lowering, boiling point elevation, freezing point depression, and osmotic pressure.

A. Vapor Pressure Lowering

At any given temperature, the vapor pressure of a solution containing a nonvolatile solute is less than that of the pure solvent (see Section 6.5B for a discussion of vapor pressure). The solid line in Figure 7.6 is a plot of the vapor pressure of pure water against temperature. The break in the curve at $0°C$ is

Figure 7.6 The vapor pressure of water is lowered by the addition of a solute.

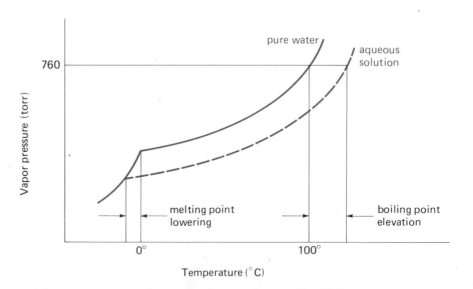

Figure 7.7 Effect of solute molecules on vapor pressure. In (a), no solute molecules interfere with the escape tendencies of the solvent molecules. In (b), the solute molecules interfere with the escape tendencies of the solvent molecules, lowering the vapor pressure.

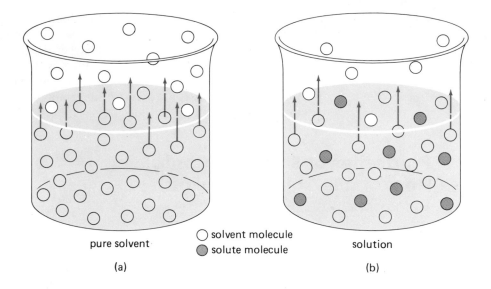

pure solvent

○ solvent molecule
● solute molecule

solution

(a)   (b)

the intersection of the vapor pressure of the solid with the vapor pressure of the liquid. The dashed line of Figure 7.6 is a plot of the vapor pressure of an aqueous solution against temperature. Notice that the vapor pressure of the solution is always less than that of the pure solvent. What causes this difference?

The surface of a pure solvent (Figure 7.7a) is populated only by solvent molecules. The surface of a solution is populated mostly with solvent molecules, but there are also some solute molecules or ions on the surface (Figure 7.7b). These molecules or ions are nonvolatile and do not contribute to the vapor pressure of the solution. Indeed, they interfere with the escape tendencies of the solvent. As a consequence, the vapor pressure of the solvent is decreased.

B. Boiling Point Elevation

The boiling point of a substance is the temperature at which the vapor pressure of the substance equals atmospheric pressure. A solution containing a nonvolatile solute, having a lower vapor pressure than the pure solvent, must be at a higher temperature before its vapor pressure equals atmospheric pressure and it boils. Thus, the boiling point of a solution containing a nonvolatile solute is higher than that of the pure solvent (see Figure 7.6).

C. Freezing Point Depression

Recall that freezing point and melting point are two terms that describe the same temperature, the temperature at which the vapor pressure of the solid equals the vapor pressure of the liquid and the solid and the liquid are in equilibrium. Remember, too, that vapor pressure decreases as the temperature decreases. The vapor pressure of a solution is lower than that of the pure solvent, so the vapor pressure of a solution will equal that of the solid at a lower temperature than in the case of the pure solvent. Thus, the freezing point will be lower for a solution than for the pure solvent (see Figure 7.6). It should be remembered that just as it is the solvent that vaporizes when a solution boils, it is the solvent, not the solution, that becomes solid when a solution freezes. When a salt solution freezes, the ice is pure water (solid); the remaining solution contains all the salt.

Application of this principle leads us to add antifreeze (a nonvolatile solute) to the water in the radiators in our cars. We thus lower the freezing point of the solvent (water) and the solution remains liquid even in subfreezing temperatures.

D. Osmosis and Osmotic Pressure

Osmosis and osmotic pressure depend on the ability of small molecules to pass through semipermeable membranes such as a thin piece of rubber, a cell membrane, or a thin piece of plastic wrap. Solvent particles are small and can easily pass through; solute particles are larger and cannot (Figure 7.8). When a semipermeable membrane separates a solution from pure solvent, solvent molecules move back and forth through the membrane, but not in equal numbers. More move from the pure solvent into the solution than from the solution into the solvent. Molecular movement continues until equal pressure is exerted by sol-

Figure 7.8 A semipermeable membrane allows small solvent molecules, such as water, to pass through, but blocks the passage of larger particles, such as sugar molecules.

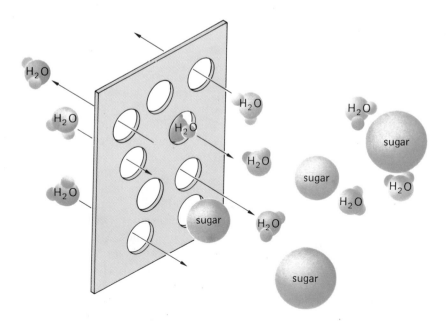

Figure 7.9 Osmosis and osmotic pressure. Differing amounts of pure solvent on either side of a semipermeable membrane (a) will, through osmosis, become equally divided on either side of the membrane (b). However, if solute molecules are added to one side (c), some of the solvent will migrate into the solution side, causing a difference in osmotic pressure (d). The difference in pressure can be counteracted by increased surface pressure on the solution side (e).

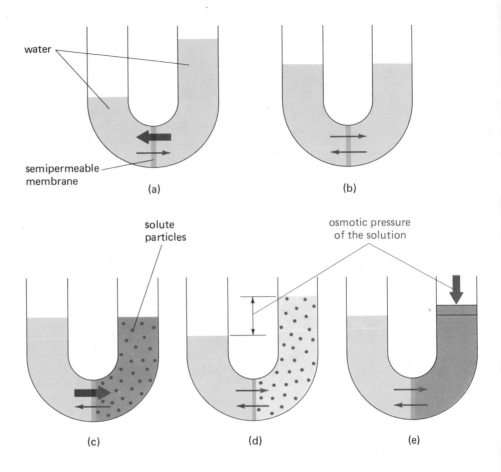

water

semipermeable membrane

(a)

(b)

solute particles

osmotic pressure of the solution

(c)

(d)

(e)

vent molecules on both sides of the membrane. The movement of particles through a semipermeable membrane in an attempt to balance the concentration of those particles on either side of the membrane is called **osmosis.**

In Figure 7.9(a), different amounts of pure solvent (water) are separated by a semipermeable membrane. Molecules from both sides move through the membrane until the pressure of solvent on both sides of the membrane is equal. This equality is indicated by the equal heights of the columns (Figure 7.9b).

In Figure 7.9(c), solute molecules have been added to the solvent, creating a solution. On the solution side, some of the molecules exerting pressure on the membrane are solute molecules. On the solvent side of the membrane all the molecules are solvent molecules. The pressure on the membrane is greater on the side of the pure solvent than on the solution side. Solvent molecules will move through the membrane into the solution until the pressure they exert on the membrane is the same on both sides. Figure 7.9(d) shows the situation when the pressures have become equal. The column of solution is higher than the column of pure solvent. The pressure exerted by this extra height is the **osmotic pressure** of the solution.

Osmotic pressure is also being measured in Figure 7.9(e). Here, pressure is being applied to the surface of the solution. The pressure needed to equalize the heights of the two columns is the osmotic pressure of the solution.

Osmosis and osmotic pressure are very important to living organisms. Blood is an aqueous solution containing many solutes and has an osmotic pressure of 7.7 atm measured against pure water. When red blood cells are put in pure water, water molecules move into the cells through the cell membrane to equate the osmotic pressure in the blood cell with that of the pure water outside. So much water may pass into the cell that the cell membrane ruptures because it cannot stretch to contain the additional water; this rupturing is called **hemolysis** (Figure 7.10).

By a similar mechanism, blood cells put in a concentrated salt solution shrink as water molecules move out of the cell into the salt solution. The water molecules move out of the cell because the concentration of water is lower outside the cell than inside; hence, the osmotic pressure is higher outside the cell. This shrinking process is called **crenation** (Figure 7.10b).

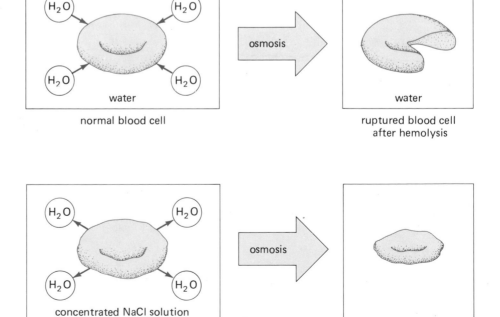

Figure 7.10(a) Hemolysis (rupturing) of red blood cells in pure water. The relatively high osmotic pressure within the cell causes water molecules to move into the cell by osmosis, eventually resulting in the rupturing of the cell.

normal blood cell

ruptured blood cell after hemolysis

Figure 7.10(b) Crenation of red blood cells in a concentrated salt solution. The higher osmotic pressure outside the cell causes water to move out of the cell, resulting in crenation.

normal blood cell

crenated blood cell

To prevent either of these destructive processes from occurring in the body, fluids fed intravenously must have the same osmotic pressure as blood itself. Solutions having the same osmotic pressure are called **isotonic.** An isotonic glucose solution contains 5.3 g glucose per 100 g solution; an isotonic saline solution contains 0.9 g sodium chloride per 100 g solution. Both solutions have an osmotic pressure of 7.7 atm.

For any solution, the amount that the vapor pressure is lowered, the freezing point depressed, or the boiling point elevated with respect to the properties of the pure solvent depends on the *number* of solute particles in solution, not on the nature of those particles. Similarly, the osmotic pressure of a solution is dependent only on the number of solute particles, not on their nature. Table 7.4 shows the melting point (freezing point), boiling point, and osmotic pressure of several glucose solutions. The number of moles of glucose (therefore, the number of glucose molecules in a given amount of water) differs among these solutions. The greater the number of molecules of solute, the greater the difference between the properties of the pure solvent and those of the solution.

The colligative properties of a solution with an ionic solute differ from those of a solution with a molecular solute. One mole of sugar, a molecular compound, dissolves to yield one mole of particles; however, one mole of sodium chloride, an ionic compound, dissolves to yield two moles of particles, one of sodium ion and one of chloride ion. One mole of sodium sulfate (Na_2SO_4) dissolves to yield three moles of particles, two of sodium ion and one of sulfate ion. Because it is the number of particles that determines the colligative properties of a solution, one mole of sodium chloride dissolved in a given amount of water causes approximately twice the change in colligative properties as does one mole of glucose dissolved in the same volume of water. One mole of sodium sulfate dissolved in the same amount of water causes approximately three times the change. Several colligative properties of solutions of glucose, sodium chloride, and sodium sulfate are shown in Table 7.4.

Table 7.4 Colligative properties of some ionic and nonionic solutions.

Solution	g solute/ 1000 g water	mp, °C	bp, °C	Osmotic Pressure at 25°C
water	0	0	100	0
glucose solutions	18 (0.1 mole)	−0.19	100.05	2.4 atm
	36 (0.2 mole)	−0.36	100.10	4.8 atm
	180 (1.0 mole)	−1.8	100.52	24 atm
	360 (2.0 moles)	−3.6	101.04	73 atm
NaCl solution	6 (0.1 mole)	−0.33	100.08	4.1 atm
Na_2SO_4 solution	14 (0.1 mole)	−0.43	100.17	5.4 atm

7.4 Expressing Concentrations of Solutions

A complete description of a solution states what the solute is, what the solvent is, and how much solute is dissolved in a given amount of solvent or solution. The quantitative relationship between solute and solvent is the **concentration of the solution.** This concentration may be expressed using several different methods, as discussed below.

A. By Weight

The concentration of a solution may be given as the weight of solute in a given amount of solution, as in the following statements: The northern part of the Pacific Ocean contains 35.9 g of salt in each 1000 g of seawater. The North Atlantic Ocean has a higher salt concentration, 37.9 g salt/1000 g seawater.

B. By Percent

The concentration of a solution is often expressed as a percent by weight or percent by volume of solute in solution. **Percent by weight** is calculated from the weight of solute in a given weight of solution. A 5% by weight aqueous solution of sodium chloride contains 5 grams of sodium chloride and 95 g of water in each 100 grams of solution.

Example 7.1

How many grams of glucose and of water are in 500 g of a 5.3% by weight glucose solution, the glucose solution that is isotonic with blood?

Solution

5.3% of the solution is glucose:

$$\frac{5.3 \text{ g glucose}}{100 \text{ g solution}} \times 500 \text{ g solution} = 26.5 \text{ g glucose}$$

The remainder of the 500 g is water.

$$500 \text{ g} - 26.5 \text{ g} = 473.5 \text{ g water}$$

Problem 7.1

How would you prepare 400 g of isotonic saline solution (0.9% by weight sodium chloride in water)?

If both solute and solvent are liquids, the concentration may be expressed as **percent by volume.** Both ethyl alcohol and water are liquids; the concentration of alcohol-water solutions is often given as percent by volume. For example, a 95% solution of ethyl alcohol contains 95 cm^3 ethyl alcohol and 5 cm^3 water in each 100 cm^3 solution.

Example 7.2

Rubbing alcohol is an aqueous solution containing 70% isopropyl alcohol. How would you prepare 250 cm^3 rubbing alcohol from pure isopropyl alcohol?

Solution

70% of the volume is isopropyl alcohol:

$$\frac{70 \text{ cm}^3 \text{ isopropyl alcohol}}{100 \text{ cm}^3 \text{ solution}} \times 250 \text{ cm}^3 \text{ solution} = 175 \text{ cm}^3 \text{ isopropyl alcohol}$$

The rest of the volume is water:

$$250 \text{ cm}^3 - 175 \text{ cm}^3 = 75 \text{ cm}^3 \text{ water}$$

Combining these two amounts will give 250 cm^3 of 70% isopropyl alcohol solution.

Problem 7.2

Using pure ethyl alcohol and water, how would you prepare 1.0 liter of a 40% ethyl alcohol solution?

Because the density of many liquids changes slightly as the temperature changes, a concentration given in percent by weight is more accurate over a range of temperatures than is a concentration given in percent by volume. Many laboratories use a combination of weight and volume to express the concentration: the weight of solute dissolved in each 100 cm^3 water. Using this method, a 5% solution of sodium chloride contains 5 g sodium chloride in each 100 cm^3 water.

C. By Parts per Million (ppm) and Parts per Billion (ppb)

The terms **parts per million (ppm)** and **parts per billion (ppb)** are encountered more and more frequently as we become more aware of the effects of those substances present in trace amounts in water and air, and as we develop instruments sensitive enough to detect compounds present in such low concentrations. Unfortunately, users of these terms do not always cite units. In discussing mass, parts per million should mean concentrations of grams per 10^6 grams, or micrograms per gram. In discussing volume, parts per million may mean milliliters per 10^3 liters, or the mixed designation of milligrams per liter. Unfortunately, current users of the terms are apt to mix metric and English units or mass and volume units fairly indiscriminately. Fortunately, there is a general movement toward the use of micrograms per liter (ppb) when discussing water contaminants, micrograms per cubic meter (ppb) for air, and micrograms per kilogram (ppb) for soil concentrations.

D. In Terms of Moles

The concentration of a solution may be stated as **molarity (M),** the number of moles of solute in a liter of solution.

$$\text{molarity } (M) = \frac{\text{moles solute}}{\text{volume (L) solution}}$$

A 6M (say "six-molar") solution of hydrochloric acid contains six moles of hydrochloric acid in one liter solution. A 0.1M solution of sodium iodide contains 0.1 mole sodium iodide in one liter solution.

The molarity of a solution gives a ratio between moles of solute and volume of solution. It can be used as a conversion factor between these two units in calculations involving solutions. Three examples of such calculations follow.

Example 7.3

How many moles of hydrochloric acid are in 200 cm^3 of 0.15M HCl?

Solution

Wanted:
? moles HCl

Given:
200 cm^3 0.15M HCl

Conversion factors:
1 liter contains 1000 cm^3
0.15M HCl contains 0.15 mole HCl in 1 liter solution

Equation:

$$? \text{ mole HCl} = 200 \text{ cm}^3 \ 0.15M \text{ HCl} \times \frac{1 \text{ L}}{1000 \text{ cm}^3} \times \frac{0.15 \text{ mole HCl}}{1 \text{ L } 0.15M \text{ HCl}}$$

Answer:
0.030 mole HCl

Example 7.4

What weight of sodium hydroxide (NaOH) is needed to prepare 100 cm^3 of 0.125M sodium hydroxide?

Solution

Wanted:
? g NaOH

Given:
100 cm^3 0.125M NaOH

Conversion factors:
1 liter contains 1000 cm^3
1 liter 0.125M NaOH contains 0.125 mole NaOH
1 mole NaOH weighs 40.0g (23.0 + 16.0 + 1.0)

Equation:

$$? \text{ g NaOH} = 100 \text{ cm}^3 \ 0.125M \text{ NaOH} \times \frac{1 \text{ liter}}{1000 \text{ cm}^3}$$

$$\times \ \frac{0.125 \text{ mole NaOH}}{1 \text{ liter solution}} \times \frac{40.0 \text{ g NaOH}}{1 \text{ mole NaOH}}$$

Answer:
0.500 g NaOH

Example 7.5

What volume of 3.25M sulfuric acid is needed to prepare 0.500 L of 0.130M H$_2$SO$_4$?

Solution

Wanted:
? L of 3.25M H$_2$SO$_4$

Given:
0.500 L 0.130M H$_2$SO$_4$

Conversion factors:
one liter 3.25M H$_2$SO$_4$ contains 3.25 moles H$_2$SO$_4$
one liter 0.130M H$_2$SO$_4$ contains 0.130 mole H$_2$SO$_4$

    The moles of sulfuric acid in 0.500 L of 0.130M H$_2$SO$_4$ must equal the moles of sulfuric acid in the volume of 3.25M sulfuric acid that is to be diluted.

Equation:
$$? \text{ L } 3.25M \text{ H}_2\text{SO}_4 = 0.500 \text{ L } 0.130M \text{ H}_2\text{SO}_4$$

$$\times \ \frac{0.130 \text{ mole H}_2\text{SO}_4}{1 \text{ L } 0.130M \text{ H}_2\text{SO}_4} \times \frac{1.00 \text{ L } 3.25M \text{ H}_2\text{SO}_4}{3.25 \text{ moles H}_2\text{SO}_4}$$

Answer:
0.0200 L 3.25M H$_2$SO$_4$

To prepare the solution, combine 20.0 cm^3 (0.0200 L) of 3.25M H$_2$SO$_4$ with 480 cm^3 water, giving a total volume of 500 cm^3 (0.500 L).

Problem 7.3   How many moles of glucose are in 450 cm^3 of 0.125M glucose?

Problem 7.4   What weight of sodium chloride is needed to prepare 1.50 L of 0.125M NaCl?

Problem 7.5   What volume of 6.0M HCl is needed to prepare 275 cm^3 of 0.255M HCl?

E. As Osmolality

Osmolality measures the moles of particles per kilogram of solvent. The most basic postulate of osmolality is that the identity of the particles is not significant. For a 2.0 osmolal solution, 1 kg solvent may contain 2 moles of glucose or 1 mole of sodium chloride (which would yield 2 moles of ions), or 1.0 mole of glucose and 0.5 mole sodium chloride (which would also yield a total of 2.0 moles of particles).

The osmolality of body fluids is closely related to health. For a discussion of the clinical importance of osmolality see the mini-essay immediately following this chapter.

Table 7.5 sums up the various methods for stating concentration. In the table, note that the value of the millimole/milliliter term is the same as that of the mole/liter term.

Table 7.5 Common units of concentration.

	Solute	Solvent	Solution	Comments
percent by weight	? g	+ ? g	$\longrightarrow$ 100 g	accurate, independent of temperature
percent by volume	? cm^3	+ ? cm^3	$\longrightarrow$ 100 cm^3	used when solute is liquid, concentration varies slightly with temperature
percent, weight/volume	? g	+ 100 cm^3 $\longrightarrow$	—	used in technical labs
molarity (M)	moles	—	1 liter	used in chemical calculations
millimole/liter	10^{-3} mole	—	1 liter ⎱	used in medical laboratories to express concen-
millimole/milliliter	10^{-3} mole	—	10^{-3} liter ⎰	trations of biological fluids
parts per million (ppm)	mg	—	kg ⎱	used in environmental studies
parts per billion (ppb)	μg	—	kg ⎰	
osmolality	mole of particles	kg	—	used for biological fluids and solutions of mixed solutes

7.5 Calculations Involving Concentrations

A. Stoichiometry

The stoichiometry calculations done in Chapter 5 involved pure substances. We determined the amount of a solid or liquid reactant by weighing the sample. In Chapter 6, we solved some problems where the reactant was a gas. In these cases, we measured its pressure, volume, and temperature and, by using the ideal gas equation and the concept of molar volume, we determined its mass. Now, by using the concentration of a solution as a conversion factor, we can extend stoichiometric calculations to include solutions.

Example 7.6

What weight of barium sulfate is precipitated by the addition of an excess of sulfuric acid to 55.6 cm^3 of 0.54M barium chloride? The balanced equation for this reaction is:

$$BaCl_2 + H_2SO_4 \longrightarrow BaSO_4 + 2\ HCl$$

Solution

Wanted:
? g BaSO$_4$

Given:
55.6 cm^3 of 0.54M barium chloride

Conversion factors:
0.54 mole BaCl$_2$ in one liter of 0.54M BaCl$_2$
1000 cm^3 = 1 liter
1 mole BaCl$_2$ forms 1 mole BaSO$_4$
1 mole BaSO$_4$ weighs 233.4 g

Equation:

$$? \text{ g BaSO}_4 = 55.6 \text{ cm}^3 \times \frac{1 \text{ liter}}{1000 \text{ cm}^3} \times \frac{0.54 \text{ mole BaCl}_2}{1 \text{ liter}}$$

$$\times \frac{1 \text{ mole BaSO}_4}{1 \text{ mole BaCl}_2} \times \frac{233.4 \text{ g BaSO}_4}{1 \text{ mole BaSO}_4}$$

Answer:
7.0 g BaSO$_4$

Example 7.7

What is the concentration of an aqueous solution of hydrochloric acid if 25 cm^3 of the acid reacts with an excess of solid calcium carbonate to yield 0.307 L carbon dioxide, measured at 0.95 atm and 25°C?

Solution

Analysis of the problem:

1. The product of the reaction is a gas whose volume was not measured at standard conditions of temperature and pressure. We can calculate the number of moles of gas produced by using the ideal gas equation: $PV = nRT$.

2. Using the number of moles of carbon dioxide produced, we can calculate the number of moles of hydrochloric acid used.

3. The moles of HCl are in 25 cm³ of solution. Molarity is a ratio between moles of solute and volume of solution. By dividing the number of moles of HCl by the volume (L) of solution in which it was dissolved, we will obtain the molarity of the acid solution.

Calculations:

1. Using the ideal gas equation, calculate the moles of CO_2 produced:

$$P \quad \times \quad V \quad = n \times \quad R \quad \times \quad T$$

$$0.95 \text{ atm} \times 0.307 \text{ L} = n \times 0.0821 \frac{\text{L atm}}{\text{mole K}} \times (25 + 273)$$

$$n = 0.012 \text{ mole } CO_2$$

2. Carry out the stoichiometric calculation:

Equation:
$$2 \text{ HCl} + CaCO_3 \longrightarrow CO_2\uparrow + CaCl_2 + H_2O$$

Wanted:

the molarity of the acid solution   or $\dfrac{\text{moles HCl}}{\text{liter solution}}$

Given:
0.012 mole carbon dioxide

Conversion factors:
2 moles HCl yields 1 mole CO_2
1 liter contains 1000 cm³

Arithmetic equation:

$$?M \text{ HCl} = \frac{? \text{ moles HCl}}{\text{liter solution}} = 0.012 \text{ mole } CO_2 \times \frac{2 \text{ moles HCl}}{1 \text{ mole } CO_2}$$

$$\times \frac{1}{25 \text{ cm}^3} \times \frac{1000 \text{ cm}^3}{1 \text{ L}}$$

Answer:
M HCl $= 0.96M$ HCl

Problem 7.6

Calculate the concentration of iodide ion in a solution of sodium iodide if 24.2 cm^3 of the solution reacts completely with 16.7 cm^3 of 0.176M silver nitrate solution. The balanced equation for the reaction is:

$$NaI + AgNO_3 \longrightarrow AgI + NaNO_3$$

Problem 7.7

What volume of carbon dioxide measured at 27°C and 0.93 atm is formed by the reaction of 22.5 cm^3 of 0.105M HCl with an excess of solid magnesium carbonate? The balanced equation for the reaction is:

$$MgCO_3 + 2\,HCl \longrightarrow MgCl_2 + CO_2 + H_2O$$

B. Titration

Laboratories, whether medical or industrial, are frequently asked to determine the exact concentration of a particular substance in a solution. For example, what is the concentration of acetic acid in a sample of vinegar? What are the concentrations of iron, calcium, and magnesium ions in a hard water sample? Such determinations may be made using a technique known as titration.

In **titration** a known volume of a solution of unknown concentration is reacted with, or titrated by, a known volume of a solution of known concentration. By knowing the titration volumes and the mole ratio in which the two solutes react, the concentration of the second solution can be calculated. The method used is similar to that used in Examples 7.6 and 7.7. The solution of unknown concentration may contain an acid (such as stomach acid), a base (such as ammonia), an ion (such as iodide ion), or any other substance whose concentration must be determined.

There are several requirements for analytical titrations:

1. The equation for the reaction must be known, so that a stoichiometric ratio can be obtained for use in calculations.

2. The reaction must be fast and complete.

3. When the reactants have combined in the stoichiometric ratio, there must be a clear-cut change in some measurable property of the reaction mixture. The occurrence of this change is called the **endpoint** of the reaction.

4. There must be a way of measuring exactly the amount of each reactant, whether that reactant is initially in solution or is a solid to be dissolved.

Let us discuss these requirements as they apply to a particular titration, that of a solution of sulfuric acid of known concentration with a sodium hydroxide solution of unknown concentration. The balanced equation for this acid-base reaction is:

$$2\ NaOH + H_2SO_4 \longrightarrow Na_2SO_4 + 2\ H_2O$$

Sodium hydroxide ionizes in water to form sodium ions and hydroxide ions; sulfuric acid ionizes to form hydrogen ions and sulfate ions. The reaction between hydroxide ions and hydrogen ions is rapid and complete; thus, the second requirement for an analytical titration is met.

What clear-cut change in property will occur when the reaction is complete? Suppose the sodium hydroxide solution is slowly added to the acid solution. As each hydroxide ion is added, it reacts with a hydrogen ion to form a water molecule. As long as there are unreacted hydrogen ions in solution, the solution is acidic. When the number of hydroxide ions added exactly equals the original number of hydrogen ions, the solution becomes neutral. As soon as any extra hydroxide ions are added, the solution becomes basic. How will the experimenter know when the solution becomes basic? In Section 4.8 you learned about those organic compounds called *indicators* which have one color in acidic solutions and another in basic solutions. If such an indicator is present in an acid-base titration, it changes color when the solution changes from acidic to basic. Phenolphthalein is an acid-base indicator that is colorless in acid solutions and pink in basic solutions. If phenolphthalein is added to the original sample of sulfuric acid, the solution will be colorless and remain so, as long as there is an excess of hydrogen ions. After enough sodium hydroxide solution has been added to react with all of the hydrogen ions, the next drop of base will provide a slight excess of hydroxide ion and the solution will turn pink. Thus, we will have a visible and clear-cut indication of the occurrence of the endpoint. Table 7.6 lists several indicators that could be used in an acid-base titration.

The requirement that the volumes of solutions used must be measured accurately is met by the use of volumetric glassware, in particular, burets, to measure the volumes of the solutions. Remember from Section 1.3C that the

Table 7.6 Common acid-base indicators.

Indicator	Color in Acid	Color in Base
phenolphthalein	colorless	pink
methyl orange	red	yellow
bromothymol blue	yellow	blue

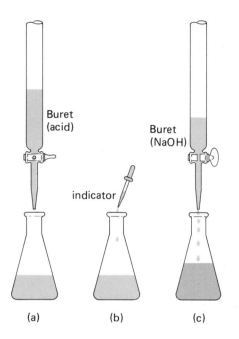

Buret
(acid)

Buret
(NaOH)

indicator

(a)   (b)   (c)

Figure 7.11 A typical acid-base titration. (a) An exact volume of an acidic solution of known concentration is measured into an Erlenmeyer flask. (b) A few drops of indicator, such as phenolphthalein, are added. (c) The basic solution of unknown concentration is added until a faint pink color becomes visible, which means that the endpoint has been reached. The volume of basic solution added is then calculated from the readings on the buret.

Table 7.7 Data from the titration of 0.108M sulfuric acid with a solution of sodium hydroxide of unknown concentration.

	Trial I	Trial II	Trial III
Data:			
Volume of 0.108M H_2SO_4	25.0 cm^3	25.0 cm^3	25.0 cm^3
NaOH solution:			
Buret readings: Finish	34.12 cm^3	39.61 cm^3	35.84 cm^3
Start	0.64 cm^3	6.15 cm^3	2.34 cm^3
Volume of NaOH to reach the endpoint:	33.48 cm^3	33.46 cm^3	33.50 cm^3
Average volume used: 33.48 cm^3			

Equation:

$$2\,NaOH + H_2SO_4 \longrightarrow Na_2SO_4 + 2\,H_2O$$

Calculation:

$$?M\ NaOH = 2.50 \times 10^{-2}\,L\ H_2SO_4 \times \frac{0.108\ \text{moles}\ H_2SO_4}{1\ L\ H_2SO_4} \times \frac{2\ \text{moles}\ NaOH}{1\ \text{mole}\ H_2SO_4}$$

$$\times \frac{1}{3.348 \times 10^{-2}\,L\ NaOH}$$

Answer:

0.161M NaOH

precision of a buret is one part per thousand. Figure 7.11 shows a typical titration set-up.

Data and calculations for a typical acid-base titration are shown in Table 7.7. Notice that three trials were run; this is a standard procedure to check the precision of the titration.

Example 7.8

What volume of 0.154M sodium hydroxide will completely react with 25.0 cm^3 of 0.0952M hydrochloric acid? The balanced equation for the reaction is:

$$NaOH + HCl \longrightarrow NaCl + H_2O$$

Solution

Wanted:
? cm^3 of 0.154M NaOH

Given:
25.0 cm^3 of 0.0952M HCl

Conversion factors:
1 L (1000 cm^3) of 0.0952M HCl contains 0.0952 mole of hydrochloric acid
1 mole of sodium hydroxide reacts with 1 mole hydrochloric acid
1 L (1000 cm^3) of 0.154M NaOH contains 0.154 mole NaOH

Calculations:

$$? \text{ cm}^3 \ 0.154M \text{ NaOH} = 25.0 \text{ cm}^3 \ 0.0952M \text{ HCl} \times \frac{0.0952 \text{ mole HCl}}{1000 \text{ cm}^3}$$

$$\times \frac{1 \text{ mole NaOH}}{1 \text{ mole HCl}} \times \frac{1000 \text{ cm}^3 \ 0.154M \text{ NaOH}}{0.154 \text{ mole NaOH}}$$

Answer:
15.5 cm^3 0.154M NaOH

Problem 7.8

Calculate the molarity of a solution of hydrochloric acid if 15.0 cm^3 of this solution reacts completely with 26.2 cm^3 of 0.126M potassium hydroxide.

Titration reactions are not always acid-base reactions. They may be oxidation-reduction reactions, precipitation reactions, or combination reactions. The endpoint may be determined by a change in color of an added indicator or by a change in color of the solution when the reaction is complete. In special titrations, the endpoint may be marked by a change in electrical conductivity of the reaction mixture, by a change in turbidity, or by a variety of other means.

7.6 Colloids

A. Properties of Colloids

Earlier in this chapter, we described a colloid as a uniform suspension of very small particles of one substance in another. Milk and mayonnaise are colloids. The suspended substance is insoluble in the second substance, as the butterfat in milk and the oil in mayonnaise are both insoluble in water. If the particles were smaller, they would dissolve to form a solution. If they were larger, they would separate out. The properties of a colloid are dependent on the small diameter (roughly 10^{-7}–10^{-4} cm) of each particle and the enormous surface area represented by a large collection of small particles. To see how the total surface area depends on the particle size, picture a cube, 1 cm on each side. Its surface area is 6 cm^2; its ratio of surface area to volume is 6 cm^2 : 1 cm^3. If the same cube is divided into 8 equal cubes, 0.5 cm on each side, the total surface area increases to 12 cm^2, but the volume remains the same. The ratio of surface area to volume is now 12 cm^2 : 1 cm^3. Further dividing these cubes increases the ratio. If the original cube of volume 1 cm^3 is divided into 10^{21} cubes, each 10^{-7} cm on a side (the approximate size of a colloidal particle), the total surface area becomes 6×10^3 cm^2, or 1.5 acre. Although the colloidal particles are spherical rather than cubic, the ratio of surface area to volume is still enormous (approximately 6000 : 1).

Figure 7.12 A schematic diagram of an electrostatic precipitator. The electrical charges on the wire and wall of the precipitator neutralize the charges on the particles in the smoke, causing them to settle out.

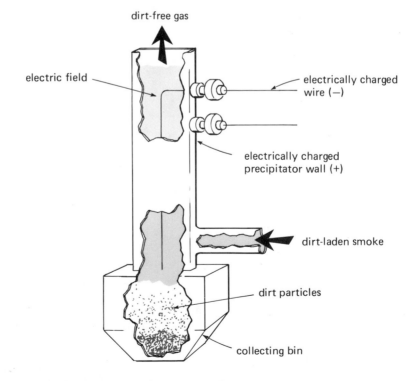

dirt-free gas

electric field

electrically charged wire (−)

electrically charged precipitator wall (+)

dirt-laden smoke

dirt particles

collecting bin

Colloidal particles are kept in constant motion by the buffeting and bumping of other molecules present. As the colloidal particles move around, they rub against one another and develop static electricity on their surface, much as you acquire static electricity in winter by scuffing across a carpet. The great stability of colloids, the fact that they do not easily separate into two separate phases, is due to these charges. Neutralizing these charges by the addition of a solution of ions can destroy the colloid. The Mississippi River picks up a great deal of finely divided soil and mud as it flows south through the prairie. By the time the river reaches New Orleans, this mud is colloidally dispersed in the river. When the river empties into the ocean, this colloid is dumped into the solution of sodium and chloride ions we call salt water. The electric charges on the surface of the tiny mud particles that kept them suspended are neutralized. The colloid separates and the mud settles out, forming the river delta.

The same phenomenon occurs when smoke (a colloidal suspension of solids in air) is released through a chimney equipped with a precipitator, shown in Figure 7.12. The precipitator consists of metal plates or wires carrying an electric charge. As the smoke passes between these plates, the static electric charges on the colloidal smoke particles are neutralized, the dirt falls to the bottom of the chimney, and no smoke is released to pollute the atmosphere. Electrostatic filters used in homes by people with severe pollen allergies work on the same principle.

B. Purification

A colloid is not always a single pure substance suspended in another pure substance. The dispersing medium may also contain dissolved ions or molecules. These ions and small molecules can be removed by a process called **dialysis.** Figure 7.13 diagrams the essential features of dialysis. The colloid is put in a thin plastic bag and the bag is suspended in a large water bath in which the

Figure 7.13 Dialysis. The colloid is placed in a bag made of semipermeable membrane and lowered into a water tank. Through osmosis, the ions and small molecules migrate out into the water tank and are removed. The colloidal particles are too large to pass through the membrane. Eventually, the colloid is purified.

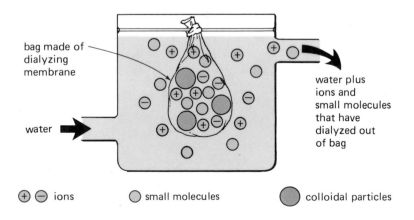

bag made of dialyzing membrane

water

water plus ions and small molecules that have dialyzed out of bag

⊕ ⊖ ions   small molecules   colloidal particles

water is being constantly changed. The plastic of the bag resembles the membrane shown in Figure 7.8, but is slightly more permeable. Small ions and molecules, as well as solvent molecules, can migrate through the bag walls, but not the larger colloidal particles. While the bag is suspended in the water bath, this migration takes place. The solvent molecules move in both directions, and the small ions and molecules move mostly outwards from the bag into the bath in a futile effort to attain equal concentration on both sides of the plastic wall. Needless to say, equilibrium is never reached because the solvent in the bath is constantly being changed. Eventually, the solution in the bag around the colloid particles becomes infinitely dilute and the colloid is said to be purified.

The kidney machine is an application of dialysis to human needs. The machine is used to purify the blood of people whose kidneys have been removed or are not functioning properly. The blood is diverted through a long tube made of dialyzing membrane. The tube is immersed in a container filled with a solution isotonic with blood in essential ions (sodium, potassium, and so on). This solution is continually changed. As the blood passes through the tube, unwanted electrolytes and small molecules migrate out through the membrane. The blood is returned to the body in a purified condition. The process is the same as that performed much more efficiently by healthy kidneys. Figure 7.14 shows a schematic diagram of a kidney machine.

Figure 7.14 Dialysis occurring in a kidney machine.

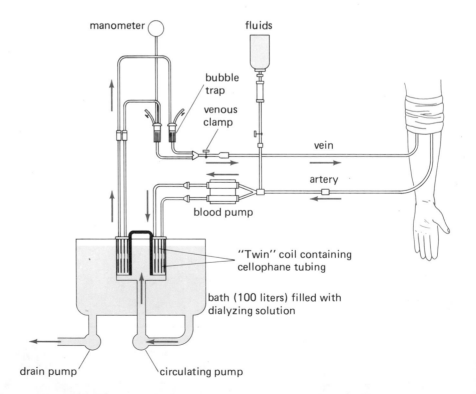

Key Terms and Concepts

Brownian movement (7.1B)

colligative properties (7.3)

colloid (7.1)

concentration (7.4)

crenation (7.3D)

dialysis (7.6B)

dispersed phase (7.1A)

dispersing medium (7.1A)

endpoint (7.5B)

equilibrium (7.2B)

filtration (7.1B)

hemolysis (7.3D)

hyperbaric (7.2C)

immiscible (7.2)

indicator (7.5B)

isotonic (7.3D)

miscible (7.2)

molarity (7.4D)

nonvolatile solute (7.3)

osmolality (7.4E)

osmosis (7.3D)

osmotic pressure (7.3D)

parts per billion (7.4C)

parts per million (7.4C)

polar solvents (7.2C1)

saturated solution (7.2B)

semi-permeable membrane (7.3D)

solubility (7.2)

solute (7.1A)

solution (7.1)

solvent (7.1A)

surface area of colloids (7.6A)

suspension (7.1)

titration (7.5B)

Tyndall effect (7.1B)

unsaturated solution (7.2B)

vapor pressure lowering (7.3A)

Problems

Colloids, Solutions, and Suspensions (Section 7.1)

7.9   How does a colloid differ from a solution?

7.10 Give several examples of colloids. Tell what two phases are present in each.

Solubility (Section 7.2)

7.11 Which of the following compounds would you expect to be quite soluble in water?
 a. potassium chloride (KCl)
 b. chloroform ($CHCl_3$)
 c. benzene (C_6H_6)
 d. hydrogen iodide (HI)
 e. sodium acetate (CH_3CO_2Na)
 f. formaldehyde (HCHO)

7.12 For each of the following, predict its relative solubility in water and in gasoline, a nonpolar solvent. Remember the solubility rules in Chapter 4.
 a. barium sulfate
 b. lithium nitrate
 c. octane (C_8H_{18})
 d. bromine
 e. carbon tetrafluoride (CF_4)

Colligative Properties (Section 7.3)

7.13 What is a colligative property? Explain why the vapor pressure of a pure liquid is greater than that of a solution in which that liquid is the solvent.

7.14 Which of the following would you expect to have the highest boiling point: $0.1M$ glucose, $1.0M$ glucose, or $10.0M$ glucose? Why?

Osmosis (Section 7.3D)

7.15 What is osmosis? Reverse osmosis is used to concentrate waste solutions and obtain fresh water from the ocean. Suggest how it might work.

Concentrations of Solutions (Section 7.4)

7.16 Carry out the following conversions:
   a. 0.90% by weight sodium chloride to molarity
   b. 5.0 g $NaHCO_3$ in 1 liter solution to molarity
   c. 12M HCl to g HCl/liter solution
   d. 0.15M sodium hydroxide to g NaOH/100 cm^3 solution
   e. 1.33 g silver nitrate/100 cm^3 solution to molarity

7.17 Describe how to prepare:
   a. 5.0 liter 0.15M sulfuric acid from 18M H_2SO_4
   b. 400 cm^3 of 0.10M KOH from solid potassium hydroxide
   c. 100 cm^3 of 0.250M HCl from 6M HCl
   d. 500 cm^3 of 50% alcohol from 95% alcohol (% by volume)
   e. 450 cm^3 of 3% (wt/volume) glucose

7.18 What is the weight of the solute in:
   a. 1.5 liter 0.10M $AgNO_3$?
   b. 0.500 liter 3.0M H_2SO_4?
   c. 25.0 cm^3 0.155M NaOH?

7.19 What volume of each of the following solutions contains 0.10 mole solute?
   a. 0.15M barium chloride   b. 0.25M copper(II) sulfate
   c. 5.0M ammonium nitrate   d. 15M nitric acid
   e. 0.30M iron(II) chloride   f. 0.55M zinc(II) nitrate

7.20 What are the differences among a saturated solution, a 1 molar solution, and a 1 osmolal solution?

7.21 The solubility of sodium bicarbonate is 6.9 g/100 g water at 25°C. What weight of sodium bicarbonate will dissolve in 250 g water at that temperature? What is the molarity of this solution?

7.22 A solution contains 2.6 g glucose in 150 cm^3 solution. Calculate the percent concentration (weight/volume) and the molarity of this solution.

7.23 Copper sulfate is obtained as a pentahydrate, $CuSO_4 \cdot 5\,H_2O$. This means that in the solid form, each unit of $CuSO_4$ is associated with five molecules of H_2O. Starting with copper sulfate pentahydrate, describe how you would prepare 500 mL of a solution that is 0.25M in copper ion.

7.24 An aqueous solution contains one part per million by weight of fluoride ion (μg/g). How would you prepare 10 liters of this solution using sodium fluoride? What is the molarity of this solution?

7.25 A solution of vitamin C, ascorbic acid (MW 176.1), contains 1.0 g per 200 mL. What is the molarity of this solution?

7.26 Beer is 12% ethyl alcohol by volume.
   a. Calculate the volume of alcohol in one liter of beer.
   b. Calculate the mass of alcohol in one liter of beer (density of alcohol = 0.789 g/cm^3).
   c. Calculate the molarity of ethyl alcohol in beer (the molecular formula of ethyl alcohol is C_2H_6O).

7.27 Phenol (C_6H_6O) is a mild antiseptic used in several nonprescription mouthwashes. In two of these, the concentration of phenol is 1.4% weight/volume. Calculate the molarity of phenol in these solutions.

7.28 A commercial liquid noncaloric sweetener contains 1.62% (weight/volume) of the calcium salt of saccharin. The molecular formula of this calcium salt is $Ca(C_7H_4NSO_3)_2$. Calculate the molarity of this solution.

**Stoichiometry
(Section 7.5A)**

7.29 What volume of hydrogen at STP is formed by the reaction of 15.0 cm^3 of 0.635M HCl with 1.0 g zinc?

7.30 Calculate the concentration of a hydrochloric acid solution if 25.0 cm^3 of it react exactly with 33.5 cm^3 of 0.1035M silver nitrate.

7.31 What volume of carbon dioxide measured at STP is formed by the reaction of 5.0 g sodium bicarbonate with 0.100 L of 0.156M hydrochloric acid?

7.32 Calculate the concentration of a solution of sulfuric acid if 15.0 cm^3 of this solution reacts completely with 26.2 cm^3 of 0.125M potassium hydroxide solution.

7.33 Calculate the molarity of a sodium hydroxide solution if 23.90 cm^3 of this solution reacts completely with 25.0 cm^3 0.215M hydrochloric acid.

7.34 What are the requirements of a titration?

**Dialysis (Section
7.6B)**

7.35 What is dialysis? How does it differ from osmosis?

The Osmolality of Body Fluids

In a healthy person, the kidneys have several functions: to purge body fluids of toxic substances; to keep the body's volume of extracellular fluid constant; and to keep the body's concentration of solutes, such as sodium and bicarbonate ions, creatinine, and glucose, within a healthy range. Disorders of kidney function are among the most common encountered in clinical medicine. Evaluation of kidney function in, for example, post-surgical patients, newborns, and patients in a comatose state must be accomplished rapidly on small samples, with minimal trauma for the patients. This can be accomplished with urinalysis or by analysis of extracellular fluid.

Traditionally, the initial evaluation of kidney function has been based on the specific gravity of urine. However, specific gravity depends on the total mass of the material dissolved in a unit volume. Specific gravity does not really tell the concentration of the urine, due to the large differences in mass among typical solutes. The mass of a mole of sodium ions (23 g) is very different from that of a mole of glucose molecules (180 g) or albumin (68,500 g). Because of these problems, tests on body fluids measuring the fluids' osmolality, not their specific gravity, are often used.

Osmolality measures the total number of particles in a kilogram of solvent. Osmolality depends only on the number of particles present, not their mass. Osmolality (O) is expressed as osmoles per kilogram of solvent (Os/kg) or as milliosmoles per kilogram of solvent (mOs/kg). The difference between the freezing point of a solution and that of the pure solvent in the solution is used to determine osmolality. Freezing-point lowering is one of the colligative properties of solutions. In aqueous solution, one mole of particles, regardless of their masses or charges, lowers the freezing point from $0.00°C$ to $-1.86°C$. Apparatus has been developed, called an **osmometer,** which can measure in a few seconds and to $0.001°C$ the freezing point of a 0.02 mL sample. An osmometer is calibrated using sodium chloride solutions of known osmolality (moles ions/kg water) and reads directly in mOs/kg water.

In a healthy person, the osmolality of serum (ECF) is $280-300$ mOs/kg fluid water. Of this, about 140 mOs/kg is contributed by sodium ions, about 100 mOs/kg by chloride ions, and the rest by other ions and by molecules such as glucose and urea. Solutions that are isotonic with body fluids, such as 5% dextrose or normal saline, are prepared to have the same osmolality as body fluids. One speaks of the **tonicity** of fluids: a hypertonic solution has a higher osmolality than normal body fluids, a hypotonic solution has lower than normal osmolality. Normal tonicity is maintained by close regulation of renal excretion of water which, in turn, is dependent on circulating levels of posterior pituitary hormone vasopressin (antidiuretic hormone). A $1-2\%$ change in plasma osmolality will suppress or stimulate thirst and vasopressin secretion. This control permits enormous variation in fluid intake ($0.5-25$ L/day) without alteration in plasma. If there are disturbances in kidney function or vasopressin secretion, the urinary diluting and concentrating mechanisms fail and abnormalities of body tonicity appear.

Serum osmolality alone can indicate problems. Hyperosmolality (>295 mOs/kg) is most often due to alcohol ingestion. Hyperosmolality may also be due to diabetes insipidus and other causes of increased sodium ion in the serum. The serum of mildly comatose patients has an osmolality greater than 365 mOs/kg. Plasma hypoosmolality indicates wholly different conditions such as stroke, concussion, and adrenal disease.

The blood levels of sodium, glucose, and blood urea nitrogen (BUN) can be used to predict serum osmolality, using the equation

$$O_{calc'd} = Na(186) + \frac{BUN}{2.8} + \frac{glucose}{18}$$

In healthy persons, the measured osmolality is approximately the same as the calculated value. If the measured osmolality is considerably higher than that calculated, severe liver failure, poisoning, or severe dehydration should be suspected. Marked elevations have also been observed in patients in traumatic shock. If the difference decreases, prognosis for recovery is good; if it continues, prognosis is ominous. The elevated osmolality in such cases is caused by the continued presence in the plasma of abnormal metabolites that are not detectable by routine laboratory tests. Failure to remove these metabolites has proven fatal to human beings and animals.

Urine osmolality by itself is not useful for diagnosis unless water intake has been controlled, for the concentration of the solute depends on the total volume of solution. However, urine osmolality combined with water restriction is a remarkably sensitive test of renal function. After a 14-hour fluid fast, normal urine osmolality should be greater than 800 mOs/kg. A range of 600–800 mOs/kg implies minimal kidney impairment, 400–600 mOs/kg implies moderate impairment, and less than 400, severe impairment. This impairment may exist even though other tests are normal. The ratio of urine osmolality to serum osmolality also gives useful information. When functioning normally, kidneys concentrate the solutes. In such cases, the urine osmolality should be greater than serum osmolality, usually by a factor of three or more. In acute renal failure, the urine osmolality approaches that of the serum. In untreated diabetes insipidus, urine osmolality is less than serum osmolality.

This mini-essay has touched only briefly on the usefulness of osmolality data in clinical evaluation of patients. It is remarkable that so much can be learned from a few freezing points measured quickly in the lab or at the patient's bedside. Osmolality is indeed a powerful tool.

References

Andreoli, T.E.; Grantham, J.J.; and Rector, F.C., Jr. 1977. *Disturbances in body fluid osmolality.* Bethesda, MD: American Physiological Society.

Duarte, C.G. 1980. *Renal function tests.* Boston: Little, Brown.

Wallach, J. 1978. *Interpretation of diagnostic tests.* 3rd ed. Boston: Little, Brown.

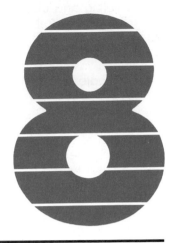

Equilibrium

Thus far we have written equations for reactions, calculated the amounts of reactants needed to form a given amount of product, and measured the enthalpy change. We have observed that some reactions are exothermic, releasing heat energy as they occur, and others are endothermic, requiring the input of heat energy. We have assumed that a reaction occurs as soon as the reactants are mixed and is rapidly completed.

Such is not always the case. Some reactions do begin immediately and are completed rapidly. Reactions between ions, such as neutralization reactions or precipitation reactions, do take place as soon as the reactants are combined. Other reactions take place slowly and require added energy to get started. We know that gasoline burns. But we also know that gasoline may be stored for years, in contact with oxygen in the air, without causing a fire. Combustion does not take place until energy, such as a spark, is added. Some reactions occur easily in one direction at some temperatures but equally easily in the opposite direction at other temperatures. Ice begins melting immediately at room temperature, but water begins freezing immediately at temperatures below 0°C. Obviously, we need to extend our study of reactions to explain these various phenomena. In this chapter we will consider the criteria of reactions: why some reactions proceed rapidly (or slowly); what makes a reaction *spontaneous*; and the special characteristics of reactions involving partially ionized substances.

8.1 Requirements for a Reaction

A. At the Molecular Level

When molecules react, bonds between the atoms in the reacting molecules break and new bonds form to combine the atoms in a different way. For this to happen, the reacting molecules must collide. This means that together they must have enough kinetic energy (energy of motion) to overcome the repulsion between the clouds of electrons that surround the molecules. As they collide, the two reacting molecules must be oriented so that those atoms that will be bonded together in the product are next to each other. Without this orientation, the molecules will retreat from the collision without reacting (Figure 8.1).

B. Energy Requirements

A reaction is **spontaneous** if it occurs with the release of free energy. **Free energy (G)** is energy that is available to do work. The **free energy change (ΔG)** associated with a reaction depends on: the enthalpy change (ΔH) of the reaction (see Section 5.8); the temperature (K) at which the reaction occurs; and the

Figure 8.1 Ineffective and effective molecular orientations at collision.

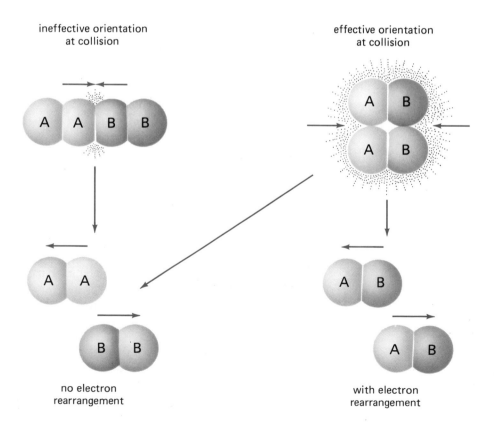

ineffective orientation at collision

effective orientation at collision

no electron rearrangement

with electron rearrangement

Figure 8.2 Order (a) and disorder (b): the concept of entropy.

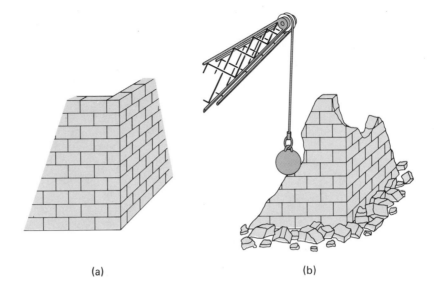

(a)   (b)

change in the amount of disorder caused by the reaction.

This last factor, involving order and disorder, is called **entropy (S)** and is, to a degree, a measure of how the event affects the orderliness of the universe (Figure 8.2). Other factors being equal, it is the nature of matter to move toward a state of maximum disorder. For example, when a tower of blocks is bumped, it falls into a disordered pile. In this process, the degree of disorder (entropy) of the blocks is increased. As another example, suppose that you have three bags of jellybeans, one of red, one of white, and one of blue, and pour them onto a tray. Before you pour them they are collected in separate bags and have some degree of order. After you pour them, they are all mixed together in one disordered pile. Mixing increases the disorder (entropy) of the jellybeans. Recall from our study of solids, liquids, and gases in Chapter 6 that the structure of a solid is usually well-ordered, the structure of a liquid is less ordered, and that of a gas is quite random. From these facts, we can deduce that the change from a solid to a liquid is accompanied by an increase in entropy (Figure 8.3a); so is a change from a liquid to a gas. When a solid dissolves in a liquid, there may be an increase in entropy (Figure 8.3b). When molecules containing many atoms break apart to form smaller molecules, there is also an increase in entropy.

An increase in entropy alone is not enough to guarantee that a chemical reaction or other event in nature will take place. The temperature and enthalpy change must also be considered. A spontaneous reaction is one that occurs with a decrease in free energy, that is, it is one for which the sign of the free energy change is negative. The equation for calculating the free energy change is:

$$\Delta G = \Delta H - T\Delta S$$

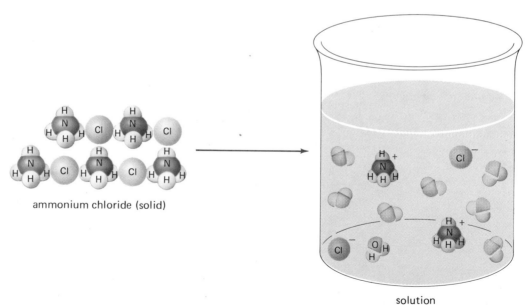

crystalline solid
(perfect order)

liquid
(less order)

gas
(considerable disorder)

(a)

ammonium chloride (solid)

solution

(b)

Figure 8.3 Order and disorder among molecules. (a) Increasing disorder as a solid melts and vaporizes. (b) Increasing disorder as ammonium chloride dissolves in water.

Table 8.1 Enthalpy, entropy and free energy.

If a reaction is:

exothermic $\Delta H < 0$	and	increases the disorder, $\Delta S > 0$	it is always spontaneous $\Delta G < 0$
exothermic $\Delta H < 0$	and	decreases the disorder, $\Delta S < 0$	spontaneity depends on the temperature
endothermic $\Delta H > 0$	and	increases the disorder, $\Delta S > 0$	spontaneity depends on the temperature
endothermic $\Delta H > 0$	and	decreases the disorder, $\Delta S < 0$	it is never spontaneous $\Delta G > 0$

This equation shows: (a) If a reaction is exothermic (ΔH is negative) and the entropy change is positive (more disorder), the free energy change is always negative and the reaction is always spontaneous. (b) If a reaction is endothermic (ΔH is positive) and the entropy change is negative (less disorder), the free energy change is always positive and the reaction is never spontaneous. (c) If both the enthalpy change and the entropy change are positive, the spontaneity of the reaction depends on the temperature. An example of the third class of reactions is the spontaneous melting of ice at room temperature (293 K). The change is endothermic but there is an increase in entropy as the water molecules lose the ordered arrangement of the ice crystals. When ΔG is calculated for melting ice at 293 K, the $T\Delta S$ factor is numerically larger than ΔH. ΔG is then negative, so the melting is spontaneous at that temperature.

Similarly, you can show that the spontaneity of a reaction that is exothermic and accompanied by a decrease in entropy also depends on temperature. The spontaneous freezing of water at temperatures below 273 K is an example.

A summary of the relationships among enthalpy, entropy, and free energy is shown in Table 8.1.

To conclude: A spontaneous event is accompanied by a release of free energy and is **exergonic.** A nonspontaneous event requires the addition of free energy in order to occur; such an event is **endergonic.**

C. Energy Changes During a Reaction

Figure 8.4 is a plot of the energy changes that take place during a reaction. The initial average energy of the reactants is indicated at the left side of each graph. Individual molecules must have much more energy than the average if they are going to collide, enough to overcome the repulsive forces between them. This added amount of energy, the **activation energy,** is the difference between the inital energy and that at the peak of each graph. Molecules having that energy can collide, and, if they are correctly oriented at collision, their bonds may break and the new bonds of the products will form. As the new

Figure 8.4 Energy changes during a reaction. (a) An endothermic reaction; (b) an exothermic reaction.

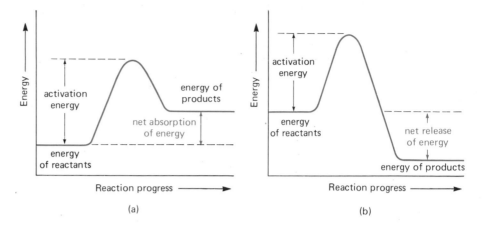

(a)

(b)

bonds form, energy is released. If the energy released is less than the activation energy, so there is a net *absorption* of energy, the reaction is endothermic. A graph of an endothermic reaction is shown in Figure 8.4(a). If the energy released is greater than the activation energy, so there is a net *release* of energy, the reaction is exothermic. A graph of an exothermic reaction is shown in Figure 8.4(b). Notice that of all the molecules present, only some will collide, and of those collisions, only some are effective and result in reactions.

This picture of a reaction is analogous to riding a bicycle over a mountain pass. The activation energy of the reaction is comparable to the energy needed to pedal to the top of the pass. The energy released by the rearrangement of bonds is comparable to that gained in coasting down from the top of the pass to the floor of the next valley. If this second valley is higher than the one you started from, the energy gained in coasting down is less than the energy expended in pedaling up. This corresponds to an endothermic reaction, where there is a net absorption of energy. If the second valley is lower than the one you started from, you gain more energy coasting down than was used pedaling up. This corresponds to an exothermic reaction, which results in a net release of energy.

8.2 The Rate of a Reaction

The **rate** of a reaction measures how fast the concentrations of the reactants decrease or how fast the concentrations of the products increase. The free energy of a reaction shows only whether the reaction will take place spontaneously; it does not show how fast the concentrations change.

A reaction will form products more rapidly if the conditions under which the reaction occurs are changed so that more molecules have enough energy to reach the peak of either of the graphs in Figure 8.4. There are three ways to increase this number, described on the following pages.

1. Increase the number of reactant molecules present. The more molecules present in the reaction vessel, the more likely is a collision. We can increase the number of molecules by increasing the concentration of the reactants. If the reactants are both gases, increasing the pressure decreases the volume and brings the molecules closer together, increasing the likelihood of collision.

2. Increase the number of molecules with enough energy to collide. The rate of a reaction will increase if the number of molecules with enough energy to provide the activation energy of the reaction increases.

Figure 8.5 shows the distribution of energies in a collection of molecules at two different temperatures. (Notice that this is the same distribution we considered in Chapter 6; see Figures 6.3 and 6.17.) In Figure 8.5, molecules with an energy greater than point A are sufficiently energetic to provide the activation energy necessary for collision. The shaded area under each graph represents the number of molecules at that temperature with an energy greater than point A. The shaded area is much larger under the higher temperature curve. Therefore, at the higher temperature, there are more collisions and the reaction proceeds faster. At lower temperatures, these results are reversed and the reaction is slower.

We store food in a refrigerator because of this effect of temperature on reaction rates. The rates of the reactions that lead to food spoilage are decreased considerably by cooling the food from room temperature to that in a refrigerator. The rates of these reactions are decreased even further by storing food in a freezer. Recent developments in low-temperature surgery have resulted from the application of this principle. By cooling the patient, metabolic reactions are slowed and the operation can be performed more deliberately, with less trauma to the patient.

Figure 8.5 Energy distribution in a collection of molecules at two different temperatures.

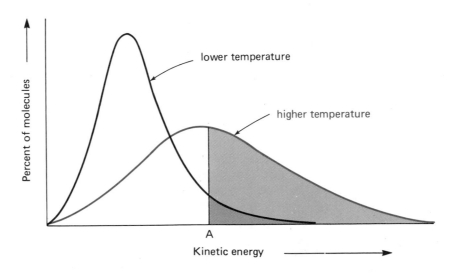

Figure 8.6 The effect of a catalyst on activation energy. The black line represents energy changes in an uncatalyzed reaction. The color line shows the energy changes for the same reaction in the presence of a catalyst.

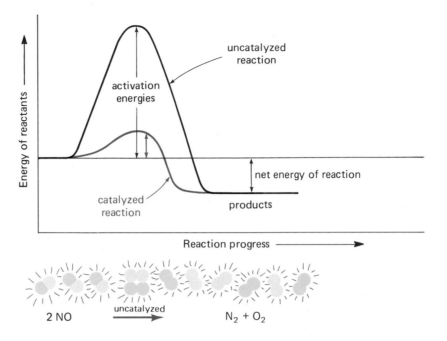

3. Lower the activation energy required for reaction. We have said that a certain amount of energy, the activation energy, is necessary for reaction. If the activation energy could be lowered (in our bicycling analogy, if the pass were not quite so far above the first valley), more molecules would be able to react. In Figure 8.6, the color line represents a lower activation energy. How can the activation energy of a reaction be lowered? Just as another pass between the valleys may be lower than that originally used, another pathway for the reaction may have a lower activation energy.

A catalyst can provide such an alternate pathway. A **catalyst** is a substance which, when added to a reaction mixture, increases the rate of the overall reaction, yet it is recovered unchanged after the reaction is complete. Suppose a substance, C, is added to a reaction mixture. If the formation of the product occurs at a faster rate in the presence of C than in its absence and C is recovered unchanged, then C is a catalyst for the reaction. The color line in Figure 8.6 shows the energy changes for the same reaction shown by the black line, but in the presence of a catalyst. There is still an activation energy, but it is less than that of the uncatalyzed reaction.

There are many examples of catalysts. Since the mid-seventies, many automobile exhaust systems have been manufactured with catalysts for the reaction

$$2 CO + O_2 \longrightarrow 2 CO_2$$

In the absence of a catalyst, this reaction requires a very high temperature and does not occur significantly at normal exhaust temperatures. The well-being of the public required that cars stop spewing out large amounts of carbon monoxide. The introduction of a catalyst to the exhaust system of the car makes possible the oxidation of carbon monoxide to carbon dioxide at lower exhaust temperatures, with a considerable improvement in air quality.

The **enzymes** that trigger biological processes are catalysts. Enzymes have enormous power to change the rates of chemical reactions. In fact, most of the reactions that occur so readily in the living cell would, in the absence of enzymes, occur too slowly to support life. For example, the enzyme carbonic anhydrase catalyzes the reaction of carbon dioxide and water to form carbonic acid:

$$CO_2 + H_2O \xrightarrow{\text{carbonic anhydrase}} H_2CO_3$$

Carbonic anhydrase increases the rate of this reaction to almost 10^7 times that of the uncatalyzed reaction. Red blood cells are especially rich in this enzyme. For this reason, they are able to absorb carbon dioxide as it is produced in the body and transport it back to the lungs, where it is released as one of the waste products of the body.

Catalysts, whether inorganic like those in automotive emission control systems, or organic like the enzymes of living systems, are so remarkably effective because they provide an alternative pathway for a reaction, one that has a lower energy of activation.

8.3 Chemical Equilibrium

A. Definition of Chemical Equilibrium

Many chemical reactions are **reversible**; that is, the products of the reaction can combine to re-form the reactants. An example of a reversible reaction is that of hydrogen with iodine to form hydrogen iodide:

$$H_2 + I_2 \rightleftharpoons 2\,HI$$

We can study this reversible reaction by placing hydrogen and iodine in a reaction vessel and then measuring the concentrations of H_2, I_2, and HI at various times after the reactants are mixed. Figure 8.7 is a plot of the concentrations of reactants and products of this equation against time. The concentration of HI increases very rapidly at first, then more slowly, and finally, after the time indicated by the vertical line marked "Equilibrium," remains constant. Similarly, the concentrations of H_2 and I_2 are large at the start of the reaction but decrease rapidly at first, and then more slowly. Finally, they, too, become constant at the equilibrium line.

If this reaction were not reversible, the concentrations of hydrogen and iodine would have continued to decrease and the concentration of hydrogen iodide to increase. This does not happen. Instead, as soon as any molecules

Figure 8.7 Concentration changes during the reversible reaction $H_2 + I_2 \rightarrow 2\ HI$, as it proceeds towards equilibrium (E).

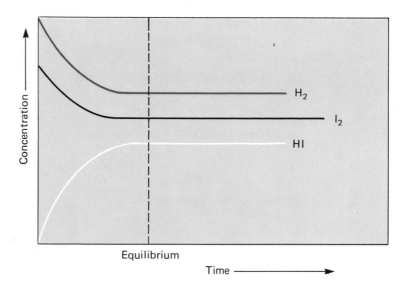

of hydrogen iodide are formed, some decompose into hydrogen and iodine. Two reactions are taking place simultaneously: the formation of hydrogen iodide, and its decomposition. When the concentrations of all these components become constant (at the equilibrium line in Figure 8.7), the rate of the forward reaction ($H_2 + I_2 \rightarrow 2\ HI$) must be equal to the rate of the reverse reaction ($2\ HI \rightarrow H_2 + I_2$). A state of **dynamic chemical equilibrium** has then been reached, one in which two opposing reactions are proceeding at equal rates, with no net changes in concentration.

B. The Characteristics of Chemical Equilibrium

1. Equal rates. At equilibrium, the rate of the forward reaction is equal to the rate of the reverse reaction.

2. Constant concentrations. At equilibrium, the concentrations of the substances participating in the equilibrium are constant. Although individual reactant molecules may be reacting to form product molecules and individual product molecules may be reacting to re-form the reactants, the concentrations of the reactants and the products remain constant.

3. Free energy requirements. At equilibrium, the free energy change is zero. Neither the forward nor the reverse reaction is spontaneous and neither is favored. Consider the ice-water change. Above 0°C, ice melts spontaneously to form liquid water; ΔG for this change is negative. Below 0°C, the change from ice to water is not spontaneous; ΔG is positive. At 0°C, the two states are in equilibrium. The rate of melting is equal to the rate of freezing; the amount of ice and the amount of liquid water present remain constant and the free energy change is zero.

C. The Equilibrium Constant

When a reaction is at equilibrium, there is a mathematical relationship between the concentrations of the components of the equilibrium that is known as the **equilibrium constant (K_{eq})**. For the reaction

$$H_2 + I_2 \rightleftharpoons 2\,HI$$

the equilibrium constant is

$$K_{eq} = \frac{[HI]^2}{[H_2][I_2]}$$

where [ ] means concentration in moles/liter. For the general equation

$$a\mathrm{A} + b\mathrm{B} \rightleftharpoons c\mathrm{C} + d\mathrm{D}$$

the equilibrium constant is

$$K_{eq} = \frac{[\mathrm{C}]^c[\mathrm{D}]^d}{[\mathrm{A}]^a[\mathrm{B}]^b}$$

In each of these expressions, the concentrations of the products of the reaction, each raised to a power equal to the coefficient of that product in the balanced equation for the reaction, are multiplied in the numerator. The concentrations of the reactants in the equation, each raised to a power equal to the coefficient of that reactant in the balanced equation, are multiplied in the denominator.

Example 8.1

Write the equilibrium constant for the reaction

$$N_2 + 3\,H_2 \rightleftharpoons 2\,NH_3$$

Solution

1. The numerator of the constant contains the product NH_3 enclosed in brackets to represent concentration and raised to the second power, because 2 is its coefficient in the equation:

$$[NH_3]^2$$

2. The denominator includes the reactants of the equation, N_2 and H_2, enclosed in brackets. The nitrogen term is at the first power, the hydrogen

term is raised to the third power:

$$[N_2][H_2]^3$$

3. The complete expression is:

$$K_{eq} = \frac{[NH_3]^2}{[N_2][H_2]^3}$$

Problem 8.1

Write the equilibrium constant for the reaction

$$2\ NO + O_2 \rightleftharpoons 2\ NO_2$$

The value of an equilibrium constant does not depend on how equilibrium was reached. Table 8.2 presents data on the hydrogen-iodine-hydrogen iodide equilibrium. It shows several different sets of initial concentrations, and the accompanying concentrations at equilibrium. The value of the equilibrium constant is given for each experiment. Notice that the value of the equilibrium constant is the same, regardless of whether the initial material was hydrogen and iodine or hydrogen iodide and whether the components were present in equal or different concentrations.

The value of the equilibrium constant does depend on how the equation for the equilibrium is written. For example, the equilibrium constants given in Table 8.2 were calculated from the expression

$$K_{eq} = \frac{[HI]^2}{[H_2][I_2]} = 45.9$$

which is the equilibrium constant for the equation

$$H_2 + I_2 \rightleftharpoons 2\ HI$$

Table 8.2 The hydrogen-iodine-hydrogen iodide equilibrium at 490°C.

Original Concentrations (mole/L)			Final Concentrations (mole/L)			Equilibrium Constant
$[H_2]$	$[I_2]$	$[HI]$	$[H_2]$	$[I_2]$	$[HI]$	$\dfrac{[HI]^2}{[H_2][I_2]}$
1.0	1.0	0	0.228	0.228	1.544	45.9
0	0	2.0	0.228	0.228	1.544	45.9
1.0	2.0	3.0	0.316	1.316	4.368	45.9

If the equation for the H_2-I_2-HI equilibrium is written

$$2\,HI \rightleftharpoons H_2 + I_2$$

then the equilibrium constant becomes

$$K'_{eq} = \frac{[H_2][I_2]}{[HI]^2} = 2.18 \times 10^{-2}$$

The two equilibrium constants are related to each other:

$$2.18 \times 10^{-2} = \frac{1}{45.9}$$

The formulations of the equilibrium constants are related in the same way:

$$K'_{eq} = \frac{[H_2][I_2]}{[HI]^2} = \frac{1}{\dfrac{[HI]^2}{[H_2][I_2]}} = \frac{1}{K_{eq}}$$

The value of an equilibrium constant does change with a change in temperature. The equilibrium constant for the $H_2 + I_2 \rightleftharpoons 2\,HI$ reaction is 45.9 only at 490°C. At 445°C, it is 64. At other temperatures, the equilibrium constant for this equation has other values, increasing as the temperature decreases, decreasing as the temperature increases.

8.4 Shifting Equilibria: Le Chatelier's Principle

A system in equilibrium is a special case, where everything is in balance. However, things rarely stay in balance; changes occur which shift the balance and the equilibrium involved. We have discussed such changes in previous chapters. In Chapter 6, we talked about physical equilibria; for example, the equilibrium between a liquid and its vapor in a closed container. At a particular temperature, the vapor exerts a given pressure. If the temperature is increased, the vapor pressure increases and has a higher value when the liquid and vapor again come into equilibrium at the higher temperature. By changing the conditions of the equilibrium, we caused the equilibrium to shift. The concentrations of the substances in equilibrium changed, but they were still in equilibrium.

Chemical equilibria can also shift in response to a change in conditions. Such changes could be changes in concentrations, changes in temperature, or, for those equilibria involving gases, changes in pressure. Such a change in conditions causes a stress on the equilibrium.

A French chemist named Le Chatelier (1850–1936) is credited with recognizing this property of equilibria. In 1888 he stated the following principle: When a stress is applied to a system at equilibrium, the system shifts to relieve that stress. In considering this statement, it is important to realize that although the system is initially at equilibrium, the stress sends it *out* of equilibrium. Concentrations then change so that the system comes back to equilibrium with a different set of concentrations of reactants and products, still related by the same constant, K_{eq}. If there is also a change in temperature, not only their concentrations but also their relationship, the equilibrium constant, changes. The system still shifts to come back to equilibrium.

A. The Effect of Concentration Changes on Equilibria

Of the various stresses that chemical equilibria are subjected to, concentration changes are most important to us, because biological reactions usually take place at constant temperature and pressure. Under normal conditions, our bodies are not subjected to more than slight changes in temperature and pressure, but they are subject to changes in concentration. Consider how a change in concentration of one of the reactants or products affects the hydrogen-iodine-hydrogen iodide equilibrium. There are two reactions proceeding simultaneously. The forward reaction

$$H_2 + I_2 \rightleftharpoons 2\,HI$$

is the combination of hydrogen and iodine, and the reverse reaction is the decomposition of hydrogen iodide. Le Chatelier's principle tells us that increasing the concentration of one of the components of an equilibrium mixture favors the reaction that consumes that component. Decreasing the concentration of a component favors the reaction that produces that component.

Suppose you have a flask containing hydrogen, iodine, and hydrogen iodide, all at equilibrium concentrations at 490°C. If we inject some iodine into the flask, the forward reaction which consumes iodine will be favored. That reaction will proceed more rapidly than the reverse reaction until the imbalance is corrected and the ratio of concentrations again matches the equilibrium constant. The new concentrations of iodine and of hydrogen iodide will be greater than the original concentrations of these substances, and the concentration of hydrogen will be less.

In summary, if the concentration of one of the components of an equilibrium is changed, the concentrations of the other components of the equilibrium will change to compensate for that change. Or, to restate Le Chatelier's principle: When a stress (such as a change in concentration) is applied to a system at equilibrium, the system shifts (all the concentrations change) to relieve that stress.

B. The Effect of Pressure Changes on Equilibria

A change in pressure is a stress to those equilibria that involve gases. Increased pressure favors the reaction that produces fewer gaseous molecules. In the equilibrium

$$N_2(g) + 3 H_2(g) \rightleftharpoons 2 NH_3(g)$$

the forward reaction produces 2 molecules of gas while the reverse reaction produces 4 molecules of gas; in terms of moles, there are 4 moles of gas on the left of this equilibrium and 2 on the right. Recall from Section 6.4B that the volume of a gas is independent of the composition of the gas and that at the same temperature and pressure, one mole of any gas occupies the same volume as one mole of any other gas. We then know that the gases on the left occupy a total of four volumes and those on the right occupy two volumes. Increased pressure decreases the volume available to this gaseous equilibrium and favors the forward reaction, because the forward reaction produces only 2 units instead of 4. An increase in pressure on this equilibrium will favor the formation of more ammonia; a decrease in pressure will favor the decomposition of ammonia.

C. The Effect of Temperature Changes on Equilibria

A change in temperature is a stress on a system in equilibrium. It changes the rate of both reactions and also changes the value of the equilibrium constant.

In each equilibrium, there are two reactions proceeding simultaneously, one forward and one reverse. One of these is endothermic ($\Delta H > 0$) and one is exothermic ($\Delta H < 0$). When the equilibrium is shown as an equation, the enthalpy term refers only to the forward reaction. For example, in the hydrogen iodide equilibrium, the forward reaction is exothermic. This is shown:

$$H_2(g) + I_2(g) \rightleftharpoons 2 HI(g) \qquad \Delta H = -6 \text{ kcal/mole}$$

When the temperature of an equilibrium mixture is increased, the rate of both reactions increases (see Section 8.2) but the rate of the endothermic reaction (the reaction that absorbs the added energy) is increased more. For the hydrogen iodide equilibrium, an increase in temperature favors the endothermic reverse reaction. When the system returns to equilibrium, the hydrogen iodide concentration will be smaller and the concentrations of hydrogen and iodine larger. The equilibrium constant will also be changed:

$$\text{at } 445°C, \ K_{eq} = \frac{[HI]^2}{[H_2][I_2]} = 64 \qquad \text{at } 490°C, \ K_{eq} = \frac{[HI]^2}{[H_2][I_2]} = 45.9$$

D. The Effect of a Catalyst on Equilibria

A catalyst changes the rate of a reaction by providing an alternate pathway with a lower energy of activation. The lower-energy pathway is available to both the forward and the reverse reactions of the equilibrium. The addition of a catalyst to a system in equilibrium does not favor one reaction over the other. Instead, it increases equally the rates of both the forward and the reverse reactions. The rate at which equilibrium is reached is increased, but the relative concentrations of reactants and products at equilibrium, hence the equilibrium constant, are unchanged.

Example 8.2

Given the equilibrium

$$PCl_3(g) + Cl_2(g) \rightleftharpoons PCl_5(g) \qquad \Delta H = -22.1 \text{ kcal/mole}$$

a. Write the equilibrium constant for this reaction.

b. How will the equilibrium shift if the temperature is increased?

c. How will the equilibrium shift if more chlorine is added to the reaction mixture? What will happen to the concentration of phosphorus trichloride (PCl_3)?

d. How will an increase in pressure affect the relative concentrations of products and reactants?

e. Is the value of K_{eq} increased, decreased, or unchanged by the changes in conditions in parts (b), (c), and (d)?

Solution

a. $K_{eq} = \dfrac{[PCl_5]}{[PCl_3][Cl_2]}$

b. The forward reaction is exothermic and the reverse reaction is endothermic. Therefore, at a higher temperature the reverse reaction is favored and the concentrations change, forming more PCl_3 and Cl_2. The concentration of PCl_5 decreases.

c. Chlorine is consumed by the forward reaction. The addition of more chlorine increases the rate of the forward reaction, forming more phosphorus pentachloride and decreasing the concentration of phosphorus trichloride.

d. In the forward reaction, two gaseous molecules combine to form one molecule of PCl_5. In the reverse reaction, one molecule of PCl_5 decomposes to produce two molecules. Therefore, if pressure is increased, the forward reaction is favored, resulting in the formation of more PCl_5.

e. In (b): Increasing the temperature results in a decrease in the value of K_{eq} for this reaction.

In (c): Adding more Cl_2 results in changes in the relative concentrations of Cl_2, PCl_3, and PCl_5 but does not change the value of K_{eq}.

In (d): Increasing the pressure changes the relative concentrations of Cl_2, PCl_3, and PCl_5 but does not change the value of K_{eq}.

Problem 8.2

Given the equilibrium

$$2\,H_2O(g) \rightleftharpoons 2\,H_2(g) + O_2(g) \qquad \Delta H = +115.6\ kcal/mole$$

a. Write the equilibrium constant for this reaction.

b. In what way will the addition of more O_2 affect this equilibrium?

c. In what way will a decrease in pressure affect this equilibrium?

d. In what way will an increase in temperature affect this equilibrium?

e. Which of the three changes in parts (b), (c), and (d) affect the value of K_{eq}?

8.5 Equilibria Involving Weak Electrolytes

A. Characteristics of Weak Electrolytes

The chemical equilibria reactions we have discussed so far have involved only molecules. Equilibria may also involve ions. Of the many types of ionic equilibria, those involving compounds only partially dissociated into ions in aqueous solution are the most important in biochemical reactions. Such compounds are called **weak electrolytes** (see Section 4.4B). **Strong electrolytes** are completely dissociated into ions in aqueous solution. Salts, metallic hydroxides, and strong mineral acids, such as nitric acid (HNO_3), are strong electrolytes. Aqueous solutions of these substances do not contain undissociated molecules (Figure 8.8a).

A solution of a weak electrolyte contains both molecules and ions. Nitrous acid, HNO_2, is a weak electrolyte. A solution of nitrous acid contains molecules of nitrous acid, nitrite ions, and hydrogen ions (Figure 8.8b). In a solution of nitrous acid, some molecules are dissociating into ions, according to the equation

$$HNO_2 \longrightarrow H^+ + NO_2^-$$

and at the same time, H^+ ions and NO_2^- ions are combining to re-form molecules:

$$H^+ + NO_2^- \longrightarrow HNO_2$$

Figure 8.8 (a) Strong and (b) weak electrolytes in solution.

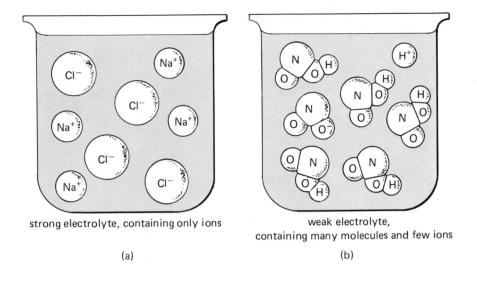

strong electrolyte, containing only ions

(a)

weak electrolyte, containing many molecules and few ions

(b)

When the rates of these two reactions are equal, the system is in dynamic equilibrium:

$$HNO_2 \rightleftharpoons H^+ + NO_2^-$$

Like equilibria involving molecules, an equilibrium involving ions and molecules has an equilibrium constant. The equilibrium constant for the ionization of nitrous acid is

$$K_{eq} = \frac{[H^+][NO_2^-]}{[HNO_2]}$$

By convention, the equations for ionic equilibria are written so that the forward reaction produces ions. The concentrations of the ions are therefore always in the numerator of the equilibrium constant. Only a very few molecules are ionized in aqueous solutions of weak electrolytes; therefore, the concentration values of the un-ionized molecules in the denominator of the equilibrium constant are much larger than those of the ions in the numerator. Consequently, values of the equilibrium constants for weak electrolytes are always much less than 1.

B. Weak Acids

Most biologically important acids and some mineral acids are weak electrolytes in aqueous solution. The equilibrium constant describing their dissociation is called the **acid dissociation** or **acid ionization constant** and is given the symbol

Table 8.3 Some common weak acids.

Weak Acid	Equilibrium	Acid Dissociation Constant	K_a
acetic acid	$CH_3CO_2H \rightleftharpoons H^+ + CH_3CO_2^-$	$\dfrac{[H^+][CH_3CO_2^-]}{[CH_3CO_2H]}$	1.8×10^{-5}
formic acid	$HCO_2H \rightleftharpoons H^+ + HCO_2^-$	$\dfrac{[H^+][HCO_2^-]}{[HCO_2H]}$	1.8×10^{-4}
nitrous acid	$HNO_2 \rightleftharpoons H^+ + NO_2^-$	$\dfrac{[H^+][NO_2^-]}{[HNO_2]}$	4.6×10^{-4}
hydrocyanic acid	$HCN \rightleftharpoons H^+ + CN^-$	$\dfrac{[H^+][CN^-]}{[HCN]}$	4.9×10^{-10}
carbonic acid	$CO_2 + H_2O \rightleftharpoons H^+ + HCO_3^-$	$\dfrac{[H^+][HCO_3^-]}{[CO_2]}$	4.3×10^{-7}
	$HCO_3^- \rightleftharpoons H^+ + CO_3^{2-}$	$\dfrac{[H^+][CO_3^{2-}]}{[HCO_3^-]}$	5.6×10^{-11}
phosphoric acid	$H_3PO_4 \rightleftharpoons H^+ + H_2PO_4^-$	$\dfrac{[H^+][H_2PO_4^-]}{[H_3PO_4]}$	7.5×10^{-3}
	$H_2PO_4^- \rightleftharpoons H^+ + HPO_4^{2-}$	$\dfrac{[H^+][HPO_4^{2-}]}{[H_2PO_4^-]}$	6.2×10^{-8}
	$HPO_4^{2-} \rightleftharpoons H^+ + PO_4^{3-}$	$\dfrac{[H^+][PO_4^{3-}]}{[HPO_4^{2-}]}$	2.2×10^{-13}
ammonium ion	$NH_4^+ \rightleftharpoons H^+ + NH_3$	$\dfrac{[H^+][NH_3]}{[NH_4^+]}$	5.5×10^{-10}

K_a. Table 8.3 lists several weak acids, equations for the ionization of each, and the values of their acid dissociation constants. Several points about Table 8.3 should be noted.

1. An acid with more than one ionizable hydrogen loses those hydrogen ions one by one. For each loss, there is an equilibrium equation and an equilibrium constant. Phosphoric acid has three ionizable hydrogens. The acid molecule ionizes to yield a hydrogen ion and a dihydrogen phosphate ion:

$$H_3PO_4 \rightleftharpoons H^+ + H_2PO_4^- \qquad K_a = 7.5 \times 10^{-3}$$

The dihydrogen phosphate ion ionizes to yield another hydrogen ion and the monohydrogen phosphate ion:

$$H_2PO_4^- \rightleftharpoons H^+ + HPO_4^{2-} \qquad K_a = 6.2 \times 10^{-8}$$

Finally, the monohydrogen phosphate ion ionizes to yield another hydrogen ion and the phosphate anion:

$$HPO_4^{2-} \rightleftharpoons H^+ + PO_4^{3-} \qquad K_a = 2.2 \times 10^{-13}$$

Notice that the acid dissociation constant becomes smaller with each ionization. For acids with more than one ionizable hydrogen, the first dissociation constant is always the largest; the value for each successive dissociation constant decreases.

2. Carbonic acid (H_2CO_3) is a solution of carbon dioxide in water. Molecules of carbonic acid are not stable. The mixture of carbon dioxide and water ionizes stepwise, as does phosphoric acid.

3. Ammonium ion acts as a weak acid, ionizing to ammonia and a hydrogen ion:

$$NH_4^+ \rightleftharpoons H^+ + NH_3$$

Example 8.3

Ascorbic acid, vitamin C, is a weak electrolyte. Its molecular formula is $C_6H_8O_6$. In aqueous solution, ascorbic acid ionizes to form H^+ and ascorbate ion, $C_6H_7O_6^-$.

a. Write an equation for the equilibrium established in this ionization.

b. Write an expression for the K_a of ascorbic acid.

Solution

a. The equation shows the loss of one hydrogen as an ion; the rest of the molecule is the ascorbate anion.

$$C_6H_8O_6 \rightleftharpoons H^+ + C_6H_7O_6^-$$

ascorbic acid   ascorbate ion

b. The acid dissociation constant has the concentrations of the ions in the numerator and that of the un-ionized acid molecule in the denominator:

$$K_a = \frac{[H^+][C_6H_7O_6^-]}{[C_6H_8O_6]}$$

Problem 8.3

Citric acid $(C_6H_8O_7)$ is found in the juice of lemons and other citrus fruits, hence its name. It is a weak electrolyte and ionizes in aqueous solution to form H^+ and citrate ion, $C_6H_7O_7^-$.

a. Write the equilibrium equation for this ionization.

b. Write the expression for the acid dissociation constant of citric acid.

C. Hydrogen Ion Concentrations in Acid Solutions

The course of a chemical reaction in solution frequently depends on the hydrogen ion concentration of that solution.

1. Hydrogen ion concentration in solutions of strong acids. Strong acids with one ionizable hydrogen are completely ionized in aqueous solution; therefore, the hydrogen ion concentration of these solutions is equal to the molar concentration of the acid.

Example 8.4

What is the hydrogen ion concentration in $1.0M$ HCl?

Solution

Hydrochloric acid is a strong acid, completely ionized in water:

$$HCl \longrightarrow H^+ + Cl^-$$

Therefore, in a solution prepared by adding 1.0 mole of HCl to enough water to make one liter of solution, the concentration of H^+ is $1.0M$, that of Cl^- is $1.0M$ and that of undissociated acid is 0.

Problem 8.4

What is the hydrogen ion concentration of $0.1M$ HNO_3?

2. In weak acids. The hydrogen ion concentration of an aqueous solution of a weak acid depends on the value of its acid dissociation constant and is always less than the concentration of the weak acid. The value of $[H^+]$ can be calculated using the value of K_a and the concentration of the weak acid.

Acetic acid is a weak acid that ionizes according to the equation:

$$CH_3CO_2H \rightleftharpoons H^+ + CH_3CO_2^-$$

Its acid dissociation constant is:

$$K_a = \frac{[H^+][CH_3CO_2^-]}{[CH_3CO_2H]} = 1.8 \times 10^{-5}$$

The hydrogen ion concentration of a $1.0M$ acetic acid solution can be calculated as follows. The solution contains 1.0 mole of acetic acid in 1.0 liter of solution. Because acetic acid is a weak electrolyte, only a small fraction of the molecules ionize to hydrogen and acetate ions; most remain as un-ionized acetic acid molecules. Let x stand for the number of moles of acetic acid that ionize. If x moles ionize, then $1.0 - x$ moles remain un-ionized. For x moles of acetic acid that ionize, x moles of H^+ and x moles of $CH_3CO_2^-$ are formed. The resulting concentrations of acetic acid, hydrogen ion, and acetate ion at equilibrium are

$$CH_3CO_2H \rightleftharpoons H^+ + CH_3CO_2^-$$
$$1.0 - x \qquad\qquad x \qquad\quad x$$

Substituting these values into the formula for the acid dissociation constant gives:

$$K_a = \frac{[H^+][CH_3CO_2^-]}{[CH_3CO_2H]} = \frac{(x)(x)}{1.0 - x} = 1.8 \times 10^{-5}$$

The tiny value of the acid dissociation constant suggests that the amount of acid dissociated is very small (less than $0.01M$). Using the rules for significant figures in addition and subtraction (Section 1.3D2), we know that the quantity $(1.0 - 0.01)$ expressed to two significant figures is 1.0. If x has a value of less than 0.01, it is appropriate to ignore x in the expression $1.0 - x$, changing the K_a equation to:

$$\frac{x^2}{1.0} = 1.8 \times 10^{-5}$$

Solving this equation gives:

$$x^2 = \sqrt{1.8 \times 10^{-5}} = \sqrt{18 \times 10^{-6}}$$
$$x = 4.2 \times 10^{-3}M = [H^+] = [CH_3CO_2^-]$$

The hydrogen ion concentration in $1.0M$ acetic acid solution is, then, 4.2×10^{-3} M, or $0.0042M$. Notice that 4.2×10^{-3} is not significant when subtracted from 1.0, so that our simplification of the original equation was justified. If the acid is very dilute, for example, 10^{-3} M, or if it is one with a large acid dissociation constant, such as 10^{-2}, this simplification would not be valid.

These calculations emphasize the difference between strong and weak acids. The hydrogen ion concentration of a $1.0M$ solution of a strong acid is $1.0M$. The hydrogen ion concentration of a $1.0M$ solution of a weak acid is much less than $1.0M$, and is calculated from the acid dissociation constant of the weak acid.

Example 8.5

Calculate the hydrogen ion concentration in $0.10M$ ascorbic acid ($C_6H_8O_6$). K_a for ascorbic acid is 8.0×10^{-5}.

Solution

In $0.10M$ $C_6H_8O_6$, the equilibrium equation is

$$C_6H_8O_6 \rightleftharpoons H^+ + C_6H_7O_6^-$$
$$0.10 - x \qquad x \qquad x$$

The K_a for this equilibrium is

$$K_a = \frac{[H^+][C_6H_7O_6^-]}{[C_6H_8O_6]} = 8.0 \times 10^{-5}$$

Let $[H^+] = x$; then $[C_6H_7O_6^-]$ also equals x and

$$[C_6H_8O_6] = 0.10 - x$$

Substituting these values in the formula for the dissociation constant gives:

$$\frac{(x)(x)}{0.10 - x} = 8.0 \times 10^{-5}$$

Assuming, as before, that $[H^+]$ is so much less than 0.10 as to be insignificant, the equation becomes

$$\frac{x^2}{0.10} = 8.0 \times 10^{-5}$$

Solving for x, we get:

$$x^2 = 8.0 \times 10^{-6}$$
$$x = 2.8 \times 10^{-3}$$

Then in $0.10M$ $C_6H_8O_6$, $[H^+] = 2.8 \times 10^{-3}$ M

Problem 8.5

Calculate the hydrogen ion concentration of $0.1M$ formic acid. (See Table 8.3 for the molecular formula of formic acid and the value of its acid dissociation constant.)

D. pH

The **pH** of a solution is the negative log of its hydrogen ion concentration. This term is used because the hydrogen ion concentration in solutions of weak acids and in many other fluids is generally much less than one; therefore, when the concentration is expressed exponentially, it contains a negative exponent. Many people find numbers with negative exponents confusing and answer with some hesitation such questions as: "Is 1.8×10^{-4} larger or smaller than 3.6×10^{-5}?" (To answer the question, state both numbers with the same exponent of 10. This changes 3.6×10^{-5} to 0.36×10^{-4}, a value that is clearly less than 1.8×10^{-4}.) To avoid confusion when dealing with small numbers, a system has been set up which describes dissociation constants and ion concentrations, not in exponential form, but as the negative logarithms of the actual values. The letter "p" has been chosen to mean "negative logarithm of." Thus, pH means the negative log of the hydrogen ion concentration and pOH means the negative log of the hydroxide ion concentration.

1. Calculation of pH. To calculate the pH of a solution, the hydrogen ion concentration must be stated in exponential form. For a solution with a hydrogen ion concentration of $0.003M$, restate that concentration as 3.0×10^{-3} M. Next, determine the logarithm (log) of that number. The log of the product of two numbers is the sum of their logs.

$$log(3.0 \times 10^{-3}) = log(3) + log(10^{-3})$$

The log of the first term can be found in Table 8.4; the log of the exponential term is its exponent.

$$log(3.0 \times 10^{-3}) = 0.477 + (-3)$$
$$= -2.523$$

The pH is the negative log of the hydrogen ion concentration. For $[H^+] = 3.0 \times 10^{-3}$, pH $= 2.523$.

log 1.0 = 0.000	log 6.0 = 0.778
log 2.0 = 0.301	log 7.0 = 0.845
log 3.0 = 0.477	log 8.0 = 0.903
log 4.0 = 0.602	log 9.0 = 0.954
log 5.0 = 0.699	log 10.0 = 1.000

Table 8.4 Logarithms of small whole numbers.

Example 8.6

a. Calculate the pH of a solution with a hydrogen ion concentration of $0.0004M$.

b. Calculate the hydrogen ion concentration of a solution of pH 8.52.

Solution

a. State $[H^+]$ in exponential form.

$$[H^+] = 0.0004 = 4 \times 10^{-4}$$

Determine the log of $[H^+]$ using Table 8.4.

$$pH = -\log(4 \times 10^{-4}) = -(\log 4 + \log 10^{-4})$$
$$= -[(0.602) + (-4)] = -0.602 + 4 = 3.398$$

b. This calculation is performed by reversing the steps in part (a).

$$pH = -\log[H^+] = 8.52 = 9 - 0.48$$
$$\log[H^+] = -9 + 0.48$$
$$[H^+] = (\text{antilog } 0.48) \times 10^{-9} = 3.0 \times 10^{-9}$$

Problem 8.6

a. Calculate the pH of a solution with a hydrogen ion concentration of 5.0×10^{-4}.

b. What is the hydrogen ion concentration of a solution of pH 3.16?

2. *The interpretation of pH values.* When the hydrogen ion concentration is stated in exponential notation, the smaller the exponent, the greater the acidity of the solution. Consequently, with pH values, the lower the pH, the more acidic the solution.

in 0.1M HCl: in 0.0001M HCl:

$[H^+] = 1 \times 10^{-1}$ $[H^+] = 1 \times 10^{-4}$

pH $= 1$ pH $= 4$

Figure 8.9 shows the pH of several familiar fluids. Many of these values are the mid-point of a range. Human blood plasma normally varies only between pH 7.35 and pH 7.45. Human gastric fluid is much more acidic; its normal range is from pH 1.0 to pH 3.0.

E. pK_a

The pK_a of an acid is the negative log of its acid dissociation constant. Just as pH can be used to describe the hydrogen ion concentration of a solution, pK_a can be used to describe the dissociation constant of a weak acid. The higher the pK_a of an acid, the weaker is the acid.

Table 8.5 lists the same weak acids that were listed in Table 8.3 and the pK_a of each. Notice that the acids with larger ionization constants have smaller pK_a's. For example, formic acid ($K_a = 1.8 \times 10^{-4}$) is a stronger acid than acetic acid ($K_a = 1.8 \times 10^{-5}$). The pK_a of formic acid is 3.7, a smaller number

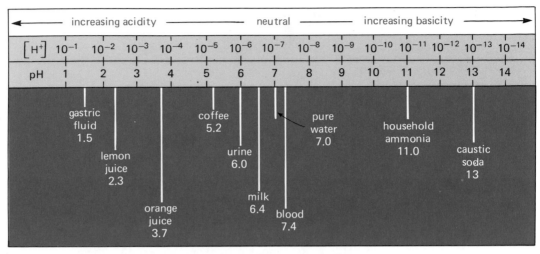

Figure 8.9 pH and hydrogen ion concentration.

than 4.7, the pK_a of acetic acid. Notice, too, that for the polyprotic acids, the pK_a increases with each ionization. For example, the pK_a for the first ionization of phosphoric acid

$$H_3PO_4 \rightleftharpoons H^+ + H_2PO_4^- \qquad pK_a = 2.12$$

is much smaller than that of the second ionization:

$$H_2PO_4^- \rightleftharpoons H^+ + HPO_4^{2-} \qquad pK_a = 7.21$$

Phosphoric acid is a much stronger acid than the dihydrogen phosphate ion.

Table 8.5 The K_a and pK_a of several common weak acids.

Weak Acid	K_a	pK_a
acetic acid	1.8×10^{-5}	4.74
formic acid	1.8×10^{-4}	3.74
nitrous acid	4.6×10^{-4}	3.34
hydrocyanic acid	4.9×10^{-10}	9.31
carbonic acid	4.3×10^{-7}	6.37
	5.6×10^{-11}	10.25
phosphoric acid	7.5×10^{-3}	2.12
	6.2×10^{-8}	7.21
	2.2×10^{-13}	12.67
ammonium ion	5.5×10^{-10}	9.26

8.6 Water as a Weak Electrolyte

To a very small but very important extent, water is a weak electrolyte that ionizes to yield hydrogen and hydroxide ions. This is an equilibrium reaction with the equation:

$$H_2O \rightleftharpoons H^+ + OH^-$$

Like all other equilibria, it has an equilibrium constant:

$$K_{eq} = \frac{[H^+][OH^-]}{[H_2O]}$$

In any aqueous solution, the concentration of water is so high (55.5 moles of water in 1000 cm^3 water) and the number of ionized water molecules is so low (1.0×10^{-7} mole in 1000 cm^3 water), that the concentration of the water molecules is a constant. The **ionization constant of water,** K_w, then includes the constant factor $[H_2O]$:

$$K_w = K_{eq}[H_2O] = 1.0 \times 10^{-14}; \ pK_w = 14$$

The constant K_w brings the types of equilibrium constants mentioned to three. They are tabulated in Table 8.6.

In pure water, the hydrogen ion concentration $[H^+]$ equals the hydroxide ion concentration $[OH^-]$. We can calculate what these concentrations are, given that the equation for the ionization of water is

$$H_2O \rightleftharpoons H^+ + OH^-$$

Let x equal the hydrogen ion concentration $[H^+]$.
Then x also equals the hydroxide ion concentration $[OH^-]$.

Table 8.6 Types of equilibrium constants.

Name of Constant	Symbol	Typical Equation	Formulation of Constant
equilibrium constant	K_{eq}	$A_2 + B_2 \rightleftharpoons 2\,AB$	$\dfrac{[AB]^2}{[A_2][B_2]}$
acid dissociation constant	K_a	$HA \rightleftharpoons H^+ + A^-$	$\dfrac{[H^+][A^-]}{[HA]}$
ionization constant of water	K_w	$H_2O \rightleftharpoons H^+ + OH^-$	$[H^+][OH^-]$

Substituting into the equation, $[H^+][OH^-] = 1.0 \times 10^{-14}$, we obtain

$$x^2 = 1.0 \times 10^{-14}$$
$$x = 1.0 \times 10^{-7}$$
$$[H^+] = [OH^-] = 1.0 \times 10^{-7}M$$

A **neutral** solution is one that is neither acidic nor basic. The hydrogen ion concentration equals the hydroxide ion concentration, and both equal $1.0 \times 10^{-7}M$. In a neutral solution, pH = pOH = 7.

An **acidic** solution is one in which the hydrogen ion concentration is greater than the hydroxide ion concentration; in other words, the hydrogen ion concentration is greater than 1.0×10^{-7} and the hydroxide ion concentration is less than 1.0×10^{-7}. In terms of pH, an acidic solution has a pH less than 7.

What is the hydroxide ion concentration in an acidic solution? To answer this question, we need to know that the following relationships exist whenever water is present:

$$[H^+][OH^-] = 1.0 \times 10^{-14} \quad \text{and} \quad pH + pOH = 14$$

If we know the hydrogen ion concentration, we can use the first relationship to calculate the hydroxide ion concentration. Suppose we know that the hydrogen ion concentration is $0.1M$. Substituting in and rearranging the first relationship gives

$$[OH^-] = \frac{1.0 \times 10^{-14}}{0.10} = 1.0 \times 10^{-13}M$$

Supposing we know that the pH of an acidic solution is 1.0. We can use the second relationship to calculate the pOH.

$$1.0 + pOH = 14; \quad pOH = 13$$

An **alkaline** or **basic** solution is one in which the hydrogen ion concentration is less than 1.0×10^{-7}. In terms of pH, an alkaline solution is one in which pH is greater than 7.0. Table 8.7 shows the relationship between pH and pOH over a scale from 0 to 14.

Table 8.7 The relationship between pOH and pH.

pH	0	1	2	3	4	5	6	7	8	9	10	11	12	13	14
pOH	14	13	12	11	10	9	8	7	6	5	4	3	2	1	0

increasingly acidic ← ——————————————— neutral ——————————→ increasingly alkaline (basic)

Example 8.7

Calculate the hydroxide ion concentration, the pH, and the pOH of $0.01M$ nitric acid (HNO_3).

Solution

Nitric acid is a strong acid, and is therefore completely ionized in solution. Thus, $[H^+]$ is $0.01M$ and pH $= 2$. The hydroxide ion concentration can be calculated using the K_w constant:

$$K_w = [H^+][OH^-] = 1.0 \times 10^{-14}$$

Substituting and rearranging gives:

$$0.01 \times [OH^-] = 1.0 \times 10^{-14}$$

$$[OH^-] = \frac{1.0 \times 10^{-14}}{0.01} = 1.0 \times 10^{-12}M$$

Since pH + pOH $= 14$, pOH $= 12$.

Problem 8.7

Calculate the hydrogen ion concentration of a solution having a hydroxide ion concentration of $1.0 \times 10^{-4}M$. What is the pH of this solution?

8.7 The Salts of Weak Electrolytes

The food that we eat is transformed in the body into a variety of substances. Many of these are weak acids or the anions of those acids. In the presence of enzymes (see Section 8.2), these substances react further to provide the energy we need, as well as other compounds. The efficiency of these reactions depends enormously on the hydrogen ion concentration in the solutions in which the reactions take place. We have shown earlier how the concentrations of all the components of an equilibrium change when the concentration of one of the components changes. Ionic equilibria follow the same rules. It is possible to change the hydrogen ion concentration in a solution of a weak acid by changing the concentration of the anion of the weak acid.

A. Hydrogen Ion Concentration in a Solution of a Weak Acid and Its Salt

Suppose we have one liter of a solution containing one mole of acetic acid and one mole of sodium acetate. The acetic acid is present as an equilibrium mixture of acetic acid molecules, hydrogen ions, and acetate ions. The sodium acetate is present only as ions (recall from Chapter 4 that salts are completely ionized in solution). The acetate ions from both the acid and the sodium acetate participate in the acetic acid equilibrium:

$$CH_3CO_2H \rightleftharpoons H^+ + CH_3CO_2^- \qquad K_a = 1.8 \times 10^{-5}$$

Although sodium ions are also present in solution, they do not participate in the equilibrium, they do not appear in the equation for the equilibrium, and they play no role in determining the concentrations of those substances whose formulas do appear in the equilibrium expression. If x equals the moles of ionized acetic acid molecules, the concentrations at equilibrium are:

$1.0 - x =$ the concentration of un-ionized acetic acid molecules

$x =$ the concentration of hydrogen ions

$1.0 + x =$ the concentration of acetate ions (the concentration of sodium acetate in the solution plus the acetate ions from the ionization of acetic acid)

In the equilibrium equation, the concentrations are:

$$CH_3CO_2H \rightleftharpoons H^+ + CH_3CO_2^-$$
$$1.0 - x \qquad x \qquad 1.0 + x$$

Our earlier studies of the effect of concentration changes on equilibria allow us to predict that, in this solution, the large concentration of acetate ions will shift the equilibrium to the left and lower the hydrogen ion concentration to a value that is not significant when added to or subtracted from 1.0. Given this assumption, $1.0 + x$ is approximately equal to 1.0 and $1.0 - x$ is also approximately equal to 1.0. Thus, the concentrations can be expressed as:

$$[CH_3CO_2H] \qquad [H^+] \qquad [CH_3CO_2^-]$$
$$1.0 \qquad\qquad x \qquad\qquad 1.0$$

Substituting these values into the formula for the acid dissociation constant for acetic acid and solving gives:

$$K_a = \frac{[H^+][CH_3CO_2^-]}{[CH_3CO_2H]} = \frac{(x)(1.0)}{(1.0)} = 1.8 \times 10^{-5}$$
$$x = 1.8 \times 10^{-5}$$
$$[H^+] = 1.8 \times 10^{-5} M$$

The addition of sodium acetate has decreased the hydrogen ion concentration from $4.2 \times 10^{-3} M$ in $1.0M$ acetic acid solution to $1.8 \times 10^{-5} M$ in a $1.0M$ acetic acid solution that is also $1.0M$ in acetate ion. This is a tremendous decrease. The addition of other concentrations of the acid or of its anion will change the hydrogen ion concentration to other values.

Example 8.8

Calculate the hydrogen ion concentration of a solution that is $1.0M$ in acetic acid and $0.2M$ in sodium acetate.

Solution

If we let x equal the moles of acetic acid that ionize, the concentrations at equilibrium will be:

$$x = \text{hydrogen ion concentration}$$
$$1.0 - x = \text{acetic acid concentration (un-ionized)}$$
$$0.2 + x = \text{acetate ion concentration}$$

We can drop x from the concentration of acetic acid and acetate ions because, as has been shown before, x is not significant when added to or subtracted from a number as large as 1.0 or 0.2. The equilibrium concentrations become:

$$CH_3CO_2H \rightleftharpoons H^+ + CH_3CO_2^-$$

1.0	x	0.2

Substituting these values in the formula for the acid dissociation constant gives:

$$K_a = \frac{[H^+][CH_3CO_2^-]}{[CH_3CO_2H]} = \frac{(x)(0.2)}{1.0} = 1.8 \times 10^{-5}$$

or

$$0.2x = 1.8 \times 10^{-5}$$
$$x = 9.0 \times 10^{-5}$$

A solution that is $1.0M$ in acetic acid and $0.2M$ in sodium acetate has a hydrogen ion concentration of $9.0 \times 10^{-5}M$.

Problem 8.8

Calculate the hydrogen ion concentration in one liter of a solution that is $0.1M$ in formic acid and $0.1M$ in sodium formate.

B. The Henderson-Hasselbalch Equation

The acid dissociation constant of a weak acid, HA, is:

$$K_a = \frac{[H^+][A^-]}{[HA]}$$

This can be stated in terms of pH and pK_a as:

$$pK_a = pH - \log \frac{[A^-]}{[HA]}$$

This rearranges to form:

$$pH = pK_a + \log \frac{[A^-]}{[HA]}$$

This last equation is known as the **Henderson-Hasselbalch equation.** Example 8.9 shows how it is used.

Example 8.9

What is the pH of $0.5M$ formic acid that contains 1.0 mole sodium formate per liter?

$$pK_a \text{ (formic acid)} = 3.74$$

Solution

The concentrations are

$$[HCO_2H] = 0.5M \qquad [HCO_2^-] = 1.0M$$

Substituting into the equation $pH = pK_a + \log \dfrac{[A^-]}{[HA]}$, we get:

$$pH = 3.74 + \log \frac{1.0}{0.5} = 3.74 + \log 2$$

$$= 3.74 + 0.30 \text{ (from Table 8.4)}$$

$$= 4.04$$

Problem 8.9

Using the Henderson-Hasselbalch equation, calculate the pH of $0.2M$ acetic acid solution that contains 0.4 mole of sodium acetate in each liter of solution.

C. The Hydrolysis of Ions

Hydrolysis means reaction with water. The extent to which ions react with water depends on the ionization of water and whether or not the ion is part of a weak electrolyte. The omnipresent aspects of the equilibria involved in the ionization of water and of weak electrolytes in water cannot be underestimated. Whenever water is present, whether in the ocean, in a raindrop, or in blood, hydrogen and hydroxide ions are present in amounts which satisfy the equation:

$$K_w = [H^+][OH^-] = 1.0 \times 10^{-14}$$

Whenever an ion is present in water and is related through ionization to either a weak acid or a weak base, that equilibrium, too, must be satisfied. For example, whenever acetate ion is present in water, the equilibrium

$$CH_3CO_2^- + H^+ \rightleftharpoons CH_3CO_2H$$

an ion   *a weak acid*

is present. The concentration of hydrogen ion, acetate ion, and acetic acid molecules must satisfy the acid dissociation constant for acetic acid:

$$K_a = \frac{[H^+][CH_3CO_2^-]}{[CH_3CO_2H]} = 1.8 \times 10^{-5}$$

The hydrogen ion concentration of a solution of sodium acetate must satisfy two equilibrium constants: the ion product of water (K_w) and the acid dissociation constant of acetic acid (K_a). The hydrogen ion concentration of water is decreased by the amount of hydrogen ions that react with acetate ion to satisfy the acid dissociation constant of acetic acid. In the equilibrium of water with its ions, this decrease in the concentration of hydrogen ions favors the reaction that produces those ions.

$$H_2O \rightleftharpoons H^+ + OH^-$$

The longer arrow to the right indicates the favored direction of the reaction. This reaction, the forward reaction as the equation is written, also produces hydroxide ions and results in a basic solution, one whose hydroxide ion concentration is greater than $1.0 \times 10^{-7} M$. Such reactions of ions with the ions of water are examples of hydrolysis.

In terms of the Brønsted-Lowry acid-base system, ion hydrolysis reactions are acid-base reactions. Recall from Section 4.8 that in this system, an acid is a proton donor and a base is a proton acceptor. The anion of a weak acid, then, becomes a base, and is described as the **conjugate base** of the weak acid. The conjugate base of acetic acid, a weak acid, is acetate ion.

$$CH_3CO_2H \rightleftharpoons CH_3CO_2^- + H^+$$

acid   *conjugate*
                          base

Similarly, hydroxide ion is the conjugate base of water, nitrite ion is the conjugate base of nitrous acid, and water is the conjugate base of the hydronium ion, H_3O^+. Using these concepts we can see that hydrolysis is an acid-base reaction in which the products are the conjugate acid and conjugate base of the reacting species. When sodium acetate is dissolved in water, acetate ion (a conjugate base) reacts with water (an acid) to form acetic acid and hydroxide ion:

$$CH_3CO_2^- + H_2O \rightleftharpoons CH_3CO_2H + OH^-$$

conjugate   *acid*   *acid*   *conjugate*
base of CH_3CO_2H *base of* H_2O

Notice that the equation produces hydroxide ion. A solution of sodium acetate in water is basic, as we predicted.

Hydrolysis of anions always produces hydroxide ions. For example, nitrite ion in water reacts according to the equation:

$$NO_2^- + H_2O \rightleftharpoons HNO_2 + OH^-$$

conjugate   *acid$_2$*   *acid$_1$*   *conjugate*
base$_1$ *base$_2$*

This reaction results in an excess of hydroxide ions and the solution is basic.

Hydrolysis of a cation always produces hydrogen ions. Ammonium ion is one of the weak acids listed in Table 8.3. It hydrolyzes in water according to the equation:

$$NH_4^+ + H_2O \rightleftharpoons NH_3 + H_3O^+$$

acid$_1$ *conjugate*   *conjugate*   *acid$_2$*
base$_2$ *base$_1$*

This reaction results in an excess of hydrogen (or hydronium) ions and the solution is acidic.

8.8 Buffers

A. Characteristics of Buffers

As we shall see in Chapters 24, 25, and 26, the metabolism of sugars, fats, and proteins results in the production of a variety of acids and bases. For example, lactic acid is produced during the metabolism of glucose; ammonia is produced during the metabolism of amino acids. These metabolically produced acids and bases are then transported to the liver and kidneys for eventual excretion. The ability of blood to transport these and other acids and bases without appreciable changes in pH requires a very effective system for pH control. A chemical system that resists change in pH is called a **buffer system** or **buffer.**

Buffers are especially important in the health and biological sciences, for virtually all chemical reactions in living systems take place in buffered solutions.

These include reactions in intracellular and intercellular fluids, as well as those in blood.

A chemical buffer contains a weak acid and the conjugate base of that acid. Such a solution exhibits the equilibrium:

$$HA \rightleftharpoons H^+ + A^-$$

$$\underset{acid}{} \qquad \underset{\substack{conjugate \\ base}}{}$$

If hydrogen ions are added to the solution, the reverse reaction, which consumes the added hydrogen ion, is favored. If hydroxide ion, carbonate ion, or some other base is added, more HA ionizes to replace the hydrogen ions that have combined with the added base. If the concentrations of the acid and of the anions are sufficiently larger than that of the added base or hydrogen ions, the pH of the solution will remain essentially constant.

Figure 8.10 shows a plot of pH against concentration for a series of solutions containing various amounts of acetate ion and un-ionized acetic acid. For all solutions shown, the sum of the concentrations of acetate ion and acetic acid equals 0.2 mole. In other words, the relative amounts of each change, but their combined concentration is unchanged:

$$[CH_3CO_2^-] + [CH_3CO_2H] = 0.2 \text{ mole}$$

The graph shows a change of only two pH units as the acetic acid concentration decreases from $0.18M$ to $0.02M$ and the acetate ion concentration increases from $0.02M$ to $0.18M$. It is within this range that the solution acts as a buffer. Let us show how this works by doing some calculations involving buffers.

Figure 8.10 pH vs concentration for a series of acetic acid-acetate solutions.

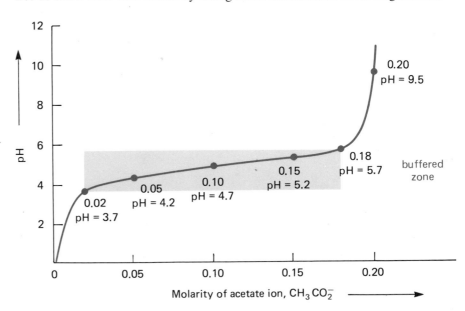

pH vs Molarity of acetate ion, $CH_3CO_2^-$

0.02 pH = 3.7
0.05 pH = 4.2
0.10 pH = 4.7
0.15 pH = 5.2
0.18 pH = 5.7
0.20 pH = 9.5

buffered zone

B. Calculations Involving Buffers

In Section 8.7A we learned how to calculate the hydrogen ion concentration in a solution of a weak acid and its salt, as for example, one containing acetic acid and sodium acetate. We can calculate the pH of buffer solutions in the same way.

Example 8.10

Calculate the following:

a. The pH of a buffer solution, one liter of which contains 0.10 mole acetic acid and 0.10 mole acetate ion. The pK_a of acetic acid is 4.74.

b. The pH of 1.0 L of the buffer solution in part (a) after the addition of 0.01 mole of hydrochloric acid to the solution.

c. The pH of 1.0 L of the buffer solution in part (a) after 0.005 mole of hydroxide ion has been added.

Solution

The equilibrium involved is

$$CH_3CO_2H \rightleftharpoons H^+ + CH_3CO_2^-$$

The original concentrations are

$$[CH_3CO_2H] = 0.10M$$
$$[CH_3CO_2^-] = 0.10M$$

a. Calculate the pH of the buffer solution using the Henderson-Hasselbalch equation.

$$pH = pK_a + \log \frac{[A^-]}{[HA]}$$

$$= 4.74 + \log \frac{0.10}{0.10} = 4.74$$

Notice that the pH of a solution containing equal amounts of a weak acid and the anion of that weak acid equals the pK_a of the acid.

b. Before the addition of HCl, 1.0 L of the solution contains 0.10 mole each of acetate ion and acetic acid:

$$[CH_3CO_2^-] = 0.10M$$
$$[CH_3CO_2H] = 0.10M$$

The 0.01 mole of hydrochloric acid added to the solution is completely ionized and therefore adds 0.01 mole of H^+ to the solution. The added hydrogen ions combine with acetate ions to form more un-ionized molecules of acetic acid. Therefore, the concentration of acetate ion decreases and the concentration of acetic acid in the solution increases:

$$[CH_3CO_2^-] = 0.10 - 0.01 = 0.09M$$
$$[CH_3CO_2H] = 0.10 + 0.01 = 0.11M$$

We can use the Henderson-Hasselbalch equation to calculate the resulting pH.

$$pH = pK_a + \log \frac{[CH_3CO_2^-]}{[CH_3CO_2H]}$$

$$pH = 4.74 + \log \frac{0.09}{0.11} = 4.74 + (-0.09) = 4.65$$

This is a change of 0.09 pH units.

c. Before addition of NaOH, 1 L of the solution contained 0.10 mole of acetic acid and 0.10 mole of acetate ion. The added NaOH (0.005 mole) reacts with the acetic acid to form 0.005 mole sodium acetate. This increases the concentration of acetate ion by 0.005 mole and decreases the concentration of acetic acid by the same amount.

$$[CH_3CO_2^-] = 0.10M + 0.005M = 0.105M$$
$$[CH_3CO_2H] = 0.10M - 0.005M = 0.095M$$

The resulting pH can be calculated using the equation, $pH = pK_a + \log \frac{[A^-]}{[HA]}$:

$$pH = 4.74 + \log \frac{0.105}{0.095} = 4.74 + 0.04 = 4.78$$

This is a change of 0.04 pH units from the original solution.

Problem 8.10

One liter of a buffer solution contains 0.10 mole of formic acid and 0.20 mole of sodium formate.

a. Calculate the pH of this solution.

b. Calculate the pH after 0.01 mole of hydrogen ion is added to one liter of the buffer.

c. Calculate the pH after 0.01 mole of hydroxide ion is added to one liter of the buffer.

The addition of small amounts of either hydrogen or hydroxide ions to a buffer does not appreciably change the pH of that solution. Let's see how different the results would be if, instead of using a solution of acetic acid, we used a solution of a strong acid with the same pH.

Example 8.11

A solution contains the same number of moles of HCl and NaCl and has a pH of 4.74. Calculate the final pH of one liter of this solution after the addition of 0.01 mole of hydrogen ion.

Solution

Because HCl is a strong acid, it is completely ionized. There is no tendency for Cl^- to combine with H^+ to form HCl molecules. At pH 4.74, the concentration of hydrogen ion is $1.8 \times 10^{-5}M$. Any addition of H^+ will simply increase this concentration. The final $[H^+]$ of the solution will be:

$$0.01 + 0.000018 = 0.010018M \quad \text{or} \quad pH = 2$$

Thus, the addition of a small amount of H^+ to a very dilute solution of a strong acid causes a large change in pH.

Problem 8.11

Calculate the change in pH when 0.01 mole of hydroxide ion is added to one liter of $0.000018M$ HCl.

C. Choosing a Buffer

1. pH. Buffering is most effective if the pH required for the reaction is close to the pK_a of the weak electrolyte of the buffer.

2. Concentration. The **concentration of a buffer** refers to the total concentration of both the weak acid and the anion of the weak acid. A $0.1M$ acetate buffer may be made up of 0.025 mole of acetic acid and 0.075 mole of sodium acetate in a liter of solution, or any other combination in which $[CH_3CO_2H] + [CH_3CO_2Na] = 0.1$ M. The buffer used in Figure 8.10 has a concentration of $0.2M$. At each point on the graph, the sum of the concentrations of acetic acid and acetate ion equals 0.2. The concentration of the buffer should be greater than the amount of H^+ or OH^- that will be produced by the reaction being buffered.

3. Capacity. The capacity of a buffer, or its effectiveness, is the amount of hydrogen or hydroxide ion that the buffer can absorb without a significant change in pH. Capacity depends on concentration and pH. An effective buffer has a pH that equals the pK_a of the weak acid, ± 1. We can see the significance of this range by using the Henderson-Hasselbalch equation. If the concentration of the acid, HA, is ten times that of the salt, A$^-$, then:

$$pH = pK_a + \log \frac{[A^-]}{[HA]}$$

$$= pK_a + \log \frac{1}{10}$$

$$= pK_a - 1$$

and the pH is one unit less than the pK_a of the weak acid.

When the concentration of the salt, A$^-$, is ten times that of the acid, HA, then:

$$pH = pK_a + \log \frac{[A^-]}{[HA]}$$

$$= pK_a + \log \frac{10}{1}$$

$$= pK_a + 1$$

and the pH of the solution is one unit greater than the pK_a of the weak acid.

For acetic acid, pK_a is 4.74; therefore, a solution of acetic acid and sodium acetate will function as a buffer within a pH range of approximately 3.74–5.74.

The most effective buffer is one with equal concentrations of a weak acid and its salt, that is, one in which the pH of the solution is equal to the

Table 8.8 Some common buffers and the pH at which each is most effective.

Name of Acid	Acid Formula	K_a	pK_a of Anion	pH of Buffer when $\frac{\text{anion}}{\text{acid}} = 1$
phosphoric	H_3PO_4	7.5×10^{-3}	$H_2PO_4^-$	2.12
formic	HCO_2H	1.8×10^{-4}	HCO_2^-	3.74
acetic	CH_3CO_2H	1.8×10^{-5}	$CH_3CO_2^-$	4.74
carbonic	$CO_2 + H_2O$	4.3×10^{-7}	HCO_3^-	6.37
dihydrogen phosphate	$H_2PO_4^-$	6.2×10^{-8}	HPO_4^{2-}	7.21
bicarbonate	HCO_3^-	5.6×10^{-11}	CO_3^{2-}	10.25
monohydrogen phosphate	HPO_4^{2-}	2.2×10^{-13}	PO_4^{3-}	12.67

pK_a of the weak acid. While a solution of acetic acid and sodium acetate will buffer within the pH range 3.74–5.74, it is most effective as a buffer at, or very near, pH 4.74. Table 8.8 lists some common buffers and the pH at which each is most effective.

8.9 Acid-Base Balance in Blood Plasma

In a healthy person, the pH of blood remains at a remarkably constant level of 7.35–7.45. The principal buffer of blood plasma is bicarbonate. Carbonic acid is a diprotic acid. Since about 99% of the carbonic acid in blood is in the form of dissolved carbon dioxide, it is correct to refer to the acid form as CO_2. In terms of buffering action in blood plasma, the first dissociation constant is most important. The first dissociation of carbonic acid can be written

$$CO_2 + H_2O \qquad HCO_3^- + H^+$$

The solubility of CO_2 in plasma is greater than that in water. In blood, the pK_a for this reaction is 6.1. By using the Henderson-Hasselbalch equation we can calculate the ratio of the acid form, CO_2, to salt form HCO_3^-, at pH 7.4.

$$pH = pK_a + \log \frac{[HCO_3^-]}{[CO_2]}$$

$$7.4 = 6.1 + \log \frac{[HCO_3^-]}{[CO_2]}$$

Solving this equation shows that in blood plasma at pH 7.4,

$$\frac{[HCO_3^-]}{[CO_2]} = \frac{20}{1}$$

Thus, in blood plasma at pH 7.4, the ratio of bicarbonate to carbonic acid is 20 to 1. The normal concentration of bicarbonate in plasma is about 0.025 mole per liter and that of carbonic acid is about 0.0012 mole per liter. Therefore, the concentration of this buffer system is approximately 0.026 mole per liter, consisting mainly of bicarbonate.

Recall from our discussion in Section 8.8 that a weak acid and its salt are most effective as an acid-base buffer in the concentration range from 10% salt/90% acid to 90% salt/10% acid, that is, in the region $pH = pK_a \pm 1$. The pK_a of carbonic acid is 1.3 units smaller than the pH of blood. At pH 7.4, this buffer is approximately 95% bicarbonate and 5% carbonic acid (CO_2). We would predict that this buffer system, being at the outer limit of the useful range, would not be effective—yet it is. Its effectiveness is enhanced by the respiratory system, which provides a means for making very rapid adjustments in the concentration of carbon dioxide in blood. In addition, the kidneys provide a means for making slower, long-term adjustments in the concentration of bicarbonate. Through the cooperative interaction of these two systems, the bicarbonate to carbon dioxide ratio can be kept very close to 20 to 1.

From the clinical standpoint, respiratory acidosis, respiratory alkalosis, metabolic acidosis, and metabolic alkalosis are the four major disturbances in acid-base balance. **Acidosis** is brought about by any abnormal condition that leads to the accumulation of excess acid in the body or excessive loss of alkali. In acidosis, the pH of blood falls below 7.30. Two apparently unrelated activities, starvation and unusually strenuous physical activity, can cause **alkalosis**: loss of acid or accumulation in the body of excess alkali. In alkalosis, the pH of blood rises above 7.50. Alkalosis may also occur after vomiting or forced breathing.

Any chronic respiratory difficulty or depression of breathing rate can increase the carbon dioxide concentration in blood plasma, causing acidosis:

$$\text{pH} = \text{p}K_a + \log \frac{[\text{HCO}_3^-]}{[\text{CO}_2]} \quad \begin{array}{l} \textit{This increases in chronic respiratory} \\ \textit{difficulty or depression of breathing; as a} \\ \textit{consequence, pH decreases.} \end{array}$$

Because of this increase, the ratio of HCO_3^- to CO_2 decreases to something less than 20 to 1, and the pH decreases. The blood becomes more acidic. Even holding your breath to get rid of the hiccups can temporarily decrease the pH of blood to 7.30 or below.

Hyperventilation, or prolonged rapid, deep breathing, results in a "blow off" or decrease in the concentration of dissolved CO_2:

$$\text{pH} = \text{p}K_a + \log \frac{[\text{HCO}_3^-]}{[\text{CO}_2]} \quad \textit{When this decreases, pH increases.}$$

Because the concentration of CO_2 decreases, the ratio of HCO_3^- to CO_2 increases to greater than 20 to 1 and the pH of blood increases. Even mild hyperventilation can increase blood pH to as high as 7.51.

Whenever respiratory alkalosis or respiratory acidosis occurs, the kidneys act to bring conditions back to normal. They will try to increase or decrease the concentration of HCO_3^- in an effort to restore the 20 to 1 ratio of bicarbonate to carbonic acid in normal blood. In the case of respiratory acidosis, the kidneys compensate by increasing the reabsorption of bicarbonate. As long as the concentration of carbon dioxide is elevated, the kidneys will stabilize the bicarbonate at an elevated level.

In respiratory alkalosis, the kidneys will increase the concentration of H^+, thereby allowing the reaction of H^+ and HCO_3^- to form more carbonic acid. Instead of excreting hydrogen ions in an acidic urine, the kidneys will excrete other cations, mainly Na^+ and K^+, and the urine will become less acidic or even slightly alkaline.

Metabolic acidosis may result from a variety of causes, including increased biosynthesis of acids such as the *ketone bodies* (see Section 25.4) and decreased excretion of hydrogen ion due to kidney failure. Metabolic alkalosis may result from impaired nitrogen metabolism or any other factor that leads to an increase in the production of bases.

Key Terms and Concepts

acid dissociation constant, K_a (8.5B)

acidosis (8.9)

activation energy (8.1c)

alkalosis (8.9)

blood pH (8.9)

buffers (8.8)

catalyst (8.2)

chemical equilibrium (8.3)

endergonic (8.1B)

entropy (S) (8.1B)

enzymes (8.2)

equilibrium (8.3)

equilibrium constant (8.2C)

exergonic (8.1B)

free energy (G) (8.1B)

Henderson-Hasselbalch equation (8.7B)

hydrolysis (8.7C)

ionization constant of water (8.6)

Le Chatelier's principle (8.4)

molecular collisions (8.1C)

pH (8.5D)

pK_a (8.5E)

rate of reaction (8.2)

reversible reactions (8.3A)

spontaneous reactions (8.1B)

strong acids (8.5C1)

weak acids (8.7A)

weak electrolytes (8.5)

Problems

The Rate of a Reaction (Section 8.2)

8.12 Plot the course of a reaction against energy for:
  a. an exothermic reaction
  b. an endothermic reaction
  c. an exothermic reaction in the presence of a catalyst

8.13 Label the activation energy and the enthalpy in each part of Problem 8.12.

8.14 Explain why increasing the concentration of the reactants may increase the rate of a reaction. Explain why increasing the pressure on a reaction involving gases increases the rate of the reaction.

8.15 Describe how a catalyst can change the rate of a reaction.

8.16 Draw a curve showing the normal distribution of energies in a collection of molecules. Draw another curve showing how this distribution of energies changes as the temperature is increased. Why does the rate of a reaction increase with increased temperature?

8.17 What role do enzymes play in biological reactions?

8.18 Consider the reaction:

$$4\,NH_3(g) + 5\,O_2(g) \longrightarrow 4\,NO(g) + 6\,H_2O(g) \qquad \Delta H = -385 \text{ kcal}$$

What changes would increase the rate of the reaction?

8.19 Calculate the equilibrium constant for the hypothetical reaction

$$AB \rightleftharpoons A + B$$

if the concentrations at equilibrium are $[AB] = 2$, $[A] = 2$, $[B] = 2$. After a stress has been absorbed by this reaction, $[A] = 8$ and $[AB] = 16$. What is the new concentration of B?

8.20 The Haber process for the industrial preparation of ammonia uses the reaction

$$N_2(g) + 3 H_2(g) \rightleftharpoons 2 NH_3(g) \qquad \Delta H = -22 \text{ kcal/mole}$$

   a. Write the equilibrium constant for this reaction.
   b. How will increased concentration of nitrogen change the concentration of ammonia?
   c. How will increased pressure change the concentration of ammonia?
   d. How will the addition of a catalyst change the equilibrium?
   e. How will increased temperature change the concentration of ammonia?

8.21 Write the equilibrium constant for each of the following equilibria.
   a. The ionization of propionic acid, $C_2H_5CO_2H$.
   b. The reaction of nitric oxide with oxygen to form nitrogen dioxide.

8.22 One of the components of smog is the brown gas nitrogen dioxide, NO_2. It participates in the equilibrium

$$2 NO_2 \rightleftharpoons N_2O_4 \qquad \Delta H = -14.7 \text{ kcal/mole}$$

Dinitrogen tetroxide (N_2O_4) is colorless. Is smog apt to be darker in winter or in summer? If the nitrogen dioxide were in a closed cylinder, would the gas get lighter or darker as the pressure is increased?

8.23 Using the data in Table 8.3, calculate the $[H^+]$ in $0.10M$ HCN. Calculate the pH of one liter of this solution to which 0.1 mole solid sodium cyanide has been added.

8.24 Using the data in Table 8.3 for the ionization of NH_4^+ as a weak acid, calculate the pH of a solution of $0.10M$ NH_4Cl.

8.25 Calculate the pH of a solution of:
   a. $1 \times 10^{-3}M$ HCl
   b. $0.10M$ $Ca(OH)_2$
   c. one liter of $0.1M$ HCl containing 0.1 mole sodium chloride

8.26 Using the Henderson-Hasselbalch equation, calculate the pH of one liter of the following solutions:
   a. $0.5M$ CH_3CO_2H and $0.1M$ CH_3CO_2Na
   b. $0.5M$ CH_3CO_2H and $0.5M$ CH_3CO_2Na
   c. $0.1M$ HCO_2H and $0.25M$ HCO_2K

8.27 Vinegar is a 5% solution (weight/volume) of acetic acid in water.
   a. Calculate the molarity of the acetic acid.
   b. Calculate the hydrogen ion concentration and pH of this solution.
   c. What volume of $0.10M$ sodium hydroxide would be necessary to react completely with 100 cm³ vinegar?

**Solutions of Acids
and Their Anions;
Buffers (Sections 8.7
and 8.8)**

8.28 Given pure acetic acid and pure sodium acetate, what weight of each would you mix in water to prepare one liter of a $0.1M$ buffer solution of pH 4.74? The total concentration of acetic acid and sodium acetate should equal 0.1 mole.

8.29 Calculate the pH change on adding 0.04 g of solid NaOH to 100 cm³ of a solution containing 0.05 mole acetic acid and 0.05 mole sodium acetate.

8.30 In acidosis, the kidneys may restore acid-base balance by excreting urine of pH 4.8. Calculate the ratio of HPO_4^{2-} to $H_2PO_4^-$ at this pH.

8.31 In alkalosis, the urine may have a pH of 8.0. What ratio of HPO_4^{2-} to $H_2PO_4^-$ would give this pH?

8.32 Calculate the ratio of bicarbonate ion to carbonate ion in a solution of pH 10.

$$HCO_3^- \rightleftharpoons CO_3^{2-} + H^+ \qquad K_a = 5.6 \times 10^{-11}$$

8.33 Calculate the pH of a solution, 2 liters of which contain 4.662 g of KH_2PO_4 and 16.595 g of K_2HPO_4.

8.34 A solution is prepared by adding 6.36 g of sodium carbonate (Na_2CO_3) and 3.36 g of sodium bicarbonate ($NaHCO_3$) to water and making the total volume of the solution 1.00 liter. What is the pH of the solution?

8.35 What weight of sodium bicarbonate and of sodium carbonate would you use to make 1.00 liter of a buffer solution of $0.1M$ concentration and pH 11.03? of pH 9.95?

8.36 a. Write the equilibrium constant, K_{a_1} for the reaction:

$$NH_3 + H_2O \rightleftharpoons NH_4^+ + OH^-$$

b. Write the equilibrium constant, K_{a_2}, for the reaction:

$$NH_4^+ \rightleftharpoons NH_3 + H^+$$

c. Show for these reactions that $K_{a_1} \times K_{a_2} = K_w$.

8.37 Calculate the $[H^+]$ and the pH of a $0.1M$ solution of nitrous acid which is also $0.1M$ in nitrite ion (K_a, $HNO_2 = 4.6 \times 10^{-4}$).

8.38 Calculate the $[H^+]$ and the pH of a $0.01M$ solution of acrylic acid whose $K_a = 5.6 \times 10^{-5}$.

8.39 The ionization constant for butyric acid is 1.5×10^{-5}. Calculate the pH of a solution which is $0.01M$ in butyric acid and $0.02M$ in sodium butyrate.

8.40 At body temperature, $pK_w = 13.6$. Calculate $[H^+]$, $[OH^-]$, pH, and pOH for pure water at body temperature.

Organic Chemistry: The Study of the Compounds of Carbon

Organic chemistry is the study of the compounds of carbon. While the term *organic* reminds us that a great many compounds of carbon are of either plant or animal origin, by no means is that the limit of organic chemistry. Certainly, organic chemistry is the study of the naturally occurring medicines such as penicillin, cortisone, and streptomycin; but it also includes the study of novocaine, the sulfa drugs, aspirin, and other man-made medicines. Organic chemistry is the study of naturally occurring textile fibers such as cotton, silk and wool. It is also the study of man-made textile fibers such as nylon, Dacron, Orlon, and rayon. Organic chemistry is the study of Saran, Teflon, Styrofoam, polyethylene, and other man-made polymers used to manufacture the films and molded plastics with which we are so familiar today. The list could go on and on.

Perhaps the most remarkable feature of organic chemistry is that it comprises the chemistry of carbon and only a few other elements: chiefly hydrogen, nitrogen, and oxygen. Let us begin this chapter with a review of how atoms of these elements combine to form molecules.

9.1 The Covalence of Carbon, Hydrogen, Oxygen, and Nitrogen

Table 9.1 Lewis structures for hydrogen, carbon, nitrogen, and oxygen.

Recall from Section 4.5 that in discussing covalent bonding in hydrogen, carbon, nitrogen, and oxygen, we can focus our attention on the electrons in the outermost or *valence* shell, for it is these electrons that participate in covalent bonding and chemical reactions. Lewis structures for H, C, N, and O are shown in Table 9.1.

The single valence electron of hydrogen belongs to the first principal energy level; this shell is completely filled with two electrons. Thus, hydrogen can form only one covalent bond with another element; hydrogen has a valence of one.

H·   ·Ċ: ·N̈:   ·Ö:

Carbon, with four valence electrons, needs four additional electrons to complete its octet; carbon has a valence of four. This valence of carbon can be satisfied by appropriate combinations of single, double, or triple bonds, as illustrated by methane, ethane, ethylene, and acetylene:

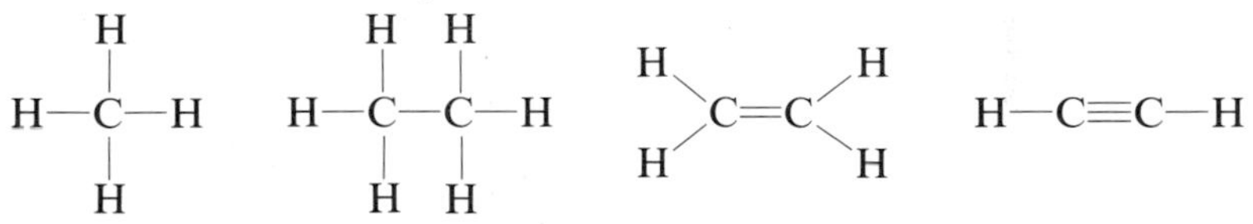

methane   ethane   ethylene   acetylene

Nitrogen, with five valence electrons, needs three additional electrons to complete its octet; nitrogen has a valence of three. This valence can be satisfied by appropriate combinations of single, double, or triple bonds, as illustrated by ammonia, nitrous acid, hydrogen cyanide, and nitrogen:

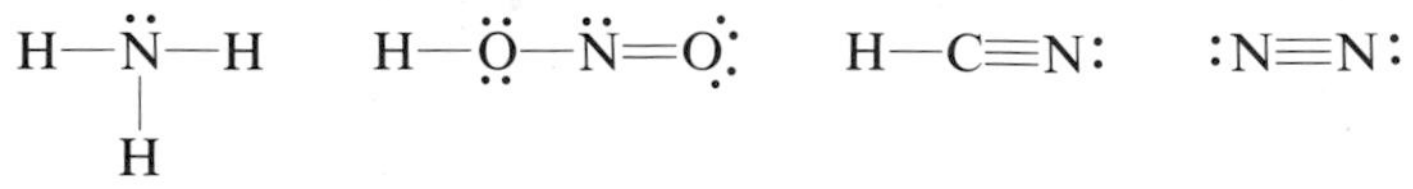

ammonia   nitrous acid   hydrogen cyanide   nitrogen molecule

In each of these neutral, uncharged molecules, nitrogen forms three bonds and

H
|
H—N⁺—H
|
H

ammonium ion

has one unshared pair of electrons. Recall from Section 4.5C that nitrogen can also form four covalent bonds, as in the ammonium ion (NH_4^+) (see margin). In NH_4^+, the nitrogen atom bears a formal charge of $+1$.

Oxygen, with six valence electrons, needs two additional electrons to complete its octet; oxygen has a valence of two. This valence can be satisfied by either two single bonds, as in water and methanol, or by one double bond, as in formaldehyde and carbonic acid.

$$H—\ddot{O}—H \qquad H—\overset{\overset{\displaystyle H}{|}}{\underset{\underset{\displaystyle H}{|}}{C}}—\ddot{O}—H \qquad \overset{H}{\underset{H}{>}}C=\ddot{O}: \qquad H—\ddot{O}—\overset{\overset{\displaystyle :O:}{\|}}{C}—\ddot{O}—H$$

| water | methanol | formaldehyde | carbonic acid |

H—Ö⁺—H
|
H

hydronium ion

In each of these neutral, uncharged molecules, oxygen forms two bonds and has two unshared pairs of electrons. Recall from Section 4.5C that oxygen can also form three covalent bonds, as in the hydronium ion (H_3O^+) (see margin). In H_3O^+, the oxygen atom bears a formal charge of $+1$.

By examining together these molecules and the others used in Chapter 4 to illustrate covalent bonding, we can make the following generalizations. For stable, uncharged molecules containing atoms of carbon, nitrogen, and oxygen:

1. Carbon forms four covalent bonds. Bond angles on carbon are approximately 109.5° for four attached groups, 120° for three attached groups, and 180° for two attached groups.

Four Attached Groups	Three Attached Groups	Two Attached Groups
$-\overset{\displaystyle |}{\underset{\displaystyle |}{C}}-$	$>C=$	$-C\equiv$ $=C=$
109.5°	120°	180°
	$-\overset{\displaystyle ..}{N}-$	$-\overset{\displaystyle ..}{N}=$
	109.5°	120°
	$-\overset{\displaystyle ..}{\underset{\displaystyle ..}{O}}-$	
	109.5°	

Table 9.2 Summary of predicted bond angles for neutral covalent compounds of carbon, nitrogen, and oxygen.

2. Nitrogen forms three covalent bonds and has one unshared pair of electrons. Bond angles on nitrogen are approximately 109.5° for three attached groups and one unshared pair of electrons; they are 120° for two attached groups and one unshared pair of electrons.
3. Oxygen forms two covalent bonds and has two unshared pairs of electrons. Bond angles on oxygen are approximately 109.5° for two attached groups.

These generalizations are summarized in Table 9.2.

9.2 Common Organic Functional Groups

In the preceding section we examined the different types of covalent bonds formed by carbon with hydrogen, nitrogen, and oxygen. These atoms combine in various ways to form characteristic structural features known as **functional groups.** The concept of a functional group is important to organic chemistry for two reasons. First, functional groups are sites of chemical reaction. A particular functional group undergoes the same characteristic reactions in whatever organic molecule it is found. Second, functional groups serve as a basis for nomenclature (naming) of organic compounds.

In this section, we will study the characteristic structural features of five functional groups: alcohols, ethers, aldehydes, ketones, and carboxylic acids.

A. Alcohols and Ethers

The characteristic structural feature of an **alcohol** is an oxygen atom bonded to atoms of carbon and hydrogen. We say that an alcohol contains a **hydroxyl (—OH) group.** The characteristic structural feature of an **ether** is an oxygen atom bonded to two carbon atoms.

Lewis structures:

an alcohol an ether

characteristic structural feature:

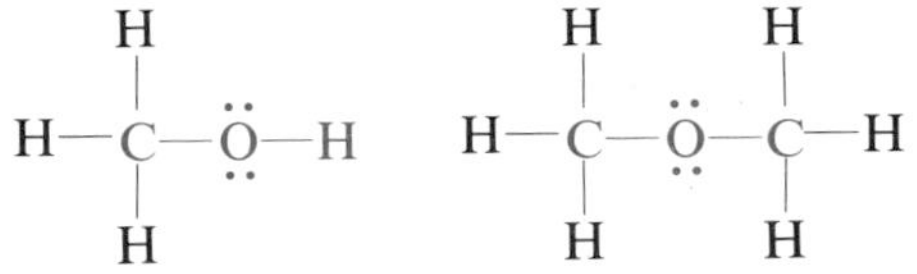

We can write organic formulas in a more abbreviated way by using what are called **condensed structural formulas.** In a condensed structural formula, CH_3— indicates a carbon with three attached hydrogens, —CH_2— indicates a carbon with two attached hydrogens, and $-\overset{|}{C}H-$ indicates a carbon with one attached hydrogen. Following are Lewis structures and condensed structural formulas for the alcohol and ether of molecular formula C_2H_6O.

Lewis structures:

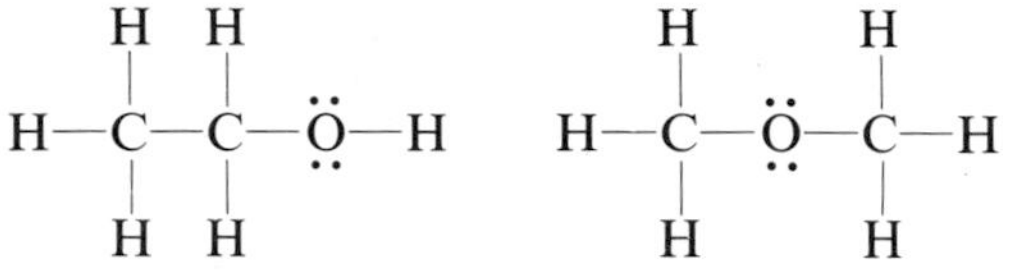

$$\begin{array}{cc} \underset{\underset{H}{|}}{\overset{\overset{H}{|}}{H-C-}} \underset{\underset{H}{|}}{\overset{\overset{H}{|}}{C-}} \ddot{O}-H \qquad & \underset{\underset{H}{|}}{\overset{\overset{H}{|}}{H-C-}} \ddot{O} \underset{\underset{H}{|}}{\overset{\overset{H}{|}}{-C-H}} \end{array}$$

condensed structural formulas: CH_3-CH_2-OH CH_3-O-CH_3

Example 9.1

Draw Lewis structures and write condensed structural formulas for two alcohols of molecular formula C_3H_8O.

Solution

The characteristic structural feature of an alcohol is an atom of oxygen bonded to one carbon and one hydrogen atom:

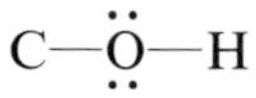

$$C-\ddot{O}-H$$

The molecular formula contains three carbon atoms. These can be bonded together in a chain with the —OH group attached to either the end carbon or the middle carbon of the chain:

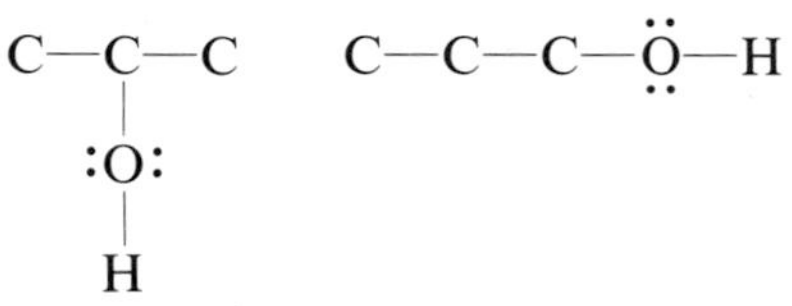

$$C-C-C \qquad C-C-C-\ddot{O}-H$$
$$\underset{\underset{H}{|}}{\overset{|}{:}O:}$$

Finally, add seven hydrogens to each structure to satisfy the tetravalence of carbon and give the correct molecular formulas.

Lewis structures:

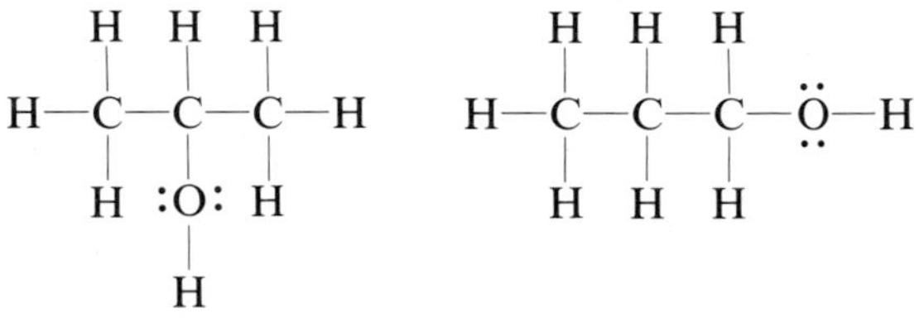

$$\begin{array}{cc} \underset{\underset{\underset{H}{|}}{:O:}}{\overset{\overset{H}{|}}{H-C-}} \underset{}{\overset{\overset{H}{|}}{C-}} \overset{\overset{H}{|}}{C-H} \qquad & \underset{\underset{H}{|}}{\overset{\overset{H}{|}}{H-C-}} \underset{\underset{H}{|}}{\overset{\overset{H}{|}}{C-}} \underset{\underset{H}{|}}{\overset{\overset{H}{|}}{C-}} \ddot{O}-H \end{array}$$

condensed structural formulas: 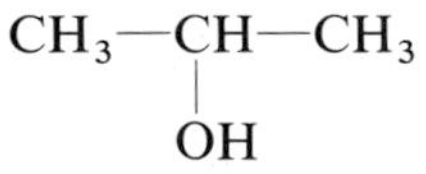 $CH_3-CH-CH_3$ $CH_3-CH_2-CH_2-OH$
                                      $\quad\quad\quad |$
                                      $\quad\quad\;\; OH$

Problem 9.1

Draw the Lewis structure and the condensed structural formula for the ether of molecular formula C_3H_8O.

B. Aldehydes and Ketones

The characteristic structural feature of an **aldehyde** is a **carbonyl group (C=O)** bonded to a hydrogen atom. Formaldehyde (Section 4.5B) is the simplest molecule containing an aldehyde functional group, and it is the only aldehyde that contains two hydrogen atoms bonded to the carbonyl group. All other aldehydes have one carbon and one hydrogen bonded to the carbonyl group. The characteristic structural feature of a **ketone** is a carbonyl group bonded to two carbon atoms.

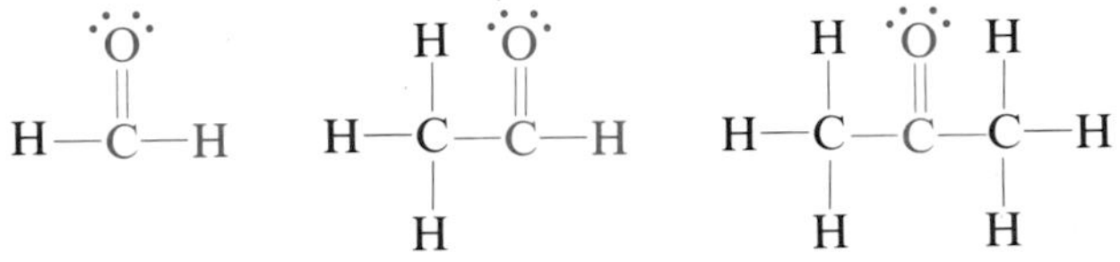

formaldehyde   an aldehyde   a ketone

characteristic structural feature:

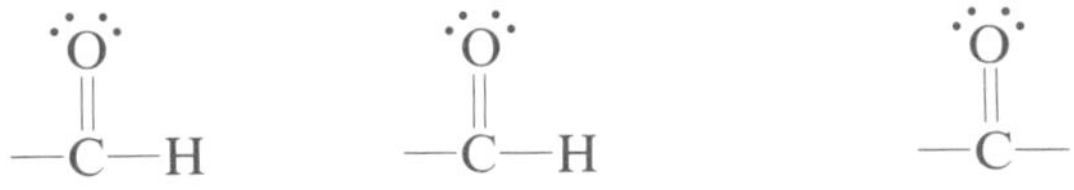

Example 9.2

Draw Lewis structures and condensed structural formulas for two aldehydes of molecular formula C_4H_8O.

Solution

First draw the characteristic structural feature of the aldehyde group and then add the remaining carbons. These may be attached in two different ways:

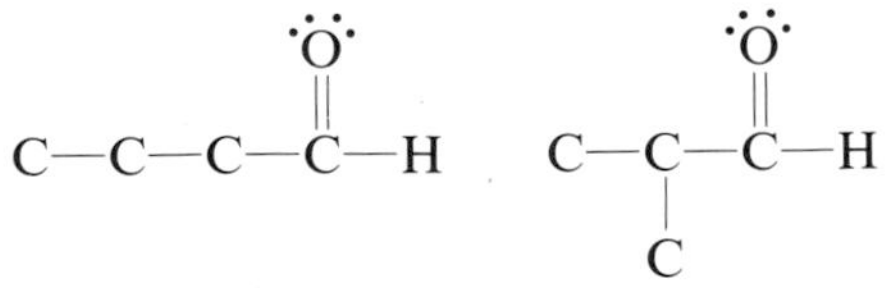

Finally, add seven hydrogens to complete the tetravalence of carbon and give the correct molecular formula.

Lewis structures:

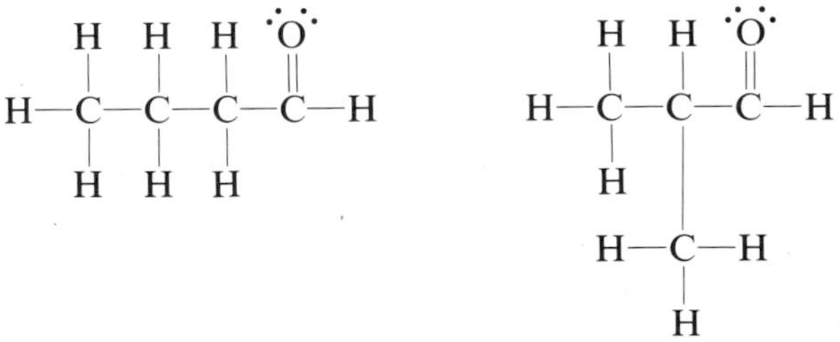

condensed structural formulas:

$$CH_3-CH_2-CH_2-\overset{\displaystyle O}{\overset{\|}{C}}-H \qquad CH_3-\underset{\underset{\displaystyle CH_3}{|}}{CH}-\overset{\displaystyle O}{\overset{\|}{C}}-H$$

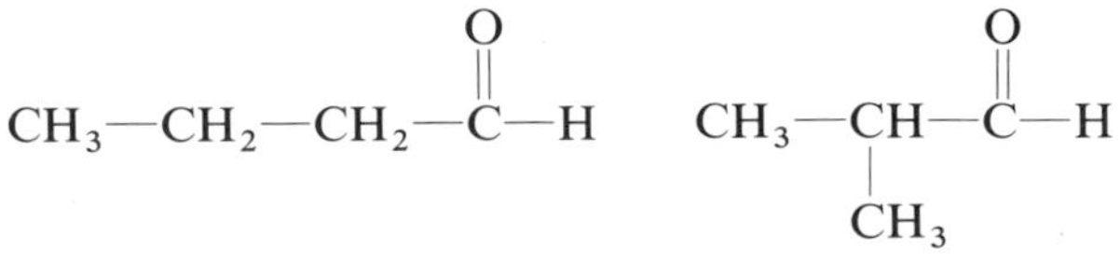

Problem 9.2

Draw Lewis structures and condensed structural formulas for the three ketones of molecular formula $C_5H_{10}O$.

C. Carboxylic Acids

The characteristic structural feature of a **carboxylic acid** is a —CO_2H group. We say that a carboxylic acid contains a **carboxyl** (*carb*onyl + hydr*oxyl*) **group.**

carboxylic acids:

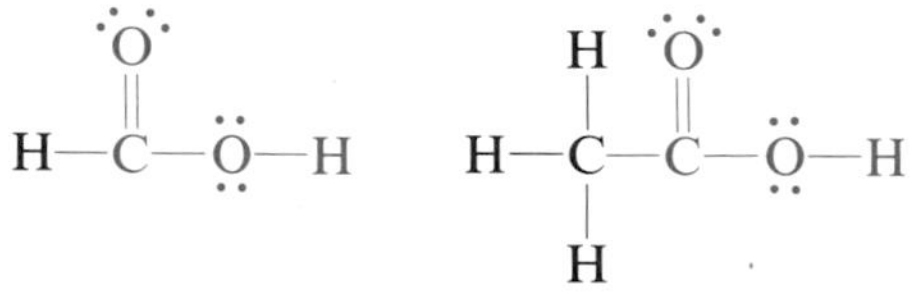

characteristic structural feature:

Example 9.3

Draw Lewis structures and condensed structural formulas for the one carboxylic acid of molecular formula $C_3H_6O_2$.

Solution

Lewis structure:

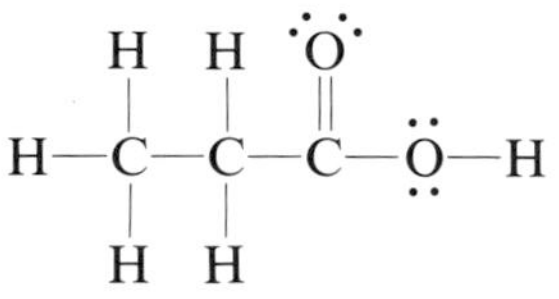

condensed structural formulas: CH_3—CH_2—$\overset{\displaystyle O}{\overset{\|}{C}}$—OH or CH_3—CH_2—CO_2H

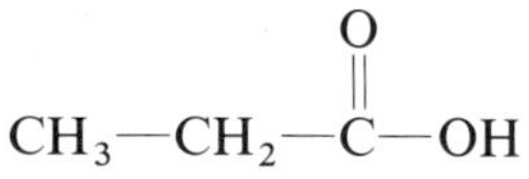

Problem 9.3

Draw Lewis structures and condensed structural formulas for the two carboxylic acids of molecular formula $C_4H_8O_2$.

9.3 Structural Isomerism

In Section 9.2, you saw that for most molecular formulas it is possible to draw more than one structural formula. For example, the following compounds have the same molecular formula, C_2H_6O, but different structural formulas.

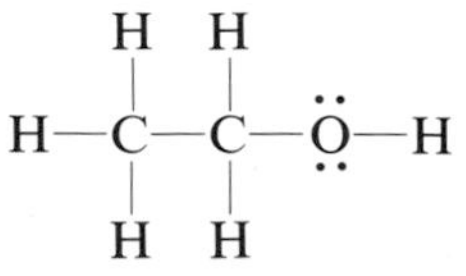

an alcohol 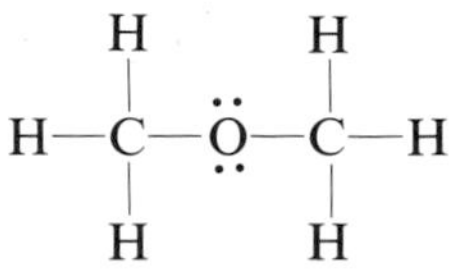   an ether

Similarly, the following alcohols have the same molecular formula, C_3H_8O, but different structural formulas.

$$CH_3-CH_2-CH_2-OH \qquad CH_3-CH-OH$$
$$| \\ CH_3$$

an alcohol   an alcohol

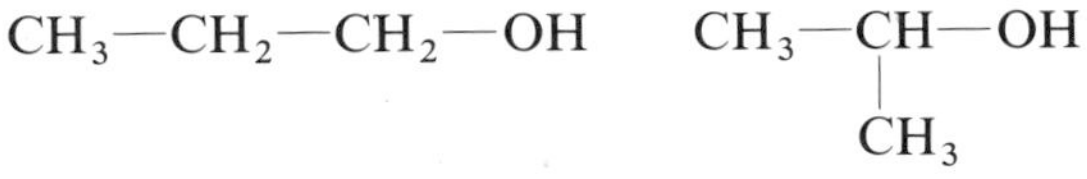

Compounds that have the same molecular formula but different structural formulas are called **structural isomers.** Structural isomers are different compounds and they have different chemical and physical properties.

Example 9.4

Divide the following compounds into groups of structural isomers.

a. $CH_3-CH_2-\overset{\overset{\displaystyle O}{\|}}{C}-O-H$

b. $CH_3-O-CH_2-\overset{\overset{\displaystyle O}{\|}}{C}-H$

c. $CH_2{=}CH-CH_2-O-H$

d. $CH_3-\underset{\underset{\displaystyle CH_3}{|}}{CH}-\overset{\overset{\displaystyle O}{\|}}{C}-H$

e. $CH_3-\overset{\overset{\displaystyle O}{\|}}{C}-O-CH_2-CH_2-CH_3$

f. $CH_3-\overset{\overset{\displaystyle O}{\|}}{C}-CH_2-CH_2-CH_2-OH$

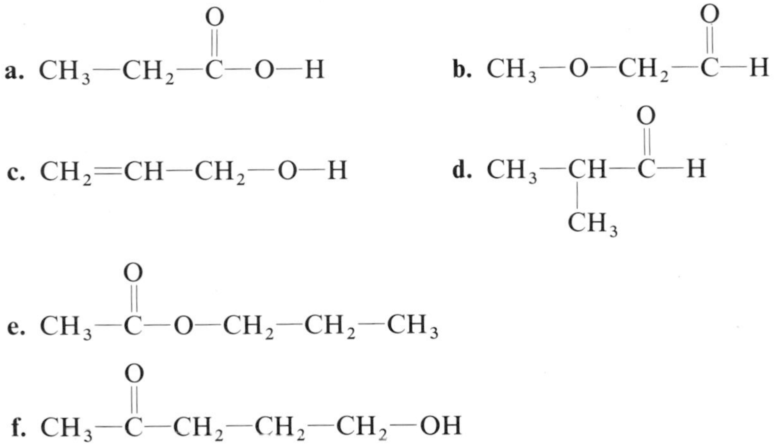

Solution

To determine which are structural isomers, first write the molecular formula of each compound and then compare them. All those that have the same molecular formula and different structural formulas are structural isomers. Compounds (a) and (b) have the same molecular formula, $C_3H_6O_2$, and different structural formulas, so they are structural isomers. Compounds (e) and (f) have the same molecular formula, $C_5H_{10}O_2$, and are structural isomers. There are no structural isomers in this problem for compounds (c) and (d).

Problem 9.4

Divide the following compounds into groups of structural isomers.

a. $CH_2{=}CH-O-CH{=}CH_2$

b. 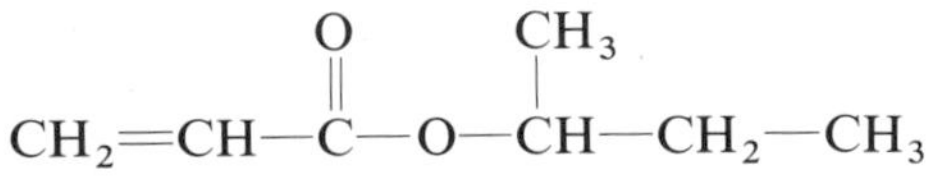 $CH_2{=}CH-\overset{\overset{\displaystyle O}{\|}}{C}-O-\underset{\underset{\displaystyle CH_3}{|}}{CH}-CH_2-CH_3$

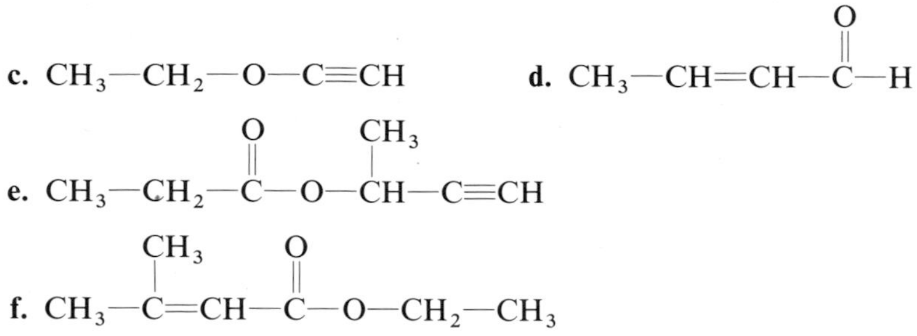

c. $CH_3-CH_2-O-C\equiv CH$ d. $CH_3-CH=CH-\overset{\overset{\displaystyle O}{\|}}{C}-H$

e. $CH_3-CH_2-\overset{\overset{\displaystyle O}{\|}}{C}-O-\overset{\overset{\displaystyle CH_3}{|}}{CH}-C\equiv CH$

f. $CH_3-\overset{\overset{\displaystyle CH_3}{|}}{C}=CH-\overset{\overset{\displaystyle O}{\|}}{C}-O-CH_2-CH_3$

9.4 The Need for Another Model of Covalent Bonding

As much as the Lewis and valence-shell electron-pair repulsion models (Section 4.6) have helped us understand covalent bonding and the geometry of organic molecules, they leave many important questions unanswered. The most important of these is the relationship between molecular structure and chemical reactivity. For example, a carbon-carbon single bond is quite different in chemical reactivity from a carbon-carbon double bond. Most carbon-carbon single bonds are quite unreactive, but carbon-carbon double bonds react with a variety of reactants under a variety of experimental conditions. The Lewis model of bonding gives us no way to account for these differences. Yet, to discuss modern organic and biochemistry, we must have an understanding of how chemists account for such differences. Therefore, let us now look at the formation of covalent bonds in terms of the overlap of atomic orbitals.

9.5 Covalent Bond Formation by the Overlap of Atomic Orbitals

According to modern bonding theory, formation of a covalent bond between two atoms amounts to bringing them together in such a way that an atomic orbital of one atom overlaps an atomic orbital of the other. For example, in forming the covalent bond in H_2, two hydrogen atoms approach each other so that their atomic $1s$ orbitals overlap (Figure 9.1). The new orbital formed by the overlap of two atomic orbitals encompasses both hydrogen nuclei and is

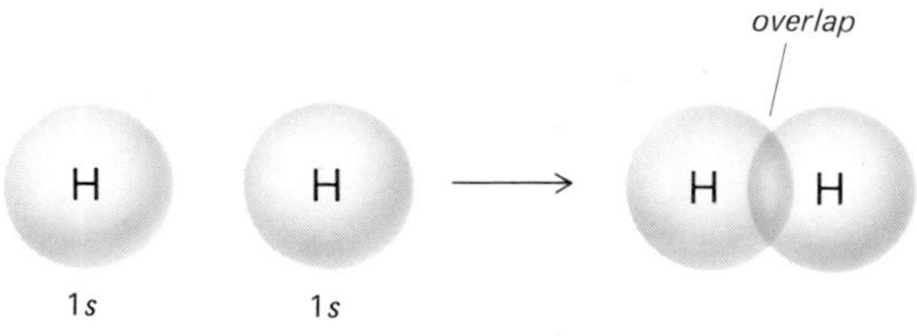

Figure 9.1 The overlap of Is orbitals of two hydrogen atoms to form a sigma bond. The atomic Is orbitals are spherical and each contains one electron. The molecular orbital formed by the overlap of the atomic orbitals contains two electrons.

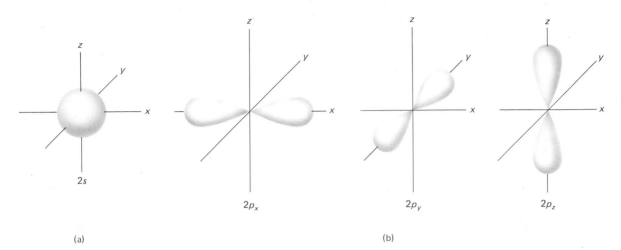

(a)

(b)

Figure 9.2 The atomic orbitals of the second principal energy level: (a) shape of the 2s orbital; (b) shapes of the three 2p orbitals; (c) orientations in space of the three 2p orbitals relative to each other.

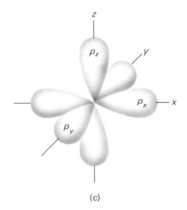

(c)

called a **molecular orbital.** Like an atomic orbital, a molecular orbital can accomodate two electrons.

The molecular orbital resulting from the overlap of two 1s orbitals has the shape of a cylinder whose central axis is the line joining the two nuclei. A chemical bond formed by a molecular orbital with a cylindrical distribution of electrons along the line joining the two nuclei is called a **sigma (σ) bond.**

Carbon, nitrogen, and oxygen form covalent bonds using atomic orbitals of the second principal energy level. The second principal energy level consists of four atomic orbitals: a single 2s orbital and three 2p orbitals. The 2p orbitals are designated $2p_x$, $2p_y$, and $2p_z$ and are oriented along the x-axis, y-axis, and z-axis, respectively. Figure 9.2 shows these atomic orbitals.

The three 2p orbitals form 90° angles with each other, and if atoms of carbon, nitrogen, or oxygen used only 2p orbitals to form covalent bonds, we would expect to find bond angles of 90° about a central atom. However, bond angles of 90° are not observed; instead, we find angles of approximately 109.5°, 120°, or 180°. According to modern bonding theory, atoms can combine

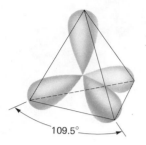

Figure 9.3 sp^3 hybrid orbitals.

Figure 9.4 Orbital overlap diagrams for methane (CH_4), ammonia (NH_3), and water (H_2O) molecules.

$2s$ and $2p$ orbitals to form new atomic orbitals that then form bonds with the angles we do observe. The combination of atomic orbitals is called **hybridization** and the new atomic orbitals formed are called **hybrid atomic orbitals.**

A. sp^3 Hybrid Orbitals

Combination of one $2s$ orbital and three $2p$ orbitals produces four equivalent orbitals called sp^3 **hybrid orbitals** (Figure 9.3). The four sp^3 hybrid orbitals are directed toward the corners of a regular tetrahedron. Thus, sp^3 hybridization always results in orbitals with angles of approximately 109.5°. Figure 9.4 shows orbital overlap diagrams for CH_4, NH_3, and H_2O. The C—H, N—H, and

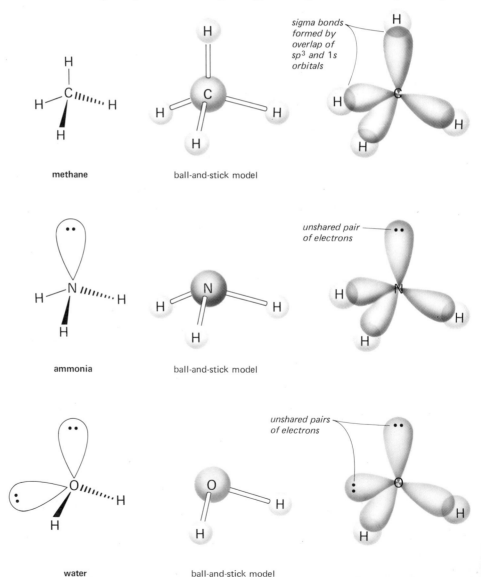

methane   ball-and-stick model

ammonia   ball-and-stick model

water   ball-and-stick model

O—H sigma bonds are formed by overlap of sp^3 orbitals and $1s$ orbitals. The unshared pairs of electrons on nitrogen and oxygen lie in sp^3 orbitals.

B. sp^2 Hybrid Orbitals

Combination of one $2s$ orbital and two $2p$ orbitals produces three equivalent orbitals called **sp^2 hybrid orbitals.** The three sp^2 hybrid orbitals lie in a plane with angles of 120° between them (Figure 9.5a). The remaining unhybridized $2p$ orbital is perpendicular to the plane created by the sp^2 orbitals (Figure 9.5b).

Figure 9.5 An atom in the sp^2 hybrid state. (a) Three sp^2 orbitals in a plane with 120° angles between them; (b) the remaining $2p$ orbital, perpendicular to the plane of the sp^2 orbitals.

(a)

(b)

The sp^2 orbitals are used to form double bonds. Consider ethylene, C_2H_4. In this compound, two carbon atoms are arranged so that they form a sigma bond by the overlap of sp^2 hybrid orbitals (Figure 9.6b). Each carbon also forms sigma bonds to two hydrogens by the overlap of sp^2 orbitals of carbon and $1s$ orbitals of hydrogen. The remaining $2p$ orbitals on adjacent carbons overlap to form a second bond called a **pi (π) bond** (Figure 9.6c). A pi bond

Figure 9.6 Covalent bond formation in ethylene. (a) Lewis structure; (b) a sigma bond between carbon atoms is formed by the overlap of sp^2 orbitals (p orbitals are shown uncombined); (c) the overlap of parallel $2p$ orbitals forms a pi bond.

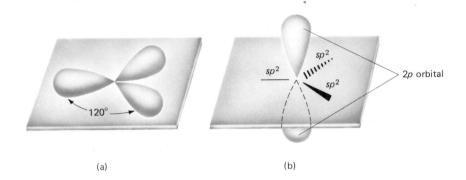

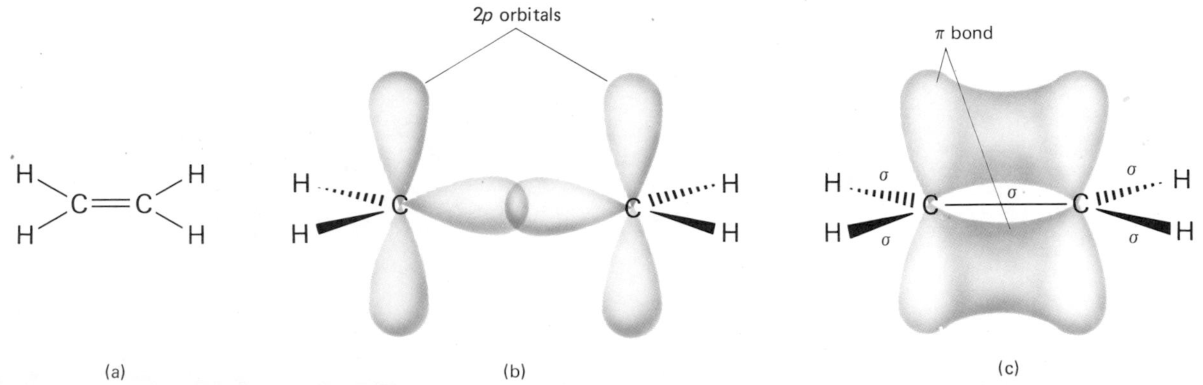

(a)

(b)

(c)

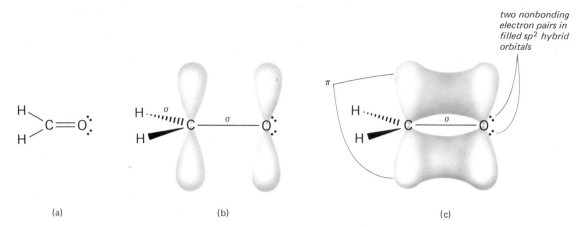

two nonbonding electron pairs in filled sp² hybrid orbitals

(a)   (b)   (c)

Figure 9.7 Formation of the carbon-oxygen double bond in formaldehyde. (a) Lewis structure; (b) sigma bonds shown without the 2p orbitals overlapping; (c) overlap of parallel 2p orbitals to form a pi bond.

consists of two sausage-shaped regions of electron density, one on either side of the plane formed by the carbon and hydrogen atoms. Because the degree of overlap is less, a pi bond joining the two carbons is weaker than a sigma bond.

A carbon-oxygen double bond can be described in the same way as a carbon-carbon double bond. In formaldehyde, the simplest organic molecule containing a carbon-oxygen double bond, carbon forms sigma bonds to two hydrogen atoms by the overlap of $1s$ orbitals of hydrogen and sp^2 orbitals of carbon (Figure 9.7). Carbon and oxygen are joined by a sigma bond (formed by the overlap of sp^2 orbitals of carbon and oxygen) and a pi bond (formed by the overlap of $2p$ orbitals).

C. *sp* Hybrid Orbitals

Combination of one $2s$ orbital and one $2p$ orbital produces two equivalent orbitals, called ***sp* hybrid orbitals,** that lie in a straight line at an angle of 180° (Figure 9.8). *sp* Hybridization results in bond angles of 180°.

Figure 9.8 An atom in the *sp* hybrid state. (a) Two *sp* hybrid orbitals; (b) the two remaining 2p orbitals are perpendicular to each other and to the linear *sp* orbitals.

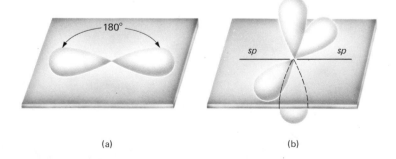

180°

sp   sp

(a)   (b)

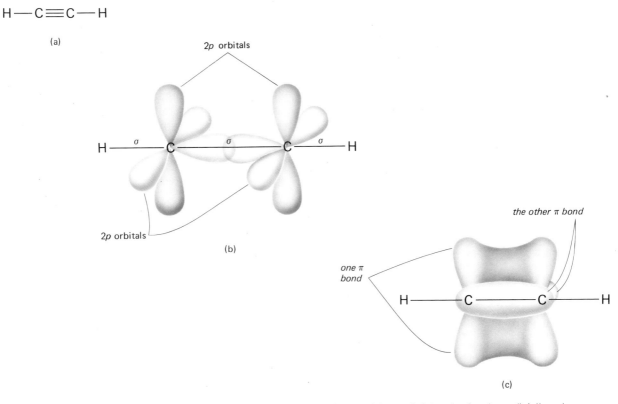

Figure 9.9 Covalent bonding in acetylene. (a) Lewis structure; (b) the sigma-bond framework, shown along with the 2p orbitals; (c) formation of two pi bonds by the overlap of two sets of 2p orbitals.

Figure 9.9 shows a Lewis structure for acetylene, C_2H_2. The carbon-carbon triple bond consists of one sigma bond and two pi bonds. The sigma bond is formed by the overlap of one *sp* hybrid orbital from each carbon. One carbon-carbon pi bond is formed by the overlap of unhybridized 2p orbitals in a plane; the other pi bond is formed by the overlap of the remaining unhybridized 2p orbitals in a second, perpendicular plane.

Example 9.5

Describe the bonding in the following molecules in terms of the orbitals involved and, based on this orbital description, predict all bond angles.

a. $H-\overset{\overset{\displaystyle H}{|}}{\underset{\underset{\displaystyle H}{|}}{C}}-\overset{..}{\underset{..}{O}}-H$

b. $H-\overset{\overset{\displaystyle H}{|}}{\underset{\underset{\displaystyle H}{|}}{C}}-\overset{\overset{\displaystyle ..\overset{.}{O}.}{\parallel}}{C}-\overset{..}{\underset{..}{O}}-H$

Solution

The problem here is how to show (1) the hybridization of each atom, (2) the orbitals involved in each covalent bond, and (3) all bond angles. This is done below in three separate diagrams. Labels on the first diagram point to atoms and show the hybridization of each atom. Labels on the second diagram point to bonds and show the type of bond, either sigma or pi. The orbitals involved in each covalent bond can be determined from the state of hybridization of each atom. Labels on the third diagram point to atoms and show predicted bond angles about each atom.

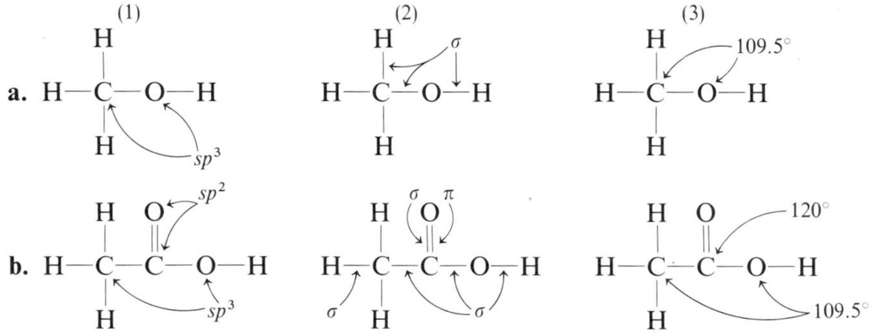

Problem 9.5

Describe the bonding in the following molecules in terms of the orbitals involved and, based on this orbital description, predict all bond angles.

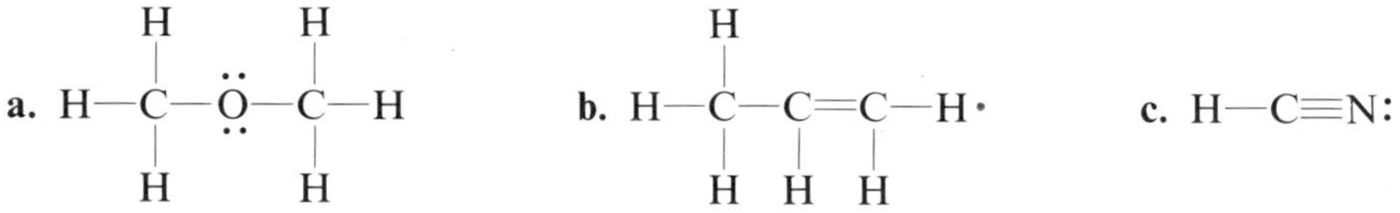

If you compare the predictions of bond angles in the small molecules we have studied in this chapter, you will see that the valence-shell electron-pair repulsion model (Section 4.6) and the orbital overlap model give equally good predictions. However, as we have seen, the orbital overlap model gives us a more useful understanding of double and triple bonds. A double bond is not a combination of two identical bonds. Rather, it consists of one sigma bond and one pi bond. The distribution of the bonding electrons in a pi bond is quite different from that in a sigma bond. Similarly, a triple bond is not a combination of three identical bonds. Rather, it consists of one sigma bond and two pi bonds. We will use the orbital overlap picture of covalent bonding in later chapters to help us understand why compounds containing double and triple bonds have quite different chemical properties from compounds containing only single bonds.

Key Terms and Concepts

alcohol (9.2A)

aldehyde (9.2B)

carbonyl group (9.2B)

carboxyl group (9.2C)

condensed structural formula (9.2A)

ether (9.2A)

functional group (9.2)

sp hybrid orbitals (9.5C)

sp^2 hybrid orbitals (9.5B)

sp^3 hybrid orbitals (9.5A)

hybridization (9.5)

hydroxyl group (9.2A)

ketone (9.2B)

molecular orbital (9.5)

pi bond (9.5B)

sigma bond (9.5)

structural isomerism (9.3)

Problems

Lewis Structures of Covalent Molecules and Ions (Section 9.1; also review Sections 4.3 and 4.5)

9.6   Write Lewis structures for the following atoms and state how many electrons each must gain to achieve the same electron configuration as that of the noble gas nearest it in the periodic table.

a. carbon

b. nitrogen

c. sulfur

d. oxygen

e. hydrogen

f. chlorine

g. phosphorus

h. bromine

i. iodine

9.7   Write Lewis structures for the following molecules. Be certain to show all valence electrons.

a. H_2O_2

b. N_2H_4

c. CH_3OH

d. CH_3SH

e. CH_3NH_2

f. CH_3Cl

g. CH_3OCH_3

h. C_2H_6

i. C_2H_4

j. C_2H_2

k. CO_2

l. H_2CO_3

m. CH_2O

n. CH_3CHO

o. CH_3COCH_3

p. HCO_2H

q. CH_3CO_2H

r. $CH_3CO_2CH_3$

9.8   Write Lewis structures for the following ions. Be certain to show all valence electrons and to assign formal charges.

a. OH^-

b. H_3O^+

c. NH_4^+

d. Cl^-

e. HCO_3^-

f. CO_3^{2-}

g. HCO_2^-

h. $CH_3CO_2^-$

9.9   The following compounds contain both ionic and covalent bonds. Write a Lewis structure for each compound. Use a dash to show electrons shared in covalent bonds and use plus and minus signs to show ionic bonds.

a. NaOH

b. NH_4Cl

c. $NaHCO_3$

d. Na_2CO_3

e. CH_3ONa

f. CH_3CO_2Na

Bond Angles and Shapes of Covalent Molecules (Section 9.1; also review Section 4.6)

9.10 Explain how the valence-shell electron-pair repulsion model is used to predict bond angles.

9.11 Following are Lewis structures for several molecules and ions. Use the valence-shell electron-pair repulsion model to predict bond angles about each circled atom.

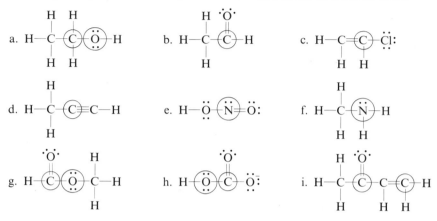

9.12 For the following molecules, show all valence electrons on each atom of nitrogen and oxygen. Then, use the valence-shell electron-pair repulsion model to predict bond angles in each molecule.

a. CH_3-CH_2-O-H

b. $CH_3-CH_2-\overset{\overset{\displaystyle O}{\|}}{C}-H$

c. $CH_3-CH=CH_2$

d. $CH_3-C\equiv C-H$

e. CH_3-NH_2

f. $CH_3-\underset{\underset{\displaystyle CH_3}{|}}{N}-CH_3$

g. $CH_3-CH_2-\overset{\overset{\displaystyle O}{\|}}{C}-O-H$

h. $CH_3-\overset{\overset{\displaystyle O}{\|}}{C}-O-CH_3$

i. $CH_3-\overset{\overset{\displaystyle O}{\|}}{C}-CH_3$

The Polarity of Covalent Bonds (review Sections 4.3B and 4.6)

9.13 Arrange the following bonds in order of increasing polarity within each set.
a. C—H, O—H, N—H
b. C—O, P—O, N—O
c. C—O, C—H, C—S
d. C—Cl, C—I, C—Br
e. Na—O, Mg—O, C—O

9.14 For each polar covalent bond in the following molecules, label the partially positive atom with δ^+ and the partially negative atom with δ^-.

a. $H-\underset{\underset{\displaystyle H}{|}}{N}-H$

b. $H-\underset{\underset{\displaystyle H}{|}}{\overset{\overset{\displaystyle H}{|}}{C}}-\underset{\underset{\displaystyle H}{|}}{\overset{\overset{\displaystyle H}{|}}{C}}-\overset{\overset{\displaystyle O}{\|}}{C}-O-H$

c. $H-\underset{\underset{\displaystyle H}{|}}{\overset{\overset{\displaystyle H}{|}}{C}}-\underset{\underset{\displaystyle H}{|}}{\overset{\overset{\displaystyle H}{|}}{C}}-S-H$

d. $H-\underset{\underset{\displaystyle H}{|}}{\overset{\overset{\displaystyle H}{|}}{C}}-\overset{\overset{\displaystyle O}{\|}}{C}-\underset{\underset{\displaystyle H}{|}}{\overset{\overset{\displaystyle H}{|}}{C}}-H$

e. $H-\underset{\underset{\displaystyle H}{|}}{\overset{\overset{\displaystyle H}{|}}{C}}-S-S-\underset{\underset{\displaystyle H}{|}}{\overset{\overset{\displaystyle H}{|}}{C}}-H$

f. $H-\underset{\underset{\displaystyle H}{|}}{\overset{\overset{\displaystyle H}{|}}{C}}-\underset{\underset{\displaystyle H}{|}}{\overset{\overset{\displaystyle H}{|}}{C}}-F$

g. $H-\underset{\underset{H}{|}}{\overset{\overset{H}{|}}{C}}-\underset{\underset{H}{|}}{\overset{\overset{H}{|}}{C}}-O-H$

h. $H-\underset{\underset{H}{|}}{\overset{\overset{H}{|}}{C}}-\underset{\underset{H}{|}}{N}-H$

i. $\underset{H}{\overset{Cl}{\diagdown}}C=C\underset{H}{\overset{H}{\diagup}}$

j. $H-\underset{\underset{H}{|}}{\overset{\overset{H}{|}}{C}}-Mg-\underset{\underset{H}{|}}{\overset{\overset{H}{|}}{C}}-H$

Functional Groups (Section 9.2)

9.15 Draw Lewis structures for the following functional groups. Be certain to show all valence electrons.
   a. carbonyl group
   b. carboxyl group
   c. hydroxyl group

9.16 Write Lewis structures and condensed structural formulas for the four carboxylic acids of molecular formula $C_5H_{10}O_2$.

9.17 Write Lewis structures and condensed structural formulas for:
   a. the four aldehydes of molecular formula $C_5H_{10}O$
   b. the six ketones of molecular formula $C_6H_{12}O$

9.18 Write condensed structural formulas for:
   a. the four alcohols of molecular formula $C_4H_{10}O$
   b. the three ethers of molecular formula $C_4H_{10}O$

9.19 Write Lewis structures and condensed structural formulas for a compound of molecular formula C_4H_8O that contains the following functional groups:
   a. a carbon-carbon double bond and an alcohol
   b. a ketone
   c. an aldehyde
   d. a carbon-carbon double bond and an ether

Structural Isomerism (Section 9.3)

9.20 Which of the following are true of structural isomers?
   a. They have the same molecular formula.
   b. They have the same molecular weight.
   c. They have the same order of attachment of atoms.
   d. They have the same physical properties.

9.21 Are the following pairs of molecules structural isomers or are they identical?

a. $CH_3-\underset{\underset{CH_3-CH-CH_3}{\overset{\overset{|}{O}}{|}}}{CH}-CH_3$ and $CH_3-\underset{\underset{H}{}}{\overset{\overset{CH_3}{|}}{CH}}-O-\overset{\overset{CH_3}{|}}{CH}-CH_3$

b. $CH_2=CH-CH_2-CH_3$ and $CH_3-CH=CH-CH_3$

c. $CH_3-O-\overset{\overset{O}{\|}}{C}-H$ and $CH_3-\overset{\overset{O}{\|}}{C}-O-H$

d. $HO-CH_2-\overset{\overset{CH_3}{|}}{CH}-OH$ and $CH_3-\overset{\overset{CH_2-OH}{|}}{CH}-OH$

e. $H-\overset{\overset{O}{\parallel}}{C}-CH_2-\overset{\overset{\displaystyle CH_3}{|}}{CH}-CH_3$ and $\overset{\displaystyle CH_3}{\underset{\displaystyle CH_3}{\diagdown}}CH-CH_2-\overset{\overset{O}{\parallel}}{C}-H$

f. $CH_3-\overset{\overset{\displaystyle }{|}}{\underset{\underset{\displaystyle CH_3}{|}}{CH}}-CH_2-CH_2-\overset{\underset{\underset{\displaystyle CH_3}{|}}{}}{CH}-CH_3$ and $CH_3-CH-CH_2-CH_3$
 $CH_2-CH-CH_3$
 CH_3

9.22 Divide the following molecules into groups of structural isomers.

a. $CH_3-\overset{\underset{\underset{\displaystyle CH_3}{|}}{}}{N}-CH_3$

b. $CH_3-\overset{\underset{\underset{\displaystyle OH}{|}}{}}{CH}-CH_2-NH_2$

c. $CH_3-CH_2-CH_2-OH$

d. $CH_3-O-CH_2-CH_3$

e. $\overset{\underset{\underset{\displaystyle OH}{|}}{}}{CH_2}-CH_2-\overset{\overset{\overset{\displaystyle H}{|}}{}}{N}-CH_3$

f. $CH_3-\overset{\overset{O}{\parallel}}{C}-CH_2-CH_3$

g. $CH_3-\overset{\overset{O}{\parallel}}{C}-O-CH_2-CH_3$

h. $CH_3-O-\overset{\overset{O}{\parallel}}{C}-CH_2-CH_3$

i. $CH_3-CH_2-CH_2-\overset{\overset{O}{\parallel}}{C}-H$

j. $CH_3-CH_2-O-CH=CH_2$

k. $CH_3-CH_2-CH_2-NH_2$

l. $HO-CH_2-CH_2-OH$

Covalent Bond Formation by the Overlap of Atomic Orbitals (Section 9.5)

9.23 Following are Lewis structures for several molecules. State the orbital hybridization of each circled atom.

a. $H-\overset{\overset{\displaystyle H}{|}}{\underset{\underset{\displaystyle H}{|}}{C}}-\underset{\underset{\underset{\displaystyle H}{|}}{\overset{\overset{\displaystyle H}{|}}{\boxed{C}}}}{}-H$

b. $\overset{\displaystyle H}{\underset{\displaystyle H}{\diagup}}C=C\overset{\displaystyle H}{\underset{\displaystyle H}{\diagdown}}$

c. $H-\boxed{C}\equiv C-H$

d. $\overset{\displaystyle H}{\underset{\displaystyle H}{\diagup}}\boxed{C}=\ddot{O}:$

e. $H-\boxed{C}-\overset{\overset{\displaystyle :O:}{\parallel}}{\underset{\underset{\displaystyle }{}}{}}\ddot{O}-H$

f. $H-\overset{\overset{\displaystyle H}{|}}{\underset{\underset{\displaystyle H}{|}}{C}}-\boxed{\ddot{O}}-H$

g. $H-\overset{\overset{\displaystyle H}{|}}{\underset{\underset{\displaystyle H}{|}}{C}}-\boxed{\ddot{N}}-H$
H

9.24 Following are Lewis structures for several molecules. Describe each circled bond in terms of the overlap of atomic orbitals.

a. 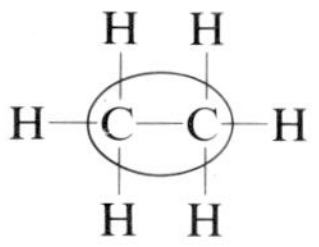

b. 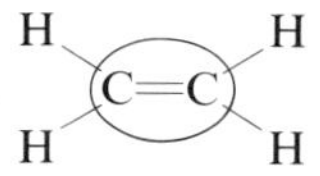

c. 

d. 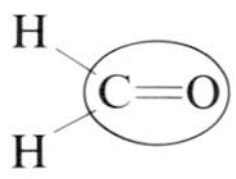

e. 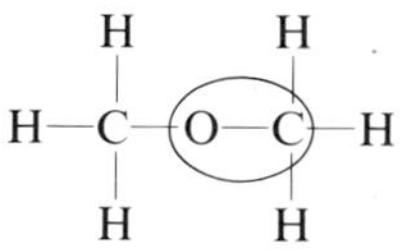

f. 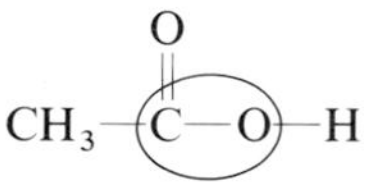

g. 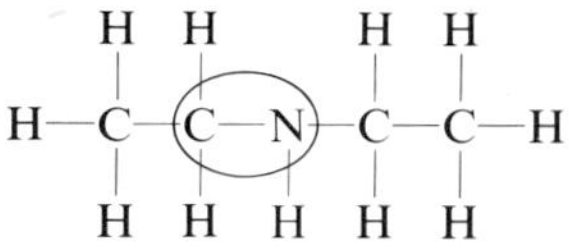

h. 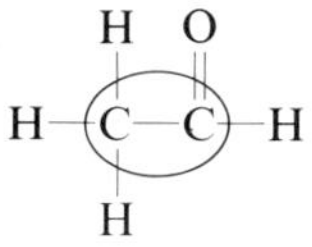

Saturated Hydrocarbons: Alkanes and Cycloalkanes

Compounds consisting solely of carbon and hydrogen are called **hydrocarbons.** If the carbon atoms in a hydrocarbon are joined together by only covalent single bonds, the hydrocarbon is called a **saturated hydrocarbon,** or **alkane.** If any of the carbon atoms in a hydrocarbon are bonded together by one or more double or triple bonds, the hydrocarbon is called an **unsaturated hydrocarbon.** In this chapter we shall discuss saturated hydrocarbons. They are the simplest organic compounds from the structural point of view, and therefore a good place to begin the study of organic chemistry.

10.1 Structure of Alkanes

Methane, CH_4, and ethane, C_2H_6, are the first members of the alkane family. As you can see from the Lewis structures in Figure 10.1, these two compounds fit the definition of an alkane, since each contains atoms of carbon and hydrogen bonded together by only covalent single bonds.

320

Figure 10.1 Alkanes: the structures of methane and ethane.
(a) Lewis structures;
(b) three-dimensional structures; (c) ball-and-stick models.

(a)

(b)

(c)

The carbon atom in methane is bonded to four hydrogens by the overlap of sp^3 hybrid orbitals of carbon with $1s$ orbitals of hydrogen. This tetrahedral shape of methane is shown in the three-dimensional drawing and the ball-and-stick model in Figure 10.1. Each carbon atom in ethane is also tetrahedral, as can be seen in Figure 10.1.

By increasing the number of carbon atoms bonded to each other in a chain, we can form the next members of the alkane series: propane, C_3H_8; butane, C_4H_{10}; and pentane, C_5H_{12}.

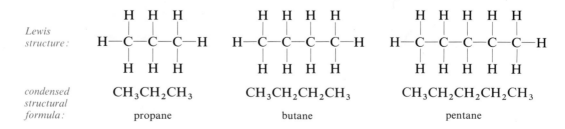

Lewis structure:

$CH_3CH_2CH_3$ — *condensed structural formula:* — propane

$CH_3CH_2CH_2CH_3$ — butane

$CH_3CH_2CH_2CH_2CH_3$ — pentane

Table 10.1 Names, molecular formulas, and condensed structural formulas of the first 10 alkanes.

Name	Molecular Formula	Condensed Structural Formula
methane	CH_4	CH_4
ethane	C_2H_6	CH_3CH_3
propane	C_3H_8	$CH_3CH_2CH_3$
butane	C_4H_{10}	$CH_3(CH_2)_2CH_3$
pentane	C_5H_{12}	$CH_3(CH_2)_3CH_3$
hexane	C_6H_{14}	$CH_3(CH_2)_4CH_3$
heptane	C_7H_{16}	$CH_3(CH_2)_5CH_3$
octane	C_8H_{18}	$CH_3(CH_2)_6CH_3$
nonane	C_9H_{20}	$CH_3(CH_2)_7CH_3$
decane	$C_{10}H_{22}$	$CH_3(CH_2)_8CH_3$

The condensed structural formulas of butane, pentane, and higher alkanes can be written in an even more abbreviated form. For example, the structural formula of pentane contains three $—CH_2—$ groups in the middle of the chain. These can be grouped together and the structural formula can be written as $CH_3(CH_2)_3CH_3$.

Table 10.1 shows the names, molecular formulas, and condensed structural formulas for the first ten alkanes. The name of each alkane in this table consists of two parts: a root to indicate the number of carbon atoms and the ending -*ane*. *But*ane contains four carbon atoms, *pent*ane contains five, *hex*ane contains six, and so on up to *dec*ane, which contains ten carbon atoms. The

Figure 10.2 The pentane molecule. (a) Lewis structure; (b) ball-and-stick model.

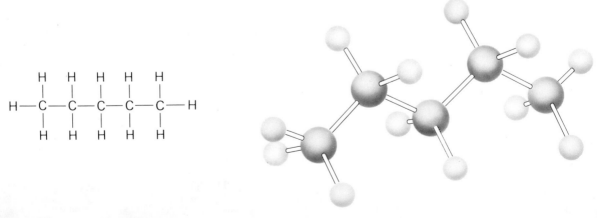

(a)   (b)

names of these first ten alkanes must be learned, for they are used to form the names of substituted alkanes as well as several other classes of organic compounds.

Alkanes have the general formula C_nH_{2n+2}. For example, pentane, with five carbon atoms ($n = 5$), has ($2n + 2 = 12$) hydrogen atoms; hexane, with six carbon atoms ($n = 6$), has ($2n + 2 = 14$) hydrogen atoms.

Although the three-dimensional shapes of propane and larger alkanes are more complex than those of methane and ethane, each individual carbon atom is still at the center of a tetrahedral structure with bond angles of approximately 109.5°. Figure 10.2 shows a Lewis structure and a ball-and-stick model of pentane, C_5H_{12}.

10.2 Structural Isomerism in Alkanes

Recall from Section 9.3 that two or more compounds with the same molecular formula but different orders of attachment of atoms are *structural isomers*. For molecular formulas CH_4, C_2H_6, and C_3H_8, there is only one possible order of attachment of carbon atoms and therefore, methane, ethane, and propane have no structural isomers. Although you may at first think that the three drawings in Figure 10.3 represent different compounds, they all represent the same compound (propane), for in each structure the three carbon atoms are in a continuous chain.

Figure 10.3 The propane molecule.
(a) Lewis structures;
(b) ball-and-stick model. Note that all Lewis structures have the three carbons in a continuous chain.

(a)

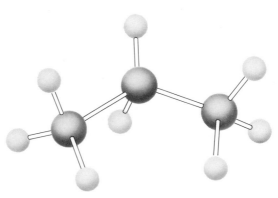

(b)

For the molecular formula C_4H_{10}, there are two possible orders of attachment of the atoms. In one of these, the four carbon atoms are attached in a chain; in the other, three are attached in a chain, with the fourth carbon as a branch on the middle carbon of the chain. These two compounds are named butane and 2-methylpropane. (We will show how to name alkanes in the next section.)

$$CH_3-CH_2-CH_2-CH_3$$

butane (C_4H_{10})
bp $-0.5°C$

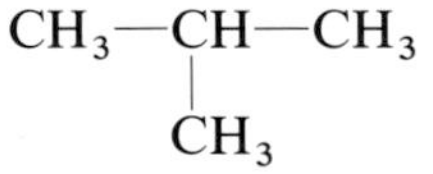

$$CH_3-CH-CH_3$$
$$|$$
$$CH_3$$

2-methylpropane (C_4H_{10})
bp $-10.2°C$

Butane and 2-methylpropane are structural isomers. They have the same molecular formula but different orders of attachment of atoms. Structural isomers are different compounds and have different physical and chemical properties. Notice that the boiling points of butane and 2-methylpropane differ by over 9°C.

There are 3 structural isomers of C_5H_{12}, 5 structural isomers of C_6H_{14}, 18 of C_8H_{18}, and 75 of $C_{10}H_{22}$. Even for a rather small number of carbon and hydrogen atoms, a very large number of structural isomers are possible.

Example 10.1

Identify the following pairs as identical compounds or as structural isomers.

a. 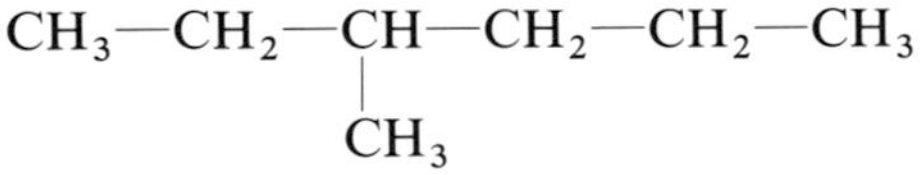 $CH_3-CH_2-CH-CH_2-CH_2-CH_3$
       $\qquad\qquad\quad |$
       $\qquad\qquad\ CH_3$

and $CH_3-CH_2-CH_2-CH-CH_3$
        $\qquad\qquad\qquad\qquad\ |$
        $\qquad\qquad\qquad\quad CH_2$
        $\qquad\qquad\qquad\qquad\ |$
        $\qquad\qquad\qquad\quad CH_3$

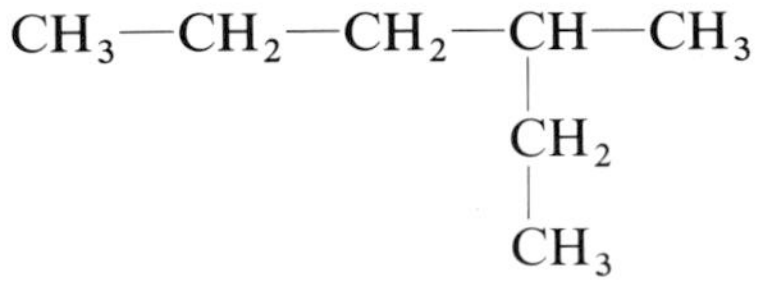

b. 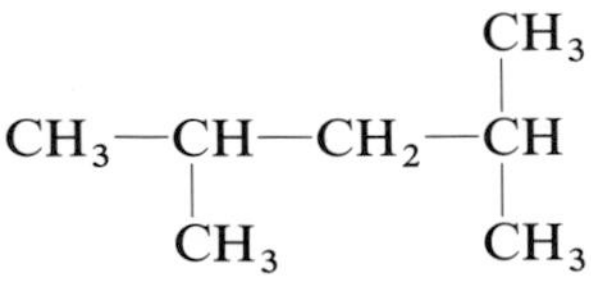 $CH_3-CH-CH_2-\overset{\displaystyle CH_3}{\overset{\displaystyle |}{CH}}$ and $CH_3-CH_2-CH-\overset{\displaystyle CH_3}{\overset{\displaystyle |}{CH}}-CH_3$
        $\qquad\ \ |\qquad\quad\ |$ $\qquad\qquad\qquad\quad |$
        $\qquad CH_3\qquad CH_3$ $\qquad\qquad\qquad\ CH_3$

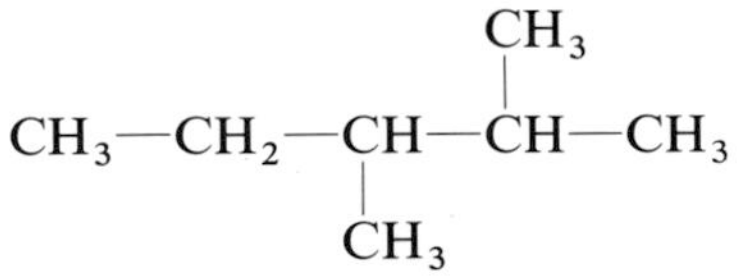

Solution

To determine if two formulas are identical or represent structural isomers, find the longest chain of carbon atoms in each and number each carbon in it from the end nearest the first branch. Then compare the lengths of each carbon chain and the size and location of any branches.

a. 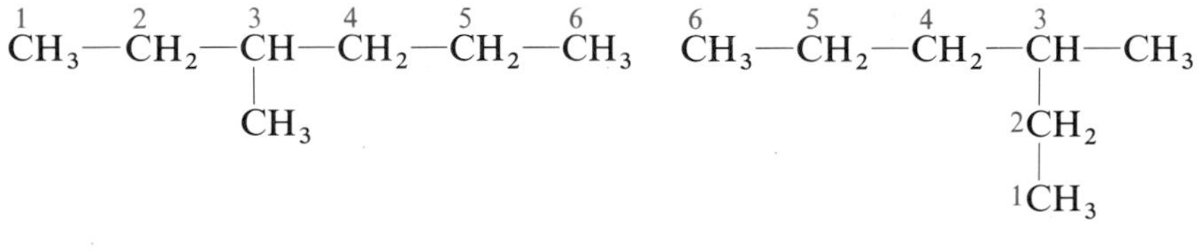
$$\overset{1}{C}H_3-\overset{2}{C}H_2-\overset{3}{C}H-\overset{4}{C}H_2-\overset{5}{C}H_2-\overset{6}{C}H_3 \qquad \overset{6}{C}H_3-\overset{5}{C}H_2-\overset{4}{C}H_2-\overset{3}{C}H-\overset{}{C}H_3$$

with CH_3 branch on carbon 3 (left), and $_2CH_2$, $_1CH_3$ branch (right).

Each structural formula has a chain of six carbon atoms with a CH_3 group on the third carbon of the chain. Therefore, they are identical and represent the same compound.

b. 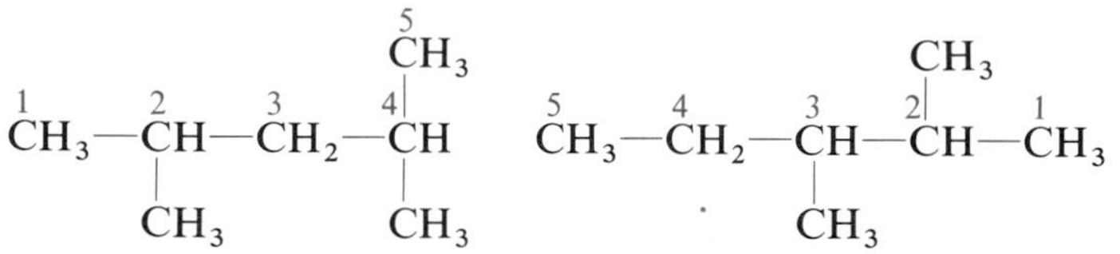
$$\overset{1}{C}H_3-\overset{2}{C}H-\overset{3}{C}H_2-\overset{4}{C}H \qquad \overset{5}{C}H_3-\overset{4}{C}H_2-\overset{3}{C}H-\overset{2}{C}H-\overset{1}{C}H_3$$

with $\overset{5}{C}H_3$ and CH_3 branches on first chain; CH_3 and CH_3 branches on second.

Both structural formulas have chains of five carbon atoms with two CH_3 branches. However, while the size of the branches is identical, they have different locations on the two chains; these two formulas are structural isomers.

Problem 10.1

Identify the following pairs as identical or as structural isomers.

a. 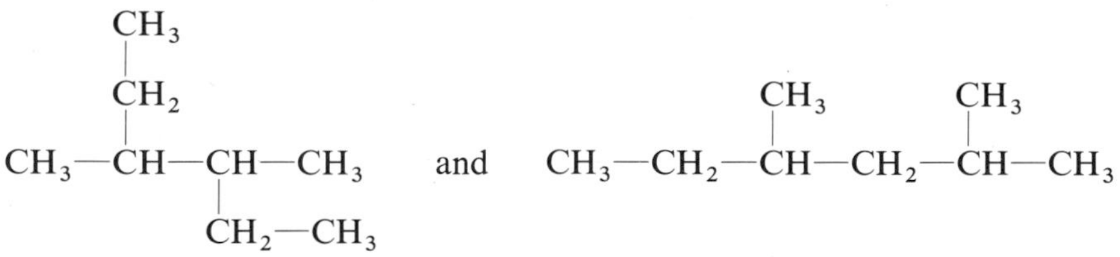
$$CH_3-CH-CH-CH_3 \qquad and \qquad CH_3-CH_2-CH-CH_2-CH-CH_3$$

with CH_3, CH_2, CH_2-CH_3 groups on the left molecule; CH_3 and CH_3 branches on the right.

b. 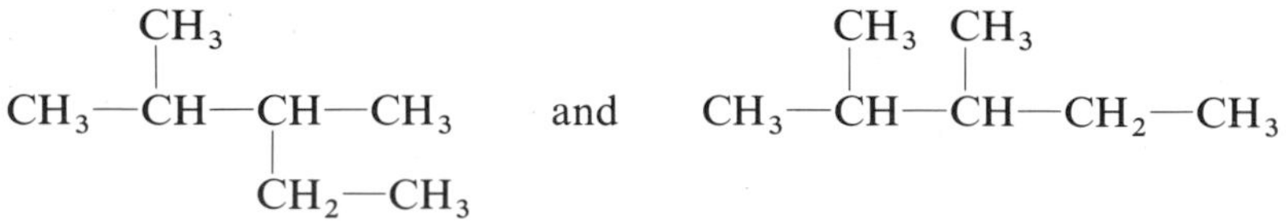
$$CH_3-CH-CH-CH_3 \qquad and \qquad CH_3-CH-CH-CH_2-CH_3$$

with CH_3 and CH_2-CH_3 groups on the left; CH_3 and CH_3 branches on the right.

Example 10.2

Draw structural formulas for the five structural isomers of molecular formula C_6H_{14}.

Solution

You should approach this type of problem (there will be more like it later) in a systematic manner. First, draw the structural isomer with six carbon atoms in an unbranched chain. Then draw all structural isomers with five carbons in a chain and one carbon as a branch on the chain. Finally, draw all structural isomers with four carbons in a chain and two carbons as branches. By working in this systematic manner, you will arrive at all possible structural isomers.

six carbons in an unbranched chain:

$$\overset{1}{C}H_3-\overset{2}{C}H_2-\overset{3}{C}H_2-\overset{4}{C}H_2-\overset{5}{C}H_2-\overset{6}{C}H_3$$

five carbons in a chain; one carbon as a branch:

$$\overset{1}{C}H_3-\overset{2}{C}H-\overset{3}{C}H_2-\overset{4}{C}H_2-\overset{5}{C}H_3$$
$$| \\ CH_3$$

$$\overset{1}{C}H_3-\overset{2}{C}H_2-\overset{3}{C}H-\overset{4}{C}H_2-\overset{5}{C}H_3$$
$$| \\ CH_3$$

four carbons in a chain; two carbons as branches:

$$\begin{array}{cc} \qquad CH_3 \qquad\qquad\qquad CH_3 \\ \qquad | \qquad\qquad\qquad\qquad | \\ \overset{1}{C}H_3-\overset{2}{C}-\overset{3}{C}H_2-\overset{4}{C}H_3 \qquad \overset{1}{C}H_3-\overset{2}{C}H-\overset{3}{C}H-\overset{4}{C}H_3 \\ \qquad | \qquad\qquad\qquad\qquad\qquad | \\ \qquad CH_3 \qquad\qquad\qquad\qquad CH_3 \end{array}$$

Problem 10.2

Draw structural formulas for the three structural isomers of molecular formula C_5H_{12}.

10.3 Nomenclature of Alkanes

A. The IUPAC System of Nomenclature

Ideally, every organic compound should have a name that clearly describes the structure of the compound and from which a structural formula can be drawn. For this purpose, chemists throughout the world have accepted a set of rules proposed by the International Union of Pure and Applied Chemistry (IUPAC). This system is known as the **IUPAC system,** or alternatively as the **Geneva system,** after the fact that the first meetings of the IUPAC were held in Geneva, Switzerland. There are two major components to the IUPAC name of an alkane: a **parent name,** which indicates the longest chain of carbon atoms in the formula, and **substituent names,** which indicate groups attached to the parent chain and their location.

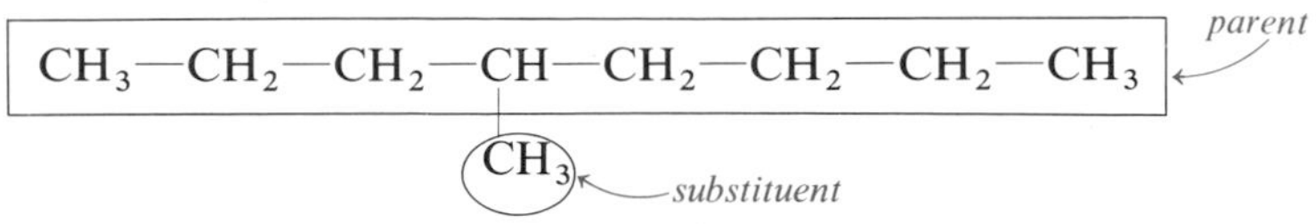

$$CH_3-CH_2-CH_2-CH-CH_2-CH_2-CH_2-CH_3 \quad \textit{parent}$$
$$\underset{\textit{substituent}}{CH_3}$$

Parent names for chains of from one to ten carbon atoms are simply the names of the first ten alkanes (Table 10.1). In Table 10.2 are listed names and structural formulas for eight of the most common alkyl groups. An **alkyl group** is named by dropping the *-ane* from the name of the parent alkane and adding the suffix *-yl*. For example:

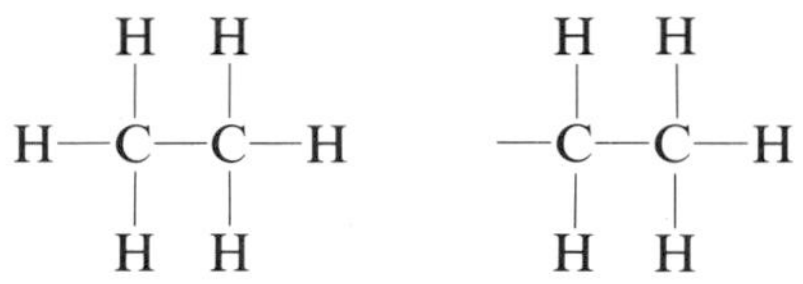

ethane
(a parent hydrocarbon)

ethyl group
(an alkyl group)

Following are IUPAC rules for naming alkanes.

1. The **root** or **stem name** is derived from the name of the parent alkane, the longest chain of carbon atoms.

2. Groups attached to the parent chain are called **substituents.** Each substituent is given a name and a number. The number shows the carbon atom of the parent chain to which the substituent is attached.

Table 10.2 Common alkyl groups.

IUPAC Name	Condensed Structural Formula	IUPAC Name	Condensed Structural Formula
methyl	$-CH_3$	butyl	$-CH_2-CH_2-CH_2-CH_3$
ethyl	$-CH_2-CH_3$	isobutyl	$-CH_2-CH-CH_3$ $\;\;\;\;\;\;\;\;\;CH_3$
propyl	$-CH_2-CH_2-CH_3$	*sec*-butyl	$-CH-CH_2-CH_3$ $\;\;CH_3$
isopropyl	$-CH-CH_3$ $\;\;CH_3$	*tert*-butyl	CH_3 $-C-CH_3$ $\;\;CH_3$

3. If the same substituent occurs more than once, the number of each carbon of the parent chain to which the substituent is attached is given. In addition, the number of times the substituent group occurs is indicated by a prefix (di-, tri-, tetra-, penta-, or hexa-).

4. If there is only one substituent, number the parent chain from the end that gives the substituent the lowest number. If there are two or more substituents, number the parent chain from the end that gives the lowest number to the substituent encountered first.

5. If there are two or more different substituents, list them in alphabetical order. When listing substituents, dimethyl- is alphabetized under d-, triethyl- under t-, and so on.

Example 10.3

Name the following compounds by the IUPAC system.

a. 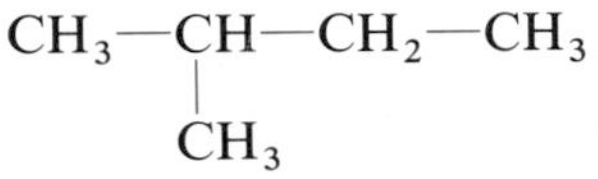 CH_3—CH—CH_2—CH_3
    |
    CH_3

b. 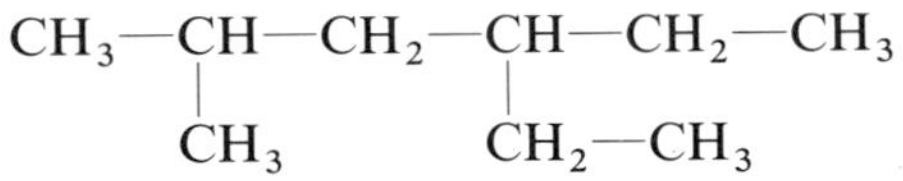 CH_3—CH—CH_2—CH—CH_2—CH_3
    | |
    CH_3 CH_2—CH_3

c. CH_3—CH_2—CH_2—C—CH_3
                        |
       CH_3 (above C) and CH_3 (below C)

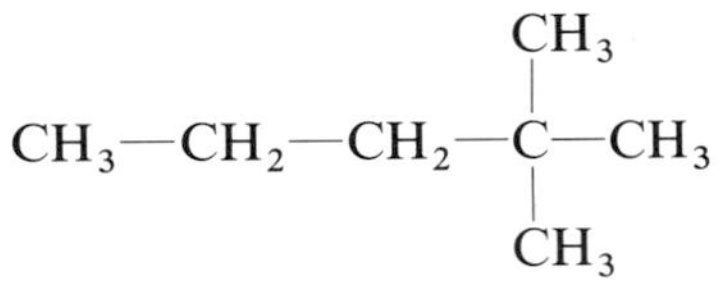

Solution

a. There are four carbon atoms in the longest chain, and the name of the parent compound is butane (rule 1). The butane chain must be numbered so that the single methyl group is on carbon 2 of the chain (rule 4). The correct name of this hydrocarbon is 2-methylbutane.

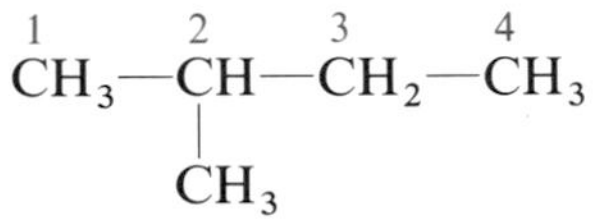

$$\overset{1}{CH_3}-\overset{2}{CH}-\overset{3}{CH_2}-\overset{4}{CH_3}$$
    |
    CH_3

2-methylbutane

b. The longest chain contains six carbon atoms and the parent compound is a hexane (rule 1). There are two substituents on the hexane chain: a methyl group and an ethyl group. The hexane chain must be numbered so that the first substituent (a methyl group) encountered is on carbon-2 of the chain (rule 4). The ethyl and methyl substituents are listed in alphabetical order to give the name 4-ethyl-2-methylhexane.

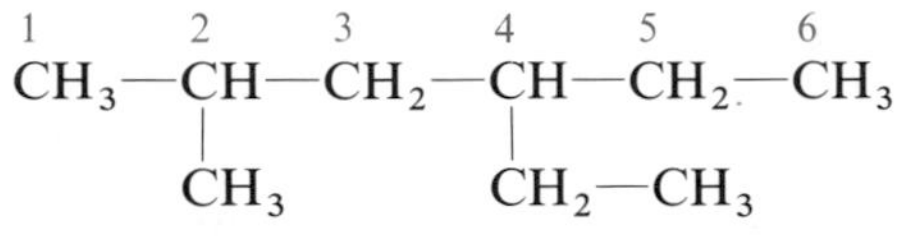

$$\underset{1}{\text{CH}_3}-\underset{2}{\text{CH}}-\underset{3}{\text{CH}_2}-\underset{4}{\text{CH}}-\underset{5}{\text{CH}_2}-\underset{6}{\text{CH}_3}$$

with CH_3 below carbon 2 and CH_2-CH_3 below carbon 4

4-ethyl-2-methylhexane

c. The longest chain contains five carbon atoms and the parent compound is a pentane (rule 1). The pentane chain must be numbered so that the substituents are on carbon-2 of the chain (rule 4). There are two substituents and each must have a name and a number (rule 3). Because the substituents are identical, they are grouped together using the prefix -di (rule 3). The correct IUPAC name is 2,2-dimethylpentane.

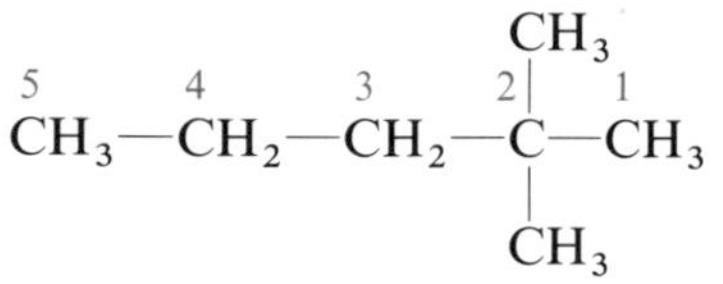

$$\underset{5}{\text{CH}_3}-\underset{4}{\text{CH}_2}-\underset{3}{\text{CH}_2}-\underset{2}{\overset{\overset{\text{CH}_3}{|}}{\underset{\underset{\text{CH}_3}{|}}{\text{C}}}}-\underset{1}{\text{CH}_3}$$

2,2-dimethylpentane

Problem 10.3

Name the following alkanes by the IUPAC system.

a.
$$\text{CH}_3-\overset{\overset{\text{CH}_3}{|}}{\text{CH}}-\text{CH}_2-\text{CH}_2-\overset{\overset{}{\underset{\underset{\underset{\underset{\text{CH}_2-\text{CH}_3}{|}}{\text{CH}_2}}{|}}{}}{\text{CH}}-\overset{\overset{\text{CH}_3}{|}}{\text{CH}}-\text{CH}_3$$

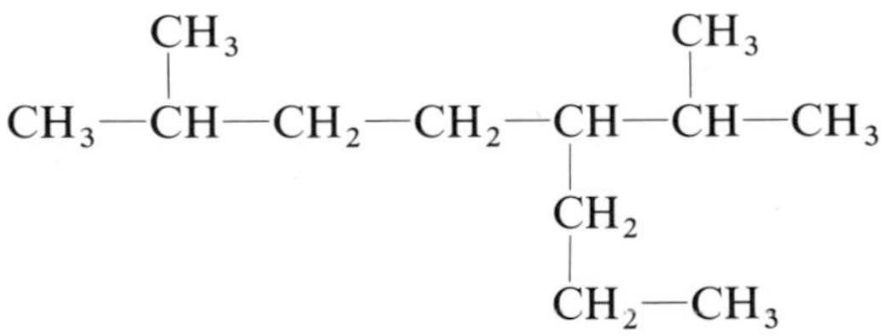

b.
$$\text{CH}_3-\text{CH}_2-\text{CH}_2-\overset{\overset{\text{CH}_2-\text{CH}_2-\text{CH}_3}{|}}{\underset{\underset{\underset{\underset{\text{CH}_3}{|}}{\text{CH}-\text{CH}_3}}{|}}{\text{C}}}-\text{CH}_2-\text{CH}_2-\text{CH}_3$$

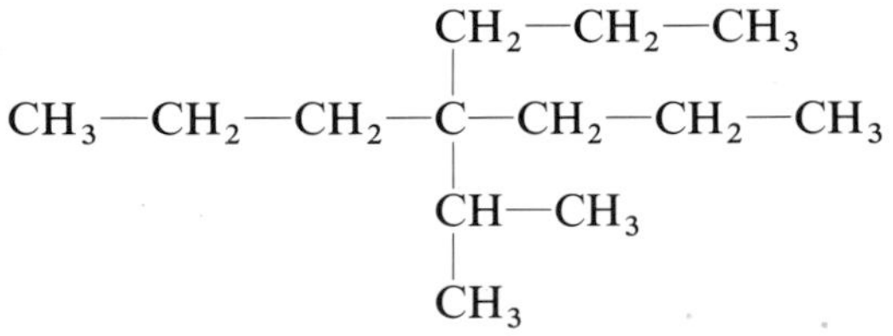

B. Other Names of Alkanes

In spite of the precision of the IUPAC system, routine communication in organic chemistry still relies on a mixture of nomenclature systems. The reasons for this situation are rooted in both convenience and historical development.

In the older, **semisystematic** nomenclature, the total number of carbon atoms in the alkane, regardless of the arrangement, determines the name. The first three alkanes are methane, ethane, and propane. All alkanes of formula

C_4H_{10} are called butanes, all alkanes of formula C_5H_{12} are called pentanes, all those of formula C_6H_{14} are hexanes, and so on. For those alkanes beyond propane, the prefix *normal* or *n-* is used to indicate that all carbons are joined in a continuous chain. The prefix *iso-* indicates that one end of an otherwise continuous chain terminates in the $(CH_3)_2CH-$ group. The prefix *neo-* indicates that one end of an otherwise continuous chain of carbon atoms terminates in the $(CH_3)_3C-$ group.

$$CH_3CH_2CH_2CH_3$$

n-butane

$$CH_3CHCH_3$$
$$|$$
$$CH_3$$

isobutane

$$CH_3CH_2CH_2CH_2CH_3$$

n-pentane

$$CH_3CHCH_2CH_3$$
$$|$$
$$CH_3$$

isopentane

$$CH_3CCH_3$$ with CH_3 above and CH_3 below

neopentane

$$CH_3CH_2CH_2CH_2CH_2CH_3$$

n-hexane

$$CH_3CHCH_2CH_2CH_3$$
$$|$$
$$CH_3$$

isohexane

$$CH_3CCH_2CH_3$$ with CH_3 above and CH_3 below

neohexane

This semisystematic system has no way of handling other branching patterns. For more complex alkanes it is necessary to use the more flexible IUPAC system of nomenclature.

We shall strive as far as possible to use IUPAC names in this text, but unavoidably, there will be some use of semisystematic names.

10.4 Cycloalkanes

Hydrocarbons that contain a ring of carbon atoms are called **cyclic hydrocarbons.** When all carbon atoms of a ring are saturated, the molecules are called **cycloalkanes.** Figure 10.4 shows structural formulas for four cycloalkanes: cyclopropane, cyclobutane, cyclopentane, and cyclohexane. As a matter of convenience, the organic chemist does not usually write out the structural formulas for cycloalkanes showing all carbons and hydrogens. Rather, the rings are represented by polygons with the same number of sides as the rings. For example, cyclopropane is represented by a triangle and cyclohexane by a hexagon.

To name cycloalkanes, prefix the name of the corresponding open-chain hydrocarbon by cyclo- and name each substituent on the ring. If there is only a

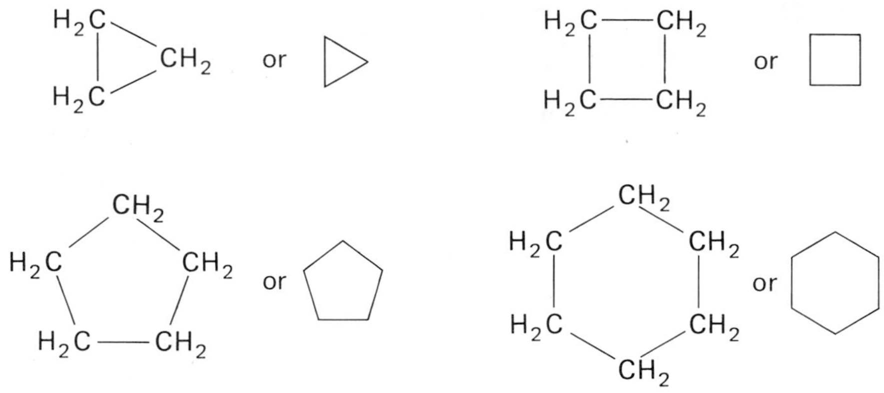

Figure 10.4 Examples of cycloalkanes.

single substituent on the cycloalkane ring, there is no need to give it a number. If there are two or more substituents, each substituent must have a number to indicate its position on the ring.

Example 10.4

Name the following cycloalkanes.

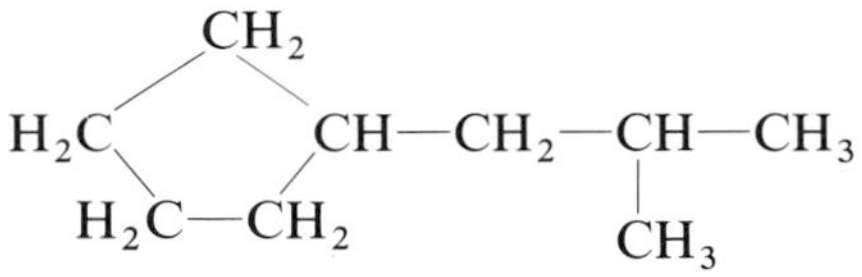

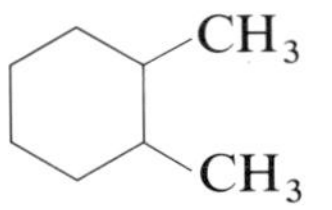

Solution

a. The ring contains five carbon atoms and is a cyclopentane. Because there is only one substituent on the ring, there is no need to number the atoms of the ring. The IUPAC name is isobutylcyclopentane.

b. The ring contains six carbon atoms and is a cyclohexane. Because there are two substituents on the ring, a numbering system must be used to locate the groups. The IUPAC name is 1,2-dimethylcyclohexane.

Problem 10.4

Name the following cycloalkanes.

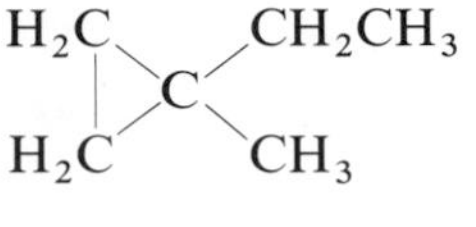

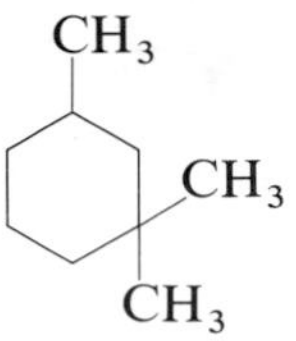

10.5 Conformations of Alkanes and Cycloalkanes

A molecule can be twisted into a number of different three-dimensional shapes without breaking any bonds. These different shapes are called **conformations.** A simple molecule such as ethane has an infinite number of conformations. You can visualize all of these conformations with the help of a ball-and-stick model by holding one CH_3 group in your hand and twisting the other CH_3 group about the single bond joining the two carbon atoms.

Figure 10.5(a) shows a ball-and-stick model of a staggered conformation of ethane. In this conformation, hydrogen atoms are staggered, meaning hydrogens on the front carbon lie between those on the back carbon and vice versa. In Figure 10.5(b), the staggered conformation is turned in space so that you are looking at it from another angle. In Figure 10.5(c), this same conformation is drawn in a form called a **Newman projection.** In a Newman projection, the molecule is viewed along the axis of the C—C bond. Atoms or groups of atoms on the front carbon are shown on lines from the center of a circle. Those on the back carbon are shown on lines from the circumference of a circle.

Figure 10.6 shows an **eclipsed conformation** of ethane. In this conformation, the hydrogen atoms on the back carbon are lined up, or eclipsed with, the hydrogen atoms on the front carbon.

An ethane molecule can be twisted into an infinite number of conformations between the staggered and eclipsed conformations. Yet any given ethane molecule spends most of its time in the staggered conformation because in this conformation, hydrogen atoms on one carbon are as far apart as possible from hydrogens on the other carbon. For this reason, the staggered conformation is the most stable or **preferred conformation.**

Figure 10.5 Staggered conformation of ethane. (a,b) Ball-and-stick models; (c) Newman projection.

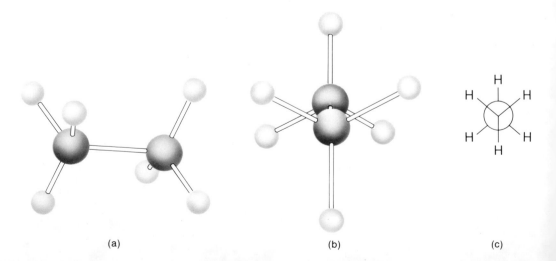

(a)   (b)   (c)

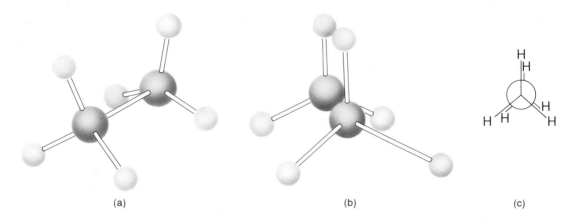

(a)   (b)   (c)

Figure 10.6 Eclipsed conformation of ethane. (a,b) Ball-and-stick models; (c) a Newman projection.

Example 10.5

Draw Newmann projections for two staggered conformations of butane. Consider only conformations along the bond between carbons 2 and 3 of the butane chain. Which of these is the more stable conformation? Which is the less stable?

Solution

The condensed structural formula of butane is:

$$\overset{1}{C}H_3 - \overset{2}{C}H_2 - \overset{3}{C}H_2 - \overset{4}{C}H_3$$

First, view the molecule along the bond between carbon atoms 2 and 3. Then, to see the possible conformations asked for, hold carbon atom 2 in place and rotate carbon 3 about the single bond joining carbons 2 and 3. Staggered conformation (a) is the more stable conformation because in it the two —CH_3 groups are as far from each other as possible.

ball-and-stick model of a staggered conformation of butane

Newman projections of two staggered conformations of butane

Problem 10.5

Draw two eclipsed conformations of butane. Consider only conformations along the bond between carbons 2 and 3. Which of these conformations is the more stable? Which is less stable?

   Next, let us look at the conformations of cycloalkanes. Figure 10.7 shows the shapes of cyclopropane, cyclobutane, and cyclopentane. For all practical purposes, these molecules are planar. While there is some puckering of the ring, it is insignificant. In each of these molecules, the C—C—C bond angles are less than the tetrahedral angle of 109.5°.

Figure 10.7 Ball-and-stick models of (a) cyclopropane; (b) cyclobutane; (c) cyclopentane.

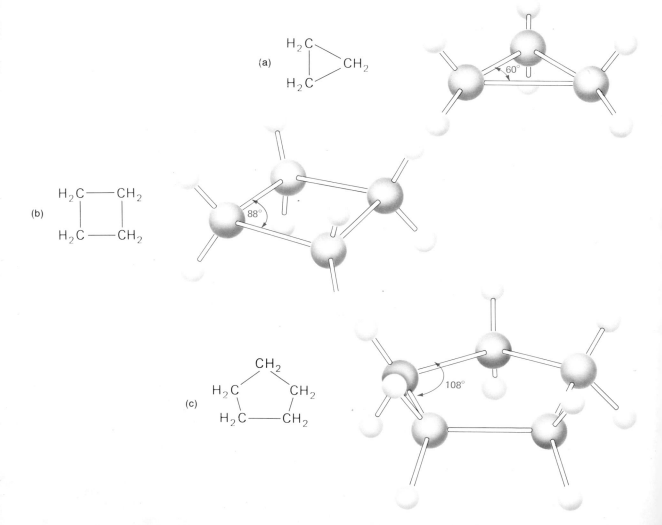

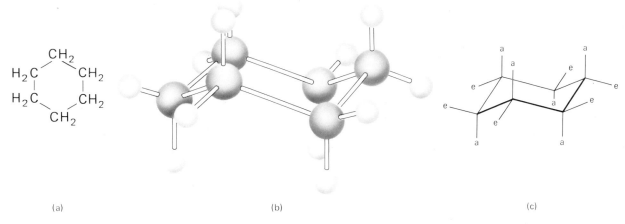

Figure 10.8 Chair conformation of cyclohexane. (a) Lewis structure; (b) ball-and-stick model; (c) three-dimensional drawing. In the chair conformation, six hydrogens are equatorial (e) and six are axial (a).

Cyclohexane and all larger cycloalkanes exist in nonplanar or puckered conformations. A ball-and-stick model of the most stable puckered conformation of cyclohexane is shown in Figure 10.8(b). This conformation is commonly called a **chair** because of its similarity in shape to a beach chair with a back rest, a seat, and a leg rest. All C—C—C bond angles in a chair conformation are approximately 109.5°. Hydrogens in a chair conformation are in two different geometrical positions (Figure 10.8c). Six, called **equatorial hydrogens,** project straight out and are roughly parallel to the plane of the ring. The other six, called **axial hydrogens,** are perpendicular to the plane of the ring.

If a hydrogen atom of cyclohexane is replaced by a methyl group or other substituent, the group may occupy either an axial position or an equatorial position.

Example 10.6

Draw a chair conformation for 1,2-dimethylcyclohexane in which (a) both methyl groups are equatorial, and (b) one methyl group is equatorial and the other axial.

Solution

Problem 10.6

Draw a chair conformation for 1,3-dimethylcyclohexane in which (a) the two methyl groups are equatorial, (b) one methyl group is equatorial and the other axial, and (c) both methyl groups are axial.

10.6 *Cis-Trans* Isomerism in Cycloalkanes

The principle of ***cis-trans* isomerism** in cycloalkanes can be illustrated by looking at 1,2-dimethylcyclopentane. In the following drawings, the cyclopentane ring is drawn as a planar pentagon as though you are looking at it through the plane of the ring. Carbon-carbon bonds on the ring projecting toward you are shown as heavy lines. Substituents attached to the ring project above and below the plane of the ring. In one isomer, both methyl groups are on the same side of the ring; in the other isomer, they are on opposite sides of the ring. The prefix *cis-* indicates that substituents are on the same side of the ring; the prefix *trans-* indicates that they are on opposite sides of the ring.

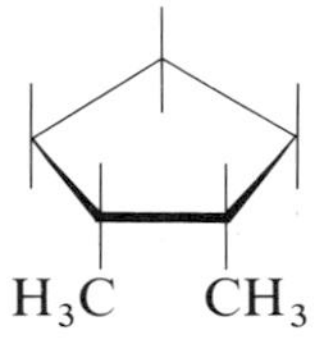

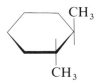

cis-1,2-dimethylcyclopentane 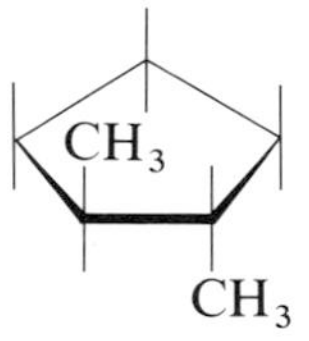   *trans*-1,2-dimethylcyclopentane

Following are drawings of the *cis* and *trans* isomers of 1,2-dimethylcyclohexane. Heavy lines show the carbon-carbon bonds of the ring projecting toward you.

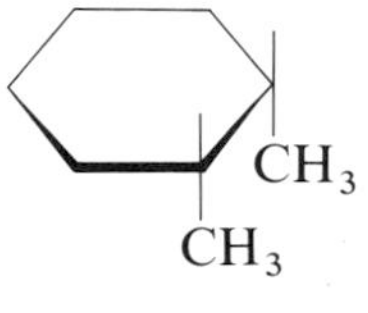

cis-1,2-dimethylcyclohexane   *trans*-1,2-dimethylcyclohexane

As illustrated by these examples, *cis-trans* isomers have the same order of attachment of atoms but different arrangements of atoms in space. In addition, these different arrangements of atoms in space cannot be converted from one to another by rotation about carbon-carbon single bonds. Cycloalkanes show *cis-trans* isomerism because there is restricted rotation about the carbon-carbon single bonds of the ring.

Example 10.7

Which of the following cycloalkanes show *cis-trans* isomerism? For each that does, draw both *cis-* and *trans-* isomers.

a. 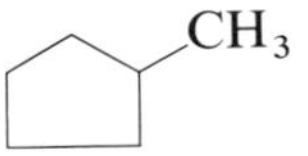 (structure: cyclopentane with CH₃)   b. 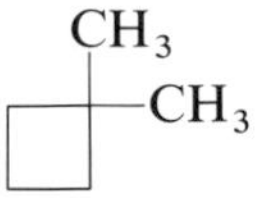 (structure: cyclobutane with CH₃ and —CH₃ at carbon-1)   c. 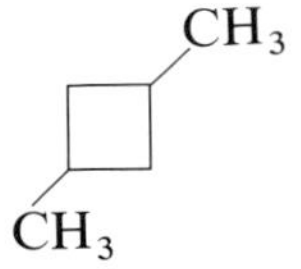 (structure: cyclobutane with CH₃ and CH₃)

Solution

a. Methylcyclopentane does not show *cis-trans* isomerism because there is only one substituent on the ring.

b. 1,1-Dimethylcyclobutane does not show *cis-trans* isomerism because there is only one possible arrangement for the two methyl groups on carbon-1 of the ring.

c. 1,3-Dimethylcyclobutane does show *cis-trans* isomerism.

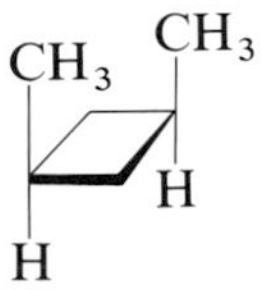

cis-1,3-dimethylcyclobutane 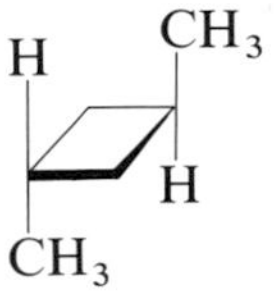   *trans*-1,3-dimethylcyclobutane

Problem 10.7

Which of the following cycloalkanes show *cis-trans* isomerism? For each that does, draw both *cis-* and *trans-* isomers.

a. 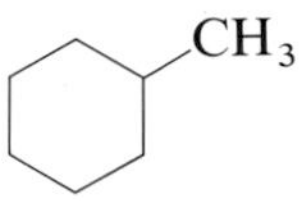 (cyclohexane with CH₃)   b. 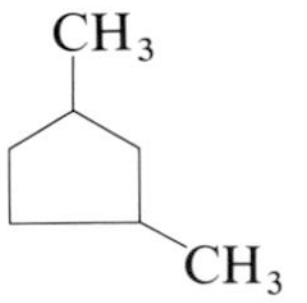 (cyclopentane with CH₃ and CH₃)   c. 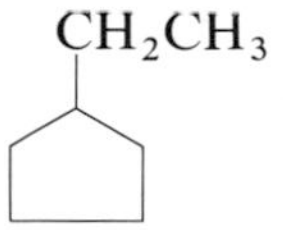 (cyclopentane with CH₂CH₃)   d. 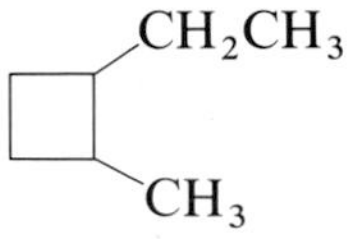 (cyclobutane with CH₂CH₃ and CH₃)

10.7 Physical Properties of Saturated Hydrocarbons

You are almost certainly familiar with some of the physical properties of alkanes and cycloalkanes from your everyday experience. Low-molecular-weight alkanes such as methane (natural gas), ethane, propane, and butane are gases at room temperature and atmospheric pressure. Higher-molecular-weight alkanes such as those in gasoline and kerosene are liquids. Very high-molecular-

Table 10.3 Physical properties of alkanes.

Name	Structural formula	mp (°C)	bp (°C)	Density (g/mL)
methane	CH_4	−182	−164	(a gas)
ethane	CH_3CH_3	−183	−88	(a gas)
propane	$CH_3CH_2CH_3$	−190	−42	(a gas)
butane	$CH_3(CH_2)_2CH_3$	−138	0	(a gas)
pentane	$CH_3(CH_2)_3CH_3$	−130	36	0.626
hexane	$CH_3(CH_2)_4CH_3$	−95	69	0.659
heptane	$CH_3(CH_2)_5CH_3$	−90	98	0.684
octane	$CH_3(CH_2)_6CH_3$	−57	126	0.703
nonane	$CH_3(CH_2)_7CH_3$	−51	151	0.718
decane	$CH_3(CH_2)_8CH_3$	−30	174	0.730

weight alkanes such as those found in paraffin wax are solids. Melting points, boiling points, and densities of the first ten alkanes are listed in Table 10.3.

To account for these trends in physical properties, we need to consider the molecular structure of alkanes. Alkanes are nonpolar molecules, and the only forces of attraction between alkane molecules are very weak dispersion forces (Section 6.2). Because interaction forces between alkane molecules are very weak, alkanes have lower boiling points than almost any other type of compound of the same molecular weight. As the number of atoms and the molecular weight of an alkane increases, the number of dispersion forces per molecule also increases. Therefore, boiling points of alkanes increase as their molecular weight increases.

Melting points of alkanes also increase with increasing molecular weight. However, the increase is not as regular as that observed for boiling points. The density of the alkanes listed in Table 10.3 is about 0.7 g/mL. That of the higher-molecular-weight hydrocarbons found in crude oil and paraffin wax is about 0.8 g/mL. Thus, all alkanes are less dense than water (1.0 g/mL).

An observation you have certainly made is that oil and water do not mix. For example, gasoline does not dissolve in water. We refer to organic compounds that are insoluble in water as **hydrophobic** (''water-hating''). Alkanes do dissolve in nonpolar liquids such as carbon tetrachloride (CCl_4), carbon disulfide (CS_2), and, of course, in other hydrocarbons.

Alkanes that are structural isomers of each other are different compounds and have different physical and chemical properties. Table 10.4 lists boiling points, melting points, and densities of the five structural isomers of molecular formula C_6H_{14}. Notice that each of the branched-chain isomers has a boiling point lower than hexane itself, and that the more branching there is, the lower

Table 10.4 Physical properties of the isomeric alkanes of molecular formula C_6H_{14}.

Name	bp (°C)	mp (°C)	Density (g/mL)
hexane	68.7	−95	0.659
2-methylpentane	60.3	−154	0.653
3-methylpentane	63.3	−118	0.664
2,3-dimethylbutane	58.0	−129	0.661
2,2-dimethylbutane	49.7	−98	0.649

the boiling point. To account for the fact that chain branching lowers the boiling point, remember that the only forces of attraction between alkane molecules are very weak dispersion forces, and that boiling point depends on the number of dispersion forces between molecules. As branching increases, the shape of an alkane molecule becomes more compact and the surface area decreases. As surface area decreases, the contact between molecules decreases, the number of dispersion forces decreases, and boiling point decreases. For any group of alkane structural isomers, the least branched isomer usually has the highest boiling point; the most branched isomer usually has the lowest boiling point. Figure 10.9 (next page) shows ball-and-stick models of hexane and its most highly branched isomer, 2,2-dimethylbutane.

Example 10.8

Arrange the following compounds in order of increasing boiling point within each group.

a. $CH_3CH_2CH_2CH_3$ $CH_3CH_2CH_2CH_2CH_2CH_3$ $CH_3(CH_2)_8CH_3$

b. $CH_3(CH_2)_6CH_3$ $\underset{\displaystyle CH_3}{\overset{\displaystyle CH_3\ \ CH_3}{CH_3CCH_2CHCH_3}}$ $\underset{\displaystyle CH_3}{CH_3CHCH_2CH_2CH_2CH_3}$

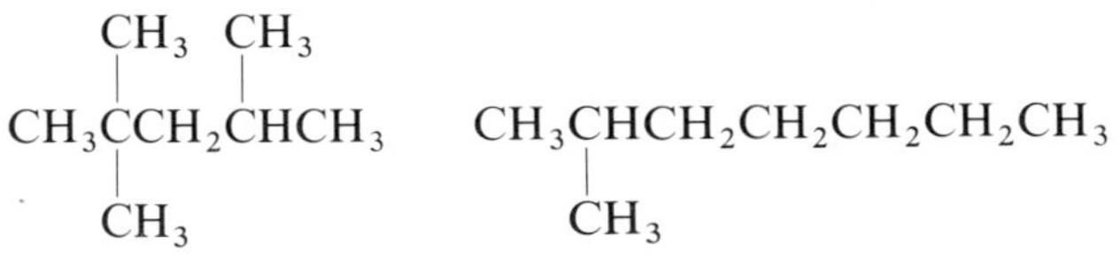

Solution

a. All are unbranched alkanes. As the number of carbon atoms in the chain increases, dispersion forces between molecules increase and boiling points increase.

$$CH_3CH_2CH_2CH_3 \qquad CH_3CH_2CH_2CH_2CH_2CH_3 \qquad CH_3(CH_2)_8CH_3$$

butane
bp −0.5°C

hexane
bp 69°C

decane
bp 174°C

Figure 10.9 Ball-and-stick models of (a) hexane, and (b) 2,2-dimethylbutane.

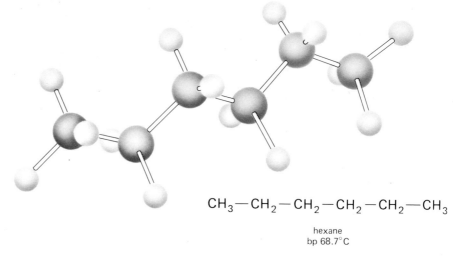

$$CH_3—CH_2—CH_2—CH_2—CH_2—CH_3$$

hexane
bp 68.7°C

(a)

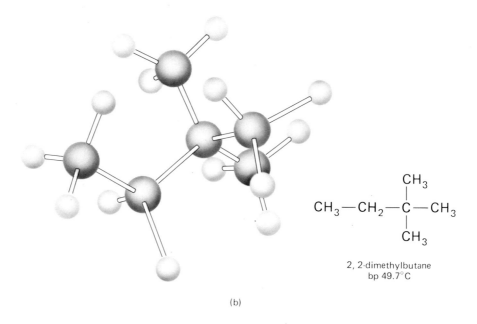

$$CH_3—CH_2—\underset{\underset{CH_3}{|}}{\overset{\overset{CH_3}{|}}{C}}—CH_3$$

2, 2-dimethylbutane
bp 49.7°C

(b)

b. These three alkanes have the same molecular formula, C_8H_{18}, and are structural isomers of each other. The most highly branched is 2,2,4-trimethylpentane, the least highly branched is octane. Because dispersion forces are least between molecules of 2,2,4-trimethylpentane, it has the lowest boiling point. Dispersion forces are greatest between molecules of octane, so it has the highest boiling point.

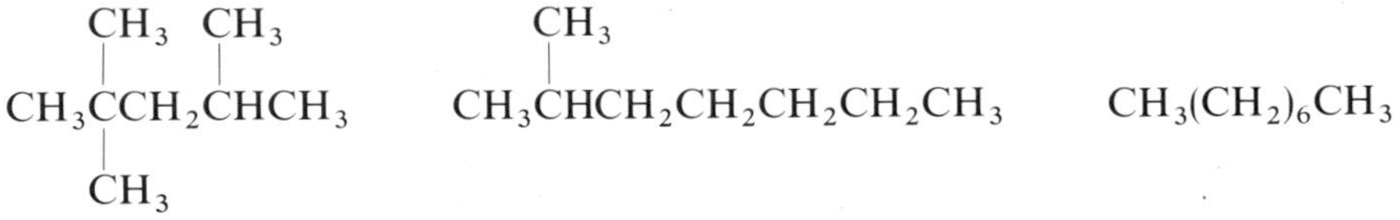

$$CH_3CCH_2CHCH_3$$

2,2,4-trimethylpentane
(bp 99 C)

$$CH_3CHCH_2CH_2CH_2CH_2CH_3$$

2-methylheptane
(bp 118 C)

$$CH_3(CH_2)_6CH_3$$

octane
(bp 126 C)

Problem 10.8

Arrange the following compounds in order of increasing boiling point.

a. $CH_3—CH_2—CH—CH_3$ $CH_3—C—CH_3$

$CH_3—CH_2—CH_2—CH_2—CH_3$

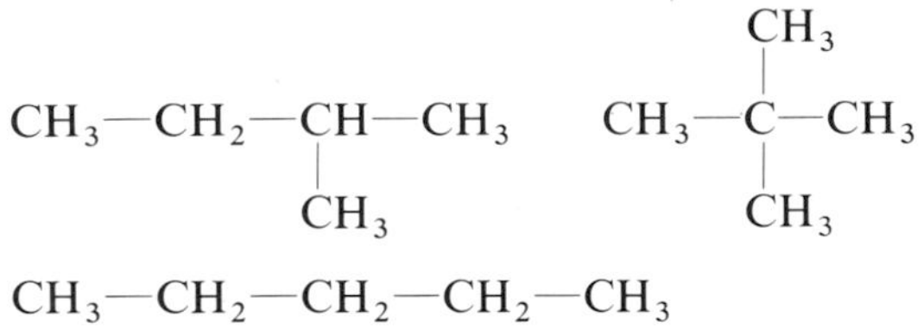

b.

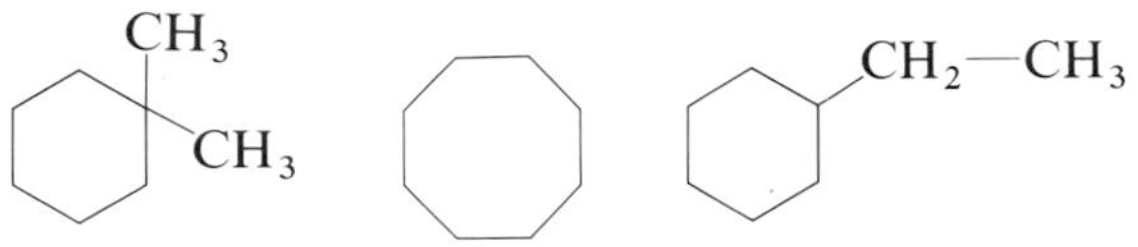

10.8 Sources of Alkanes

The two major sources of alkanes throughout the world are natural gas and petroleum. Natural gas consists of approximately 80% methane, 10% ethane, and a mixture of other relatively low-boiling alkanes such as propane, butane, and 2-methylpropane.

Petroleum is a liquid mixture of thousands of compounds, most of them hydrocarbons, formed from the decomposition of marine plants and animals. Petroleum and petroleum-derived products fuel automobiles, aircraft, and trains. They provide heat for buildings and fuel foe electric generating plants. They provide most of the greases and lubricants required for the machinery of our highly industrialized society. Furthermore, petroleum, along with natural gas, provides close to 90% of the organic raw materials for the synthesis and manufacture of synthetic fibers, plastics, detergents, and a multitude of other products.

The fundamental process in refining petroleum is **distillation.** Practically all crude oil that enters a refinery today goes to distillation units (called *columns*) where it is heated to temperatures as high as 370°C to 425°C and separated into **fractions.** Each fraction contains a mixture of hydrocarbons that boils within a particular range. Following are names of several of these fractions, along with the major uses of each.

1. Gases boiling below 20°C are taken off at the top of the distillation column. This fraction is a mixture of low-molecular-weight hydrocarbons, predominantly propane, butane, and 2-methylpropane. These three hydrocarbons can be liquefied under pressure at room temperature. The liquefied mixture, known as liquefied petroleum gas (LPG), can be stored and shipped in metal tanks and is a convenient source of gaseous fuel.

2. Naphthas, bp 20–200°C, are a mixture of C_4 to C_{10} alkanes and cyclo-alkanes. Naphthas also contain some aromatic hydrocarbons (Section 11.7) such as benzene, toluene, and xylene. The light naphtha fraction, bp 20–150°C, is the source of what is known as *straight run gasoline*, and averages approximately 25% of crude petroleum. Naphthas are also a major source of raw materials for the organic chemical industry.

3. Kerosene, bp 175–275°C, is a mixture of C_9 to C_{15} hydrocarbons.

4. Gas oil, bp 200–400°C, is a mixture of C_{15} to C_{25} hydrocarbons. It is from this fraction that diesel fuel is obtained.

5. Lubricating oil and heavy fuel oil distill from the column at temperatures over 350°C.

6. Asphalt is the name given to the black, tarry residue remaining after the removal of the other volatile fractions.

Gasoline is a mixture of C_4 to C_{10} hydrocarbons. The quality of gasoline as a fuel for an internal combustion engine is measured by **octane rating.** A high-octane fuel is one that burns smoothly and delivers power smoothly to the piston. A low-octane fuel tends to "explode" in the cylinder; that is, it undergoes too rapid combustion and leads to engine "knocking." When the scale of octane ratings was set up in 1929, 2,2,4-trimethylpentane (common name, isooctane), which has excellent knock properties, was assigned an octane rating of 100. Heptane, which causes a great deal of knocking, was assigned an octane rating of 0. The octane rating of a particular gasoline is equal to the percent of isooctane in a mixture of isooctane and heptane that has equivalent knock properties. For example, the knock properties of 2-methylhexane are the same as those of a mixture of 42% isooctane and 58% heptane; therefore, the octane rating of 2-methylhexane is 42. Octane itself has an octane rating of -20, which indicates that it produces even more engine knocking than heptane.

10.9 Reactions of Alkanes

Alkanes and cycloalkanes are resistant to attack by most acids, bases, and oxidizing and reducing agents because (1) they contain only nonpolar carbon-carbon and carbon-hydrogen bonds and (2) they have no unshared pairs of electrons and therefore are not susceptible to attack by electron-deficient reactants. However, all alkanes and cycloalkanes do react with the halogens and with oxygen.

A. Halogenation of Alkanes

Halogenation, the reaction of an alkane with chlorine or bromine, is the first step in the synthesis of many useful and important compounds. If a mixture of methane and chlorine gas is kept in the dark at room temperature, no reaction occurs. However, if the mixture is heated or exposed to light, a reaction begins almost at once. The products are chloromethane and hydrogen chloride. What occurs is a **substitution reaction:** the replacement of one hydrogen atom of methane by a chlorine atom and the production of an equivalent amount of hydrogen chloride.

$$
\underset{\text{methane}}{H-\overset{\displaystyle H}{\underset{\displaystyle H}{C}}-H} + Cl-Cl \xrightarrow{\text{light}} \underset{\substack{\text{chloromethane}\\\text{(methyl chloride)}}}{H-\overset{\displaystyle H}{\underset{\displaystyle H}{C}}-Cl} + H-Cl
$$

If chloromethane is reacted with more chlorine, further chlorination produces a mixture of dichloromethane (methylene chloride), trichloromethane (chloroform), and tetrachloromethane (carbon tetrachloride), each of which is widely used as a solvent.

$$
CH_3Cl \xrightarrow{Cl_2} \underset{\substack{\text{dichloromethane}\\\text{(methylene chloride)}}}{CH_2Cl_2} \xrightarrow{Cl_2} \underset{\substack{\text{trichloromethane}\\\text{(chloroform)}}}{CHCl_3} \xrightarrow{Cl_2} \underset{\substack{\text{tetrachloromethane}\\\text{(carbon tetrachloride)}}}{CCl_4}
$$

Reaction of ethane with chlorine gives chloroethane:

$$
\underset{\text{ethane}}{H-\overset{\displaystyle H}{\underset{\displaystyle H}{C}}-\overset{\displaystyle H}{\underset{\displaystyle H}{C}}-H} + Cl-Cl \xrightarrow{\text{light}} \underset{\substack{\text{chloroethane}\\\text{(ethyl chloride)}}}{H-\overset{\displaystyle H}{\underset{\displaystyle H}{C}}-\overset{\displaystyle H}{\underset{\displaystyle H}{C}}-Cl} + H-Cl
$$

When propane reacts with chlorine, two different chloropropanes are produced.

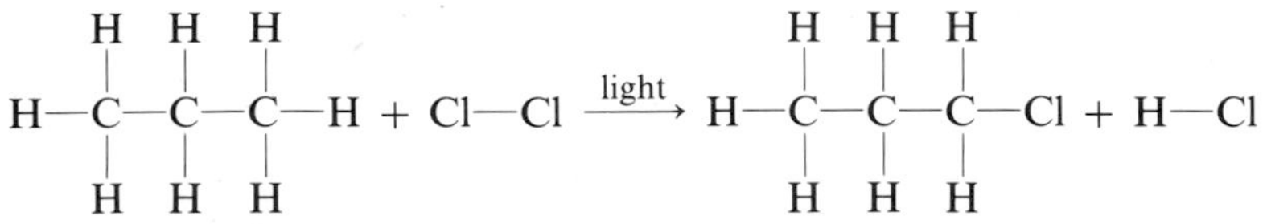

$$\underset{\text{propane}}{H-\overset{\overset{\displaystyle H}{|}}{\underset{\underset{\displaystyle H}{|}}{C}}-\overset{\overset{\displaystyle H}{|}}{\underset{\underset{\displaystyle H}{|}}{C}}-\overset{\overset{\displaystyle H}{|}}{\underset{\underset{\displaystyle H}{|}}{C}}-H} + Cl-Cl \xrightarrow{\text{light}} \underset{\substack{\text{1-chloropropane}\\(n\text{-propyl chloride})}}{H-\overset{\overset{\displaystyle H}{|}}{\underset{\underset{\displaystyle H}{|}}{C}}-\overset{\overset{\displaystyle H}{|}}{\underset{\underset{\displaystyle H}{|}}{C}}-\overset{\overset{\displaystyle H}{|}}{\underset{\underset{\displaystyle H}{|}}{C}}-Cl} + H-Cl$$

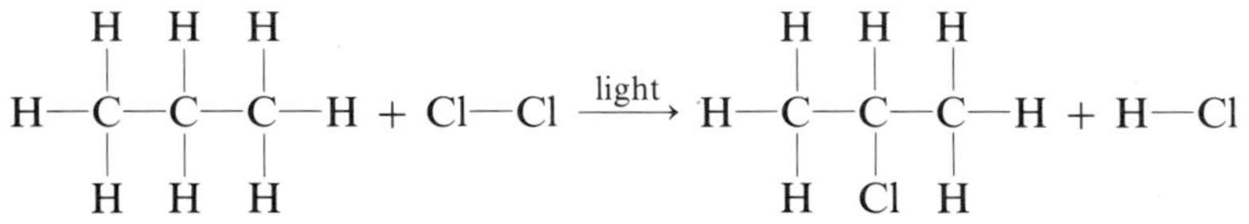

$$\underset{\text{propane}}{H-\overset{\overset{\displaystyle H}{|}}{\underset{\underset{\displaystyle H}{|}}{C}}-\overset{\overset{\displaystyle H}{|}}{\underset{\underset{\displaystyle H}{|}}{C}}-\overset{\overset{\displaystyle H}{|}}{\underset{\underset{\displaystyle H}{|}}{C}}-H} + Cl-Cl \xrightarrow{\text{light}} \underset{\substack{\text{2-chloropropane}\\(\text{isopropyl chloride})}}{H-\overset{\overset{\displaystyle H}{|}}{\underset{\underset{\displaystyle H}{|}}{C}}-\overset{\overset{\displaystyle H}{|}}{\underset{\underset{\displaystyle Cl}{|}}{C}}-\overset{\overset{\displaystyle H}{|}}{\underset{\underset{\displaystyle H}{|}}{C}}-H} + H-Cl$$

These are different reactions and we have written them separately. However, it is more common to show all possible products in one equation. The single equation below shows that the reaction of one mole of propane with one mole of chlorine produces one mole of HCl and a mixture totalling one mole of 1-chloropropane and 2-chloropropane combined.

$$CH_3-CH_2-CH_3 + Cl_2 \xrightarrow{\text{light}} CH_3-CH_2-CH_2-Cl + CH_3-\underset{\underset{\displaystyle Cl}{|}}{CH}-CH_3 + HCl$$

Haloalkanes (halogenated alkanes) are named according to the rules we have already used for naming alkanes. The parent chain is numbered from the direction that gives the alkyl substituent encountered first the lowest number. Halogen substituents are not taken into consideration in numbering the parent alkane. Halogen substituents are indicated by the prefixes fluoro-, chloro-, bromo-, and iodo-. In the common system of names, haloalkanes are named as alkyl halides; the name of the alkyl group is given first, followed by the name of the halide. For example, the IUPAC name for CH_3CH_2Cl is chloroethane; the common name for this compound is ethyl chloride.

Example 10.9

Name and draw structural formulas for all monochlorination products of the following reactions.

a. $CH_3-\overset{\overset{\displaystyle CH_3}{|}}{CH}-CH_3 + Cl_2 \xrightarrow{\text{light}}$

b. $CH_3-\overset{\overset{\displaystyle CH_3}{|}}{CH}-CH_2-CH_2-CH_3 + Cl_2 \xrightarrow{\text{light}}$

Solution

a. As we did when drawing all possible structural isomers of a given molecular formula, it is best to devise a system and then follow it. The most direct way is to start at one end of the carbon chain and substitute —Cl for —H. Then do the same thing on each carbon until you come to the other end of the chain. There are only two monochlorination derivatives of 2-methylpropane:

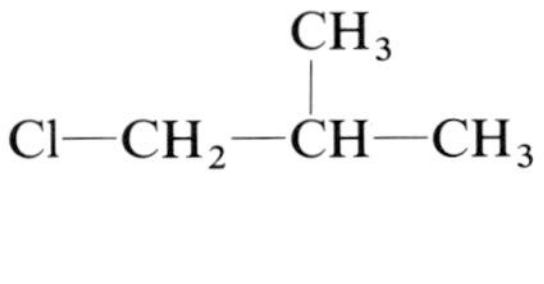

$$Cl—CH_2—\overset{\overset{\displaystyle CH_3}{|}}{CH}—CH_3 \qquad CH_3—\overset{\overset{\displaystyle CH_3}{|}}{\underset{\underset{\displaystyle Cl}{|}}{C}}—CH_3$$

1-chloro-2-methylpropane   2-chloro-2-methylpropane
(isobutyl chloride)   (*t*-butyl chloride)

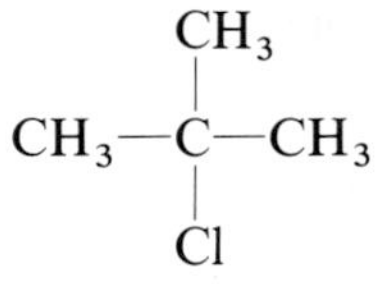

b. There are five monochlorination derivatives of 2-methylpentane:

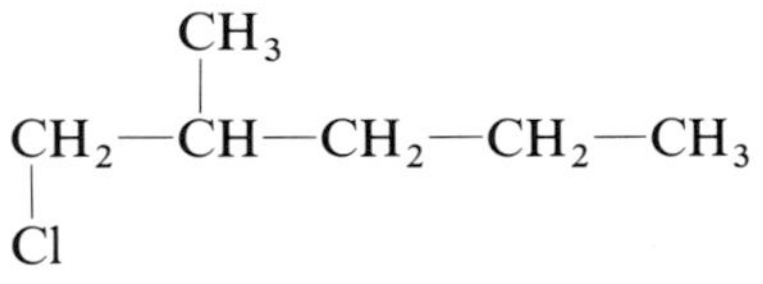

$$\underset{\underset{\displaystyle Cl}{|}}{CH_2}—\overset{\overset{\displaystyle CH_3}{|}}{CH}—CH_2—CH_2—CH_3 \qquad CH_3—\overset{\overset{\displaystyle CH_3}{|}}{\underset{\underset{\displaystyle Cl}{|}}{C}}—CH_2—CH_2—CH_3$$

1-chloro-2-methylpentane 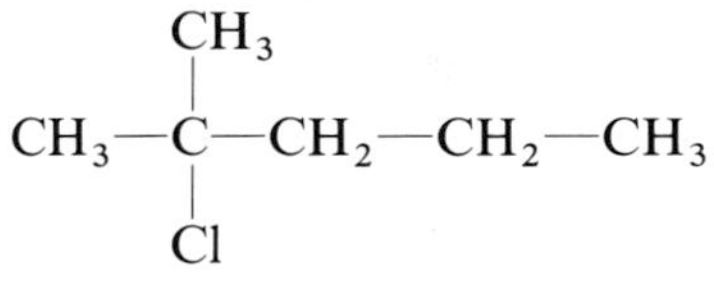   2-chloro-2-methylpentane

$$CH_3—\overset{\overset{\displaystyle CH_3}{|}}{CH}—\underset{\underset{\displaystyle Cl}{|}}{CH}—CH_2—CH_3 \qquad CH_3—\overset{\overset{\displaystyle CH_3}{|}}{CH}—CH_2—\underset{\underset{\displaystyle Cl}{|}}{CH}—CH_3$$

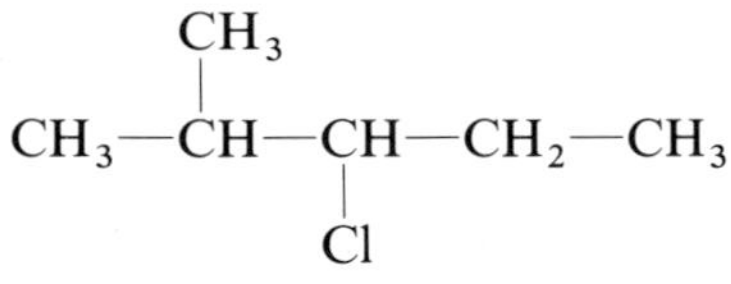

3-chloro-2-methylpentane 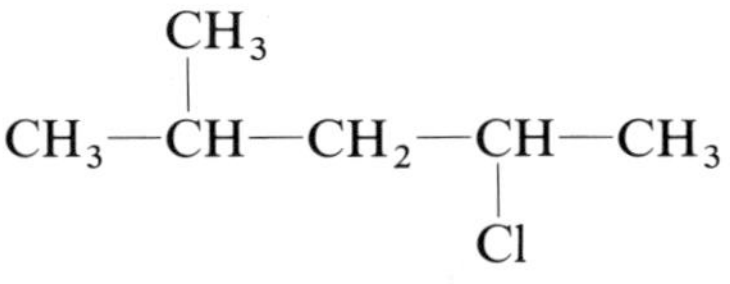   4-chloro-2-methylpentane

$$CH_3—\overset{\overset{\displaystyle CH_3}{|}}{CH}—CH_2—CH_2—\underset{\underset{\displaystyle Cl}{|}}{CH_2}$$

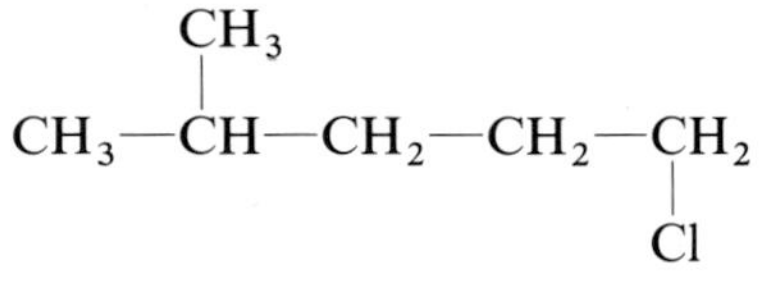

5-chloro-2-methylpentane

Problem 10.9

Name and draw structural formulas for all monohalogenation products of the following reactions:

a. 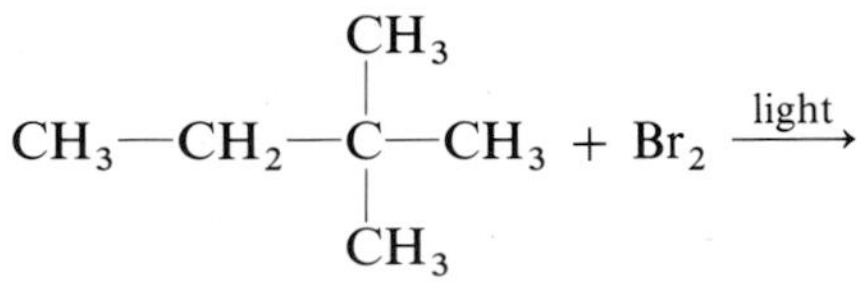 $CH_3—CH_2—\overset{\overset{\displaystyle CH_3}{|}}{\underset{\underset{\displaystyle CH_3}{|}}{C}}—CH_3 + Br_2 \xrightarrow{\text{light}}$

b. 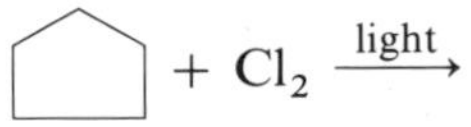 ⬠ $+ Cl_2 \xrightarrow{\text{light}}$

Because of their physical and chemical properties, many halogenated hydrocarbons have found wide commercial use. These include applications as commercial solvents, refrigerants, dry-cleaning agents, local and inhalation anesthetics, and insecticides.

Carbon tetrachloride, CCl_4, is a colorless, dense, nonflammable liquid, bp 77°C, that is inert to most common reagents and laboratory conditions. It is immiscible with water but is a good solvent for oils and greases, and at one time found wide use in the dry-cleaning industry. It is somewhat toxic, readily absorbed through the skin, and like all organic solvents should be used only with adequate ventilation. Prolonged exposure to carbon tetrachloride vapors results in liver and kidney damage.

Chloroform, $CHCl_3$, is a colorless, dense, sweet-smelling liquid, bp 62°C. It, too, is a widely used solvent for organic compounds.

Ethyl chloride, CH_3CH_2Cl, is used as a fast-acting, topical anesthetic. This haloalkane owes its anesthetic property more to its physical properties than to its chemical properties. Ethyl chloride boils at 12°C and, unless under pressure, is a gas at room temperature. When sprayed on the skin, it evaporates rapidly and cools the skin surface and underlying nerve endings. Skin and underlying nerve endings become anesthetized when skin temperature drops to about 15°C. Halothane is widely used as an inhalation anesthetic.

$$\begin{array}{c} \quad\;\; F \quad\; H \\ \quad\;\; | \qquad | \\ F\!-\!C\!-\!C\!-\!Br \\ \quad\;\; | \qquad | \\ \quad\;\; F \quad\; Cl \end{array}$$

halothane

Of all the fluoroalkanes, those manufactured under the trade name Freon are the most widely known and used. Freons were developed in a search for a new refrigerant, a compound that would be nontoxic, nonflammable, odorless, and noncorrosive. In 1930, General Motors Corp. announced the discovery of just such a compound, dichlorodifluoromethane, which was marketed under the trade name Freon-12. The Freons are manufactured by reacting a chlorinated hydrocarbon with hydrofluoric acid in the presence of an antimony pentafluoride or antimony chlorofluoride catalyst. Freon-11 is prepared from carbon tetrachloride. Freon-12 is prepared from Freon-11. Freon-22, monochlorodifluoromethane, is made from chloroform.

$$CCl_4 + HF \xrightarrow{SbF_5} \underset{\text{Freon-11}}{CCl_3F} + HCl$$

$$CCl_3F + HF \xrightarrow{SbF_5} \underset{\text{Freon-12}}{CCl_2F_2} + HCl$$

By 1974, U.S. production of Freons had grown to more than 1.1 billion pounds annually, almost one-half of world production. Worldwide production of these substances is now at an all-time high, reflecting their use in refrigeration and air-conditioning systems, aerosols (outside the U.S.), and as solvents.

Concern about the environmental impact of chlorofluorocarbons like Freon-11 and Freon-12 arose in 1974 when Drs. Sherwood Rowland and Mario Molina of the University of California, Irvine, announced their theory of ozone destruction by these compounds. When used as aerosol propellants and refrigerants, chlorofluorocarbons escape to the lower atmosphere but because of their general inertness, they do not decompose there. Slowly, they find their way to the stratosphere where they absorb ultraviolet radiation from the sun and then decompose. As they decompose, they set up a chemical reaction that may also lead to the destruction of the stratospheric ozone layer. What makes this a serious problem is that the stratospheric ozone layer acts as a shield for the Earth against excess ultraviolet radiation from the sun. An increase in ultraviolet radiation reaching the Earth may lead to the destruction of certain crops and agricultural species, and even increased incidence of skin cancer in sensitive individuals. Controversy continues over the potential for ozone depletion and its impact on the environment. In the meantime, both government and the chemical industry in the United States have taken steps to limit sharply the production and use of chlorofluorocarbons.

B. Combustion of Alkanes

The reaction of alkanes with oxygen to form carbon dioxide and water—and, most importantly, energy—is the basis for the use of hydrocarbons in our modern industrial and technological society as a source of heat (natural gas and fuel oil) and power (gasoline, diesel, and aviation fuel). An average of 170 kcal of heat energy is liberated from an alkane per mole of carbon atoms burned.

$$CH_4 \quad + 2\,O_2 \longrightarrow CO_2 + 2\,H_2O \qquad \Delta H = -212 \text{ kcal/mole}$$

methane
(natural gas)

$$CH_3 \overset{\underset{\textstyle CH_3}{|}}{\underset{\underset{\textstyle CH_3}{|}}{C}} CH_2 \overset{\underset{\textstyle CH_3}{|}}{CH} CH_3 + \tfrac{25}{2}\,O_2 \longrightarrow 8\,CO_2 + 9\,H_2O \qquad \Delta H = -1304 \text{ kcal/mole}$$

2,2,4-trimethylpentane
(a hydrocarbon with an octane rating of 100)

Example 10.10   Write a balanced equation for the combustion of butane and estimate the heat energy released (a) per mole of butane and (b) per gram of butane.

Solution   First write the balanced equation for the combustion of butane:

$$2\ CH_3—CH_2—CH_2—CH_3 + 13\ O_2 \longrightarrow 8\ CO_2 + 10\ H_2O$$

Butane has four carbons and releases approximately 4×170 kcal or 680 kcal of heat energy per mole. The heat energy per gram is:

$$\frac{680\ \text{kcal}}{\text{mole}} \times \frac{1\ \text{mole}}{58\ \text{g}} = 11.7\ \text{kcal/gram}$$

Problem 10.10   Write a balanced equation for the combustion of 2-methylpentane and estimate the amount of heat energy released (a) per mole of 2-methylpentane and (b) per gram of 2-methylpentane.

Key Terms and Concepts

alkane (10.1)   equatorial position (10.5)
alkyl group (10.3A)   halogenation of alkanes (10.9A)
axial position (10.5)   hydrocarbon (10.1)
chair conformation (10.5)   IUPAC system of nomenclature (10.3)
cis-trans isomerism in cycloalkanes (10.6)   Newman projection (10.5)
combustion of alkanes (10.9B)   saturated hydrocarbon (10.1)
conformation (10.5)   staggered conformation (10.5)
cycloalkane (10.4)   structural isomerism (10.2)
eclipsed conformation (10.5)   .   unsaturated hydrocarbon (10.1)

Key Reactions

1. Halogenation (Section 10.9A):

$$CH_4 + Cl_2 \xrightarrow{\text{light}} CH_3—Cl + H—Cl$$

2. Oxidation (Section 10.9B):

$$CH_4 + 2\ O_2 \longrightarrow CO_2 + 2\ H_2O \qquad \Delta H = -212\ \text{kcal/mole}$$

Problems

Names and
Structural Formulas
of Alkanes and
Cycloalkanes
(Sections 10.1, 10.3,
and 10.4)

10.11 Name the following alkyl groups.

a. CH_3-

b. CH_3-CH_2-

c. $CH_3-\overset{\overset{\displaystyle CH_3}{|}}{CH}-CH_2-$

d. $CH_3-CH_2-CH_2-$

e. $CH_3-\overset{\overset{\displaystyle CH_3}{|}}{CH}-$

f. $\overset{\displaystyle CH_2}{\underset{\displaystyle CH_2}{|}}CH-$

g. $CH_3-\overset{\overset{\displaystyle CH_3}{|}}{\underset{\underset{\displaystyle CH_3}{|}}{C}}-$

h. (cyclohexyl group)

i. $CH_3-CH_2-CH_2-CH_2-$

j. (cyclobutyl group)

k. $CH_3-\overset{\overset{\displaystyle CH_3}{|}}{CH}-CH_2-CH_2-$

10.12 Write IUPAC names for the following structural formulas.

a. $CH_3\overset{}{\underset{\underset{\displaystyle CH_3}{|}}{CH}}CH_2CH_2CH_3$

b. $CH_3\overset{}{\underset{\underset{\displaystyle CH_3}{|}}{CH}}CH_2CH_2\overset{}{\underset{\underset{\displaystyle CH_3}{|}}{CH}}CH_3$

c. $CH_3CH_2\overset{}{\underset{\underset{\displaystyle CH_3}{|}}{CH}}CH_2\overset{}{\underset{\underset{\displaystyle CH_2CH_3}{|}}{CH}}CH_3$

d. $(CH_3)_3CH$

e. $CH_3CH_2\overset{}{\underset{\underset{\displaystyle CH_3CHCH_3}{|}}{CH}}CH_2CH_2CH_2CH_3$

f. $CH_3CH_2CH_2\overset{}{\underset{\underset{\displaystyle CH_2CH_2CH_3}{|}}{CH}}CH_3$

g. $CH_3(CH_2)_8CH_3$

h. $(CH_3)_2CHCH_2CH_2CH(CH_3)_2$

i. (cyclopropane with) $\overset{\displaystyle CH_3}{\underset{\displaystyle CH_3}{}}$

j. (cyclopentane with) $CH_2-\overset{\overset{}{}}{\underset{\underset{\displaystyle CH_3}{|}}{CH}}-CH_3$

k. (cyclohexane with CH_2CH_3 and H_3C)

l. (cyclobutane with two CH_3)

m. (cyclohexane with $\overset{\overset{\displaystyle CH_3}{|}}{CH}-CH_3$)

n. (cyclohexane with cyclopropane)

o. (cyclohexane)$-CH_2-CH_2-\overset{\overset{}{}}{\underset{\underset{\displaystyle CH_3}{|}}{CH}}-CH_3$

10.13 Write structural formulas for the following compounds.

a. 2,2,4-trimethylhexane
b. 1-isopropyl-4-methylcyclohexane
c. 2,2-dimethylpropane
d. 2,4,5-trimethyl-3-ethyloctane
e. 2,4,6-trimethyloctane
f. 2,4-dimethyl-5-butylnonane
g. 4-isopropyloctane
h. 3,3-dimethylpentane
i. pentylcyclohexane
j. 1,3-dimethylcyclopentane
k. 1,2-diethylcyclobutane
l. 1-methyl-2-propylcyclopentane

10.14 Explain why each of the following names is incorrect. Write a correct IUPAC name.

a. 1,3-dimethylbutane
b. 4-methylpentane
c. 2,2-diethylbutane
d. 2-ethyl-3-methylpentane
e. 4,4-dimethylhexane
f. 2-propylpentane
g. 2,2-diethylheptane
h. 5-butyloctane
i. 2-dimethylpropane
j. 2-*sec*-butyloctane
k. 4-isopentylheptane
l. 1,3-dimethyl-6-ethylcyclohexane

Structural Isomerism in Alkanes and Cycloalkanes (Sections 10.2 and 10.4)

10.15 Are the following molecules pairs of structural isomers or are they identical?

a.

$$CH_3-\underset{\underset{\displaystyle CH_3}{|}}{CH}-\underset{\underset{\displaystyle CH_3}{|}}{CH}-CH_2-CH_3 \quad \text{and} \quad CH_3-CH-CH_2-CH_3$$
$$\underset{\displaystyle CH_3 \quad CH_3}{\overset{\displaystyle |}{CH}}$$

b.

$$CH_3-CH_2-CH-\underset{\underset{\displaystyle CH_3}{|}}{CH}-CH_2-CH_3 \quad \text{and} \quad CH_3-CH_2-CH-CH_3$$
$$\underset{\displaystyle \overset{\displaystyle |}{CH_3}-\underset{\underset{\displaystyle CH_3}{|}}{C}-CH_3}{}$$

(with CH_3 above the third carbon on the left molecule)

c.

$$CH_3-CH_2-CH-\underset{\underset{\displaystyle CH_2}{|}}{CH}-CH_3 \quad \text{and} \quad$$
$$\underset{\displaystyle CH_3}{}$$

with CH_3 above the CH on the left; and on the right:

$$\overset{\displaystyle CH_3}{\underset{\displaystyle CH_3}{\diagdown}}CH-CH\overset{\displaystyle CH_2-CH_3}{\underset{\displaystyle CH_2-CH_3}{\diagup}}$$

10.16 Which of the following structural formulas represent structural isomers and which represent identical compounds?

a.

$$CH_3-CH_2-\underset{\underset{\displaystyle Cl}{|}}{CH}-CH_3$$

b.

$$CH_3-\underset{\underset{\displaystyle Cl}{|}}{CH}-CH_3$$

c.

$$CH_3-\underset{\underset{\displaystyle Cl}{|}}{CH}-CH\ -CH_3$$

d.

$$\overset{\displaystyle Cl}{\underset{\displaystyle |}{CH_2}}-CH_2-CH_2-CH_3$$

e. 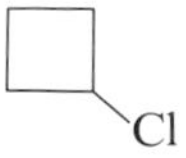 ⬚ Cl

f. 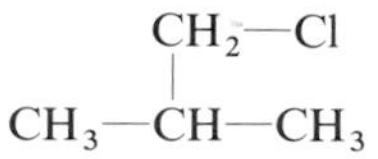 CH$_3$—CH—CH$_3$ with CH$_2$—Cl above CH

g. CH$_3$—CH$_2$—CH$_2$—CH$_2$—Cl

h. 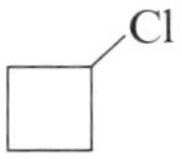 ⬚ Cl

i. 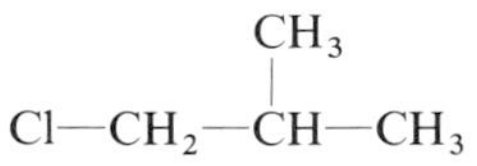 Cl—CH$_2$—CH—CH$_3$ with CH$_3$ above CH

j. 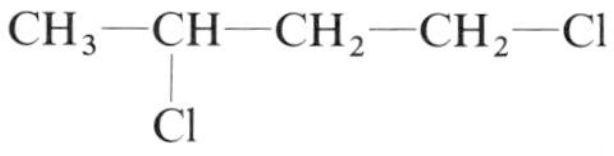 CH$_3$—CH—CH$_2$—CH$_2$—Cl with Cl below CH

k. 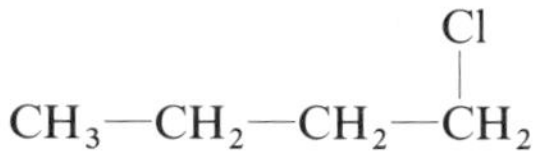 CH$_3$—CH$_2$—CH$_2$—CH$_2$ with Cl above terminal CH$_2$

l. 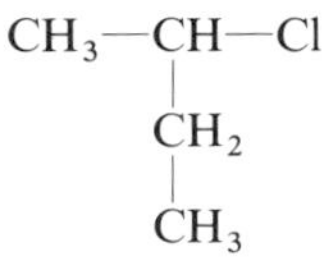 CH$_3$—CH—Cl with CH$_2$ and CH$_3$ below

m. 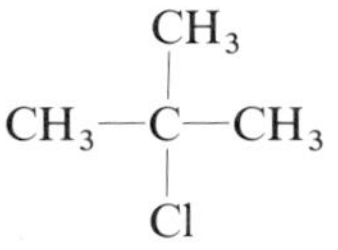 CH$_3$—C—CH$_3$ with CH$_3$ above C and Cl below C

Conformations of Alkanes and Cycloalkanes (Section 10.5)

10.17 Given 1-bromo-2-chloroethane, draw structural formulas for:
a. a conformation where the Br and Cl atoms are eclipsed by each other.
b. a conformation where the Br and Cl atoms are eclipsed by hydrogen atoms.
c. two different staggered conformations.

10.18 Draw a chair conformation for 1,4-dimethylcyclohexane in which:
a. both methyl groups are equatorial.
b. one methyl group is equatorial and the other is axial.
c. both methyl groups are axial.

10.19 Below on the left is a derivative of cyclohexane with substituent groups on carbons 1–5 of the ring. On the right is a drawing of a chair conformation of cyclohexane with the carbon atoms numbered 1–6. On this chair conformation show:
a. all substituent groups equatorial.
b. the —OH group on carbon-1 axial and the rest equatorial.
c. the —OH group on carbon-2 axial and the rest equatorial.
d. the —OH group on carbon-4 axial and the rest equatorial.

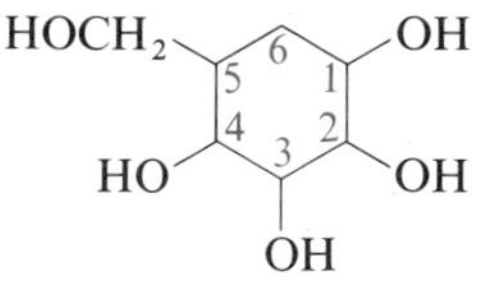

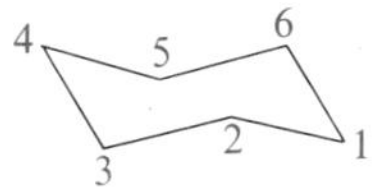

Cis-Trans Isomerism in Cycloalkanes (Section 10.6)

10.20 Draw structural formulas for the *cis-* and *trans-* isomers of 1,3-dimethylcyclohexane and 1,4-dimethylcyclohexane. Use a planar hexagon representation for the cyclohexane ring.

10.21 There are four *cis-trans* isomers of 2-isopropyl-5-methylcyclohexanol.

2-isopropyl-5-methylcyclohexanol

a. Draw each of these *cis-trans* isomers. Show the cyclohexane ring as a planar hexagon.

b. Draw a chair conformation with all three substituent groups in equatorial positions. If you have drawn this correctly you have drawn the structural formula of menthol.

10.22 "Benzene hexachloride," more properly named 1,2,3,4,5,6-hexachlorocyclohexane, is a mixture of various *cis-trans* isomers. The crude mixture is sold as the insecticide benzene hexachloride (BHC). The insecticidal properties of this mixture arise from one isomer known as the γ-isomer (gamma isomer), which is marketed under the trade names Lindane and Gammexane. The γ-isomer is *cis*-1,2,4,5-*trans*-3,6-hexachlorocyclohexane. Using a planar hexagon representation for the cyclohexane ring, draw a structural formula for the γ-isomer.

benzene hexachloride (BHC)

10.23 a. For 2,3-dimethylbutane, as viewed along the bond between carbon atoms 2 and 3, draw two different staggered conformations and two different eclipsed conformations.

b. Draw the *cis*- and *trans*- isomers of 1,2-dimethylcyclobutane.

c. Explain why the staggered and eclipsed conformations of 2,3-dimethylbutane drawn in part (a) do not represent different compounds, whereas *cis*- and *trans*-1,2-dimethylcyclobutane drawn in part (b) do represent different compounds.

Physical Properties of Alkanes and Cycloalkanes (Section 10.7)

10.24 What generalization can you make about the densities of alkanes relative to that of water?

10.25 Which unbranched alkane has about the same boiling point as water? (Refer to Table 10.3 for data on the physical properties of alkanes.) Calculate the molecular weight of this alkane and compare it with that of water.

10.26 a. Name and draw structural formulas for all isomeric alkanes of molecular formula C_7H_{16}.

b. Predict which isomer in part (a) has the lowest boiling point; the highest boiling point.

10.27 Name and draw structural formulas for all monochlorination products formed in the following reactions:

a. $CH_3-CH-CH_2-CH_3 + Cl_2 \xrightarrow{light}$
   (with CH_3 on the CH)

b. $CH_3-CH_2-CH_2-CH_2-CH_3 + Cl_2 \xrightarrow{light}$

c. $CH_3-CH-CH-CH_3 + Cl_2 \xrightarrow{light}$
   (with CH_3 and CH_3 on the two CH groups)

d. 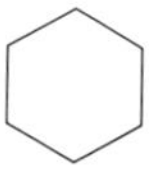 (hexagon) $+ Cl_2 \xrightarrow{light}$

10.28 There are three isomeric alkanes of molecular formula C_5H_{12}. When reacted with chlorine gas at 300°C, isomer A gives a mixture of four monochlorination products. Under the same conditions, isomer B gives a mixture of three monochlorination products, while isomer C gives only one monochlorination product. From this information, assign structural formulas to isomers A, B, and C.

10.29 Draw structural formulas for Freon-11, Freon-12, and Freon-22. Explain why Freons such as these have become so widely used in refrigeration and air-conditioning systems.

10.30 In a handbook of chemistry or other suitable reference, look up the densities of methylene chloride, chloroform, and carbon tetrachloride. Which are more dense than water? Which are less dense?

10.31 Complete and balance the following combustion reactions. Assume that each hydrocarbon is converted completely to carbon dioxide and water. In addition, estimate the heat energy liberated per mole of hydrocarbon.
   a. propane $+ O_2 \longrightarrow$ b. octane $+ O_2 \longrightarrow$
   c. cyclohexane $+ O_2 \longrightarrow$ d. 3-methylpentane $+ O_2 \longrightarrow$
   e. methylcyclopentane $+ O_2 \longrightarrow$ f. cyclobutane $+ O_2 \longrightarrow$

10.32 What is the major component of natural gas? Of bottled gas or LPG?

10.33 Calculate the heat of combustion per gram of methane, propane, and 2,2,4-trimethylpentane. When compared on a gram-for-gram basis, which of these hydrocarbons is the best source of heat energy?

Hydrocarbon	A Major Component of:	Heat of Combustion
CH_4	natural gas	212 kcal/mole
$CH_3CH_2CH_3$	LPG	531 kcal/mole
$CH_3CCH_2CHCH_3$ (with CH_3, CH_3 on C and CH_3 below)	gasoline	1304 kcal/mole

Unsaturated Hydrocarbons

Unsaturated hydrocarbons are hydrocarbons that contain one or more carbon-carbon double bonds or triple bonds. There are three classes of unsaturated hydrocarbons: alkenes, alkynes, and aromatic hydrocarbons. **Alkenes** contain one or more carbon-carbon double bonds and **alkynes** contain one or more carbon-carbon triple bonds. The structural formulas of ethene, the simplest alkene, and ethyne, the simplest alkyne, are:

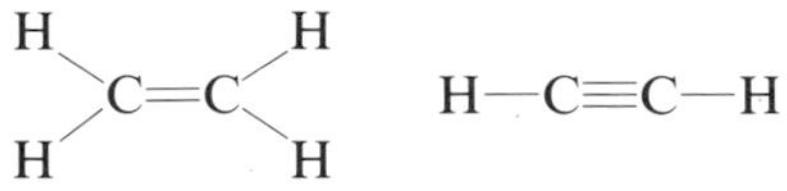

$$\begin{array}{ccc} H & & H \\ & \diagdown C = C \diagup & \\ H & & H \end{array} \qquad H - C \equiv C - H$$

ethene
(an alkene)

ethyne
(an alkyne)

Because alkenes and alkynes have fewer hydrogen atoms than alkanes with the same number of carbon atoms, they are commonly referred to as unsaturated hydrocarbons; they are unsaturated with respect to hydrogen atoms.

The third class of unsaturated hydrocarbons is the **aromatic hydrocarbons.** The simplest aromatic hydrocarbon, benzene, has three carbon-carbon double bonds in a six-member ring.

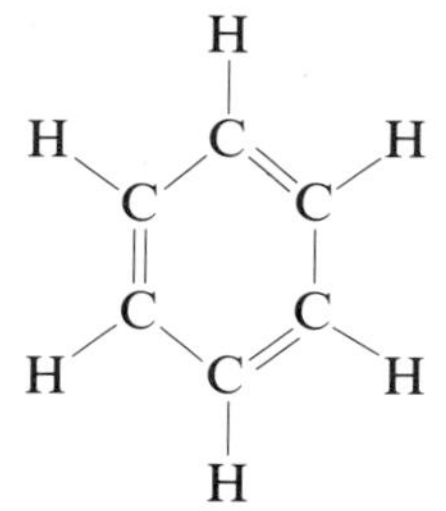

benzene
(an aromatic hydrocarbon)

Benzene and other aromatic hydrocarbons have chemical properties quite different from those of alkenes and alkynes.

The structural feature that relates alkenes, alkynes, and aromatic hydrocarbons is the presence of one or more pi bonds between carbon atoms. Consequently, this chapter is the study of carbon-carbon pi bonds.

11.1 Structure of Alkenes and Alkynes

A. Alkenes

Ethylene, C_2H_4, is the first member of the alkene family.

$$CH_2{=}CH_2$$

ethylene

The second member of the alkene family is propene, C_3H_6.

$$CH_3{-}CH{=}CH_2$$

propene

There are three alkenes of formula C_4H_8.

$$CH_3{-}CH_2{-}CH{=}CH_2 \qquad CH_3{-}CH{=}CH{-}CH_3 \qquad CH_3{-}\underset{\underset{\displaystyle CH_3}{|}}{C}{=}CH_2$$

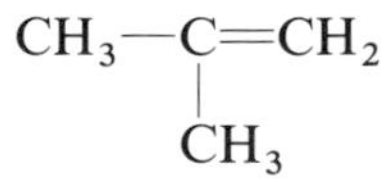

The number of alkene structural isomers increases rapidly as the number of carbons increases, for in addition to variations in chain length and branching, there are variations in the position of the double bond. The general formula of an alkene is C_nH_{2n}.

Example 11.1

Draw all structural isomers for alkenes of molecular formula C_5H_{10}.

Solution

Approach this type of problem systematically. First, draw possible carbon skeletons for molecules of five carbon atoms. There are three such skeletons:

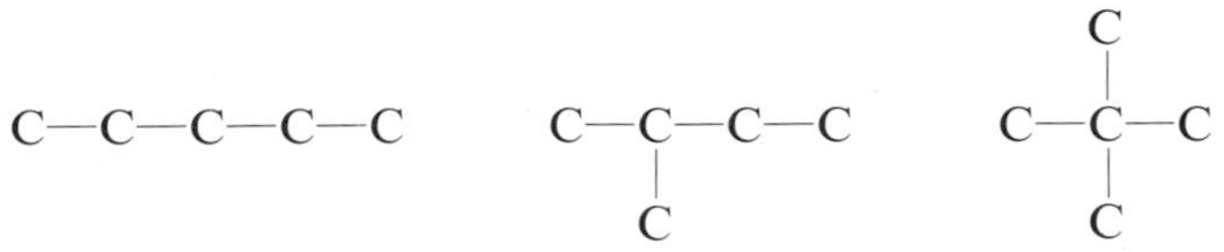

$$C-C-C-C-C \qquad \begin{matrix} C-C-C-C \\ | \\ C \end{matrix} \qquad \begin{matrix} C \\ | \\ C-C-C \\ | \\ C \end{matrix}$$

(a)   (b)   (c)

Note that skeleton (c) cannot contain a carbon-carbon double bond because the central carbon atom is already bonded to four other carbon atoms. Therefore, we need consider only skeletons (a) and (b) when drawing structural isomers for alkenes of molecular formula C_5H_{10}. Next, locate the double bond between carbon atoms along the chain. For carbon skeleton (a), there are two possible locations for the carbon-carbon double bond; for carbon skeleton (b), there are three possible locations for the carbon-carbon double bond:

for (a): $C=C-C-C-C$ and $C-C=C-C-C$

for (b): $\begin{matrix} C=C-C-C \\ | \\ C \end{matrix}$ and $\begin{matrix} C-C=C-C \\ | \\ C \end{matrix}$ and $\begin{matrix} C-C-C=C \\ | \\ C \end{matrix}$

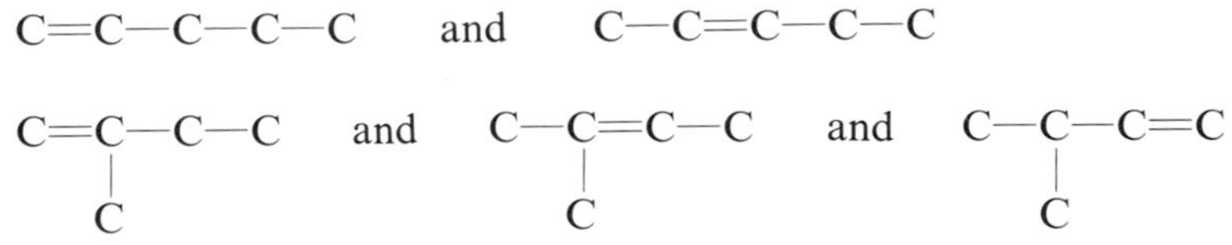

Finally, add hydrogen atoms to complete the tetravalence of carbon and give the correct molecular formula.

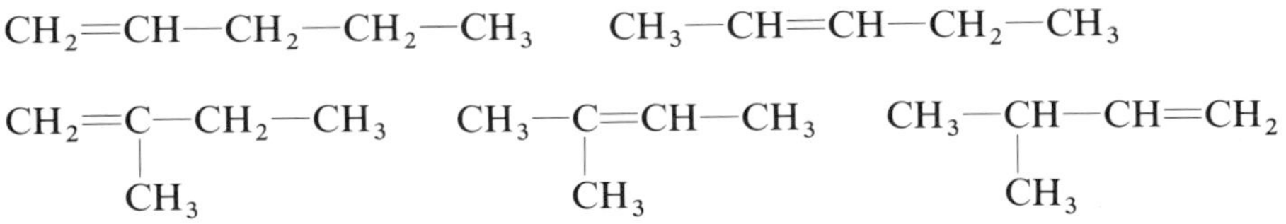

$$CH_2=CH-CH_2-CH_2-CH_3 \qquad CH_3-CH=CH-CH_2-CH_3$$

$$\begin{matrix} CH_2=C-CH_2-CH_3 \\ | \\ CH_3 \end{matrix} \qquad \begin{matrix} CH_3-C=CH-CH_3 \\ | \\ CH_3 \end{matrix} \qquad \begin{matrix} CH_3-CH-CH=CH_2 \\ | \\ CH_3 \end{matrix}$$

Problem 11.1

Draw structural isomers for all alkenes of molecular formula C_6H_{12} that have the following carbon skeletons:

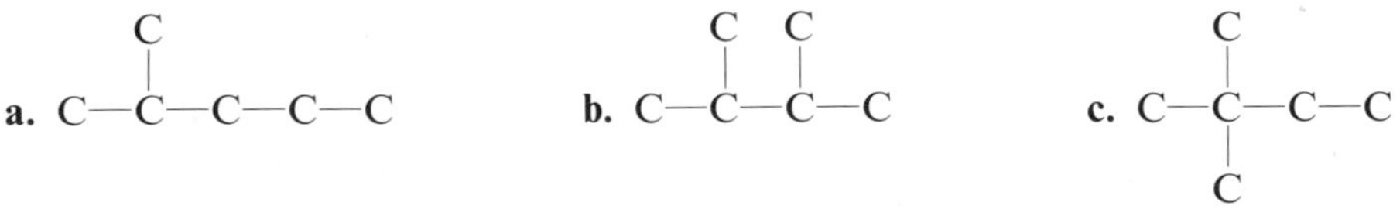

a. $\begin{matrix} C \\ | \\ C-C-C-C-C \end{matrix}$ **b.** $\begin{matrix} C \quad C \\ | \quad | \\ C-C-C-C \end{matrix}$ **c.** $\begin{matrix} C \\ | \\ C-C-C-C \\ | \\ C \end{matrix}$

B. Alkynes

Ethyne (acetylene) is the first member of the alkyne family. Propyne is the second.

$$H-C\equiv C-H \qquad CH_3-C\equiv C-H$$

ethyne · · · · · · · · · · propyne
(acetylene)

Alkynes of four or more carbons that can have different locations for the triple bond show structural isomerism. For instance, there are two alkynes of molecular formula C_4H_6:

$$CH_3-CH_2-C\equiv CH \qquad CH_3-C\equiv C-CH_3$$

Alkynes form a series of compounds with the general formula C_nH_{2n-2}.

Example 11.2

There are three alkynes of molecular formula C_5H_8. Draw structural formulas for each.

Solution

First draw all possible carbon skeletons that can contain a carbon-carbon triple bond. Next, locate the triple bond, and finally, add hydrogens to complete the tetravalence of carbon. There are only two carbon skeletons for C_5H_8 that can contain a triple bond:

$$\begin{array}{c} \\ C-C-C-C-C \end{array} \qquad \begin{array}{c} C \\ | \\ C-C-C-C \end{array}$$

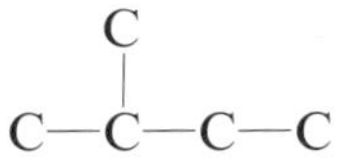

There are two possible locations for the triple bond in the first carbon skeleton and only one in the second. Structural formulas for the three possible alkynes are:

$$CH_3-CH_2-CH_2-C\equiv CH \qquad CH_3-CH_2-C\equiv C-CH_3 \qquad \overset{\displaystyle CH_3}{\underset{\displaystyle |}{CH_3-CH-C\equiv CH}}$$

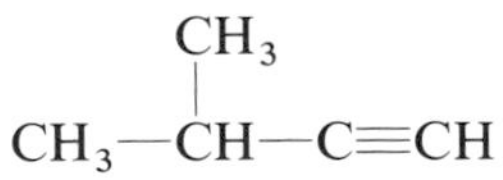

Problem 11.2

Draw structural formulas for the seven alkynes of molecular formula C_6H_{10}.

11.2 Nomenclature of Alkenes and Alkynes

A. Alkenes

IUPAC names of alkenes are formed by changing the -ane of the parent alkane to -ene. The molecule $CH_2=CH_2$ is named ethene and $CH_3-CH=CH_2$ is named propene. The IUPAC system retains the common name ethylene and therefore there are two acceptable IUPAC names for this alkene: ethene and ethylene. Ethylene and propene can contain a double bond in only one position. In butene and all higher alkenes, there are isomers that differ in the location of the double bond and therefore a numbering system must be used to locate the double bond. According to the IUPAC system, the longest carbon chain that contains the double bond is numbered to give the double-bonded carbons the lowest possible numbers and the position of the double bond is indicated by the number of the first carbon of the double bond. Branched or substituted alkenes are named in a manner similar to that used for alkanes. The carbon atoms are numbered, substituent groups are located and named, the double bonds are located, and the main chain is named.

$$\overset{6}{CH_3}-\overset{5}{CH_2}-\overset{4}{CH_2}-\overset{3}{CH}=\overset{2}{CH}-\overset{1}{CH_3} \qquad \overset{6}{CH_3}-\overset{5}{CH_2}-\overset{4}{CH}-\overset{3}{CH}=\overset{2}{CH}-\overset{1}{CH_3}$$

$$\underset{CH_3}{|}$$

2-hexene   4-methyl-2-hexene

Example 11.3

In Example 11.1, you drew formulas for alkenes of molecular formula C_6H_{10}. Give each alkene an IUPAC name.

Solution

First, number the carbon chain from the end that gives the double-bonded carbons the lowest numbers. In writing the name, be certain that each substituent is named and numbered and that the location of the double bond is given.

$$\overset{1}{CH_2}=\overset{2}{CH}-\overset{3}{CH_2}-\overset{4}{CH_2}-\overset{5}{CH_3} \qquad \overset{1}{CH_3}-\overset{2}{CH}=\overset{3}{CH}-\overset{4}{CH_2}-\overset{5}{CH_3}$$

1-pentene   2-pentene

$$\overset{1}{CH_2}=\overset{2}{C}-\overset{3}{CH_2}-\overset{4}{CH_3} \qquad \overset{1}{CH_3}-\overset{2}{C}=\overset{3}{CH}-\overset{4}{CH_3} \qquad \overset{4}{CH_3}-\overset{3}{CH}-\overset{2}{CH}=\overset{1}{CH_2}$$

$$\underset{CH_3}{|} \qquad\qquad \underset{CH_3}{|} \qquad\qquad \underset{CH_3}{|}$$

2-methyl-1-butene   2-methyl-2-butene   3-methyl-1-butene

Problem 11.3

In Problem 11.1, you drew formulas for alkenes having three different carbon skeletons. Give each alkene an IUPAC name.

In naming cycloalkenes, carbon atoms of the double bond are numbered 1 and 2 in the direction that gives substituents the smallest numbers possible.

3-methylcyclopentene   1,4-dimethylcyclohexene
                              (*not* 2,5-dimethylcyclohexene)

Alkenes that contain more than one double bond are called dienes, trienes, and so on. Following are structural formulas for two dienes of molecular formula C_5H_8.

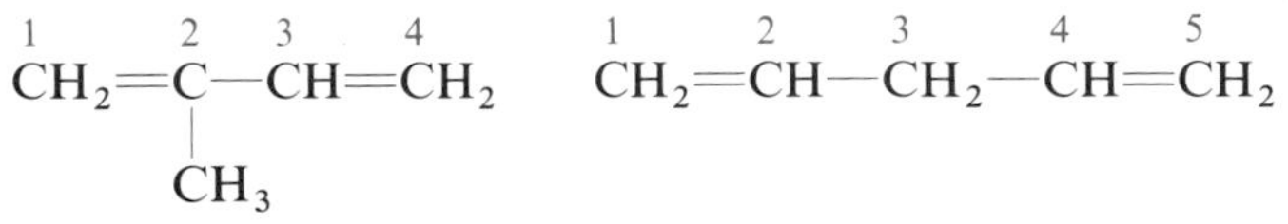

$$\overset{1}{C}H_2{=}\overset{2}{C}{-}\overset{3}{C}H{=}\overset{4}{C}H_2 \qquad \overset{1}{C}H_2{=}\overset{2}{C}H{-}\overset{3}{C}H_2{-}\overset{4}{C}H{=}\overset{5}{C}H_2$$
$$\qquad\; |$$
$$\qquad CH_3$$

2-methyl-1,3-butadiene   1,4-pentadiene
(isoprene)

Many alkenes, particularly the lower-molecular-weight ones, are known almost exclusively by their common names. For example, the common name of propene is propylene; that of 2-methylpropene is isobutylene.

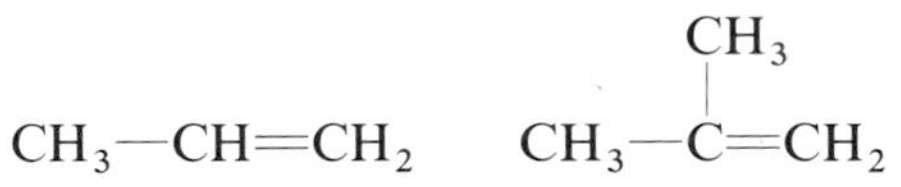

$$\qquad\qquad\qquad\qquad\qquad\qquad CH_3$$
$$\qquad\qquad\qquad\qquad\qquad\qquad |$$
$$CH_3{-}CH{=}CH_2 \qquad CH_3{-}C{=}CH_2$$

IUPAC name: propene   2-methylpropene
common name: propylene   isobutylene

The use of common names is generally avoided for alkenes of more than four carbons because of the large number of structural isomers possible and the lack of suitable prefixes to distinguish one structural isomer from another.

B. Alkynes

In the IUPAC nomenclature, the ending *-yne* indicates the presence of a triple bond. Hence, $HC{\equiv}CH$ is named ethyne and $CH_3{-}C{\equiv}CH$ is named propyne. The IUPAC system retains the name acetylene, and therefore there are two acceptable IUPAC names for this alkyne: acetylene and ethyne. For longer-chain alkynes, the longest carbon chain that contains the triple bond is numbered in such a manner as to give triple-bonded carbons the lowest possible numbers. Its location is indicated by the number of the first carbon of the triple bond.

$$CH_3{-}CH_2{-}C{\equiv}CH \qquad CH_3{-}C{\equiv}C{-}CH_3$$

1-butyne   2-butyne

Example 11.4

Give IUPAC names for the following alkynes:

a. $CH_3—CH—C\equiv C—CH_3$
    |
    CH_3

b. $CH_3—CH_2—C\equiv C—CH_2—CH—CH_3$
                            |
                            CH_3

Solution

a. Number the carbon chain so that the triple bond is between carbons 2 and 3 of the chain. The correct IUPAC name is 4-methyl-2-pentyne.

$$\overset{5}{C}H_3—\overset{4}{C}H—\overset{3}{C}\equiv\overset{2}{C}—\overset{1}{C}H_3$$
                |
                CH_3

4-methyl-2-pentyne

b. The correct IUPAC name is 6-methyl-3-heptyne.

$$\overset{1}{C}H_3—\overset{2}{C}H_2—\overset{3}{C}\equiv\overset{4}{C}—\overset{5}{C}H_2—\overset{6}{C}H—\overset{7}{C}H_3$$
                                        |
                                        CH_3

6-methyl-3-heptyne

Problem 11.4

Write structural formulas for the following alkynes:

a. 1-hexyne   b. 2-hexyne   c. 3-ethyl-4-methyl-1-pentyne

11.3 *Cis-Trans* Isomerism in Alkenes

Stereoisomers have the same structural formula but different arrangements of atoms in space. *Cis-trans* isomerism is one type of stereoisomerism. In cyclo-alkanes, *cis-trans* isomerism depends on the arrangement of substituent groups on a ring (Section 10.6). In alkenes, it depends on the arrangement of groups on a double bond.

There are two important features of a carbon-carbon double bond that make *cis-trans* isomerism possible. First, as shown in Figure 11.1(a), the two carbons of a double bond and the four atoms attached to them all lie in a plane. Second, there is restricted rotation about the two carbons of the double bond. Recall from Section 9.5 that a double bond consists of one sigma bond formed by overlap of sp^2 hybrid orbitals and one pi bond formed by overlap of un-hybridized $2p$ orbitals. The sigma electrons lie along the line connecting the two atoms bonded together (Figure 11.1b) and the pi electrons lie above and

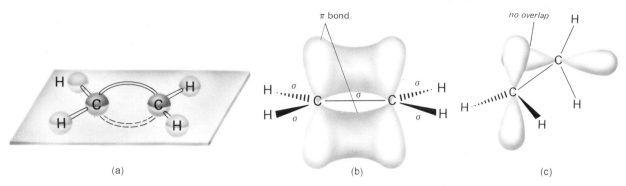

Figure 11.1 Restricted rotation about the carbon-carbon double bond. (a) Ball-and-stick model; (b) orbital overlap model showing the pi bond; (c) the pi bond must be broken for the orbitals to rotate.

below the plane created by the sigma-bond skeleton. In alkenes, the double-bonded carbon atoms cannot be twisted with respect to each other, for to do so would break the pi bond (Figure 11.1c).

Ethylene, propene, 1-butene, and 2-methylpropene have only one possible arrangement of atoms. 2-Butene has two possible arrangements. In one of them, the two methyl groups are on the same side of the double bond: this isomer is called *cis*-2-butene. In the other arrangement, the two methyl groups are on opposite sides of the double bond: this isomer is called *trans*-2-butene. Shown in Figure 11.2 are Lewis structures of the *cis*- and *trans*- isomers of 2-butene.

Cis- and *trans*- isomers are different compounds and have different physical and chemical properties. Note that the *cis*- and *trans*- isomers of 2-butene differ in melting points by 33°C and in boiling points by 3°C.

Figure 11.2 The *cis*- and *trans*- isomers of 2-butene.

cis-2-butene
mp: −139°C
bp: 4°C

trans-2-butene
mp: −106°C
bp: 1°C

Example 11.5

Which of the following alkenes show *cis-trans* isomerism?

a. $CH_2=CHCH_2CH_2CH_3$ **b.** $CH_3CH=CHCH_2CH_3$

c. $CH_2=CCH_2CH_3$
       $|$
       CH_3

Solution

a. Begin by drawing the carbon-carbon double bond and showing the bond angles of 120° about it.

$$\text{C}=\text{C}$$

Next, complete the structural formula and show all four groups attached to the double bond.

$$\begin{array}{c}\text{H}\diagdown\quad\diagup\text{H}\\\text{C}=\text{C}\\\text{H}\diagup\quad\diagdown\text{CH}_2\text{CH}_2\text{CH}_3\end{array}$$

To determine if this molecule shows *cis-trans* isomerism, switch the positions of the —H and —$\text{CH}_2\text{CH}_2\text{CH}_3$ groups and compare the two structural formulas:

$$\begin{array}{c}\text{H}\diagdown\quad\diagup\text{H}\\\text{C}=\text{C}\\\text{H}\diagup\quad\diagdown\text{CH}_2\text{CH}_2\text{CH}_3\end{array}\qquad\begin{array}{c}\text{H}\diagdown\quad\diagup\text{CH}_2\text{CH}_2\text{CH}_3\\\text{C}=\text{C}\\\text{H}\diagup\quad\diagdown\text{H}\end{array}$$

In this case, they represent the same molecule. To see this, imagine that you could pick up either structure and turn it over as you would turn your hand from palm down to palm up. If you do this correctly, you will see that one structural formula fits exactly on top of the other (that is, it can be superimposed on the other). Therefore, they are identical; 1-pentene does not show *cis-trans* isomerism.

b. Draw structural formulas for possible *cis-trans* isomers as you did in part (a), and then examine them to see if the drawings represent the same molecule or *cis-trans* isomers.

$$\begin{array}{c}\text{H}_3\text{C}\diagdown\quad\diagup\text{H}\\\text{C}=\text{C}\\\text{H}\diagup\quad\diagdown\text{CH}_2\text{CH}_3\end{array}\qquad\begin{array}{c}\text{H}_3\text{C}\diagdown\quad\diagup\text{CH}_2\text{CH}_3\\\text{C}=\text{C}\\\text{H}\diagup\quad\diagdown\text{H}\end{array}$$

trans-2-pentene   cis-2-pentene

The orientation of the four groups attached to the double bond of the molecule on the left is different from that of the molecule on the right. Therefore, the structures represent *cis-trans* isomers.

$$\begin{array}{c}
\text{c.} \quad \underset{H}{\overset{H}{\diagdown}}C=C\underset{CH_2CH_3}{\overset{CH_3}{\diagup}} \quad \text{and} \quad \underset{H}{\overset{H}{\diagdown}}C=C\underset{CH_3}{\overset{CH_2CH_3}{\diagup}}
\end{array}$$

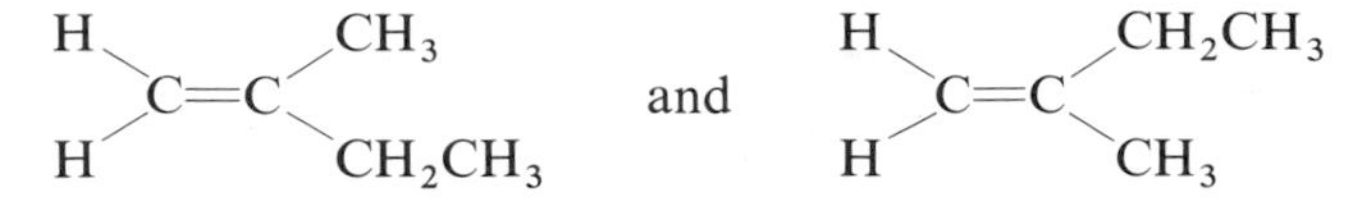

These two structural formulas are superimposable; 2-methyl-1-butene does not show *cis-trans* isomerism. These three examples show that an alkene shows *cis-trans* isomerism only if each of the carbon atoms of the double bond has two different groups attached to it.

Problem 11.5

Which of the following alkenes show *cis-trans* isomerism? For each that does, draw structural formulas for the isomers.

a. $CH_2{=}C{-}CH_2{-}CH_2{-}CH_3$
       |
      CH_3

b. $CH_3{-}C{=}CH{-}CH_2{-}CH_3$
          |
         CH_3

c. $CH_3{-}CH{-}CH{=}CH{-}CH_3$
          |
         CH_3

d. $CH_3{-}CH{-}CH_2{-}CH{=}CH_2$
          |
         CH_3

Vitamin A is an example of a biologically important molecule that can exist as a number of *cis-trans* isomers.

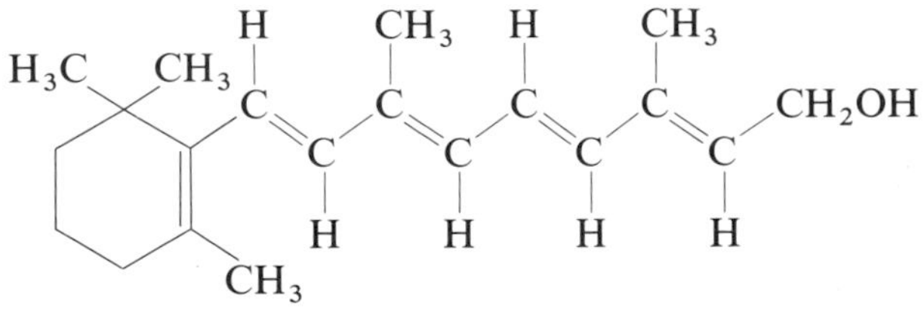

vitamin A

There are four carbon-carbon double bonds in the chain of atoms attached to the cyclohexene ring, and each of the four can show *cis-trans* isomerism. Thus, there are 2 × 2 × 2 × 2 = 16 possible *cis-trans* isomers of vitamin A.

11.4 Physical Properties of Alkenes and Alkynes

Alkenes and alkynes are nonpolar compounds and their physical properties are much the same as those of alkanes with similar carbon skeletons. Alkenes and alkynes of 2, 3, and 4 carbon atoms are gases at room temperature. Those of five or more carbons are colorless liquids, less dense than water. Alkenes and alkynes are insoluble in water, but they are soluble in each other, in ethanol,

Table 11.1 Physical properties of some alkenes and alkynes.

IUPAC Name	Structural Formula	bp (°C)	mp (°C)
ethene or ethylene	$CH_2{=}CH_2$	-104	-169
propene	$CH_3CH{=}CH_2$	-47	-185
1-butene	$CH_3CH_2CH{=}CH_2$	-6	-185
1-pentene	$CH_3CH_2CH_2CH{=}CH_2$	30	-138
cis-2-pentene	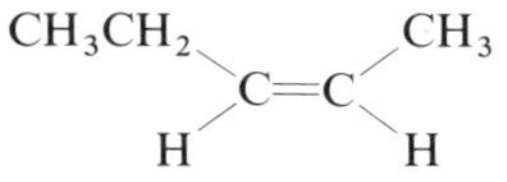 $\begin{array}{c} CH_3CH_2 \diagdown\quad\diagup CH_3 \\ C{=}C \\ H \diagup\quad\diagdown H \end{array}$	37	-151
trans-2-pentene	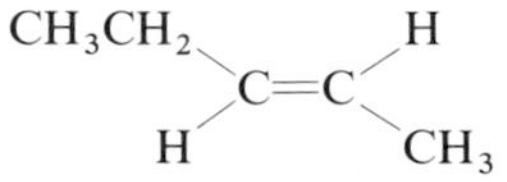 $\begin{array}{c} CH_3CH_2 \diagdown\quad\diagup H \\ C{=}C \\ H \diagup\quad\diagdown CH_3 \end{array}$	36	-136
2-methyl-2-butene	$\begin{array}{c} CH_3 \\ \mid \\ CH_3C{=}CHCH_3 \end{array}$	39	-134
ethyne or acetylene	$HC{\equiv}CH$	-84	-81
propyne	$CH_3C{\equiv}CH$	-23	-101
1-butyne	$CH_3CH_2C{\equiv}CH$	8	-126
1-pentyne	$CH_3CH_2CH_2C{\equiv}CH$	40	-90

and in other nonpolar organic solvents. Table 11.1 lists the physical properties of some alkenes and alkynes.

11.5 Reactions of Alkenes

In contrast to alkanes, alkenes react readily with halogens, certain strong acids, a variety of oxidizing and reducing agents, and even water in the presence of concentrated sulfuric acid. The characteristic reaction of alkenes is addition to the carbon-carbon double bond.

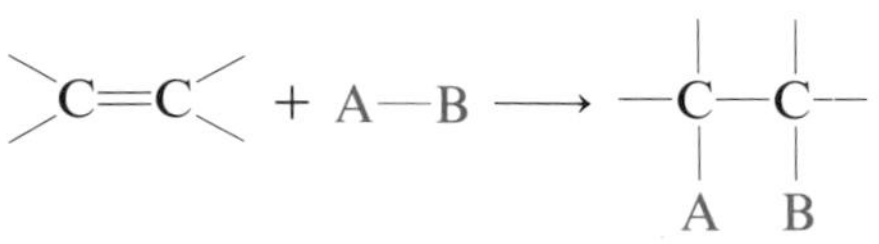

$$\diagup\!\!\!\!C{=}C\!\!\!\!\diagdown \;+\; A{-}B \;\longrightarrow\; \begin{array}{c} \mid\quad\mid \\ {-}C{-}C{-} \\ \mid\quad\mid \\ A\quad B \end{array}$$

In addition reactions, one sigma bond (A—B) and one pi bond (C=C) are broken and two new sigma bonds are formed. Consequently, addition reactions to the double bond are almost always energetically favorable because there is net conversion of one pi bond to a sigma bond.

A. Addition of Hydrogen: Reduction

Addition of hydrogen to a carbon-carbon double bond is called **hydrogenation** and reduces an alkene to an alkane. The process requires the presence of a metal catalyst.

$$CH_2{=}CH_2 + H{-}H \xrightarrow[\text{catalyst}]{\text{metal}} \underset{\underset{H\ \ \ H}{|\ \ \ |}}{\overset{\overset{H\ \ \ H}{|\ \ \ |}}{H{-}C{-}C{-}H}}$$

In catalytic hydrogenation, both hydrogen atoms are added to the same side of the alkene molecule. For example, addition of hydrogen to 1,2-dimethylcyclopentene yields *cis*-1,2-dimethylcyclopentane.

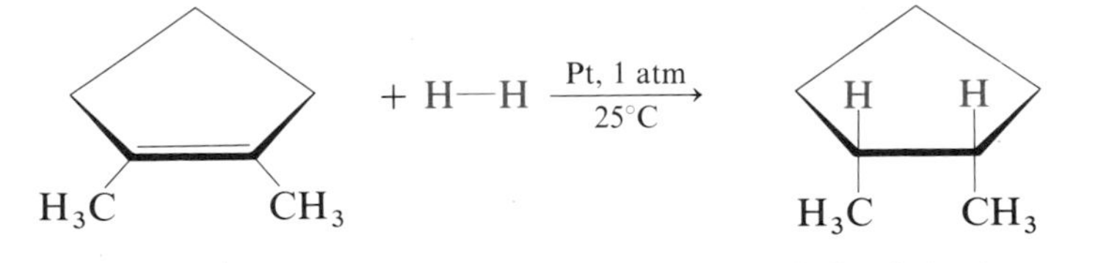

1,2-dimethylcyclopentene   *cis*-1,2-dimethylcyclopentane

Although addition of hydrogen to an alkene is exothermic, these reactions are immeasurably slow at moderate temperatures in the gas phase. However, they take place rapidly in the presence of finely divided platinum, palladium, or nickel. Separate experiments have shown that these and other transition metals near the center of the periodic table are able to *adsorb* (take in on their surfaces) large quantities of hydrogen gas. Adsorption on these metals weakens the bond between hydrogen atoms (Figure 11.3a). Alkenes also interact with

Figure 11.3 Addition of hydrogen to ethylene in the presence of a metal catalyst.

(a)   (b)   (c)

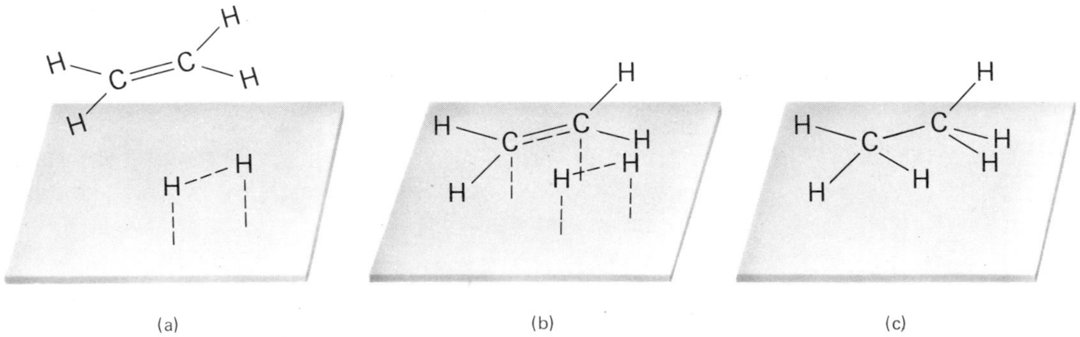

metal surfaces, forming reactive intermediates in which the $2p$ orbitals of carbon interact with the metal surface (Figure 11.3b). If hydrogen and the adsorbed alkene are positioned properly on the metal surface, hydrogen adds to the double bond to give a saturated hydrocarbon (Figure 11.3c) which is then *desorbed* (released from the surface).

B. Addition of Hydrogen Halides

HCl, HBr, and HI add readily to alkenes. These reactions may be carried out either with the pure reagents or in the presence of a polar solvent such as acetic acid.

$$\underset{\text{ethylene}}{\overset{\displaystyle H}{\underset{\displaystyle H}{\Large >}}C=C\overset{\displaystyle H}{\underset{\displaystyle H}{\Large <}} + H-Cl \longrightarrow \underset{\text{chloroethane}}{H-\overset{\overset{\displaystyle H}{|}}{\underset{\underset{\displaystyle H}{|}}{C}}-\overset{\overset{\displaystyle H}{|}}{\underset{\underset{\displaystyle Cl}{|}}{C}}-H}$$

In the addition of HCl to ethylene, hydrogen bonds to one of the carbon atoms and chlorine to the other; only one product, chloroethane, is formed. In the addition of HBr to propene, there are two possible products, 2-bromopropane and 1-bromopropane.

$$CH_3-CH=CH_2 + H-Br \longrightarrow CH_3-\underset{\underset{\displaystyle Br}{|}}{CH}-\underset{\underset{\displaystyle H}{|}}{CH_2} + CH_3-\underset{\underset{\displaystyle H}{|}}{CH}-\underset{\underset{\displaystyle Br}{|}}{CH_2}$$

propene	2-bromopropane	1-bromopropane
	(*major product*)	(*minor product*)

Both isomers are formed, but 2-bromopropane is the major product. Only minor amounts of 1-bromopropane are formed.

After studying a large number of reactions involving the addition of HCl, HBr, and HI to alkenes, the Russian chemist Vladimir Markovnikov proposed the following rule: In addition of HCl, HBr, or HI to an alkene, the major product is the one formed by addition of hydrogen to the carbon atom of the double bond with the greater number of hydrogen atoms already attached to it.

Example 11.6

Draw structural formulas for the products of the following alkene addition reactions. Use Markovnikov's rule to predict which is the major product and which is the minor product.

a. $CH_3-\overset{\overset{\displaystyle CH_3}{|}}{C}=CH_2 + HI \longrightarrow$ b. $+ HCl \longrightarrow$

Solution

a. HI adds to 2-methylpropene to form two products:

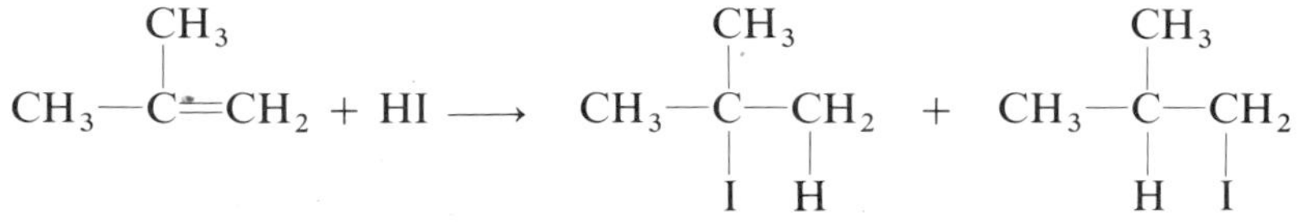

2-methylpropene   2-iodo-2-methylpropane   1-iodo-2-methylpropane
(isobutylene)   (*major product*)   (*minor product*)

To form 2-iodo-2-methylpropane (*t*-butyl iodide), hydrogen adds to the carbon of the double bond bearing two hydrogens; to form 1-iodo-2-methylpropane, hydrogen adds to the carbon of the double bond bearing no hydrogens. Markovnikov's rule predicts that 2-iodo-2-methylpropane will be the major product and 1-iodo-2-methylpropane the minor product.

b. Addition of HCl to 1-methylcyclopentene forms two products:

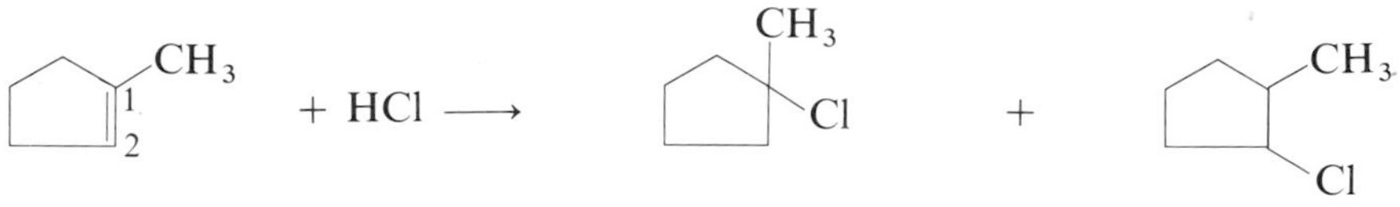

1-methylcyclopentene   1-chloro-1-methylcyclopentane   2-chloro-1-methylcyclopentane
                     (*major product*)   (*minor product*)

Carbon-1 of this cycloalkene has no attached hydrogens and carbon-2 has only one attached hydrogen. According to Markovnikov's rule, hydrogen tends to add to carbon-2 and 1-chloro-1-methylcyclopentane is the major product. 2-Chloro-1-methylcyclopentane is the minor product.

Problem 11.6

Draw structural formulas for the products of the following alkene addition reactions. Use Markovnikov's rule to predict which is the major product and which the minor product.

$$CH_3$$
$$|$$
a. 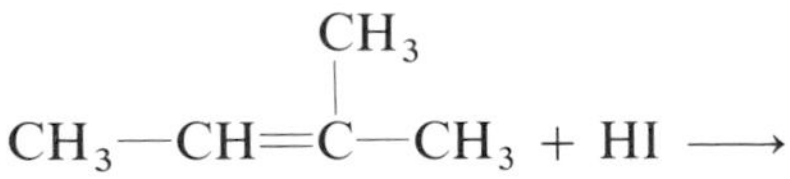 $CH_3-CH=C-CH_3 + HI \longrightarrow$

b. $CH_3-CH=CH-CH_2-CH_3 + HI \longrightarrow$

It is important to remember that while Markovnikov's rule helps us to predict the major and minor products of alkene addition reactions, it does not explain why one product predominates over the other.

C. Addition of Water: Hydration

In the presence of an acid catalyst, often 60% aqueous sulfuric acid, water adds to alkenes to produce alcohols. Hydrogen adds to the carbon of the double bond with the greater number of hydrogens; OH adds to the carbon of the double bond with fewer hydrogens. Thus, H—OH adds to alkenes in accordance with Markovnikov's rule.

$$CH_3-CH=CH_2 + H-OH \xrightarrow{H^+} CH_3-CH-CH_2 + CH_3-CH-CH_2$$

$$\quad\quad\quad\quad\quad\quad\quad\quad\quad\quad\quad\quad | \quad | \quad\quad\quad\quad | \quad |$$
$$\quad\quad\quad\quad\quad\quad\quad\quad\quad\quad\quad\quad OH \quad H \quad\quad\quad\quad H \quad OH$$

propene	2-propanol	1-propanol
	(major product)	*(minor product)*

$$CH_3 \quad\quad\quad\quad\quad\quad\quad\quad\quad CH_3 \quad\quad\quad\quad CH_3$$
$$|\quad\quad\quad\quad\quad\quad\quad\quad\quad\quad\quad\quad |\quad\quad\quad\quad\quad\quad\quad |$$
$$CH_3-C=CH_2 + H-OH \xrightarrow{H^+} CH_3-C-CH_2 + CH_3-C-CH_2$$
$$\quad\quad\quad\quad\quad\quad\quad\quad\quad\quad\quad\quad\quad\quad | \quad |\quad\quad\quad\quad | \quad |$$
$$\quad\quad\quad\quad\quad\quad\quad\quad\quad\quad\quad\quad\quad\quad OH \quad H \quad\quad\quad\quad H \quad OH$$

2-methylpropene	2-methyl-2-propanol	2-methyl-1-propanol
	(major product)	*(minor product)*

The addition of water to an alkene is called **hydration.** Hydration of alkenes is an important reaction in the laboratory as well as in biological systems. For example, one of the reactions in the metabolism of fatty acids (Chapter 25) involves the following hydration:

$$CH_3(CH_2)_{14}CH=CH-\overset{\overset{\displaystyle O}{\|}}{C}-OH + H_2O \xrightarrow{enzyme}$$

$$CH_3(CH_2)_{14}\overset{\overset{\displaystyle OH}{|}}{C}HCH_2-\overset{\overset{\displaystyle O}{\|}}{C}-OH$$

Example 11.7

Draw structural formulas for the products of the following alkene addition reactions. Use Markovnikov's rule to predict which is the major product and which is the minor product.

a.
$$\overset{\overset{\displaystyle CH_3}{|}}{CH_3-CH}-CH=CH_2 + H_2O \xrightarrow{H_2SO_4}$$

b. (cyclohexene with CH₃ substituent) + $H_2O \xrightarrow{H_2SO_4}$

Solution

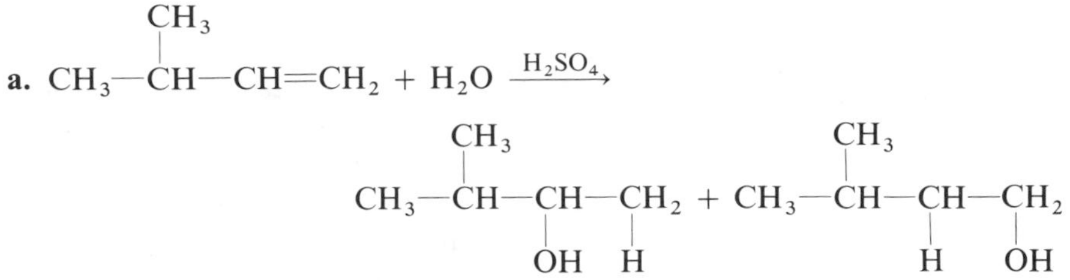

a.
$$CH_3-\overset{\overset{\displaystyle CH_3}{|}}{CH}-CH=CH_2 + H_2O \xrightarrow{H_2SO_4}$$

$$CH_3-\overset{\overset{\displaystyle CH_3}{|}}{CH}-\underset{\underset{\displaystyle OH}{|}}{CH}-\underset{\underset{\displaystyle H}{|}}{CH_2} + CH_3-\overset{\overset{\displaystyle CH_3}{|}}{CH}-\underset{\underset{\displaystyle H}{|}}{CH}-\underset{\underset{\displaystyle OH}{|}}{CH_2}$$

                  (*major product*)   (*minor product*)

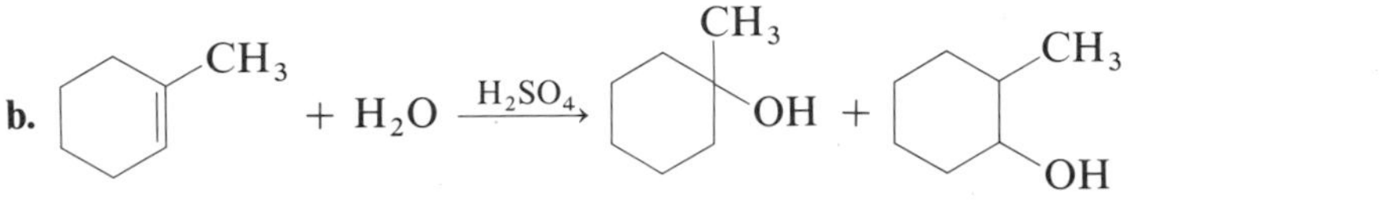

b.

            (*major product*)   (*minor product*)

Problem 11.7

Draw structural formulas for the products of the following alkene addition reactions. Use Markovnikov's rule to predict which is the major product and which is the minor product.

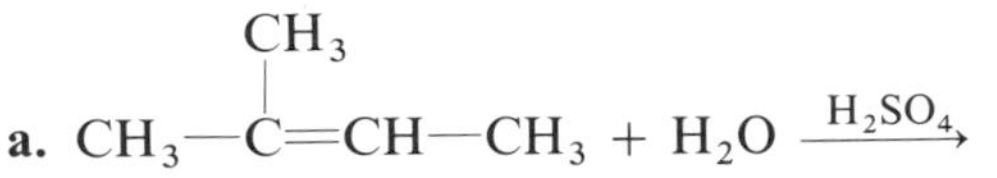

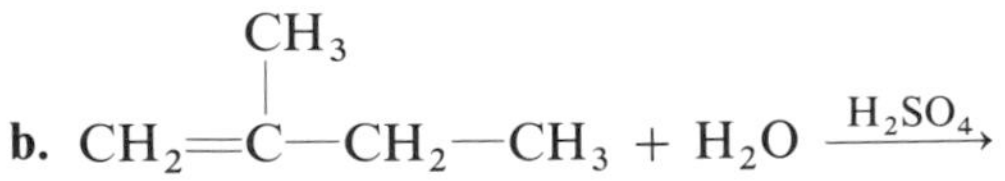

a.
$$CH_3-\overset{\overset{\displaystyle CH_3}{|}}{C}=CH-CH_3 + H_2O \xrightarrow{H_2SO_4}$$

b.
$$CH_2=\overset{\overset{\displaystyle CH_3}{|}}{C}-CH_2-CH_3 + H_2O \xrightarrow{H_2SO_4}$$

D. Mechanism of Addition to Alkenes: Electrophilic Attack

In studying chemical reactions, chemists try to understand how each occurs. In the terminology of chemistry, we ask: What is the **mechanism** of the reaction? Quite literally, a **reaction mechanism** is a step-by-step description of how a reaction occurs. It is a description of which covalent bonds are broken and which new ones are formed, the role of catalysts, and the rates at which the various steps take place. Further, it is a description of the role of **reactive intermediates,** species formed during the reaction that have an atom or atoms in an abnormal valence state. In this chapter we will encounter one type of reactive intermediate, *carbocations*. In Chapter 13 we will encounter another type of reactive intermediate, *carbanions*. To understand how a reaction mechanism works, let us look at the mechanism proposed for the addition of H—X to an alkene, where X stands for a halogen.

The first step in the addition of H—X to an alkene is ionization of H—X to form H$^+$ and X$^-$. In writing equations for steps in a reaction mechanism, we often use curved arrows to show the movement of electron pairs. A curved arrow shows that a pair of electrons moves from the position indicated by the tail of the arrow to the position indicated by the head of the arrow. The curved arrow in Step 1 below shows that the pair of electrons shared by the two atoms in H—X shifts and in the products of the equation belongs entirely to the halogen atom, thus giving it a negative charge. The other product is the hydrogen ion. H$^+$ is electron-deficient; it needs two electrons to complete its valence shell. Any atom, molecule, or ion that seeks electrons to complete its valence shell is called an **electrophile** (an electron-seeking reagent). In Step 2, two electrons of the pi bond move to form a new sigma bond between carbon and H$^+$. In this way, hydrogen acquires two electrons and completes its valence shell. Step 2 leaves a carbon atom with only six electrons in its valence shell. A carbon with six electrons in its valence shell bears a positive charge and is called a **carbocation.** (The older term "carbonium ion" means the same thing as carbocation.) Carbocations are classified as primary (1°), secondary (2°), or tertiary (3°), according to the number of alkyl groups attached to the carbon bearing the positive charge. Because they are electron-deficient, all carbocations are electrophiles.

Step 1: $\quad$ H$-$Ẍ: $\longrightarrow$ H$^+$ + :Ẍ:$^-$

Step 2:

$$
\begin{array}{cc}
\text{H} & \text{H} \\
| & | \\
\text{H}-\text{C}=\text{C}-\text{H} + \text{H}^+ & \longrightarrow
\end{array}
\qquad
\begin{array}{cc}
\text{H} & \text{H} \\
| & | \\
\text{H}-\text{C}-\overset{+}{\text{C}}-\text{H} \\
| \\
\text{H}
\end{array}
$$

pi electrons are used to form a new sigma bond

—*a carbocation*

In the final step, the carbocation reacts with a pair of electrons of the halide ion to form a new sigma bond between carbon and halogen.

Step 3:

$$
\begin{array}{cc}
\text{H} & \text{H} \\
| & | \\
\text{H}-\text{C}-\overset{+}{\text{C}}-\text{H} + :\text{Ẍ}:^- & \longrightarrow
\end{array}
\qquad
\begin{array}{cc}
\text{H} & \text{H} \\
| & | \\
\text{H}-\text{C}-\text{C}-\text{H} \\
| & | \\
\text{H} & :\text{Ẍ}:
\end{array}
$$

an electron pair of X$^-$ used to form a new sigma bond

To account for the fact that H—X adds to alkenes to give a major product and a minor product, the carbocation mechanism proposes that a tertiary carbocation is formed more easily than a secondary carbocation, which is, in turn, formed more easily than a primary carbocation. Following are examples of primary, secondary, and tertiary carbocations.

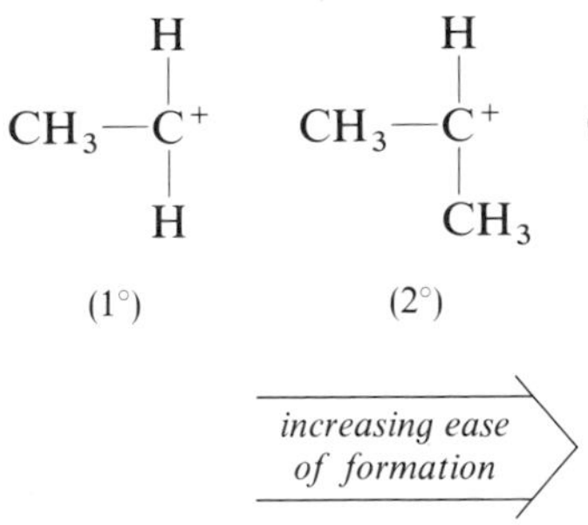

$$
\begin{array}{ccc}
\underset{\overset{|}{H}}{\overset{\overset{\textstyle H}{|}}{CH_3-C^+}} & \underset{\overset{|}{CH_3}}{\overset{\overset{\textstyle H}{|}}{CH_3-C^+}} & \underset{\overset{|}{CH_3}}{\overset{\overset{\textstyle CH_3}{|}}{CH_3-C^+}} \\
(1°) & (2°) & (3°)
\end{array}
$$

increasing ease of formation ⟩

Addition of a proton to propene gives a propyl carbocation (a primary carbocation) or an isopropyl carbocation (a secondary carbocation).

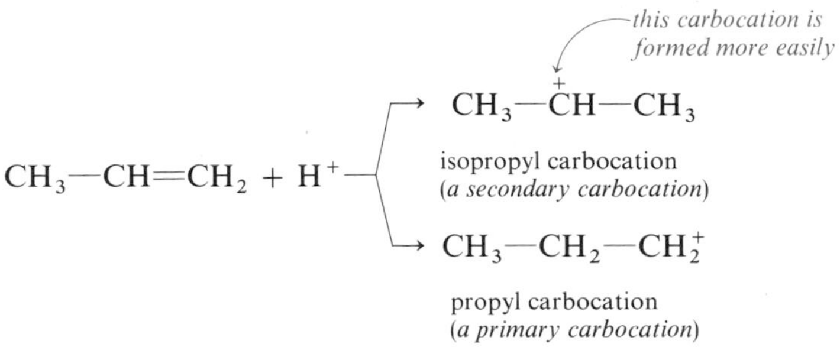

—*this carbocation is formed more easily*

$$CH_3-\overset{+}{C}H-CH_3$$

isopropyl carbocation
(*a secondary carbocation*)

$$CH_3-CH=CH_2 + H^+$$

$$CH_3-CH_2-CH_2^+$$

propyl carbocation
(*a primary carbocation*)

The isopropyl carbocation is formed more easily than the propyl carbocation. The isopropyl carbocation reacts with halide ion to give the major product of the reaction and the propyl carbocation reacts with halide ion to give the minor product.

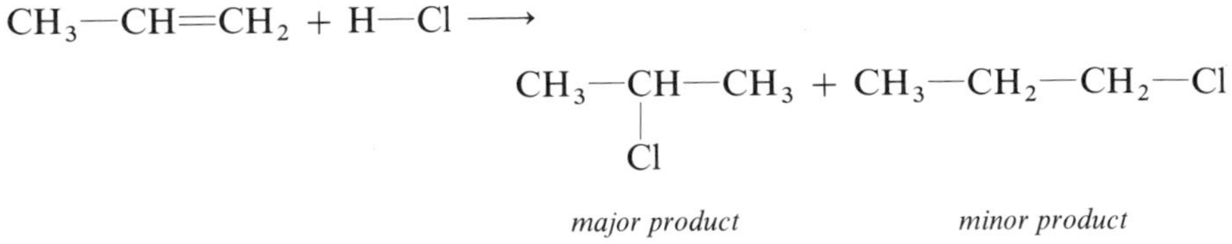

$$CH_3-CH=CH_2 + H-Cl \longrightarrow$$

$$\underset{\overset{|}{Cl}}{CH_3-CH-CH_3} + CH_3-CH_2-CH_2-Cl$$

major product   *minor product*

As another example, reaction of 2-methylpropene with a proton gives the isobutyl carbocation (a primary carbocation) and the *tert*-butyl carbocation (a tertiary carbocation). The tertiary carbocation is formed more easily than the primary carbocation. The tertiary carbocation reacts with halide ion to form the major product of the reaction, and the primary carbocation reacts with halide to form the minor product.

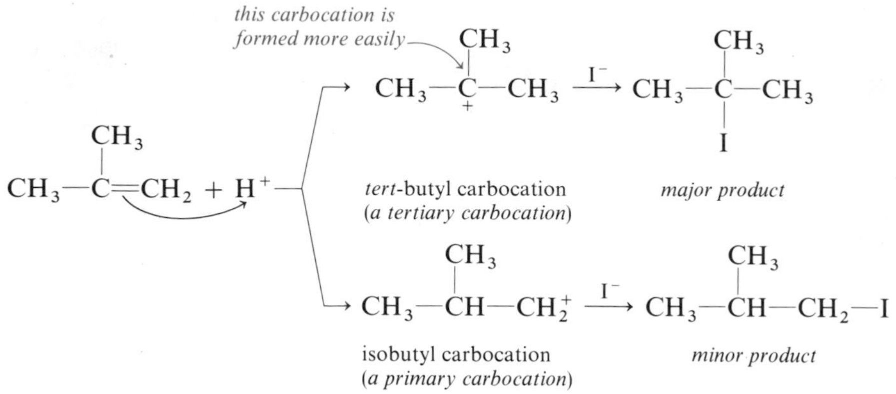

this carbocation is formed more easily

$$CH_3-C=CH_2 + H^+ \left\{ \begin{array}{l} \longrightarrow CH_3-\overset{CH_3}{\underset{+}{C}}-CH_3 \xrightarrow{I^-} CH_3-\overset{CH_3}{\underset{I}{C}}-CH_3 \\ \\ \longrightarrow CH_3-\overset{CH_3}{CH}-CH_2^+ \xrightarrow{I^-} CH_3-\overset{CH_3}{CH}-CH_2-I \end{array} \right.$$

tert-butyl carbocation   *major product*
(*a tertiary carbocation*)

isobutyl carbocation   *minor product*
(*a primary carbocation*)

Acid-catalyzed hydration of alkenes follows a carbocation mechanism much the same as that for the addition of hydrogen halides. In the presence of an acid catalyst, ethylene adds a molecule of water to form ethanol. The first step involves reaction of the pi bond of the alkene with a proton to form a carbocation. In Step 2, this reactive intermediate completes its valence shell by forming a new covalent bond with an unshared electron pair on the oxygen of a water molecule. Finally, loss of a proton in Step 3 results in formation of an alcohol and regeneration of a proton.

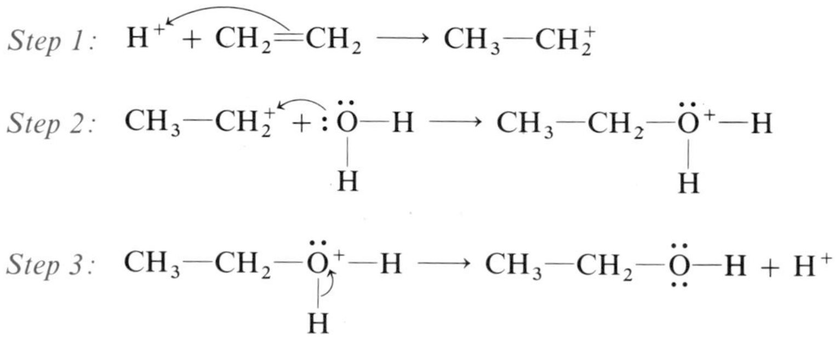

Step 1: $\quad H^+ + CH_2=CH_2 \longrightarrow CH_3-CH_2^+$

Step 2: $\quad CH_3-CH_2^+ + :\ddot{O}-H \longrightarrow CH_3-CH_2-\overset{+}{\ddot{O}}-H$
$$\qquad\qquad\qquad\quad\; H \qquad\qquad\qquad\qquad\quad H$$

Step 3: $\quad CH_3-CH_2-\overset{+}{\ddot{O}}-H \longrightarrow CH_3-CH_2-\ddot{\ddot{O}}-H + H^+$
$$\qquad\qquad\qquad\quad\; H$$

Given the concept of the relative ease of formation of primary, secondary, and tertiary carbocations, this mechanism explains why hydration of propene gives mostly 2-propanol and very little 1-propanol.

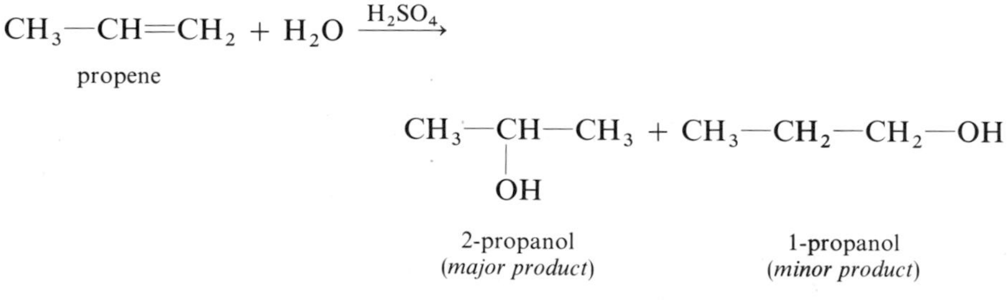

$$CH_3-CH=CH_2 + H_2O \xrightarrow{H_2SO_4}$$
propene

$$CH_3-\overset{}{\underset{OH}{CH}}-CH_3 + CH_3-CH_2-CH_2-OH$$

2-propanol   1-propanol
(*major product*)   (*minor product*)

Example 11.8

Write a mechanism for formation of the major product of the following reaction. Identify all electrophiles in your mechanism.

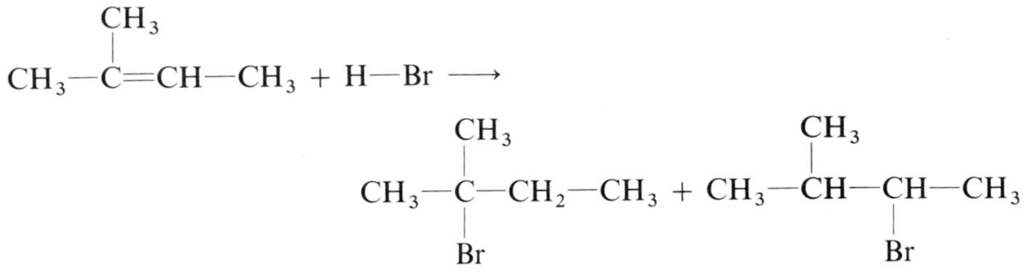

$$CH_3-\underset{\underset{CH_3}{|}}{C}=CH-CH_3 + H-Br \longrightarrow$$

$$CH_3-\underset{\underset{Br}{|}}{\overset{\overset{CH_3}{|}}{C}}-CH_2-CH_3 + CH_3-\underset{\underset{Br}{|}}{\overset{\overset{CH_3}{|}}{CH}}-CH-CH_3$$

major product   *minor product*

Solution

Propose a three-step mechanism involving a carbocation reactive intermediate.

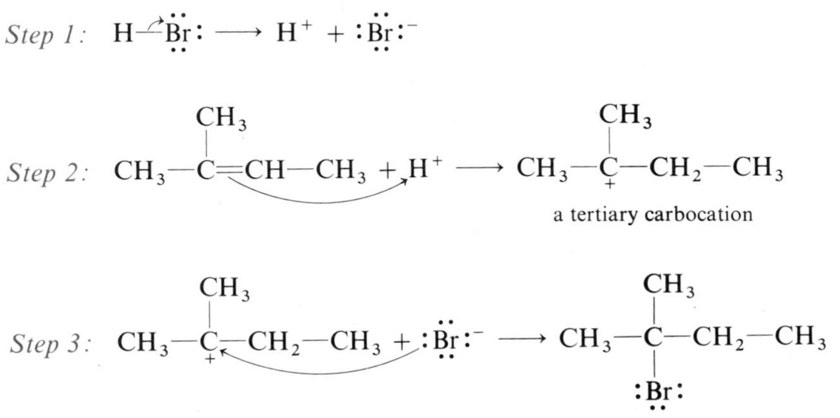

Step 1: $H-\overset{..}{\underset{..}{Br}}: \longrightarrow H^+ + :\overset{..}{\underset{..}{Br}}:^-$

Step 2: $CH_3-\underset{\underset{CH_3}{|}}{C}=CH-CH_3 + H^+ \longrightarrow CH_3-\underset{\underset{CH_3}{|}}{\overset{+}{C}}-CH_2-CH_3$

a tertiary carbocation

Step 3: $CH_3-\underset{\underset{CH_3}{|}}{\overset{+}{C}}-CH_2-CH_3 + :\overset{..}{\underset{..}{Br}}:^- \longrightarrow CH_3-\underset{\underset{:\overset{..}{\underset{..}{Br}}:}{|}}{\overset{\overset{CH_3}{|}}{C}}-CH_2-CH_3$

In this mechanism, both H^+ and the tertiary carbocation are electrophiles.

Problem 11.8

Write a mechanism for the formation of the major product of the following reaction. Identify all electrophiles in your mechanism.

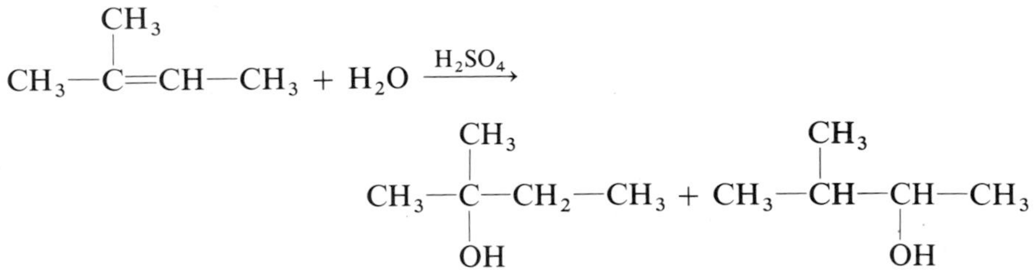

$$CH_3-\underset{\underset{CH_3}{|}}{C}=CH-CH_3 + H_2O \xrightarrow{H_2SO_4}$$

$$CH_3-\underset{\underset{OH}{|}}{\overset{\overset{CH_3}{|}}{C}}-CH_2-CH_3 + CH_3-\underset{\underset{OH}{|}}{\overset{\overset{CH_3}{|}}{CH}}-CH-CH_3$$

major product   *minor product*

E. Addition of Halogens: Halogenation

Bromine (Br_2) and chlorine (Cl_2) add readily to alkenes to form single covalent bonds on adjacent carbons. Halogenation with bromine or chlorine is carried out either with the pure reagents or by mixing the reagents in CCl_4 or some other inert solvent.

$$CH_3-CH=CH-CH_3 + Br-Br \longrightarrow CH_3-\underset{\underset{Br}{|}}{C}H-\underset{\underset{Br}{|}}{C}H-CH_3$$

2-butene   2,3-dibromobutane

cyclopentene   1,2-dichlorocyclopentane

Addition of Br_2 or Cl_2 to an alkene occurs by a two-step mechanism and a carbocation intermediate as illustrated by the reaction of bromine and propene to form 1,2-dibromopropane. The nonpolar bromine molecule becomes partially polarized as it approaches the carbon-carbon double bond, a region of high electron density. The polarized bromine molecule is then capable of acting as an electrophile and initiating attack on the double bond. In Step 1, the bromine atom which first bonds with carbon leaves behind the pair of electrons that formerly bonded it to the other bromine. The other bromine atom acquires a negative charge and becomes a bromide ion. The carbocation formed in Step 1 reacts with the bromide ion in Step 2 to give the product, 1,2-dibromopropane.

Step 1: $:\overset{..}{\underset{..}{Br}}-\overset{..}{\underset{..}{Br}}: + CH_2=CH-CH_3 \longrightarrow :\overset{..}{\underset{..}{Br}}:^- + Br-CH_2-\overset{+}{C}H-CH_3$

                                                           (a 2° carbocation)

Step 2: $:\overset{..}{\underset{..}{Br}}:^- + Br-CH_2-\overset{+}{C}H-CH_3 \longrightarrow Br-CH_2-\underset{\underset{Br}{|}}{C}H-CH_3$

Reaction with bromine is a particularly useful qualitative test for alkenes. A solution of bromine in carbon tetrachloride is red, whereas alkenes and dibromoalkanes are usually colorless. If a few drops of bromine in carbon tetrachloride is added to an alkene, the red color of the test solution is discharged. Discharge of the red color is a sensitive test for the presence of alkenes.

We might pause a moment to compare and contrast halogenation of alkenes with that of alkanes (Section 10.9A). Recall that chlorine and bromine do not react with alkanes unless the halogen-alkane mixture is exposed to light or heated to temperatures of 250–400°C. The reaction that occurs is substitution of halogen for hydrogen and the formation of an equivalent amount of HCl or HBr. Halogenation of most higher alkanes gives a complex mixture of monosubstitution products. In contrast, alkenes react with chlorine and bromine at room temperature in the dark to give a high yield of a single addition product.

Example 11.9

Name and draw structural formulas for the products of the following reactions.

a. 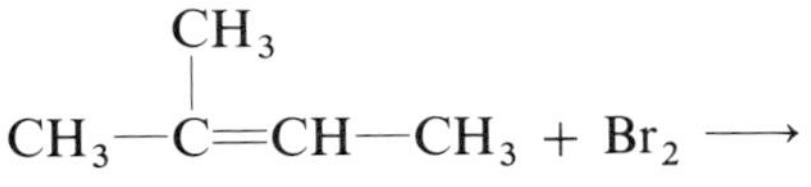 $CH_3-\overset{\overset{\displaystyle CH_3}{|}}{C}=CH-CH_3 + Br_2 \longrightarrow$

b. 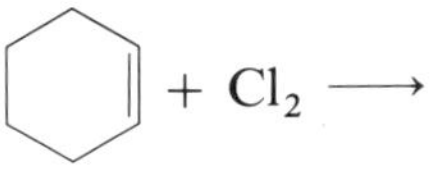 ⬡ $+ Cl_2 \longrightarrow$

Solution

a. 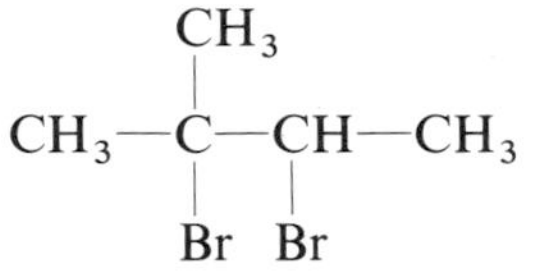 $CH_3-\overset{\overset{\displaystyle CH_3}{|}}{\underset{\underset{\displaystyle Br}{|}}{C}}-\overset{\overset{}{}}{\underset{\underset{\displaystyle Br}{|}}{CH}}-CH_3$

2,3-dibromo-2-methylbutane

b. 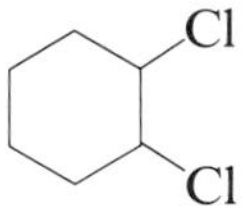 (cyclohexane ring with Cl at C1 and Cl at C2)

1,2-dichlorocyclohexane

Problem 11.9

Name and draw structural formulas for the alkene that reacts with Br_2 to give:

a. 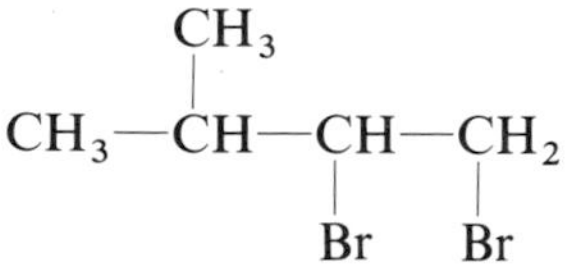 $CH_3-\overset{\overset{\displaystyle CH_3}{|}}{CH}-\overset{\overset{}{}}{\underset{\underset{\displaystyle Br}{|}}{CH}}-CH_2$
   (with Br below first CH₂)

b. 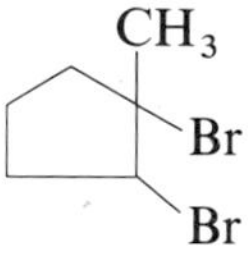 (cyclopentane ring with CH₃ and Br on one carbon, Br on adjacent carbon)

F. Oxidation of Alkenes

Pi bonds of alkenes are readily attacked by a variety of oxidizing agents, including oxygen (O_2), ozone (O_3), potassium permanganate ($KMnO_4$), and potassium dichromate ($K_2Cr_2O_7$). In these oxidations, the carbon-carbon double bond is cleaved.

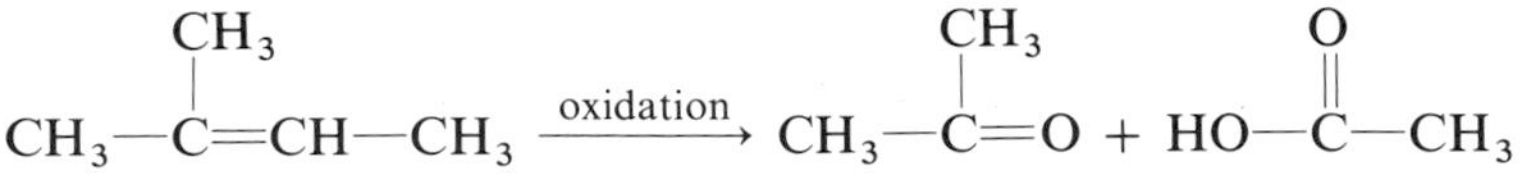

$$CH_3-\overset{\overset{\displaystyle CH_3}{|}}{C}=CH-CH_3 \xrightarrow{\text{oxidation}} CH_3-\overset{\overset{\displaystyle CH_3}{|}}{C}=O + HO-\overset{\overset{\displaystyle O}{||}}{C}-CH_3$$

a ketone   a carboxylic acid

The products of alkene oxidation depend on the number of hydrogen atoms attached to each carbon atom of the double bond. If there are two hydrogens attached, the product is carbonic acid; if there is one hydrogen attached, the product is a carboxylic acid; if there are no hydrogens attached, the product is a ketone.

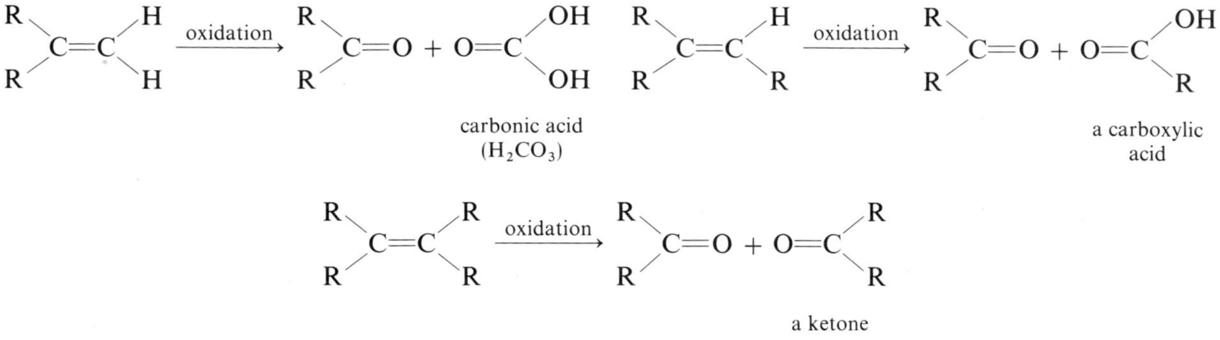

carbonic acid
(H_2CO_3)

a carboxylic
acid

a ketone

Example 11.10

Draw structural formulas for the products of oxidation by $K_2Cr_2O_7$ of the following alkenes. Name the new functional group formed in each oxidation product.

$$\text{a. } CH_3-CH_2-CH=CH-\underset{\underset{CH_3}{|}}{CH}-CH_3$$

b.

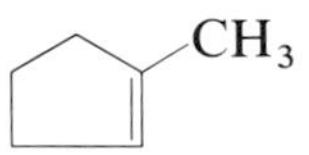

Solution

$$\text{a. } CH_3-CH_2-\overset{\overset{O}{||}}{C}-OH + HO-\overset{\overset{O}{||}}{C}-\underset{\underset{CH_3}{|}}{CH}-CH_3$$

a carboxylic acid   *a carboxylic acid*

b.

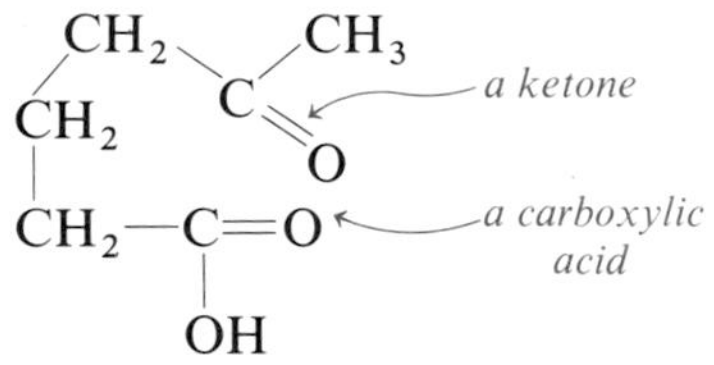

a ketone

a carboxylic acid

In the oxidation of 1-methylcyclopentene the carbon-carbon double bond is broken to give a single molecule that contains a ketone and a carboxylic acid. The product of this oxidation can also be written:

$$CH_3-\overset{\overset{O}{||}}{C}-CH_2-CH_2-CH_2-\overset{\overset{O}{||}}{C}-OH$$

Problem 11.10

Draw structural formulas for the products of the oxidation of the following alkenes. Name the new functional group formed in each oxidation product.

a. CH_3—CH=CH—CH=C—CH$_3$
  with CH_3 branch on the carbon

b. (cyclohexene ring with CH_3 substituent and CH attached to CH_3 CH_3)

G. Polymerization of Substituted Ethylenes

From the perspective of the chemical industry, the single most important reaction of alkenes is **polymerization.** Polymerization is the joining together of many small units known as **monomers** (Greek: *mono- + meros*, "single part") into very large, high-molecular-weight **polymers** (Greek: *poly + meros*, "many parts"). In **addition polymerization,** monomer units are joined together without loss of any atoms. An example of addition polymerization is the formation of polyethylene from ethylene monomers:

$$\cdots + \underset{\underset{H}{|}}{\overset{\overset{H}{|}}{C}} = \underset{\underset{H}{|}}{\overset{\overset{H}{|}}{C}} + \underset{\underset{H}{|}}{\overset{\overset{H}{|}}{C}} = \underset{\underset{H}{|}}{\overset{\overset{H}{|}}{C}} + \underset{\underset{H}{|}}{\overset{\overset{H}{|}}{C}} = \underset{\underset{H}{|}}{\overset{\overset{H}{|}}{C}} + \cdots \xrightarrow{\text{polymerization}} $$

pi bonds break and the pairs of electrons form new C—C sigma bonds

new sigma bonds formed

In the polymerization of ethylene, pi bonds are broken and the electrons are used to form new sigma bonds between the monomer units. The preceding example shows the addition polymerization of three ethylene units. In practice, hundreds of monomer units polymerize and molecular weights of polyethylene range from 50,000 to over 1,500,000. Polymerization reactions are more generally written in the following way:

$$n \;\underset{\underset{H}{|}}{\overset{\overset{H}{|}}{C}} = \underset{\underset{H}{|}}{\overset{\overset{H}{|}}{C}} \xrightarrow{\text{polymerization}} \left[\underset{\underset{H}{|}}{\overset{\overset{H}{|}}{C}} - \underset{\underset{H}{|}}{\overset{\overset{H}{|}}{C}} \right]_n$$

repeating unit

monomer

Table 11.2 Polymers derived from substituted ethylene monomers.

Monomer	Monomer Name	Polymer Name or Trade Name
$CH_2{=}CH_2$	ethylene	polyethylene, Polythene, for unbreakable containers and tubing
$CH_2{=}CHCH_3$	propylene	polypropylene, Herculon, fibers for carpeting and clothes
$CH_2{=}CHCl$	vinyl chloride	polyvinyl chloride, PVC, Koroseal
$CH_2{=}CCl_2$	1,1-dichloroethylene	Saran, food wrappings
$CH_2{=}CHCN$	acrylonitrile	polyacrylonitrile, Orlon, Acrylics
$CF_2{=}CF_2$	tetrafluoroethylene	polytetrafluoroethylene, Teflon
$CH_2{=}CHC_6H_5$	styrene	polystyrene, Styrofoam, for insulation
$CH_2{=}CCO_2CH_3$ $\quad\mid$ $\quad CH_3$	methyl methacrylate	polymethyl methacrylate, Lucite, Plexiglas, for glass substitutes
$CH_2{=}CHCO_2CH_3$	methyl acrylate	polymethyl acrylate, Acrylics, for latex paints

The subscript, n, in the formula of polyethylene indicates that the monomer-derived unit, $-CH_2CH_2-$, repeats n times in the polymer. Table 11.2 lists several important polymers of substituted ethylenes, along with their common names and uses. The alkene polymers, mainly polyethylene, polypropylene, and polystyrene, are the plastics produced in the largest quantities.

Much of the early interest in polymers arose from a need to make synthetic rubber and relieve the near total dependence of the industrial world on natural rubber. This dependence became a crisis at the beginning of World War II, when Japan seized control of almost all the natural-rubber producing areas in the world. In the mid-1920s, several companies, including du Pont in the United States and I.G. Farbenindustrie in Germany, began research programs in this area. By the mid-1930s, du Pont was producing the first synthetic rubber, Neoprene, on a commercial basis. Neoprene is a polymer of chloroprene and its chemical name is polychloroprene.

$$n CH_2{=}\overset{\overset{\textstyle Cl}{\textstyle |}}{C}{-}CH{=}CH_2 \xrightarrow{\text{polymerization}} -(CH_2{-}\overset{\overset{\textstyle Cl}{\textstyle |}}{C}{=}CH{-}CH_2)_n{-}$$

chloroprene
(2-chlorobutadiene)

Neoprene
(polychloroprene)

The years since the 1930s have seen extensive research and development in polymer chemistry and physics, and an almost explosive growth in plastics, coatings, and rubber technology has created a worldwide multibillion-dollar industry. A few basic factors account for this phenomenal growth. First, the raw materials for plastics are derived mainly from petroleum and natural gas, raw materials whose continued plentiful supply has been taken for granted, at least until recently. Second, scientists have learned how to make polymers with specific chemical and physical properties. Third, many plastics can be fabricated more cheaply than competing materials. For example, plastic technology created the water-based latex paints that have revolutionized the coatings industry; plastic films and foams have done the same for the packaging industry. The list could go on and on as we think of the manufactured items that surround us in our daily lives.

The tetrafluoroethylene polymer Teflon provides an interesting example of how a chance observation followed by creative research led to the development of a revolutionary new product. Teflon was discovered accidentally in 1938 by du Pont chemists during a search for new refrigerants (the Freons described in Section 10.9A). One morning, a cylinder of tetrafluoroethylene appeared to be empty (no gas escaped when the valve was opened), yet the weight of the cylinder indicated that it was full. The cylinder was opened and inside was found a waxy solid. The solid proved to have very unusual properties: extraordinary chemical inertness, outstanding heat resistance, very high melting point, and unusual frictional properties. In 1941, du Pont began limited production of polytetrafluoroethylene and gave it the trade name Teflon. The small amount of Teflon polymer from initial production was preempted at once by the wartime Manhattan Project (the project that developed the atomic bomb), where it was used in equipment to contain the highly corrosive UF_6 during the gaseous diffusion separation of the isotopes of uranium. In 1948, du Pont built the first commercial Teflon plant and the product was used to make gaskets, bearings for automobiles, nonstick equipment for candy manufacturers and commercial bakers, seals for rotating equipment, and a number of other items. Teflon became a household word in 1961 with the introduction of nonstick frying pans in the U.S. market.

11.6 Reactions of Alkynes

Alkynes undergo addition reactions similar to those of alkenes. Reagents such as HCl, Cl_2, and Br_2 add to alkynes, though somewhat more slowly than to alkenes. Acetylene reacts with one mole of HCl to form vinyl chloride and with two moles of HCl to form 1,1-dichloroethane:

$$H-C\equiv C-H + HCl \longrightarrow \overset{H}{\underset{H}{>}}C=C\overset{Cl}{\underset{H}{<}}$$

vinyl chloride

$$H-C\equiv C-H + 2\,HCl \longrightarrow H-\overset{\overset{\displaystyle H}{|}}{\underset{\underset{\displaystyle H}{|}}{C}}-\overset{\overset{\displaystyle Cl}{|}}{\underset{\underset{\displaystyle Cl}{|}}{C}}-H$$

1,1-dichloroethane

Propyne adds one mole of HCl to form 2-chloropropene. It adds two moles of HCl to form 2,2-dichloropropane:

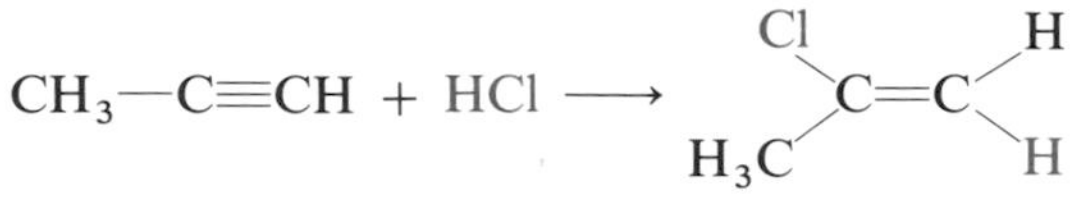

$$CH_3-C\equiv CH + HCl \longrightarrow$$

Cl   H

C=C

H_3C   H

2-chloropropene

$$CH_3-C\equiv CH + 2\ HCl \longrightarrow CH_3-\overset{\overset{\displaystyle Cl}{|}}{\underset{\underset{\displaystyle Cl}{|}}{C}}-\overset{\overset{\displaystyle H}{|}}{\underset{\underset{\displaystyle H}{|}}{C}}-H$$

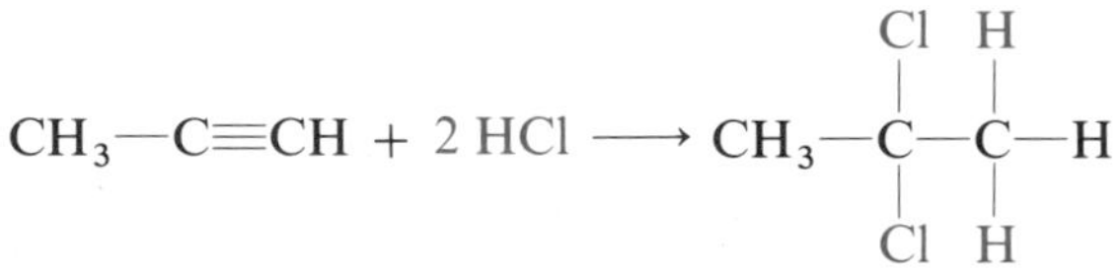

2,2-dichloropropane

Similarly, alkynes add one mole of bromine to form dibromoalkenes and two moles of bromine to form tetrabromoalkanes:

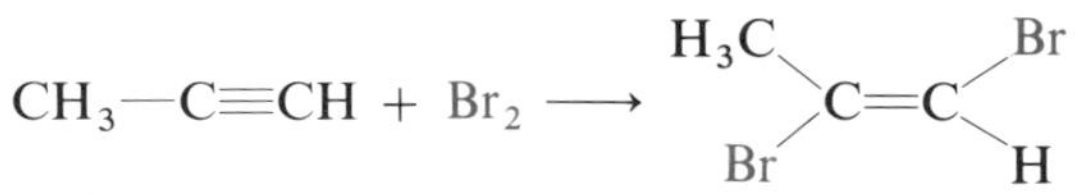

$$CH_3-C\equiv CH + Br_2 \longrightarrow$$

H_3C   Br

C=C

Br   H

1,2-dibromopropene

$$CH_3-C\equiv CH + 2\ Br_2 \longrightarrow CH_3-\overset{\overset{\displaystyle Br}{|}}{\underset{\underset{\displaystyle Br}{|}}{C}}-\overset{\overset{\displaystyle Br}{|}}{\underset{\underset{\displaystyle Br}{|}}{C}}-H$$

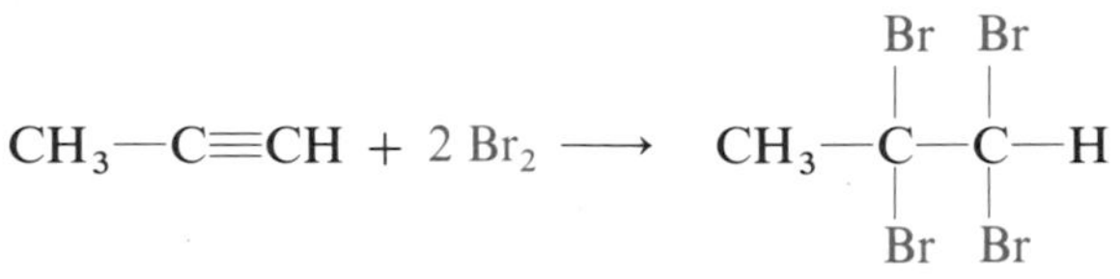

1,1,2,2-tetrabromopropane

11.7 Aromatic Hydrocarbons

Benzene is a colorless liquid with a boiling point of $80°C$. It was first isolated by Michael Faraday in 1825 from the oily liquid that collected in the illuminating gas lines of London. Its molecular formula, C_6H_6, suggests a high degree of unsaturation. (Remember that a saturated alkane of six carbons has the formula C_6H_{14} and that a saturated cycloalkane of six carbons has the formula C_6H_{12}.) Considering the fact that benzene has six fewer hydrogens than cyclohexane, it was expected that benzene would be highly reactive and show reactions characteristic of alkenes and alkynes. Surprisingly, benzene does not undergo characteristic alkene reactions. For example, benzene does not react with bromine, hydrogen chloride, hydrogen bromide, or other reagents that add to double and triple bonds. When it does react, benzene does so by substitution, in which one of its hydrogen atoms is replaced by another atom or group of atoms. For example, benzene reacts with bromine in the presence of ferric bromide to form bromobenzene and hydrogen bromide.

$$C_6H_6 + Br\!-\!Br \xrightarrow{FeBr_3} C_6H_5Br + HBr$$

benzene   bromobenzene

The terms **aromatic** and **aromatic compounds** were used to classify benzene and a number of its derivatives because many of them have distinctive odors. However, after a time it became clear that a sounder classification for these compounds should be based, not on aroma, but on chemical reactivities. Currently, the term *aromatic* is used to refer to the unusual chemical properties of benzene and its derivatives. They are resistant to typical alkene addition reactions and, when they do react, they do so by substitution.

A. The Structure of Benzene

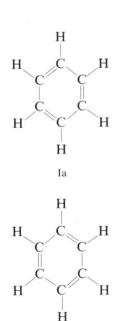

Ia

Ib

There are two common ways to describe the structure of benzene: (1) in terms of Lewis structures and (2) in terms of the overlap of atomic orbitals. Let us look at each of these descriptions.

The six carbon atoms of benzene form a regular hexagon with bond angles of 120°. One hydrogen atom is bonded to each carbon. Two Lewis structures, Ia and Ib, can be drawn for this arrangement of atoms. They differ only in the positions of the double bonds. Lewis structures Ia and Ib do account for the fact that benzene is a cyclic, unsaturated hydrocarbon. However, they do not account for the fact that benzene does not undergo typical alkene addition reactions. If benzene contains three double bonds, why doesn't it add three molecules of bromine to form hexabromocyclohexane? The resonance theory of Linus Pauling provided the first adequate description of the structure and unusual chemical properties of benzene.

Recall from Section 4.5D that when a molecule can be represented by two or more contributing structures that differ only in the position of valence electrons, the actual molecule is best represented as a *resonance hybrid*. The two major contributing structures for the benzene resonance hybrid are:

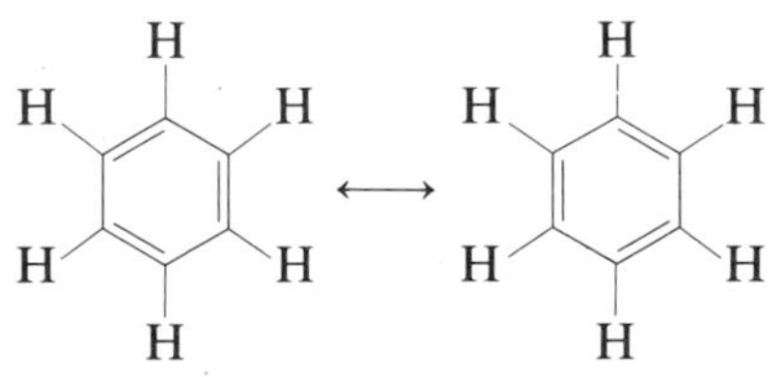

One of the consequences of resonance is a marked increase in stability of the hybrid over that of any one of the contributing structures. Resonance stabilization is particularly large and important in benzene and other aromatic hydrocarbon. Because of their stability, benzene and other aromatic hydrocarbons do not undergo typical alkene addition reactions.

We can also describe the structure of benzene in terms of the overlap of atomic orbitals (Figure 11.4). To bond with three other atoms, carbon uses sp^2 hybrid orbitals. The six carbon atoms are joined together by sigma bonds formed by the overlap of sp^2 hybrid orbitals. Each carbon is further joined to one hydrogen by a sigma bond formed from the overlap of sp^2-$1s$ orbitals. These twelve C—C and C—H sigma bonds form the skeletal framework of the ring. Each carbon also has a single $2p$ orbital containing one electron. Overlap of these six $2p$ orbitals forms a pi molecular orbital. This pi orbital is in two parts: half in the "upper lobe" and half in the "lower lobe."

The pi cloud of a benzene ring or other aromatic hydrocarbon is considerably more stable and less reactive than the pi cloud of a simple carbon-carbon or carbon-oxygen double bond. This completely symmetrical distribution of pi electrons in benzene is shown by a regular hexagon with a circle inside:

Figure 11.4 (a) The sigma bonds of benzene formed by overlap of sp^2-sp^2 and sp^2-$1s$ orbitals. The six $2p$ orbitals are uncombined. (b) Overlap of six $2p$ orbitals to form the pi orbital, split into the upper and lower lobes.

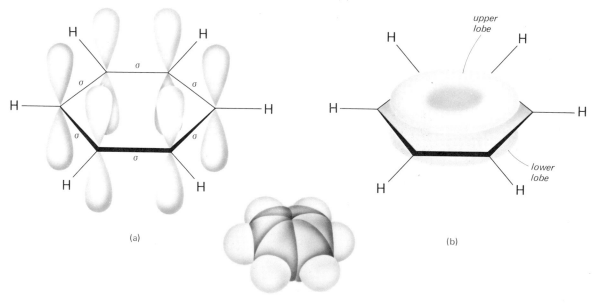

Figure 11.5 Symbols to represent the structure of benzene.

(a) resonance model   (b) orbital overlap model

Each of these approaches to the structure of benzene uses a particular symbol to represent benzene. The resonance approach uses two Lewis contributing structures connected by a double-headed arrow (Figure 11.5a). The overlap of atomic orbitals approach uses a regular hexagon with a circle inside (Figure 11.5b). As you might expect, there are very strongly held opinions about which symbol is the best description of the structure of benzene and which symbol should be used. Throughout the remainder of the book, we will use a single Lewis structure to represent benzene.

B. Nomenclature of Aromatic Hydrocarbons

In the IUPAC system, monosubstituted alkyl benzenes are named as derivatives of benzene; for example, ethylbenzene. The IUPAC system retains common names for many of the simpler monosubstituted benzenes. Examples are toluene, styrene, and cumene.

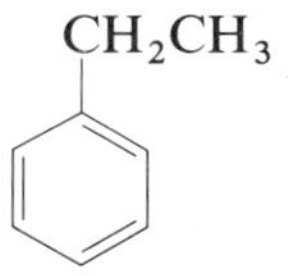

ethylbenzene 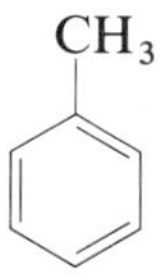   toluene 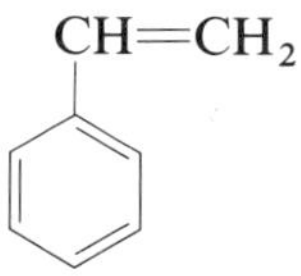   styrene 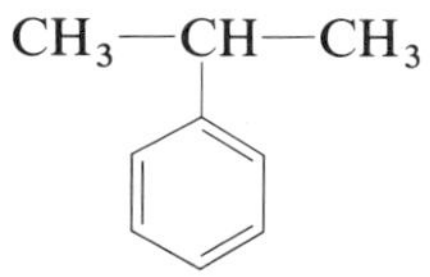   cumene
(isopropylbenzene)

When there are two substituents on a benzene ring, three structural isomers are possible. The prefixes *ortho-* (*o*), *meta-* (*m*), and *para-* (*p*) are used to locate the substituents. *Ortho-* substituents are on carbons 1 and 2 of the benzene ring; *meta-* substituents are on carbons 1 and 3; and *para-* substituents are on carbons 1 and 4. The IUPAC system retains the name xylene for the dimethylbenzenes:

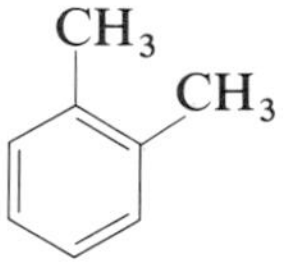

ortho-xylene 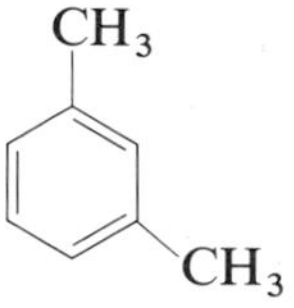   *meta*-xylene 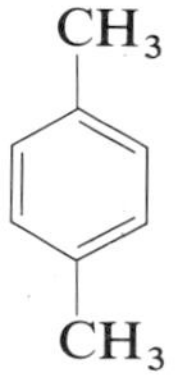   *para*-xylene

With three or more substituents, a numbering system must be used.

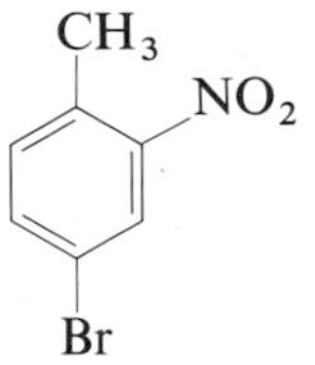

4-bromo-2-nitrotoluene

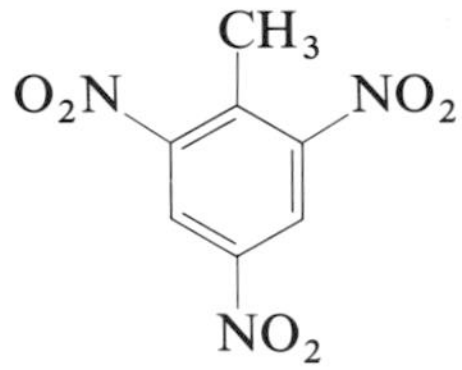

2,4,6-trinitrotoluene
(TNT)

In naming more complex molecules, the benzene ring is often named as a substituent on a parent chain. In this case, C_6H_5— is called a **phenyl group.**

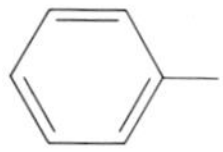

$$CH_3-CH_2-CH_2-CH-CH_3$$

phenyl group

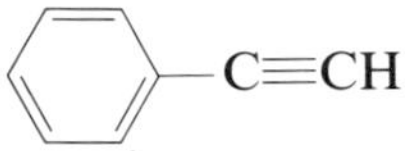

phenylacetylene

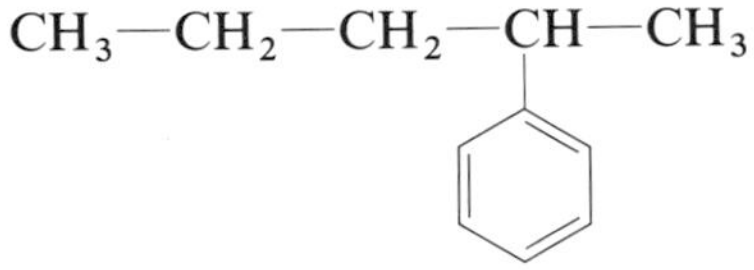

2-phenylpentane

Closely related to benzene are numerous aromatic hydrocarbons having two or more six-member rings joined together.

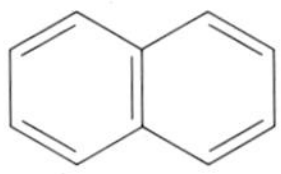

naphthalene
mp 80°C

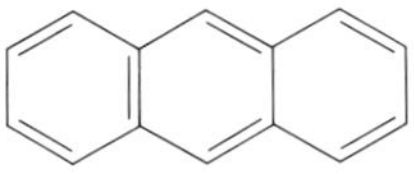

anthracene
mp 217°C

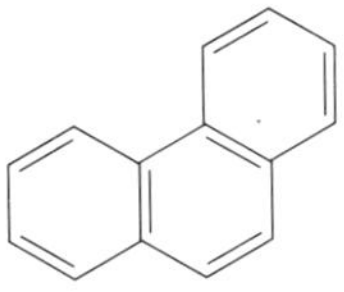

phenanthrene
mp 99°C

C. Reactions of Aromatic Hydrocarbons — Electrophilic Aromatic Substitution

The most striking characteristic of aromatic hydrocarbons is their tendency to react by substitution rather than addition. Following are examples of a few of the more common aromatic substitution reactions.

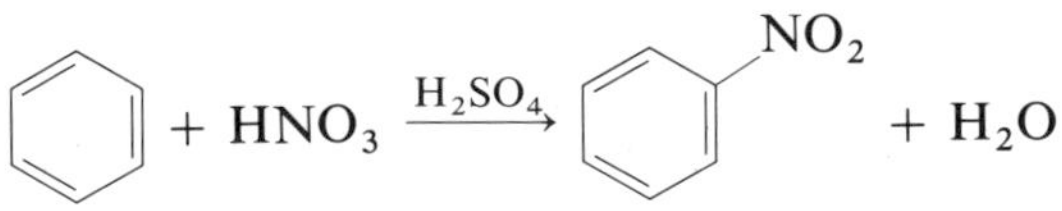

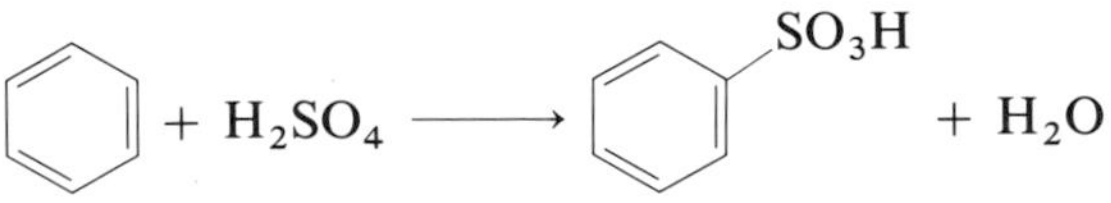

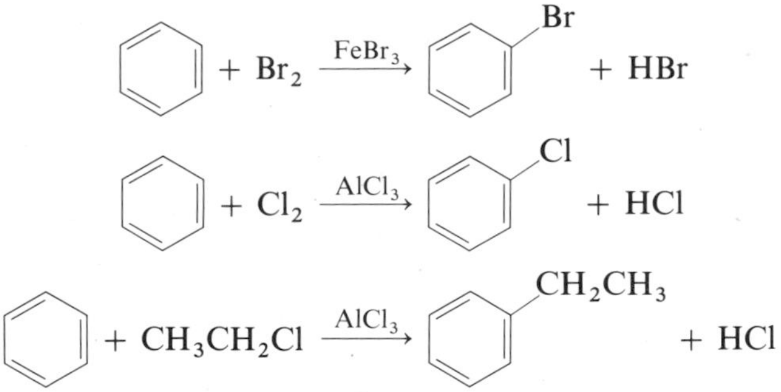

When studying these reactions, we will deal with the question of how they occur and why benzene, in contrast to alkenes, typically undergoes substitution rather than additions.

The characteristic feature of nearly all aromatic substitution reactions is that the benzene ring is attacked by electrophiles. We can illustrate this principle by studying the mechanism for the bromination of benzene. Benzene and most substituted benzenes react very slowly with bromine. However, in the presence of $AlBr_3$, $FeBr_3$, or some other catalyst, reaction proceeds rapidly. The function of the catalyst is to react with Br_2 and break the bond between the two bromine atoms as shown by the curved arrow in Step 1 of Figure 11.6. One of the bromine atoms now has only six electrons in its valence shell and bears a positive charge. The name for Br^+ is **bromonium ion.** Br^+ is an electrophile and in Step 2, reaction of a pair of pi electrons of the benzene ring with Br^+ produces a carbocation. Note that there are six hydrogens on this carbo-

Figure 11.6 A mechanism for the bromination of benzene. An example of electrophilic aromatic substitution.

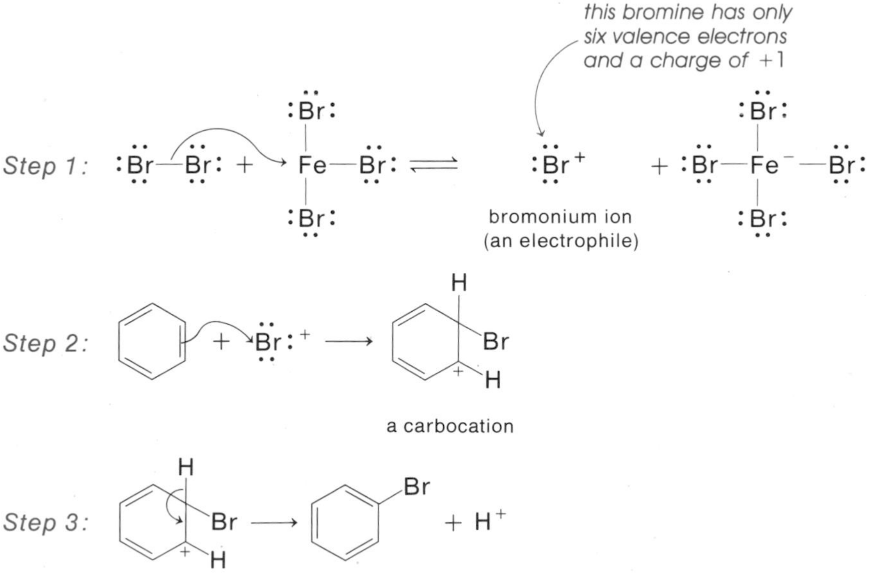

cation. Two of the hydrogens are shown to remind you (1) that the carbon bearing the positive charge has a hydrogen bonded to it, and (2) that the carbon bearing the bromine also has a hydrogen bonded to it. In the final step of the reaction, the carbocation loses H^+ to give bromobenzene.

The major difference between halogen addition to an alkene and halogen substitution on an aromatic ring centers on the fate of the carbocation intermediate formed in the first step. Recall from Section 11.5E that addition of bromine to propene is a two-step process:

Step 1: $\quad CH_3-CH=CH_2 + \overset{..}{\underset{..}{Br}}-\overset{..}{\underset{..}{Br}} \longrightarrow CH_3-\overset{+}{CH}-CH_2-\overset{..}{\underset{..}{Br}} + \overset{..}{\underset{..}{Br}}\overset{-}{:}$

Step 2: $\quad CH_3-\overset{+}{CH}-CH_2-\overset{..}{\underset{..}{Br}} + \overset{..}{\underset{..}{Br}}\overset{-}{:} \longrightarrow CH_3-\underset{\underset{\overset{|}{\underset{..}{Br}}}{}}{CH}-CH_2-\overset{..}{\underset{..}{Br}}$

In the case of an alkene, the carbocation intermediate reacts with bromide ion to complete the addition. In the case of an aromatic hydrocarbon, the carbocation intermediate loses a proton to regenerate the aromatic ring and regain the large resonance stabilization. There is no such resonance stabilization to be regained in the case of an alkene.

Chlorination, nitration, sulfonation, and alkylation of benzene can be formulated in much the same terms. In nitration, sulfuric acid facilitates formation of the nitronium ion, NO_2^+, which then reacts with the aromatic ring.

$$H^+ + H-\overset{..}{\underset{..}{O}}-\overset{+}{N}\overset{\overset{..}{\underset{..}{O}}{:}^-}{\underset{\overset{..}{O}{\cdot}}{}} \rightleftharpoons \overset{H}{\underset{H}{}}\overset{+}{\underset{}{O}}-\overset{+}{N}\overset{\overset{..}{O}{:}^-}{\underset{\overset{..}{O}{\cdot}}{}} \rightleftharpoons H_2\overset{..}{\underset{..}{O}}: + \overset{..}{\underset{..}{O}}=\overset{+}{N}=\overset{..}{\underset{..}{O}}$$

nitric
acid

nitronium
ion, NO_2^+

Reaction of an alkyl halide with benzene or other aromatic hydrocarbon in the presence of a catalyst is known as the **Friedel-Crafts reaction.**

⬡ $+ CH_3-Cl \xrightarrow{AlCl_3}$ ⬡$-CH_3$ $+ HCl$

The function of $AlCl_3$ in this reaction is to react with the alkyl chloride and form an alkyl carbocation. The alkyl carbocation is the electrophile that then attacks the aromatic ring.

$$CH_3-\overset{..}{\underset{..}{Cl}}: + \overset{\overset{:\overset{..}{Cl}:}{|}}{\underset{\underset{:\overset{..}{Cl}:}{|}}{Al}}-\overset{..}{\underset{..}{Cl}}: \rightleftharpoons CH_3^+ + :\overset{..}{\underset{..}{Cl}}-\overset{\overset{:\overset{..}{Cl}:}{|}}{\underset{\underset{:\overset{..}{Cl}:}{|}}{Al}}^--\overset{..}{\underset{..}{Cl}}:$$

a carbocation
(an electrophile)

Key Terms and Concepts

alkene (introduction)

alkyne (introduction)

aromatic hydrocarbon (11.7)

carbocation (11.5D)

cis-trans isomerism in alkenes (11.3)

electrophile (11.5D)

electrophilic aromatic substitution
 (11.7C)

halogenation of alkenes (11.5F)

hydration of alkenes (11.5C)

Markovnikov's rule (11.5B)

monomer (11.5G)

oxidation of alkenes (11.5F)

polymer (11.5G)

polymerization (11.5G)

reduction of alkenes (11.5A)

Key Reactions

1. Addition of hydrogen to alkenes (Section 11.5A):

$$CH_3-CH_2-CH=CH_2 + H_2 \xrightarrow{Pt} CH_3-CH_2-CH_2-CH_3$$

2. Addition of HCl, HBr, and HI to alkenes (Section 11.5B):

$$CH_3-CH=CH_2 + HBr \longrightarrow CH_3-\underset{\underset{Br}{|}}{CH}-CH_3 + CH_3-CH_2-\underset{\underset{Br}{|}}{CH_2}$$

   major product   *minor product*

3. Addition of H_2O to alkenes (Section 11.5C):

$$CH_3-CH=CH_2 + H_2O \xrightarrow{H_2SO_4} CH_3-\underset{\underset{OH}{|}}{CH}-CH_3 + CH_3-CH_2-\underset{\underset{OH}{|}}{CH_2}$$

   major product   *minor product*

4. Addition of Cl_2 and Br_2 to alkenes (Section 11.5E):

$$CH_3-CH=CH_2 + Br_2 \longrightarrow CH_3-\underset{\underset{Br}{|}}{CH}-\underset{\underset{Br}{|}}{CH_2}$$

5. Oxidation of alkenes (Section 11.5F):

$$CH_3-CH=\overset{\overset{CH_3}{|}}{C}-CH_3 \xrightarrow{oxidation} CH_3-\overset{\overset{O}{||}}{C}-OH + CH_3-\overset{\overset{O}{||}}{C}-CH_3$$

   Oxidizing agents described are $K_2Cr_2O_7$ and ozone.

6. Addition polymerization of ethylene and substituted ethylenes (Section 11.5G):

$$n CH_2=CH-Cl \xrightarrow{polymerization} -(CH_2-\overset{\overset{Cl}{|}}{CH})_n-$$

   vinyl chloride   polyvinylchloride
   (PVC)

7. Addition of HCl and HBr to alkynes (Section 11.6):

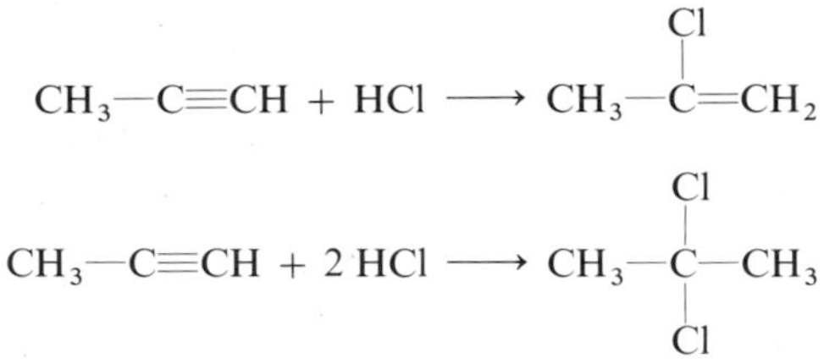

$$CH_3—C≡CH + HCl \longrightarrow CH_3—\overset{\overset{\displaystyle Cl}{|}}{C}=CH_2$$

$$CH_3—C≡CH + 2 HCl \longrightarrow CH_3—\overset{\overset{\displaystyle Cl}{|}}{\underset{\underset{\displaystyle Cl}{|}}{C}}—CH_3$$

8. Addition of Cl_2 and Br_2 to alkynes (Section 11.6):

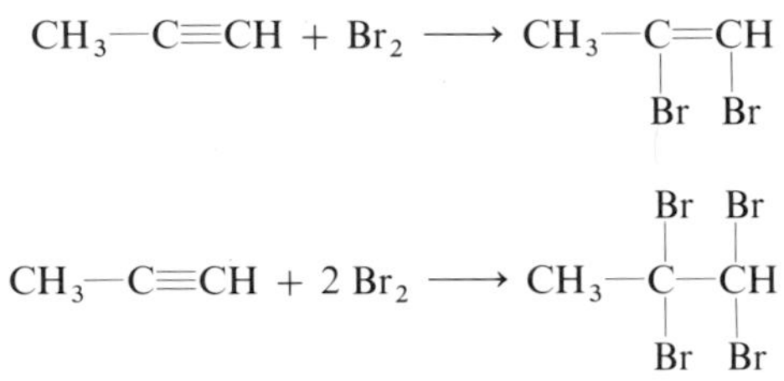

$$CH_3—C≡CH + Br_2 \longrightarrow CH_3—\underset{\underset{\displaystyle Br}{|}}{C}=\underset{\underset{\displaystyle Br}{|}}{CH}$$

$$CH_3—C≡CH + 2 Br_2 \longrightarrow CH_3—\overset{\overset{\displaystyle Br}{|}}{\underset{\underset{\displaystyle Br}{|}}{C}}—\overset{\overset{\displaystyle Br}{|}}{\underset{\underset{\displaystyle Br}{|}}{CH}}$$

9. Electrophilic aromatic substitution (11.7C):

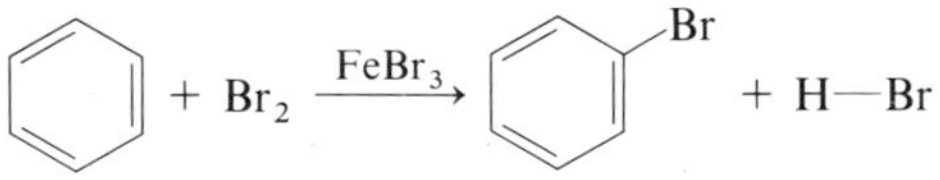

Problems

Structure of Alkenes and Alkynes (Section 11.1)

11.11 Predict all bond angles about each circled carbon atom. To make these predictions, use the valence-shell electron-pair repulsion model. (Review Section 4.6)

a.

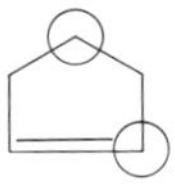

b.  $CH_3—C≡\text{Ⓒ}—\text{Ⓒ}H_2—CH_3$

c.

$—CH_2—O—H$

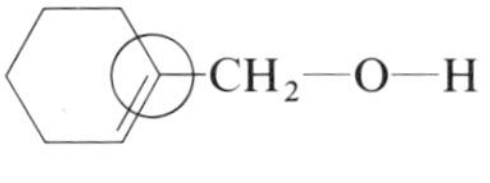

d.

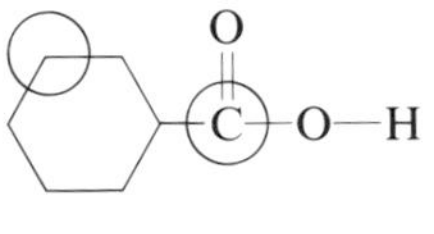

e.

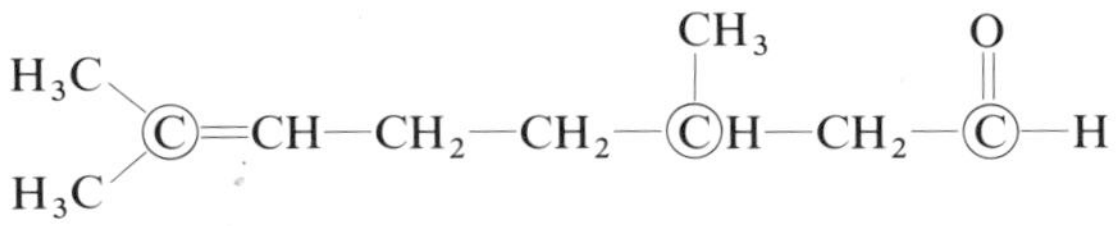

11.12 For each circled carbon atom in the previous problem, identify which atomic orbitals are used to form each sigma bond and which are used to form each pi bond.

Nomenclature of Alkenes and Alkynes (Section 11.2)

11.13 Name the following compounds:

a. 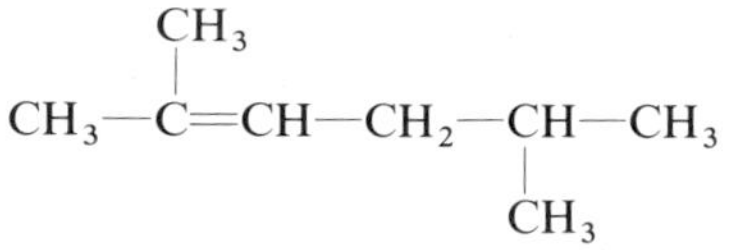 $CH_3-\overset{\overset{\displaystyle CH_3}{|}}{C}=CH-CH_2-\overset{\overset{\displaystyle }{|}}{\underset{\underset{\displaystyle CH_3}{}}{CH}}-CH_3$

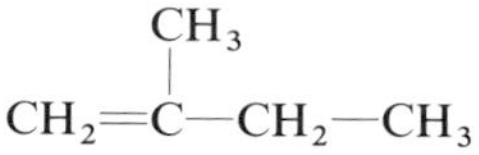

b. $CH_2=\overset{\overset{\displaystyle CH_3}{|}}{C}-CH_2-CH_3$

c. $CH_2=\overset{\overset{\displaystyle CH_3}{|}}{C}-CH=CH_2$

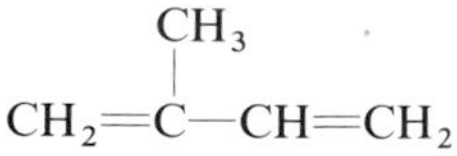

d. $Cl-CH=CH-Cl$

e.

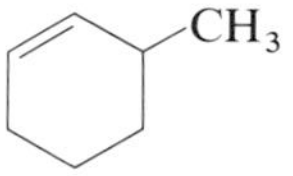

f. 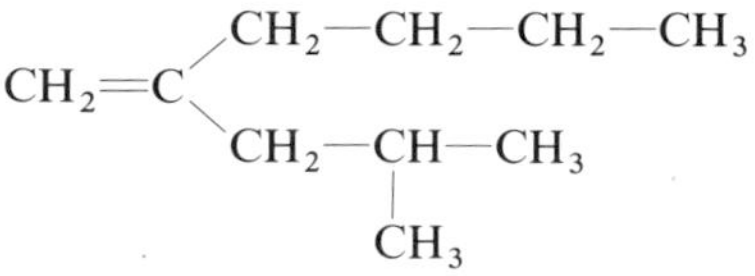 $CH_2=C\overset{\displaystyle \diagup CH_2-CH_2-CH_2-CH_3}{\diagdown CH_2-\underset{\underset{\displaystyle CH_3}{|}}{CH}-CH_3}$

g.

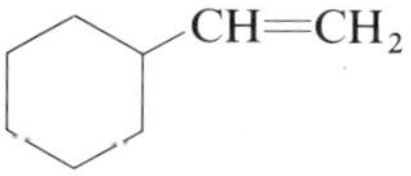

h.

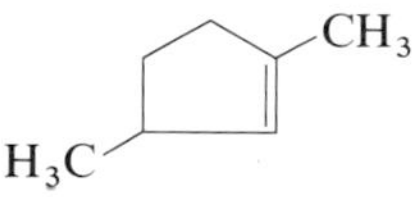

i. $(CH_3)_2CHCH=C(CH_3)_2$

j. $(CH_3)_3CCH_2CH=CH_2$

k. $CH_2=CH-CH=CH_2$

l. $CH_2=CH-Cl$

m. $CH_3-CH_2-C\equiv CH$

n. 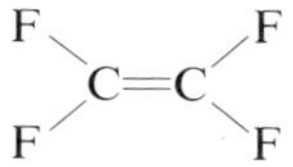 $\overset{F}{\underset{F}{\diagup}}C=C\overset{\diagup F}{\underset{\diagdown F}{}}$

o. 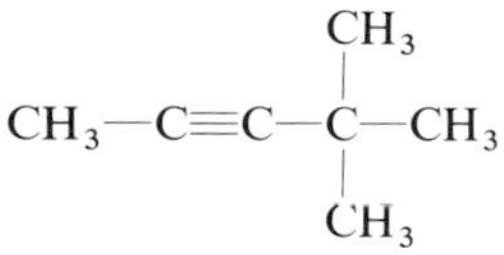 $CH_3-C\equiv C-\overset{\overset{\displaystyle CH_3}{|}}{\underset{\underset{\displaystyle CH_3}{|}}{C}}-CH_3$

p. $CH_2=CH-CH_2Cl$

11.14 Draw structural formulas for the following compounds.
   a. 2-methyl-3-hexene
   b. 2-methyl-2-hexene
   c. 2-methyl-1-butene
   d. 3-ethyl-3-methyl-1-pentene
   e. 2,3-dimethyl-2-butene
   f. 1-pentene
   g. 2-pentene
   h. 1-chloropropene
   i. 2-chloropropene
   j. 3-methylcyclohexene
   k. 1-isopropyl-4-methylcyclohexene
   l. 1-phenylcyclohexene
   m. 3-hexyne
   n. 5-isopropyl-3-octyne
   o. 3-penyl-1-butyne
   p. tetrachloroethylene

Cis-Trans Isomerism in Alkenes (Section 11.3)

11.15 Which of the molecules in Problem 11.14 show *cis-trans* isomerism? For each that does, draw both *cis*- and *trans*-isomers.

11.16 Which of the following molecules show *cis-trans* isomerism?

a. [structure: cyclohexane with two CH₃ groups on same carbon, top CH₃ and CH₃]

b. [structure: cyclohexane with CH₃ and CH₃ on adjacent carbons]

c. [structure: cyclohexene with CH₃ and CH₃]

d. ClCH=CHCl

e. $CH_3(CH_2)_5CH=CH(CH_2)_7\overset{\overset{\displaystyle O}{\|}}{C}OH$

f. $HO\overset{\overset{\displaystyle O}{\|}}{C}-CH=CH-\overset{\overset{\displaystyle O}{\|}}{C}OH$

11.17 Draw structural formulas for all compounds of molecular formula C_5H_{10} that are:
  a. alkenes that do not show *cis-trans* isomerism.
  b. alkenes that do show *cis-trans* isomerism.
  c. cycloalkanes that do not show *cis-trans* isomerism.
  d. cycloalkanes that do show *cis-trans* isomerism.

11.18 Draw structural formulas for the four isomeric chloropropenes (C_3H_5Cl).

Reactions of Alkenes – Addition (Sections 11.5A–11.5E)

11.19 Draw structural formulas for the following alkene addition reactions. Where two products are possible, state which is the major product and which is the minor product.

a. $CH_3-\underset{\underset{\displaystyle CH_3}{|}}{C}=CH-CH_3 + H_2O \xrightarrow{H_2SO_4}$

b. $CH_3-\underset{\underset{\displaystyle CH_3}{|}}{C}=CH-CH_3 + HBr \longrightarrow$

c. $CH_3-\underset{\underset{\displaystyle CH_3}{|}}{C}=CH-CH_3 + Br_2 \longrightarrow$

d. $CH_3-\underset{\underset{\displaystyle CH_3}{|}}{C}=CH-CH_3 + H_2 \xrightarrow{Pt}$

e. [cyclohexene structure] $+ H_2O \xrightarrow{H_2SO_4}$

f. $CH_2=\underset{\underset{\displaystyle CH_3}{|}}{C}-CH_2-CH_3 + H_2O \xrightarrow{H_2SO_4}$

g. [benzene ring with CH=CH₂ substituent] $+ Cl_2 \longrightarrow$

h. [cyclopentene structure] $+ HCl \longrightarrow$

i. $\underset{H}{\overset{H_3C}{>}}C=C\underset{H}{\overset{CH_2CH_3}{<}} + H_2O \xrightarrow{H_2SO_4}$

j. $\underset{H}{\overset{H_3C}{>}}C=C\underset{CH_2CH_3}{\overset{H}{<}} + H_2O \xrightarrow{H_2SO_4}$

k. $CH_2=\underset{\underset{\displaystyle CH_3}{|}}{C}-CH=CH_2 + 2H_2 \xrightarrow{Pt}$

1. $CH_3-\overset{\overset{\displaystyle H_3C}{|}}{C}=\overset{\overset{\displaystyle CH_3}{|}}{C}-CH_3 + Br_2 \longrightarrow$

m. 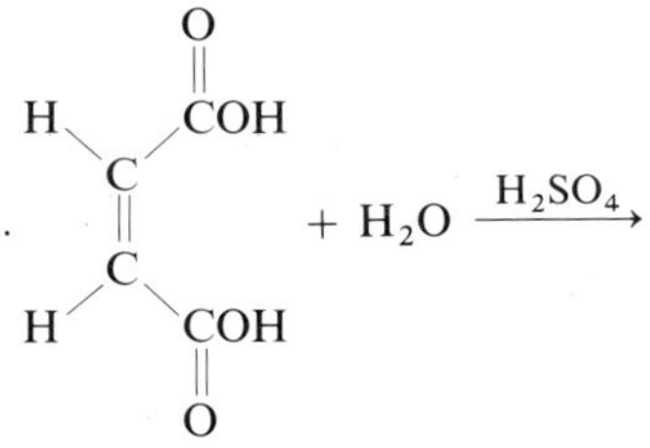
$$
\begin{array}{c}
\overset{H}{\underset{H}{}} \diagdown \overset{\overset{\displaystyle O}{\|}}{C} \diagup \overset{COH}{\underset{\displaystyle}{}} \\
\| \\
\underset{C}{} \\
COH \\
\|\\
O
\end{array}
+ H_2O \xrightarrow{H_2SO_4}
$$

11.20 Draw the structural formula for an alkene or alkenes of molecular formula C_5H_{10} that will react to give the indicated compound as the major product. Note that in several parts of this problem (for example, part a), more than one alkene will give the same substance as the major product.

a. $C_5H_{10} + H_2O \xrightarrow{H_2SO_4} CH_3-\overset{\overset{\displaystyle CH_3}{|}}{\underset{\underset{\displaystyle OH}{|}}{C}}-CH_2CH_3$

b. $C_5H_{10} + Br_2 \longrightarrow CH_3-\overset{\overset{\displaystyle CH_3}{|}}{CH}-\overset{\underset{\displaystyle Br}{|}}{CH}-\overset{\underset{\displaystyle Br}{|}}{CH_2}$

c. $C_5H_{10} + H_2 \xrightarrow{Pt} CH_3-CH_2-CH_2-CH_2-CH_3$

d. $C_5H_{10} + H_2O \xrightarrow{H_2SO_4} CH_3-\overset{\underset{\displaystyle OH}{|}}{CH}-CH_2-CH_2-CH_3$

e. $C_5H_{10} + HCl \longrightarrow CH_3-\overset{\overset{\displaystyle CH_3}{|}}{\underset{\underset{\displaystyle Cl}{|}}{C}}-CH_2-CH_3$

11.21 Draw structural formulas for the carbocations formed by the reaction of H^+ with the following alkenes. Where two different carbocations are possible, state which is the more stable.

a. $CH_3-CH_2-\overset{\overset{\displaystyle CH_3}{|}}{C}=CH-CH_3 + H^+ \longrightarrow$

b. $CH_3-CH_2-CH=CH-CH_3 + H^+ \longrightarrow$

c. 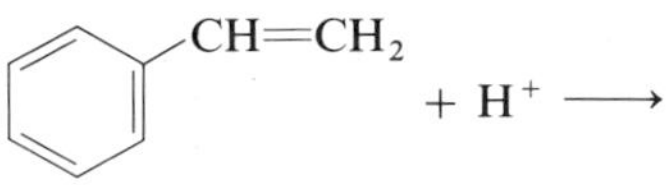 (phenyl)$-CH=CH_2$ $+ H^+ \longrightarrow$ d. 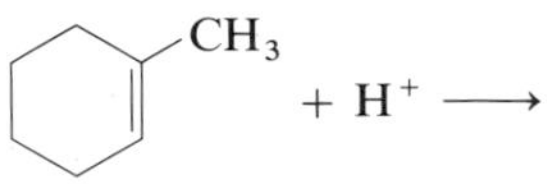 (cyclohexenyl)$-CH_3$ $+ H^+ \longrightarrow$

11.22 Write a reaction mechanism for the following alkene addition reactions. For each mechanism, identify all electrophiles and reactive intermediates.

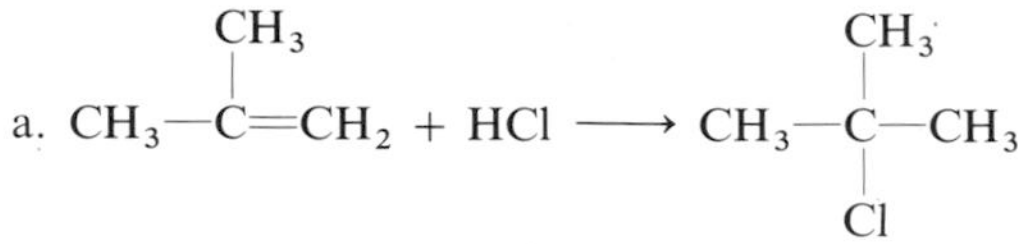

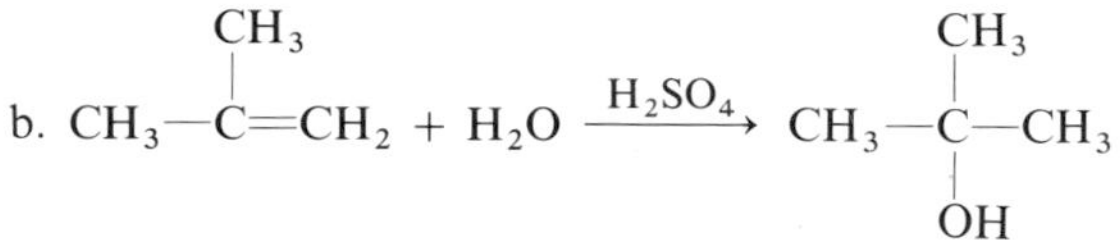

a. $CH_3-\underset{\underset{\displaystyle CH_3}{|}}{C}=CH_2 + HCl \longrightarrow CH_3-\underset{\underset{\displaystyle Cl}{|}}{\overset{\overset{\displaystyle CH_3}{|}}{C}}-CH_3$

b. $CH_3-\underset{\underset{\displaystyle CH_3}{|}}{C}=CH_2 + H_2O \xrightarrow{H_2SO_4} CH_3-\underset{\underset{\displaystyle OH}{|}}{\overset{\overset{\displaystyle CH_3}{|}}{C}}-CH_3$

11.23 Terpin hydrate is prepared commercially by the addition of two moles of water to limonene in the presence of dilute sulfuric acid. Limonene is one of the main components of lemon, orange, caraway, dill, bergamot, and some other oils. Terpin hydrate is used medicinally as an expectorant for coughs. It may be given as terpin hydrate and codeine. Propose a structure for terpin hydrate and a reasonable mechanism to account for the formation of the product you have predicted.

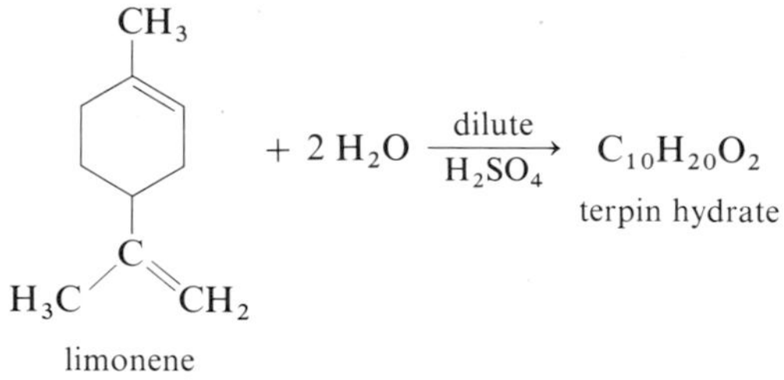

$$+ \ 2\,H_2O \xrightarrow[H_2SO_4]{dilute} C_{10}H_{20}O_2$$

terpin hydrate

limonene

11.24 Reaction of 2-methylpropene with methanol in the presence of concentrated H_2SO_4 yields a product of molecular formula $C_5H_{12}O$.

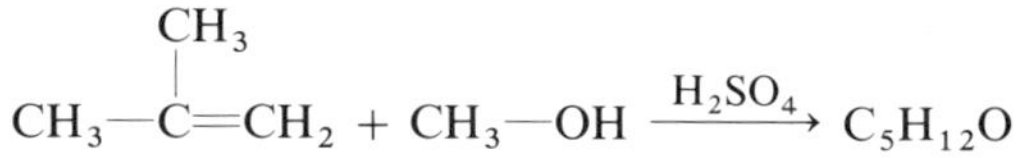

$$CH_3-\underset{\underset{\displaystyle CH_3}{|}}{C}=CH_2 + CH_3-OH \xrightarrow{H_2SO_4} C_5H_{12}O$$

a. Propose a structural formula for $C_5H_{12}O$.
b. Propose a reaction mechanism for the formation of this product.

11.25 Show how you might distinguish between the members of the following pairs of compounds by a simple chemical test. In each case, tell what test you would perform, what you would expect to observe, and write an equation for each positive test.
a. cyclohexane and 1-hexene
b. 1-hexene and 2-chlorohexane
c. 2,3-dimethyl-2-butene and 1,1-dimethylcyclopentane

Reactions of Alkenes – Oxidation (Section 11.5F)

11.26 Define oxidation (review Section 5.4).

11.27 Show by writing a balanced half-reaction that the following reactions are oxidations.

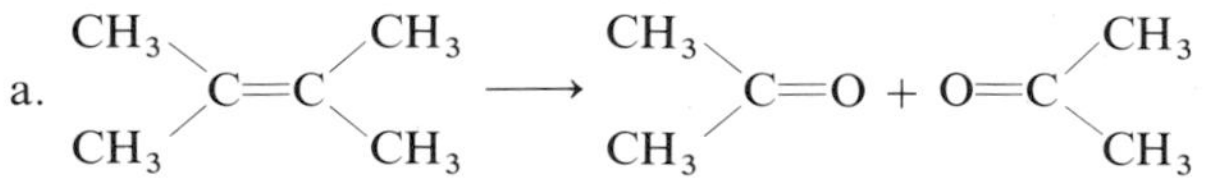

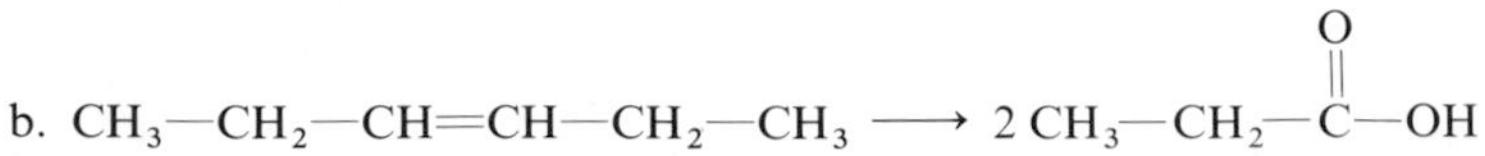

a. $\underset{\displaystyle CH_3}{\overset{\displaystyle CH_3}{>}}C=C\underset{\displaystyle CH_3}{\overset{\displaystyle CH_3}{<}} \longrightarrow \underset{\displaystyle CH_3}{\overset{\displaystyle CH_3}{>}}C=O + O=C\underset{\displaystyle CH_3}{\overset{\displaystyle CH_3}{<}}$

b. $CH_3-CH_2-CH=CH-CH_2-CH_3 \longrightarrow 2\,CH_3-CH_2-\overset{\overset{\displaystyle O}{||}}{C}-OH$

c. $\longrightarrow$ $CH_3-\overset{\overset{\displaystyle O}{\|}}{C}-CH_2-CH_2-CH_2-\overset{\overset{\displaystyle O}{\|}}{C}-OH$

11.28 Draw structural formulas for the products of oxidation of the following alkenes.

a. $CH_3-CH_2-CH=\overset{\overset{\displaystyle CH_3}{|}}{C}-CH_2-CH_3$ $\xrightarrow{\text{oxidation}}$

b. $\xrightarrow{\text{oxidation}}$

c. $\xrightarrow{\text{oxidation}}$

d. $CH_3CH_2CH=CH-CH_2CH_2-\overset{\overset{\displaystyle O}{\|}}{C}OH$ $\xrightarrow{\text{oxidation}}$

11.29 Draw the structural formula of an alkene of given molecular formula that can be oxidized by hot potassium permanganate to give the indicated products.

a. C_6H_{12} $\xrightarrow{\text{oxidation}}$ $CH_3-CH_2-\overset{\overset{\displaystyle O}{\|}}{C}-OH + CH_3-\overset{\overset{\displaystyle O}{\|}}{C}-CH_3$

b. C_6H_{12} $\xrightarrow{\text{oxidation}}$ $CH_3-\overset{\overset{\displaystyle O}{\|}}{C}-OH + CH_3-CH_2-CH_2-\overset{\overset{\displaystyle O}{\|}}{C}-OH$

c. C_7H_{12} $\xrightarrow{\text{oxidation}}$ $CH_3-\overset{\overset{\displaystyle O}{\|}}{C}-CH_2-CH_2-CH_2-\overset{\overset{\displaystyle O}{\|}}{C}-CH_3$

d. C_6H_{10} $\xrightarrow{\text{oxidation}}$ $HO-\overset{\overset{\displaystyle O}{\|}}{C}-CH_2-CH_2-CH_2-CH_2-\overset{\overset{\displaystyle O}{\|}}{C}-OH$

e. C_9H_{16} $\xrightarrow{\text{oxidation}}$ $=O + CH_3-\overset{\overset{\displaystyle O}{\|}}{C}-CH_3$

f. C_9H_{16} $\xrightarrow{\text{oxidation}}$ $-\overset{\overset{\displaystyle O}{\|}}{C}-OH + CH_3-\overset{\overset{\displaystyle O}{\|}}{C}-OH$

Reactions of Alkenes – Polymerization (Section 11.5G)

11.30 Following is a structural formula for a section of polypropylene derived from three propylene monomers.

$$-CH_2-\overset{\overset{\displaystyle CH_3}{|}}{CH}-CH_2-\overset{\overset{\displaystyle CH_3}{|}}{CH}-CH_2-\overset{\overset{\displaystyle CH_3}{|}}{CH}-$$

Draw structural formulas for comparable three-unit sections of:
a. polyvinyl chloride   b. Saran
c. Teflon   d. Orlon
e. Styrofoam   f. Plexiglas

11.31 Natural rubber is a polymer. The repeating unit in this polymer is:

$$-(CH_2-\underset{\underset{CH_3}{|}}{C}=CH-CH_2)_n-$$

a. Draw the structural formula for a section of natural rubber, showing three repeating units.

b. Draw the structural formula of the product of oxidation in natural rubber by ozone and name the two new functional groups in the product.

c. The smog prevalent in Los Angeles contains oxidizing agents. How might you account for the fact that this type of smog attacks natural rubber (automobile tires, etc.) but does not affect polyethylene or polyvinyl chloride?

Structure of Benzene (Section 11.7A)

11.32 For each circled atom, predict bond angles formed by the attached atoms. Use the valence-shell electron-pair repulsion model to make these predictions.

a. 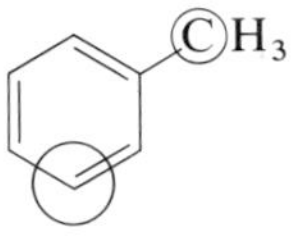 b. 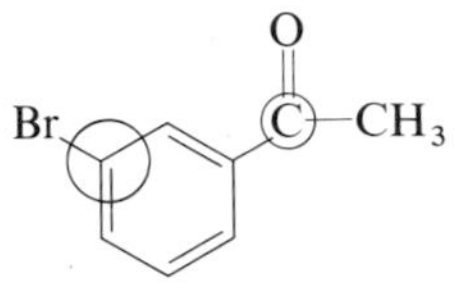 c.

11.33 For each circled atom in the previous problem, identify which atomic orbitals are used to form each sigma bond and which are used to form each pi bond.

Nomenclature of Aromatic Hydrocarbons (Section 11.7B)

11.34 Name the following compounds.

a.  b. 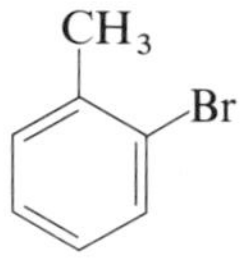 c.

d. 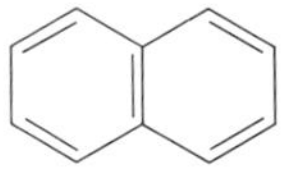 e. 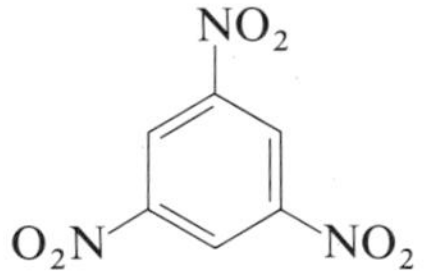 f.

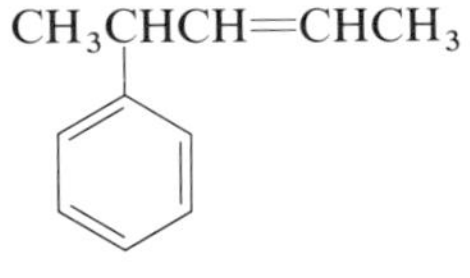

g. 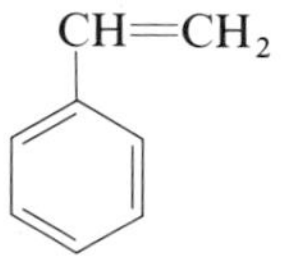 h. 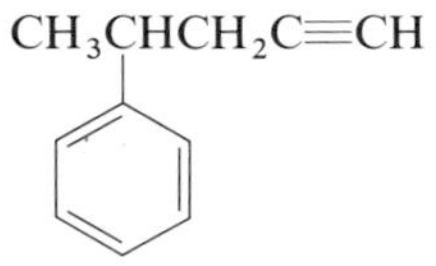 i.

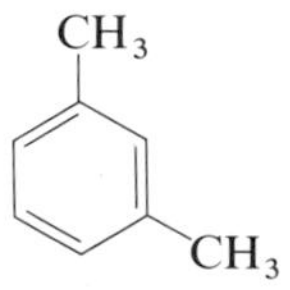

j.  k. 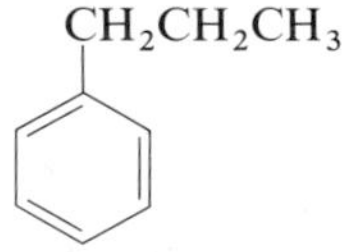 l.

11.35 Draw structural formulas for the following molecules.
    a. *m*-dibromobenzene   b. 2,4,6-trinitrotoluene (TNT)
    c. *p*-chloroiodobenzene   d. 2-ethyl-4-isopropyltoluene
    e. *p*-xylene   f. 2-ethylnaphthalene
    g. *p*-diiodobenzene   h. 2-phenyl-2-pentene
    i. phenanthrene   j. anthracene
    k. isopropylbenzene (cumene)

11.36 Name and draw structural formulas for all derivatives of benzene having the following molecular formulas:
    a. $C_6H_3Br_3$ b. C_8H_{10} c. C_9H_{12}

Reactions of Aromatic Hydrocarbons (Section 11.7C)

11.37 Draw structural formulas for the major product of the following reactions. Where you predict no reaction, write N.R.

    a. benzene + Br_2 $\xrightarrow{\text{room temp.}}$
    b. benzene + Br_2 $\xrightarrow{FeBr_3}$
    c. cyclohexene + Br_2 $\xrightarrow{\text{dark}}$
    d. cyclohexane + Br_2 $\xrightarrow{\text{dark}}$
    e. cyclohexane + Br_2 $\xrightarrow{\text{UV light}}$
    f. benzene + HBr $\longrightarrow$
    g. cyclohexene + HBr $\longrightarrow$
    h. benzene + H_2O $\xrightarrow{H_2SO_4}$
    i. cyclohexene + H_2O $\xrightarrow{H_2SO_4}$
    j. cyclohexane + H_2O $\xrightarrow{H_2SO_4}$
    k. benzene + $KMnO_4$ $\xrightarrow{H_2O, \text{ NaOH}}$
    l. cyclohexene + $KMnO_4$ $\xrightarrow{H_2O, \text{ NaOH}}$
    m. cyclohexane + $KMnO_4$ $\xrightarrow{H_2O, \text{ NaOH}}$

11.38 Draw structural formulas for the products of the following electrophilic aromatic substitution reactions.

    a.  (benzene) + Cl_2 $\xrightarrow{AlCl_3}$

    b.  (benzene) + $CH_3-CH-CH_3$ (with Cl) $\xrightarrow{AlCl_3}$

    c. 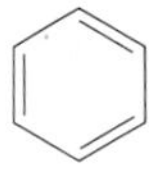 (benzene) + HNO_3 $\xrightarrow{H_2SO_4}$

    d.  (benzene) + $CH_3-\overset{\overset{\displaystyle CH_3}{|}}{\underset{\underset{\displaystyle CH_3}{|}}{C}}-Cl$ $\xrightarrow{AlCl_3}$

Terpenes

A wide variety of compounds in the plant and animal worlds contain one or more carbon-carbon double bonds. In this essay, we will focus our attention on one group of naturally occurring alkenes — the **terpene hydrocarbons.** The characteristic structural feature of a terpene is a carbon skeleton that can be divided into two or more units that are identical with the carbon skeleton of isoprene. In discussing terpenes, it is common to refer to the **head** and **tail** of an isoprene unit. The head of an isoprene unit is the carbon atom nearer the methyl branch. The tail is the carbon atom farther from the methyl branch.

$$CH_2=C-CH=CH_2$$
$$\ \ \ \ |$$
$$\ CH_3$$

isoprene

an isoprene unit

There are several important reasons for looking at this group of organic compounds. First, the number of terpenes found in bacteria, plants, and animals is staggering. Sec-ond, terpenes provide a glimpse of the wondrous diversity that nature generates from even a relatively simple carbon skeleton. Third, terpenes illustrate an important principle of the molecular logic of living systems: In building what might seem to be complex molecules, living systems piece together small subunits to produce complex but logically designed skeletal frameworks. In this mini-essay, we will show how to identify the skeletal framework of terpenes.

Probably, the terpenes most familiar to you, at least by odor, are components of the so-called **essential oils** obtained by steam distillation or ether extraction from various parts of plants. Essential oils contain relatively low-molecular-weight compounds which are, in large part, responsible for characteristic plant fragrances. Many essential oils, particularly those from flowers, are used in perfumes.

An example of a terpene obtained from an essential oil is myrcene, $C_{10}H_{16}$, obtained from bayberry wax and from oils of bay and verbena. Its parent chain of eight carbon atoms contains three double bonds and two one-carbon branches (see Figure 1a). Fig-

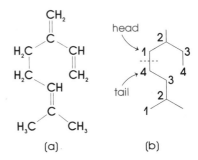

Figure 1 (a) The structure and (b) the carbon skeleton of myrcene.

ure 1(b) shows only the carbon skeleton of myrcene. As you can see, it can be divided into two isoprene units linked head-to-tail. Head-to-tail linkages of isoprene units are vastly more common in nature than are the alternative head-to-head or tail-to-tail patterns.

Figure 2 shows structural formulas for six more terpenes. Geraniol and the aggregating pheromone of the bark beetle (see Mini-Essay 8, "Pheromones") have the same carbon skeleton as myrcene but the carbon-carbon double bonds are located differently. In addition, each has an —OH group. In the last four terpenes shown in Figure 2, the framework of carbon atoms present in myrcene, geraniol, and the bark beetle pheromone is

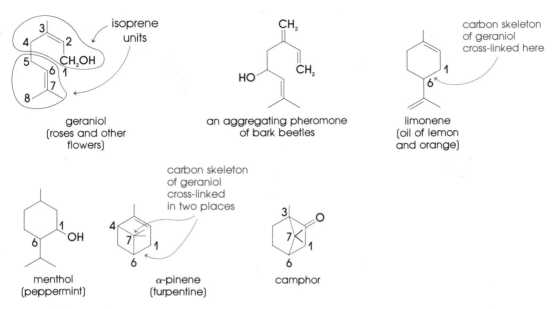

Figure 2 Several terpenes of 10-carbon atoms.

cross-linked to form cyclic structures. To help you identify the points of cross-linkage and ring formation, the carbon atoms of the geraniol skeleton are numbered 1 through 8. Bond formation between carbon atoms 1 and 6 of the geraniol skeleton gives the carbon skeletons of limonene and menthol; formation of bonds between carbons 1,6 and 4,7 gives the carbon skel-

eton of α-pinene; and between 1,6 and 3,7 gives the carbon skeleton of camphor. Shown in Figure 3 are structural formulas of several terpenes of 15 carbon atoms (sesquiterpenes). For reference, the carbon atoms of the parent chain of farnesol are numbered 1 through 12. A bond between carbon atoms 1 and 6 of this skeleton gives the carbon skeleton of zingiberene. You

should try to discover for yourself what patterns of cross-linking give the carbon skeletons of β-selinene and caryophyllene.

Structural formulas for vitamin A, a terpene of molecular formula $C_{20}H_{30}O$, and β-carotene, a terpene of molecular formula $C_{40}H_{56}$ are shown in Figure 4 (next page). Vitamin A consists of four isoprene units linked head-to-tail and

Figure 3 Several terpenes of 15 carbon atoms.

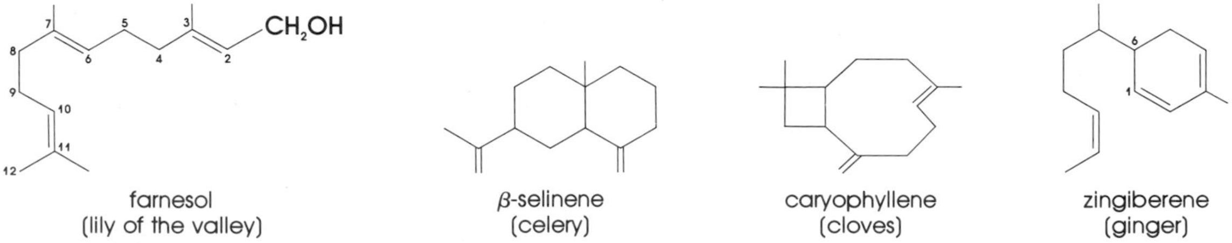

Figure 4 Vitamin A, a 20-carbon terpene, and β-carotene, a 40-carbon terpene.

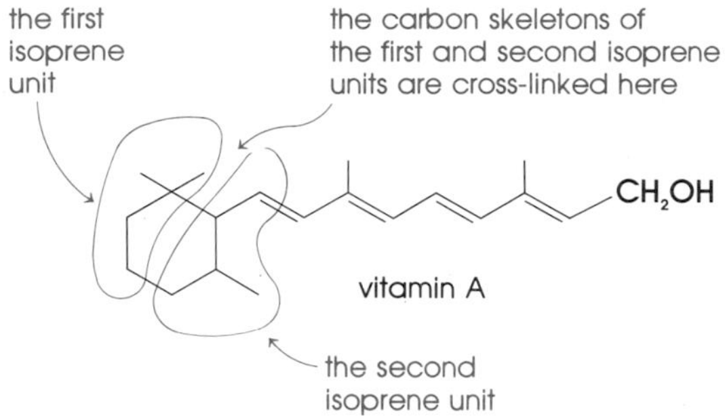

the first isoprene unit

the carbon skeletons of the first and second isoprene units are cross-linked here

CH_2OH

vitamin A

the second isoprene unit

a tail-to-tail bond joining two diterpenes

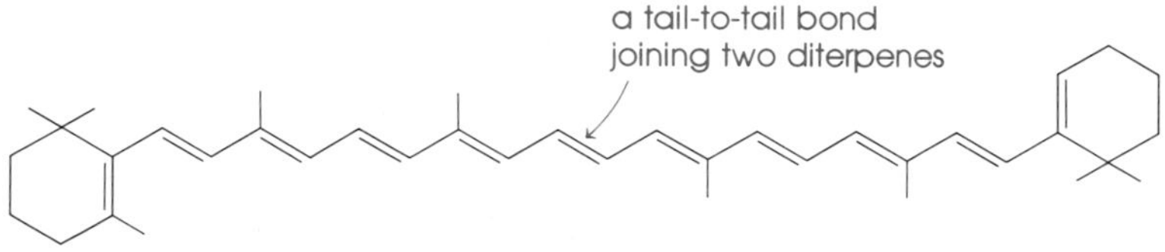

β-carotene

cross-linked at one point to form a six-member ring. The function of vitamin A is discussed in Section 18.5A. β-Carotene can be divided into two diterpenes (20-carbon terpenes), each joined tail-to-tail. The function of β-carotene is also discussed in Section 18.5A. Figure 5 shows another

use for terpenes.

We have presented only a few of the terpenes that abound in nature, but these examples should be enough to suggest their widespread distribution in living systems, the biological individuality that plants and animals achieve through terpene syn-

thesis, and the structural pattern (the isoprene unit) that underlies this apparent diversity in structural formula. In the future, when you encounter molecules of 10, 15, 20, or more carbon atoms derived from living systems, you might study their structural formulas to see if they are terpenes.

Ethylene

The U.S. chemical industry produces more ethylene, on a pound-per-pound basis, than any other organic chemical. Reports on this and other key chemicals can be found regularly in the weekly publication *Chemical and Engineering News (C&EN)*. One such report is given in Figure 1. According to *C&EN*, ethylene production in 1981 totalled almost 31 billion pounds with a commercial value of approximately $6.75 billion.

Of the estimated 250 billion pounds of organic chemicals produced each year in the United States, approximately 100 billion pounds are derived from ethylene. Clearly, in terms of its volume and the volume of the chemicals derived from it, ethylene is the organic chemical industry's most important building block. This mini-essay will focus on how this vital starting material is produced, its major derivatives, its major end uses, and consumer products made from ethylene.

First, how do we obtain ethylene? More than 90% of all organic chemicals used by the chemical industry are derived from petroleum and natural gas. However, ethylene is not found in either of these resources. If we do not find ethylene in nature, then how do we make it from the raw materials available to us? The answer is not an easy one. In fact, the answer differs from one part of the world to another, depending on avail-

Figure 1 Ethylene $CH_2{=}CH_2$

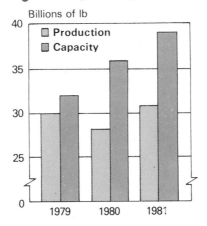

Billions of lb

How it's made: thermal (steam) or catalytic cracking of hydrocarbons ranging from natural gas-derived ethane to oil-derived gas oil (fuel oil).

Major derivatives: polyethylenes 45%, ethylene oxide 20%, vinyl chloride 15%, styrene 10%.

Major end uses: fabricated plastics 65%, antifreeze 10%, fibers 5%, solvents 5%.

Commercial value: $6.75 billion total production in 1981.

ability of raw materials and economic demand. As we shall see presently, how we make ethylene today may be quite different from how we will be forced to make it in the future.

Ethylene is produced by the **cracking** (breaking down) of hydrocarbons. In the United States, where there are vast reserves of natural gas, the major process for ethylene production has been **thermal cracking,** in the presence of steam, of the small quantities of ethane, propane, and butane that can be recovered from natural gas. For this reason, ethylene generating plants constructed in the past have been concentrated near sites of natural gas reserves. Of the 34 ethylene plants in operation in the United States in 1979, 23 were located on the Gulf Coast of Texas and Louisiana and they accounted for over 80% of total U.S. ethylene production capacity.

When written as a balanced equation, thermal (steam) cracking of ethane seems simple enough:

$$CH_3{-}CH_3$$
$$\downarrow \text{ thermal cracking}$$
$$CH_2{=}CH_2 + H_2$$

Actually, the reaction is very complicated and several other substances are produced along with ethylene. For every billion pounds of ethylene produced from ethane, 36 million pounds of propylene and 35 million pounds of butadiene are also obtained as co-products. These quantities of starting materials and products may seem enormous to you, but they reflect the scale on which the U.S. chemical industry operates. Although other low-molecular-weight alkanes can be cracked to give ethylene, thermal cracking of ethane gives the highest-percentage and highest-purity ethylene. Figure 2 shows a cracking facility in Texas.

In the United States, natural gas is currently the major source of the raw materials for the manufacture of ethylene. Approximately 10% of the natural gas consumed each year is used for this purpose. In Europe and Japan, however, supplies of natural gas are much more limited. As a result, these countries depend almost entirely on **catalytic cracking** of petroleum-derived naphtha for their ethylene. Thus, these countries produce not only aromatic hydrocarbons such as benzene, toluene, and xylene from naphtha (just as we do), but they also produce ethylene from it. The problem is that naphtha is in heavy demand as a source of straight-run gasoline, as a feedstock for catalytic cracking and reforming to produce higher-octane gasolines, and as a source of starting materials for the organic chemical industry.

The price of naphtha has increased steadily over the years, in step with increased worldwide demand. This upward spiral of demand and price has provided incentive in Europe and Japan, and more recently in the United States, to develop an ethylene-generating technology using higher-boiling petroleum fractions, particularly gas oil. Thus, while the world has depended in the past on natural gas and naphtha as raw materials for ethylene manufacture, it seems almost certain that in the future we will depend more and more on gas oil to meet this need. It

Figure 2 Located along the Gulf Coast of Louisiana and Texas near the sites of natural gas reserves, catalytic cracking facilities, such as this refinery in Port Arthur, Texas, account for more than 80% of total ethylene production in the United States. (Gulf Oil Corp.)

Table 1 Major derivatives and end uses of ethylene.

Major derivatives of ethylene	Structural formula	1981 production (billions of lbs)	Major end uses
polyethylene	$-(CH_2-CH_2)_n-$	12.0	fabricated plastics
ethylene oxide/ ethylene glycol	CH_2-CH_2, CH_2-CH_2 (with O bridge, and OH OH)	9.17	antifreeze, polyester textile fibers, solvents
vinyl chloride	$CH_2=CHCl$	6.72	polyvinyl chloride fabricated plastics
Styrene	⬡$-CH=CH_2$	6.61	polystyrene, fabricated plastics, and synthetic rubbers

seems certain that we will also come to depend on coal as coal gasification and liquefaction technologies advance.

Now that we have seen how we make ethylene, let us turn to the second question: How do we use it? Each year, ethylene is the starting material for the synthesis of almost 100 billion pounds of chemicals and polymers. As you can see from Table 1, its major derivatives are polyethylene, ethylene oxide and ethylene glycol, vinyl chloride, and styrene.

We will concentrate on just one major derivative of ethylene: the fabricated polyethylene plastics that account for approximately 45% of all ethylene used in this country. The first commercial process for ethylene polymerization used peroxide catalysts at temperatures of 500°C and a pressure of 1000 atmospheres and produced a polymer known as **low-density poly-ethylene (LDPE).** Low-density polyethylene is a soft, tough plastic. It has a density between 0.91 and 0.94 g/cm^3 and a melting point of about 115°C. Since LDPE's melting point is only slightly above 100°, it is not used for products that will be exposed to boiling water. LDPE is about 50–60% crystalline. Although polymers do not crystallize in the conventional sense, they often have regions where their chains are precisely ordered with respect to each other and interact by noncovalent forces. Such regions are said to be **crystalline.** Because it is cheap, the major use of LDPE is as film for throwaway packaging of such consumer items as baked goods, vegetables and other produce; as coatings for cardboard and paper; and, perhaps most importantly, as trash bags.

An alternative method of ethylene polymerization uses catalysts composed of titanium chloride and organoaluminum compounds. With catalysts of this type, ethylene can be polymerized under conditions as low as 60°C and 20 atmospheres pressure. Polyethylene produced in this manner has a density of 0.96 g/cm^3 and is called **high-density polyethylene (HDPE).** HDPE has a higher degree of crystallinity (90%) and a higher melting point (135°C) than LDPE. It is best described as a hard, tough plastic. The physical properties and cost of HDPE relative to other materials make it ideal for the production of plastic bottles, lids, caps, and so on. It is also molded into housewares such as mixing bowls and refrigerator and freezer containers.

Information from *Chemical and Engineering News* on high-density polyethylene and low-density polyethylene is summarized in Figure 3 (shown on next page).

Approximately 45% of all high-density polyethylene used in the United States is

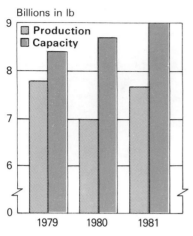

How it's made: polymerization of ethylene at high temperatures and low pressures aided by initiators.

Major fabricated forms: film, largely for packaging, 65%; injection molding 10%; coatings 10%; extrusions 5%.

Commercial value: $3.2 billion for total production in 1981.

Figure 3(a) Low-density polyethylene $\sim$[CH$_2$—CH$_2$]$_n\sim$ (Photo courtesy of Sharon A. Bazarian)

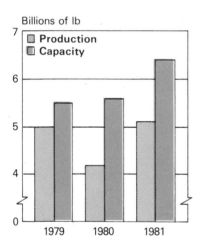

How it's made: polymerization of ethylene under moderate pressure catalyzed by metal salts and alkyls.

Major fabricated forms: Blow molding, mostly containers, 45%; injection molding 20%; extrusions 10%; film 5%.

Commercial value: $2.0 billion for total production in 1981.

Figure 3(b) High-density polyethylene $\sim$[CH$_2$—CH$_2$]$_n\sim$ (Photo courtesy of Michael E. Katin)

blow molded, mostly into containers. The process of blow molding a HDPE bottle is illustrated in Figure 4.

Approximately 65% of all low-density polyethylene is used for the manufacture of films. Fabrication of LDPE into films is done by a variation of the blow-molding technique illustrated in Figure 4. A tube of LDPE, along with a jet of compressed air, is forced through an opening and blown into a giant, thin-walled bubble (Figure 5). The film is then cooled and taken up onto a roller. This double-walled film can be slit down the side to give LDPE film or it can be sealed at points along its length to give LDPE trash bags.

References

American Chemical Society. 1973. *Chemistry in the economy.* Washington, D.C.

Bilmeyer, F.W., Jr. 1971. *Textbook of polymer science.* 2nd ed. New York: John Wiley & Sons.

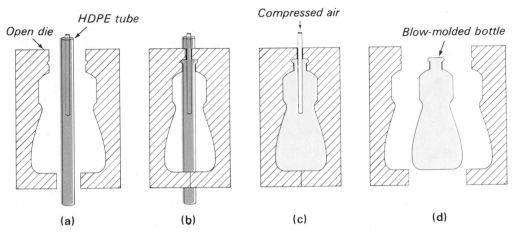

Figure 4 Blow molding of a high-density polyethylene bottle. (a) A short length of HDPE tubing is placed in an open die. (b) The die is closed, sealing the bottom of the tube. (c) Compressed air is forced into the warm polyethylene/die assembly, and the tubing is literally blown up to take the shape of the mold. (d) The die is opened, and there is a bottle!

Figure 5 Fabrication of LDPE film.

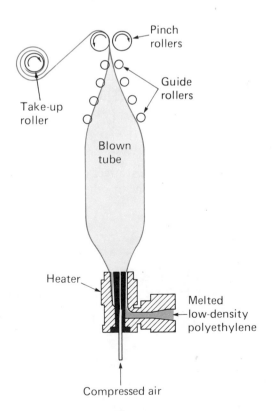

Fernelius, W.C.; Wittcoff, H.A.; and Varnerid, R.E.; eds. 1979. Ethylene: the organic chemical industry's most important building block. *J. Chem. Ed.* 56: 385–387.

Layman, P.L. C&EN's top 50 chemical products. *Chem. and Eng. News,* 3 May 1982.

Stinson, S.C. Ethylene technology moves to liquid fuels. *Chem. and Eng. News,* 28 May 1979.

Wittcoff, H.A.; and Reuben, B.G. 1980. *Industrial organic chemicals in perspective.* New York: John Wiley & Sons.

Alcohols, Ethers, Phenols, and Thiols

Alcohols and **ethers** are derivatives of water in which one or both hydrogens are replaced by alkyl groups (indicated by R in the following structures):

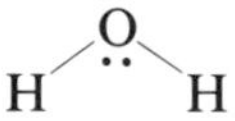

$$\overset{\cdot\cdot}{\underset{\cdot\cdot}{O}}$$

H   H   R   H   R   R

water    an alcohol 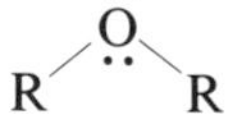   an ether

Phenols are derivatives of water in which one hydrogen is replaced by an aryl group (a benzene or other aromatic ring) indicated by the symbol Ar-:

Ar-   Ar   H

an aryl group   a phenol

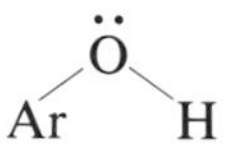

Closely related to alcohols are **thiols,** derivatives of hydrogen sulfide in which one hydrogen is replaced by an alkyl group (represented by R):

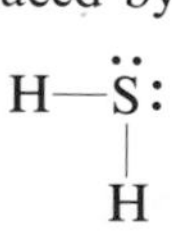

$$H\!-\!\overset{\cdot\cdot}{\underset{|}{S}}\!:\qquad\qquad R\!-\!\overset{\cdot\cdot}{\underset{|}{S}}\!:$$

H   H

hydrogen sulfide 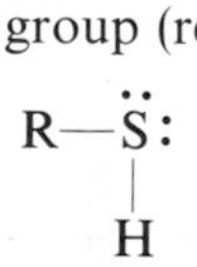   a thiol
(a thialcohol)

12.1 Structure of Alcohols and Ethers

The characteristic structural feature of an alcohol is a hydroxyl (—OH) group bonded to a hydrocarbon chain. Figure 12.1 shows the Lewis structure of methanol, CH_3OH, the simplest alcohol. As we have already seen in Section 9.5, the oxygen atom of an alcohol is sp^3 hybridized. Two sp^3 hybrid orbitals form sigma bonds to hydrogen and carbon, and the remaining sp^3 hybrid orbitals hold the two unshared pairs of electrons. Because oxygen is sp^3 hybridized, we predict 109.5° for the C—O—H bond angle. The measured bond angle for methanol is 108.9°, which is very close to the predicted value.

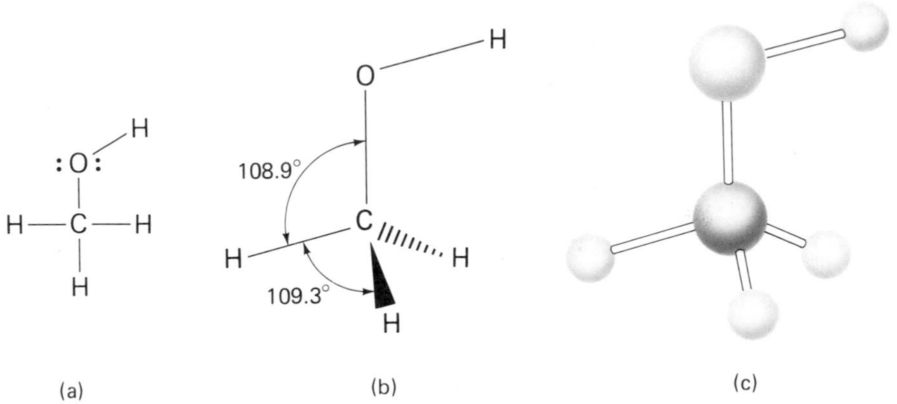

Figure 12.1 Structure of methanol, CH_3OH. (a) Lewis structure; (b) three-dimensional shape; (c) ball-and-stick model.

(a)   (b)   (c)

The characteristic structural feature of an ether is an oxygen atom bonded to two hydrocarbon chains. Figure 12.2 shows a Lewis structure of dimethyl ether, CH_3OCH_3, the simplest ether. In an ether, the oxygen atom uses sp^3 hybrid orbitals to form sigma bonds to the two hydrocarbon chains. The two unshared pairs of electrons are in the remaining sp^3 hybrid orbitals. The C—O—C bond angle in dimethyl ether is 110.3°, close to the predicted angle of 109.5°.

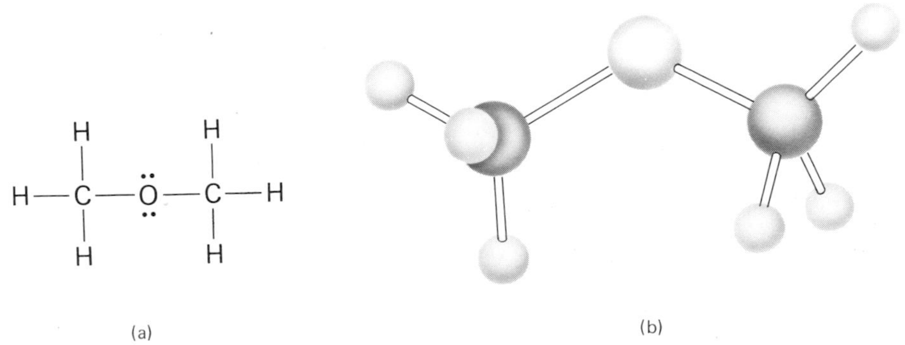

Figure 12.2 The structure of dimethyl ether, CH_3OCH_3. (a) Lewis structure; (b) ball-and-stick model.

(a)   (b)

12.2 Nomenclature of Alcohols and Ethers

A. Alcohols

In naming alcohols, the IUPAC system selects the longest carbon chain that contains the —OH group as the parent compound. The alcohol is then named by replacing the -e of the parent compound by -ol and adding a number to show the location of the —OH group. The chain is numbered in the direction that will give the —OH group the lowest possible number. Common names for simple alcohols are derived by naming the alkyl group attached to oxygen and then adding the word alcohol. Figure 12.3 lists IUPAC names and, in parentheses, common names for a variety of simple alcohols.

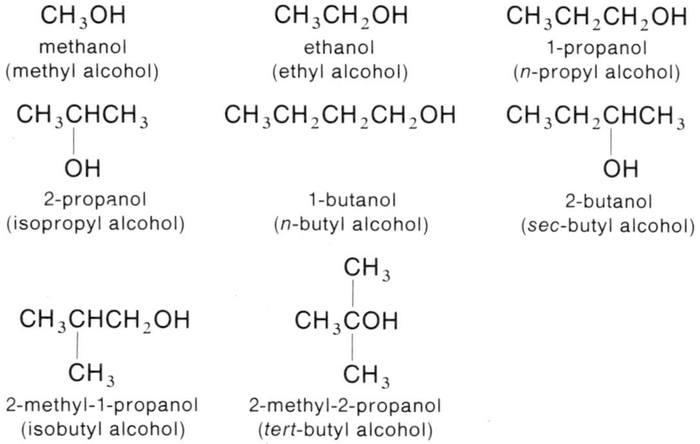

Figure 12.3 Names and structural formulas of some simple alcohols.

CH_3OH
methanol
(methyl alcohol)

CH_3CH_2OH
ethanol
(ethyl alcohol)

$CH_3CH_2CH_2OH$
1-propanol
(n-propyl alcohol)

CH_3CHCH_3
|
OH
2-propanol
(isopropyl alcohol)

$CH_3CH_2CH_2CH_2OH$
1-butanol
(n-butyl alcohol)

$CH_3CH_2CHCH_3$
|
OH
2-butanol
(sec-butyl alcohol)

CH_3CHCH_2OH
|
CH_3
2-methyl-1-propanol
(isobutyl alcohol)

CH_3
|
CH_3COH
|
CH_3
2-methyl-2-propanol
(tert-butyl alcohol)

Example 12.1

Give IUPAC names for the following alcohols:

a. 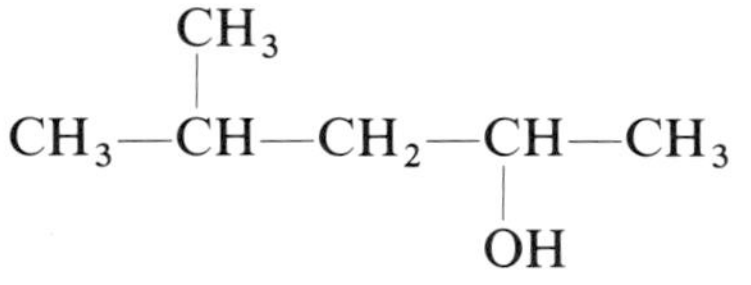
$$CH_3 \atop |$$
$$CH_3—CH—CH_2—CH—CH_3$$
$$\underset{OH}{|}$$

b. 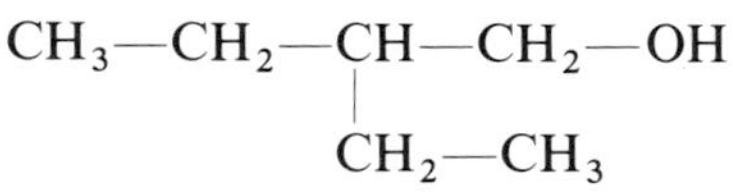 $CH_3—CH_2—CH—CH_2—OH$
$$\underset{CH_2—CH_3}{|}$$

c.

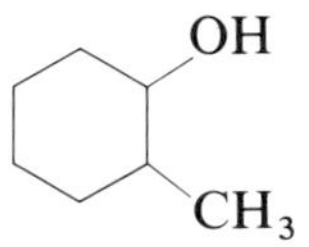

Solution

a. First select the longest chain of carbon atoms that contains the —OH group and number it to give the —OH group the lowest possible number. In this example, the longest chain is five carbon atoms and the —OH is on carbon-2. Therefore, this alcohol is a derivative of 2-pentanol. Finally, give the —CH_3 substituent a name and a number.

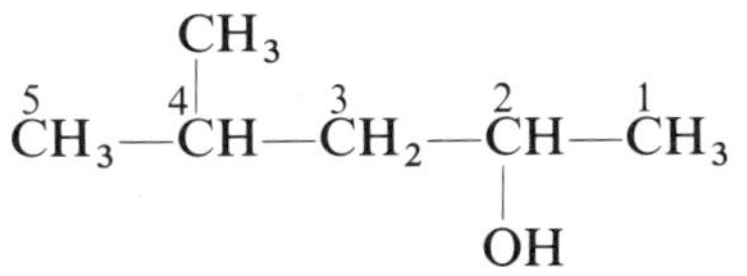

$$\overset{5}{C}H_3-\overset{4}{\underset{\underset{CH_3}{|}}{C}}H-\overset{3}{C}H_2-\overset{2}{\underset{\underset{OH}{|}}{C}}H-\overset{1}{C}H_3$$ 4-methyl-2-pentanol

b. In this molecule, the longest chain that contains the —OH group is four carbons. Note that there is a longer chain of five carbon atoms, but it does not contain the —OH group.

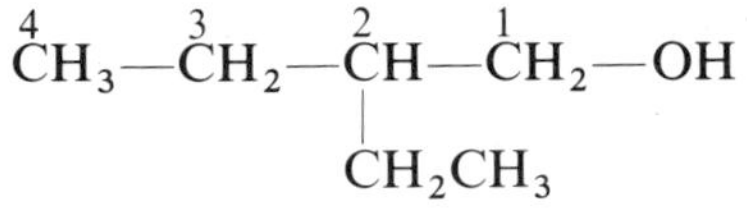

$$\overset{4}{C}H_3-\overset{3}{C}H_2-\overset{2}{\underset{\underset{CH_2CH_3}{|}}{C}}H-\overset{1}{C}H_2-OH$$ 2-ethyl-1-butanol

c. In cyclic alcohols, the numbering of the carbon atoms of the ring starts with the carbon bearing the —OH group. Because the —OH is automatically on carbon-1, there is no need to give a number for its location. The name of this compound is 2-methylcyclohexanol.

Problem 12.1

Give IUPAC names for the following alcohols:

a. 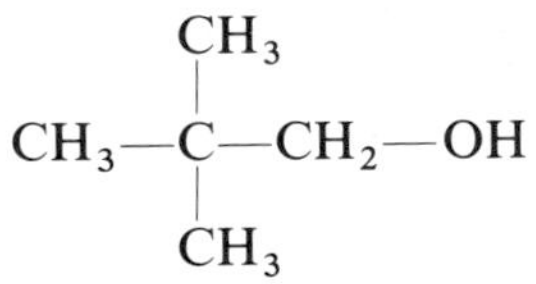 $CH_3-\underset{\underset{CH_3}{|}}{\overset{\overset{CH_3}{|}}{C}}-CH_2-OH$

b. $CH_3-CH_2-\overset{\overset{CH_3}{|}}{C}H-\underset{\underset{OH}{|}}{C}H-CH_2-CH_3$

c. (cyclopentane with OH and CH_3)

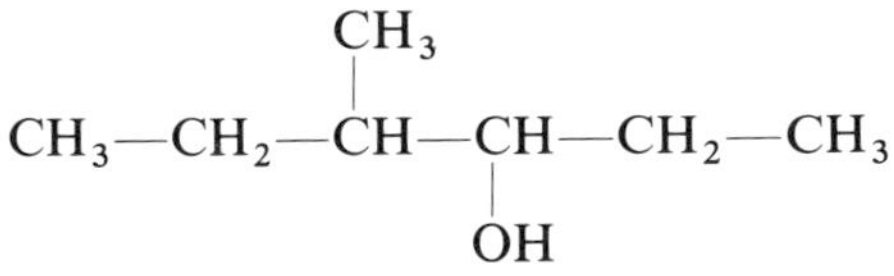

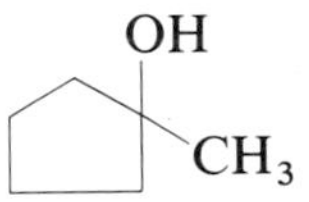

We often refer to alcohols as being **primary (1°), secondary (2°), or tertiary (3°),** depending on whether there are one, two, or three alkyl or aryl groups attached to the carbon atom bearing the —OH group (Figure 12.4).

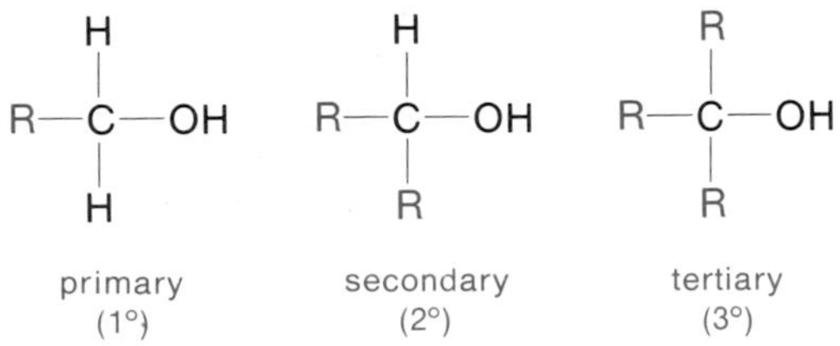

Figure 12.4 Classification of alcohols: primary, secondary, and tertiary.

$$R-\underset{\underset{H}{|}}{\overset{\overset{H}{|}}{C}}-OH \qquad R-\underset{\underset{R}{|}}{\overset{\overset{H}{|}}{C}}-OH \qquad R-\underset{\underset{R}{|}}{\overset{\overset{R}{|}}{C}}-OH$$

primary   secondary   tertiary
(1°)   (2°)   (3°)

Example 12.2

Classify the following alcohols as primary, secondary, or tertiary:

a. 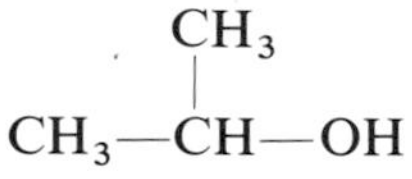
$$CH_3-\overset{\displaystyle CH_3}{\underset{\displaystyle |}{CH}}-OH$$

b. 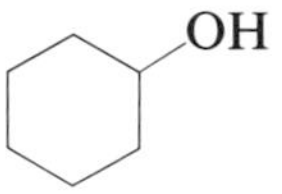 (cyclohexane with OH)

c. 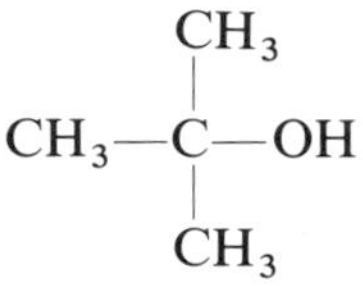
$$CH_3-\overset{\displaystyle CH_3}{\underset{\displaystyle CH_3}{\overset{\displaystyle |}{\underset{\displaystyle |}{C}}}}-OH$$

d. 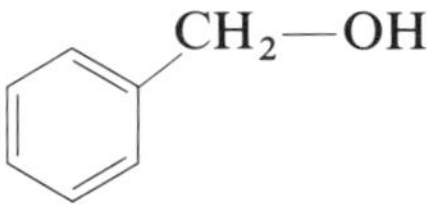 (benzene ring with CH_2-OH)

Solution

a. A secondary alcohol; because it has two attached alkyl groups, the carbon bearing the —OH group is a secondary carbon.

b. A secondary alcohol; the carbon bearing the —OH group is a secondary carbon.

c. A tertiary alcohol; the carbon bearing the —OH group has three attached alkyl groups and is a tertiary carbon.

d. A primary alcohol; the carbon bearing the —OH group has one attached aryl group and is a primary carbon.

Problem 12.2

Classify the following alcohols as primary, secondary, or tertiary:

a. 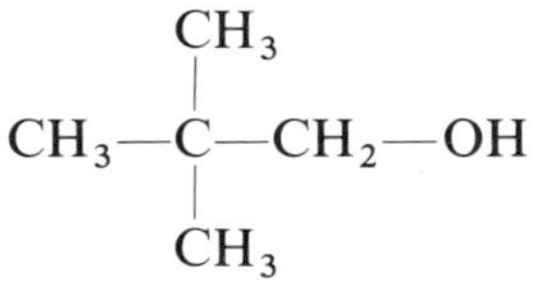
$$CH_3-\overset{\displaystyle CH_3}{\underset{\displaystyle CH_3}{\overset{\displaystyle |}{\underset{\displaystyle |}{C}}}}-CH_2-OH$$

b. 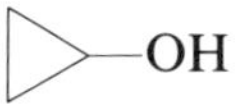 (cyclopropane with —OH)

c. $CH_2{=}CH-CH_2-OH$

d. 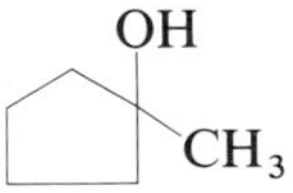 (cyclobutane with OH and CH_3)

Compounds containing two hydroxyl groups are called **diols,** those containing three hydroxyl groups are called **triols,** and so on. In the common system of nomenclature, diols are often called **glycols.** Under each of the following molecules is given its IUPAC name and, in parentheses, its common name, where appropriate.

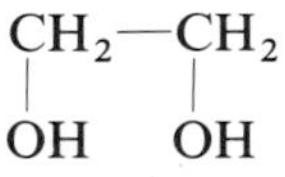

$$\underset{\displaystyle \underset{\displaystyle OH\quad OH}{|\qquad |}}{CH_2-CH_2}$$

1,2-ethanediol
(ethylene glycol)

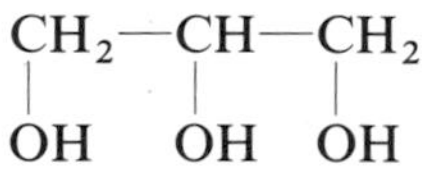

$$\underset{\displaystyle \underset{\displaystyle OH\quad OH\quad OH}{|\qquad |\qquad |}}{CH_2-CH-CH_2}$$

1,2,3-propanetriol
(glycerol, glycerine)

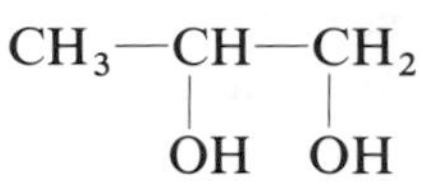

$$CH_3-CH-CH_2$$
$$\qquad\quad|\qquad\;\;|$$
$$\qquad\quad OH\;\;OH$$

1,2-propanediol
(propylene glycol)

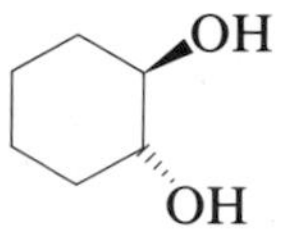

trans-1,2-cyclohexanediol

For compounds containing two or more different functional groups, the IUPAC name must show the presence and location of each. Compounds containing a hydroxyl group and a carbon-carbon double bond are named as alcohols and the carbon chain is numbered to give the —OH group the lowest number. The presence of the double bond is shown by changing the -an- of the parent compound to -en-. A number is used to locate the first carbon of the double bond.

Example 12.3

Name the following unsaturated alcohols:

a. $CH_2=CH-CH_2-OH$

b. $CH_3-CH_2-CH=CH-CH_2-CH_2-OH$

Solution

a. The parent chain contains three carbons, so this compound is named as a derivative of propane. The —OH group is on carbon-1 and the double bond is between carbons 2 and 3. The correct name is 2-propen-1-ol.

presence of carbon-carbon double bond
location of carbon-carbon double bond
2-propen-1-ol
presence of —OH *group*
location of —OH *group*

The common name of this compound is allyl alcohol.

b. The parent chain contains six carbons and this substance is named as a derivative of hexane. The correct name is 3-hexen-1-ol.

presence of carbon-carbon double bond
location of carbon-carbon double bond
3-hexen-1-ol
presence of —OH *group*
location of —OH *group*

Problem 12.3

Name the following alcohols:

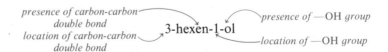

a. $CH_3-CH=CH-CH_2OH$ **b.**

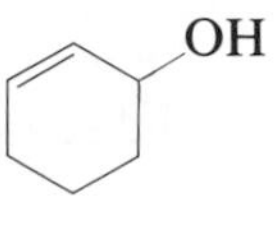

One last comment on the nomenclature of alcohols. The suffix -ol is generic to alcohols, and although names such as glycerol, menthol, and cholesterol contain no clues to their carbon skeleton, they do indicate that each compound contains at least one hydroxyl group.

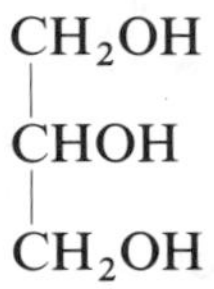

CH_2OH
$CHOH$
CH_2OH

glycerol

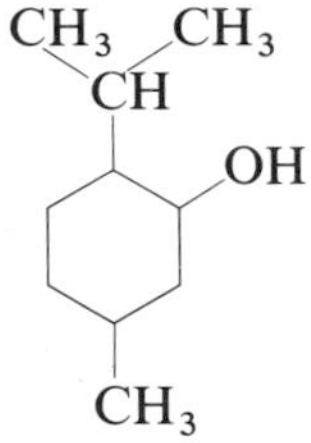

menthol

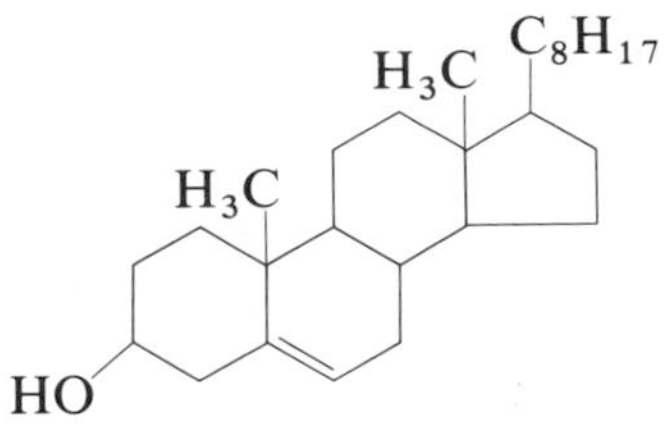

cholesterol

B. Ethers

To name ethers, list the names of the alkyl or aryl groups attached to oxygen in alphabetical order before the name *ether*. In naming simple ethers, the prefix di- is sometimes not used.

$CH_3—O—CH_3$

$CH_3CH_2—O—CH_2CH_3$

$—O—CH_3$

dimethyl ether
(methyl ether)

diethyl ether
(ethyl ether)

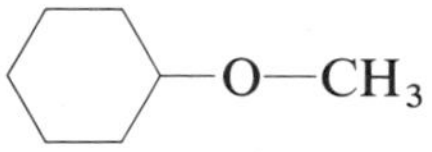

cyclohexyl
methyl ether

Heterocyclic ethers, cyclic compounds in which the ether oxygen is one of the atoms in a ring, are given special names.

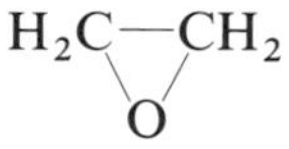

ethylene
oxide

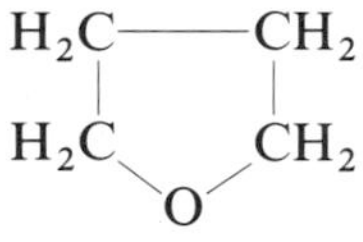

tetrahydrofuran

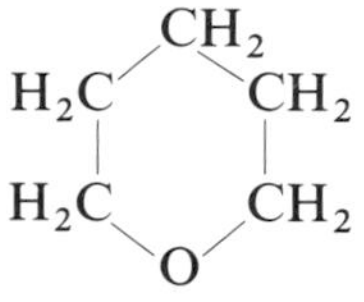

tetrahydropyran

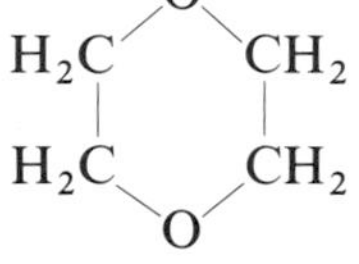

1,4-dioxane

For ethers where there is no simple name for one of the attached alkyl groups or where the ether is only one of several functional groups, —OR is named as an alkoxy group: methoxy- for $—OCH_3$, ethoxy- for $—OCH_2CH_3$, and so on.

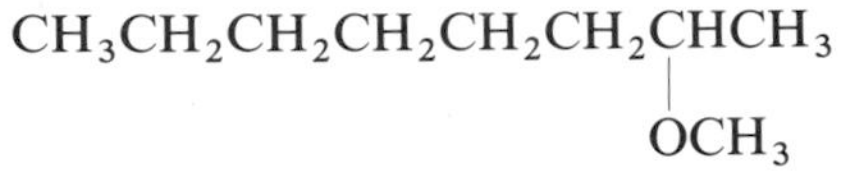

$$CH_3CH_2CH_2CH_2CH_2CH_2CHCH_3$$
$$\underset{\displaystyle OCH_3}{|}$$

2-methoxyoctane

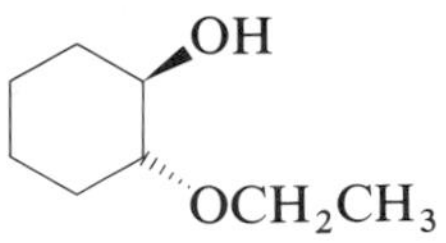

trans-2-ethoxycyclohexanol

Example 12.4

Name the following ethers:

a. 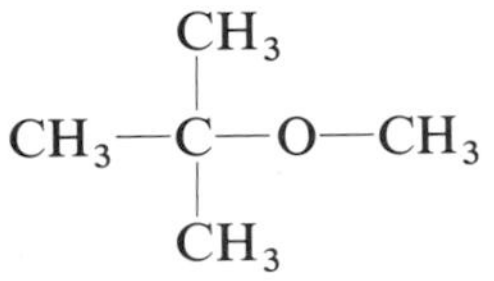 $CH_3\!\!-\!\!\underset{\displaystyle \underset{|}{CH_3}}{\overset{\displaystyle \overset{|}{CH_3}}{C}}\!\!-\!\!O\!\!-\!\!CH_3$

b. 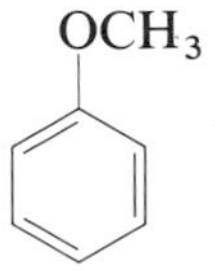 [structure: benzene ring with OCH₃ substituent]

Solution

a. Both alkyl groups have IUPAC names, and this ether is named *tert*-butyl methyl ether.

b. In this ether, methyl and phenyl groups are attached to oxygen, and it can be named methyl phenyl ether. Alternatively, —OCH_3 can be called a methoxy group and the ether named methoxybenzene. The common name of this ether is anisole.

Problem 12.4

Name the following ethers:

a. 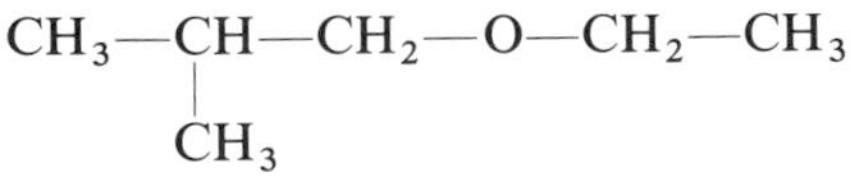 $CH_3\!\!-\!\!\underset{\displaystyle \underset{|}{CH_3}}{CH}\!\!-\!\!CH_2\!\!-\!\!O\!\!-\!\!CH_2\!\!-\!\!CH_3$

b. 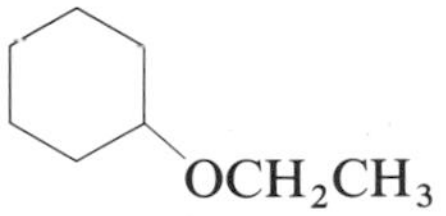 [structure: cyclohexane ring with OCH_2CH_3 substituent]

12.3 Physical Properties of Alcohols and Ethers

A. Polarity of Alcohols and Ethers

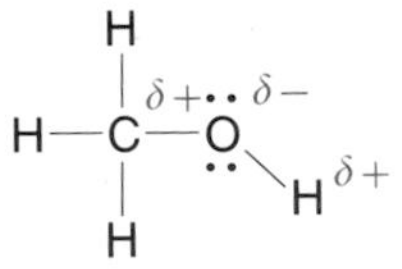

Because of the presence of the C—O—H group, alcohols are polar compounds. Oxygen is more electronegative than either carbon or hydrogen; therefore, there are partial positive charges on carbon and hydrogen and a partial negative charge on oxygen in this group (Figure 12.5). In the pure state, there is extensive hydrogen bonding between partially negative oxygen atoms and partially positive hydrogen atoms. Figure 12.6 (next page) shows the association of ethanol molecules in the liquid state by hydrogen bonding.

Figure 12.5 Polarity of the C—O—H bonds in alcohols.

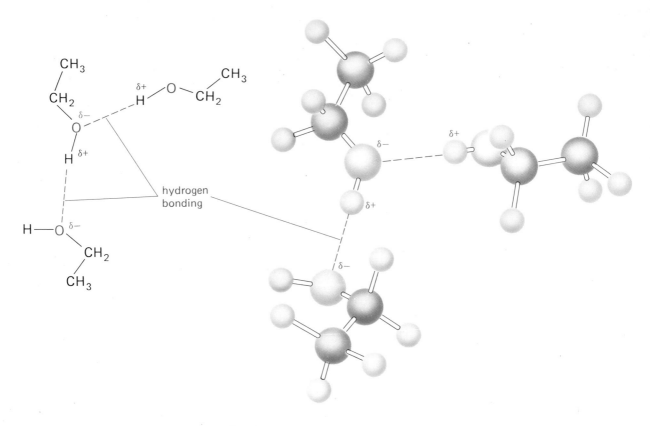

Figure 12.6 The association of ethanol by hydrogen bonding in the liquid state. (a) Lewis structures; (b) ball-and-stick models.

Ethers are also polar molecules in which the oxygen atom bears a partial negative charge and each of the attached carbon atoms bears a partial positive charge. Association by hydrogen bonding is not possible for ethers because there is no partially positive hydrogen atom attached to the oxygen to participate in the formation of hydrogen bonds (Figure 12.7).

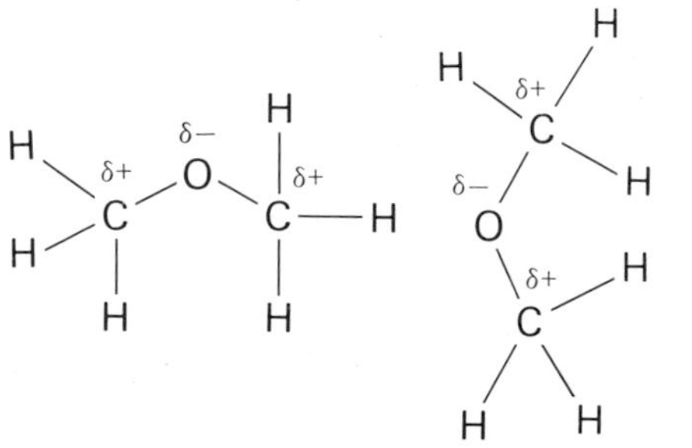

Figure 12.7 Hydrogen bonding is not possible in dimethyl ether because there are no partially positive hydrogens attached to oxygen.

B. Boiling Points of Alcohols and Ethers

Table 12.1 lists the boiling points and solubilities in water for several groups of alcohols, ethers, and hydrocarbons of similar molecular weights. Of the three classes of compounds compared in this table, alcohols have the highest boiling points because of hydrogen bonding between —OH groups. Because there is very little interaction between molecules of ether in the pure state, the boiling points of ethers are close to those of nonpolar hydrocarbons of comparable molecular weight. The effect of hydrogen bonding is dramatically illustrated by comparing the boiling points of ethanol (bp 78°C) and its structural isomer dimethyl ether (bp −24°C). The difference in boiling point between

Table 12.1 Boiling points and solubilities in water of several groups of alcohols, ethers, and hydrocarbons of similar molecular weight. The symbol ∞ means "infinity"; "soluble (∞)" means "soluble in all proportions."

Structural Formula	Name	Molecular Weight	bp (C°)	Solubility in Water (g/100 g H$_2$O)
CH_3OH	methanol	32	65	soluble (∞)
CH_3CH_3	ethane	30	−89	insoluble
CH_3CH_2OH	ethanol	46	78	soluble (∞)
CH_3OCH_3	dimethyl ether	46	−24	soluble (7 g/100 g)
$CH_3CH_2CH_3$	propane	44	−42	insoluble
$CH_3CH_2CH_2OH$	1-propanol	60	97	soluble (∞)
$CH_3CH_2OCH_3$	ethyl methyl ether	60	11	soluble
$CH_3CH_2CH_2CH_3$	butane	58	0	insoluble
$CH_3CH_2CH_2CH_2OH$	1-butanol	74	117	soluble (8 g/100 g)
$CH_3CH_2OCH_2CH_3$	diethyl ether	74	35	soluble (8 g/100 g)
$CH_3CH_2CH_2CH_2CH_3$	pentane	72	36	insoluble
$CH_3CH_2CH_2CH_2CH_2OH$	1-pentanol	88	138	slightly soluble (2.3 g/100 g)
$CH_3CH_2CH_2CH_2OCH_3$	butyl methyl ether	88	71	slightly soluble
$HOCH_2CH_2CH_2CH_2OH$	1,4-butanediol	90	230	soluble (∞)
$CH_3OCH_2CH_2OCH_3$	ethylene glycol dimethyl ether	90	84	soluble (∞)
$CH_3CH_2CH_2CH_2CH_2CH_3$	hexane	88	69	insoluble

them is due to the presence of a polar O—H group in the alcohol. Alcohol molecules interact by hydrogen bonding; ether molecules do not. Compare also the boiling points of 1-propanol (97°C) and ethyl methyl ether (11°C); and 1-butanol (117°C) and diethyl ether (35°C).

The presence of additional hydroxyl groups in a molecule further increases the effects of hydrogen bonding, as you can see by comparing the boiling points of hexane (bp 69°C), 1-pentanol (bp 138°C), and 1,4-butanediol (bp 230°C). Note that all three of these compounds have approximately the same molecular weight, but very different boiling points.

Boiling points of alcohols and ethers increase with increasing molecular weight, due to the increased dispersion forces among the hydrocarbon portions of the molecules. To see this, compare the boiling points of ethanol, 1-propanol, and 1-butanol.

Example 12.5

Arrange the following compounds in order of increasing boiling point:

$$CH_3—CH_2—OH \qquad CH_3—CH_2—Cl \qquad CH_3—CH_2—CH_3$$

Solution

Propane and ethanol have similar molecular weights. However, propane is a nonpolar hydrocarbon, and the only interactions between molecules in the pure liquid are very weak dispersion forces (review Section 10.7). Ethanol is a polar compound, and there is extensive hydrogen bonding between alcohol molecules in the pure liquid. For this reason, ethanol has a considerably higher boiling point than propane. Chloroethane has a higher molecular weight than ethanol and is a polar molecule. However, it cannot associate by hydrogen bonding so it has a lower boiling point than ethanol. In order of increasing boiling point, the three compounds are:

$$CH_3CH_2CH_3 \qquad CH_3CH_2Cl \qquad CH_3CH_2OH$$

| bp -87°C | bp 12°C | bp 78°C |
| (mw 44.1) | (mw 64.5) | (mw 46.1) |

Problem 12.5

Arrange the following compounds in order of increasing boiling point:

$$CH_3OCH_2CH_2OCH_3 \qquad HOCH_2CH_2OH \qquad CH_3OCH_2CH_2OH$$

C. Solubilities of Alcohols and Ethers in Water

Because they can form hydrogen bonds with water, alcohols and ethers are considerably more soluble in water than are alkanes of comparable molecular weight (Figure 12.8). Methanol, ethanol, and 1-propanol are soluble in water in all proportions.

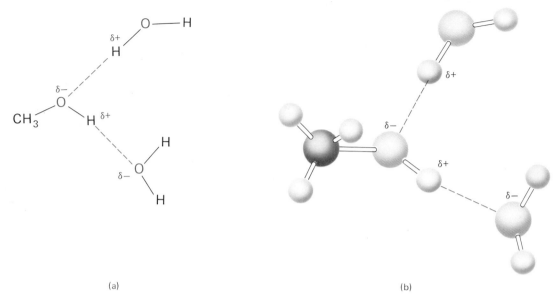

Figure 12.8 Hydrogen bonding between water and methanol. (a) Lewis structures; (b) ball-and-stick models.

As molecular weight increases, the physical properties of alcohols and ethers become more like those of hydrocarbons of comparable molecular weight. Alcohols of higher molecular weight are much less soluble in water because of the increase in size of the hydrocarbon portion of the molecule. For example, 1-decanol is insoluble in water.

Example 12.6

Arrange the following compounds in order of increasing solubility in water:

$$CH_3CH_2CH_2CH_2CH_2CH_3 \qquad CH_3OCH_2CH_2OCH_3 \qquad CH_3CH_2OCH_2CH_3$$

Solution

Water is a polar solvent. Hexane, C_6H_{14}, is a nonpolar hydrocarbon and has the lowest solubility in water. Both diethyl ether and ethylene glycol dimethyl ether contain polar C—O—C bonds, and each can interact with water molecules by hydrogen bonding. Ethylene glycol dimethyl ether is more soluble in water than diethyl ether because it has more sites within the molecule for hydrogen bonding with water molecules. The solubilities of these compounds in water are given in Table 12.1.

$$CH_3CH_2CH_2CH_2CH_2CH_3 \qquad CH_3CH_2OCH_2CH_3 \qquad CH_3OCH_2CH_2OCH_3$$

insoluble   8 g/100 g water   soluble in all
                                                        proportions

Problem 12.6　Arrange the following compounds in order of increasing solubility in water:

$$Cl—CH_2—CH_2—Cl \quad CH_3—CH_2—CH_2—OH \quad CH_3—CH_2—O—CH_2—CH_3$$

Example 12.7　Arrange the following compounds in order of increasing solubility in benzene:

$$CH_3(CH_2)_{14}CH_2OH \quad CH_3(CH_2)_2CH_2OH$$

Solution　Benzene is a nonpolar hydrocarbon solvent. Both 1-butanol and 1-hexadecanol contain polar —OH groups and nonpolar hydrocarbon chains. Because the nonpolar hydrocarbon portion of 1-hexadecanol is considerably larger than that of 1-butanol, 1-hexadecanol is much more soluble in benzene than is 1-butanol.

Problem 12.7　Arrange the following compounds in order of increasing solubility in hexane:

$$CH_3(CH_2)_6CH{=\!=}CH_2 \quad \underset{\underset{OH}{|}}{CH_3CHCH_3} \quad \underset{\underset{OH}{|}}{CH_3(CH_2)_6CHCH_3}$$

12.4 Preparation of Alcohols

Methanol, commonly called wood alcohol, was prepared by the **destructive distillation** of wood, at least until 1923. When wood is heated to temperatures above 250°C in the absence of air, it decomposes to charcoal, methanol, acetic acid, and traces of acetone. Today, methanol is made on an industrial scale by high-pressure catalytic hydrogenation of carbon monoxide:

$$CO + 2\ H_2 \xrightarrow[\text{350°C, 200 atm}]{\text{ZnO-Cr}_2\text{O}_3} CH_3OH$$
$$\text{methanol}$$

Ethanol, or simply "alcohol" in nonscientific language, has been prepared since antiquity by the fermentation of sugars and starches by yeast.

$$C_6H_{12}O_6 \xrightarrow{\text{enzymes}} 2\ CH_3CH_2OH + 2\ CO_2$$
$$\text{glucose} \qquad\qquad \text{ethanol}$$

The sugars for the fermentation process come from a variety of sources, including blackstrap molasses (a residue from the refining of cane sugar), grains (hence the name *grain alcohol*), grape juice, and various vegetables. The immediate product of fermentation is a water solution containing up to about 15% alcohol. This alcohol may be concentrated by distillation. Beverage alcohol may contain traces of flavor derived from the source (grapes in brandy, grains in whiskey) or may be essentially flavorless (like vodka). The most important industrial method for the preparation of ethanol is the hydration of ethylene (Section 11.5C).

Ordinary commercial ethanol is a mixture of 95% alcohol and 5% water. Absolute or 100% ethanol is prepared from 95% ethanol by techniques that remove water from the mixture.

Ethylene glycol is prepared commercially from ethylene by air oxidation at high temperature over a silver catalyst to give ethylene oxide, followed by acid-catalyzed hydrolysis.

$$CH_2\!=\!CH_2 + \tfrac{1}{2}O_2 \xrightarrow[300°C]{Ag} \underset{\underset{\displaystyle O}{\diagdown\diagup}}{CH_2\!-\!CH_2} \xrightarrow[H^+]{H_2O} \underset{\underset{\displaystyle OH \quad OH}{|\qquad|}}{CH_2\!-\!CH_2}$$

ethylene   ethylene   ethylene
oxide   glycol

Ethylene glycol is used as a permanent antifreeze in automobile cooling systems because it is soluble in water in all proportions and because a 50% solution with water freezes at $-34°C$. Among its many other important commercial uses, ethylene glycol is a starting material for the production of Dacron polyester.

Glycerol, a sweet-tasting, syrupy liquid, is a by-product of the manufacture of soaps (Section 16.7). Because of the presence of the three hydroxyl groups, glycerol has a very high boiling point (290°C) and is soluble in water and alcohol in all proportions. It is widely used in the manufacture of resins and nitroglycerine, in the pharmaceutical and cosmetic industries, and as a moistening agent.

Alcohols can also be prepared by acid-catalyzed hydration of alkenes (Section 11.5C). In this reaction, water adds to carbon-carbon double bonds according to Markovnikov's rule:

$$CH_3\!-\!CH\!=\!CH_2 + H_2O \xrightarrow{H_2SO_4} \underset{\underset{\displaystyle CH_3}{|}}{\overset{\overset{\displaystyle OH}{|}}{CH_3\!-\!CH\!-\!CH_3}}$$

12.5 Reactions of Alcohols

Alcohols undergo a variety of important reactions, including conversion to alkyl halides, dehydration, and oxidation to aldehydes, ketones, and carboxylic acids. Thus, alcohols are valuable starting materials for the synthesis of other compounds.

A. Conversion to Alkyl Halides

An alcohol can be converted to an alkyl halide by several reagents, the most common of which are the hydrogen halides, HCl, HBr, or HI. In the reaction between an alcohol and H—X, the C—OH bond is broken and a new C—X bond is formed. The —OH of the alcohol and the —H of H—X combine to form H—OH.

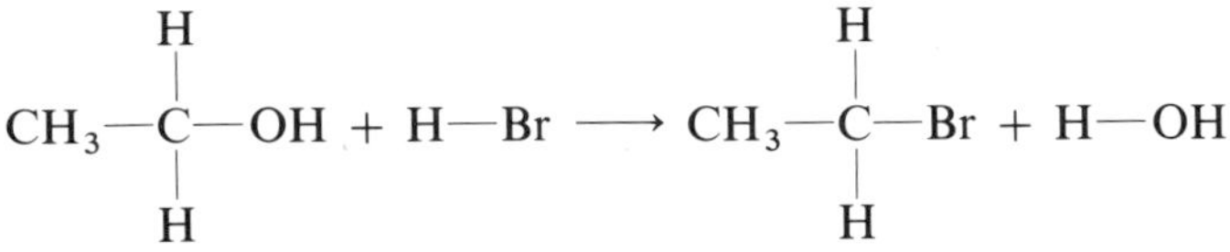

$$CH_3-\underset{\underset{\displaystyle H}{|}}{\overset{\overset{\displaystyle H}{|}}{C}}-OH + H-Br \longrightarrow CH_3-\underset{\underset{\displaystyle H}{|}}{\overset{\overset{\displaystyle H}{|}}{C}}-Br + H-OH$$

Because an —OH group is replaced by —Br, conversion of an alcohol to an alkyl halide is classified as a substitution reaction.

It has been found that tertiary alcohols react much more rapidly than secondary alcohols, and that secondary alcohols react much more rapidly than primary alcohols. For example, just mixing 2-methyl-2-propanol (a tertiary alcohol) with HCl for a few minutes at room temperature converts the alcohol to 2-chloro-2-methylpropane. On the other hand, 1-butanol (a primary alcohol) must be heated with concentrated HCl in the presence of a zinc chloride catalyst to convert it to 1-chlorobutane.

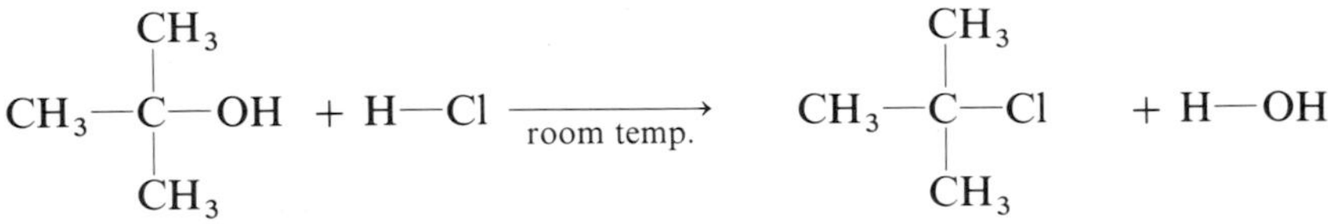

$$CH_3-\underset{\underset{\displaystyle CH_3}{|}}{\overset{\overset{\displaystyle CH_3}{|}}{C}}-OH + H-Cl \xrightarrow[\text{room temp.}]{} CH_3-\underset{\underset{\displaystyle CH_3}{|}}{\overset{\overset{\displaystyle CH_3}{|}}{C}}-Cl + H-OH$$

2-methyl-2-propanol   2-chloro-2-methylpropane
(*tert*-butyl alcohol)   (*tert*-butyl chloride)

$$CH_3-CH_2-CH_2-CH_2-OH + H-Cl \xrightarrow[\text{reflux}]{ZnCl_2} CH_3-CH_2-CH_2-CH_2-Cl + H-OH$$

1-butanol   1-chlorobutane
(*n*-butyl alcohol)   (*n*-butyl chloride)

Example 12.8

Complete the following substitution reactions and predict which reaction occurs more rapidly.

a. 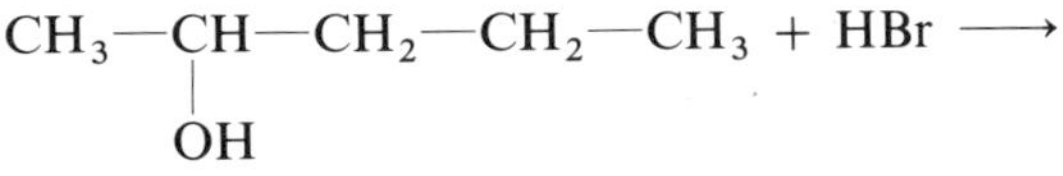 $CH_3-\underset{\underset{\displaystyle OH}{|}}{CH}-CH_2-CH_2-CH_3 + HBr \longrightarrow$

b. 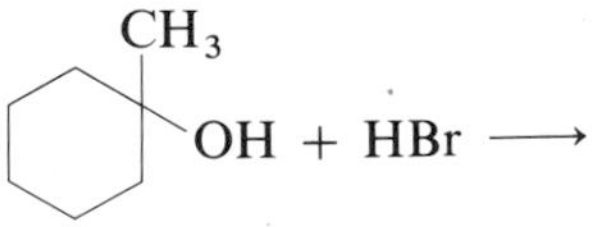 $OH + HBr \longrightarrow$

Solution

a. $CH_3-CH-CH_2-CH_2-CH_3 + HBr \longrightarrow CH_3-CH-CH_2-CH_2-CH_3 + H_2O$
         | |
         OH   Br

b. 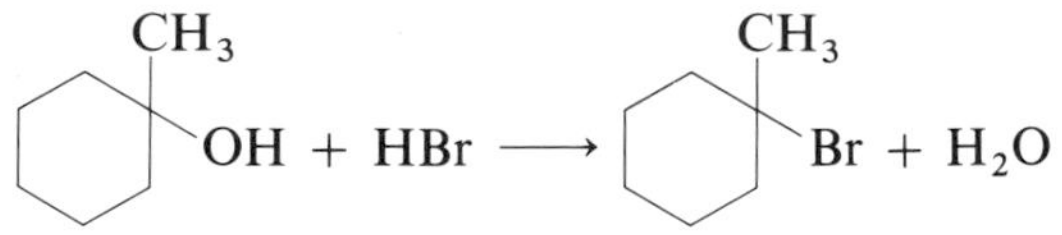 (structure) $-OH + HBr \longrightarrow$ (structure) $-Br + H_2O$

with CH_3 on each cyclohexane ring

2-Pentanol in part (a) is a secondary alcohol, while 1-methylcyclohexanol in part (b) is a tertiary alcohol. Because tertiary alcohols react more rapidly with HBr than secondary alcohols, we can predict that reaction (b) occurs more rapidly than reaction (a).

Problem 12.8

Complete the following substitution reactions and predict which reaction occurs more rapidly.

        CH_3
        |
a. $CH_3-C-CH_2-CH_3 + HI \longrightarrow$
        |
        OH

          CH_3
          |
b. $CH_3-CH-CH-CH_3 + HI \longrightarrow$
              |
              OH

B. Acid-Catalyzed Dehydration of Alcohols

An alcohol may be converted to an alkene by elimination of a molecule of water from adjacent carbon atoms. Elimination of water from an alcohol is called **dehydration.** In the laboratory, dehydration of an alcohol to an alkene is usually brought about by heating the alcohol with 85% phosphoric acid or concentrated sulfuric acid. For example, acid-catalyzed dehydration of ethanol in the presence of sulfuric acid yields ethylene:

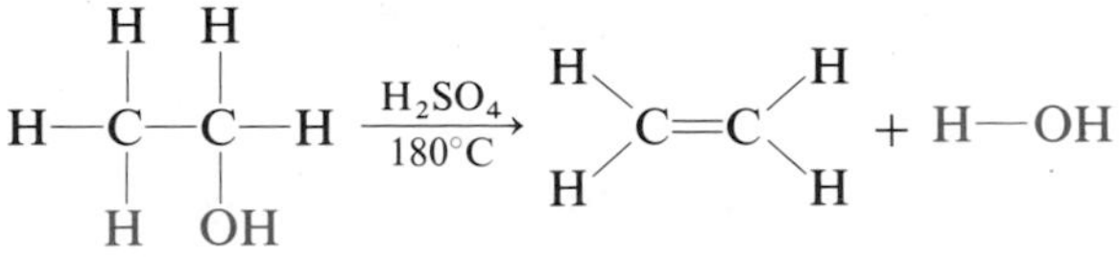

      H H
      | | $\xrightarrow[180°C]{H_2SO_4}$ $H\diagdown\;\;\;\;\diagup H$
$H-C-C-H$ $C=C$ $+ H-OH$
      | | $H\diagup\;\;\;\;\diagdown H$
      H OH

       ethanol   ethylene

In this process, two sigma bonds (C—H and C—OH) are broken and one new pi bond (C=C) and one new sigma bond (H—OH) are formed. Dehydration of cyclohexanol in the presence of 85% phosphoric acid yields cyclohexene:

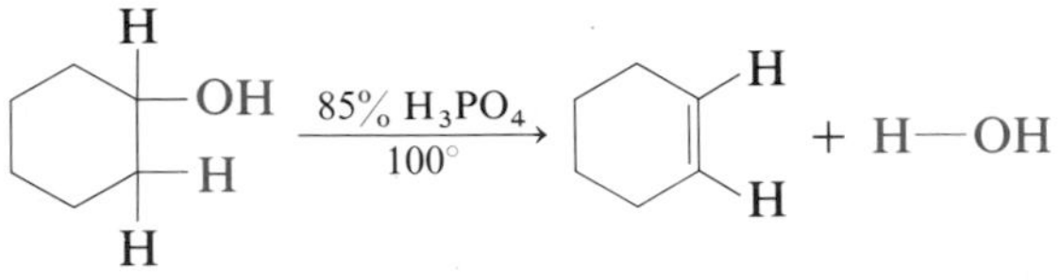

$$\text{cyclohexanol} \xrightarrow[100°]{85\% \ H_3PO_4} \text{cyclohexene} + H—OH$$

cyclohexanol   cyclohexene

Often, acid-catalyzed dehydration of an alcohol gives a mixture of isomeric alkenes. In the dehydration of 2-butanol, removal of —H and —OH from carbons 1 and 2 gives 1-butene. Removal of —H and —OH from carbons 2 and 3 gives 2-butene. In practice, acid-catalyzed dehydration of 2-butanol yields 80% 2-butene and 20% 1-butene.

$$\overset{1}{C}H_3\overset{2}{C}H\overset{3}{C}H_2\overset{4}{C}H_3 \xrightarrow[\text{heat}]{85\% \ H_3PO_4} CH_3CH{=}CHCH_3 + CH_2{=}CHCH_2CH_3 + H_2O$$

$$\underset{\text{OH}}{|}$$

2-butanol

2-butene 80%   1-butene 20%

Where it is possible to obtain isomeric alkenes from the acid-catalyzed dehydration of an alcohol, the alkene having more substituents on the double bond generally predominates.

Example 12.9

Draw structural formulas for the alkenes formed from acid-catalyzed dehydration of the following alcohols. Predict which alkene is the major product and which is the minor product.

a.
$$\begin{array}{c} CH_3 \\ | \\ CH_3-CH-CH-CH_3 \\ | \\ OH \end{array}$$

3-methyl-2-butanol

b. 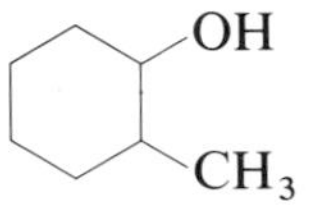
(structure of 2-methylcyclohexanol with OH and CH₃)

2-methylcyclohexanol

Solution

a. Dehydration of 3-methyl-2-butanol can result in elimination of —H and —OH, from carbons 1 and 2 to produce 3-methyl-1-butene, or from carbons 2 and 3 to produce 2-methyl-2-butene:

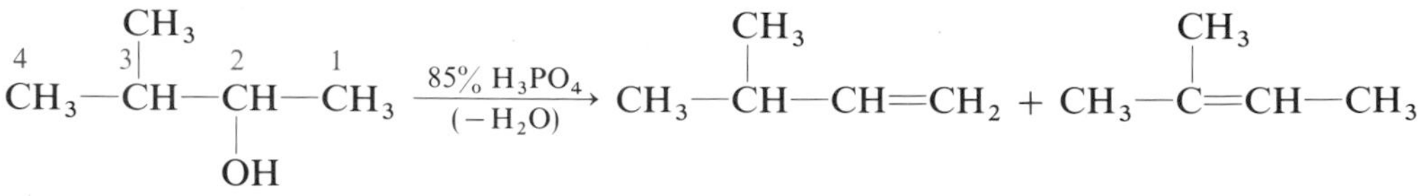

$$\underset{\substack{\text{3-methyl-2-butanol}}}{\overset{\substack{CH_3 \\ | \\ CH_3-\overset{3}{CH}-\overset{2}{CH}-\overset{1}{CH_3} \\ | \\ OH}}{}} \xrightarrow[(-H_2O)]{85\% \ H_3PO_4} \underset{\substack{\text{3-methyl-1-butene} \\ (\textit{minor product})}}{\overset{\substack{CH_3 \\ | \\ CH_3-CH-CH=CH_2}}{}} + \underset{\substack{\text{2-methyl-2-butene} \\ (\textit{major product})}}{\overset{\substack{CH_3 \\ | \\ CH_3-C=CH-CH_3}}{}}$$

3-Methyl-1-butene has one alkyl substituent (an isopropyl group) on the double bond. 2-Methyl-2-butene has three alkyl substituents (three methyl groups) on the double bond. Therefore, we predict that 2-methyl-2-butene is the major product and 3-methyl-1-butene is the minor product.

b.

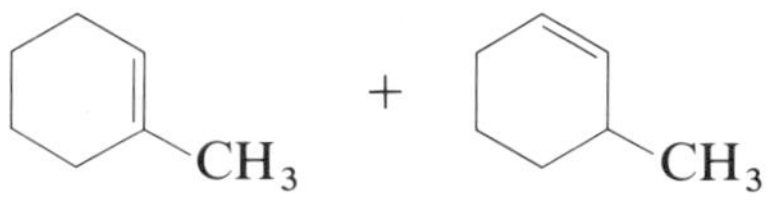

1-methylcyclohexene   3-methylcyclohexene
(*major product*)   (*minor product*)

The major product, 1-methylcyclohexene, has three alkyl substituents on the carbon-carbon double bond. The minor product, 3-methylcyclohexene, has only two.

Problem 12.9

Draw structural formulas for the alkenes formed from acid-catalyzed dehydration of the following alcohols. Where two alkenes are possible, predict which is the major product and which is the minor product.

a. $CH_3-CH_2-CH_2-CH_2-CH_2-CH_2-OH$

b. 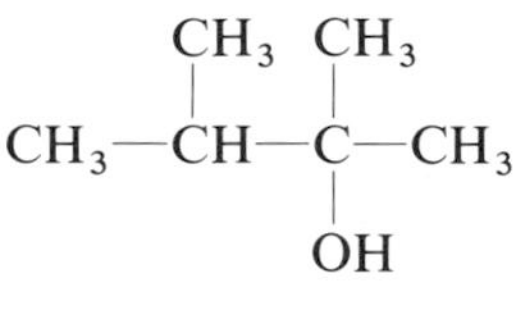 $CH_3-\underset{\substack{| \\ OH}}{\overset{\substack{CH_3 \ CH_3 \\ | \ \ \ \ | }}{C}}H-C-CH_3$

c.

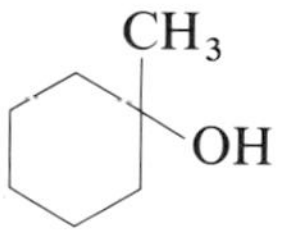

   We have seen in this section that alcohols can be dehydrated in the presence of an acid catalyst to alkenes, and in Section 11.5C we saw that alkenes can be hydrated in the presence of an acid catalyst to alcohols.

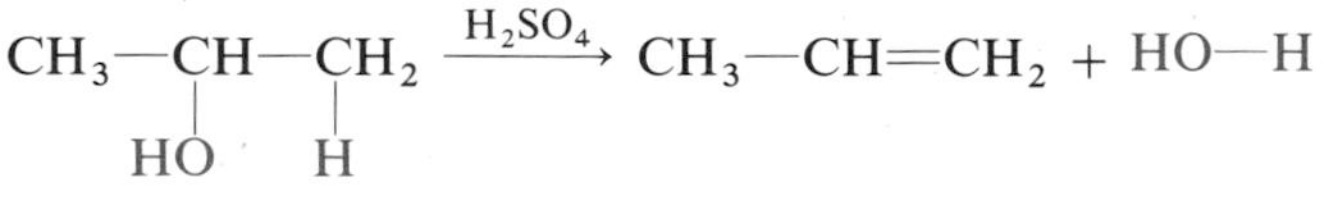

$$CH_3-\underset{\substack{| \\ HO}}{C}H-\underset{\substack{| \\ H}}{C}H_2 \xrightarrow{H_2SO_4} CH_3-CH=CH_2 + HO-H \qquad \textit{a dehydration reaction}$$

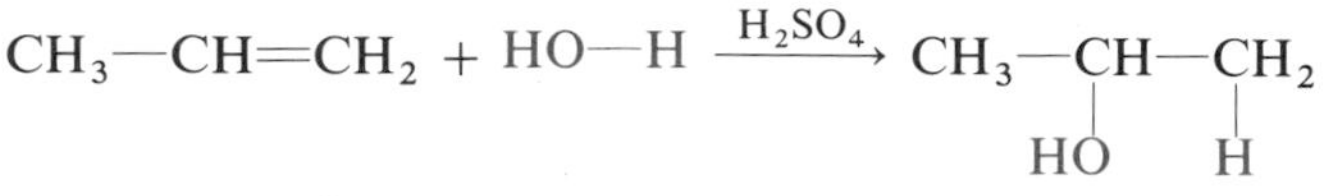

$$CH_3-CH=CH_2 + HO-H \xrightarrow{H_2SO_4} CH_3-\underset{\substack{| \\ HO}}{C}H-\underset{\substack{| \\ H}}{C}H_2 \qquad \textit{a hydration reaction}$$

In Section 11.5D we proposed a mechanism for acid-catalyzed hydration of alkenes to alcohols. This mechanism involved electrophilic attack on the pi bond of the alkene by H^+ (an electrophile) to generate a carbocation intermediate (another electrophile) and then reaction of the carbocation with a molecule of water to give the alcohol. The mechanism for acid-catalyzed dehydration of an alcohol to an alkene is the reverse of this mechanism. In step 1, the electrophile H^+ reacts with an unshared pair of electrons on oxygen to form a new sigma bond between oxygen and hydrogen. The intermediate produced is called an oxonium ion.

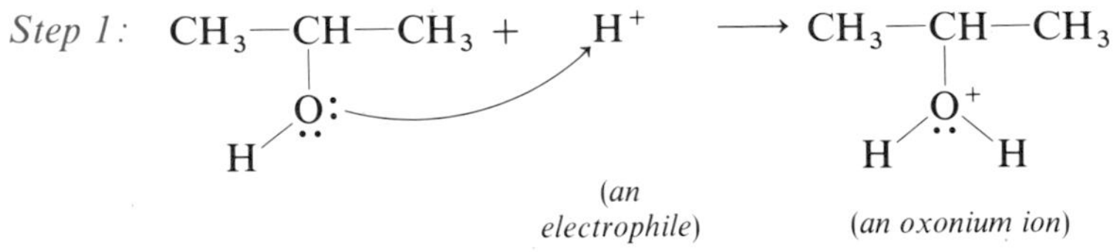

Step 1:

$$CH_3-CH-CH_3 + \quad H^+ \quad \longrightarrow \quad CH_3-CH-CH_3$$

(an electrophile)   (an oxonium ion)

In step 2, the C—O bond breaks to give a water molecule and a carbocation:

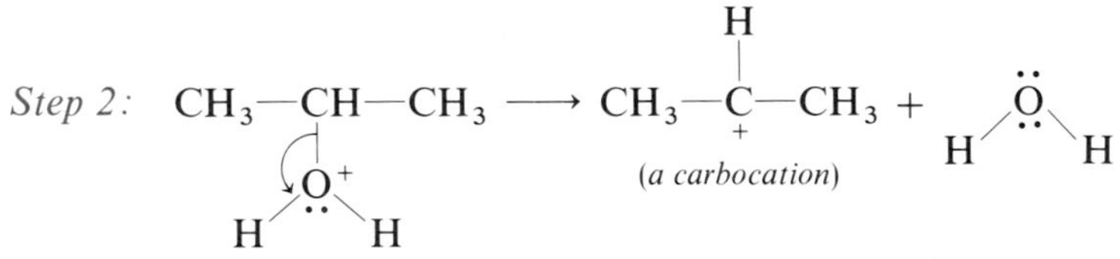

Step 2:

$$CH_3-CH-CH_3 \longrightarrow CH_3-\overset{H}{\underset{+}{C}}-CH_3 + \quad \overset{..}{\underset{..}{O}}$$

(a carbocation)

Finally, in step 3, a C—H bond breaks to form H^+ and the pair of electrons of the C—H sigma bond is used to form the pi bond of the alkene.

Step 3:

$$H-\overset{H}{\underset{H}{C}}-\overset{H}{\underset{+}{C}}-CH_3 \longrightarrow \overset{H}{\underset{H}{}}C=C\overset{CH_3}{\underset{H}{}} + H^+$$

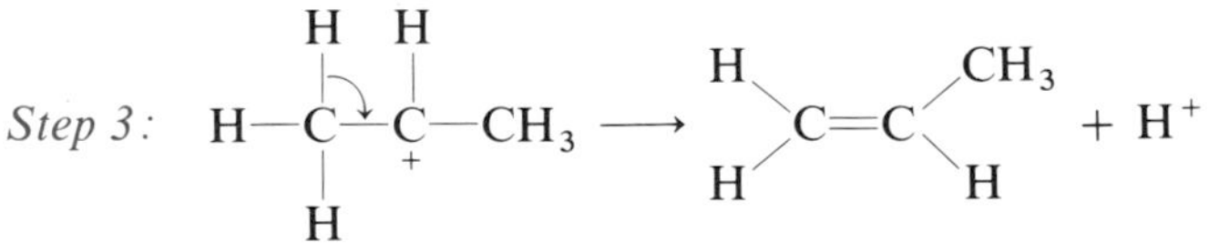

Since the reactive intermediate in this mechanism is a carbocation, the relative ease of dehydration of alcohols is the same as the ease of formation of carbocations. Tertiary alcohols dehydrate more readily than secondary alcohols because a tertiary carbocation is more readily formed than a secondary carbocation. For the same reason, a secondary alcohol undergoes acid-catalyzed dehydration more readily than a primary alcohol. Thus, the relative ease of dehydration of alcohols is:

$3°$ alcohols $> 2°$ alcohols $> 1°$ alcohols

C. Oxidation to Aldehydes, Ketones, and Carboxylic Acids

Primary alcohols are oxidized to aldehydes by oxidizing agents such as potassium dichromate ($K_2Cr_2O_7$) and potassium permanganate ($KMnO_4$). Oxidation of a primary alcohol to an aldehyde involves removal of two hydrogen atoms from the alcohol (one from oxygen and the other from carbon) and the formation of a carbon-oxygen double bond.

$$CH_3-CH_2-\overset{\displaystyle OH}{\underset{\displaystyle H}{C}}-H + Cr_2O_7^{2-} \xrightarrow[\text{water}]{H^+} CH_3-CH_2-\overset{\displaystyle O}{C}-H + Cr^{3+}$$

1-propanol   (orange)   propanal   (green)
(a primary alcohol)   (an aldehyde)

Aldehydes are very easily oxidized to carboxylic acids in a reaction that involves conversion of the aldehyde C—H bond to a C—OH bond.

$$CH_3-\overset{\displaystyle O}{C}-H + Cr_2O_7^{2-} \xrightarrow{H^+} CH_3-\overset{\displaystyle O}{C}-OH + Cr^{3+}$$

acetaldehyde   acetic acid
(an aldehyde)   (a carboxylic acid)

Under carefully controlled reaction conditions, it is possible to oxidize primary alcohols to aldehydes in good yield. However, unless special precautions are taken, primary alcohols are oxidized directly to carboxylic acids. In this oxidation, the aldehyde is an intermediate.

$$CH_3CH_2OH + Cr_2O_7^{2-} \xrightarrow[\text{oxidation}]{H^+} \left[CH_3-\overset{\displaystyle O}{C}-H \right] \xrightarrow[\text{oxidation}]{H^+} CH_3-\overset{\displaystyle O}{C}-OH + Cr^{3+}$$

ethanol   acetic acid

Oxidation of a secondary alcohol yields a ketone. This oxidation involves removal of hydrogen atoms from carbon and oxygen and formation of a carbon-oxygen double bond.

$$CH_3-\overset{\displaystyle OH}{CH}-CH_3 + Cr_2O_7^{2-} \xrightarrow{H^+} CH_3-\overset{\displaystyle O}{C}-CH_3 + Cr^{3+}$$

2-propanol   acetone
(a secondary   (a ketone)
alcohol)

For example, oxidation of menthol yields menthone:

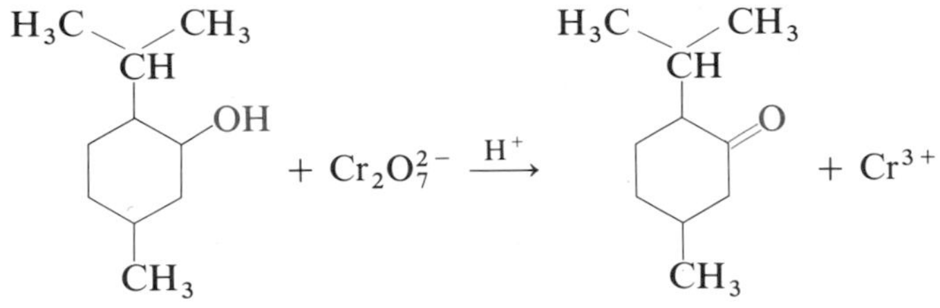

$$+ Cr_2O_7^{2-} \xrightarrow{H^+} \qquad + Cr^{3+}$$

menthol   menthone

Tertiary alcohols are not oxidized under these conditions because the carbon bearing the —OH has no hydrogen atom on it; it is already bonded to three other carbon atoms, and cannot form a carbon-oxygen double bond.

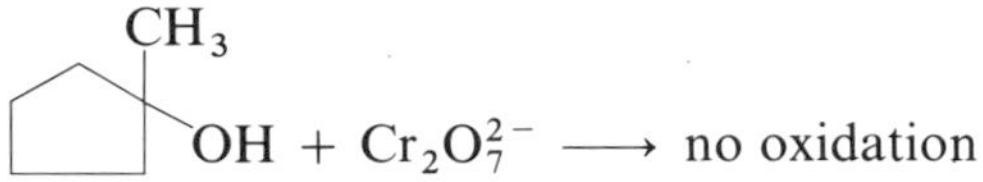

$$OH + Cr_2O_7^{2-} \longrightarrow \text{no oxidation}$$

D. How to Recognize an Oxidation-Reduction Reaction

Because of the importance of oxidation and reduction in the laboratory and the biological world, it is important that you be able to recognize reactions that involve oxidation, those that involve reduction, and those that involve neither oxidation nor reduction. As a specific example, pyruvic acid is converted by one enzyme-catalyzed reaction to acetaldehyde and carbon dioxide, and by another enzyme-catalyzed reaction to acetic acid and carbon dioxide.

$$CH_3-\overset{O}{\underset{}{C}}-\overset{O}{\underset{}{C}}-OH$$

pyruvic acid

$$\xrightarrow{\text{enzyme}} CH_3CHO + CO_2$$
acetaldehyde

$$\xrightarrow{\text{enzyme}} CH_3CO_2H + CO_2$$
acetic acid

One of these conversions involves oxidation and requires an oxidizing agent; the other involves neither oxidation nor reduction. You can tell which is which by looking at balanced half-reactions (Section 5.4C). To do this, follow these three steps:

1. Write a half-reaction showing the organic starting material and product(s).
2. Balance the number of atoms on each side of the equation. To balance the number of oxygen and hydrogen atoms, use H^+ and H_2O.
3. Balance the charge for the half-reaction by adding an appropriate number of electrons to one side or the other.

Oxidation is defined as the loss of electrons. If electrons appear on the right side of the balanced half-reaction, the organic starting material has given up electrons and has been oxidized. *Reduction* is defined as the gain of electrons. If electrons appear on the left side of the balanced half-reaction, the organic starting material has gained electrons and has been reduced. If no electrons appear in the balanced half-reaction, the conversion involves neither oxidation nor reduction.

Let us apply these steps to the conversion of ethanol to acetic acid. First write a half-reaction showing ethanol as the starting material and acetic acid as the product.

$$CH_3CH_2OH \longrightarrow CH_3CO_2H \tag{1}$$

To balance the number of atoms on each side of the half-reaction, add $4 H^+$ to the right side and $1 H_2O$ to the left side.

$$CH_3CH_2OH + H_2O \longrightarrow CH_3CO_2H + 4 H^+ \tag{2}$$

Finally, add $4 e^-$ to the right side of the half-reaction to balance charges:

$$CH_3CH_2OH + H_2O \longrightarrow CH_3CO_2H + 4 H^+ + 4 e^- \tag{3}$$

Equation (3) is a balanced half-reaction, and from it you can see that during the conversion of ethanol to acetic acid, ethanol loses four electrons. Thus, conversion of ethanol to acetic acid is a four-electron oxidation. To carry out this reaction, you would need an oxidizing agent.

Example 12.10

State whether the following reactions are oxidations, reductions, or neither.

$$\text{a. } CH_3-\overset{\overset{\displaystyle O}{\|}}{C}-H \longrightarrow CH_3-\overset{\overset{\displaystyle O}{\|}}{C}-OH$$

$$\text{b. } CH_3-\overset{\overset{\displaystyle O}{\|}}{C}-H \longrightarrow CH_3-CH_2-OH$$

$$\text{c. } CH_3-CH=CH-\overset{\overset{\displaystyle O}{\|}}{C}-OH \longrightarrow CH_3-\overset{\overset{\displaystyle OH}{|}}{CH}-CH_2-\overset{\overset{\displaystyle O}{\|}}{C}-OH$$

Solution

$$\text{a. } CH_3-\overset{\overset{\displaystyle O}{\|}}{C}-H + H_2O \longrightarrow CH_3-\overset{\overset{\displaystyle O}{\|}}{C}-OH + 2 H^+ + 2 e^-$$

This reaction is a two-electron oxidation.

$$\text{b. } CH_3 - \overset{\overset{\displaystyle O}{\|}}{C} - H + 2\,H^+ + 2\,e^- \longrightarrow CH_3 - CH_2 - OH$$

This reaction is a two-electron reduction.

$$\text{c. } CH_3 - CH = CH - \overset{\overset{\displaystyle O}{\|}}{C} - OH + H_2O \longrightarrow CH_3 - \overset{\overset{\displaystyle OH}{|}}{CH} - CH_2 - \overset{\overset{\displaystyle O}{\|}}{C} - OH$$

This reaction is neither oxidation nor reduction. It is an example of hydration of an alkene.

Problem 12.10

State whether the following are oxidations, reductions, or neither.

$$\text{a. } CH_3 - \overset{\overset{\displaystyle OH}{|}}{CH} - CH_2 - \overset{\overset{\displaystyle O}{\|}}{C} - OH \longrightarrow CH_3 - \overset{\overset{\displaystyle O}{\|}}{C} - CH_2 - \overset{\overset{\displaystyle O}{\|}}{C} - OH$$

$$\text{b. } CH_3 - \overset{\overset{\displaystyle OH}{|}}{CH} - CH_2 - \overset{\overset{\displaystyle O}{\|}}{C} - OH \longrightarrow CH_3 - CH = CH - \overset{\overset{\displaystyle O}{\|}}{C} - OH$$

$$\text{c. } CH_3 - \overset{\overset{\displaystyle O}{\|}}{C} - CH_2 - \overset{\overset{\displaystyle O}{\|}}{C} - OH \longrightarrow CH_3 - \overset{\overset{\displaystyle O}{\|}}{C} - CH_3 + CO_2$$

12.6 Reactions of Ethers

Ethers resemble hydrocarbons in their resistance to chemical reaction. They do not react with oxidizing agents such as potassium dichromate or potassium permanganate. They are not affected by most acids or strong bases at moderate temperatures. It is precisely this general resistance to chemical reaction on the one hand, and good solvent characteristics on the other, that make ethers excellent solvents.

12.7 Phenols

A. Structure

The characteristic structural feature of phenols is the presence of a hydroxyl group bonded to a benzene ring. The Lewis structure of phenol, the simplest member of this class of compounds, is shown in Figure 12.9.

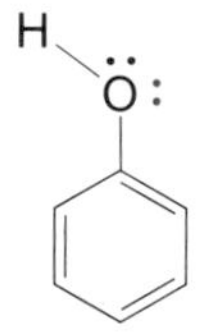

B. Nomenclature

Phenols are named either as derivatives of the parent hydrocarbon or by common names.

Figure 12.9 Lewis structure of phenol.

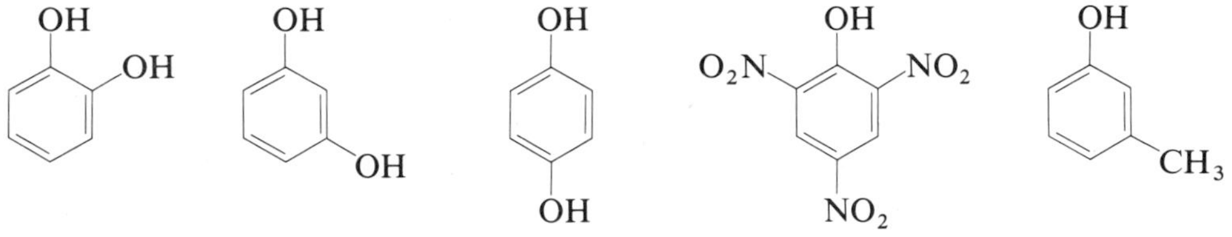

catechol   resorcinol   hydroquinone   2,4,6-trinitrophenol (picric acid)   *m*-cresol

C. Acidity of Phenols

Phenols and alcohols both contain —OH groups. However, phenols are grouped together as a separate class of compounds because they have different chemical properties from alcohols. One of the most important of these differences is that phenols are significantly more acidic than alcohols. Just how large this difference in acidity is can be seen by comparing the acid dissociation constants for phenol and ethanol.

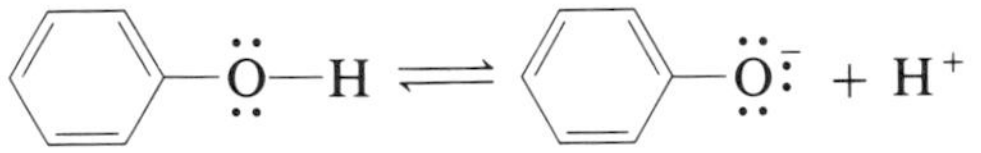

$$K_a = \frac{[C_6H_5O^-][H^+]}{[C_6H_5OH]} = 1 \times 10^{-10}$$

$$CH_3CH_2-\overset{..}{\underset{..}{O}}-H \rightleftharpoons CH_3CH_2-\overset{..}{\underset{..}{O}}\!\!:^- + H^+ \qquad K_a = \frac{[CH_3CH_2O^-][H^+]}{[CH_3CH_2OH]} = 10^{-16}$$

The acid dissociation constant for phenol is approximately one million, or 10^6, times larger than that of ethanol.

     Another way to compare the relative acid strengths of alcohols and phenols is to look at the hydrogen ion concentration and pH of aqueous solutions of ethanol, phenol, and hydrochloric acid. The hydrogen ion concentration of $0.1M$ ethanol in water is the same as that of pure water. Phenol is only slightly dissociated in water and is classified as a weak acid. A $0.1M$ solution of phenol has a pH of 5.4. Hydrochloric acid is a strong acid (Section 8.5C). The hydrogen ion concentration of $0.1M$ HCl is $0.1M$ and the pH of this solution is 1.0. This information is summarized in Table 12.2.

Table 12.2 Relative acidities of 0.1*M* solutions of ethanol, phenol, and hydrochloric acid.

Dissociation Equation	$[H^+]$	pH
$CH_3CH_2OH \rightleftharpoons CH_3CH_2O^- + H^+$	10^{-7}	7.0
$C_6H_5OH \rightleftharpoons C_6H_5O^- + H^+$	3.6×10^{-6}	5.4
$HCl \longrightarrow Cl^- + H^+$	0.1	1.0

Phenols react with strong bases such as sodium hydroxide to form salts. However, they do not react with weaker bases such as sodium bicarbonate.

weak acid strong base

weak acid weak base

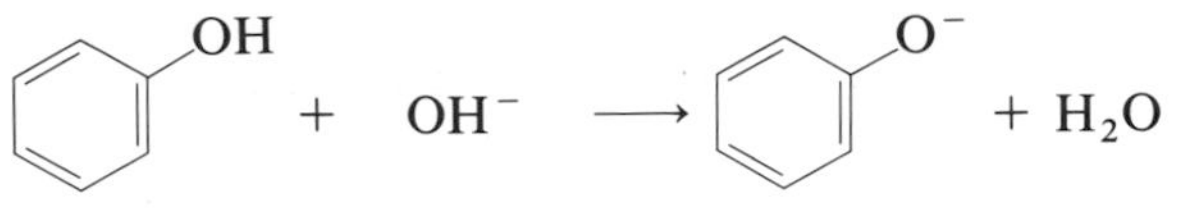

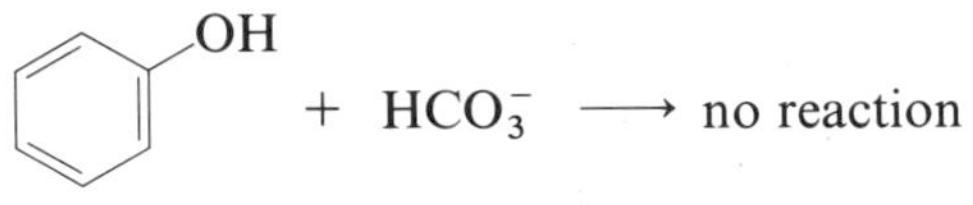

The fact that phenols are weak acids whereas alcohols are neutral provides a very convenient way to separate phenols from water-insoluble alcohols. Suppose we want to separate phenol from cyclohexanol, each of which is only slightly soluble in water.

cyclohexanol

phenol

sodium phenoxide (a water-soluble salt)

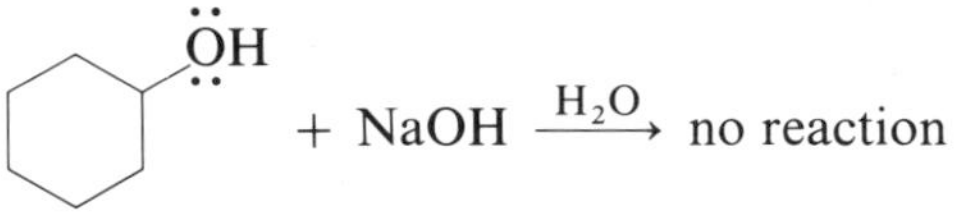

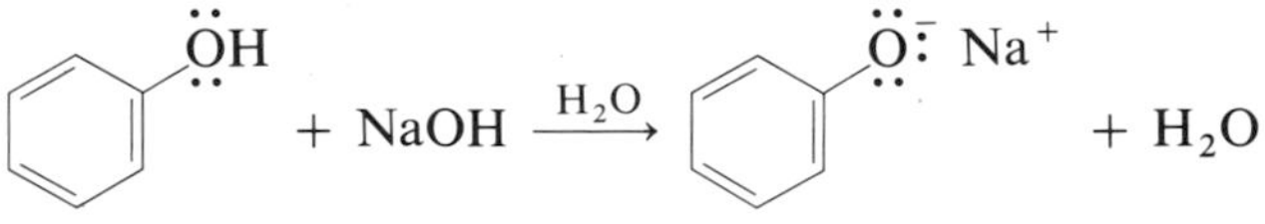

Phenol reacts with aqueous sodium hydroxide to form the water-soluble salt sodium phenoxide. Cyclohexanol does not react with NaOH and remains as a water-insoluble layer. The water layer containing sodium phenoxide can be separated from the water-insoluble cyclohexanol. Addition of a strong acid (for example, aqueous hydrochloric acid) converts sodium phenoxide to phenol.

sodium phenoxide (salt) — hydrochloric acid (strong acid) — phenol (weaker acid)

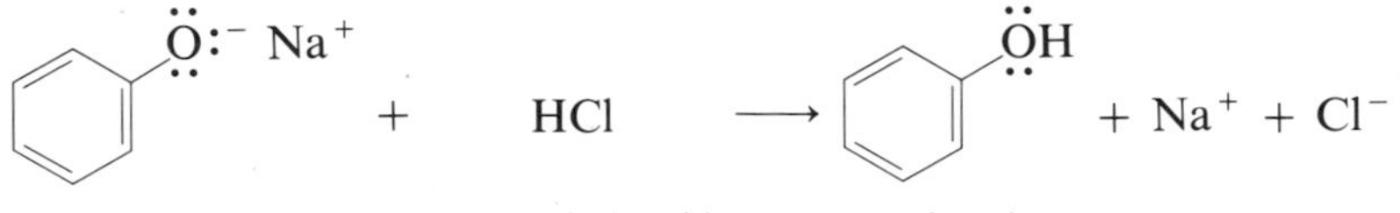

D. Some Naturally Occurring Phenols

Phenols are widely distributed in nature. Phenol itself and the isomeric cresols (*ortho*-, *meta*-, and *para*-cresol) are found in coal tar and petroleum. Thymol and vanillin are important constituents of thyme and vanilla beans.

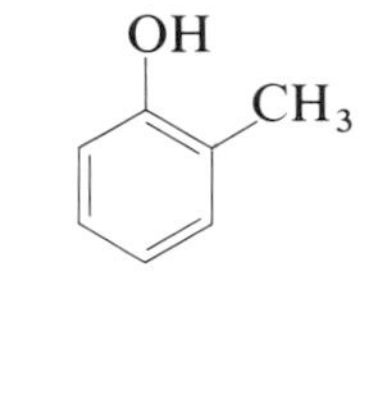

OH

o-cresol

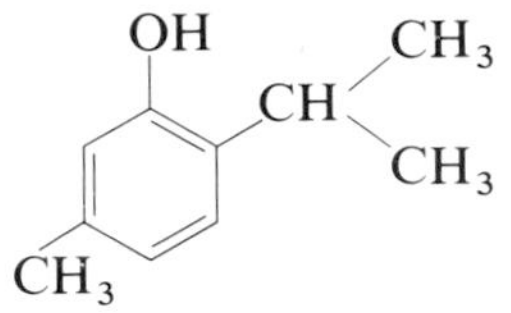

OH CH₃

thymol

OH OCH₃

CHO

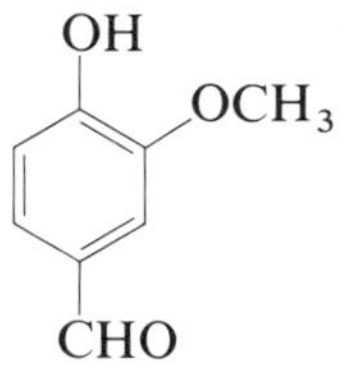

vanillin

The amino acid tyrosine contains a phenolic hydroxyl group as one of its three functional groups. This compound is known as an **amino acid** because it contains an **amino group** ($-NH_2$) and a carboxyl group ($-CO_2H$). Tyrosine is one of the 20 amino acids that are essential building blocks of proteins (Chapter 20).

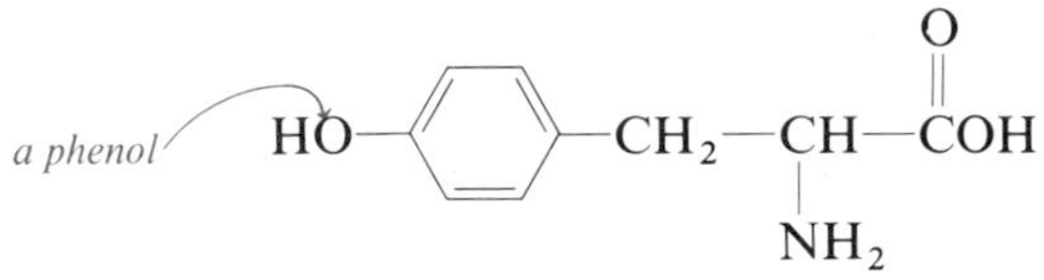

a phenol

$$HO-\langle\ \rangle-CH_2-CH-\overset{\overset{\displaystyle O}{\|}}{C}OH$$

$$NH_2$$

tyrosine

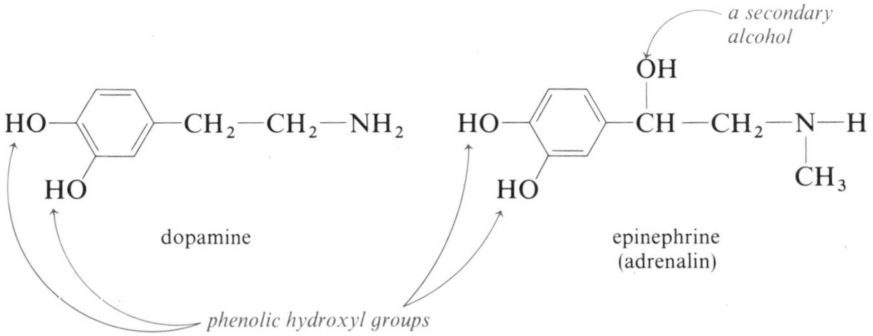

a secondary alcohol

$$HO-\langle\ \rangle-CH_2-CH_2-NH_2$$

HO

dopamine

$$HO-\langle\ \rangle-\overset{\overset{\displaystyle OH}{|}}{CH}-CH_2-N-H$$

HO

CH₃

epinephrine
(adrenalin)

phenolic hydroxyl groups

Tyrosine is used by the body as a raw material for the synthesis of dopamine and epinephrine. Dopamine is a neurotransmitter in the central nervous system; a deficiency in dopamine is one of the underlying biochemical defects in Parkinson's Disease. Dopamine contains two phenolic —OH groups. Epinephrine, more commonly known as adrenalin, is a hormone secreted by the adrenal glands. Epinephrine contains two phenolic —OH groups and one secondary alcohol group.

E. Phenols as Antiseptics

Phenol, or carbolic acid as it was once called, is a low-melting solid which is only slightly soluble in water. In sufficiently high concentrations, it is corrosive to all kinds of cells. In dilute solutions, it has some antiseptic properties and was first used in the 19th century by Joseph Lister for antiseptic surgery. It has been replaced by other antiseptics that are more powerful and have fewer undesirable side effects. Among these is *n*-hexylresorcinol (in Sucrets brand lozenges and mouthwashes), a compound used in household preparations as a mild antiseptic and disinfectant.

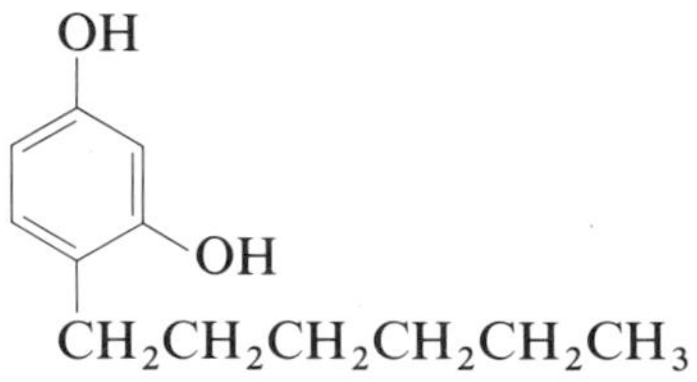

n-hexylresorcinol

12.8 Thiols, Sulfides, and Disulfides

A. Structure

Sulfur analogs of alcohols are called **thioalcohols,** or more simply, **thiols.** Shown in Figure 12.10 are Lewis structures of methanethiol (the simplest thiol) and methanol, along with the three-dimensional shape of methanethiol.

Oxygen and sulfur are both in Column VI of the periodic table and each has six electrons in its valence shell. For oxygen, these valence electrons are in the second principal energy level; for sulfur, they are in the third principal energy level. Lewis structures of methanethiol and methanol show both sulfur and oxygen forming two covalent bonds; each atom also has two unshared pairs of electrons. The C—O—H bond angle in methanol is 108.9°, very close to the predicted tetrahedral angle of 109.5°. The C—S—H bond angle in methanethiol is 100.3°.

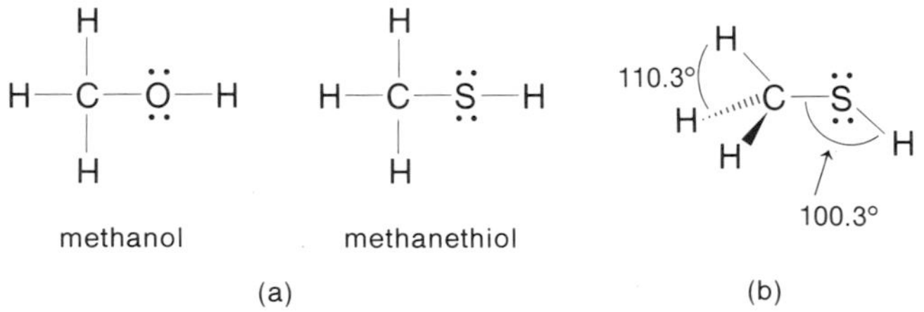

Figure 12.10 (a) Lewis structures of methanol and methanethiol; (b) the three-dimensional shape of methanethiol.

B. Nomenclature

In naming thiols, the IUPAC system selects the longest chain containing the —SH group as the parent compound. The thiol is named by adding the suffix *-thiol* to the name of the parent compound. In the common system of nomenclature, thiols are known as **mercaptans.** Common names for simple thiols are

Table 12.3 Names and structural formulas for several low-molecular-weight thiols.

CH_3-CH_2-SH	$CH_3-CH_2-CH_2-CH_2-SH$	$CH_3-CH-CH_2-SH$
ethanethiol	1-butanethiol	$\quad\quad\mid$
(ethyl mercaptan)	(*n*-butyl mercaptan)	$\quad\quad CH_3$
		2-methyl-1-propanethiol
		(isobutyl mercaptan)
	$CH_3-CH_2-CH-CH_3$	$HS-CH_2-CH_2-SH$
	$\quad\quad\quad\mid$	1,2-ethanedithiol
	$\quad\quad\quad SH$	(ethylene mercaptan)
	2-butanethiol	
	(*sec*-butyl mercaptan)	

derived by naming the alkyl group attached to —SH and then adding the word *mercaptan*. Table 12.3 lists IUPAC names and, in parentheses, common names for several low-molecular-weight thiols. For molecules containing other functional groups, the presence of an —SH group is indicated by the prefix *mercapto-*.

$$HS-CH_2-CH_2-OH$$

2-mercaptoethanol
(β-mercaptoethanol)

Sulfur analogs of ethers are known as **thioethers** and are most commonly named by using the word *sulfide* to show the presence of the —S— group. Following are common names of two thioethers.

$$CH_3-S-CH_3 \quad\quad CH_3-CH_2-S-CH-CH_3$$
$$\quad\quad\quad\quad\quad\quad\quad\quad\quad\quad\quad\quad\quad\quad\quad\mid$$
$$\quad\quad\quad\quad\quad\quad\quad\quad\quad\quad\quad\quad\quad\quad CH_3$$

dimethyl sulfide   ethyl isopropyl sulfide

The characteristic structural feature of a disulfide is the presence of an —S—S— group. Disulfides are named by listing the names of the alkyl or aryl groups attached to sulfur and adding the word *disulfide*.

$$CH_3-S-S-CH_3 \quad\quad CH_3-CH_2-S-S-CH_2-CH_3$$

dimethyl disulfide   diethyl disulfide

C. Physical Properties

The physical properties of thiols are quite different from those of alcohols, primarily because of the difference in electronegativity between sulfur and oxygen. Because the electronegativities of sulfur and hydrogen are identical, the S—H bond is nonpolar covalent. By comparison, the electronegativity

Table 12.4 Boiling
points of several thiols
and alcohols with the
same number of car-
bon atoms.

Thiol	bp (°C)	Alcohol	bp (°C)
methanethiol	6.2	methanol	65
ethanethiol	35	ethanol	78
1-butanethiol	98.5	1-butanol	117

difference between oxygen and hydrogen is 0.9 unit (3.0 − 2.1), so the O—H
bond is polar covalent.

$$CH_3-S \overset{\longarrow}{} H \qquad CH_3 \overset{\delta-}{-O} \overset{\delta+}{-H}$$

a nonpolar covalent   *a polar covalent*
bond   *bond*

Thiols show little association by hydrogen bonding. Consequently, thiols have
lower boiling points and are less soluble in water and other polar solvents than
alcohols of comparable molecular weights. Table 12.4 gives structural formulas
and boiling points for several low-molecular-weight thiols. Shown for comparison are boiling points of alcohols with the same number of carbon atoms.
Recall that in Section 12.3 we illustrated the importance of hydrogen bonding
by comparing the boiling points of ethanol (bp 78°C) and its structural isomer
dimethyl ether (bp −24°C). The difference in boiling point between these
isomers is due to the presence of a polar —OH group in ethanol and association
of alcohol molecules in the pure state by hydrogen bonding. By comparison,
the boiling point of ethanethiol is 35°C and that of its structural isomer dimethyl
sulfide is 37°.

$$CH_3-CH_2-SH \qquad CH_3-S-CH_3$$

ethanethiol   dimethyl sulfide
bp 35°C   bp 37°C

The fact that the boiling points of these two compounds are almost identical
indicates that there is little or no association by hydrogen bonding of thiols in
the pure liquid.

D. Acidity of Thiols

Thiols are stronger acids than alcohols. Compare, for example, the acid
dissociation constants of ethanol and ethanethiol.

$$CH_3-CH_2-OH \rightleftharpoons CH_3-CH_2-O^- + H^+ \qquad pK_a = 16$$
$$CH_3-CH_2-SH \rightleftharpoons CH_3-CH_2-S^- + H^+ \qquad pK_a = 8.5$$

Thiols are sufficiently strong acids that, when dissolved in aqueous sodium hydroxide, they are converted completely to alkyl-sulfide salts.

$$CH_3-CH_2-SH + Na^+OH^- \longrightarrow CH_3-CH_2-S^-Na^+ + H-OH$$

(stronger acid)   sodium   (weaker
                                                ethyl sulfide   acid)

To name metal salts of thiols, give the name of the metal cation first, followed by the name of the alkyl group to which is attached the suffix *-sulfide*. For example, the sodium salt derived from ethanethiol is named sodium ethylsulfide.

Like hydrogen sulfide and hydrosulfide salts, thiols form water-insoluble salts with most heavy metals. In fact, the name *mercaptan* is derived from the Latin *mercurium captans*, which means "mercury-capturing."

$$Hg^{2+} + 2\,RSH \qquad Hg(SR)_2 + 2\,H^+$$

Reaction with Pb(II) is often used as a qualitative test for the presence of a sulfhydryl group. Treatment of a mercaptan with a saturated solution of lead(II) acetate, $Pb(OAc)_2$, usually gives a yellow solid that constitutes a positive test for the presence of a thiol.

$$R-SH + Pb^{2+} \longrightarrow Pb(S-R)_2 + 2\,H^+$$

a yellow
precipitate

E. Oxidation of Thiols

Many of the chemical properties of thiols are related to the fact that a divalent sulfur atom is a reducing agent; it is easily oxidized to several higher oxidation states. Thiols are oxidized by mild oxidizing agents such as iodine, I_2, to disulfides.

$$2\,CH_3-S-H + I_2 \longrightarrow CH_3-S-S-CH_3 + 2\,H^+ + 2\,I^-$$

(a thiol)   (a disulfide)

They are also oxidized to disulfides by molecular oxygen. In fact, thiols are so susceptible to oxidation by molecular oxygen that they must be protected from contact with air during storage.

$$2\,CH_3-S-H + \tfrac{1}{2}O_2 \longrightarrow CH_3-S-S-CH_3 + H_2O$$

methanethiol   dimethyl disulfide

The **disulfide bond** is an important structural feature of many biomolecules, including proteins (Chapter 20). Most proteins contain intermolecular or intramolecular disulfide bonds that help hold the protein chain or chains in a unique three-dimensional shape. Following is a representation of two protein chains, P_1 and P_2, joined by a disulfide bond.

$$P_1-CH_2-S-S-CH_2-P_2$$

Often, in studying the structural and biological properties of a protein it is necessary to cleave any disulfide bonds. One common way to do this is to react the protein with hydrogen peroxide in the presence of aqueous acetic acid (CH_3CO_2H). Under these conditions, disulfides are oxidized to two sulfonic acids.

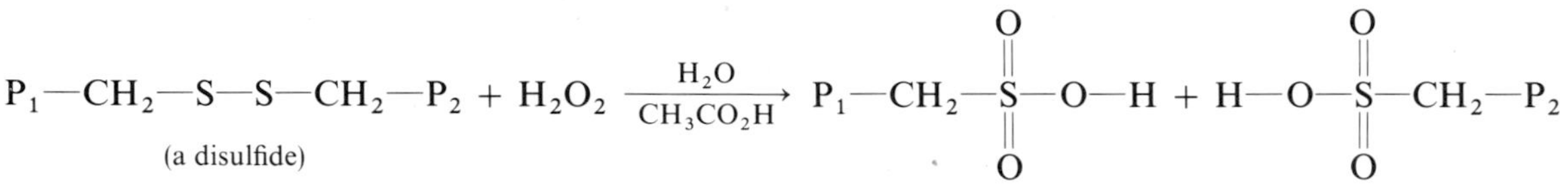

$$P_1-CH_2-S-S-CH_2-P_2 + H_2O_2 \xrightarrow[CH_3CO_2H]{H_2O} P_1-CH_2-\overset{\overset{\displaystyle O}{\|}}{\underset{\underset{\displaystyle O}{\|}}{S}}-O-H + H-O-\overset{\overset{\displaystyle O}{\|}}{\underset{\underset{\displaystyle O}{\|}}{S}}-CH_2-P_2$$

(a disulfide)   a sulfonic acid   a sulfonic acid

The characteristic structural feature of a sulfonic acid is the presence of an —SO_3H group. Thiols are also oxidized to sulfonic acids by hydrogen peroxide in aqueous acetic acid. A second way to cleave a disulfide bond is to reduce it to two thiols. A common reducing agent is hydrogen in the presence of a heavy metal catalyst.

$$CH_3-CH_2-S-S-CH_2-CH_3 + H_2 \xrightarrow{Pt} 2\ CH_3-CH_2-S-H$$

a disulfide   two thiols

Key Terms and Concepts

absolute ethanol (12.4)

acidity of phenols (12.7C)

acidity of thiols (12.8D)

alcohol (12.1)

dehydration of alcohols (12.5B)

diol (12.2A)

disulfide (12.8B)

ether (12.2B)

glycol (12.2A)

hydrogen bonding in alcohols (12.3A)

hydroxyl group (12.1)

mercaptan (12.8B)

oxidation of alcohols (12.5C)

oxidation of thiols (12.8E)

phenols (12.7A)

primary alcohol (12.2A)

secondary alcohol (12.2A)

sulfide (12.8B)

tertiary alcohol (12.2A)

thiol (12.8A)

Key Reactions

1. Conversion of alcohols to alkyl halides (Section 12.5A):

$$CH_3-CH_2-\underset{\underset{OH}{|}}{CH}-CH_3 + HCl \longrightarrow CH_3-CH_2-\underset{\underset{Cl}{|}}{CH}-CH_3 + H_2O$$

2. Acid-catalyzed dehydration of alcohols (Section 12.5B):

$$CH_3-CH_2-\underset{\underset{OH}{|}}{CH}-CH_3 \xrightarrow{H_2SO_4}$$

$$CH_3-CH=CH-CH_3 + CH_3-CH_2-CH=CH_2 + H_2O$$

major product   *minor product*

3. Oxidation of alcohols (Section 12.5C):
   a. Primary alcohols are oxidized to aldehydes or carboxylic acids, depending on the experimental conditions:

$$CH_3-CH_2-OH + Cr_2O_7^{2-} \xrightarrow{H^+} CH_3-\overset{\overset{O}{||}}{C}-H + Cr^{3+}$$

$$CH_3-CH_2-OH + Cr_2O_7^{2-} \xrightarrow{H^+} CH_3-\overset{\overset{O}{||}}{C}-OH + Cr^{3+}$$

   b. Secondary alcohols are oxidized to ketones:

$$CH_3-\underset{\underset{OH}{|}}{CH}-CH_3 + Cr_2O_7^{2-} \xrightarrow{H^+} CH_3-\overset{\overset{O}{||}}{C}-CH_3 + Cr^{3+}$$

   c. Tertiary alcohols have no hydrogen atom attached to the carbon bearing the —OH groups, and therefore are resistant to oxidation.

4. Reaction of phenols with strong bases to form salts (Section 12.7C):

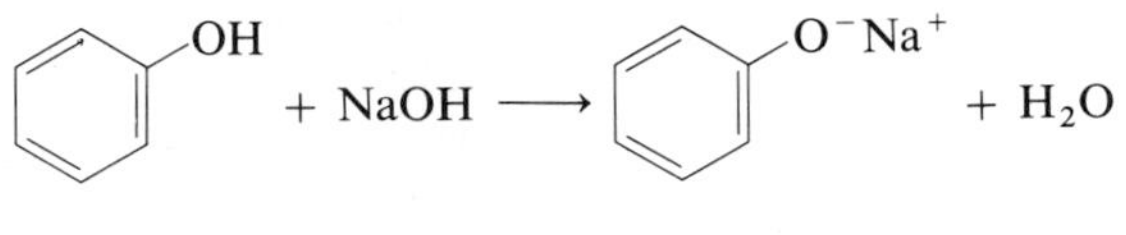

$$+ \text{ NaOH} \longrightarrow \qquad + H_2O$$

5. Reaction of thiols with heavy-metal cations to form salts (Section 12.8D):

$$2\ CH_3CH_2SH + PbCl_2 \longrightarrow (CH_3CH_2S)_2Pb + 2\ HCl$$

6. Oxidation of thiols to disulfides (Section 12.8E):

$$2\ CH_3CH_2CH_2SH + H_2O_2 \longrightarrow$$

$$CH_3CH_2CH_2-S-S-CH_2CH_2CH_3 + 2\ H_2O$$

7. Reduction of disulfides to thiols (Section 12.8E):

$$CH_3CH_2CH_2-S-S-CH_2CH_2CH_3 + H_2 \xrightarrow{Pt} 2\ CH_3CH_2CH_2-SH$$

Problems

Structure and Nomenclature of Alcohols and Ethers (Sections 12.1 and 12.2)

12.11 Name each of the following compounds.

a. $CH_3{-}CH_2{-}CH_2{-}CH_2OH$

b. $HO{-}CH_2{-}CH_2{-}CH_2{-}CH_2{-}OH$

c. 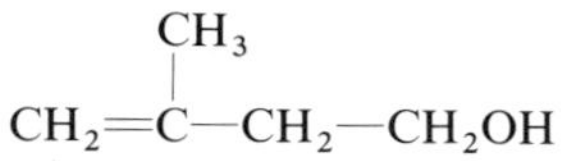

d. 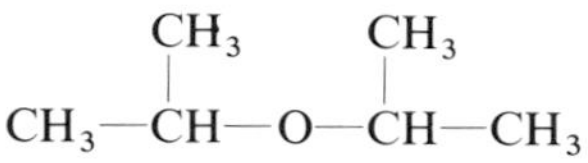

e. $CH_3{-}O{-}CH_2{-}CH_2OH$

f. 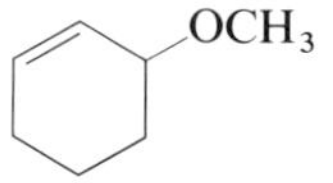

g. 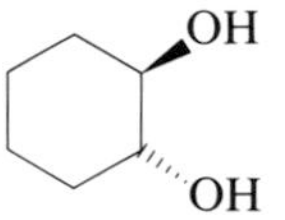

h. 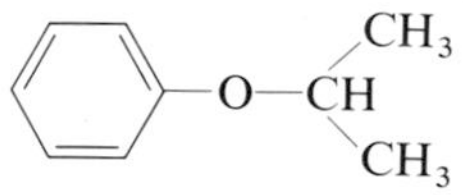

i. 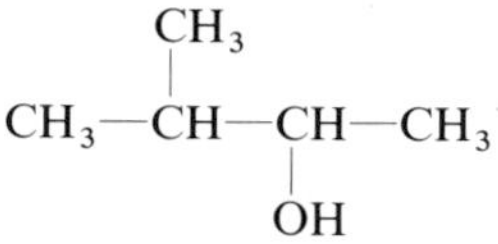

j. 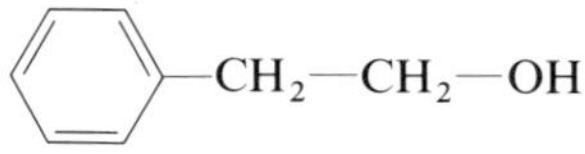

k. $CH_3(CH_2)_8CH_2OH$

l. 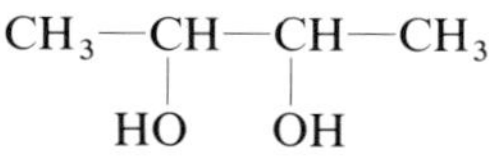

m. 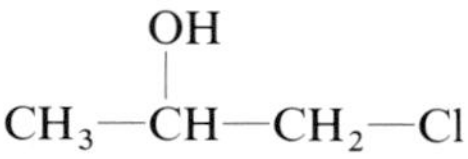

n. 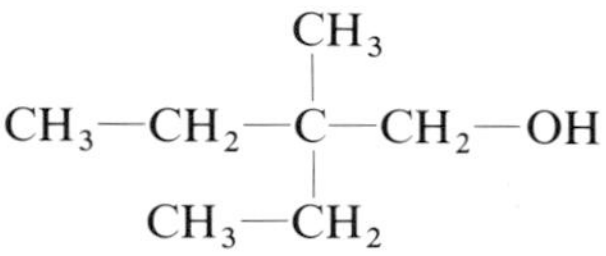

12.12 Write structural formulas for each of the following compounds:

a. isopropyl methyl ether
b. propylene glycol
c. 2-methyl-2-propylpropane-1,3-diol
d. 1-chloro-2-hexanol
e. 5-methyl-2-hexanol
f. 2-isopropyl-5-methylcyclohexanol
g. 2,2-dimethyl-1-propanol
h. *tert*-butyl alcohol
i. methyl cyclopropyl ether
j. ethylene glycol
k. methyl phenyl ether
l. *trans*-2-ethylcyclohexanol

12.13 Name and draw structural formulas for the eight isomeric alcohols of molecular formula $C_5H_{12}O$. Classify each as primary, secondary, or tertiary.

12.14 Name and draw structural formulas for the six isomeric ethers of molecular formula $C_5H_{12}O$.

Physical Properties of Alcohols and Ethers (Section 12.3)

12.15 In the following compounds: (i) circle each hydrogen atom that is capable of hydrogen-bonding, and (ii) put a square around each atom capable of hydrogen bonding to a partially positive hydrogen atom.

a.
```
        H
        |
    H—C—O—H
        |
    H—C—O—H
        |
    H—C—O—H
        |
        H
```

b.
```
        H  H
        |  |
    H—O—C—C—S—H
        |  |
        H  H
```

c.
(benzene ring with O—H and below)
```
        O—H
    ⬡
        H
        |
    O—C—H
        |
        H
```

d.
```
        H      H  H
        |      |  |
    H—C—O—C—C—O—H
        |      |  |
        H      H  H
```

12.16 Arrange the following sets of compounds in order of decreasing boiling points. Explain your reasoning.

a. $CH_3CH_2CH_3$ $CH_3CH_2CH_2CH_2CH_2CH_2CH_3$ $CH_3CH_2CH_2CH_2CH_3$

b. N_2H_4 H_2O_2 CH_3CH_3

c. CH_3CO_2H CH_3CH_2OH CH_3OCH_3

d.
```
    CH3CHCH3        CH3—CH—CH2      CH2—CH—CH2
       |               |   |          |   |   |
       OH              OH  OH         OH  OH  OH
```

12.17 Arrange the following compounds in order of decreasing solubility in water. Explain your reasoning.
a. ethanol, butane, diethyl ether   b. 1-hexanol, 1,2-hexanediol, hexane

12.18 Diethyl ether has a much lower boiling point than 1-butanol, yet each of these compounds has about the same solubility in water (8 grams per 100 mL of water). How do you account for these characteristics?

$CH_3CH_2CH_2CH_2OH$ $CH_3CH_2OCH_2CH_3$

1-butanol   diethyl ether
bp 117°C   bp 35°C

12.19 How do you account for the fact that ethanol (molecular weight 46, bp 78°C) has a boiling point over 43° higher than that of ethanethiol (molecular weight 62, bp 35°C)?

12.20 Both propanoic acid and methyl acetate (see margin) have the molecular formula $C_3H_6O_2$. Both are liquids at room temperature. The boiling point of one of these liquids is 57°C, the boiling point of the other is 141°C.

a. Which of these compounds has the boiling point of 141°C? The boiling point of 57°C? Explain your reasoning.

b. Which of these two compounds is more soluble in water? Explain your reasoning.

12.21 Compounds that contain N—H bonds show association by hydrogen bonding. Would you expect this association to be stronger or weaker than that in compounds containing O—H bonds? (Hint: remember the table of relative electronegativities, Figure 4.1).

```
           O
           ||
    CH3—CH2—C—OH
```
propanoic acid

```
        O
        ||
    CH3—C—O—CH3
```
methyl acetate

12.22 What is meant by the following names?
a. wood alcohol   b. grain alcohol   c. absolute ethanol

12.23 Write structural formulas for the alkenes of the given molecular formula that undergo acid-catalyzed hydration to give the alcohol shown as the major product.

a. 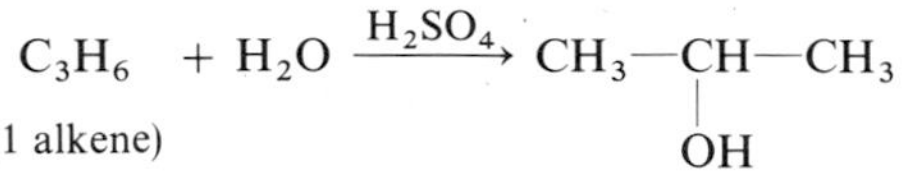 C_3H_6 + H_2O $\xrightarrow{H_2SO_4}$ $CH_3-CH-CH_3$
(1 alkene) |
                                            OH

b. 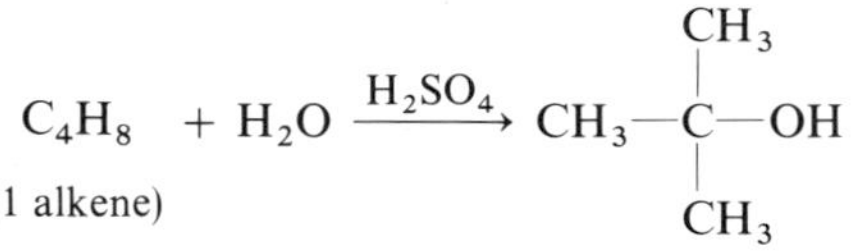 C_4H_8 + H_2O $\xrightarrow{H_2SO_4}$ $CH_3-\overset{\displaystyle CH_3}{\underset{\displaystyle CH_3}{C}}-OH$
(1 alkene)

c. 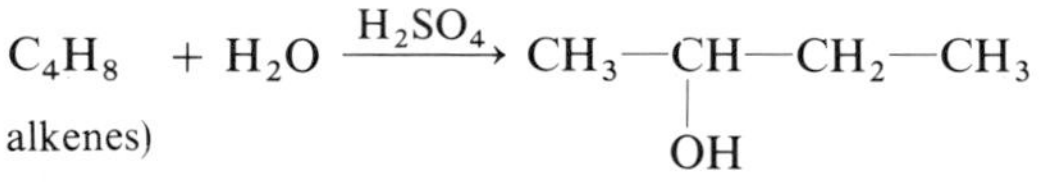 C_4H_8 + H_2O $\xrightarrow{H_2SO_4}$ $CH_3-CH-CH_2-CH_3$
(3 alkenes) |
                                            OH

d. 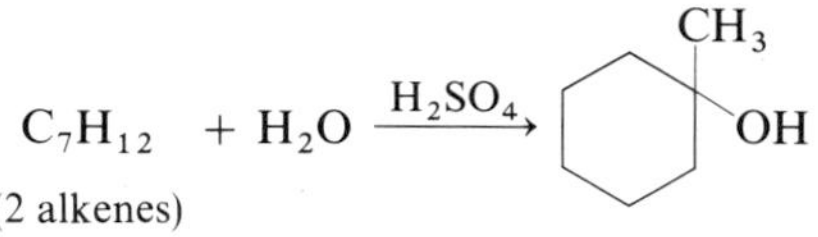 C_7H_{12} + H_2O $\xrightarrow{H_2SO_4}$
(2 alkenes)

12.24 Propose a mechanism for the following acid-catalyzed hydration.

$CH_3-\overset{\displaystyle CH_3}{C}=CH-CH_3 + H_2O \xrightarrow{H_2SO_4} CH_3-\overset{\displaystyle CH_3}{\underset{\displaystyle OH}{C}}-CH_2-CH_3$

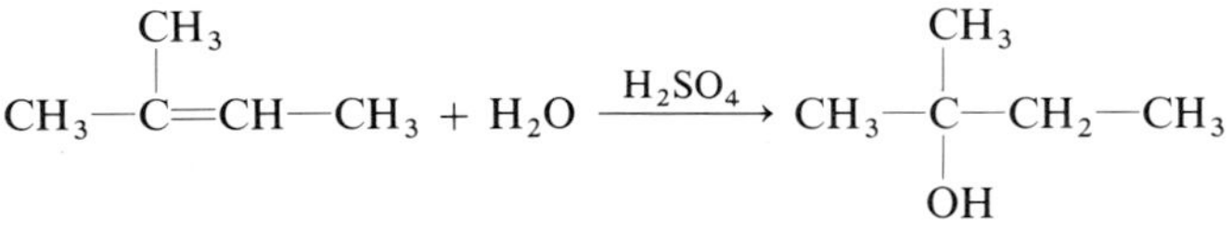

12.25 Hydration of fumaric acid, catalyzed by aqueous sulfuric acid, produces malic acid. Propose a mechanism for this reaction.

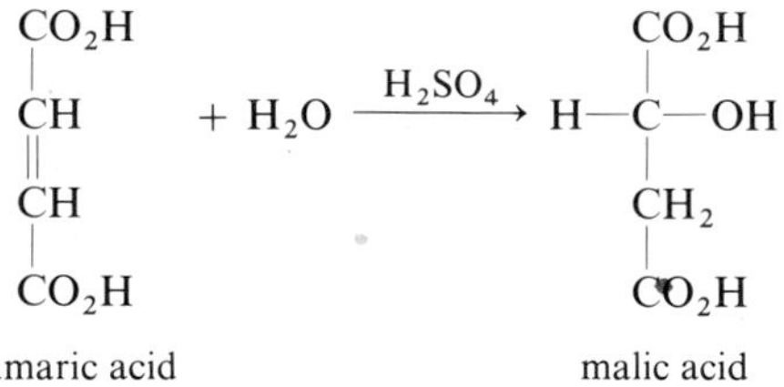

CO_2H ... fumaric acid + $H_2O \xrightarrow{H_2SO_4}$ malic acid

12.26 Draw structural formulas for the alkene or alkenes formed from acid-catalyzed dehydration of the following alcohols. Where two or more alkenes are formed, predict which is the major product and which is the minor product.
a. $CH_3-CH_2-CH_2-OH$ b. 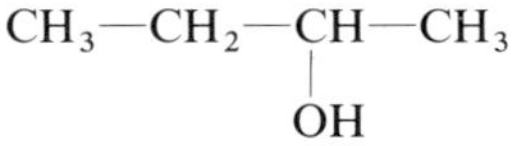 $CH_3-CH_2-CH-CH_3$ (OH)
c. 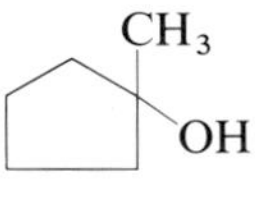 (cyclopentane with CH_3 and OH)   d. 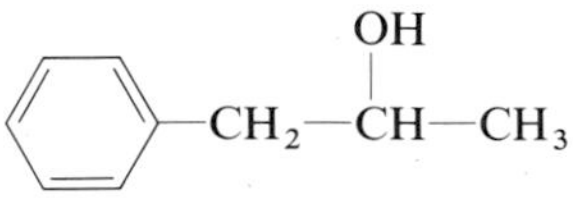 (benzene)$-CH_2-CH-CH_3$ (OH)

e. $HO-CH_2-CH_2-CH_2-CH_2-OH$

f. 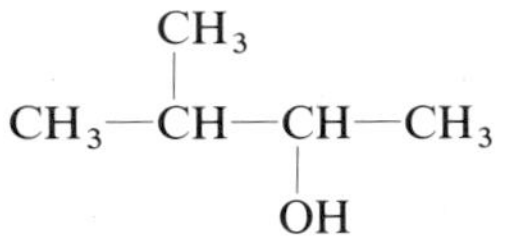 $CH_3-\underset{\underset{OH}{|}}{\overset{\overset{CH_3}{|}}{CH}}-CH-CH_3$

g. 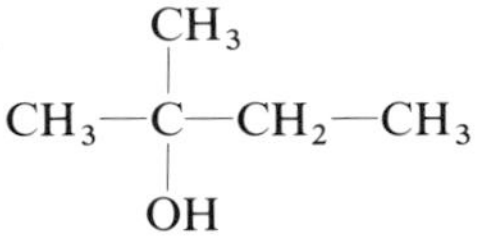 $CH_3-\underset{\underset{OH}{|}}{\overset{\overset{CH_3}{|}}{C}}-CH_2-CH_3$

12.27 Propose a mechanism for the acid-catalyzed dehydration of cyclohexanol to give cyclohexene and water.

12.28 Predict the relative ease with which the following alcohols undergo acid-catalyzed dehydration. Draw a structural formula for the major product of each dehydration.

a. 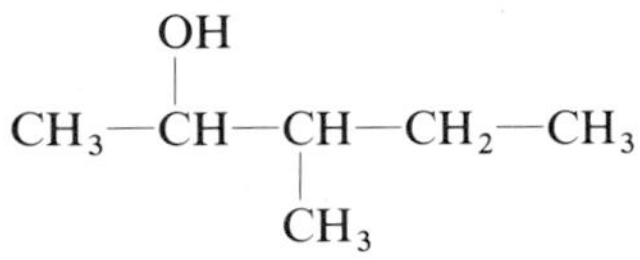 $CH_3-\underset{\underset{CH_3}{|}}{\overset{\overset{OH}{|}}{CH}}-CH-CH_2-CH_3$

b. 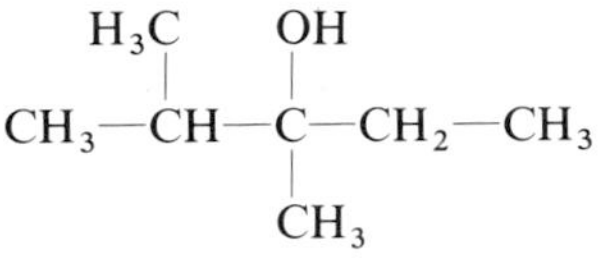 $CH_3-\underset{H_3C}{\overset{}{CH}}-\underset{\underset{CH_3}{|}}{\overset{\overset{OH}{|}}{C}}-CH_2-CH_3$

c. ⬠$-CH_2-CH_2OH$

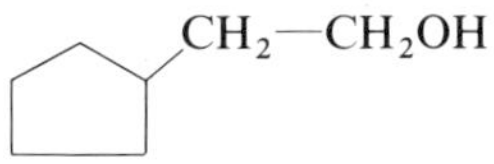

12.29 One of the reactions in the metabolism of glucose is the isomerization of citric acid to isocitric acid. The isomerization is catalyzed by the enzyme aconitase.

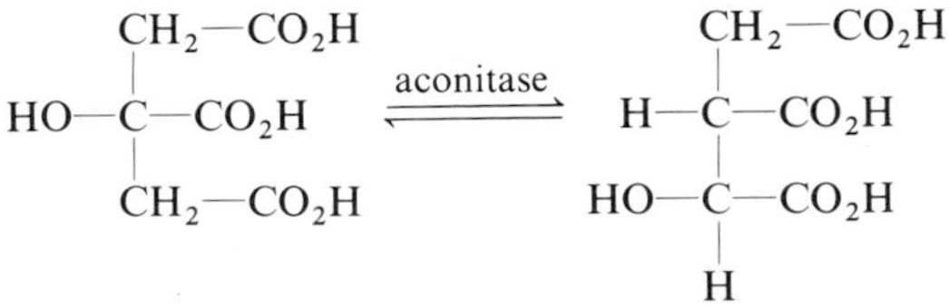

$$HO-\underset{\underset{CH_2-CO_2H}{|}}{\overset{\overset{CH_2-CO_2H}{|}}{C}}-CO_2H \underset{\text{aconitase}}{\rightleftharpoons} \underset{\underset{HO-C-CO_2H}{|}\atop\underset{H}{|}}{\overset{\overset{CH_2-CO_2H}{|}}{H-C-CO_2H}}$$

citric acid   isocitric acid

Propose a reasonable mechanism to account for this isomerization. (*Hint:* Within its structure, aconitase has groups that can function as acids.)

12.30 Write structural formulas for the major organic product(s) of the following oxidations.

a. 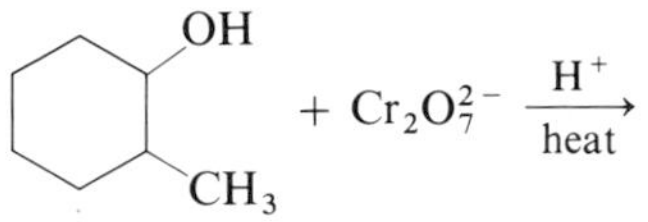 $\underset{CH_3}{\overset{OH}{⬡}} + Cr_2O_7^{2-} \xrightarrow[\text{heat}]{H^+}$

b. $HO-CH_2-CH_2-CH_2-CH_2-OH + Cr_2O_7^{2-} \xrightarrow[\text{heat}]{H^+}$

c. $CH_3-CH_2-OH + O_2 \text{ (excess)} \xrightarrow{\text{(combustion)}}$

d. 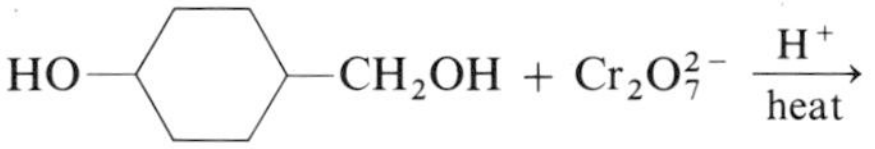 $HO-⬡-CH_2OH + Cr_2O_7^{2-} \xrightarrow[\text{heat}]{H^+}$

e. 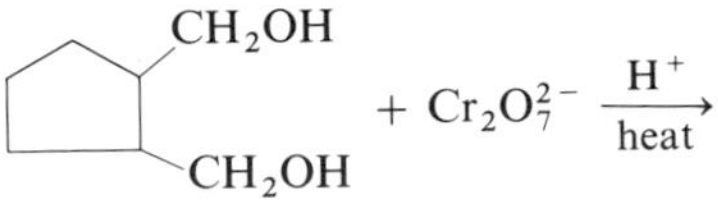 $\underset{CH_2OH}{\overset{CH_2OH}{⬠}} + Cr_2O_7^{2-} \xrightarrow[\text{heat}]{H^+}$

12.31 Look again at the two reactions of pyruvic acid shown in Section 12.5D. By completing a half-reaction, show that:

a. Conversion of pyruvic acid to acetic acid and carbon dioxide is a two-electron oxidation.

b. Conversion of pyruvic acid to acetaldehyde and carbon dioxide is neither oxidation nor reduction.

12.32 The following reactions are important in the metabolism of either fats or carbohydrates. State which are oxidations, which are reductions, and which are neither oxidation nor reduction.

a. 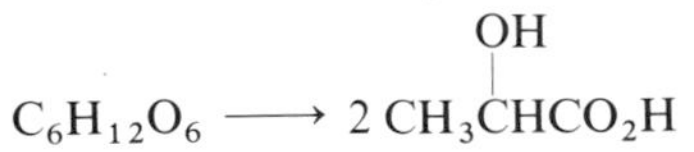 $C_6H_{12}O_6 \longrightarrow$ 2 $CH_3\overset{\overset{\displaystyle OH}{|}}{C}HCO_2H$

glucose   lactic acid

b. $CH_3(CH_2)_{12}CH_2CH_2CO_2H \longrightarrow CH_3(CH_2)_{12}CH{=}CHCO_2H$

c. 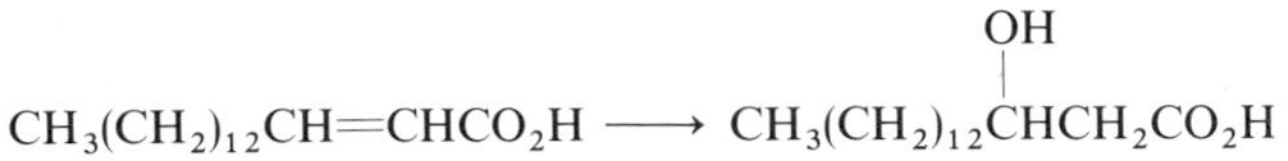 $CH_3(CH_2)_{12}CH{=}CHCO_2H \longrightarrow CH_3(CH_2)_{12}\overset{\overset{\displaystyle OH}{|}}{C}HCH_2CO_2H$

d. 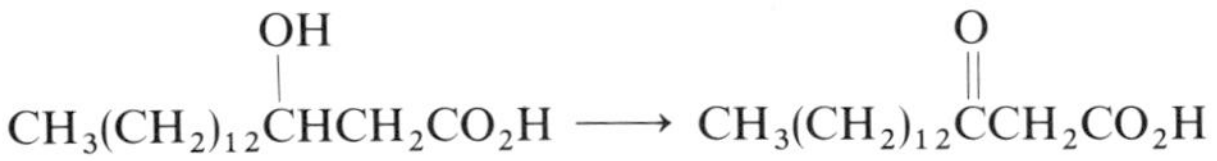 $CH_3(CH_2)_{12}\overset{\overset{\displaystyle OH}{|}}{C}HCH_2CO_2H \longrightarrow CH_3(CH_2)_{12}\overset{\overset{\displaystyle O}{\|}}{C}CH_2CO_2H$

e. 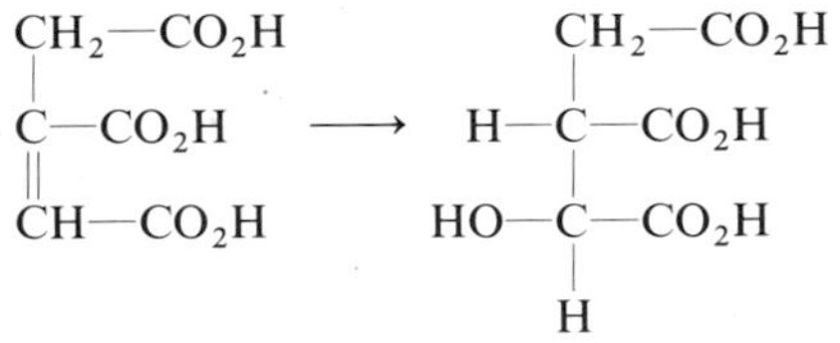 $\begin{array}{c} CH_2{-}CO_2H \\ | \\ C{-}CO_2H \\ \| \\ CH{-}CO_2H \end{array} \longrightarrow \begin{array}{c} CH_2{-}CO_2H \\ | \\ H{-}C{-}CO_2H \\ | \\ HO{-}C{-}CO_2H \\ | \\ H \end{array}$

f. 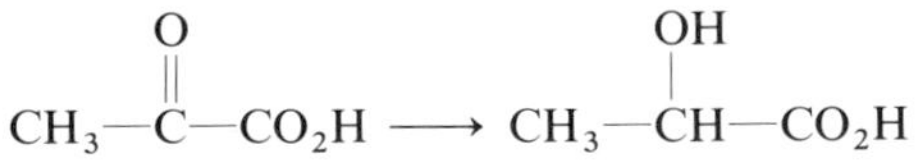 $CH_3{-}\overset{\overset{\displaystyle O}{\|}}{C}{-}CO_2H \longrightarrow CH_3{-}\overset{\overset{\displaystyle OH}{|}}{C}H{-}CO_2H$

g. 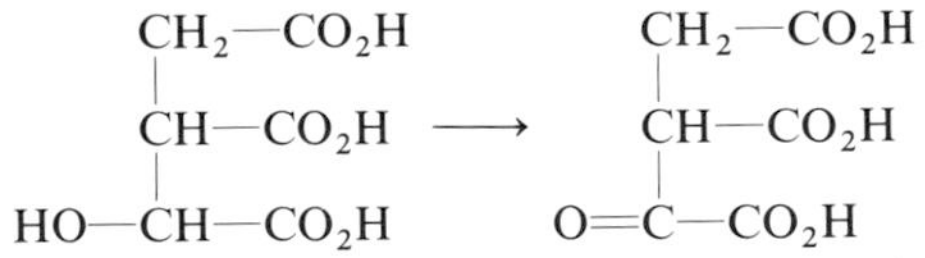 $\begin{array}{c} CH_2{-}CO_2H \\ | \\ CH{-}CO_2H \\ | \\ HO{-}CH{-}CO_2H \end{array} \longrightarrow \begin{array}{c} CH_2{-}CO_2H \\ | \\ CH{-}CO_2H \\ | \\ O{=}C{-}CO_2H \end{array}$

h.

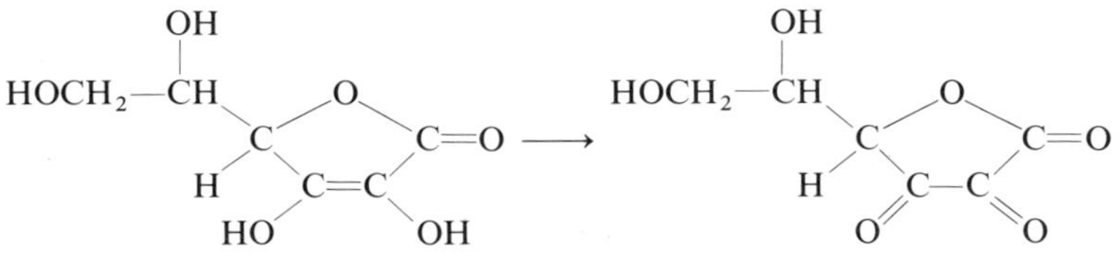

ascorbic acid
(vitamin C)   dehydroascorbic
acid

12.33 Write structural formulas for the haloalkanes formed in the following reactions.

a.

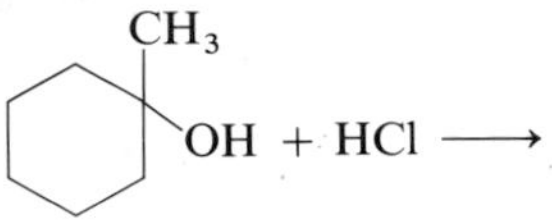

$$\text{(cyclohexane ring with } CH_3 \text{ and OH)} + HCl \longrightarrow$$

b. $CH_3—CH_2—CH_2—CH_2—OH + H—Br \longrightarrow$

c.

$$\text{(cyclopentane ring with } CH_2OH \text{ and OH)} + 2\, HI \longrightarrow$$

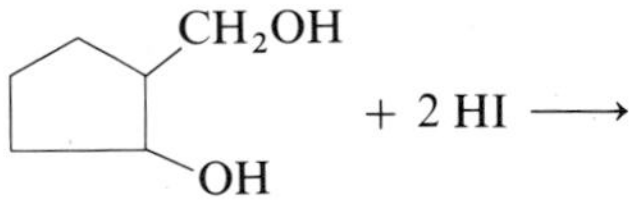

d. $CH_2{=}CH—CH_2—Cl + HCl \longrightarrow$

e. $CH_2{=}CH—CH_2—OH + Cl_2 \longrightarrow$

12.34 The following conversions can be carried out in one step. Show the reagent you would use to bring about each conversion.

a.

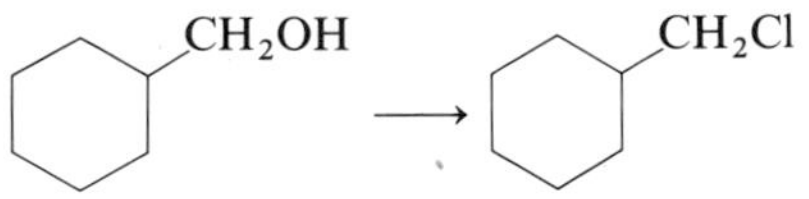

$$\text{(cyclohexane ring with } CH_2OH) \longrightarrow \text{(cyclohexane ring with } CH_2Cl)$$

b.

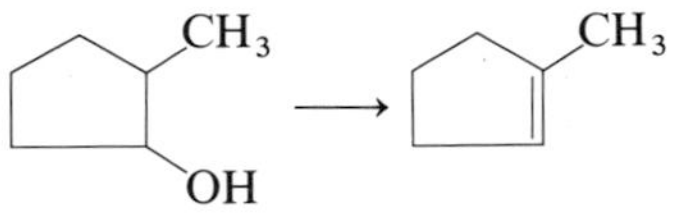

$$\text{(cyclopentane ring with } CH_3 \text{ and OH)} \longrightarrow \text{(cyclopentene ring with } CH_3)$$

c.

$$\text{(cyclopentene ring with } CH_3) \longrightarrow \text{(cyclopentane ring with } CH_3 \text{ and OH)}$$

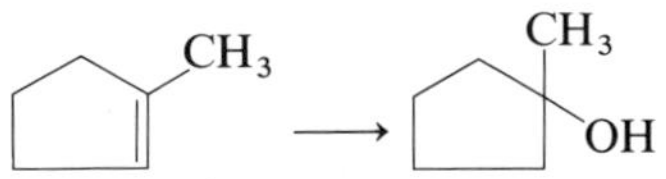

d. $CH_3—\underset{\underset{CH_3}{|}}{CH}—CH_2OH \longrightarrow CH_3—\underset{\underset{CH_3}{|}}{CH}—\overset{\overset{O}{\|}}{C}—OH$

e.

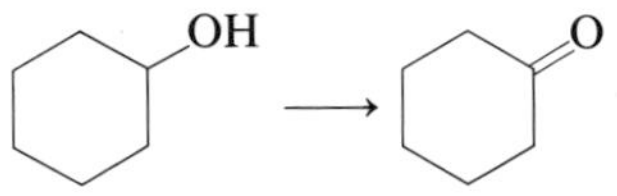

$$\text{(cyclohexane ring with OH)} \longrightarrow \text{(cyclohexanone ring with =O)}$$

f.

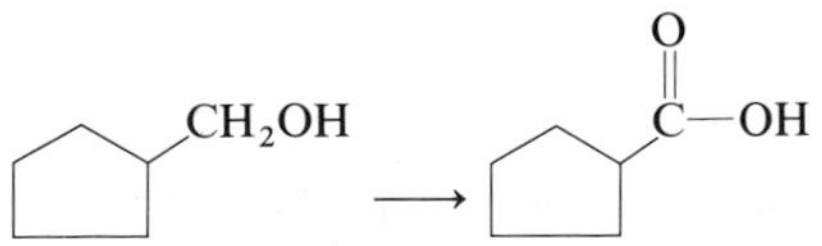

$$\text{(cyclopentane ring with } CH_2OH) \longrightarrow \text{(cyclopentane ring with } \overset{\overset{O}{\|}}{C}—OH)$$

g.

$$\text{(benzene ring with } \overset{\overset{O}{\|}}{C}—H) \longrightarrow \text{(benzene ring with } \overset{\overset{O}{\|}}{C}—OH)$$

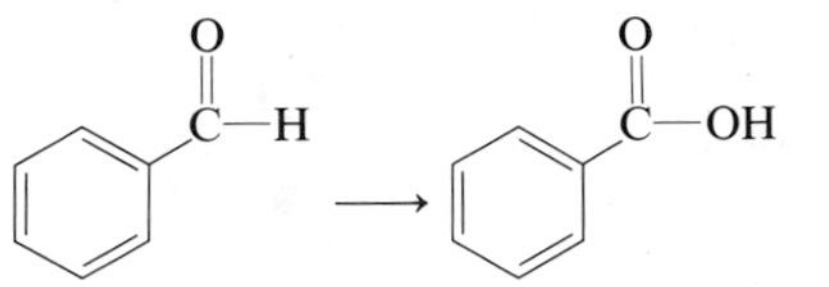

12.35 Following are a series of conversions in which a starting material is converted to the indicated product in two steps. State the reagent or reagents you would use to bring about each conversion.

a. 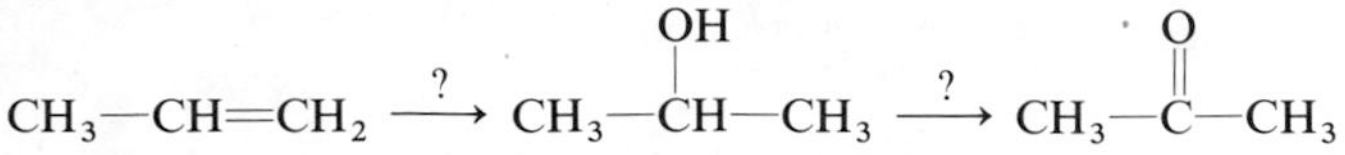 $CH_3-CH=CH_2 \xrightarrow{?} CH_3-\underset{\underset{OH}{|}}{CH}-CH_3 \xrightarrow{?} CH_3-\underset{\underset{\|}{O}}{C}-CH_3$

b. [cyclopentanol] $\xrightarrow{?}$ [cyclopentene] $\xrightarrow{?}$ [1,2-dichlorocyclopentane]

c. [methylcyclopentanol] $\xrightarrow{?}$ [methylcyclopentene] $\xrightarrow{?}$ [1-methylcyclopentanol]

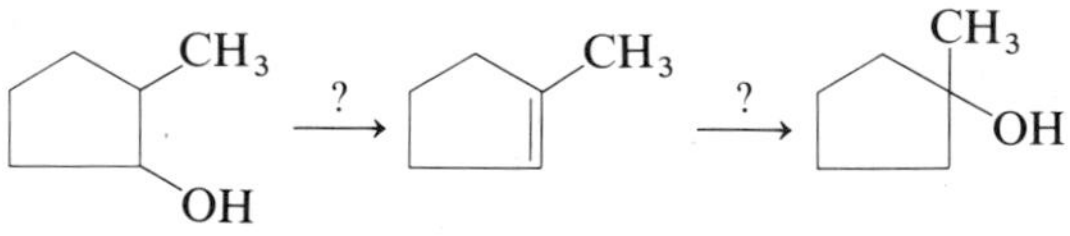

d. 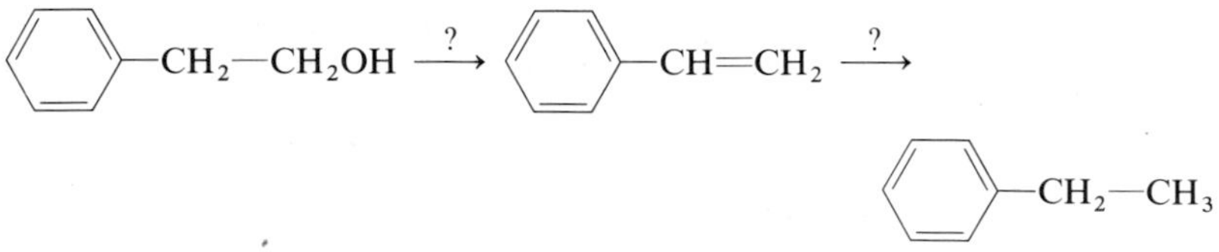 [benzene]$-CH_2-CH_2OH \xrightarrow{?}$ [benzene]$-CH=CH_2 \xrightarrow{?}$

[benzene]$-CH_2-CH_3$

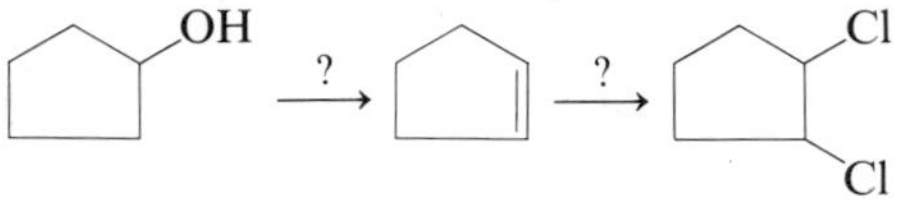

12.36 The following conversions can be carried out in two steps. Show reagents you would use and the structural formula of the intermediate formed in each conversion.

a. 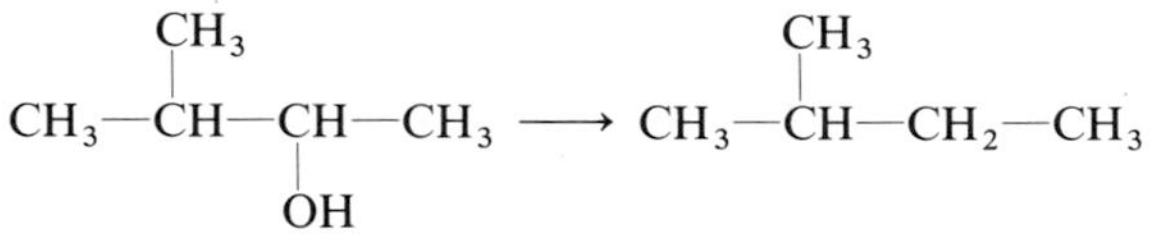 $CH_3-\underset{\underset{OH}{|}}{\underset{|}{CH}}-\underset{\overset{|}{CH_3}}{CH}-CH_3 \longrightarrow CH_3-\underset{\overset{|}{CH_3}}{CH}-CH_2-CH_3$

b. 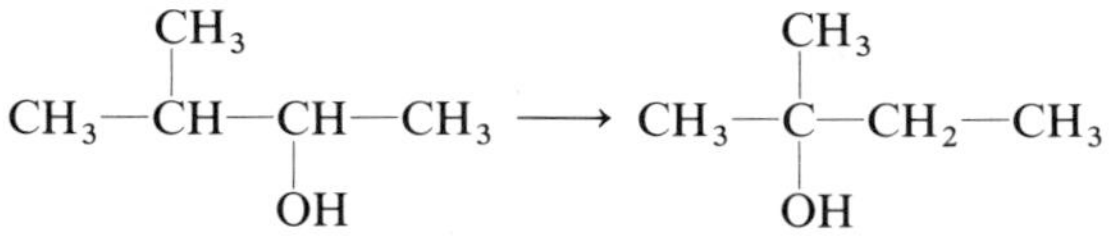 $CH_3-\underset{\underset{OH}{|}}{\underset{|}{CH}}-\underset{\overset{|}{CH_3}}{CH}-CH_3 \longrightarrow CH_3-\underset{\underset{OH}{|}}{\underset{|}{C}}-CH_2-CH_3$ (with CH_3 top)

c. $CH_3-CH_2-CH_2-OH \longrightarrow CH_3-\underset{\underset{Cl}{|}}{CH}-\underset{\underset{Cl}{|}}{CH_2}$

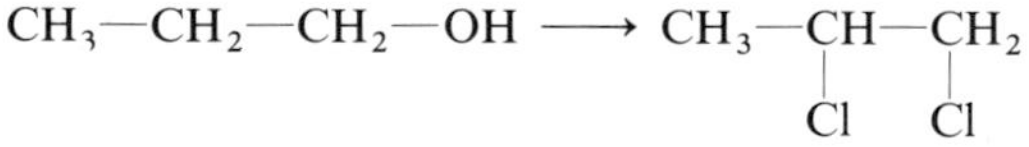

d. 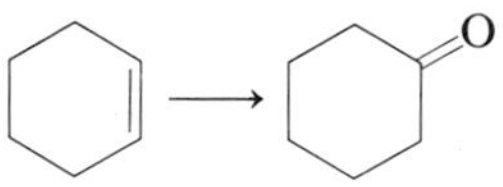 [cyclohexene] $\longrightarrow$ [cyclohexanone]

**Phenols
(Section 12.7)**

12.37 Name the following compounds.

a. 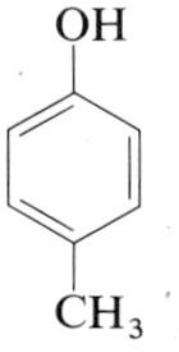 [4-methylphenol]

b. C_6H_5-[phenol with OH]

c. [2,4,6-trinitrophenol, with OH, O_2N, NO_2, NO_2]

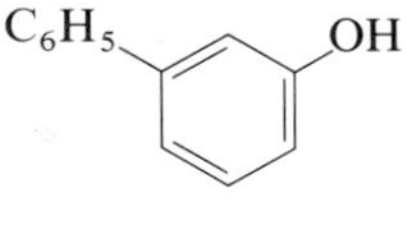

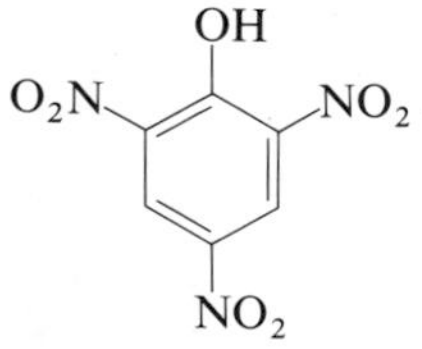

12.38 Write structural formulas for the following compounds.
    a. 2,4-dimethoxyphenol   b. sodium phenoxide
    c. 2-isopropyl-4-methylphenol   d. *m*-cresol

12.39 Identify all functional groups in the following compounds.

a. 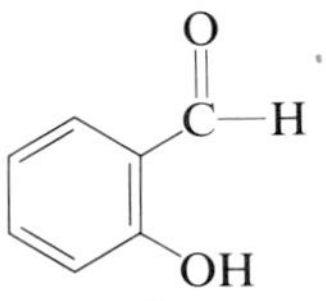 salicylaldehyde   b. 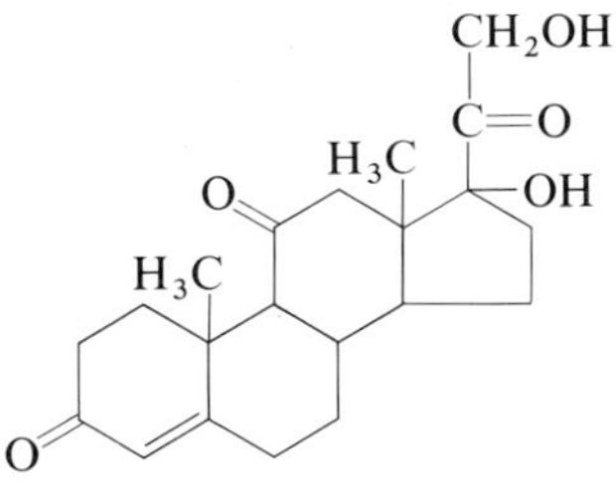 cortisone   c. 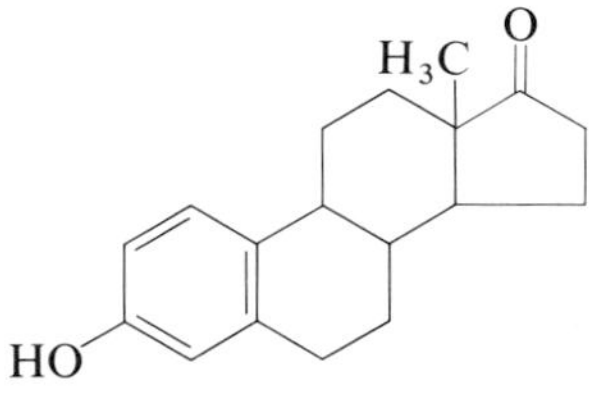 estrone (a female sex hormone)

d. 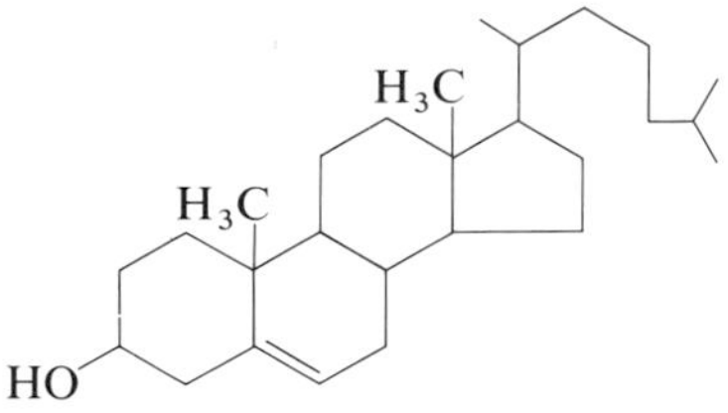 cholesterol   e. 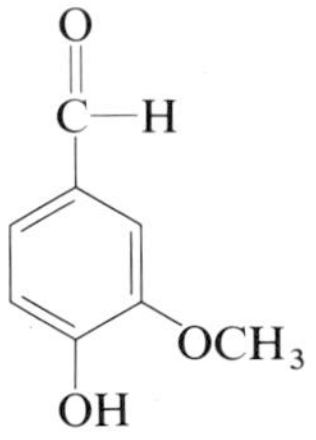 vanillin

12.40 Complete the following reactions. Where you predict no reaction, write N.R.

a. 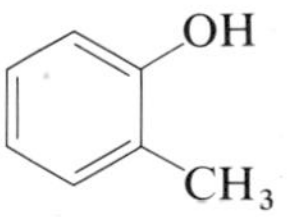 + NaOH $\longrightarrow$ b. 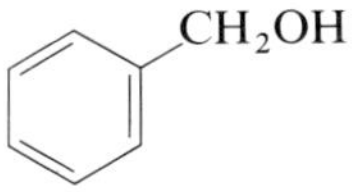 + NaOH $\longrightarrow$

c. 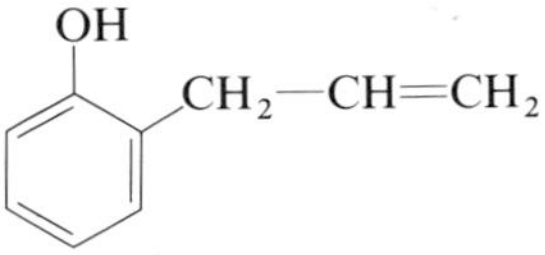 $+ H_2 \xrightarrow{Pt}$

12.41 Show how you might distinguish between the following pairs of compounds by a simple chemical test. In each case, tell what test you would perform, what you would expect to observe, and write an equation for each positive test.

a. 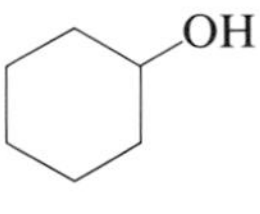 and 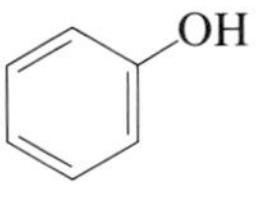   b. 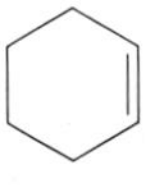 and

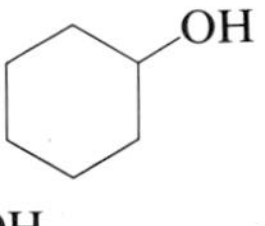

c.  and 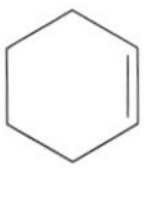   d. 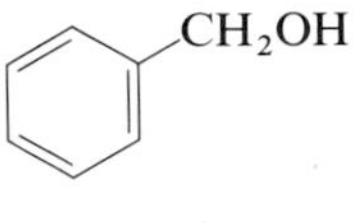 and

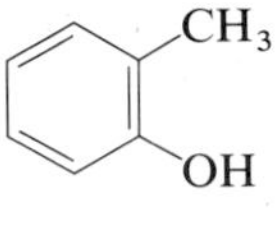

Thiols (Section 12.8)

12.42 Name the following:

a. CH_3—CH_2—SH

b. 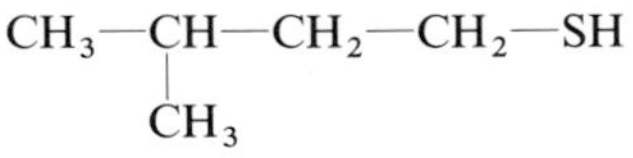 CH_3—$\underset{\underset{CH_3}{|}}{CH}$—$CH_2$—$CH_2$—$SH$

c. 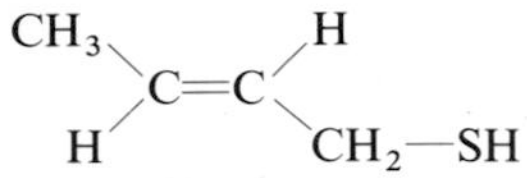 $\underset{H}{\overset{CH_3}{\diagdown}}C=C\underset{CH_2-SH}{\overset{H}{\diagup}}$

d. HO—CH_2—CH_2—SH

12.43 Draw structural formulas for the following compounds.

a. 2-pentanethiol
b. cyclopentanethiol
c. 1,2-ethanedithiol
d. diisobutyl sulfide
e. diisobutyl disulfide
f. 2,3-dimercapto-1-propanol

12.44 Write a balanced half-reaction for the conversion of two molecules of a thiol to a disulfide and show that this conversion is a two-electron oxidation.

12.45 Write a balanced half-reaction for the conversion of a disulfide to two sulfonic acids and show that this conversion is a ten-electron oxidation.

12.46 Draw structural formulas for the major organic products of the following reactions.

a. $CH_3(CH_2)_6CH_2SH + NaOH \longrightarrow$

b. HS—CH_2—CH_2—CH_2—$SH + \frac{1}{2}O_2 \longrightarrow$

c. CH_3—CH=CH—CH_2—$SH + H_2 \xrightarrow{Pt}$

d. 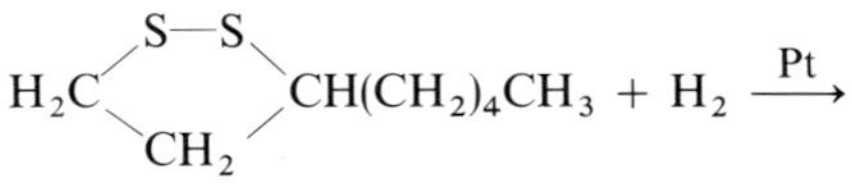 $H_2C\underset{CH_2}{\overset{S-S}{\diagdown}}CH(CH_2)_4CH_3 + H_2 \xrightarrow{Pt}$

e. 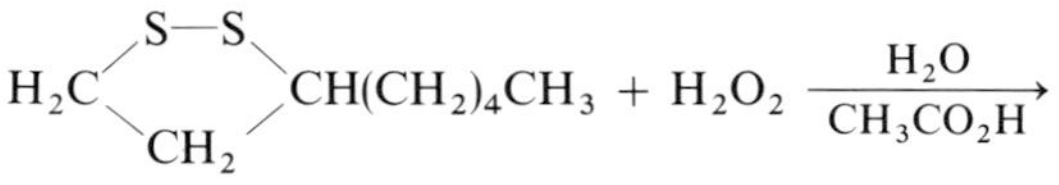 $H_2C\underset{CH_2}{\overset{S-S}{\diagdown}}CH(CH_2)_4CH_3 + H_2O_2 \xrightarrow[CH_3CO_2H]{H_2O}$

12.47 Penicillamine can be used to treat lead poisoning. Write an equation for the reaction of penicillamine with Pb^{2+} and explain how penicillamine might be used to counter-act lead poisoning.

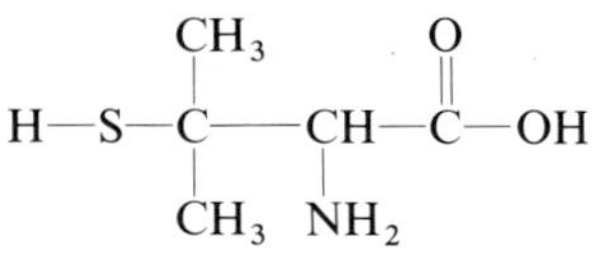

$$H-S-\underset{\underset{CH_3}{|}}{\overset{\overset{CH_3}{|}}{C}}-\underset{\underset{NH_2}{|}}{CH}-\overset{\overset{O}{||}}{C}-OH$$

penicillamine

12.48 One treatment for mercury poisoning uses 2,3-dimercapto-1-propanol, a substance that forms a water-soluble complex with Hg(II) ion and is excreted in the urine. Draw the structural formulas of this water-soluble complex.

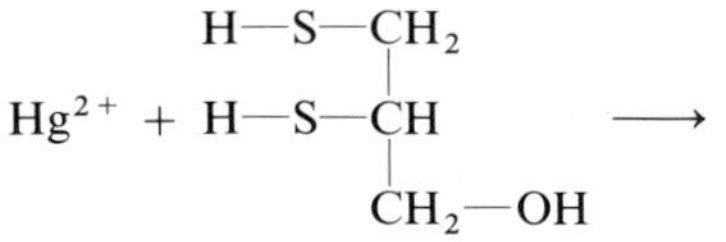

$$Hg^{2+} + \begin{array}{l} H-S-CH_2 \\ \quad\quad | \\ H-S-CH \\ \quad\quad | \\ \quad\quad CH_2-OH \end{array} \longrightarrow$$

2,3-dimercapto-
1-propanol

Aldehydes and Ketones

In Chapter 11 we studied the physical and chemical properties of compounds containing carbon-carbon double bonds. In the next several chapters, we will study the physical and chemical properties of compounds containing carbon-oxygen double bonds. The **C═O (carbonyl) group** is the central structural feature of aldehydes, ketones, carboxylic acids, esters, and amides. The reactions of the carbonyl group are quite simple, and an understanding of the few reaction themes leads very quickly to an understanding of a wide variety of reactions in organic chemistry and biochemistry. In this chapter we will study the reactions of aldehydes and ketones.

13.1 Structure of Aldehydes and Ketones

The characteristic structural feature of an aldehyde is a carbonyl group bonded to a hydrogen atom. In the simplest aldehyde, formaldehyde, a carbonyl group is bonded to two hydrogen atoms; in more complex aldehydes, the carbonyl group is bonded to one hydrogen and one carbon, as in acetaldehyde. Shown on the next page are Lewis structures of formaldehyde and acetaldehyde.

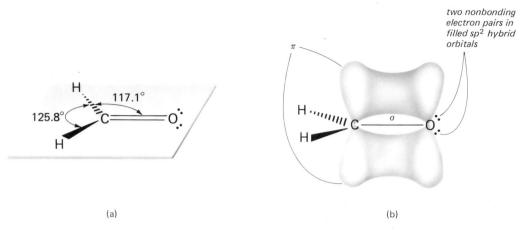

Figure 13.1 Bonding in formaldehyde. (a) Lewis structure showing all bond angles; (b) covalent bond formation by the overlap of atomic orbitals.

formaldehyde   acetaldehyde

The characteristic structural feature of a ketone is a carbonyl group bonded to two carbon atoms. A Lewis structure of acetone, the simplest ketone, is:

acetone

A carbon-oxygen double bond consists of: one sigma bond formed by overlap of sp^2 hybrid orbitals of carbon and oxygen; and one pi bond formed by overlap of $2p$ orbitals (Section 9.5B). Figure 13.1 shows an orbital overlap diagram of the bonding in formaldehyde. The H—C—O bond angle is 117.1°, close to the predicted value of 120° about an sp^2 hybridized carbon atom.

13.2 Nomenclature of Aldehydes and Ketones

The IUPAC system of nomenclature for aldehydes follows the familiar pattern of selecting as the parent compound the longest chain of carbon atoms that contains the functional group. The presence of the aldehyde group is shown by changing the -e of the parent compound to -al. The aldehyde functional group can only occur on the end of a parent chain, and therefore there is no need to use the number 1 to locate its position. Following are structural formulas and IUPAC names for methanal, ethanal, and 4-methylpentanal.

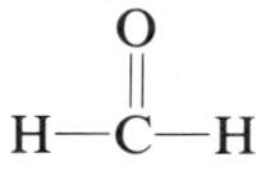

methanal
(formaldehyde)

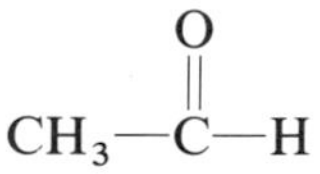

ethanal
(acetaldehyde)

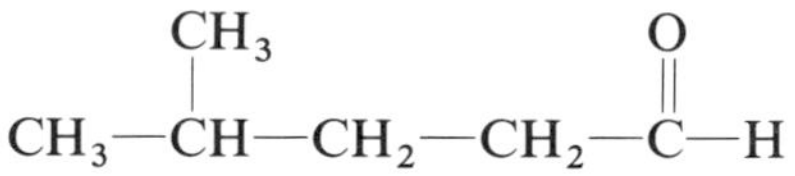

4-methylpentanal

The IUPAC system retains certain common names. For example, formaldehyde and acetaldehyde are retained as IUPAC names for methanal and ethanal. Other common names retained in the IUPAC system are benzaldehyde and cinnamaldehyde.

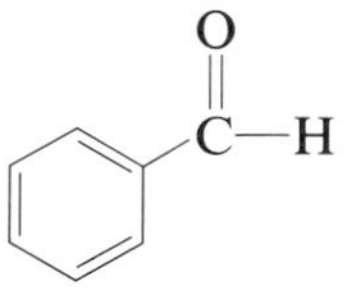

benzaldehyde

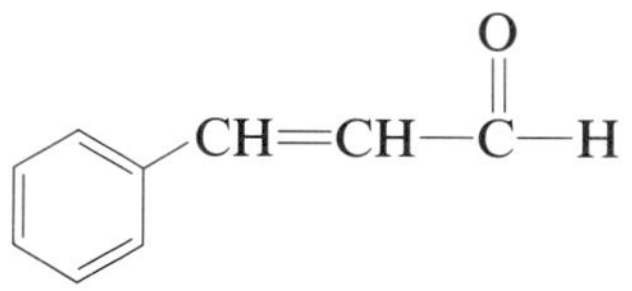

cinnamaldehyde

In the IUPAC system, ketones are named by selecting as the parent compound the longest chain that contains the carbonyl group, and then showing the presence of the carbonyl group by changing the -e of the parent compound to -one. The parent chain is numbered in the direction that gives the carbonyl group the lowest possible number. The IUPAC system retains the common names acetone and acetophenone.

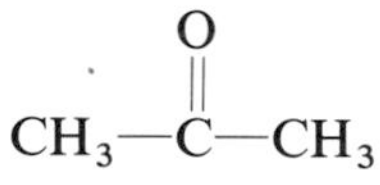

2-propanone
(acetone)

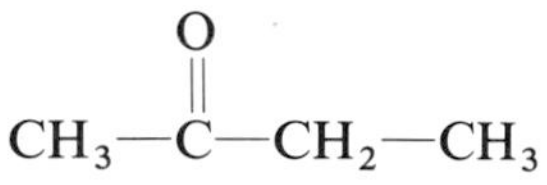

2-butanone

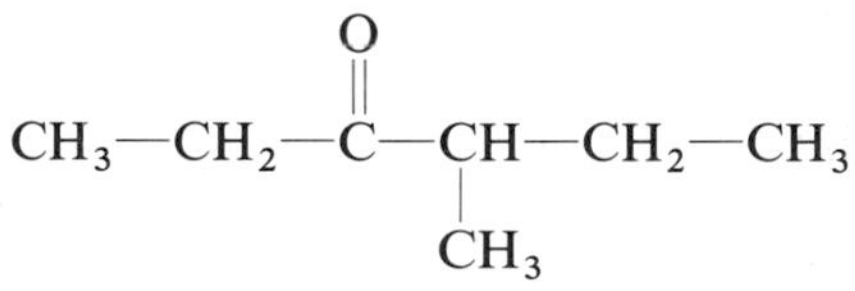

4-methyl-3-hexanone

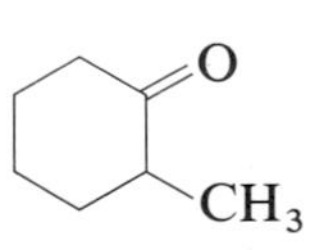

2-methylcyclohexanone

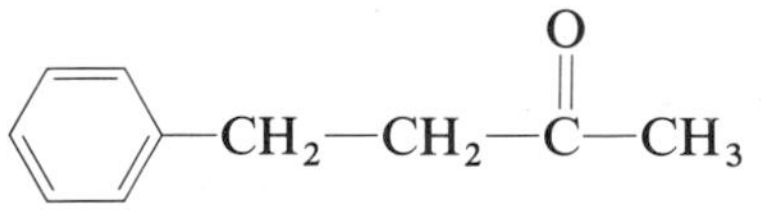

4-phenyl-2-butanone

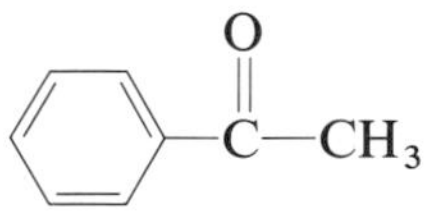

acetophenone

An older system of naming ketones lists the names of the two hydrocarbon groups attached to the carbonyl group, followed by the word "ketone." In the following examples, the name using this older system is listed in parentheses below the IUPAC name.

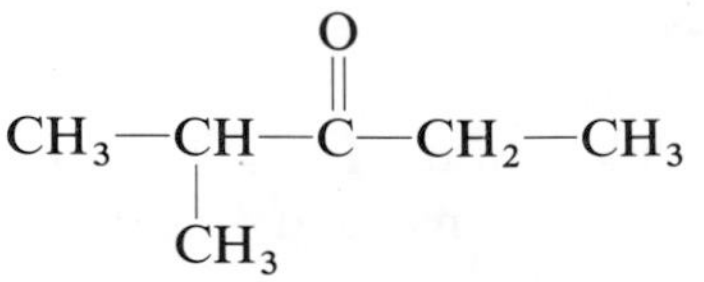

$$CH_3-\overset{\displaystyle CH}{\underset{\displaystyle CH_3}{|}}-\overset{\displaystyle O}{\overset{\displaystyle \|}{C}}-CH_2-CH_3 \qquad CH_3-CH_2-\overset{\displaystyle O}{\overset{\displaystyle \|}{C}}-CH_2-CH_3 \qquad CH_3-\overset{\displaystyle O}{\overset{\displaystyle \|}{C}}-CH_2-CH_3$$

2-methyl-3-pentanone   3-pentanone   2-butanone
(ethyl isopropyl ketone)   (diethyl ketone)   (methyl ethyl ketone)

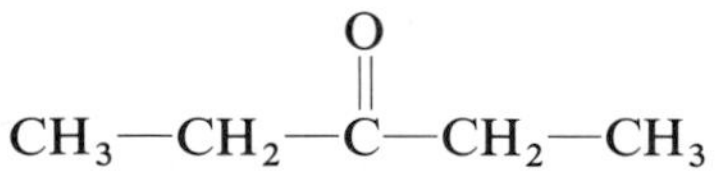

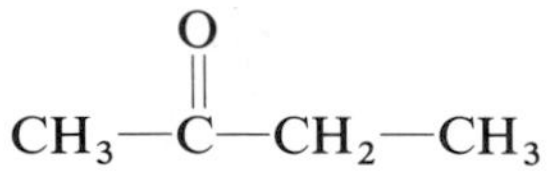

Example 13.1

Give IUPAC names for the following compounds.

a. 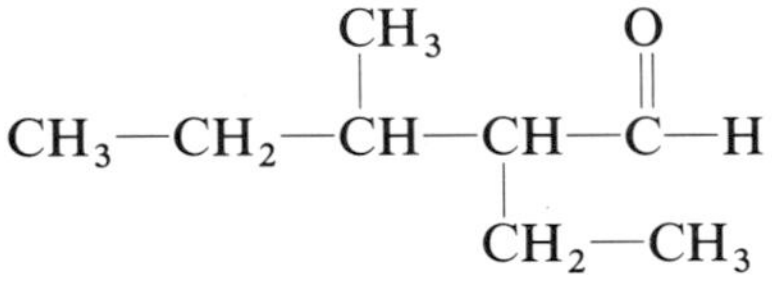 $CH_3-CH_2-\overset{\displaystyle CH_3}{\underset{\displaystyle CH_2-CH_3}{|}}{\overset{}{C}H}-\overset{\displaystyle O}{\overset{\displaystyle \|}{C}}-H$ **b.** 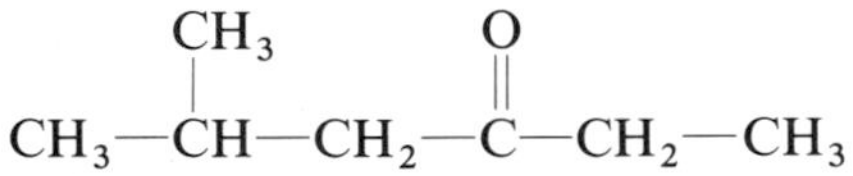 $CH_3-\overset{\displaystyle CH_3}{\overset{\displaystyle |}{C}H}-CH_2-\overset{\displaystyle O}{\overset{\displaystyle \|}{C}}-CH_2-CH_3$

c.

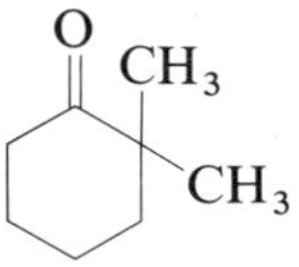

d.

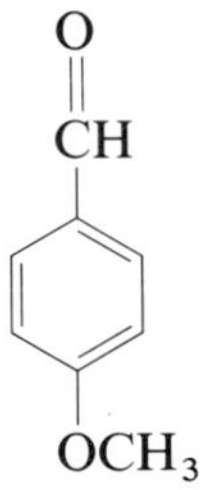

Solution

a. Number the longest chain that contains the aldehyde group, and then give each substituent a name and a number. Note that the longest carbon chain in this molecule is 6 carbons, but the longest chain that contains the aldehyde is 5 carbons:

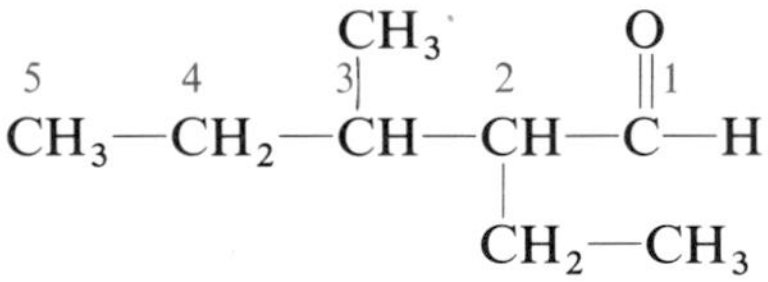

$$\underset{5}{CH_3}-\underset{4}{CH_2}-\underset{3}{\overset{\displaystyle CH_3}{\overset{\displaystyle |}{CH}}}-\underset{2}{\underset{\displaystyle |}{CH}}-\underset{1}{\overset{\displaystyle O}{\overset{\displaystyle \|}{C}}}-H$$
$$\underset{\displaystyle CH_2-CH_3}{}$$

2-ethyl-3-methylpentanal

b. Number the 6-carbon chain to give the carbonyl group the lowest possible number.

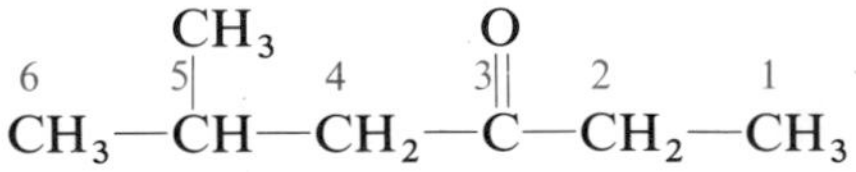

$$\underset{6}{CH_3}-\underset{5}{\overset{\displaystyle CH_3}{\overset{\displaystyle |}{CH}}}-\underset{4}{CH_2}-\underset{3}{\overset{\displaystyle O}{\overset{\displaystyle \|}{C}}}-\underset{2}{CH_2}-\underset{1}{CH_3}$$

5-methyl-3-hexanone

This molecule can also be named ethyl isobutyl ketone.

c. Number the six-membered ring beginning with the carbon bearing the carbonyl group. The name is 2,2-dimethylcyclohexanone.

d. This molecule is a derivative of benzaldehyde. The methoxy substituent can be located either by numbering the carbons of the ring or by using the *ortho-*, *meta-*, *para-*system. The name is either 4-methoxybenzaldehyde or *p*-methoxybenzaldehyde.

Problem 13.1

Give an acceptable name for the following compounds:

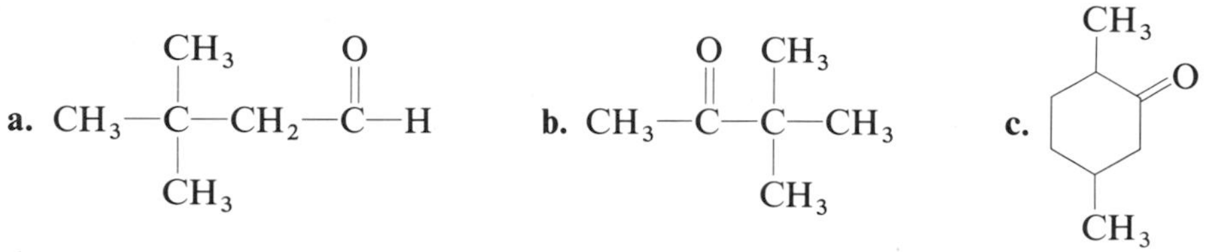

13.3 Some Naturally Occurring Aldehydes and Ketones

A great variety of aldehydes and ketones have been isolated from natural sources and are best known by their common or trivial names. Such names are usually derived from that of a source from which the compound can be isolated, or they may refer to a characteristic property of the compound. Figure 13.2 shows structural formulas for several naturally occurring aldehydes and ketones, together with their common names.

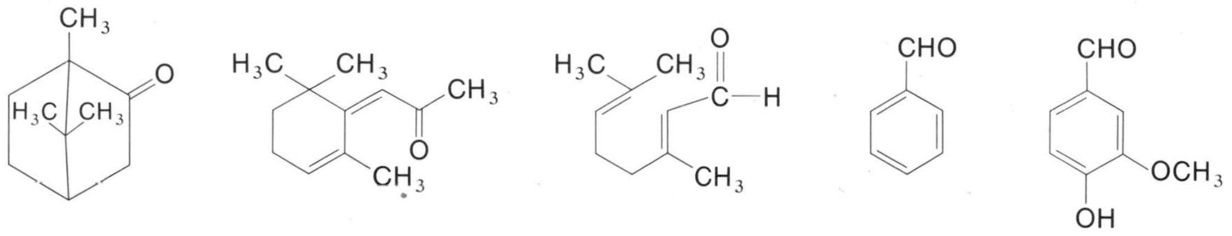

camphor
(camphor tree)

irone
(violet)

citral
(lemon grass oil)

benzaldehyde
(bitter almond)

vanillin
(vanilla bean)

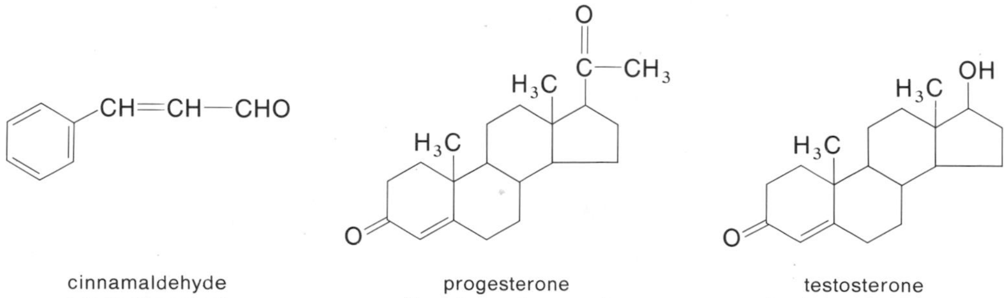

cinnamaldehyde
(oil of cinnamon)

progesterone
(female sex hormone)

testosterone
(male sex hormone)

Figure 13.2 Common names of some naturally occurring aldehydes and ketones.

13.4 Physical Properties of Aldehydes and Ketones

A. Polarity of the Carbonyl Group

Oxygen is more electronegative than carbon (Section 4.3B), and therefore the carbon-oxygen double bond is polar covalent. Oxygen bears a partial negative charge and carbon a partial positive charge, as illustrated for formaldehyde:

$$\begin{array}{c} H \\ \diagdown {\scriptstyle \delta+} {\scriptstyle \delta-} \\ C{=}\ddot{\underset{\displaystyle ..}{O}}: \\ \diagup \\ H \end{array}$$

B. Boiling Points

Because they are polar compounds and interact in the pure state by dipole-dipole interaction, aldehydes and ketones have higher boiling points than nonpolar compounds of comparable molecular weight. Table 13.1 lists boiling points of six compounds of comparable molecular weight. Pentane, a nonpolar hydrocarbon, and methyl propyl ether have the lowest boiling points. While methyl propyl ether is a polar compound, there is little association between molecules in the liquid state; hence, its boiling point is only slightly higher than that of pentane. Both butanal and 2-butanone are polar compounds, and because of the association between the partially positive carbon of one molecule and a partially negative oxygen of another molecule, their boiling points are higher than those of pentane and methyl propyl ether. Because aldehydes and ketones have no partially positive hydrogen atom attached to oxygen, they cannot associate by hydrogen bonding. Therefore, their boiling points are

Table 13.1 Boiling points of compounds of similar molecular weight.

Structural Formula	IUPAC Name	Molecular Weight	bp (°C)
$CH_3CH_2CH_2CH_2CH_3$	pentane	72	36
$CH_3CH_2CH_2OCH_3$	methyl propyl ether	74	39
$CH_3CH_2CH_2\overset{\displaystyle O}{\overset{\|}{C}}H$	butanal	72	76
$CH_3CH_2\overset{\displaystyle O}{\overset{\|}{C}}CH_3$	2-butanone	72	80
$CH_3CH_2CH_2CH_2OH$	1-butanol	74	117
$CH_3CH_2\overset{\displaystyle O}{\overset{\|}{C}}OH$	propanoic acid	74	141

lower than those of alcohols and carboxylic acids, compounds that can associate by hydrogen bonding.

C. Solubility in Water

Aldehydes and ketones can interact with water molecules as hydrogen-bond acceptors (Figure 13.3). For this reason, low-molecular-weight aldehydes and ketones are more soluble in water than nonpolar compounds of comparable molecular weight.

Table 13.2 lists boiling points and solubility in water for several aldehydes and ketones.

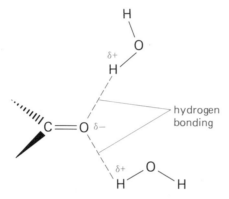

Figure 13.3 Intermolecular interaction between a carbonyl group and water molecules.

Table 13.2 Physical properties of aldehydes and ketones.

Structural Formula	IUPAC Name	Common Name	bp (°C)	Solubility (g/100 g H$_2$O)
HCHO	methanal	formaldehyde	−21	very
CH$_3$CHO	ethanal	acetaldehyde	20	∞
CH$_3$CH$_2$CHO	propanal	propionaldehyde	49	16
CH$_3$(CH$_2$)$_2$CHO	butanal	butyraldehyde	76	7
CH$_3$(CH$_2$)$_3$CHO	pentanal	valeraldehyde	103	slightly
CH$_3$(CH$_2$)$_4$CHO	hexanal	caproaldehyde	129	slightly
CH$_3$COCH$_3$	2-propanone	acetone	56	∞
CH$_3$COCH$_2$CH$_3$	2-butanone	methyl ethyl ketone	80	26
CH$_3$COCH$_2$CH$_2$CH$_3$	2-pentanone	methyl propyl ketone	102	6
CH$_3$CH$_2$COCH$_2$CH$_3$	3-pentanone	diethyl ketone	101	5

13.5 Reactions of the Carbonyl Group

Reactions of aldehydes and ketones may be divided into two classes: (1) *nucleophilic* addition to the carbonyl group, and (2) reaction at the carbon atom adjacent to the carbonyl group. These two classes of reactions are among the most important in organic and biochemistry for the formation of new carbon-carbon bonds; for this reason, they deserve special attention.

A **nucleophile** is any atom or group of atoms with an unshared pair of electrons that can be shared with another atom or group of atoms to form a new covalent bond. In nucleophilic additions to a carbonyl group, the nucleophile adds to the carbonyl carbon to produce a tetrahedral carbonyl addition compound. In the following general reaction, the nucleophilic reagent is indicated by the symbol H—Nu: to emphasize the presence of the unshared pair of electrons in the nucleophile.

general reaction:

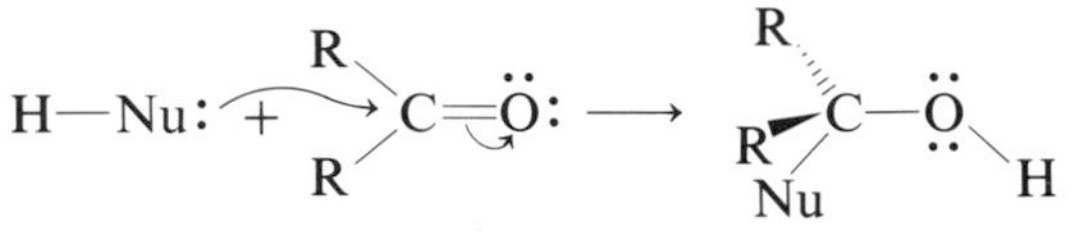

nucleophile *tetrahedral carbonyl addition compound*

As an example of nucleophilic addition, water adds to the carbonyl group of formaldehyde. In this example, the oxygen of water adds to the carbonyl carbon and one hydrogen adds to the carbonyl oxygen. In water at 20°C, formaldehyde is more than 99% hydrated.

reaction with a nucleophile:

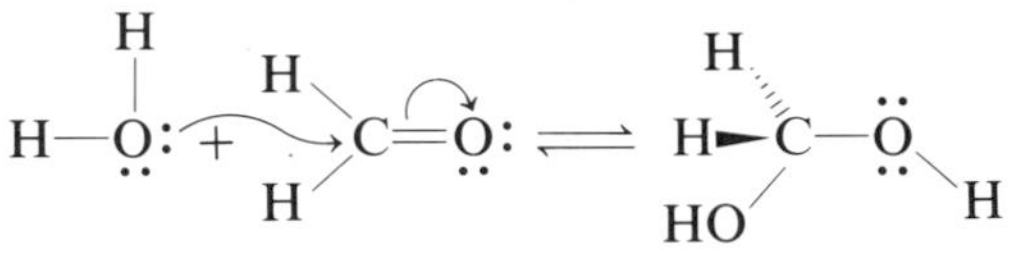

0.01%   99.99%

Under the same conditions, acetaldehyde is about 58% hydrated. The extent of hydration of most ketones is very small.

A. Addition of Alcohols: Formation of Acetals and Ketals

Alcohols add to aldehydes and ketones in the same manner as described for the addition of water. Addition of one molecule of alcohol to an aldehyde forms a **hemiacetal.** The comparable reaction of a ketone forms a **hemiketal.**

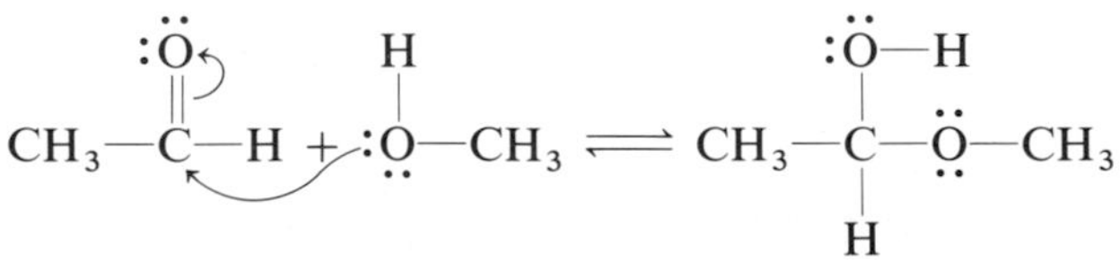

a hemiacetal

$$CH_3-\overset{\overset{\displaystyle :\ddot{O}}{\|}}{C}-CH_3 + :\overset{\displaystyle H}{\underset{\displaystyle ..}{O}}-CH_2-CH_3 \rightleftharpoons CH_3-\overset{\overset{\displaystyle :\ddot{O}-H}{}}{\underset{\underset{\displaystyle CH_3}{|}}{C}}-\overset{..}{\underset{..}{O}}-CH_2-CH_3$$

a hemiketal

Hemiacetals and hemiketals are only minor components of an equilibrium mixture except in one very important case. When the hydroxyl group is a part of the same molecule that contains the carbonyl group and a five- or six-member ring can form, a cyclic hemiacetal or hemiketal is frequently the major component of the equilibrium mixture.

$$CH_3-\underset{\underset{\displaystyle H-\overset{..}{\underset{..}{O}}:}{|}}{CH}-CH_2-CH_2-\overset{\overset{\displaystyle :\ddot{O}}{\|}}{C}-H \rightleftharpoons$$

4-hydroxypentanal
(*minor*)

a cyclic hemiacetal
(*major*)

Hemiacetals and hemiketals react further with alcohols to form **acetals** and **ketals.** These reactions are catalyzed by acids.

$$CH_3-\overset{\overset{\displaystyle OH}{|}}{\underset{\underset{\displaystyle H}{|}}{C}}-OCH_3 + CH_3O-H \underset{}{\overset{H^+}{\rightleftharpoons}} CH_3-\overset{\overset{\displaystyle OCH_3}{|}}{\underset{\underset{\displaystyle H}{|}}{C}}-OCH_3 + H-OH$$

a hemiacetal   an acetal

$$CH_3-\overset{\overset{\displaystyle OH}{|}}{\underset{\underset{\displaystyle CH_3}{|}}{C}}-OCH_2CH_3 + CH_3CH_2O-H \underset{}{\overset{H^+}{\rightleftharpoons}} CH_3-\overset{\overset{\displaystyle OCH_2CH_3}{|}}{\underset{\underset{\displaystyle CH_3}{|}}{C}}-OCH_2CH_3 + H-OH$$

a hemiketal   a ketal

Acetal and ketal formation are equilibrium reactions and in order to obtain high yields of acetal or ketal, water must be removed from the reaction mixture. In the presence of H^+ and water, acetals and ketals are hydrolyzed to aldehydes or ketones and alcohols.

$$CH_3-\overset{\overset{\displaystyle OCH_3}{|}}{\underset{\underset{\displaystyle OCH_3}{|}}{C}}-H + H_2O \underset{}{\overset{H^+}{\rightleftharpoons}} CH_3-\overset{\overset{\displaystyle O}{\|}}{C}-H + 2\ CH_3OH$$

Example 13.2

For the following, show the reaction of each carbonyl group with one molecule of alcohol to form a hemiacetal or hemiketal and then with a second molecule of alcohol to form an acetal or ketal.

a. $CH_3CH_2\overset{\displaystyle O}{\overset{\|}{C}}CH_3 + 2\ CH_3CH_2OH\ \xrightarrow{H^+}$

b. 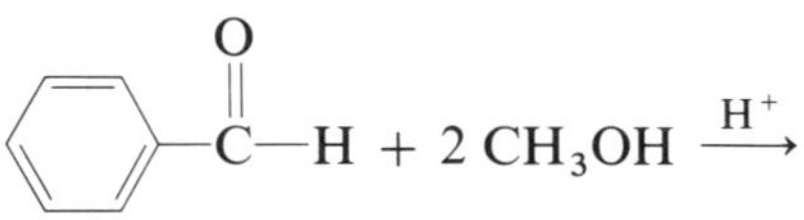 $\overset{\displaystyle O}{\overset{\|}{C}}-H + 2\ CH_3OH\ \xrightarrow{H^+}$

Solution

a. 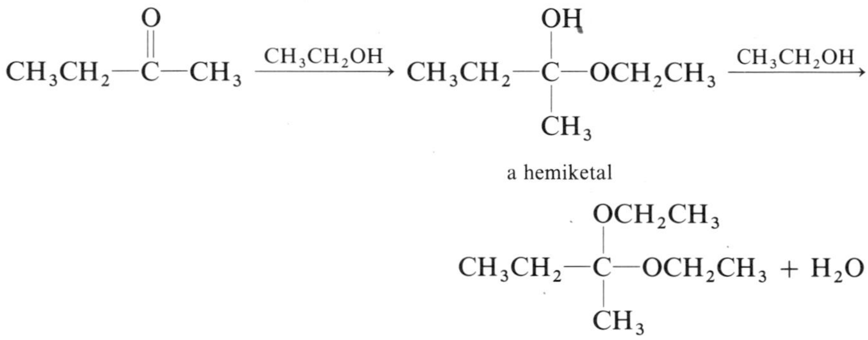 $CH_3CH_2\overset{\displaystyle O}{\underset{}{\overset{\|}{C}}}-CH_3\ \xrightarrow{CH_3CH_2OH}\ CH_3CH_2\overset{\displaystyle OH}{\underset{\displaystyle CH_3}{\overset{|}{\underset{|}{C}}}}-OCH_2CH_3\ \xrightarrow{CH_3CH_2OH}$

a hemiketal

$CH_3CH_2\overset{\displaystyle OCH_2CH_3}{\underset{\displaystyle CH_3}{\overset{|}{\underset{|}{C}}}}-OCH_2CH_3 + H_2O$

a ketal

b. 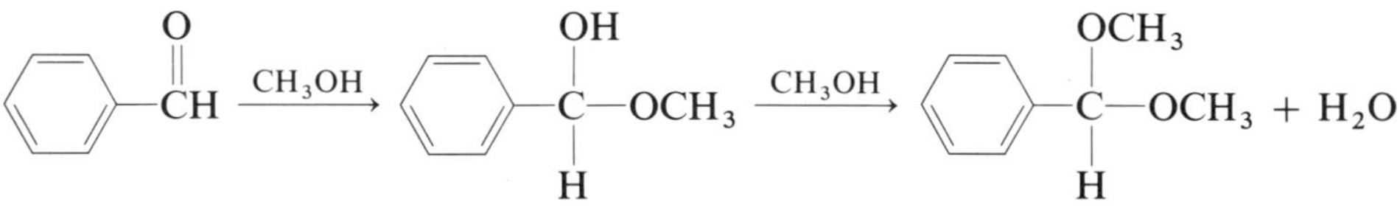 $\overset{\displaystyle O}{\overset{\|}{C}}H\ \xrightarrow{CH_3OH}$ $\overset{\displaystyle OH}{\underset{\displaystyle H}{\overset{|}{\underset{|}{C}}}}-OCH_3\ \xrightarrow{CH_3OH}$ $\overset{\displaystyle OCH_3}{\underset{\displaystyle H}{\overset{|}{\underset{|}{C}}}}-OCH_3 + H_2O$

a hemiacetal   an acetal

Problem 13.2

Following are structural formulas for one ketal and two acetals. Draw structural formulas for the products of the reaction of each with H_2O in the presence of HCl.

a.

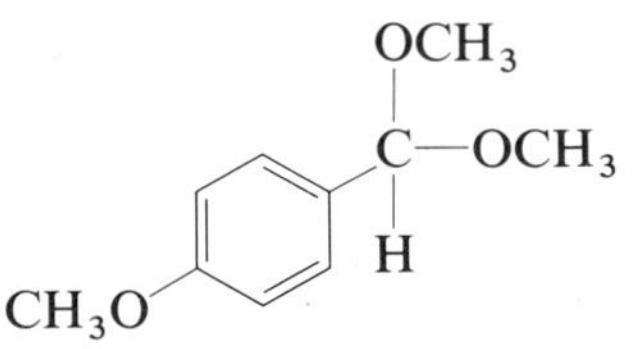

an acetal

b.

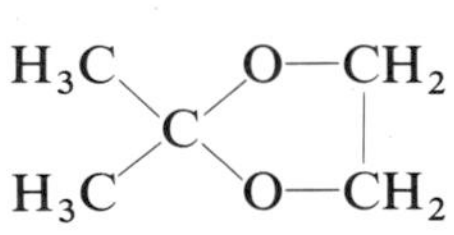

a ketal

c. 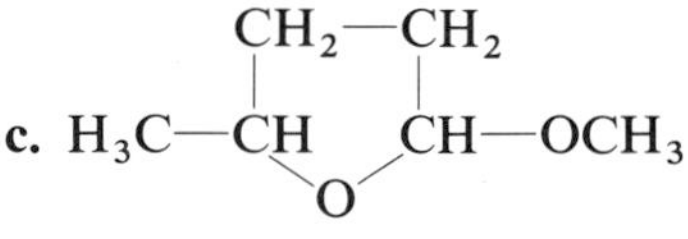 H_3C

an acetal

A mechanism for the acid-catalyzed conversion of acetaldehyde hemi-acetal into acetaldehyde dimethylacetal follows. In step 1, H$^+$ (an electrophile) reacts with the —OH group of the hemiacetal to form an oxonium ion. In step 2, the oxonium ion loses a molecule of water to form a carbocation.

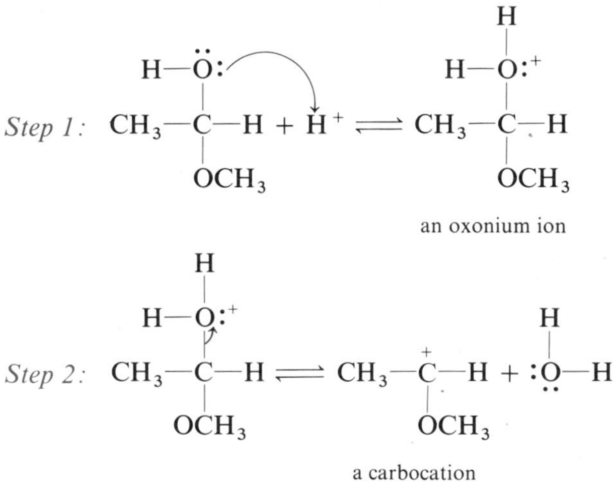

Step 1:

an oxonium ion

Step 2:

a carbocation

Note that these two steps are similar to the first two steps in the acid-catalyzed dehydration of alcohols (Section 12.5B). A carbocation is an electrophile and completes its valence shell by reacting with a nucleophile to form a new co-valent bond. In acetal formation, the nucleophile is another molecule of alcohol.

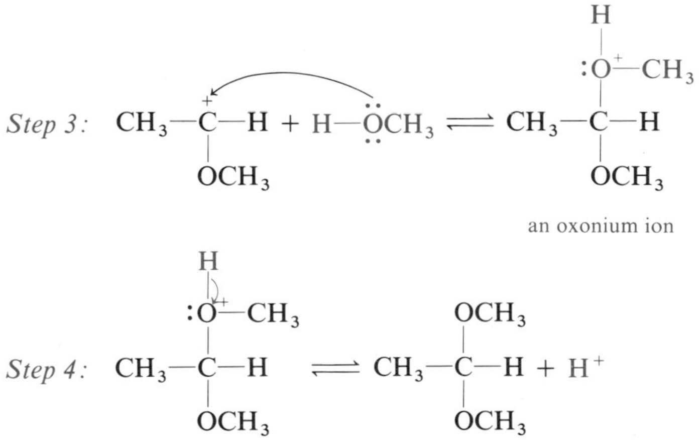

Step 3:

an oxonium ion

Step 4:

Steps 3 and 4 of this mechanism are similar to the steps in the acid-catalyzed hydration of an alkene to form an alcohol (Section 11.5D). In each case, a carbocation reacts with a pair of electrons on an oxygen atom to form a new covalent bond and release H$^+$.

B. Addition of Ammonia and its Derivatives: Formation of Schiff Bases

Ammonia, primary amines ($R-NH_2$), and certain monosubstituted amines such as hydroxylamine ($HO-NH_2$) add to the carbonyl groups of aldehydes and ketones to form **tetrahedral carbonyl addition intermediates.** In these additions, the amine nitrogen adds to the carbonyl carbon and an amine hydrogen adds to the carbonyl oxygen. These tetrahedral carbonyl addition intermediates lose a molecule of water to form compounds called **Schiff bases.** The characteristic structural feature of a Schiff base is the presence of a carbon-nitrogen double bond.

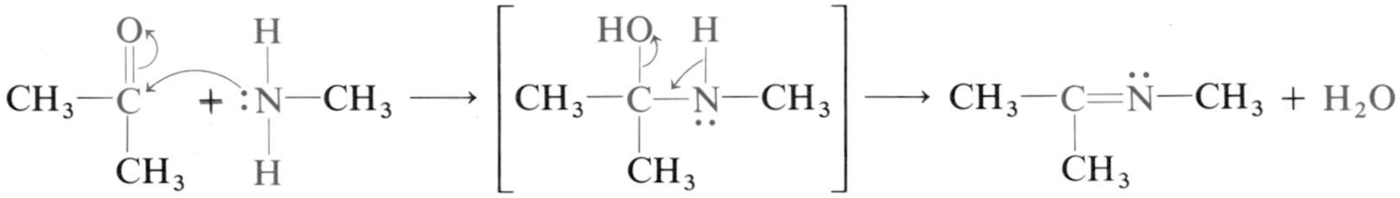

tetrahedral carbonyl addition compound

Schiff base

Example 13.3

Write structural formulas for the tetrahedral carbonyl addition intermediates and Schiff bases formed in the following reactions.

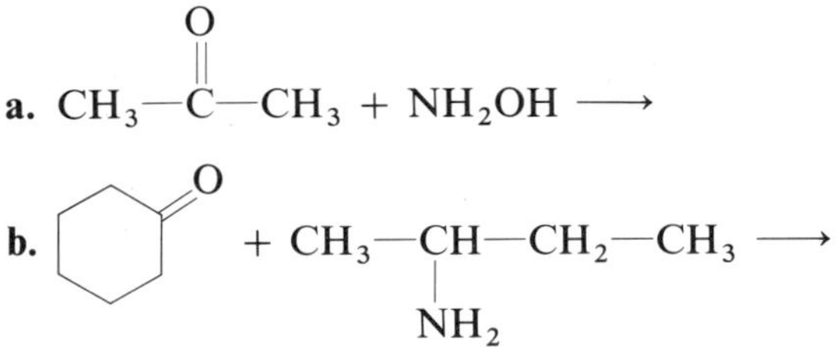

Solution

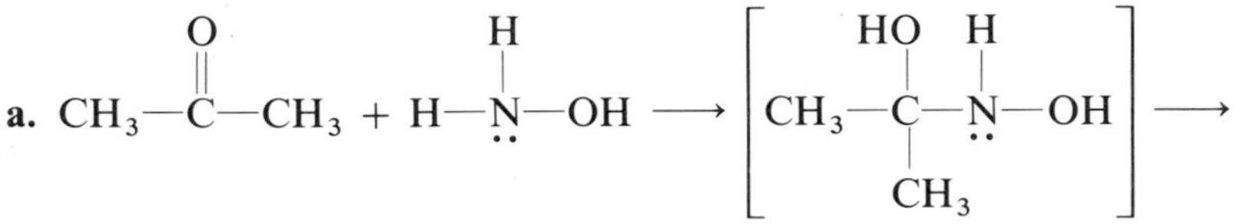

tetrahedral carbonyl addition intermediate

$$CH_3-C=\overset{..}{N}-OH + H_2O$$
$$\overset{|}{CH_3}$$

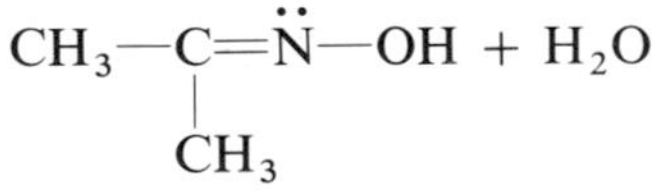

Schiff base

b.

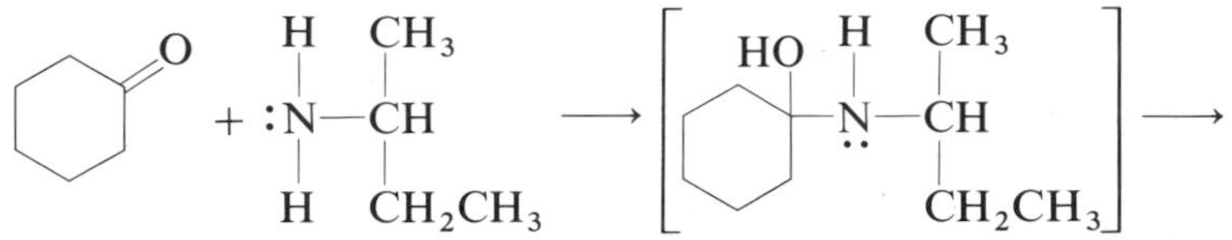

tetrahedral carbonyl
addition intermediate

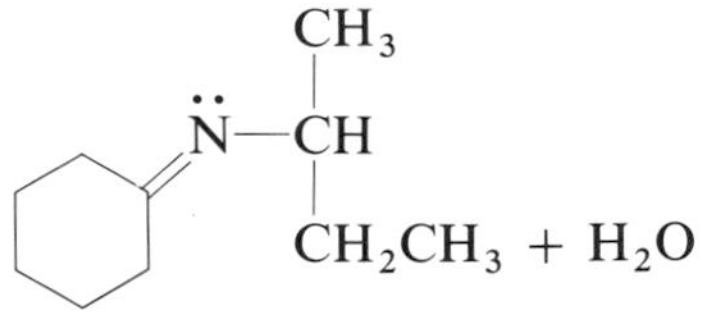

Schiff base

Problem 13.3

Write structural formulas for the amine and the aldehyde or ketone formed when each of the following Schiff bases is reacted with water. (The reaction of a Schiff base with water to form an amine and aldehyde or ketone is called **hydrolysis.**)

a.

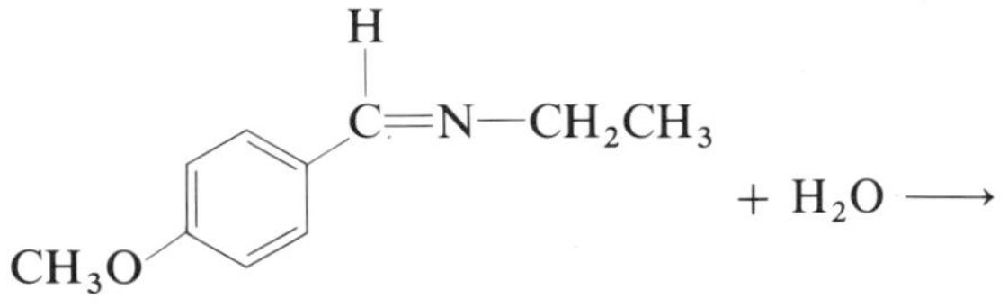

$+ H_2O \longrightarrow$

b. 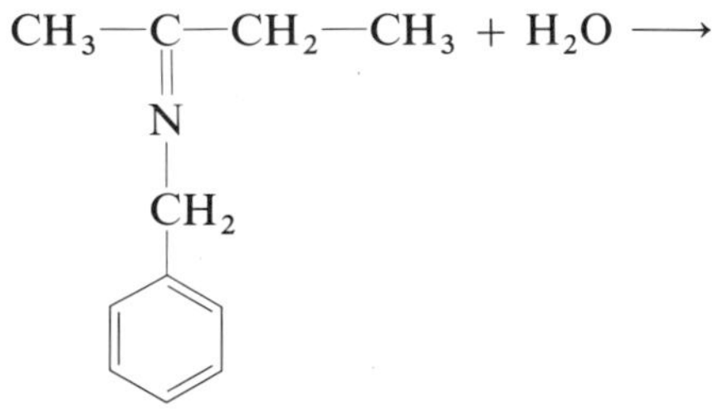 $CH_3-\underset{\underset{\underset{\underset{\bigcirc}{|}}{CH_2}}{\overset{\displaystyle \| }{N}}}{C}-CH_2-CH_3 + H_2O \longrightarrow$

C. Oxidation-Reduction of Aldehydes and Ketones

We have seen in Section 12.5C that aldehydes are oxidized to carboxylic acids by oxidizing agents such as $K_2Cr_2O_7$. Ketones are not oxidized by these reagents. Aldehydes are also oxidized to carboxylic acids by a solution of silver nitrate in ammonium hydroxide. This reagent, known as **Tollens' test,** is selective for the oxidation of aldehydes; it does not oxidize alkenes, alkynes, alcohols, or ketones.

The use of Ag^+ in NH_4OH is a convenient way to distinguish between aldehydes and other compounds that contain a carbonyl group. In Tollens' test, a solution of silver nitrate in ammonium hydroxide is added to the compound suspected of being an aldehyde. Within a few minutes at room temperature, silver ion is reduced to metallic silver and the aldehyde is oxidized to a carboxylic acid. In the alkaline medium necessary for this test, the carboxylic acid reacts with ammonium hydroxide to form a water-soluble ammonium salt.

$$\underset{\substack{\text{an} \\ \text{aldehyde}}}{R-\overset{\displaystyle O}{\overset{\|}{C}}-H} + 2\,Ag^+ + 3\,OH^- \xrightarrow{NH_4OH} \underset{\substack{\text{anion of a} \\ \text{carboxylic} \\ \text{acid}}}{R-\overset{\displaystyle O}{\overset{\|}{C}}-O^-} + 2\,Ag + 2\,H_2O$$

If the test is done properly, metallic silver deposits as a mirror on a glass surface. For this reason, the test is commonly called the **silver mirror test.**

Aldehydes are reduced to primary alcohols, and ketones to secondary alcohols, by hydrogen in the presence of a heavy metal catalyst. The most commonly used catalysts are palladium, platinum, or nickel.

Catalytic hydrogenation of an aldehyde or ketone is generally more difficult than hydrogenation of an alkene. Consequently, if a molecule contains both a carbon-carbon double bond and an aldehyde or ketone, the conditions necessary to reduce the carbonyl group may also saturate the alkene.

A second and more common method for the reduction of aldehydes and ketones uses either of two metal hydrides, lithium aluminum hydride ($LiAlH_4$) or sodium borohydride ($NaBH_4$).

$$\underset{\substack{\text{lithium aluminum} \\ \text{hydride}}}{Li^+H-\overset{\displaystyle H}{\underset{\displaystyle H}{Al}}\!=\!H} \qquad \underset{\substack{\text{sodium} \\ \text{borohydride}}}{Na^+H-\overset{\displaystyle H}{\underset{\displaystyle H}{B}}\!=\!H}$$

Reaction of either with an aldehyde or ketone gives a metal salt. Reaction of the metal salt with water gives the alcohol.

$$4\,CH_3\overset{\displaystyle O}{\overset{\|}{C}}-H + NaBH_4 \longrightarrow \underset{\substack{\text{a metal salt}}}{(CH_3CH_2O)_4B^-Na^+} \xrightarrow{H_2O} 4\,CH_3CH_2OH$$

This rearrangement of an α-hydrogen to a carbonyl oxygen is called **tautomerism.** For most aldehydes and ketones, the keto form predominates over the enol form at equilibrium by factors of well over 1000 to 1. However, when an α-hydrogen is flanked by a second carbonyl group, as in 2,4-pentanedione, the equilibrium shifts toward the enol form. Liquid 2,4-pentanedione is an equilibrium mixture of which approximately 80% is the enol form.

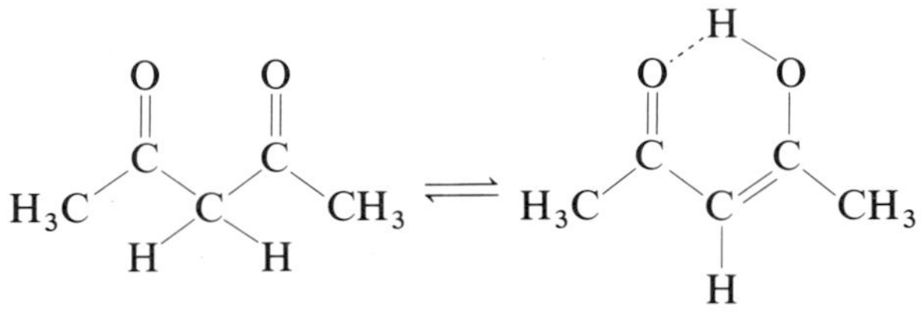

keto form (20%)   enol form (80%)

2,4-pentanedione
(acetyl acetone)

Note that the enol form of 2,4-pentanedione is stabilized by hydrogen bonding between the carbonyl oxygen and the O—H of the enol.

Example 13.5

Write the indicated number of enols for the following compounds:

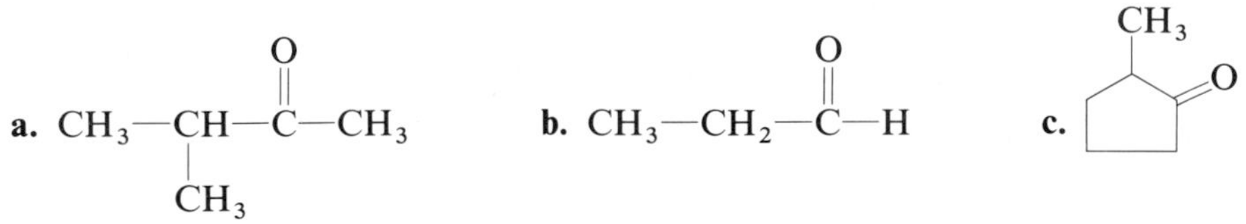

a. $CH_3-CH-C-CH_3$
      | ||
      CH_3 O

(2 enol forms)

b. CH_3-CH_2-C-H
                ||
                O

(1 enol form)

c.

(2 enol forms)

Solution

a. 3-Methyl-2-butanone has two α-carbons and forms two different enols.

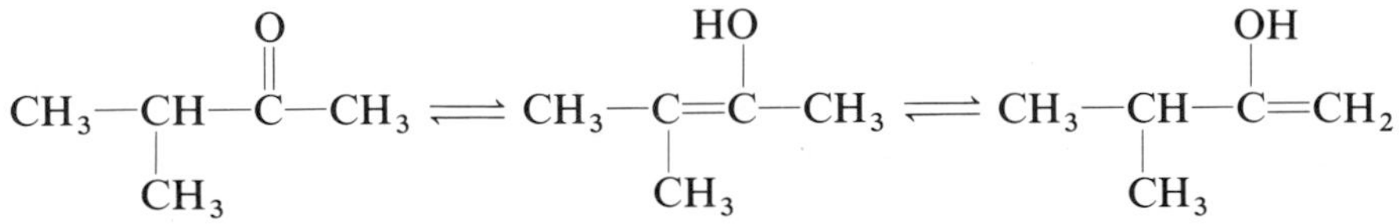

b. Propanal has only one α-carbon and forms only one enol.

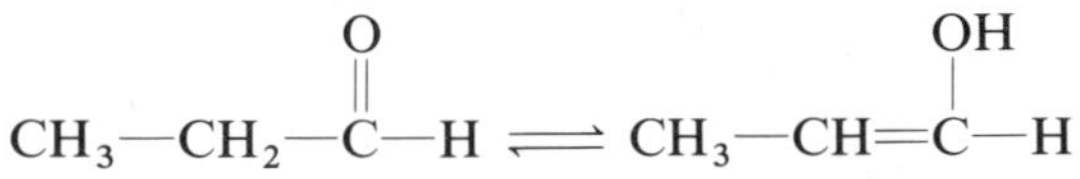

c. 2-Methylcyclopentanone has two α-carbons and forms two different enols.

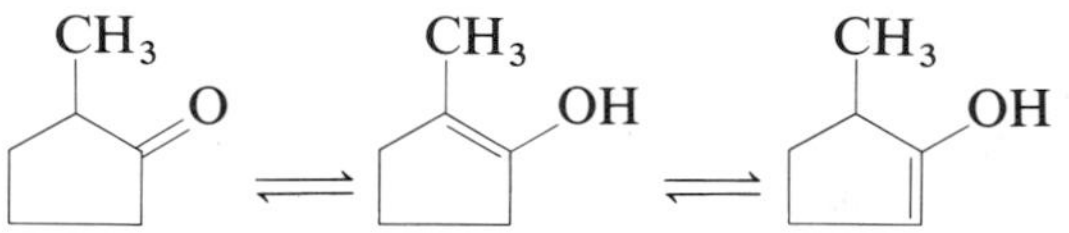

Problem 13.5

Following are structural formulas for three enols. Draw the structural formula of the keto form of each.

a. $CH_3-CH_2-CH_2-CH=CH$
    with OH on the terminal CH

b. 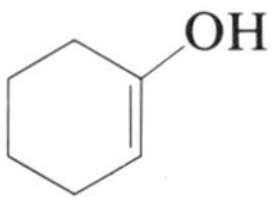 (cyclohexene with OH)

c. 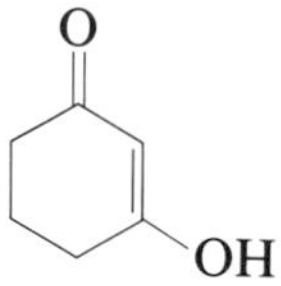 (cyclohexenone-type structure with O and OH)

E. The Aldol Condensation

Because hydrogen and carbon have similar electronegativities, there is normally no appreciable polarity to C—H bonds and no tendency for them to ionize or to show acidic properties. However, in the case of hydrogens alpha to a carbonyl group, the situation is different. The reaction of an aldehyde or ketone containing an α-hydrogen and a strong base such as hydroxide ion forms a carbon-containing anion and a molecule of water.

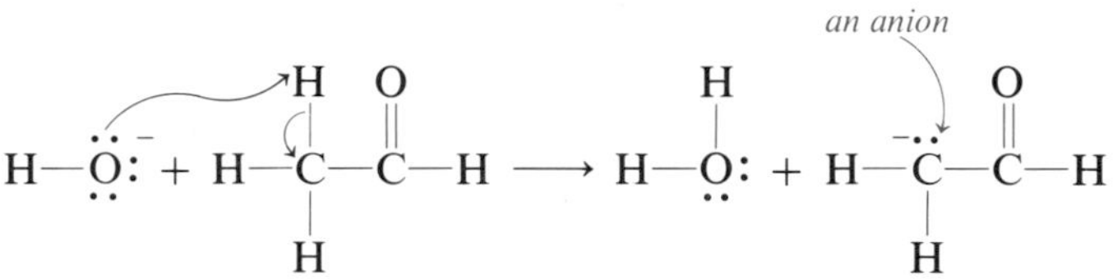

In this reaction, an α-hydrogen is removed as H^+ and the pair of electrons of the C—H bond remain on the α-carbon, giving it a formal charge of −1.

The most important reaction of an anion derived from an aldehyde or ketone is addition to the carbonyl group of another aldehyde or ketone. In this reaction, a new carbon-carbon single bond is formed between the α-carbon of one molecule and the carbonyl carbon of the other molecule. For example, reaction of two molecules of acetaldehyde in aqueous NaOH yields 3-hydroxy-butanal, commonly known as **aldol**. The name aldol is derived from the structural features of 3-hydroxybutanal: it is both an *ald*ehyde and an alcoh*ol*.

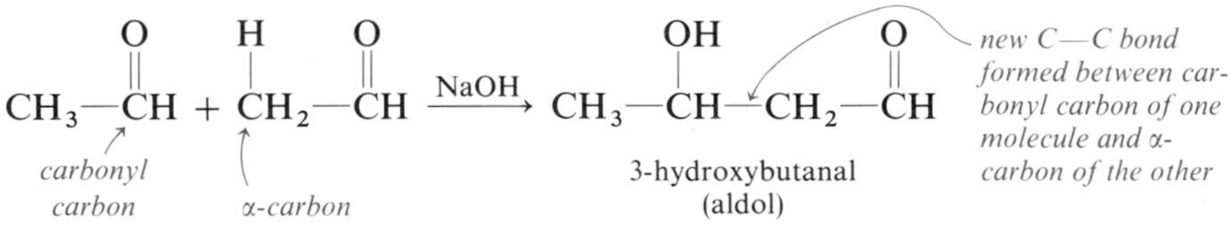

This reaction is known as an **aldol condensation.** Aldol condensation between two molecules of propanal yields 2-methyl-3-hydroxypentanal.

$$CH_3-CH_2-\overset{\overset{\displaystyle O}{\|}}{C}H + CH_2-\overset{\overset{\displaystyle O}{\|}}{C}H \xrightarrow{\text{NaOH, } H_2O} CH_3-CH_2-\overset{\overset{\displaystyle OH}{|}}{C}H-\underset{\underset{\displaystyle CH_3}{|}}{C}H-\overset{\overset{\displaystyle O}{\|}}{C}H$$

$$\underset{\underset{\displaystyle CH_3}{|}}{}$$

   propanal   propanal   2-methyl-3-hydroxypentanal

Ketones also undergo aldol condensation, as illustrated by the condensation of acetone in the presence of barium hydroxide.

$$2\ CH_3-\overset{\overset{\displaystyle O}{\|}}{C}-CH_3 \underset{}{\overset{Ba(OH)_2}{\rightleftharpoons}} CH_3-\underset{\underset{\displaystyle CH_3}{|}}{\overset{\overset{\displaystyle OH}{|}}{C}}-CH_2-\overset{\overset{\displaystyle O}{\|}}{C}-CH_3$$

4-hydroxy-4-methyl-2-pentanone
(diacetone alcohol)

The characteristic structural feature of a product of an aldol condensation is a β-hydroxyaldehyde or β-hydroxyketone.

$$\overset{\beta}{CH_3}-\overset{\overset{\displaystyle OH}{|}}{C}H-CH_2-\overset{\overset{\displaystyle O}{\|}}{C}-H \qquad CH_3-CH_2-\overset{\overset{\displaystyle OH}{|}}{C}H-\underset{\underset{\displaystyle CH_3}{|}}{C}H-\overset{\overset{\displaystyle O}{\|}}{C}-H \qquad CH_3-\underset{\underset{\displaystyle CH_3}{|}}{\overset{\overset{\displaystyle OH}{|}}{C}}-CH_2-\overset{\overset{\displaystyle O}{\|}}{C}-CH_3$$

  a β-hydroxyaldehyde   a β-hydroxyaldehyde   a β-hydroxyketone

Chemists have proposed a three-step mechanism for aldol condensation. This mechanism is illustrated for the aldol condensation of acetaldehyde to form 3-hydroxybutanal. In step 1, hydroxide ion (a strong base) forms the anion of acetaldehyde. Nucleophilic addition of this anion to the carbonyl carbon of another molecule of acetaldehyde in step 2 gives a tetrahedral carbonyl addition compound, which reacts with water in step 3 to give the final product and to regenerate hydroxide ion.

Step 1: $$H-\overset{..}{\underset{..}{O}}{:}^- + H-\overset{\overset{\displaystyle H}{|}}{\underset{\underset{\displaystyle H}{|}}{C}}-\overset{\overset{\displaystyle O}{\|}}{C}-H \rightleftharpoons H-\overset{..}{\underset{..}{O}}{:} + H-\overset{\overset{\displaystyle H}{|}}{\underset{\underset{\displaystyle H}{|}}{\overset{..}{C}}}-\overset{\overset{\displaystyle O}{\|}}{C}-H$$

Step 2: $$CH_3-\overset{\overset{\displaystyle :\overset{..}{O}}{\|}}{C}-H + H-\overset{\overset{\displaystyle ..}{}}{\underset{\underset{\displaystyle H}{|}}{C}}-\overset{\overset{\displaystyle O}{\|}}{C}-H \rightleftharpoons CH_3-\underset{\underset{\displaystyle H}{|}}{\overset{\overset{\displaystyle :\overset{..}{O}:^-}{|}}{C}}-CH_2-\overset{\overset{\displaystyle O}{\|}}{C}-H$$

Step 3: $H\overset{..}{\underset{..}{O}}: + CH_3-\underset{\underset{H}{|}}{\overset{\overset{\displaystyle :\overset{..}{\underset{..}{O}}:^-}{|}}{C}}-CH_2-\overset{\overset{\displaystyle O}{||}}{C}-H \rightleftharpoons H-\overset{..}{\underset{..}{O}}:^- + CH_3-\underset{\underset{H}{|}}{\overset{\overset{\displaystyle OH}{|}}{C}}-CH_2-\overset{\overset{\displaystyle O}{||}}{C}-H$

The ingredients in the key step of an aldol condensation are an anion and a carbonyl group. In a **self-condensation,** both roles are played by one kind of molecule. Mixed aldol condensations are also possible. Consider the mixed aldol condensation of acetone and formaldehyde. Formaldehyde cannot function as an anion because it contains no α-hydrogen, but it can function as an anion acceptor. Acetone forms an anion because it has α-hydrogens. Mixed aldol condensation of acetone and formaldehyde gives 4-hydroxy-2-butanone.

$$CH_3-\overset{\overset{\displaystyle O}{||}}{C}-CH_3 + H-\overset{\overset{\displaystyle O}{||}}{C}-H \xrightarrow{OH^-} CH_3-\overset{\overset{\displaystyle O}{||}}{C}-CH_2-\overset{\overset{\displaystyle OH}{|}}{C}H_2$$

4-hydroxy-2-butanone

In the case of mixed aldol condensations where there is no appreciable difference in reactivity between the two compounds, mixtures of products result. For example, in the condensation of equimolar quantities of propanal and butanal, both α-carbons are alike and the carbonyls are alike. As a consequence, aldol condensation in which a mixture of these two aldehydes is used results in a mixture of four possible products.

β-Hydroxyaldehydes and β-hydroxyketones are very easily dehydrated to compounds with a carbon-carbon double bond α-β to the remaining carbonyl group. Often the conditions necessary to bring about the condensation also cause dehydration. Alternatively, warming the aldol product in dilute mineral acid leads to dehydration.

$$CH_3-\overset{\overset{\displaystyle OH}{|}}{C}H-CH_2-\overset{\overset{\displaystyle O}{||}}{C}-H \xrightarrow[\text{warm}]{\text{dilute HCl}} CH_3-\overset{\beta}{C}H=\overset{\alpha}{C}H-\overset{\overset{\displaystyle O}{||}}{C}-H + H_2O$$

3-hydroxybutanal   2-butenal
(aldol)   (crotonaldehyde)

Example 13.6

Write structural formulas for the products of the following aldol condensations and of the unsaturated compounds formed by loss of water.

a. $2 \, CH_3-\overset{\overset{\displaystyle O}{||}}{C}-CH_3 \xrightarrow{\text{base}}$

b. $\langle\!\!\bigcirc\!\!\rangle-\overset{\overset{\displaystyle O}{||}}{C}H + CH_3-\overset{\overset{\displaystyle O}{||}}{C}-CH_3 \xrightarrow{\text{base}}$

Solution

a. $2\ CH_3{-}\overset{\displaystyle O}{\overset{\|}{C}}{-}CH_3 \xrightarrow[\text{condensation}]{\text{aldol}} CH_3{-}\underset{\underset{\displaystyle CH_3}{|}}{\overset{\displaystyle OH}{\overset{|}{C}}}{-}CH_2{-}\overset{\displaystyle O}{\overset{\|}{C}}{-}CH_3 \xrightarrow[-H_2O]{}$

$$CH_3{-}\underset{\underset{\displaystyle CH_3}{|}}{C}{=}CH{-}\overset{\displaystyle O}{\overset{\|}{C}}{-}CH_3$$

b. (phenyl)$-\overset{\displaystyle O}{\overset{\|}{C}}H + CH_3{-}\overset{\displaystyle O}{\overset{\|}{C}}{-}CH_3 \xrightarrow[]{\text{aldol condensation}}$

(phenyl)$-\overset{\displaystyle OH}{\overset{|}{C}}H{-}CH_2{-}\overset{\displaystyle O}{\overset{\|}{C}}{-}CH_3 \xrightarrow[-H_2O]{}$ (phenyl)$-CH{=}CH{-}\overset{\displaystyle O}{\overset{\|}{C}}{-}CH_3$

Problem 13.6

Draw structural formulas for the products of the following aldol condensations and for the unsaturated compounds produced by loss of water.

a. (phenyl)$-\overset{\displaystyle O}{\overset{\|}{C}}{-}CH_3 \xrightarrow{\text{base}}$

b. (phenyl)$-\overset{\displaystyle O}{\overset{\|}{C}}{-}H + CH_3{-}\overset{\displaystyle O}{\overset{\|}{C}}{-}H \xrightarrow{\text{base}}$

The double bonds of alkenes, aldehydes, and ketones can be reduced by catalytic hydrogenation or by chemical means. Hence, aldol condensation is often used for the preparation of saturated alcohols. For example, acetaldehyde is converted into 1-butanol by first making 2-butenal.

$2\ CH_3{-}\overset{\displaystyle O}{\overset{\|}{C}}{-}H \xrightarrow[\text{dehydration}]{\substack{\text{aldol}\\ \text{condensation}\\ \text{then}}} \underset{\text{2-butenal}}{CH_3{-}CH{=}CH{-}\overset{\displaystyle O}{\overset{\|}{C}}{-}H} \xrightarrow{2\ H_2/Pt} \underset{\text{1-butanol}}{CH_3{-}CH_2{-}CH_2{-}CH_2{-}OH}$

Alternatively, if the β-hydroxyaldehyde is isolated, selective oxidation of the aldehyde with silver nitrate in ammonia solution produces a β-hydroxy-carboxylic acid.

$\underset{\text{aldol}}{CH_3{-}\overset{\displaystyle OH}{\overset{|}{C}}H{-}CH_2{-}\overset{\displaystyle O}{\overset{\|}{C}}{-}H} + 2\ Ag^+ \xrightarrow{NH_4OH} \underset{\substack{\text{3-hydroxybutanoic acid}\\ (\beta\text{-hydroxybutyric acid})}}{CH_3{-}\overset{\displaystyle OH}{\overset{|}{C}}H{-}CH_2{-}\overset{\displaystyle O}{\overset{\|}{C}}{-}OH} + 2\ Ag$

Example 13.7   Show reagents and conditions to illustrate how the following products can be synthesized from the indicated starting materials by way of aldol condensation reactions.

a. $CH_3-\overset{\overset{\displaystyle O}{\|}}{C}-CH_3 \longrightarrow CH_3-\underset{\underset{\displaystyle CH_3}{|}}{CH}-CH_2-\overset{\overset{\displaystyle OH}{|}}{CH}-CH_3$

b. $\langle\!\!\!\bigcirc\!\!\!\rangle-\overset{\overset{\displaystyle O}{\|}}{C}-H + CH_3-\overset{\overset{\displaystyle O}{\|}}{C}-H \longrightarrow \langle\!\!\!\bigcirc\!\!\!\rangle-CH=CH-\overset{\overset{\displaystyle O}{\|}}{C}-OH$

Solution

a. $CH_3-\overset{\overset{\displaystyle O}{\|}}{C}-CH_3 \xrightarrow[\substack{\text{aldol} \\ \text{condensation}}]{OH^-} CH_3-\underset{\underset{\displaystyle CH_3}{|}}{\overset{\overset{\displaystyle OH}{|}}{C}}-CH_2-\overset{\overset{\displaystyle O}{\|}}{C}-CH_3 \xrightarrow[-H_2O]{H^+,\ \text{warm}}$

$CH_3-\underset{\underset{\displaystyle CH_3}{|}}{C}=CH-\overset{\overset{\displaystyle O}{\|}}{C}-CH_3 \xrightarrow{2\ H_2/Pt} CH_3-\underset{\underset{\displaystyle CH_3}{|}}{CH}-CH_2-\overset{\overset{\displaystyle OH}{|}}{CH}-CH_3$

Aldol condensation of acetone yields 4-methyl-4-hydroxy-2-pentanone. Warming this β-hydroxyketone in acid leads to dehydration and formation of an α,β-unsaturated ketone. Catalytic hydrogenation reduces both the carbon-carbon double bond and the ketone to form the desired alcohol.

b. $\langle\!\!\!\bigcirc\!\!\!\rangle-\overset{\overset{\displaystyle O}{\|}}{C}-H + CH_3-\overset{\overset{\displaystyle O}{\|}}{C}-H \xrightarrow[\text{NaOH}]{\substack{\text{aldol} \\ \text{condensation}}}$

$\langle\!\!\!\bigcirc\!\!\!\rangle-\overset{\overset{\displaystyle OH}{|}}{CH}-CH_2-\overset{\overset{\displaystyle O}{\|}}{C}-H \xrightarrow[-H_2O]{\text{warm}} \langle\!\!\!\bigcirc\!\!\!\rangle-CH=CH-\overset{\overset{\displaystyle O}{\|}}{C}-H \xrightarrow{Ag(NH_3)_2^+}$

$\langle\!\!\!\bigcirc\!\!\!\rangle-CH=CH-\overset{\overset{\displaystyle O}{\|}}{C}-OH$

Mixed aldol condensation between benzaldehyde and acetaldehyde produces a β-hydroxyaldehyde that undergoes dehydration under the conditions used for the aldol condensation. Oxidation of the aldehyde using silver ion in NH_4OH (Tollens' test) forms the desired cinnamic acid.

Problem 13.7

Show reagents and conditions to illustrate how the following products can be synthesized from the indicated starting materials by way of aldol condensations.

a. 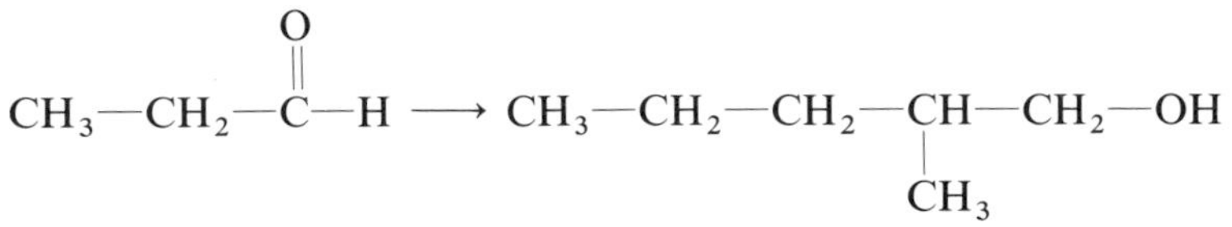 $CH_3-CH_2-\overset{\overset{\displaystyle O}{\|}}{C}-H \longrightarrow CH_3-CH_2-CH_2-\underset{\underset{\displaystyle CH_3}{|}}{CH}-CH_2-OH$

b. 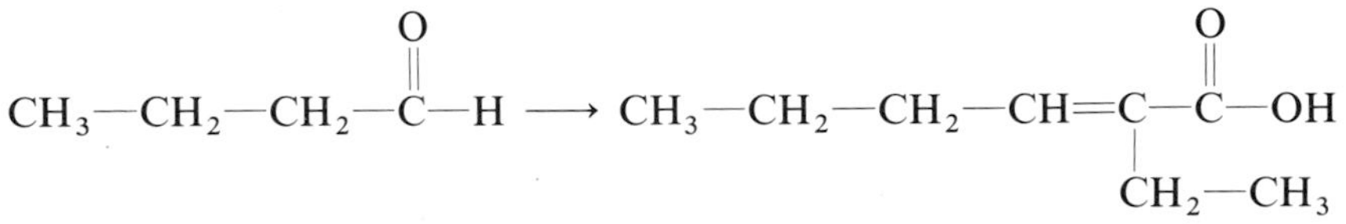 $CH_3-CH_2-CH_2-\overset{\overset{\displaystyle O}{\|}}{C}-H \longrightarrow CH_3-CH_2-CH_2-CH=\underset{\underset{\displaystyle CH_2-CH_3}{|}}{C}-\overset{\overset{\displaystyle O}{\|}}{C}-OH$

Key Terms and Concepts

acetal (13.5A)
aldehyde (13.1)
aldol condensation (13.5E)
carbonyl group (introduction)
enol (13.5D)
hemiacetal (13.5A)
hemiketal (13.5A)
ketal (13.5A)
ketone (13.1)

nucleophile (13.5)
nucleophilic addition to a carbonyl
  group (13.4)
polarity of a carbonyl group (13.4)
Schiff base (13.5B)
tautomerism (13.5D)
tetrahedral carbonyl addition
  compound (13.5)
Tollens' test (13.5C)

Key Reactions

1. Addition of water (hydration) of aldehydes and ketones (Section 13.5):

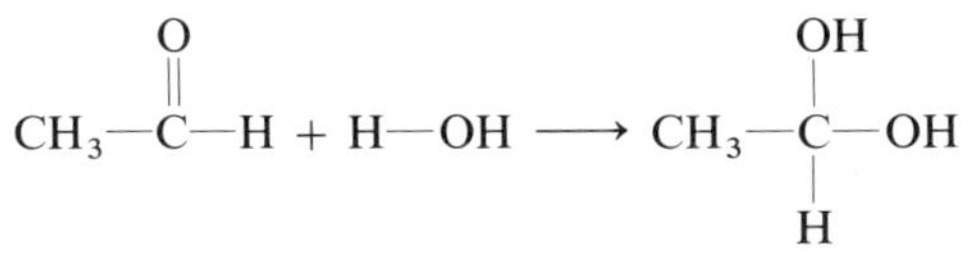

$$CH_3-\overset{\overset{\displaystyle O}{\|}}{C}-H + H-OH \longrightarrow CH_3-\underset{\underset{\displaystyle H}{|}}{\overset{\overset{\displaystyle OH}{|}}{C}}-OH$$

2. Addition of alcohols: formation of acetals and ketals (Section 13.5A):

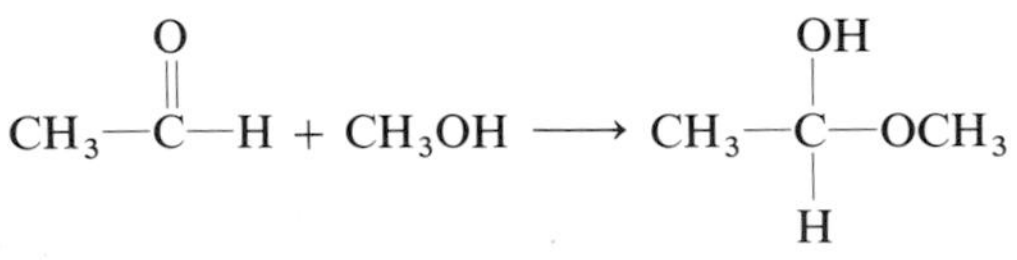

$$CH_3-\overset{\overset{\displaystyle O}{\|}}{C}-H + CH_3OH \longrightarrow CH_3-\underset{\underset{\displaystyle H}{|}}{\overset{\overset{\displaystyle OH}{|}}{C}}-OCH_3$$

a hemiacetal

$$CH_3-\underset{\underset{H}{|}}{\overset{\overset{OH}{|}}{C}}-OCH_3 + CH_3OH \xrightarrow{H^+} CH_3-\underset{\underset{H}{|}}{\overset{\overset{OCH_3}{|}}{C}}-OCH_3 + H_2O$$

an acetal

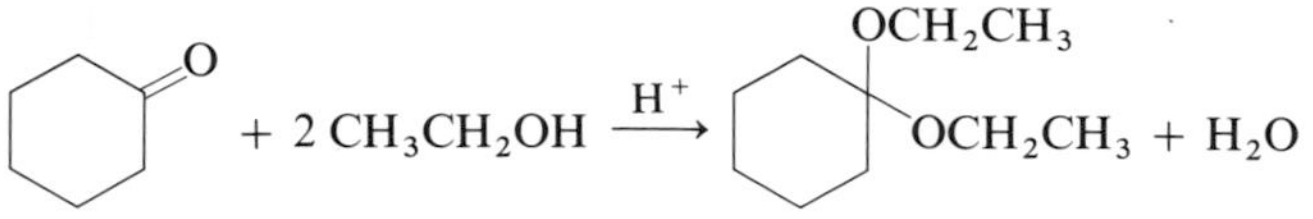

$$\text{(cyclohexanone)} + 2\ CH_3CH_2OH \xrightarrow{H^+} \text{(cyclohexane with } OCH_2CH_3 \text{ and } OCH_2CH_3) + H_2O$$

a ketal

3. Addition of ammonia and its derivatives to form Schiff bases (Section 13.5B):

$$\underset{H_3C}{\overset{H_3C}{>}}C=O + H_2N-CH_2CH_2CH_3 \longrightarrow \underset{H_3C}{\overset{H_3C}{>}}C=N-CH_2CH_2CH_3 + H_2O$$

a Schiff base

4. Oxidation of aldehydes to carboxylic acids (Section 13.5C):

$$CH_3-\overset{\overset{O}{\|}}{C}-H + Ag^+ \xrightarrow{NH_4OH} CH_3-\overset{\overset{O}{\|}}{C}-OH + Ag$$

$$CH_3-\overset{\overset{O}{\|}}{C}-H + Cr_2O_7^{2-} \xrightarrow{H^+} CH_3-\overset{\overset{O}{\|}}{C}-OH + Cr^{3+}$$

5. Reduction of aldehydes and ketones to alcohols (Section 13.5C):

$$CH_3-CH_2-CH_2-\overset{\overset{O}{\|}}{C}-H + H_2 \xrightarrow{Pt} CH_3-CH_2-CH_2-CH_2-OH$$

$$CH_3-CH_2-\overset{\overset{O}{\|}}{C}-CH_3 \xrightarrow{LiAlH_4} \xrightarrow{H_2O} CH_3-CH_2-\underset{\underset{}{\overset{\overset{OH}{|}}{C}}}{C}H-CH_3$$

6. Tautomerism (Section 13.5D):

$$CH_3-\overset{\overset{O}{\|}}{C}-CH_3 \rightleftharpoons CH_3-\overset{\overset{OH}{|}}{C}=CH_2$$

keto form   enol form

7. Aldol condensation (Section 13.5E):

$$2\ CH_3-\overset{\overset{O}{\|}}{C}-H \xrightarrow{NaOH} CH_3-\overset{\overset{OH}{|}}{C}H-CH_2-\overset{\overset{O}{\|}}{C}-H$$

$$2\ CH_3-\overset{\overset{O}{\|}}{C}-CH_3 \xrightarrow{Ba(OH)_2} CH_3-\underset{\underset{CH_3}{|}}{\overset{\overset{OH}{|}}{C}}-CH_2-\overset{\overset{O}{\|}}{C}-CH_3$$

8. Acid-catalyzed dehydration of the products of aldol condensations (Section 13.5E):

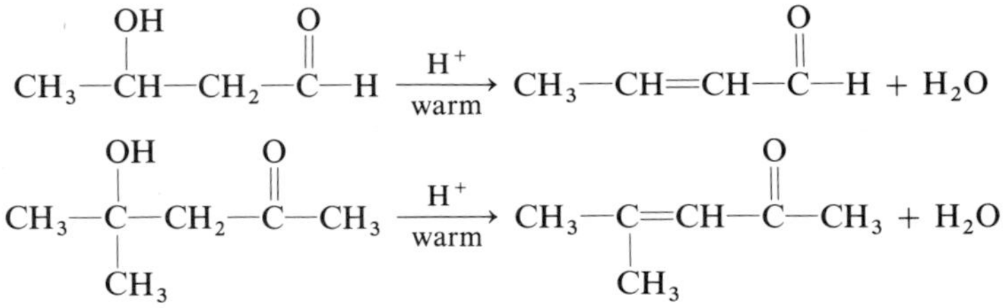

$$CH_3-\underset{\underset{\displaystyle OH}{|}}{CH}-CH_2-\overset{\overset{\displaystyle O}{\|}}{C}-H \xrightarrow[\text{warm}]{H^+} CH_3-CH=CH-\overset{\overset{\displaystyle O}{\|}}{C}-H + H_2O$$

$$CH_3-\underset{\underset{\displaystyle CH_3}{|}}{\overset{\overset{\displaystyle OH}{|}}{C}}-CH_2-\overset{\overset{\displaystyle O}{\|}}{C}-CH_3 \xrightarrow[\text{warm}]{H^+} CH_3-\underset{\underset{\displaystyle CH_3}{|}}{C}=CH-\overset{\overset{\displaystyle O}{\|}}{C}-CH_3 + H_2O$$

Problems

Structure and Nomenclature of Aldehydes and Ketones (Sections 13.1 and 13.2)

13.8   Name each of the following compounds.

a. 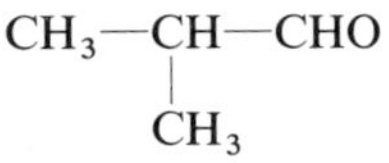 $CH_3-\underset{\underset{\displaystyle CH_3}{|}}{CH}-CHO$

b. 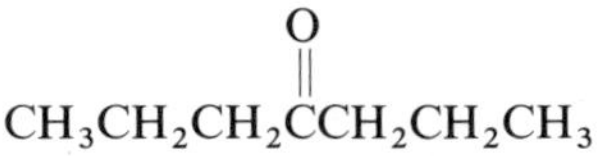 $CH_3CH_2CH_2\overset{\overset{\displaystyle O}{\|}}{C}CH_2CH_2CH_3$

c. (cyclopentanone with CH_3)

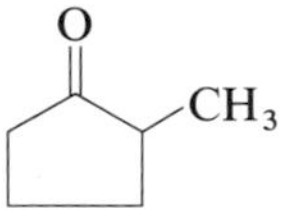

d. 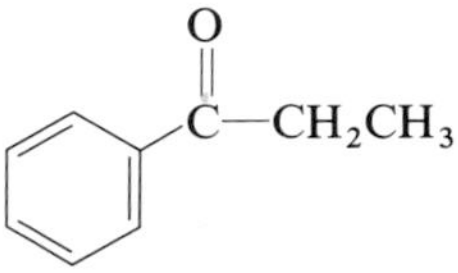 $\overset{\overset{\displaystyle O}{\|}}{C}-CH_2CH_3$ (phenyl)

e. 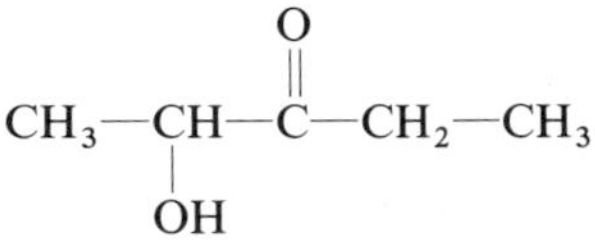 $CH_3-\underset{\underset{\displaystyle OH}{|}}{CH}-\overset{\overset{\displaystyle O}{\|}}{C}-CH_2-CH_3$

f. 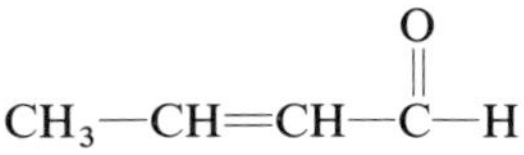 $CH_3-CH=CH-\overset{\overset{\displaystyle O}{\|}}{C}-H$

g. 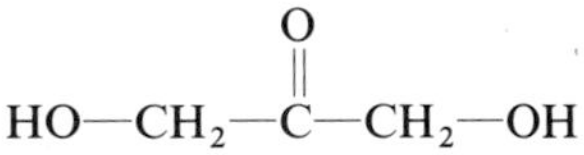 $HO-CH_2-\overset{\overset{\displaystyle O}{\|}}{C}-CH_2-OH$

h. $\underset{\underset{\displaystyle HO}{|}}{CH_2}-\underset{\underset{\displaystyle OH}{|}}{CH}-\overset{\overset{\displaystyle O}{\|}}{C}-H$

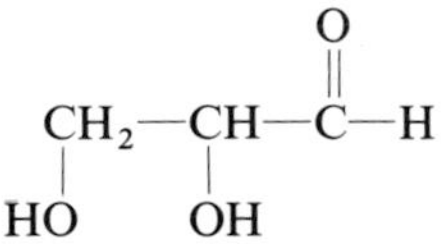

i. 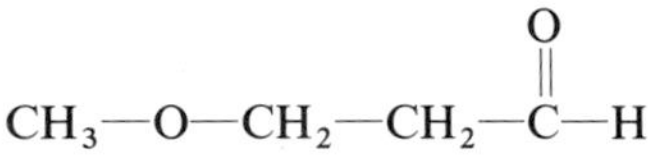 $CH_3-O-CH_2-CH_2-\overset{\overset{\displaystyle O}{\|}}{C}-H$

j. 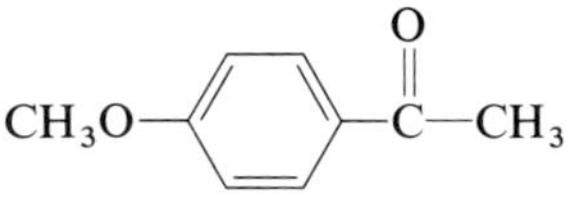 CH_3O- $-\overset{\overset{\displaystyle O}{\|}}{C}-CH_3$

k. 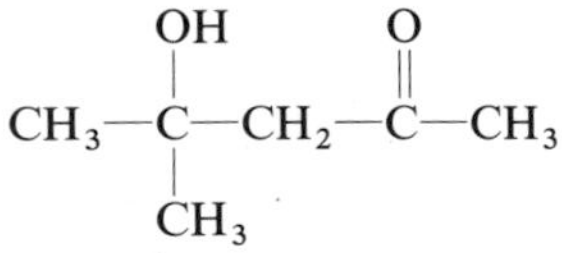 $CH_3-\underset{\underset{\displaystyle CH_3}{|}}{\overset{\overset{\displaystyle OH}{|}}{C}}-CH_2-\overset{\overset{\displaystyle O}{\|}}{C}-CH_3$

l. 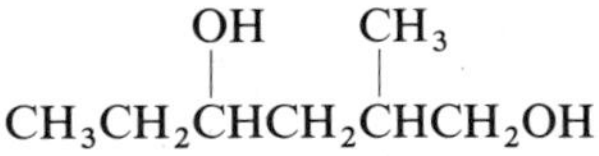 $CH_3CH_2\underset{\underset{\displaystyle OH}{|}}{CH}CH_2\underset{\underset{\displaystyle CH_3}{|}}{CH}CH_2OH$

m. 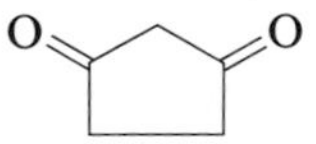 (cyclopentane-1,3-dione)

n. 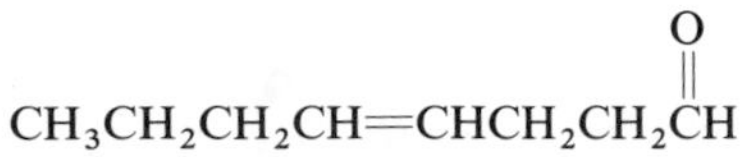 $CH_3CH_2CH_2CH=CHCH_2CH_2\overset{\overset{\displaystyle O}{\|}}{CH}$

13.9 Write structural formulas for the following compounds.

a. cycloheptanone
b. propanal
c. 2-methylpropanal
d. benzaldehyde
e. 3,3-dimethyl-2-butanone
f. 2,4-pentanedione
g. hexanal
h. 2-decanone
i. propenal
j. *p*-bromocinnamaldehyde
k. 3-methoxy-4-hydroxybenzaldehyde (vanillin from the vanilla bean)
l. 3-phenyl-2-propenal (from oil of cinnamon)
m. 3,7-dimethyl-2,6-octadienal (from orange blossom oil)

13.10 Draw the structures for all aldehydes of formula C_4H_8O; of formula $C_5H_{10}O$. Give each an IUPAC name.

13.11 Draw the structures of all ketones of formula C_4H_8O; of formula $C_5H_{10}O$. Give each an IUPAC name.

Reactions – Acetals and Ketals; Schiff Bases; Oxidation-Reduction (Sections 13.5A–C)

13.12 Complete the following reactions. Where there is no reaction, write N.R.

a. $CH_3CH_2CH_2\overset{\overset{\displaystyle O}{\|}}{C}H + H_2 \xrightarrow{Pt}$

b. 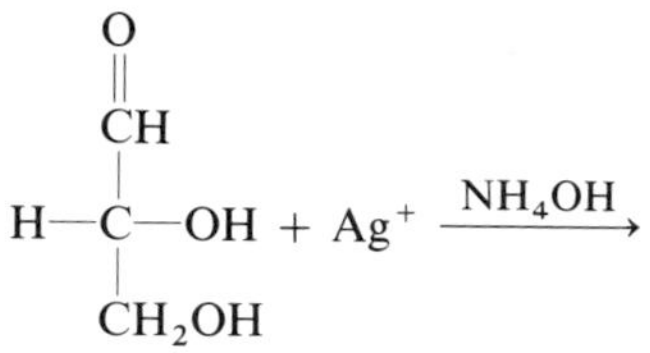 $H-\overset{\overset{\displaystyle \overset{\overset{O}{\|}}{CH}}{|}}{\underset{\underset{\displaystyle CH_2OH}{|}}{C}}-OH + Ag^+ \xrightarrow{NH_4OH}$

c. $CH_3-\overset{\overset{\displaystyle OH}{|}}{C}H-CH_2-\overset{\overset{\displaystyle O}{\|}}{C}-CH_3 + K_2Cr_2O_7 \xrightarrow{H^+}$

d. $CH_3-\overset{\overset{\displaystyle OH}{|}}{C}H-CH_2-\overset{\overset{\displaystyle O}{\|}}{C}-CH_3 + H_2 \xrightarrow{Pt}$

e. 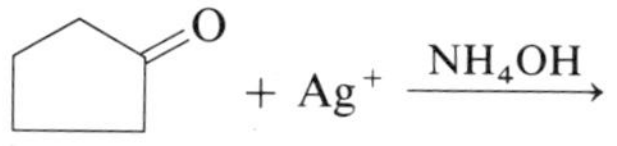 (cyclopentanone) $+ Ag^+ \xrightarrow{NH_4OH}$

f. 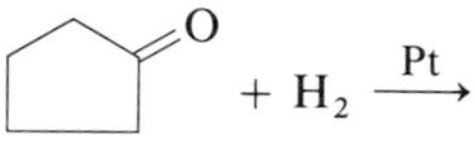 (cyclopentanone) $+ H_2 \xrightarrow{Pt}$

g. $CH_3CH_2CH_2\overset{\overset{\displaystyle O}{\|}}{C}H + 1\ CH_3CH_2OH \longrightarrow$

h. $CH_3CH_2CH_2\overset{\overset{\displaystyle O}{\|}}{C}H + 2\ CH_3CH_2OH \xrightarrow{H^+}$

i. 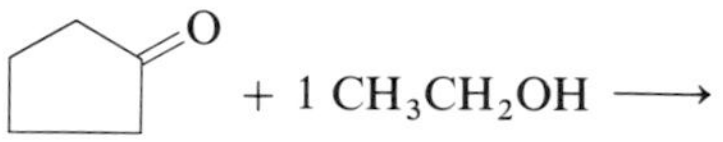 (cyclopentanone) $+ 1\ CH_3CH_2OH \longrightarrow$

j. 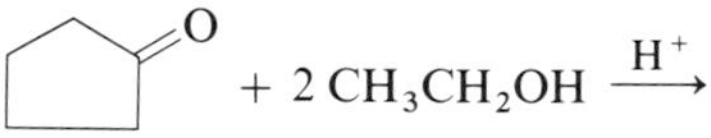 (cyclopentanone) $+ 2\ CH_3CH_2OH \xrightarrow{H^+}$

k. $\underset{H_3C}{\overset{H_3C}{>}}C\underset{OCH_3}{\overset{OCH_3}{<}} + H_2O \xrightarrow{H^+}$

l. 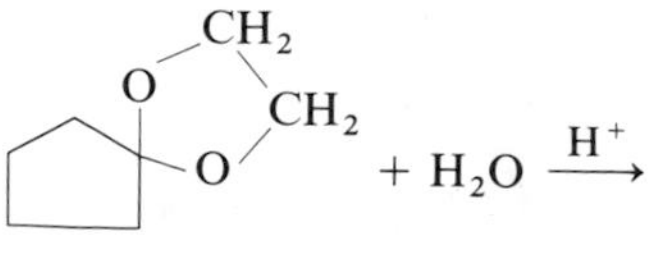 (cyclopentane with dioxolane-type ring, $O-CH_2$, $O-CH_2$) $+ H_2O \xrightarrow{H^+}$

m. 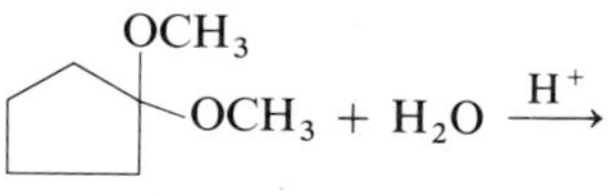 (cyclopentane with OCH_3, OCH_3) $+ H_2O \xrightarrow{H^+}$

n. $\xrightarrow[\text{(2) H}_2\text{O}]{\text{(1) LiAlH}_4}$ o. $CH_3-\underset{\underset{CH_3}{|}}{C}=CH-\overset{\overset{O}{\|}}{C}-CH_3 \xrightarrow[\text{(2) H}_2\text{O}]{\text{(1) NaBH}_4}$

13.13 Show how you might distinguish between the following pairs of compounds by a simple chemical test. In each case, tell what test you would perform, what you would expect to observe, and write an equation for each positive test.
a. benzaldehyde and acetophenone
b. benzaldehyde and benzyl alcohol

c. CH_2OH CHO
    $CHOH$ $CHOH$
    CH_2OH CH_2OH

    glycerol   glyceraldehyde

13.14 Propose a mechanism to account for the formation of cyclic acetal from 4-hydroxy-pentanal and one molecule of methyl alcohol:

$$CH_3CHCH_2CH_2\overset{\overset{O}{\|}}{C}H + CH_3OH \xrightarrow{H^+} H_3C\text{—}O\text{—}OCH_3 + H_2O$$
$$\underset{OH}{|}$$

If the carbonyl oxygen of 4-hydroxypentanal were enriched with oxygen-18, would you predict that the oxygen-18 would appear in the cyclic acetal or in the water?

13.15 5-Hydroxyhexanal readily forms a six-member hemiacetal:

$$CH_3-\underset{\underset{OH}{|}}{CH}-CH_2-CH_2-CH_2-\overset{\overset{O}{\|}}{C}-H \longrightarrow \text{a cyclic hemiacetal}$$

5-hydroxyhexanal

a. Draw a structural formula for this cyclic hemiacetal.
b. How many cis-trans isomers are possible for this cyclic hemiacetal?
c. Draw planar hexagon representations for each cis-trans isomer.
d. Draw a chair conformation for the cyclic hemiacetal in which the —OH and —CH_3 groups are equatorial. Is this a cis- or a trans- isomer?

13.16 Acetaldehyde reacts with ethylene glycol in the presence of a trace of sulfuric acid to give a cyclic acetal of formula $C_4H_8O_2$. Draw the structural formula of this acetal and propose a mechanism for its formation.

13.17 The following conversions can be carried out in either one or two steps. Show the reagents you would use for each step and draw structural formulas for the inter-mediate formed in any conversion that requires two steps.

a. $CH_3-\underset{\underset{OH}{|}}{CH}-CO_2H \longrightarrow CH_3-\overset{\overset{O}{\|}}{C}-CO_2H$

b. 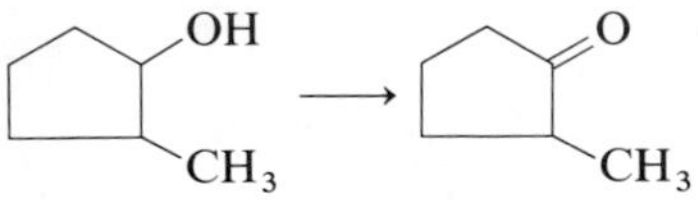 [cyclopentane with OH and CH₃] ⟶ [cyclopentanone with CH₃]

c. 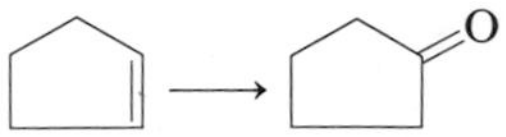 [cyclopentene] ⟶ [cyclopentanone]

d. $CH_3-\overset{\displaystyle OH}{\underset{\displaystyle \underset{CH_3}{|}}{CH}}-CH-CH_2-\overset{\displaystyle O}{\overset{\|}{C}}-H \longrightarrow CH_3-\overset{\displaystyle O}{\overset{\|}{C}}-\underset{\underset{CH_3}{|}}{CH}-CH_2-\overset{\displaystyle O}{\overset{\|}{C}}-OH$

e. 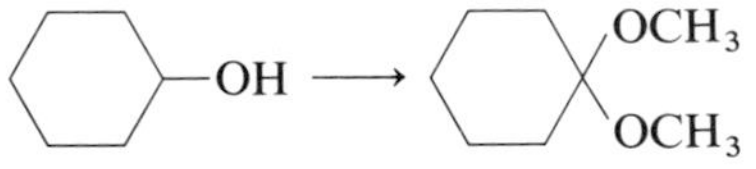 [cyclohexane]—OH ⟶ [cyclohexane with OCH₃, OCH₃]

f. 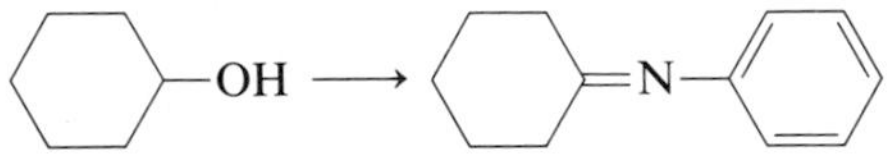 [cyclohexane]—OH ⟶ [cyclohexane]=N—[phenyl]

g. $CH_3-CH{=}CH_2 \longrightarrow CH_3-\overset{\displaystyle O}{\overset{\|}{C}}-CH_3$

Reactions of Aldehydes and Ketones – Tautomerism (Section 13.5D)

13.18 What is meant by the term *keto-enol tautomerism?*

13.19 Draw the indicated number of enol structures for the following aldehydes and ketones.

a. $CH_3CH_2CH_2\overset{\displaystyle O}{\overset{\|}{C}}H$

(1 enol form)

b. $CH_3CH_2\overset{\displaystyle O}{\overset{\|}{C}}CH_3$

(2 enol forms)

c. 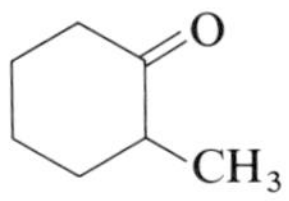 [cyclohexanone with CH₃]

(2 enol forms)

d. 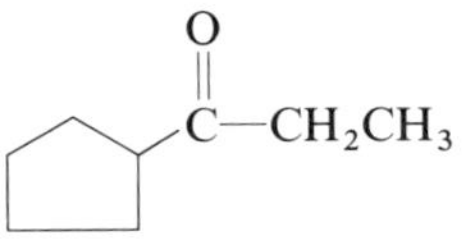 [cyclopentane ring]—$\overset{\displaystyle O}{\overset{\|}{C}}$—$CH_2CH_3$

(2 enol forms)

13.20 The following are enols. Draw structural formulas for the keto form of each.

a. $CH_2{=}\overset{\underset{\displaystyle |}{OH}}{CH}$

b. 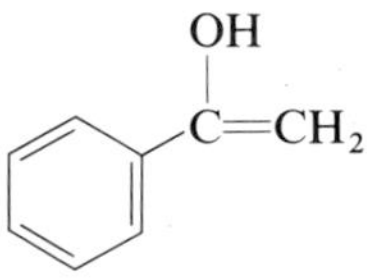 [phenyl]—$\overset{\underset{\displaystyle |}{OH}}{C}{=}CH_2$

c. [cyclopentane ring]$=\overset{\underset{\displaystyle |}{OH}}{C}-CH_3$

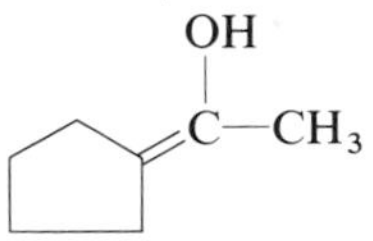

d. $CH_3-\overset{\displaystyle O}{\overset{\|}{C}}-CH{=}\overset{\underset{\displaystyle |}{OH}}{C}-CH_3$

e. $CH_3-\overset{\displaystyle O}{\overset{\|}{C}}-\overset{\underset{\displaystyle |}{OH}}{C}{=}CH-CH_3$

13.21 The following compound is an **enediol,** a compound with two hydroxyl groups on a carbon-carbon double bond. Draw structural formulas for the two carbonyl-containing compounds with which this enediol is in equilibrium.

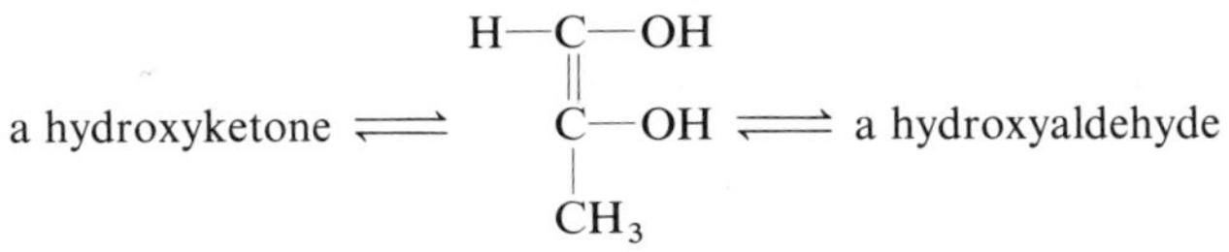

$$
\text{H—C—OH}
$$

a hydroxyketone $\rightleftharpoons$
$$
\begin{array}{c}
\text{C—OH} \\
| \\
\text{CH}_3
\end{array}
$$
 $\rightleftharpoons$ a hydroxyaldehyde

an enediol

13.22 How would you account for the fact that in dilute aqueous NaOH, glyceraldehyde is converted into an equilibrium mixture of glyceraldehyde and dihydroxyacetone?

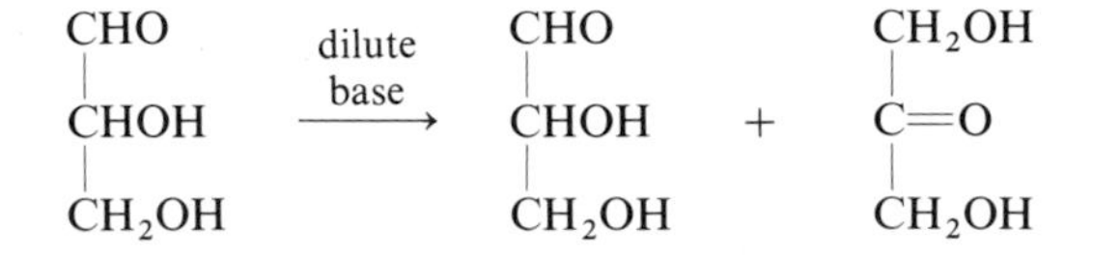

$$
\begin{array}{c}
\text{CHO} \\
| \\
\text{CHOH} \\
| \\
\text{CH}_2\text{OH}
\end{array}
\quad
\xrightarrow[\text{base}]{\text{dilute}}
\quad
\begin{array}{c}
\text{CHO} \\
| \\
\text{CHOH} \\
| \\
\text{CH}_2\text{OH}
\end{array}
\quad + \quad
\begin{array}{c}
\text{CH}_2\text{OH} \\
| \\
\text{C=O} \\
| \\
\text{CH}_2\text{OH}
\end{array}
$$

glyceraldehyde   glyceraldehyde   dihydroxyacetone

Reactions of Aldehydes and Ketones – Aldol Condensation (Section 13.5E)

13.23 Draw structural formulas for the products of the following aldol condensations and of the unsaturated compounds formed by loss of water from the aldol product.

a. 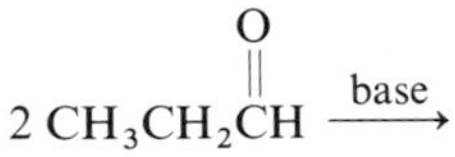 2 CH$_3$CH$_2$CH (O) $\xrightarrow{\text{base}}$

b. 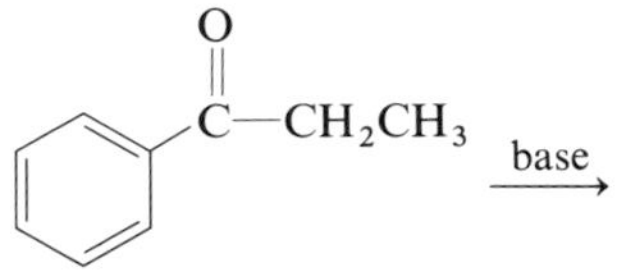 phenyl-C(=O)-CH$_2$CH$_3$ $\xrightarrow{\text{base}}$

c. cyclohexanone $\xrightarrow{\text{base}}$

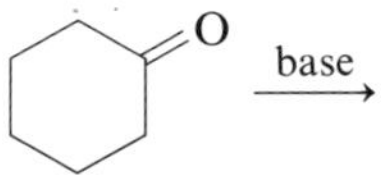

d. 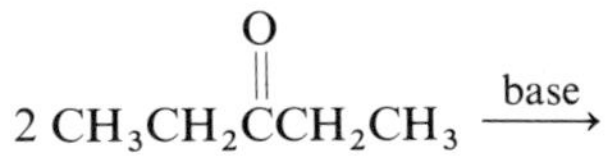 2 CH$_3$CH$_2$CCH$_2$CH$_3$ (O) $\xrightarrow{\text{base}}$

13.24 Draw structural formulas for the products of the following mixed aldol condensations and of the unsaturated compounds formed by loss of water from the aldol product.

a. 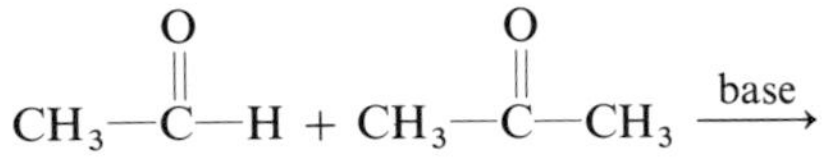 CH$_3$—C—H + CH$_3$—C—CH$_3$ $\xrightarrow{\text{base}}$

b. phenyl-C(=O)-H + CH$_3$—C-phenyl $\xrightarrow{\text{base}}$

c. cyclohexanone + H—C—H $\xrightarrow{\text{base}}$

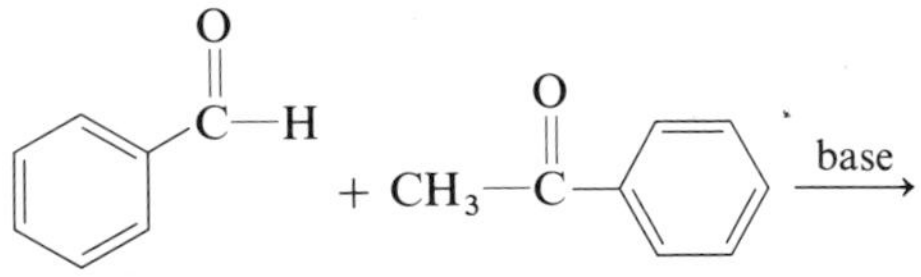

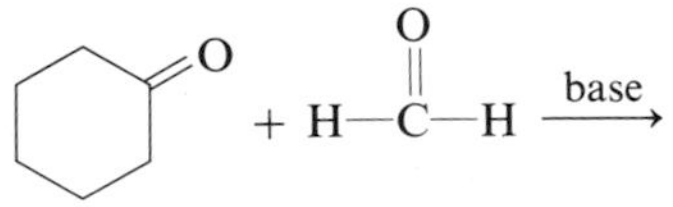

13.25 Show reagents and conditions to illustrate how the following products can be synthesized from the indicated starting materials by way of aldol condensation reactions.

a. $CH_3CHO \longrightarrow CH_3-CH_2-CH_2-CH_2OH$

b. $CH_3CHO \longrightarrow CH_3-CH=CH-CO_2H$

c. 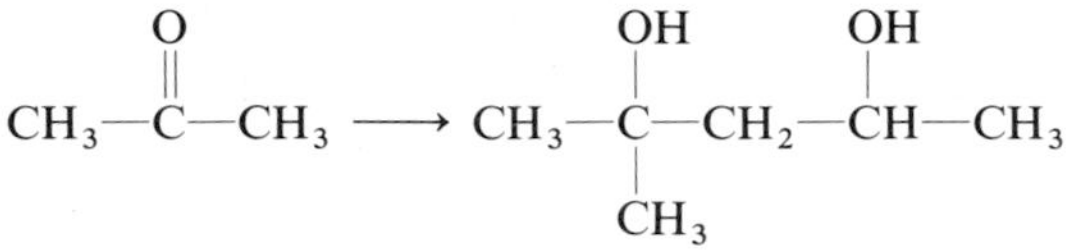
$$CH_3-\overset{\overset{\displaystyle O}{\|}}{C}-CH_3 \longrightarrow CH_3-\underset{\underset{\displaystyle CH_3}{|}}{\overset{\overset{\displaystyle OH}{|}}{C}}-CH_2-\overset{\overset{\displaystyle OH}{|}}{CH}-CH_3$$

d. 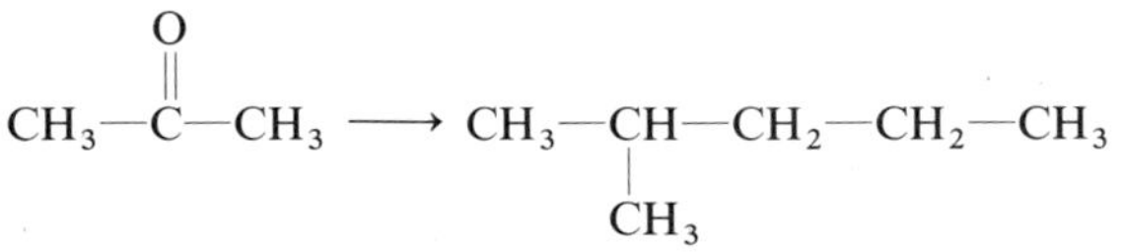
$$CH_3-\overset{\overset{\displaystyle O}{\|}}{C}-CH_3 \longrightarrow CH_3-\underset{\underset{\displaystyle CH_3}{|}}{CH}-CH_2-CH_2-CH_3$$

e. 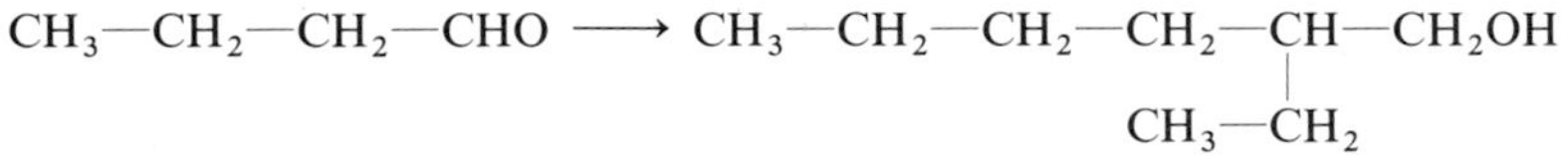 $CH_3-CH_2-CH_2-CHO \longrightarrow CH_3-CH_2-CH_2-CH_2-\underset{\underset{\displaystyle CH_3-CH_2}{|}}{CH}-CH_2OH$

f. 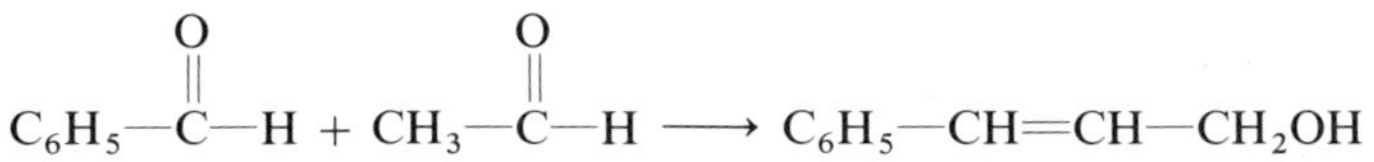 $C_6H_5-\overset{\overset{\displaystyle O}{\|}}{C}-H + CH_3-\overset{\overset{\displaystyle O}{\|}}{C}-H \longrightarrow C_6H_5-CH=CH-CH_2OH$

g. 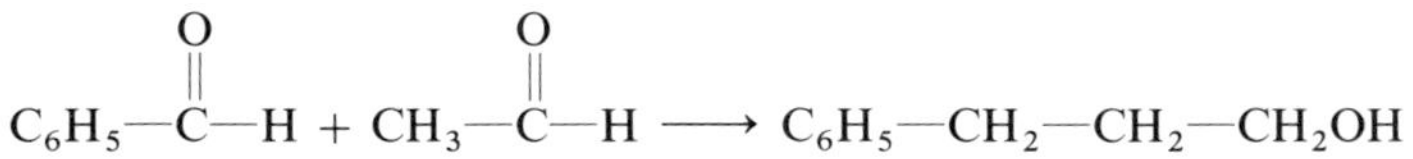 $C_6H_5-\overset{\overset{\displaystyle O}{\|}}{C}-H + CH_3-\overset{\overset{\displaystyle O}{\|}}{C}-H \longrightarrow C_6H_5-CH_2-CH_2-CH_2OH$

Pheromones

Chemical communication abounds in nature: the clinging, penetrating odor of the skunk's defensive spray; the hound, nose to the ground, in pursuit of prey; the female dog making known her sexual availability; the female moth attracting males from great distances for mating. As biologists and chemists cooperate to extend our knowledge of other animals, it is becoming increasingly clear that chemical communication is the primary mode of communication for the vast majority of species in the animal world.

Prior to 1950, isolation of enough biologically active material to permit us to decipher any chemical communications seemed an insurmountable task. For example, obtaining a mere 12 milligrams of gypsy moth sex attractant required processing of 500,000 virgin female moths, each yielding only 0.02 microgram of attractant. However,

rapid progress in instrumental techniques, particularly spectroscopy and chromatography, has now made it possible to isolate and carry out structural determinations on as little as a few micrograms of material. However, even with these advances, the isolation and identification of the components of insect chemical communication signals remain a major challenge to technical and experimental expertise.

The term **pheromone** (from the Greek *pherein,* "to carry," and *horman,* "to excite") is the accepted name for chemicals secreted by an organism of one species to evoke a response in another member of the same species. In this essay, we will look at insect pheromones, for these have been the most widely studied.

Pheromones are generally divided into two classes: **primer** and **releaser**

pheromones. Primer pheromones cause important physiological changes that affect the organism's development and later behavior. The most clearly understood primer pheromones are those involved in regulation of caste systems in social insects (bees, ants, and termites). A typical colony of honey bees (*Apis mellifera*) consists of one queen, several hundred drones (males) and thousands of workers (sterile females). The queen bee is the only fully developed female in the colony. She secretes a "queen substance" that prevents the development of workers' ovaries and promotes the construction of royal colony cells for the rearing of new queens. One of the components of the primer pheromone in the queen substance has been identified as 9-keto-*trans*-2-decenoic acid (Figure 1). In addition, this same compound serves as a sex pheromone,

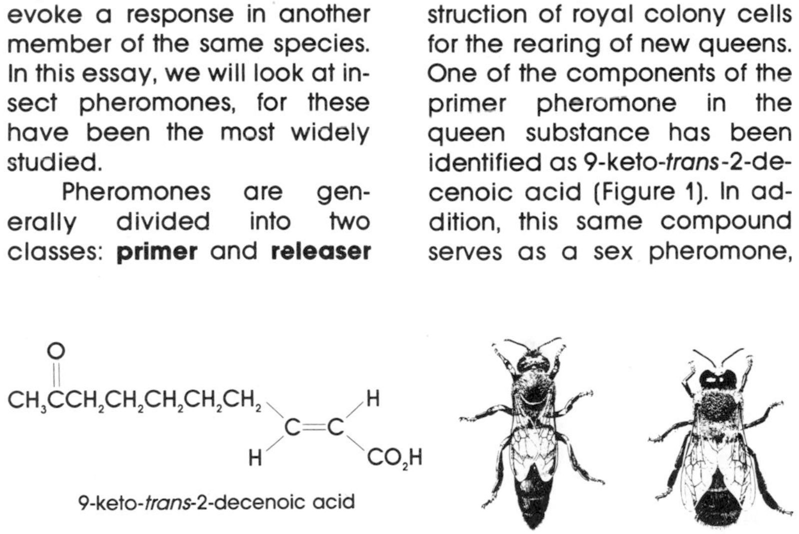

Figure 1 (a) 9-Keto-*trans*-2-decenoic acid, a component of the queen substance. (b) *Apis mellifera,* a queen bee, and (c) a drone. (Courtesy of USDA)

9-keto-*trans*-2-decenoic acid

(a)

(b)

(c)

466a

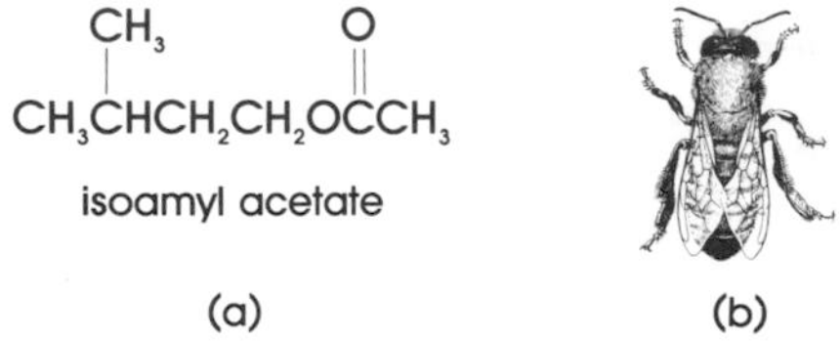

Figure 2 (a) Isoamyl acetate, a component of the alarm pheromone of the honey bee; (b) a worker (female). (Courtesy of USDA)

isoamyl acetate

(a)   (b)

attracting drones to the queen during her mating flight.

Releaser pheromones produce rapid, reversible changes in behavior such as sexual attraction and stimulation, aggregation, trail marking, territorial and home range marking, and other social behaviors. Some of the earliest observations of releaser pheromones were recorded on the alarm pheromones of the honey bee. Bee keepers, and perhaps some of the rest of us too, are well aware that the sting of one bee often causes swarms of angry workers to attack the same spot. When a worker stings an intruder, it discharges, along with venom, an alarm pheromone that evokes aggressive attack by other bees. One component of this alarm pheromone is isoamyl acetate, a compound with an odor similar to that of banana oil (Figure 2).

Of all classes of pheromones, sex and aggregating pheromones have received the greatest attention in both the scientific community and the popular press. Larvae of certain insects that release these classes of pheromones, particularly the moths and beetles, are among the world's most serious agricultural and forestland pests.

Sex pheromones are commonly referred to as "sex attractants" but this term is misleading because it implies only attraction. Actually, the behavior elicited by sex pheromones is considerably more complex. Low levels of sex pheromone stimulation cause orientation and flight of a male toward a female (or in some species, flight of a female toward a male). If the level of stimulation is high enough, copulation follows.

One of the aggregating pheromone systems recently identified and studied is that of the Ips family of bark beetles. *Ips paraconfusus,* an insect especially destructive to ponderosa pines in the Sierra Nevada range in California, lives in the soil during the winter. In early spring, when the temperature begins to rise, a few males emerge from the ground and seek out ponderosa pines in which to construct breeding chambers. The few males bore into trees and during this process, a pheromone produced in the hind gut is emitted and triggers a massive secondary invasion of both males and females. As they bore into ponderosa pines, these bark beetles infect the tree with fungal spores, and it is the fungal spore infection that actually kills the tree. After fertilization and hatching, Ips larva grow and develop behind the bark. In autumn, they leave the tree and return to the soil to begin another life cycle.

Investigation of the aggregating pheromone of *Ips paraconfusus* led to the isolation and identification of three components (Figure 3). All are terpene alcohols. Ipsenol and ipsdienol have carbon skeletons identical to that of myrcene (see the mini-essay "Terpenes") and differ only in the presence of a carbon-carbon

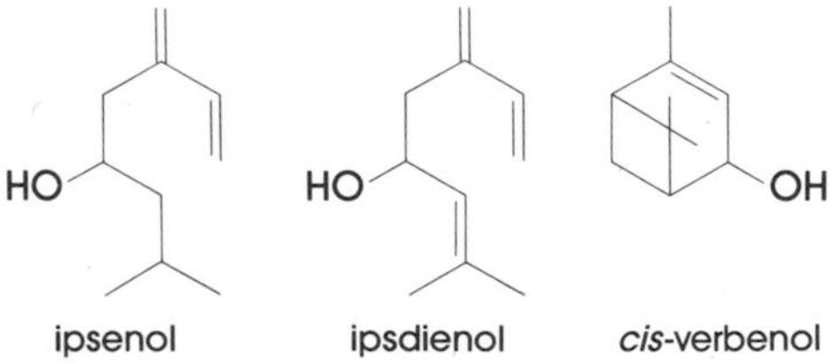

Figure 3 Components of the aggregating pheromone of the bark beetle *Ips paraconfusus.* A mixture of all three is necessary for attraction of males and females in the field.

ipsenol   ipsdienol   *cis*-verbenol

cis-11-tetradecenyl acetate

trans-11-tetradecenyl acetate

(a)   (b)

Figure 4 (a) Tetradecenyl acetate, a component of the sex phero-mone of the European corn borer. (b) Corn infested with European corn borers (courtesy of USDA).

double bond. The third component, verbenol, has a carbon skeleton identical to that of α-pinene (see the mini-essay, "Terpenes"). In fact, it has been shown that *Ips paraconfusus* can synthesize verbenol from α-pinene it encounters in the thick resin that flows from an injured ponderosa pine.

Several groups of scientists have studied the components of the sex pheromone of both Iowa and New York strains of the European corn borer. Females of these closely related species secrete the same sex attractant, 11-tetradecenyl acetate (Figure 4). Males of the Iowa strain show maximum response to a mixture containing about 96% of the *cis*-isomer and 4% of the *trans*-isomer. When the pure *cis*-isomer is used alone, males are only weakly attracted. Males of the New York strain show an entirely different response pattern. They respond maximally to a reversed ratio of isomers: 3% *cis*-isomer and 97% *trans*-isomer. There is increasing evidence that optimum response to a narrow range of stereoisomers as we see here, or to a mixture of components as we saw in the case of *Ips paraconfusus*, is widespread in nature and that at least some species maintain species isolation (at least for the purposes of mating and reproduction) by the very nature of their pheromone systems. Figure 5 shows the terrible devastation caused by corn borers.

Figure 5 A field of corn devastated by European corn borers. The larvae damage corn by boring into the stalks and then feeding on the ears (courtesy of USDA).

Figure 6 Components of the aggregating pheromone of the European elm bark beetle. α-Cubebene is synthesized by the Dutch elm and is a necessary component of the elm bark beetle aggregating pheromone.

$$CH_3CH_2CHCHCH_2CH_2CH_3$$
$$OH$$
$$CH_3$$

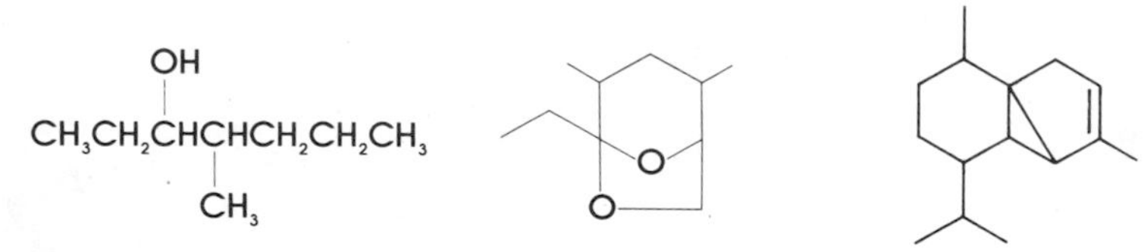

4-methyl-3-heptanol

multistriatin

α-cubebene

The isolation and identification of the components of the aggregating pheromone of the European elm bark beetle (*Scolytus multistriatus*) provides an unusual example of a unique relationship between predator and host. This beetle is responsible for the spread of Dutch elm disease, an insect-borne blight which has wiped out elm trees in more than half of the United States. The fungus, which destroys the trees by interrupting the flow of sap, is transported on the bodies of the attacking beetles. The aggregating pheromone of the elm bark beetle consists of three components (Figure 6). Two of these, 4-methyl-3-heptanol and multistriatin (a name derived from that of the beetle) are synthesized by the female and excreted as it bores into an elm tree. The third component, α-cubebene, is produced by the elm tree! A mixture of all three compounds is required for aggregation. In this regard, the Dutch elm tree itself affects the behavior of the attacking insect. Note that α-cubebene is a terpene. Its carbon skeleton can be derived from that of farnesol (see the mini-essay "Terpenes") by cross-linking between carbons 1 and 6, 1 and 10, and 2 and 6.

Within the last decade, scientists have developed a number of practical applications of pheromone systems for the monitoring and control of selected insect pests. First, pheromone-baited traps can be placed in the field to monitor populations of selected insect pests. In this way, changes in population levels can be determined and major or potential areas of infestation defined. The great value of this type of information is that the use of large-scale spraying of conventional insecticides can be drastically reduced or even avoided in areas where populations are below threshold levels. Several companies here and abroad now market pheromone-baited traps to monitor population levels of such insect pests as the Japanese beetle, the gypsy moth, and the boll weevil.

Second, pheromones can be used for mass trapping and population suppression of particular insect pests. Probably the largest single effort to date involving trapping and population suppression was that undertaken in Norway and Sweden to prevent a potentially catastrophic infestation of *Ips typographus*, a bark beetle largely confined to the coniferous forests of Europe and Asia and particularly attracted to the commercially valuable Norway spruce. The aggregating pheromone of *Ips typographus* consists of three components (Figure 7).

Figure 7 The three components of the aggregating pheromone of the bark beetle, *Ips typographus*.

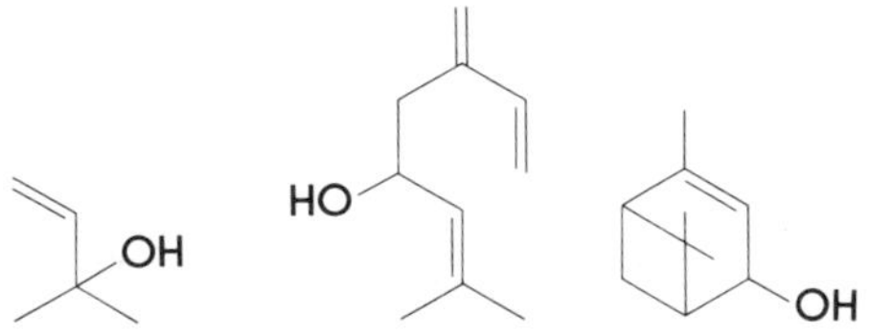

2-methyl-3-buten-2-ol

ipsdienol

verbenol

In the three years prior to 1979, severe drought in Norway and Sweden affected huge numbers of spruce trees and in that year bark beetles killed or severely damaged an estimated 5 million trees. The governments of Norway and Sweden initiated a large-scale program to control *Ips typographus;* in the summer of 1979, they placed about 1 million pheromone-baited traps in infested forests. The beetle catch in 1979 was estimated at 2.9 billion. The program was repeated in 1980 with an estimated catch in Norway alone of 4.5 billion beetles. Although tree mortality during these years was still high, the feared catastrophic infestation of *Ips typographus* did not occur.

In a third use of pheromones in the field, insects can be lured to traps and treated there with insecticides, insect juvenile hormones or juvenile hormone analogs, or species-specific pathogenic organisms, all of which are then spread throughout the local population of that particular pest when the insects are released.

In a fourth use, specific pheromones can be spread through the air to disrupt mating or aggregation. Clearly, this means of population control and suppression requires an understanding of the growth and behavior patterns of the particular insect pest and good timing.

Based on the information in this mini-essay, it should be clear that pheromones are becoming an integral part of environmentally sound means of insect pest control.

References

Birch, M.C., ed. 1974. Pheromones. *Frontiers of biology.* New York: Elsevier North-Holland.

Klun, J.A. et al. 1973. Insect sex pheromones: minor amounts of opposite geometrical isomer critical to attraction. *Science* 181: 661.

O'Sullivan, D.A. Pheromone lures help control bark beetles. *Chem. and Eng. News* 30 July 1979.

Sanders, H.J. New weapons against insects: a special report. *Chem. and Eng. News* 28 July 1975.

Shorey, H.H. 1976. *Animal communication by pheromones.* New York: Academic Press.

Silverstein, R.M. 1981. Pheromones: background and potential for use in pest control. *Science* 213: 1326.

Optical Isomerism

All compounds that have the same structural formula but different orientations of their atoms in space are classified as **stereoisomers.** *Cis-trans* isomerism is one type of stereoisomerism; **optical isomerism** is a second type of stereoisomerism. In this chapter we will deal with optical isomers known as *enantiomers*, *diastereomers*, and *meso compounds*.

We shall first examine the structural features that give rise to optical isomerism, how it can be detected in the laboratory, and finally its significance in the biological world.

14.1 Molecules with One Chiral Center

A. Molecules and Their Mirror Images

A molecule, like any other object in nature, has a mirror image. Shown in Figure 14.1 (next page) is a **stereorepresentation** of a lactic acid molecule and its mirror image. In these stereorepresentations, the—CO_2H groups are in the plane of the page and parallel to the mirror; the—OH groups are also in the plane of the page, but point away from the mirror. The—H groups are behind the plane of the page (shown by ⅲⅲⅲⅲ) and point toward the mirror, and the —CH_3 groups are in front of the plane of the page (shown by ►) and point toward the mirror.

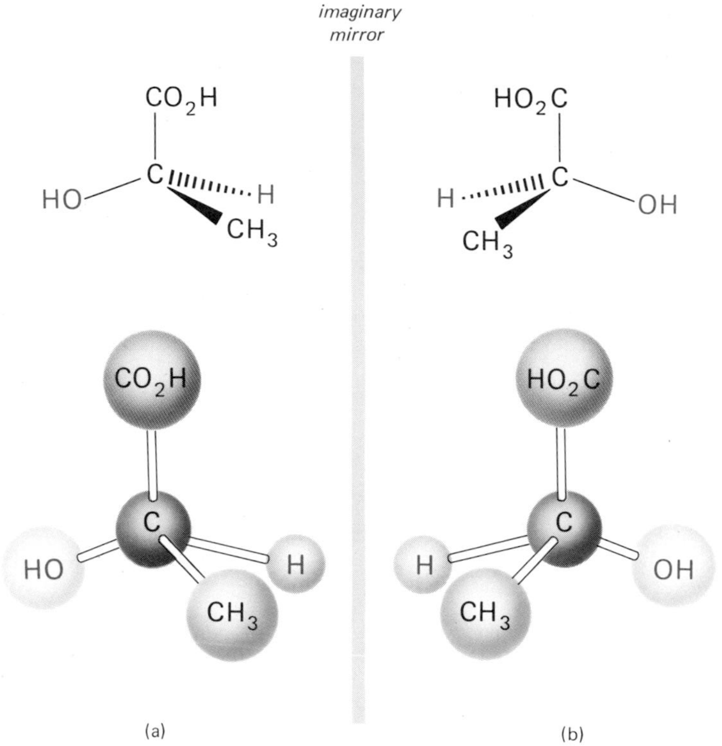

Figure 14.1 Stereorepresentations and ball-and-stick models of (a) a lactic acid molecule and (b) its mirror image. In these formulas, ▶ represents a bond projecting in front of the plane of the page, — represents a bond in the plane of the page, and ····· represents a bond projecting behind the plane of the page.

The question is: What is the relationship between these two structural formulas? Do they represent the same or different compounds? The answer is that they represent different compounds. The lactic acid molecule drawn in Figure 14.1(a) can be turned any direction in space, but as long as no bonds are broken, only two of the four groups attached to the central carbon in (a) can be made to coincide with those in (b), as shown in Figure 14.2. The mirror images of lactic acid differ from each other just as a right hand differs from a left hand. They are related by reflection, but they are not superimposable on each other.

Objects that are not identical with their mirror images are said to be **chiral** (pronounced *ki-ral*, to rhyme with *spiral;* from the Greek *cheir*, "hand"). In a chiral molecule, a carbon that has four different groups attached to it is called a **chiral carbon,** or an **asymmetric carbon.** It is important to realize that chirality is a property of an object, the property of being nonsuperimposable on its mirror image.

B. Enantiomers

Molecules that are nonsuperimposable mirror images of each other are called **enantiomers** (Greek: *enantio*, "opposite" + *meros*, "parts"). Many properties

Figure 14.2 A molecule of lactic acid and its mirror image are not superimposable on each other. No matter how the mirror image is turned in space, only two of the four groups attached to the central carbon can be made to coincide.

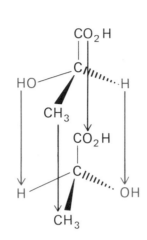

of enantiomers are identical: they have the same melting points, the same boiling points, the same solubilities in various solvents. Yet they are isomers and we can expect them to show some differences in their properties. One important difference, as we shall see in the next section, is optical activity.

Example 14.1

Draw stereorepresentations for the enantiomers of:

a. CH_3—CH—CH_2- CH_3
        |
        OH

b. CH_3—CH_2—CH—CH_2—OH
                    |
                    CH_3

Solution

First, locate and draw the chiral (asymmetric) carbon atom and the four bonds from it arranged to show the tetrahedral geometry. Next, draw the four groups attached to the chiral carbon. Finally, draw the nonsuperimposable mirror image.

a. CH_3—$\overset{*}{C}H$—CH_2—CH_3
            |
            OH

mirror ⟶

b. 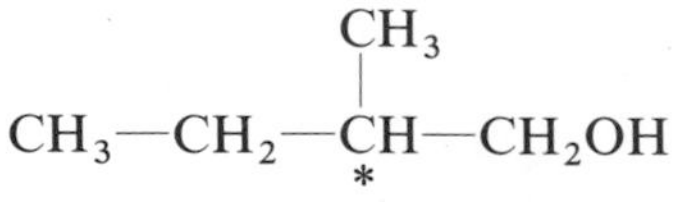 CH_3—CH_2—$\overset{\overset{\displaystyle CH_3}{|}}{\underset{*}{CH}}$—$CH_2OH$

the chiral carbon
atom marked by an
asterisk

tetrahedral
geometry of
the chiral
carbon

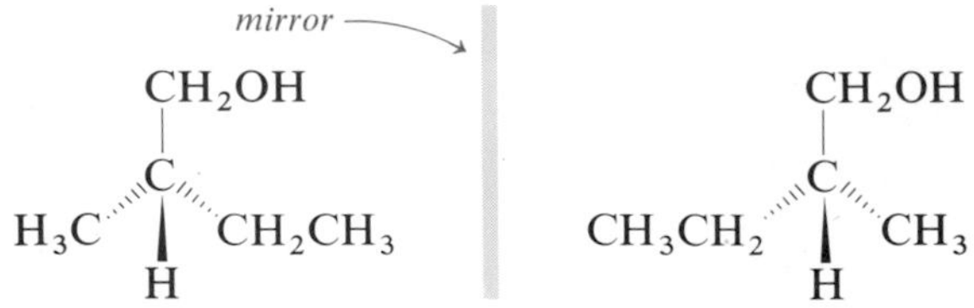

mirror

a pair of enantiomers
(nonsuperimposable mirror images)

Problem 14.1

Draw stereorepresentations for the enantiomers of:

a. 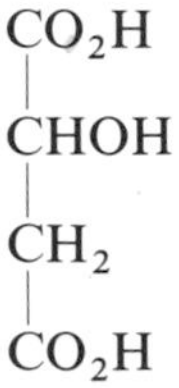 $\begin{array}{l} CO_2H \\ | \\ CHOH \\ | \\ CH_2 \\ | \\ CO_2H \end{array}$ malic acid

b. 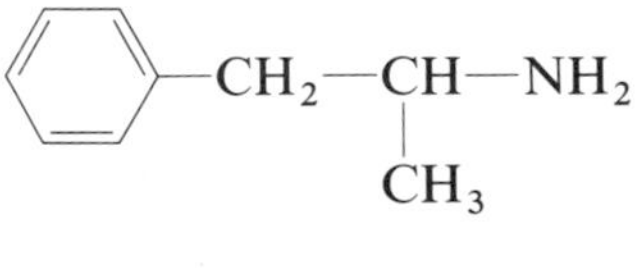 [benzene ring]—CH_2—$\overset{\overset{\displaystyle \ }{|}}{\underset{\underset{\displaystyle CH_3}{|}}{CH}}$—$NH_2$

amphetamine

C. Achiral Molecules

For all molecules that contain a single chiral carbon atom, the mirror images
are nonsuperimposable and the compounds show enantiomerism. For many
other compounds, however, the molecule and its mirror image are superimposable. Consider, for example, the amino acid glycine. Figure 14.3 shows
stereorepresentations of (a) glycine and (b) its mirror image. Are they nonsuperimposable or superimposable? The answer is that they are superimposable
mirror images. If (b) is turned by 180° about the C—CO_2H bond, it is possible
to superimpose (b) on (a). Molecules that are superimposable on their mirror
images are identical and therefore cannot show optical isomerism. They are
achiral ("without handedness").

D. Plane of Symmetry

A **plane of symmetry** is defined as a plane (often visualized as a mirror) cleaving
a molecule in such a way that one side of the molecule is the mirror image of

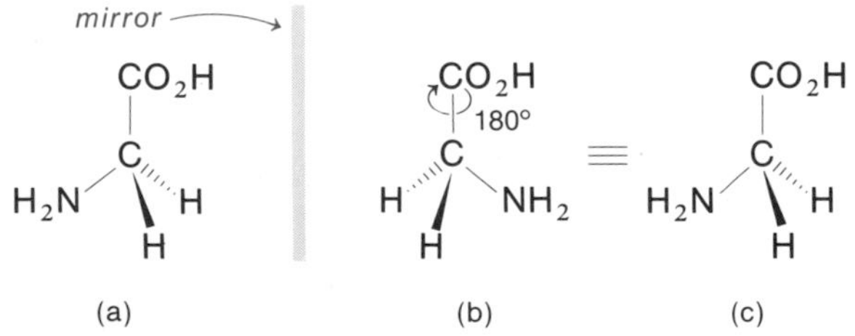

Figure 14.3 Stereore-
presentations of (a)
glycine and (b) its mir-
ror image. If (b) is ro-
tated by 180° about
the C — CO_2H bond, it
can be seen that (b) is
superimposable on,
hence identical to, (a).

mirror

(a)   (b)   (c)

Figure 14.4 Two stereorepresentations of glycine. A plane of symmetry runs through the axis of the C — C — N bond.

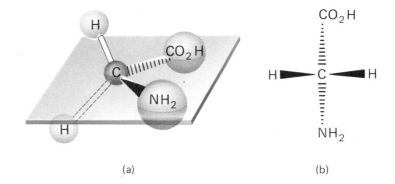

(a)   (b)

the other. Any molecule that has a plane of symmetry is superimposable on its mirror image and does not show optical isomerism.

Inspection of the stereorepresentation of glycine (Figure 14.4a) shows that it has a plane of symmetry running through the molecule on the axis of the C—C—N bonds. Alternatively, we might rotate the molecule in space into a different stereorepresentation (b), and see the plane of symmetry again running through the axis of the C—C—N bonds, this time perpendicular to the plane of the page.

14.2 How to Predict Optical Isomerism

There are three methods you can use to determine whether a molecule shows optical isomerism. The first and most direct test is to build a model of the molecule and one of its mirror image. If the two are superimposable, the molecule and its mirror image are identical and do not show optical isomerism. If the two are not superimposable, then the molecule shows optical isomerism.

A second method is to look for a plane of symmetry. If the molecule has a plane of symmetry, then the mirror images are superimposable and the molecule does not show optical isomerism.

Third, you can look for the presence or absence of a chiral carbon atom.

Example 14.2

Which of the following compounds show optical isomerism? For each that does, label the chiral carbon and draw stereorepresentations for both enantiomers. For each that does not, draw a stereorepresentation to show a plane of symmetry.

a. $CH_2\!=\!CH\!-\!CH\!-\!CH_2\!-\!CH_3$
         $\quad\quad\quad\quad\;|$
         $\quad\quad\quad\quad OH$

b. $CH_3\!-\!CH_2\!-\!CH\!-\!CH_2\!-\!CH_3$
         $\quad\quad\quad\quad\quad\quad\;|$
         $\quad\quad\quad\quad\quad\quad OH$

c. $CH_3\!-\!CH_2\!-\!CH\!-\!CH_2\!-\!CH_2OH$
         $\quad\quad\quad\quad\quad\;|$
         $\quad\quad\quad\quad\quad OH$

Solution

Both (a) and (c) have chiral carbon atoms and show optical isomerism. Compound (b) has no chiral carbon and does not show optical isomerism.

a.

$$CH_2{=}CH \overset{\displaystyle H}{\underset{\displaystyle OH}{\overset{|}{\underset{|}{C}}}} CH_2CH_3 \qquad mirror \longrightarrow \qquad CH_3CH_2 \overset{\displaystyle H}{\underset{\displaystyle OH}{\overset{|}{\underset{|}{C}}}} CH{=}CH_2 \quad \text{— chiral carbon atom}$$

b.

$$CH_3CH_2 \overset{\displaystyle H}{\underset{\displaystyle OH}{\overset{|}{\underset{|}{C}}}} CH_2CH_3 \qquad \textit{the plane of symmetry bisects the molecule through the O—C—H bonds}$$

c.

$$HO \overset{\displaystyle CH_2CH_3}{\underset{\displaystyle H}{\overset{|}{\underset{|}{C}}}} CH_2CH_2OH \qquad HOCH_2CH_2 \overset{\displaystyle CH_2CH_3}{\underset{\displaystyle H}{\overset{|}{\underset{|}{C}}}} OH \quad \text{— chiral carbon atom}$$

Problem 14.2

Which of the following compounds show optical isomerism? For each that does, label the chiral carbon and draw stereorepresentations for both enantiomers. For each that does not, draw a stereorepresentation to show a plane of symmetry.

a. $CH_3{-}CH{-}CO_2H$
$\qquad\qquad |$
$\qquad\quad\;\; NH_2$

b. (cyclopentane)$-CH{-}CH_3$ with OH on the CH

c. (cyclopentane) with CH_3 and OH

14.3 Detection of Optical Isomerism in the Laboratory

A. Plane Polarized Light

Ordinary light consists of waves vibrating in all possible planes perpendicular to its path (Figure 14.5). Certain materials such as *Polaroid sheet*, a plastic film containing properly oriented crystals of an organic substance embedded in it, selectively transmit light waves vibrating in one specific plane. Light which vibrates in only one specific plane is said to be **plane polarized.** No doubt you are familiar with the effect of polarizing sheets on light from experiences with sunglasses or camera filters made of this material. If two polarizing discs are placed in a light path in such a way that their polarizing axes are parallel to

Figure 14.5 Schematic diagram of a polarimeter with the sample tube empty.

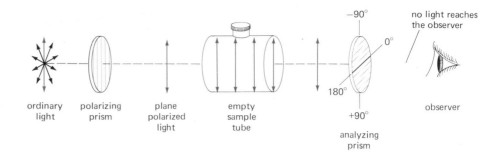

each other, a maximum intensity of light passes through the second disc to you. However, if their polarizing axes are perpendicular to each other (that is, at an angle of 90°), no light passes through the second polarizing disc.

B. The Polarimeter

A **polarimeter** is an instrument used to measure the effect of an optically active compound on the plane of polarized light. This device consists of a light source, a polarizing prism, a sample tube, and an analyzing prism (Figure 14.5). If the sample tube is empty, the intensity of light reaching the observer is a maximum when the polarizing axes of the two prisms are parallel. If the analyzing prism is turned either clockwise or counterclockwise, less light is transmitted. When the axis of the analyzing prism is at right angles to that of the polarizing prism, the field of view in the instrument is dark. We take this as the **zero point,** or 0° on the optical scale. If a solution of an optically active compound is placed in the sample tube, the plane of polarized light is rotated and a certain amount of light now passes through the analyzing prism to the observer. Turning the analyzing prism a few degrees clockwise or counterclockwise restores darkness to the field of view (Figure 14.6). The number of degrees, α, that the analyzing

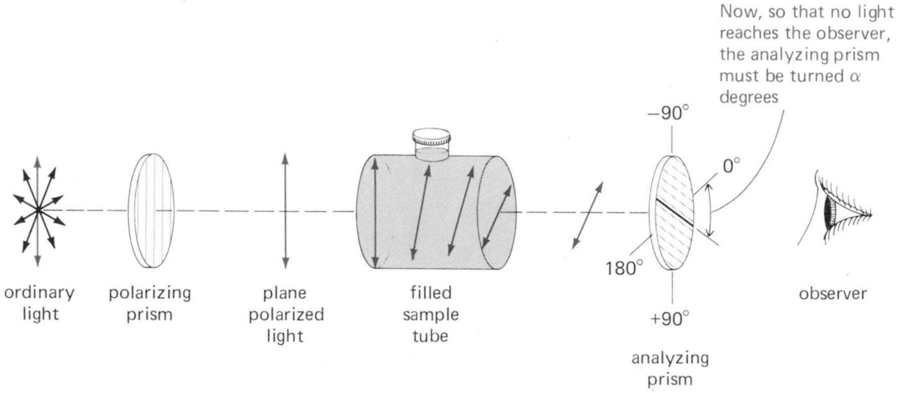

Figure 14.6 Schematic diagram of a polarimeter with the sample tube containing an optically active compound. To restore the dark field of view, the analyzer has been rotated clockwise by α degrees.

Now, so that no light reaches the observer, the analyzing prism must be turned α degrees

ordinary light   polarizing prism   plane polarized light   filled sample tube   analyzing prism   observer

−90°   0°   180°   +90°

prism must be turned to restore darkness to the field of view is called the **observed rotation.** If the analyzing prism must be turned to the right (clockwise) to restore darkness, we say the compound is **dextrorotatory.** If the analyzing prism must be turned to the left (counterclockwise) to restore darkness, we say the compound is **levorotatory.** In either case, the compound is **optically active.** It is because of the effect of chiral compounds on the plane of polarized light that this type of isomerism is called **optical isomerism.**

C. Specific Rotation

Specific rotation, $[\alpha]$, is defined as the rotation caused by a compound at a concentration of one gram per cubic centimeter in a sample tube ten centimeters long. Specific rotation depends on the temperature and the wavelength of the light source, and these values must be reported as part of the measurement. The most commonly used light source is the D line of sodium, the same line that is responsible for the yellow color of excited sodium vapor. In reporting either observed or specific rotation, it is common to indicate a dextrorotatory compound by a positive sign ($+$) and a levorotatory compound by a negative sign ($-$). Using these conventions, the specific rotation of sucrose (table sugar) dissolved in water at 25°C, using the D line of sodium as the light source, is reported as follows:

$$[\alpha]_D^{25°C} = +66.5(H_2O)$$

As we have already indicated, enantiomers are different compounds and have different properties. One important difference is optical activity. One member of a pair of enantiomers is dextrorotatory and the other member is levorotatory. For each, the number of degrees of the specific rotation is the same, but the rotations differ in sign, as illustrated below by the dextro- and levorotatory isomers of 2-butanol.

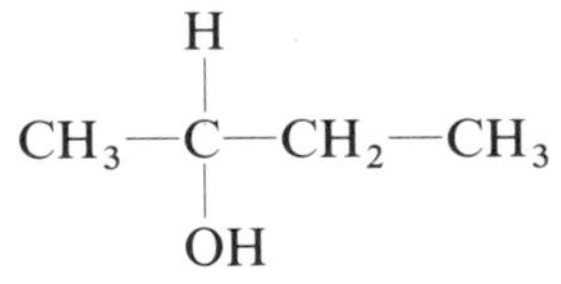

$$
\begin{array}{c}
H \\
| \\
CH_3-C-CH_2-CH_3 \\
| \\
OH
\end{array}
$$

(+)-2-butanol $[\alpha]_D^{20°C} = +13.9°$
(−)-2-butanol $[\alpha]_D^{20°C} = -13.9°$

14.4 Racemic Mixtures and Resolution of Racemic Mixtures

Lactic acid was one of the first optically active compounds discovered. It was originally isolated in 1780 from sour milk and found to be optically inactive, that is, its specific rotation was 0°. In 1807, lactic acid was isolated from muscle tissue. Whereas lactic acid from sour milk is optically inactive, that from muscle tissue is dextrorotatory and designated (+)-lactic acid. We have already demonstrated that lactic acid and its mirror image are nonsuperimposable (it is a chiral molecule), and that it shows optical isomerism. How is it that lactic acid from fermentation in milk is optically inactive while that from muscle is optically active? The answer is that the inactive form of lactic acid is a mixture of equal numbers of (+)-lactic acid and (−)-lactic acid molecules. Because the mixture contains equal numbers of molecules that rotate the plane of polarized light to the right and to the left, the mixture does not rotate the plane of polarized light to the right or to the left; it is optically inactive. A mixture containing equal amounts of a pair of enantiomers is called a **racemic mixture.**

The process of separating a pair of enantiomers into (+)- and (−)-isomers is called **resolution.** The first demonstration of this process was the resolution of racemic tartaric acid by Louis Pasteur in 1848. Below 25°C, the (+)-enantiomer of sodium ammonium tartrate crystallizes into one type of crystal, while the (−)-enantiomer crystallizes into a mirror image of it. By carefully hand-picking the crystals, Pasteur separated the mixture into two types and examined their solutions separately in a polarimeter. He discovered that a solution of one type rotates the plane of polarized light to the right and the other rotates it to the left. When equal weights of the two types of crystals were dissolved in water, the solution of the mixture, like the starting material, had no effect on the plane of polarized light. This initial resolution by Pasteur is particularly remarkable, for since that time very few additional examples have been found in which crystallization produces enantiomorphic (mirror-image) crystals large enough so that hand separation is possible. Fortunately, other methods of resolution are now known.

14.5 Molecules with Multiple Chiral Centers

A. Enantiomers and Diastereomers

Compounds that contain two or more chiral centers can exist in more than two optical isomers. The maximum number of optical isomers is 2^n, where n is the number of chiral carbon atoms.

As an example of a molecule with two chiral centers, consider 2,3,4-trihydroxybutanal.

$$HO-CH_2-\overset{*}{C}H-\overset{*}{C}H-\overset{\overset{\displaystyle O}{\|}}{C}-H$$
$$\hspace{2.5cm}| \hspace{0.6cm} |$$
$$\hspace{2.5cm}OH \hspace{0.3cm} OH$$

2,3,4-trihydroxybutanal

Figure 14.7 Stereoisomers of 2,3,4-trihydroxybutanal, a substance with two chiral carbon atoms.

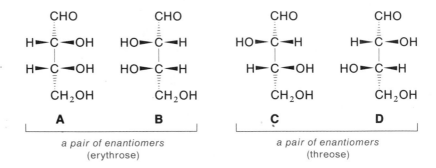

$$
\begin{array}{cc}
\text{CHO} & \text{CHO} \\
\text{H—C—OH} & \text{HO—C—H} \\
\text{H—C—OH} & \text{HO—C—H} \\
\text{CH}_2\text{OH} & \text{CH}_2\text{OH} \\
\textbf{A} & \textbf{B}
\end{array}
$$

a pair of enantiomers
(erythrose)

a pair of enantiomers
(threose)

The two chiral carbons are marked by asterisks. There are four optical isomers ($2^2 = 4$) of this molecule, each of which is shown in Figure 14.7. A and B are nonsuperimposable mirror images and are a pair of enantiomers. C and D are a second pair of nonsuperimposable mirror images and are a second pair of enantiomers. What is the relationship of A to C or A to D? The answer is that they are optical isomers of each other but are not mirror images. Optical isomers that are not mirror images of each other are called **diastereomers.** Of the optical isomers drawn in Figure 14.7, AB and CD are pairs of enantiomers; AC, AD, BC, and BD are pairs of diastereomers. Diastereomers have different chemical and physical properties, and sometimes they are even given different names. The diastereomers of trihydroxybutanal are given the common names erythrose and threose.

B. *Meso* **Compounds**

Some molecules have special symmetry properties which reduce the number of optical isomers below 2^n, where n is the number of chiral carbons. Tartaric acid is an example of a molecule possessing two chiral carbon atoms. The 2^n rule predicts four optical isomers; in fact, only three are known (Figure 14.8). Structures E and F are mirror images, and because they are not superimposable on

Figure 14.8 Optical isomers of tartaric acid.

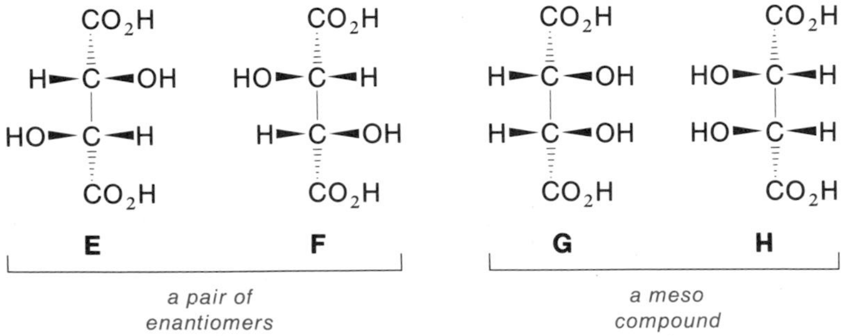

$$
\begin{array}{cccc}
\text{CO}_2\text{H} & \text{CO}_2\text{H} & \text{CO}_2\text{H} & \text{CO}_2\text{H} \\
\text{H—C—OH} & \text{HO—C—H} & \text{H—C—OH} & \text{HO—C—H} \\
\text{HO—C—H} & \text{H—C—OH} & \text{H—C—OH} & \text{HO—C—H} \\
\text{CO}_2\text{H} & \text{CO}_2\text{H} & \text{CO}_2\text{H} & \text{CO}_2\text{H} \\
\textbf{E} & \textbf{F} & \textbf{G} & \textbf{H}
\end{array}
$$

a pair of
enantiomers

a meso
compound

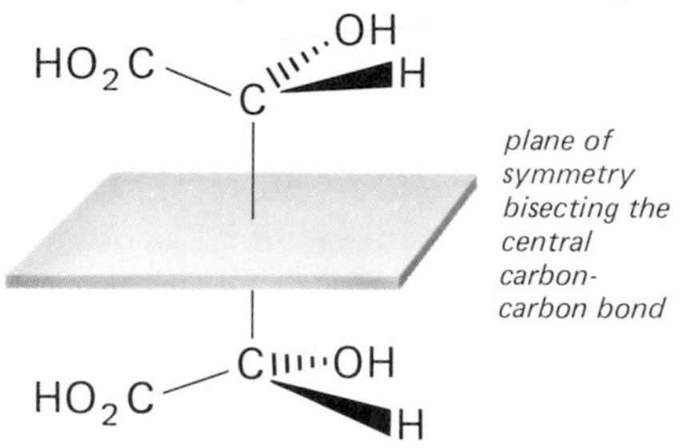

Figure 14.9 Plane of symmetry in *meso*-tartaric acid.

plane of symmetry bisecting the central carbon-carbon bond

each other, they are a pair of enantiomers. G and H are mirror images, but they are superimposable. To see how, turn H by 180° in the plane of the page and you will see that it fits exactly on G. In Section 14.1, we said that any molecule superimposable on its mirror image must have a plane of symmetry. The plane of symmetry in G is shown in Figure 14.9.

Symmetrical molecules that contain two or more chiral carbon atoms are called ***meso* compounds.** *meso*-Tartaric acid is a diastereomer of (−)-tartaric acid and of (+)-tartaric acid. Physical properties of the three optical isomers of tartaric acid are given in Table 14.1.

Table 14.1 Physical properties of the tartaric acids.

Acid	mp (°C)	$[\alpha]_D^{25°C}$
dextro-	170	+12°
levo-	170	−12°
meso-	146	inactive

Example 14.3

1,2-cyclopentanediol

Shown in the margin is the structural formula of 1,2-cyclopentanediol. This compound has two chiral carbon atoms, marked by asterisks.

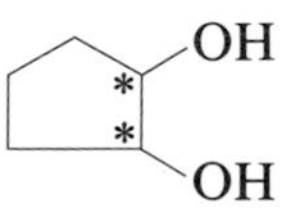

a. Draw a stereorepresentation for the *cis*-isomer and its mirror image. Are they superimposable?

b. Draw a stereorepresentation for the *trans*-isomer and its mirror image. Are they superimposable?

c. State the total number and kind of stereoisomers of 1,2-cyclopentanediol.

Solution

a. To show *cis-trans* isomerism in derivatives of cyclopentane, draw the ring as a planar pentagon with the substituent groups above and below the plane of the ring. The *cis-* isomer and its mirror image are superimposable on each other. Because *cis*-1,2-cyclopentanediol has two chiral carbon atoms yet it and its mirror image are superimposable, the *cis-* isomer is a *meso* compound. The plane of symmetry in this molecule bisects the —CH—CH— bond and cuts through the ring.

$$\begin{array}{ccc} & | & | \\ & OH & OH \end{array}$$

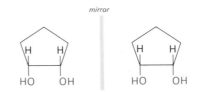

cis-1,2-cyclopentanediol (a *meso* compound)

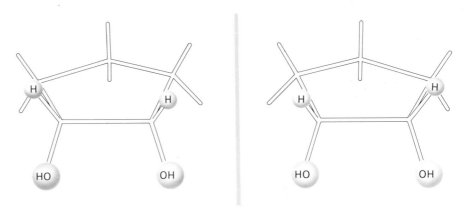

ball-and-stick models of *cis*-1,2-cyclopentanediol

b. The *trans-* isomer and its mirror image are nonsuperimposable; therefore, the *trans-* isomer shows enantiomerism—it exists as a pair of enantiomers.

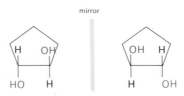

trans-1,2-cyclopentanediol (a pair of enantiomers)

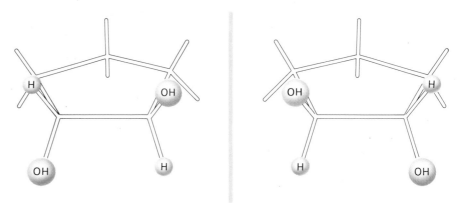

ball-and-stick models for the enantiomers of *trans*-1, 2-cyclopentanediol

c. There are three stereoisomers of 1,2-cyclopentanediol: one pair of enantiomers and one *meso* compound.

Problem 14.3

Following is the structural formula for 2-methylcyclopentanol:

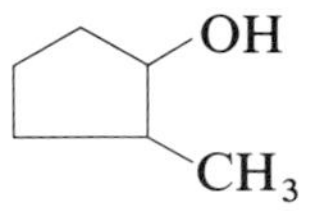

OH

CH$_3$

2-methylcyclopentanol

a. Draw a stereorepresentation for the *cis*- isomer and its mirror image. Are they superimposable?

b. Draw a stereorepresentation for the *trans*- isomer and its mirror image. Are they superimposable?

c. State the total number and kind of stereoisomers of 2-methylcyclopentanol.

14.6 The Significance of Chirality in the Biological World

Except for inorganic salts and a relatively few low-molecular-weight organic compounds, most molecules of living organisms, both plant and animal, are chiral. Although these molecules can exist as a mixture of stereoisomers, almost invariably only one stereoisomer is found in nature. There are, of course, instances where both enantiomers can be found in nature, but they rarely exist together in the same biological system. We can make the further generalization that only one enantiomer can be used or assimilated by an organism. In fact, this latter observation is the basis for a sometimes-used resolution technique.

A. Microbiological Resolution

Pasteur discovered in 1858–1860 that when *Penicillium glaucum*, a green mold found in aging cheese and rotting fruit, is grown in a solution containing racemic tartaric acid, the solution slowly becomes levorotatory. The microorganism preferentially consumes (metabolizes) (+)-tartaric acid. If the process is interrupted at the right time, (−)-tartaric acid can be crystallized from solution in pure form. If the process is allowed to continue, the microorganism eventually consumes the (−)-tartaric acid as well. Thus, while both enantiomers of tartaric acid are metabolized by the microorganism, the (+)-enantiomer is metabolized more rapidly. As another example, when racemic mevalonic acid is fed to rats, one enantiomer is metabolized, while almost all of the other enantiomer is excreted in the urine.

mevalonic
acid
(metabolized)

mevalonic
acid
(excreted)

B. Chirality in the Biological World

The general observations that only one enantiomer is found in a given biological system and that only one enantiomer can be metabolized at a time should be enough to convince us that it is a chiral world in which we live. At least it is chiral at the molecular level. Essentially all chemical reactions in the biological world take place in a chiral environment. Perhaps the most conspicuous examples of chirality among biological molecules are enzymes, all of which have a very high degree of chirality. To illustrate this point, consider chymotrypsin, an enzyme that functions so efficiently in the intestine of animals at pH 7–8 to catalyze the digestion of proteins. This enzyme contains 251 chiral carbon atoms. The number of possible optical isomers of chymotrypsin is 2^{251}, a number beyond comprehension. Fortunately, nature does not squander its precious resources and energies unnecessarily: only one of these stereoisomers is produced by any given organism.

Enzymes catalyze biological reactions by first adsorbing on their surface the molecule or molecules about to undergo reaction. These molecules may be bound by an accumulation of hydrogen bonds, ionic bonds, dispersion forces, or by covalent bond formation. Thus, whether the molecules are chiral or not, they are held for reaction in a chiral environment.

C. How an Enzyme Distinguishes Between a Molecule and its Enantiomer

It is generally agreed that an enzyme with specific binding sites for three of the four substituents on a chiral center can distinguish between a pair of enantiomers. Assume, for example, that an enzyme involved in the catalysis of a glyceraldehyde reaction has three binding sites, one specific for —H, another for —OH, and a third for —CHO, and that the three sites are arranged on the enzyme surface as shown in Figure 14.10. The enzyme can "recognize" (+)-glyceraldehyde (the natural or biologically active form) from (−)-glyceraldehyde because the correct enantiomer can be adsorbed with three groups attached to the appropriate binding sites, while the other enantiomer can, at best, bind to only two of these sites.

Given that interactions between molecules in living systems take place in a chiral environment, such events as the selective metabolism of (+)-tartaric acid by *Penicillium glaucum* or the selective metabolism of one enantiomer of mevalonic acid by rats should be no surprise. Further, it should be no surprise to discover that a molecule and its enantiomer have quite different psychological or physiological properties; after all, a molecule and its enantiomer are different compounds. For example, (+)-leucine tastes sweet, while its enantiomer, (−)-leucine, is bitter.

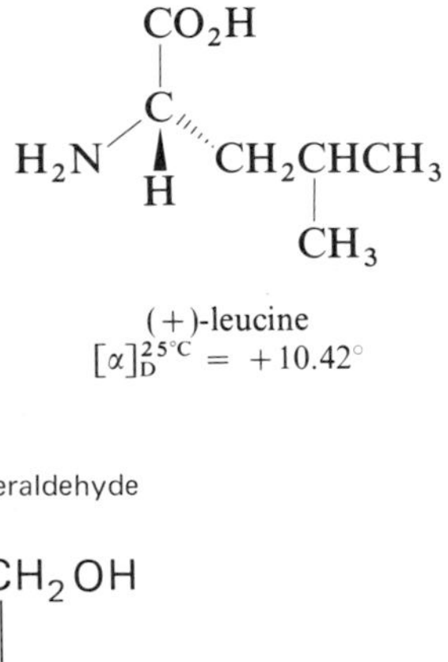

(+)-leucine
$[\alpha]_D^{25°C} = +10.42°$

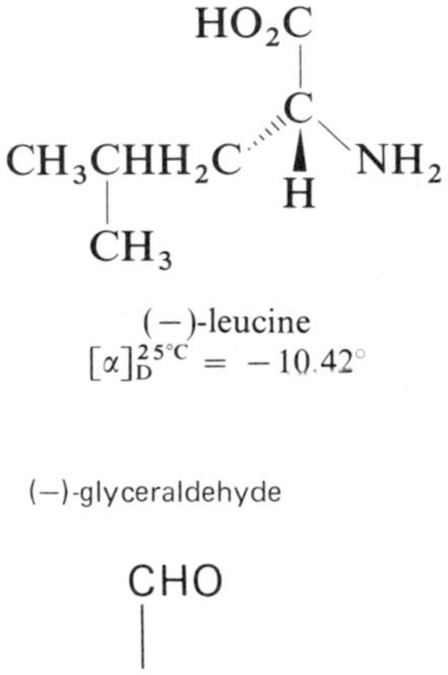

(−)-leucine
$[\alpha]_D^{25°C} = -10.42°$

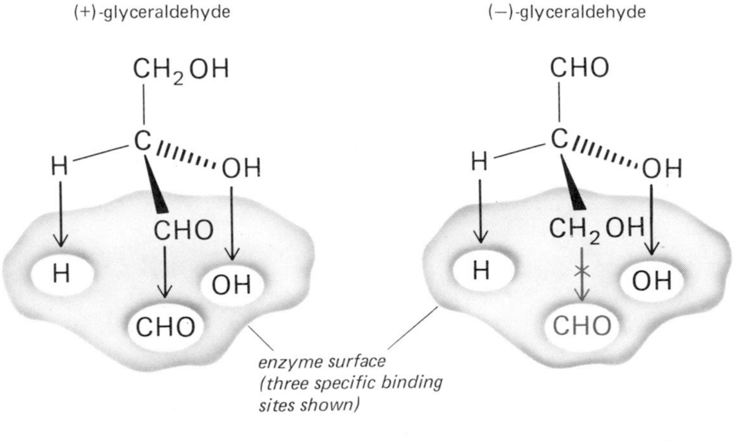

Figure 14.10 A schematic diagram of an enzyme surface capable of interacting with (+)-glyceraldehyde at three binding sites but with (−)-glyceraldehyde at only two of the three binding sites.

(+)-glyceraldehyde

(−)-glyceraldehyde

enzyme surface
(three specific binding sites shown)

The fact that the interactions of molecules in the biological world are so very specific in geometry and chirality is not surprising, but just how these interactions are accomplished with such precision and efficiency is one of the great puzzles that modern science has only recently begun to solve.

Key Terms and Concepts

asymmetric carbon (14.1A)

chiral (14.1A)

dextrorotatory (14.3B)

diastereomer (14.5A)

enantiomer (14.1B)

levorotatory (14.3B)

meso compound (14.5B)

mirror image (14.1A)

nonsuperimposable mirror image (14.1A)

observed rotation (14.3B)

optical activity (14.3B)

optical isomerism (introduction)

plane of symmetry (14.1D)

plane-polarized light (14.3A)

polarimeter (14.3A)

racemic mixture (14.4)

resolution (14.4)

specific rotation (14.3C)

Problems

Molecules with One Chiral Center (Section 14.1)

14.4   Draw mirror images for the following:

a.

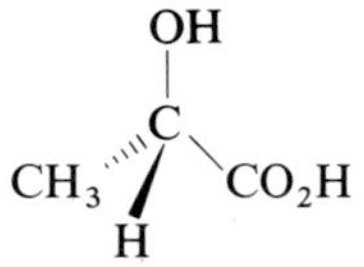

b.

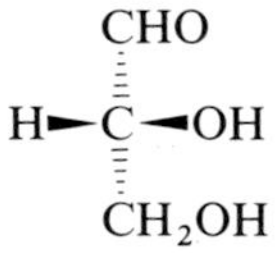

c.

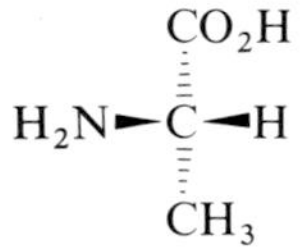

d.

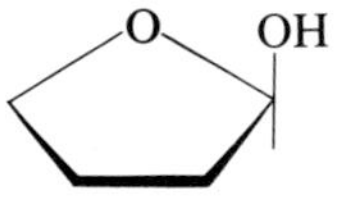

e.

f.

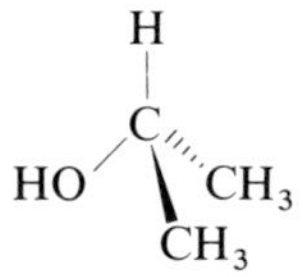

14.5   Drawn here are several stereorepresentations of lactic acid. Take (a) as a reference structure. Which of the stereorepresentations are identical to (a) and which are mirror images of (a)?

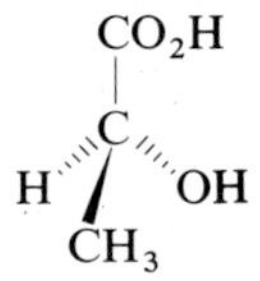

a. 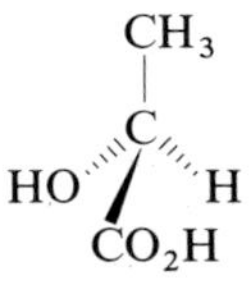   b. 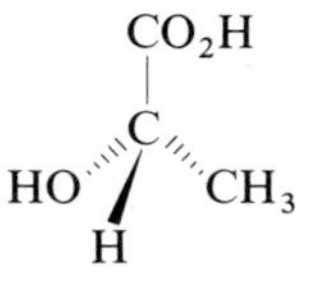   c. 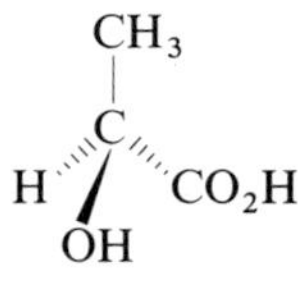   d. 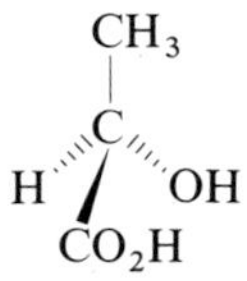   e.

14.6   Which of the following molecules contain a chiral carbon? For each that does, draw stereorepresentations for both enantiomers.

a. CH_3—$\overset{\displaystyle CH_3}{\underset{\displaystyle OH}{\overset{|}{\underset{|}{C}}}}$—$CH=CH_2$

b. H—$\overset{\displaystyle CO_2H}{\underset{\displaystyle CH_2OH}{\overset{|}{\underset{|}{C}}}}$—$OH$

c. CH_3—CH—$\overset{\displaystyle CH_3}{\overset{|}{C}}$—$CH$—$\underset{\displaystyle NH_2}{\underset{|}{CH}}$—$CO_2H$

d. CH_3—$\overset{\displaystyle O}{\overset{\|}{C}}$—$CH_2$—$CH_3$

e. CH_3—$\overset{\displaystyle O}{\overset{\|}{C}}$—$CO_2H$

f. $\underset{\displaystyle H}{\overset{\displaystyle H_3C}{}}C=C\underset{\displaystyle CH_3}{\overset{\displaystyle CO_2H}{}}$

g. H—$\overset{\displaystyle CH_2OH}{\underset{\displaystyle CH_2OH}{\overset{|}{\underset{|}{C}}}}$—$OH$

h. CH_3—CH—$\overset{\displaystyle CH_3}{\overset{|}{CH}}$—$\underset{\displaystyle OH}{\underset{|}{CH}}$—$CH_3$

i. HO—$\overset{\displaystyle CH_2—CO_2H}{\underset{\displaystyle CH_2—CO_2H}{\overset{|}{\underset{|}{C}}}}$—$CO_2H$

j. HO—CH_2—$\underset{\displaystyle NH_2}{\underset{|}{CH}}$—$CO_2H$

k. HS—CH_2—$\underset{\displaystyle NH_2}{\overset{\displaystyle O}{\overset{\|}{\underset{|}{CH}}}}$—$C$—$OH$

l. CH_3—CH_2—$\underset{\displaystyle OCH_3}{\underset{|}{CH}}$—$CH_3$

14.7   Draw the structural formula of at least one alkene of molecular formula C_5H_9Br that shows:
   a. neither *cis-trans* isomerism nor optical isomerism.
   b. *cis-trans* isomerism but not optical isomerism.
   c. optical isomerism but not *cis-trans* isomerism.
   d. both *cis-trans* isomerism and optical isomerism.

Molecules with Two or More Chiral Carbons (Section 14.5)

14.8   Mark each chiral carbon in the following molecules with an asterisk.

a.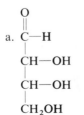

b. $\underset{\displaystyle HO—CH—CO_2H}{\overset{\displaystyle CH_2—CO_2H}{H—\overset{|}{\underset{|}{C}}—CO_2H}}$

c. 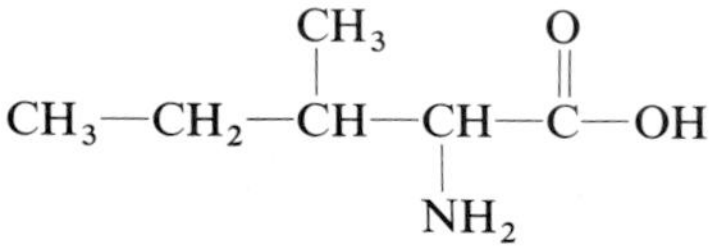 $CH_3-CH_2-\overset{\overset{\displaystyle CH_3}{|}}{CH}-\overset{\overset{\displaystyle}{|}}{\underset{\underset{\displaystyle NH_2}{|}}{CH}}-\overset{\overset{\displaystyle O}{||}}{C}-OH$

d.

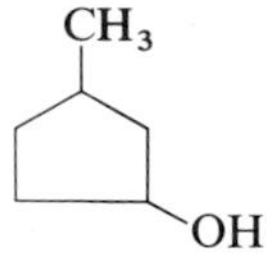

e.

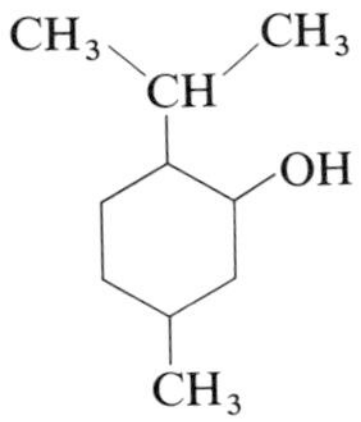

f.

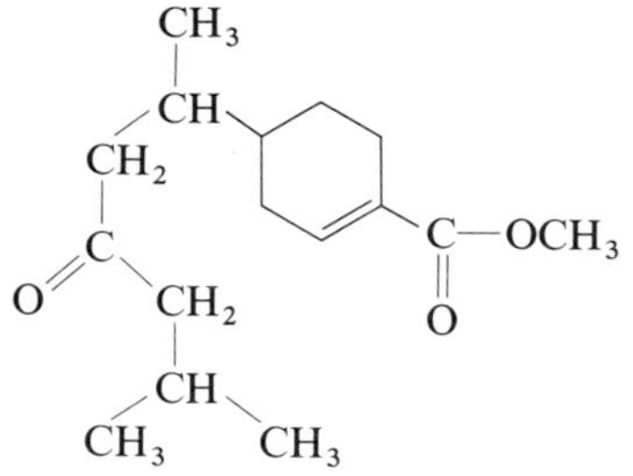

g.

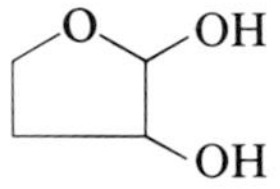

h.

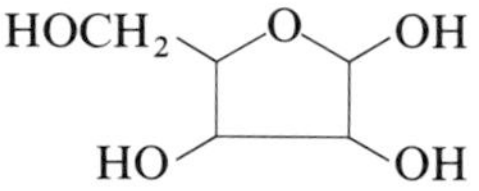

i.

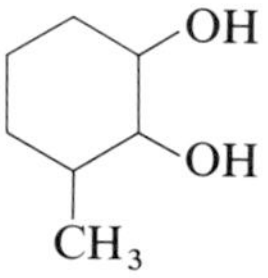

j. 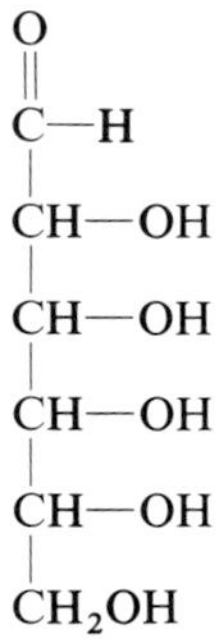
$$\overset{\overset{\displaystyle O}{||}}{C}-H$$
$$CH-OH$$
$$CH-OH$$
$$CH-OH$$
$$CH-OH$$
$$CH_2OH$$

k. 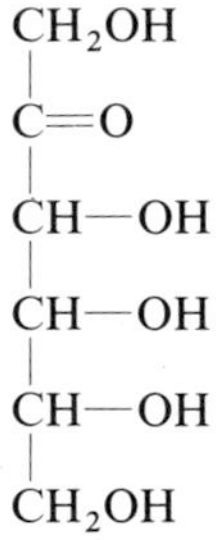
$$CH_2OH$$
$$C=O$$
$$CH-OH$$
$$CH-OH$$
$$CH-OH$$
$$CH_2OH$$

14.9 a. How many optical isomers are possible for each molecule in Problem 14.8?

b. How many pairs of enantiomers are possible for each molecule in Problem 14.8?

14.10 In Section 13.5A you saw that 4-hydroxypentanal forms a five-member cyclic hemiacetal.

$$CH_3-\overset{\overset{\displaystyle CH_2-CH_2}{|\qquad|}}{\underset{\underset{\displaystyle OH}{|}}{CH}}\quad CHO \longrightarrow \overset{\displaystyle CH_2-CH_2}{\underset{\displaystyle CH_3\ \ O\quad OH}{CH\quad CH}}$$

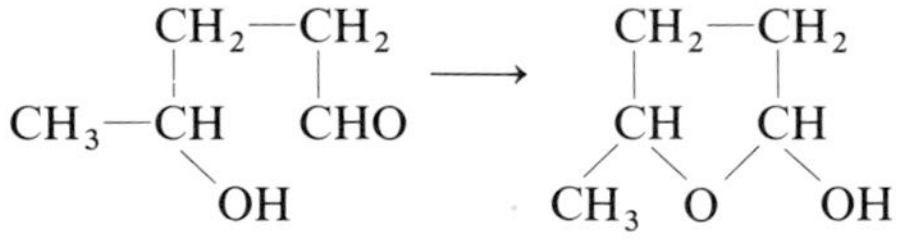

4-hydroxypentanal   a cyclic hemiacetal

How many stereoisomers are possible for this cyclic hemiacetal? Draw stereorepresentations of each.

14.11 5-Hydroxyhexanal readily forms a six-member cyclic hemiacetal.

$$CH_3-\overset{\overset{\displaystyle OH}{|}}{CH}-CH_2-CH_2-CH_2-\overset{\overset{\displaystyle O}{||}}{CH} \longrightarrow \text{a cyclic hemiacetal}$$

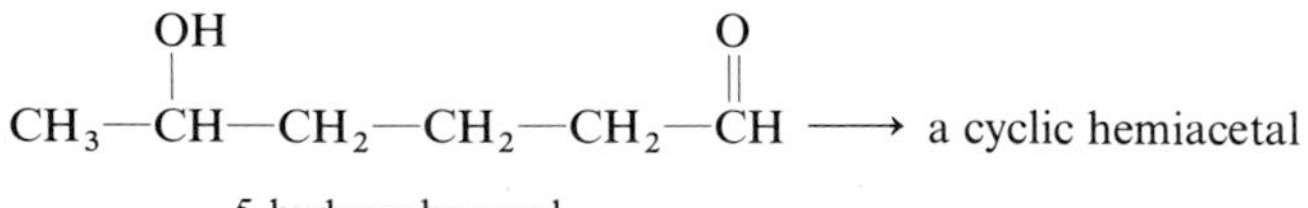

5-hydroxyhexanal

a. Draw a structural formula for the cyclic hemiacetal.
b. How many stereoisomers are possible for 5-hydroxyhexanal?
c. How many stereoisomers are possible for the cyclic hemiacetal?
d. Draw planar hexagon representations for each stereoisomer of the cyclic hemiacetal.

14.12 Glucose, a polyhydroxyaldehyde, forms a six-member cyclic hemiacetal in which the oxygen on carbon-5 of the chain reacts with the aldehyde on carbon-1.

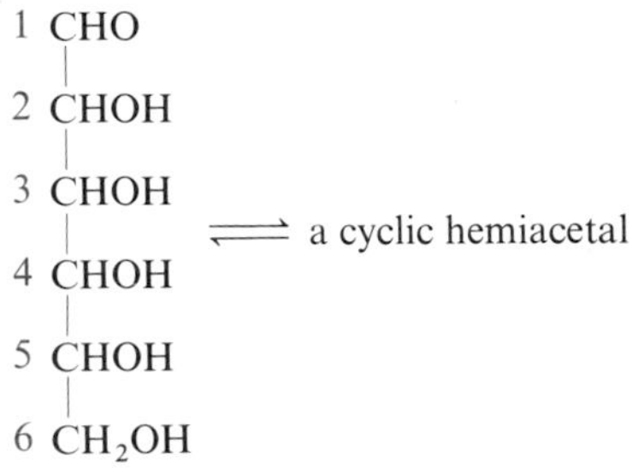

$$1 \ CHO$$
$$|$$
$$2 \ CHOH$$
$$|$$
$$3 \ CHOH$$
$$|$$
$$4 \ CHOH \quad \rightleftharpoons \ \text{a cyclic hemiacetal}$$
$$|$$
$$5 \ CHOH$$
$$|$$
$$6 \ CH_2OH$$

glucose

a. How many chiral carbon atoms are there in glucose? How many stereoisomers are possible for a molecule of this structure?
b. Draw a structural formula for the cyclic hemiacetal of glucose (do not worry about showing stereochemistry).
c. How many chiral carbon atoms are there in the cyclic hemiacetal formed by glucose? How many stereoisomers are possible for the cyclic hemiacetal?

14.13 Explain the difference in molecular structure between *meso*-tartaric acid and racemic tartaric acid.

14.14 Which of the following are *meso* compounds?

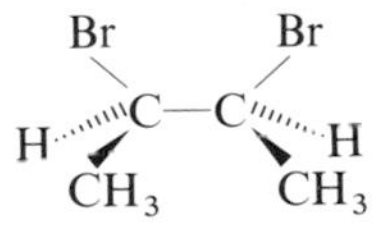

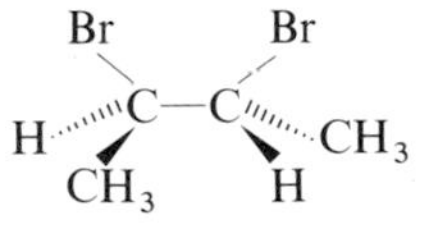

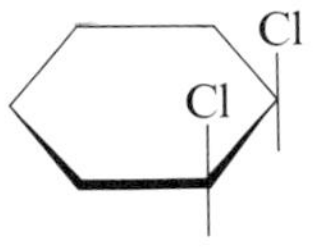

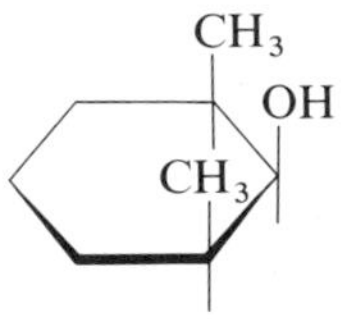

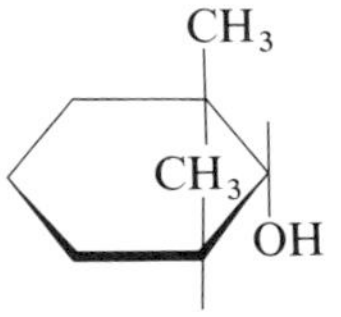

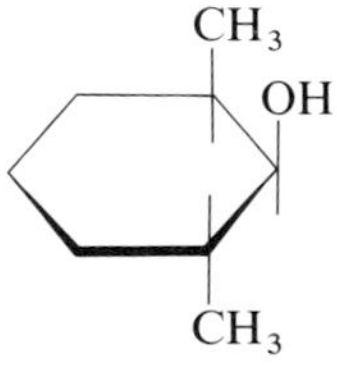

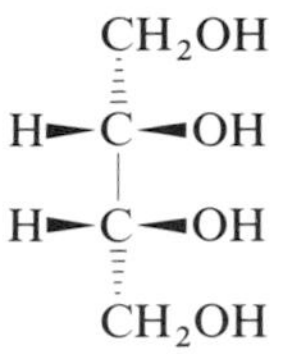

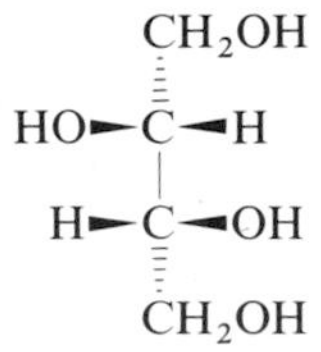

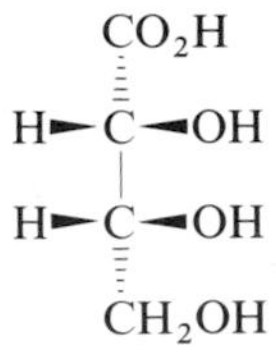

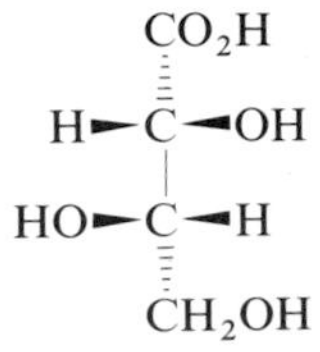

14.15 Below the structural formula of each of the following molecules is given the number of stereoisomers, and, in parentheses, the number of these stereoisomers that are *meso* compounds. For each *meso* compound, draw a stereorepresentation to show the plane of symmetry.

a.

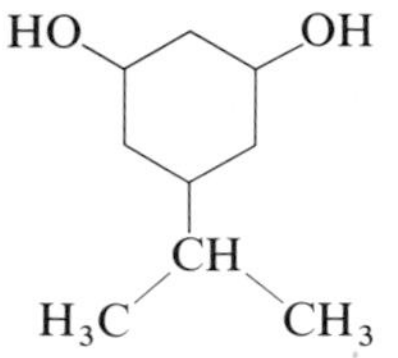

4(2 *meso*)

b. 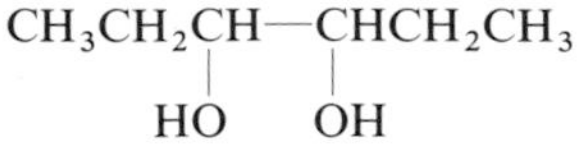 $CH_3CH_2CH{-}CHCH_2CH_3$

   with HO and OH substituents

3(1 *meso*)

c.

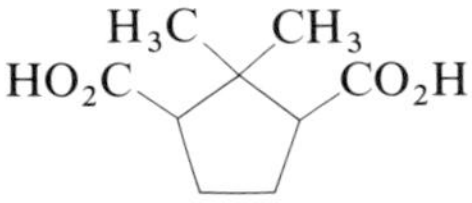

3(1 *meso*)

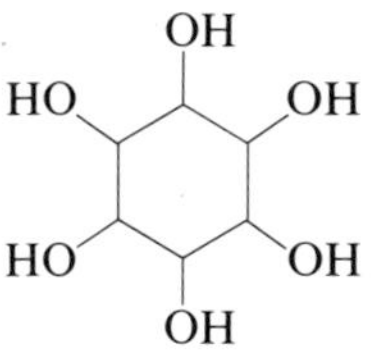

14.16 Inositol is a growth factor for animals and microorganisms. It is used medically for treatment of cirrhosis of the liver, hepatitis, and fatty infiltration of the liver. The most prevalent natural form is *cis*-1,2,3,5-*trans*-4,6-cyclohexanehexol. Draw a stereorepresentation of the natural isomer (show the cyclohexane ring as a planar hexagon) and determine whether it shows enantiomerism or is a *meso* compound.

inositol

14.17 Draw all stereoisomers for the following compounds. Classify them into pairs of enantiomers or *meso* compounds.

a. 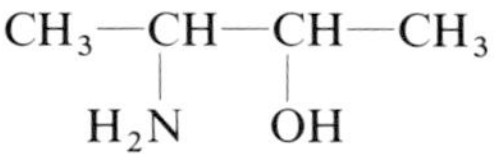 $CH_3{-}CH{-}CH{-}CH_3$
   with H_2N and OH substituents

b. 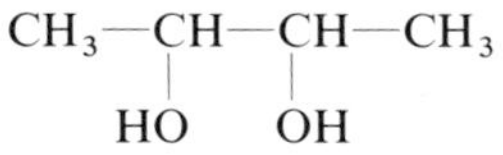 $CH_3{-}CH{-}CH{-}CH_3$
   with HO and OH substituents

c.

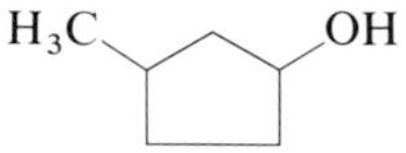

d.

14.18 Draw the four stereoisomers of grandisol, a sex hormone, secreted by the hind gut of the male boll weevil (*Anthonomus grandis*).

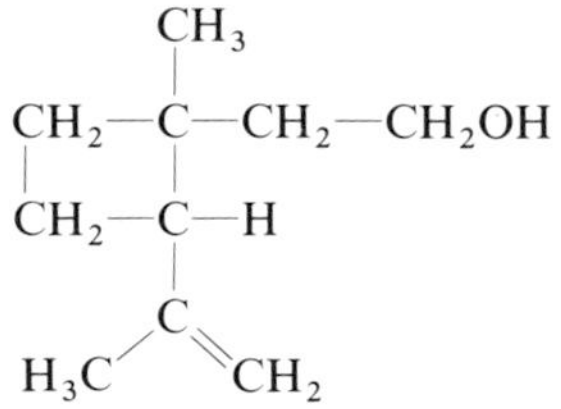

grandisol

The Significance of Chirality in the Biological World (Section 14.6)

14.19 How might you explain the following observations:

a. An enzyme is able to distinguish between a pair of enantiomers and catalyze a biochemical reaction of one enantiomer but not of its mirror image.

b. The microorganism *Penicillium glaucum* preferentially metabolizes $(+)$-tartaric acid rather than $(-)$-tartaric acid.

Carbohydrates

Carbohydrates are among the most abundant constituents of the plant and animal worlds. They serve many vital functions, such as: storehouses of chemical energy (glucose, starch, glycogen); supportive structural components in plants (cellulose); and essential components in the mechanisms for the genetic control of development and growth of living cells (D-ribose and 2-deoxy-D-ribose).

The name *carbohydrate* was derived from early observations that many members of this class of compounds have the formula $C_n(H_2O)_m$; hence, they were termed "hydrates of carbon." For example, the molecular formula of glucose, $C_6H_{12}O_6$, can also be written $C_6(H_2O)_6$. The formula of sucrose, $C_{12}H_{22}O_{11}$, can also be written $C_{12}(H_2O)_{11}$. It soon became clear, however, that not all carbohydrates have this general formula. For example, some carbohydrates contain nitrogen in addition to carbon, oxygen, and hydrogen. Although the term *carbohydrate* is not fully descriptive, it is firmly rooted in chemical nomenclature and has persisted as the name of this class of compounds.

Carbohydrates are often referred to as **saccharides** because of the sweet taste of the simpler members of the family, the **sugars** (Latin, *saccharum*, "sugar").

Carbohydrates are polyhydroxyaldehydes, polyhydroxyketones, or substances that yield either polyhydroxyaldehydes or polyhydroxyketones after hydrolysis. Therefore, the chemistry of carbohydrates is essentially the chemistry of two functional groups: hydroxyl groups and carbonyl groups.

15.1 Monosaccharides

A. Structure

Monosaccharides are carbohydrates that cannot be hydrolyzed to simpler compounds. They have the general formula $C_nH_{2n}O_n$, where n varies from three to eight. The terms *triose*, *tetrose*, *pentose*, and so on refer to the number of carbon atoms in the monosaccharide. For example, a triose contains three carbon atoms; a tetrose contains four carbon atoms.

$C_3H_6O_3$	triose	$C_6H_{12}O_6$	hexose
$C_4H_8O_4$	tetrose	$C_7H_{14}O_7$	heptose
$C_5H_{10}O_5$	pentose	$C_8H_{16}O_8$	octose

There are only two trioses: glyceraldehyde and dihydroxyacetone. Glyceraldehyde contains two —OH groups and an aldehyde; dihydroxyacetone contains two —OH groups and a ketone.

glyceraldehyde (*an aldose*)　　dihydroxyacetone (*a ketose*)

All monosaccharides contain a carbonyl group. Those with a carbonyl group in the form of an aldehyde are known as **aldoses** (*ald*ehyde + *-ose*). Glyceraldehyde is the simplest aldose. Those with a carbonyl group in the form of a ketone are known as **ketoses** (*ket*one + *-ose*). Dihydroxyacetone is the simplest ketose.

Dihydroxyacetone has no chiral carbon, and therefore cannot show stereoisomerism. Glyceraldehyde, however, has one chiral carbon and shows stereoisomerism. Shown in Figure 15.1 are structural formulas for the enantiomers of glyceraldehyde. The compound of structural formula Ia is named D-glyceraldehyde and has a specific rotation of $+13.5°$. Its enantiomer, Ib, is named L-glyceraldehyde and has a specific rotation of $-13.5°$. Note that the magnitude of the specific rotation for the enantiomers of glyceraldehyde is equal but opposite in sign. Also shown in Figure 15.1 are ball-and-stick models of the enantiomers of glyceraldehyde.

In the three-dimensional formulas in Figure 15.1, the configuration of each atom or group of atoms bonded to the chiral carbon atom is indicated by a combination of dashed wedges and solid wedges. **Fischer projection for-**

Figure 15.1 The enantiomers of glyceraldehyde (Ia and Ib) and ball-and-stick models of each.

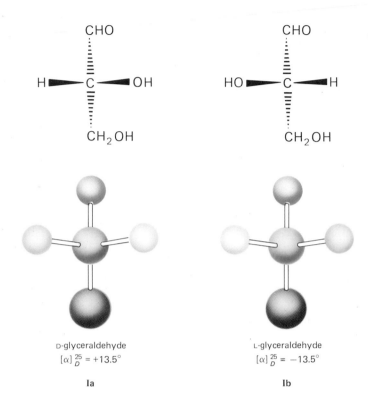

D-glyceraldehyde
$[\alpha]_D^{25} = +13.5°$

Ia

L-glyceraldehyde
$[\alpha]_D^{25} = -13.5°$

Ib

mulas are a simplified way to show the configuration about such chiral carbon atoms. According to this convention, the carbon chain is written vertically, with the most highly oxidized carbon atom at the top. Horizontal lines show groups projecting above the plane of the page; vertical lines show groups projecting behind the plane of the page. Applying the Fischer projection rules to the three-dimensional drawings of the enantiomers of glyceraldehyde gives formulas Ia′ and Ib′.

Ia′
D-glyceraldehyde

Ib′
L-glyceraldehyde

The configurations of D-glyceraldehyde and L-glyceraldehyde serve as reference points for the assignment of configuration to all other aldoses and ketoses. All those that have the same configuration as D-glyceraldehyde about the chiral carbon farthest from the aldehyde or ketone group are called D-monosaccharides; all those that have the same configuration as L-glycer-

Table 15.1 The isomeric D-aldotetroses, D-aldopentoses, and D-aldohexoses derived from D-glyceraldehyde.

D-aldotriose			

$$\begin{array}{c} \text{CHO} \\ | \\ \text{H—C—OH} \\ | \\ \text{CH}_2\text{OH} \end{array}$$

D-glyceraldehyde

D-aldotetroses

$$\begin{array}{c} \text{CHO} \\ | \\ \text{H—C—OH} \\ | \\ \text{H—C—OH} \\ | \\ \text{CH}_2\text{OH} \end{array} \qquad \begin{array}{c} \text{CHO} \\ | \\ \text{HO—C—H} \\ | \\ \text{H—C—OH} \\ | \\ \text{CH}_2\text{OH} \end{array}$$

D-erythrose   D-threose

D-aldopentoses

$$\begin{array}{c} \text{CHO} \\ | \\ \text{H—C—OH} \\ | \\ \text{H—C—OH} \\ | \\ \text{H—C—OH} \\ | \\ \text{CH}_2\text{OH} \end{array} \quad \begin{array}{c} \text{CHO} \\ | \\ \text{HO—C—H} \\ | \\ \text{H—C—OH} \\ | \\ \text{H—C—OH} \\ | \\ \text{CH}_2\text{OH} \end{array} \quad \begin{array}{c} \text{CHO} \\ | \\ \text{H—C—OH} \\ | \\ \text{HO—C—H} \\ | \\ \text{H—C—OH} \\ | \\ \text{CH}_2\text{OH} \end{array} \quad \begin{array}{c} \text{CHO} \\ | \\ \text{HO—C—H} \\ | \\ \text{HO—C—H} \\ | \\ \text{H—C—OH} \\ | \\ \text{CH}_2\text{OH} \end{array}$$

D-ribose   D-arabinose   D-xylose   D-lyxose

D-aldohexoses

$$\begin{array}{c} \text{CHO} \\ | \\ \text{H—C—OH} \\ | \\ \text{H—C—OH} \\ | \\ \text{H—C—OH} \\ | \\ \text{H—C—OH} \\ | \\ \text{CH}_2\text{OH} \end{array} \quad \begin{array}{c} \text{CHO} \\ | \\ \text{HO—C—H} \\ | \\ \text{H—C—OH} \\ | \\ \text{H—C—OH} \\ | \\ \text{H—C—OH} \\ | \\ \text{CH}_2\text{OH} \end{array} \quad \begin{array}{c} \text{CHO} \\ | \\ \text{H—C—OH} \\ | \\ \text{HO—C—H} \\ | \\ \text{H—C—OH} \\ | \\ \text{H—C—OH} \\ | \\ \text{CH}_2\text{OH} \end{array} \quad \begin{array}{c} \text{CHO} \\ | \\ \text{HO—C—H} \\ | \\ \text{HO—C—H} \\ | \\ \text{H—C—OH} \\ | \\ \text{H—C—OH} \\ | \\ \text{CH}_2\text{OH} \end{array}$$

D-allose   D-altrose   D-glucose   D-mannose

$$\begin{array}{c} \text{CHO} \\ | \\ \text{H—C—OH} \\ | \\ \text{H—C—OH} \\ | \\ \text{HO—C—H} \\ | \\ \text{H—C—OH} \\ | \\ \text{CH}_2\text{OH} \end{array} \quad \begin{array}{c} \text{CHO} \\ | \\ \text{HO—C—H} \\ | \\ \text{H—C—OH} \\ | \\ \text{HO—C—H} \\ | \\ \text{H—C—OH} \\ | \\ \text{CH}_2\text{OH} \end{array} \quad \begin{array}{c} \text{CHO} \\ | \\ \text{H—C—OH} \\ | \\ \text{HO—C—H} \\ | \\ \text{HO—C—H} \\ | \\ \text{H—C—OH} \\ | \\ \text{CH}_2\text{OH} \end{array} \quad \begin{array}{c} \text{CHO} \\ | \\ \text{HO—C—H} \\ | \\ \text{HO—C—H} \\ | \\ \text{HO—C—H} \\ | \\ \text{H—C—OH} \\ | \\ \text{CH}_2\text{OH} \end{array}$$

D-gulose   D-idose   D-galactose   D-talose

Table 15.2 The isomeric D-ketopentoses and D-ketohexoses derived from dihydroxyacetone and D-erythrulose.

ketotriose	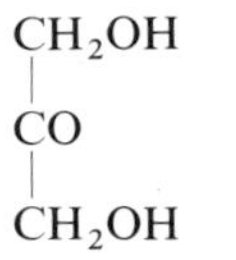 CH$_2$OH | CO | CH$_2$OH dihydroxyacetone

D-ketotetrose	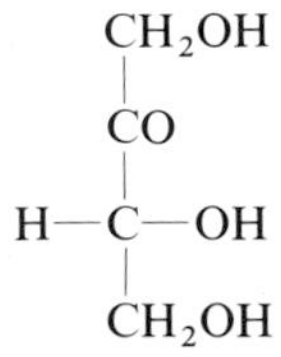 CH$_2$OH | CO | H—C—OH | CH$_2$OH D-erythrulose

D-ketopentoses	CH$_2$OH | CO | H—C—OH | H—C—OH | CH$_2$OH D-ribulose	CH$_2$OH | CO | HO—C—H | H—C—OH | CH$_2$OH D-xylulose

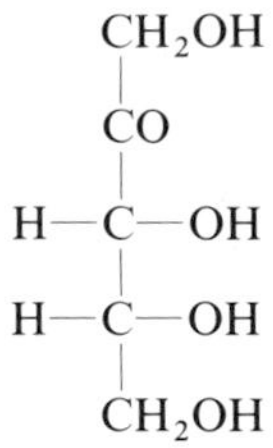

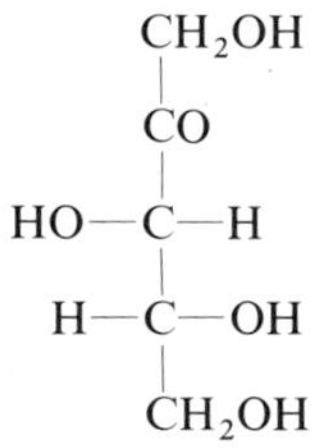

D-ketohexoses	CH$_2$OH | CO | H—C—OH | H—C—OH | H—C—OH | CH$_2$OH D-psicose	CH$_2$OH | CO | HO—C—H | H—C—OH | H—C—OH | CH$_2$OH D-fructose	CH$_2$OH | CO | H—C—OH | HO—C—H | H—C—OH | CH$_2$OH D-sorbose	CH$_2$OH | CO | HO—C—H | HO—C—H | H—C—OH | CH$_2$OH D-tagatose

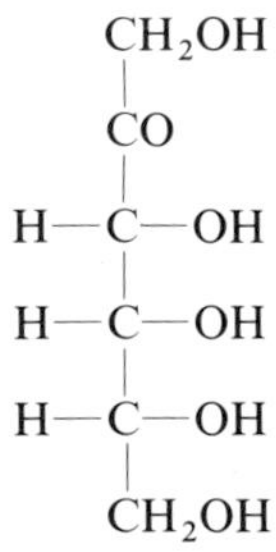

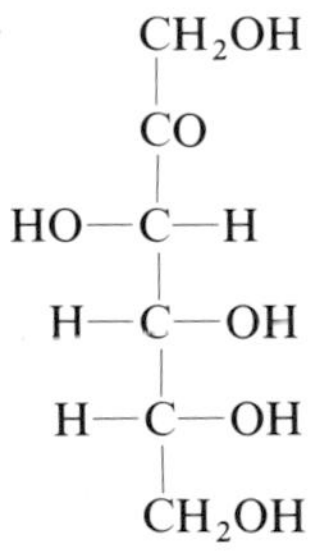

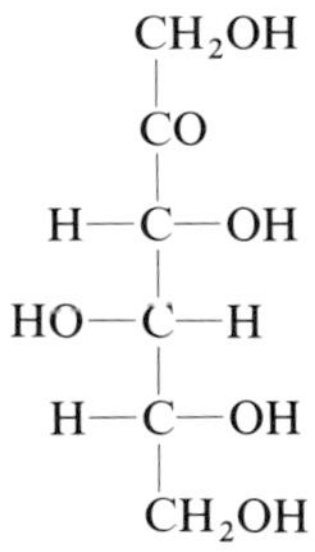

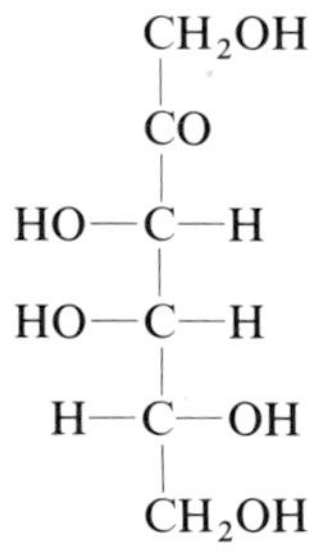

aldehyde about the chiral carbon farthest from the aldehyde or ketone group are called L-monosaccharides.

Tables 15.1 and 15.2 show the names and structural formulas for all trioses, tetroses, pentoses, and hexoses of the D-series.

D-Ribose and 2-deoxy-D-ribose, the most important pentoses, are building blocks of nucleic acids: D-ribose in ribonucleic acids (RNA) and 2-deoxy-D-ribose in deoxyribonucleic acids (DNA). The structures of D-ribose and 2-deoxy-D-ribose are shown on the next page.

$$
\begin{array}{c}
\text{CHO} \\
\mid \\
\text{H—C—OH} \\
\mid \\
\text{H—C—OH} \\
\mid \\
\text{H—C—OH} \\
\mid \\
\text{CH}_2\text{OH}
\end{array}
\qquad
\begin{array}{c}
\text{CHO} \\
\mid \\
\text{H—C—H} \\
\mid \\
\text{H—C—OH} \\
\mid \\
\text{H—C—OH} \\
\mid \\
\text{CH}_2\text{OH}
\end{array}
$$

D-ribose   2-deoxy-D-ribose

The most abundant hexoses in the biological world are D-glucose, D-galactose, and D-fructose. The first two of these are D-aldohexoses; the third, fructose, is a D-ketohexose.

$$
\begin{array}{c}
\text{CHO} \\
\mid \\
\text{H—C—OH} \\
\mid \\
\text{HO—C—H} \\
\mid \\
\text{H—C—OH} \\
\mid \\
\text{H—C—OH} \\
\mid \\
\text{CH}_2\text{OH}
\end{array}
\qquad
\begin{array}{c}
\text{CHO} \\
\mid \\
\text{H—C—OH} \\
\mid \\
\text{HO—C—H} \\
\mid \\
\text{HO—C—H} \\
\mid \\
\text{H—C—OH} \\
\mid \\
\text{CH}_2\text{OH}
\end{array}
\qquad
\begin{array}{c}
\text{CH}_2\text{OH} \\
\mid \\
\text{C=O} \\
\mid \\
\text{HO—C—H} \\
\mid \\
\text{H—C—OH} \\
\mid \\
\text{H—C—OH} \\
\mid \\
\text{CH}_2\text{OH}
\end{array}
$$

D-glucose   D-galactose   D-fructose

Glucose is by far the most common hexose monosaccharide. It is also known as dextrose because it is dextrorotatory. Other names for this monosaccharide are grape sugar, blood sugar, and corn sugar, names that clearly indicate its sources in an uncombined state. Human blood normally contains 65–110 mg of glucose per 100 mL.

Fructose is found combined with glucose in the disaccharide sucrose, or as it is more commonly known, table sugar (Section 15.3C). D-Galactose is found combined with glucose in the disaccharide lactose (Section 15.3B). Lactose appears nowhere else except in milk.

Example 15.1

a. Draw Fischer projection formulas for all aldoses of four carbon atoms.

b. Label those substances in part (a) that are D-monosaccharides and those that are L-monosaccharides.

c. Indicate which are pairs of enantiomers.

d. Refer to Table 15.1 and give names to the aldoses whose formulas you have drawn.

Solution

Aldotetroses have an aldehyde group on carbon-1 and —OH groups on carbons 2, 3, and 4. Following are Fischer projection formulas for the four aldotetroses. D- and L- refer to the arrangement of groups attached to the next-to-last carbon. In aldotetroses, the next-to-last carbon is number 3. In the Fischer projection of a D-aldotetrose, the —OH on carbon-3 is on the right; in an L-aldotetrose, it is on the left.

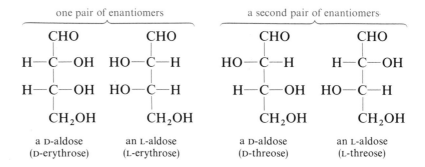

Problem 15.1

a. Draw Fischer projection formulas for all 2-ketoses with five carbon atoms.

b. Label those that are D-ketopentoses and those that are L-ketopentoses.

c. Indicate any pairs of enantiomers.

d. Refer to Table 15.2 and give names to the ketoses you have drawn.

B. Amino Sugars

Amino sugars contain an —NH₂ group in place of an —OH group. Only three amino sugars are common in nature: D-glucosamine, D-mannosamine, and D-galactosamine.

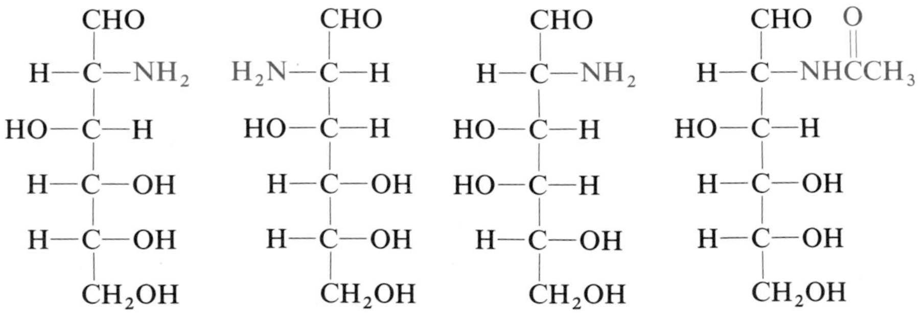

$$\begin{array}{cccc}
\text{CHO} & \text{CHO} & \text{CHO} & \text{CHO} \quad \text{O} \\
\text{H—C—NH}_2 & \text{H}_2\text{N—C—H} & \text{H—C—NH}_2 & \text{H—C—NHCCH}_3 \\
\text{HO—C—H} & \text{HO—C—H} & \text{HO—C—H} & \text{HO—C—H} \\
\text{H—C—OH} & \text{H—C—OH} & \text{HO—C—H} & \text{H—C—OH} \\
\text{H—C—OH} & \text{H—C—OH} & \text{H—C—OH} & \text{H—C—OH} \\
\text{CH}_2\text{OH} & \text{CH}_2\text{OH} & \text{CH}_2\text{OH} & \text{CH}_2\text{OH}
\end{array}$$

D-glucosamine   D-mannosamine   D-galactosamine   N-acetyl-D-glucosamine (NAG)

In most cases where these monosaccharides occur in nature, the —NH_2 group is acetylated. *N*-Acetyl-D-glucosamine, a derivative of D-glucosamine, is a component of many polysaccharides including *chitin*, the hard, shell-like exoskeleton of lobsters, crabs, shrimps, and other crustaceans.

C. The Cyclic Structure of Monosaccharides

As we saw in Section 13.5A, aldehydes and ketones react with alcohols to form hemiacetals and hemiketals.

$$R{-}\overset{\overset{\displaystyle O}{\|}}{C}{-}H + CH_3O{-}H \rightleftharpoons R{-}\overset{\overset{\displaystyle OH}{|}}{\underset{\underset{\displaystyle H}{|}}{C}}{-}OCH_3$$

an aldehyde   a hemiacetal

$$R{-}\overset{\overset{\displaystyle O}{\|}}{C}{-}R + CH_3O{-}H \rightleftharpoons R{-}\overset{\overset{\displaystyle OH}{|}}{\underset{\underset{\displaystyle R}{|}}{C}}{-}OCH_3$$

a ketone   a hemiketal

We also saw that cyclic hemiacetals and hemiketals form when hydroxyl and carbonyl groups are part of the same molecule. For example, 4-hydroxypentanal forms a five-member cyclic hemiacetal.

4-hydroxypentanal   a cyclic hemiacetal
(*minor form*)   (*major form*)

Monosaccharides have hydroxyl and carbonyl groups in the same molecule, and they, too, can form cyclic hemiacetals and hemiketals. One such monosaccharide is glucose.

In the cyclic hemiacetal of glucose, the —OH on carbon-5 bonds to carbon-1 to form a six-member cyclic hemiacetal. When this cyclic hemiacetal forms, carbon-1 becomes a chiral center and is called an **anomeric carbon.** Because of this new chiral carbon, D-glucose exists as a pair of diastereomers (Section 14.5A), each of which has different chemical and physical properties. Monosaccharides that differ from each other only in the configuration at the anomeric carbon are called **anomers.** One anomer of D-glucose, called α-D-glucose, has a specific rotation of $+112°$. The other anomer, called β-D-glucose, has a specific rotation of $+19°$. Following are Fischer projection formulas for the α- and β-anomers of D-glucose, along with the formula for the open-chain form.

α-D-glucose
mp 146°C
[α] = +112°

open-chain or
free aldehyde form
of D-glucose

β-D-glucose
mp 150°C
[α] = +19°

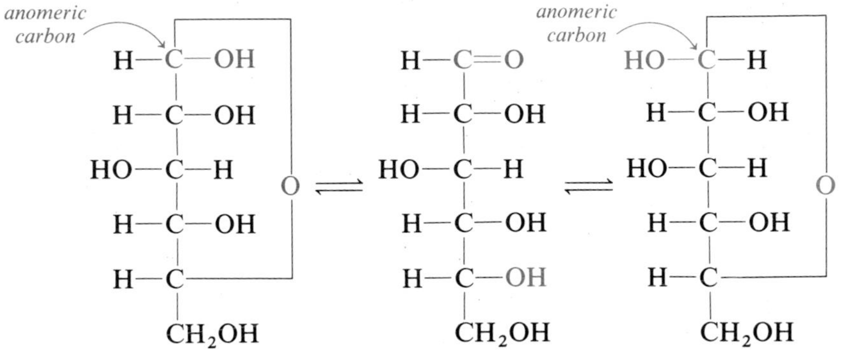

In Fischer projection formulas, the —OH on the anomeric carbon is on the right in α-D-glucose and on the left in β-D-glucose. We can also draw structural formulas showing these cyclic hemiacetals as planar hexagons. Such representations are called **Haworth structures.** Haworth structures for α-D-glucose and β-D-glucose are

α-D-glucose

β-D-glucose

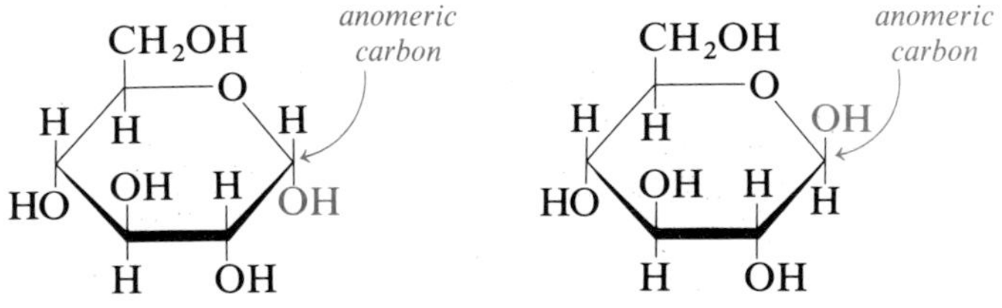

In Haworth structures, the —OH on the anomeric carbon is below the plane of the ring in the α-anomer and above the plane of the ring in the β-anomer.

Six-member cyclic hemiacetals may also be drawn as strain-free chair conformations. Following are chair conformations for α-D-glucose and β-D-glucose, along with the open-chain form of D-glucose. In these drawings, only substituents on the rings are shown; hydrogen atoms are not shown.

α-D-glucose
[α] = +112°

open-chain form

β-D-glucose
[α] = +19°

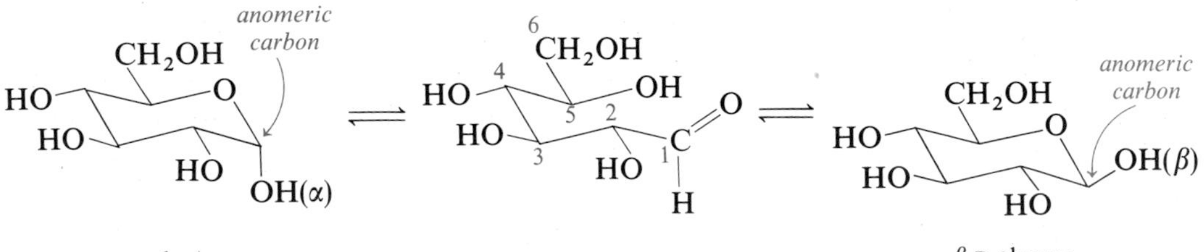

Notice that in the chair conformations of α- and β-D-glucose, substituents on carbons 2, 3, and 4 of each ring are equatorial. The —OH on the anomeric carbon of α-D-glucose is axial and that on the anomeric carbon of β-D-glucose is equatorial.

Other monosaccharides also form cyclic hemiacetals and hemiketals. Following are Haworth structures for the cyclic hemiketals formed by D-fructose. Note that these cyclic hemiketals contain five-member rings.

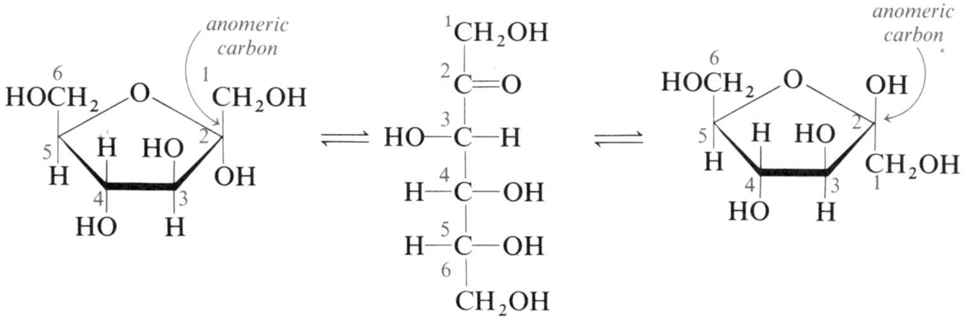

α-D-fructose
$[\alpha] = +21°$

open-chain or
free ketone form
of D-fructose

β-D-fructose
$[\alpha] = -133°$

In the cyclic hemiketals of fructose, the —OH on the anomeric carbon (carbon-2 of fructose) is below the plane of the ring in the α-anomer and above the plane of the ring in the β-anomer.

Example 15.2

D-Galactose forms a cyclic hemiacetal containing a six-member ring. Draw Haworth structures for α-D-galactose and β-D-galactose. Label the anomeric carbon in each cyclic hemiacetal.

Solution

One way to draw Haworth structures for six-member cyclic hemiacetals is to use the α- and β-forms of D-glucose as reference points. D-Galactose differs from D-glucose in the configuration of carbon-4. Therefore, the α- and β-forms of D-galactose differ from the α- and β-forms of D-glucose only in the orientation of the —OH group on carbon-4.

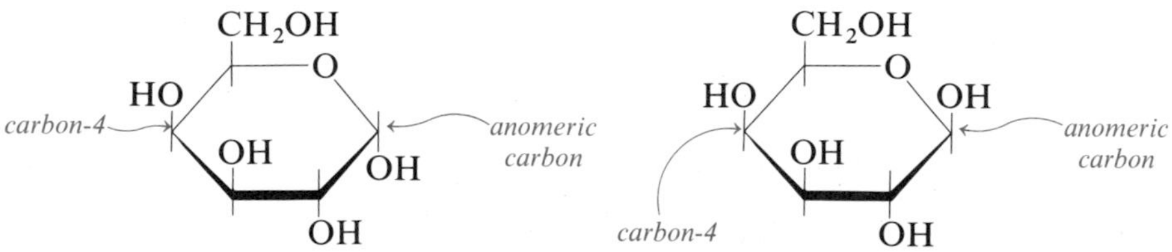

α-D-galactose

β-D-galactose

Problem 15.2

D-Mannose forms a cyclic hemiacetal containing a six-member ring. Draw Haworth structures for α-D-mannose and β-D-mannose. Label the anomeric carbon atom in each of these cyclic hemiacetals.

D. Mutarotation

The α- and β-anomers of monosaccharides are interconvertible in aqueous solution. The change in specific rotation that accompanies this interconversion is known as **mutarotation.** For example, a freshly prepared solution of α-D-glucose shows an initial rotation of +112°, which gradually decreases to +52° as α-D-glucose reaches an equilibrium with β-D-glucose. A solution of β-D-glucose also undergoes mutarotation, during which the specific rotation changes from an initial value of +19° to the same equilibrium value of +52°.

Mutarotation is common to all monosaccharides that exist in α- and β-forms. Shown in Table 15.3 are specific rotations for α- and β-forms of D-galactose and D-fructose, with equilibrium values for the specific rotation of each after mutarotation.

Table 15.3 Specific rotations of α-and β-forms of D-galactose and D-fructose before and after mutarotation.

Monosaccharide	Specific Rotation	Specific Rotation After Mutarotation
α-D-galactose	+151°	+84°
β-D-galactose	−53°	
α-D-fructose	+21°	−92°
β-D-fructose	−133°	

E. Physical Properties of Monosaccharides

Monosaccharides are colorless crystalline solids. Because of the possibility of hydrogen bonding between polar —OH groups and water, all monosaccharides are very soluble in water. They are only slightly soluble in alcohol and are insoluble in nonhydroxylic solvents such as ether, chloroform, and benzene.

Although all monosaccharides are sweet to the taste, some are sweeter than others (Table 15.4, next page). Of the monosaccharides, D-fructose tastes the sweetest, even sweeter than sucrose (table sugar). The sweet taste of honey is due largely to the presence of D-fructose, while that of corn syrup is due largely to the presence of glucose. Molasses is a by-product of table sugar manufacture. In the production of table sugar, sugar cane is boiled with water and then cooled. As the mixture cools, both sucrose crystals and light molasses

Table 15.4 Relative sweetness of some monosaccharides, disaccharides, and other sweetening agents.

Monosaccharides		Disaccharides		Other Sweetening Agents	
D-fructose	174	sucrose		honey	97
D-glucose	74	(table sugar)	100	molasses	74
D-xylose	0.40	lactose		corn syrup	74
D-galactose	0.22	(milk sugar)	0.16		

separate and are collected. Subsequent boilings and coolings yield a dark, thick syrup known as blackstrap molasses.

15.2 Reactions of Monosaccharides

A. Oxidation

Monosaccharides (and carbohydrates in general) are classified as **reducing** or **nonreducing sugars** according to their behavior toward Cu^{2+} **(Benedict's solution, Fehling's solution)** or toward Ag^+ in ammonium hydroxide **(Tollens' solution).** To understand the chemical basis for this classification, you need to remember two things: First, all monosaccharides contain either an aldehyde or an α-hydroxyketone; second, in dilute base (the conditions of Benedict's, Tollens', and Fehling's tests) ketoses are in equilibrium with aldoses via an enediol intermediate (Problems 13.21 and 13.22).

$$
\begin{array}{ccc}
\text{H} & & \\
| & & \\
\text{H}-\text{C}-\text{OH} & \text{H}-\text{C}-\text{OH} & \text{H}-\text{C}=\text{O} \\
| & || & | \\
\text{C}=\text{O} \rightleftharpoons & \text{C}-\text{OH} \rightleftharpoons & \text{H}-\text{C}-\text{OH} \\
| & | & | \\
\text{CH}_2\text{OH} & \text{CH}_2\text{OH} & \text{CH}_2\text{OH} \\
\text{a ketose} & \text{an enediol} & \text{an aldose}
\end{array}
$$

Both Ag(I) and Cu(II) oxidize aldehydes to carboxylic acids.

$$
\begin{array}{c}
\text{O} \\
|| \\
\text{R}-\text{C}-\text{H} + \text{Ag}^+ \longrightarrow \text{R}-\text{C}-\text{O}^- + \text{Ag}
\end{array}
$$

silver mirror

$$
\begin{array}{c}
\text{O} \\
|| \\
\text{R}-\text{C}-\text{H} + \text{Cu}^{2+} \longrightarrow \text{R}-\text{C}-\text{O}^- + \text{Cu}_2\text{O}
\end{array}
$$

brick-red precipitate

In Tollens' test, silver ion is reduced to metallic silver; if the reaction is done properly, silver precipitates as a mirror-like coating on the surface of the container. In Benedict's or Fehling's test, copper(II) is reduced to copper(I) which precipitates as Cu_2O, a brick-red solid. Any carbohydrate that reduces silver ion to metallic silver or copper(II) ion to Cu_2O is classified as a reducing sugar. Sugars that do not reduce these reagents are classified as nonreducing sugars.

Even though monosaccharides exist predominantly in the cyclic hemiacetal form, the cyclic forms are in equilibrium with the free aldehyde and are therefore susceptible to oxidation by the solutions mentioned. All monosaccharides are reducing sugars.

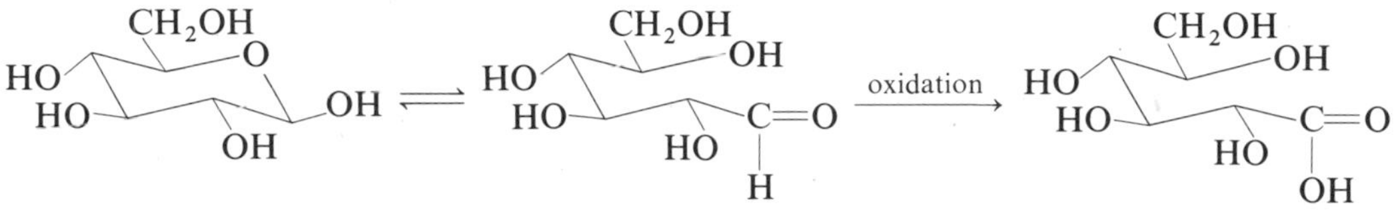

hemiacetal form   open-chain form   carboxylic acid

Oxidation of aldoses by Fehling's, Benedict's, or Tollens' solutions forms monocarboxylic acids known as **aldonic acids.** For example, D-glucose forms D-gluconic acid.

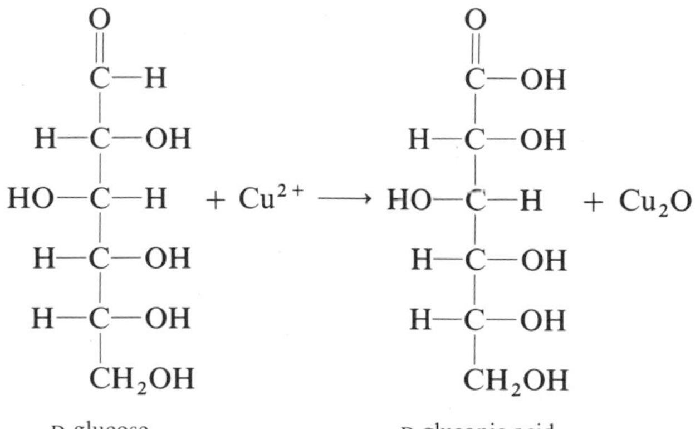

D-glucose   D-gluconic acid

B. Reduction

The carbonyl group of a monosaccharide is reduced to an alcohol by a variety of reducing agents, including $NaBH_4$ and hydrogen in the presence of a metal catalyst. These reduction products are known as **alditols.** Reduction of D-glucose gives D-glucitol, more commonly known as sorbitol (see next page).

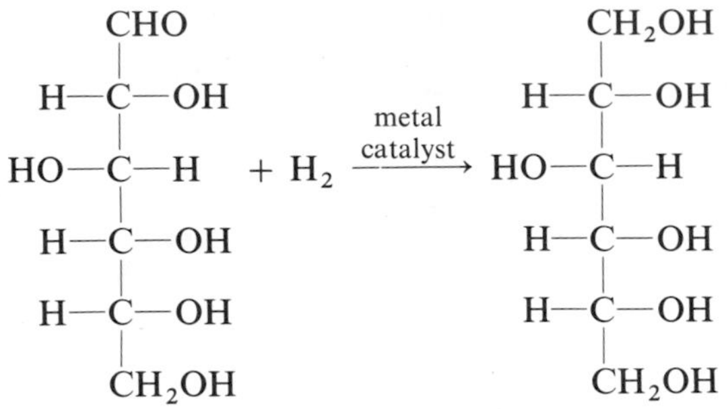

D-glucose + H₂ --metal catalyst--> D-glucitol (sorbitol)

C. Formation of Glycosides (Acetals)

We have already seen in Section 15.1C that monosaccharides of five or more carbon atoms form cyclic hemiacetals or hemiketals. Reaction of a monosaccharide hemiacetal or hemiketal with a second molecule of alcohol forms an acetal or ketal. Following is a Haworth structure for the acetal formed by reaction of β-D-glucose with methanol.

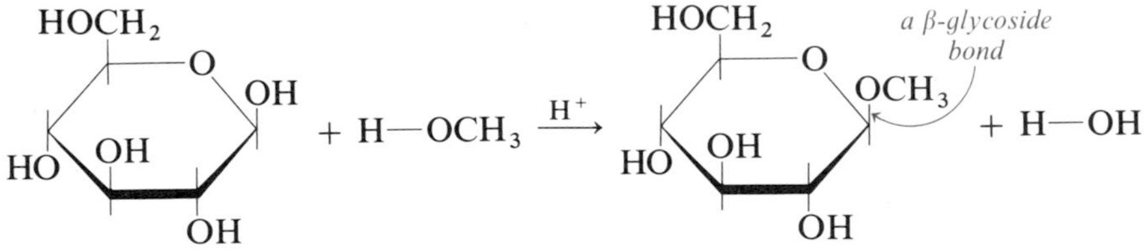

β-D-glucose + methanol → a methyl glycoside (methyl β-D-glucoside) + H—OH

In this reaction, the —OH on the anomeric carbon is replaced by an —OR group. A cyclic acetal or ketal derived from a monosaccharide is called a **glycoside,** and the bond from the anomeric carbon to the —OR group is called a **glycoside bond.** The name of the glycoside is derived by dropping the terminal -e from the name of the monosaccharide and adding -ide. For example, glycosides derived from D-glucose are named D-glucosides; those derived from D-ribose are named D-ribosides. In Haworth structures, the —OR on the anomeric carbon of the glycoside is below the plane of the ring in the α-anomer and above the plane of the ring in the β-anomer. In chair conformations, the —OR on the anomeric carbon is axial in the α-anomer and equatorial in the β-anomer. Glycosides of all monosaccharides are nonreducing sugars.

Example 15.3

Draw structural formulas for the following glycosides.

a. β-methyl-D-riboside

b. α-methyl-D-galactoside

Solution

a. D-Ribose forms a five-member cyclic hemiacetal that reacts with methanol to form a cyclic acetal. The —OCH_3 group on the anomeric carbon is above the plane of the ring in a β-riboside.

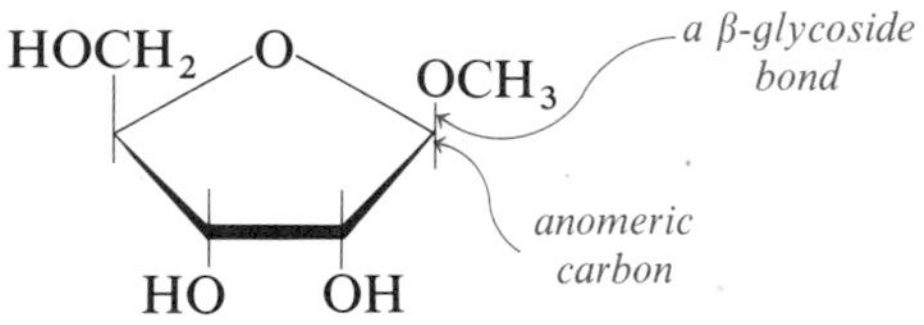

b. D-Galactose forms a six-member cyclic hemiacetal that reacts with methanol to form a cyclic acetal. The —OCH_3 group is below the plane of the ring in an α-galactoside. Following are Haworth and chair structures for α-methyl-D-galactoside.

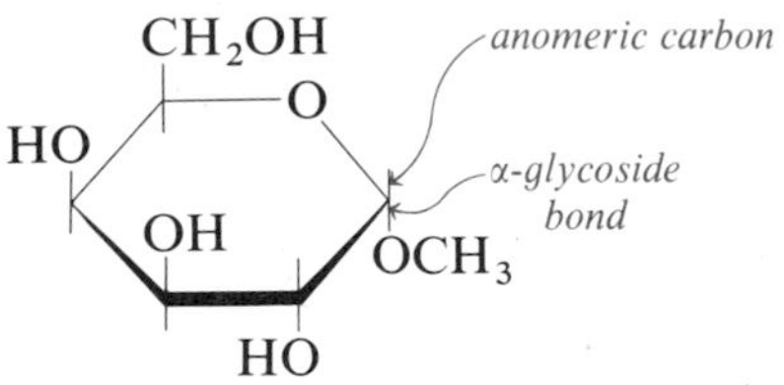

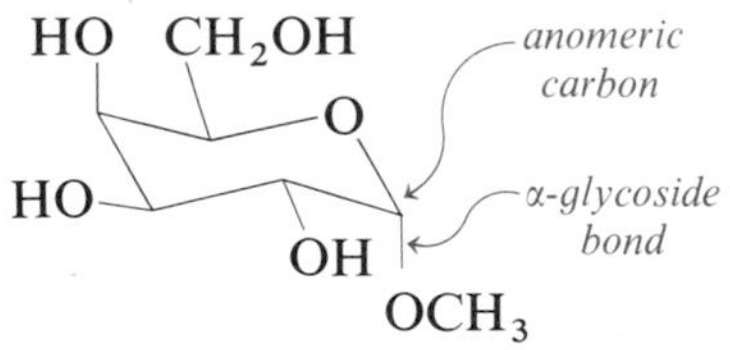

Problem 15.3

Draw structural formulas for the following glycosides. In each, label the anomeric carbon and the glycoside bond.

a. β-methyl-D-fructoside   **b.** α-methyl-D-mannoside

Just as the anomeric carbon of a cyclic hemiacetal or hemiketal can react with R—OH to form a glycoside, it can also react with an N- -H group to form an **N-glycoside**. Especially important in the biological world are the N-glycosides formed between D-ribose and 2-deoxy-D-ribose and the heterocyclic aromatic amines uracil, cytosine, thymine, adenine, and guanine (Figure 15.2). N-Glycosides of these purine and pyrimidine bases are structural units of DNA and RNA (Chapter 22).

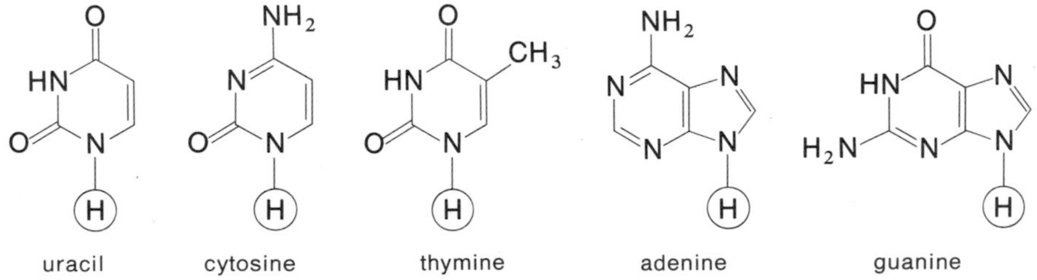

Figure 15.2 Structural formulas of the most important purine and pyrimidine bases found in DNA and RNA. The circled hydrogen atom is lost in forming an N-glycoside.

uracil   cytosine   thymine   adenine   guanine

Example 15.4

Draw structural formulas for the N-glycosides formed between the following compounds. In each, label the anomeric carbon and the N-glycoside bond.

a. β-D-ribose and cytosine **b.** β-2-deoxy-D-ribose and adenine

Solution

a.

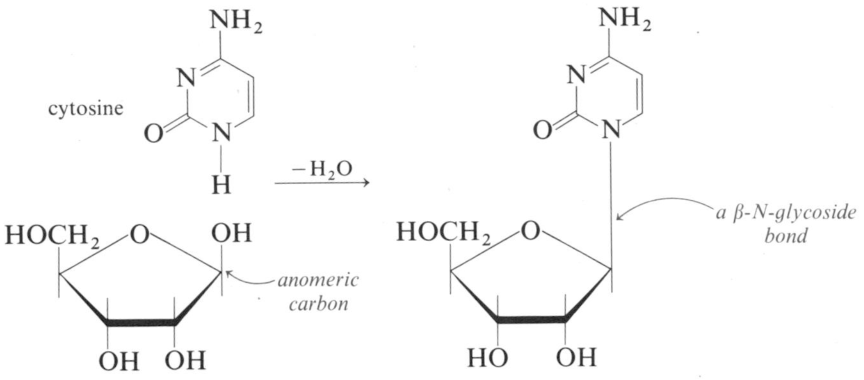

β-D-ribose

b.

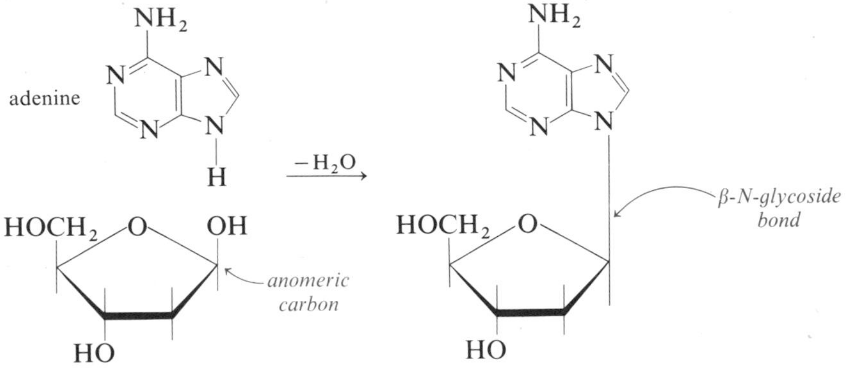

β-2-deoxy-D-ribose

Problem 15.4

Draw structural formulas for the N-glycosides formed between the following compounds. In each, label the anomeric carbon and the N-glycoside bond.

a. β-2-deoxy-D-ribose and uracil **b.** β-D-ribose and guanine

15.3 Disaccharides

Most carbohydrates in nature contain more than one monosaccharide. Those that contain two monosaccharide units are called **disaccharides,** those that contain three monosaccharide units are called **trisaccharides**, and those that contain many monosaccharide units are called **polysaccharides.** In a disaccharide, two monosaccharide units are joined together by a glycoside bond between the anomeric carbon of one unit and an —OH of the other. In this section, we will look at the structure and natural occurrence of three disaccharides: maltose, lactose, and sucrose.

A. Maltose

Maltose derives its name from the fact that it occurs in malt liquors, the juice from sprouted barley and other cereal grains. Maltose consists of two molecules of glucose joined by a glycoside bond between carbon-1 (the anomeric carbon) of one glucose unit and carbon-4 of the second glucose unit. The oxygen atom on the anomeric carbon of the first glucose unit is α (below the plane of the ring); therefore, the bond joining the two glucose units is called an α-1,4-glycoside bond. Following are Haworth and chair formulas for β-maltose. This form of maltose is called β-maltose because the —OH on the anomeric carbon of the rightmost glucose unit is β.

α-D-glucose unit | β-D-glucose unit | α-D-glucose unit | β-D-glucose unit

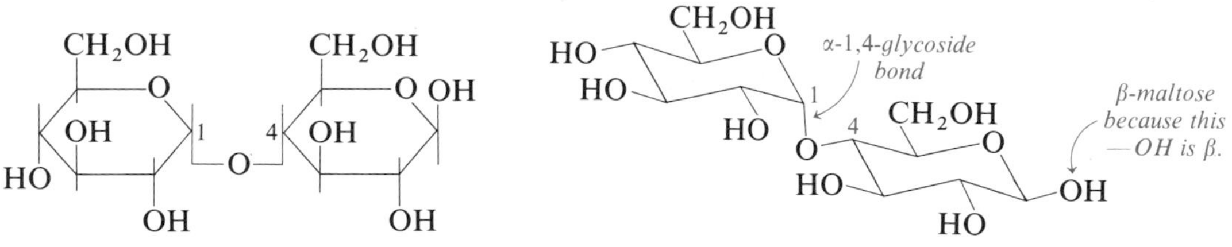

β-maltose

Hydrolysis of maltose yields two molecules of glucose. Maltose is a reducing sugar because the anomeric carbon on the right unit of D-glucose is in equiiibrium with the free aldehyde that can be oxidized to a carboxylic acid.

B. Lactose

Lactose is the major sugar present in milk. It makes up about 5–8% of human milk and 4–6% of cow's milk. Hydrolysis of lactose yields D-glucose and D-galactose. In lactose, galactose is joined to glucose by a β-1,4-glycoside bond.

β-D-galactose unit | β-D-glucose unit | β-D-galactose unit | β-D-glucose unit

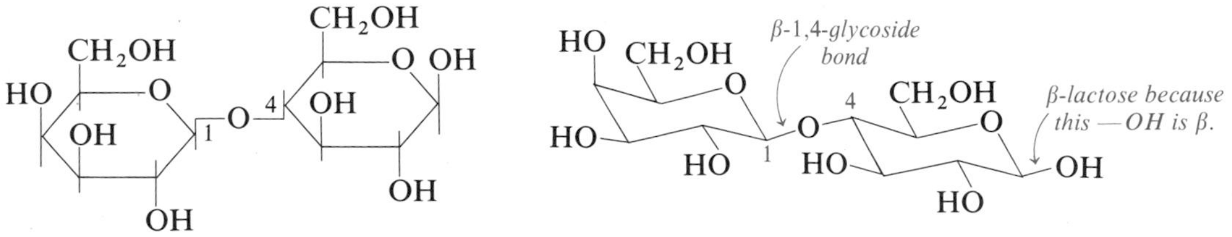

β-lactose (from the milk of mammals)

C. Sucrose

Sucrose (table sugar) is the most abundant disaccharide. It is obtained from the juice of sugar cane and sugar beets.

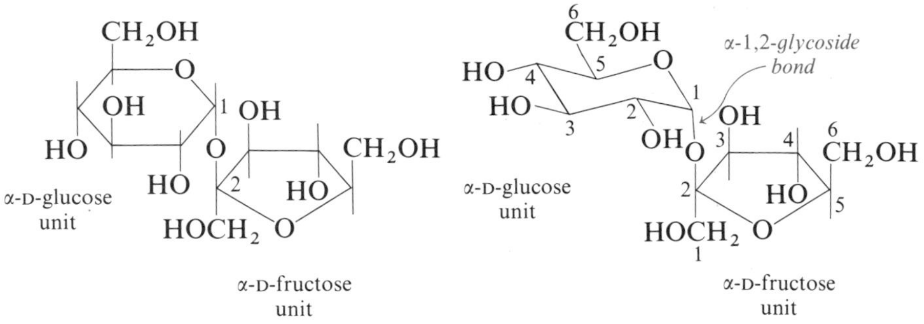

sucrose (cane or beet sugar)

In sucrose, carbon-1 of glucose is joined to carbon-2 of fructose by an α-1,2-glycoside bond. Glucose is in a six-member ring form while fructose is in a five-member ring form. Sucrose is a nonreducing sugar because the anomeric hemiacetal carbons of both glucose and fructose are involved in the formation of the glycoside bond.

Example 15.5

Draw Haworth and chair formulas of a disaccharide in which two units of D-glucose are joined by an α-1,6-glycoside bond.

Solution

First draw the structural formulas of α-D-glucose. Then connect the anomeric carbon of this monosaccharide to carbon-6 of the second glucose unit by an α-glycoside bond.

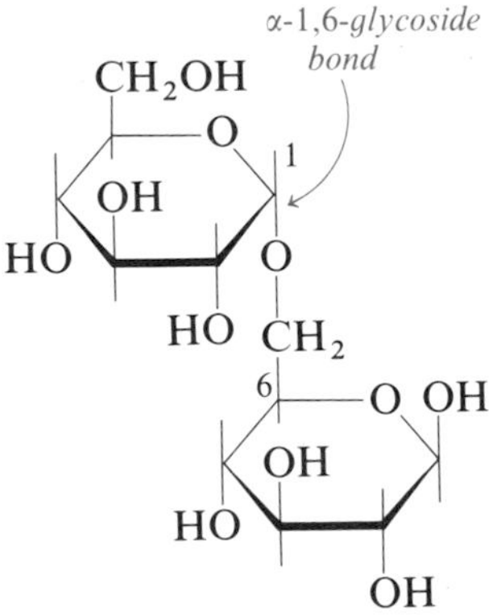

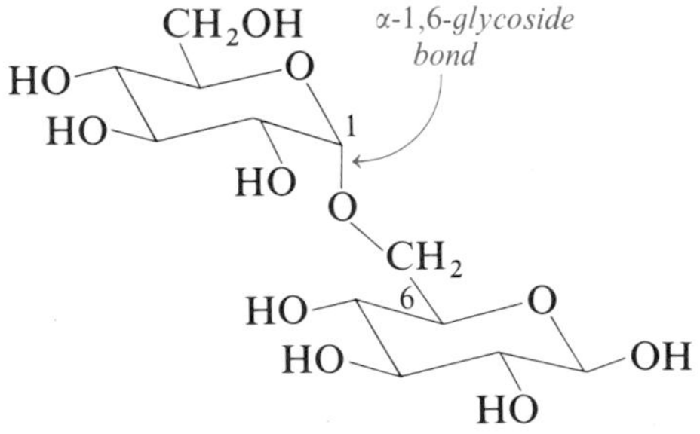

Problem 15.5

Draw Haworth and chair formulas for a disaccharide in which two units of D-glucose are joined by a β-1,6-glycoside bond.

15.4 Polysaccharides

A. Starch

Starch is the reserve carbohydrate for plants. It is found in all plant seeds and tubers and is the form in which glucose is stored for later use by plants. Starch can be separated into two fractions by making a paste with water and warming it to 60–80°C. One fraction, amylose, comprising about 20% of starch, is soluble in water. The water-insoluble fraction is amylopectin. Amylose has a molecular weight range of 10,000–50,000 (60 to 300 glucose units), and amylopectin has a molecular weight range of 50,000–1,000,000 (300 to 600 glucose units). X-Ray diffraction studies of amylose show a continuous, unbranched chain of glucose units joined by α-1,4-glycoside bonds.

Amylopectin has a highly branched structure. It contains the same type of repetitive sequence of α-1,4-glycoside bonds as does amylose, but chain lengths vary from about 24 to 30 units. In addition, there is considerable branching from this linear network. At branch points, a new chain is started by an α-1,6-glycoside linkage between carbon-1 of one glucose unit and carbon-6 of another glucose unit.

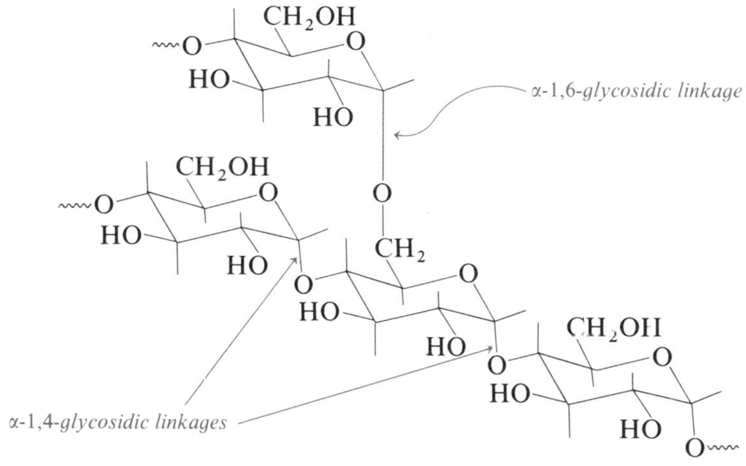

amylopectin

Why are carbohydrates stored in plants as polysaccharides rather than as monosaccharides? The answer has to do with osmotic pressure (Section 7.3D), which is proportional to the molar concentration, not the molecular weight, of a solute. If we assume that 1000 molecules of glucose are assembled into one starch macromolecule, then we can predict that a solution containing 1 gram of starch per 10 mL will have only 1/1000 the osmotic pressure of a solution of 1 gram of glucose in the same volume of solution. This feat of packaging is of tremendous advantage because it reduces the strain on various membranes enclosing such macromolecules.

B. Glycogen

Glycogen is the reserve carbohydrate for animals. Like amylopectin, glycogen is a nonlinear polymer of glucose units joined by α-1,4- and α-1,6-glycoside bonds, but it has a lower molecular weight and a more highly branched structure. The total amount of glycogen in the body of a well-nourished adult is about 350 grams, divided almost equally between liver and muscle. Figure 15.3 illustrates the highly branched structure of glycogen.

C. Cellulose

Cellulose is the most widely distributed skeletal polysaccharide. It constitutes almost half of the cell wall material of wood. Cotton is almost pure cellulose. Cellulose is a linear polymer of glucose units joined together by β-1,4-glycoside linkages. It has a molecular weight of approximately 400,000, corresponding to 2800 glucose units. Cellulose fibers consist of bundles of parallel polysaccharide chains held together by hydrogen bonding between hydroxyl groups on adjacent chains. This arrangement of parallel chains in bundles and the resulting hydrogen bonding gives cellulose fibers their high mechanical strength.

β-1,4-glycoside linkages

cellulose chain

Figure 15.3 Glycogen.

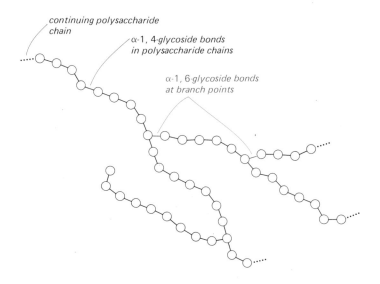

continuing polysaccharide chain

α-1, 4-glycoside bonds in polysaccharide chains

α-1, 6-glycoside bonds at branch points

Humans and other animals cannot use cellulose as a food because our digestive systems do not contain β-glycosidases, enzymes that catalyze hydrolysis of the β-glycoside bonds. Our digestive systems contain only α-glycosidases; hence, the polysaccharides we use as sources of glucose are starch and glycogen. On the other hand, many bacteria and microorganisms do contain β-glycosidases and are able to digest cellulose. Termites are fortunate to have such bacteria in their intestine and can use wood as their principal food. Ruminants (cud-chewing animals) can also digest grasses and wood because of the presence of β-glycosidase-containing microorganisms within their specially constructed alimentary systems.

Key Terms and Concepts

aldose (15.1A)

amino sugar (15.2)

anomeric carbon (15.2B)

Benedict's solution (15.2A)

carbohydrate (introduction)

disaccharide (15.3)

Fehling's solution (15.2A)

Fischer projection formula (15.1A)

α-D-glucose (15.2C)

β-D-glucose (15.2C)

glycoside (15.2C)

N-glycoside (15.2C)

Haworth structure (15.2C)

hexose (15.1A)

ketose (15.1A)

D-monosaccharide (15.1A)

L-monosaccharide (15.1A)

mutarotation (15.2D)

nonreducing sugar (15.2A)

pentose (15.1A)

polysaccharide (15.4)

reducing sugar (15.2A)

saccharide (introduction)

tetrose (15.1A)

Tollens' solution (15.2A)

triose (15.1A)

Key Reactions

1. Formation of cyclic hemiacetals and hemiketals (Section 15.1C)

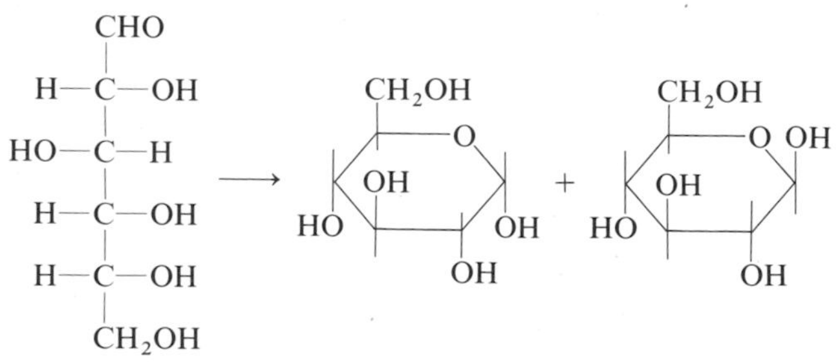

D-glucose

α-D-glucose
$[\alpha] = +112°$

β-D-glucose
$[\alpha] = +19°$

2. Oxidation to aldonic acids (Section 15.2A).

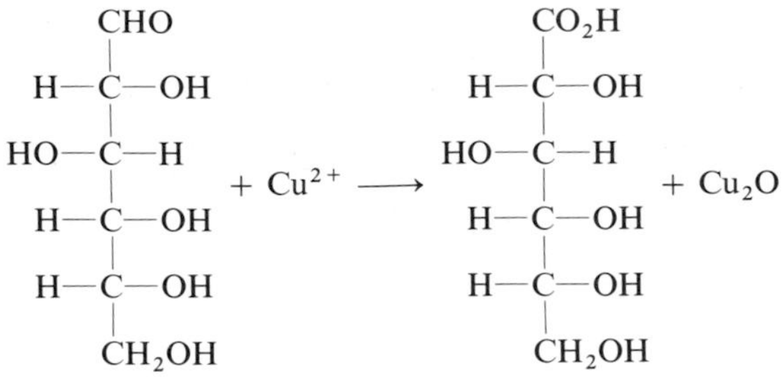

$$CHO \quad H-C-OH \quad HO-C-H \quad H-C-OH \quad H-C-OH \quad CH_2OH \quad + Cu^{2+} \longrightarrow CO_2H \quad H-C-OH \quad HO-C-H \quad H-C-OH \quad H-C-OH \quad CH_2OH \quad + Cu_2O$$

D-glucose   D-gluconic acid

3. Reduction to alditols (Section 15.2B).

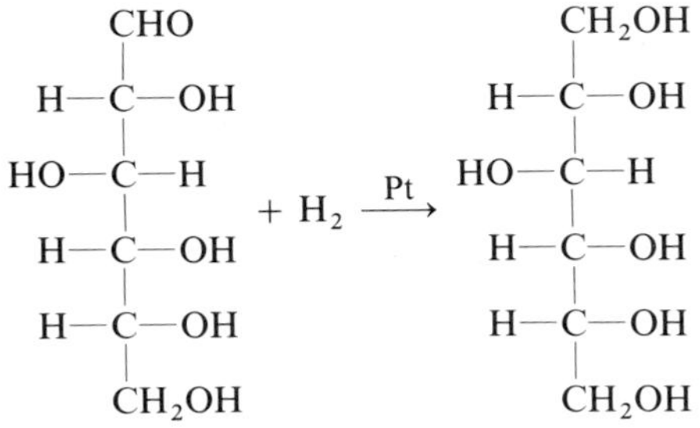

$$CHO \quad H-C-OH \quad HO-C-H \quad H-C-OH \quad H-C-OH \quad CH_2OH \quad + H_2 \xrightarrow{Pt} CH_2OH \quad H-C-OH \quad HO-C-H \quad H-C-OH \quad H-C-OH \quad CH_2OH$$

D-glucose   D-glucitol

4. Formation of glycosides (Section 15.2C)

$$CHO \quad H-C-OH \quad HO-C-H \quad H-C-OH \quad H-C-OH \quad CH_2OH \quad + CH_3OH \xrightarrow{H^+}$$

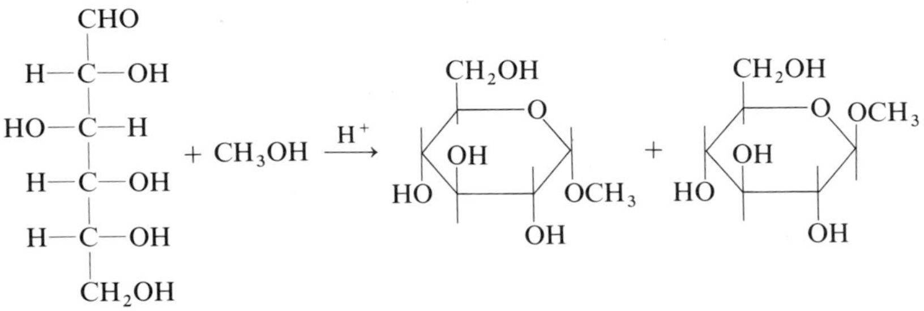

D-glucose

α-methyl-D-glucoside
$[\alpha] = +159°$

β-methyl-D-glucoside
$[\alpha] = -34°$

5. Formation of N-glycosides (Section 15.2C).

$$CHO \quad H-C-OH \quad H-C-OH \quad H-C-OH \quad CH_2OH \quad + \quad \text{(uracil)} \quad \longrightarrow \quad + H_2O$$

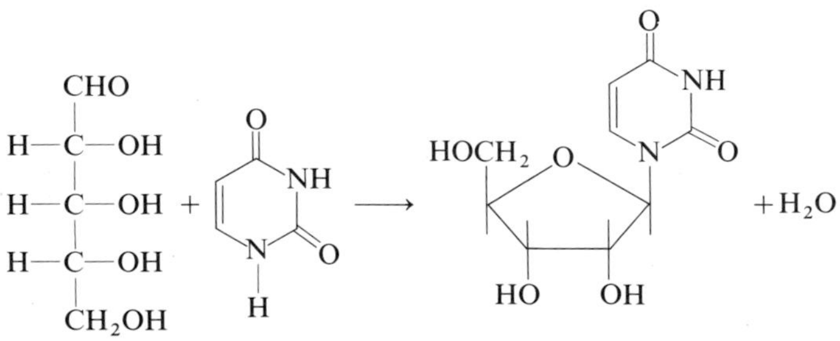

D-ribose   uracil

Problems

15.6   The term *carbohydrate* is derived from "hydrates of carbon." Show the origin of this term by reference to the molecular formulas of D-ribose, D-fructose, and lactose.

15.7   Explain the meaning of the designations D- and L- as used to specify the stereochemistry of monosaccharides.

15.8   List the rules for drawing Fischer projection formulas.

15.9   Draw the four stereoisomers of 2,3,4-trihydroxybutanal. Label them A, B, C, and D. Which are pairs of enantiomers?

15.10 What is the difference in structure between D-ribose and 2-deoxy-D-ribose?

15.11 Both D-ribose and 2-deoxy-D-ribose form five-member cyclic hemiacetals. Draw structural formulas for the α- and β-forms of each.

15.12 Draw a Fischer projection formula of D-glucose. State the total number of chiral carbons and the total number of stereoisomers possible for this structural formula. Of these, only D-glucose, D-galactose, and D-mannose are common in nature. Draw Fischer projection formulas for D-galactose and D-mannose.

15.13 Table 15.1 shows a Fischer projection formula of D-arabinose. Draw a Fischer projection formula of L-arabinose, a naturally occurring aldopentose of the "unnatural" L-configuration.

15.14 There are three common conventions for representing the stereochemistry of carbohydrates. They are (1) Fischer projections, (2) Haworth structures, and (3) chair structures. Draw α-D-glucose according to the rules of each of these conventions. Do the same for β-D-glucose.

15.15 a. Build a molecular model of D-glucose in the open-chain form.
       b. Using this molecular model, show the reaction of the —OH on carbon-5 with the aldehyde of carbon-1 to form a cyclic hemiacetal. Show that either α-D-glucose or β-D-glucose can be formed, depending on the direction from which the —OH group interacts with the aldehyde group.

15.16 Explain the convention α- and β- as used to designate the stereochemistry of cyclic forms of monosaccharides.

15.17 Draw structural formulas for the open-chain and cyclic forms of D-fructose.

15.18 D-Glucosamine, D-mannosamine, and D-galactosamine form six-member cyclic hemiacetals. Draw Haworth and chair structures for the α- and β-forms of each.

15.19 Explain the phenomenon of mutarotation with reference to carbohydrates. By what means is it detected?

15.20 A solution of α-D-glucose has a specific rotation of $+112°$; one of β-D-glucose has a specific rotation of $+19°$. On mutarotation, the specific rotation of each solution changes to an equilibrium value of $+52°$. Calculate the percentage of α-D-glucose in the equilibrium mixture.

15.21 Fischer attempted to convert D-glucose into its dimethyl acetal according to the following reaction:

$$\text{(i) } C_6H_{12}O_6 + 2\,CH_3OH \xrightarrow{H^+} C_8H_{18}O_7 + H_2O$$

D-glucose   D-glucose
                          dimethyl
                          acetal

However, the reaction that actually takes place is:

(ii) $C_6H_{12}O_6 + CH_3OH \xrightarrow{H^+} C_7H_{14}O_6 + H_2O$

    D-glucose   methyl
                                  D-glucoside

Draw a structural formula for $C_8H_{18}O_7$, the expected product of reaction (i). Reaction (ii) gives isomeric methyl glucosides, designated α-methyl-D-glucoside and β-methyl-D-glucoside. Draw Haworth and chair projections for each of these glucosides.

15.22 There are four isomeric D-aldopentoses (Table 15.1). Suppose that each of these is reduced with $NaBH_4$. Which of the four will yield optically inactive D-alditols? Which will yield optically active alditols?

15.23 An important technique for establishing relative configurations among isomeric aldoses is to convert both terminal carbon atoms into the same functional group. This can be done by either selective oxidation or selective reduction. As a specific example, nitric acid oxidation of D-erythrose gives *meso*-tartaric acid. Oxidation of D-threose under similar conditions gives D-tartaric acid.

$$D\text{-threose} \xrightarrow[\text{oxidation}]{HNO_3} D\text{-tartaric acid}$$

$$D\text{-erythrose} \xrightarrow[\text{oxidation}]{HNO_3} meso\text{-tartaric acid}$$

Using this information, show which of the following structural formulas is D-erythrose and which is D-threose. Check your answer by referring to Table 15.1.

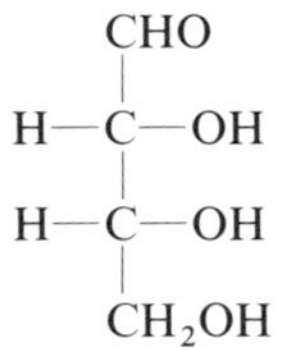

      CHO   CHO
  H—C—OH   HO—C—H
  H—C—OH   H—C—OH
    CH_2OH CH_2OH

      (a) 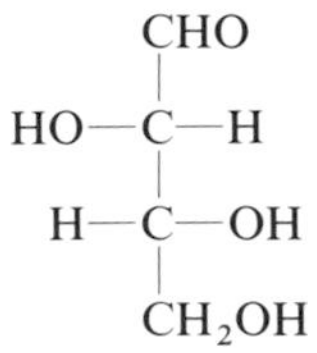   (b)

15.24 Classify the following as reducing or nonreducing sugars.
    a. α-D-glucose   b. β-D-ribose
    c. 2-deoxy-D-ribose   d. α-methyl-D-glucoside

15.25 Treatment of D-glucose in dilute aqueous base at room temperature yields an equilibrium mixture of D-glucose, D-mannose, and D-fructose. How might you account for this conversion? (*Hint:* review Section 13.5D and your answers to problems 13.21 and 13.22.)

15.26 Ketones are not oxidized by mild oxidizing agents. However, both dihydroxyacetone and fructose give a positive Benedict's test and are classified as reducing sugars. How might you account for the fact that these ketoses are reducing sugars? (*Hint:* these tests are done in dilute aqueous base.)

15.27 L-Fucose is one of several monosaccharides commonly found in the surface polysaccharides of animal cells. This 6-deoxyaldohexose is synthesized from D-mannose in a series of eight steps, shown below.

a. Describe the type of reaction (oxidation, reduction, hydration, and so on) involved in each step.

b. Explain why this monosaccharide belongs to the L-series even though it is derived biochemically from a D-sugar.

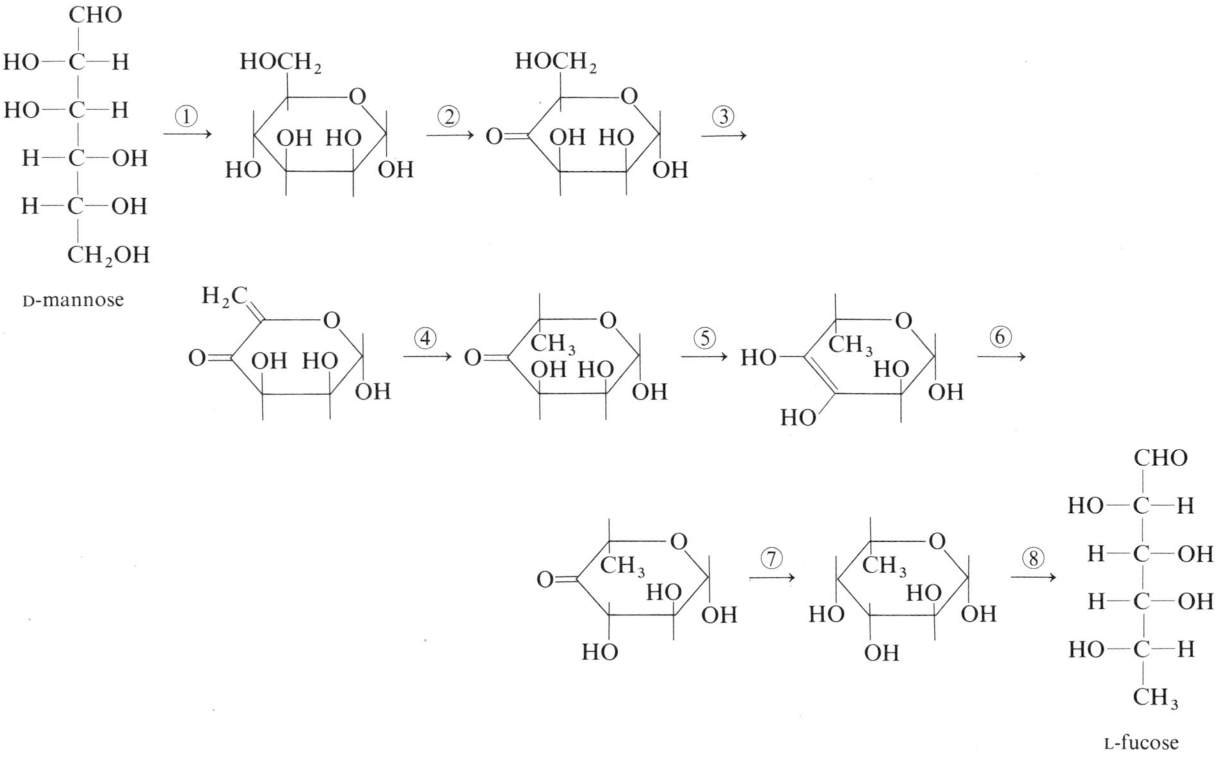

D-mannose

L-fucose

Disaccharides and Trisaccharides (Section 15.3)

15.28 Classify the following as reducing or nonreducing sugars.
   a. sucrose
   b. lactose
   c. α-methyl-lactoside
   d. maltose
   e. β-methyl-maltoside

15.29 Trehalose, a disaccharide consisting of two glucose units joined by an α-1,1-glycoside bond, is found in young mushrooms and is the chief carbohydrate in the blood of certain insects.

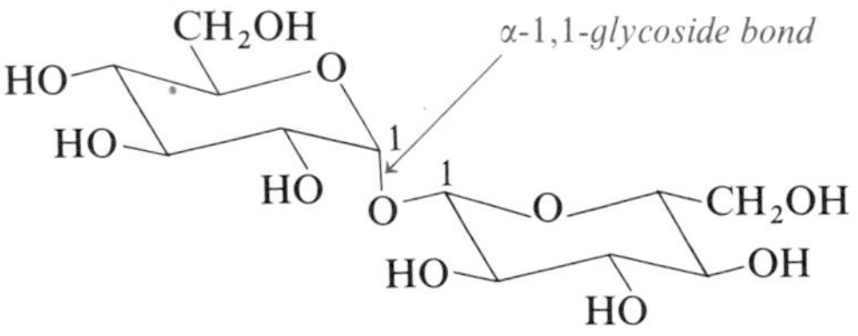

On the basis of its structural formula, would you expect trehalose (a) to be a reducing sugar? (b) To undergo mutarotation?

15.30 Raffinose is the most abundant trisaccharide in nature.
  a. Name the three monosaccharide units in raffinose.
  b. There are two glycoside bonds in raffinose. Describe each as you have already done for other disaccharides (for example, an α-1,2-glycoside bond in sucrose).
  c. Would you expect raffinose to be a reducing sugar?
  d. Would you expect raffinose to undergo mutarotation?

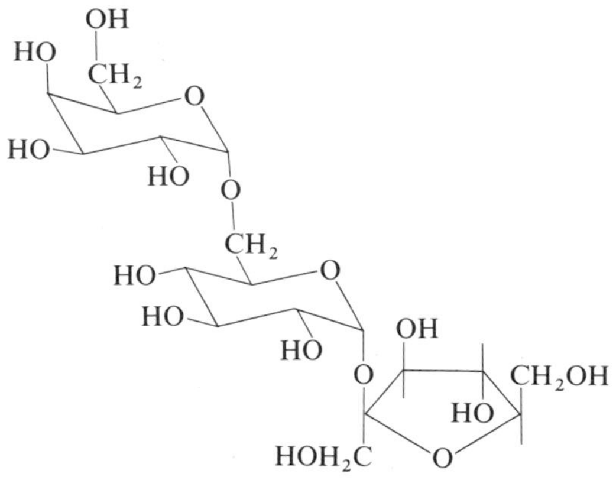

raffinose

15.31 Following is the Fischer projection formula for N-acetyl-D-glucosamine. This substance forms a six-member cyclic hemiacetal.

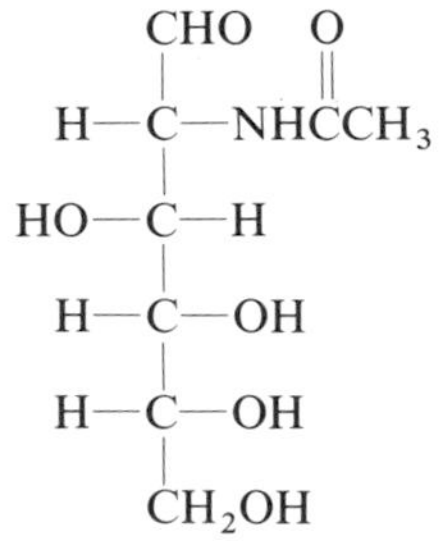

N-acetyl-D-glucosamine

  a. Draw Haworth and chair structures for the α- and β-forms of this monosaccharide.
  b. Draw Haworth and chair structures for the disaccharide formed by joining two units of N-acetyl-D-glucosamine by a β-1,4-glycoside bond. (If you have done this correctly, you have drawn the structural formula of the repeating dimer of chitin, the polysaccharide component of the shells of lobster and other crustacea.)

Polysaccharides (Section 15.4)

15.32 What is the major difference in structure between cellulose and starch? Why are humans unable to digest cellulose?

15.33 Propose a likely structure for the following polysaccharides.
  a. Alginic acid, isolated from seaweed, is used as a thickening agent in ice cream and other foods. Alginic acid is a polymer of D-mannuronic acid units joined together by β-1,4-glycoside bonds.

b. Pectic acid is the main constituent of pectin, which is responsible for the formation of jellies from fruits and berries. Pectic acid is a polymer of D-galacturonic acid units joined together by α-1,4-glycoside bonds.

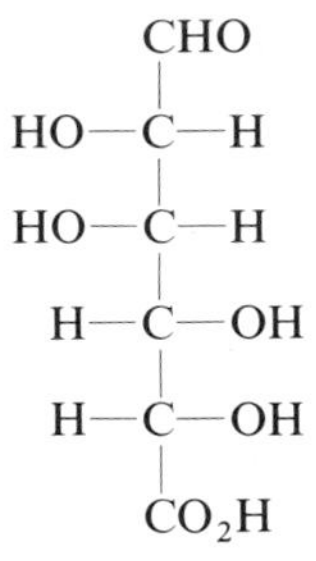

$$
\begin{array}{ccc}
 & \text{CHO} & & \text{CHO} \\
\text{HO}-&\text{C}&-\text{H} & \text{H}-&\text{C}&-\text{OH} \\
\text{HO}-&\text{C}&-\text{H} & \text{HO}-&\text{C}&-\text{H} \\
\text{H}-&\text{C}&-\text{OH} & \text{HO}-&\text{C}&-\text{H} \\
\text{H}-&\text{C}&-\text{OH} & \text{H}-&\text{C}&-\text{OH} \\
 & \text{CO}_2\text{H} & & \text{CO}_2\text{H}
\end{array}
$$

  D-mannuronic acid 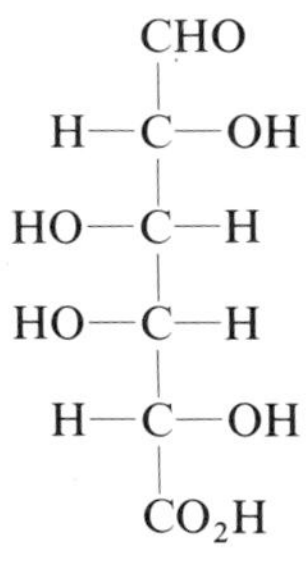   D-galacturonic acid

Clinical Chemistry: The Search for Specificity

The analytical procedure most often performed in the clinical chemistry laboratory is the determination of glucose in blood, urine, or other biological fluids. The need for a rapid and reliable test for blood glucose stems from the high incidence of diabetes meilitus. There are approximately 2 million known diabetics in the United States and it is estimated that another 2 million are undiagnosed.

Diabetes mellitus is characterized by insufficient blood levels of the polypeptide hormone insulin (Section 20.3F).

Deficiency of insulin results in the inability of muscle and liver cells to absorb glucose, leading to increased levels of blood glucose; impaired metabolism of fats and proteins; ketosis; and, possibly, diabetic coma. Thus, it is critical for the early diagnosis and effective management of this disease to have a rapid and reliable procedure for the determination of blood glucose.

Over the past 70 years or more, a great many tests have been developed. We will discuss four of these, each chosen to illustrate some of the

problems involved in developing suitable clinical laboratory tests. These tests will demonstrate the use of both chemical and enzymatic techniques in the modern clinical chemistry laboratory.

The first widely used glucose test was based on the fact that glucose is a reducing sugar. Specifically, the aldehyde group of glucose is oxidized by ferricyanide ion to a carboxyl group. In the process, Fe^{3+} in ferricyanide ion is reduced to Fe^{2+} in ferrocyanide ion (Figure 1). The reaction is carried out in the presence

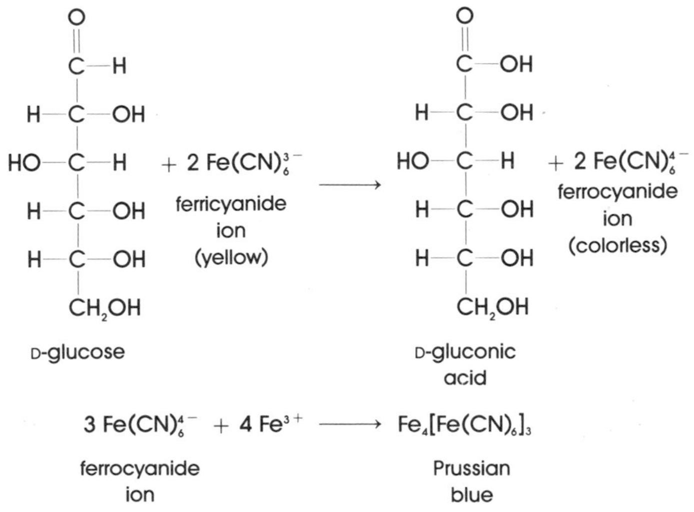

Figure 1 The first glucose test was based on the fact that glucose reduces ferricyanide ion to ferrocyanide ion which, in turn, reacts with Fe^{3+} to form Prussian blue.

$$3 Fe(CN)_6^{4-} + 4 Fe^{3+} \longrightarrow Fe_4[Fe(CN)_6]_3$$

ferrocyanide ion   Prussian blue

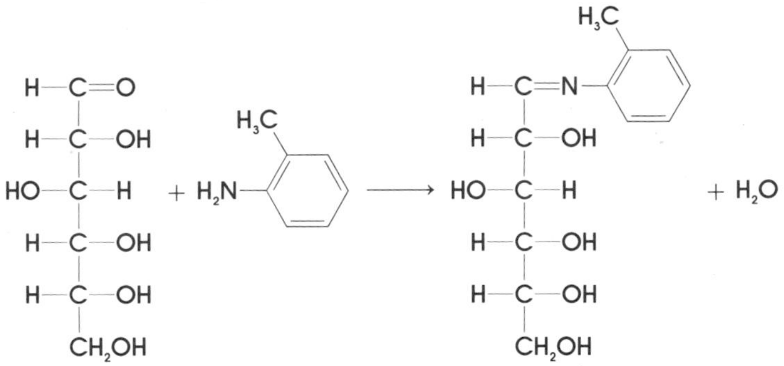

Figure 2 For many years, the *o*-toluidine test has been the standard clinical chemistry laboratory test for D-glucose.

D-glucose   *o*-toluidine   a Schiff base
                                          (blue-green)

of excess ferric ion. Under these conditions, ferrocyanide ion reacts further to form ferric ferrocyanide, or, as it is more commonly known, Prussian blue. The concentration of Prussian blue in the test sample is measured spectrophotometrically. In this test, the absorbance of Prussian blue is directly proportional to the concentration of glucose in the test sample.

Although this method can be used to measure glucose concentration, it has the disadvantage that ferricyanide also oxidizes several other reducing substances found in blood, including ascorbic acid, uric acid, certain amino acids, and phenols. In addition, any other aldoses present in blood will also reduce ferricyanide. All of these substances are said to give false positive results. The ferricyanide and other early oxidative tests often gave values that were 30% or more higher than the so-called true glucose value.

A more satisfactory approach in the search for an accurate blood glucose test lay in attacking the problem in a completely different way; namely, by taking advantage of a chemical reactivity of glucose other than its properties as a reducing sugar. One of the most successful and widely used of these nonoxidative methods involves reaction of glucose with o-toluidine to form a blue-green Schiff base. The absorbance of this Schiff base can be measured spectrophotometrically at a wavelength of 625 nm and is directly proportional to glucose concentration (Figure 2).

The o-toluidine test can be applied directly to serum, plasma, cerebro-spinal fluid, and urine, and will work on samples as small as 20 microliters (20×10^{-6} liter). In addition, it does not give false positive results with other reducing substances because the procedure does not involve oxidation. However, galactose and mannose, and, to a lesser

extent, lactose and xylose are all potential sources of false positive results, for they also react with o-toluidine to give colored Schiff bases. This is generally not a problem because these mono- and disaccharides are normally present in serum and plasma only in very low concentrations.

In recent years, the search for even greater specificity in glucose determinations has led to the introduction of enzyme-based glucose assay procedures. What was needed was an enzyme that catalyzes a specific reaction of glucose but does not catalyze comparable reactions of any other substance normally present in biological fluids. The enzyme glucose oxidase meets these requirements. It catalyzes the oxidation of D-glucose to D-gluconic acid (Figure 3, next page).

Oxygen, O_2, is the oxidizing agent; in this process, oxygen is reduced to hydrogen peroxide, H_2O_2. Hydrogen peroxide is, in turn, used to oxidize

Figure 3 The glucose oxidase method is the most highly specific test yet developed for measurement of the concentration of D-glucose in biological fluids.

$$
\begin{array}{c}
\text{O} \\
\| \\
\text{C—H} \\
| \\
\text{H—C—OH} \\
| \\
\text{HO—C—H} \\
| \\
\text{H—C—OH} \\
| \\
\text{H—C—OH} \\
| \\
\text{CH}_2\text{OH}
\end{array}
\; + \; \text{O}_2 + \text{H}_2\text{O}
\; \xrightarrow{\text{glucose oxidase}} \;
\begin{array}{c}
\text{O} \\
\| \\
\text{C—OH} \\
| \\
\text{H—C—OH} \\
| \\
\text{HO—C—H} \\
| \\
\text{H—C—OH} \\
| \\
\text{H—C—OH} \\
| \\
\text{CH}_2\text{OH}
\end{array}
\; + \; \text{H}_2\text{O}_2
$$

D-glucose   D-gluconic   hydrogen
                        acid   peroxide

Figure 4 The insulin pump worn by this diabetes patient mimics the normal pancreas by continuously injecting insulin under the skin in tiny, predetermined amounts. However, the patient must measure his own glucose levels and then adjust the pump so it releases the correct amount of insulin. (Joslin Diabetes Center, Boston, Massachusetts.)

another substance whose concentration can then be determined spectrophotometrically. In one procedure, H_2O_2 is reacted with iodide ion to form molecular iodine, I_2.

$$2\,I^- + H_2O_2 + 2\,H^+$$

$$\downarrow$$

$$I_2 + 2\,H_2O$$

The absorbance at 420 nm is used to calculate iodine concentration, which is then used to calculate glucose concentration. In another procedure, H_2O_2 is used to oxidize the aromatic amine o-toluidine to a colored product. The enzyme peroxidase catalyzes this second oxidation. The concentration of the colored oxidation product is determined spectrophotometrically.

o-toluidine + H_2O_2

$$\begin{array}{c} \text{peroxidase} \\ \big| \\ \downarrow \end{array}$$

oxidized o-toluidine + H_2O
(colored)

A number of commercially available test kits employ the glucose oxidase reaction for qualitative determination of glucose in urine. One of these, Clinistix (produced by the Ames Co., Elkhart, Indiana), consists of a filter paper strip impregnated with glucose oxidase, peroxidase, and o-toluidine. The test end of the paper is dipped in the urine sample, removed, and examined after 10 seconds. A blue color develops if the concentration of glucose in the urine exceeds about 1 mg/mL.

Any determination of glucose in blood or a 24-hour urine sample reflects glucose levels only during the sampling period. There is now a simple, convenient laboratory method that can be used to monitor long-term blood glucose levels. This method depends on the measurement of the relative amounts of hemoglobin and certain hemoglobin derivatives normally present in the blood. Hemoglobin A (HbA) is the major type of hemoglobin present in normal red blood cells. In addition, there are a number of minor components, including glycosylated hemoglobins (HbA_1). Glycosylated hemoglobins are synthesized within red blood cells in a two-step process. In step 1, D-glucose reacts with the free $-NH_2$ group of a beta chain of hemoglobin to form a Schiff base. (See Section 20.3D and

| D-glucose | the terminal —NH₂ group of a beta chain of normal hemoglobin | a Schiff base (unstable) | glycosylated hemoglobin (stable) |

the mini-essay "Abnormal Human Hemoglobins" for details of the structure of hemoglobin.) Step 1 is reversible and the Schiff base is in equilibrium with D-glucose and hemoglobin. In a slower, irreversible second step, a portion of the Schiff base undergoes a type of keto-enediol tautomerism (Section 14.5D) to form glycosylated hemoglobin, a stable molecule. The molecular structure of glycosylated hemoglobin is illustrated schematically in Figure 5. Because this slow, irreversible second step occurs continuously throughout the 120-day life span of a red blood cell, the level of glycosylated hemoglobins within the red blood cell population reflects the average blood glucose level during that period.

Normal levels of glycosylated hemoglobins fall within the range 4.5–8.5% of total hemoglobin. In cases of uncontrolled or poorly controlled diabetes, the percent of glycosylated hemoglobins may rise to two or three times these values. Conversely, once long-term blood glucose control is achieved, glycosylated hemoglobins gradually fall to within normal range. Thus, the level of glycosylated hemoglobin can be used to give a picture of the average blood glucose levels over the previous 8 to 10 weeks.

References

Bunn, H.F. 1981. Evaluation of glycosylated hemoglobin in diabetic patients. *Diabetes* 30:613.

Garel, M.C.; Blouquit, Y.; Molko, F.; and Rosa, J. 1979. HbA$_1$ — a review of its structure, biosynthesis, clinical significance and methods of assay. *Biomedicine* 30:234.

Figure 5 Schematic diagram of glycosylated hemoglobin.

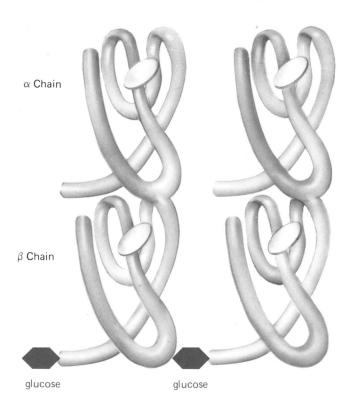

α Chain

β Chain

glucose   glucose

Hammons, G.T. 1981. Glycosylated hemoglobins and
diabetes mellitus. *Lab.
Med.* 12:213.

Henry, R.J.; Cannon, D.O.; and
Winkleman, J.W., eds.
1974. *Clinical chemistry,
principles and techniques.*
2nd ed. New York: Harper
and Row.

Rahbar, S. 1980. Glycosylated
hemoglobins. *New York
State J. Med.* 80:553.

Tietz, N., ed. 1976. *Fundamentals
of clinical chemistry.* New
York: W.B. Saunders.

Carboxylic Acids

In this chapter we shall discuss the structure and acidity of carboxylic acids. These acids and their functional derivatives are widespread, both in the biological world and, thanks to research and technology, in the world of man-made materials as well. In addition, we shall describe a special group of carboxylic acids known as *fatty acids*. The discussion of fatty acids will lead us into the chemistry of soaps and synthetic detergents.

16.1 Structure of Carboxylic Acids

The characteristic structural feature of a carboxylic acid is the presence of a carboxyl group (Section 9.2C). Shown in Figure 16.1(a) is the Lewis structure

Figure 16.1 The structure of formic acid. (a) Lewis structure; (b) ball-and-stick model.

(a)   (b)

514

of formic acid, HCO_2H, the simplest compound containing a carboxyl group. Predicted bond angles of the carboxyl group are 120° about the carbonyl carbon and 109.5° about the hydroxyl oxygen. The observed angles in formic acid are quite close to those predicted.

16.2 Nomenclature of Carboxylic Acids

A. IUPAC Names

The IUPAC system of nomenclature for carboxylic acids selects as the parent compound the longest chain of carbon atoms that contains the $—CO_2H$ group. The presence of the carboxyl group is indicated by changing the *-e* of the parent compound to *-oic acid*. Because the carbon of the carboxyl group is always carbon-1 of the parent compound, there is no need to give it a number. Following are structural formulas and IUPAC names for several carboxylic acids:

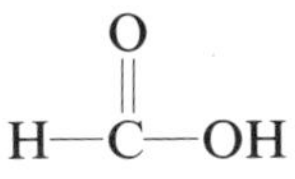

methanoic acid
(formic acid)

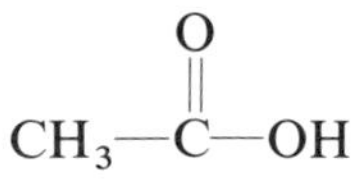

ethanoic acid
(acetic acid)

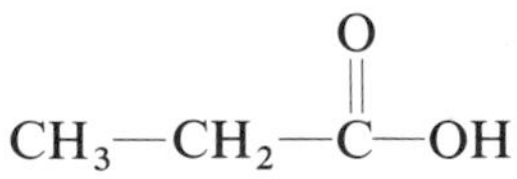

propanoic acid

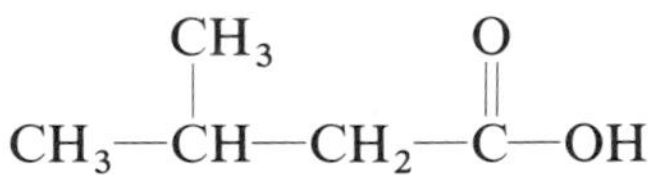

3-methylbutanoic
acid

The IUPAC system retains the common names formic acid and acetic acid.
   Dicarboxylic acids have the ending *-dioic acid* to indicate the presence of two $—CO_2H$ groups. Following are IUPAC names for three dicarboxylic acids, along with the common names of each in parentheses. Most dicarboxylic acids are known almost exclusively by their common names.

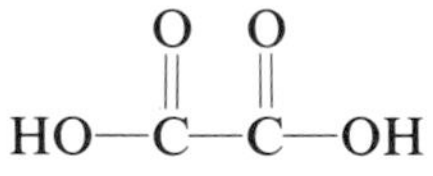

ethanedioic acid
(oxalic acid)

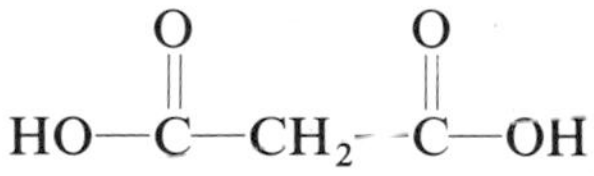

propanedioic acid
(malonic acid)

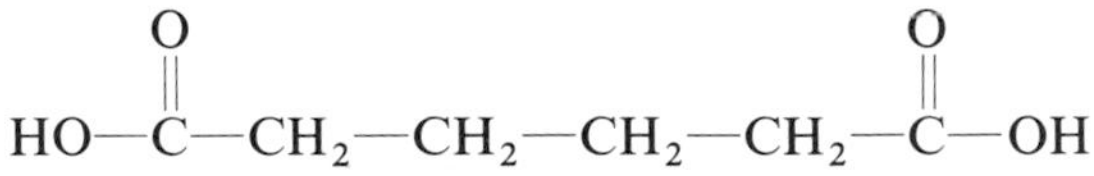

hexanedioic acid
(adipic acid)

Aromatic carboxylic acids are named by changing the ending of the parent hydrocarbon name to *-oic acid*. For example, the carboxylic acid derived from benzene is named benzoic acid.

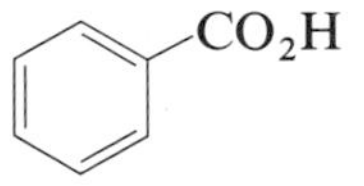

benzoic acid

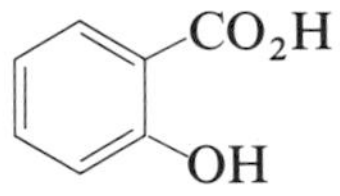

2-hydroxybenzoic acid
(salicylic acid)

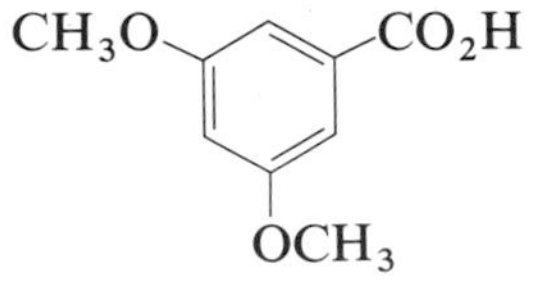

3,5-dimethoxybenzoic acid

In more complex structural formulas, the carboxyl group may be named by adding the words -*carboxylic acid* to the name of the parent hydrocarbon structure:

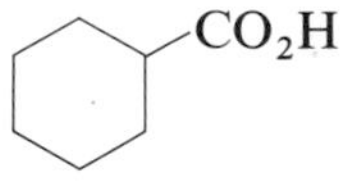

$$\text{cyclohexane—CO}_2\text{H}$$

cyclohexanecarboxylic acid

B. Common Names

Many lower-molecular-weight carboxylic acids are still known by their common names. When using common names, Greek letters (α, β, γ, δ) are used to locate substituents. An α-carbon is the one next to the carboxyl group, and an α-substituent in a common name is equivalent to a 2-substituent in the IUPAC name.

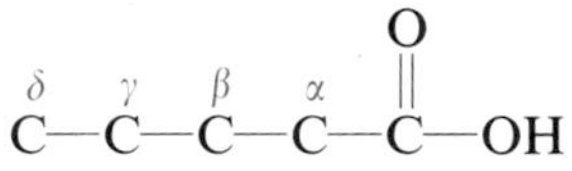

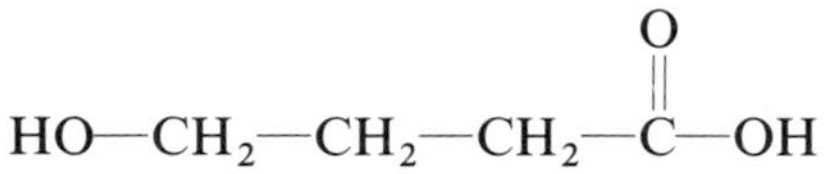

$$\overset{\delta}{C}-\overset{\gamma}{C}-\overset{\beta}{C}-\overset{\alpha}{C}-\overset{O}{\overset{\|}{C}}-OH \qquad HO-CH_2-CH_2-CH_2-\overset{O}{\overset{\|}{C}}-OH \qquad CH_3-\overset{Cl}{\overset{|}{CH}}-\overset{O}{\overset{\|}{C}}-OH$$

γ-hydroxybutyric acid 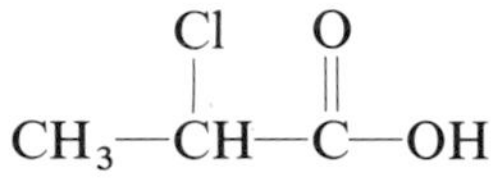 α-chloropropionic acid

Table 16.1 lists IUPAC and common names for several straight-chain monocarboxylic acids and for dicarboxylic acids of up to six carbons.

Table 16.1 IUPAC and common names of some mono- and di-carboxylic acids.

Structural Formula	IUPAC Name	Common Name
HCO_2H	methanoic, formic	formic
CH_3CO_2H	ethanoic, acetic	acetic
$CH_3CH_2CO_2H$	propanoic	propionic
$CH_3(CH_2)_2CO_2H$	butanoic	butyric
$CH_3(CH_2)_3CO_2H$	pentanoic	valeric
$CH_3(CH_2)_4CO_2H$	hexanoic	caproic
$CH_3(CH_2)_{10}CO_2H$	dodecanoic	lauric
$CH_3(CH_2)_{12}CO_2H$	tetradecanoic	myristic
$CH_3(CH_2)_{14}CO_2H$	hexadecanoic	palmitic
$CH_3(CH_2)_{16}CO_2H$	octadecanoic	stearic
$CH_3(CH_2)_{18}CO_2H$	eicosanoic	arachidic
HO_2CCO_2H	ethanedioic	oxalic
$HO_2CCH_2CO_2H$	propanedioic	malonic
$HO_2C(CH_2)_2CO_2H$	butanedioic	succinic
$HO_2C(CH_2)_3CO_2H$	pentanedioic	glutaric
$HO_2C(CH_2)_4CO_2H$	hexanedioic	adipic

Example 16.1

Give IUPAC names for the following carboxylic acids:

a. 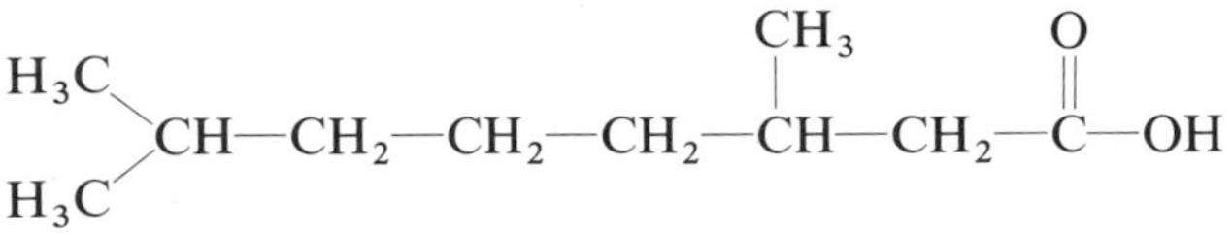
$$\underset{H_3C}{\overset{H_3C}{\diagdown}}CH\!-\!CH_2\!-\!CH_2\!-\!CH_2\!-\!\underset{\underset{CH_3}{|}}{CH}\!-\!CH_2\!-\!\overset{\overset{O}{\|}}{C}\!-\!OH$$

b. 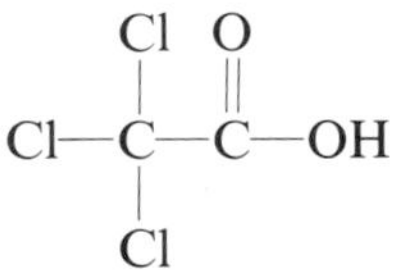 $Cl\!-\!\underset{\underset{Cl}{|}}{\overset{\overset{Cl}{|}}{C}}\!-\!\overset{\overset{O}{\|}}{C}\!-\!OH$

c.

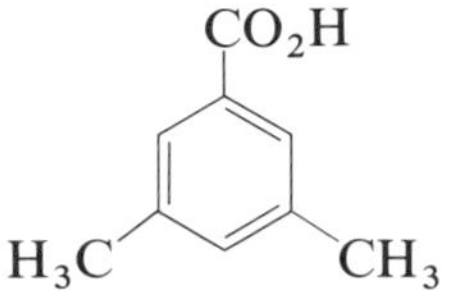

d. 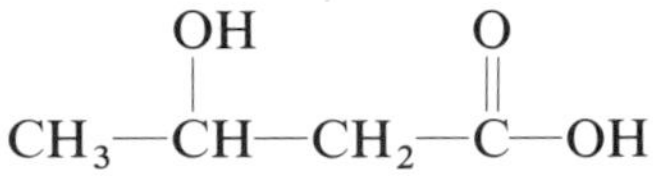 $CH_3\!-\!\underset{\underset{OH}{|}}{CH}\!-\!CH_2\!-\!\overset{\overset{O}{\|}}{C}\!-\!OH$

Solution

a. 3,7-Dimethyloctanoic acid. The longest chain containing the carboxyl group is eight carbons; therefore, this is a disubstituted octanoic acid.

b. Trichloroacetic acid. Acetic acid is accepted as the IUPAC name for CH_3CO_2H. It is not necessary to use a numbering system to show the location of the three chlorine substituents for they can only be on the second carbon of the acid.

c. 3,5-Dimethylbenzoic acid.

d. 3-Hydroxybutanoic acid. The common name of this acid is β-hydroxybutyric acid.

Problem 16.1

Give IUPAC names for the following carboxylic acids.

a.

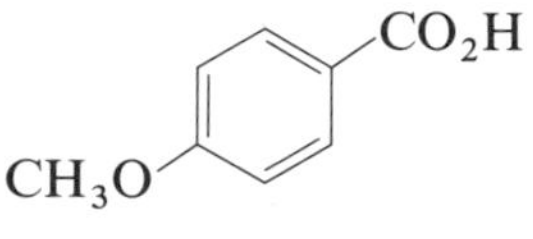

b. 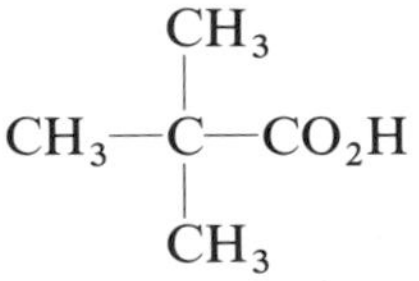 $CH_3\!-\!\underset{\underset{CH_3}{|}}{\overset{\overset{CH_3}{|}}{C}}\!-\!CO_2H$

c. $HO_2C\!-\!CH_2\!-\!\underset{\underset{CH_3}{|}}{CH}\!-\!CO_2H$

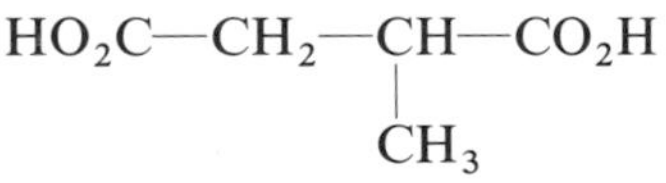

d. 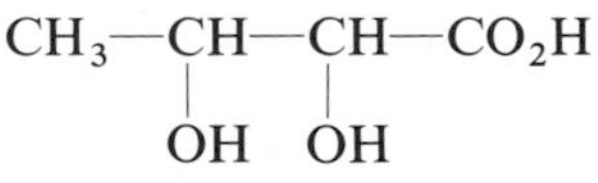 $CH_3\!-\!\underset{\underset{OH}{|}}{CH}\!-\!\underset{\underset{OH}{|}}{CH}\!-\!CO_2H$

Salts of carboxylic acids are named in much the same manner as salts of inorganic acids: the cation is named first and then the anion. The anion derived from a carboxylic acid is named by dropping the terminal *-ic* from the name of the acid and adding *-ate*.

Example 16.2

Name the following salts.

a. CH_3CO_2Na b. $(CH_3CH_2CO_2)_2Ca$ c. 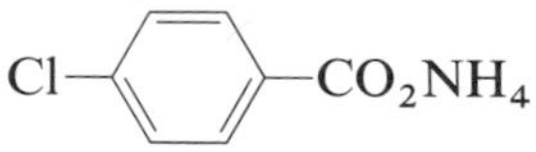 $Cl-\langle\bigcirc\rangle-CO_2NH_4$

Solution

In the following answers, the name and structural formula of the carboxylic acid are listed, followed by the name and structural formula of the anion derived from the acid. Finally, the salt is named.

a. 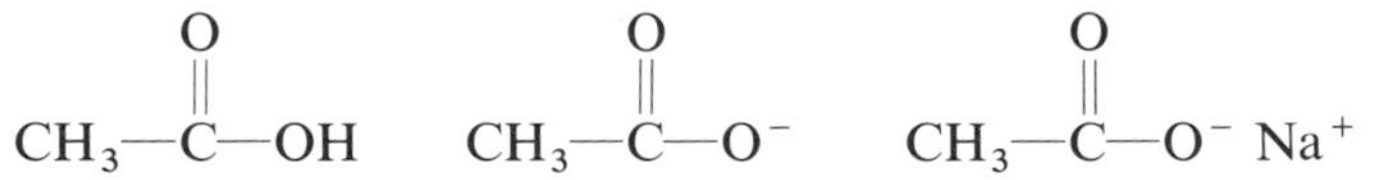 $CH_3-\overset{\displaystyle O}{\overset{\|}{C}}-OH$ $CH_3-\overset{\displaystyle O}{\overset{\|}{C}}-O^-$ $CH_3-\overset{\displaystyle O}{\overset{\|}{C}}-O^-\ Na^+$

   acetic   acetate   sodium
   acid   anion   acetate

b. 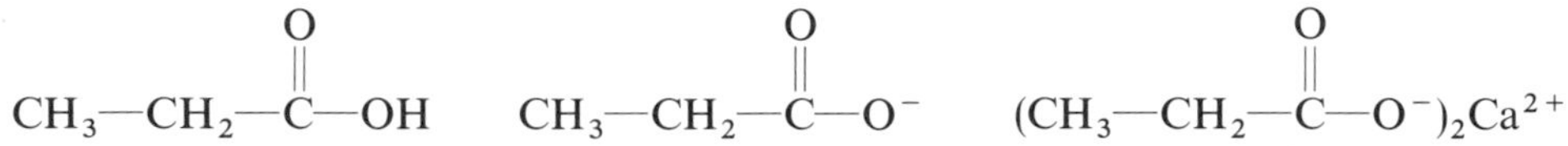 $CH_3-CH_2-\overset{\displaystyle O}{\overset{\|}{C}}-OH$ $CH_3-CH_2-\overset{\displaystyle O}{\overset{\|}{C}}-O^-$ $(CH_3-CH_2-\overset{\displaystyle O}{\overset{\|}{C}}-O^-)_2Ca^{2+}$

   propanoic   propanoate   calcium
   acid   anion   propanoate

c. 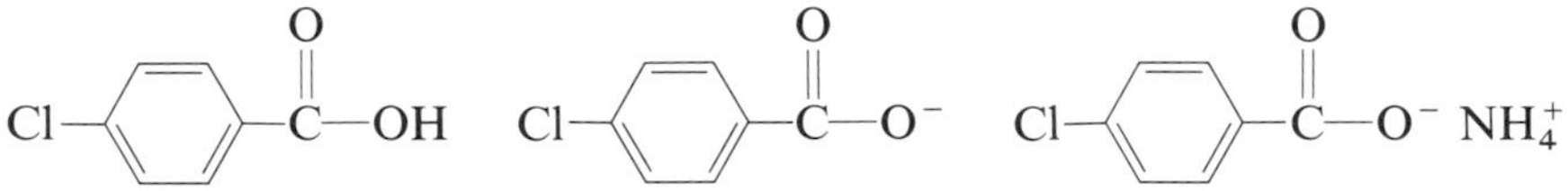 $Cl-\langle\bigcirc\rangle-\overset{\displaystyle O}{\overset{\|}{C}}-OH$ $Cl-\langle\bigcirc\rangle-\overset{\displaystyle O}{\overset{\|}{C}}-O^-$ $Cl-\langle\bigcirc\rangle-\overset{\displaystyle O}{\overset{\|}{C}}-O^-\ NH_4^+$

   p-chlorobenzoic   *p*-chlorobenzoate   ammonium
   acid   anion   *p*-chlorobenzoate

Problem 16.2

Name the following salts.

a. $CH_3\overset{\displaystyle }{\underset{\displaystyle CH_3}{CH}}CO_2K$ b. $CH_3(CH_2)_8CO_2Na$ c. $CH_3CO_2NH_4$

16.3 Physical Properties of Carboxylic Acids

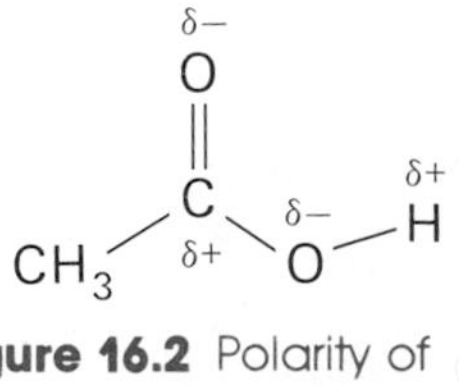

Figure 16.2 Polarity of the carboxyl group.

Because the carboxyl group contains three polar covalent bonds, carboxylic acids are polar compounds. The carbonyl oxygen and the hydroxyl oxygen each bear partial negative charges, and the carbonyl carbon and the hydroxyl hydrogen bear partial positive charges, as shown in Figure 16.2. Carboxylic acids can participate in hydrogen bonding through both the C=O and O—H groups, as shown in Figure 16.3 for acetic acid. Because carboxylic acids are even more extensively hydrogen bonded than alcohols, their boiling points are higher than alcohols of comparable molecular weight. For example, propanoic acid and 1-butanol have almost identical molecular weights, but the boiling point of propanoic acid is over 20° higher than that of 1-butanol.

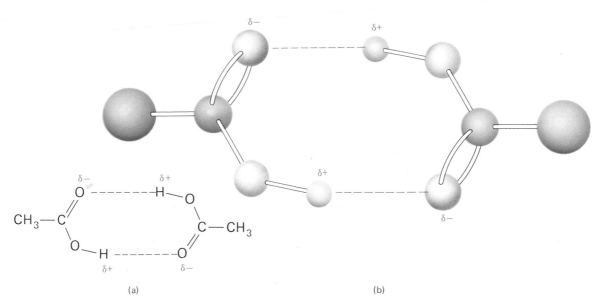

Figure 16.3 Hydrogen bonding between acetic acid molecules in pure liquid acetic acid. (a) Lewis structure; (b) ball-and-stick model.

$$CH_3—CH_2—\overset{\overset{\textstyle O}{\|}}{C}—OH \qquad CH_3—CH_2—CH_2—CH_2—OH$$

propanoic acid   1-butanol
mw 74, bp 141°C   mw 74, bp 117°C

Carboxylic acids also interact with water molecules by hydrogen bonding through both the carbonyl oxygen and the hydroxyl group.

Because of their interaction with water molecules, carboxylic acids are more soluble in water than are alkanes, ethers, alcohols, aldehydes, or ketones. For example, propanoic acid is soluble in water in all proportions, while the solubility of 1-butanol is only 8 g/100 g water.

The first four carboxylic acids (formic, acetic, propanoic, and butanoic) are soluble in water in all proportions. The higher-molecular-weight carboxylic acids have longer nonpolar hydrocarbon chains. For this reason, the solubility of carboxylic acids in water decreases as molecular weight increases.

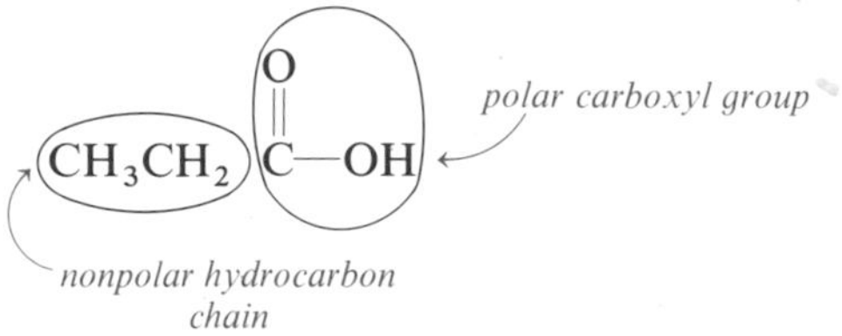

$$CH_3CH_2 \quad C-OH$$

polar carboxyl group

nonpolar hydrocarbon chain

propanoic acid
(soluble in water in all proportions)

$$CH_3CH_2CH_2CH_2CH_2CH_2CH_2CH_2CH_2CH_2CH_2CH_2CH_2CH_2 \quad C-OH$$

nonpolar hydrocarbon chain

polar carboxyl group

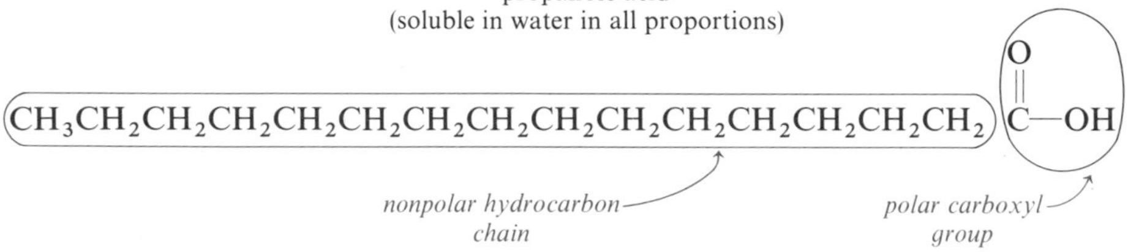

hexadecanoic acid
(insoluble in water)

The polar (hydrophilic) carboxyl group of propanoic acid is large compared to the nonpolar (hydrophobic) hydrocarbon chain, and this acid is soluble in water in all proportions. In hexadecanoic acid, the polar carboxyl group is small in comparison to the nonpolar hydrocarbon chain, and this acid is insoluble in water.

Table 16.2 shows the physical properties of some monocarboxylic acids.

Table 16.2 Physical properties of some monocarboxylic acids.

Name	Structural Formula	mp (°C)	bp (°C)	Solubility in Water (g/100g H_2O)	
formic acid	HCO_2H	8	100	∞	
acetic acid	CH_3CO_2H	16	118	∞	soluble in all proportions
propanoic acid	$CH_3CH_2CO_2H$	−22	141	∞	
butanoic acid	$CH_3(CH_2)_2CO_2H$	−6	164	∞	
hexanoic acid	$CH_3(CH_2)_4CO_2H$	−3	205	1.0	slightly soluble
decanoic acid	$CH_3(CH_2)_8CO_2H$	32	—	—	insoluble

Example 16.3

Arrange the following compounds in order of increasing boiling point:

$$CH_3CH_2CH_2\overset{\overset{\displaystyle O}{\|}}{C}OH \qquad CH_3CH_2CH_2CH_2\overset{\overset{\displaystyle O}{\|}}{C}H \qquad CH_3CH_2CH_2CH_2CH_2OH$$

butanoic acid   pentanal   1-pentanol

Solution

All three compounds are polar molecules of comparable molecular weight. Pentanal has no polar —OH group, cannot participate in hydrogen bonding, and has the lowest boiling point. 1-Pentanol participates in hydrogen bonding through the —OH group and is next in boiling point. The carboxyl group of butanoic acid participates in hydrogen bonding through both the —OH group and carbonyl oxygen and has the highest boiling point.

pentanal	1-pentanol	butanoic acid
mw 82	mw 84	mw 86
bp 103°	bp 137°	bp 164°

Problem 16.3

Arrange the following compounds in order of increasing solubility in water:

$$CH_3CH_2OCH_2CH_3 \qquad CH_3CH_2CH_2\overset{\overset{\displaystyle O}{\|}}{C}OH \qquad CH_3(CH_2)_8\overset{\overset{\displaystyle O}{\|}}{C}OH$$

16.4 Preparation of Carboxylic Acids

A. Oxidation of Primary Alcohols

Oxidation of primary alcohols (Section 12.5C) yields carboxylic acids. In the laboratory, the most commonly used oxidizing agents are potassium dichromate and potassium permanganate. (Note that the following oxidation equations show reactants and products, but are not balanced by charge or mass.)

$$CH_3(CH_2)_5CH_2OH + Cr_2O_7^{2-} \xrightarrow{H_3O^+} CH_3(CH_2)_5\overset{\overset{\displaystyle O}{\|}}{C}OH + Cr^{3+}$$

1-heptanol   heptanoic acid

$$HOCH_2CH_2CH_2CH_2OH + MnO_4^- \xrightarrow{H_3O^+} HO\overset{\overset{\displaystyle O}{\|}}{C}CH_2CH_2\overset{\overset{\displaystyle O}{\|}}{C}OH + Mn^{2+}$$

1,4-butanediol   butanedioic acid
                                      (succinic acid)

B. Oxidation of Aldehydes

Aldehydes are oxidized to carboxylic acids by $KMnO_4$, $K_2Cr_2O_7$, and even by weak oxidizing agents such as $Ag(NH_3)_2^+$. For example:

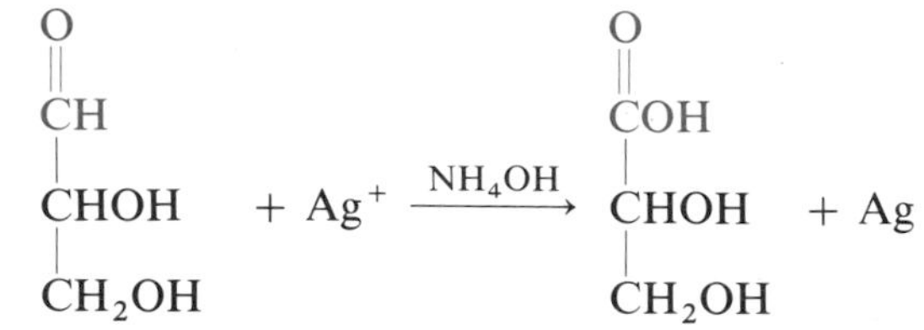

$$\underset{\text{glyceraldehyde}}{\overset{\overset{\displaystyle O}{\underset{\displaystyle |}{\overset{\displaystyle \|}{C}H}}}{\underset{\overset{\displaystyle |}{CH_2OH}}{\underset{\displaystyle |}{CHOH}}}} + Ag^+ \xrightarrow{NH_4OH} \underset{\text{glyceric acid}}{\overset{\overset{\displaystyle O}{\underset{\displaystyle |}{\overset{\displaystyle \|}{C}OH}}}{\underset{\overset{\displaystyle |}{CH_2OH}}{\underset{\displaystyle |}{CHOH}}}} + Ag$$

During the oxidation shown above, silver is reduced from the $+1$ oxidation state to silver metal.

C. Oxidation of Alkenes

Oxidation of disubstituted alkenes of the $RCH{=}CHR$ type by $K_2Cr_2O_7$ or $KMnO_4$ results in cleavage of the carbon-carbon double bond and formation of two carboxylic acids.

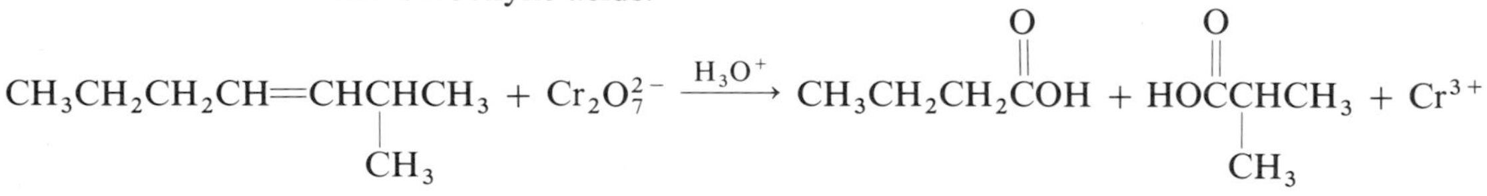

$$\underset{\text{2-methyl-3-heptene}}{CH_3CH_2CH_2CH{=}\underset{\underset{CH_3}{|}}{C}HCHCH_3} + Cr_2O_7^{2-} \xrightarrow{H_3O^+} \underset{\text{butanoic acid}}{CH_3CH_2CH_2\overset{\overset{\displaystyle O}{\|}}{C}OH} + \underset{\underset{\text{propanoic acid}}{\text{2-methyl-}}}{HO\overset{\overset{\displaystyle O}{\|}}{C}\underset{\underset{CH_3}{|}}{C}HCH_3} + Cr^{3+}$$

Oxidation of a cycloalkene such as cyclohexene cleaves the carbon-carbon double bond and yields a dicarboxylic acid.

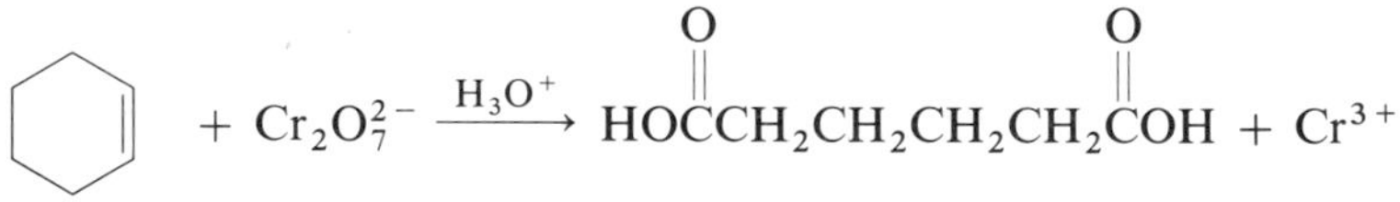

$$\underset{\text{cyclohexene}}{\bighexagon} + Cr_2O_7^{2-} \xrightarrow{H_3O^+} \underset{\underset{\text{(adipic acid)}}{\text{hexanedioic acid}}}{HO\overset{\overset{\displaystyle O}{\|}}{C}CH_2CH_2CH_2CH_2\overset{\overset{\displaystyle O}{\|}}{C}OH} + Cr^{3+}$$

Potassium dichromate oxidation converts a trisubstituted alkene of the $R_2C{=}CHR$ type to a ketone and a carboxylic acid. Under these conditions, 2-methyl-2-pentene is oxidized to acetone and propanoic acid:

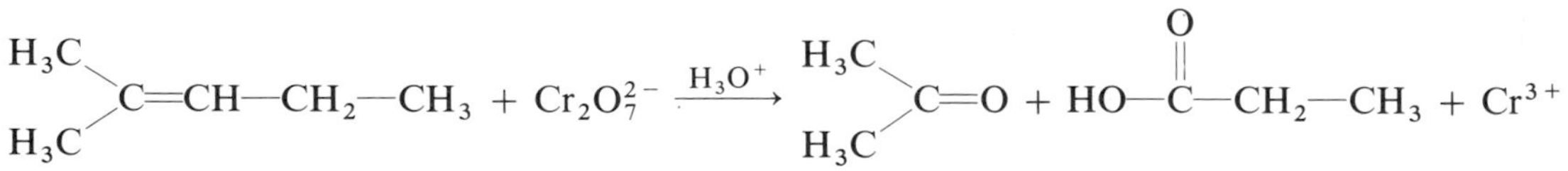

$$\underset{\text{2-methyl-2-pentene}}{\overset{H_3C}{\underset{H_3C}{>}}C{=}CH{-}CH_2{-}CH_3} + Cr_2O_7^{2-} \xrightarrow{H_3O^+} \underset{\text{acetone}}{\overset{H_3C}{\underset{H_3C}{>}}C{=}O} + \underset{\text{propanoic acid}}{HO{-}\overset{\overset{\displaystyle O}{\|}}{C}{-}CH_2{-}CH_3} + Cr^{3+}$$

Example 16.4

Draw structural formulas for the products of the following oxidations:

a.

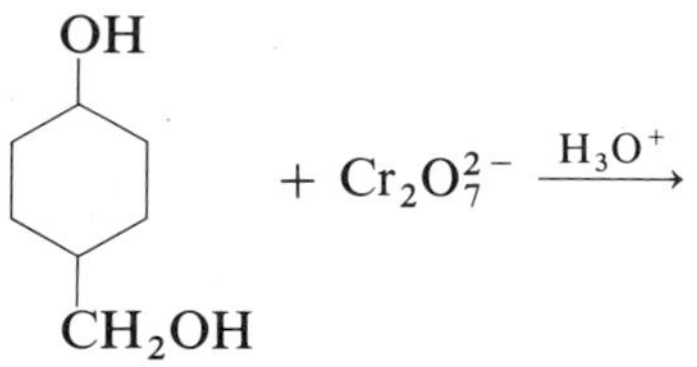

OH

(cyclohexane ring with OH at top and CH₂OH at bottom) + $Cr_2O_7^{2-}$ $\xrightarrow{H_3O^+}$

ĊH₂OH

b.

(cyclohexene ring with CH₃) + $Cr_2O_7^{2-}$ $\xrightarrow{H_3O^+}$

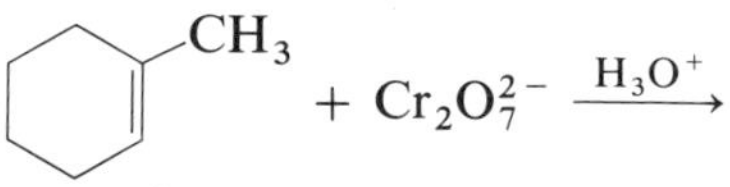

Solution

a. The starting material contains both a primary alcohol and a secondary alcohol. In the presence of $Cr_2O_7^{2-}$, the primary alcohol is oxidized to a carboxylic acid and the secondary alcohol is oxidized to a ketone. The product is:

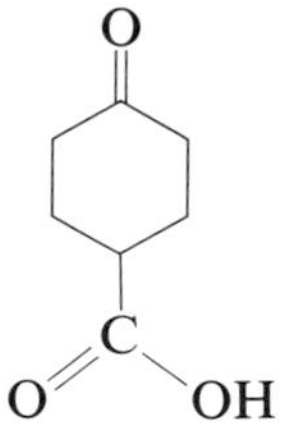

O

(cyclohexane ring with =O at top and C attached to O and OH at bottom)

O   OH

b. Oxidation of a trisubstituted alkene yields a ketone and a carboxylic acid. The structural formula of the product is drawn below in two different ways: the first to emphasize where the ring is cleaved, the second to show the molecule as a chain of seven carbon atoms.

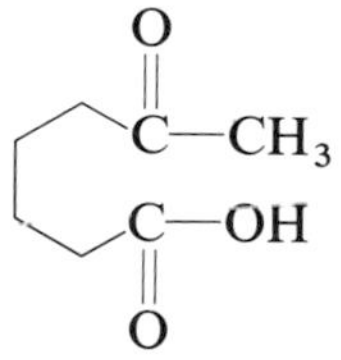

O
‖
C—CH₃
(ring)
C—OH
‖
O

$$CH_3-\overset{O}{\overset{\|}{C}}-CH_2-CH_2-CH_2-CH_2-\overset{O}{\overset{\|}{C}}-OH$$

Problem 16.4

Draw structural formulas for the products of the following oxidations:

a.

H₃C
    \
     C=CH—CH₂—CH₂—CH=C
    /   \
H₃C   CH₃

with CH₃ + $Cr_2O_7^{2-}$ $\xrightarrow{H_3O^+}$

b. $HO-CH_2-CH_2-CH_2-\overset{O}{\overset{\|}{C}}-OH$ + $Cr_2O_7^{2-}$ $\xrightarrow{H_3O^+}$

16.5 Reactions of Carboxylic Acids

A. Acidity of Carboxylic Acids

Carboxylic acids ionize in water to give acidic solutions. However, carboxylic acids are quite different from inorganic acids such as HCl, HBr, HNO_3, and H_2SO_4. These inorganic acids are 100% ionized in aqueous solution and are classified as strong acids.

$$HCl \longrightarrow H^+ + Cl^-$$

Carboxylic acids are only slightly ionized in aqueous solution and are classified as weak acids. When a carboxylic acid is dissolved in water, an equilibrium is established between the carboxylic acid, the carboxylate anion, and H^+.

$$CH_3-\overset{\overset{\displaystyle O}{\|}}{C}-OH \rightleftharpoons CH_3-\overset{\overset{\displaystyle O}{\|}}{C}-O^- + H^+$$

The difference in acidity between a strong acid such as HCl and weak acids such as acetic acid and phenol can be seen by comparing the hydrogen ion concentration of a $0.1M$ solution of each of these acids in water (Table 16.3). Because they are only partially dissociated in water solution, both carboxylic acids and phenols are classed as weak acids. Note, however, that carboxylic acids are much stronger acids than phenols.

Sodium, potassium, and ammonium salts of carboxylic acids are ionic compounds and are much more soluble in water than their parent carboxylic acids. For example, benzoic acid is insoluble in water. When reacted with sodium hydroxide, a strong base, it forms sodium benzoate, a water-soluble salt.

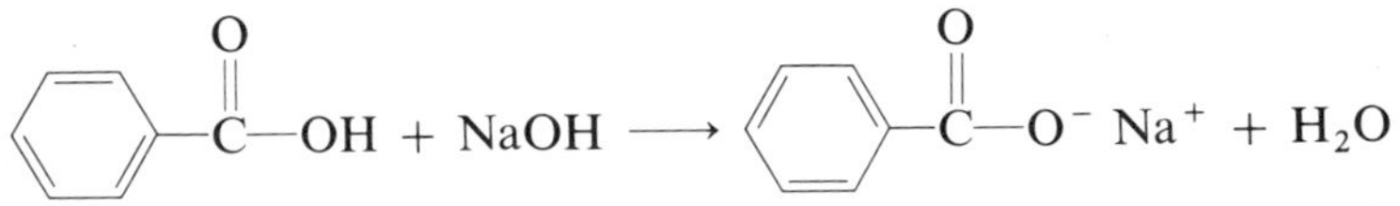

benzoic acid
(insoluble in water)

sodium benzoate
(soluble in water)

Table 16.3 Relative acid strengths of hydrochloric acid, acetic acid, and phenol.

Acid	K_a	Ionization in Water	$[H^+]$ of $0.1M$ solution
HCl	very large	100%	$0.1M$
$CH_3-\overset{\overset{\displaystyle O}{\|}}{C}-OH$	1.8×10^{-5}	1.3%	$0.0013M$
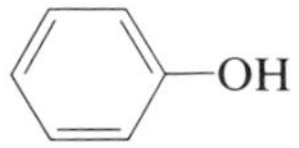 ⬡—OH	1.3×10^{-10}	0.0036%	$3.6 \times 10^{-6}M$

Carboxylic acids also react with weaker bases, such as bicarbonate and carbonate ions. In these reactions, the carboxylic acid is converted to a sodium salt. Bicarbonate and carbonate ions react with the acid to produce carbonic acid, H_2CO_3, which breaks down spontaneously to form CO_2 and H_2O.

$$2\ CH_3-\overset{\displaystyle O}{\overset{\|}{C}}-OH + Na_2CO_3 \longrightarrow 2\ CH_3-\overset{\displaystyle O}{\overset{\|}{C}}-O^-\ Na^+ + CO_2 + H_2O$$

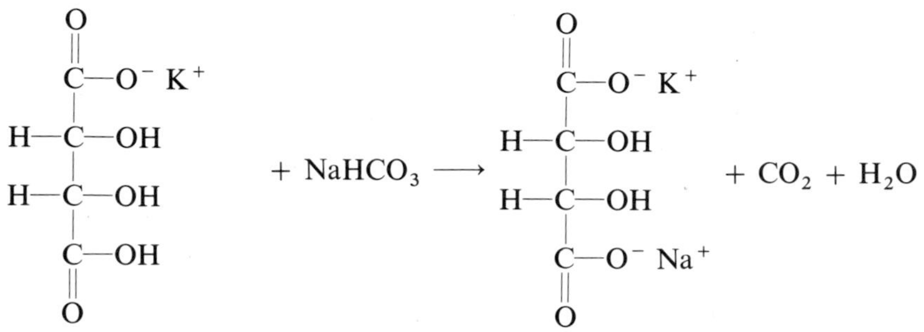

potassium hydrogen tartrate
(cream of tartar)

potassium sodium
tartrate

It is this last reaction that explains why sodium carbonate and sodium bicarbonate are used in baking. Baking powder is a combination of sodium carbonate or sodium bicarbonate and potassium hydrogen tartrate. When water is added to baking powder, the acid and base react to liberate CO_2, which forms bubbles in the batter or dough and causes it to rise. Baking soda or sodium bicarbonate can also be mixed with vinegar (a solution of acetic acid in water) or lemon juice (a solution containing citric acid) to produce CO_2. Some recipes call for mixing baking soda and sour cream (containing lactic acid), which also react to produce carbon dioxide.

Example 16.5

Write equations for the following acid-base reactions.

a. $CH_3-\underset{\underset{\displaystyle OH}{|}}{CH}-\overset{\displaystyle O}{\overset{\|}{C}}-OH + NaOH \xrightarrow{H_2O}$

b. $CH_3-\overset{\displaystyle O}{\overset{\|}{C}}-CH_2-CH_2-CH_2-\overset{\displaystyle O}{\overset{\|}{C}}-OH + NaHCO_3 \xrightarrow{H_2O}$

c. $HO-\overset{\displaystyle O}{\overset{\|}{C}}-CH_2-CH_2-\overset{\displaystyle O}{\overset{\|}{C}}-OH + Na_2CO_3 \xrightarrow{H_2O}$

Solution

a. 2-Hydroxypropanoic acid (lactic acid) is a monocarboxylic acid and reacts in a 1:1 ratio with sodium hydroxide:

$$\underset{\text{lactic acid}}{CH_3-\underset{\underset{\displaystyle OH}{|}}{CH}-\overset{\overset{\displaystyle O}{\|}}{C}-OH} + NaOH \longrightarrow \underset{\text{sodium lactate}}{CH_3-\underset{\underset{\displaystyle OH}{|}}{CH}-\overset{\overset{\displaystyle O}{\|}}{C}-O^-\,Na^+} + H_2O$$

b. 5-Ketohexanoic acid is a monocarboxylic acid and reacts in a 1:1 ratio with sodium bicarbonate:

$$\underset{\text{5-ketohexanoic acid}}{CH_3\overset{\overset{\displaystyle O}{\|}}{C}CH_2CH_2CH_2\overset{\overset{\displaystyle O}{\|}}{C}OH} + NaHCO_3 \longrightarrow \underset{\text{sodium 5-ketohexanoate}}{CH_3\overset{\overset{\displaystyle O}{\|}}{C}CH_2CH_2CH_2\overset{\overset{\displaystyle O}{\|}}{C}O^-Na^+} + CO_2 + H_2O$$

c. Butanedioic acid (succinic acid) is a dicarboxylic acid. It reacts with sodium carbonate in a 1:1 ratio:

$$\underset{\text{succinic acid}}{HO\overset{\overset{\displaystyle O}{\|}}{C}CH_2CH_2\overset{\overset{\displaystyle O}{\|}}{C}OH} + Na_2CO_3 \longrightarrow \underset{\text{sodium succinate}}{Na^+{}^-O\overset{\overset{\displaystyle O}{\|}}{C}CH_2CH_2\overset{\overset{\displaystyle O}{\|}}{C}O^-Na^+} + CO_2 + H_2O$$

Problem 16.5

Write equations for the following acid-base reactions.

a. $CH_3-CH_2-CH{=}CH-CO_2H + NaOH \xrightarrow{H_2O}$

b. $2\,C_6H_5CO_2H + Na_2CO_3 \xrightarrow{H_2O}$

c. $HO-\underset{\underset{\displaystyle CH_2-CO_2H}{|}}{\overset{\overset{\displaystyle CH_2-CO_2H}{|}}{C}}-CO_2H \quad + 3\,NaHCO_3 \xrightarrow{H_2O}$

     citric acid

B. Reduction of Carboxylic Acids

Carboxylic acids are reduced to primary alcohols by hydrogen in the presence of a metal catalyst:

$$CH_3CH_2CH_2CH_2CH_2\overset{\overset{\displaystyle O}{\|}}{C}OH + 2\,H_2 \xrightarrow[\text{high pressure}]{\text{catalyst}}$$

hexanoic acid

$$CH_3CH_2CH_2CH_2CH_2CH_2OH + H_2O$$

1-hexanol

The reduction of carboxylic acids to primary alcohols requires much higher pressures (up to 100 atmospheres) and temperatures than are required for the reduction of alkenes, aldehydes, or ketones. Therefore, it is possible to reduce alkenes, aldehydes, and ketones without affecting the carboxyl group.

$$CH_3-\overset{\overset{\displaystyle O}{\|}}{C}-CH_2-\overset{\overset{\displaystyle O}{\|}}{C}-OH + H_2 \xrightarrow{\text{Pt}} CH_3-\overset{\overset{\displaystyle OH}{|}}{C}H-CH_2-\overset{\overset{\displaystyle O}{\|}}{C}-OH$$

3-ketobutanoic acid   3-hydroxybutanoic acid

Carboxylic acids are also reduced to primary alcohols by $LiAlH_4$. Recall from Section 13.5C that this reagent also reduces aldehydes and ketones but not alkenes, alkynes, or aromatic rings.

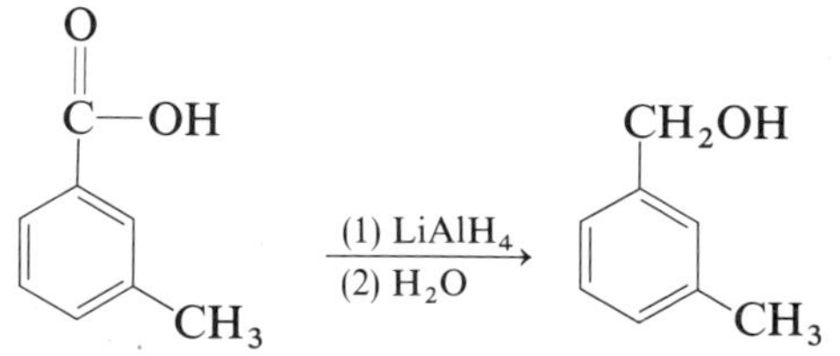

3-methylbenzoic acid
(*m*-toluic acid)

(1) $LiAlH_4$
(2) H_2O

m-methylbenzyl
alcohol

C. Decarboxylation of β-Keto Acids

Carboxylic acids that have a carbonyl group on the carbon atom beta to the carboxyl group lose CO_2 on heating. This reaction results in loss of the carboxyl group as CO_2 and is called **decarboxylation.** For example, when heated, 3-ketobutanoic acid decarboxylates to yield acetone and carbon dioxide.

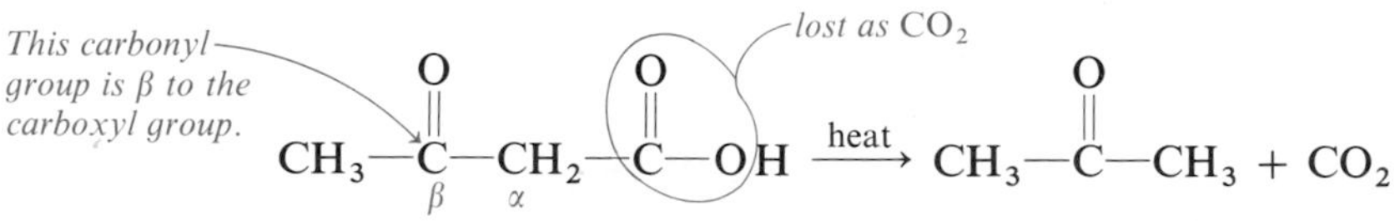

This carbonyl group is β to the carboxyl group.

lost as CO_2

$$CH_3-\overset{\overset{\displaystyle O}{\|}}{\underset{\beta}{C}}-CH_2-\overset{\overset{\displaystyle O}{\|}}{\underset{\alpha}{C}}-OH \xrightarrow{\text{heat}} CH_3-\overset{\overset{\displaystyle O}{\|}}{C}-CH_3 + CO_2$$

3-ketobutanoic
acid

acetone

Decarboxylation on heating is a unique property of β-ketocarboxylic acids.

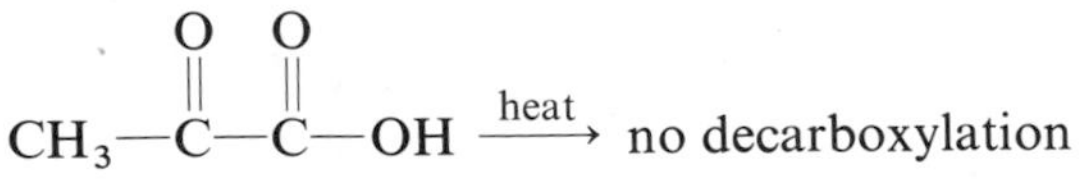

$$\underset{\substack{\text{(an alpha-keto-} \\ \text{carboxylic acid)}}}{CH_3-\overset{\overset{\displaystyle O}{\|}}{C}-\overset{\overset{\displaystyle O}{\|}}{C}-OH} \xrightarrow{\text{heat}} \text{no decarboxylation}$$

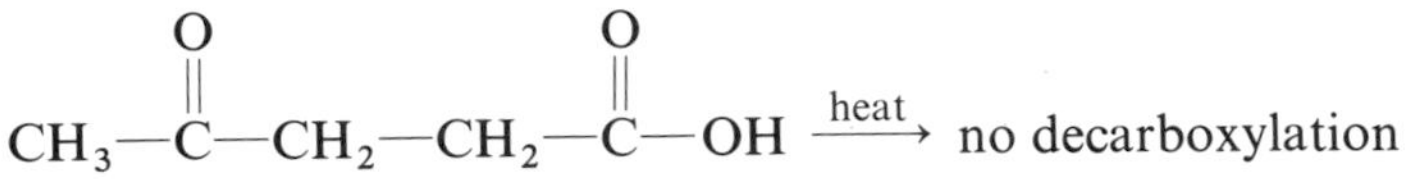

$$\underset{\substack{\text{(a gamma-keto-} \\ \text{carboxylic acid)}}}{CH_3-\overset{\overset{\displaystyle O}{\|}}{C}-CH_2-CH_2-\overset{\overset{\displaystyle O}{\|}}{C}-OH} \xrightarrow{\text{heat}} \text{no decarboxylation}$$

The mechanism for decarboxylation of β-ketoacids is illustrated by the decarboxylation of 3-ketobutanoic acid. The reaction involves a cyclic six-member transition state which, by rearrangement of electrons, leads to the enol form of acetone and carbon dioxide. The enol form of acetone is in equilibrium with the keto form.

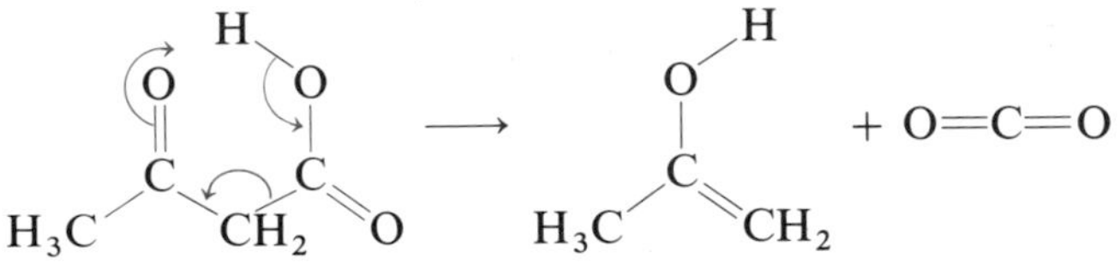

3-ketobutanoic acid   enol form
of acetone

An important example of decarboxylation of a β-ketoacid occurs during the oxidation of foodstuffs in the tricarboxylic acid cycle (Section 24.5). Oxalo-succinic acid, a tricarboxylic acid, undergoes decarboxylation to produce α-ketoglutaric acid. Only one of the three carboxyl groups of oxalosuccinic acid has a carbonyl group in the β-position to it. It is this one that is lost as CO_2.

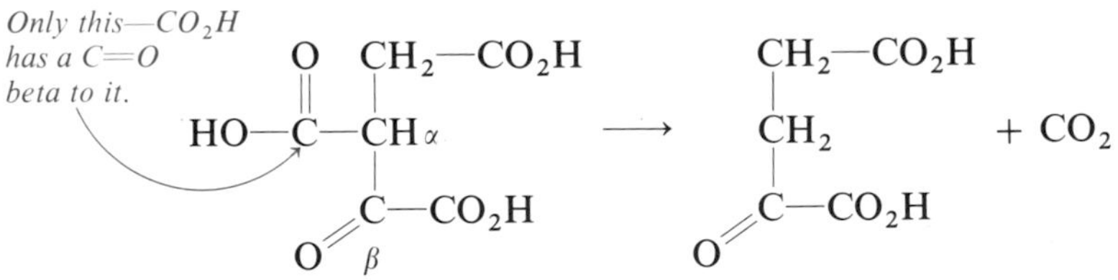

oxalosuccinic acid α-ketoglutaric acid

16.6 Fatty Acids

A. Structure

Fatty acids are monocarboxylic acids obtained from the hydrolysis of neutral fats and oils. Neutral fats and oils (Chapter 18) are triesters of glycerol, and their hydrolysis yields one molecule of glycerol and three molecules of fatty acid.

$$
\begin{array}{c}
\underset{\displaystyle |}{CH_2-O-\overset{\displaystyle O}{\overset{\displaystyle \|}{C}}-R} \\[2mm]
\underset{\displaystyle |}{CH-O-\overset{\displaystyle O}{\overset{\displaystyle \|}{C}}-R'} \;+\; 3\,H_2O \;\xrightarrow[\substack{\text{or enzyme}\\ \text{catalysis}}]{H^+ \text{ or } OH^-}\;
\underset{\displaystyle |}{CH-OH} \;+\; R'-CO_2H \\[2mm]
CH_2-O-\overset{\displaystyle O}{\overset{\displaystyle \|}{C}}-R''
\end{array}
\qquad
\begin{array}{l}
CH_2-OH \quad R-CO_2H \\[2mm]
CH-OH \\[2mm]
CH_2-OH \quad R''-CO_2H
\end{array}
$$

a neutral fat   glycerol   fatty acids

Over 70 fatty acids have been isolated from various cells and tissues. Table 16.4 gives common names and structural formulas for several of the most abundant fatty acids.

We can make certain generalizations about the more abundant fatty acid components of higher plants and animals.

1. Nearly all fatty acids have an even number of carbon atoms, usually between 14 and 22 carbons, in an unbranched chain. Those having 16 or 18 carbon atoms are by far the most abundant in nature.

2. Unsaturated fatty acids have lower melting points than their saturated counterparts.

Table 16.4 Some naturally occurring fatty acids.

Carbon Atoms	Structural Formula	Common Name	mp (°C)
Saturated fatty acids			
12	$CH_3(CH_2)_{10}COOH$	lauric	44
14	$CH_3(CH_2)_{12}COOH$	myristic	58
16	$CH_3(CH_2)_{14}COOH$	palmitic	63
18	$CH_3(CH_2)_{16}COOH$	stearic	70
20	$CH_3(CH_2)_{18}COOH$	arachidic	77
Unsaturated fatty acids			
16	$CH_3(CH_2)_5CH{=}CH(CH_2)_7COOH$	palmitoleic	-1
18	$CH_3(CH_2)_7CH{=}CH(CH_2)_7COOH$	oleic	16
18	$CH_3(CH_2)_4CH{=}CHCH_2CH{=}CH(CH_2)_7COOH$	linoleic	-5
18	$CH_3CH_2(CH{=}CHCH_2)_3(CH_2)_6COOH$	linolenic	-11
20	$CH_3(CH_2)_3(CH_2CH{=}CH)_4(CH_2)_3COOH$	arachidonic	-49

Figure 16.4 Space-filling models of (a) palmitic acid, a saturated fatty acid, and (b) oleic acid, an unsaturated fatty acid. Note that the *cis* configuration of the double bond in oleic acid produces a bend in the hydrocarbon chain.

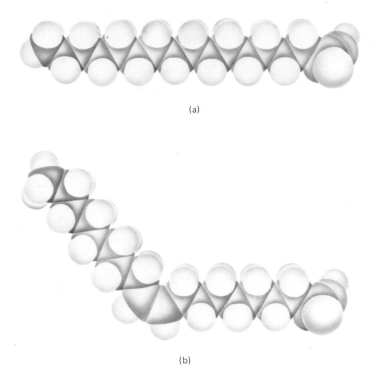

(a)

(b)

3. In most of the unsaturated fatty acids of higher organisms, there is a double bond between carbon atoms 9 and 10. In these unsaturated fatty acids, *cis* isomers predominate; the *trans* configuration is very rare (see Figure 16.4).

B. Physical Properties

Because of their long hydrophobic hydrocarbon chains, fatty acids are insoluble in water. However, they do interact with water in a particular way. If a drop of fatty acid is placed on the surface of water, it spreads out to form a thin film one molecule thick (a monomolecular layer), with the polar carboxyl groups dissolved in the water and the nonpolar hydrocarbon chains forming a hydrocarbon layer on the surface of the water (Figure 16.5).

C. Essential Fatty Acids

If fatty acids (in the form of fats) are withheld entirely from the diet of rats, they soon begin to suffer from retarded growth, scaly skin, and kidney damage. Addition of unsaturated fatty acids (linoleic, linolenic, and arachidonic acids) to their diet cures this condition. Humans and higher animals produce the enzymes necessary to catalyze the synthesis of fatty acids with double bonds between carbons 9 and 10. However, they lack the enzymes necessary to synthesize fatty acids with unsaturation beyond carbon-10, so these fatty acids must be supplied in the diet. Strictly speaking, linoleic acid is the critical fatty

Figure 16.5 The interaction of a drop of fatty acid with water to form a monomolecular layer.

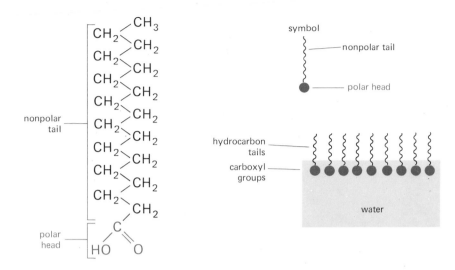

acid, because it can be converted within the body to linolenic and arachidonic acids. Because linoleic acid must be obtained in the diet for normal growth and well-being of humans and higher animals, it is classified as an **essential fatty acid.**

As you can see from Table 16.5, most animal fats are rich in saturated fatty acids, and though the percentage of unsaturated fatty acids is also high,

Table 16.5 Distribution of saturated and unsaturated fatty acids in some foods.

	% Fat in Edible Portion of Food	% Total Fat		
		Saturated	Oleic	Linoleic
Animal fats				
beef	5–37	43–48	43	0.5–3.0
butter	81	57	33	3
eggs	11.5	35	44	8.7
fish (tuna)	4.1	24.4	24.6	0.5
milk (whole, pasteurized)	3.7	57	33	3
pork	52	36.5	42	9.6
Vegetable fats				
coconut oil	100	85	6	0.5
corn oil	100	10	28	53
margarine	81	22.2	58	17.3
peanut oil	100	18	47	29
soybean oil	100	15	20	52
cottonseed oil	100	25	21	50

this unsaturation is due mostly to oleic acid. Vegetable fats generally have a lower content of saturated fatty acids and a higher content of unsaturated fatty acids, including linoleic acid. Corn, cottonseed, soybean, and wheat-germ oils are especially rich in linoleic acid.

There is no set minimum requirement for linoleic acid, but the Food and Nutrition Board suggests an intake of about 6 grams per day for adults. For infants and premature babies, the requirements are higher. Human milk and commercially prepared infant formulas provide a generous allowance of linoleic acid.

16.7 Soaps

A. Structure and Preparation of Soaps

Natural **soaps** are sodium or potassium salts of fatty acids (Section 16.6). One of the most ancient organic reactions known to man is the preparation of soaps by boiling lard or other animal fat with potash (potassium hydroxide). The reaction that takes place is hydrolysis of naturally occurring fats and oils. This reaction is called **saponification.**

$$
\begin{array}{l}
CH_2-O-\overset{\displaystyle O}{\overset{\|}{C}}(CH_2)_{14}CH_3 \\
\quad\quad\quad\quad\quad\overset{\displaystyle O}{\overset{\|}{} } \\
CH-O-\overset{\|}{C}(CH_2)_{14}CH_3 \; + 3\,NaOH \xrightarrow{\;\text{saponification}\;} \\
\quad\quad\quad\quad\quad\overset{\displaystyle O}{\overset{\|}{} } \\
CH_2-O-\overset{\|}{C}(CH_2)_{14}CH_3
\end{array}
$$

a fat

$$
\begin{array}{l}
CH_2-OH \\
CH-OH \; + 3\,CH_3(CH_2)_{14}CO_2Na \\
CH_2-OH
\end{array}
$$

glycerol   a soap

In the present-day industrial manufacture of soap, tallow (the fat of cattle, sheep, and so on) is heated with sodium hydroxide. After saponification is complete, sodium chloride is added to precipitate the soap as thick curds. The water layer is drawn off and glycerol is recovered from it by vacuum distillation.

The crude soap contains sodium chloride, sodium hydroxide, and glycerol as impurities. These are removed by boiling the curds in water and reprecipitating with salt. After several such purifications, the soap may be used without further processing as an inexpensive industrial soap. Fillers such as sand or pumice may be added to make a scouring soap. Other treatments transform the crude soap into laundry soaps, medicated soaps, cosmetic soaps, liquid soaps, and others.

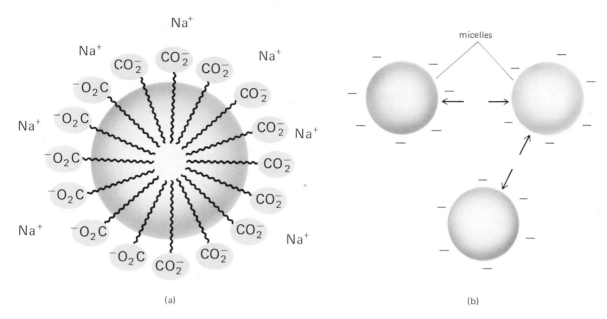

Figure 16.6 Soap micelles. (a) Diagram of a soap micelle. (b) Because of the negative charges on their surfaces, soap micelles repel each other.

B. How Soaps Clean

Soap owes its remarkable cleansing properties to its ability to act as an emulsifying agent. Regarded from one end, the organic portion of a natural soap is a polar, negatively charged, hydrophilic (water-seeking) carboxylate group, $—CO_2^-$, which interacts with surrounding water molecules by hydrogen bonding and ion-dipole interactions. Regarded from the other end, it is a long, nonpolar hydrophobic (water-repelling) hydrocarbon chain, which does not interact at all with surrounding water molecules. Because the long hydrocarbon chains of natural soaps are insoluble in water, they cluster together, attracted to each other by dispersion forces. These clusters are called **micelles.** In soap micelles, the charged carboxylate groups form a negatively charged surface and the nonpolar, water-insoluble hydrocarbon chains lie buried within the center (Figure 16.6). Soap micelles have net negative charges and remain suspended or dispersed because of mutual repulsion.

Most of the things we commonly think of as dirt, such as grease, oil, fat stains, and so on, are nonpolar and insoluble in water. When soap and dirt are mixed together, as in a washing machine, the nonpolar hydrocarbon ends of soap micelles "dissolve" the nonpolar dirt molecules. In effect, new soap micelles form, this time with nonpolar dirt molecules in the center. In this way, nonpolar organic oil, grease, and dirt are dissolved and washed away in the polar wash water (Figure 16.7, next page).

Figure 16.7 Soap micelle with a "dissolved" oil droplet.

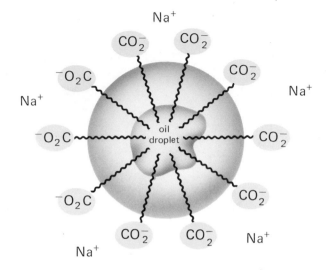

Soaps are not without their disadvantages. First, they are salts of weak acids and are converted by mineral acids into free fatty acids.

$$CH_3(CH_2)_{16}CO_2^- Na^+ + HCl \longrightarrow CH_3(CH_2)_{16}CO_2H + Na^+ + Cl^-$$

(soluble in water)   (insoluble in water)

Fatty acids are far less soluble than their potassium or sodium salts and precipitate, forming a scum. For this reason, soaps cannot be used in acidic solution. Second, soaps form insoluble salts when used in water containing calcium, magnesium, or iron(III) ions (hard water).

$$2\ CH_3(CH_2)_{16}CO_2^- Na^+ + Ca^{2+} \longrightarrow [CH_3(CH_2)_{16}CO_2^-]_2Ca^{2+} + 2\ Na^+$$

(water soluble)   (water insoluble)

These insoluble salts create problems, including rings around the bathtub, films that spoil the luster of hair, and grayness and roughness that build up on textiles after repeated washing.

Given these limitations on the use of natural soap, the problem for chemists was to create a new type of cleansing agent that would be soluble in both acidic and alkaline solutions and would not form insoluble precipitates when used in hard water. Despite considerable effort, there was no significant progress in solving this problem until the early 1940s.

16.8 Synthetic Detergents

Given an understanding of the cleansing action of soaps, the design criteria for a synthetic detergent were as follows: A molecule with a long hydrocarbon chain (12 to 18 carbon atoms) and a highly polar group or groups at one end of the molecule that do not form insoluble salts with Ca^{2+}, Mg^{2+}, or other ions present in hard water. Chemists recognized that the essential characteristics of a soap could be produced in a molecule containing a sulfate group rather than a carboxylate group. Such compounds, known as alkyl acid sulfate esters, are strong acids, comparable in strength to sulfuric acid. Furthermore, the calcium, magnesium, and iron(III) salts of alkyl acid sulfate esters are soluble in water.

In the earliest method of detergent production, a long-chain alcohol was reacted with sulfuric acid to form an alkyl acid sulfate ester. Neutralization with sodium hydroxide formed a synthetic **detergent.** The first synthetic detergent was produced from 1-dodecanol (lauryl alcohol).

$$CH_3(CH_2)_{10}CH_2OH + H_2SO_4 \longrightarrow CH_3(CH_2)_{10}CH_2O-SO_3H + H_2O$$

1-dodecanol
(lauryl alcohol)

dodecyl hydrogen sulfate
(lauryl hydrogen sulfate)

$$\downarrow NaOH$$

$$CH_3(CH_2)_{10}CH_2O-SO_3^- \ Na^+ + H_2O$$

sodium dodecyl sulfate (SDS)
(sodium lauryl sulfate)

The physical resemblances between this synthetic detergent and natural soaps are obvious: a long nonpolar hydrocarbon chain and a highly polar end.

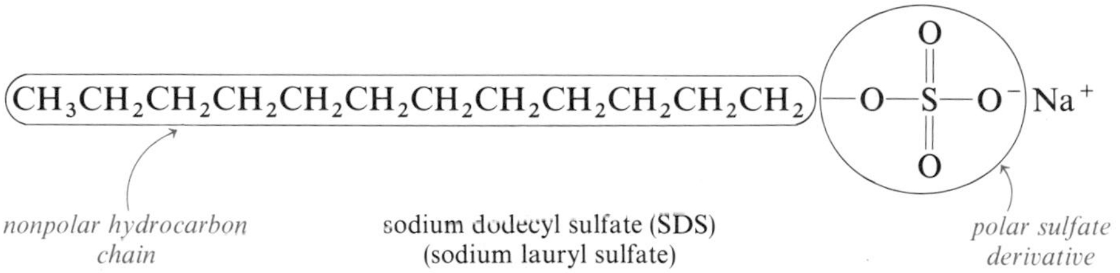

nonpolar hydrocarbon chain

sodium dodecyl sulfate (SDS)
(sodium lauryl sulfate)

polar sulfate derivative

Large-scale commercial production of SDS was not possible because of the lack of bulk quantities of 1-dodecanol.

A second major advance in detergents came in the late 1940s when technology was developed to make alkylbenzene sulfonate detergents. Ethylene and benzene, the starting materials for this synthesis, were readily available from refining of petroleum (benzene) and natural gas (ethylene).

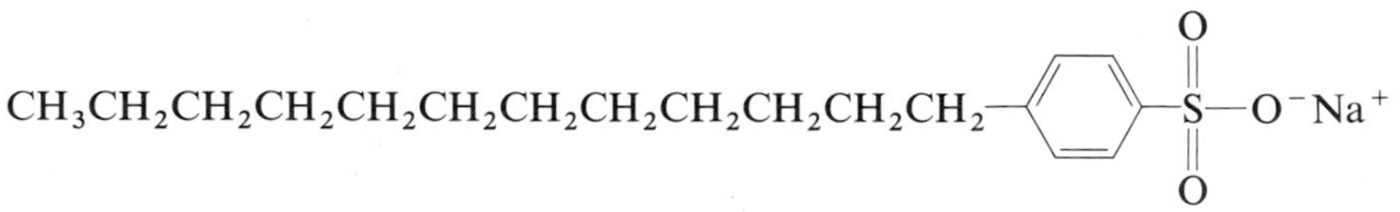

a sodium alkylbenzene sulfonate detergent

Alkylbenzene sulfonate detergents were introduced in the 1950s and today they command close to 90% of the market once held by soaps.

Key Terms and Concepts

acidity of carboxylic acids (16.5A)

carboxyl group (16.8)

decarboxylation (16.5C)

detergent (16.8)

essential fatty acid (16.6C)

fatty acid (16.6A)

hydrophilic (16.3)

hydrophobic (16.3)

β-ketoacid (16.5C)

micelle (16.7B)

soap (16.7)

Key Reactions

1. Preparation of carboxylic acids (Section 16.4):
   a. Oxidation of primary alcohols:

   $$CH_3(CH_2)_5CH_2OH + Cr_2O_7^{2-} \xrightarrow{H_3O^+} CH_3(CH_2)_5\overset{\overset{\displaystyle O}{\|}}{C}OH + Cr^{3+}$$

   b. Oxidation of aldehydes:

   $$\underset{CH_2OH}{\overset{\overset{\displaystyle O}{\|}}{\underset{|}{\overset{C-H}{|}}}}H-\overset{|}{\underset{|}{C}}-OH + Ag^+ \xrightarrow{NH_4OH} \underset{CH_2OH}{\overset{\overset{\displaystyle O}{\|}}{\underset{|}{\overset{C-OH}{|}}}}H-\overset{|}{\underset{|}{C}}-OH + Ag$$

   c. Oxidation of disubstituted alkenes of the type RCH=CHR:

   $$CH_3CH_2CH_2CH=\underset{\underset{CH_3}{|}}{C}HCHCH_3 + Cr_2O_7^{2-} \xrightarrow{H_3O^+} CH_3CH_2CH_2\overset{\overset{\displaystyle O}{\|}}{C}OH + HO\overset{\overset{\displaystyle O}{\|}}{C}\underset{\underset{CH_3}{|}}{C}HCH_3 + Cr^{3+}$$

2. Acidity of carboxylic acids (Section 16.5A):

   $$CH_3-\overset{\overset{\displaystyle O}{\|}}{C}-OH \rightleftharpoons CH_3-\overset{\overset{\displaystyle O}{\|}}{C}-O^- + H^+$$

3. Reduction of carboxylic acids to primary alcohols (Section 16.5B):

   $$CH_3(CH_2)_{10}\overset{\overset{\displaystyle O}{\|}}{C}OH + 2\,H_2 \xrightarrow[\substack{heat \\ high\ pressure}]{catalyst} CH_3(CH_2)_{10}CH_2OH + H_2O$$

4. Decarboxylation of β-ketocarboxylic acids (Section 16.5C):

   $$CH_3-\underset{\beta}{\overset{\overset{\displaystyle O}{\|}}{C}}-\underset{\alpha}{CH_2}-\overset{\overset{\displaystyle O}{\|}}{C}-OH \xrightarrow{heat} CH_3-\overset{\overset{\displaystyle O}{\|}}{C}-CH_3 + CO_2$$

$$2\ CH_3(CH_2)_{16}CO_2^- + Mg^{2+} \longrightarrow [CH_3(CH_2)_{16}CO_2]_2Mg$$

Problems

Structure and Nomenclature of Carboxylic Acids (Sections 16.1 and 16.2)

16.6   Name each of the following molecules.

a. $CH_3CHCH_2CH_2CO_2H$
     $|$
     OH

b.

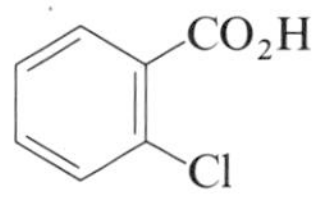

c. $ClCH_2CO_2H$

d. CH_3CHCO_2H
   $|$
   CH_3

e. $C_6H_5CH_2CH_2CH_2CO_2H$

f.

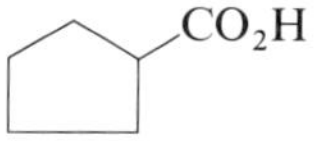

g. $C_6H_5CO_2Na$

h. $CH_3CH_2CH_2CH_2CO_2NH_4$

i. $HO_2CCH_2CH_2CH_2CH_2CO_2H$

j. CF_3CO_2H

k. $(CH_3CH_2CO_2)_2Mg$

l. $CH_3CH_2CH_2CH_2CH_2CH_2CH_2CH_2CH{=}CHCO_2H$

16.7   Draw structural formulas for each of the following molecules.
a. 3-hydroxybutanoic acid
b. sodium oxalate
c. trichloroacetic acid
d. 4-aminobutanoic acid
e. sodium hexadecanoate
f. calcium octanoate
g. potassium phenylacetate
h. octanoic acid
i. 2-hydroxypropanoic acid (lactic acid)
j. 2-aminopropanoic acid (alanine)
k. *p*-methoxybenzoic acid
l. potassium 2,4-hexadienoate (the food preservative, potassium sorbate)

Physical Properties of Carboxylic Acids (Section 16.3)

16.8   Draw structural formulas to illustrate hydrogen bonding between the circled atoms.

a. 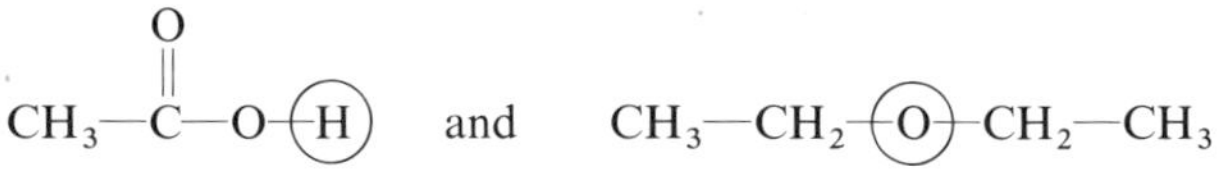 $CH_3{-}\overset{\overset{\displaystyle O}{\|}}{C}{-}O{-}\boxed{H}$ and $CH_3{-}CH_2{-}\boxed{O}{-}CH_2{-}CH_3$

b. 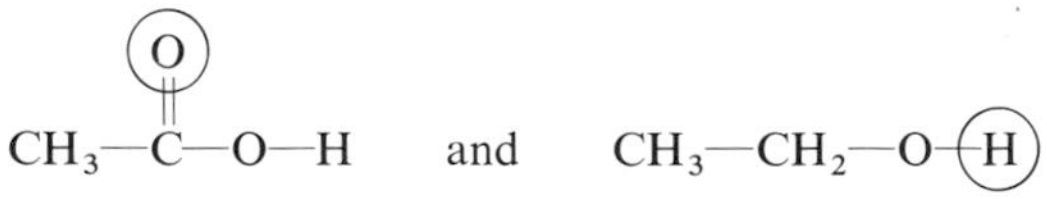 $CH_3{-}\overset{\overset{\displaystyle \boxed{O}}{\|}}{C}{-}O{-}H$ and $CH_3{-}CH_2{-}O{-}\boxed{H}$

c.

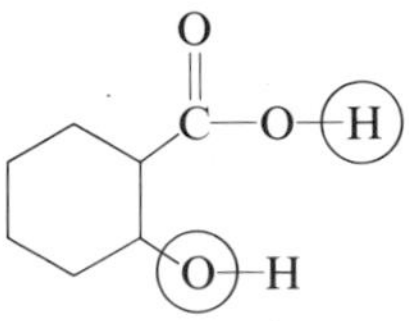

d.

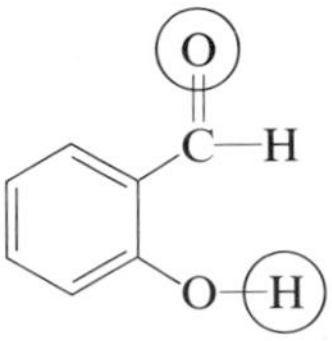

e. 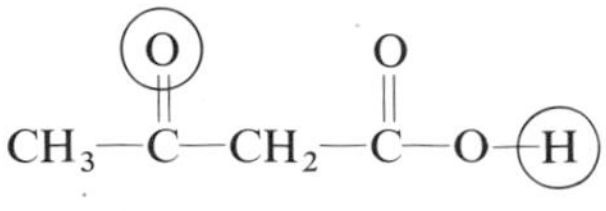 $CH_3{-}\overset{\overset{\displaystyle \boxed{O}}{\|}}{C}{-}CH_2{-}\overset{\overset{\displaystyle O}{\|}}{C}{-}O{-}\boxed{H}$

16.9 Arrange the following compounds in order of increasing boiling points.

a. $CH_3CH_2\overset{\displaystyle O}{\overset{\displaystyle \|}{C}}OH$ $CH_3CH_2CH_2CH_2OH$ $CH_3CH_2OCH_2CH_3$

b. $CH_3(CH_2)_8CO_2H$ ⬡—OH $CH_3(CH_2)_4CO_2H$

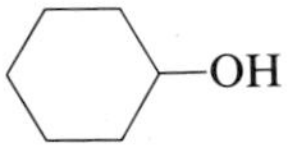

Preparation of Carboxylic Acids (Section 16.4)

16.10 Complete the following reactions:

a. $CH_3(CH_2)_4CH_2OH + Cr_2O_7^{2-} \xrightarrow[\text{heat}]{H_3O^+}$

b. 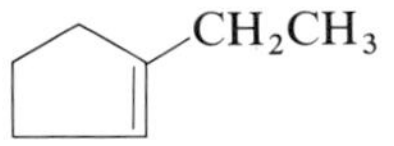 [cyclopentene with CH_2CH_3 substituent] $+ Cr_2O_7^{2-} \xrightarrow[\text{heat}]{H_3O^+}$

c. 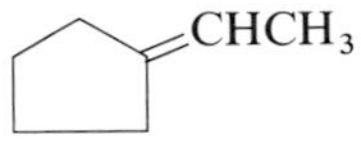 [cyclopentane with $=CHCH_3$ substituent] $+ Cr_2O_7^{2-} \xrightarrow[\text{heat}]{H_3O^+}$

d. $CH_3(CH_2)_7CH{=}CH(CH_2)_7\overset{\displaystyle O}{\overset{\displaystyle \|}{C}}OH + Cr_2O_7^{2-} \xrightarrow[\text{heat}]{H_3O^+}$

   oleic acid

e. 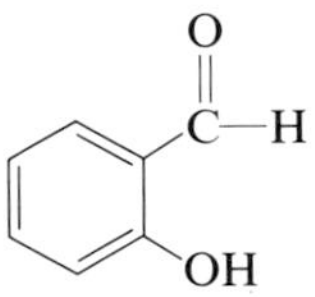 [benzene ring with $\overset{\displaystyle O}{\overset{\displaystyle \|}{C}}{-}H$ and OH substituents] $+ Ag^+ \xrightarrow{NH_4OH}$

   salicylaldehyde

f. 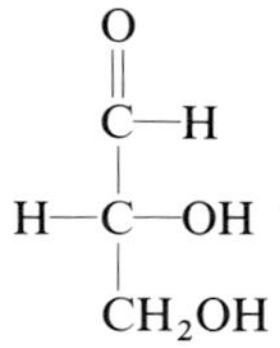 $H{-}\overset{\displaystyle \overset{\textstyle O}{\|}}{\underset{\displaystyle CH_2OH}{\overset{\displaystyle |}{C}}}{-}OH + Cu^{2+} \longrightarrow$

   with the $C{-}H$ group on top

   D-glyceral-dehyde

16.11 Draw the structural formula of a molecule of the given molecular formula that on oxidation gives the carboxylic acid or dicarboxylic acid shown.

a. $C_6H_{14}O \xrightarrow{\text{oxidation}} CH_3(CH_2)_4\overset{\displaystyle O}{\overset{\displaystyle \|}{C}}OH$

b. $C_6H_{12}O \xrightarrow{\text{oxidation}} CH_3(CH_2)_4\overset{\displaystyle O}{\overset{\displaystyle \|}{C}}OH$

c. $C_6H_{14}O_2 \xrightarrow{\text{oxidation}} HO\overset{\displaystyle O}{\overset{\displaystyle \|}{C}}CH_2CH_2CH_2CH_2\overset{\displaystyle O}{\overset{\displaystyle \|}{C}}OH$

d. $C_{13}H_{22}O \xrightarrow{\text{oxidation}}$ [cyclohexanone with H_3C and CH_3 groups at one position and CH_3 at another] $+ CH_3{-}\overset{\displaystyle O}{\overset{\displaystyle \|}{C}}{-}CH_2{-}\overset{\displaystyle O}{\overset{\displaystyle \|}{C}}{-}OH$

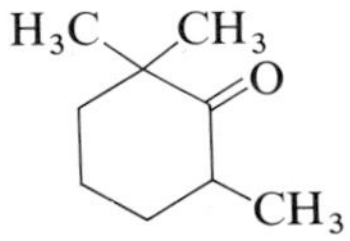

**Reactions of
Carboxylic Acids
(Section 16.5)**

16.12 Arrange the following compounds in order of increasing acidity.

a. 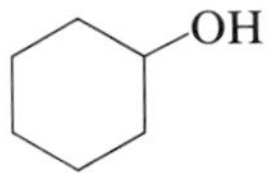 (cyclohexyl)—OH   b. 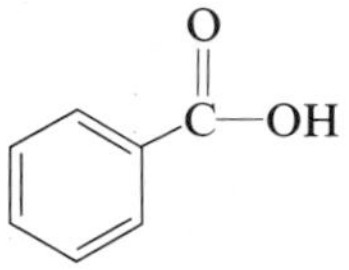 (phenyl)—$\overset{\overset{\displaystyle O}{\|}}{C}$—OH   c. (phenyl)—OH

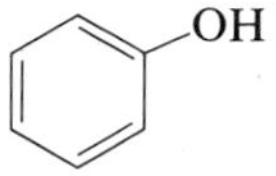

16.13 Complete the following reactions. Where there is no reaction, write NR.

a. $CH_3CO_2H + NaOH \longrightarrow$ b. 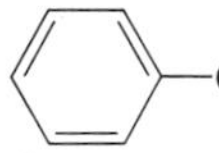 (phenyl)—$CO_2H + NH_3 \longrightarrow$

c. $CH_3(CH_2)_{14}CO_2Na + H_2SO_4 \longrightarrow$

d. $CH_3CH_2CH_2CO_2H + NaHCO_3 \longrightarrow$

e. 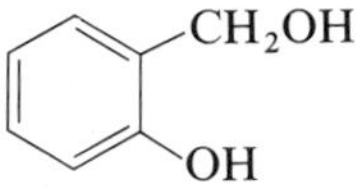 (phenyl with CH₂OH and OH) $+ NaHCO_3 \longrightarrow$ f. 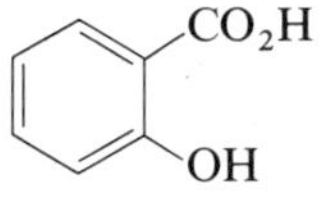 (phenyl with CO₂H and OH) $+ NaHCO_3 \longrightarrow$

g. $CH_3-\overset{\overset{\displaystyle OH}{|}}{CH}-CO_2H + 2\,H_2 \xrightarrow[\text{high pressure}]{\text{catalyst}}$

h. $CH_3-CH_2-\overset{\overset{\displaystyle O}{\|}}{C}-\overset{\overset{\displaystyle}{\underset{\underset{\displaystyle CH_3}{|}}{CH}}}-\overset{\overset{\displaystyle O}{\|}}{C}-OH \xrightarrow{\text{heat}}$

i. $HO-\overset{\overset{\displaystyle O}{\|}}{C}-CH_2-\overset{\overset{\displaystyle O}{\|}}{C}-\overset{\overset{\displaystyle O}{\|}}{C}-OH \xrightarrow{\text{heat}}$

16.14 Show how you might distinguish between the following pairs of compounds by a simple chemical test. In each case, tell what test you would perform, what you would expect to observe, and write an equation for each positive test.
a. acetic acid and acetaldehyde
b. hexanoic acid and 1-hexanol
c. benzoic acid and phenol
d. sodium salicylate and salicylic acid
e. oleic acid and stearic acid (see Table 16.4 for structural formulas of these fatty acids)
f. phenylacetic acid and acetophenone (methyl phenyl ketone)
g. sodium lauryl sulfate (a synthetic detergent, Section 16.8) and sodium stearate (a natural soap)

16.15 Decarboxylation is a general reaction for any molecule that has a carbonyl group on the carbon atom beta to a carboxylic acid.

$$HO-\overset{\overset{\displaystyle O}{\|}}{C}-CH_2-\overset{\overset{\displaystyle O}{\|}}{C}-OH \xrightarrow{\text{heat}} \text{a carboxylic acid} + CO_2$$

malonic acid

a. Draw the structural formula of the carboxylic acid formed in decarboxylation of malonic acid.

b. Would you expect succinic acid to undergo the same type of decarboxylation?

16.16 β-Ketoacids and 1,3-dicarboxylic acids undergo decarboxylation. The reverse of this process, carboxylation, is an important step in several metabolic pathways. Two of these are:

$$CH_3-\overset{\overset{\displaystyle O}{\|}}{C}-CO_2H + CO_2 \longrightarrow \text{a dicarboxylic acid} \qquad CH_3-CH_2-CO_2H + CO_2 \longrightarrow \text{a dicarboxylic acid}$$

pyruvic acid   propanoic acid

The carboxylation of pyruvic acid is one means by which a cell is able to maintain an adequate supply of intermediates in the Krebs or tricarboxylic acid cycle during certain kinds of metabolic stress. The carboxylation of propanoic acid is a key step in the degradation of fatty acids with odd numbers of carbon atoms and of certain amino acids.

a. Draw structural formulas for the dicarboxylic acids produced in these carboxylations.

b. The name of one of these dicarboxylic acids is α-ketosuccinic acid, that of the other is methylmalonic acid. Which dicarboxylic acid has which name?

16.17 The following conversions can be carried out in either one or two steps. Show the reactants you would use and draw structural formulas for the intermediate formed in any conversion that requires two steps.

a. $CH_3(CH_2)_6\overset{\overset{\displaystyle O}{\|}}{C}H \longrightarrow CH_3(CH_2)_6CO_2H$

b. $CH_3(CH_2)_6CH_2OH \longrightarrow CH_3(CH_2)_5CO_2H$

c. $\longrightarrow HO\overset{\overset{\displaystyle O}{\|}}{C}CH_2CH_2CH_2\overset{\overset{\displaystyle O}{\|}}{C}OH$

glutaric acid

d. $\longrightarrow HO\overset{\overset{\displaystyle O}{\|}}{C}CH_2CH_2CH_2CH_2\overset{\overset{\displaystyle O}{\|}}{C}OH$

adipic acid

e. $HO\overset{\overset{\displaystyle O}{\|}}{C}CH_2CH_2\overset{\overset{\displaystyle O}{\|}}{C}OH \longrightarrow Na^+{}^-O\overset{\overset{\displaystyle O}{\|}}{C}CH_2CH_2\overset{\overset{\displaystyle O}{\|}}{C}O^-Na^+$

succinic acid   sodium succinate

f. $\longrightarrow$
$$\begin{array}{c} CO_2H \\ | \\ CH_2 \\ | \\ CH_2 \\ | \\ CO_2H \end{array}$$

fumaric acid   succinic acid

g. $HO_2CCH_2CH_2CO_2H \longrightarrow HOCH_2CH_2CH_2CH_2OH$

h. 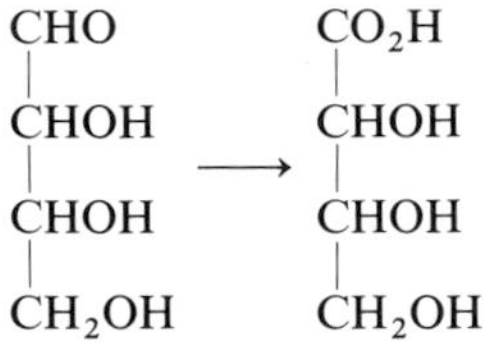
$$
\begin{array}{ccc}
CHO & & CO_2H \\
| & & | \\
CHOH & & CHOH \\
| & \longrightarrow & | \\
CHOH & & CHOH \\
| & & | \\
CH_2OH & & CH_2OH
\end{array}
$$

i. $CH_3(CH_2)_6CH_2OH \longrightarrow CH_3(CH_2)_6CO_2^- \ NH_4^+$

Fatty Acids (Section 16.6)

16.18 Examine the structural formulas for lauric, palmitic, stearic, oleic, linoleic, and arachidonic acids. For each that shows *cis-trans* isomerism, state the total number of such isomers possible.

16.19 What does it mean to say that linoleic acid is an "essential" fatty acid? Name several dietary sources of linoleic acid.

16.20 By using structural formulas, illustrate how fatty acid molecules interact with water to form a monomolecular layer on the surface of water.

16.21 Draw structural formulas for the products of the following reactions.

a.
$$
\overset{\displaystyle O}{\overset{\displaystyle \|}{CH_3(CH_2)_7CH{=}CH(CH_2)_7C\text{---}OH}} + Br_2 \longrightarrow
$$

       oleic acid

b. oleic acid + H_2 $\xrightarrow[\substack{3 \text{ atm pressure} \\ 25°C}]{Pt}$
      c. oleic acid + $3 H_2$ $\xrightarrow[\substack{\text{high pressure} \\ \text{high temperature}}]{Pt}$

d. oleic acid + NaOH $\longrightarrow$
      e. oleic acid + NH_3 $\longrightarrow$

Soaps and Detergents (Sections 16.7 and 16.8)

16.22 By using structural formulas, show how a soap "dissolves" fats, oils, and grease.

16.23 Show by balanced equations the reaction of a soap with (a) hard water, and (b) acidic solution.

16.24 Characterize the structural features necessary to make a good detergent. Illustrate by structural formulas two different classes of synthetic detergents. Name each example.

16.25 Following are structural formulas for a cationic detergent and a nonionic detergent. How would you account for the detergent properties of each?

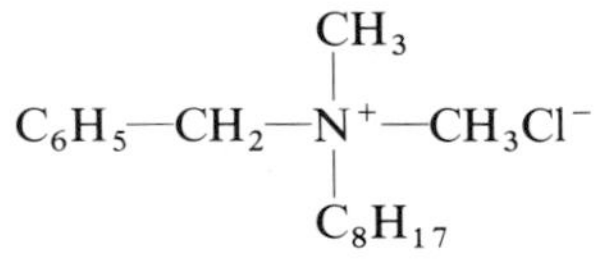

$$
\underset{\begin{array}{c}\text{benzyldimethyloctylammonium} \\ \text{chloride} \\ \text{(a cationic detergent)}\end{array}}{C_6H_5\text{---}CH_2\overset{\displaystyle CH_3}{\underset{\displaystyle C_8H_{17}}{\text{---}N^+\text{---}CH_3Cl^-}}}
\qquad
\underset{\begin{array}{c}\text{pentaerythrityl palmitate} \\ \text{(a nonionic detergent)}\end{array}}{CH_3(CH_2)_{14}\overset{\displaystyle O}{\overset{\displaystyle \|}{C}}\text{---}O\text{---}CH_2\overset{\displaystyle CH_2OH}{\underset{\displaystyle CH_2OH}{\text{---}C\text{---}CH_2OH}}}
$$

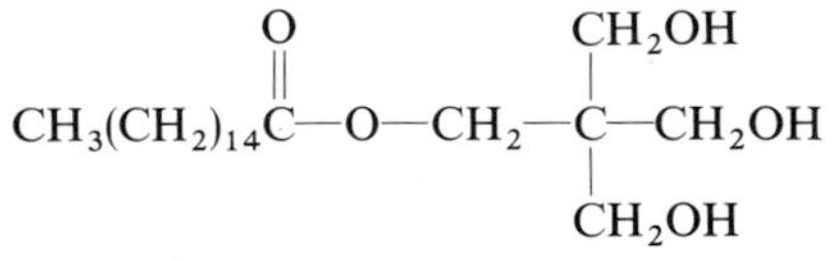

Prostaglandins

Prostaglandins are a group of naturally occurring substances, all of which have the 20-carbon skeleton of prostanoic acid.

prostanoic acid

Prostaglandins and prostaglandin-derived materials have been found in virtually all human tissues examined thus far and are intimately involved in a host of bodily processes. For example, they are involved in both the induction of the inflammatory response and in its relief. The medical significance of these facts becomes obvious when we realize that more than five million Americans suffer from rheumatoid arthritis, an inflammatory disease. Prostaglandins are also involved in almost every phase of reproductive physiology.

The discovery and structural determination of prostaglandins began in 1930, when Raphael Kurzrok and Charles Lieb, both gynecologists practicing in New York, observed that human seminal fluid stimulates contraction of isolated human uterine muscle. A few years later, in Sweden, Ulf von Euler confirmed this report and noted that human seminal fluid also produces contraction of intestinal smooth muscle and lowers blood pressure when injected into the blood stream. Von Euler proposed the name *prostaglandin* for the mysterious substance or substances responsible for such diverse effects, because at that time it was believed that the substances originated in the prostate gland. We now know that prostaglandin production is by no means limited to the prostate gland. However, the name has stuck. By 1960, several prostaglandins had been isolated in pure crystalline form and their structural formulas had been determined. Structural formulas for three common prostaglandins are given in Figure 1.

Prostaglandins are abbreviated **PG**, with an additional letter and numerical subscript to indicate the type and series. The various types differ in the functional groups present in the five-member ring. Those of the A-type are α,β-unsaturated ketones; those of the B-type are β-hydroxyketones; and those of the F-type are 1,3-diols. The subscript α in the F-type indicates that the hydroxyl group at carbon-9 is below the plane of the five-member ring and on the same side as the hydroxyl at carbon-11. The various series of prostaglandins differ in the number of double bonds on the two side chains. Those of the 1-series have only one double bond; those of the 2-series have two double bonds; and those of the 3-series have three double bonds.

At the same time investigations of the chemical structure of prostaglandins were going on, clinical scientists began to study the biochemistry of these remarkable substances and their potential as drugs. Initially, research was hampered by the high cost and great difficulty of isolating and purifying the substances. If they could not be isolated easily, could they be synthesized instead? The first totally synthetic prostaglandins became available in 1968 when Dr. John Pike of the Upjohn Company and Professor E. J. Corey of Harvard University each announced laboratory syntheses of several prostaglandins and prostaglandin analogs (Figure 2). However, costs were still high. Then, in 1969, the price of prostaglandins dropped dramatically with the discovery that the gor-

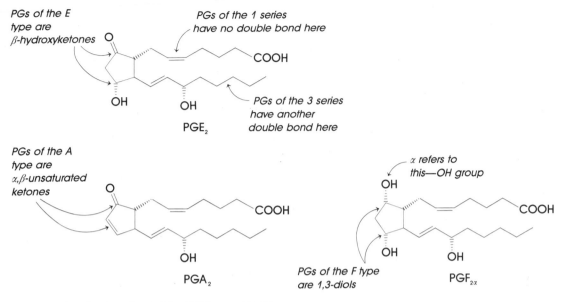

PGs of the E type are β-hydroxyketones

PGs of the 1 series have no double bond here

COOH

OH   OH

PGE₂

PGs of the 3 series have another double bond here

PGs of the A type are α,β-unsaturated ketones

COOH

OH

PGA₂

α refers to this—OH group

OH

COOH

OH   OH

PGs of the F type are 1,3-diols

PGF₂α

Figure 1 Prostaglandins PGA₂, PGE₂, and PGF₂α.

Figure 2 A molecular model of prostaglandin E₁ is examined by Dr. John Pike, Head of Experimental Chemistry Research, The Upjohn Company. (Courtesy of The Upjohn Company)

Figure 3 Found in the Caribbean Sea, *Plexaura homomalla*, known as the sea whip or gorgonian, contains the highest concentration of prostaglandin-like substances so far found in nature. Before synthetic production of prostaglandins, Upjohn extracted the rare substances from this soft coral. (Courtesy of The Upjohn Company)

gonian sea whip or sea fan, *Plexaura homomalla*, which grows on the coral reefs off the coast of Florida and in the Caribbean, is a rich source of prostaglandin-like materials (Figure 3). The concentration of PG-like substances in this marine organism is about 100 times the normal concentration found in most mammalian sources. In the laboratory, the PG-like compounds were extracted and then transformed into prostaglandins and prostaglandin analogs. At the present time, however, there is no need to depend on this natural source, for chemists have developed highly effective and stereospecific laboratory schemes for the synthesis of almost any prostaglandin or prostaglandin-like substance.

Prostaglandins are not stored as such in tissues; instead, they are synthesized in response to specific environmental or physiological triggers. Starting materials for prostaglandin synthesis are unsaturated fatty acids of twenty carbon atoms. Prostaglandins of the 2-series are derived from arachidonic acid (5,8,11,14-eicosatetraenoic acid), an unsaturated fatty acid containing four carbon-carbon double bonds. Steps in the biochemical pathways by which arachidonic acid is converted to several key prostaglandins are summarized in Figure 4. Arachidonic acid is drawn in this figure in such a way as to show the relationship between its structural formula and that of the prostaglandins derived from it.

A key step in the biosynthesis of prostaglandins of the 2-series is reaction of arachidonic acid with two molecules of oxygen, O_2, to form PGG_2. This complex reaction, catalyzed by the enzyme cyclooxygenase, is shown in de-

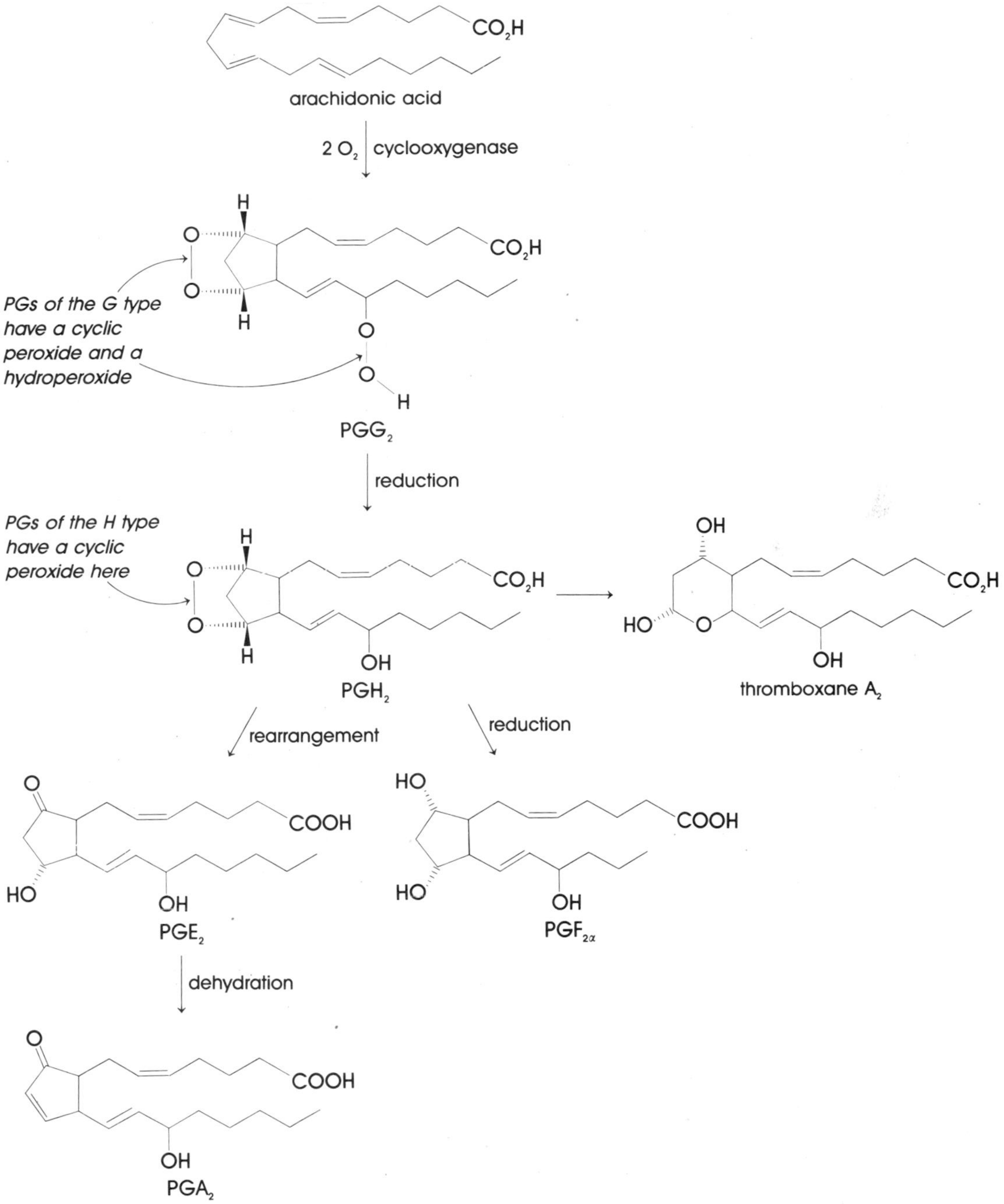

Figure 4 Biosynthesis of several prostaglandins from arachidonic acid.

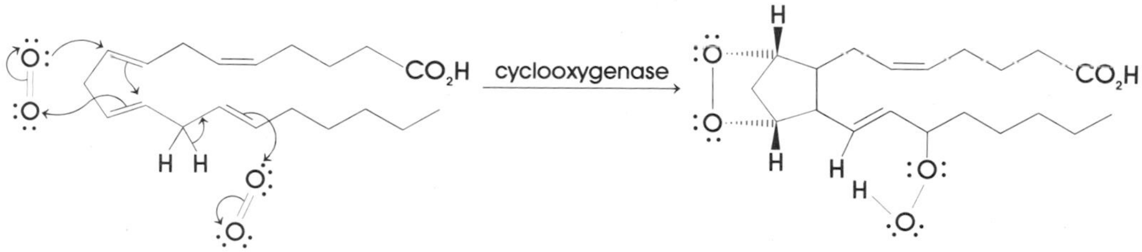

Figure 5 A schematic drawing illustrating the transformations catalyzed by cyclooxygenase.

tail in Figure 5, with arrows showing where bonds are broken and new ones formed. The actual mechanism of the reaction catalyzed by cyclooxygenase is considerably more complicated than is indicated by the arrows.

Enzyme-catalyzed reduction of PGG_2 gives PGH_2. This substance is a key intermediate from which all other prostaglandins of the 2-series are synthesized. Within minutes it is converted into other prostaglandins and prostaglandin-derived substances. Figure 4 shows the biosynthesis of types A, E, F, G, and H. There are other types, also derived from the key intermediate PGH_2. Precisely which prostaglandins or prostaglandin-like substances are produced depends on the enzymes present in a particular tissue. In blood platelets, for example, primarily thromboxane A_2 is synthesized.

Now that we have seen the types of transformations by which the body synthesizes prostaglandins, let us look at several functions of these substances within the body.

First is the participation of prostaglandins in blood clotting. There are three distinct phases to the physiological mechanisms that the body uses to stop bleeding from a ruptured blood vessel. The first phase, called **platelet aggregation,** is initiated by agents such as thrombin. During platelet aggregation, blood platelets become sticky and form a platelet plug at the site of the injury. If damage is minor and the blood vessel is small, a platelet plug may be sufficient to stop loss of blood from the vessel. If it is not sufficient, the platelets are stimulated to release a group of substances (the **platelet release reaction**) which, in turn, promote a second wave of platelet aggregation and constriction of the injured vessel. The third phase is the triggering of the **blood coagulation process.**

We have learned within the past few years that among the substances released in platelet release reactions is thromboxane A_2. This prostaglandin-derived molecule is a very potent vasoconstrictor and the key substance in triggering platelet aggregation. Just as we have known for some time that thrombin stimu-

lates the second and irreversible phase of platelet aggregation, we have also known that aspirin and aspirin-like drugs such as indomethacin inhibit this second phase. How these drugs are able to do this remained a mystery until it was discovered that aspirin inhibits cyclooxygenase, the enzyme that initiates the synthesis of thromboxane A_2.

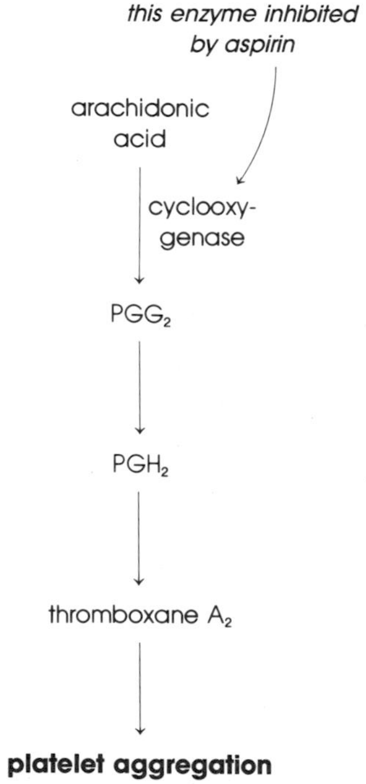

this enzyme inhibited by aspirin

arachidonic acid

↓

cyclooxy-genase

↓

PGG_2

↓

PGH_2

↓

thromboxane A_2

↓

platelet aggregation

There is now good evidence that the ability of aspirin to reduce inflammation is also related to its ability to inhibit prostaglandin synthesis. Further research on prostaglandins may help us understand even more about inflammatory diseases such as rheumatoid arthritis, asthma, and other allergic responses.

As indicated in the introduction to this essay, the first recorded observations of the biological activity of prostaglandins were those of gynecologists Kurzrok and Lieb. The first widespread clinical application of these substances was also by gynecologists and obstetricians. The observation that prostaglandins stimulate contraction of uterine smooth muscle led to the suggestion that these substances might be used for termination of second-trimester pregnancy. One problem with the use of naturally occurring prostaglandins for this purpose is that they are very rapidly degraded within the body. Therefore, their use required repeated administration over a period of hours. In the search for less rapidly degraded prostaglandins, a number of semi-synthetic prostaglandin analogs were synthesized. One of the most effective of these was 15-methyl prostaglandin $F_{2\alpha}$ which is longer acting and has 10 to 20 times the potency of $PGF_{2\alpha}$.

The potential clinical use of prostaglandins and prostaglandin analogs for termination of second trimester pregnancy was explored in a

an extra methyl group at carbon 15

15-methyl prostaglandin $F_{2\alpha}$

study designed and conducted by the World Health Organization Task Force on the Use of Prostaglandins for the Regulation of Fertility. This multicenter, multinational study, entitled "Prostaglandins and Abortion," is described in the *American Journal of Obstetrics and Gynecology* (1977) and concludes that a single intraamniotic injection of 15-methyl $PGF_{2\alpha}$ is a safe and effective means for termination of second-trimester pregnancy.

In this mini-essay we have looked at only a few aspects of the biosynthesis of prostaglandins and their importance in human physiology. From even this brief encounter, it should be clear that we are only beginning to understand the chemistry and biochemistry of these substances. It should also be clear that the enormous prostaglandin research effort now under way offers great promise for even deeper insight into human physiology and for the development of new and highly effective drugs for use in clinical medicine.

References

Bergstrom, S. 1967. Prostaglandins: members of a new hormonal system. *Science* 157: 382.

Bergstrom, S.; Carlson, L.A.; and Weeks, J.R. 1968. The prostaglandins: a family of biologically active lipids. *Pharmacological Review* 20: 1.

Goodman, L.S. and Gilman, A. 1975. *The pharmacological basis of therapeutics.* 5th ed. New York: Macmillan.

Kuehl, F.A., and Egan, R.W. 1980. Prostaglandins, arachidonic acid, and inflammation. *Science* 210: 978–984.

Needleman, P., et al. 1976. Prostaglandins and abortion. *Nature* 261: 558–560.

Ramwell, P.W., ed. 1973. *The Prostaglandins.* Vol. I. New York: Plenum Press.

Samuellson, B., et al. 1975. Prostaglandins. *Annual Review of Biochemistry* 44: 669–695.

World Health Organization. 1977. Prostaglandins and abortion. *The American Journal of Obstetrics and Gynecology* 129: 593–606.

Functional Derivatives of Carboxylic Acids

In this chapter we shall describe the structures and chemical properties of organic *esters*, *amides*, and *anhydrides*—all functional derivatives of carboxylic acids in which the —OH of the carboxyl group has been replaced by —OR, —NH_2, or OCOR.

$$\underset{\text{ester}}{R-\overset{\displaystyle O}{\overset{\|}{C}}-OR} \qquad \underset{\text{amide}}{R-\overset{\displaystyle O}{\overset{\|}{C}}-NH_2} \qquad \underset{\text{anhydride}}{R-\overset{\displaystyle O}{\overset{\|}{C}}-O-\overset{\displaystyle O}{\overset{\|}{C}}-R}$$

17.1 Nomenclature

In the IUPAC system of nomenclature, **esters** are named as derivatives of carboxylic acids. The alkyl or aryl group attached to oxygen is named first, then the acid from which the ester is derived. The parent acid is indicated by dropping the suffix *-ic* from the IUPAC or common name of the acid and adding *-ate*.

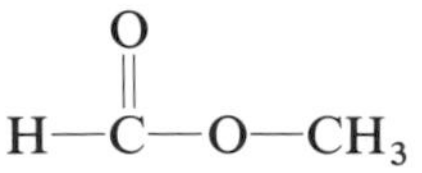

$$H-\overset{\overset{\displaystyle O}{\|}}{C}-O-CH_3 \qquad CH_3-\overset{\overset{\displaystyle O}{\|}}{C}-O-CH_2-CH_3$$

methyl formate 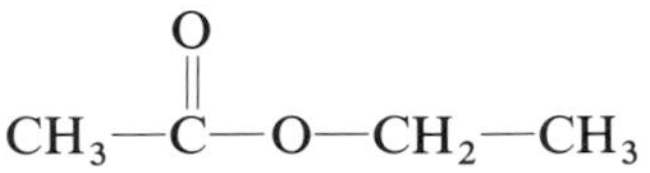   ethyl acetate

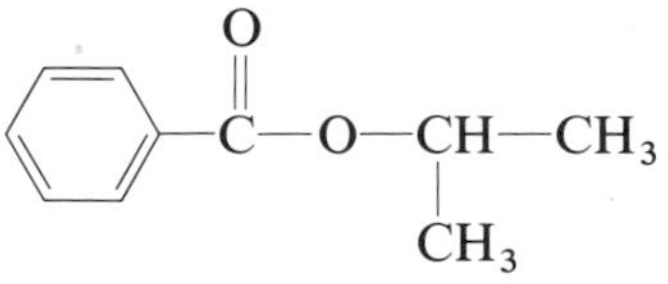

$$\text{(phenyl)}-\overset{\overset{\displaystyle O}{\|}}{C}-O-\underset{\underset{\displaystyle CH_3}{|}}{CH}-CH_3 \qquad CH_3-\overset{\overset{\displaystyle O}{\|}}{C}-O-\text{(phenyl)}$$

isopropyl benzoate 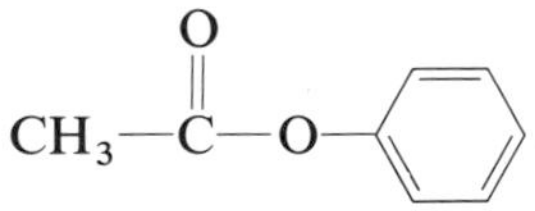   phenyl acetate

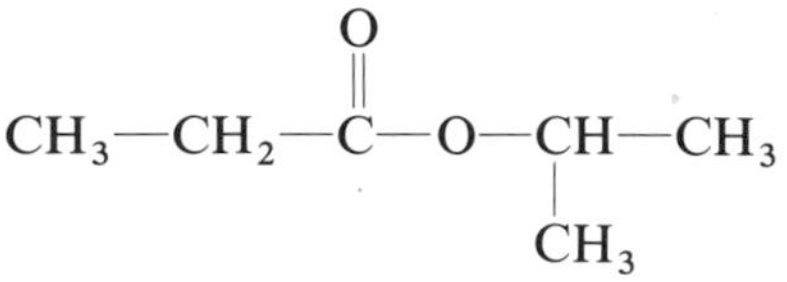

$$CH_3-CH_2-\overset{\overset{\displaystyle O}{\|}}{C}-O-\underset{\underset{\displaystyle CH_3}{|}}{CH}-CH_3 \qquad \underset{CH_2-\overset{\overset{\displaystyle O}{\|}}{C}-O-CH_2CH_3}{\overset{CH_2-\overset{\overset{\displaystyle O}{\|}}{C}-O-CH_2CH_3}{|}}$$

isopropyl propanoate 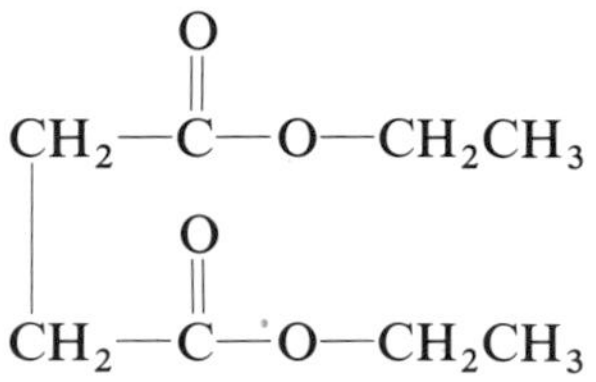   diethyl butanedioate
                                     (diethyl succinate)

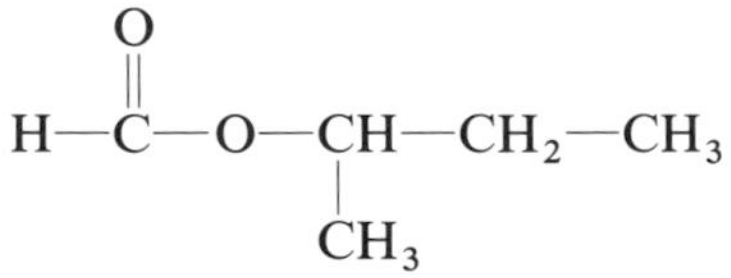

Example 17.1

Name the following esters.

a. 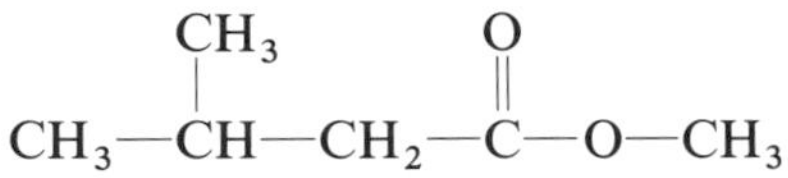 $CH_3-\underset{\underset{\displaystyle CH_3}{|}}{CH}-CH_2-\overset{\overset{\displaystyle O}{\|}}{C}-O-CH_3$

b. $H-\overset{\overset{\displaystyle O}{\|}}{C}-O-\underset{\underset{\displaystyle CH_3}{|}}{CH}-CH_2-CH_3$ c. $CH_3(CH_2)_6\overset{\overset{\displaystyle O}{\|}}{C}OCH_3$

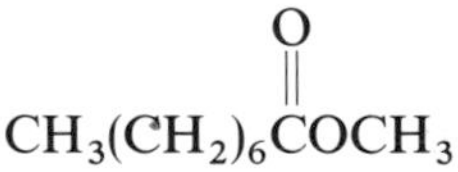

Solution

Names of these esters are derived below in a stepwise manner. First is given the name of the alkyl group attached to oxygen; then the IUPAC name of the carboxylic acid from which the ester is derived; and finally, the name of the ester.

Alkyl Group Attached to Oxygen	IUPAC Name of Carboxylic Acid	Name of Ester
a. methyl	3-methylbutanoic acid	methyl 3-methylbutanoate
b. *sec*-butyl	formic acid	*sec*-butyl formate
c. methyl	octanoic acid	methyl octanoate

Problem 17.1

Name the following esters.

a. 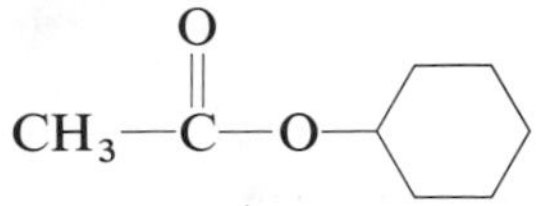 $CH_3-\overset{\displaystyle O}{\overset{\displaystyle \|}{C}}-O-\langle\text{cyclohexyl}\rangle$

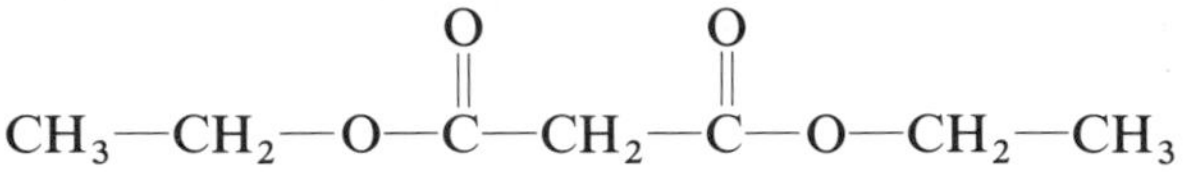

b. $CH_3-CH_2-O-\overset{\displaystyle O}{\overset{\displaystyle \|}{C}}-CH_2-\overset{\displaystyle O}{\overset{\displaystyle \|}{C}}-O-CH_2-CH_3$

c. $CH_3-CH_2-CH_2-CH_2-\overset{\displaystyle O}{\overset{\displaystyle \|}{C}}-O-\overset{\displaystyle CH_3}{\underset{\displaystyle CH_3}{C}}-CH_3$

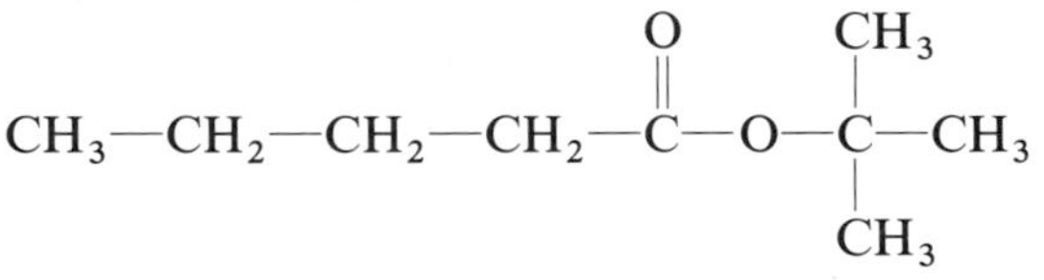

Esters of thiols are named by adding the prefix *thio-* to the name of the parent carboxylic acid to indicate the presence of a sulfur atom.

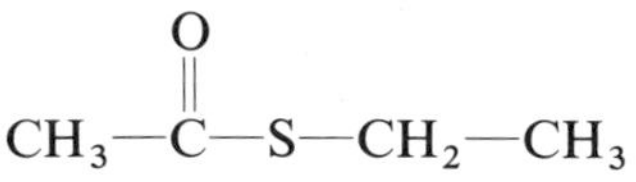

$$CH_3-\overset{\displaystyle O}{\overset{\displaystyle \|}{C}}-S-CH_2-CH_3$$

ethyl thioacetate

Amides are named as derivatives of carboxylic acids by dropping the suffix *-oic* from the IUPAC name of the acid, or the suffix *-ic* from the common name of the acid, and adding *-amide*.

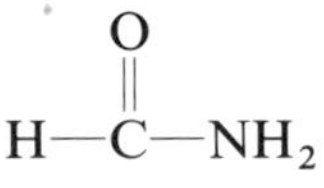

$H-\overset{\displaystyle O}{\overset{\displaystyle \|}{C}}-NH_2$

formamide

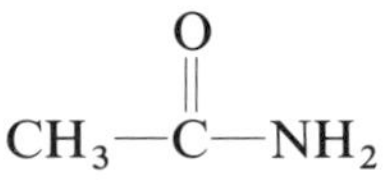

$CH_3-\overset{\displaystyle O}{\overset{\displaystyle \|}{C}}-NH_2$

acetamide

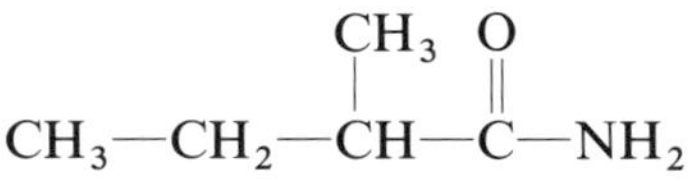

$CH_3-CH_2-\overset{\displaystyle CH_3}{\underset{\displaystyle |}{CH}}-\overset{\displaystyle O}{\overset{\displaystyle \|}{C}}-NH_2$

2-methylbutanamide

$\langle\text{phenyl}\rangle-\overset{\displaystyle O}{\overset{\displaystyle \|}{C}}-NH_2$

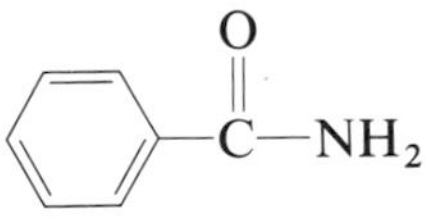

benzamide

If the nitrogen atom is substituted with an alkyl or aryl group, the substituent is named and its location on nitrogen is indicated by a capital N.

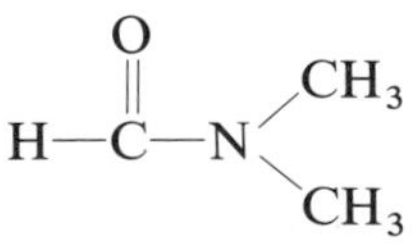

$H-\overset{\displaystyle O}{\overset{\displaystyle \|}{C}}-N\overset{\displaystyle CH_3}{\underset{\displaystyle CH_3}{\big\langle}}$

N,N-dimethylformamide

$CH_3CH_2CH_2\overset{\displaystyle O}{\overset{\displaystyle \|}{C}}-NH-\langle\text{phenyl}\rangle$

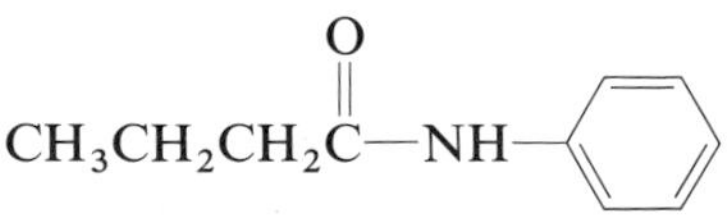

N-phenylbutanamide

Example 17.2

Name the following amides.

a. $CH_3-\overset{\displaystyle O}{\overset{\|}{C}}-NH-CH_3$

b. $\begin{array}{c} H_3C \\ \diagdown \\ \diagup \\ H_3C \end{array} CH-CH_2-\overset{\displaystyle O}{\overset{\|}{C}}-NH_2$

c. $CH_3-\overset{\displaystyle \overset{H_3C}{|}}{\underset{\underset{\displaystyle CH_3}{|}}{C}}-\overset{\displaystyle O}{\overset{\|}{C}}-NH_2$

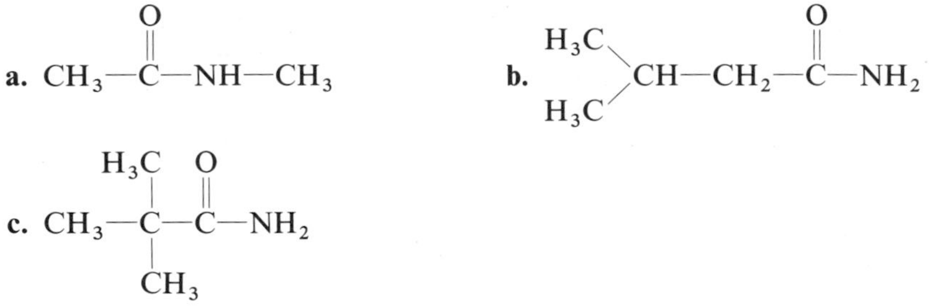

Solution

Names of these amides are derived below in a stepwise manner, just as we did those of the esters in Example 17.1. First is given the name of the alkyl or aryl group (if any) attached to nitrogen; then the name of the carboxylic acid from which the amide is derived; and finally, the name of the amide.

Alkyl or Aryl Group Attached to Nitrogen	IUPAC Name of Carboxylic Acid	Name of Amide
a. N-methyl	acetic acid	N-methylacetamide
b. —	3-methylbutanoic acid	3-methylbutanamide
c. —	2,2-dimethylpropanoic acid	2,2-dimethylpropanamide

Problem 17.2

Name the following amides.

a. $CH_3-\overset{\displaystyle O}{\overset{\|}{C}}-NH-\bigcirc$

b. $CH_3(CH_2)_4\overset{\displaystyle O}{\overset{\|}{C}}NH_2$

c. (structure: benzene ring with $\overset{\displaystyle O}{\overset{\|}{C}}-NH_2$ and OH substituents)

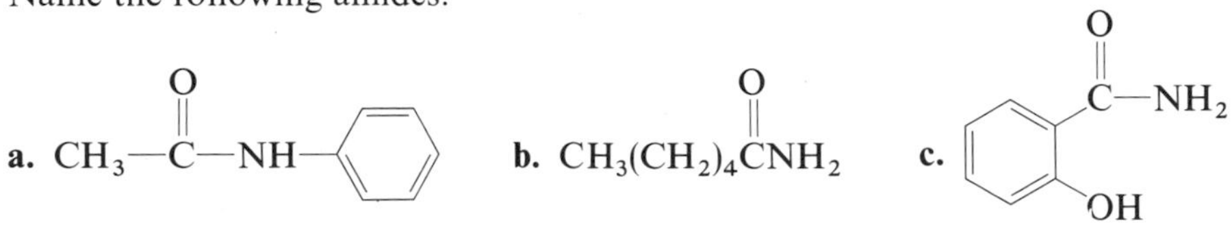

Anhydrides are named by adding the word *anhydride* to the name of the parent acid. For our purposes, the most important organic anhydride is acetic anhydride.

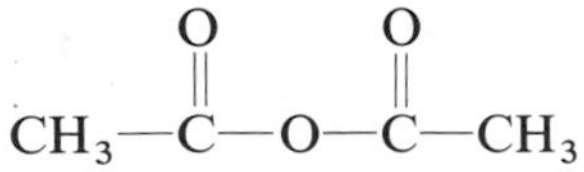

$$CH_3-\overset{\displaystyle O}{\overset{\|}{C}}-O-\overset{\displaystyle O}{\overset{\|}{C}}-CH_3$$
acetic anhydride

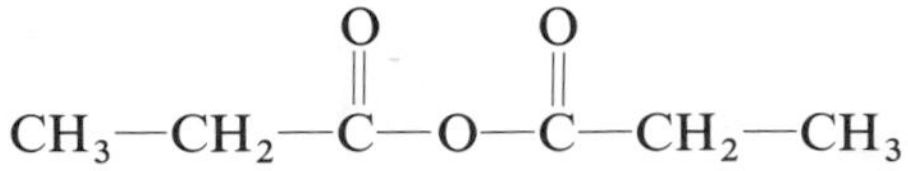

$$CH_3-CH_2-\overset{\displaystyle O}{\overset{\|}{C}}-O-\overset{\displaystyle O}{\overset{\|}{C}}-CH_2-CH_3$$
propanoic anhydride

17.2 Physical Properties of Esters, Amides, and Anhydrides

Esters are polar molecules and are attracted to each other in the pure state by a combination of dipole-dipole interactions between polar $-CO_2-$ groups and dispersion forces between nonpolar hydrocarbon portions of molecules. Most esters are insoluble in water due to the hydrophobic character of the hydrocarbon portion of the molecule. Esters are soluble in polar organic solvents such as ether and acetone.

Low-molecular-weight esters have rather pleasant odors. The characteristic fragrances of many flowers and fruits, for example, are due to the presence of esters, either singly or in mixtures. Some of the more familiar esters are ethyl formate (artificial rum flavor), methyl butanoate (apples), octyl acetate (oranges), and ethyl butanoate (pineapples).

Amides are polar molecules and there is the possibility for hydrogen bonding between the partially positive hydrogen atom of one amide group and the partially negative oxygen atom of another amide.

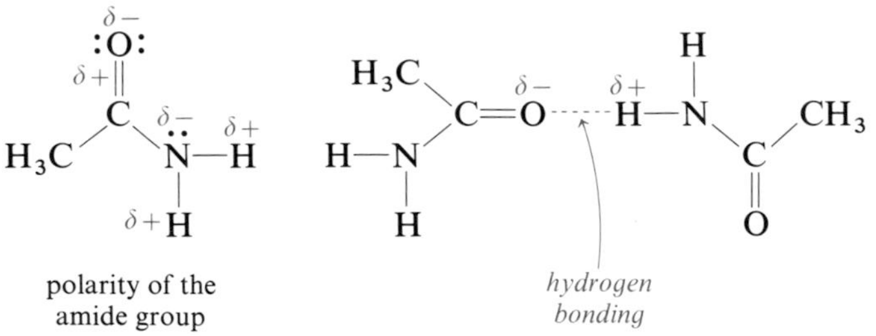

polarity of the amide group   hydrogen bonding

Because of this polarity and the resulting hydrogen bonding, amides have higher boiling points and are more soluble in water than esters of comparable molecular weight. Except for formamide (mp 3°C), all amides are solids at room temperature.

17.3 Nucleophilic Substitution at a Carbonyl Carbon

The reaction scheme common to the carbonyl group of aldehydes, ketones, carboxylic acids, esters, amides, and anhydrides is **nucleophilic addition to the carbonyl group.** In the case of aldehydes and ketones, the carbonyl addition product is sometimes the final product of the reaction. For example, in aldol condensations (Section 13.5E), the carbonyl group undergoing reaction is transformed into an alcohol, and this addition product is the final product of the reaction.

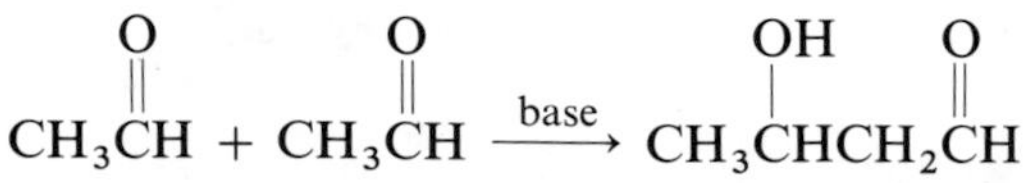

In other reactions of aldehydes and ketones, a carbonyl addition product is formed but then undergoes loss of H_2O or another small molecule to yield a new functional group. For example, the reaction of an aldehyde or ketone with a primary amine forms a carbonyl addition product which then loses a molecule of H_2O to form a Schiff base; in this reaction, a $>C=O$ group is transformed into a $>C=N-$ group.

$$H_3C \atop H_3C \Big\rangle C=O + H_2N-\underset{\underset{CH_3}{|}}{\overset{\overset{CH_3}{|}}{CH}} \longrightarrow \left[CH_3 - \underset{\underset{CH_3}{|}}{\overset{\overset{OH}{|}}{C}} - \underset{H}{N} - \underset{\underset{CH_3}{|}}{\overset{\overset{CH_3}{|}}{CH}} \right] \longrightarrow {H_3C \atop H_3C} \Big\rangle C=N-\underset{\underset{CH_3}{|}}{\overset{\overset{CH_3}{|}}{CH}} + H_2O$$

a Schiff base

With the new functional groups to be studied in this chapter, the carbonyl product collapses to regenerate the carbonyl group.

$$\underset{R}{\overset{O}{\underset{\|}{C}}}{\diagdown}_Y + H-Z \longrightarrow \left[\overset{HO}{\underset{R}{\diagup}} \overset{Z}{\underset{Y}{C}} \right] \longrightarrow \underset{R}{\overset{O}{\underset{\|}{C}}}{\diagdown}_Z + H-Y$$

tetrahedral carbonyl
addition intermediate

general reaction

The effect of this reaction is to substitute a new atom or group of atoms for one already attached to the carbonyl group. For this reason, we characterize these reactions as **nucleophilic substitution at a carbonyl carbon.**

17.4 Preparation of Esters – Fischer Esterification

A carboxylic acid can be converted into an ester by heating with an alcohol in the presence of an acid catalyst, usually dry hydrogen chloride, concentrated sulfuric acid, or an ion-exchange resin in the hydrogen ion form. Direct esterification of alcohols and acids in this manner is called **Fischer esterification.** As an example, reaction of acetic acid and ethanol in the presence of concentrated sulfuric acid produces ethyl acetate and water. Fischer esterification is an example of the general reaction cited in the previous section. In the conversion of acetic acid and ethanol to ethyl acetate and water, one group of atoms ($-OCH_2CH_3$) is substituted at the carbonyl carbon for another group of atoms ($-OH$).

$$\underset{\text{acetic acid}}{CH_3-\overset{\overset{O}{\|}}{C}-OH} + \underset{\text{ethanol}}{HO-CH_2-CH_3} \overset{H^+}{\rightleftharpoons} \underset{\text{ethyl acetate}}{CH_3-\overset{\overset{O}{\|}}{C}-O-CH_2-CH_3} + H-O-H$$

Acid-catalyzed esterification is reversible, and generally at equilibrium there are appreciable quantities of both ester and acid present. If 60.6 g (one mole) of acetic acid and 60.0 g (one mole) of 1-propanol are refluxed for a short time in the presence of a few drops of concentrated sulfuric acid, the reaction mixture at equilibrium contains about 68.0 g (0.67 mole) of propyl acetate, 12.0 g (0.67 mole) of water, and 20.0 g (0.33 mole) each of acetic acid and 1-propanol. At equilibrium there is about 67% conversion of acid and alcohol into ester.

$$CH_3\overset{\overset{\displaystyle O}{\|}}{C}-OH + HOCH_2CH_2CH_3 \underset{}{\overset{H^+}{\rightleftharpoons}} CH_3\overset{\overset{\displaystyle O}{\|}}{C}-OCH_2CH_2CH_3 + H_2O$$

initial:	1.00 mole	1.00 mole	0.00 mole	0.00 mole
equilibrium:	0.33 mole	0.33 mole	0.67 mole	0.67 mole

By careful control of reaction conditions, direct esterification can be used to prepare esters in high yield. For example, if the alcohol is inexpensive, a large excess of it can be used to drive the reaction to the right and achieve a high conversion of acid into ester. Or, it may be possible to take advantage of a situation in which the boiling points of the reactants and ester are higher than that of water. Heating the reaction mixture above 100°C removes water as the product is formed and shifts the equilibrium toward the production of higher yields of ester.

Example 17.3

Name and draw structural formulas for the esters produced in the following reactions.

a. $HCO_2H + CH_3CH_2OH \xrightarrow{H^+}$ **b.** 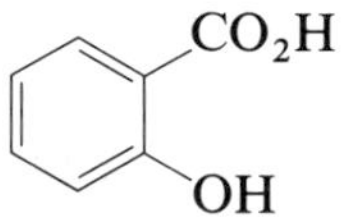 [benzene ring with CO_2H and OH substituents] $+ CH_3OH \xrightarrow{H^+}$

c. $HO\overset{\overset{\displaystyle O}{\|}}{C}CH_2CH_2CH_2\overset{\overset{\displaystyle O}{\|}}{C}OH + 2\ CH_3CH_2OH \xrightarrow{H^+}$

Solution

Ester formation follows the general reaction shown in Section 17.3:

$$\textit{general reaction:}\quad R-\overset{\overset{\displaystyle O}{\|}}{C}-Y + H-Z \longrightarrow R-\overset{\overset{\displaystyle O}{\|}}{C}-Z + H-Y$$

a. $H-\overset{\overset{\displaystyle O}{\|}}{C}-OH + H-OCH_2CH_3 \longrightarrow H-\overset{\overset{\displaystyle O}{\|}}{C}-OCH_2CH_3 + H-OH$

formic   ethanol   ethyl
acid   formate

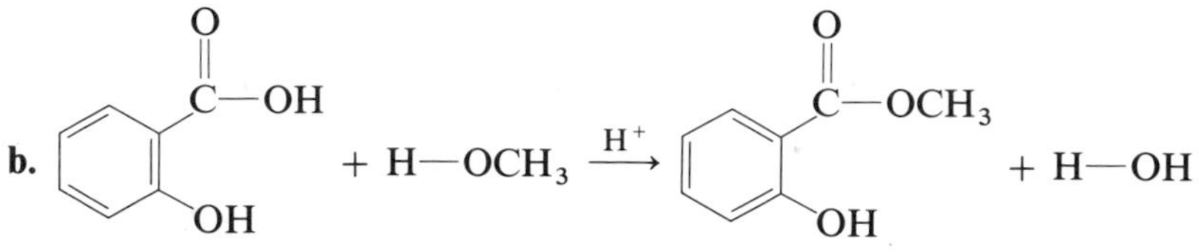

b.

salicylic acid + H—OCH$_3$ $\xrightarrow{\text{H}^+}$ methyl salicylate + H—OH

salicylic
acid   methanol   methyl
salicylate

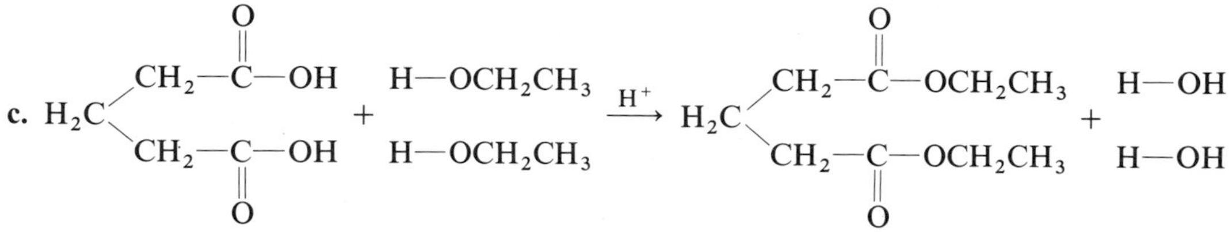

c. H$_2$C

pentanedioic acid
(glutaric acid)

+ H—OCH$_2$CH$_3$
    H—OCH$_2$CH$_3$

ethanol

$\xrightarrow{\text{H}^+}$ H$_2$C

diethyl pentanedioate
(diethyl glutarate)

+ H—OH
    H—OH

Problem 17.3

Name and draw structural formulas for the esters produced in the following reactions.

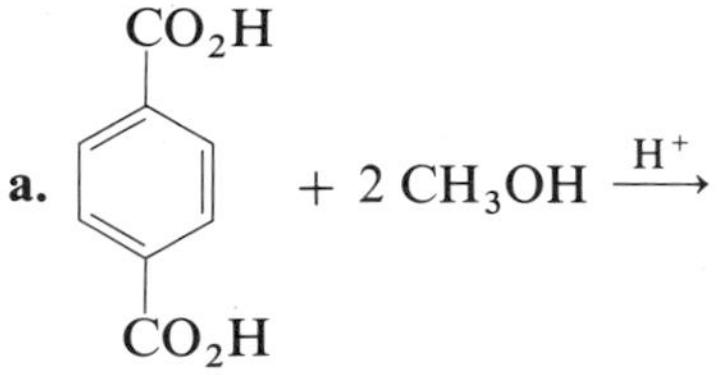

a.

CO$_2$H

CO$_2$H

+ 2 CH$_3$OH $\xrightarrow{\text{H}^+}$

b. CH$_3$CO$_2$H + CH$_3$CH$_2$CH$_2$CH$_2$CH$_2$OH $\xrightarrow{\text{H}^+}$

c. CH$_3$CH$_2$CO$_2$H + CH$_3$CHCH$_2$CH$_2$OH $\xrightarrow{\text{H}^+}$
                                 |
                                CH$_3$

A key step in the mechanism of Fischer esterification is formation of a tetrahedral carbonyl addition intermediate.

CH$_3$—C—OH + HO—CH$_3$ $\xrightleftharpoons{\text{H}^+}$ $\left[\text{CH}_3-\underset{\text{OH}}{\overset{\text{OH}}{\text{C}}}-\text{O}-\text{CH}_3 \right]$ $\xrightleftharpoons{\text{H}^+}$ CH$_3$—C—O—CH$_3$ + H—OH

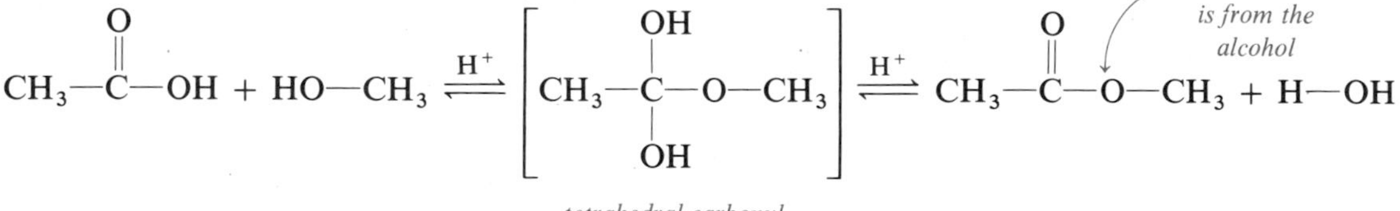

this oxygen is from the alcohol

*tetrahedral carbonyl
addition intermediate*

This mechanism predicts that the molecule of water formed during the reaction is derived from the —OH of the carboxylic acid and the —H of the alcohol.

This prediction has been tested in the following way. Oxygen in nature is a mixture of three isotopes (Section 2.6D): 99.7% ^{16}O, 0.04% ^{17}O, and 0.20% ^{18}O. Through the use of modern techniques of isotope separation, it is possible to prepare oxygen-containing compounds significantly enriched in oxygen-18. One of these is H—$^{18}OCH_3$. When methanol enriched with oxygen-18 reacts with acetic acid containing only ordinary oxygen, all of the oxygen-18 is found in the ester. The water formed in the reaction contains none of the oxygen-18.

$$CH_3-\overset{\overset{\displaystyle O}{\|}}{C}-OH + H-^{18}OCH_3 \longrightarrow CH_3-\overset{\overset{\displaystyle O}{\|}}{C}-^{18}OCH_3 + H_2O$$

Example 17.4

Draw structural formulas for the tetrahedral carbonyl addition intermediates formed in the following acid-catalyzed esterifications.

a. $CH_3CH_2\overset{\overset{\displaystyle O}{\|}}{C}OH + HOCH_2CH_3 \xrightarrow{H^+} CH_3CH_2\overset{\overset{\displaystyle O}{\|}}{C}OCH_2CH_3 + H_2O$

b. $\langle\!\!\!\bigcirc\!\!\!\rangle-\overset{\overset{\displaystyle O}{\|}}{C}OH + HOCH_3 \xrightarrow{H^+} \langle\!\!\!\bigcirc\!\!\!\rangle-\overset{\overset{\displaystyle O}{\|}}{C}OCH_3 + H_2O$

Solution

In the solutions below, the starting materials are drawn to emphasize that the oxygen of the —OH group adds to the carbonyl carbon and hydrogen adds to the carbonyl oxygen.

a. $CH_3-CH_2-\overset{\overset{\displaystyle O}{\|}}{\underset{\underset{\displaystyle OH}{|}}{C}} + \overset{\overset{\displaystyle H}{|}}{O}-CH_2CH_3 \longrightarrow \left[CH_3-CH_2-\overset{\overset{\displaystyle O-H}{|}}{\underset{\underset{\displaystyle OH}{|}}{C}}-O-CH_2CH_3 \right]$

b. $\langle\!\!\!\bigcirc\!\!\!\rangle-\overset{\overset{\displaystyle O}{\|}}{\underset{\underset{\displaystyle OH}{|}}{C}} + \overset{\overset{\displaystyle H}{|}}{O}-CH_3 \longrightarrow \left[\langle\!\!\!\bigcirc\!\!\!\rangle-\overset{\overset{\displaystyle O-H}{|}}{\underset{\underset{\displaystyle OH}{|}}{C}}-O-CH_3 \right]$

Problem 17.4

Draw structural formulas for the tetrahedral carbonyl addition intermediates formed in the following acid-catalyzed esterifications.

a. $H\overset{\overset{\displaystyle O}{\|}}{C}OH + HO\underset{\underset{\displaystyle CH_3}{|}}{C}HCH_3 \xrightarrow{H^+} H\overset{\overset{\displaystyle O}{\|}}{C}O\underset{\underset{\displaystyle CH_3}{|}}{C}HCH_3 + H_2O$

b.

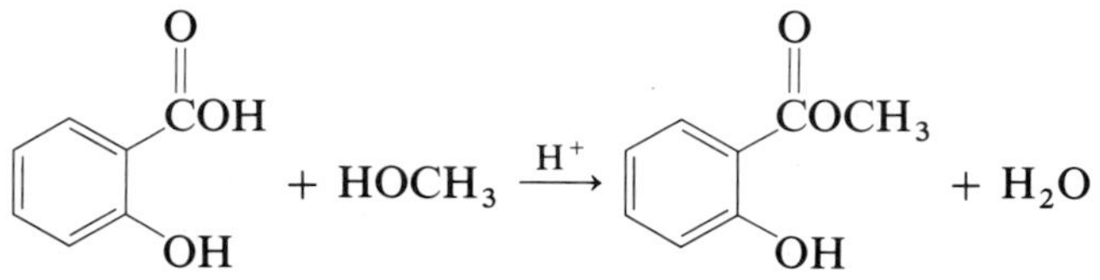

$$\text{(2-COH on benzene ring with OH)} + HOCH_3 \xrightarrow{H^+} \text{(2-COCH}_3 \text{ on benzene ring with OH)} + H_2O$$

17.5 Reactions of Esters

A. Reduction

Esters can be reduced by hydrogen in the presence of a catalyst; however, as with carboxylic acids, catalytic reduction of esters requires high temperatures and pressures of hydrogen. Esters are reduced easily at room temperature by $LiAlH_4$. Reduction of an ester gives two alcohols, one derived from the carboxylic acid portion of the ester, the other derived from the alkyl group on oxygen.

$$CH_3CH_2O\overset{O}{\overset{||}{C}}(CH_2)_4\overset{O}{\overset{||}{C}}OCH_2CH_3 \xrightarrow[\text{(2) } H_2O]{\text{(1) } LiAlH_4} HOCH_2(CH_2)_4CH_2OH + 2\ CH_3CH_2OH$$

diethyl adipate   1,6-hexanediol   ethanol

B. Hydrolysis

Esters are converted to the acids and alcohols by hydrolysis in either aqueous acid or base.

$$CH_3-\overset{O}{\overset{||}{C}}-O-CH_2CH_3 + H-O-H \underset{}{\overset{H^+}{\rightleftharpoons}} CH_3-\overset{O}{\overset{||}{C}}-OH + HO-CH_2CH_3$$

Since the mechanism proposed in Section 17.4 for acid-catalyzed esterification is reversible, the formation of the same tetrahedral carbonyl addition intermediate also occurs in acid-catalyzed hydrolysis.

$$CH_3-\overset{O}{\overset{||}{C}}-OCH_2CH_3 + H-OH \underset{}{\overset{H^+}{\rightleftharpoons}}$$

$$\left[CH_3-\overset{OH}{\underset{OH}{\overset{|}{\underset{|}{C}}}}-OCH_2CH_3 \right] \rightleftharpoons CH_3-\overset{O}{\overset{||}{C}}-OH + H-OCH_2CH_3$$

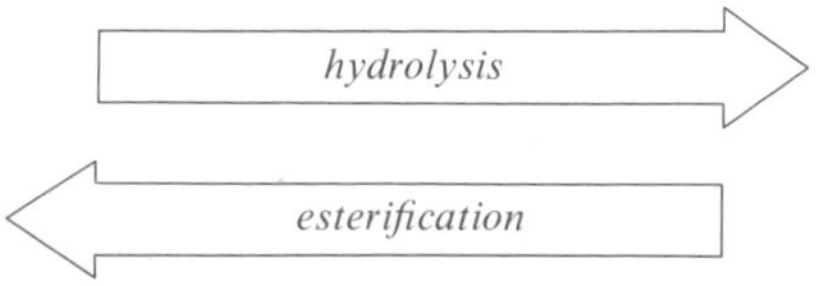

hydrolysis →

← *esterification*

By carrying out acid-catalyzed hydrolysis in a large excess of water, the position of equilibrium is shifted to favor formation of the carboxylic acid and alcohol.

In alkaline hydrolysis of esters, hydroxide ion adds to the carbonyl carbon to form a tetrahedral carbonyl addition intermediate. This intermediate then eliminates a molecule of alcohol to form the carboxylate anion, $RCOO^-$.

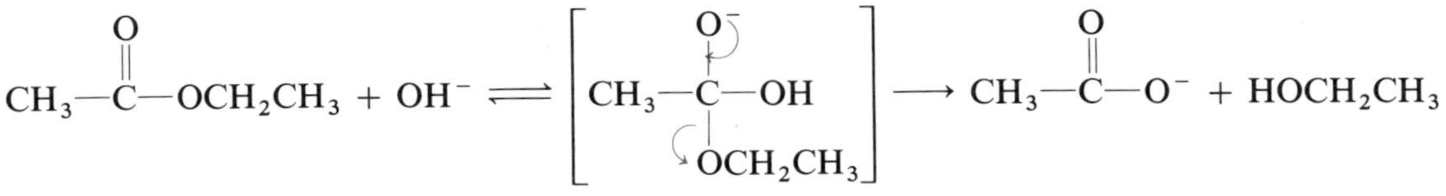

$$CH_3-\overset{\overset{\displaystyle O}{\|}}{C}-OCH_2CH_3 + OH^- \rightleftharpoons \left[CH_3-\overset{\overset{\displaystyle O^-}{|}}{\underset{\underset{\displaystyle OCH_2CH_3}{|}}{C}}-OH\right] \longrightarrow CH_3-\overset{\overset{\displaystyle O}{\|}}{C}-O^- + HOCH_2CH_3$$

tetrahedral carbonyl
addition intermediate

This mechanism, like that for acid-catalyzed esterification and hydrolysis, involves cleavage of the C—O bond of the carboxylic acid portion of the molecule rather than the C—O bond of the alcohol. For practical purposes, alkaline hydrolysis of an ester is irreversible because a carboxylate anion, the final product, shows no tendency to react with alcohol.

Example 17.5

Write equations for acid-catalyzed hydrolysis of the following esters.

a. 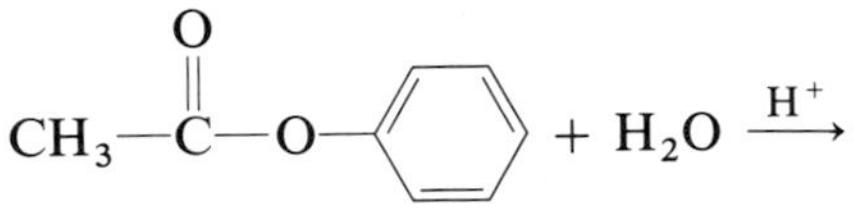 $CH_3-\overset{\overset{\displaystyle O}{\|}}{C}-O-\langle\bigcirc\rangle + H_2O \overset{H^+}{\longrightarrow}$

b. $CH_3-\overset{\overset{\displaystyle O}{\|}}{C}-O-CH_2-CH_2-O-\overset{\overset{\displaystyle O}{\|}}{C}-CH_3 + 2H_2O \overset{H^+}{\longrightarrow}$

Solution

a. 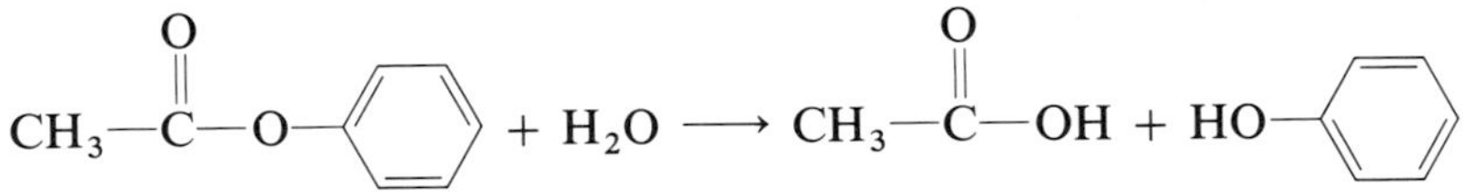 $CH_3-\overset{\overset{\displaystyle O}{\|}}{C}-O-\langle\bigcirc\rangle + H_2O \longrightarrow CH_3-\overset{\overset{\displaystyle O}{\|}}{C}-OH + HO-\langle\bigcirc\rangle$

b. $CH_3-\overset{\overset{\displaystyle O}{\|}}{C}-O-CH_2-CH_2-O-\overset{\overset{\displaystyle O}{\|}}{C}-CH_3 + 2H_2O \longrightarrow$

$$2\,CH_3-\overset{\overset{\displaystyle O}{\|}}{C}-OH + HO-CH_2-CH_2-OH$$

This substance is a diester and on hydrolysis yields two molecules of acetic acid and one of ethylene glycol.

Problem 17.5   Write equations for alkaline hydrolysis of the following esters.

a.

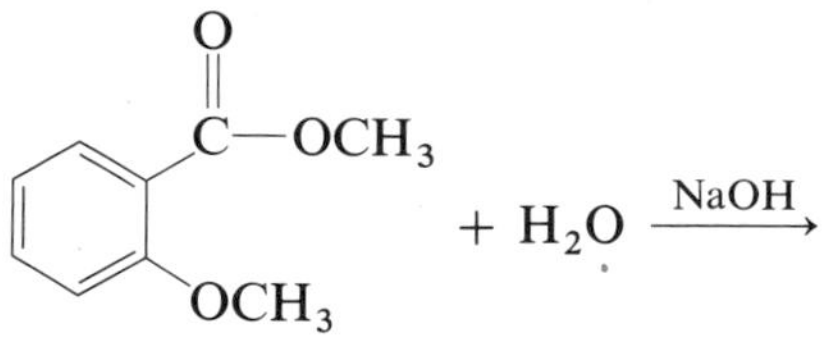

$+ H_2O \xrightarrow{\text{NaOH}}$

b.

$+ 3 H_2O \xrightarrow{\text{NaOH}}$

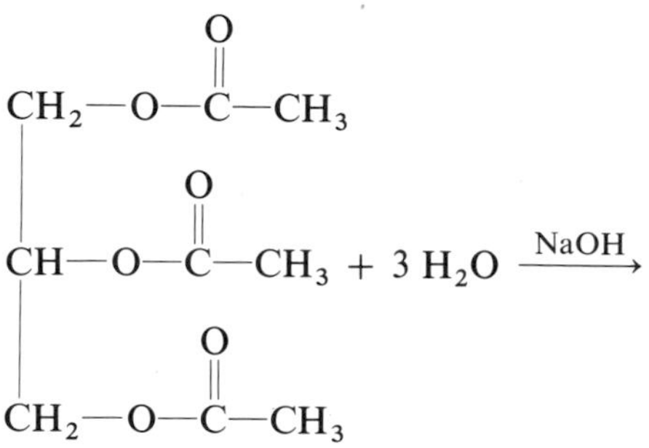

C. Ammonolysis

Reaction with ammonia converts an ester into an amide. This reaction is similar to hydrolysis and is called **ammonolysis.** Ammonia is a strong nucleophile and adds directly to the carbonyl carbon; no catalyst is necessary.

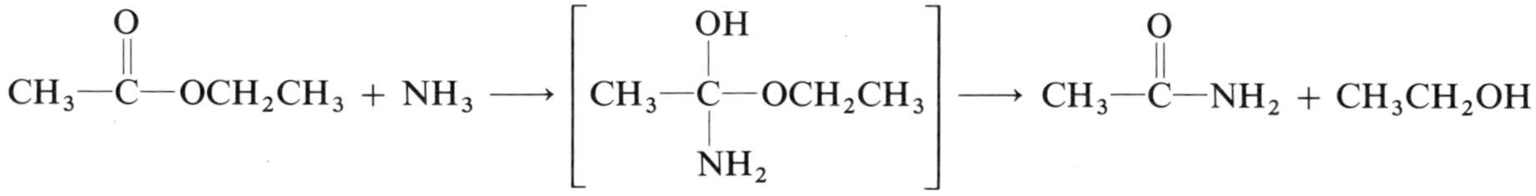

ethyl acetate   *tetrahedral carbonyl*   acetamide
                                 addition intermediate

Another example of ammonolysis of an ester is the laboratory synthesis of barbituric acid and barbiturates. Heating urea and diethyl malonate at 110°C in the presence of sodium ethoxide yields barbituric acid.

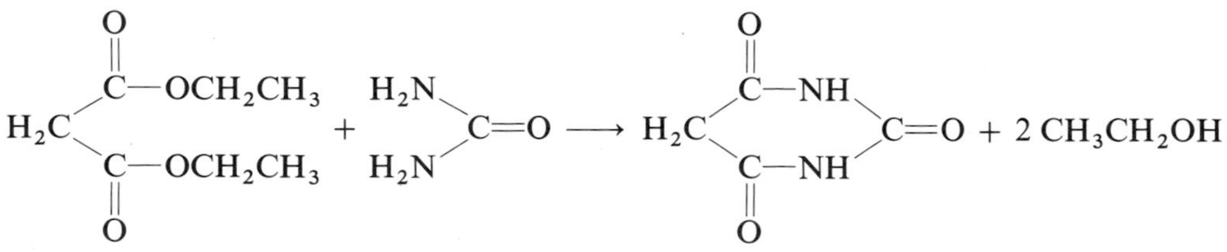

diethyl malonate   urea   barbituric acid

Mono- and disubstituted malonic esters yield substituted barbituric acids known as **barbiturates.**

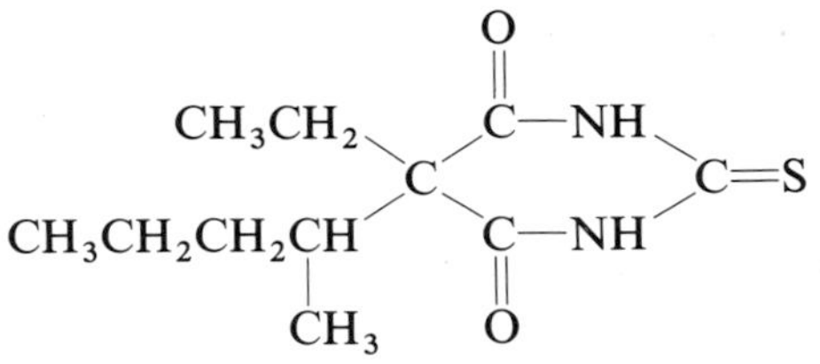

thiopental
(Penthothal)

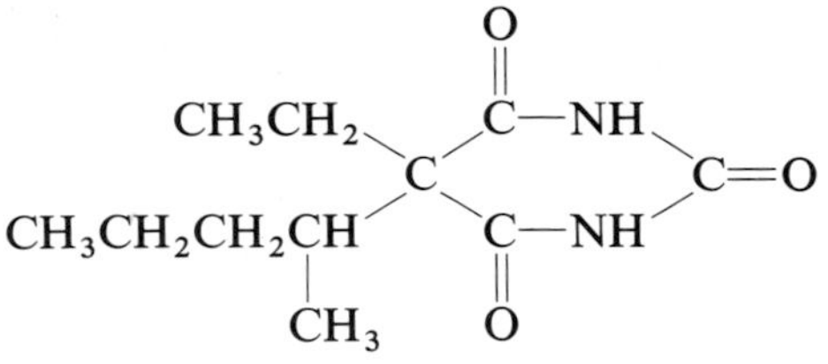

pentobarbital
(Nembutal)

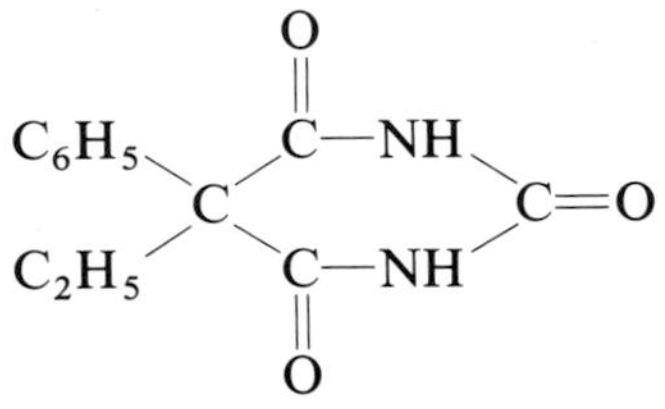

phenobarbital
(Luminal)

Barbiturates produce effects ranging from mild sedation to deep anesthesia and even death, depending on the dose and the particular barbiturate. Sedation, long- or short-acting, depends on the structure of the barbiturate. Phenobarbital is long-acting while pentobarbital acts for a shorter time, about three hours. Thiopental is very fast-acting and is used as an anesthetic for producing deep sedation quickly. With barbiturates in general, sleep can be produced with as little as 0.1 g and toxic symptoms and even death can result from 1.5 g.

Example 17.6

Write equations for the following ammonolysis reactions.

a. $\underset{\text{O}}{\overset{\text{O}}{\text{HC}}}-\text{OCH}_2\text{CH}_3 + \text{NH}_3 \longrightarrow$

b. $\text{CH}_3\text{CH}_2\text{O}-\overset{\text{O}}{\underset{}{\text{C}}}-\text{OCH}_2\text{CH}_3 + 1\ \text{NH}_3 \longrightarrow$

c. $\text{CH}_3\text{CH}_2\text{O}-\overset{\text{O}}{\underset{}{\text{C}}}-\text{OCH}_2\text{CH}_3 + 2\ \text{NH}_3 \longrightarrow$

Solution

a. $\text{HC}-\text{OCH}_2\text{CH}_3 + \text{NH}_3 \longrightarrow \text{HC}-\text{NH}_2 + \text{HOCH}_2\text{CH}_3$

ethyl formate   formamide

b.
$$CH_3CH_2O{-}\overset{\displaystyle O}{\overset{\|}{C}}{-}OCH_2CH_3 + 1\ NH_3 \longrightarrow CH_3CH_2O{-}\overset{\displaystyle O}{\overset{\|}{C}}{-}NH_2 + HOCH_2CH_3$$

diethyl carbonate   ethyl carbamate

Diethyl carbonate is a diester, and reacts with one mole of ammonia to form a compound that contains both an amide and an ester.

c.
$$CH_3CH_2O{-}\overset{\displaystyle O}{\overset{\|}{C}}{-}OCH_2CH_3 + 2\ NH_3 \longrightarrow H_2N{-}\overset{\displaystyle O}{\overset{\|}{C}}{-}NH_2 + 2\ HOCH_2CH_3$$

diethyl carbonate   urea

Problem 17.6

Write equations for the following ammonolysis reactions.

a. 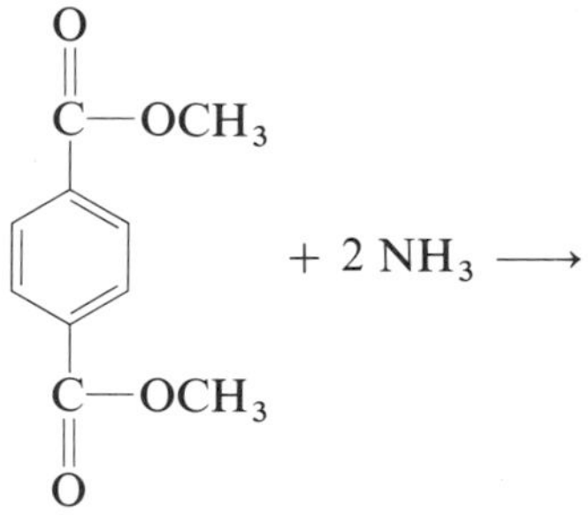 + 2 NH$_3$ ⟶   b. 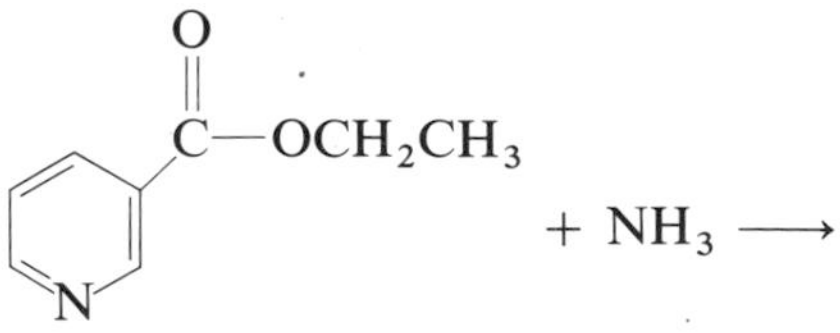 + NH$_3$ ⟶

c.
$$CH_3CH_2\overset{\displaystyle O}{\overset{\|}{C}}OCHCH_3 + NH_3 \longrightarrow$$
$$\hspace{3.5cm}|$$
$$\hspace{3.5cm}CH_3$$

D. Transesterification

Reaction of an ester with an alcohol in the presence of an acid catalyst results in the interchange of alkyl groups on the carboxyl oxygen.

$$CH_3{-}\overset{\displaystyle O}{\overset{\|}{C}}{-}OCH_3 + HOCH_2CH_3 \overset{H^+}{\rightleftharpoons} CH_3{-}\overset{\displaystyle O}{\overset{\|}{C}}{-}OCH_2CH_3 + HOCH_3$$

methyl acetate   ethanol   ethyl acetate   methanol

Because the reaction of an ester with an alcohol results in the formation of a different ester, the process is called **transesterification.** Transesterification is an equilibrium reaction. An excess of alcohol can be used to drive the equilibrium to the right.

Example 17.7

Complete the following transesterification reactions.

a. $CH_3\overset{O}{\overset{\|}{C}}-OCH_2(CH_2)_8CH_3 + CH_3OH \xrightarrow{H^+}$

b. $CH_3\overset{O}{\overset{\|}{C}}-OCH_2CH_2O-\overset{O}{\overset{\|}{C}}CH_3 + 2\ CH_3OH \xrightarrow{H^+}$

Solution

a. $CH_3\overset{O}{\overset{\|}{C}}-OCH_3 + HOCH_2(CH_2)_8CH_3$ b. $2\ CH_3\overset{O}{\overset{\|}{C}}-OCH_3 + HOCH_2CH_2OH$

Problem 17.7

Complete the following transesterification reactions.

a. 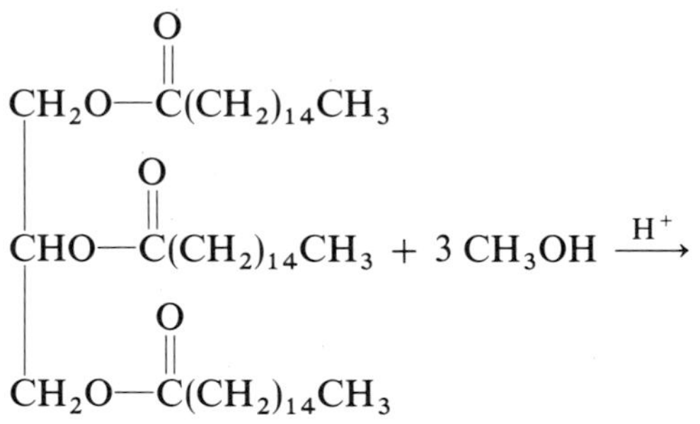 $CH_2O-\overset{O}{\overset{\|}{C}}(CH_2)_{14}CH_3$

$\quad\ CHO-\overset{O}{\overset{\|}{C}}(CH_2)_{14}CH_3 + 3\ CH_3OH \xrightarrow{H^+}$

$\quad\ CH_2O-\overset{O}{\overset{\|}{C}}(CH_2)_{14}CH_3$

b. $2\ \langle\!\!\langle\ \rangle\!\!\rangle-\overset{O}{\overset{\|}{C}}-OCH_3 + HOCH_2CH_2OH \xrightarrow{H^+}$

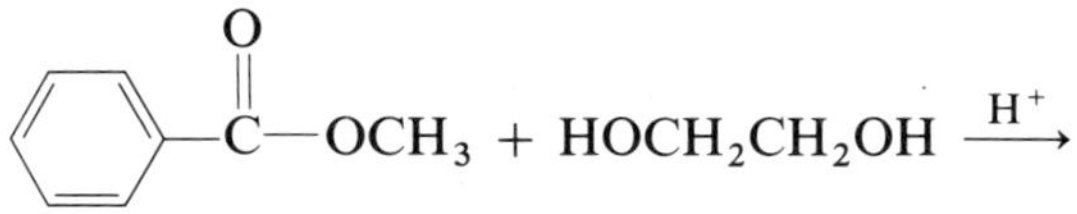

17.6 Reactions of Acid Anhydrides

The most important uses of anhydrides are reaction with water to form carboxylic acids, reaction with alcohols to form esters, and reaction with amines to form amides.

A. Hydrolysis

Anhydrides of carboxylic acids are converted to the corresponding acids by hydrolysis in either aqueous acid or base. For example, hydrolysis of acetic anhydride gives two molecules of acetic acid.

$CH_3-\overset{O}{\overset{\|}{C}}-O-\overset{O}{\overset{\|}{C}}-CH_3 + H-OH \longrightarrow CH_3-\overset{O}{\overset{\|}{C}}-OH + HO-\overset{O}{\overset{\|}{C}}-CH_3$

acetic anhydride

Acetic anhydride reacts so readily with water that it must be protected from moisture during storage.

B. Reaction with Alcohols

Anhydrides react with alcohols to form an ester and a carboxylic acid.

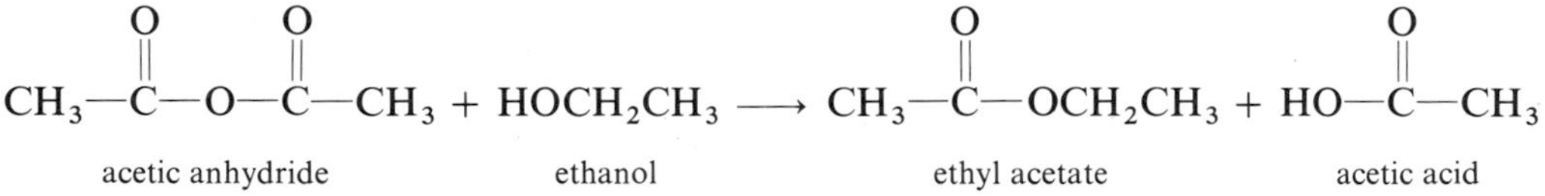

$$CH_3-\overset{\overset{O}{\|}}{C}-O-\overset{\overset{O}{\|}}{C}-CH_3 + HOCH_2CH_3 \longrightarrow CH_3-\overset{\overset{O}{\|}}{C}-OCH_2CH_3 + HO-\overset{\overset{O}{\|}}{C}-CH_3$$

acetic anhydride · · · · · · · · · · ethanol · · · · · · · · · · · ethyl acetate · · · · · · · · · acetic acid

Aspirin is prepared by the reaction of acetic anhydride and the —OH group of salicylic acid.

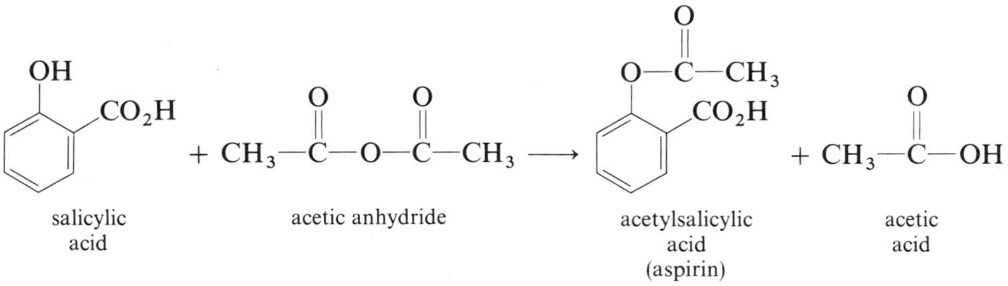

salicylic acid · · · · · · · · · · · acetic anhydride · · · · · · · · · · · acetylsalicylic acid (aspirin) · · · · · · · · · acetic acid

C. Reaction with Ammonia and Amines

Anhydrides react very rapidly with ammonia to give one molecule of amide and one of a carboxylate salt. For example, acetic anhydride reacts with ammonia to form acetamide and ammonium acetate. In this reaction, the anhydride and ammonia first react to form a tetrahedral carbonyl addition intermediate, which then breaks apart to give an amide and a carboxylic acid. The carboxylic acid then reacts with a second molecule of ammonia to form the ammonium salt.

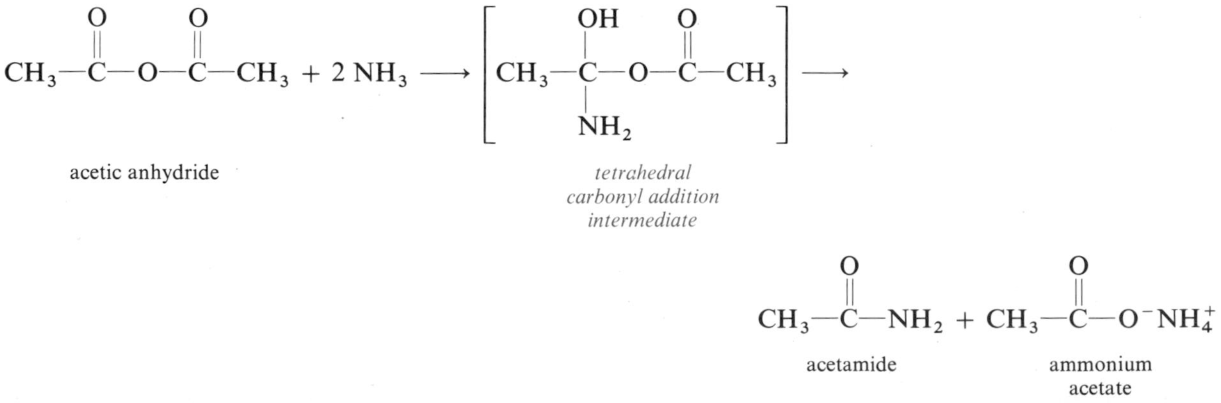

$$CH_3-\overset{\overset{O}{\|}}{C}-O-\overset{\overset{O}{\|}}{C}-CH_3 + 2\ NH_3 \longrightarrow \left[CH_3-\overset{\overset{OH}{|}}{\underset{\underset{NH_2}{|}}{C}}-O-\overset{\overset{O}{\|}}{C}-CH_3 \right] \longrightarrow$$

acetic anhydride

tetrahedral carbonyl addition intermediate

$$CH_3-\overset{\overset{O}{\|}}{C}-NH_2 + CH_3-\overset{\overset{O}{\|}}{C}-O^-NH_4^+$$

acetamide · · · · · · · · · · · ammonium acetate

Example 17.8

Complete the following reactions. (The stoichiometry of each is given in the example.)

a. 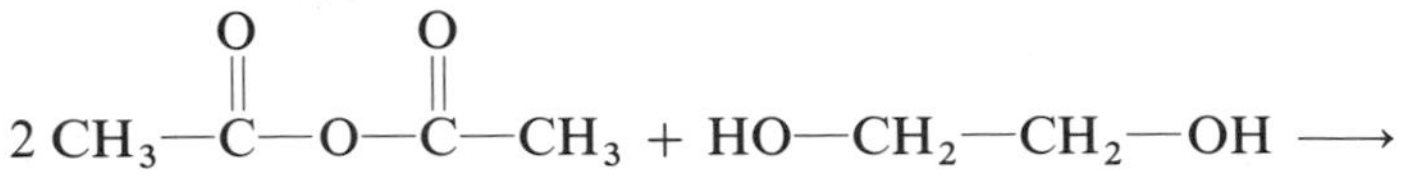 $2 \ CH_3-\overset{\overset{\displaystyle O}{\|}}{C}-O-\overset{\overset{\displaystyle O}{\|}}{C}-CH_3 + HO-CH_2-CH_2-OH \longrightarrow$

b. 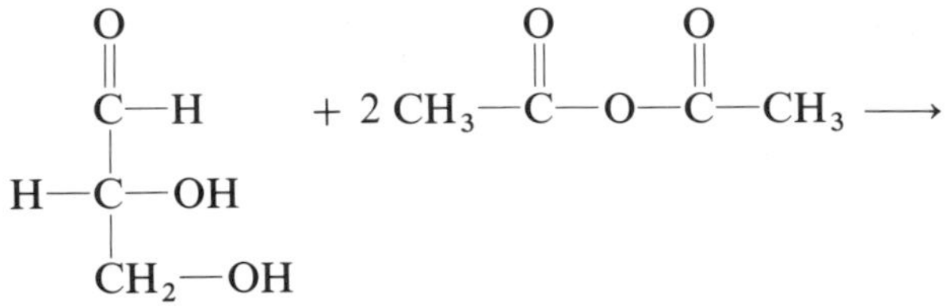
$$\begin{array}{l} \overset{\overset{\displaystyle O}{\|}}{C}-H \\ | \\ H-C-OH \\ | \\ CH_2-OH \end{array} \quad + \ 2 \ CH_3-\overset{\overset{\displaystyle O}{\|}}{C}-O-\overset{\overset{\displaystyle O}{\|}}{C}-CH_3 \longrightarrow$$

Solution

a. 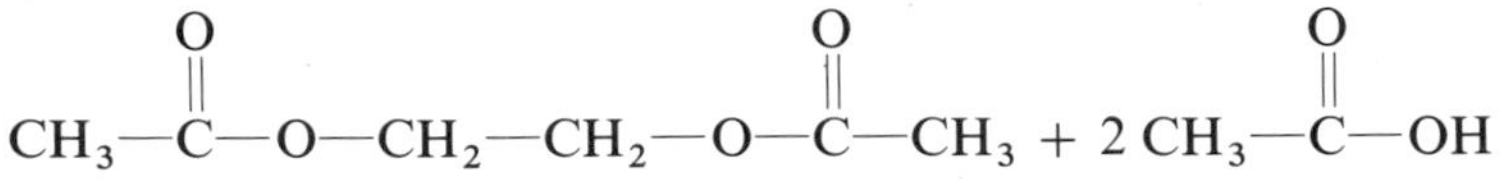 $CH_3-\overset{\overset{\displaystyle O}{\|}}{C}-O-CH_2-CH_2-O-\overset{\overset{\displaystyle O}{\|}}{C}-CH_3 + 2 \ CH_3-\overset{\overset{\displaystyle O}{\|}}{C}-OH$

The starting material, ethylene glycol, is a diol and reacts with two moles of acetic anhydride to produce a diester and two moles of acetic acid.

b. 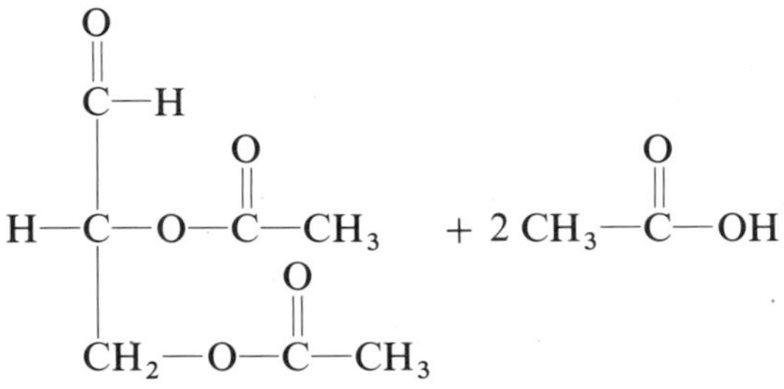
$$\begin{array}{l} \overset{\overset{\displaystyle O}{\|}}{C}-H \\ | \quad\quad\quad \overset{\displaystyle O}{\|} \\ H-C-O-C-CH_3 \\ | \quad\quad\quad \overset{\displaystyle O}{\|} \\ CH_2-O-C-CH_3 \end{array} \quad + \ 2 \ CH_3-\overset{\overset{\displaystyle O}{\|}}{C}-OH$$

Glyceraldehyde is a diol and reacts with two moles of acetic anhydride to give a diester and two moles of acetic acid.

Problem 17.8

Complete the following reactions.

a. 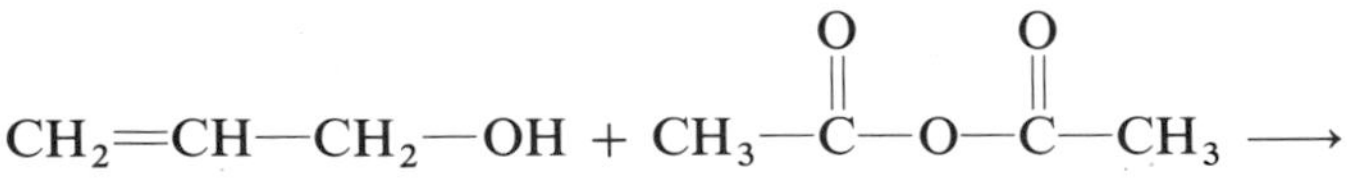 $CH_2{=}CH-CH_2-OH + CH_3-\overset{\overset{\displaystyle O}{\|}}{C}-O-\overset{\overset{\displaystyle O}{\|}}{C}-CH_3 \longrightarrow$

b. 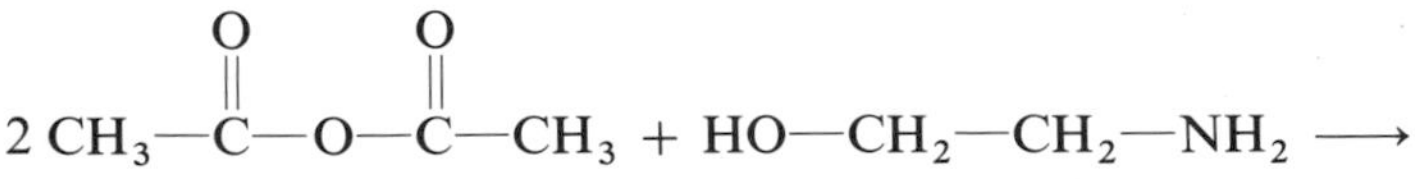 $2 \ CH_3-\overset{\overset{\displaystyle O}{\|}}{C}-O-\overset{\overset{\displaystyle O}{\|}}{C}-CH_3 + HO-CH_2-CH_2-NH_2 \longrightarrow$

17.7 Preparation of Amides

Amides are most commonly synthesized in the laboratory by reaction of an anhydride with ammonia (Section 17.6C), or reaction of an ester with ammonia (Section 17.5C). We have already seen several examples of each of these reactions.

$$CH_3-\overset{\overset{\displaystyle O}{\|}}{C}-O-CH_2CH_3 + NH_3 \longrightarrow CH_3-\overset{\overset{\displaystyle O}{\|}}{C}-NH_2 + CH_3CH_2OH$$

$$CH_3-\overset{\overset{\displaystyle O}{\|}}{C}-O-\overset{\overset{\displaystyle O}{\|}}{C}-CH_3 + 2\,NH_3 \longrightarrow CH_3-\overset{\overset{\displaystyle O}{\|}}{C}-NH_2 + CH_3-\overset{\overset{\displaystyle O}{\|}}{C}-O^-NH_4^+$$

Amides can also be prepared by heating the ammonium salt of a carboxylic acid above its melting point.

$$CH_3-\overset{\overset{\displaystyle O}{\|}}{C}-O^-NH_4^+ \xrightarrow{\text{heat}} CH_3-\overset{\overset{\displaystyle O}{\|}}{C}-NH_2 + H_2O$$

ammonium
acetate   acetamide

17.8 Hydrolysis of Amides

Amides are very resistant to hydrolysis. However, in the presence of concentrated aqueous acid or base, hydrolysis does occur, though not as rapidly as in the case of esters. Often, it is necessary to reflux an amide for several hours with concentrated hydrochloric acid to bring about hydrolysis.

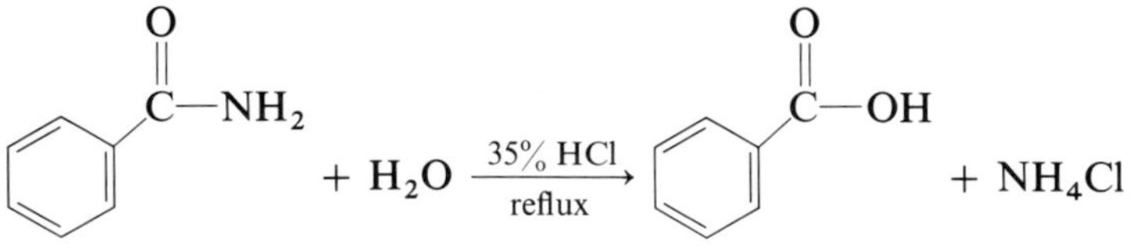

benzamide   benzoic acid

17.9 Relative Reactivities of Esters, Amides, and Anhydrides

The three common functional derivatives of carboxylic acids described in this chapter show marked differences in reactivities. For example, consider the ease of hydrolysis of an anhydride, an ester, and an amide. Acetic anhydride hydrolyzes so readily with water that it must be protected from atmospheric

Table 17.1 Interconversion of functional derivatives of carboxylic acids.

Acid Derivative	Can be Converted to	By Reaction with	Name of Process
acid anhydride ⟶	carboxylic acid	water	hydrolysis
	ester	alcohol or phenol	alcoholysis
	amide	NH_3, primary or secondary amine	ammonolysis
ester ⟶	carboxylic acid	water	hydrolysis
	amide	NH_3, primary or secondary amine	ammonolysis
amide ⟶	carboxylic acid	water	hydrolysis

increasing reactivity ↑

moisture during storage. Ethyl acetate hydrolyzes slowly with water at room temperature, but reacts readily on heating in the presence of an acid or base catalyst. Acetamide is very resistant to hydrolysis except in the presence of strong acid or base and heat. The reactivity of these functional derivatives of carboxylic acids decreases in the following order:

$$\underset{\text{acid anhydride}}{CH_3\overset{O}{\overset{\|}{C}}-O-\overset{O}{\overset{\|}{C}}CH_3} > \underset{\text{ester}}{CH_3\overset{O}{\overset{\|}{C}}-OCH_2CH_3} > \underset{\text{amide}}{CH_3\overset{O}{\overset{\|}{C}}-NH_2}$$

Any less reactive derivative of a carboxylic acid may be prepared directly from a more reactive derivative, but not vice versa. For example, an ester can be synthesized from an acid anhydride plus an alcohol or phenol. However, an ester cannot be synthesized from an amide plus an alcohol. These inter-conversions are summarized in Table 17.1.

17.10 Some Esters, Amides, and Anhydrides of Inorganic Acids

A. Phosphoric Acid

In esters of carboxylic acids, the —OH of the carboxyl group is replaced by —OR. Inorganic acids also form esters in which the —OH of the acid is replaced by —OR. The most common inorganic esters in biological chemistry are those of phosphoric acid. Phosphoric acid, H_3PO_4, has three —OH groups and forms mono-, di-, and triesters. Following are examples of each:

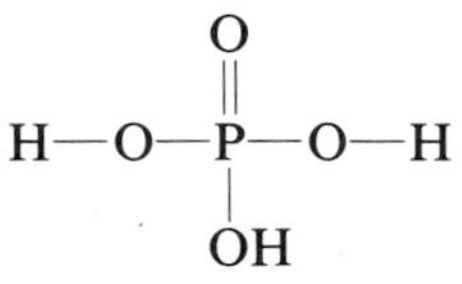

phosphoric acid 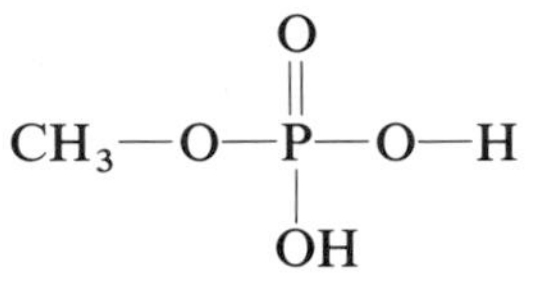   methyl phosphate   dimethyl phosphate   trimethyl phosphate
                          (a monoester)   (a diester)   (a triester)

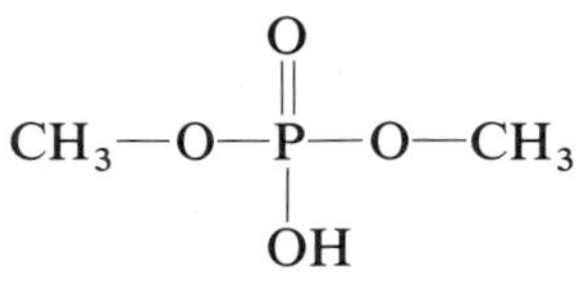

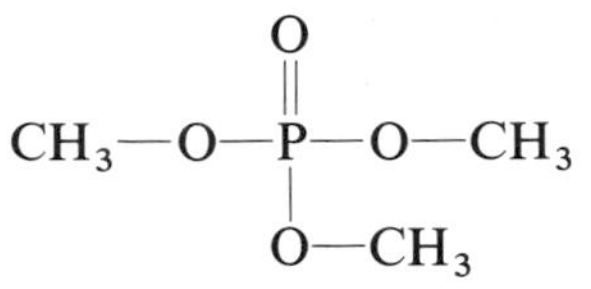

Esters of phosphoric acid are especially important in biological chemistry because many organic molecules, such as glyceraldehyde and dihydroxyacetone, can be metabolized only as phosphate esters. Vitamin B$_6$, or pyridoxal, is metabolically active only after it is converted to pyridoxal phosphate.

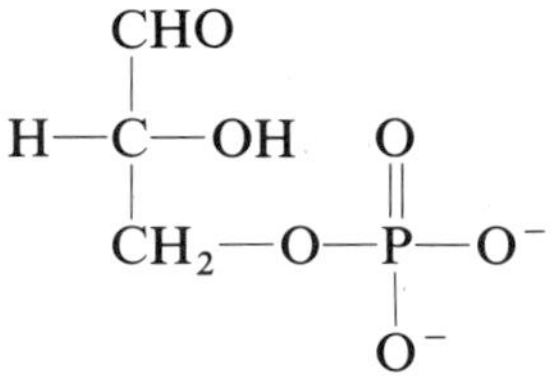

glyceraldehyde   dihydroxyacetone   pyridoxal phosphate
3-phosphate   phosphate

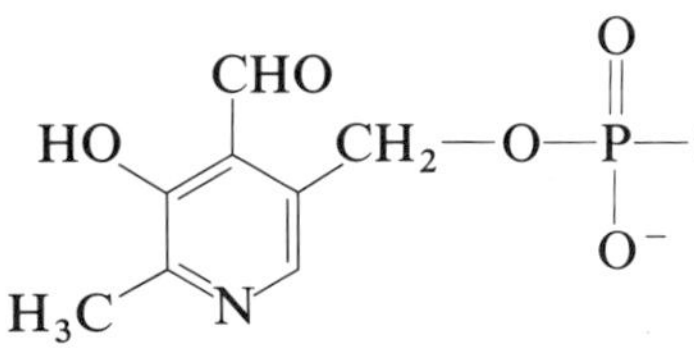

In drawing structural formulas for phosphate esters and anhydrides found in living systems, we will show the state of ionization at pH 7.4. Phosphoric acid is a triprotic acid; at physiological pH (7.4), two protons of H_3PO_4 are completely ionized. Glyceraldehyde 3-phosphate is a monoester of phosphoric acid; at pH 7.4, the remaining two protons of the phosphate group are ionized, giving this substance a net charge of −2.

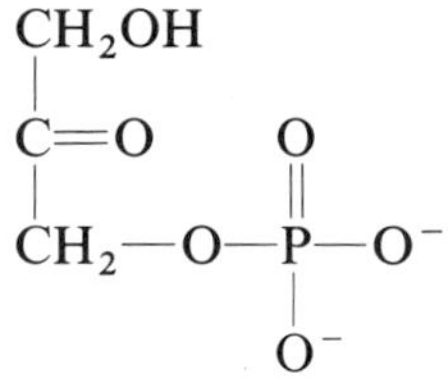

Hydrolysis of a monoester of phosphoric acid requires one molecule of water, hydrolysis of a diester requires 2 H$_2$O, and hydrolysis of a triester requires 3 II$_2$O. For example, hydrolysis of diethyl phosphate requires 2 H$_2$O and yields phosphoric acid and two molecules of ethanol.

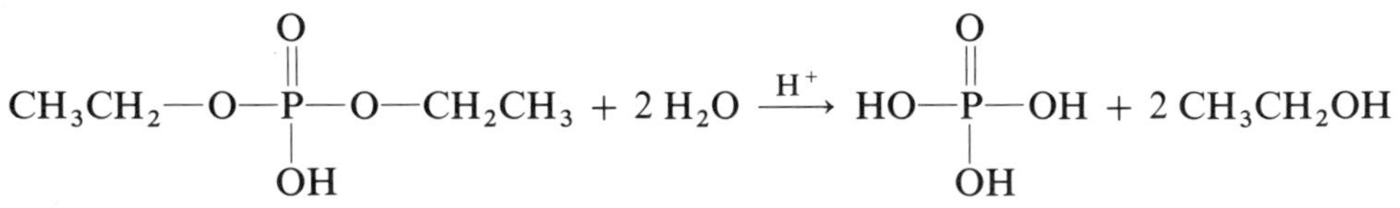

diethyl phosphate

Example 17.9

Draw structural formulas for the following phosphate esters. Calculate the net charge on each at pH 7.4.

a. α-D-glucose-6-phosphate   **b.** β-D-fructose-1,6-diphosphate

Solution

a. Draw D-glucose as a six-member cyclic hemiacetal with the —OH on the anomeric carbon atom below the plane of the ring. Show a phosphate ester bond between the oxygen atom of carbon-6 and phosphoric acid.

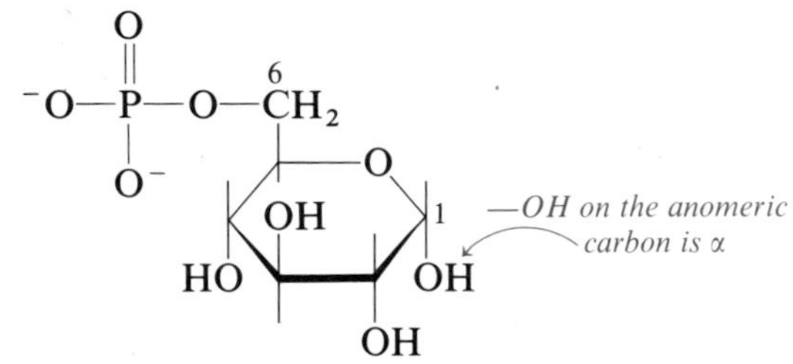

The net charge at pH 7.4 is −2.

b. Draw D-fructose in the five-member cyclic hemiketal form. Then draw one phosphate ester between the —OH on carbon-6 and phosphate and another between the —OH on carbon-1 and phosphate.

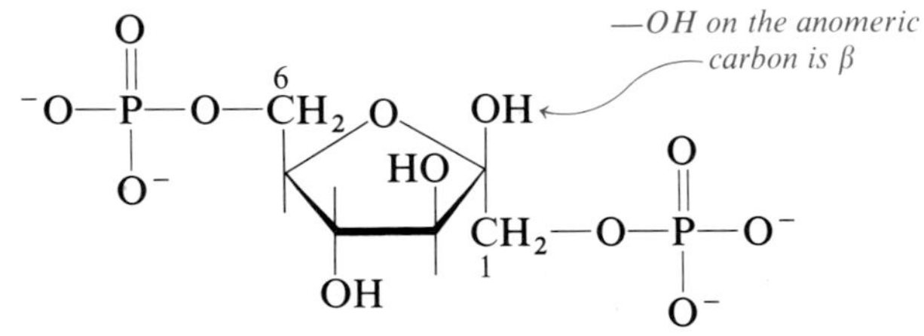

The net charge at pH 7.4 is −4.

Problem 17.9

Draw structural formulas for the following phosphate esters. Calculate the net charge on each at pH 7.4.

a. α-D-glucose-1-phosphate   **b.** β-D-ribose-5-phosphate

Phosphoric acid also forms anhydrides. Following is the structural formula of phosphoric acid anhydride. The common name for this compound is pyrophosphoric acid. Phosphoric acid anhydride has a net charge of −4 at pH 7.4.

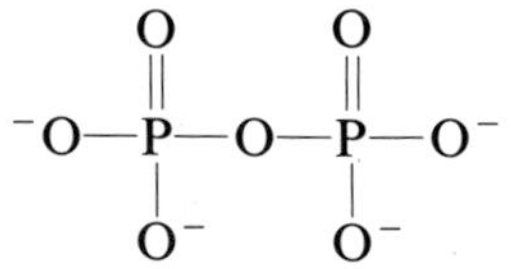

phosphoric acid anhydride
(pyrophosphoric acid)

Phosphoric acid forms mixed anhydrides with carboxylic acids. Following is the structural formula of acetyl phosphate, an anhydride derived from one molecule of phosphoric acid and one molecule of acetic acid. At pH 7.4, acetyl phosphate has a net charge of -2.

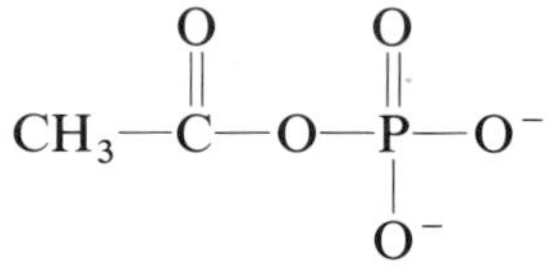

$$CH_3-\overset{\overset{\displaystyle O}{\|}}{C}-O-\underset{\underset{\displaystyle O^-}{|}}{\overset{\overset{\displaystyle O}{\|}}{P}}-O^-$$

acetyl phosphate

Phosphoric acid also forms amides in which an $-OH$ group is replaced by $-NH_2$ or a derivative of $-NH_2$. Following is the structural formula of creatine phosphate, an amide of phosphoric acid.

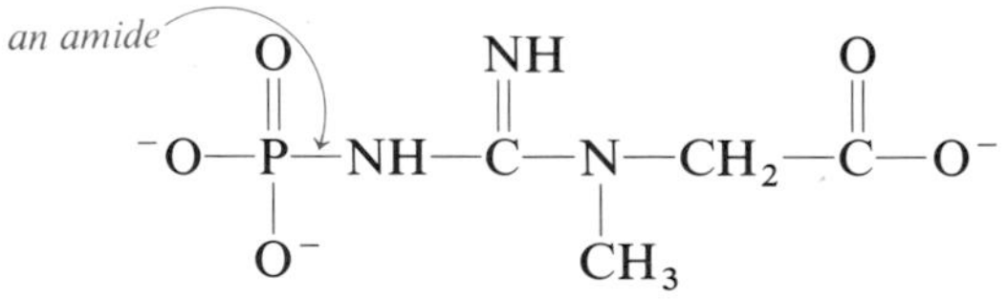

$$\overset{an\ amide}{}\quad ^-O-\underset{\underset{\displaystyle O^-}{|}}{\overset{\overset{\displaystyle O}{\|}}{P}}-NH-\overset{\overset{\displaystyle NH}{\|}}{C}-\underset{\underset{\displaystyle CH_3}{|}}{N}-CH_2-\overset{\overset{\displaystyle O}{\|}}{C}-O^-$$

creatine phosphate

Creatine phosphate is found in muscle tissue and is a source of energy to drive muscle contraction (Section 23.6C).

B. Sulfonic Acids and Sulfa Drugs

Sulfonic acids form amides in which the $-OH$ is replaced by $-NH_2$. Following are structural formulas of benzenesulfonic acid, benzenesulfonamide, and p-aminobenzenesulfonamide.

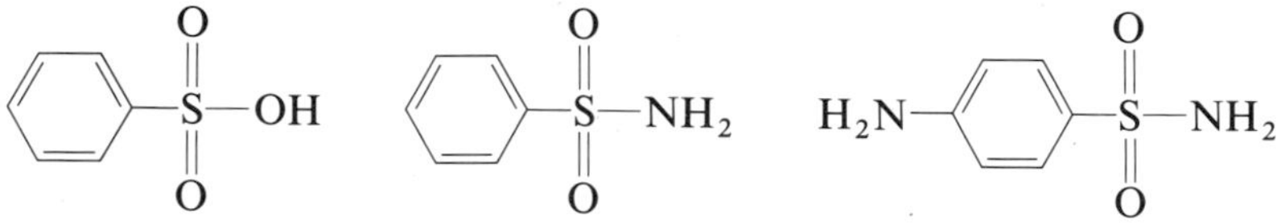

benzenesulfonic acid	benzenesulfonamide	p-aminobenzenesulfonamide (sulfanilamide)

The discovery of the medicinal uses of sulfanilamide and its derivatives was a milestone in the history of chemotherapy, because it represents one of the first rational investigations of synthetic organic molecules as potential drugs to fight infection. Sulfanilamide was first prepared in 1908 in Germany, but it was not until 1932 that its possible therapeutic value was realized. In that year, the dye Protonsil was prepared. During research over the next two years, the German scientist G. Domagk observed Protonsil's remarkable effectiveness in curing streptococcal and staphylococcal infections in mice and other experimental animals. Domagk further discovered that Protonsil is rapidly reduced in cells to sulfanilamide, and that it is sulfanilamide, not Protonsil, that is the actual antibiotic. His discoveries were honored in 1939 by a Nobel Prize in Medicine. Sulfanilamide is a member of a group of drugs called **sulfa drugs.**

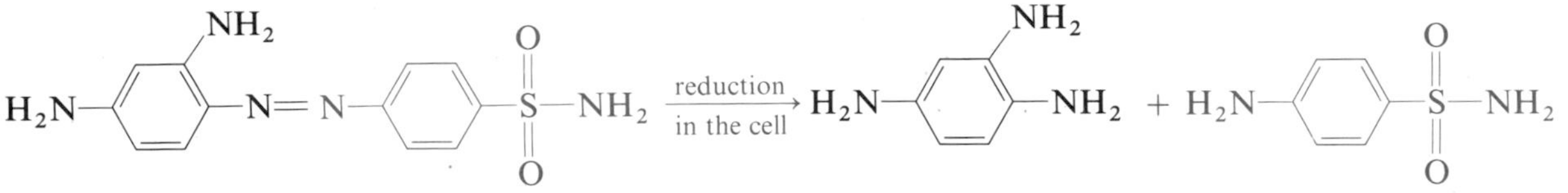

Protonsil   sulfanilamide

The key to understanding the action of sulfanilamide came in 1940 with the observation that inhibition of bacterial growth caused by sulfanilamide can be reversed by adding large amounts of p-aminobenzoic acid (PABA). From this experiment, it was recognized that p-aminobenzoic acid is a growth factor for certain bacteria, and that in some way not then understood, sulfanilamide interferes with bacteria's ability to use PABA.

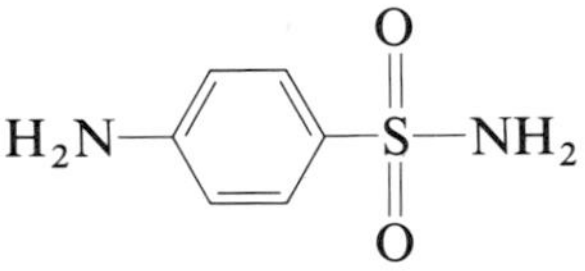

There are obvious structural similarities between p-aminobenzoic acid and sulfanilamide.

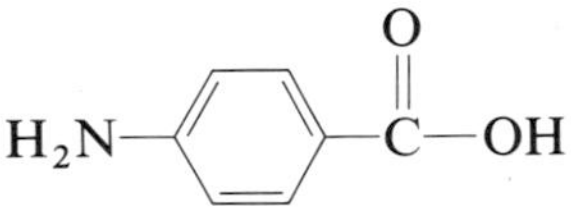

p-aminobenzoic acid   sulfanilamide
(PABA)

It now appears that sulfanilamide drugs inhibit one or more enzyme-catalyzed steps in the synthesis of folic acid from p-aminobenzoic acid. The ability of sulfanilamide to combat infections in humans without harming the patient depends on the fact that humans also require folic acid but do not make it from p-aminobenzoic acid. For humans, folic acid is a vitamin and must be supplied in the diet.

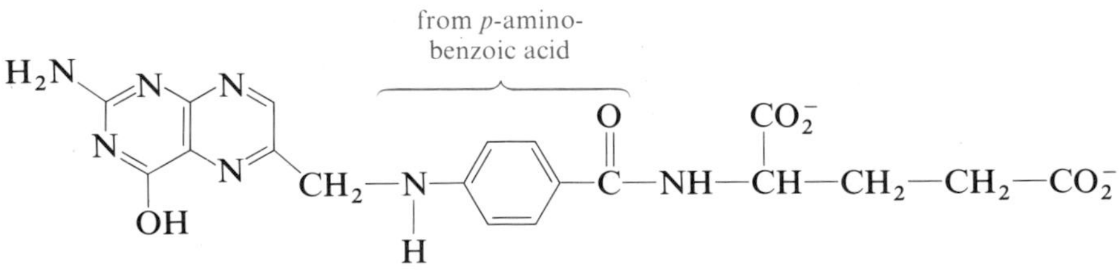

folic acid

In the search for even better sulfa drugs, literally thousands of derivatives of sulfanilamide have been synthesized. Two of the more effective sulfa drugs are:

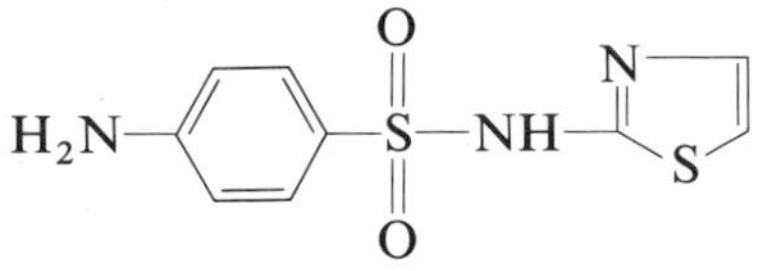

sulfathiazole 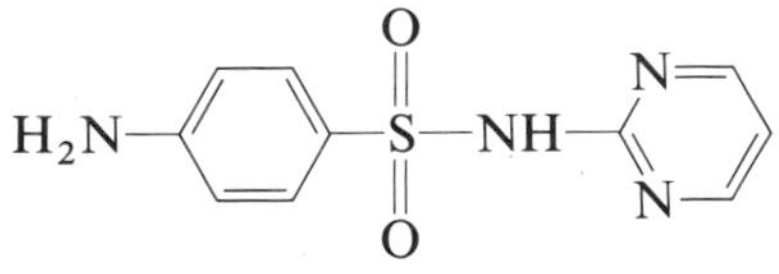   sulfadiazine

Sulfa drugs were found to be effective in the treatment of tuberculosis, pneumonia, and diphtheria, and they helped usher in a new era in public health in the United States in the 1930s. During World War II, they were routinely sprinkled on wounds to prevent infection. These drugs were among the first of the new "wonder drugs." As an historical footnote, the use of sulfa drugs to fight bacterial infection has been largely supplanted by an even newer wonder drug, the penicillins. (See the mini-essay, "The Penicillins.")

C. Nitric and Nitrous Acids

Several organic esters of nitric acid and nitrous acid have been used as drugs for more than 100 years. Two of these are glyceryl trinitrate (more commonly known as nitroglycerine) and 3-methylbutyl nitrite (more commonly called isopentyl nitrite or isoamyl nitrite).

$$\text{H—O—NO}_2$$

$$
\begin{array}{l}
\text{CH}_2\text{—O—NO}_2 \\
| \\
\text{CH—O—NO}_2 \\
| \\
\text{CH}_2\text{—O—NO}_2
\end{array}
$$

nitric acid

glyceryl trinitrate
(nitroglycerine)

$$\text{H—O—NO}$$

$$\text{CH}_3\text{—CH—CH}_2\text{—CH}_2\text{—O—NO}$$ with CH_3 branch

nitrous acid

3-methylbutyl nitrite
(isopentyl nitrite)
(isoamyl nitrite)

Because nitroglycerine and isoamyl nitrite produce rapid relaxation of most smooth muscles of the body, they are called **vasodilators.** Their most important medical use is relaxation of the smooth muscle of blood vessels and dilation of all large and small arteries of the heart. Both esters are used for the treatment of angina pectoris, a heart disease characterized by agonizing chest pains.

17.11 The Claisen Condensation: β-Ketoesters

The **Claisen condensation** involves condensation of the α-carbon of one molecule of ester with the carbonyl carbon of a second molecule of ester. For example, when ethyl acetate is heated with sodium ethoxide in ethanol, the α-carbon of one molecule of ethyl acetate condenses with the carbonyl carbon of a second molecule to form a new carbon-carbon bond.

$$
\underset{\substack{\uparrow \\ \textit{carbonyl} \\ \textit{carbon}}}{CH_3\overset{O}{\overset{\|}{C}}{-}OCH_2CH_3} + \underset{\substack{\uparrow \\ \textit{α-carbon}}}{CH_3\overset{O}{\overset{\|}{C}}OCH_2CH_3} \xrightarrow[CH_3CH_2OH]{CH_3CH_2O^-Na^+} CH_3\overset{O}{\overset{\|}{C}}{-}CH_2\overset{O}{\overset{\|}{C}}OCH_2CH_3 + CH_3CH_2OH
$$

new C—C bond formed

ethyl 3-ketobutanoate
(ethyl acetoacetate)

The characteristic structural feature of the product of a Claisen condensation is a ketone on carbon-3 of an ester chain. In the common name system of nomenclature, carbon-3 of a carboxylic acid is called a beta-carbon (β-carbon), and, for this reason, the products of Claisen condensation reactions are often called **β-ketoesters.** In naming β-ketoesters, the prefix *keto-* is used to indicate the presence of the ketone group on the ester chain.

$$
CH_3-\overset{O}{\overset{\|}{\underset{\beta}{C}}}-\overset{}{\underset{\alpha}{CH_2}}-\overset{O}{\overset{\|}{C}}-O-CH_2CH_3
$$

a β-ketoester

Claisen condensation of ethyl propanoate yields the following β-ketoester.

$$
\underset{\substack{\textit{carbonyl carbon}}}{CH_3CH_2\overset{O}{\overset{\|}{C}}\underset{\nearrow}{OCH_2CH_3}} + \underset{\substack{\uparrow \\ \textit{α-carbon}}}{CH_2\overset{O}{\overset{\|}{C}}OCH_2CH_3 \atop CH_3} \xrightarrow{CH_3CH_2O^-Na^+} CH_3CH_2\overset{O}{\overset{\|}{C}}{-}\underset{CH_3}{CH}\overset{O}{\overset{\|}{C}}OCH_2CH_3 + CH_3CH_2OH
$$

new C—C bond

ethyl 2-methyl-3-ketopentanoate
(a β-ketoester)

The mechanism of a Claisen condensation is similar to the three-step mechanism for aldol condensations (Section 13.5E) in that it begins with formation of an anion on the carbon atom alpha to the carbonyl group. Recall from Section 13.5E that hydrogen atoms on alpha carbons show acidity in the presence of strong bases. The anion of ethyl acetate is a nucleophile, and in Step 2 it attacks the carbonyl carbon of a second molecule of ethyl acetate to form a tetrahedral carbonyl addition intermediate. Elimination of ethoxide ion in Step 3 gives ethyl acetoacetate.

Step 1: $CH_3CH_2O^-$ + H—CH_2—$\overset{\overset{\displaystyle O}{\|}}{C}$—$OCH_2CH_3$ $\rightleftharpoons$ CH_3CH_2OH + $^-{:}CH_2$—$\overset{\overset{\displaystyle O}{\|}}{C}$—$OCH_2CH_3$

Step 2: CH_3—$\overset{\overset{\displaystyle \ddot{O}:}{\|}}{C}$—$OCH_2CH_3$ + $^-{:}CH_2$—$\overset{\overset{\displaystyle O}{\|}}{C}$—$OCH_2CH_3$ $\rightleftharpoons$ CH_3—$\overset{\overset{\displaystyle :\ddot{O}:^-}{|}}{\underset{\underset{\displaystyle OCH_2CH_3}{|}}{C}}$—$CH_2$—$\overset{\overset{\displaystyle O}{\|}}{C}$—$OCH_2CH_3$

tetrahedral carbonyl addition intermediate

Step 3: CH_3—$\overset{\overset{\displaystyle :\ddot{O}:^-}{|}}{\underset{\underset{\displaystyle OCH_2CH_3}{|}}{C}}$—$CH_2$—$\overset{\overset{\displaystyle O}{\|}}{C}$—$OCH_2CH_3$ $\rightleftharpoons$ CH_3—$\overset{\overset{\displaystyle O}{\|}}{C}$—$CH_2$—$\overset{\overset{\displaystyle O}{\|}}{C}$—$OCH_2CH_3$ + $CH_3CH_2O^-$

Example 17.10

Draw structural formulas for the products of the following Claisen condensations.

a. C_6H_5—$\overset{\overset{\displaystyle O}{\|}}{C}$—$OCH_2CH_3$ + $CH_3CH_2\overset{\overset{\displaystyle O}{\|}}{C}$—$OCH_2CH_3$ $\xrightarrow{\text{base}}$

b. (cyclohexanone) + H—$\overset{\overset{\displaystyle O}{\|}}{C}$—$OCH_2CH_3$ $\xrightarrow{\text{base}}$

Solution

a. Ethyl benzoate has no α-hydrogens and in a Claisen condensation can function only as an anion acceptor. Ethyl propanoate has an α-hydrogen and reacts with base to form an anion which then reacts with the carbonyl group of ethyl benzoate.

new C—C bond

C_6H_5—$\overset{\overset{\displaystyle O}{\|}}{\underset{\underset{\displaystyle OCH_2CH_3}{|}}{C}}$ + $CH_2\overset{\overset{\displaystyle O}{\|}}{C}OCH_2CH_3$ $\xrightarrow{\text{base}}$ C_6H_5—$\overset{\overset{\displaystyle O}{\|}}{C}$—$\overset{}{\underset{\underset{\displaystyle CH_3}{|}}{CH}}$$\overset{\overset{\displaystyle O}{\|}}{C}OCH_2CH_3$ + CH_3CH_2OH

with CH_3 substituent

ethyl 2-methyl-3-keto-3-phenylpropanoate

b. Only cyclohexanone has an α-hydrogen that can be removed by base to form an anion. The carbonyl group of ethyl formate functions as the anion acceptor.

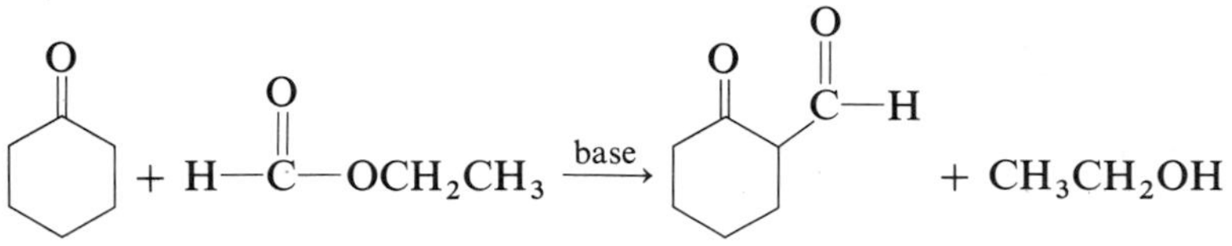

Problem 17.10

Draw structural formulas for the products of the following Claisen condensations.

a. $CH_3CH_2O-\overset{O}{\overset{||}{C}}(CH_2)_5\overset{O}{\overset{||}{C}}-OCH_2CH_3 \xrightarrow{\text{base}}$

b. 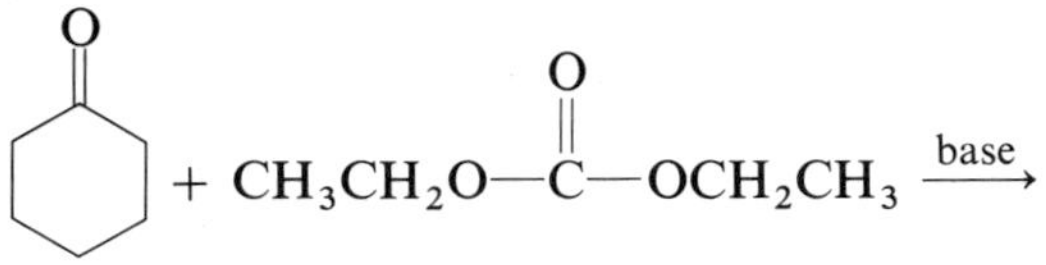 $+ CH_3CH_2O-\overset{O}{\overset{||}{C}}-OCH_2CH_3 \xrightarrow{\text{base}}$

Hydrolysis of a β-ketoester in aqueous acid gives the corresponding β-ketoacid.

$$CH_3\overset{O}{\overset{||}{C}}CH_2\overset{O}{\overset{||}{C}}OCH_2CH_3 + H_2O \xrightarrow[\text{heat}]{H^+} CH_3\overset{O}{\overset{||}{C}}CH_2\overset{O}{\overset{||}{C}}OH + CH_3CH_2OH$$

3-ketobutanoic acid
(β-ketobutyric acid)

β-Ketoacids lose carbon dioxide on heating to give ketones (Section 16.5C).

$$CH_3\overset{O}{\overset{||}{C}}CH_2\overset{O}{\overset{||}{C}}OH \xrightarrow{\text{heat}} CH_3\overset{O}{\overset{||}{C}}CH_3 + O=C=O$$

In the more usual reaction, ester hydrolysis and decarboxylation occur together, and only the ketone is isolated.

$$CH_3CH_2\overset{O}{\overset{||}{C}}\underset{\underset{CH_3}{|}}{CH}\overset{O}{\overset{||}{C}}OCH_2CH_3 + H_2O \xrightarrow[\text{heat}]{H^+} CH_3CH_2\overset{O}{\overset{||}{C}}CH_2CH_3 + CO_2 + CH_3CH_2OH$$

ethyl 2-methyl-3-keto-
pentanoate   3-pentanone

Key Terms and Concepts

amide (17.1)

anhydride (17.1)

Claisen condensation (17.11)

ester (17.1)

Fischer esterification (17.4)

β-ketoester (17.11)

nucleophilic substitution at a carbonyl
carbon (17.3)

transesterification (17.5D)

Key Reactions

1. Fischer esterification (Section 17.4):

$$CH_3\overset{O}{\overset{\|}{C}}-OH + H-OCH_2CH_2CH_3 \underset{\longleftarrow}{\overset{H^+}{\rightleftharpoons}} CH_3\overset{O}{\overset{\|}{C}}-OCH_2CH_2CH_3 + H_2O$$

2. Hydrolysis
   a. of esters (Section 17.5B):

$$CH_3\overset{O}{\overset{\|}{C}}-OCH_2CH_2CH_3 + H_2O \overset{H^+}{\rightleftharpoons} CH_3\overset{O}{\overset{\|}{C}}-OH + HOCH_2CH_2CH_3$$

$$CH_3\overset{O}{\overset{\|}{C}}-OCH_2CH_2CH_3 + NaOH \longrightarrow CH_3\overset{O}{\overset{\|}{C}}-O^-Na^+ + HOCH_2CH_2CH_3$$

$$CH_3CH_2O-\underset{\underset{OCH_2CH_3}{|}}{\overset{O}{\overset{\|}{P}}}-OCH_2CH_3 + 3\,H_2O \overset{H_3O^+}{\rightleftharpoons} 3\,CH_3CH_2OH + HO-\underset{\underset{OH}{|}}{\overset{O}{\overset{\|}{P}}}-OH$$

   b. of amides (Section 17.8):

$$CH_3\overset{O}{\overset{\|}{C}}-NH_2 + H_2O + HCl \longrightarrow CH_3\overset{O}{\overset{\|}{C}}-OH + NH_4Cl$$

   c. of anhydrides (Section 17.6A):

$$CH_3\overset{O}{\overset{\|}{C}}-O-\overset{O}{\overset{\|}{C}}CH_3 + H_2O \longrightarrow 2\,CH_3\overset{O}{\overset{\|}{C}}-OH$$

3. *Ammonolysis
   a. of esters (Section 17.5C):

$$CH_3\overset{O}{\overset{\|}{C}}-OCH_2CH_3 + NH_3 \longrightarrow CH_3\overset{O}{\overset{\|}{C}}-NH_2 + CH_3CH_2OH$$

   b. of anhydrides (Section 17.6C):

$$CH_3\overset{O}{\overset{\|}{C}}-O-\overset{O}{\overset{\|}{C}}CH_3 + 2\,NH_3 \longrightarrow CH_3\overset{O}{\overset{\|}{C}}-NH_2 + CH_3\overset{O}{\overset{\|}{C}}-O^-NH_4^+$$

4. Alcoholysis
   a. of esters (Section 17.5D):

$$CH_3\overset{O}{\underset{\|}{C}}-OCH_3 + HOCH_2CH_2CH_3 \xrightarrow{H^+} CH_3\overset{O}{\underset{\|}{C}}-OCH_2CH_2CH_3 + CH_3OH$$

The reaction of an ester with an alcohol is also called transesterification.
   b. of anhydrides (Section 17.6B):

$$CH_3\overset{O}{\underset{\|}{C}}-O-\overset{O}{\underset{\|}{C}}CH_3 + CH_3CH_2OH \longrightarrow CH_3\overset{O}{\underset{\|}{C}}-OCH_2CH_3 + HO-\overset{O}{\underset{\|}{C}}CH_3$$

5. Reduction of esters (Section 17.5A):

$$C_6H_5-\overset{O}{\underset{\|}{C}}-O-CH_2CH_3 \xrightarrow[(2) \, H_2O]{(1) \, LiAlH_4} C_6H_5-CH_2OH + HOCH_2CH_3$$

6. Claisen condensation (Section 17.11):

$$2\, CH_3\overset{O}{\underset{\|}{C}}-OCH_2CH_3 \xrightarrow{CH_3CH_2ONa} CH_3\overset{O}{\underset{\|}{C}}-\underset{\beta}{CH_2}-\underset{\alpha}{\overset{O}{\underset{\|}{C}}}-OCH_2CH_3 + CH_3CH_2OH$$

The product of a Claisen condensation is a β-ketoester.

Problems

Structure and Nomenclature of Esters, Amides, and Anhydrides (Section 17.1)

17.11 Name the following compounds.

a. $CH_3-CH_2-\overset{O}{\underset{\|}{C}}-O-\underset{\underset{CH_3}{|}}{CH}-CH_3$

b. $CH_3-\overset{O}{\underset{\|}{C}}-NH_2$

c. 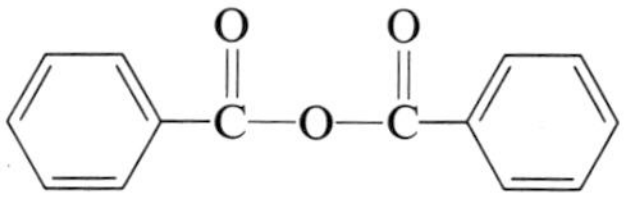 benzene ring—$\overset{O}{\underset{\|}{C}}-O-\overset{O}{\underset{\|}{C}}$—benzene ring

d. $CH_2{=}CH-\overset{O}{\underset{\|}{C}}-O-CH_3$

e. $CH_3-CH_2-CH_2-CH_2-\overset{O}{\underset{\|}{C}}-NH-CH_3$

f. $H_2N-\overset{O}{\underset{\|}{C}}-NH_2$

g. $CH_3-CH_2-\overset{O}{\underset{\|}{C}}-S-CH_3$

h. $CH_3-O-\overset{O}{\underset{\|}{C}}-CH_2-CH_2-\overset{O}{\underset{\|}{C}}-O-CH_3$

i. 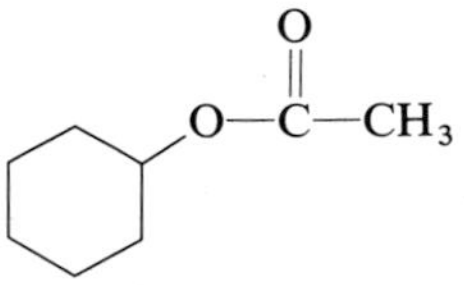 cyclohexyl—$O-\overset{O}{\underset{\|}{C}}-CH_3$

j. 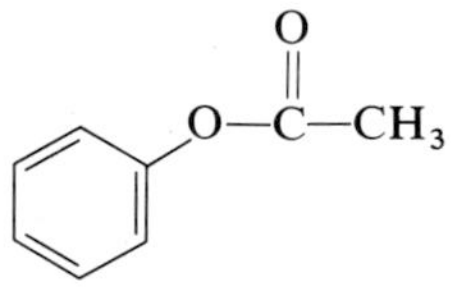 phenyl—$O-\overset{O}{\underset{\|}{C}}-CH_3$

17.12 Draw structural formulas for the following compounds:

a. phenyl benzoate
b. diethyl carbonate
c. benzamide
d. cyclobutyl butanoate
e. methyl 3-methylbutanoate
f. isopropyl 3-methylhexanoate
g. diethyl oxalate
h. ethyl *cis*-2-pentenoate
i. acetamide
j. N,N-dimethylacetamide
k. acetic anhydride
l. N-phenylbutanamide
m. diethyl malonate
n. formamide
o. ethyl 3-hydroxybutanoate
p. methyl formate
q. trimethyl citrate
r. *p*-nitrophenyl acetate
s. ethyl *p*-hydroxybenzoate

17.13 Draw structural formulas for the nine isomeric esters of molecular formula $C_5H_{10}O_2$. Give each an IUPAC name.

Physical Properties of Esters, Amides, and Anhydrides (Section 17.2)

17.14 Both acetic acid and methyl formate have the same molecular formula, $C_2H_4O_2$. Both are liquids. One has a boiling point of 32°C, the other a boiling point of 118°C. Which of the two compounds would you predict to have the boiling point of 118°C? Which the boiling point of 32°C? Explain your reasoning.

$$CH_3-\overset{\overset{\displaystyle O}{\|}}{C}-OH \qquad H-\overset{\overset{\displaystyle O}{\|}}{C}-OCH_3$$

acetic acid   methyl formate

17.15 Draw structural formulas to show hydrogen bonding between the circled atoms.

$$CH_3-\overset{\overset{\displaystyle \textcircled{O}}{\|}}{C}-\underset{\underset{\displaystyle H}{|}}{N}-H \qquad \text{and} \qquad CH_3-\overset{\overset{\displaystyle O}{\|}}{C}-\underset{\underset{\displaystyle H}{|}}{N}-\textcircled{H}$$

17.16 Following are melting and boiling points for acetamide and ethyl acetate.

a. What is the physical state (solid, liquid, or gas) of each at room temperature?
b. How might you account for the fact that the boiling point of acetamide is considerably higher than that of ethyl acetate?

$$CH_3-\overset{\overset{\displaystyle O}{\|}}{C}-NH_2 \qquad CH_3-\overset{\overset{\displaystyle O}{\|}}{C}-O-CH_2-CH_3$$

acetamide   ethyl acetate
mp 82.3°C   mp −83.6°C
bp 221.2°C   bp 77.1°C

Preparation of Esters (Section 17.4)

17.17 a. Write an equation for the equilibrium established when acetic acid and 1-propanol are refluxed in the presence of a few drops of concentrated sulfuric acid.
b. Using the data in Section 17.4, calculate the equilibrium constant for this reaction.

17.18 If 15 grams of salicylic acid is reacted with excess methanol, how many grams of methyl salicylate (oil of wintergreen) could be formed?

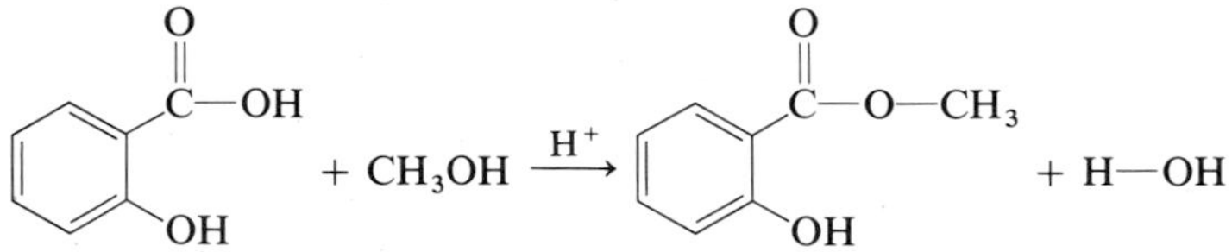

$$\text{salicylic acid} + CH_3OH \xrightarrow{H^+} \text{methyl salicylate} + H-OH$$

salicylic
acid

methyl
salicylate

17.19 Carboxylic acids and alcohols may be converted to esters by a variety of chemical methods. When the acid contains more than one COOH group and the alcohol contains more than one —OH group, then under appropriate experimental conditions, hundreds of molecules may be linked together to give a polyester. Dacron is a polyester of terephthalic acid and ethylene glycol.

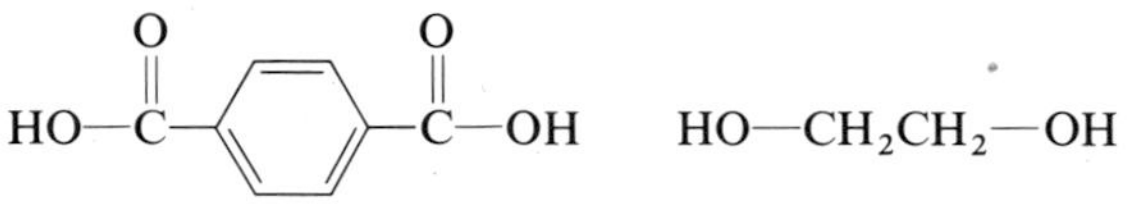

$$HO-C-\bigcirc-C-OH \qquad HO-CH_2CH_2-OH$$

terephthalic acid   ethylene glycol

a. Formulate a structure for Dacron polyester. Be certain to show in principle how several hundred molecules can be hooked together to form the polyester.
b. Write an equation for the chemistry involved when a drop of concentrated hydrochloric acid makes a hole in a Dacron polyester shirt or blouse.
c. From what starting materials do you think the condensation fiber Kodel Polyester is made?

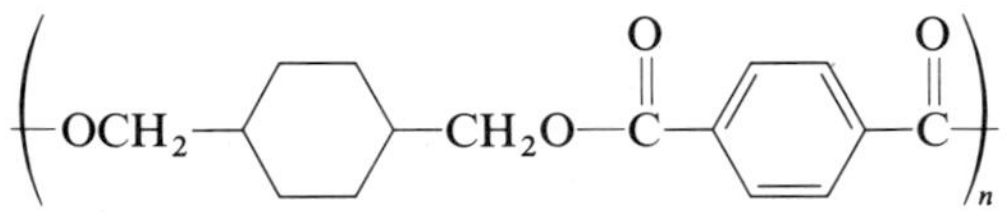

$$\left(OCH_2-\bigcirc-CH_2O-C-\bigcirc-C\right)_n$$

Kodel Polyester

Reactions of Esters, Amides, and Anhydrides (Sections 17.5–17.9)

17.20 Complete the equations for the hydrolysis of the following esters, amides, and anhydrides.

a. $CH_3-C-OCH_2CH_3 + H_2O \longrightarrow$

b. 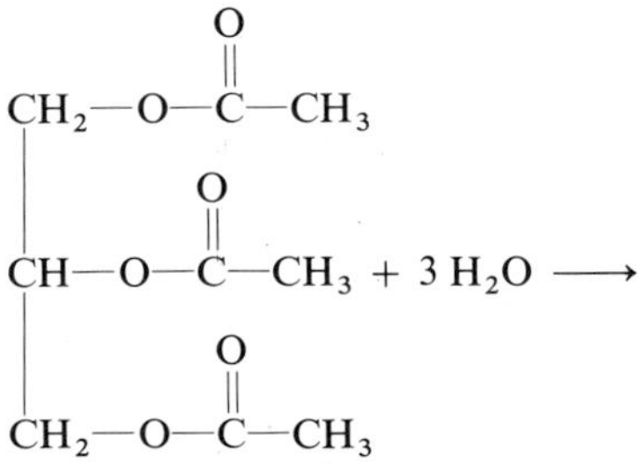
$$\begin{array}{l} CH_2-O-C-CH_3 \\ | \\ CH-O-C-CH_3 + 3\,H_2O \longrightarrow \\ | \\ CH_2-O-C-CH_3 \end{array}$$

c. $CH_3CH_2-\overset{\overset{\displaystyle O}{\|}}{C}-O-CH_2CH_2-O-\overset{\overset{\displaystyle O}{\|}}{C}-CH_2CH_3 + 2\,H_2O \longrightarrow$

d. $CH_3CH_2O-\overset{\overset{\displaystyle O}{\|}}{C}CH_2CH_2\overset{\overset{\displaystyle O}{\|}}{C}-OCH_2CH_3 + 2\,H_2O \longrightarrow$

e. $CH_3\overset{\overset{\displaystyle O}{\|}}{C}-SCH_2CH_3 + H_2O \longrightarrow$

f. $CH_3CH_2O-\overset{\overset{\displaystyle O}{\|}}{C}-OCH_2CH_3 + 2\,H_2O \longrightarrow$

g. 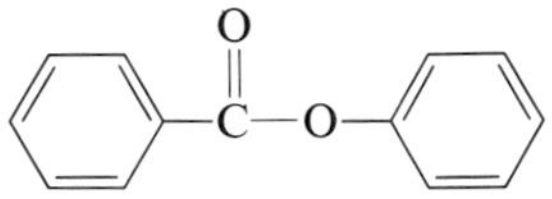 $+\ H_2O \longrightarrow$

h. $H_2N-\overset{\overset{\displaystyle O}{\|}}{C}-NH_2 + 2\,H_2O \longrightarrow$

i. 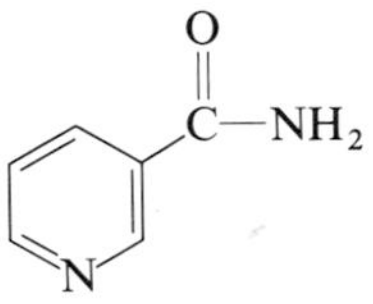 $+\ H_2O \longrightarrow$

j. 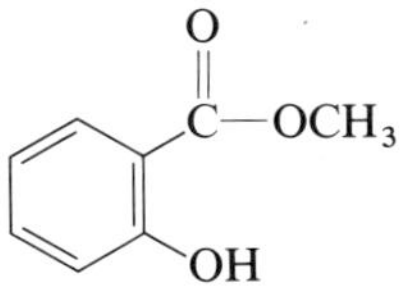 $+\ H_2O \longrightarrow$

k. $CH_3-\overset{\overset{\displaystyle O}{\|}}{C}-NH-CH_2-CH_2-CH_3 + H_2O \longrightarrow$

l. $CH_3CH_2\overset{\overset{\displaystyle O}{\|}}{C}-O-\overset{\overset{\displaystyle O}{\|}}{C}CH_2CH_3 + H_2O \longrightarrow$

m. 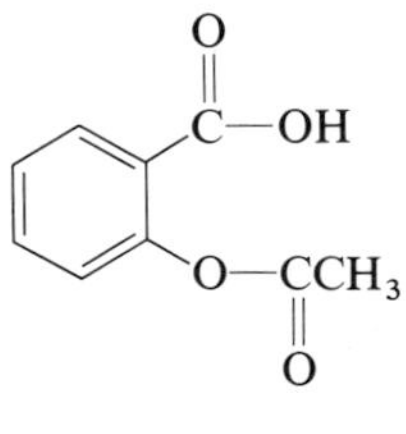 $+\ H_2O \longrightarrow$

17.21 Following are structural formulas of two drugs widely used in clinical medicine. The first is a tranquilizer. Miltown is one of the several trade names for this substance. The second drug, phenobarbital, is a long-acting sedative, hypnotic, and a central nervous system depressant. Phenobarbital is used to treat mild hypertension and temporary emotional strain.

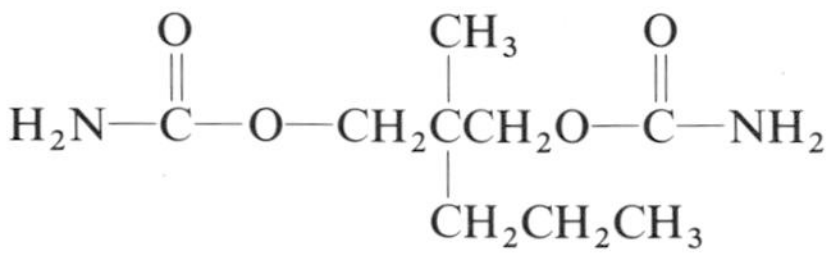

Miltown 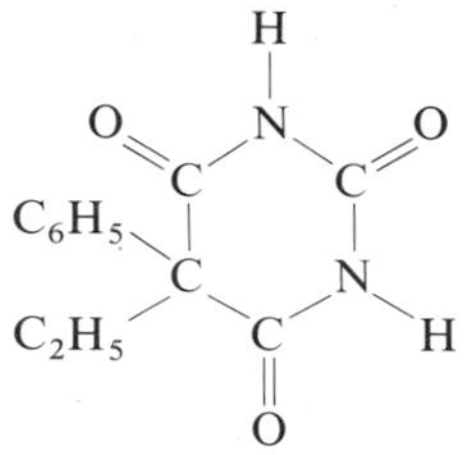   phenobarbital

Predict the products of hydrolysis in aqueous acid of each of these compounds.

17.22 Explain why it is preferable to hydrolyze esters in aqueous base rather than in aqueous acid.

17.23 The following compounds all contain carbon-oxygen-carbon (C—O—C) linkages. Compare the reactivity of each with water.
   a. diethyl ether   b. acetic anhydride   c. ethyl acetate

17.24 The following are derivatives of acetic acid. Compare the reactivities of each with water.
   a. ethyl acetate   b. acetic anhydride   c. acetamide

17.25 Complete the following reactions:

a. $CH_3CH_2\overset{\overset{O}{\|}}{C}-OCH_3 + NH_3 \longrightarrow$

b. 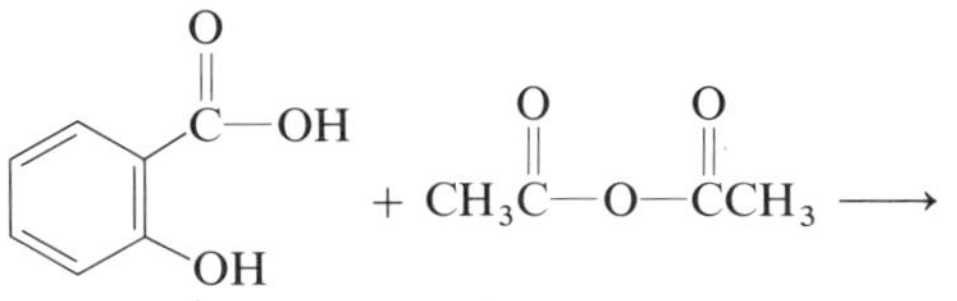 (salicylic acid) $+ CH_3\overset{\overset{O}{\|}}{C}-O-\overset{\overset{O}{\|}}{C}CH_3 \longrightarrow$

c. $CH_3CH_2O-\!\!\!\langle\ \rangle\!\!\!-NH_2 + CH_3\overset{\overset{O}{\|}}{C}-O-\overset{\overset{O}{\|}}{C}CH_3 \longrightarrow$ (phenacetin, a pain reliever)

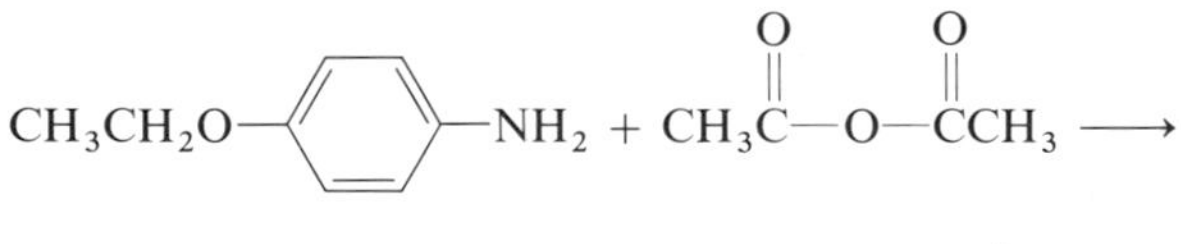

d. 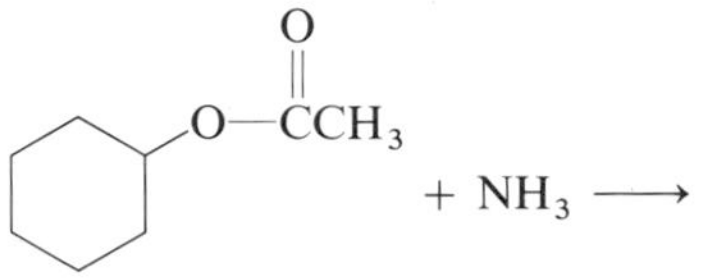 (cyclohexyl acetate) $+ NH_3 \longrightarrow$ e. 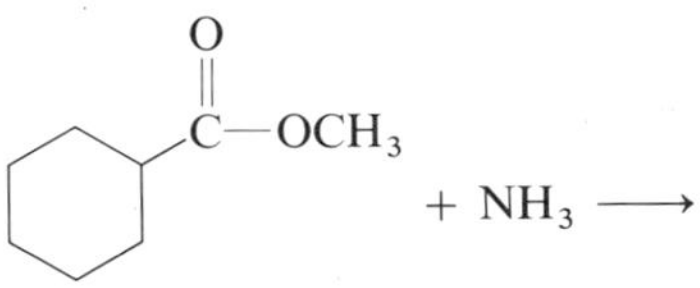 (methyl cyclohexanecarboxylate) $+ NH_3 \longrightarrow$

f. 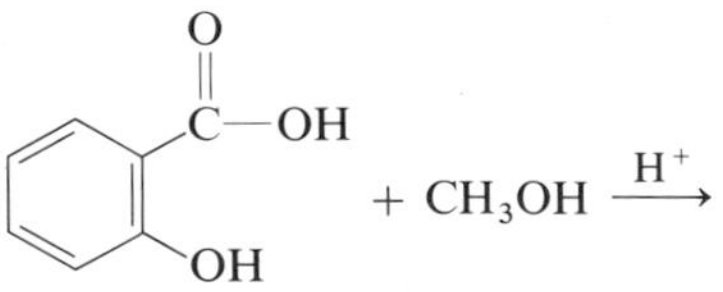 (salicylic acid) $+ CH_3OH \xrightarrow{H^+}$ (oil of wintergreen)

g. $2\ CH_3\overset{\overset{O}{\|}}{C}-O-\overset{\overset{O}{\|}}{C}CH_3 + HOCH_2CH_2OH \longrightarrow$

$$\text{h. } CH_3\overset{\overset{\displaystyle O}{\|}}{C}-OCH_3 + CH_3-\underset{\underset{\displaystyle H}{|}}{N}-H \longrightarrow$$

$$\text{i. } CH_3-\overset{\overset{\displaystyle O}{\|}}{C}-OCH_3 + CH_3-\underset{\underset{\displaystyle CH_3}{|}}{N}-H \longrightarrow$$

17.26 Following is an example of transesterification.

$$CH_3-\overset{\overset{\displaystyle O}{\|}}{C}-O-(CH_2)_6CH_3 + CH_3-OH \xrightarrow{H^+} CH_3-\overset{\overset{\displaystyle O}{\|}}{C}-O-CH_3 + CH_3(CH_2)_6OH$$

a. Draw a structural formula for the tetrahedral carbonyl addition intermediate formed in this reaction.

b. If the methanol used in this reaction is enriched with oxygen-18, would you expect the oxygen-18 label to appear in methyl acetate or 1-heptanol? Explain.

17.27 Show how you might distinguish between the following pairs of compounds by a simple chemical test. In each case, tell what test you would perform, what you would expect to observe, and write an equation for each positive test.

a. isopropyl formate and 2-methylpropanoic acid (isobutyric acid)

b. butanoic acid and butanamide

c. methyl benzoate and acetophenone (methyl phenyl ketone)

d. methyl hexanoate and hexanal

e.

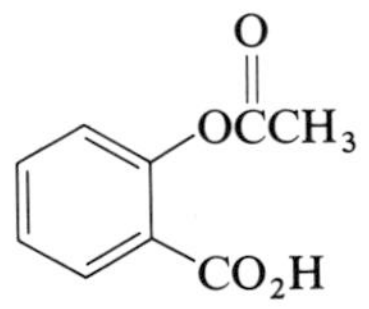

aspirin   and   phenacetin

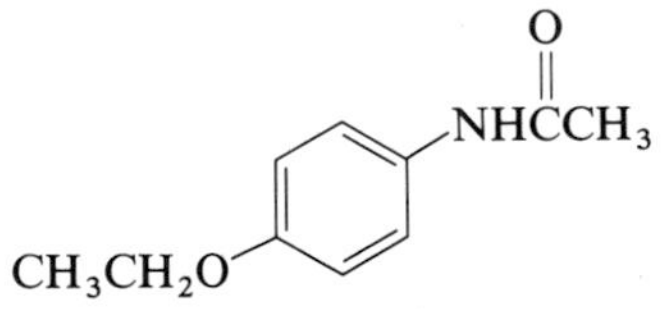

f.

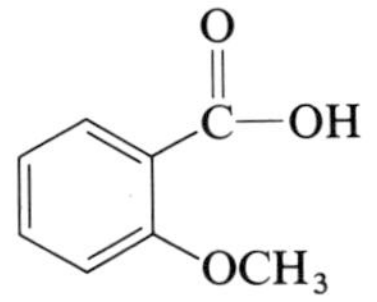

    and

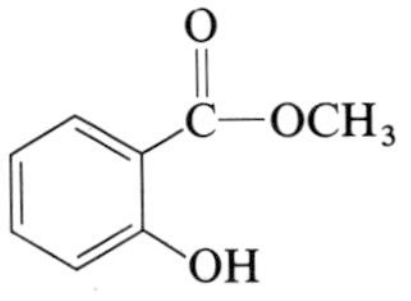

17.28 The following conversions can be done in either one or two steps. Show the reagents you would use and draw structural formulas for any intermediate involved in a two-step reaction. In addition to the indicated starting material, use any necessary inorganic and organic compounds.

$$\text{a. } CH_3CH_2CH_2CH_2CH_2OH \longrightarrow CH_3CH_2CH_2CH_2\overset{\overset{\displaystyle O}{\|}}{C}OH$$

$$\text{b. } CH_3CH_2CH_2CH_2OH \longrightarrow CH_3CH_2CH_2\overset{\overset{\displaystyle O}{\|}}{C}OCH_2CH_2CH_2CH_3$$

c. $CH_3CH_2CH_2CH_2OH \longrightarrow CH_3CH_2CH=CH_2$

d. 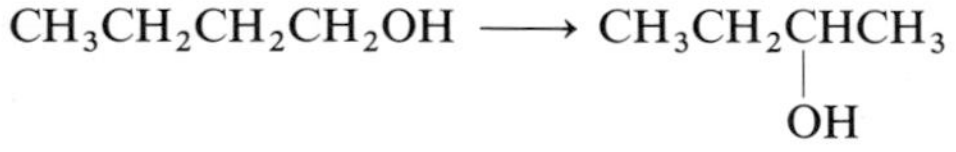 $CH_3CH_2CH_2CH_2OH \longrightarrow CH_3CH_2\underset{\underset{OH}{|}}{C}HCH_3$

e. 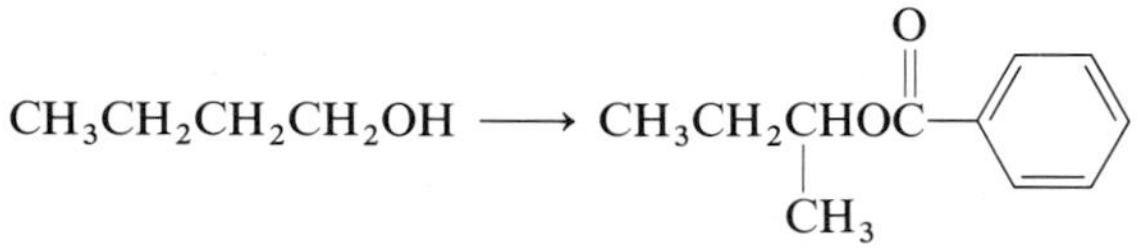 $CH_3CH_2CH_2CH_2OH \longrightarrow CH_3CH_2\underset{\underset{CH_3}{|}}{C}H\overset{\overset{O}{\parallel}}{O}\overset{\overset{O}{\parallel}}{C}\text{—}C_6H_5$

f.

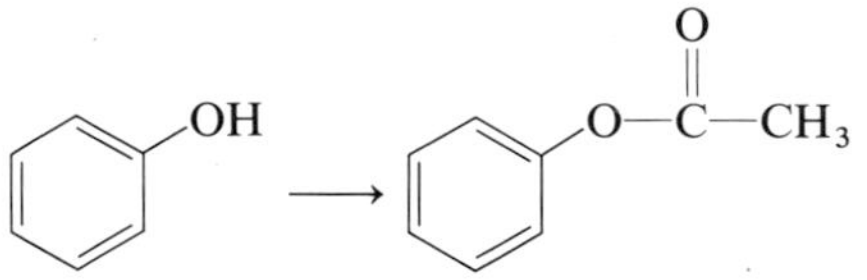

g. 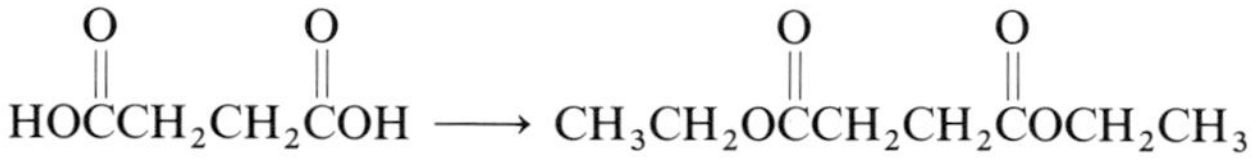 $HO\overset{\overset{O}{\parallel}}{C}CH_2CH_2\overset{\overset{O}{\parallel}}{C}OH \longrightarrow CH_3CH_2O\overset{\overset{O}{\parallel}}{C}CH_2CH_2\overset{\overset{O}{\parallel}}{C}OCH_2CH_3$

h.

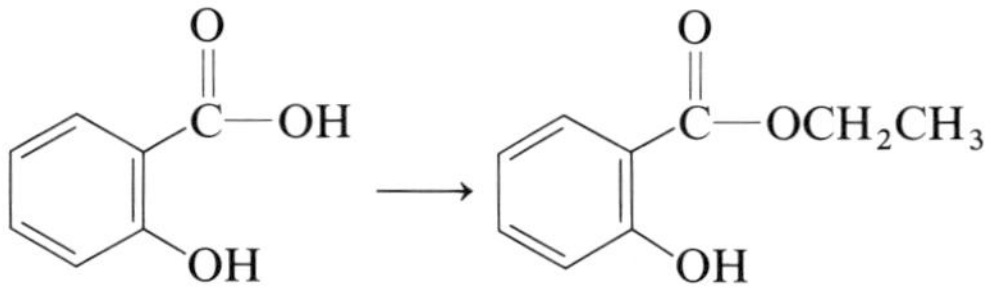

i.

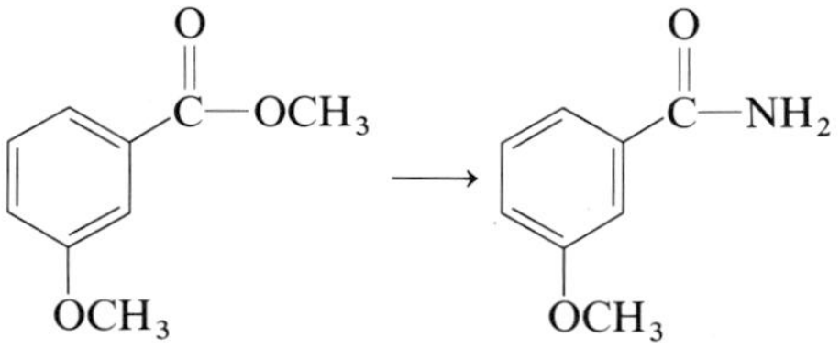

j. 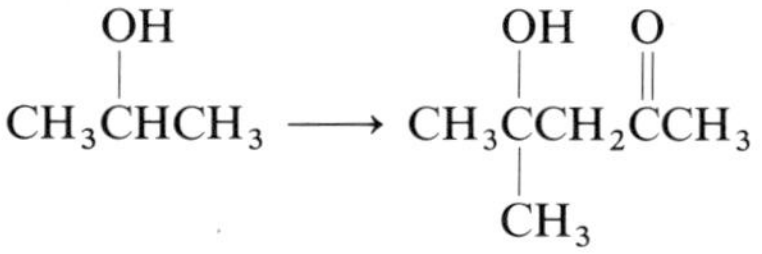 $CH_3\underset{\underset{OH}{|}}{C}HCH_3 \longrightarrow CH_3\underset{\underset{CH_3}{|}}{\overset{\overset{OH}{|}}{C}}CH_2\overset{\overset{O}{\parallel}}{C}CH_3$

k.

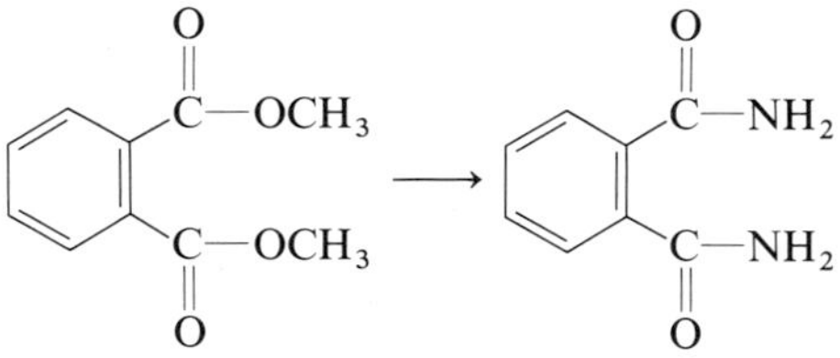

l.

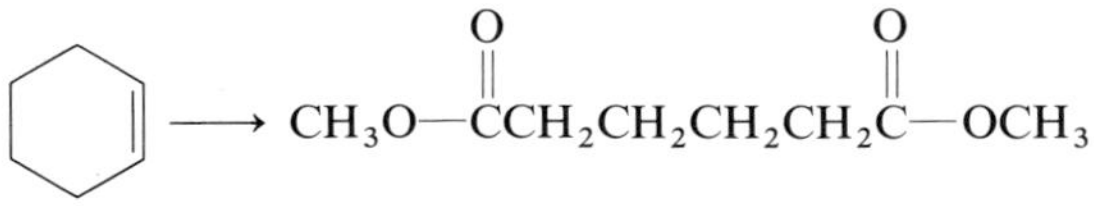

m. 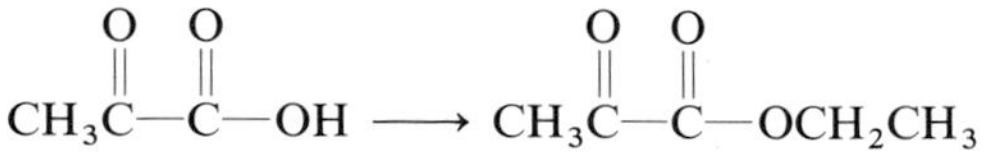 $CH_3\overset{\overset{O}{\parallel}}{C}\text{—}\overset{\overset{O}{\parallel}}{C}\text{—}OH \longrightarrow CH_3\overset{\overset{O}{\parallel}}{C}\text{—}\overset{\overset{O}{\parallel}}{C}\text{—}OCH_2CH_3$

n. 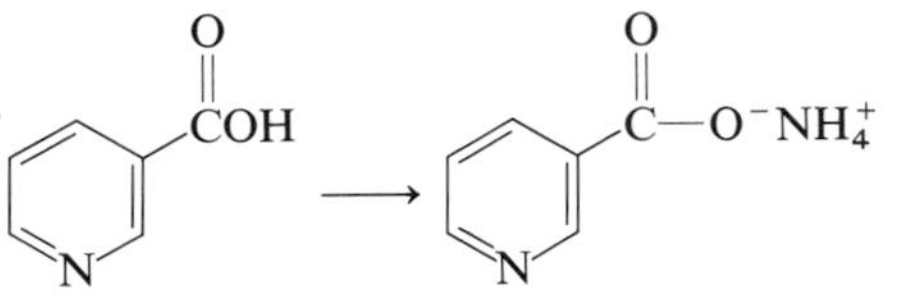 (structure: pyridine-3-carboxylic acid) $\underset{}{\longrightarrow}$ (structure: pyridine-3-carboxylate ammonium salt, $C-O^-NH_4^+$)

Esters, Amides, and Anhydrides of Inorganic Acids (Section 17.10)

17.29 Draw structural formulas for the following phosphate esters.
a. β-D-galactose-6-phosphate
b. β-D-ribose-3-phosphate
c. β-2-deoxy-D-ribose-5-phosphate
d. β-D-ribose-1,3-diphosphate
e. glycerol 1-phosphate

17.30 The backbone of ribonucleic acid (RNA) consists of units of β-D-ribose joined together by phosphate ester bonds between the hydroxyl on carbon-3 of one ribose and the hydroxyl on carbon-5 of another ribose. Draw the structural formula of two units of β-D-ribose joined in this manner.

$$(\beta\text{-D-ribose})\text{-}(\text{phosphate})\text{-}(\beta\text{-D-ribose})$$

17.31 Draw structural formulas for the products of hydrolysis of the following phosphate esters, amides, and anhydrides.

a. 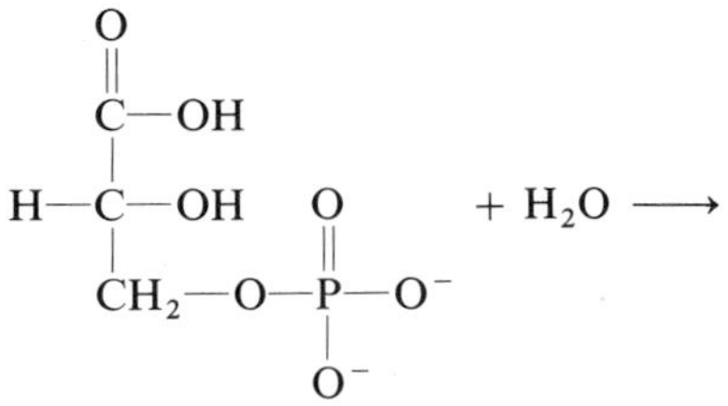 (structure) $\text{H}-\overset{\overset{\displaystyle \text{C}-\text{OH}}{|}}{\underset{\underset{\displaystyle \text{CH}_2-\text{O}-\text{P}-\text{O}^-}{}}{\text{C}}}-\text{OH}$ $+\,H_2O \longrightarrow$

b. $CH_3-\overset{O}{\overset{||}{C}}-O-\overset{O}{\overset{||}{\underset{\underset{O^-}{|}}{P}}}-O^- + H_2O \longrightarrow$

c. $^-O-\overset{O}{\overset{||}{\underset{\underset{O^-}{|}}{P}}}-O-\overset{O}{\overset{||}{\underset{\underset{O^-}{|}}{P}}}-O^- + H_2O \longrightarrow$

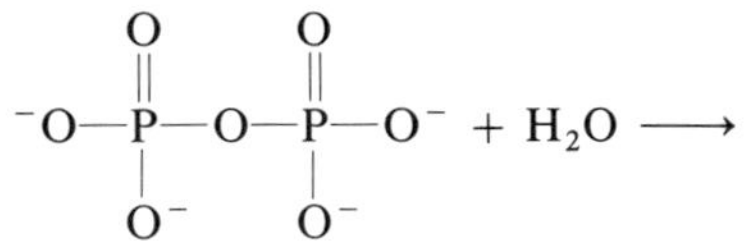

d. 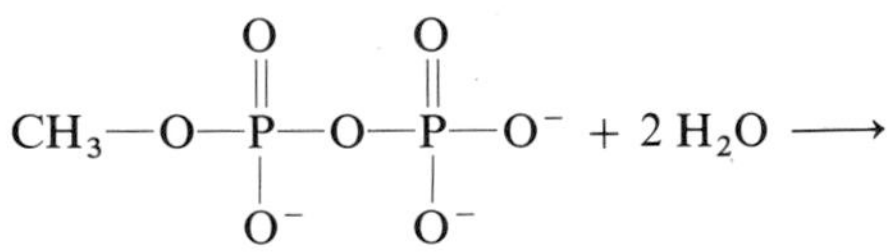 $CH_3-O-\overset{O}{\overset{||}{\underset{\underset{O^-}{|}}{P}}}-O-\overset{O}{\overset{||}{\underset{\underset{O^-}{|}}{P}}}-O^- + 2\,H_2O \longrightarrow$

e. 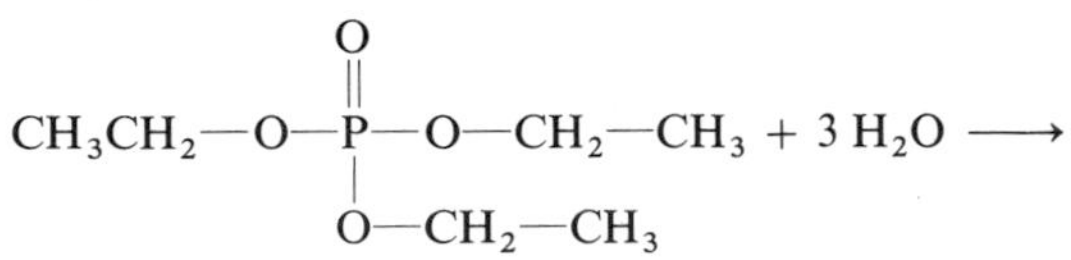 $CH_3CH_2-O-\overset{O}{\overset{||}{\underset{\underset{O-CH_2-CH_3}{|}}{P}}}-O-CH_2-CH_3 + 3\,H_2O \longrightarrow$

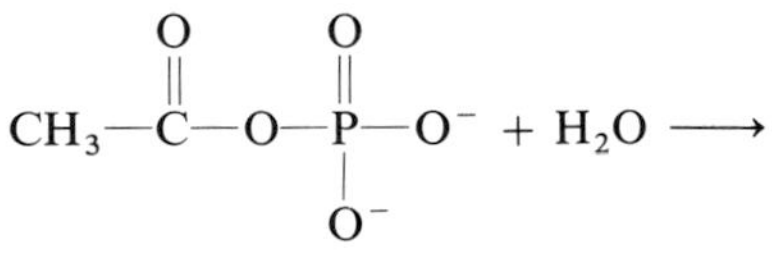

f.

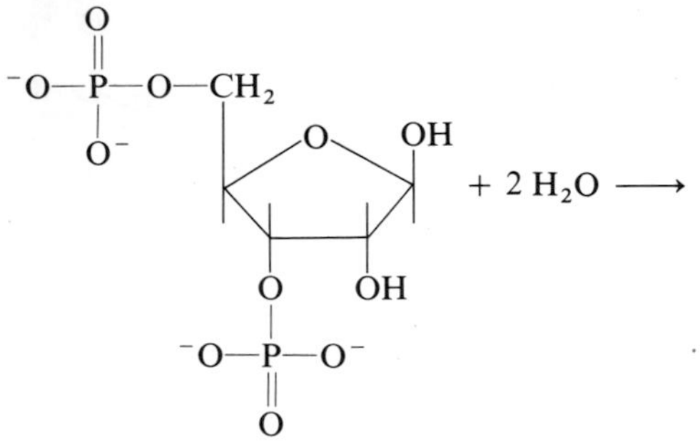

$+ 2 H_2O \longrightarrow$

17.32 Following is the structural formula of β-D-ribose 5-triphosphate.

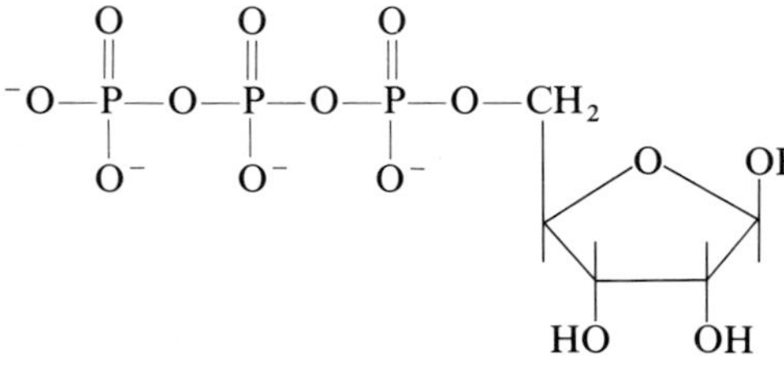

a. Label all phosphate ester and phosphate anhydride bonds.
b. Calculate the net charge on this molecule at pH 7.4.
c. Draw structural formulas for the products of complete hydrolysis of this compound. Show products as they would exist at pH 7.4.

The Claisen Condensation (Section 17.11)

17.33 Draw structural formulas for the products of the following Claisen condensations.

a. $CH_3-CH_2-\overset{\displaystyle O}{\overset{\|}{C}}-O-CH_2-CH_3 \xrightarrow[CH_3CH_2OH]{CH_3CH_2O^-Na^+}$

b. 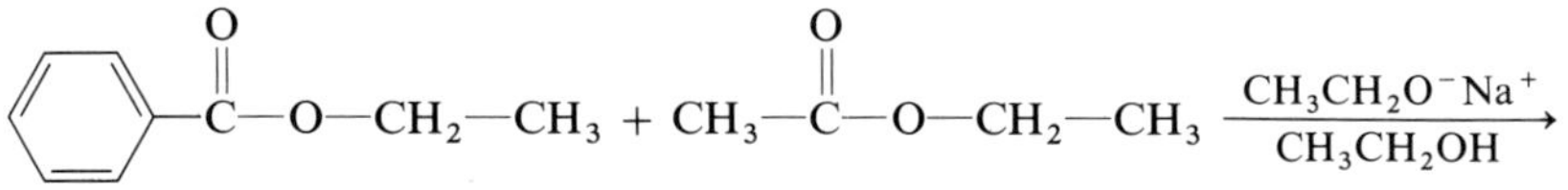 $+ CH_3-\overset{\displaystyle O}{\overset{\|}{C}}-O-CH_2-CH_3 \xrightarrow[CH_3CH_2OH]{CH_3CH_2O^-Na^+}$

17.34 What is the characteristic structural feature of the product of a Claisen condensation?

17.35 What is the characteristic structural feature of the product of an aldol condensation?

17.36 Write equations to show how ethyl propanoate could be converted into the following compounds.

a. $CH_3-CH_2-\overset{\displaystyle O}{\overset{\|}{C}}-\underset{\underset{\displaystyle CH_3}{|}}{CH}-\overset{\displaystyle O}{\overset{\|}{C}}-OCH_2CH_3$

b. $CH_3-CH_2-\overset{\displaystyle O}{\overset{\|}{C}}-\underset{\underset{\displaystyle CH_3}{|}}{CH}-\overset{\displaystyle O}{\overset{\|}{C}}-OH$

c. 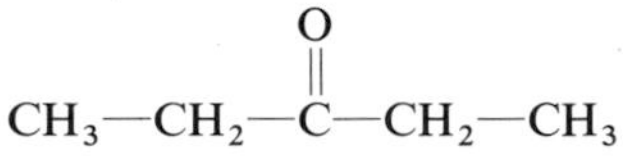 $CH_3-CH_2-\overset{\overset{\displaystyle O}{\|}}{C}-CH_2-CH_3$

d. 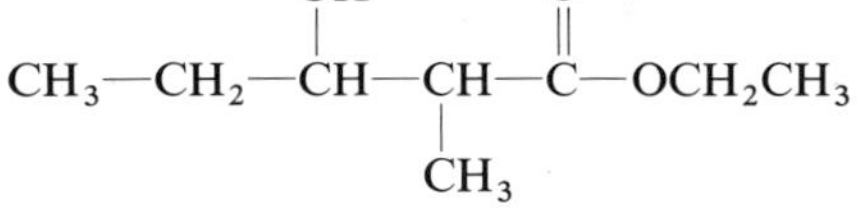 $CH_3-CH_2-\overset{\overset{\displaystyle OH}{|}}{CH}-\underset{\underset{\displaystyle CH_3}{|}}{CH}-\overset{\overset{\displaystyle O}{\|}}{C}-OCH_2CH_3$

e. 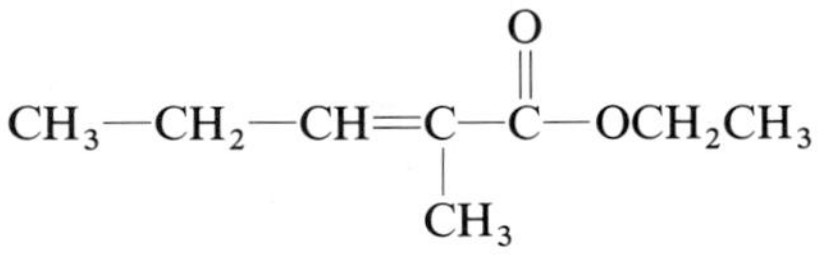 $CH_3-CH_2-CH{=}\underset{\underset{\displaystyle CH_3}{|}}{C}-\overset{\overset{\displaystyle O}{\|}}{C}-OCH_2CH_3$

f. 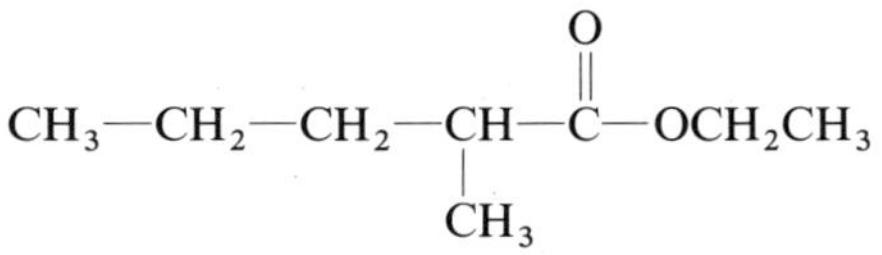 $CH_3-CH_2-CH_2-\underset{\underset{\displaystyle CH_3}{|}}{CH}-\overset{\overset{\displaystyle O}{\|}}{C}-OCH_2CH_3$

g. 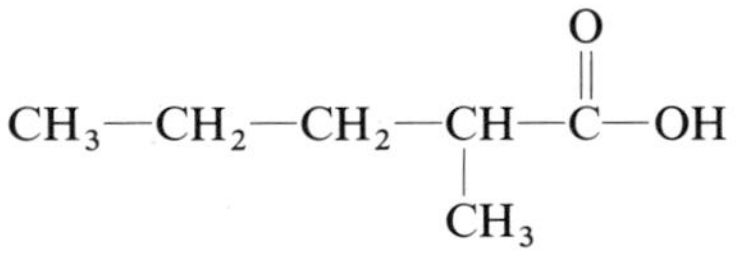 $CH_3-CH_2-CH_2-\underset{\underset{\displaystyle CH_3}{|}}{CH}-\overset{\overset{\displaystyle O}{\|}}{C}-OH$

h. 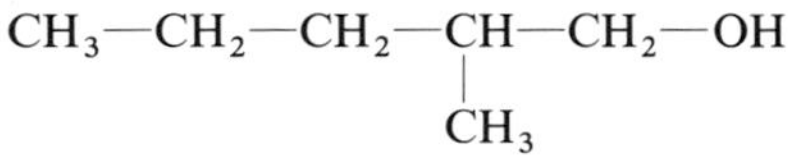 $CH_3-CH_2-CH_2-\underset{\underset{\displaystyle CH_3}{|}}{CH}-CH_2-OH$

Nylon and Dacron

In the years following World War I, a number of chemists recognized the need to develop a basic knowledge of polymer chemistry. One of the most creative of these pioneers was Wallace M. Carothers of E. I. du Pont de Nemours & Co., Inc. In the early 1930s, Carothers and his associates began fundamental research in the reactions of aliphatic dicarboxylic acids with diols. From adipic acid and ethylene glycol, they obtained a polyester of high molecular weight which could be drawn into fibers. However, the melting points of the first polyester fibers obtained by Carothers were too low for them to be used as textile fibers and they were not further investigated until a decade later. Carothers then turned his attention to the reactions of dicarboxylic

acids and diamines and in 1934 synthesized Nylon 66, the first purely synthetic fiber. Nylon 66 is so named because it is synthesized from two different organic starting materials, each having six carbon atoms.

In the synthesis of Nylon 66, adipic acid and hexamethylenediamine (HMDA) dissolved in aqueous alcohol are reacted to form a one-to-one salt called **nylon salt.** Nylon salt is then heated in an autoclave to 250°C. As the temperature increases in the closed system, the internal pressure rises to about 15 atmospheres. Under these conditions, $-CO_2^-$ groups from adipic acid and $-NH_3^+$ groups from hexamethylenediamine react to form amides. Water is formed as a by-product (see equation below). As polymerization proceeds

and more water is formed, steam and alcohol vapors are continuously bled from the autoclave to maintain a constant internal pressure. The temperature is gradually raised to about 275°C and when all water vapor is removed, the internal pressure of the reaction vessel falls to one atmosphere. Nylon 66 formed under these conditions melts at 250–260°C and has a molecular weight range of 10,000–20,000.

In the first stage of fiber production, crude Nylon 66 is melted, spun into fibers, and cooled to room temperature. In the second stage of fiber production, the melt-spun fiber is drawn at room temperature (cold-drawn) to about four times its original length. As the fiber is drawn, chains within the polymer are oriented in the direction of the fiber axis

$$\underset{\text{adipic acid}}{HOC(CH_2)_4COH} + \underset{\substack{\text{hexamethylene} \\ \text{diamine} \\ \text{(HMDA)}}}{H_2N(CH_2)_6NH_2} \longrightarrow \underset{\text{nylon salt}}{{}^-OC(CH_2)_4CO^- \, H_3\overset{+}{N}(CH_2)_6NH_3^+} \xrightarrow{\text{heat}}$$

$$\underset{\text{Nylon 66}}{\left[C(CH_2)_4C-NH(CH_2)_6NH \right]_n} + H_2O$$

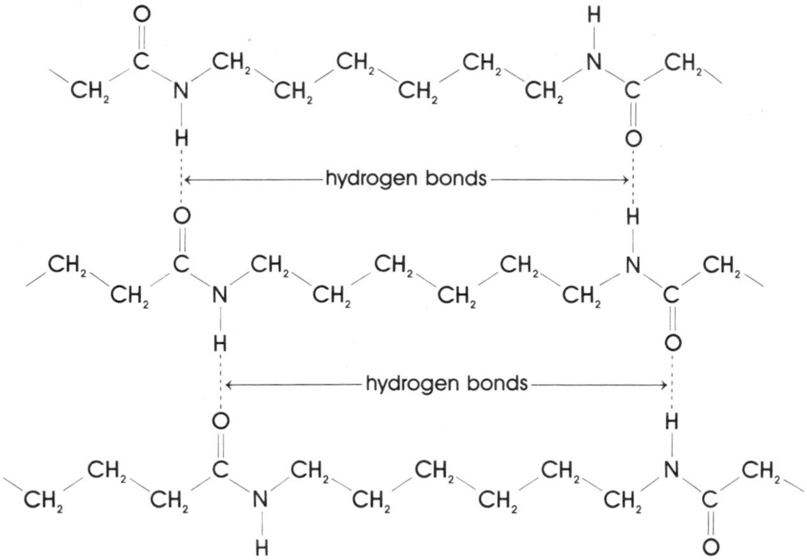

Figure 1 The structure of cold-drawn Nylon 66 fiber. Hydrogen bonds between adjacent chains hold molecules together.

and hydrogen bonds are formed between the carbonyl oxygens of one chain and amide hydrogens of adjacent chains (Figure 1). The effects of orientation of polymer chains on the physical properties of the polymer are dramatic; both tensile strength and stiffness are increased dramatically. Cold-drawing is an important step in the production of all synthetic fibers.

At the time du Pont's management decided to begin production of Nylon 66, of the two reactants, only adipic acid was commercially available. The raw material base for the production of adipic acid at this time was benzene. Catalytic reduction of benzene to cyclohexane followed by air oxidation yields a mixture of cyclohexanol and cyclohexanone. Nitric acid oxi-

dation of this mixture gives adipic acid (see equation below). Adipic acid was, in turn, the starting material for the synthesis of hexamethylenediamine. Reaction of adipic acid with ammonia gives a diammonium salt which when heated gives adipamide. Catalytic reduction of adipamide gives hexamethylenediamine (see equation below). At this time, benzene

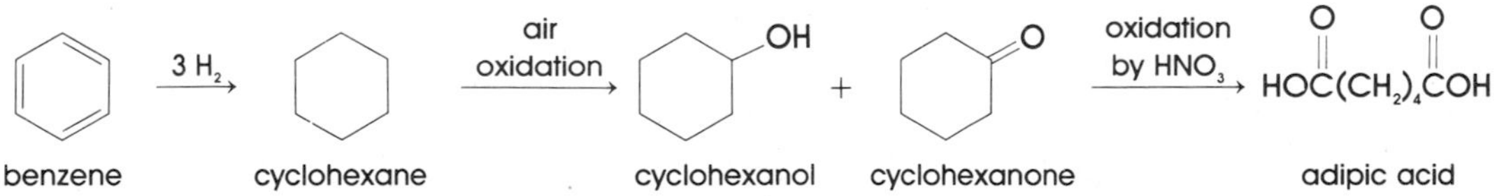

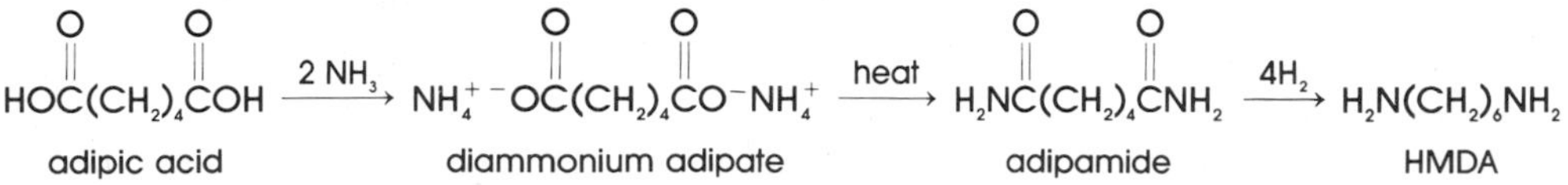

$$CH_2=CH-CH=CH_2 \xrightarrow{Cl_2} Cl-CH_2-CH=CH-CH_2-Cl \xrightarrow{NaCN} N\equiv C-CH_2-CH=CH-CH_2-C\equiv N$$

butadiene   1,4-dichloro-2-butene   1,4-dicyano-2-butene

$$\xrightarrow{5\ H_2} H_2N(CH_2)_6NH_2$$

hexamethylenediamine

was derived largely from coal and both nitric acid and ammonia were derived from nitrogen in the air. Thus, du Pont could rightly claim that Nylon 66 was derived from coal, air, and water.

Following World War II, du Pont embarked on a major expansion of its Nylon 66 capacity, a move that demanded a major expansion in facilities for the production of hexamethylenediamine. The problem facing management was whether the company should continue to make this key starting material from benzene or whether it should convert to a more economical raw material base and a more economical synthesis. Within a few years du Pont developed a new synthesis of HMDA using butadiene as a starting material. Recall that butadiene is obtained as a co-product of thermal (steam) cracking of ethane and other light hydrocarbons extracted from natural gas. It is also obtained from catalytic cracking and reforming of naphthas and other petroleum fractions. Reaction of butadiene with chlorine under carefully controlled conditions forms 1,4-dichloro-2-butene, which is, in turn, reacted with two moles of sodium cyanide to form 1,4-dicyano-2-butene. Reaction of

this intermediate with hydrogen in the presence of a catalyst reduces both the carbon-carbon double bond and the carbon-nitrogen triple bonds (see equation above).

By the early 1950s, almost all hexamethylenediamine was being made from butadiene, derived from petroleum and natural gas. The starting material for the production of adipic acid was and still is benzene. Today, however, benzene is derived almost entirely from catalytic cracking and reforming of petroleum. Thus, the raw material base for the synthesis of Nylon 66 has shifted from coal, air, and water to petroleum and natural gas.

Nylon 66 has been the primary nylon fiber synthesized in the United States and Canada. The primary nylon fiber synthesized in many other parts of the world, particularly Germany, Italy, and Japan, is Nylon 6. The manufacture of Nylon 6 uses only one starting material, caprolactam, a six-carbon cyclic amide.

The raw material for the synthesis of caprolactam is benzene. During synthesis of Nylon 6, caprolactam is partially hydrolyzed and then heated to 250°C to drive off water and bring about polymerization.

Why is Nylon 66 the primary nylon textile fiber produced in the United States and Canada, while Nylon 6 is the primary nylon fiber produced in Germany, Italy, and Japan? The answer lies chiefly in the availability of raw materials. In the United States and Canada, butadiene is readily available from the thermal cracking of ethane and other low-molecular-weight hydrocarbons extracted from natural gas, itself a major natural resource. Because natural gas is not as plentiful in Europe and Japan, these countries have been forced to depend on petroleum as a raw material base. It is more economical for them to synthesize Nylon 6 from caprolactam than it is to synthesize Nylon 66 from adipic acid and hexamethylenediamine.

By the 1940s, scientists

caprolactam   Nylon 6

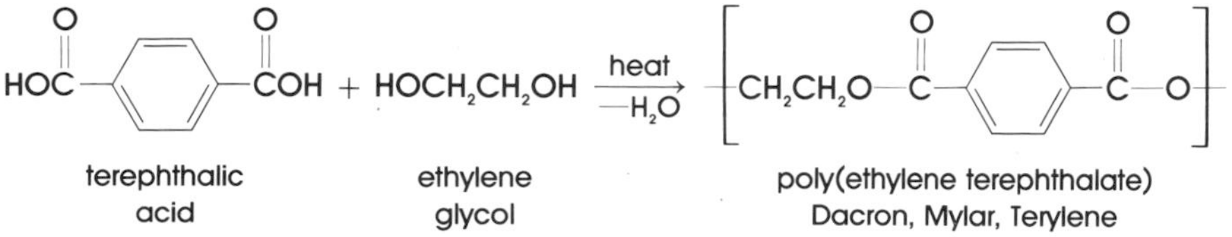

Figure 2 The synthesis of Dacron polyester.

were beginning to understand some of the fundamental relationships between molecular structure and bulk physical properties, and polyester condensations were re-examined. Recall that Carothers and others had already concluded that polyester fibers from aliphatic dicarboxylic acids and diols were not suitable for textile fibers because they were too low-melting. Winfield and Dickson at the Calico Printers Association in England reasoned, quite correctly it turned out, that a greater resistance to rotation in the polymer backbone would stiffen the polymer, raise its melting point, and thereby make it more acceptable as a polyester fiber. To create stiffness in the polyester chain, they used terephthalic acid, an aromatic dicarboxylic acid (Figure 2).

The crude polyester is first spun into fibers and then cold-drawn to form a textile fiber with the trade name *Dacron polyester*. The outstanding features of Dacron polyester are its stiffness (about four times that of Nylon 66), very high strength, and a remarkable resistance to creasing and wrinkling. Because Dacron polyesters are harsh to the touch (due to their stiffness), they are usually blended with cotton or wool to make acceptable textile fibers. Crude polyester is also fabricated into films and tapes and marketed under the trade name *Mylar*.

The sources for ethylene glycol and terephthalic acid are petroleum and natural gas. Ethylene glycol is prepared by air oxidation of eth-

ylene, followed by hydrolysis. Terephthalic acid is obtained by air oxidation of p-xylene, an aromatic hydrocarbon derived along with benzene and toluene from catalytic cracking and reforming of naphthas and other petroleum fractions (Figure 3).

In 1981, the production of man-made textile fibers in the United States exceeded 10 billion pounds. Heading the list were polyester fibers (4.8 billion pounds) and polyamide fibers (2.7 billion pounds).

The fruits of research into the relationships between molecular structure and physical properties of polymers and advances in fabrication techniques are nowhere better illustrated than by the polyaromatic amides, or aramids, introduced by du Pont in the

Figure 3 The starting materials for the synthesis of Dacron and Mylar are derived from petroleum and natural gas.

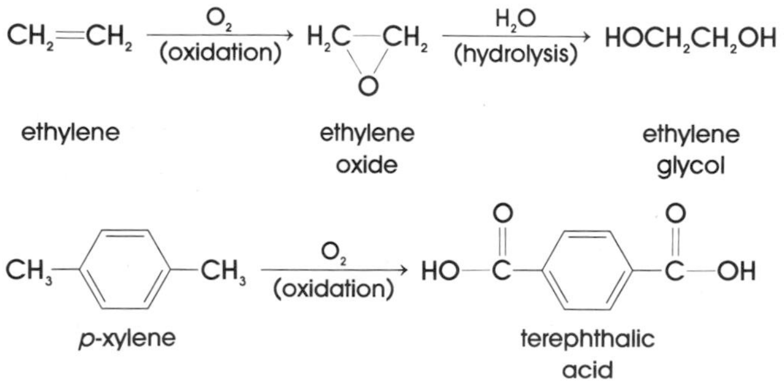

Figure 4 Kevlar, an aramid polymer.

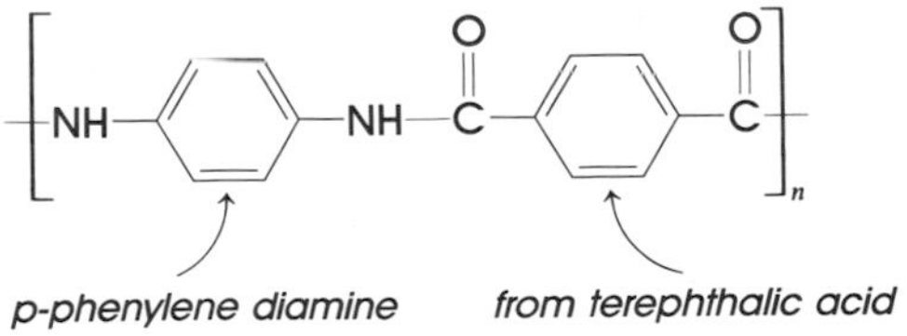

from p-phenylene diamine   *from terephthalic acid*

early 1950s. Researchers reasoned that a polyamide composed of aromatic rings would be stiffer and stronger than a polyamide such as Nylon 66 or Nylon 6, composed of amide bonds connected by hydrocarbon chains. In early 1960, du Pont introduced a polyaromatic amide fiber synthesized from terephthalic acid and 1,4-diaminobenzene (*p*-phenylene diamine). This polymer is marketed under the trade name Kevlar (Figure 4). One of the remarkable features of Kevlar is its extremely light weight, hence, the weight that can be saved by using it as a replacement for other materials. For example, a 3-inch cable woven of Kevlar has a strength equal to that of a similarly woven 3″ steel cable. But whereas the steel cable weighs about 20 pounds per foot, a comparable Kevlar cable weighs only 4 pounds per foot. Kevlar now finds use in such things as anchor cables for offshore drilling rigs and reinforcement fibers for automobile and truck tires. Kevlar can also be woven into a fabric that when struck by a projectile stretches almost like a trampoline and absorbs the impact. Today there is a rapidly growing market among VIPs for Kevlar-lined vests, jackets, and raincoats (Figure 5).

References

Anderson, B.C.; Barton, L.R.; and Collette, J.W. 1980. Trends in polymer development. *Science* 208: 807−812.

Deanin, R.D. 1979. *New industrial polymers*. Washington, D.C.: Am. Chemical Society.

Encyclopedia of polymer science and technology. 1976. New York: Wiley.

Stille, J.K. 1968. *Industrial organic chemistry*. Englewood Cliffs, N.J.: Prentice-Hall.

Wittcoff, H.A., and Reuben, B.G. 1980. *Industrial organic chemicals in perspective*. New York: Wiley.

Figure 5 A Kevlar-lined bulletproof vest. (Courtesy of Peter Garfield)

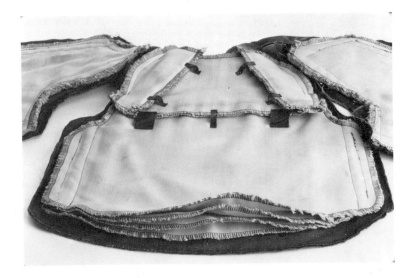

The Penicillins

The most successful of all antibiotics are the penicillins, the first of the so-called miracle drugs. These truly remarkable drugs are almost completely harmless to all living organisms except for certain classes of bacteria. As a result of extensive research on the structure, chemistry, and mechanism of antibacterial activity of penicillins, it is safe to say that we have a clearer understanding of the penicillins than of almost any other class of antibiotics.

The discovery of penicillin was purely accidental. In 1928, the Scottish bacteriologist Alexander Fleming (later Sir Alexander Fleming) reporting the following:

"While working with staphylococcal variants, a number of culture plates were set aside on the laboratory bench and examined from time to time. In the examinations, these plates were necessarily exposed to the air and they became contaminated with various microorganisms. It was noticed that around a large colony of contaminating mold, the staphylococcal colonies became transparent and were obviously undergoing lysis. . . . I was sufficiently interested in the antibacterial substance produced by the mold to pursue the subject."

Because the contaminating mold was *Penicillium notatum,* Fleming named the antibacterial substance *penicillin.* Despite several attempts, he was unable to isolate and purify an active form of it. Nonetheless, he continued to maintain his cultures of the mold.

The outbreak of war in Europe in 1939 stimulated an intensive search for new drugs, and in Great Britain the potential of Fleming's penicillin was reinvestigated. Howard Florey, an Australian experimental pathologist, and Ernst Chain, a Jewish chemist who had fled Nazi Germany, worked together on the project. Using the newly discovered technique of freeze-drying (lyophilization), they succeeded in isolating penicillin from Fleming's cultures. Within a few months, larger quantities of purified penicillin were available and many of its physical, chemical, and antibacterial properties determined. In 1940, Florey and Chain published a report of treatment of bacterial infection in mice and two years later the drug was tested on humans. By 1943, pharmaceutical companies in Britain and the United States were producing penicillin on a large scale and it was authorized for use by the military. The following year, penicillin became available for civilian use, and in 1945 Fleming, Florey, and Chain were awarded the Nobel Prize in Medicine and Physiology. Thus, in less than two decades, penicillin progressed from a chance observation in a research laboratory to a drug that has been recognized as one of the greatest contributions of medical science in the service of humanity.

Preliminary investigations of the structure of penicillin presented a confusing picture until it was discovered that *P. notatum* produces different kinds of penicillin, depending on the composition of the medium in which the mold is grown. Initially, six different penicillins were recognized. All proved to be derivatives of 6-aminopenicillanic acid (Figure 1, next page). Of the six, penicillin G (benzyl penicillin) became the most widely used and the standard against which others were judged.

The structural formula of penicillins consists of a five-member ring fused to a four-member ring. The four-member ring is a cyclic amide, the special name for which is *lac-*

Figure 1 The penicillins.

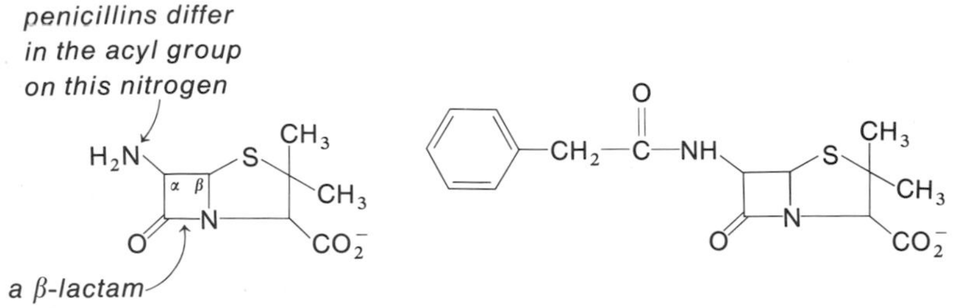

penicillins differ
in the acyl group
on this nitrogen

a β-lactam

6-aminopenicillanic acid

penicillin G or
benzylpenicillin
$(R— = C_6H_5—CH_2—)$

tam. Because the nitrogen atom of this lactam is on the carbon atom beta to the carbonyl group, the general name given to this type of ring is β-lactam.

The penicillins undergo a variety of chemical reactions, some of which are very complex. We will concentrate on just two of these, each chosen because it has important consequences for the medical use of these antibiotics. Treatment of penicillin with aqueous HCl brings about hydrolysis of the amide bond of the highly strained β-lactam and cleavage of the five-member ring as well (Figure 2). Both penaldic acid and penicillamine are devoid of antibacterial activity. Because penicillin G is rapidly inactivated by this type of hydrolysis in the acid

conditions of the stomach, it cannot be administered orally; it must be given by injection.

The second important reaction of the penicillins is selective hydrolysis of the β-lactam ring catalyzed by a group of enzymes known as β-lactamases (Figure 3). The product of β-lactamase-catalyzed hydrolysis is penicilloic acid. Like penaldic acid and penicillamine, penicilloic acid possesses no antibacterial activity. The main basis for the natural resistance of certain bacteria to penicillins is the ability of these resistant strains to synthesize β-lactamases and thereby inactive penicillins with which they come in contact.

The penicillins owe their antibacterial activity to a

common mechanism that inhibits the biosynthesis of a vital part of bacterial cell walls. Within a bacterial cell, the concentration of lower-molecular-weight substances is often considerably higher than that in the surrounding medium, causing a high osmotic pressure within the cell. The osmotic pressure within certain types of bacterial cells is as high as 10–20 atmospheres and without the support of a cell wall, the bacterial cell would rupture. One of the simplest types of bacterial cell wall is made up of long polysaccharide chains to which are attached short polypeptide chains (Figure 4).

The final sequence of reactions in the construction of a bacterial cell wall is formation of amide bonds be-

Figure 2 Acid-catalyzed hydrolysis of penicillin. Neither penaldic acid nor penicillamine have any antibiotic activity.

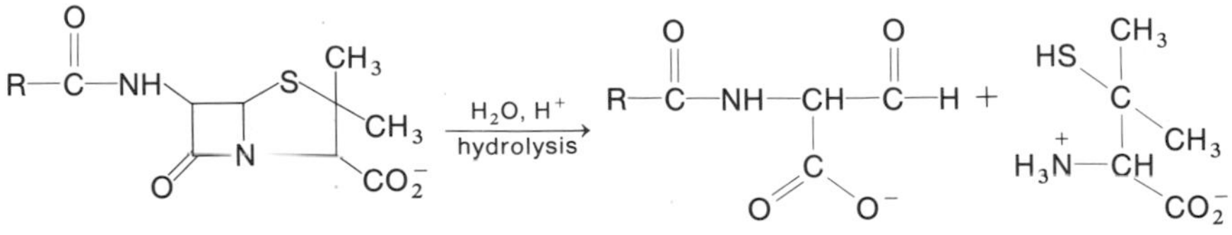

penicillin

penaldic acid

penicillamine

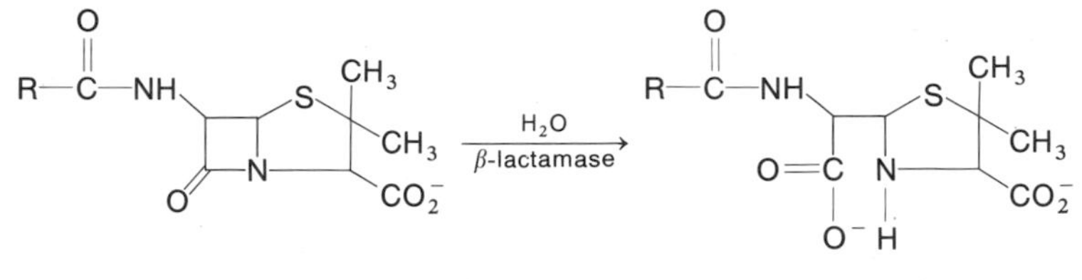

Figure 3 Selective hydrolysis of the β-lactam ring catalyzed by β-lactamases.

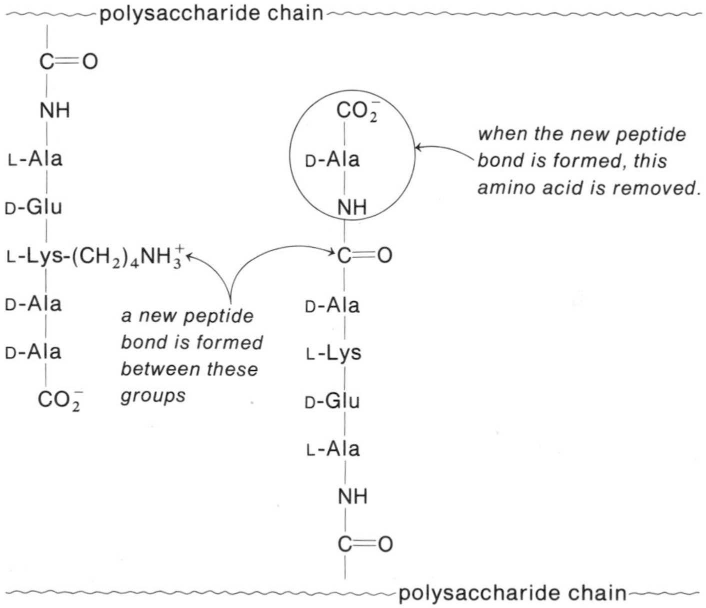

penicillin   penicilloic acid

Figure 4 Bacterial cell walls are constructed of long polysaccharide chains to which are attached short polypeptide chains. In the final stage of cell wall construction, cross-linking of the peptide chains creates an enormous bag-shaped macromolecule.

when the new peptide bond is formed, this amino acid is removed.

a new peptide bond is formed between these groups

tween adjacent polypeptide chains (Figure 4). Through this type of polypeptide bond interchange, polysaccharide chains are cross-linked to form one enormous polymer, a literally bag-shaped macromolecule.

It is the formation of the final cross-linked macromolecule that is inhibited by the penicillins; a number of hypotheses have been proposed to explain how they do this. One hypothesis is that the penicillins are similar in structure to D-alanyl-D-alanine, the terminal amino acids of the short peptide chains that must be cross-linked. This similarity is illustrated in Figure 5. Penicillin is thought to bind selectively to the active site of the enzyme complex that catalyzes peptide bond interchange in the final step of cell wall construction. Hence, at the molecular level, penicillin's antibacterial activity appears to be one of selective enzyme inhibition.

The fact that this pattern of cell wall construction is unique

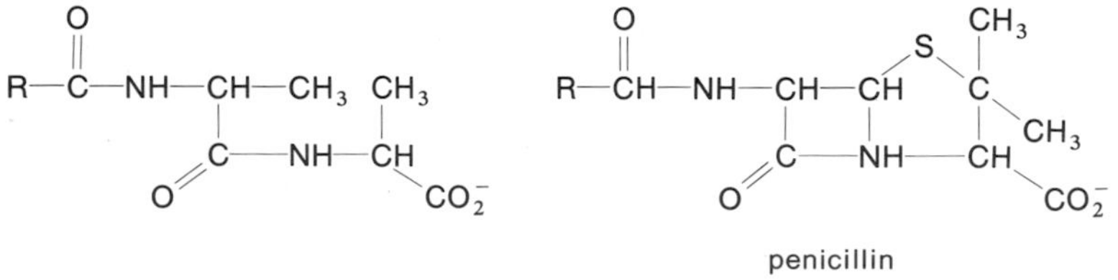

Figure 5 Structural formulas of D-alanyl-D-alanine and penicillin drawn to suggest a structural similarity between the two.

penicillin

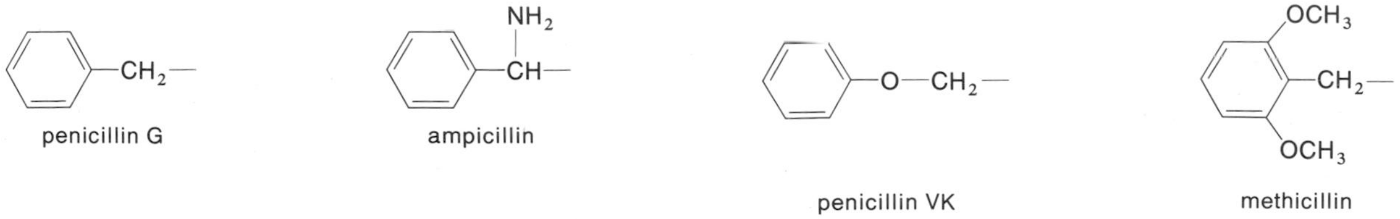

penicillin G

ampicillin

penicillin VK

methicillin

Figure 6 Structural relationships between penicillin G and three widely used semisynthetic penicillins.

to bacteria and that it is not found in mammalian cells no doubt accounts for the lack of toxicity of penicillins to humans. However, the use of penicillins does have its problems. A significant percentage of the population has become hypersensitive to the drug and develops severe allergic reactions. The factor responsible for the allergic reactions is not penicillin itself but rather certain degradation products, particularly 6-aminopenicillanic acid (Figure 1).

The susceptibility of penicillin G to hydrolysis by acid and the emergence of β-lactamase-producing strains of bacteria have provided incentive for scientists in the medical professions and pharmaceutical industries to develop newer, more effec-

tive penicillins. As a result of this effort, by 1974 over 20,000 semisynthetic penicillins had been prepared. At the present time, the three most widely prescribed penicillins are ampicillin, penicillin G, and penicillin VK. The side chains of each of these, along with that of methicillin, are shown in Figure 6.

Penicillin G is most effective against gram-negative bacteria. Ampicillin is a broader range antibiotic and attacks both gram-negative and gram-positive bacteria. Penicillin VK is more resistant to acid hydrolysis than penicillin G and can be given orally. Methicillin is about 100 times more resistant to the action of β-lactamases than penicillin G and is used to treat infections caused by "pen-

icillin-resistant" organisms.

Soon after the penicillins were introduced into medical practice, resistant strains began to appear and have proliferated ever since. Many argue that widespread, unnecessary use of penicillins is the primary cause for the emergence of resistant strains. One of the most serious is *Neisseria gonorrhoeae,* a strain of which causes a gonorrhea that is very difficult to treat. One approach to the problem of resistant strains is to synthesize newer, more effective β-lactam antibiotics. The penicillins were the first of the so-called β-lactam antibiotics. A newer class of β-lactam antibiotics are the **cephalosporins** (Figure 7). The first cephalosporin was isolated from the fungus *Cephalosporium*

Figure 7 Cephalothin, one of the first of a new generation of β-lactam antibiotics.

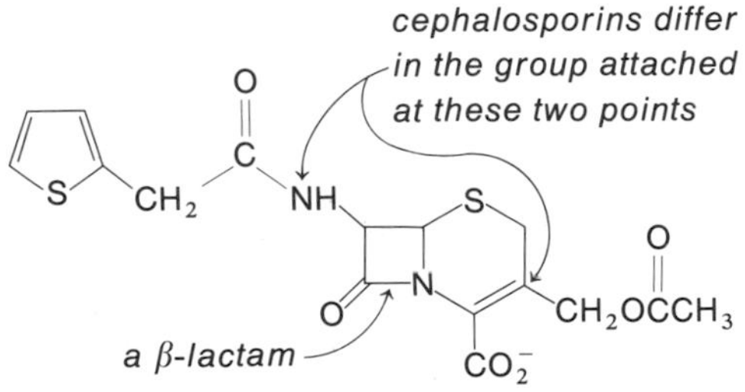

cephalosporins differ in the group attached at these two points

a β-lactam

acremonium. Several cephalosporins have already been approved for clinical use and approval of at least a dozen others is pending. Clinical studies indicate that this class of β-lactam antibiotics has a broader spectrum of antibacterial activity than the penicillins and at the same time has a greater resistance to β-lactamases. Cephalosporins now account for approximately 35% of all antibiotic use.

The cephalosporins contain a β-lactam ring fused to a six-member ring containing atoms of sulfur and nitrogen and a carbon-carbon double bond. As a family, cephalosporins differ from one another in the acyl group attached to the —NH$_2$ group of the β-lactam ring and in the substituent on carbon-3 of the six-member ring (Figure 7).

An entirely different approach to the treatment of β-lactam resistant infections is to use either a penicillin or a cephalosporin in combination with a compound that inhibits the activity of β-lactamases. One such compound, clavulinic acid, has virtually no antibiotic activity by itself but it is a powerful, irreversible inhibitor of β-lactamases. Note that clavulinic acid itself is a β-lactam (Figure 8). With this combination of drugs, it now appears possible to interfere not only with an infectious organism's essential biochemistry but also with its first line of defense against attack by drugs.

The penicillins and cephalosporins are and almost certainly will continue to be

a β-lactam

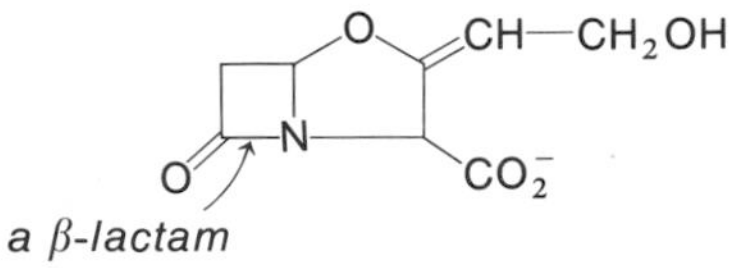

Figure 8 Clavulinic acid, a powerful inhibitor of β-lactamases.

among the most important anti-infective agents. Because the possibilities for substituting different groups on the essential ring structures of each are manifold, it is probable that newer, even more effective β-lactam antibiotics are yet to be discovered. So, too, is it probable that, in response to these drugs, even newer, drug-resistant strains of bacteria will emerge.

Lipids

Lipids are a heterogeneous class of naturally occurring organic compounds, grouped together not by the presence of a distinguishing functional group or structural feature, but rather on the basis of common solubility properties. Lipids are all insoluble in water and very soluble in one or more organic solvents, including ether, chloroform, benzene, and acetone. In fact, these four solvents are often referred to as *lipid-solvents* or *fat-solvents*. Proteins, carbohydrates, and nucleic acids are largely insoluble in these solvents.

In this chapter we will describe the structure and biological function of representative members of the five major types of lipids: fats and oils, waxes, phospholipids, the fat-soluble vitamins, and steroids. In addition, we will describe the structure of biological membranes and the function of a group of membrane-bound polysaccharides.

18.1 Fats and Oils

A. Structure

You certainly are familiar with fats and oils since you encounter them every day in such things as milk, butter, oleomargarine, corn oil, and other liquid vegetable oils, as well as in many other foods. Fats and oils are triesters of glycerol and are called **triglycerides.** Triglycerides are the most abundant naturally occurring lipids. Complete hydrolysis of a triglyceride yields one molecule of glycerol and three molecules of fatty acid.

$$
\begin{array}{l}
CH_2-O-\overset{\overset{\displaystyle O}{\|}}{C}-R \\[2ex]
CH-O-\overset{\overset{\displaystyle O}{\|}}{C}-R \quad +\ 3\,H_2O \ \longrightarrow \\[2ex]
CH_2-O-\overset{\overset{\displaystyle O}{\|}}{C}-R
\end{array}
\qquad
\begin{array}{l}
CH_2-OH \\[2ex]
CH-OH \quad +\ 3\,R-\overset{\overset{\displaystyle O}{\|}}{C}-OH \\[2ex]
CH_2-OH
\end{array}
$$

a triglyceride   glycerol   fatty acids

A triglyceride in which all three fatty acids are identical is called a *simple* triglyceride; an example is tristearin. Simple triglycerides are rare in nature, with mixed triglycerides much more common. Following is a mixed triglyceride formed from glycerol and molecules of stearic acid, oleic acid, and linoleic acid, three of the most abundant fatty acids.

$$
\begin{array}{l}
CH_2-O-\overset{\overset{\displaystyle O}{\|}}{C}-(CH_2)_{16}CH_3 \\[2ex]
CH-O-\overset{\overset{\displaystyle O}{\|}}{C}-(CH_2)_{16}CH_3 \\[2ex]
CH_2-O-\overset{\overset{\displaystyle O}{\|}}{C}-(CH_2)_{16}CH_3
\end{array}
$$

a simple triglyceride
(tristearin)

$$
\begin{array}{l}
CH_2-O-\overset{\overset{\displaystyle O}{\|}}{C}-(CH_2)_{16}CH_3 \\[2ex]
CH-O-\overset{\overset{\displaystyle O}{\|}}{C}-(CH_2)_7CH=CH(CH_2)_7CH_3 \\[2ex]
CH_2-O-\overset{\overset{\displaystyle O}{\|}}{C}-(CH_2)_7CH=CHCH_2CH=CH(CH_2)_4CH_3
\end{array}
$$

an ester of stearic acid
an ester of oleic acid
an ester of linoleic acid

a mixed triglyceride

B. Physical Properties

The physical properties of a triglyceride depend on its fatty acid components. In general, the melting point of a triglyceride increases as the number of carbons in its hydrocarbon chains increases and decreases as the degree of unsaturation increases. Triglycerides rich in oleic acid, linoleic acid, and other unsaturated fatty acids are generally liquid at room temperature and are called **oils.** Triglycerides rich in palmitic, stearic, and other saturated fatty acids are generally semisolids or solids at room temperature and are called **fats.** Table 18.1 (next page) lists the percent composition in grams of fatty acid per 100 grams of triglyceride for several common fats and oils. Notice that beef tallow is approximately 41.5% saturated and 53.1% unsaturated fatty acids by weight.

Table 18.1 Fatty acid composition by weight of several triglycerides. Percentages are given for the most abundant fatty acids; other fatty acids are present in lesser amounts.

| Fat or Oil | mp (°C) | Saturated Fatty Acids | | Unsaturated Fatty Acids | | |
		Palmitic	Stearic	Palmitoleic	Oleic	Linoleic
depot fat (human)	15	24.0	8.4	5.0	46.9	10.2
beef tallow	—	27.4	14.1	—	49.6	2.5
corn oil	−20	10.2	3.0	1.5	49.6	34.3
peanut oil	3.0	8.3	3.1	—	56.0	26.0
soybean oil	−16	9.8	2.4	0.4	28.9	50.7
wheat germ oil	—	{	16.0 }	—	28.1	52.3
butter fat	32	29.0	9.2	4.6	26.7	3.6

Figure 18.1 A saturated triglyceride, tri-palmitin. (a) Structural formula; (b) space-filling model.

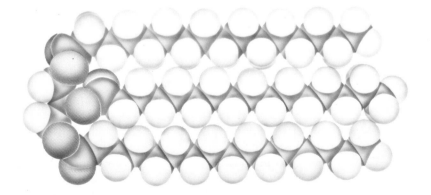

(a)

(b)

Vegetable oils such as corn oil, soybean oil, and wheat germ oil, are all approximately 80% by weight unsaturated fatty acids. Butter fat is distinctive in that it contains significant amounts of lower-molecular-weight fatty acids.

The lower melting points of triglycerides rich in unsaturated fatty acids are related to differences in three-dimensional shape between the hydrocarbon chains of unsaturated and saturated fatty acid components. Shown in Figure 18.1 is the structural formula of tripalmitin and a space-filling model of this saturated triglyceride. Notice that the three saturated hydrocarbon chains of tripalmitin lie parallel to each other and that the molecule has an ordered, compact shape. Dispersion forces between these hydrocarbon chains are strong. Because of their compact nature and interaction of dispersion forces, triglycerides rich in saturated fatty acids have melting points above room temperature.

The three-dimensional shape of an unsaturated fatty acid is quite different from that of a saturated fatty acid. Recall from Section 16.6 that in unsaturated fatty acids of higher organisms, *cis*- isomers predominate, and *trans*- isomers are very rare. Figure 18.2 (next page) shows the structural formula of a triglyceride derived from one molecule each of palmitic acid, oleic acid, and linoleic acid. Notice the *cis* configuration about the double bonds in the hydrocarbon chains of oleic and linoleic acids. Also shown in Figure 18.2 is a space-filling model of this unsaturated triglyceride.

Compared to saturated triglycerides, molecules of unsaturated triglycerides are more bulky, pack together less well, and dispersion forces between them are weaker. For these reasons, unsaturated triglycerides have lower melting points than saturated triglycerides.

Example 18.1

Following is the fatty acid composition by percent of two triglycerides. Predict which triglyceride has the lower melting point.

	Palmitic Acid	Stearic Acid	Palmitoleic Acid	Oleic Acid	Linoleic Acid
triglyceride A	24.0	8.4	5.0	46.9	10.2
triglyceride B	9.8	2.4	0.4	28.9	50.7

Solution

Triglyceride A is composed of approximately 32% saturated fatty acids and 62% unsaturated fatty acids. Triglyceride B is composed of 12% saturated fatty acids and 80% unsaturated fatty acids. Of the unsaturated fatty acids in B, more than 50% are linoleic acid, a fatty acid with two double bonds. Because of its higher degree of unsaturation, triglyceride B has a lower melting point. Refer to Table 18.1 and you will see that triglyceride A is human depot fat (mp 15°C) and triglyceride B is soybean oil (mp −16°C).

Figure 18.2 An unsaturated triglyceride. (a) Structural formula; (b) space-filling model.

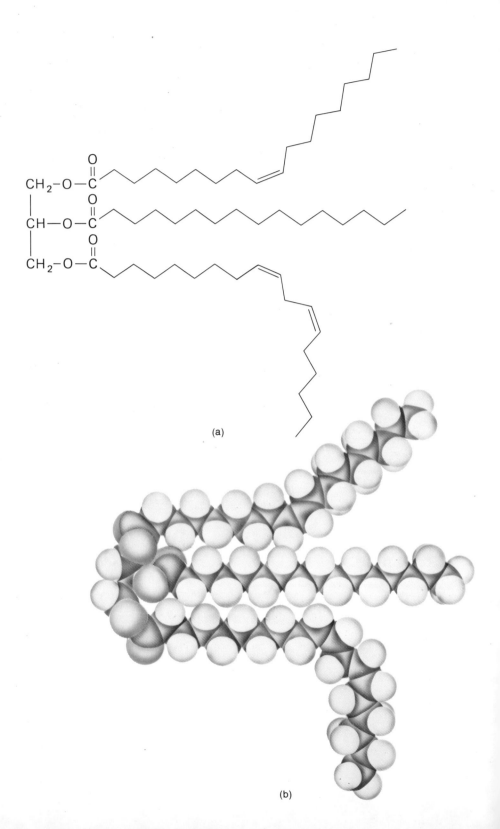

(a)

(b)

Problem 18.1   How do you account for the fact that both beef tallow and corn oil are composed of approximately 50% oleic acid, yet they have such different melting points?

C. Rancidity

On exposure to air, most triglycerides develop an unpleasant odor and flavor, and are said to become **rancid.** In part, rancidity is the result of slight hydrolysis of the fat and oil, causing production of low-molecular-weight fatty acids. The odor of rancid butter is due largely to the presence of butanoic acid formed by the hydrolysis of butterfat. These same low-molecular-weight fatty acids can be formed by air oxidation of unsaturated fatty-acid side chains. The rate of rancidification varies with individual triglycerides, largely because of the presence of certain naturally occurring substances called **antioxidants,** which inhibit the process. One of the most common lipid antioxidants is vitamin E (Section 18.5C).

D. Reduction of Fats and Oils

For a variety of reasons, in part convenience and in part dietary preference, conversion of oils to fats has become a major industry. The process is called *hardening* and involves reaction of an oil with hydrogen in the presence of a catalyst and reduction of some or all of the carbon-carbon double bonds of a triglyceride. If all double bonds are reduced (saturated with hydrogen), the resulting triglyceride is hard and brittle. In practice, the degree of hardening is carefully controlled to produce fat of a desired consistency. The resulting fats are sold for kitchen use (Crisco, Spry, and others). Oleomargarine and other butter substitutes are prepared by hydrogenation of cottonseed, soybean, corn, or peanut oils. The resulting product is often churned with milk and artificially colored to give it a flavor and consistency resembling those of butter.

18.2 Waxes

Waxes are esters of fatty acids and alcohols, each having from 16 to 34 carbon atoms. Carnauba wax, which coats the leaves of the carnauba palm native to Brazil, is largely myricyl cerotate, $C_{25}H_{51}CO_2C_{30}H_{61}$. Beeswax, secreted from the wax glands of the bee, is largely myricyl palmitate, $C_{15}H_{31}CO_2C_{30}H_{61}$.

$$CH_3(CH_2)_{14}\overset{\overset{\textstyle O}{\|}}{C}-O(CH_2)_{29}CH_3$$

myricyl palmitate
(a major component of beeswax)

Waxes are harder, more brittle, and less greasy to the touch than fats. Applications are found in polishes, cosmetics, ointments, and other pharmaceutical preparations.

18.3 Phospholipids

Phospholipids are the second most abundant kind of naturally occurring lipids. They are found almost exclusively in plant and animal membranes, which typically consist of about 40–50% phospholipids and 50–60% protein.

   The most abundant phospholipids contain glycerol and fatty acids, as do the simple fats. In addition, they also contain phosphoric acid and a low-molecular-weight alcohol. The most common of these low-molecular-weight alcohols are choline, ethanolamine, serine, and inositol.

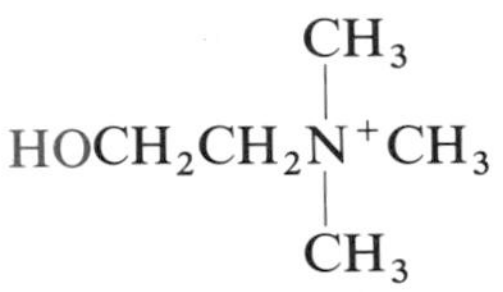

$$HOCH_2CH_2\overset{\overset{\displaystyle CH_3}{|}}{\underset{\underset{\displaystyle CH_3}{|}}{N}}{}^+CH_3 \qquad HOCH_2CH_2NH_3^+ \qquad HOCH_2\underset{\underset{\displaystyle NH_3^+}{|}}{CH}CO_2^-$$

choline   ethanolamine   serine 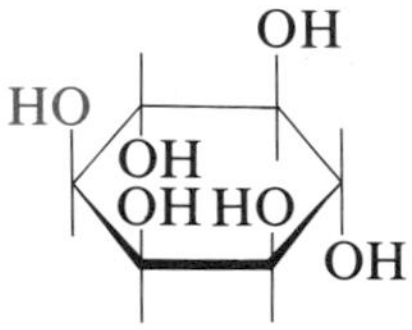   inositol

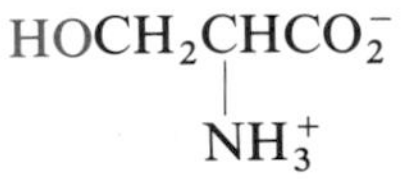

   The most abundant phospholipids in higher plants and animals are the lecithins and the cephalins.

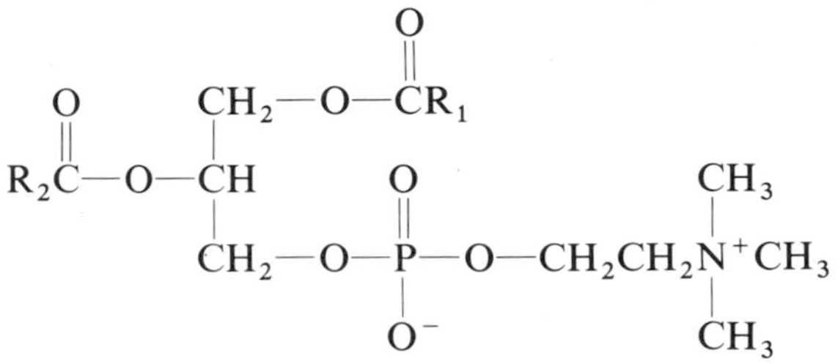

a lecithin
(*a phospholipid containing choline*)

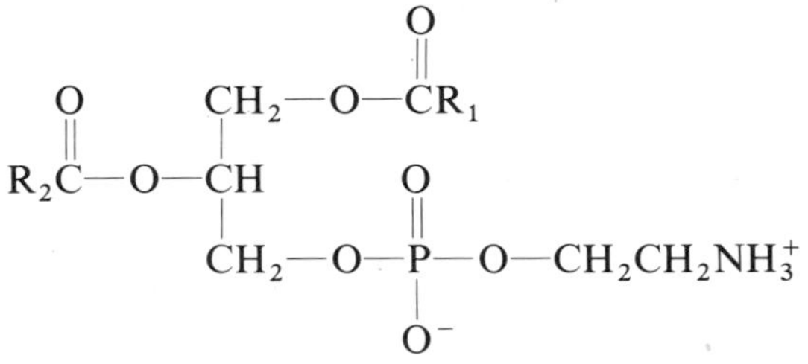

a cephalin
(*a phospholipid containing ethanolamine*)

Lecithins are phosphate esters of choline and **cephalins** are phosphate esters of ethanolamine. Lecithin and cephalin are shown as they would be ionized at pH 7.4. The fatty acids most common in these membrane phospholipids are palmitic and stearic acids (both fully saturated) and oleic acid (one double bond in the hydrocarbon chain).

18.4 Steroids

Steroids are a group of lipids that contain as a characteristic structural feature four fused carbon rings: 3 six-member rings and 1 five-member ring. These seventeen carbon atoms make up a structural unit known as the **steroid nucleus.**

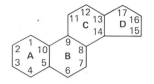

Figure 18.3 The steroid nucleus.

Figure 18.3 shows both the numbering system and the letter designations for the steroid nucleus.

The steroid nucleus is found in a number of extremely important bio-molecules. For our discussion, we will divide these into four groups: cholesterol, adrenocorticoid hormones, sex hormones, and bile acids.

A. Cholesterol

Cholesterol (Figure 18.4) is a white, water-insoluble compound found in varying amounts in practically all living organisms except bacteria. In animal cells, it serves (1) as a component of biological membranes, and (2) as the precursor of bile acids, steroid hormones, and vitamin D. In humans, the central and peripheral nervous systems have a very high cholesterol content (about 10% of dry brain weight). Human plasma contains an average of 50 mg of free cholesterol per 100 mL and about 170 mg of cholesterol esterified with fatty acids. Gallstones are almost pure cholesterol.

Since it is relatively easy to measure the concentration of cholesterol in serum, a great deal of information has been collected in attempts to correlate serum cholesterol levels with various diseases. One of these diseases, arterio-sclerosis or hardening of the arteries, is among the most common diseases of aging. With increasing age, humans normally develop decreased capacity to metabolize fat, and therefore cholesterol concentration in tissues increases. When arteriosclerosis is accompanied by a build-up of cholesterol and other lipids on the inner surfaces of arteries, the condition is known as **atherosclerosis** and results in a decrease in the diameter of the channels through which blood must flow. This decreased diameter, together with increased turbulence, leads to a greater probability of clot formation. If a blood vessel is blocked by a clot, cells may be deprived of oxygen and die. Death of tissue in this way is called *infarction*.

Infarction can occur in any tissue and the clinical symptoms depend upon which vessels and tissues are involved. Myocardial infarction involves the myocardium, or heart muscle tissue. For a discussion of an important laboratory method for diagnosis of myocardial infarction, see the mini-essay "Clinical Enzymology—the Search for Specificity."

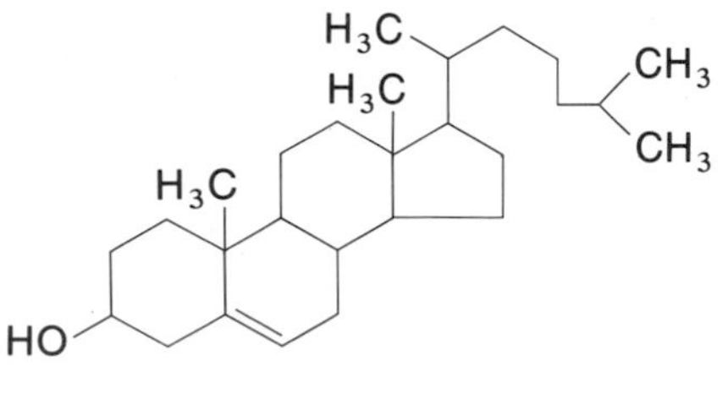

Figure 18.4 The structure of cholesterol.

B. Adrenocorticoid Hormones

The cortex of the adrenal gland synthesizes several hormones that affect (1) water and electrolyte balance, and (2) carbohydrate and protein metabolism. These are the **adrenocorticoid hormones.** Those that control mineral balance are called *mineralocorticoid* hormones; those that control glucose and carbohydrate balance are called *glucocorticoid* hormones.

Aldosterone (Figure 18.5) is the most effective mineralocorticoid hormone secreted by the adrenal cortex. This hormone acts on kidney tubules to stimulate resorption of sodium ions, thus regulating water and electrolyte metabolism. An adult on a diet with normal sodium content secretes about 0.1 mg of aldosterone per day.

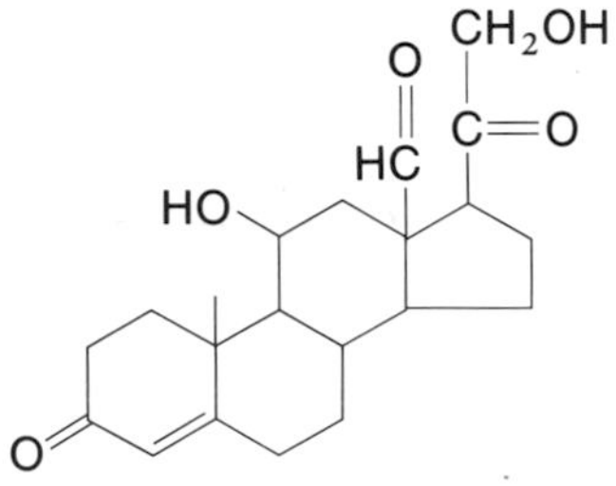

Figure 18.5 The structure of the mineralocorticoid hormone aldosterone.

Cortisol (Figure 18.6) is the principal glucocorticoid hormone of the adrenal cortex, which secretes about 25 mg of this substance per day. Cortisol affects: the metabolism of carbohydrates, proteins, and fats; water and electrolyte balance; and inflammatory processes within the body. In the presence of cortisol, the synthesis of protein in muscle tissue is depressed, protein degradation is increased, and there is an increase in the supply of free amino acids in both muscle cells and blood plasma. The liver, in turn, is stimulated to use the carbon skeletons of certain amino acids for the synthesis of glucose and glycogen. Thus, cortisol and other glucocorticoid hormones act to increase the supply of glucose and liver glycogen at the expense of body protein. Cortisol also has some mineralocorticoid action: it promotes resorption of sodium ions

cortisol   cortisone   prednisolone

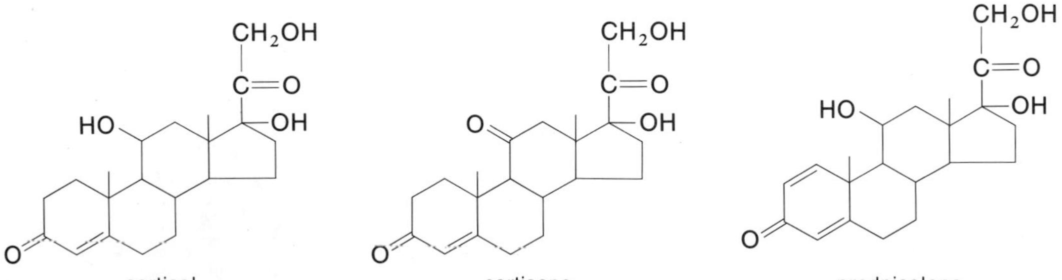

Figure 18.6 Some glucocorticoid hormones.

Bile acids have several important functions. First, they are products of the breakdown of cholesterol and thus are a major pathway for the elimination of cholesterol from the body via the feces. Second, because they are able to emulsify fats in the intestine, bile acids aid in the digestion and absorption of dietary fats. Third, they can dissolve cholesterol by the formation of cholesterol-bile salt micelles or cholesterol-lecithin-bile salt micelles. In this way cholesterol, whether it is from the diet, synthesized in the liver, or removed from circulation by the liver, can be made soluble.

18.5 Fat-Soluble Vitamins

Vitamins are divided into two broad classes on the basis of solubility: those that are fat-soluble and those that are water-soluble. The **fat-soluble vitamins** include vitamins A, D, E, and K. At the present time, the molecular basis of their function is poorly understood. The water-soluble vitamins serve as coenzymes or as building blocks for coenzymes (Section 21.3). The role of coenzymes is quite well understood.

A. Vitamin A

Vitamin A, or retinol, is a primary alcohol of molecular formula $C_{20}H_{30}O$. Vitamin A alcohol occurs only in the animal world, where the best sources are cod-liver oil and other fish-liver oils, animal liver, and dairy products. Vitamin A in the form of a precursor, or provitamin, is found in the plant world in pigments called carotenes. The most common of these, β-carotene ($C_{40}H_{56}$), has an orange-red color and is used as a food coloring. The carotenes have no vitamin A activity; however, after ingestion, β-carotene is cleaved at the central carbon-carbon double bond to give a vitamin-A-related molecule.

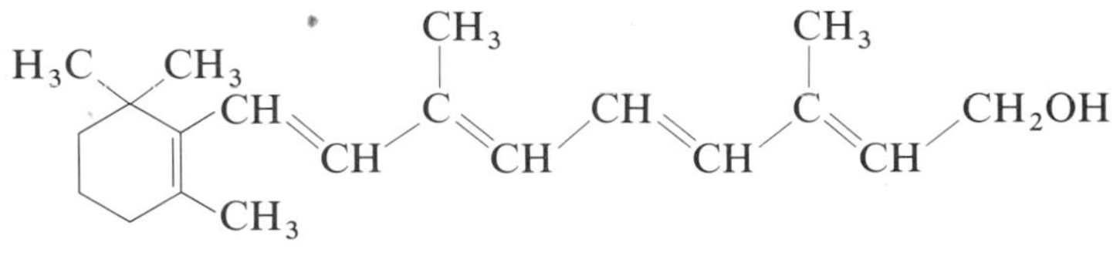

vitamin A
(retinol)

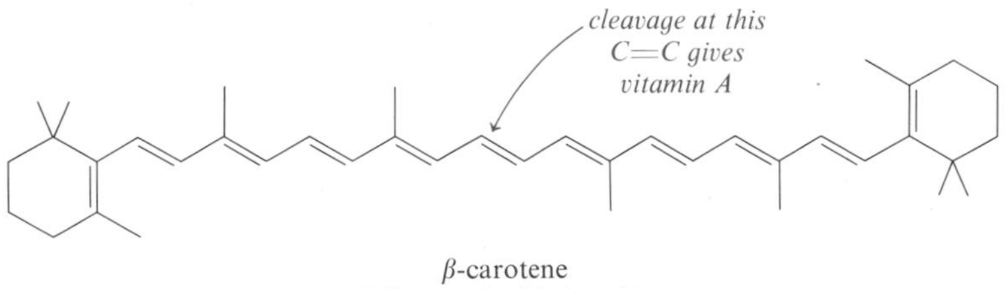

cleavage at this C=C gives vitamin A

β-carotene
(all *trans* double bonds)

A deficiency of vitamin A or vitamin-A precursors leads to a slowing or stopping of growth. Probably the major action of this vitamin is on epithelial cells, particularly those of the mucous membranes of the eye, respiratory tract, and genitourinary tract. Without adequate supplies of vitamin A, these mucous membranes become hard and dry, a process known as keratinization. One of the first and most obvious effects of vitamin-A deficiency is on the eyes. As cells of the tear glands become keratinized, they stop secreting tears and the external surfaces of the eyes become dry, dull, and often scaly. Without tears to remove bacteria, the eye is much more susceptible to serious infection. If this condition is not treated in time, blindness results. The mucous membranes of the respiratory, digestive, and urinary tracts also become keratinized in vitamin-A deficiency and become susceptible to infection.

A less serious condition, one frequently seen in humans whose diets contain insufficient vitamin A, is night blindness. This is the inability to see in dim light or to adapt to a decrease in light intensity.

B. Vitamin D

The term **vitamin D** is a generic name for a group of structurally related compounds produced by the action of ultraviolet light on certain provitamins. Vitamin D_3, also called cholecalciferol, is produced in the skin of mammals by the action of sunlight on 7-dehydrocholesterol. Sunlight causes opening of one of the six-member rings and the formation of a triene. With normal exposure to sunlight, enough 7-dehydrocholesterol is converted to vitamin D_3 so that no dietary vitamin D is necessary. Only when skin manufacture of vitamin D_3 is inadequate is there a need to supplement the diet with artificially fortified foods or multivitamins. Vitamin D_3 has little or no biological activity, but must be metabolically activated before it can function in its target tissues. In the liver, vitamin D_3 undergoes a selective enzyme-catalyzed oxidation at carbon-25 of the side chain to form 25-hydroxyvitamin D_3. This oxidation corresponds to conversion of a C—H bond to a C—OH group. Although 25-hydroxyvitamin D_3 is the most abundant form in the circulatory system, it has only modest biological activity and undergoes further oxidation in the kidneys to form 1,25-dihydroxyvitamin D_3, the hormonally active form of the vitamin. Notice that the first oxidation of the activation process takes place in the liver; the second oxidation takes place in the kidneys.

The principal function of vitamin D metabolites is to regulate calcium metabolism. 1,25-Dihydroxyvitamin D_3 acts in the small intestine to facilitate absorption of calcium and phosphate ions; it acts in the kidneys to stimulate reabsorption of filtered calcium ions; and it acts in bone to stimulate demineralization and release calcium and phosphate ions into the blood stream. A deficiency of vitamin D in childhood is associated with rickets, a mineral-metabolism disease that leads to bowlegs, knock-knees, and enlarged joints.

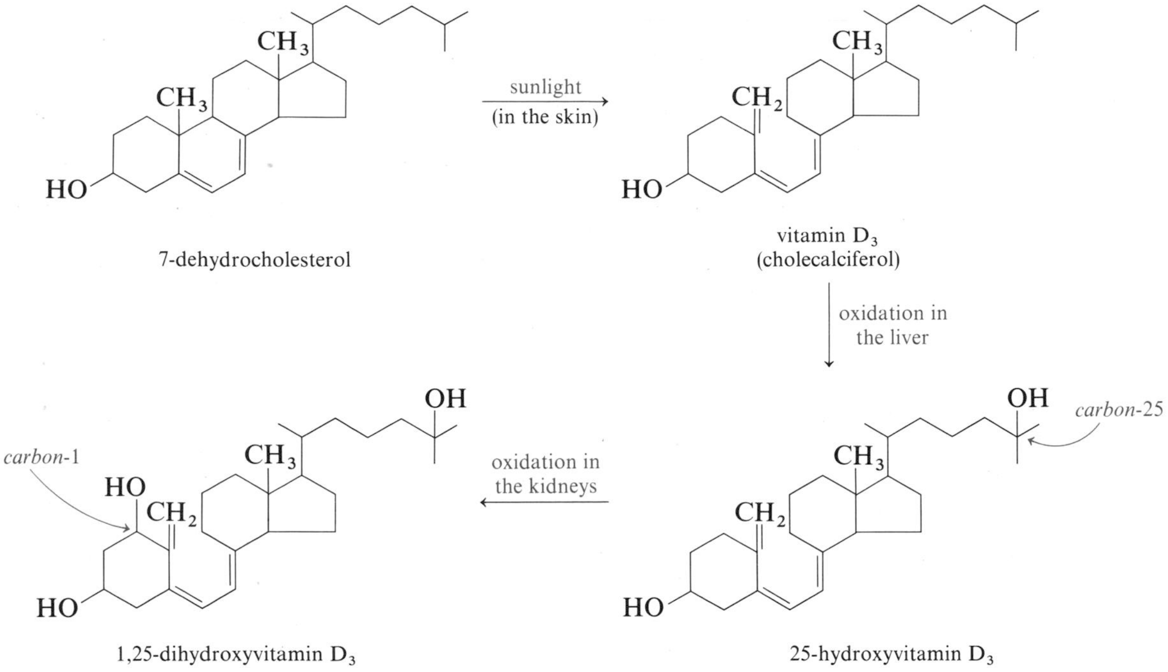

7-dehydrocholesterol

vitamin D$_3$
(cholecalciferol)

1,25-dihydroxyvitamin D$_3$

25-hydroxyvitamin D$_3$

C. Vitamin E

Vitamin E is a group of about seven compounds of similar structure. Of these, α-tocopherol has the greatest potency.

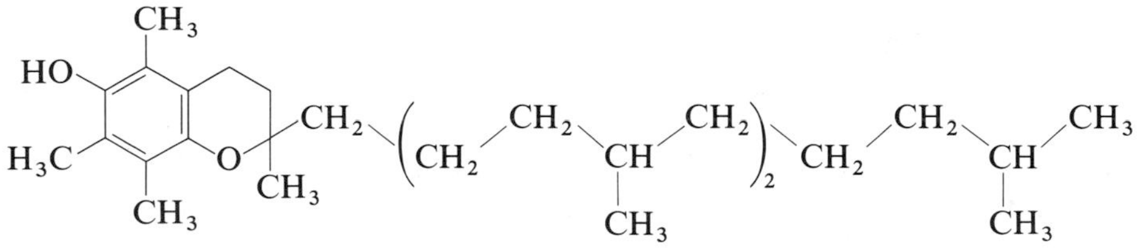

vitamin E (α-tocopherol)

Vitamin E was first recognized in 1922 as a dietary factor essential for normal reproduction in rats. Its name comes from the Greek, *tocopherol*, "promoter of childbirth." Vitamin E occurs in fish oil, in other oils such as cottonseed and peanut oil, and in leafy green vegetables. The richest source of vitamin E is wheat germ oil. In the body, vitamin E functions as an antioxidant in that it inhibits the oxidation of unsaturated lipids by molecular oxygen. In addition, it is necessary for the proper development and function of membranes in red blood cells, muscle cells, and so on.

D. Vitamin K

Vitamin K was discovered in 1935 as a result of a study of newly hatched chicks that had a fatal disease in which the blood was slow to clot. This condition could be prevented and cured by the administration of a substance found in hog liver and in alfalfa. It was later discovered that the delayed clotting time of the blood was caused by a deficiency of prothrombin, and it is now known that vitamin K is essential for the synthesis of prothrombin in the liver. The natural vitamin has a long, branched alkyl chain of usually 20 to 30 carbon atoms.

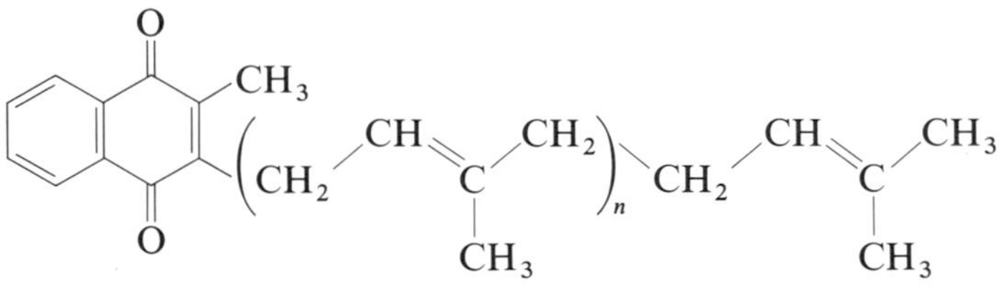

vitamin K (*n* may be 5, 6, or 8)

The natural vitamins of the K family have for the most part been replaced by synthetic preparations. Menadione, one such synthetic material with vitamin-K activity, has only hydrogen in the place of the alkyl chain.

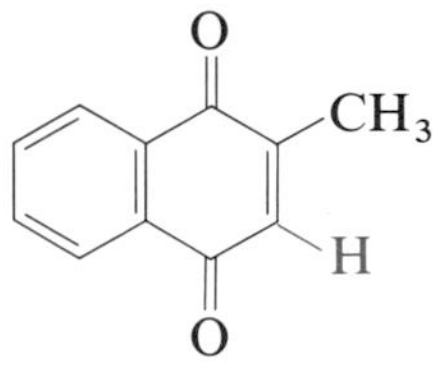

menadione

18.6 Biological Membranes

A. Composition

Membranes are an important feature of cell structure and are vital for all living organisms. Some of the most important functions of membranes can be illustrated by considering the cell membrane. First, a cell membrane is a mechanical barrier which separates the contents of a cell from its environment. Second, a cell membrane controls the passage of molecules and ions into and out of the cell. For example, essential nutrients are transported into the cell and metabolic wastes out of the cell through the membrane. Cell membranes also help to regulate the concentrations of molecules and ions within the cell. Third, a cell membrane provides structural support for certain proteins. Some of these proteins are "receptors" for hormone-carried "messages"; others are

Figure 18.11 Diagram of a cell.

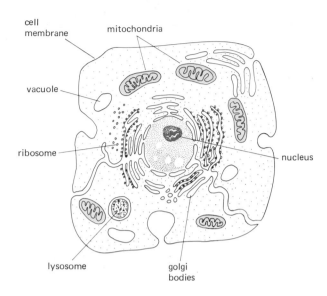

cell membrane

mitochondria

vacuole

ribosome

nucleus

lysosome

golgi bodies

specific enzyme complexes. The cell membrane is but one of the membranes common to most cells. Several other types of membranes found within a typical cell are shown in Figure 18.11.

The functions of the subcellular membranes shown in Figure 18.11 are similar to those of the cell membrane itself. For example, the nuclear membrane separates the nucleus from the rest of the cell. The inner membranes of the mitochondria contain the enzymes that catalyze the reactions of the final states of respiration (Chapter 23). The endoplasmic reticulum contains enzymes that carry out a number of synthetic reactions. The rough endoplasmic reticulum supports ribosomes and the enzymes that catalyze the synthesis of proteins from amino acids. The smooth endoplasmic reticulum contains hydroxylation enzymes, steroid synthesis enzymes, and enzymes for drug metabolism. Lysosomes contain enzymes that digest substances brought into the cell. Membranes are more than impervious, mechanical barriers separating the cell and its organelles from the environment. They are highly specialized structures that perform a multitude of tasks with great precision and accuracy.

B. Structure

The determination of the detailed molecular structure of membranes is one of the most challenging problems in biochemistry today. Despite intensive research, many aspects of membrane structure and activity still are not understood. Before we discuss a model for membrane structure, let us first consider the shapes of phospholipid molecules and the organization of phospholipid molecules in aqueous solution.

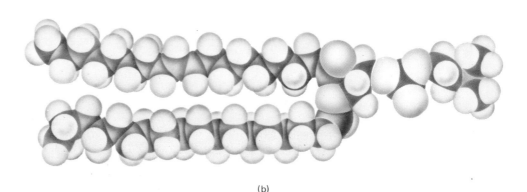

Figure 18.12 A lecithin. (a) Structural formula; (b) space-filling model.

Shown in Figure 18.12 is a structural formula and space-filling model of a lecithin, a major type of membrane phospholipid. Lecithin and other phospholipids are elongated, almost rod-like molecules, with the nonpolar (hydrophobic) hydrocarbon chains lying essentially parallel to one another and with the polar (hydrophilic) phosphate ester group pointing in the opposite direction.

To understand what happens when phospholipid molecules are placed in an aqueous medium, recall from Section 16.7 that soap molecules placed in water form micelles in which polar head groups interact with water molecules and nonpolar hydrocarbon tails cluster within the micelle and are removed from contact with water. One possible arrangement for phospholipids in water also is micelle formation (Figure 18.13).

Another arrangement that satisfies the requirement that polar groups interact with water and nonpolar groups cluster together to exclude water is a **lipid bilayer.** A schematic diagram of a lipid bilayer is shown in Figure 18.14. The favored structure for phospholipids in aqueous solution is a lipid bilayer rather than a micelle, because micelles can only grow to a limited size before holes begin to appear in the outer polar surface. Lipid bilayers can grow to almost infinite extent and provide a boundary surface for a cell or organelle, whatever its size.

It is important to realize that self-assembly of phospholipid molecules into a bilayer is a spontaneous process driven by two types of noncovalent forces: (1) hydrophobic interactions, which result when nonpolar hydrocarbon chains cluster together and exclude water molecules, and (2) electrostatic

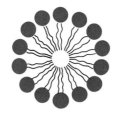

Figure 18.13 Proposed micelle formation of phospholipids in an aqueous medium.

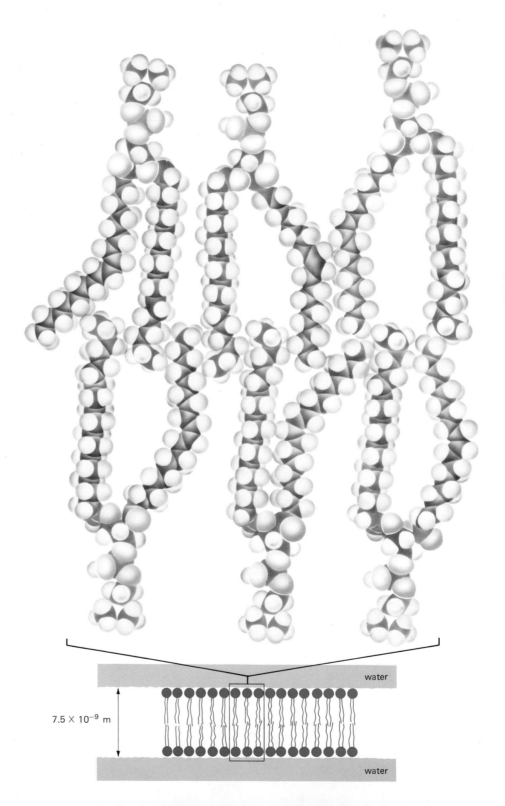

Figure 18.14 A section of lipid bilayer (lower part). Enlarged (upper part) is a section of six phospholipid molecules in the bilayer. Note in the enlargement that 50% of the hydrocarbon chains are unsaturated.

7.5×10^{-9} m

water

water

Figure 18.15 Fluid-mosaic model of a membrane, showing the lipid bilayer and membrane proteins oriented (a) on the outer surface of the membrane, (b) penetrating the entire thickness of the membrane, (c) embedded within the membrane, and (d) on the inner surface of the membrane.

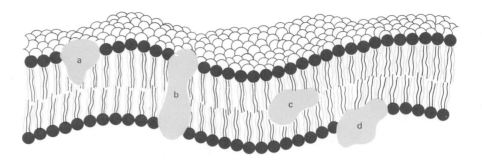

interactions and hydrogen bonding, which result when polar head groups interact with water molecules.

As you might expect from their structural characteristics, lipid bilayers are highly impermeable to ions and most polar molecules, for it would take a great deal of energy to transport an ion or a polar molecule through the nonpolar interior of the bilayer. However, water readily passes in and out of the lipid bilayer. Glucose passes through lipid bilayers 1×10^4 times more slowly than water, and sodium ion 1×10^9 times more slowly than water.

The most satisfactory current model for the arrangement of proteins and phospholipids in plant and animal membranes is the **fluid-mosaic model.** According to this model, membrane phospholipids form a lipid bilayer and membrane proteins are imbedded in this bilayer. Some proteins are exposed to the aqueous environment on the outer surface of the membrane; others provide channels that penetrate from the outer to the inner surface of the membrane; while still others are imbedded within the lipid bilayer. Four possible protein arrangements are shown schematically in Figure 18.15.

The fluid-mosaic model is consistent with the evidence provided by chemical analysis and electron microscope pictures of cell membranes. However, this model does not explain just how membrane proteins act as pumps and gates for the transport of ions and molecules across the membrane, or how they act as receptors for hormone-borne messages and communications between one cell and another. Nor does it explain how enzymes bound on membrane surfaces catalyze reactions. All of these questions are very active areas of research today.

18.7 Blood Group Substances

Plasma membranes of animal cells have large numbers of relatively small carbohydrates bound to them. In fact, it now appears that the outsides of most plasma membranes are literally sugar-coated. Typically, these membrane-bound sugars contain from 4 to 15 monosaccharides and are built from just a few monosaccharides, chiefly D-galactose, D-mannose, L-fucose, N-acetyl-D-glucosamine, and N-acetyl-D-galactosamine. Following is the structural formula of L-fucose.

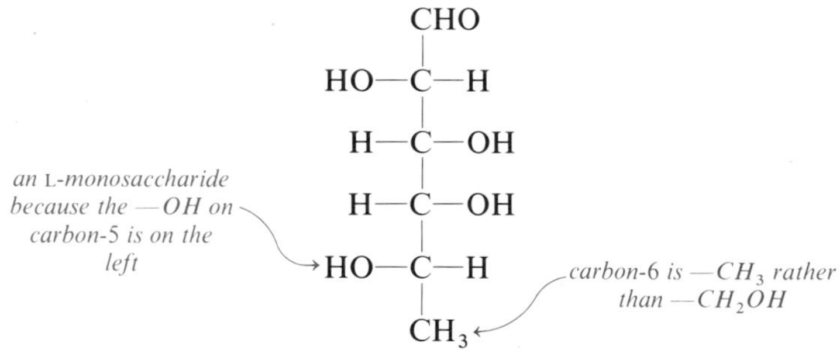

$$
\begin{array}{c}
\text{CHO} \\
\text{HO}-\text{C}-\text{H} \\
\text{H}-\text{C}-\text{OH} \\
\text{H}-\text{C}-\text{OH} \\
\text{HO}-\text{C}-\text{H} \\
\text{CH}_3
\end{array}
$$

an L-monosaccharide because the —OH on carbon-5 is on the left

carbon-6 is —CH₃ rather than —CH₂OH

L-fucose

One of the first discovered and best understood classes of these membrane-bound carbohydrates are the so-called **blood group substances.** Although blood group substances, or "markers," are found chiefly on the surface of erythrocytes (red blood cells), they are also found on proteins and lipids in other parts of the body. In the ABO system, first described in 1900, individuals are classified according to four blood types: A, B, AB, and O. Blood from individuals of the same type can be mixed without clumping of erythrocytes. However, if blood from a type A individual is mixed with type B blood, or vice versa, the erythrocytes clump. Blood from a type O individual does not cause clumping of either type A or type B blood. At the cellular level, the chemical basis for ABO classification involves three relatively small, membrane-bound carbohydrates. Following is the terminal tetrasaccharide portion of the sugar found on erythrocytes of individuals with type A blood. The tetrasaccharide is shown by using planar hexagons for the monosaccharide units. The stereochemistry of each glycoside bond is indicated by an α or β written over the glycoside bond itself.

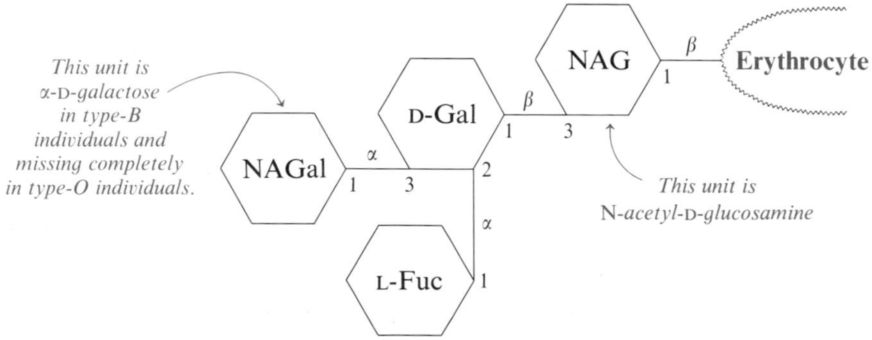

This unit is α-D-galactose in type-B individuals and missing completely in type-O individuals.

NAGal D-Gal NAG Erythrocyte

This unit is N-acetyl-D-glucosamine

L-Fuc

There are several distinctive features of this tetrasaccharide. First, it contains a monosaccharide of the "unnatural" or L-series, namely L-fucose. Second, it contains D-galactose to which are bonded two other monosaccharides, one by an α-1,2-glycoside bond, the other by an α-1,3-glycoside bond.

It is the last monosaccharide mentioned that determines the ABO classification. In type-A individuals, the chain terminates in N-acetyl-D-galactosamine (NAGal); in type-B individuals it terminates instead in D-galactose (D-Gal); and in type-O individuals, the last monosaccharide is missing completely. The saccharides of type-AB individuals contain both kinds of carbohydrate chains.

Key Terms and Concepts

adrenocorticoid hormone (18.4B)

atherosclerosis (18.4A)

bile acid (18.4D)

biological membrane (18.6)

blood group substances (18.7)

fat (18.1)

fat-soluble vitamin (18.5)

fluid-mosaic model (18.6)

hardening of oils (18.1D)

infarction (18.4A)

lecithin (18.3)

lipid (introduction)

lipid bilayer (18.6)

oil (18.1)

phospholipid (18.3)

rancidity (18.1C)

sex hormone (18.4C)

triglyceride (18.1)

vitamin A (18.5A)

vitamin D (18.5B)

vitamin E (18.5C)

vitamin K (18.5D)

wax (18.2)

Problems

**Fats and Oils
(Section 18.1)**

18.2   List six major functions of lipids in the human body. Name and draw a structural formula for a lipid representing each function.

18.3   How many isomers (including stereoisomers) are possible for a triglyceride containing one molecule each of palmitic, stearic, and oleic acid?

18.4   What is meant by the term "hardening" as applied to fats and oils?

18.5   Saponification is the alkaline hydrolysis of naturally occurring fats and oils. A saponification number is the number of mg of potassium hydroxide required to saponify 1 g of a fat or oil. Calculate the saponification number of tristearin, molecular weight 890.

18.6   The saponification number of butter is approximately 230; that of oleomargarine is approximately 195. Calculate the average molecular weight of butter fat and of oleomargarine.

18.7   The percentage of unsaturated fatty acids in butter fat is approximately 35%. Compare this with the percentage of unsaturated fatty acids in corn oil, soybean oil, and wheat germ oil.

**Phospholipids
(Section 18.3)**

18.8   Draw a structural formula for a phospholipid containing serine; a phospholipid containing inositol.

18.9   Draw structural formulas for the products of complete hydrolysis of a lecithin; a cephalin.

Steroids (Section 18.4)

18.10 Draw the structural formula of cholesterol; label all chiral carbons and state the total number of stereoisomers possible for this structural formula.

18.11 Esters of cholesterol and fatty acids are normal constituents of blood plasma. The fatty acids esterified with cholesterol are generally unsaturated. Draw the structural formula for cholesteryl oleate.

18.12 Cholesterol is an important component of the lipid fraction of cell membranes. How do you think a cholesterol molecule might be oriented in a biological membrane?

18.13 Name the six functional groups in cortisol; in aldosterone.

18.14 Examine the structural formulas of testosterone, a male sex hormone, and progesterone, a female sex hormone. What are the similarities in structure between the two? The differences?

18.15 Why are testosterone, progesterone and the estrogens called *sex hormones*?

18.16 Describe how a combination of progesterone and estrogen analogs functions as an oral contraceptive.

18.17 Examine the structural formula of cholic acid and account for the fact that this and other bile acids are able to emulsify fats and oils.

**Fat-Soluble Vitamins
(Section 18.5)**

18.18 Examine the structural formula of vitamin A and state the number of *cis-trans* isomers possible for this molecule.

18.19 In fish-liver oils, vitamin A is present as esters of fatty acids. The most common of these esters is vitamin A palmitate. Draw a structural formula for this substance.

18.20 Describe the symptoms of severe vitamin-A deficiency.

18.21 Examine the structural formulas of vitamins A, D_3, E, and K_2. Based on their structural formulas, would you expect them to be more soluble in water or in olive oil? Would you expect them to be soluble in blood plasma?

18.22 Explain why vitamin E is added to some processed foods.

**Biological
Membranes (Section 18.6)**

18.23 Two of the major noncovalent forces directing the organization of biomolecules in aqueous solution are the tendencies to (1) arrange polar groups so that they interact with water by hydrogen bonding, and (2) arrange nonpolar groups so that they are shielded from water. Show how these forces direct micelle formation by soap molecules and lipid bilayer formation by phospholipids.

18.24 Describe the major features of the fluid-mosaic model of the structure of biological membranes.

Amines

In Chapter 12 we discussed the structure of alcohols and ethers—compounds in which first one and then both hydrogens of H_2O are replaced by alkyl or aromatic groups. In this chapter, we will discuss **amines,** derivatives of ammonia in which one, two, or three hydrogens of NH_3 are replaced by alkyl or aromatic groups. The most important chemical property of amines is their basicity.

19.1 Structure of Amines

The structure of simple alkyl amines is much like that of ammonia, NH_3. Shown in Figure 19.1 are structural formulas and ball-and-stick models of methylamine, dimethylamine, and trimethylamine.

In amines, nitrogen uses three sp^3 hybrid orbitals to form sigma bonds with three other atoms. The unshared pair of electrons on the nitrogen atom is in the remaining sp^3 hybrid orbital. Therefore, we predict bond angles of 109.5° about the nitrogen atom (Section 9.5).

Amines are classified as primary (1°), secondary (2°), or tertiary (3°), depending on the number of hydrogen atoms that have been replaced by carbon atoms. In a **primary amine,** one hydrogen is replaced by a carbon; in a **secondary amine,** two hydrogens are replaced; in a **tertiary amine,** three hydrogens are replaced.

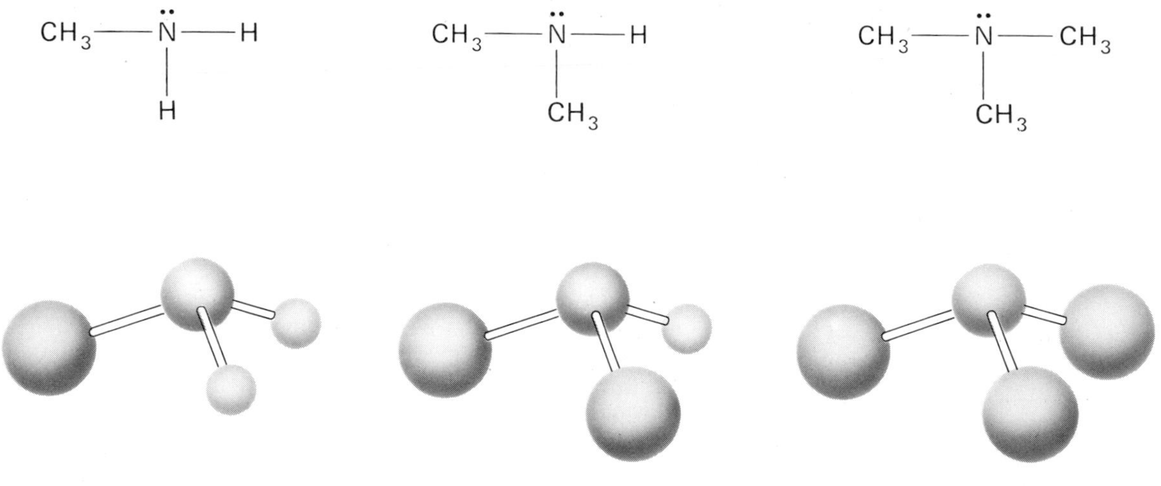

Figure 19.1 Structural formulas and ball-and-stick models of (a) methylamine (a primary amine); (b) dimethylamine (a secondary amine); and (c) trimethylamine (a tertiary amine).

19.2 Nomenclature of Amines

Common names of simple aliphatic amines are derived by listing the names of the alkyl groups attached to nitrogen and adding the ending *-amine*.

isobutylamine
(*primary*)

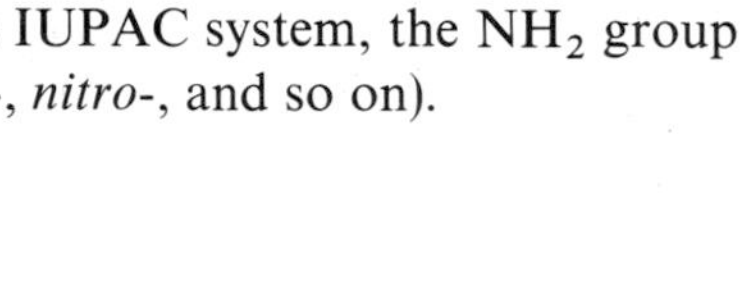

ethylisopropylamine
(*secondary*)

triethylamine
(*tertiary*)

In the IUPAC system, the NH_2 group is named as an amino substituent (like *chloro-*, *nitro-*, and so on).

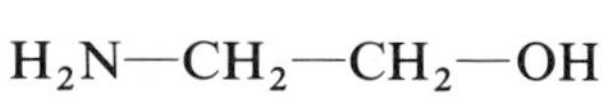

2-aminoethanol
(ethanolamine)

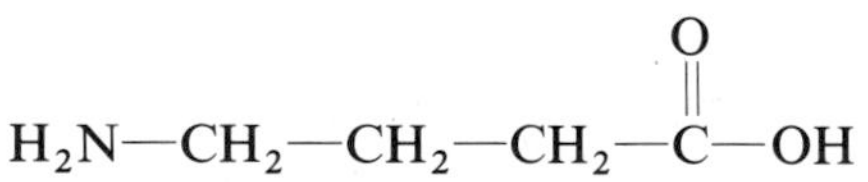

4-aminobutanoic acid
(γ-aminobutyric acid)

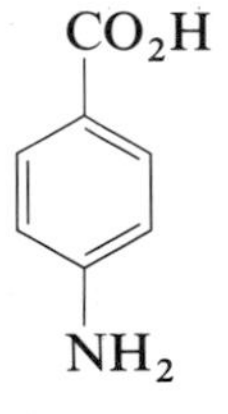

p-aminobenzoic acid

A nitrogen with four hydrocarbon groups attached is positively charged and is known as a **quaternary ammonium ion,** an example of which is tetramethylammonium ion, $(CH_3)_4N^+$.

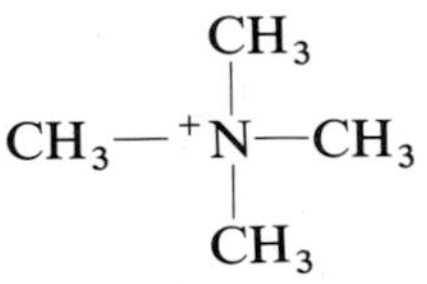

CH₃—⁺N(CH₃)(CH₃)—CH₃   CH₃—⁺N(CH₃)(CH₃)—CH₃ Cl⁻

tetramethylammonium ion 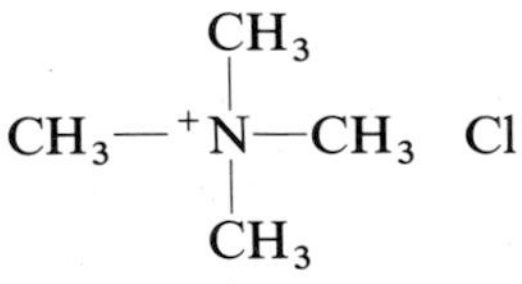   tetramethylammonium chloride

Compounds containing an —NH₂ group on a benzene ring are named as derivatives of aniline. Many simple substitution derivatives are also known by common names; these include anisidine (methoxyaniline) and toluidine (methylaniline).

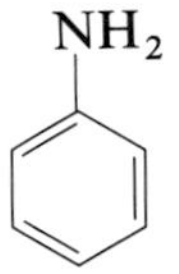

aniline 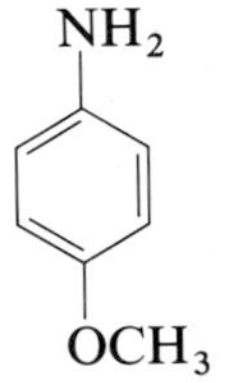   p-methoxyaniline   o-methylaniline   m-nitroaniline
            (p-anisidine)   (o-toluidine)

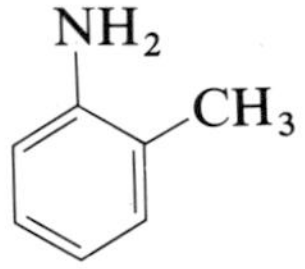

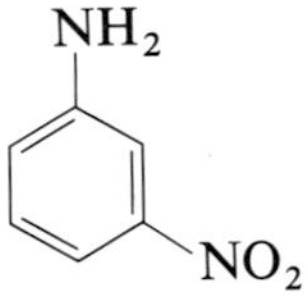

Heterocyclic amines, cyclic compounds in which the amine nitrogen is one of the atoms of a ring, are given special names. Following are structural formulas for piperidine and pyrrolidine, both secondary amines. N-Methylpyrrolidine is an example of a heterocyclic amine in which the amine is tertiary.

piperidine   pyrrolidine   N-methylpyrrolidine
(a secondary   (a secondary   (a tertiary
amine)   amine)   amine)

The most important six-member heterocyclic aromatic amines are pyridine and pyrimidine. Following are structural formulas for these amines, along with the numbering systems used to locate substituents on each ring:

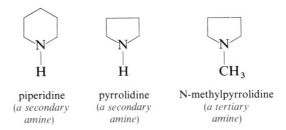

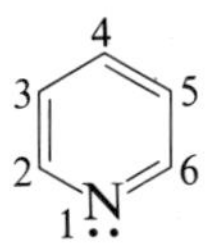

pyridine 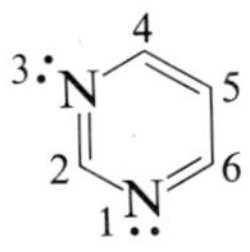   pyrimidine

Pyridine is a derivative of benzene in which one C—H group is replaced by a nitrogen atom. In pyrimidine, two C—H groups are replaced by nitrogen atoms. These heterocyclic amines are classified as aromatic because they tend to undergo ring substitution reactions like benzene rather than addition reactions like alkenes.

The pyridine ring is a building block of a great many compounds in the biological world. Two especially important pyridine derivatives are nicotinamide adenine dinucleotide (NAD^+) and its phosphorylated derivative, nicotinamide adenine dinucleotide phosphate ($NADP^+$), the structural formulas of which are shown in Figure 23.10. These molecules, along with appropriate enzymes, function in oxidation-reduction reactions in both plants and animals. Pyrimidine itself does not occur in nature, but pyrimidine derivatives with substituents at positions 2, 4, and 5 are essential building blocks of nucleic acids.

The structures of two five-member heterocyclic aromatic amines, pyrrole and imidazole, are shown below. The imidazole ring is part of the essential amino acid histidine.

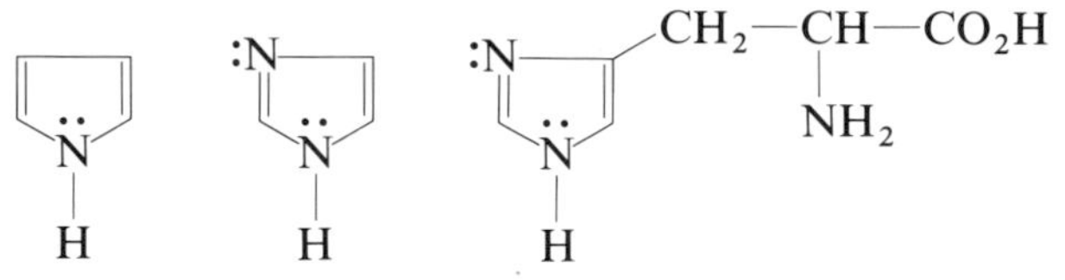

pyrrole   imidazole   histidine (an amino acid)

Two other heterocyclic aromatic amines that are especially important in the biological world are purine and indole.

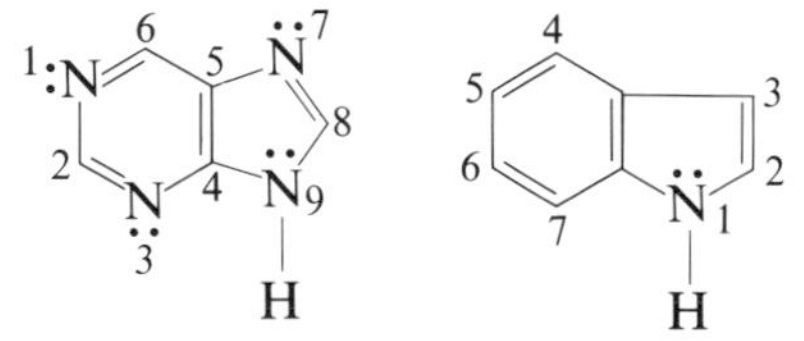

purine   indole

Purine contains a six-member pyrimidine ring and a five-member imidazole ring that share two atoms in common. Purine and pyrimidine rings are building blocks of deoxyribonucleic acid (DNA) and ribonucleic acid (RNA). (The structure and function of DNA and RNA are discussed in detail in Chapter 22.)

Indole contains two fused rings, a six-member benzene ring and a five-member pyrrole ring. The indole nucleus is found in nature in the amino acid tryptophan, in the neurotransmitter serotonin, and in many other substances of plant and animal origin.

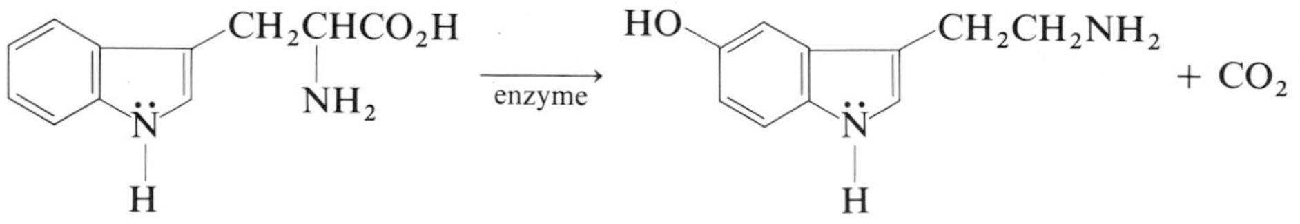

tryptophan   serotonin

Example 19.1

Name the following compounds.

a. $CH_3—CH_2—NH—CH_3$

b. 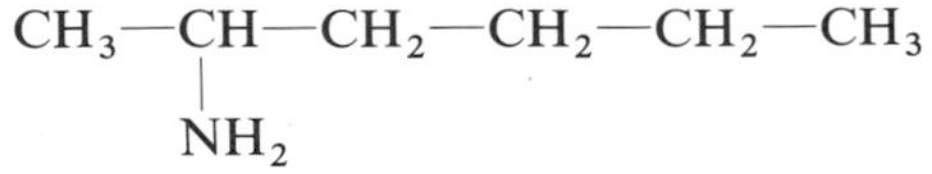 $CH_3—CH—CH_2—CH_2—CH_2—CH_3$
      $|$
     NH_2

c. 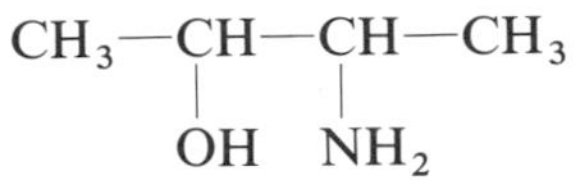 $CH_3—CH—CH—CH_3$
      $|$ $|$
     OH NH_2

d. 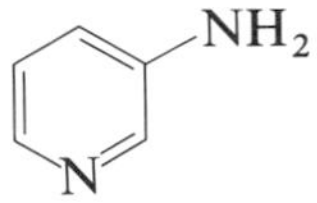 [pyridine ring structure with NH_2 substituent]

Solution

a. Name both of the alkyl groups attached to nitrogen and add the ending -*amine* to give the name ethylmethylamine.

b. There is no simple name for this six-carbon alkyl group. Therefore, show the —NH_2 group as a substituent. The IUPAC name is 2-aminohexane.

c. Name this substance as an alcohol and show the amine group as a substituent: 3-amino-2-butanol.

d. Name this substance as a derivative of pyridine with an amino substituent on carbon-3 of the ring: 3-aminopyridine.

Problem 19.1

Name the following compounds.

a. $H_2NCH_2CH_2CH_2CH_2CH_2CH_2NH_2$

b. 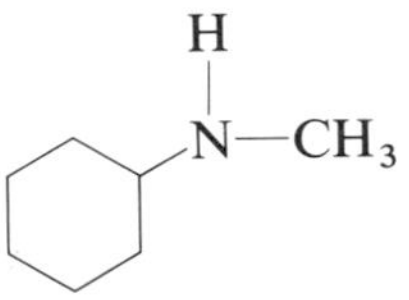 [cyclohexane ring with N(H)—CH$_3$ substituent]

c. 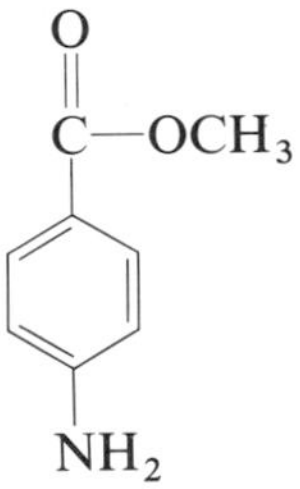 [benzene ring with $C(=O)—OCH_3$ and NH_2 substituents]

19.3 Physical Properties of Amines

Amines are polar compounds and both primary and secondary amines can form intermolecular hydrogen bonds.

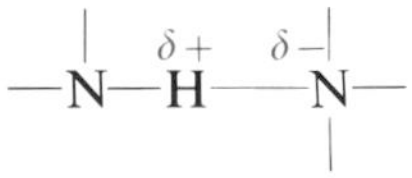

$$—N—H\overset{\delta+}{}\overset{\delta-}{}N—$$

However, because the difference in electronegativity between nitrogen and hydrogen $(3.0 - 2.1 = 0.9)$ is not as great as that between oxygen and hydrogen

Table 19.1 Physical properties of amines.

Name	bp (°C)	Solubility (g/100g H_2O)	K_b	pK_b	pK_a
ammonia	−33	90	1.8×10^{-5}	4.74	9.26
methylamine	−7	∞	4.4×10^{-4}	3.36	10.64
ethylamine	17	∞	6.3×10^{-4}	3.20	10.80
diethylamine	55	∞	3.1×10^{-4}	3.51	10.49
triethylamine	89	1.5	1.0×10^{-3}	3.00	11.00
cyclohexylamine	134	slightly	5.5×10^{-4}	3.26	10.74
benzylamine	185	∞	2.1×10^{-5}	4.67	9.33
aniline	184	3.7	4.2×10^{-10}	9.37	4.63
pyridine	116	∞	1.8×10^{-9}	8.74	5.26
imidazole	257	∞	8.9×10^{-8}	7.05	6.95

(3.5 − 2.1 = 1.4), the N—H---N hydrogen bond is not nearly as strong as the O—H---O hydrogen bond. The boiling points of ethane, methylamine, and methanol, all compounds of comparable molecular weight, are:

$$CH_3—CH_3 \qquad CH_3—NH_2 \qquad CH_3—OH$$

mol. wt.	30	31	32
bp (°C)	−88	−7	65

Ethane is a nonpolar hydrocarbon, and the only interactions between its molecules in the pure liquid are very weak dispersion forces. Therefore, it has the lowest boiling point of the three. Both methylamine and methanol are polar molecules and interact in the pure liquid by hydrogen bonding. Hydrogen bonding is weaker in methylamine than in methanol, and therefore methylamine has a lower boiling point than methanol.

All classes of amines form hydrogen bonds with water and therefore are more soluble in water than hydrocarbons of comparable molecular weight. Most low-molecular-weight amines are completely soluble in water. The higher-molecular-weight amines are only slightly soluble in water. Boiling points and solubilities in water for several amines are listed in Table 19.1.

19.4 Reactions of Amines

A. Basicity

Like ammonia, all primary, secondary, and tertiary amines are weak bases, and aqueous solutions of amines are basic.

$$NH_3 + H_2O \rightleftharpoons NH_4^+ + OH \qquad CH_3NH_2 + H_2O \rightleftharpoons CH_3NH_3^+ + OH^-$$

The equilibrium constant for the reaction of methylamine and water is called a **base dissociation constant** (K_b) and is given by the expression

$$K_b = \frac{[CH_3NH_3^+][OH^-]}{[CH_3NH_2]}$$

Values for K_b for some primary, secondary, tertiary, and aromatic amines are given in Table 19.1. All aliphatic amines, whether primary, secondary, tertiary, or saturated heterocyclic amines, have about the same base strength as ammonia.

Aromatic amines such as aniline are significantly less basic than aliphatic amines. The K_b of aniline is less than that of cyclohexylamine by a factor of 10^6.

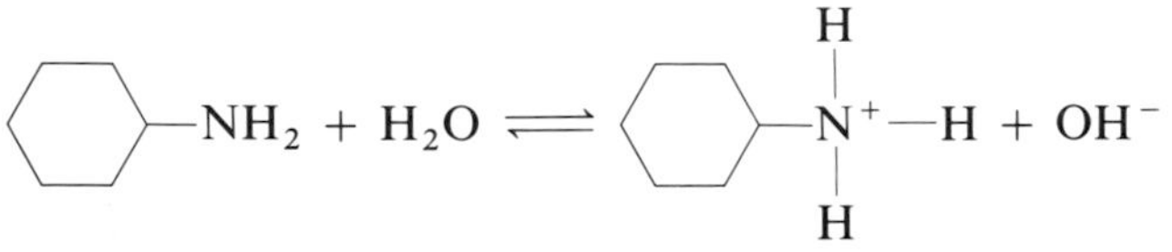

cyclohexylamine   cyclohexylammonium $K_b = 5.5 \times 10^{-4}$
                         ion

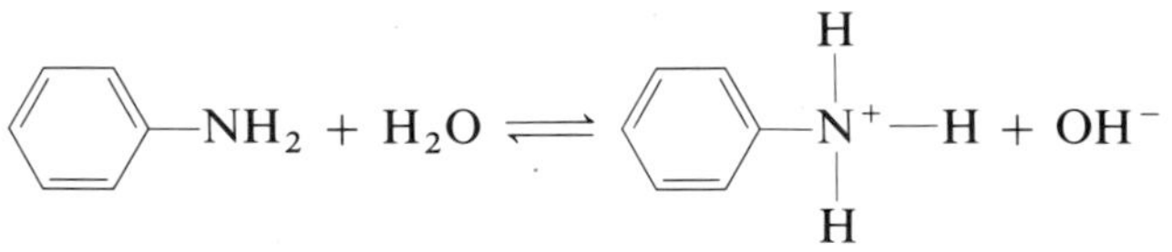

aniline   anilinium $K_b = 4.2 \times 10^{-10}$
                 ion

To extend our discussion of the basicity of the —NH_2 group one step further, let us compare the basicities of cyclohexylamine, aniline, and acetamide, which all contain the —NH_2 group.

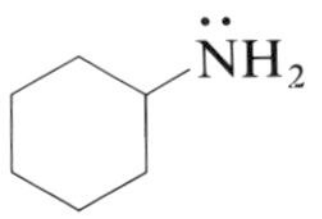

cyclohexylamine   aniline   acetamide
($K_b = 5.5 \times 10^{-4}$)   ($K_b = 4.2 \times 10^{-10}$)   ($K_b = 10^{-16}$ (approx.))

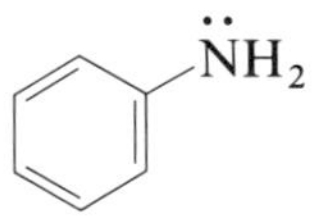

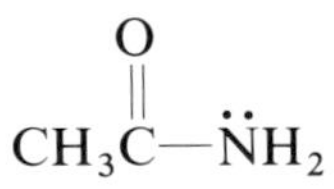

Of the three classes of compounds containing the —NH_2 group, aliphatic amines are the strongest bases. Aromatic amines such as aniline are also bases, but they are considerably weaker than aliphatic amines. The —NH_2 group of amides shows no basicity at all, therefore amides are neutral substances.

One final note about the basicity of amines. Until recently it was common practice to list only K_b or pK_b values for amines. Now it is becoming more and more common to list only K_a and pK_a values for amines. For example, only

K_a and pK_a values are used in discussing the acid-base properties of the amino group of amino acids (Chapter 20). For this reason, Table 19.1 also lists pK_a values for a variety of common amines. Note that pK_a + pK_b = 14.

To illustrate the difference between pK_b and pK_a for an amine, consider methylamine:

$$CH_3NH_2 + H_2O \rightleftharpoons CH_3NH_3^+ + OH^- \quad K_b = 4.4 \times 10^{-4} \quad pK_b = 3.36$$
$$CH_3NH_3^+ \rightleftharpoons CH_3NH_2 + H^+ \quad K_a = 2.7 \times 10^{-11} \quad pK_a = 10.64$$

pK_b directly measures the strength of CH_3NH_2 as a base while pK_a directly measures the strength of $CH_3NH_3^+$ as an acid. For perspective, you might compare the pK_a values for acetic acid and methylamine.

$$CH_3CO_2H \rightleftharpoons CH_3CO_2^- + H^+ \quad K_a = 1.8 \times 10^{-5} \quad pK_a = 4.74$$
$$CH_3NH_3^+ \rightleftharpoons CH_3NH_2 + H^+ \quad K_a = 2.3 \times 10^{-11} \quad pK_a = 10.64$$

By using K_a or pK_a values for carboxylic acids and amines, we can compare them directly, for in each case we are looking at the dissociation of an acid to form a base and a proton. It is obvious that acetic acid is a much stronger acid than methylammonium ion.

All amines, whether soluble or insoluble in water, react quantitatively with acids to form substituted ammonium ions.

$$CH_3-NH_2 + HCl \longrightarrow \quad CH_3-NH_3^+Cl^-$$

methylamine   methylammonium chloride   cyclohexylamine   cyclohexylammonium chloride

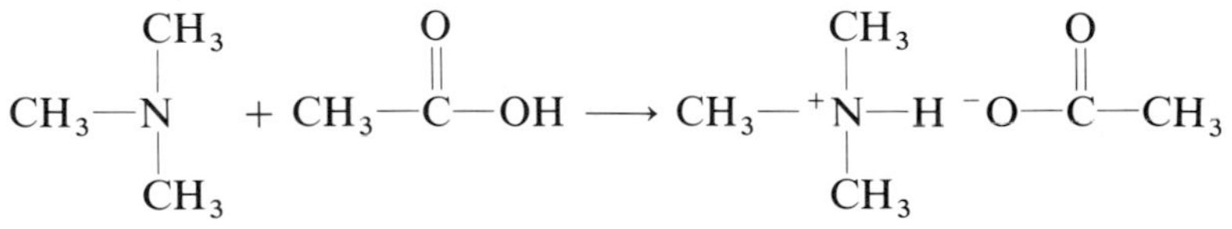

trimethylamine   trimethylammonium acetate

Example 19.2

Complete the following acid-base reactions.

a. $(CH_3CH_2)_2NH + HCl \longrightarrow$ **b.** 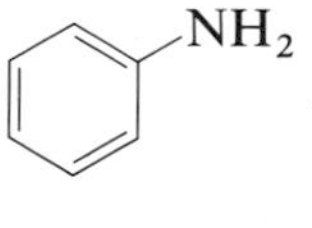 [structure] $+ HCl \longrightarrow$

c. 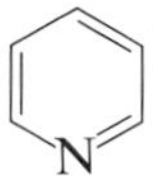 [structure] $+ CH_3CO_2H \longrightarrow$

Solution

a. $(CH_3CH_2)_2\overset{..}{N}H + HCl \longrightarrow CH_3-CH_2-\overset{\overset{\displaystyle H}{|}}{\underset{\underset{\displaystyle H}{|}}{N^+}}-CH_2-CH_3 \ Cl^-$

Diethylamine, a secondary amine, reacts with HCl to form the salt diethyl-ammonium chloride.

b. 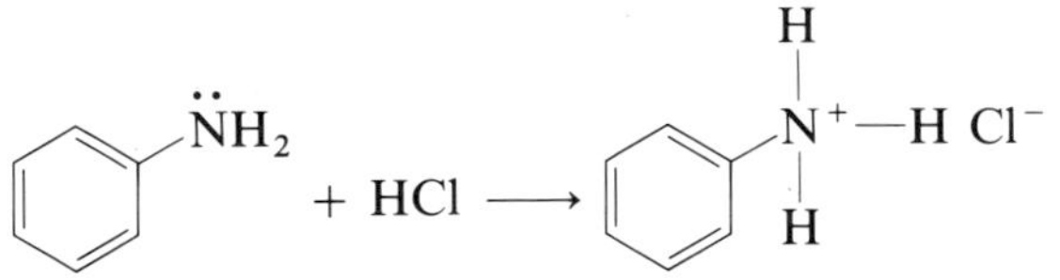 ⬡—$\overset{..}{N}H_2$ + HCl ⟶ ⬡—$\overset{\overset{\displaystyle H}{|}}{\underset{\underset{\displaystyle H}{|}}{N^+}}$—H Cl⁻

Aniline, a primary aromatic amine, reacts with hydrochloric acid to form the salt anilinium chloride. Another name for this salt is aniline hydrochloride.

c. 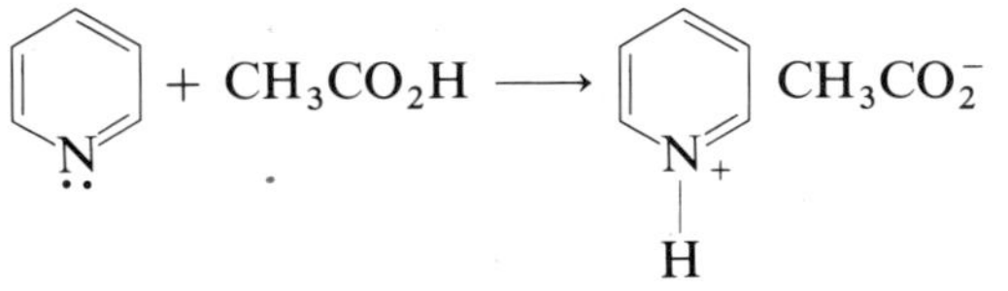 ⬡(N) + CH_3CO_2H ⟶ ⬡($\overset{}{\underset{\underset{\displaystyle H}{|}}{N^+}}$) $CH_3CO_2^-$

Pyridine, a heterocyclic aromatic amine, reacts with acetic acid to form the salt pyridinium acetate.

Problem 19.2

Complete the following acid-base reactions.

a. $HO-CH_2-CH_2-\underset{\underset{\displaystyle CH_3}{|}}{N}-CH_3 + HCl \longrightarrow$

b. $(CH_3CH_2)_3N + C_6H_5CO_2H \longrightarrow$

c. 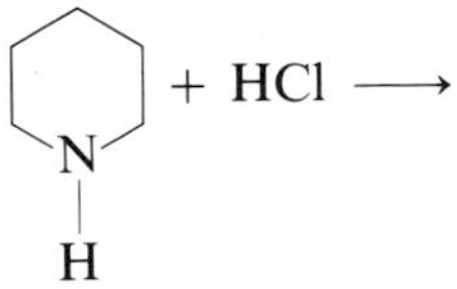 ⬡($\underset{\underset{\displaystyle H}{|}}{N}$) + HCl ⟶

The basicity of amines and the solubility of amine salts in water can be used to distinguish and separate amines and nonbasic, water-insoluble compounds. Following is a flowchart for the separation of aniline from methyl benzoate.

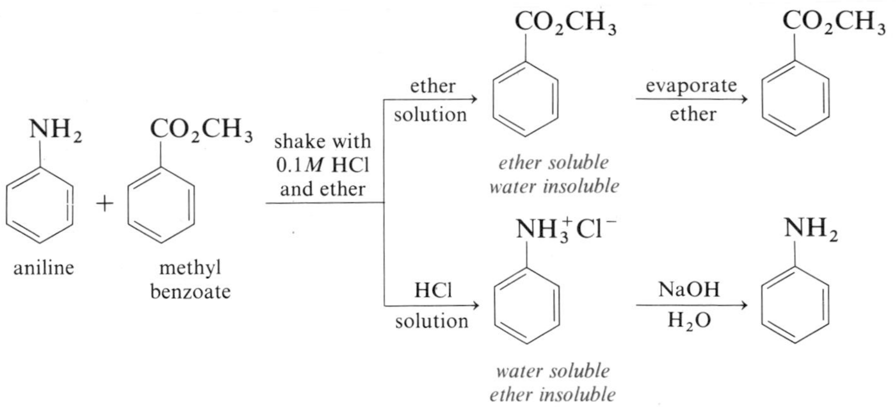

Aniline and methyl benzoate are only slightly soluble in water and therefore cannot be separated on the basis of their solubility in water. However, both dissolve in ether. When an ether solution of the two compounds is shaken with dilute aqueous hydrochloric acid, aniline reacts to form a water-soluble salt. Methyl benzoate remains in the ether layer. Separation of the ether layer and evaporation of the ether yields methyl benzoate. Treatment of the aqueous solution with sodium hydroxide converts the water-soluble salt to free aniline, which then separates as a water-insoluble layer.

B. Reaction with Carboxylic Acid Derivatives to Give Amides

As we have already seen in Chapter 17, primary and secondary amines react with esters and acid anhydrides to give amides. Following are examples of each of these reactions.

$$CH_3-\overset{\overset{\displaystyle O}{\|}}{C}-O-CH_2-CH_3 + CH_3-NH_2 \longrightarrow$$

$$CH_3-\overset{\overset{\displaystyle O}{\|}}{C}-NH-CH_3 + CH_3-CH_2-OH$$

N-methylacetamide

$$CH_3-\overset{\overset{\displaystyle O}{\|}}{C}-O-\overset{\overset{\displaystyle O}{\|}}{C}-CH_3 + CH_3-NH_2 \longrightarrow$$

$$CH_3-\overset{\overset{\displaystyle O}{\|}}{C}-NH-CH_3 + CH_3-\overset{\overset{\displaystyle O}{\|}}{C}-OH$$

N-methylacetamide

We also saw in Chapter 17 that amides can be prepared by heating ammonium salts of carboxylic acids.

$$CH_3-\overset{\overset{\textstyle O}{\|}}{C}-O^-CH_3-\overset{\overset{\textstyle H}{|}}{\underset{\underset{\textstyle CH_3}{|}}{N^+}}-H \xrightarrow[200°C]{heat} CH_3-\overset{\overset{\textstyle O}{\|}}{C}-\overset{\underset{\textstyle CH_3}{|}}{N}-CH_3 + H_2O$$

dimethylammonium acetate   N,N-dimethylacetamide

C. Reaction with Nitrous Acid

Tertiary amines react with nitrous acid in a typical acid-base reaction to form ammonium salts.

$$CH_3-\overset{\overset{\textstyle CH_3}{|}}{\underset{\underset{\textstyle CH_3}{|}}{N:}} + H-O-N=O \longrightarrow CH_3-\overset{\overset{\textstyle CH_3}{|}}{\underset{\underset{\textstyle CH_3}{|}}{{}^+N}}-H^-O-N=O$$

trimethylamine   trimethylammonium nitrite

With secondary amines, the reaction is somewhat more complex and leads to the formation of N-nitrosoamines, commonly called nitrosamines. For example, the reaction of dimethylamine and nitrous acid gives N-nitrosodimethylamine (dimethylnitrosamine).

$$CH_3-\overset{\underset{\textstyle CH_3}{|}}{N}-H + H-O-N=O \longrightarrow CH_3-\overset{\underset{\textstyle CH_3}{|}}{N}-N=O + H-OH$$

dimethylamine   N-nitrosodimethylamine (dimethylnitrosamine)

Sodium nitrite, a compound that reacts with acids to give nitrous acid, is used in many industrial processes and commercial products. For example, it was common in years past, especially before the development of adequate refrigeration for the transport and storage of meats, to add $NaNO_2$ to retard spoilage, prevent botulism food poisoning, and to preserve the red color of processed meats. Thus, there are many potential sources of nitrite in the environment. However, for humans, the environment is not the only source of nitrite. Many foods contain nitrate as a natural constituent. Nitrate is reduced to nitrite in the saliva and by bacterial action in the intestine. Thus, there is the potential for internally generated nitrites to react with amines and produce nitrosoamines.

Nitrosoamines are also formed by the reaction of amines with certain oxides of nitrogen, such as N_2O_3.

$$CH_3-\overset{\underset{\displaystyle CH_3}{|}}{N}-H + O=N-\overset{\overset{\displaystyle O}{\nearrow}}{\underset{\displaystyle O^-}{N}} \longrightarrow CH_3-\overset{\underset{\displaystyle CH_3}{|}}{N}-N=O + H-O-N=O$$

Because of the prevalence of low-molecular-weight secondary amines, nitrites, and oxides of nitrogen in the environment, N-nitrosoamines have been found in the air in such places as tannery plants, rubber tire plants, iron foundries, and new car interiors, and in such consumer products as beer, whiskey, lotions, shampoos, and cooked, nitrite-cured bacon.

What makes N-nitrosoamines of special concern is that more than 130 of them have been tested as chemical carcinogens in animals and over 100 have produced tumors in one or more species and in one or more organs. Most studies have been done on rodents. The possible link between nitroso-amines and cancer in humans is being studied in several ways. First, improvements are being made in analytical techniques to detect these compounds in the environment. Second, expanded studies of the occurrence and distribution of various forms of cancer are being carried out by both government and private research groups. Third, the chemistry and biochemistry of N-nitrosoamines are being studied intensively.

Primary amines react with nitrous acid to form diazonium ions. In the case of a primary aliphatic amine, the diazonium ion is unstable and decomposes to give nitrogen gas and a carbocation. The carbocation then reacts with water to form an alcohol or loses a proton to form an alkene.

$$CH_3CH_2NH_2 + HNO_2 \xrightarrow[\text{water}]{HCl} [CH_3CH_2N_2^+] \longrightarrow CH_3CH_2^+ + N_2$$

diazonium
ion

$$-H^+ \qquad\qquad +H_2O$$

$$CH_2=CH_2 \qquad CH_3CH_2OH + H^+$$

For each mole of primary amine, one mole of N_2 is formed, and therefore the reaction of a primary amine with nitrous acid can be used as a quantitative measure of the number of moles of primary amine present in a sample of organic material.

Primary aromatic amines also react with cold nitrous acid to give diazonium salts. Aromatic diazonium salts are stable at 0°C, but they, too, decompose to produce nitrogen gas when they are warmed to room temperature. In water, phenol is the organic product of the decomposition.

19.5 Some Naturally Occurring Amines

Structural formulas of a variety of naturally occurring amines of both plant and animal origin are shown in Figure 19.2. These are chosen to illustrate something of the structural diversity and range of physiological activity of amines, their value as drugs, and their importance in nutrition. Coniine from the water hemlock is highly toxic. It can cause weakness, labored respiration, paralysis, and eventually death. The levorotatory isomer of coniine is the toxic

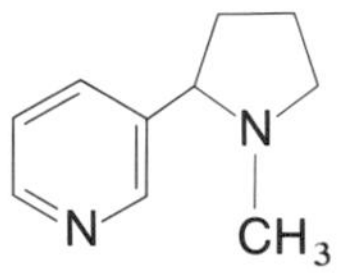

nicotine
(from tobacco)

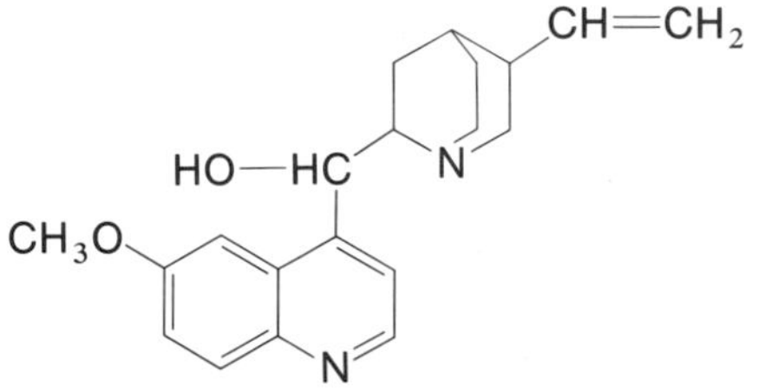

quinine

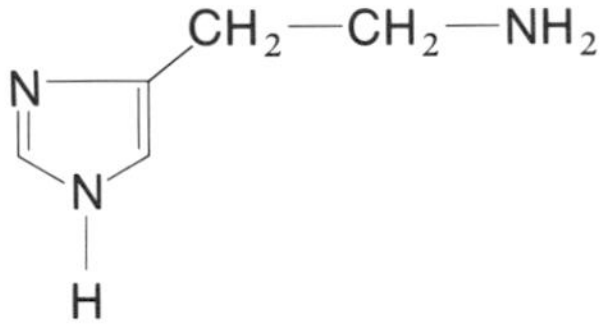

histamine

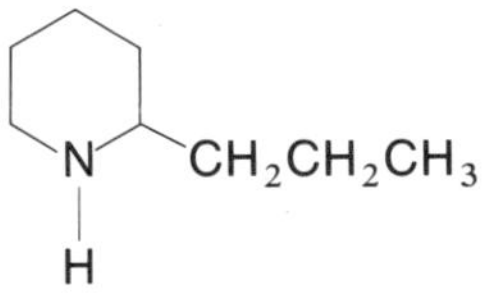

coniine
(from poison hemlock)

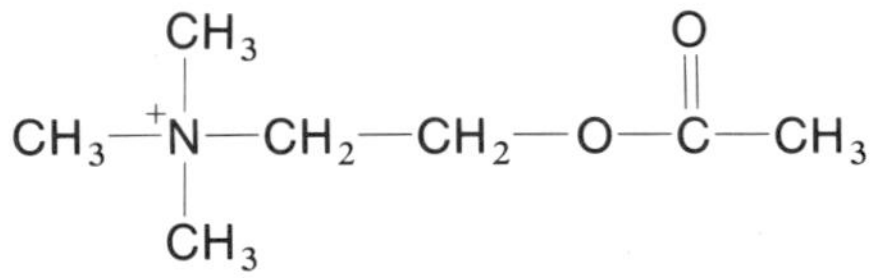

acetylcholine

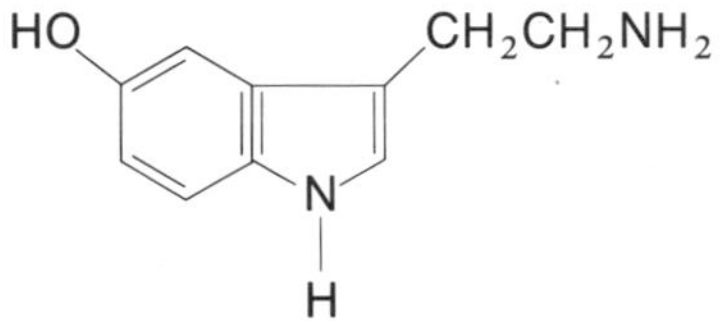

serotonin
(5-hydroxytryptamine)

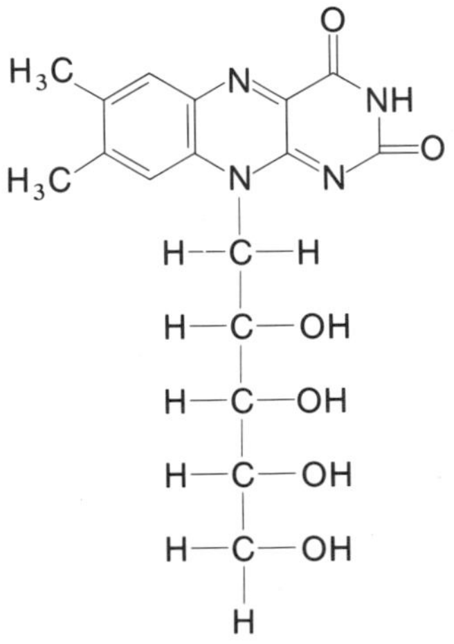

riboflavin
or vitamin B$_2$

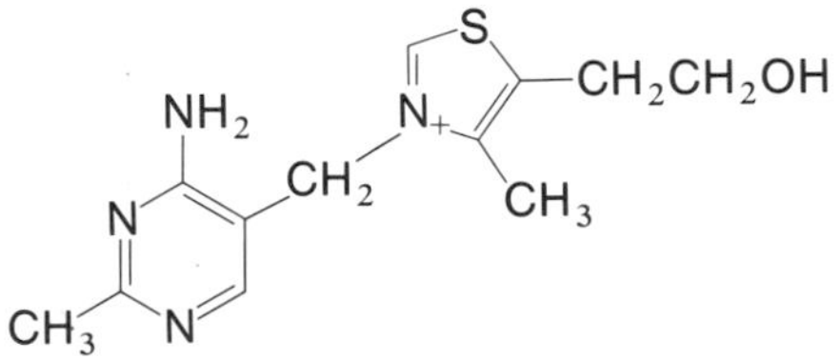

thiamine
or vitamin B$_1$

Figure 19.2 Several naturally occurring amines of plant and animal origin.

compound in "poison hemlock," the potion used by Socrates to commit suicide. Nicotine is one of the chief heterocyclic aromatic amines in tobacco. In small doses, nicotine is a stimulant; in larger doses, it causes depression, nausea, and vomiting. In still larger doses, it is a poison. Solutions of nicotine are often used as insecticides. Note that nicotinic acid, an oxidation product of nicotine, is one of the water-soluble vitamins required by humans for proper nutrition. You should also note that ingested or inhaled nicotine does not give rise to nicotinic acid in the body, for humans have no enzyme systems capable of catalyzing this conversion. Smoking does not supply any vitamins!

Quinine, isolated from the bark of the cinchona tree in South America, has long been used as a medicine for the treatment of malaria. Histamine, formed by the decarboxylation of the amino acid histidine (Table 20.1) is present in all tissues of the body, combined in some manner with proteins. Extensive production of histamine occurs during hypersensitive allergic reactions, and the symptoms of this production are, unfortunately, familiar to most of us, in particular those who suffer from hay fever. The search for antihistamines—drugs that inhibit the effects of histamine—has led to the synthesis of several drugs whose trade names are well known. Structural formulas for three of the more widely used antihistamines are shown in Figure 19.3.

Observe the structural similarity in the three antihistamines: each has two aromatic rings and a dimethylaminoethyl ($-CH_2CH_2N(CH_3)_2$) group. Dexbrompheniramine is one of the most potent of the available antihistamines. Notice that it has one chiral carbon atom. Pharmacological studies have shown that the ($+$)-isomer is nearly twice as potent as the racemic mixture and about thirty times as potent as the ($-$)-isomer.

Both serotonin and acetylcholine are neurotransmitters. Serotonin works in parts of the central nervous system mediating affective behavior; acetylcholine works in certain motor neurons responsible for causing contraction of voluntary muscles. Acetylcholine is stored in synaptic vesicles and released in response to electrical activity in the neuron. It then diffuses across the

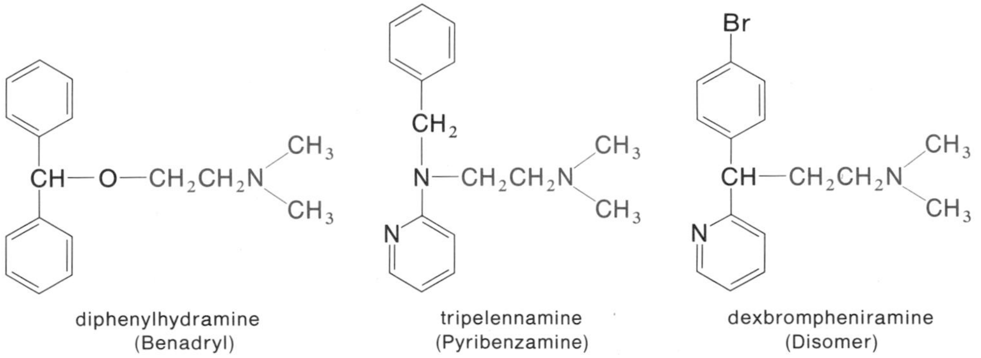

diphenylhydramine (Benadryl)

tripelennamine (Pyribenzamine)

dexbrompheniramine (Disomer)

Figure 19.3 Three synthetic antihistamines.

synapse and interacts with receptor sites on a neighboring neuron and causes transmission of a nerve impulse. After interacting with receptor sites, acetylcholine is deactivated by hydrolysis catalyzed by the enzyme acetylcholinesterase.

$$CH_3-\overset{\overset{\displaystyle CH_3}{|}}{\underset{\underset{\displaystyle CH_3}{|}}{N^+}}-CH_2-CH_2-O-\overset{\overset{\displaystyle O}{||}}{C}-CH_3 + H_2O \xrightarrow{\text{acetylcholin-esterase}}$$

acetylcholine

$$CH_3-\overset{\overset{\displaystyle CH_3}{|}}{\underset{\underset{\displaystyle CH_3}{|}}{N^+}}-CH_2-CH_2-OH + {}^-O-\overset{\overset{\displaystyle O}{||}}{C}-CH_3$$

choline   acetate

Inhibition of acetylcholinesterase leads to increased concentration of acetylcholine in synapses and increased stimulation of particular neurons. In response to this overload, neurons cease to respond, leading to loss of response in voluntary muscles.

Several water-soluble vitamins contain cyclic amines. Riboflavin (Figure 19.2) contains a fused three-ring system known as flavin. One of the nitrogen atoms of this ring system is substituted with a five-carbon chain derived from the sugar ribose, hence the name *riboflavin*. Thiamine, or vitamin B_1 (Figure 19.2), contains a substituted pyrimidine ring as well as a five-member ring including one atom each of nitrogen and sulfur. You have already seen the structural formula of pyridoxal phosphate in Section 17.10B. This substance, derived from pyridoxine, or vitamin B_6, contains a substituted pyridine ring.

Key Terms and Concepts

amino group (19.2)

base dissociation constant (19.4A)

basicity of amines (19.4A)

heterocyclic amine (19.2)

heterocyclic aromatic amine (19.2)

hydrogen bonding in 1° and 2° amines (19.3)

primary amine (19.1)

quaternary ammonium ion (19.2)

secondary amine (19.1)

tertiary amine (19.1)

Key Reactions

1. Basicity of amines (Section 19.4A):

$$CH_3NH_2 + H_2O \qquad CH_3NH_3^+ + OH^- \qquad K_b = \frac{[CH_3NH_3^+][OH^-]}{[CH_3NH_2]} = 4.4 \times 10^{-4} \qquad pK_b = 3.36$$

2. Formation of salts with acids (Section 19.4A):

$$CH_3NH_2 + HCl \longrightarrow CH_3NH_3^+Cl^- \quad\quad CH_3NH_2 + CH_3\overset{\displaystyle O}{\overset{\|}{C}}OH \longrightarrow CH_3NH_3^+ CH_3\overset{\displaystyle O}{\overset{\|}{C}}O^-$$

3. Ammonolysis of esters to form amides (Section 19.4B):

$$CH_3\overset{\displaystyle O}{\overset{\|}{C}}-OCH_2CH_3 + CH_3NH_2 \longrightarrow CH_3\overset{\displaystyle O}{\overset{\|}{C}}-NHCH_3 + CH_3CH_2OH$$

4. Ammonolysis of anhydrides to form amides and the ammonium salt of a carboxylic acid (Section 19.4C):

$$CH_3\overset{\displaystyle O}{\overset{\|}{C}}-O-\overset{\displaystyle O}{\overset{\|}{C}}CH_3 + 2\ CH_3NH_2 \longrightarrow CH_3\overset{\displaystyle O}{\overset{\|}{C}}-NHCH_3 + CH_3\overset{\displaystyle O}{\overset{\|}{C}}O^-\ CH_3NH_3^+$$

5. Reaction with nitrous acid (Section 19.4C):

a. Tertiary amines:

$$CH_3-\underset{\underset{\displaystyle CH_3}{|}}{\overset{\overset{\displaystyle CH_3}{|}}{N}}: \ + HONO \longrightarrow CH_3-\underset{\underset{\displaystyle CH_3}{|}}{\overset{\overset{\displaystyle CH_3}{|}}{{}^+N}}-H \ \ NO_2$$

a nitrite salt

b. Secondary amines:

$$CH_3CH_2\underset{\underset{\displaystyle CH_3}{|}}{N}H + HONO \longrightarrow CH_3CH_2\underset{\underset{\displaystyle CH_3}{|}}{N}-N{=}O + H_2O$$

an N-nitrosoamine

c. Primary amines:

$$CH_3CH_2NH_2 + HONO \longrightarrow CH_3CH_2OH + H_2O + N_2$$

Problems

Structure and Nomenclature of Amines (Sections 19.1 and 19.2)

19.3 Write structural formulas for the following compounds. Classify each as a primary amine, a secondary amine, a tertiary amine, an aromatic amine, a heterocyclic aromatic amine, or an ammonium salt.

a. diethylamine
b. aniline
c. cyclohexylamine
d. pyrrole
e. pyridine
f. tetraethylammonium iodide
g. 2-aminoethanol (ethanolamine)
h. 2-aminopropanoic acid (alanine)
i. pyrimidine
j. trimethylammonium benzoate
k. p-methoxyaniline
l. N, N-dimethylacetamide
m. N-methylaniline
n. acetylcholine
o. choline acetate
p. ethyl p-aminobenzoate
q. pyridine 3-carboxylic acid (nicotinic acid)
r. 3,5-dichloroaniline
s. 3,5-diaminochlorobenzene
t. p-aminobenzenesulfonic acid

19.4   Give an acceptable name for each of the following compounds.

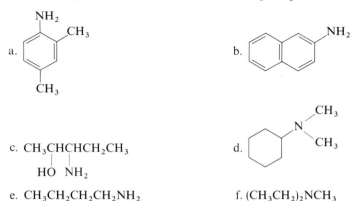

a.

b.

c. $CH_3CHCHCH_2CH_3$
       | |
      HO NH_2

d.

e. $CH_3CH_2CH_2CH_2NH_2$

f. $(CH_3CH_2)_2NCH_3$

19.5   Draw structural formulas for the eight isomeric amines of molecular formula $C_4H_{11}N$. Name each; label each as primary, secondary, or tertiary.

Physical Properties of Amines (Section 19.3)

19.6   Draw structural formulas to illustrate hydrogen bonding between the circled atoms.

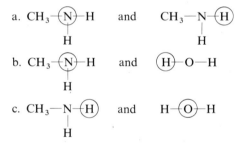

a. CH_3—Ⓝ—H   and CH_3—N—Ⓗ
         | |
         H   H

b. CH_3—Ⓝ—H   and   Ⓗ—O—H
         |
         H

c. CH_3—N—Ⓗ   and   H—Ⓞ—H
         |
         H

19.7   Both 1-aminobutane and 1-butanol are liquids at room temperature. One of these compounds has a boiling point of 117°C, the other a boiling point of 78°C. Which compound has the boiling point of 78°C? Explain your reasoning.

19.8   Arrange the following compounds in order of increasing boiling points.

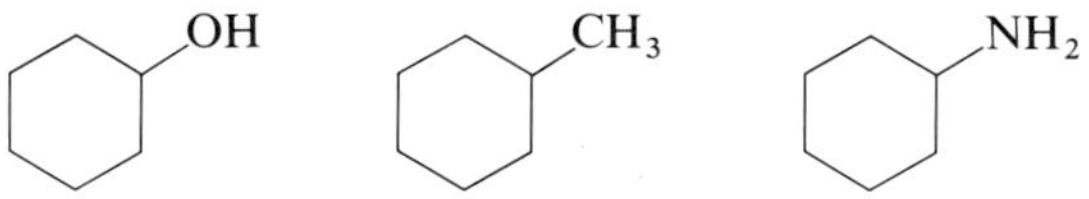

OH CH_3 NH_2

cyclohexanol   methylcyclohexane   cyclohexylamine

Reactions of Amines (Section 19.4)

19.9   Select the stronger acid in each pair.

a. $CH_3CH_2NH_3^+$ or   ⬡—NH_3^+

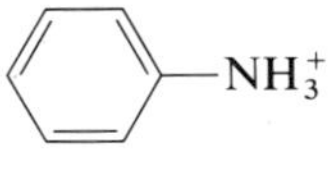

b. NH_4^+ or $CH_3NH_3^+$

c. 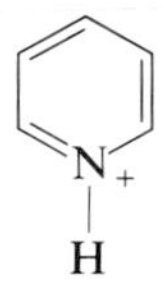   or

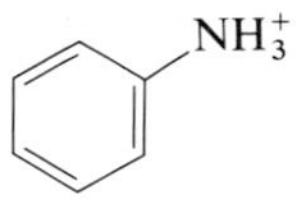

d. 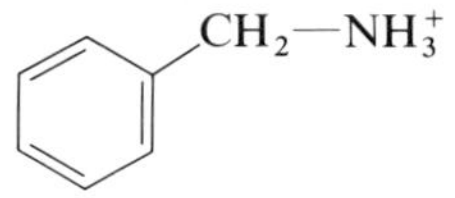   or

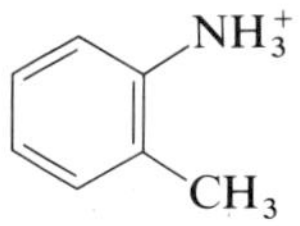

19.10 Select the stronger base in each pair.

a. $CH_3CH_2NH_2$ or $C_6H_5NH_2$
b. CH_3NH_2 or $CH_3CO_2^-$
c. $C_6H_5CH_2NH_2$ or $C_6H_5NH_2$

d. 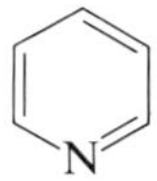   or

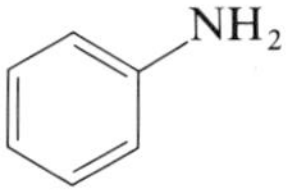

19.11 Arrange the following compounds in order of increasing basicity.

a.

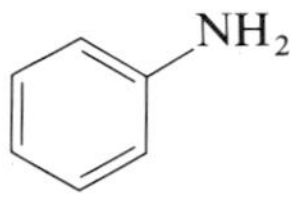

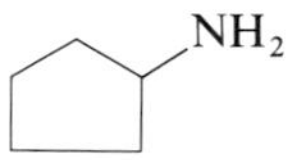

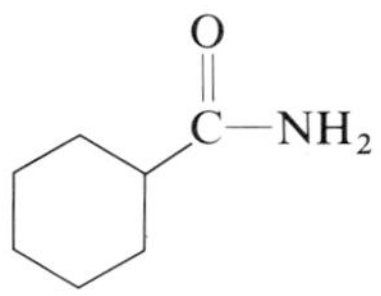

    (i)   (ii)   (iii)

b.

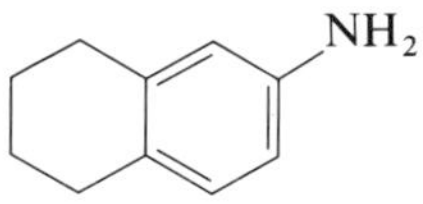

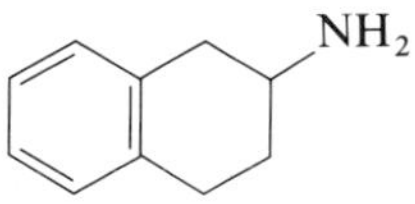

    (i)   (ii)

19.12 Suppose you are given a mixture of the following three compounds. Describe a procedure you could use to separate and isolate each in a pure state.

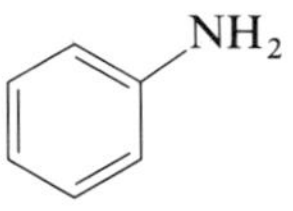

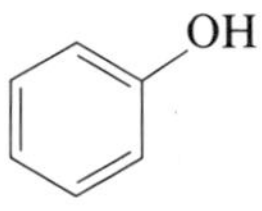

$CH_3CH_2CH_2CH_2CH_2CH_2OH$

19.13 Alanine (2-aminopropanoic acid) is one of the important amino acids found in proteins. Would you expect the structural formula of alanine to be represented better by (I) or (II)? Explain.

$$CH_3{-}CH{-}CO_2H \qquad CH_3{-}CH{-}CO_2^-$$
$$\qquad\;\; | \qquad\qquad\qquad\quad |$$
$$\qquad\;\; NH_2 \qquad\qquad\qquad NH_3^+$$

        (I)   (II)

19.14 Describe a simple chemical test by which you could distinguish between the following pairs of compounds. In each case, state what test you would perform, what you would expect to observe, and write a balanced equation for all positive tests.

a. 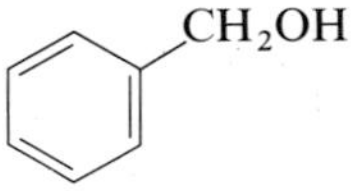   (benzene ring)—CH_2OH and   (benzene ring)—CH_2NH_2

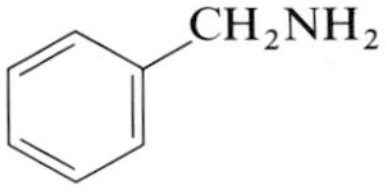

b. 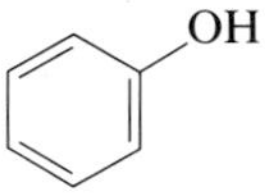   (benzene ring)—OH and   (benzene ring)—NH_2

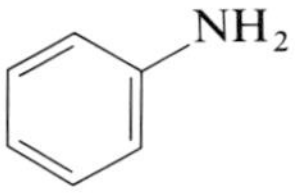

c. cyclohexanol and cyclohexylamine
d. trimethylacetic acid and 2,2-dimethylpropylamine

19.15 Would you expect aniline to be more soluble in water or in 0.1M HCl? Explain.

19.16 Draw structural formulas for the products of the following reactions.

a. 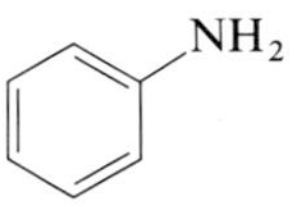   (benzene ring)—NH_2 $+ CH_3\overset{O}{\underset{||}{C}}-O-\overset{O}{\underset{||}{C}}CH_3 \longrightarrow$

b. 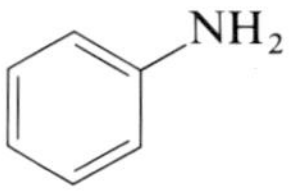   (benzene ring)—NH_2 $+ CH_3\overset{O}{\underset{||}{C}}OH \longrightarrow$

c. $CH_3\underset{\underset{NH_2}{|}}{\overset{\overset{O}{||}}{C}H}COH + CH_3CH_2OH \xrightarrow[\text{esterification}]{\overset{H^+}{\text{Fischer}}}$

d. $CH_3\underset{\underset{NH_2}{|}}{C}HCOCH_2CH_3 + CH_3\overset{O\;\;O}{\underset{||\;\;||}{C}}OCCH_3 \longrightarrow$

(with C=O indicated: $CH_3\overset{O}{\underset{|}{\underset{}{}}}CHCOCH_2CH_3$)

e. $CH_3\underset{\underset{\underset{\underset{O}{||}}{C}CH_3}{HN}}{\overset{\overset{O}{||}}{C}H}COCH_2CH_3 + H_2NCH_2CH_2CH_3 \longrightarrow$

19.17 Draw structural formulas for the products of the following reactions.

a. 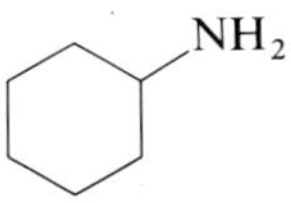   (cyclohexane ring)—NH_2 $+ HNO_2 \longrightarrow$

b. 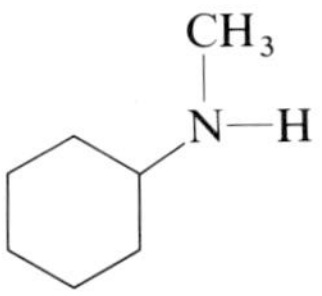   (cyclohexane ring)—$\overset{CH_3}{\underset{}{N}}-H$ $+ HNO_2 \longrightarrow$

c.

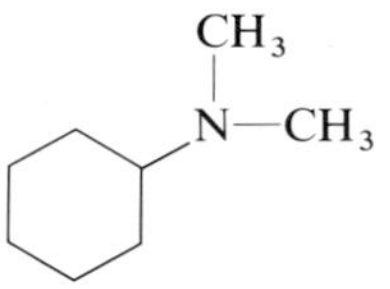

CH$_3$
|
N—CH$_3$

(cyclohexyl) + HNO$_2$ ⟶

d. 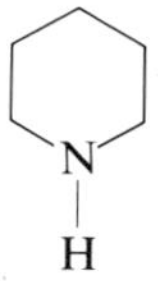 (piperidine ring)

N
|
H

+ HNO$_2$ ⟶

19.18 Name the following compounds.

a. 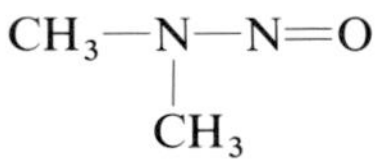 CH$_3$—N—N=O
|
CH$_3$

b. 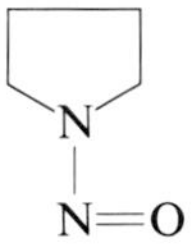 (pyrrolidine ring)

N
|
N=O

19.19 Pyridoxal phosphate is one of the metabolically active forms of vitamin B$_6$. Draw structural formulas for the Schiff bases formed by reaction of pyridoxal phosphate with (a) tyrosine and (b) glutamic acid.

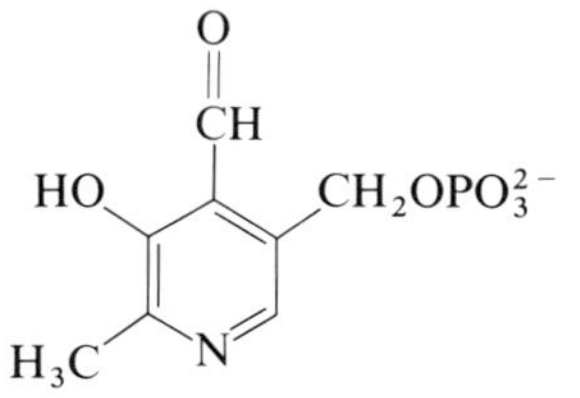

O
||
CH

HO   CH$_2$OPO$_3^{2-}$

H$_3$C   N

pyridoxal phosphate

HO—⟨benzene ring⟩—CH$_2$CHCOH
                              |
                              NH$_2$
                              O
                              ||

O
||
HOCCH$_2$CH$_2$CHCOH
                  |
                  NH$_2$
                  O
                  ||

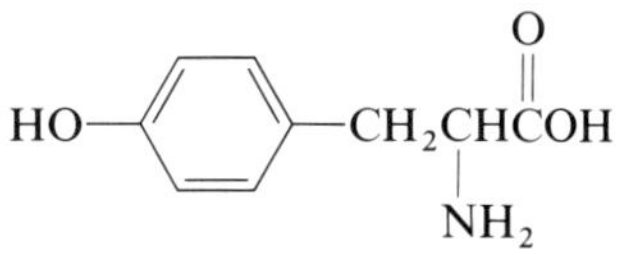

tyrosine 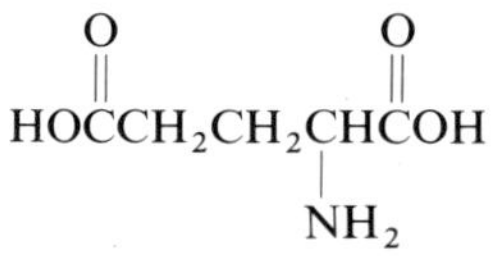   glutamic acid

19.20 Another of the metabolically active forms of vitamin B$_6$ is pyridoxamine phosphate. Draw structural formulas for the Schiff bases formed by reaction of pyridoxamine phosphate with (a) pyruvic acid and (b) oxaloacetic acid.

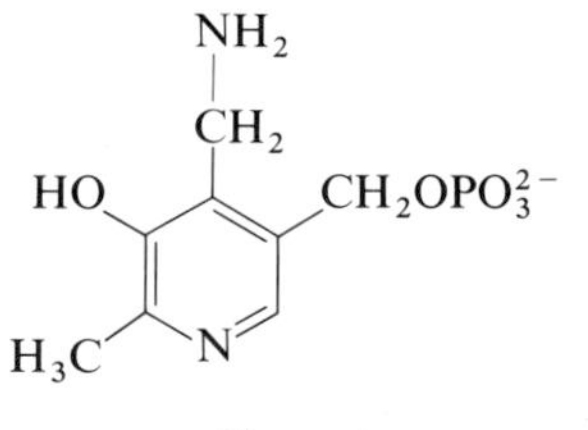

NH$_2$
|
CH$_2$

HO   CH$_2$OPO$_3^{2-}$

H$_3$C   N

pyridoxamine
phosphate

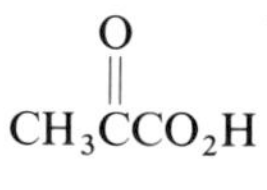

O
||
CH$_3$CCO$_2$H

pyruvic
acid

O
||
HO$_2$CCH$_2$CCO$_2$H

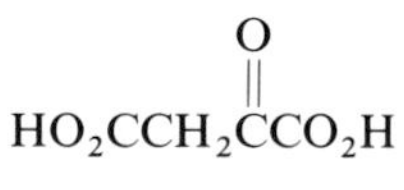

oxaloacetic
acid

Amino Acids and Proteins

Proteins, as much as any other class of compounds, are inseparable from life itself. These remarkable molecules are classified into two broad categories, depending on their physical properties: *fibrous proteins* and *globular proteins*. Fibrous proteins are stringy, physically tough, and generally insoluble in water and most solvents. There are three major classes of fibrous proteins: the keratins of skin, wool, claws, horn, scales, and feathers; the silks; and the collagens of skin, bone, and tendons. Globular proteins, the second broad category, are generally spherical in shape and soluble in water. Nearly all enzymes, antibodies, hormones, and transport proteins are globular. We will begin our study of proteins with an examination of amino acids, the building blocks of proteins.

20.1 Amino Acids

A. Structure

Amino acids are compounds that contain both a carboxyl group and an amino group. While many types of amino acids are known, the α-amino acids are the most significant in the biological world because they are the units from

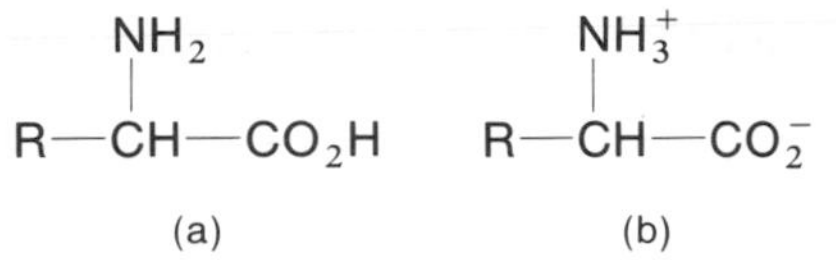

$$\begin{array}{cc} NH_2 & NH_3^+ \\ | & | \\ R-CH-CO_2H & R-CH-CO_2^- \\ (a) & (b) \end{array}$$

Figure 20.1 General formula for an α-amino acid: (a) un-ionized form; (b) zwitterion form.

which proteins are constructed. The general formula of an α-amino acid is shown in Figure 20.1. Although Figure 20.1(a) is a common way of writing structural formulas for amino acids, it is not accurate because it shows an acid ($-CO_2H$) and a base ($-NH_2$) within the same molecule. These acidic and basic groups react with each other to form an internal salt called a **zwitterion** (Figure 20.1b). Note that the zwitterion has no net charge; it contains one positive and one negative charge.

B. Acid-Base Properties and the Zwitterion

Although pK_a values for carboxyl and amino groups of particular amino acids vary, the average value of pK_a for an α-CO_2H group is 2.2; that for an α-NH_3^+ group is 9.5. Using these values, we can calculate the ratio of [α-CO_2H] to [α-CO_2^-] and of [α-NH_3^+] to [α-NH_2] at any given pH. As an example, let us calculate these ratios at pH 7.4, physiological pH. Consider first the ionization of the weak acid α-CO_2H to form H^+ and its conjugate base, α-CO_2^-.

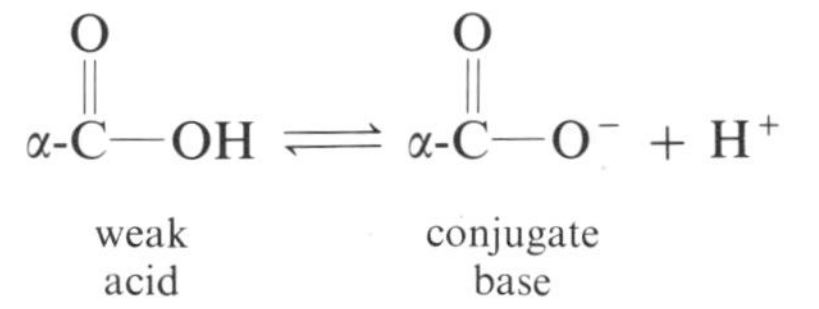

$$\begin{array}{ccc} O & & O \\ \| & & \| \\ \alpha\text{-C}-OH & \rightleftharpoons & \alpha\text{-C}-O^- + H^+ \qquad K_a = 6.3 \times 10^{-3} \qquad pK_a = 2.2 \\ \text{weak} & & \text{conjugate} \\ \text{acid} & & \text{base} \end{array}$$

The equilibrium constant for this ionization is called an acid dissociation constant and is given by the expression:

$$K_a = \frac{[H^+][\alpha\text{-}CO_2^-]}{[\alpha\text{-}CO_2H]}$$

Rearranging this expression gives

$$\frac{[\alpha\text{-}CO_2^-]}{[\alpha\text{-}CO_2H]} = \frac{K_a}{[H^+]}$$

It is from this equation that the Henderson-Hasselbalch equation was derived in Section 8.7B. Substituting the average value of K_a for an α-CO_2H group

(6.3×10^{-3}) and the hydrogen ion concentration at pH 7.4 (4.0×10^{-8}) in this equation gives

$$\frac{[\alpha\text{-}CO_2^-]}{[\alpha\text{-}CO_2H]} = \frac{6.3 \times 10^{-3}}{4.0 \times 10^{-8}} = 1.6 \times 10^5$$

Thus, we see that at pH 7.4, the physiological pH, the ratio of $[\alpha\text{-}CO_2^-]$ to $[\alpha\text{-}CO_2H]$ is over 160,000 to 1. It is clear that at pH 7.4 an α-carboxyl group is virtually 100% in the conjugate base or ionized form and has a charge of -1.

We can also calculate the ratio of acid to conjugate base for an α-amino group. The average value of pK_a for α-amino groups is 9.5.

$$\underset{\substack{\text{weak}\\\text{acid}}}{\alpha\text{-}NH_3^+} \rightleftharpoons \underset{\substack{\text{conjugate}\\\text{base}}}{\alpha\text{-}NH_2} + H^+ \qquad K_a = 3.2 \times 10^{-10} \qquad pK_a = 9.5$$

The acid dissociation constant for the ionization of $\alpha\text{-}NH_3^+$ can be rearranged to give the following expression.

$$\frac{[\alpha\text{-}NH_2]}{[\alpha\text{-}NH_3^+]} = \frac{K_a}{[H^+]}$$

Substituting the average value of K_a for an $\alpha\text{-}NH_3^+$ group (3.2×10^{-10}) and the hydrogen ion concentration at pH 7.4 (4.0×10^{-8}) gives

$$\frac{[\alpha\text{-}NH_2]}{[\alpha\text{-}NH_3^+]} = \frac{3.2 \times 10^{-10}}{4.0 \times 10^{-8}} = 8 \times 10^{-3}$$

Thus, the ratio of $\alpha\text{-}NH_2$ to $\alpha\text{-}NH_3^+$ is less than 1 to 100, and at pH 7.4, an α-amino group is almost completely in the acid or protonated form and has a charge of $+1$.

We can see from these calculations that the zwitterion form of an amino acid predominates at pH 7.4. In the remainder of the text, we shall use the zwitterion form for amino acids.

C. Chirality of Amino Acids

Protein-derived amino acids, except for glycine, have a chiral carbon atom adjacent to the carboxyl group and show stereoisomerism. Figure 20.2 shows stereorepresentations and Fischer projection formulas for the enantiomers of serine.

Using D-glyceraldehyde as a standard, it has been established that all amino acids occurring naturally in proteins have an L-configuration about the chiral carbon. D-Amino acids are not found in proteins and are not part of the metabolism of higher organisms. However, several D-amino acids are

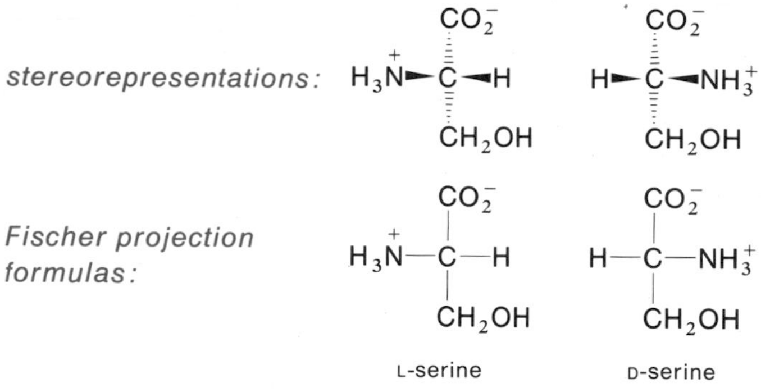

Figure 20.2 The enantiomers of serine.

stereorepresentations:

$$
\begin{array}{cc}
CO_2^- & CO_2^- \\
H_3\overset{+}{N}\!-\!C\!-\!H & H\!-\!C\!-\!NH_3^+ \\
CH_2OH & CH_2OH
\end{array}
$$

Fischer projection formulas:

$$
\begin{array}{cc}
CO_2^- & CO_2^- \\
H_3\overset{+}{N}\!-\!C\!-\!H & H\!-\!C\!-\!NH_3^+ \\
CH_2OH & CH_2OH \\
\text{L-serine} & \text{D-serine}
\end{array}
$$

important in the structure and metabolism of lower forms of life. As an example, both D-alanine and D-glutamic acid are structural components of the cell walls of certain bacteria. See the mini-essay, "The Penicillins."

D. The 20 Common Protein-Derived Amino Acids

Table 20.1 (next page) shows names, structural formulas, and standard three-letter abbreviations for the 20 common amino acids found in proteins. Amino acids in this table are grouped into three categories according to the nature of their side chain.

1. Nonpolar side chains. The nonpolar-side-chain category includes eight amino acids. Of these, glycine, alanine, and proline have small nonpolar side chains and are weakly hydrophobic. The other five amino acids in this category (phenylalanine, valine, leucine, isoleucine, and methionine) have larger side chains and are more strongly hydrophobic.

2. Polar but uncharged side chains. The polar uncharged side chain category includes eight amino acids: serine and threonine with hydroxyl groups; asparagine and glutamine with amide groups; tyrosine with a phenolic side chain; tryptophan and histidine with heterocyclic aromatic amine side chains; and cysteine with a sulfhydryl group. Three amino acids included in the polar uncharged category have side chains that show some degree of ionization, depending on the pH. These are the sulfhydryl group of cysteine, the imidazole group of histidine, and the phenolic hydroxyl of tyrosine.

3. Charged side chains. The charged side chain category contains four amino acids: aspartic acid, glutamic acid, lysine, and arginine. Aspartic and glutamic acids have carboxyl groups on the side chain. The pK_a values for these side-chain $-CO_2H$ groups are approximately 4.0, therefore each acid is completely ionized at pH 7.4. The side chains of lysine and arginine have amino groups.

Table 20.1 The 20 common amino acids found in proteins, grouped by categories. Each is shown as it would be ionized at pH 7.4.

Nonpolar Side Chains

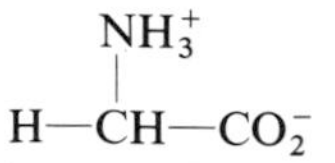

glycine (gly)

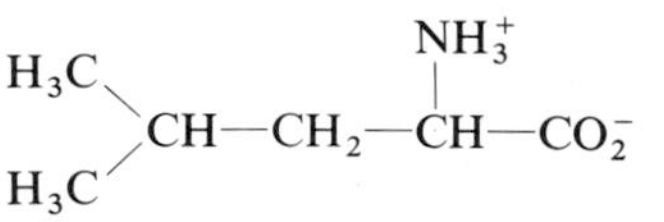

L-leucine (leu)

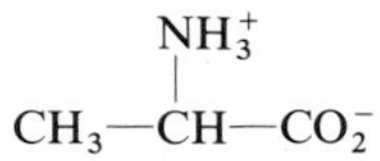

L-alanine (ala)

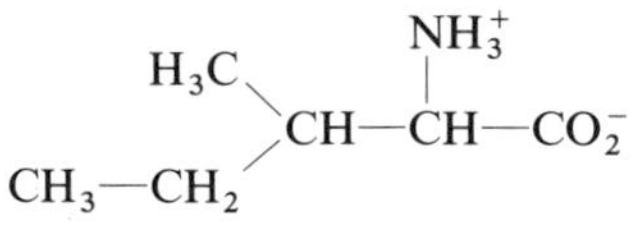

L-isoleucine (ile)

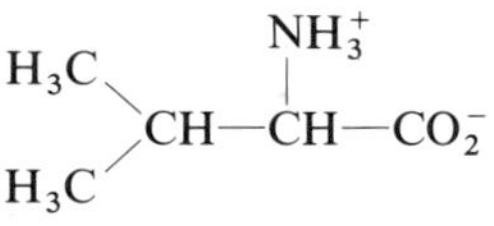

L-valine (val)

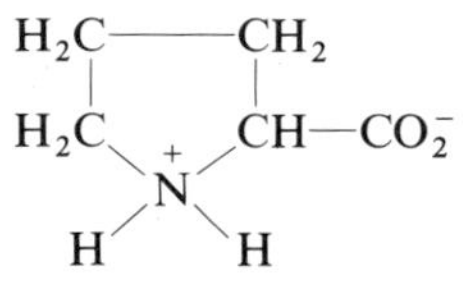

L-proline (pro)

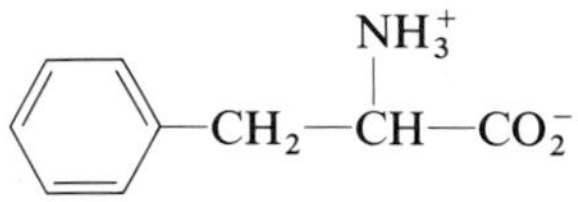

L-phenylalanine (phe)

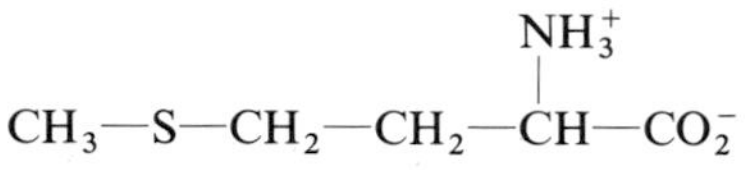

L-methionine (met)

Polar Uncharged Side Chains

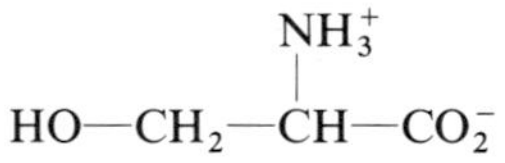

L-serine (ser)

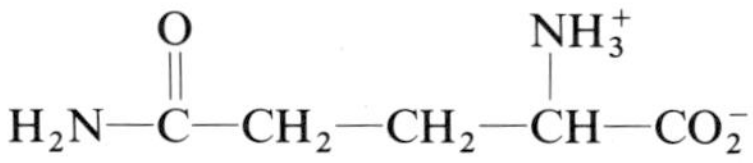

L-glutamine (gln)

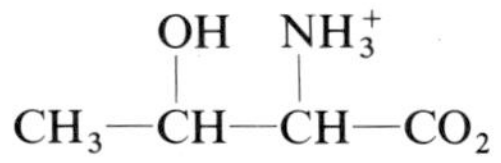

L-threonine (thr)

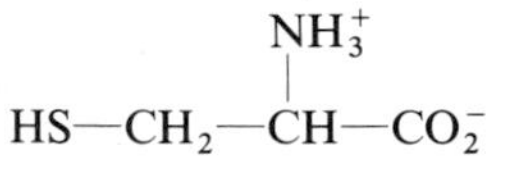

L-cysteine (cys)

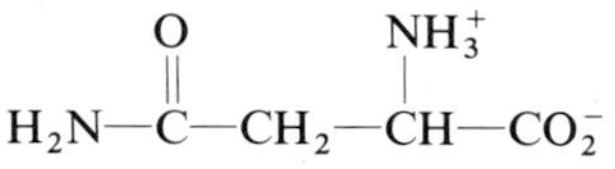

L-asparagine (asn)

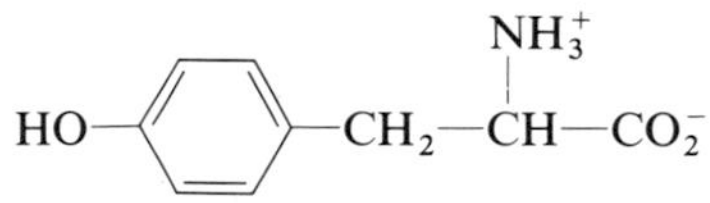

L-tyrosine (tyr)

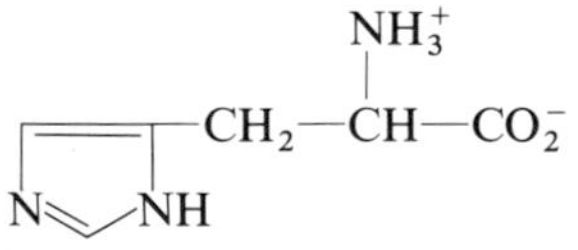

L-histidine (his)

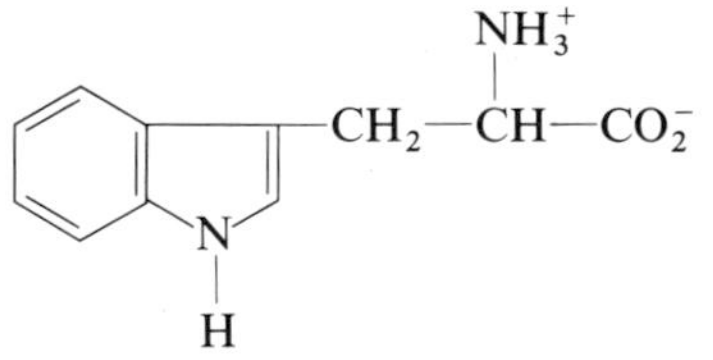

L-tryptophan (trp)

Polar Charged Side Chains

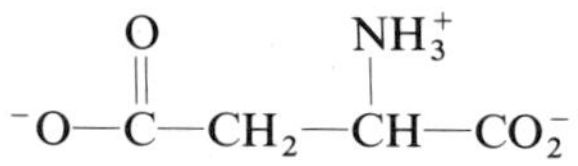

L-aspartic acid (asp)

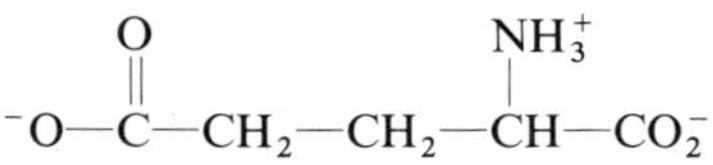

L-glutamic acid (glu)

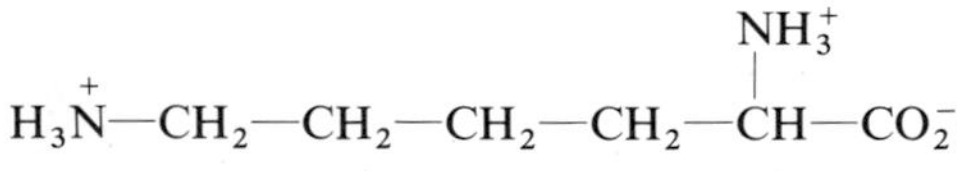

L-lysine (lys)

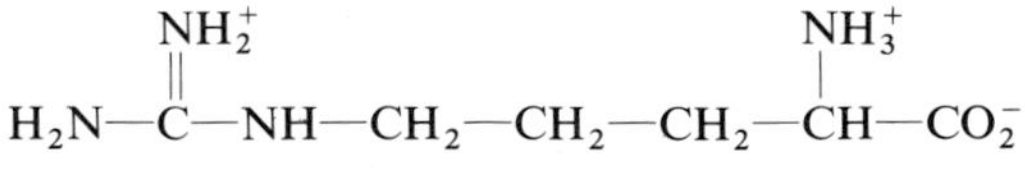

L-arginine (arg)

Table 20.2 pK$_a$ values of ionizable amino acid side chains.

Amino Acid	Side Chain (Acid Form)	pK$_a$ of Side Chain	Predominant Form at pH 7.4
aspartic acid	$-CO_2H$	3.86	$-CO_2^-$
glutamic acid	$-CO_2H$	4.07	$-CO_2^-$
histidine	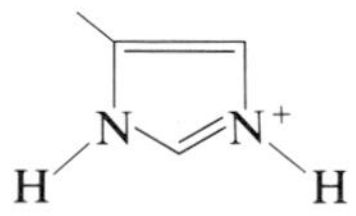	6.10	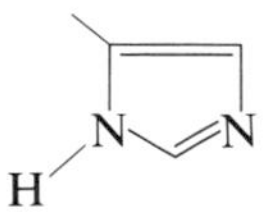
cysteine	$-SH$	8.00	$-SH$
tyrosine	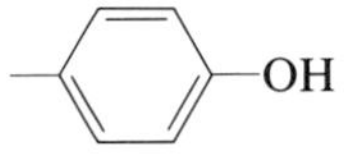 OH	10.07	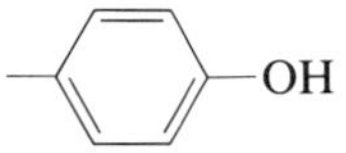 OH
lysine	$-NH_3^+$	10.53	$-NH_3^+$
arginine	$-NH-\overset{\overset{\displaystyle NH_2^+}{\|}}{C}-NH_2$	12.48	$-NH-\overset{\overset{\displaystyle NH_2^+}{\|}}{C}-NH_2$

The pK$_a$ of the lysine side chain is 10.5 and that of arginine is 12.5. Therefore, these side chains are fully protonated at pH 7.4 and lysine and arginine have a net charge of $+1$ at pH 7.4. The pK$_a$ values for the seven amino acids with ionizable side chains are given in Table 20.2.

The following guidelines can be used to estimate the degree of ionization of an alpha-carboxyl group, alpha-amino group, or ionizable side chain at any pH.

1. If the pH of the solution is 2.0 or more units greater (more basic) than the pK$_a$ of the ionizable group, then the group is almost entirely in the conjugate base form.

2. If the pH of the solution is 1.0 unit greater (more basic) than the pK$_a$ of the ionizable group, then the group is 90% in the conjugate base form and 10% in the acid form.

3. If the pH of the solution is equal to the pK$_a$ of the ionizable group, then the group is 50% in the conjugate base form and 50% in the acid form.

4. If the pH of the ionizable group is 1.0 unit less (more acidic) than the pK$_a$ of the ionizable group, then the group is 10% in the conjugate base form and 90% in the acid form.

5. If the pH of the solution is 2.0 or more units less (more acidic) than the pK$_a$ of the ionizable group, then the group is almost entirely in the acid form.

Example 20.1

Draw structural formulas for the following amino acids and estimate the net charge on each at pH 1.0, 6.0, and 12.0.

a. serine   **b.** glutamic acid

Solution

a. The pK_a for the α-carboxyl group of serine is approximately 2.2 and that of the α-amino group is 9.5. A pH of 1.0 is 1.2 units below the pK_a of the α-CO_2H group, thus the group is more than 90% in the acid form. The α-amino group is also completely in the acid form at this pH, therefore serine has a net charge of +1. At pH 6.0, the α-carboxyl group is fully ionized to —CO_2^- and the α-amino group is fully protonated to α-NH_3^+. Therefore, at pH 6.0, serine has a net charge of zero. At pH 12.0, both the α-carboxyl and α-amino groups are fully ionized to α-CO_2^- and α-NH_2, and serine has a net charge of −1.

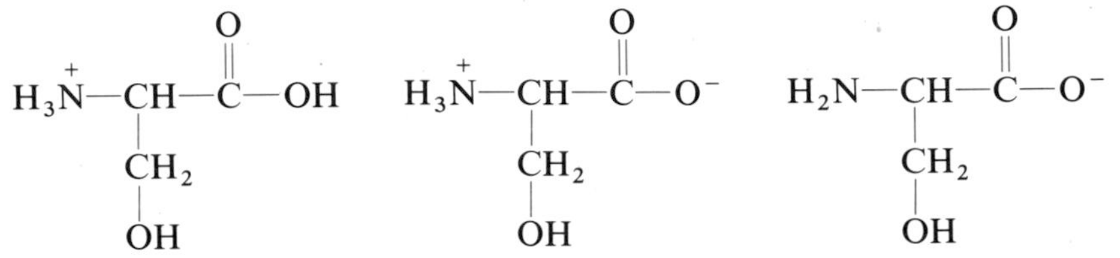

pH = 1.0	pH = 6.0	pH = 12.0
net charge = +1	net charge = 0	net charge = −1

b. The degree of ionization and charged character of the α-carboxyl and α-amino groups in glutamic acid are the same as those in serine. The pK_a of the side-chain carboxyl group is 4.07 (Table 20.2). A pH of 1.0 is more than 2 units less (more acidic) than this pK_a and pH 6.0 and 12.0 are each more than 2 units greater (more basic) than this pK_a. Therefore, the side-chain carboxyl group of glutamic acid is fully protonated at pH 1.0 and fully ionized at pH 6.0 and 12.0.

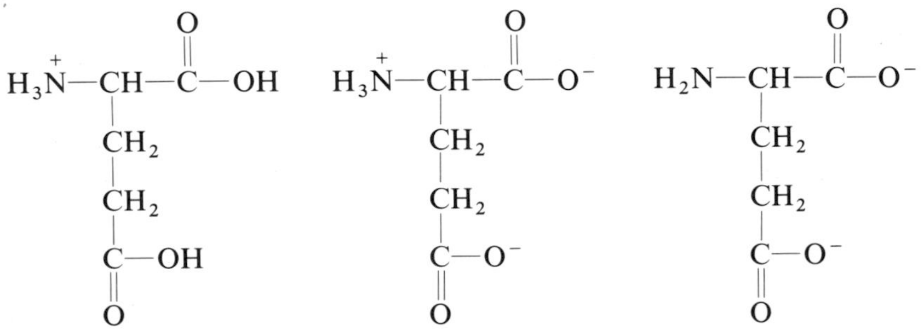

pH = 1.0	pH = 6.0	pH = 12.0
net charge = +1	net charge = −1	net charge = −2

Problem 20.1

Draw structural formulas for the following amino acids and estimate the net charge on each at pH 1.0, 6.0, and 12.0.

a. histidine   **b.** lysine

E. Isoelectric Point and Isoelectric Precipitation

As you can see from the previous discussion of amino acids and from your solutions to Example 20.1 and Problem 20.1, the net charge on an amino acid in solution becomes more positive as the pH decreases (the solution becomes more acidic), and conversely, the net charge becomes more negative as the pH increases (the solution becomes more basic). The pH at which an amino acid has a net charge of zero is called its **isoelectric point.** This pH is given the symbol **pI.**

Amino acids with non-ionizing side chains have isoelectric points in the pH range 5.5–6.5 (for example, the pIs of glycine and serine are 6.0). The isoelectric points for amino acids with ionizable side chains are given in Table 20.3.

Given values for the isoelectric point of these amino acids, you can estimate the charge on each at any pH. For example, the charge on tyrosine is zero at pH 5.63. A small fraction of tyrosine molecules are positively charged at pH 5.0. Virtually all are positively charged at pH 3.63 (2 units less than pI). As another example, the net charge on lysine is zero at pH 9.74. At pH values smaller than 9.74, some fraction of lysine molecules are positively charged.

An understanding of isoelectric point is important for two reasons. First, it helps us to understand the solubility of amino acids and proteins as a function of pH. The solubility of these molecules is a minimum at the isoelectric

Table 20.3 Isoelectric points for amino acids with ionizable side-chain groups.

Amino Acid	Ionizable Side-Chain Group	pI (Isoelectric Point)
aspartic acid	carboxyl	2.98
glutamic acid	carboxyl	3.08
histidine	imidazole	7.64
cysteine	thiol	5.02
tyrosine	phenol	5.63
lysine	amino	9.74
arginine	amino	10.76

point and increases as pH is increased or decreased from this point. In order to crystallize an amino acid or protein, the pH of an aqueous solution is adjusted to the pI of the acid or protein and the compound is precipitated, filtered, and collected. This process is called **isoelectric precipitation.** Second, an understanding of isoelectric points enables us to predict the way components of mixtures of amino acids or proteins migrate in an electrical field during electrophoresis.

F. Electrophoresis

Electrophoresis is the process of separating compounds on the basis of their electrical charge. In paper electrophoresis, a paper strip saturated with an aqueous buffer of predetermined pH serves as a bridge between two electrode vessels. A sample of amino acid or protein is applied as a spot. When an electrical potential is applied to the electrode vessels, amino acid or protein molecules migrate toward the electrode carrying the charge opposite to their own. Molecules having a high charge density move more rapidly than those with a low charge density. Any molecule already at its isoelectric point remains at the origin. After separation is complete, the strip is dried and sprayed with a dye to make the separated components visible. The dye most commonly used for amino acids is ninhydrin.

Ninhydrin consists of a benzene ring fused to a five-member ring. In aqueous solution, the middle ketone of the five-member ring reacts with a molecule of water (undergoes hydration) to give a compound called ninhydrin hydrate.

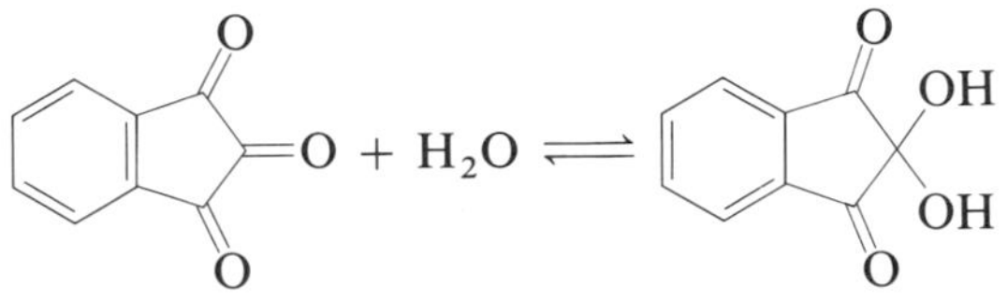

ninhydrin   ninhydrin hydrate

Ninhydrin hydrate reacts with α-amino acids to produce a purple-colored anion, an aldehyde, and carbon dioxide.

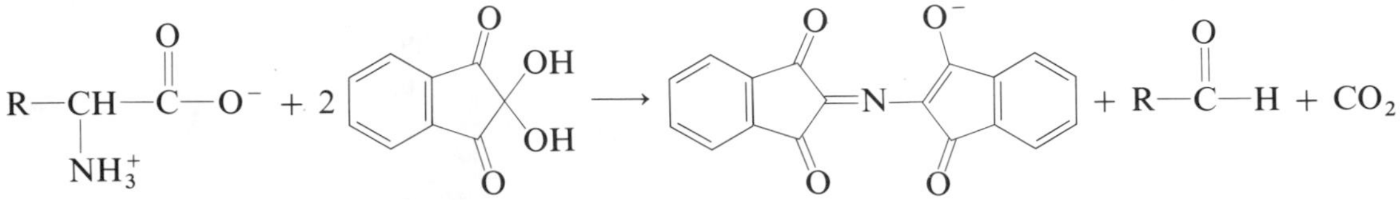

an α-amino   ninhydrin   anion
acid   hydrate   (purple-colored)

The intensity of the purple color is directly proportional to the concentration of the anion, and for this reason, the ninhydrin reaction can be used as both a qualitative and a quantitative test for the presence of amino acids.

Example 20.2

Electrophoresis of a mixture of lysine, histidine, and cysteine is carried out at pH 7.64. Describe the behavior of each of these amino acids.

Solution

The isoelectric point of histidine is 7.64. At this pH, histidine has a net charge of zero and does not move from the origin. The pI of cysteine is 5.02; at pH 7.64, cysteine has a net negative charge and moves toward the positive electrode. The pI of lysine is 9.74; at pH 7.64, lysine has a net positive charge and moves toward the negative electrode. See Figure 20.3.

Figure 20.3 Electro-phoresis of a mixture of histidine, lysine, and cysteine at pH 7.64.

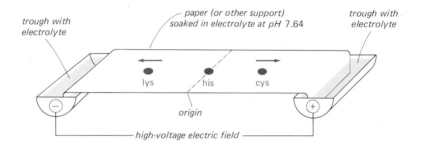

Problem 20.2

Describe the behavior of a mixture of glutamic acid, arginine, and valine on paper electrophoresis at pH 6.0.

Electrophoretic separations also can be carried out using starch, agar, certain plastics, and cellulose acetate as solid supports. This technique is extremely important in biochemical research and is also an invaluable tool in the clinical chemistry laboratory. For a discussion of the electrophoretic screening of blood samples for sickle-cell anemia, see the mini-essay, "Abnormal Human Hemoglobins," immediately following this chapter.

G. Essential Amino Acids

Of the twenty amino acids required by the human body for the synthesis of proteins, adequate amounts of twelve can be synthesized by enzyme-catalyzed reactions starting from carbohydrates or lipids and a source of nitrogen. For the remaining amino acids, either there are no biochemical pathways available for their synthesis, or the available pathways do not provide adequate amounts

Table 20.4 Estimated amino acid requirements for humans.

Amino Acid	Requirement, mg/kg Body Weight/Day		
	Infant (4–6 months)	Child (10–12 years)	Adult
histidine	33	—	—
isoleucine	83	28	12
leucine	135	42	16
lysine	99	44	12
total S-containing amino acids (methionine and cysteine)	49	22	10
total aromatic amino acids (phenylalanine and tyrosine)	141	22	16
threonine	68	28	8
tryptophan	21	4	3
valine	92	25	14

for proper nutrition. Therefore, these eight amino acids must be supplied in the diet and are called **essential amino acids.** In reality, all amino acids are essential for normal tissue growth and development. However, the term *essential* is reserved for those that must be supplied in the diet.

The estimated daily requirements of eight essential amino acids are given in Table 20.4. Anyone who consumes about 40–55 grams of protein daily in the form of meat, fish, cheese, milk, or eggs satisfies daily needs for these essential amino acids.

Tyrosine is synthesized from phenylalanine in the body. Therefore, the requirements for these two aromatic amino acids are combined in Table 20.4. Similarly, the sulfur-containing amino acids methionine and cysteine are combined.

Histidine is essential for growth in infants and it may be needed by adults as well. Arginine is synthesized by adults, but the rate of internal synthesis is not adequate to meet the needs of the body during periods of rapid growth and protein synthesis. Therefore, depending on age and state of health, either eight, nine, or ten amino acids may be *essential* for humans.

H. Biological Value of Dietary Proteins

The **biological value of a dietary protein** is a measure of the percentage that is absorbed and used to build body tissue. Some of the first information on the biological value of dietary proteins came from studies on rats. In one series of experiments, young rats were fed diets containing protein in the form of either casein (a milk protein), gliadin (a wheat protein), or zein (a corn protein). With casein as the sole source of protein, the rats remained healthy and grew normally. Those fed gliadin maintained their weight but did not grow much. Those fed zein not only failed to grow but lost weight and, if kept on this diet, eventually died. Since casein evidently supplies all required amino acids in the correct proportions needed for growth, it is called a **complete protein.** Analysis revealed that gliadin contains too little lysine, and that zein is low in both lysine and tryptophan. When the gliadin diet was supplemented with lysine, or the zein diet with lysine and tryptophan, the test animals grew normally.

Table 20.5 shows the biological value for rats of some common dietary sources of protein. The proteins in egg are the best-quality natural protein. The proteins of milk rank 84, those of meats and soybeans about 74. The legumes, vegetables, and cereal grains are in the range 50–70.

Plant proteins generally vary more from the amino acid pattern required by humans than do animal proteins. Fortunately, however, not all plant proteins are deficient in the same amino acids. For example, beans are low in the sulfur-containing amino acids cysteine and methionine, yet are high in lysine. Wheat has just the opposite pattern. By eating wheat and beans together, it is possible to increase by 33% the usable protein you would get by eating either of these foods separately.

The provision of a diet adequate in protein and essential amino acids is a grave problem in the world today, especially in areas of Asia, Africa, and South America. The overriding dimension of this problem is poverty and an inability to select foods of adequate protein and caloric content. The best overall sources of calories are the cereal grains, which provide not only calories but proteins as well. When these are supplemented with animal protein or a proper selection of plant protein, the diet is adequate for even the most vulnerable. However, as income decreases, there is less animal protein in the diet, and

Table 20.5 The biological value for rats of some common sources of dietary protein.

Food	Protein as % of Dry Solid	Biological Value of Protein, %
hen's egg, whole	48	94
cow's milk, whole	27	84
fish	72	83
beef	45	74
soybeans	41	73
rice, brown	9	73
potato, white	9	67
wheat, whole grain	14	65
corn, whole grain	11	59
dry beans, common	25	58

even cereal grains are often replaced by cheaper sources of calories such as starches or tubers, foods that have either very little or no protein. The poorest 25% of the world's population consumes diets with caloric and protein content that fall below, often dangerously below, the calculated minimum daily requirements.

Those most apt to show symptoms of too little food, too little protein, or both, are young children in the years immediately following weaning. There is a failure to grow properly and a wasting of tissue. This sickness is called marasmus, a name derived from a Greek word meaning "to waste away." The muscles become atrophied and the face develops a wizened "old man" look. Another disease, kwashiorkor, leads to tragically high death rates among children. As long as a child is breast-fed, it is healthy. At weaning (often forced when a second child is born), the first child's diet suddenly is switched to starch and inadequate sources of protein. Such children develop bloated bellies and patchy, discolored skin, and are often doomed to short lives.

One solution to the problem of quantity and quality of protein is to breed new varieties of cereal grains with higher protein content, better protein quality, or both. Alternatively, cereals and their derived products can be supplemented (fortified) with amino acids in which they are deficient: principally lysine for wheat; lysine or lysine plus threonine for rice; and lysine plus tryptophan for corn. New methods of synthesis and fermentation now provide a cheap source of these amino acids, thus making the economics of food fortification entirely practical. In another attack on the problem of protein malnutrition, several high-protein, low-cost infant foods have been developed by nutritionists. Clearly, advances in food chemistry and technology have provided the means to eradicate most hunger and malnutrition. What remains is for political and social systems to put this knowledge into practice.

I. Some Other Common Amino Acids

Thus far we have concentrated on the 20 amino acids from which the body synthesizes proteins. A few proteins contain special amino acids. For example, L-hydroxyproline and L-5-hydroxylysine are important components of collagen, but are found in very few other proteins. These and all other special amino acids are formed after proteins are constructed by modification of one of the 20 common amino acids already incorporated into the protein.

L-hydroxyproline   L-5-hydroxylysine

Table 20.6 Several amino acids not found in proteins.

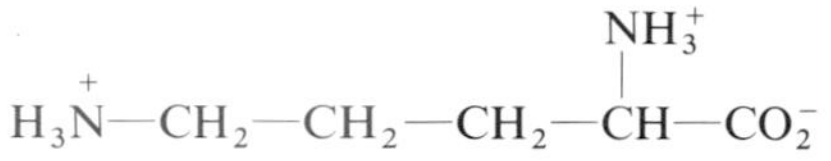

$$\overset{+}{H_3N}-CH_2-CH_2-CH_2-\overset{\overset{\displaystyle NH_3^+}{|}}{CH}-CO_2^-$$

ornithine (orn)

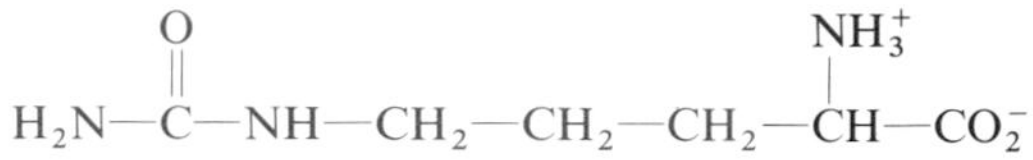

$$H_2N-\overset{\overset{\displaystyle O}{\|}}{C}-NH-CH_2-CH_2-CH_2-\overset{\overset{\displaystyle NH_3^+}{|}}{CH}-CO_2^-$$

citrulline

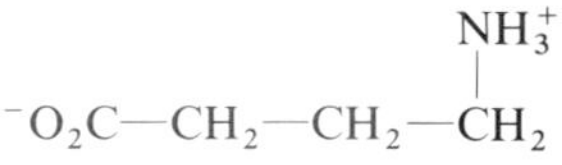

$$^-O_2C-CH_2-CH_2-\overset{\overset{\displaystyle NH_3^+}{|}}{CH_2}$$

γ-aminobutyric acid
(GABA)

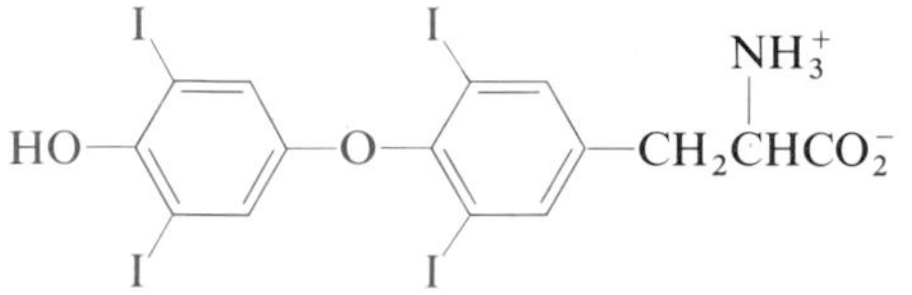

thyroxine
tetraiodothyronine or T_4

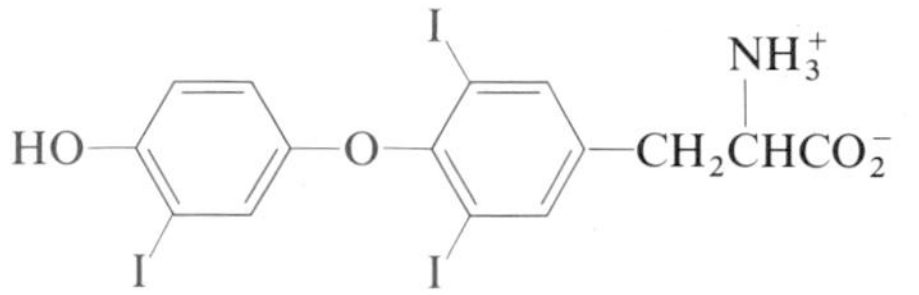

triiodothyronine *or* T_3

In addition to those amino acids listed in Table 20.1, there are a number of important nonprotein-derived amino acids, many of which are either metabolic intermediates or parts of nonprotein biomolecules. Several of these are shown in Table 20.6. Ornithine and citrulline are part of the urea cycle (Section 26.2C), a metabolic pathway that converts ammonia to urea. Gamma-amino butyric acid (GABA) is present in brain tissue. Its function there is, as yet, largely unknown.

Thyroxine, one of several hormones derived from the amino acid tyrosine, was first isolated from thyroid tissue in 1914. In 1952, triiodothyronine, a compound identical to thyroxine except that is contains only three atoms of iodine, was also discovered in the thyroid. Triiodothyronine is even more potent than thyroxine. Although the exact mechanisms of action of these thyroid hormones are not known, they are essential for the proper regulation of cellular metabolism. The levorotatory isomer of each is significantly more active than the dextrorotatory isomer.

20.2 The Peptide Bond

A. Structure of the Peptide Bond

In 1902, Emil Fischer proposed that proteins are long chains of amino acids joined together by amide bonds between the α-carboxyl group of one amino acid and the α-amino group of another. For these amide bonds, Fischer pro-

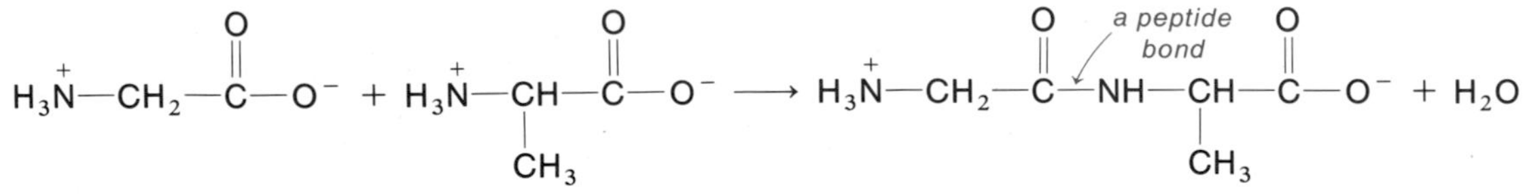

Figure 20.4 The peptide bond in glycylalanine.

posed the special name **peptide bond.** Figure 20.4 shows the peptide bond formed between glycine and alanine in the peptide glycylalanine.

A molecule containing two amino acids joined by an amide bond is called a dipeptide. Those containing larger numbers of amino acids are called tripeptides, tetrapeptides, pentapeptides, and so on. Molecules containing 10 or more amino acids are generally called polypeptides. Proteins are biological macromolecules of molecular weight 5,000 or greater, consisting of one or more polypeptide chains.

By convention, polypeptides are written from the left, beginning with the amino acid having the free H_3N^+— group and proceeding to the right toward the amino acid with the free —CO_2^- group. The amino acid with the free H_3N^+— group is called the N-terminal amino acid and that with the free —CO_2^- group is called the C-terminal amino acid. The structural formula for a polypeptide sequence may be written out in full, or the sequence of amino acids may be indicated using the standard abbreviation for each.

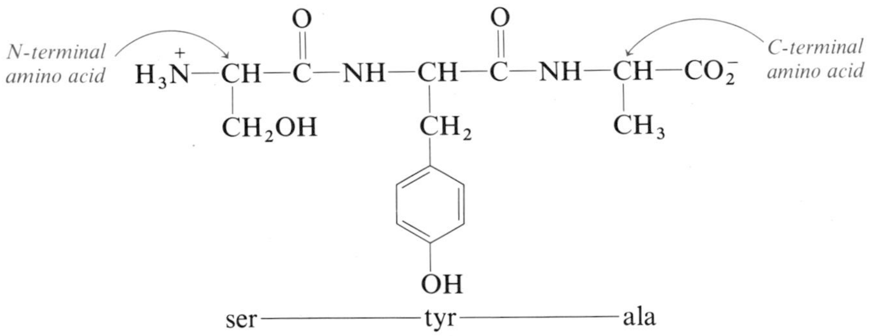

Polypeptides are named by listing each amino acid in order, from the N-terminal amino end of the chain to the C-terminal end. The name of the C-terminal amino acid is given in full. The name of each other amino acid in the chain is derived by dropping the suffix -*ine* and adding -*yl*. For example, if the order of amino acids from the N-terminal end is serine-tyrosine-alanine (ser-tyr-ala), the name of the tripeptide is seryltyrosylalanine.

Example 20.3

Draw a structural formula for the tripeptide gly-ser-asp. Label the N-terminal amino acid and the C-terminal amino acid. What is the net charge on this tripeptide at pH 6.0?

Solution

In writing the formula for this tripeptide, begin with glycine on the left. Then connect the α-carboxyl of glycine to the α-amine of serine by a peptide bond. Finally, connect the α-carboxyl of serine to the α-amine of aspartic acid by another peptide bond. The net charge on this tripeptide at pH 6.0 is −1.

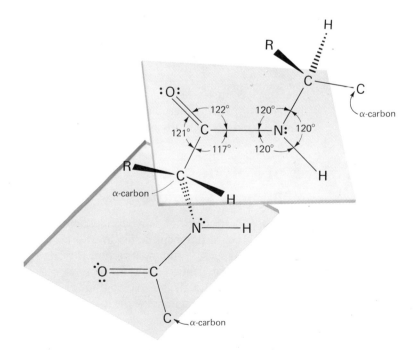

Problem 20.3

Draw a structural formula for lys-phe-ala. Label the N-terminal amino acid and the C-terminal amino acid. What is the net charge on this tripeptide at pH 6.0?

B. Geometry of a Peptide Bond

In the late 1930s, Linus Pauling began a series of studies to determine the geometry of a peptide bond. One of his first discoveries was that a peptide bond itself is planar. As shown in Figure 20.5, the four atoms of a peptide bond and the two alpha carbons joined to it all lie in the same plane.

Figure 20.5 Planarity of the peptide bond. Bond angles about the carbonyl carbon and the amide nitrogen are approximately 120°.

Had you been asked in Chapter 9 to describe the geometry of a peptide bond, you probably would have predicted bond angles of 120° about the carbonyl carbon and 109.5° about the amide nitrogen. This prediction agrees with the observed bond angles of approximately 120° about the carbonyl carbon. However, a bond angle of 120° about the amide nitrogen is unexpected. To account for this observed geometry, Pauling proposed that a peptide bond is more accurately represented as a resonance hybrid of two important contributing structures:

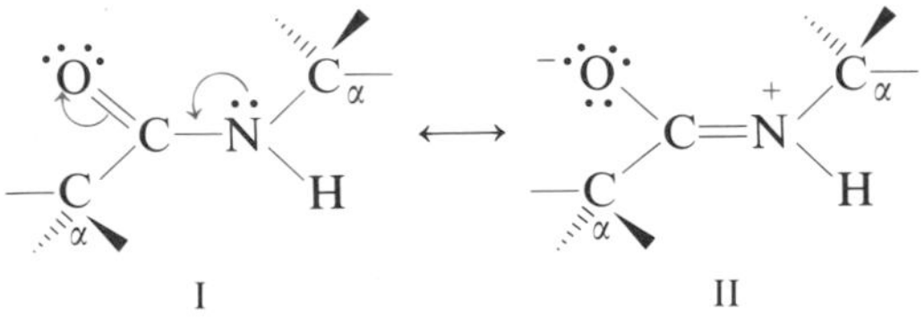

Contributing structure I shows a carbon-oxygen double bond, while structure II shows a double bond between the carbonyl carbon and the nitrogen atom of the peptide bond. The hybrid, of course, is neither of these; in the real structure, the carbon-nitrogen bond has at least partial double-bond character. Accordingly, in the hybrid, the six-atom group is planar.

There are two possible configurations for the atoms of a planar peptide bond. In one configuration, the two α-carbons are *cis* to each other; in the other, they are *trans* to each other:

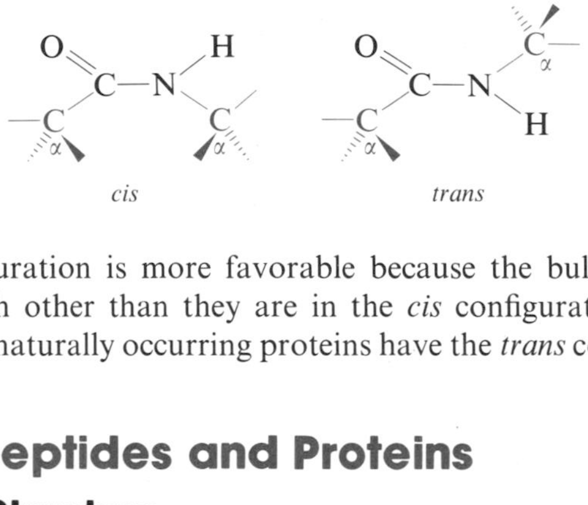

The *trans* configuration is more favorable because the bulky α-carbons are farther from each other than they are in the *cis* configuration. Virtually all peptide bonds in naturally occurring proteins have the *trans* configuration.

20.3 The Structure of Polypeptides and Proteins

A. Primary Structure

Primary (1°) structure of polypeptides and proteins refers to the sequence of amino acids in a polypeptide chain and also to the location of any disulfide bonds. In this sense, primary structure is a complete description of all covalent bonding in a polypeptide chain or protein.

To appreciate the problem of deciphering the primary structure of a polypeptide chain, just imagine the incredibly large number of different chemical words (polypeptides) that can be constructed with a 20-letter alphabet, where words can range from under ten letters to over a hundred letters. With only three amino acids, there are 27 different tripeptides possible. For glycine, alanine, and serine, the 27 tripeptides are:

gly-gly-gly	ser-ser-ser	ala-ala-ala
gly-gly-ser	ser-ser-gly	ala-ala-gly
gly-gly-ala	ser-ser-ala	ala-ala-ser
gly-ser-gly	ser-gly-ser	ala-gly-ala
gly-ala-gly	ser-ala-ser	ala-ser-ala
gly-ser-ala	ser-gly-ala	ala-gly-ser
gly-ala-ser	ser-ala-gly	ala-ser-gly
gly-ser-ser	ser-gly-gly	ala-gly-gly
gly-ala-ala	ser-ala-ala	ala-ser-ser

For a polypeptide containing one each of the 20 different amino acids, the number of possible polypeptides is $20 \times 19 \times 18 \times \cdots \times 2 \times 1$, or about 2×10^{18}. With larger polypeptides and proteins, the number of possible arrangements becomes truly countless!

1. Edman degradation. Of the various chemical methods developed for determining the amino acid sequence of a polypeptide chain, the one most widely used today is **Edman degradation,** introduced in 1950 by Pehr Edman of the University of Lund, Sweden. In this procedure, a polypeptide chain is reacted with phenylisothiocyanate, C_6H_5—N=C=S, a compound that reacts selectively with the —NH_3^+ group of the N-terminal amino acid. The effect of Edman degradation is to cleave the N-terminal amino acid as a substituted phenylthiohydantoin (Figure 20.6), which is then separated and identified.

The special value of Edman degradation is that it cleaves the N-terminal amino acid from a polypeptide chain without affecting any other bonds in the chain. Furthermore, Edman degradation can be repeated on the shortened polypeptide chain, causing the next amino acid in the sequence to be cleaved and identified. In practice, it is now possible to sequence as many as the first 40 or so amino acids in a polypeptide chain by this method.

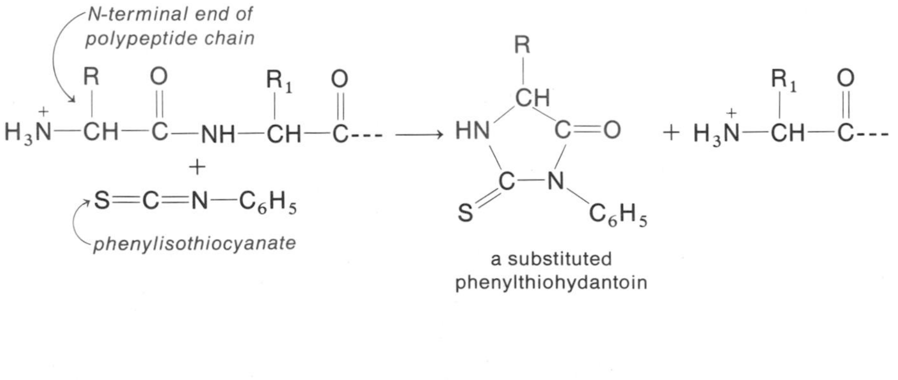

Figure 20.6 Edman degradation. Reaction of a polypeptide chain with phenylisothiocyanate selectively removes the N-terminal amino acid as a substituted phenylthiohydantoin.

Table 20.7 Specific cleavage of peptide bonds catalyzed by trypsin and chymotrypsin.

Enzyme	Catalyzes the Hydrolysis of Peptide Bonds Formed by the Carboxyl Group of	Side Chain (R-Group) of the Amino Acid Undergoing Selective Cleavage
trypsin	arginine	$-CH_2-CH_2-CH_2-NH-\overset{\overset{\displaystyle NH_2^+}{\|}}{C}-NH_2$
	lysine	$-CH_2-CH_2-CH_2-CH_2-NH_3^+$
chymotrypsin	phenylalanine	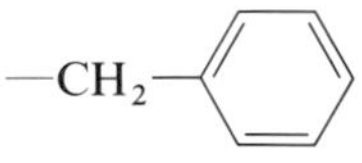 $-CH_2-\langle\bigcirc\rangle$
	tyrosine	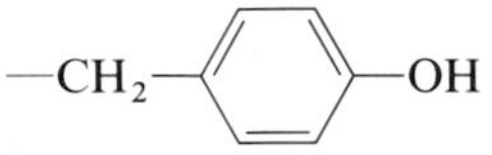 $-CH_2-\langle\bigcirc\rangle-OH$
	tryptophan	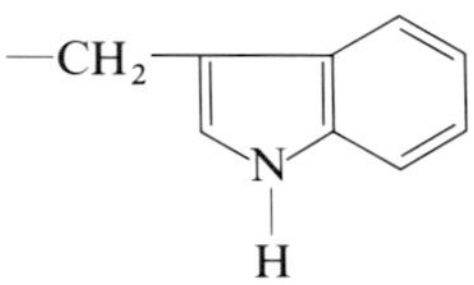 indole ring

2. Selective enzyme-catalyzed hydrolysis. If it is not possible to sequence an entire polypeptide chain by repeated Edman degradation, the chain is partially hydrolyzed to yield a series of smaller fragments and each fragment is sequenced separately. **Selective hydrolysis** is most often carried out using enzymes that catalyze the hydrolysis of specific peptide bonds.

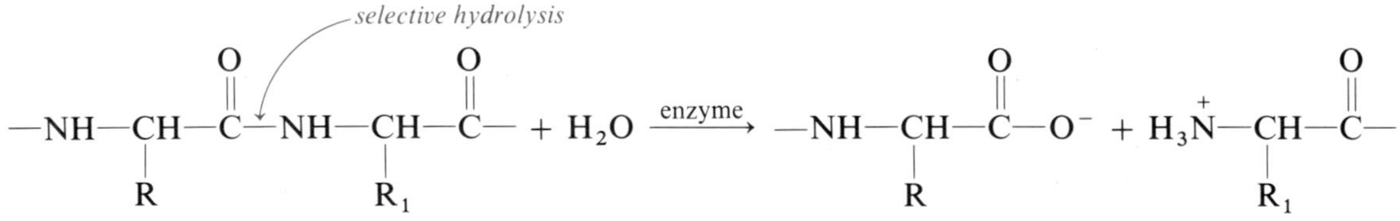

selective hydrolysis

$$-NH-\underset{R}{CH}-\overset{O}{\overset{\|}{C}}-NH-\underset{R_1}{CH}-\overset{O}{\overset{\|}{C}}- + H_2O \xrightarrow{enzyme} -NH-\underset{R}{CH}-\overset{O}{\overset{\|}{C}}-O^- + H_3\overset{+}{N}-\underset{R_1}{CH}-\overset{O}{\overset{\|}{C}}-$$

Two commonly used enzymes for selective hydrolysis are trypsin and chymotrypsin. Trypsin catalyzes the hydrolysis of peptide bonds in which the carboxyl group is contributed by either arginine or lysine; chymotrypsin catalyzes the hydrolysis of peptide bonds in which the carboxyl group is contributed by either phenylalanine, tyrosine, or tryptophan (Table 20.7).

Example 20.4

Following are amino acid sequences for several tripeptides. Which are hydrolyzed by trypsin? Which by chymotrypsin?

a. arg-glu-ser   **b.** phe-gly-lys   **c.** phe-lys-met

Solution

a. Trypsin catalyzes hydrolysis of peptide bonds between the carboxyl groups of lysine and arginine and the α-amino group of other amino acids. Therefore, the peptide bond between arginine and glutamic acid is hydrolyzed in the presence of trypsin.

$$\text{arg-glu-ser} + H_2O \xrightarrow{\text{trypsin}} \text{arg} + \text{glu-ser}$$

Chymotrypsin catalyzes the hydrolysis of peptide bonds between the carboxyl groups of phenylalanine, tyrosine, and tryptophan and other amino acids. Because none of these amino acids is present, tripeptide (a) is not affected by chymotrypsin.

b. Tripeptide (b) is not affected by trypsin. While there is a lysine present, its carboxyl group is at the C-terminal end and is not involved in peptide bond formation. Tripeptide (b) is hydrolyzed in the presence of chymotrypsin.

$$\text{phe-gly-lys} + H_2O \xrightarrow{\text{chymotrypsin}} \text{phe} + \text{gly-lys}$$

c. Tripeptide (c) is hydrolyzed by both trypsin and chymotrypsin:

$$\text{phe-lys-met} + H_2O \xrightarrow{\text{trypsin}} \text{phe-lys} + \text{met}$$

$$\text{phe-lys-met} + H_2O \xrightarrow{\text{chymotrypsin}} \text{phe} + \text{lys-met}$$

Problem 20.4

Following are amino acid sequences for three tripeptides. Which are hydrolyzed by trypsin? Which by chymotrypsin?

a. tyr-gln-val   **b.** thr-phe-ser   **c.** thr-ser-phe

Example 20.5

Deduce the amino acid sequence of a pentapeptide from the following experimental results.

amino acid composition: arg, glu, his, phe, ser
Edman degradation: glu
hydrolysis catalyzed by chymotrypsin: Fragment A: glu, his, phe
Fragment B: arg, ser
hydrolysis catalyzed by trypsin: Fragment C: arg, glu, his, phe
Fragment D: ser

Solution

Edman degradation cleaves glu from the pentapeptide chain; therefore, glutamic acid must be the N-terminal amino acid:

glu-(arg, his, phe, ser)

Next, hydrolysis catalyzed by chymotrypsin gives fragments A and B. Fragment A contains phe. Because of the specificity of chymotrypsin, phe must be the C-terminal amino acid of A. Fragment A also contains glu, which you already know is the N-terminal amino acid. From this, we conclude that the first three amino acids in the chain must be glu-his-phe, and now write the following fuller partial sequence.

glu-his-phe-(arg, ser)

The fact that trypsin cleaves the pentapeptide means that arg must be within the pentapeptide chain; it cannot be the C-terminal amino acid. Therefore, the complete sequence must be:

glu-his-phe-arg-ser

Problem 20.5

Deduce the amino acid sequence of a hexapeptide from the following experimental results:

amino acid composition: lys, pro, pro, tyr, val, val
Edman degradation: pro
hydrolysis catalyzed by trypsin: Fragment E: pro, lys, val
Fragment F: pro, tyr, val
hydrolysis catalyzed by chymotrypsin: Fragment G: lys, pro, tyr, val, val
Fragment H: pro

B. Secondary Structure

Secondary (2°) structure refers to ordered arrangements (conformations) of amino acids in localized regions of a polypeptide or protein molecule. The first studies of polypeptide conformations were carried out by Linus Pauling and Robert Corey, beginning in 1939. They assumed that in conformations of greatest stability, (1) all atoms in a peptide bond lie in the same plane, and (2) each amide group is hydrogen bonded between the N—H of one peptide bond and the C=O of another. Pauling was the first to recognize the importance of hydrogen bonding in stabilizing folding patterns. There is appreciable polarity to the C=O and N—H bonds and when two amide groups lie close to each other, the two peptide bonds interact by hydrogen bonding, as shown in Figure 20.7.

On the basis of model building, Pauling proposed that two folding patterns should be particularly stable: the α-helix and the antiparallel β-pleated sheet.

1. The α-helix. In the **α-helix** pattern shown in Figure 20.8, a polypeptide chain is coiled in a spiral. As you study this figure, note the following:

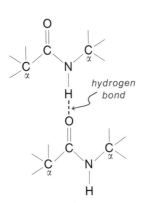

Figure 20.7 Hydrogen bonding between amide groups.

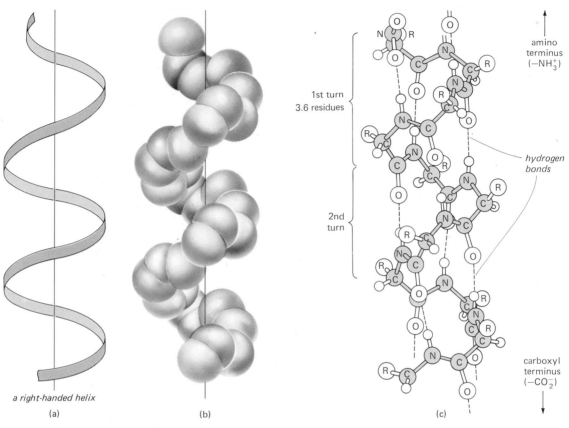

Figure 20.8 (a) A right-handed helix. (b) Space-filling model of the carbon-nitrogen backbone of an α-helix. (c) Ball-and-stick model of the α-helix showing intrachain hydrogen bonding. There are 3.6 amino acid residues per turn.

1. The helix is coiled in a clockwise or right-handed manner. Right-handed means that if you turn the helix clockwise, it twists away from you. In this sense, a right-handed helix is analogous to the right-hand thread of a common wood or machine screw.

2. There are 3.6 amino acids per turn of the helix.

3. Each peptide bond is *trans* and planar.

4. The N—H group of each peptide bond points roughly upward, parallel to the axis of the helix, and the C=O of each peptide bond points roughly downward.

5. The carbonyl group of each peptide bond is hydrogen bonded to the N—H group of the peptide bond four amino acid units away from it. Hydrogen bonds are shown as dotted lines.

6. All R— groups point outward from the helix.

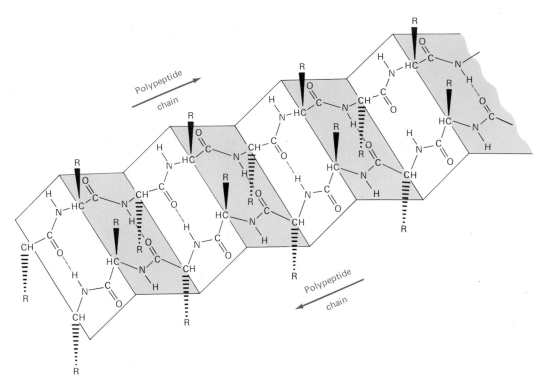

Figure 20.9 β-Pleated sheet conformation with two polypeptide chains running in opposite (antiparallel) directions. Hydrogen bonding between chains is indicated by dotted lines.

Almost immediately after Pauling proposed the α-helix structure, other researchers proved the presence of α-helix in keratin, the protein of hair and wool. It soon became obvious that the α-helix is one of the fundamental folding patterns of polypeptide chains.

2. The β-pleated sheet. **β-Pleated sheets** consist of extended polypeptide chains with neighboring chains running in opposite (antiparallel) directions. The C$=$O group of each peptide bond is hydrogen bonded to the N—H group of a peptide bond of a neighboring chain. As you study the section of β-pleated sheet shown in Figure 20.9, note the following:

1. The two polypeptide chains lie parallel to each other and run in opposite (antiparallel) directions.

2. Each peptide bond is *trans* and planar.

3. The polypeptide is a chain of flat or planar sections connected together at amino acid α-carbons.

4. The C=O and N—H groups of peptide bonds point at each other and in the same plane so that hydrogen bonding is possible between adjacent polypeptide chains.

5. The R— groups on any one chain alternate, first above the plane of the sheet and then below the plane of the sheet.

The pleated sheet conformation is stabilized by hydrogen bonding between —NH groups of one chain and C=O groups of an adjacent chain. By comparison, the α-helix is stabilized by hydrogen bonding between —NH and C=O groups within the same polypeptide chain.

The term *secondary structure* is used to describe α-helix, β-pleated sheet, and other types of periodic conformations in localized regions of polypeptide or protein molecules.

C. Tertiary Structure

Tertiary (3°) structure refers to the overall folding pattern and three-dimensional arrangement of atoms in a single polypeptide chain. Actually, there is no sharp dividing line between secondary structure and tertiary structure. Secondary structure refers to the spatial arrangement of amino acids close to one another on a polypeptide chain, while tertiary structure refers to the three-dimensional arrangement of *all* atoms of a polypeptide chain.

Disulfide bonds play an important role in maintaining tertiary structure. Disulfide bonds are formed between side chains of cysteine by oxidation of two thiol groups (—SH) to form a disulfide bond (—S—S—), as shown.

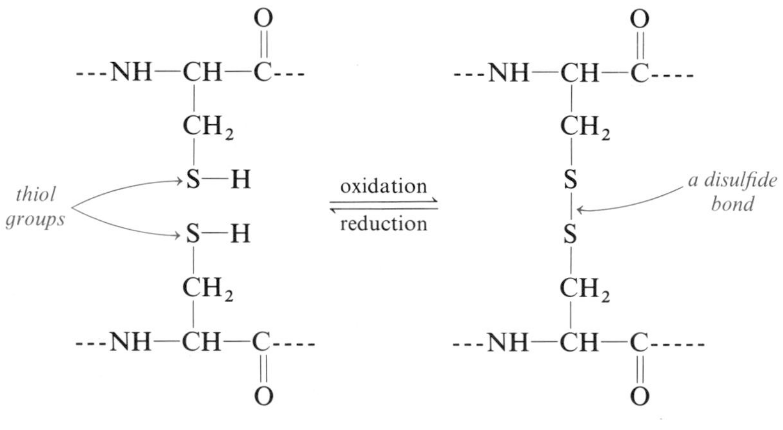

Treatment of disulfide bonds with a reducing agent regenerates the thiol groups.

We now know the primary structure for several hundred polypeptides and proteins and we know the secondary and tertiary structure of scores of these. As an example, let us look at the three-dimensional structure of myoglobin, a protein found in skeletal muscle and particularly abundant in diving

Figure 20.10 The three-dimensional structure of myoglobin. The heme group is shown in color. The N-terminal amino acid (indicated by —NH$_3^+$) is at the lower left and the wc-terminal amino acid (indicated by —CO$_2^-$) is at the upper left. Reproduced from R.E. Dickerson, in H. Neurath, ed., *The Proteins*, Vol. II (New York: Academic Press, 1964).

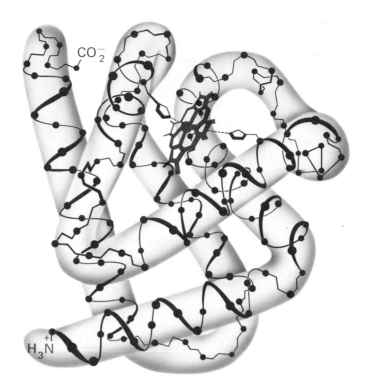

mammals such as seals, whales, and porpoises. Myoglobin and its structural relative, hemoglobin (Section 20.3D), are the oxygen transport and storage molecules of vertebrates. Hemoglobin binds molecular oxygen in the lungs and transports it to myoglobin in muscles. Myoglobin stores molecular oxygen until it is required for metabolic oxidation.

Myoglobin consists of a single polypeptide chain of 153 amino acids. The complete amino acid sequence (primary structure) of the chain is known. Myoglobin also contains a single heme unit (Figure 21.2). Determination of the three-dimensional structure of myoglobin represented a milestone in the study of molecular architecture. For his contribution to this research, J.C. Kendrew shared the Nobel Prize in Chemistry in 1963. The secondary and tertiary structure of myoglobin are shown in Figure 20.10. The single polypeptide chain is folded into a complex, almost box-like shape.

A more detailed analysis has revealed the exact location of all atoms of the peptide backbone and also the location of all side chains. The important structural features of myoglobin are:

1. The backbone consists of eight relatively straight sections of α-helix, each separated by a bend in the polypeptide chain. The longest section of α-helix has 23 amino acids, the shortest has 7. Some 75% of the amino acids are found in these eight regions of α-helix.

2. Hydrophobic side chains such as those of phenylalanine, alanine, valine, leucine, isoleucine, and methionine are clustered in the interior of the

molecule, where they are shielded from contact with water. Hydrophobic interactions between nonpolar side chains are a major factor in directing the folding of the polypeptide chain of myoglobin into this compact, three-dimensional shape.

3. The surface of myoglobin is coated with hydrophilic side chains, such as those of lysine, arginine, serine, glutamic acid, histidine, and glutamine, which interact with the aqueous environment by hydrogen bonding. The only polar side chains that point to the interior of the myoglobin molecule are those of two histidines. These side chains can be seen in Figure 20.10 as five-member rings pointing inward toward the heme group.

4. Oppositely charged amino acids close to each other in the three-dimensional structure interact by electrostatic attractions called **salt linkages.** An example of a salt linkage is the attraction of the side chains of lysine (NH_3^+) and glutamic acid ($-CO_2^-$).

The three-dimensional structures of several other globular proteins have also been determined and their secondary and tertiary structures analyzed. It is clear that globular proteins contain α-helix and β-pleated sheet structure, but that there is wide variation in the relative amounts of each. Lysozyme, with 129 amino acids in a single polypeptide chain, has only about 25% of its amino acids in α-helix regions. Cytochrome, with 104 amino acids in a single polypeptide chain, has no α-helix structure but does contain several regions of β-pleated sheet. Yet, whatever the proportions of α-helix, β-pleated sheet, or other periodic structure, virtually all nonpolar side chains of globular proteins are directed toward the interior of the molecule, while polar side chains are on the surface of the molecule and are in contact with the aqueous environment. Note that this arrangement of polar and nonpolar groups in globular proteins very much resembles the arrangement of polar and nonpolar groups of soap molecules in micelles (Figure 16.6).

Example 20.6

With which of the following amino acid side chains can the side chain of threonine form hydrogen bonds?

a. valine   **b.** phenylalanine   **c.** tyrosine

d. asparagine   **e.** histidine   **f.** alanine

Solution

The side chain of threonine contains a hydroxyl group that can participate in hydrogen bonding in two ways: the negatively charged oxygen can function as a hydrogen bond acceptor and the positively charged hydrogen can function as a hydrogen bond donor. Therefore, the side chain of threonine can function as a hydrogen bond acceptor for the side chains of tyrosine, asparagine, and histidine. The side chains of threonine can also function as hydrogen bond donors for the side chains of tyrosine, asparagine, and histidine.

Problem 20.6

At pH 7.4, with what amino acid side chains can the side chain of lysine form salt linkages?

D. Quaternary Structure

Most proteins of molecular weight greater than 50,000 consist of two or more noncovalently linked polypeptide chains. The arrangement of protein monomers into an aggregation is known as **quaternary (4°) structure.** A good example is hemoglobin, a protein that consists of four separate protein monomers: two α-chains of 141 amino acids each and two β-chains of 146 amino acids each. The quaternary structure of hemoglobin is shown in Figure 20.11.

The major factor stabilizing the aggregation of protein subunits is hydrophobic interaction. When separate monomers fold into compact three-dimensional shapes to expose polar side chains to the aqueous environment and shield nonpolar side chains from water, there are still hydrophobic "patches" on the surface, in contact with water. These patches can be shielded from water if two or more monomers assemble so that their hydrophobic patches are in contact. The molecular weights, numbers of subunits, and biological functions of several proteins with quaternary structure are shown in Table 20.8.

Figure 20.11 The quaternary structure of hemoglobin, showing the four subunits packed together. The flat disks represent four heme units. From R.E. Dickerson and I. Geis, *The Structure and Action of Proteins* (Menlo Park, Ca.: W.A. Benjamin, Inc., 1969). © Copyright, 1969. All rights reserved. Used by permission.

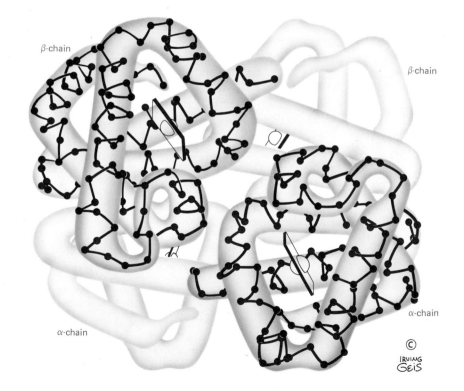

β-chain

β-chain

α-chain

α-chain

©

IRVING GEIS

Table 20.8 Quaternary structure of selected proteins.

Protein	Mol. Wt.	Number of Subunits	Subunit Mol. Wt.	Biological Function
insulin	11,466	2	5,733	a hormone regulating glucose metabolism
hemoglobin	64,500	4	16,100	oxygen transport in blood plasma
alcohol dehydrogenase	80,000	4	20,000	an enzyme of alcoholic fermentation
lactate dehydrogenase	134,000	4	33,500	an enzyme of anaerobic glycolysis
aldolase	150,000	4	37,500	an enzyme of anaerobic glycolysis
fumarase	194,000	4	48,500	an enzyme of the tri-carboxylic acid cycle
tobacco mosaic virus	40,000,000	2200	17,500	plant virus coat

E. Denaturation

Globular proteins found in living organisms are remarkably sensitive to changes in their environment. Relatively small changes in pH, temperature, or solvent composition, even for only a short period of time, may cause them to **denature.** Denaturation is a physical change, the most observable result of which is loss of biological activity. With the exception of cleavage of disulfide bonds, denaturation stems from changes in secondary, tertiary, or quaternary structure through disruption of noncovalent interactions, such as hydrogen bonds, salt linkages, and hydrophobic interactions. Common denaturing agents include the following:

1. **Heat.** Most globular proteins denature when heated above 50–60°C. For example, boiling or frying an egg causes egg-white protein to denature and form an insoluble mass.

2. **Large changes in pH.** Adding concentrated acid or alkali to a protein in aqueous solution causes changes in the charged character of ionizable side chains and interferes with salt linkages. For example, in certain clinical chemistry tests where it is necessary to first remove any protein material, trichloroacetic acid (a strong organic acid) is added to denature and precipitate any protein present.

3. **Detergents.** Treatment of a protein with sodium dodecylsulfate (SDS), a detergent, causes the native conformation to unfold and exposes the

nonpolar protein side chains. These side chains are then stabilized by hydrophobic interaction with hydrocarbon chains of the detergent.

4. **Organic solvents** such as alcohols, acetone, or ether.

5. **Mechanical treatment.** Most globular proteins denature in aqueous solution if they are stirred or shaken vigorously. An example is the whipping of egg whites to make a meringue.

6. **Urea and guanidine hydrochloride.**

$$
\underset{\text{urea}}{H_2N-\overset{\overset{\displaystyle O}{\|}}{C}-NH_2}
\qquad
\underset{\substack{\text{guanidine} \\ \text{hydrochloride}}}{H_2N-\overset{\overset{\displaystyle NH_2^+Cl^-}{\|}}{C}-NH_2}
$$

These reagents cause disruption of protein hydrogen bonding and hydrophobic interactions.

Denaturation may be partial or complete. It may also be reversible or irreversible. For example, the hormone insulin can be denatured with $8M$ urea and the three disulfide bonds reduced to —SH groups. If urea is then removed and the disulfide bonds reformed, the resulting molecule has less than 1% of its former biological activity. In this case, denaturation is both complete and irreversible. As another example, consider ribonuclease, an enzyme that consists of a single polypeptide chain of 124 amino acids folded into a compact, three-dimensional structure stabilized in part by four disulfide bonds. Treatment of ribonuclease with urea causes the molecule to unfold and the disulfide bonds are then reduced to thiol groups. At this point, the protein is completely denatured—it has no biological activity. If urea is removed from solution and the thiol groups reoxidized to disulfide bonds, the protein regains its full biological activity. In this instance, denaturation has been complete but reversible.

F. 2°, 3°, and 4° Structure Are Determined by 1° Structure

The primary structure of a protein is determined by information coded within genes. Once the primary structure of a polypeptide is established, it then directs folding of the polypeptide chain into a three-dimensional structure. In other words, information inherent in the primary structure of a protein determines its secondary, tertiary, and quaternary structures.

If the three-dimensional shape of a polypeptide or protein is determined by its primary structure, how can we account for the observation that denaturation of some proteins is reversible while that of others is irreversible?

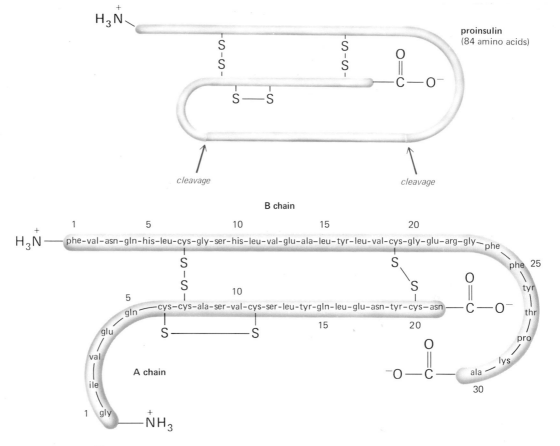

Figure 20.12 (*Upper*) A schematic diagram of proinsulin, a single polypeptide chain of 84 amino acids. (*Lower*) The amino acid sequence of bovine insulin.

The reason for this difference in behavior from one protein to another is that some proteins, like ribonuclease, are synthesized as single polypeptide chains which then fold into unique three-dimensional structures with full biological activity. Others, like insulin, are synthesized as larger molecules with no biological activity, but are "activated" at some later point by specific enzyme-catalyzed peptide bond cleavage. Insulin is synthesized in the β-cells of the pancreas as a single polypeptide chain of 84 amino acids. This molecule is called proinsulin and has no biological activity. When insulin is needed, a section of 33 amino acids is hydrolyzed from proinsulin in an enzyme-catalyzed reaction to produce the active hormone (Figure 20.12). Bovine insulin contains 51 amino acids in two polypeptide chains. The A chain contains 21 amino acids and has glycine (gly) at the $-NH_3^+$ terminus and asparagine (asn) at the $-CO_2^-$ terminus. The B chain contains 30 amino acids with phenylalanine (phe) at the $-NH_3^+$ terminus and alanine (ala) at the $-CO_2^-$ terminus.

The information directing the original folding of the single polypeptide chain of proinsulin is no longer present in the A and B chains of the active hormone. For this reason, refolding of the denatured protein is irregular and denaturation is irreversible.

The process of producing a protein in an inactive storage form is quite common. For example, the digestive enzymes trypsin and chymotrypsin and the blood-clotting enzyme thrombin are produced as inactive proteins. Enzymes produced as inactive proteins that are then activated by cleavage of one or more polypeptide bonds are called **zymogens.**

20.4 Fibrous Proteins

Fibrous proteins are stringy, physically tough macromolecules composed of rod-like polypeptide chains joined together by several types of cross-linkages to form stable, insoluble structures. There are two major classes of fibrous proteins: the keratins of skin, wool, claws, horn, scales, and feathers; and the collagens of tendons and hides.

A. The α-Keratins

Hair and wool are very flexible and they also stretch. They are elastic, so that when tension is released, the fibers revert to their original length. At the molecular level, the fundamental structural unit of hair is a polypeptide chain wound into an α-helix conformation (Figure 20.13). Several levels of structural organization are built from the simple α-helix. First, three strands of α-helix are twisted together to form a larger cable called a protofibril. Protofibrils are then wound in bundles to form an 11-strand cable called a microfibril. These, in turn, are imbedded in a larger matrix that ultimately forms a hair fiber (Figure 20.13).

Figure 20.13 The detailed structure of hair fiber. From R.E. Dickerson and I. Geis, *The Structure and Action of Proteins* (Menlo Park, Ca.: W.A. Benjamin, Inc., 1969). © Copyright 1969. All rights reserved. Used by permission.

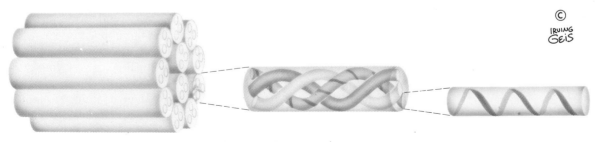

microfibril   protofibril   α-helix

When hair is stretched, hydrogen bonds along turns of each α-helix are elongated. The major force causing stretched hair fibers to return to their original length is reformation of hydrogen bonds in the α-helices.

The α-keratins of horns and claws have essentially the same structure as hair but with a much higher content of cysteine and a greater degree of disulfide bridge cross-linking between the helices. These additional disulfide bonds greatly increase the stretch resistance and produce the hard keratins of horn and claw.

B. Collagen

The second major class of fibrous proteins are the **collagens.** Collagens are constituents of skin, bone, teeth, blood vessels, tendons, cartilage, and connective tissue. In fact, they are the most abundant protein in higher vertebrates and make up almost 30% of total body protein mass in humans. Table 20.9 lists the collagen content of several tissues. Note that bone, Achilles' tendon, skin, and cornea of the eye are largely collagen.

Because of its abundance and wide distribution in vertebrates and because it is associated with a variety of diseases and problems of aging, more is known about collagen than about probably any other fibrous protein. Collagen molecules are very large and have a distinctive amino acid composition. One-third of all amino acids in collagen are glycine, and another 20% are proline and hydroxyproline. Tyrosine is present in very small amounts and the essential amino acid tryptophan is absent entirely. Because cysteine is also absent, there are no disulfide cross-links in collagen. When collagen fibers are boiled in water, they are converted into soluble gelatins. Gelatin itself has no biological food value because it lacks the essential amino acid tryptophan.

The polypeptide chains of collagen fold into a conformation that is particularly stable and unique to collagen. In this conformation, three protein strands wrap around each other to form a left-handed superhelix called the

Table 20.9 Collagen content of some body tissues.

Tissue	Collagen (% Dry Weight)
bone, mineral-free	88
Achilles' tendon	86
skin	72
cornea	68
cartilage	46–63
ligament	17
aorta	12–24

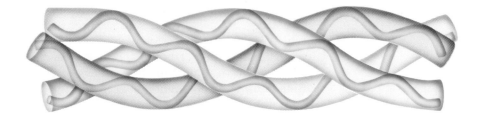

collagen triple helix. This unit is called tropocollagen and looks much like a three-stranded rope (Figure 20.14).

Collagen fibers are formed when many tropocollagen molecules line up side by side in a regular pattern and are then cross-linked by the formation of new covalent bonds. One of the effects of severe ascorbic acid deficiency is impaired synthesis of collagen. Without adequate supplies of vitamin C, cross-linking of tropocollagen strands is inhibited, with the result that they do not unite to form stable, physically tough fibers.

The extent and type of cross-linking vary with age and physiological conditions. For example, the collagen of rat Achilles' tendon is highly cross-linked, while that of the more flexible tendon of rat tail is much less highly cross-linked. Further, it is not clear when, if ever, the process of cross-linking is completed. Some believe it is a process that continues throughout life, producing increasingly stiffer skin, blood vessels, and other tissues, which then contribute to the medical problems of aging and the aged.

20.5 Plasma Proteins: Examples of Globular Proteins

Human blood consists of a fluid portion (plasma) and cellular components. The cellular components, which make up 40–45% of the volume of whole blood, consist of red blood cells (erythrocytes), white blood cells (leukocytes), and blood platelets. Human plasma consists largely of water (90–92%) in which are dissolved various inorganic ions and a heterogeneous mixture of organic molecules, the largest groups of which are the **plasma proteins.** The earliest method of separating plasma proteins into fractions used ammonium sulfate to "salt-out" different types of proteins. The fraction precipitated from plasma 50% saturated with ammonium sulfate was called globulin. The fraction not precipitated at this salt concentration but precipitated from plasma saturated with ammonium sulfate was called albumin. Today, electrophoresis is the most common method for separating proteins of biological fluids into fractions, especially in the clinical laboratory, where it is used routinely to measure proteins in human plasma, urine, and cerebrospinal fluid. It is estimated that between 15 and 20 million plasma protein electrophoretic analyses are carried out each year in the United States and Canada.

Figure 20.15 Separation of serum proteins by electrophoresis. (a) A sample is applied as a narrow line at the origin. After electrophoresis at pH 8.8, the paper is dried and stained. (b) A plot of color intensity of each spot.

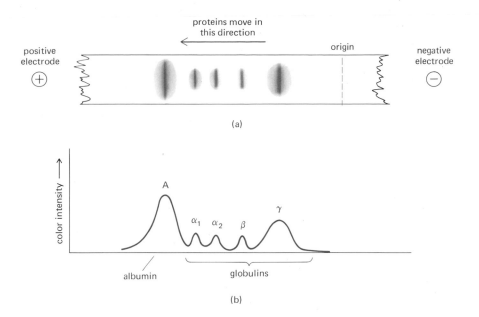

In plasma protein electrophoresis, a sample of plasma is applied as a narrow line to a cellulose acetate strip. The ends of the strip are then immersed in a buffer of pH 8.8 and a voltage is applied to the strip. At pH 8.8, plasma proteins have net negative charges and migrate toward the positive electrode. After a predetermined time, the cellulose acetate strip is removed, dried, and sprayed with a dye that selectively stains proteins. The separated protein fractions then appear on the developed strip as spots (Figure 20.15a). The amount of protein in each spot is determined using a densitometer to measure the intensity of the color in each spot. Shown in Figure 20.15(b) is a plot of intensity versus width of the spots in part (a) of the figure. The concentration of protein in each spot is proportional to the area under each peak.

Electrophoresis on cellulose acetate separates serum proteins into five major fractions: one albumin fraction and four globulin fractions. The four globulin fractions are arbitrarily designated α_1, α_2, β, and γ according to their electrophoretic mobilities. Serum albumin has an isoelectric point of about 4.9 and migrates farthest toward the positive electrode. γ-Globulin has an isoelectric point of about 7.4 and migrates the shortest distance. Shown in Table 20.10 are the concentrations of the five major protein fractions of human serum.

The primary function of albumins is to regulate the osmotic pressure of blood. In addition, albumins are important in transporting fatty acids and certain drugs such as aspirin and digitalis. The α_1 and α_2 fractions transport other biomolecules, such as fats, steroids, phospholipids, and various other lipids. The α_1 fraction also contains antitrypsin, a protein that inhibits the protein-digesting enzyme trypsin. The α_2 fraction contains haptoglobulin,

Fraction	grams/100 mL	% of Total Protein
albumin	3.5–5.0	52–67
globulins		
α_1	0.1–0.4	2.5–4.5
α_2	0.5–1.1	6.6–13.6
β	0.6–1.2	9.1–14.7
γ	0.5–1.5	9.0–20.6

Table 20.10 Concentrations of the major types of human serum proteins as determined by electrophoresis.

which binds any hemoglobin released from destroyed red blood cells, and ceruloplasmin, the principal copper-containing protein of the body. The α_2 fraction also contains prothrombin, an inactive form of the blood-clotting enzyme thrombin. The β fraction contains a variety of specific transport proteins, as well as substances involved in blood clotting.

The γ-globulin fraction consists primarily of **antibodies** (immunoglobulins) whose function is to combat **antigens** (foreign proteins) introduced into the body. Specific antibodies are formed by the immune system in response to specific antigens. This response is the basis for immunization against such

Figure 20.16 The three-dimensional shape of an antibody.

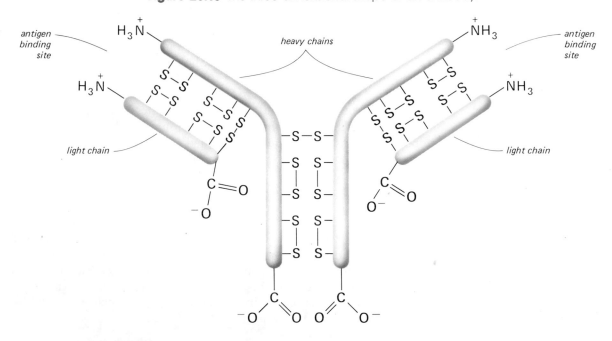

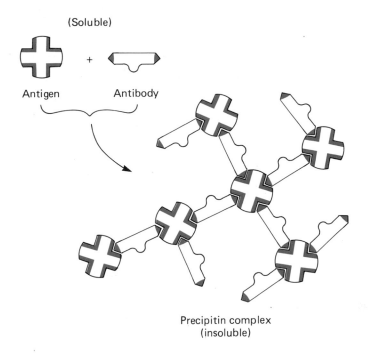

Figure 20.17 The action of an antibody and its specific antigen to form an inactive precipitin complex. The precipitated antigen-antibody complex is ingested and broken down by white blood cells.

(Soluble)

Antigen + Antibody

Precipitin complex (insoluble)

infectious diseases as polio, tetanus, and diphtheria. An antibody consists of a combination of heavy (higher-molecular-weight) and light (lower-molecular-weight) polypeptide chains held together by disulfide bonds (Figure 20.16). Each antibody has two identical binding sites that react with specific antigens to form an insoluble complex called precipitin (Figure 20.17). Formation of precipitin deactivates the antigen and permits its removal and breakdown by white blood cells.

Key Terms and Concepts

amino acid (20.1A)

antibody (20.5)

antigen (20.5)

biological value of dietary protein (20.1H)

chirality of amino acids (20.1C)

collagen (20.4C)

collagen triple helix (20.4B)

denaturation (20.3E)

Edman degradation (20.3A)

electrophoresis (20.1F)

essential amino acids (20.1G)

fibrous protein (20.4)

geometry of a peptide bond (20.2B)

globular protein (introduction)

α-helix (20.3B)

hydrophilic side chains (20.3C)

hydrophobic side chains (20.3C)

immunoglobulin (20.5)

isoelectric point (20.1E)

isoelectric precipitation (20.1E)

α-keratin (20.4A)

peptide bond (20.2A)

plasma proteins (20.5)

β-pleated sheet (20.3B)

primary structure (20.3A)

quaternary structure (20.3D)

salt linkage (20.3C)

secondary structure (20.3B)

C-terminal amino acid (20.2A)

N-terminal amino acid (20.2A)

tertiary structure (20.3C)

zwitterion (20.1A)

zymogen (20.3F)

Problems

**Amino Acids
(Section 20.1)**

20.7   Explain the meaning of the designation L- as it is used to indicate the stereochemistry of amino acids found in proteins.

20.8   What is the relationship in terms of configuration about the chiral carbon between L-serine and D-glyceraldehyde? Between L-serine and L-glyceraldehyde?

20.9   a. Which amino acid found in proteins has no chiral carbon?
b. Which amino acids found in proteins have two chiral carbons?

20.10   For amino acids with nonionizable side chains, the value of pI (the isoelectric point) can be calculated from the equation

$$pI = \tfrac{1}{2}(pK_a \text{ of } \alpha\text{-}CO_2H + pK_a \text{ of } \alpha\text{-}NH_3^+)$$

Given the following values of pK_a, calculate the isoelectric point of each amino acid listed.

	pK_a of α-CO_2H	pK_a of α-NH_3^+
glycine	2.34	9.60
serine	2.21	9.15
threonine	2.63	10.39
asparagine	2.02	8.80

20.11   For amino acids with a side-chain —CO_2H group, the value of pI can be calculated from the equation

$$pI = \tfrac{1}{2}(pK_a \text{ of } \alpha\text{-}CO_2H + pK_a \text{ of side-chain } \text{—}CO_2H)$$

Given the following values of pK_a, calculate the pI of the following amino acids. Compare these values with those given in Table 20.3.

	pK_a of α-CO_2H	pK_a of α-NH_3^+	pK_a of side-chain —CO_2H
aspartic acid	2.09	9.82	3.86
glutamic acid	2.10	9.67	4.06

20.12   For amino acids with side-chain amino groups, the value of pI can be calculated from the equation

$$pI = \tfrac{1}{2}(pK_a \text{ of } \alpha\text{-}NH_3^+ + pK_a \text{ of side-chain amino group})$$

Given the following values of pK_a, calculate the pI of each amino acid. Compare these values with those given in Table 20.3.

	pK_a of α-CO_2H	pK_a of α-NH_3^+	pK_a of side-chain amino group
lysine	2.18	8.95	10.53
arginine	2.01	9.04	12.48

20.13 For the following amino acids, draw a structural formula for the form you would expect to predominate at pH 1.0, 6.0, and 12.0. Refer to Section 20.1E and Problems 20.10–20.12 for values of pI.
a. alanine   b. tyrosine
c. aspartic acid   d. arginine

20.14 Would you expect an aqueous solution of lysine to be acidic, basic, or neutral? Explain your reasoning. (*Hint:* In thinking about this problem, first consider the effect on pH of the carboxyl and the α-amino groups together in the zwitterion form and then the effect of the terminal amino group.)

20.15 a. Estimate the pH at which the solubility of alanine in water is at a minimum.
b. Why does the solubility of alanine increase as the pH is increased?
c. Why does the solubility of alanine increase as the pH is decreased?

20.16 Which of the following amino acids will migrate toward the (+) electrode and which will migrate toward the (−) electrode on electrophoresis at pH 6.0? At pH 8.6?
a. tyrosine   b. arginine
c. cysteine   d. aspartic acid
e. asparagine   f. histidine

20.17 Write an equation for the oxidation of two molecules of cysteine by O_2 to form a disulfide bond.

20.18 Name the essential amino acids for humans. Why are they termed *essential*? Compare meat, fish, and the cereal grains in terms of their ability to supply the essential amino acids.

20.19 How many stereoisomers are possible for:
a. L-hydroxyproline   b. L-5-hydroxylysine
c. ornithine   d. citrulline
e. thyroxine

The Peptide Bond (Section 20.2)

20.20 What is the characteristic structural feature of a peptide bond?

20.21 Using structural formulas, show how the theory of resonance accounts for the fact that a peptide bond is planar.

20.22 Write a structural formula for the tripeptide glycylserylaspartic acid. Write a structural formula for an isomeric tripeptide. Calculate the net charge on each at pH 6.0.

20.23 Write the structural formula for the tripeptide lys-asp-val. Calculate the net charge on this tripeptide at pH 6.0.

20.24 How many tetrapeptides can be constructed from the 20 amino acids
a. if each of the amino acids is used only once in the tetrapeptide?
b. if each amino acid can be used up to four times in the tetrapeptide?

20.25 Following is the structural formula of glutathione (GSH), one of the most common small polypeptides in animals, plants, and bacteria.

$$H_3\overset{+}{N}-CH-CH_2-CH_2-\overset{\displaystyle O}{\overset{\|}{C}}-NH-CH-\overset{\displaystyle O}{\overset{\|}{C}}-NH-CH_2-\overset{\displaystyle O}{\overset{\|}{C}}-O^-$$

$$\begin{array}{ccc} & CO_2^- & CH_2 \\ & & | \\ & & SH \end{array}$$

glutathione (GSH)

a. Name the amino acids in this tripeptide.

b. What is unusual about the peptide bond formed between the first two amino acids in this tripeptide?

c. Write the primary structure of glutathione using three-letter abbreviations for the amino acids present.

d. Write an equation for the reaction of two molecules of glutathione with O_2 to form a disulfide bond.

The Structure of Polypeptides and Proteins (Section 20.3)

20.26 Following are amino acid sequences for several tripeptides. Indicate which are hydrolyzed by trypsin, which by chymotrypsin.

a. ala-asp-lys   b. glu-arg-ser   c. phe-met-trp

d. lys-ala-asp   e. arg-tyr-gly   f. asp-lys-phe

20.27 Deduce the amino acid sequence of an octapeptide from the following experimental results.

amino acid composition: ala, arg, leu, lys, met, phe, ser, tyr

Edman degradation: ala

hydrolysis catalyzed by trypsin: Fragment I: ala, arg

Fragment J: leu, met, tyr

Fragment K: lys, phe, ser

hydrolysis catalyzed by chymotrypsin: Fragment L: ala, arg, phe, ser

Fragment M: leu

Fragment N: lys, met, tyr

20.28 In constructing models of polypeptide chains, Linus Pauling assumed that for maximum stability, (1) all amide bonds are *trans* and coplanar, and (2) there is a maximum of hydrogen bonding between amide groups. Examine the α-helix (Figure 20.8) and the β-pleated sheet (Figure 20.9) and convince yourself that in each conformation, amide bonds are planar and that each carbonyl oxygen is hydrogen bonded to an amide hydrogen.

20.29 Examine the α-helix conformation. Are the amino acid side chains arranged all inside the helix, all outside the helix, or randomly oriented?

20.30 Draw structural formulas to illustrate the noncovalent interactions indicated.

a. Hydrogen bonding between the side chains of thr and asn.

b. Salt linkage between the side chains of lys and glu.

c. Hydrophobic interactions between the side chains of two phenylalanines.

20.31 Consider a typical globular protein in an aqueous medium of pH 6.0. Which of the following amino acid side chains would you expect to find on the outside and in contact with water? Which on the inside and shielded from contact with water?

a. glutamic acid   b. glutamine

c. arginine   d. serine

e. valine   f. phenylalanine

g. lysine   h. isoleucine

i. threonine

20.32 Which of the following peptides and proteins migrate to the (+) electrode and which migrate to the (−) electrode on electrophoresis at pH 6.0? At pH 8.6?

a. ala-glu-ile   b. gly-asp-lys

c. val-ala-leu   d. tyr-trp-arg

e. thyroglobulin (pI 4.6)   f. hemoglobin (pI 6.8)

20.33 The following proteins have approximately the same molecular weight and size. The pI of each is given.

carboxypeptidase	pI 6.0
pepsin	pI 1.0
human growth hormone	pI 6.9
ovalbumin	pI 4.6

a. State whether each is positively charged, negatively charged, or uncharged at pH 6.0.

b. Draw a diagram showing the results of electrophoresis of a mixture of the four at pH 6.0.

20.34 Account for the fact that the solubility of globular proteins is a function of pH, and that solubility is at a minimum when the pH of the solution equals the isoelectric point (pI) of the protein.

20.35 Examine the primary structure of bovine insulin (Figure 20.12) and list all asp, glu, lys, arg, and his in the molecule. Would you predict the isoelectric point of insulin to be nearer that of the acidic amino acids (pI 2.0–3.0), the neutral amino acids (pI 5.5–6.5) or the basic amino acids (pI 9.5–11.0)?

20.36 Following is the primary structure of glucagon, a polypeptide hormone that helps to regulate glycogen metabolism. Glucagon is secreted by the alpha cells of the pancreas during the fasting state, when blood glucose levels are decreasing. This hormone stimulates the enzymes that catalyze the hydrolysis of glycogen to glucose and thus helps to maintain blood glucose levels within a normal concentration range. Glucagon contains 29 amino acids and has a molecular weight of approximately 3500.

```
1           5           10          15          20          25        29
his-ser-glu-gly-thr-phe-thr-ser-asp-tyr-ser-lys-tyr-leu-asp-ser-arg-arg-ala-gln-asp-phe-val-gln-trp-leu-met-asn-thr
```

glucagon

a. Estimate the net charge on glucagon at pH 6.0.

b. Would you predict the isoelectric point of glucagon to be nearer that of the acidic amino acids (pI 2.0–3.0), the neutral amino acids (5.5–6.5), or the basic amino acids (9.5–11.0)?

20.37 Suppose insulin (Figure 20.12) is treated with a disulfide reducing agent and the A and B chains separated.

a. How many polypeptide fragments are formed when the A chain is treated with chymotrypsin?

b. How many polypeptide fragments are formed when the B chain is treated with trypsin?

c. Estimate the pI of the A chain; of the B chain. How might it be possible to separate the A and B chains by electrophoresis?

20.38 Myoglobin and hemoglobin are globular proteins. Myoglobin consists of a single polypeptide chain of 153 amino acids. Hemoglobin is composed of four polypeptide chains, two of 141 amino acids and two of 146 amino acids. The three-dimensional structures of myoglobin and hemoglobin polypeptide chains are very similar, yet myoglobin exists as a monomer in aqueous solution, while the four polypeptide chains of hemoglobin self-assemble to form a tetramer. Which polypeptide chains, those of myoglobin or hemoglobin, would you predict to have a higher percentage of nonpolar amino acids?

20.39 What is *irreversible denaturation* and how does it differ from *reversible denaturation?*

20.40 After water, proteins are the major constituents of most tissues. Often in the analysis of lesser constituents, it is necessary first to remove all protein material. The reagents most commonly employed for this purpose are $0.5M$ trichloroacetic acid, ethanol, or acetone. Explain the basis for the use of these reagents to deproteinize a solution.

20.41 Is the following statement true or false? Explain your answer. "The major factor directing the folding of globular proteins in an aqueous environment is hydrogen bonding between polar side chains and water molecules."

20.42 Suppose proinsulin is treated with a disulfide reducing agent in the presence of $8M$ urea. These reagents are then removed from solution and the denatured protein is allowed to refold in the presence of an oxidizing agent that converts thiols to disulfides. Would you expect that proinsulin after this treatment would have the same conformation as the original molecule or a different conformation? Explain your answer.

Fibrous Proteins (Section 20.4)

20.43 What is the most characteristic type of secondary structure in the protein of:
a. hair   b. hooves   c. collagen

20.44 What is meant by the following statement? "An α-helix is flexible and it is also elastic."

20.45 What is the function of collagen? Describe:
a. the macroscopic physical properties
b. the molecular structure of collagen

Serum Proteins (Section 20.5)

20.46 Of the five major types of serum proteins, which has the highest isoelectric point? Which has the lowest isoelectric point?

20.47 Explain the process of "salting out".

20.48 What is the primary function of serum albumin?

20.49 What is the major function of the proteins of the γ-globulin fraction?

Abnormal Human Hemoglobins

There are an estimated 5 billion red blood cells (erythrocytes) in the blood stream of an adult, each packed with about 270 million molecules of hemoglobin. In terms of sheer numbers, hemoglobin is one of the most plentiful proteins in the body. Hemoglobin's role is to pick up molecular oxygen in the lungs and deliver it to all parts of the body for metabolic oxidation. Normal adult hemoglobin (hemoglobin A, or Hb A) is composed of four polypeptide chains: two α-chains, each of 141 amino acids; and two β-chains, each of 146 amino acids. Each polypeptide chain surrounds one iron porphyrin or heme group (Section 21.3C), which reversibly binds oxygen. The tetrameric structure of hemoglobin is stabilized principally by hydrophobic interactions. Shown in Figure 1 is the three-dimensional shape of a single β-chain. The N-terminal amino acid is indicated by $-NH_3^+$ and the C-terminal amino acid is indicated by $-CO_2^-$.

The three-dimensional structure of hemoglobin A was determined by Max Perutz. For

Figure 1 β-chain of hemoglobin.

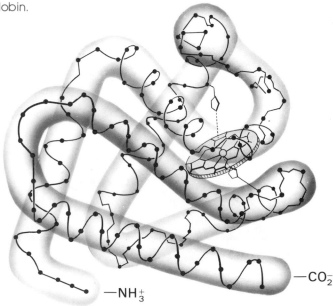

$-NH_3^+$

$-CO_2^-$

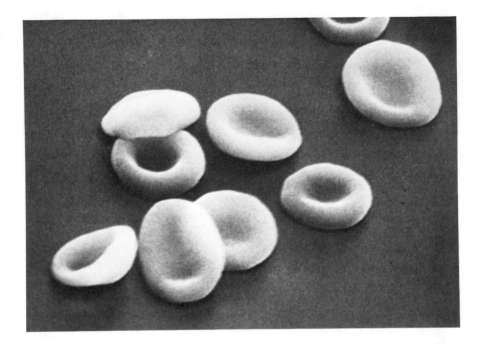

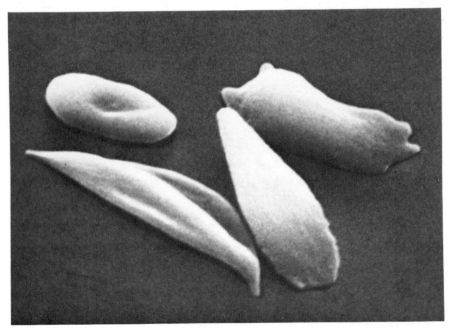

Figure 2 Normal red blood cells (top) magnified × 6750 and cells that have sickled (bottom) after discharging oxygen, magnified × 8700. (Courtesy of Dr. Marion I. Barnhart, Wayne State University School of Medicine.)

this pioneering work, he shared in the Nobel Prize in Chemistry in 1963.

It is the so-called abnormal human hemoglobins that have attracted particular attention because of the diseases associated with them. The best known of these diseases is sickle-cell anemia, a name derived from the characteristic sickle shape of affected red blood cells when they are deoxygenated. When combined with oxygen, red blood cells of persons with sickle-cell anemia have the flat, disk-like conformation of normal erythrocytes. However, when oxygen pressure is reduced, affected cells become distorted and considerably more rigid and inflexible than normal cells (Figure 2).

Because they are larger than some of the blood channels through which they must pass, sickled cells tend to become wedged in capillaries, blocking the flow of blood. Surprisingly little is known about the following aspects of sickle-cell anemia: the reason some organs and tissues are affected more than others by the disease; the normal age of onset of the disease; and male versus female susceptibility. Some persons afflicted with sickle-cell anemia die at an early age, often due to childhood infections complicated by the disease. Others lead long, productive lives.

In 1949, Linus Pauling made a discovery that opened the way to an understanding of this disease at the molecular level. He observed that there is a significant difference between normal adult hemoglobin (Hb A) and sickle-cell hemoglobin (Hb S). At pH 6.9, Hb A has a net negative charge and Hb S has a net positive charge; on paper electrophoresis at this pH, Hb A moves toward the positive electrode and Hb S toward the negative electrode. Vernon Ingram pursued this discovery and in 1956 showed that sickle-cell hemoglobin differs from normal hemoglobin only in the amino acid at the sixth position of the β-chain. Alpha-chains of both are identical, but glutamic acid at position 6 of each β-chain of Hb A is replaced by valine in Hb S. A result of the valine-glutamic acid substitution is replacement of two negatively charged side chains by two nonpolar, hydrophobic side chains.

How is the substitution of Hb S for Hb A in red blood cells related to the process of sickling? We know that hemoglobin S functions perfectly normally in transporting molecular oxygen from the lungs to cells. In this regard, it is indistinguishable from hemoglobin A. However, when it gives up its oxygen, Hb S tends to form polymers that separate from solution in crystalline form. There is now good evidence that the basic unit of crystalline Hb S polymer is a double-stranded fiber stabilized by hydrophobic interactions, including that between the valine at position 6 of one β-chain and a hydrophobic patch on other Hb S mole-cules. The double-stranded Hb S polymer molecules interact to form multi-stranded, cable-like structures. This is a remarkable phenomenon: polymerization of Hb S is facilitated by the presence of valine at β-6, but comparable polymerization of Hb A is prevented by the presence of glutamic acid at β-6.

Now that we have an understanding of sickle-cell anemia at the molecular level, the challenge is to devise specific medical treatments to prevent or at least inhibit the sickling process. One strategy being actively pursued is a search for substances that will inhibit the polymerization of Hb S by disrupting or preventing hydrophobic interactions of β-6 valines.

Sickle-cell anemia is a genetic disease. Persons with an Hb S gene from only one parent are said to have **sickle-cell trait.** About 40% of the hemoglobin in these individuals is Hb S. There are generally no ill effects associated with sickle-cell trait except under extreme conditions. Persons with Hb S genes from both parents are said to have **sickle-cell disease,** and all of their hemoglobin is Hb S. The mutant gene coding for Hb S occurs in about 10% of American blacks and in about 20% of African blacks. The gene is also present in significant numbers in the populations of countries bordering the Mediterranean Sea and parts of India.

The fact that there seems to be so much natural selec-

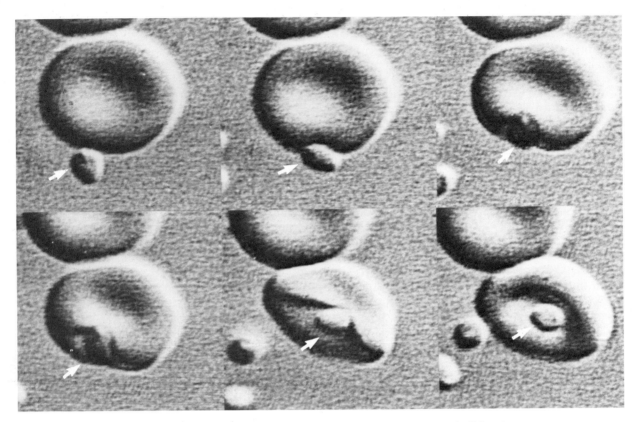

Figure 3 From top left to bottom right, the arrows in this microscope sequence follow the invasion of a red blood cell by a malaria parasite. (Courtesy of National Institute of Allergy and Infectious Diseases)

tion pressure against the Hb S gene raises two questions: Why has it persisted so long in the gene pool? Why is sickle-cell trait so common in populations of specific parts of the world? Several explanations have been offered, but the most likely, first advanced in 1949, is that sickle-cell trait provides some protection against *Plasmodium falciparum,* the parasite responsible for the most severe form of malaria.

The *falciparum* parasite lives part of its life cycle in red blood cells and grows equally well in oxygenated cells containing either Hb S or Hb A (Figure 3). However, when infected cells containing Hb S are deoxygenated and sickle, the parasites living in them are killed. Not all infected Hb S cells sickle at any one time, but the approximately 40% that do in a sickle-cell trait individual sufficiently reduce

the parasite population to reduce the severity of the malaria and prevent death.

Dramatic success in discovering the genetic and molecular basis for sickle-cell anemia spurred interest in searching for other abnormal hemoglobins. To date, several hundred have been isolated and the changes in primary structure have been determined. In the vast majority, there is but a single amino

Table 1 Abnormal human hemoglobins. Many of these names are derived from the location of their discovery.

Hemoglobin variant	Amino acid substitution		
	Position	From	To
alpha-chain			
J-Paris	12	ala	asp
G-Philadelphia	68	asn	lys
M-Boston	58	his	tyr
Dakar	112	his	gln
beta-chain			
S	6	glu	val
J-Trinidad	16	gly	asp
E	26	glu	lys
M-Hamburg	63	his	tyr

acid change in either the alpha- or the beta-chain, and each substitution is consistent with the change of a single nucleotide in one DNA codon (Section 22.11). Several abnormal hemoglobins are listed in Table 1.

Although most of the abnormal hemoglobins differ from Hb A by only a single amino acid substitution, several have been discovered in which there are either insertions or deletions of amino acids. For example, in hemoglobin-Leiden, discovered in 1968, glutamic acid at position 6 in each beta-chain is missing altogether. In hemoglobin-Gun Hill, discovered in 1967 in a 41-year-old man and one of his three daughters, there is a deletion of five amino acids in each beta-chain. Thus, each beta-chain is shortened to 141 amino acid residues (Figure 4a).

In hemoglobin-Grady, discovered in 1974 in a 25-year-old woman and her father, there is insertion of three amino acids in each alpha-chain. Thus, each alpha-chain is elongated to 144 amino acid residues (Figure 4b).

References

Beale, D., and Lehmann, H. 1965. Abnormal hemoglobins and the genetic code. *Nature* 207: 259.

Figure 4 Some abnormal human hemoglobins correspond to deletion mutations (a), others to insertion mutations (b).

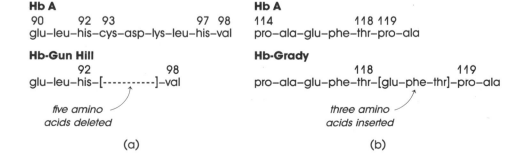

Hb A
90   92 93   97 98
glu–leu–his–cys–asp–lys–leu–his–val

Hb-Gun Hill
            92   98
glu–leu–his–[-----------]–val

five amino acids deleted

(a)

Hb A
114   118 119
pro–ala–glu–phe–thr–pro–ala

Hb-Grady
                        118   119
pro–ala–glu–phe–thr–[glu–phe–thr]–pro–ala

three amino acids inserted

(b)

Dayhoff, M.O. 1972. *Atlas of protein sequence and structure.* Vol. 5. Washington, D.C.: National Biomedical Research Foundation.

Harkness, D.R. 1976. Trends. *Biochemical Science* 1: 73–76.

Ingram, B. 1957. Gene mutations in human hemoglobin: the chemical difference between normal and sickle-cell hemoglobin. *Nature* 180: 326.

Maugh, T.H. 1981. A new understanding of sickle cell emerges. *Science* 211: 265–267.

Morimoto, H.; Lehmann, H.; and Perutz, M.F. Molecular pathology of human hemoglobins. *Nature* 232: 408.

Pauling, L., et al. 1949. Sickle-cell anemia, a molecular disease. *Science* 110: 543.

Perutz, M.F. Nov. 1964. The hemoglobin molecule. *Scientific American.*

Enzymes

One of the unique characteristics of a living cell is its ability to carry out complex reactions rapidly and with remarkable specificity. The agents responsible for these transformations are a group of protein biocatalysts called **enzymes,** each designed to catalyze a specific reaction. James Sumner, in 1926, was the first to isolate an enzyme in pure crystalline form. The enzyme was urease, which catalyzes the hydrolysis of urea to ammonia and carbon dioxide:

$$\underset{\text{urea}}{H_2N-\overset{\displaystyle O}{\overset{\|}{C}}-NH_2} + H_2O \xrightarrow{\text{urease}} 2\,NH_3 + CO_2$$

It is now clear that all enzymes are proteins and that the one feature distinguishing them from other proteins is that they are catalysts.

Practically speaking, enzyme technology has been with us for centuries. The use of enzymes for fermentation of fruit juices and grains to make alcoholic beverages is an art long practiced by humankind. Cheese was and still is made by treating milk with rennin, an enzyme obtained from the lining of a calf's stomach. The active ingredient in commercially available meat tenderizers is an enzyme extracted from the papaya plant.

Currently, there is a rapidly expanding field of research called **clinical enzymology.** This research has shown: the enzyme content of human blood changes significantly under certain pathological conditions; and information about these changes in enzyme concentration can be an important diagnostic tool for the physician.

In many ways, the study of enzyme structure and function has provided a bridge between the biological and physical sciences. As we shall see in this and subsequent chapters, an understanding of enzymes and their properties is essential to an understanding of metabolism, metabolic diseases, and therapies for metabolic diseases.

21.1 Characteristics of Enzyme Catalysis

Enzymes function as catalysts in much the same way as common inorganic or organic laboratory catalysts do. A catalyst, whatever the kind, combines with a reactant or reactants to "activate" it. In the case of enzyme-catalyzed reactions, reactants are referred to as *substrates* (S). Enzyme and substrate(s) combine to form an activated complex called an **enzyme-substrate complex (ES).** This complex then undergoes a chemical change to form product(s) and regenerates the enzyme.

$$E + S \rightleftharpoons ES$$
$$ES \longrightarrow E + P$$

As catalysts, enzymes are far superior to their nonbiological laboratory counterparts in three major ways: they have enormous catalytic power; they are highly specific in the reactions they catalyze; and the activity of many enzymes can be regulated. Let us discuss each of these characteristics in detail.

A. Catalytic Efficiency

Enzymes are able to bring about enormous increases in the rates of chemical reactions. In fact, most reactions that occur readily in living cells would occur too slowly to support life in the absence of these biocatalysts. An example is peroxidase, an enzyme that catalyzes the decomposition of hydrogen peroxide to oxygen and water.

$$2\,H_2O_2 \xrightarrow{\text{peroxidase}} 2\,H_2O + O_2$$

This is a very important reaction because many biological oxidations use molecular oxygen as the oxidizing agent and in a number of these, oxygen is reduced to hydrogen peroxide. Hydrogen peroxide is very toxic and must be decomposed rapidly to prevent damage to cellular components. Figure 21.1 shows the energy changes for the catalyzed and uncatalyzed decomposition of hydrogen peroxide. The energy of activation for the uncatalyzed decomposition is 18 kcal/mole. Peroxidase provides an alternate pathway with an energy of activation of only

Figure 21.1 Energy changes for the uncatalyzed and peroxidase-catalyzed decomposition of hydrogen peroxide.

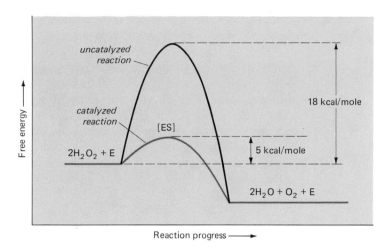

5.0 kcal/mole. This seemingly small decrease in energy of activation corresponds to an increase in rate of reaction of almost 10^{10}!

As a second example, carbonic anhydrase catalyzes the reaction of carbon dioxide and water to produce carbonic acid.

$$CO_2 + H_2O \underset{\text{anhydrase}}{\overset{\text{carbonic}}{\rightleftharpoons}} H_2CO_3 \rightleftharpoons HCO_3^- + H^+$$

Carbonic anhydrase increases the rate of hydration of carbon dioxide by almost 10^7 times compared to the uncatalyzed reaction. Red blood cells are especially rich in this enzyme and are able to promote the rapid interconversion of carbon dioxide and bicarbonate ion.

B. Specificity

The second unique property of enzymes is their high specificity in the reactions they catalyze. Most enzymes catalyze only one reaction or a single type of reaction. Competing reactions and by-products such as we find under laboratory conditions are not observed in enzyme-catalyzed reactions. We have already discussed in Chapter 14 how an enzyme might catalyze a reaction of (+)-glyceraldehyde but not of its enantiomer, (−)-glyceraldehyde. This is but one example of enzyme specificity. We will see many more examples later.

C. Potential for Regulation

Third among their unique properties is the fact that the activities of many enzymes can be regulated. Most often, regulation mechanisms are in the form of signals from certain small molecules that bind to the enzyme; in so doing, they either increase or decrease the enzyme's activity. We shall discuss regulation of enzyme activity in detail in Section 21.6.

The fact that enzymes have enormous catalytic power, are highly specific in the reactions they catalyze, and can be regulated has important consequences for living cells. Each living cell contains literally thousands of different molecules, and there is an almost infinite number of chemical reactions that are possible in this mix. Yet a cell, by virtue of its enzymes, selects which chemical reactions take place; by regulating the activities of key enzymes, the cell also controls the rates of these reactions and how much of any given product is formed. In this regard, enzymes are truly remarkable catalysts without which the living state would be impossible!

21.2 Naming Enzymes

The naming of enzymes has been systematized by a set of rules proposed by the **International Enzyme Commission (IEC).** According to these rules, each enzyme is given a name that specifies the substrate(s) upon which it acts, the functional group(s) acted upon, and the type of reaction catalyzed. All IEC names end in -*ase*. An example of an IEC name is that of the enzyme that catalyzes the hydrolysis of urea:

$$H_2N-\overset{\overset{\displaystyle O}{\|}}{C}-NH_2 + H_2O \xrightarrow{\text{enzyme}} 2\,NH_3 + CO_2$$

substrate: urea
functional group: amide
type of reaction: hydrolysis
IEC name: **urea amidohydrolase**

Enzymes are also given common names that have the form substrate-*ase* or substrate-reaction type-*ase*. An example of a common name is that given to urea amidohydrolase.

substrate: urea
common name: *urea* + *ase* = **urease**

Note that the name *urease* specifies the substrate upon which the enzyme acts but not the type of reaction catalyzed.

Another example of a common name is that of the enzyme that catalyzes the oxidation of lactate to pyruvate. Oxidations of this type in which hydrogen atoms are removed from adjacent atoms are often called **dehydrogenations.** In this enzyme-catalyzed reaction, the oxidizing agent is nicotinamide adenine dinucleotide (NAD^+).

$$\underset{\text{lactate}}{CH_3-\overset{\overset{\displaystyle OH}{|}}{CH}-CO_2^-} + NAD^+ \xrightarrow{\text{LDH}} \underset{\text{pyruvate}}{CH_3-\overset{\overset{\displaystyle O}{\|}}{C}-CO_2^-} + NADH + H^+$$

substrate: lactate
reaction type: dehydrogenation
common name: *lactate dehydrogen*ation + *ase* =
                 lactate dehydrogenase (LDH)

Throughout the text, we will refer to enzymes by their accepted common names.

21.3 Enzyme Proteins and Cofactors

Among the enzymes that act as biocatalysts, there is considerable diversity of structure. Many enzymes are *simple proteins*, which means that the protein itself is the true catalyst. Other enzymes catalyze reactions of their substrates only in the presence of specific nonprotein molecules or metal ions. Nonprotein molecules or metal ions required for enzyme activity are called **cofactors.** Cofactors are divided into three groups: *metal ions*, *coenzymes*, and *prosthetic groups*.

A. Metal Ion Cofactors

Metal ion cofactors function primarily by forming complexes with the enzyme itself or with other nonprotein groups required by the enzyme for catalytic activity. In some cases, metal ions appear to be only loosely associated with the active enzyme and can be removed easily from the enzyme. In other instances, metal ions are integral parts of the enzyme structure and are retained throughout isolation and purification procedures. As an example, virtually all reactions that involve adenosine triphosphate (ATP) and the hydrolysis of phosphate anhydride bonds require Mg^{2+} as a cofactor. In these reactions, the positively charged cation coordinates with the negatively charged oxygens of the phosphate groups. As another example, carbonic anhydrase requires one Zn^{2+} per molecule of enzyme for activity.

In recent years, we have come to realize that many other metals, some of them present in trace amounts only, are also essential for proper enzyme function in humans; hence, they are required for good health. In many instances, we have little or no understanding of what the role of these metal ions is or why they are essential. For further discussion, see Mini-Essay 1, "Elements in the Body."

B. Coenzymes and Vitamins

A **coenzyme** is a small organic molecule that binds reversibly to an enzyme and is required for activity of the enzyme. Many coenzymes are second substrates for the enzyme. For example, the enzyme lactate dehydrogenase (LDH) requires nicotinamide adenine dinucleotide (NAD^+) for activity.

$$\underset{\text{lactate}}{CH_3-\underset{\overset{|}{OH}}{CH}-CO_2^-} + NAD^+ \xrightarrow{\text{lactate}\atop\text{dehydrogenase}} \underset{\text{pyruvate}}{CH_3-\underset{\overset{\|}{O}}{C}-CO_2^-} + NADH + H^+$$

It should be obvious why NAD^+ is required: NAD^+ is the organic molecule that oxidizes lactate to pyruvate and is reduced to NADH.

Table 21.1 Twelve essential coenzymes, their vitamin precursors, and biological functions.

Coenzyme	Vitamin Precursor	Function
nicotinamide adenine dinucleotide (NAD^+)	nicotinic acid (niacin)	oxidation-reduction
nicotinamide adenine dinucleotide phosphate ($NADP^+$)	nicotinic acid (niacin)	oxidation-reduction
flavin adenine dinucleotide (FAD)	riboflavin (vitamin B_2)	oxidation-reduction
flavin mononucleotide (FMN)	riboflavin (vitamin B_2)	oxidation-reduction
ascorbic acid (vitamin C)	none	oxidation-reduction
lipoic acid	none	oxidation-reduction
thiamine pyrophosphate (TPP)	thiamine (vitamin B_1)	oxidative decarboxylation and nonoxidative decarboxylation
pyridoxal phosphate	pyridoxine (vitamin B_6)	transaminations
coenzyme A	pantothenic acid	transfer of $CH_3-\overset{\overset{\textstyle O}{\|}}{C}-$ groups
tetrahydrofolic acid (THF)	folic acid	transfer of $-CH_3$, $-CH_2OH$, $-CHO$ groups
biotin	none	carboxylation reactions
cobamide (B_{12})	cobalamine	transfer of $-\overset{\overset{\textstyle O}{\|}}{C}-S-CoA$ groups

Humans and many other organisms cannot synthesize certain coenzymes, so they must obtain from their diet either the coenzyme itself or a substance from which the coenzyme can be synthesized. These so-called essential coenzymes or coenzyme precursors are vitamins. Vitamins are divided into two classes, based on their physical properties: those that are water-soluble and those that are fat-soluble. As might be expected, water-soluble vitamins are highly polar substances, while fat-soluble vitamins are nonpolar. Most water-soluble vitamins are either coenzymes themselves or are small molecules from which coenzymes are synthesized within the body. The function of fat-soluble vitamins is less well understood.

Table 21.1 lists twelve essential coenzymes, their vitamin precursors, and the function of each coenzyme.

C. Prosthetic Groups

Coenzymes and **prosthetic groups** are similar in that both are organic molecules that bind to enzymes and are necessary for biological activity of the enzyme. The difference between the two is one of degree. Organic molecules that bind reversibly to enzymes are classified as coenzymes. Those that are strongly bound and are an integral part of an enzyme protein are called **prosthetic groups.** Metal ions that are an integral part of the protein may also be referred to as prosthetic groups. One important class of prosthetic groups are the **hemes.** The structure of heme consists of four substituted pyrrole rings joined by one-carbon bridges into a larger ring called porphyrin (Figure 21.2a). Shown in Figure 21.2(b) is the heme group found in hemoglobin and myoglobin. Note that there is an atom of iron in the center of the heme group. In hemoglobin and myoglobin, the iron atom occurs as Fe^{2+}. A magnesium ion, Mg^{2+}, embedded in a porphyrin ring is an integral part of chlorophyll.

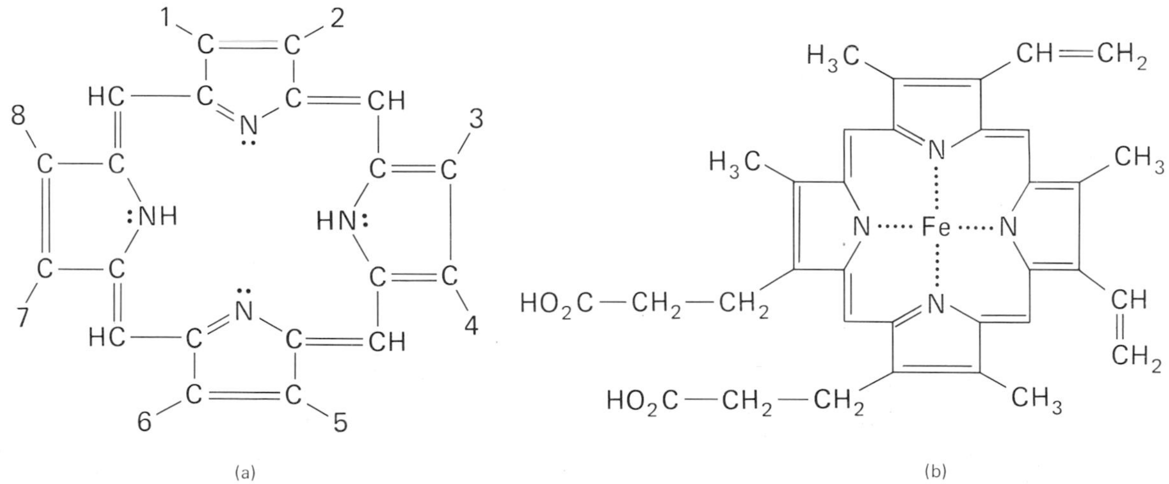

(a)   (b)

Figure 21.2 Heme coenzymes: (a) the porphyrin ring system; (b) the heme prosthetic group of hemoglobin, myoglobin, and certain enzymes.

21.4 Mechanism of Enzyme Catalysis

The formation of an enzyme-substrate complex (Section 21.1) is the first and crucial step in enzyme catalysis. Virtually all enzymes are globular proteins and even the simplest have molecular weights ranging from 12,000 to 40,000, meaning that they consist of 100–400 amino acids. Because enzymes are so large compared to molecules whose reactions they catalyze, it has been proposed that substrate and enzyme interact over only a small region of the enzyme surface. We call this region of interaction the **active site.**

To date, a large number of enzymes have been isolated in pure crystalline form and studied by X-ray crystallography. In all cases where enzyme three-dimensional structures have been determined and the interactions between enzyme and substrate studied, the active site has been found to be a portion of the enzyme surface with a unique arrangement of amino acid side chains. Often these side chains are contributed by amino acids quite far apart in the primary structure of the polypeptide chain.

In 1890, Emil Fischer likened the binding of an enzyme and its substrate to the interaction of a lock and key. According to this lock-and-key model, shown schematically in Figure 21.3, enzyme and substrate have complementary shapes and fit together. Recall from our discussion of the significance of chirality in the biological world (Section 14.6) that we accounted for the remarkable ability of enzymes to distinguish between enantiomers by proposing that an enzyme and its substrate must interact through at least three specific binding sites on the surface of the enzyme. In Figure 21.3, these three binding sites are labeled a, b, and c. The complementary regions on the substrate are labeled a', b', and c'. Groups on the enzyme surface that participate in binding enzyme and substrate to form an enzyme-substrate complex are called **binding groups.**

Once a substrate molecule is recognized and bound to the active site of the enzyme, certain functional groups in the active site participate directly in the making and breaking of chemical bonds. These are called **catalytic groups.** In a sense, then, the active site on an enzyme is a unique combination of binding groups and catalytic groups.

Figure 21.3 Lock-and-key model of the interaction of enzyme and substrate.

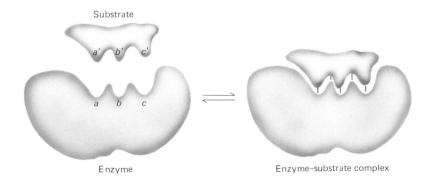

Substrate

a' b' c'

a b c

Enzyme

Enzyme–substrate complex

21.5 Factors that Affect the Rate of Enzyme-Catalyzed Reactions

The rate of an enzyme-catalyzed reaction depends on many factors, the most important of which are concentration of the enzyme, concentration of the substrate, pH, temperature, and presence of enzyme inhibitors. Let us look at each of these factors in some detail.

A. Concentration of the Enzyme

In an enzyme-catalyzed reaction, the concentration of enzyme is very small compared to the concentration of substrate(s); under these conditions, rate of reaction is directly proportional to concentration of enzyme. For example, if the concentration of enzyme is doubled, the rate of conversion of substrate(s) to product(s) is also doubled. The effect of enzyme concentration on the rate of reaction is shown in Figure 21.4.

Figure 21.4 The effect of enzyme concentration on the rate of conversion of substrate to product.

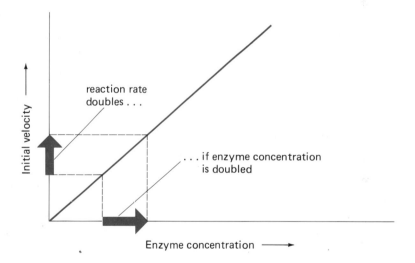

reaction rate doubles . . .

. . . if enzyme concentration is doubled

Initial velocity

Enzyme concentration

B. Concentration of the Substrate

To understand the effect of substrate concentration on the rate of an enzyme-catalyzed reaction, we need to consider a series of experiments, each using the same concentration of enzyme but a different initial concentration of substrate.

Suppose we mix a given concentration of enzyme and substrate and then measure the initial velocity of the reaction. In a second experiment, we mix the same concentration of enzyme but increase the concentration of substrate and again measure the initial reaction velocity. This can be repeated with a constant

Figure 21.5 Dependence of initial velocity of an enzyme-catalyzed reaction on the concentration of substrate.

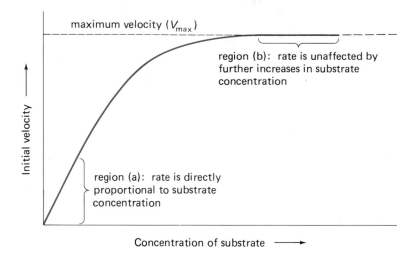

maximum velocity (V_{max})

region (b): rate is unaffected by further increases in substrate concentration

region (a): rate is directly proportional to substrate concentration

Initial velocity

Concentration of substrate ⟶

enzyme concentration and a range of substrate concentrations. Figure 21.5 shows the relationship between initial reaction velocity and concentration of substrate. In region (a) of the figure, an increase in initial substrate concentration results in a direct increase in reaction velocity. Beyond region (a), an increase in initial substrate concentration also increases the reaction velocity, but the effect becomes progressively smaller. Finally, in region (b), an increase in initial substrate concentration has no effect whatsoever on the reaction velocity. The reason for this is that in region (b), all enzyme molecules have formed enzyme-substrate complexes and are in continuous operation, catalyzing the conversion of substrate to product. At this point we say that the enzyme is saturated. The maximum velocity of an enzyme-catalyzed reaction is given by the symbol V_{max}. At V_{max}, all enzyme molecules have substrate bound to them.

The substrate concentration required to convert one-half of all enzyme molecules to enzyme-substrate complexes is a measure of the affinity of enzyme

Figure 21.6 K_m is the substrate concentration required to give an initial reaction velocity of $(1/2)V_{max}$.

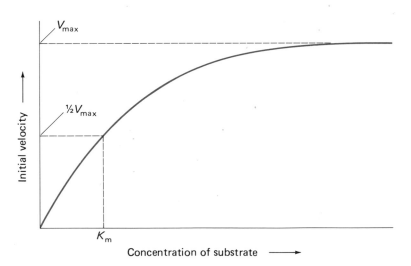

V_{max}

$\frac{1}{2}V_{max}$

Initial velocity

K_m

Concentration of substrate ⟶

for substrate. This concentration is represented by the symbol K_m. A small value of K_m means that an enzyme has a high affinity for its substrate; a large value of K_m means that the enzyme has a low affinity for its substrate. The value of K_m can be determined graphically—it is the substrate concentration required to give an initial velocity of $\frac{1}{2}V_{max}$ (Figure 21.6).

Example 21.1

Following are data for five reactions, each using the same concentration of enzyme but a different initial concentration of substrate. Also given is the initial reaction velocity for each substrate concentration. In these reactions, velocity is measured as mg of substrate per mL of solution reacting per minute.

Initial Concentration of Substrate	Initial Reaction Velocity (mg/mL/min)
$15 \ \times 10^{-5}\,M$	0.38
$10 \ \times 10^{-5}\,M$	0.33
$5.0 \times 10^{-5}\,M$	0.25
$3.3 \times 10^{-5}\,M$	0.20
$2.5 \times 10^{-5}\,M$	0.166

a. Prepare a graph of initial velocity versus substrate concentration and estimate V_{max}, $\frac{1}{2}V_{max}$, and K_m from it.

b. For this enzyme-catalyzed reaction, the initial velocity is 0.166 mg/mL/min when the substrate concentration is $2.5 \times 10^{-5}M$. What will be the initial velocity at this substrate concentration if the enzyme concentration is doubled?

Solution

a. Following is a graph plotting substrate concentration on the horizontal axis and initial reaction velocity on the vertical axis.

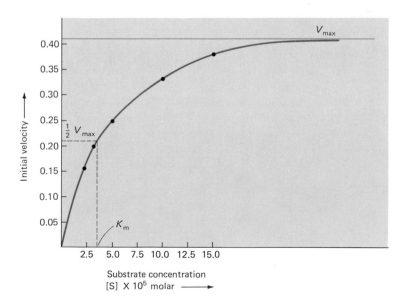

From this graph, we can estimate that V_{max} is approximately 0.41 milligram of substrate reacting per milliliter of solution per minute. Given this value, $\frac{1}{2}V_{max}$ is approximately 0.21 mg/mL/min and K_m is approximately $3.5 \times 10^{-5}M$. For this enzyme, a substrate concentration of 3.5×10^{-5} mole per liter is required to convert one-half of all enzyme molecules to enzyme-substrate complexes.

b. The velocity of an enzyme-catalyzed reaction is directly proportional to the enzyme concentration. If the enzyme concentration is doubled, the initial velocity doubles and the new initial velocity becomes 0.332 mg/mL/min.

Problem 21.1

Following are the initial velocities of an enzyme-catalyzed reaction at five different substrate concentrations [S]. Velocity is given in units of micromoles of substrate reacting per minute.

Initial Substrate Concentration	Initial Velocity (μmoles/min)
2.0×10^{-3}	14
3.0×10^{-3}	18
4.0×10^{-3}	21
10.0×10^{-3}	31
12.0×10^{-3}	33

a. Prepare a graph of initial velocity versus initial substrate concentration and estimate V_{max}, $\frac{1}{2}V_{max}$, and K_m from it.

b. From your graph, estimate the initial velocity when the initial substrate concentration is $6 \times 10^{-3}M$.

c. If the enzyme concentration is doubled, what will be the new initial velocity when $[S] = 6 \times 10^{-3}M$?

C. pH

The catalytic activity of all enzymes is affected by the pH of the solution in which the reaction occurs. Often, small changes in pH cause large changes in the ability of a given enzyme to function as a biocatalyst. Shown in Figure 21.7

Figure 21.7 The dependence of enzyme activity on pH.

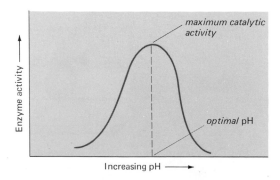

Table 21.2 Optimal pH values for several enzymes.

Enzyme	Optimal pH
pepsin	1.5
acid phosphatase	4.7
α-glucosidase	5.4
urease	6.7
α-amylase (pancreatic)	7.0
carboxypeptidase	7.5
succinic dehydrogenase	7.6
trypsin	7.8
alkaline phosphatase	9.5
arginase	9.7

is a typical plot of enzyme activity versus pH. Notice that enzyme activity is at a maximum in a narrow pH range and decreases at both higher and lower pHs. The pH corresponding to maximum enzyme activity is called the *optimal* pH. A majority of enzymes have maximum catalytic activity around pH 7, the pH of most biological fluids. Some, however, have maximum activity at considerably higher or lower pHs. For example, pepsin, a digestive enzyme of the stomach, has maximum activity around pH 1.5, the pH of gastric fluids. Table 21.2 lists optimal pH values for several enzymes.

Variations in enzyme activity with changes in pH depend on a number of factors, some or all of which may occur at the same time. First, changes in pH may cause partial denaturation of the enzyme protein. Second, catalytic activity may be possible only when certain amino acid side chains are in the correct states of ionization. For example, suppose that catalysis by a particular enzyme requires that a lysine and a glutamic acid at the active site be fully ionized (lys-NH_3^+ and glu-CO_2^-), as shown in Figure 21.8(b). This combination

Figure 21.8 The dependence of catalytic activity on the ionization of amino acid side chains: (a) below the optimal pH; (b) at the optimal pH; (c) above the optimal pH.

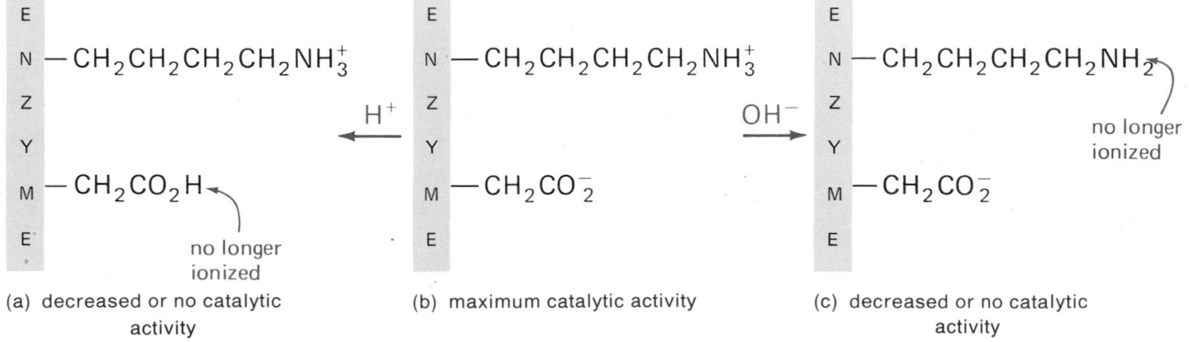

(a) decreased or no catalytic activity

(b) maximum catalytic activity

(c) decreased or no catalytic activity

of ionization states is possible only in a particular pH range. At lower pH values (more acidic), catalytic activity decreases because the carboxyl of glutamate is protonated to $-CO_2H$ (Figure 21.8a). At higher pH values (more basic), catalytic activity decreases because the side chain of lysine is deprotonated to $-NH_2$ (Figure 21.8c). The effect of pH on catalytic activity may be considerably more complex than that shown in Figure 21.7, particularly if the state of ionization of several amino acid side chains and of the substrate itself are important.

Example 21.2

Assume that one of the interactions binding a particular substrate to an enzyme is hydrogen bonding between a carbonyl group of the substrate and the $-OH$ group of tyrosine.

How might you account for the fact that the velocity of this enzyme-catalyzed reaction is at a maximum at pH 8.1 and decreases as the pH increases? (The pK_a of the side chain of tyrosine is 10.1.)

Solution

For bonding to occur between a carbonyl group of the substrate and the $-OH$ group of tyrosine, it is essential to have the side chain of tyrosine in the un-ionized form. At pH 8.1 (two pH units more acidic than the pK_a of this group), the side chain is completely in the acid ($-OH$) form, and the rate is at a maximum. As pH increases, the side chain becomes partially ionized, and at pH 10.1 the side chain is 50% ionized. The ionized form cannot participate in hydrogen bonding to bind substrate and therefore the rate of reaction decreases.

Problem 21.2

For a particular enzyme-catalyzed reaction involving the side chain of histidine, the optimal pH is 4.0. Activity of the enzyme decreases above pH 4.0. Does the active form of this enzyme require the side chain of histidine to be in the acid or conjugate base form? Explain.

D. Temperature

A fourth factor affecting the velocity of an enzyme-catalyzed reaction is temperature. Just as there is an optimal pH for a reaction, there is also an optimal temperature. Most enzymes have optimal temperatures in the range of 25–37°C. Figure 21.9 shows a typical plot of enzyme activity as a function of temperature. Enzyme activity first increases with temperature because of an increase in the number of collisions between enzyme and substrate and an increase in the

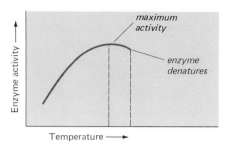

Figure 21.9 Dependence of enzyme activity on temperature.

energy of these collisions. At higher temperatures, enzyme activity decreases rapidly due to heat denaturation of the enzyme protein.

E. Inhibition of Enzyme Activity

An **inhibitor** is any compound that has the ability to decrease the rate of an enzyme-catalyzed reaction. An understanding of enzyme inhibition is important for several reasons. First, many poisons, as well as many medicines, act at the molecular level by inhibiting one or more enzymes and thereby decreasing the rates of the reactions these enzymes catalyze. Second, certain molecules present in the cell during metabolism inhibit specific enzyme-catalyzed reactions and thereby provide a means for the internal regulation of cellular metabolism.

Enzyme inhibitors can be divided into two categories, competitive and noncompetitive, depending on how the particular inhibitor acts at the molecular level. As we shall see in this section, the way to decide whether a particular inhibition is competitive or noncompetitive is by graphical analysis.

1. Competitive inhibition. A **competitive inhibitor** is one that binds to the active site of an enzyme and thus "competes" with substrate for the active site. Most often, competitive inhibitors are very closely related in structure to the enzyme substrate. As an example, consider the enzyme fumarase, which catalyzes the hydration of fumarate to malate.

$$^-O-C \overset{O}{\underset{H}{\|}} C=C \overset{H}{\underset{C-O^-}{\underset{\|}{O}}} + H_2O \xrightarrow{\text{fumarase}} \quad ^-O-C \overset{O}{\|} \overset{OH}{\underset{H}{C-C}} \overset{H}{\underset{C-O^-}{\underset{\|}{O}}}$$

fumarate   L-malate

Shown in Figure 21.10 (next page) are structural formulas for several competitive inhibitors of fumarase. Note the structural similarities between these inhibitors and fumarate.

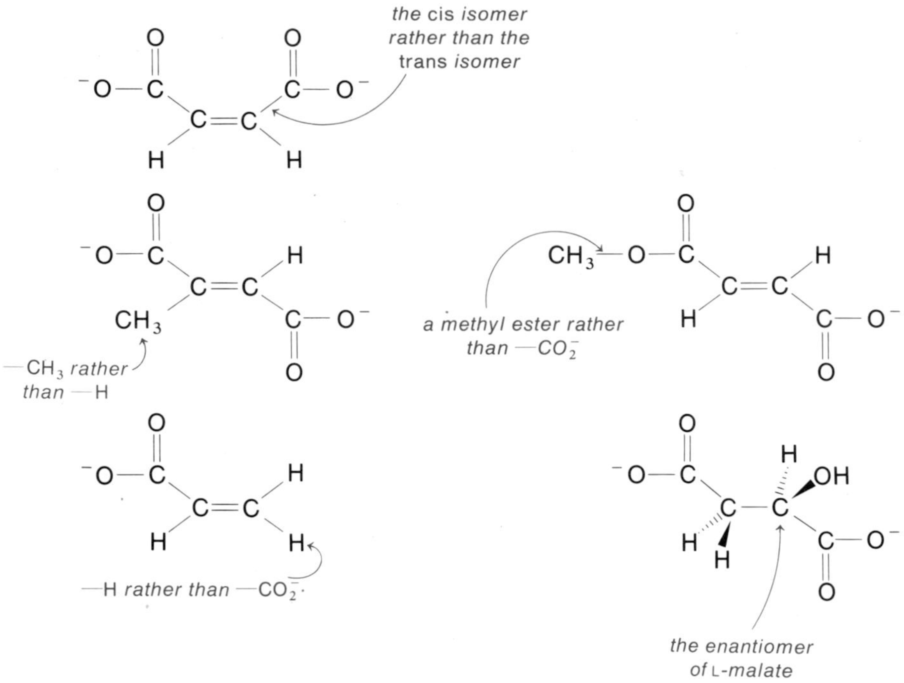

Figure 21.10 Several competitive inhibitors of the enzyme fumarase.

The nature of competitive inhibition is shown schematically in Figure 21.11. Part (a) of this figure shows an enzyme in the presence of both inhibitor and substrate. There is competition between the two for the active site and the inhibitor is shown forming an enzyme-inhibitor complex that blocks the active site for further catalytic activity.

A characteristic feature of competitive inhibition is that it can be reversed by increasing the concentration of substrate. We can account for this by looking at the two equilibria involved in a solution containing enzyme, substrate, and inhibitor. The first is the equilibrium between enzyme + substrate and enzyme-substrate complex. The second is that between enzyme + inhibitor and enzyme-inhibitor complex. If we increase the concentration of substrate, then according to Le Chatelier's principle (Section 8.4), we also increase the concentration of enzyme-substrate complex at the expense of enzyme-inhibitor complex.

$$\text{equilibrium 1:} \quad E + S \rightleftharpoons ES \qquad \textit{increases as [S] increases}$$

$$\text{equilibrium 2:} \quad E + I \rightleftharpoons EI \qquad \textit{decreases as [S] increases}$$

The reversal of competitive inhibition is shown schematically in Figure 21.11(b).

Competitive inhibition can be detected by studying the rate of an enzyme-catalyzed reaction in the presence and absence of a competitive inhibitor (Figure 21.12). In this figure, V_{max} is unchanged in the presence of a competitive

Figure 21.11 Competitive inhibition. (a) Formation of an enzyme-inhibitor complex. (b) In the presence of increased substrate, ES is increased and EI is decreased, thus reversing the effect of the inhibitor.

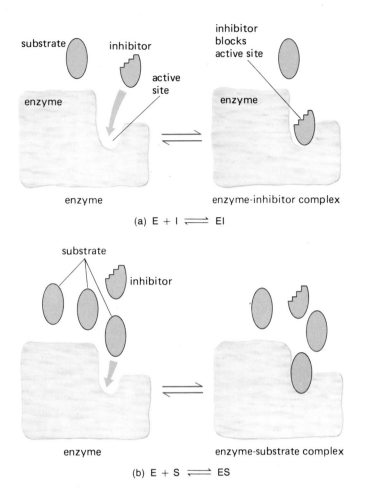

substrate inhibitor

inhibitor blocks active site

active site

enzyme enzyme

enzyme enzyme-inhibitor complex

(a) E + I $\rightleftharpoons$ EI

substrate inhibitor

enzyme enzyme-substrate complex

(b) E + S $\rightleftharpoons$ ES

Figure 21.12 An example of competitive inhibition. In the presence of the inhibitor, V_{max} is unchanged but K_m is increased.

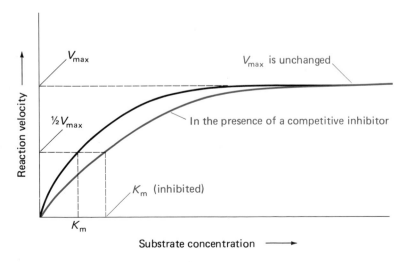

V_{max}

V_{max} is unchanged

$\frac{1}{2}V_{max}$

In the presence of a competitive inhibitor

Reaction velocity

K_m (inhibited)

K_m

Substrate concentration

Figure 21.13 In non-competitive inhibition, inhibitor binds at a site other than the active site.

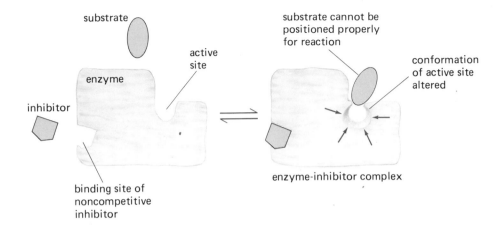

inhibitor but K_m is increased, because a greater concentration of substrate is required to achieve one-half enzyme saturation in the presence of the inhibitor.

2. Noncompetitive inhibition. **Noncompetitive inhibition** occurs when an inhibitor binds to the surface of an enzyme at a site different from the catalytically active site (Figure 21.13). The interaction of enzyme and noncompetitive inhibitor causes a change in the three-dimensional shape of the enzyme and of the active site. In the presence of a noncompetitive inhibitor: either (1) the enzyme is no longer able to bind substrate at the active site; or (2) the enzyme binds substrate but due to improper positioning of the substrate, the catalytic groups are no longer able to participate in the making and breaking of chemical bonds required for catalysis.

Unlike competitive inhibition, noncompetitive inhibition cannot be reversed by addition of more substrate because additional substrate has no effect on the equilibrium between enzyme + inhibitor and enzyme-inhibitor complex (Figure 21.14). Noncompetitive inhibition can be detected by studying the rate of enzyme-catalyzed reaction in the presence and absence of a noncom-

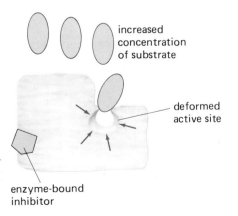

Figure 21.14 Noncompetitive inhibition is unaffected by an increase in the concentration of substrate.

Figure 21.15 In the presence of a non-competitive inhibitor, K_m is unchanged but V_{max} is decreased.

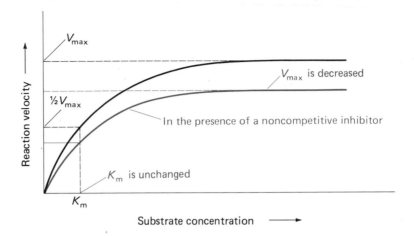

petitive inhibitor (Figure 21.15). In noncompetitive inhibition, K_m is unchanged because the substrate is still free to bind to the active site (even though no catalysis takes place). V_{max}, however, is decreased in proportion to the fraction of enzyme molecules in the form of enzyme-inhibitor complex.

21.6 Regulation of Enzyme Activity

Within the past 20 years, it has become clear that regulation of enzyme activity is a natural phenomenon common to all living systems. There are two biological mechanisms by which this is accomplished. In the first, **feedback control,** the activity of key enzymes is altered by interaction of the enzyme with molecules produced within the cell itself. The result of this interaction may be either inhibition of activity or stimulation of activity. The second biological mechanism, **genetic control,** regulates the concentration of key enzymes by regulating the rate of enzyme (protein) synthesis.

A. Feedback Control

The first example of regulation of enzyme activity by feedback control was discovered in 1957 and involves the synthesis of isoleucine in the bacterium *E. coli*. This synthesis begins with threonine and in a series of five sequential steps, each catalyzed by a different enzyme ($E_1, E_2, \ldots, E_5$), gives isoleucine.

feedback inhibition

The concentration of isoleucine within the cell is regulated by the fact that isoleucine is an inhibitor of threonine deaminase, the first enzyme in this multistep synthesis. Enzymes in a multistep synthesis that are subject to regulation by the final product of the sequence are called **regulatory enzymes.** Because the role of isoleucine in the regulation of its concentration is that of an inhibitor, it is called a **negative modifier.** Molecules or ions that increase the catalytic activity of a regulatory enzyme are called **positive modifiers.**

Regulation through alteration of the activity of key enzymes is of immense benefit to an organism because through it, the concentration of a final product can be maintained within very narrow limits. In this way, cells prevent unnecessary accumulation not only of the final product of a metabolic sequence but also of intermediates along the pathway. The evolution of regulatory enzymes was an essential step in achieving efficient use of cellular resources.

B. Genetic Control

Genetic control of enzyme activity involves regulation of the rate of protein biosynthesis by **induction** and **repression.** The first demonstrated example of enzyme induction grew out of studies in the 1950s on the metabolism of the bacterium *E. coli.* β-Galactosidase, an enzyme required for the utilization of lactose, catalyzes hydrolysis of lactose to D-galactose and D-glucose.

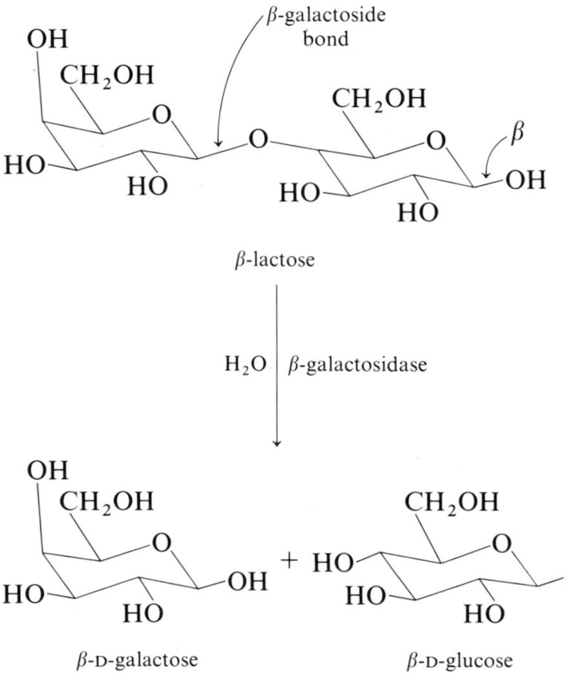

In the absence of lactose in the growth medium, no β-galactosidase is present

Figure 21.16 β-Galactosidase, an inducible enzyme. Synthesis of β-galactosidase begins within minutes after the addition of lactose to the growth medium and ceases shortly after lactose is removed from the growth medium.

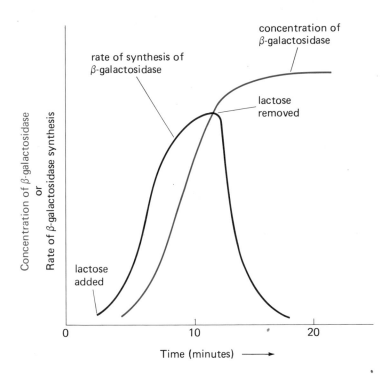

in *E. coli.* If, however, lactose is added to the medium, within minutes the bacterium begins to produce this enzyme (Figure 21.16); this is the induction step. If lactose is then removed from the growth medium, production of β-galactosidase stops. Hence, β-galactosidase is an example of an inducible enzyme.

The biological significance of genetic control of enzyme activity is that it allows an organism to adapt to changes in its environment.

21.7 Enzymes in Clinical Diagnosis

Certain enzymes, such as those involved in blood coagulation, are normal constituents of plasma, and their concentration in plasma is high compared to their concentration in cells. Other enzymes are normally present almost exclusively in cells and are released into the blood and other biological fluids only as a result of routine destruction of cells. Plasma levels of these enzymes are low, often up to a million times lower than their cell levels. However, the plasma concentration of cellular enzymes may be elevated significantly in cases of excessive cell injury and destruction or excess growth, as in cancer. By choosing appropriate enzymes for examination, changes in plasma concentrations of these enzymes can be used to detect cell damage and to suggest the site of damage or uncontrolled cell growth and proliferation. Further, the

Table 21.3 Some enzymes used in diagnostic enzymology.

Enzyme	Principal Clinical Condition in which the Enzyme Determination is Used
lactate dehydrogenase (LDH)	heart or skeletal muscle damage, myocardial infarction
alkaline phosphatase	liver and bone disease
acid phosphatase	cancer of the prostate
serum glutamate oxaloacetate transaminase (SGOT)	liver and heart disease
creatine phosphokinase (CPK)	myocardial infarction and muscle disease
α-amylase	pancreatitis

degree of elevation of plasma concentration can often be used to determine the extent of cellular damage. For these reasons, measurement of enzyme concentrations in blood plasma and other biological fluids has become a major diagnostic tool, particularly for diseases of the heart, liver, pancreas, skeletal muscle, bone, and for malignant diseases. In fact, certain enzyme determinations are performed so often that they have become routine in the clinical chemistry laboratory. Several of these enzymes are listed in Table 21.3. For a discussion of enzyme assays and diagnosis of myocardial infraction, see the mini-eassay, "Clinical Enzymology—The Search for Specificity."

Key Terms and Concepts

active site (21.4)

binding groups (21.4)

catalytic groups (21.4)

clinical enzymology (21.7)

coenzyme (21.3B)

cofactor (21.3A)

competitive inhibition (21.5E)

dehydrogenation (21.2)

enzymes as biocatalysts (21.1)

enzyme specificity (21.1B)

enzyme-substrate complex (21.1)

feedback control (21.6A)

genetic control (21.6B)

heme (21.3C)

induction (21.6B)

inhibitor (21.5E)

International Enzyme Commission (21.2)

K_m (21.5B)

metal ion cofactor (21.3A)

negative modifier (21.6A)

noncompetitive inhibition (21.5E)

positive modifier (21.6A)

prosthetic group (21.3C)

regulatory enzyme (21.6A)

repression (21.6B)

V_{max} (21.5B)

vitamin (21.3B)

Problems

Characteristics of Enzyme Catalysis (Section 21.1)

21.3 In what way does an enzyme increase the velocity of a reaction?

21.4 Is the following statement true or false? "Enzymes increase the rate at which a reaction reaches equilibrium but they do not change the position of equilibrium for the reactions they catalyze." Explain.

21.5 Compare the following for an uncatalyzed reaction and an enzyme-catalyzed reaction:
a. rate of formation of products
b. the position of equilibrium (equilibrium constant)

21.6 List three characteristics of enzymes that make them superior catalysts compared to their nonbiological laboratory counterparts.

Enzyme Proteins and Cofactors (Section 21.3)

21.7 Name three groups of enzyme cofactors.

21.8 What is the name given to cofactors that are permanently bound to an enzyme?

21.9 What is the relationship between coenzymes and water-soluble vitamins?

21.10 Of the water-soluble vitamins:
a. Which are coenzymes themselves?
b. Which are precursors from which coenzymes are synthesized in the body?

Mechanism of Enzyme Catalysis (Section 21.4)

21.11 Binding of substrate to the surface of an enzyme can involve a combination of ionic interactions, hydrogen bonding, and dispersion forces. By what type of interaction(s) might the side chains of the following amino acids bind a lecithin to form an enzyme-substrate complex?
a. phenylalanine   b. serine   c. glutamic acid
d. lysine   e. valine

21.12 Of the 20 protein-derived amino acids, the side chains of the following are most often involved as catalytic groups at the active site of an enzyme:
a. cys   b. his   c. ser
d. asp   e. glu   f. lys
What is the net charge on the side chain of each amino acid at pH 7.4?

Factors that Affect the Rate of Enzyme-Catalyzed Reactions (Section 21.5)

21.13 When storing and handling solutions of enzymes, the following precautions are commonly observed. Explain the importance of each.
a. Enzyme solutions are stored at a low temperature, usually at 0°C.
b. The pH of most enzyme solutions is kept near 7.0.
c. Enzyme solutions are prepared by dissolving the enzyme in water distilled from all-glass apparatus; ordinary tap water is never used.
d. When an enzyme is being dissolved in aqueous solution, it is dissolved with as little stirring as possible. Vigorous stirring or shaking is never used to hasten the process of dissolving.

21.14 Refer to Figure 21.5 and account for the fact that in region (a), an increase in substrate concentration results in a direct increase in reaction velocity. Also account for the fact that in region (b), an increase in substrate concentration has no effect on the reaction velocity.

21.15 What does it mean to say that an enzyme is "saturated"?

21.16 V_{max} for an enzyme-catalyzed reaction is 3 mg/mL/min. At what rate is product formed when only one-third of the enzyme molecules have substrate bound to them?

21.17 Following are the initial velocities of an enzyme-catalyzed reaction at six different substrate concentrations. Velocity is given in units of milligrams of product formed per milliliter of solution per minute.

[S]	Velocity (mg/mL/min)
0.5×10^{-3} M	1.5×10^{-3}
1.0×10^{-3} M	3.0×10^{-3}
1.5×10^{-3} M	4.4×10^{-3}
2.0×10^{-3} M	5.0×10^{-3}
3.0×10^{-3} M	5.8×10^{-3}
4.0×10^{-3} M	6.2×10^{-3}

a. Prepare a graph of initial velocity versus substrate concentration and from your graph, estimate V_{max}, $\frac{1}{2}V_{max}$, and K_m.

b. At what substrate concentration will the initial velocity be 2.5×10^{-3} mg/mL/min?

c. What percentage of enzyme is in the form of an enzyme-substrate complex when the initial velocity is 2.0×10^{-3} mg/mL/min?

21.18 Lysozyme catalyzes the hydrolysis of glycoside bonds of the polysaccharide components of certain types of bacterial cell walls. The catalytic activity of this enzyme is at a maximum at pH 5.0. The active site of lysozyme contains the side chains of asp and glu, and for maximum catalytic activity, the side chain of glu must be in the acid form and the side chain of asp in the conjugate base or deprotonated form. Explain why the velocity of lysozyme-catalyzed reactions decreases as the pH becomes more acidic than the optimal pH. Also explain why reaction velocity decreases as the pH becomes more basic than the optimal pH.

21.19 An enzyme isolated from yeast has an optimal temperature of 40°C. How would you account for the fact that the velocity of the reaction catalyzed by this enzyme:
a. decreases as the temperature is lowered to 0°C?
b. decreases as the temperature is increased above 40°C?

21.20 Following are equations for the ionization of the side chains of histidine and cysteine, along with the pK_a of each.

$$-CH_2-S-H \rightleftharpoons \quad -CH_2-S^- \quad + H^+ \qquad pK_a = 8.0$$

(this form required for maximum activity)

Assume that both histidine and cysteine are catalytic groups for a particular enzyme and that for maximum activity, the side chain of histidine must be in the protonated form and that of cysteine must be in the deprotonated form. Estimate the pH at which the catalytic activity of this enzyme is a maximum and sketch a pH-activity graph.

21.21 Various enzymes and their K_m values follow. Which enzyme has the highest affinity for substrate?

Enzyme	K_m
sucrase	$0.016M$
β-glucosidase	$6 \times 10^{-3}M$
enolase	$7 \times 10^{-5}M$
catalase	$1.17M$

21.22 How does a noncompetitive inhibitor slow an enzyme-catalyzed reaction even though the inhibitor does not bind to the active site?

21.23 Following is the initial velocity versus substrate concentration for an enzyme-catalyzed reaction and for the same reaction in the presence of an inhibitor. Inhibitor concentration is constant.

[S]	Velocity (mM/min)	Velocity (mM/min inhibited)
$1.0 \times 10^{-4}M$	0.14	0.088
$1.25 \times 10^{-4}M$	0.16	0.105
$1.67 \times 10^{-4}M$	0.19	0.13
$2.5 \times 10^{-4}M$	0.24	0.18
$5.0 \times 10^{-4}M$	0.31	0.26
$10.0 \times 10^{-4}M$	0.38	0.36

a. Prepare a graph of initial velocity versus substrate concentration for the uninhibited reaction.
b. On the same graph, plot initial velocity versus substrate concentration for the inhibited reaction.
c. Is this inhibition competitive or noncompetitive?
d. Can the effects of this inhibitor be overcome? If so, how?

Regulation of Enzyme Activity (Section 21.6)

21.24 Name two biological mechanisms for the regulation of enzyme activity.

21.25 What is meant by the term *regulatory enzyme*?

21.26 Is the inhibition of threonine deaminase by isoleucine (Section 21.6A) an example of competitive inhibition or noncompetitive inhibition? Explain.

21.27 What is meant by the term *enzyme induction*?

21.28 Below is illustrated a metabolic pathway in which substance A is converted into B and then B may be converted into substance D or substance F, depending on the needs of the cell at any particular time.

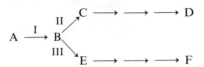

a. Which of these steps commits B to the synthesis of D?

b. Which step commits B to the synthesis of F?

c. Assume that D is a noncompetitive inhibitor of the enzyme-catalyzing reaction B → C; explain how this relationship can regulate the concentration of D.

d. Assume that both D and F are competitive inhibitors of the enzyme-catalyzed reaction A → B; show how this relationship can regulate the concentrations of B, C, and E.

Nucleic Acids and the Synthesis of Proteins

Nucleic acids are a third broad class of biopolymers which, like proteins and polysaccharides, are vital components of living materials. In this chapter we will look at the structure of nucleosides and nucleotides and the manner in which these small building blocks are bonded together to form nucleic acid molecules. Then, we will consider the three-dimensional structure of nucleic acids. Finally, we will examine the manner in which genetic information coded on deoxyribonucleic acids is expressed in protein biosynthesis.

22.1 The Components of Deoxyribonucleic Acid (DNA)

Controlled hydrolysis breaks DNA molecules into three components: (1) phosphoric acid; (2) 2-deoxy-D-ribose; and (3) heterocyclic aromatic amine bases. The heterocyclic bases fall into two classes: those derived from pyrimidine and those derived from purine. Following are structural formulas of the four most abundant bases found in DNA, along with formulas of their parent substances.

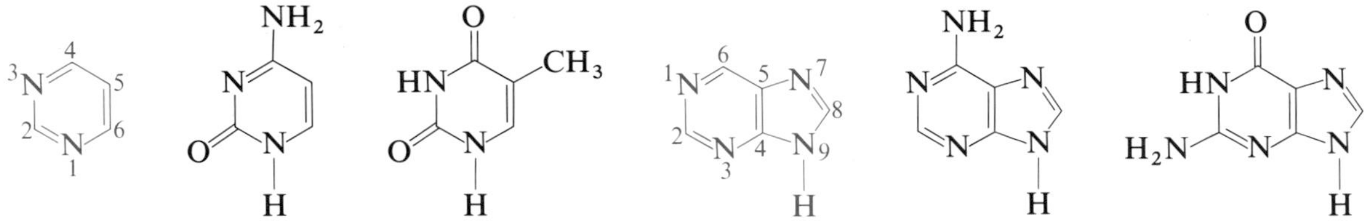

pyrimidine   cytosine (C)   thymine (T)   purine   adenine (A)   guanine (G)

22.2 Nucleosides

A **nucleoside** is a glycoside in which nitrogen-9 of a purine base or nitrogen-1 of a pyrimidine base is bonded to 2-deoxy-D-ribose by a β-N-glycoside bond (Section 15.2C). Two nucleosides, 2'-deoxyadenosine and 2'-deoxycytidine, are shown in Figure 22.1. The other two nucleosides found in DNA are 2'-deoxythymidine and 2'-deoxyguanosine.

Figure 22.1 Nucleosides; 2'-deoxyadenosine and 2'-deoxycytidine. Unprimed numbers are used to designate atoms of the purine and pyrimidine bases; primed numbers are used to designate atoms of 2-deoxy-D-ribose.

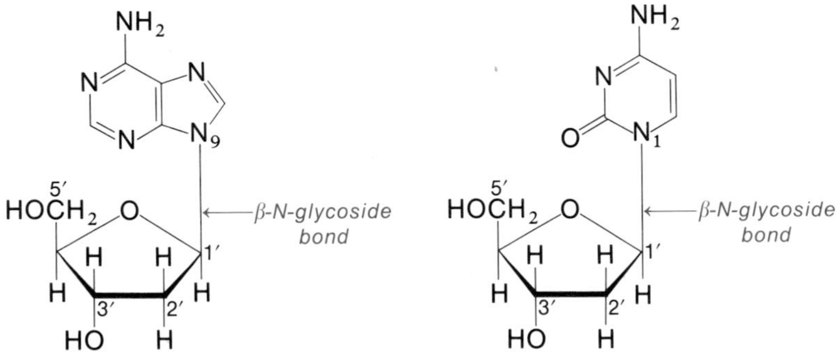

2'-deoxyadenosine   2'-deoxycytidine

22.3 Nucleotides

A **nucleotide** is a nucleoside monophosphate ester in which a molecule of phosphoric acid is esterified with a free hydroxyl group of 2-deoxy-D-ribose. Nucleoside monophosphates are illustrated in Figure 22.2 by the 5'-monophosphate ester of 2'-deoxyadenosine and the 3'-monophosphate ester of 2'-deoxycytidine. Note that at pH 7.0, the two protons of a monophosphate ester are ionized, giving this group a net charge of −2.

Mononucleotides are named as phosphate esters (for example, deoxyadenosine 5'-monophosphate); as acids (for example, deoxyadenylic acid); or by using four-letter abbreviations (for example, dAMP). In the four-letter abbreviations for mononucleotides, the letter **d** indicates 2-deoxy-D-ribose,

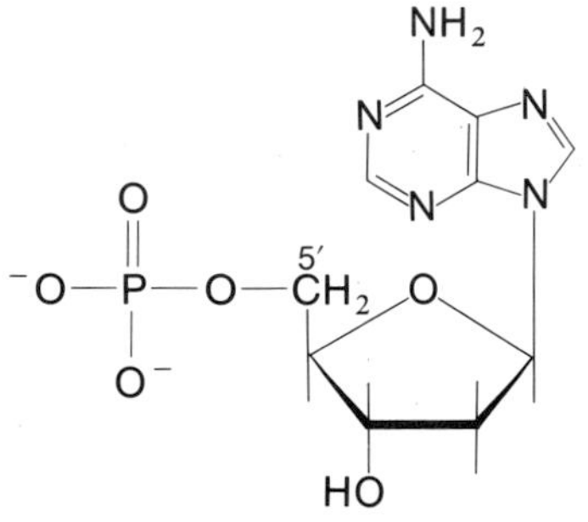

Figure 22.2 Nucleotides (nucleoside monophosphate esters).

2′-deoxyadenosine 5′-monophosphate
(dAMP)

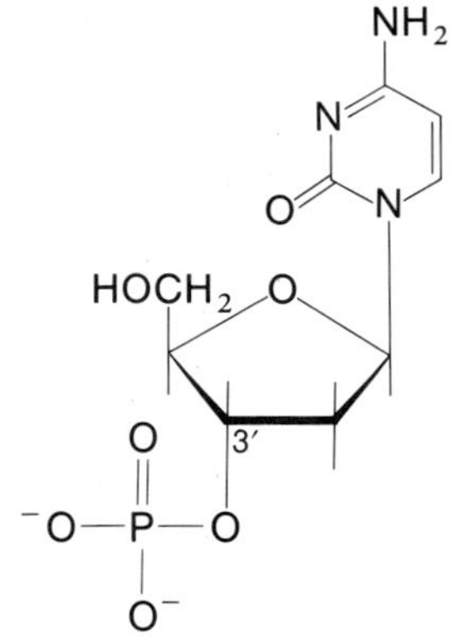

2′-deoxycytidine 3′-monophosphate
(dCMP)

Table 22.1 The major mononucleotides derived from DNA. Each is named as a monophosphate; as an acid; and by a four-letter abbreviation.

Monophosphate	Acid	Four-Letter Abbreviation
deoxyadenosine monophosphate	deoxyadenylic acid	dAMP
deoxyguanosine monophosphate	deoxyguanylic acid	dGMP
deoxycytidine monophosphate	deoxycytidylic acid	dCMP
deoxythymidine monophosphate	deoxythymidylic acid	dTMP

the second letter indicates the nucleoside, and the third and fourth letters indicate that the molecule is a monophosphate (MP) ester. Table 22.1 lists names of the major mononucleotides derived from DNA.

All nucleoside monophosphates may be further phosphorylated to form nucleoside diphosphates and nucleoside triphosphates. In the case of the diphosphates and triphosphates, the second and third phosphate groups are joined by anhydride bonds.

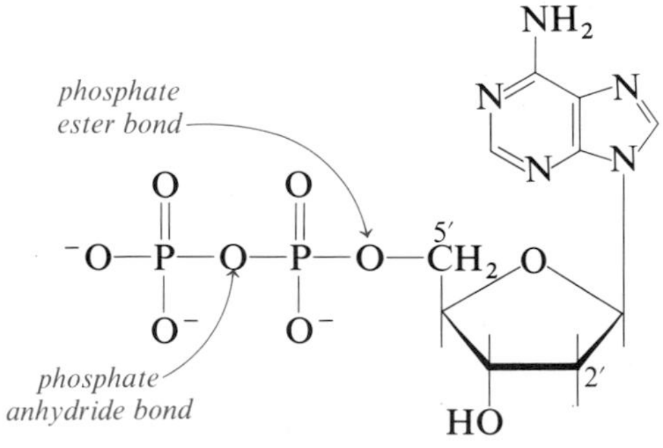

2′-deoxyadenosine 5′-diphosphate
(dADP)

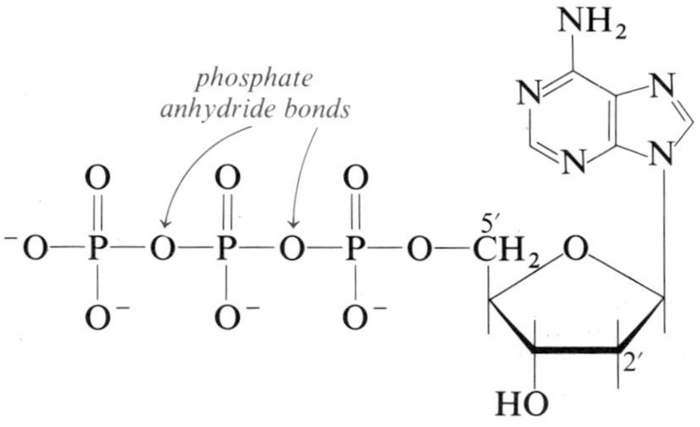

2′-deoxyadenosine 5′-triphosphate
(dATP)

At pH 7.0, all protons of diphosphate and triphosphate groups are fully ionized, giving them net charges of -3 and -4, respectively.

Example 22.1

Draw structural formulas for the following mononucleotides.

a. 2′-deoxycytidine 5′-monophosphate (dCMP)

b. 2′-deoxyguanosine 5′-triphosphate (dGTP)

Solution

a. Cytosine is joined by a β-N-glycoside bond between N-1 of cytosine and carbon-1 of the cyclic hemiacetal form of 2-deoxy-D-ribose. The 5′-hydroxyl of the pentose is bonded to phosphate by an ester bond.

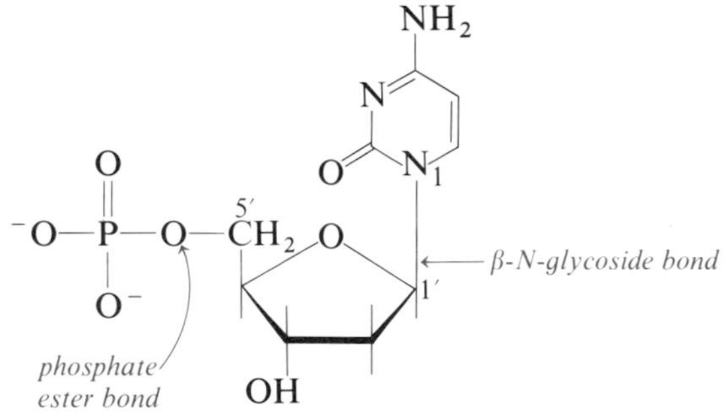

b. Guanine is joined by a β-N-glycoside bond between N-9 of guanine and carbon-1 of the cyclic hemiacetal form of 2-deoxy-D-ribose. The 5′-hydroxyl group of the pentose is joined to three phosphate groups by a combination of one ester bond and two anhydride bonds.

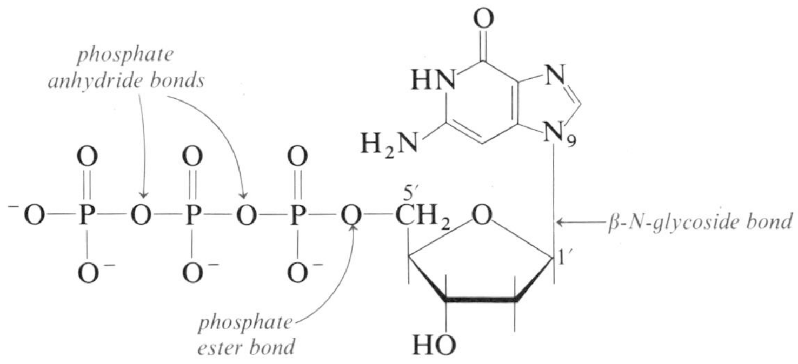

Problem 22.1

Draw structural formulas for:

a. dTPP   **b.** dGMP

22.4 The Structure of DNA

A. Primary Structure: The Covalent Backbone

Deoxyribonucleic acid (DNA) consists of a backbone of alternating units of deoxyribose and phosphate, in which the 3′-hydroxyl of one deoxyribose is joined to the 5′-hydroxyl of the next deoxyribose by a phosphodiester bond (Figure 22.3). This backbone is constant throughout the entire DNA molecule. A heterocyclic base, either adenine, guanine, thymine, or cytosine, is attached to each deoxyribose by a β-N-glycoside bond.

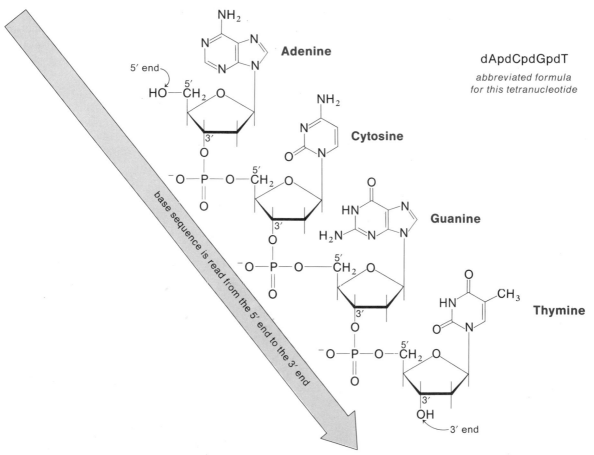

Figure 22.3 Partial structural formula of deoxyribonucleic acid (DNA), showing a tetranucleotide sequence. In the abbreviated sequence, the bases of the tetranucleotide are read from the 5′ end of the chain to the 3′ end, as indicated by the arrow.

The sequence of bases in a DNA molecule is indicated using single-letter abbreviations for each base, beginning from the free 5'-hydroxyl end of the chain. According to this convention, the base sequence of the section of DNA shown in Figure 22.3 is written as dApdCpdGpdT, where the letter **d** indicates that each nucleoside monomer is derived from deoxyribose and the letter **p** indicates a phosphodiester bond in the backbone of the molecule. Alternatively, the base sequence can be written as ACGT, a notation that emphasizes the order of the heterocyclic amine bases in the molecule.

Example 22.2

Draw a complete structural formula for a section of DNA containing the base sequence pdApdC.

Solution

The first letter in the shorthand formula of this dinucleotide is p, indicating that the 5'-hydroxyl is bonded to phosphate by an ester bond. The last letter is C, which shows that the 3'-hydroxyl is free, that is, it is not esterified with phosphate.

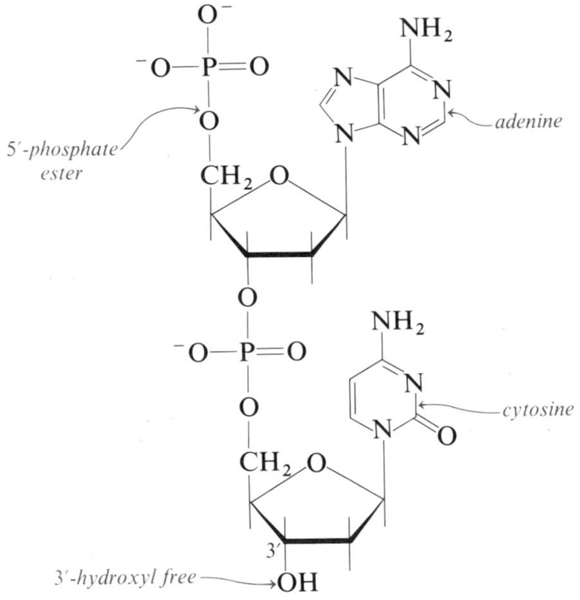

Problem 22.2

Draw a complete structural formula for a section of DNA containing the base sequence dCpdTpdGp.

B. Base Composition

By 1950, it was clear that DNA molecules consist of chains of alternating units of deoxyribose and phosphate linked by phosphodiester bonds, with a base

Table 22.2 Comparison of base composition (in mole percent) of DNA from several organisms.

Organism	A	G	C	T	$\dfrac{A}{T}$	$\dfrac{G}{C}$	$\dfrac{\text{Purines}}{\text{Pyrimidines}}$
human	30.9	19.9	19.8	29.4	1.05	1.00	1.04
sheep	29.3	21.4	21.0	28.3	1.03	1.02	1.03
sea urchin	32.8	17.7	17.3	32.1	1.02	1.02	1.02
marine crab	47.3	2.7	2.7	47.3	1.00	1.00	1.00
yeast	31.3	18.7	17.1	32.9	0.95	1.09	1.00
E. coli	24.7	26.0	26.0	23.6	1.04	1.01	1.03

attached to each deoxyribose by a β-N-glycoside bond. However, the precise sequence of bases along the chain of any particular DNA molecule was completely unknown. At one time, it was thought that the four major bases occurred in equal ratios and perhaps repeated in a regular pattern along the pentose-phosphate backbone of the molecule. However, more precise determinations of base sequence by Erwin Chargaff (Table 22.2) revealed that the bases do not occur in equal ratios. From consideration of data such as these, the following conclusions emerged.

1. The mole-percent base composition of DNA in any organism is the same in all cells and is characteristic of the organism.

2. The mole-percent of adenine equals that of thymine, and the mole-percent of guanine equals that of cytosine.

$$\%[\text{adenine}] = \%[\text{thymine}]$$
$$\%[\text{cytosine}] = \%[\text{guanine}]$$

3. The mole-percent of purine bases (A + G) equals that of pyrimidine bases (C + T).

$$\%[\text{purines}] = \%[\text{pyrimidines}]$$

C. The Molecular Dimensions of DNA

Additional information on the structure of DNA emerged from analysis of X-ray diffraction photographs of DNA fibers, taken by Rosalind Franklin and Maurice Wilkins. These photographs showed that DNA molecules are long, fairly straight, and not more than a dozen atoms thick. Furthermore, despite the fact that the base composition of DNA isolated from different organisms varies over a rather wide range, DNA molecules themselves are remarkably uniform in thickness. Herein lay one of the major problems to be solved. How could the molecular dimensions of DNA be so regular even though the relative percentages of the various bases differ so widely?

D. The Double Helix

With this accumulated information, the stage was set for the development of a hypothesis about DNA conformation. In 1953, F.H.C. Crick, a British physicist, and James D. Watson, an American biologist, postulated a precise model of the three-dimensional structure of DNA. The model not only accounted for many of the observed physical and chemical properties of DNA but also suggested a mechanism by which genetic information could be repeatedly and accurately replicated. Watson, Crick, and Wilkins shared the 1962 Nobel Prize in Physiology and Medicine for "their discoveries concerning the molecular structure of nucleic acids, and its significance for information transfer in living material."

The heart of the Watson-Crick model is the postulate that a molecule of DNA consists of two antiparallel polynucleotide strands coiled in a right-handed manner about the same axis to form a **double helix.** To account for the observed base ratios and the constant thickness of DNA, Watson and Crick postulated that purine and pyrimidine bases project inward toward the axis of the helix and are always paired in a very specific manner.

According to scale models, the dimensions of a thymine-adenine base pair are identical to those of a cytosine-guanine base pair, and the length of each pair is consistent with the thickness of a DNA strand (Figure 22.4). This fact gives rise to the principle of **complementarity.** In DNA, adenine is always paired by hydrogen bonding with thymine; hence, adenine and thymine are complementary bases. Similarly, guanine and cytosine are complementary bases. A significant fact arising from Watson and Crick's model building is that no other base pairing is consistent with the observed thickness of a DNA molecule. A pair of pyrimidine bases is too small to account for the observed thickness of a DNA molecule, while a pair of purine bases is too large. Thus, according to the Watson-Crick model, the repeating units in a double-stranded

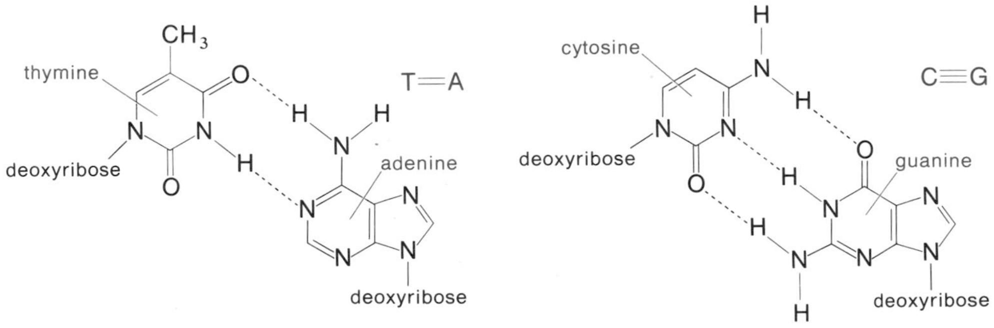

Figure 22.4 Hydrogen-bonded interaction between thymine and adenine and between cytosine and guanine. The first pair is abbreviated as T≡A (showing two hydrogen bonds) and the second pair is abbreviated as C≡G (showing three hydrogen bonds).

DNA molecule are not single bases of differing dimensions, but base pairs of identical dimensions.

To account for the periodicity observed from X-ray data, Watson and Crick postulated that base pairs are stacked one on top of the other with a distance of 3.4×10^{-8} cm between base pairs. Exactly ten base pairs are stacked in one complete turn of the helix. There is one complete turn of the helix every 34×10^{-8} cm (Figure 22.5).

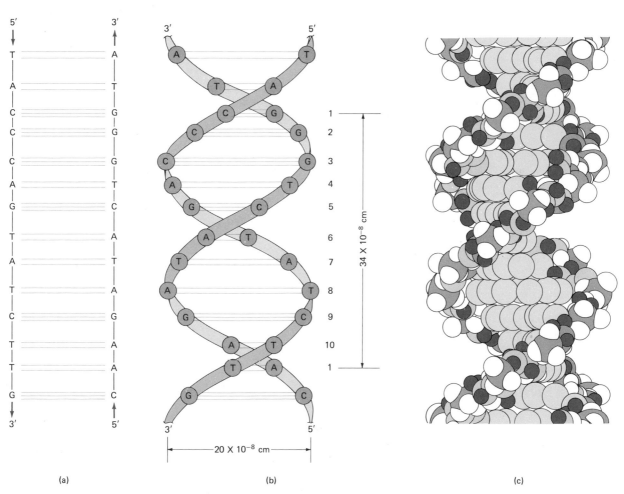

Figure 22.5 Abbreviated representation of the Watson-Crick double-helix model of DNA. On the left are shown two complementary antiparallel polynucleotide strands and the hydrogen bonds between complementary base pairs. In the middle, the strands are twisted in a double helix of thickness 20×10^{-8} cm and a repeat distance of 34×10^{-8} cm along the axis of the double helix. There are 10 base pairs per complete turn of the helix. On the right is a space-filling model of a section of DNA double helix.

Example 22.3

One strand of a DNA molecule has a base sequence of -ACTTGCCA-. Write the base sequence for the complementary strand.

Solution

Remember that base sequence is always written from the 5′ end of the strand to the 3′ end, that A is always paired by hydrogen bonding with its complement T, and that G is always paired by hydrogen bonding with its complement C. In double-stranded DNA, the strands run in opposite (antiparallel) directions so that the 5′ end of one strand is associated with the 3′ end of the other strand. Hydrogen bonds between base pairs are shown by dashed lines.

The complement of 5′-ACTTGCCA-3′ is shown under it in the solution. Writing this strand poses a communication problem. DNA strands are always written from the 5′ to 3′ end. Therefore, if the original strand is 5′-ACTTGCCA-3′, its complement is 5′-TGGCAAGT-3′.

Problem 22.3

Write the complementary base sequence for:

5′-C—C—G—T—A—C—G—A-3′

22.5 DNA Replication

At the time Watson and Crick proposed a model for the conformation of DNA, biologists had already amassed a great deal of evidence that DNA is, in fact, the hereditary or genetic material. Detailed studies revealed that during cell division, there is an exact duplication of DNA. The challenge posed to molecular biologists was: How does the genetic material duplicate itself with such unerring fidelity?

One of the exciting things about the double-helix model is that it immediately suggested how DNA might produce an exact copy of itself. The double helix consists of two parts, one the complement of the other. If the two strands separate and each serves as a template for the construction of its own complement, then each new double strand will be an exact replica of the original DNA. Because each new double-stranded DNA molecule contains one strand from the parent molecule and one newly synthesized **(daughter)** strand, the process is called **semiconservative replication** (Figure 22.6).

DNA replication is considerably more complicated than shown in Figure 22.6. In *E. coli*, replication is thought to proceed by four major steps. While there are variations from species to species, replication in other organisms follows a similar process.

Figure 22.6 Schematic diagram of semiconservative replication. The double helix uncoils, and each chain of the parent serves as a template for the synthesis of its complement. Each "daughter" DNA contains one strand from the original DNA and one newly synthesized strand.

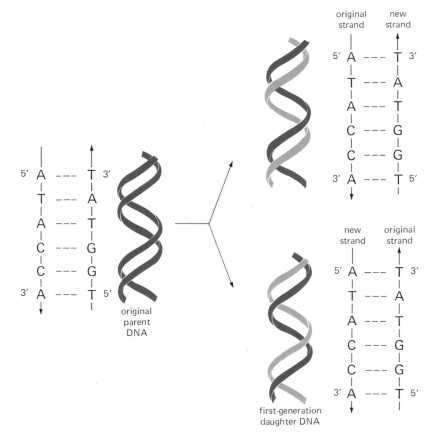

original parent DNA

first-generation daughter DNA

1. Initiation of replication. Replication starts at a specific point on a chromosome, where unwinding proteins catalyze uncoiling of the DNA helix. In the process of unwinding, hydrogen bonds between complementary base pairs are broken and purine and pyrimidine bases in the center of double-stranded DNA are exposed (Figure 22.7). The point of unwinding is called a **replication fork.**

Figure 22.7 Unwinding of DNA, creating a replication fork.

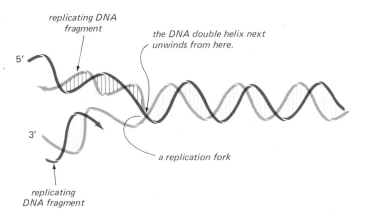

replicating DNA fragment

the DNA double helix next unwinds from here.

a replication fork

replicating DNA fragment

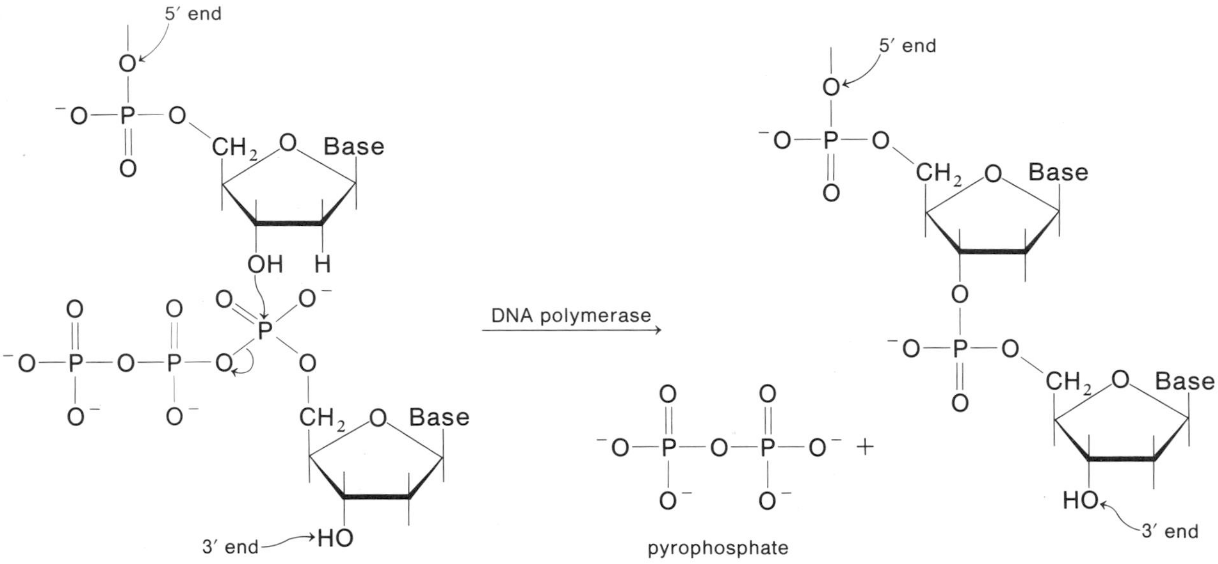

Figure 22.8 Formation of a phosphodiester bond and elongation of a daughter strand by one nucleotide.

2. Formation of DNA segments. In step 2, DNA replication proceeds along both branches of the exposed DNA template from the 3′ end toward the 5′ end. Because the two unwound DNA strands run in opposite directions, DNA synthesis proceeds toward the replication fork on one strand and away from the replication fork on the other strand. Addition of mononucleotides to growing DNA daughter strands is catalyzed by DNA polymerase, an enzyme that recognizes the 3′-hydroxyl end of the growing daughter strand and positions the proper complementary deoxynucleoside triphosphate to pair by hydrogen bonding with the next base on the template. Finally, the 3′-hydroxyl of the daughter strand displaces pyrophosphate from the 5′ end of the next nucleoside triphosphate to form a phosphodiester bond, and the daughter strand becomes elongated by one nucleotide (Figure 22.8).

3. Creation of a new replication fork and continuation of DNA synthesis. In step 3, a new section of DNA is unwound, creating a second replication fork, and the replication process is repeated. According to this mechanism, one daughter strand is synthesized as a continuous strand from the 3′ end of the DNA template toward the first replication fork and then toward each newly created replication fork as further sections of double helix are unwound. The second daughter strand is synthesized as a series of fragments, each as long as the distance from one replication fork to the next. The breaks in the second daughter strand created by the replication forks are called *nicks* (Figure 22.9) and the DNA fragments separated by nicks are called **Okazaki**

Figure 22.9 Synthesis of DNA daughter strands toward and away from a replication fork.

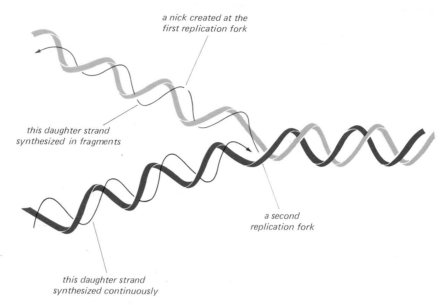

a nick created at the first replication fork

this daughter strand synthesized in fragments

a second replication fork

this daughter strand synthesized continuously

fragments. The isolation of Okazaki fragments is evidence that replication of DNA is not a continuous process.

4. Completion of the DNA strand. In step 4, an enzyme called DNA ligase closes the nicks to form a completed daughter strand.

22.6 Ribonucleic Acid (RNA)

Ribonucleic acids (RNA) are similar to deoxyribonucleic acids in that they, too, consist of long, unbranched chains of nucleotides joined by phosphodiester bonds between the 3'-hydroxyl of one pentose and the 5'-hydroxyl of the next. Thus, their structure is much the same as that of DNA shown in Figure 22.3. However, there are three major structural differences between RNA and DNA: (1) the pentose unit in RNA is D-ribose rather than 2-deoxy-D-ribose; (2) the pyrimidine bases in RNA are uracil and cytosine rather than thymine and cytosine; and (3) RNA is single-stranded rather than double-stranded. Following are structural formulas of β-D-ribose and uracil.

HOCH$_2$ O OH

HO OH

β-D-ribose

uracil
(U)

RNA is distributed throughout the cell; it is present in the nucleus, the cytoplasm, and in subcellular particles called mitochondria. Furthermore, cells contain three types of RNA: ribosomal RNA, transfer RNA, and messenger RNA. These three types of RNA differ in molecular weight and, as their names imply, they perform different functions within the cell.

A. Ribosomal RNA

Ribosomal RNA (rRNA) molecules have molecular weights of 0.5–1.0 million and comprise up to 85–90% of total cellular ribonucleic acid. rRNA is found in the cytoplasm in subcellular particles called **ribosomes,** which contain about 60% rRNA and 40% protein. Complete ribosomes (referred to as 70S ribosomes) can be dissociated into two subunits of unequal size, known as 50S subunits and 30S subunits (Figure 22.10). The designation S stands for Svedberg units. Values of S are derived from rates of sedimentation during centrifugation and are used to estimate molecular weight and compactness of ribosomal particles. A large value of S indicates a high molecular weight; conversely, a small value of S indicates a low molecular weight. The 50S ribosomal subunit is about twice the size of the 30S subunit and further dissociates into 23S and 5S subunits and approximately 30 protein molecules. The smaller 30S subunit dissociates into a single 16S subunit and about 20 different protein molecules.

Many of the proteins bound to ribosomes have a high percentage of lysine and arginine. At the pH of cells, the side chains of these amino acids have net positive charges. It is likely that interactions between positively charged amino acid side chains and negatively charged phosphate groups of RNA are an important factor stabilizing larger ribosomal particles.

Figure 22.10 Dissociation of a complete ribosome into subunits.

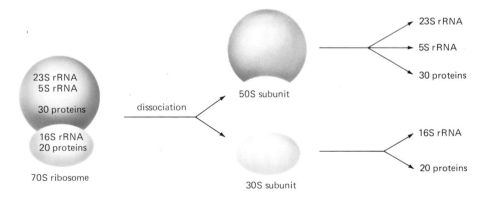

B. Transfer RNA

Transfer RNA (tRNA) molecules have the lowest molecular weight of all nucleic acids. They consist of from 75 to 80 nucleotides in a single chain that is folded into a three-dimensional structure stabilized by hydrogen bonding

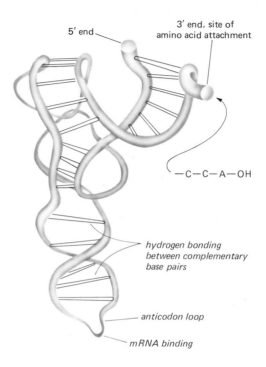

Figure 22.11 The three-dimensional shape of yeast phenylalanine tRNA.

5′ end

3′ end, site of amino acid attachment

—C—C—A—OH

hydrogen bonding between complementary base pairs

anticodon loop

mRNA binding

between complementary base pairs. Nearly all tRNAs have G at the 5′ end of the chain and the sequence CCA at the 3′ end of the chain. The three-dimensional shapes of a number of tRNAs have been determined. Shown in Figure 22.11 is a representation of the three-dimensional shape of yeast phenylalanine tRNA.

The function of tRNA is to carry amino acids to the sites of protein synthesis on ribosomes. For transportation to a ribosome, an amino acid is joined to the 3′ end of its specific tRNA by an ester bond formed between the α-carboxyl group of the amino acid and the 3′-hydroxyl group of ribose. An amino acid thus bound to tRNA is said to be "activated" because it is prepared for the synthesis of a peptide bond.

C. Messenger RNA

Messenger RNA (mRNA) is present in cells in relatively small amounts and is very short-lived. It is single-stranded, has an average molecular weight of several hundred thousand, and has a base composition much like that of the DNA of the organism from which it is isolated. The name *messenger* RNA derives from the fact that this type of RNA is made in the cell nucleus on a DNA template and carries coded genetic information to the ribosomes for the synthesis of new proteins (Section 22.9).

22.7 Transcription of Genetic Information: RNA Biosynthesis

RNA is synthesized from DNA in a manner similar to the replication of DNA. Double-stranded DNA is unwound and a complementary strand of RNA is synthesized along each strand of the DNA template, beginning from the 3′ end. The synthesis of RNA from a DNA template is called **transcription,** a term that refers to the fact that genetic information contained in a sequence of bases of DNA is transcribed into a complementary sequence of bases in RNA.

Example 22.4

Following is a base sequence from a portion of DNA. Write the sequence of bases of the RNA synthesized using this section of DNA as a template.

3′-A—G—C—C—A—T—G—T—G—A—C—C-5′

Solution

RNA synthesis begins at the 3′-end of the DNA template and proceeds toward the 5′-end. The complementary RNA strand is formed using the bases C, G, A, and U. Uracil (U) is the complement of adenine (A) on the DNA template.

direction of strand ←

3′-A—G—C—C—A—T—G—T—G—A—C—C ← *DNA template*

5′-U—C—G—G—U—A—C—A—C—U—G—G ← *daughter RNA*

→ *direction of strand*

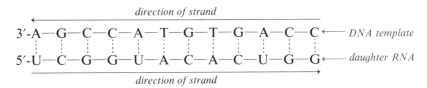

To write the base sequence of the DNA template, start from the 5′ end. If the DNA template is

5′-C—C—A—G—T—G—T—A—C—C—G—A-3′

then the complementary RNA strand is

5′-U—C—G—G—U—A—C—A—C—U—G—G-3′

Problem 22.4   Following is a base sequence from a portion of DNA. Write the sequence of bases in the RNA synthesized using this section of DNA as a template.

5′-T—C—G—G—T—A—C—A—C—T—G—G-3′

22.8 The Genetic Code

A. The Triplet Nature of the Code

It was clear by the late 1950s that the sequence of bases in DNA molecules constitutes a store of genetic information, and that the sequence of bases serves to direct the synthesis of RNA and of proteins. However, the statement that the sequence of bases in DNA directs the synthesis of proteins presented the following problem: How can a molecule containing only four variable units (adenine, cytosine, guanine, and thymine) direct the synthesis of molecules containing up to twenty variable units (the 20 common protein-derived amino acids)? How can a four-letter alphabet code for the order of letters in the 20-letter alphabet as occurs in proteins?

An obvious answer is that it is not one base but a combination of bases that codes for each amino acid. If the code consists of nucleotide pairs, there are $4^2 = 16$ combinations, a more extensive code, but still not extensive enough to code for 20 amino acids. If the code consists of nucleotides in groups of three, there are $4^3 = 64$ possible combinations, more than enough to specify the primary sequence of a protein. This appears to be a very simple solution to a system that must have taken eons of evolutionary trial-and-error to develop. Yet, there is now proof from comparison of gene (DNA) and protein (amino acid) sequences that nature does indeed use this simple 3-letter or triplet code to store genetic information. A triplet of nucleotides is called a **codon.**

B. Deciphering the Genetic Code

The next question is: Which triplets code for which amino acids? In 1961, Marshall Nirenberg provided a simple experimental approach to the problem. It was based on the observation that synthetic polynucleotides direct polypeptide synthesis in much the same manner as mRNAs direct polypeptide synthesis. Nirenberg incubated ribosomes, amino acids, tRNAs, and the appropriate protein-synthesizing enzymes. With only these components, there was no polypeptide synthesis. However, when he added synthetic polyuridylic acid (poly U), a polypeptide of high molecular weight was formed. The exciting result of this experiment was that the polypeptide synthesized contained only phenylalanine. With this discovery, the first element of the genetic code had been deciphered: the triplet UUU codes for phenylalanine.

Table 22.3 The genetic code: mRNA codons and the amino acid whose incorporation each codon directs.

UUU	Phe	UCU	Ser	UAU	Tyr	UGU	Cys
UUC	Phe	UCC	Ser	UAC	Tyr	UGC	Cys
UUA	Leu	UCA	Ser	UAA	Stop	UGA	Stop
UUG	Leu	UCG	Ser	UAG	Stop	UGG	Trp
CUU	Leu	CCU	Pro	CAU	His	CGU	Arg
CUC	Leu	CCC	Pro	CAC	His	CGC	Arg
CUA	Leu	CCA	Pro	CAA	Gln	CGA	Arg
CUG	Leu	CCG	Pro	CAG	Gln	CGG	Arg
AUU	Ile	ACU	Thr	AAU	Asn	AGU	Ser
AUC	Ile	ACC	Thr	AAC	Asn	AGC	Ser
AUA	Ile	ACA	Thr	AAA	Lys	AGA	Arg
AUG	Met	ACG	Thr	AAG	Lys	AGG	Arg
GUU	Val	GCU	Ala	GAU	Asp	GGU	Gly
GUC	Val	GCC	Ala	GAC	Asp	GGC	Gly
GUA	Val	GCA	Ala	GAA	Glu	GGA	Gly
GUG	Val	GCG	Ala	GAG	Glu	GGG	Gly

This same type of experiment was carried out with different polyribonucleotides. It was found that polyadenylic acid (poly A) leads to the synthesis of polylysine and that polycytidylic acid (poly C) leads to the synthesis of polyproline.

codon on mRNA	amino acid
UUU	phenylalanine
AAA	lysine
CCC	proline

By 1966, all 64 codons had been deciphered (Table 22.3).

C. Properties of the Genetic Code

A number of features of the genetic code are evident from Table 22.3.

1. Only 61 triplets code for amino acids. The remaining three triplets (UAA, UAG, and UGA) are signals for chain terminations, that is, they are signals to the protein-synthesizing machinery of the cell that the

primary structure of the protein is complete. The three chain termination triplets are indicated in Table 22.3 by *Stop*.

2. The code is *degenerate*, which means that many amino acids are coded for by more than one triplet. If you count the number of triplets coding for each amino acid, you will find that only methionine and tryptophan are coded for by just one triplet. Leucine, serine, and arginine are coded for by six triplets, and the remaining 15 amino acids are coded for by two, three, or four triplets.

3. For the 15 amino acids coded for by two, three, or four triplets, the degeneracy is only in the last base of the triplet. In other words, in the codons for these 15 amino acids, it is only the third letter of the code that varies. For example, glycine is coded by the triplets GGA, GGG, GGC, and GGU.

4. Finally, there is no ambiguity in the code. Each triplet codes for one, and only one, amino acid.

We must ask one last question about the genetic code: Is the code universal—is it the same for all organisms? Every bit of experimental evidence available today from the study of viruses, bacteria, and higher animals, including humans, indicates that the code is universal. Furthermore, the fact that it is the same in all these organisms means that it has been the same over billions of years of evolution.

Example 22.5

During transcription, a portion of mRNA is synthesized with the following base sequence.

$$5'\text{-AUG-GUA-CCA-CAU-UUG-UGA-}3'$$

a. Write the base sequence of the DNA from which this portion of mRNA was synthesized.

b. Write the primary structure of the polypeptide coded for by this section of mRNA.

Solution

During transcription, mRNA is synthesized from a DNA strand beginning from the 3′ end of the DNA template. The DNA strand must be complementary to the newly synthesized mRNA strand.

b. The sequence of amino acids is shown below the mRNA strand.

$$5'\text{-AUG-GUA-CCA-CAU-UUG-UGA-}3'$$
$$\text{met}\quad\text{val}\quad\text{pro}\quad\text{his}\quad\text{leu}\quad\text{stop}$$

The codon UGA codes for termination of the growing polypeptide chain; therefore, the sequence given in this problem codes for a pentapeptide only.

Problem 22.5

The following section of DNA codes for oxytocin, a polypeptide hormone.

*mRNA synthesis
begins here*

$$\rightarrow 3'\text{-ACG-ATA-TAA-GTT-TTA-ACG-GGA-GAA-CCA-ACT-}5'$$

a. Write the base sequence for the mRNA synthesized from this section of DNA.

b. Given the sequence of bases in part (a), write the amino acid sequence of oxytocin.

22.9 Translation of Genetic Information: Biosynthesis of Polypeptides

The biosynthesis of polypeptides is usually described in terms of three major processes: initiation of the polypeptide chain, elongation of the polypeptide chain, and termination of the completed polypeptide chain. These processes, along with the substances required for each, are summarized in Table 22.4.

Table 22.4 Major processes in polypeptide biosynthesis.

Process	Substances Required
initiation	tRNA carrying N-formylmethionine, mRNA, 30S and 50S ribosomal subunits, GTP, protein-initiating factors
elongation	amino acyl tRNAs, protein elongation factors, GTP
termination	termination codon on mRNA, protein termination factors

A. Initiation of Polypeptide Biosynthesis

In bacteria, all polypeptide chains are initiated with the amino acid N-formyl-methionine (f Met).

$$H-\overset{\overset{\displaystyle O}{\|}}{C}-NH-\overset{\overset{\displaystyle }{|}}{\underset{\underset{\displaystyle CH_2-S-CH_3}{|}}{\underset{\underset{\displaystyle }{CH_2}}{CH}}}-\overset{\overset{\displaystyle O}{\|}}{C}-O^-$$

formyl group

N-formylmethionine
(f Met)

Many bacterial polypeptides do have N-formylmethionine as the N-terminal amino acid. However, for most bacterial proteins, N-formylmethionine, by itself or with several other amino acids at the N-terminal end of the polypeptide chain, is cleaved to give the native polypeptide. N-Formylmethionine is bound to a specific tRNA molecule and given the symbol tRNA$_{fMet}$.

The first step in the initiation process is alignment of mRNA on a 30S ribosomal subunit so that the initiating codon is located at a specific site on the ribosome called the **P site.** The initiating codon is most commonly AUG, the one for methionine. Next, tRNA carrying N-formylmethionine (f Met) binds to the initiating codon and this complex, in turn, binds a 50S ribosomal subunit to give a unit called an **initiation complex** (Figure 22.12).

B. Elongation of the Polypeptide Chain

Elongation of a polypeptide chain consists of three steps which are repeated over and over until the entire polypeptide chain is synthesized. In the first step, a "charged" tRNA (one carrying an amino acid esterified at the 3' end of a

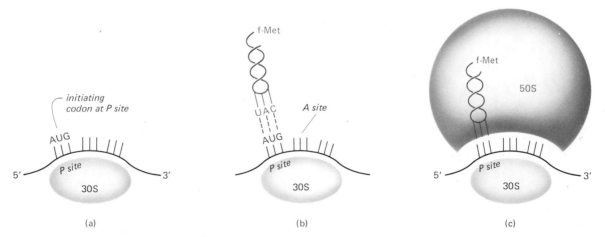

Figure 22.12 Formation of an initiation complex. (a) Alignment of mRNA on a 30S ribosomal subunit so that AUG, the initiating codon, is located at the P site. (b) Binding of tRNA carrying N-formylmethionine (fMet) to the initiating codon. (c) Association of the 50S ribosomal subunit to give an initiation complex.

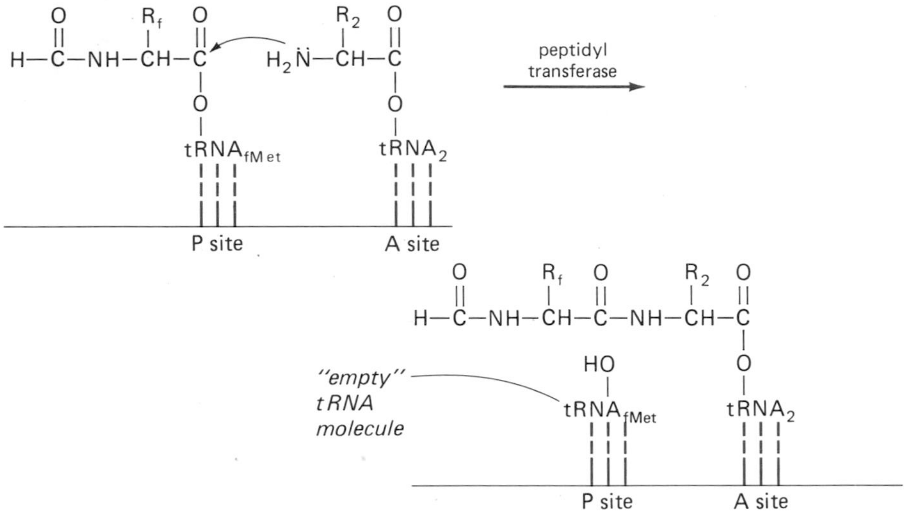

Figure 22.13 Formation of a peptide bond between the amino acid at the P site and the amino acid at the A site.

tRNA chain) binds to the **A site** of the initiating complex (see Figure 22.13). The second step is formation of a peptide bond between the carboxyl group of the tRNA-bound amino acid at the P site and the amino group of the tRNA-bound amino acid at the A site. Peptide bond formation is catalyzed by the enzyme peptidyl transferase. After formation of a new peptide bond, the tRNA bound to the P site is "empty" and the growing polypeptide chain is now attached to the tRNA bound to the A site.

The third step in the elongation cycle involves release of the "empty" tRNA from the P site and translocation of the growing polypeptide chain from the A site to the P site.

The three steps in the elongation cycle are shown schematically in Figure 22.14 for the synthesis of the tripeptide fmet-arg-phe from fmet-arg.

C. Termination of Polypeptide Biosynthesis

Polypeptide synthesis continues through the chain elongation cycle until the ribosome complex reaches a stop codon (UAA, UAG, or UGA) on mRNA. There, a specific protein called a **termination factor** binds to the stop codon and catalyzes hydrolysis of the completed polypeptide chain from tRNA. The "empty" ribosome then dissociates, ready for binding to another strand of mRNA and fMet-tRNA to form another initiation complex.

Figure 22.15 shows several ribosome complexes moving along a single strand of mRNA and illustrates the fact that several identical polypeptide chains can be synthesized simultaneously from a single mRNA molecule. Figure 22.15 also illustrates the fact that as a polypeptide chain grows, it extends out from the ribosome into the cytoplasm of the cell and folds spontaneously into its native three-dimensional conformation.

Figure 22.14 Chain elongation. (a) The growing peptide chain bound to arg-tRNA is aligned at the P site and phe-tRNA is aligned at the A site. Peptidyl transferase catalyzes peptide bond formation (b) between the carboxyl group of arginine and the amino group of phenylalanine and the growing polypeptide chain is transferred to phe-tRNA. As a result of translocation (c), phe-tRNA is moved to the P site and the next amino acid in the primary sequence, lys-tRNA, is aligned at the A site.

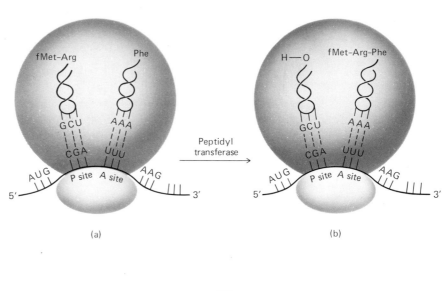

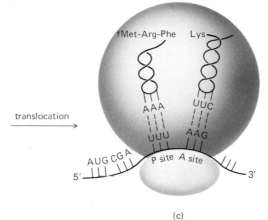

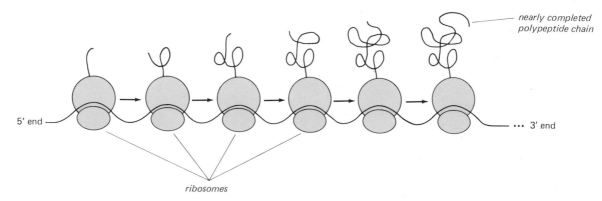

Figure 22.15 Simultaneous elongation of several identical polypeptide chains on a single strand of mRNA. The growing polypeptide chains spontaneously assume the native three-dimensional conformation.

22.10 Inhibition of Protein Synthesis and the Action of Antibiotics

Several widely used antibiotics, including tetracycline, streptomycin, chloramphenicol, and puromycin (Figure 22.16), act by inhibition of protein synthesis in bacteria at the ribosomal level. Although the general process of protein synthesis described in the previous section operates universally, there are differences in detail between the processes in bacteria and animals. Some of these differences, particularly those dealing with chain initiation and chain elongation, are quite marked. Because of these differences, many antibiotics

Figure 22.16 Structural formulas of four antibiotics and their effects on protein synthesis in bacteria.

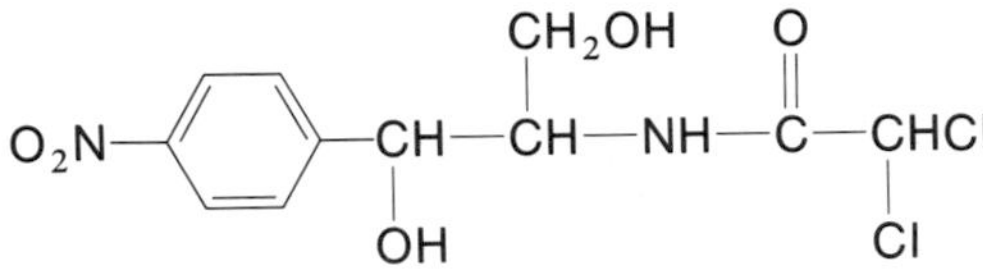

chloramphenicol (binds to the
A site and inhibits binding of
charged tRNAs)

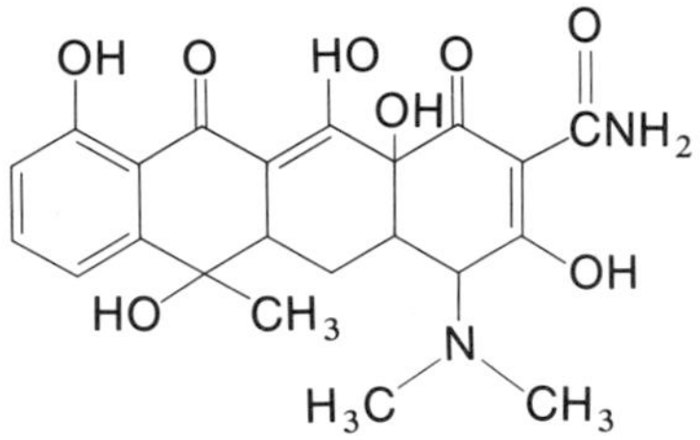

tetracycline (inhibits binding
of charged tRNAs to the 30S
ribosomal subunit)

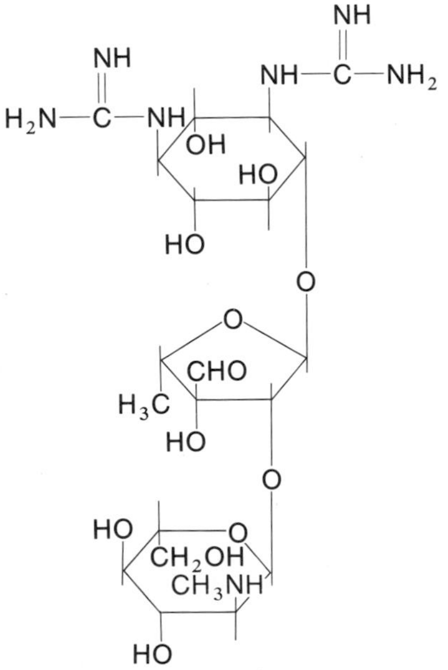

streptomycin (binds to proteins
of the 30S ribosomal subunit and
causes misreading of mRNA code)

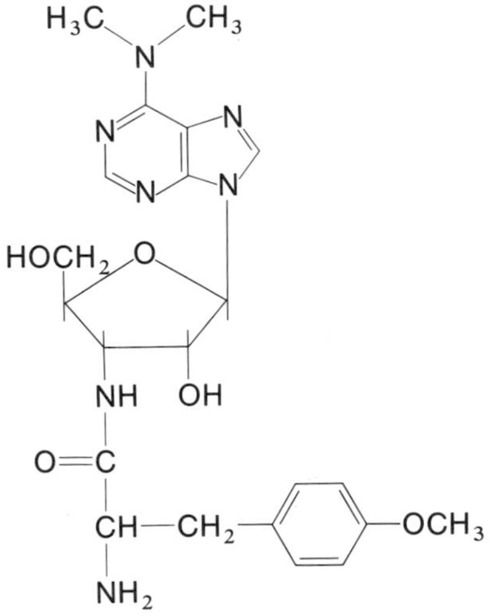

puromycin (is inserted in the
growing polypeptide chain and
causes premature termination
of polypeptide synthesis)

are able to inhibit protein synthesis in bacteria but, at the same time, have little or no effect on host cells.

Chloramphenicol is a broad-spectrum antibiotic. However, in some persons it causes serious, often toxic side effects. For this reason, the use of chloramphenicol is restricted largely to treatment of acute infections for which other antibiotics are ineffective or for medical reasons cannot be used.

Puromycin is a structural analog of a charged tRNA molecule (one bearing an amino acid esterified to the 3′-hydroxyl of the terminal nucleotide) and binds to the A site during the chain elongation phase of polypeptide synthesis. There, the enzyme peptidyl transferase catalyzes formation of a peptide bond between the growing polypeptide chain and the amino group of puromycin, at which point, further chain elongation ceases. Thus, puromycin causes premature termination of polypeptide synthesis.

Streptomycin binds with proteins of the 30S ribosomal subunit and interferes with interactions between mRNA codons and tRNAs. This interference gives rise to errors in reading the mRNA code and results in insertion of incorrect amino acids in the growing polypeptide chain.

Chloramphenicol binds specifically to the A site of the 50S ribosomal subunit and thereby prevents charged tRNAs from binding to the A site. Tetracycline prevents binding of charged tRNAs to the 30S ribosomal subunit.

22.11 Mutation

A **mutation** is any change or alteration in the sequence of heterocyclic aromatic amine bases on DNA molecules. Mutations can occur naturally during DNA replication, or they can be induced by environmental factors such as ionizing radiation or specific chemicals. At the molecular level, mutation in its simplest form is a change in a single base pair in a DNA molecule. Such mutations are called **point mutations** and can be divided into three groups: (1) base pair substitutions, (2) base-pair insertions, and (3) base-pair deletions. A point mutation in DNA is transmitted to mRNA during transcription and is ultimately expressed in the form of a specific protein with an altered primary structure.

A. Base-Pair Substitution

A point mutation leading to the substitution of one base pair for another in a section of double-stranded DNA affects only one codon. Figure 22.17 shows an example of base-pair substitution.

Figure 22.17 A portion of DNA (a) before and (b) after base-pair substitution.

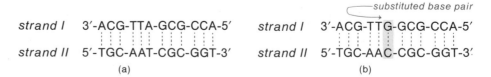

strand I   3′-ACG-TTA-GCG-CCA-5′

strand II   5′-TGC-AAT-CGC-GGT-3′

(a)

substituted base pair

strand I   3′-ACG-TTG-GCG-CCA-5′

strand II   5′-TGC-AAC-CGC-GGT-3′

(b)

Example 22.6

Refer to the DNA sequences given in Figure 22.17.

a. Write the sequence of bases in mRNA transcribed from the 3′ end of strand I of the original DNA. Also write the sequence of amino acids coded for by this section of mRNA. Remember that protein synthesis begins from the 5′ end of mRNA.

b. Do the same for strand I after mutation and compare the two amino acid sequences.

Solution

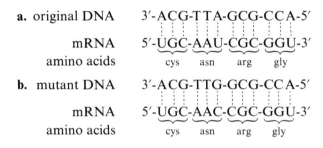

a. original DNA   3′-ACG-TTA-GCG-CCA-5′

        mRNA   5′-UGC-AAU-CGC-GGU-3′

        amino acids   cys   asn   arg   gly

b. mutant DNA   3′-ACG-TTG-GCG-CCA-5′

        mRNA   5′-UGC-AAC-CGC-GGU-3′

        amino acids   cys   asn   arg   gly

This point mutation does not cause any change in the amino acid sequence. Both AAU and AAC code for asparagine.

Problem 22.6

Following is a segment of DNA showing five triplets.

        3′-GAC-TCC-GAT-CGC-GAT-5′

a. Write the sequence of bases in the mRNA transcribed from the 3′ end of this section of DNA; write the amino acid sequence coded for by the mRNA.

b. Assume a point mutation changes the fifth base from the 3′ end of the DNA strand from C to T. Write the mRNA sequence transcribed after this mutation and the amino acid sequence it codes for.

B. Base-Pair Insertion and Deletion

An insertion mutation involves addition of one or more base pairs to a DNA strand and a deletion mutation involves removal of one or more base pairs. Both change the reading frame of all bases after the mutation point; thus, all amino acids in the primary structure following the mutation point are affected. Shown in Figure 22.18 is a section of (double-stranded) DNA and the same section after insertion of one base pair.

Figure 22.18 Insertion mutation. A section of double-stranded DNA (a) before and (b) after base-pair insertion.

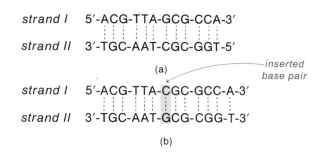

strand I   5′-ACG-TTA-GCG-CCA-3′

strand II   3′-TGC-AAT-CGC-GGT-5′

(a)

strand I   5′-ACG-TTA-CGC-GCC-A-3′

strand II   3′-TGC-AAT-GCG-CGG-T-3′

(b)

inserted base pair

Example 22.7

Following is a section of a single DNA strand showing four triplets, the section of mRNA transcribed from the 3′ end of the DNA template, and the amino acid sequence coded for by this section of mRNA. Insert A after TTA in the DNA template, write the sequence of bases in the mRNA produced by transcription from the mutant DNA, and write the amino acid sequence the new mRNA codes for.

DNA template: 3′-TTA-GGT-TGT-TGG-5′

mRNA: 5′-AAU-CCA-ACA-ACC-3′

amino acids: asn   pro   thr   thr

Solution

Insertion of A at the position indicated gives the following DNA template, mRNA, and amino acid sequence. Except for the first amino acid, the entire sequence is different.

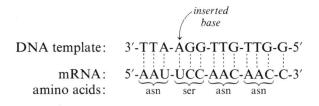

inserted base

DNA template: 3′-TTA-AGG-TTG-TTG-G-5′

mRNA: 5′-AAU-UCC-AAC-AAC-C-3′

amino acids: asn   ser   asn   asn

Problem 22.7

Given the following DNA strand sequence:

3′-GAT-GGG-ATG-TCT-5′

a. Write the base sequence of the complementary strand of mRNA and the amino acid sequence it codes for.

b. Delete T from GAT in the DNA strand and write the new mRNA sequence and the amino acid sequence it codes for.

Key Terms and Concepts

A site (22.9B)	Okazaki fragments (22.5)
base composition of DNA (22.4B)	point mutations (22.11)
codon (22.8B)	polypeptide biosynthesis (22.9)
complementarity (22.4D)	P site (22.9A)
daughter strand (22.5)	replication (22.5)
deoxyribonucleic acid (DNA) (22.4)	replication fork (22.5)
double helix (22.4D)	ribonucleic acid (RNA) (22.6)
genetic code (22.8)	ribosomal RNA (rRNA) (22.6A)
initiation complex (22.9A)	ribosome (22.6A)
messenger RNA (mRNA) (22.6C)	semiconservative replication (22.5)
mutation (22.11)	termination factor (22.9C)
nucleoside (22.2)	transcription (22.7)
nucleotide (22.3)	transfer RNA (tRNA) (22.6B)

Problems

Nucleosides and Nucleotides (Sections 22.2 and 22.3)

22.8 Examine the structure of purine. Would you predict this molecule to be planar or puckered? To exist as a number of interconvertible conformations (as in the case of cyclohexane) or to be rigid and inflexible? Explain the basis for your answer.

22.9 An important drug in the chemotherapy of leukemia is 6-mercaptopurine, a sulfur analog of adenine. Draw the structural formula of 6-mercaptopurine.

22.10 Explain the difference in structure between a nucleoside and a nucleotide.

22.11 Name and draw structural formulas for the following. In each label the N-glycoside bond.
a. a nucleoside composed of β-D-ribose and adenine.
b. a nucleoside composed of β-D-ribose and uracil.
c. a nucleoside composed of β-2-deoxy-D-ribose and cytosine.

22.12 Name and draw structural formulas for the following. Label all N-glycoside bonds, ester bonds, and anhydride bonds.
a. ADP   b. dGMP   c. GTP

22.13 Calculate the net charge on the following at pH 7.4.
a. ATP   b. 2'-deoxyadenosine   c. GMP

22.14 Cyclic-AMP (adenosine-3',5'-cyclic monophosphate), first isolated in 1959, is involved in many diverse biological processes as a regulator of metabolic and physiological activity. In it, a single phosphate group is esterified with both the 3'- and 5'-hydroxyls of adenosine. Draw the structural formula for this substance.

22.15 Following are sequences for several polynucleotides. Write structural formulas for each. Calculate the net charge on each at pH 7.4.
a. dApdGpdA   b. pppdCpdT   c. pdGpdCpdCpdTpdA

The Structure of DNA (Section 22.4)

22.16 Show by structural formulas the hydrogen bonding between thymine and adenine; between uracil and adenine.

22.17 Compare and contrast the α-helix found in proteins with the double helix of DNA in regard to the following points.
   a. The units that repeat in the backbone of the chain.
   b. The projection in space of the backbone substituents (R— groups in the case of amino acids, purine and pyrimidine bases in the case of DNA) relative to the axis of the helix.

22.18 List the postulates of the Watson-Crick model of DNA structure. This model is based on certain experimental observations of base composition and molecular dimensions. Describe these observations and show how the model accounts for each.

22.19 Explain the role of hydrophobic interaction in stabilizing
   a. soap micelles   b. lipid bilayers   c. double-stranded DNA

22.20 What type of bond or interaction holds monomers together in
   a. proteins   b. nucleic acids   c. polysaccharides

22.21 In terms of hydrogen bonding, which is more stable, an A-T base pair or a G-C base pair?

22.22 In the presence of high temperature, nucleic acids denature, that is, they unwind into disordered single strands. Account for the fact that, the higher the content of G-C base pairs, the higher the temperature required to denature a given molecule of DNA.

22.23 Given the DNA triplet ATC, is its complement TAG or GAT? Explain.

DNA Replication (Section 22.5)

22.24 What is the meaning of the adjective *semiconservative* in the term *semiconservative replication*?

22.25 From what direction is a DNA strand read during formation of its complement?

22.26 What is an Okazaki fragment? Explain how the isolation and identification of Okazaki fragments provides evidence that the synthesis of DNA is not a continuous process.

Ribonucleic Acids (RNA) (Section 22.6)

22.27 Compare and contrast DNA and RNA in regard to the following points.
   a. monosaccharide units   b. major purine and pyrimidine bases
   c. primary structure   d. location in the cell
   e. function in the cell

22.28 Compare and contrast ribosomal RNA, messenger RNA, and transfer RNA in regard to the following points.
   a. molecular weight   b. function in protein synthesis

Transcription of Genetic Information: RNA Biosynthesis (Section 22.7)

22.29 Draw a diagram of a mRNA-ribosome initiation complex and label the following:
   a. 30S subunit   b. 50S subunit
   c. 5′ and 3′ ends of mRNA

22.30 Given the following DNA strand sequence

$$\text{5′-ACC-GTT-GCC-AAT-G-3′}$$

   a. Write the sequence of its DNA complement.
   b. Write the sequence of its mRNA complement.

**The Genetic Code
(Section 22.8)**

22.31 Following is a section of mRNA.

$$5'-AGG-UCC-CAG-3'$$

a. What tripeptide is synthesized if the code is read from the 5′ end to the 3′ end?
b. What tripeptide is synthesized if the code is read from the 3′ end to the 5′ end?
c. Calculate the net charge on each tripeptide at pH 7.4.
d. Which way is the code read in the cell and which tripeptide is synthesized?

22.32 What peptide sequences are coded for by the following mRNA sequences? (Each is written from the 5′ → 3′ direction.)

a. GCU-GAA-UGG   b. UCA-GCA-AUC
c. GUC-GAG-GUG   d. GCU-UCU-UAA

22.33 Complete the following table.

DNA	DNA Complement	mRNA Complement	Amino Acid Coded For
T-G-C	_____	_____	
C-A-G	_____	_____	_____
_____	A-C-G	_____	
_____	G-T-A	_____	_____
_____	_____	G-U-C	_____
_____	_____	U-G-C	_____
_____	_____	C-A-C	_____

22.34 The α-chain of human hemoglobin has 141 amino acids in a single polypeptide chain.
a. Calculate the minimum number of bases on DNA necessary to code for the α-chain. Include in your calculation the bases necessary to specify termination of polypeptide synthesis.
b. Calculate the length in centimeters of DNA containing this number of bases.

**Translation of
Genetic Information:
Synthesis of
Polypeptides
(Section 22.9)**

22.35 a. Draw the structural formula of N-formylmethionine.
b. Draw structural formulas for the products of hydrolysis of the amide bond in N-formylmethionine. Show each product as it would be ionized at pH 7.0.

22.36 Each of the following reactions involves ammonolysis of an ester. Draw the structural formula of the amide produced in each reaction.

a.

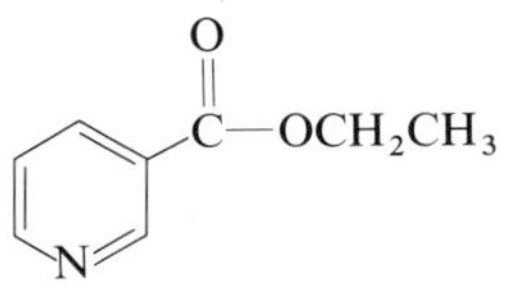

$$\text{(pyridine ring)} \overset{O}{\underset{\|}{C}}-OCH_2CH_3 + NH_3 \longrightarrow \text{(nicotinamide)} + CH_3CH_2OH$$

b. $CH_3CH_2O-\overset{O}{\underset{\|}{C}}-OCH_2CH_3 + 2\,NH_3 \longrightarrow \text{(urea)} + 2\,CH_3CH_2OH$

$$
\text{c. } H_2C \underset{\underset{O}{\overset{\|}{C}-OCH_2CH_3}}{\overset{\overset{O}{\overset{\|}{C}-OCH_2CH_3}}{}} + H_2N-\overset{\overset{O}{\|}}{C}-NH_2 \longrightarrow
$$

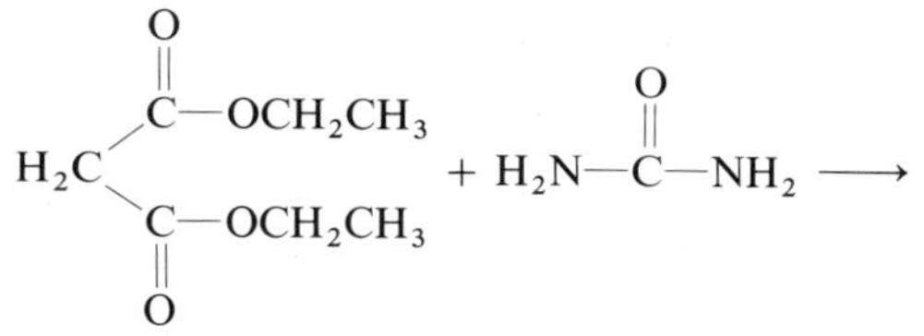

(barbituric acid) + 2 CH_3CH_2OH

22.37 Show that the reaction catalyzed by peptidyl transferase is an example of ammonolysis of an ester.

22.38 Are polypeptide chains synthesized from the N-terminal amino acid toward the C-terminal amino acid or vice versa?

Mutations (Section 22.11)

22.39 Following is a mRNA sequence written from the 5′ end to the 3′ end. Below it are three substitution mutations, one insertion mutation, and one deletion mutation. For what polypeptide sequence does the normal mRNA code, and what is the effect of each mutation on the resulting polypeptide?
  a. normal: 5′-UCC-CAG-GCU-UAC-AAA-GUA-3′
  b. substitution of A for C: 5′-UCC-AAG-GCU-UAC-AAA-GUA-3′
  c. substitution of A for C: 5′-UCC-CAG-GCU-UAA-AAA-GUA-3′
  d. substitution of A for C: 5′-UCA-CAG-GCU-UAC-AAA-GUA-3′
  e. insertion of A: 5′-UCC-CAG-GCU-AUA-CAA-AGU-A-3′
  f. deletion of A: 5′-UCC-CAG-GCU-UCA-AAG-UA-3′

22.40 In HbS, the abnormal human hemoglobin found in individuals with sickle-cell anemia (see the mini-essay "Abnormal Human Hemoglobins"), glutamic acid at position 6 of the β-chain is replaced by valine.
  a. List the two codons for glutamic acid and the four codons for valine.
  b. Show that a glutamic acid codon can be converted into a valine codon by a single substitution mutation.

The Flow of Energy in the Biological World

All living organisms need a constant supply of energy to support maintenance of cell structure and growth; in this sense, energy is the key to life itself. In this chapter, we will learn how cells extract energy from foodstuffs and how this energy is used in the performance of chemical, mechanical, and osmotic work.

23.1 Metabolism and Metabolic Pathways

Metabolism is defined as the sum of all chemical reactions used by an organism to grow, feed, move, excrete wastes, and communicate. Metabolism has two major components, **catabolism** and **anabolism.** Catabolism includes all reactions leading to the breakdown of biomolecules. Anabolism includes all reactions leading to the synthesis of biomolecules. In general, catabolism produces energy while anabolism consumes energy.

All reactions of a cell or organism are organized into orderly, carefully regulated sequences known as **metabolic pathways.** Each metabolic pathway consists of a series of consecutive steps that converts a starting material into an end product. The range of metabolic pathways in even a one-celled organism such as the bacterium *E. coli* is enormous. To support growth, a culture medium for *E. coli* needs to contain only glucose as a source of carbon atoms and energy, inorganic salts as sources of nitrogen and phosphorus, and a few other simple substances. The fact that *E. coli* can grow under these conditions means that each cell has the metabolic pathways needed to extract energy from the culture medium and use it to synthesize all of the proteins, enzymes, lipids, coenzymes, nucleic acids, carbohydrates, and other biomolecules necessary for maintenance and development. The ability of *E. coli* to grow under these conditions is truly remarkable, especially when you consider the complexity of some of the biomolecules found in living systems.

Fortunately for those who study the biochemistry of living systems, there are a great many similarities among the major metabolic pathways in humans, *E. coli*, and, for that matter, most other organisms. The number of individual reactions is large, but the number of different kinds of reactions is small. For example, the basic features of how different cells extract energy from foodstuffs and use it to synthesize other biomolecules, and even the means of self-regulation, are surprisingly similar. Because of these similarities, scientists can study the metabolism of simple organisms and then use these results to help understand the corresponding metabolic pathways in humans and other more complex organisms. We shall concentrate on human biochemistry, but you should realize that much of what we say about human metabolism can be applied equally well to the metabolism of most other organisms.

23.2 The Flow of Energy in the Biosphere

The uniqueness of living systems rests in their ability to capture energy from the environment, to store it, at least temporarily, and to use it to power the vast number of biological processes vital to life. The energy for all biological processes comes ultimately from the sun, whose enormous energy is derived from the fusion of hydrogen atoms into helium atoms.

$$4 \, H \cdot \xrightarrow{\text{nuclear fusion}} 2 \, He \colon + \text{energy}$$

A portion of this energy streams toward us as sunlight and is absorbed by chlorophyll pigments in plants. This energy drives the process of **photosynthesis,** the immediate product of which is glucose.

$$6 \, CO_2 + 6 \, H_2O + \text{energy} \xrightarrow{\text{photosynthesis}} C_6H_{12}O_6 + 6 \, O_2$$

glucose

In secondary steps, plants convert glucose into other carbohydrates, triglycerides, and proteins, all chemical storage forms of energy. Animals obtain these energy-rich molecules either directly or indirectly from plants.

During **respiration,** both plants and animals oxidize these energy-rich compounds to carbon dioxide and water. Respiration is accompanied by the release of energy.

$$\begin{matrix} \text{glucose and other} \\ \text{storage forms of} \\ \text{energy} \end{matrix} + O_2 \xrightarrow{\text{respiration}} CO_2 + H_2O + \text{energy}$$

A portion of the energy derived from respiration is transformed into **adenosine triphosphate (ATP),** a carrier of energy, which can be used directly for the performance of biological work. The remainder of the energy of respiration is liberated as heat. Steps in the flow of energy in the biosphere are summarized in Figure 23.1.

Figure 23.1 The flow of energy in the biosphere. (Adapted from David S. Page, *Principles of Biological Chemistry,* 2nd ed. Boston: Willard Grant Press, 1981.)

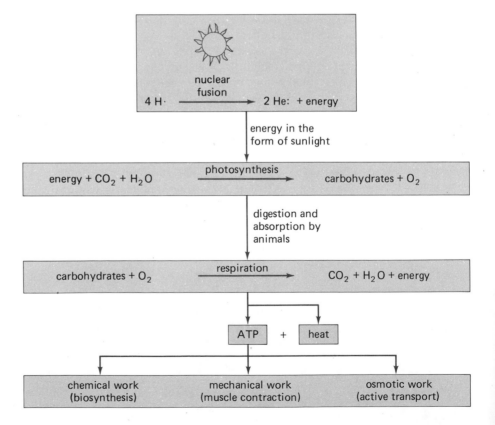

23.3 ATP: The Central Carrier of Energy

A. The Concept of Free Energy

Energy changes for reactions taking place in biological systems are commonly reported as **changes in free energy, ΔG.** A change in free energy measures the maximum work that can be obtained from a given reaction or process. ΔG^0 stands for the **change in free energy under standard conditions.**

Reactions that result in a decrease in free energy are said to be *exergonic*. Exergonic reactions include the oxidation of carbohydrates, fats, and proteins. The following equation shows that oxidation of glucose to carbon dioxide and water results in a decrease in free energy of 686,000 calories per mole of glucose.

$$C_6H_{12}O_6 + 6\,O_2 \longrightarrow 6\,CO_2 + 6\,H_2O \qquad \Delta G^0 = -686,000 \text{ cal/mole}$$
$$\text{glucose}$$

Because they result in a decrease in free energy, exergonic reactions are said to be *spontaneous*, which means that they proceed to the right as written and, at equilibrium, the concentration of products is greater than that of reactants. The more negative the value of ΔG^0, the greater the concentration of products relative to reactants. It is important to remember that although a negative value of ΔG^0 means that a reaction is spontaneous as written, it gives us no indication of the rate at which the reaction occurs. For example, the fact that the sign of ΔG^0 is negative for conversion of glucose and oxygen to carbon dioxide and water tells us that the reaction is spontaneous as written. Yet we know that in the absence of heat or appropriate catalysts, no reaction occurs.

Reactions that occur with an increase in free energy are said to be *endergonic*, which means that the reaction proceeds to the left as written and at equilibrium, the concentration of reactants is greater than that of products. The more positive the value of ΔG^0, the greater the concentration of reactants relative to products. The following equation shows that photosynthesis occurs with an increase in free energy.

$$6\,CO_2 + 6\,H_2O \longrightarrow \text{glucose} + 6\,O_2 \qquad \Delta G^0 = +686,000 \text{ cal/mole}$$

Because they occur with an increase in free energy, endergonic reactions are not spontaneous; at equilibrium there is very little product formed unless energy is supplied to drive the reaction to the right. As shown in Figure 23.1, the energy to drive photosynthesis is supplied in the form of sunlight. The relationships between sign of ΔG^0 and spontaneity are summarized in Figure 23.2 (see next page).

B. ATP, A High-Energy Compound

The central role of adenosine triphosphate (ATP) in the transfer of energy in the biological world depends on the triphosphate end of the molecule. The

Figure 23.2 Conversion of glucose and oxygen to carbon dioxide and water occurs with a decrease in free energy; it is a spontaneous reaction. Conversion of carbon dioxide and water to glucose and oxygen occurs with an increase in free energy; it is not a spontaneous process.

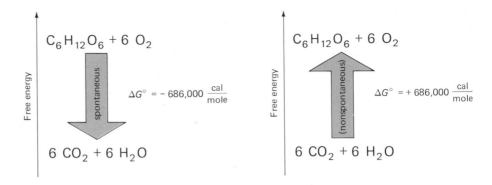

structure of ATP is shown in Figure 23.3. At pH 7.4, all protons of the triphosphate group are ionized, giving ATP a charge of -4. In the cell, ATP is present as a 1:1 complex with Mg^{2+}. The 1:1 complex has a charge of -2. Figure 23.4 shows abbreviated structural formulas of ATP, adenosine diphosphate (ADP), and adenosine monophosphate (AMP). In these formulas, only phosphate ester and anhydride bonds are shown.

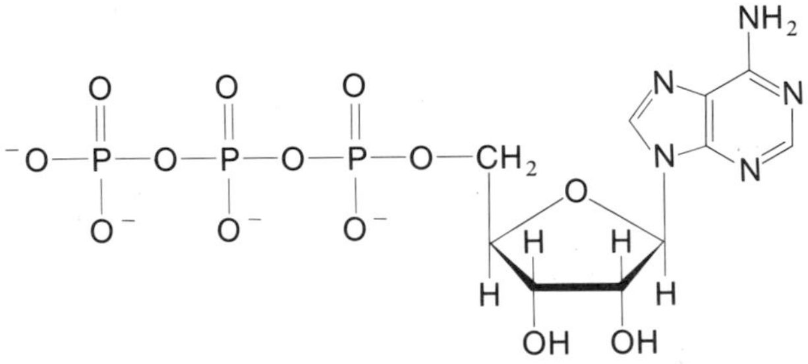

Figure 23.3 The structure of adenosine triphosphate (ATP). In the cell, ATP and Mg^{2+} form a 1:1 complex with a net charge of -2.

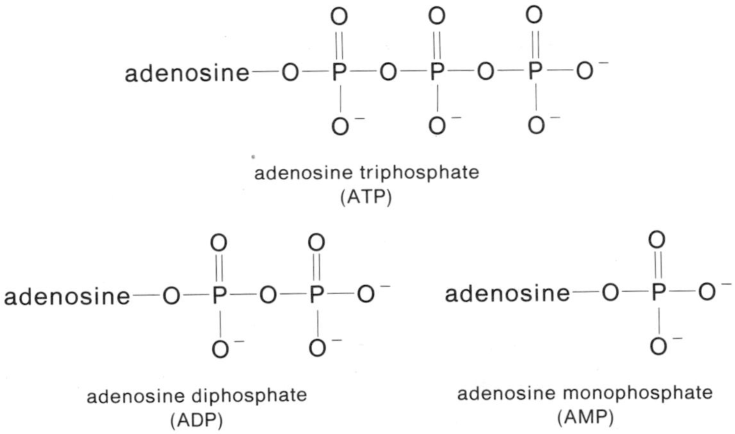

Figure 23.4 Abbreviated structural formulas of ATP, adenosine diphosphate (ADP), and adenosine monophosphate (AMP).

adenosine triphosphate
(ATP)

adenosine diphosphate
(ADP)

adenosine monophosphate
(AMP)

The key to understanding the role of ATP in the flow of energy in the biological world is knowing that it can transfer a phosphoryl group, $-PO_3^{2-}$, to another molecule. For example, during hydrolysis of ATP in water, a phosphoryl group is transferred from ATP to water. The products of this hydrolysis are ADP and phosphate ion.

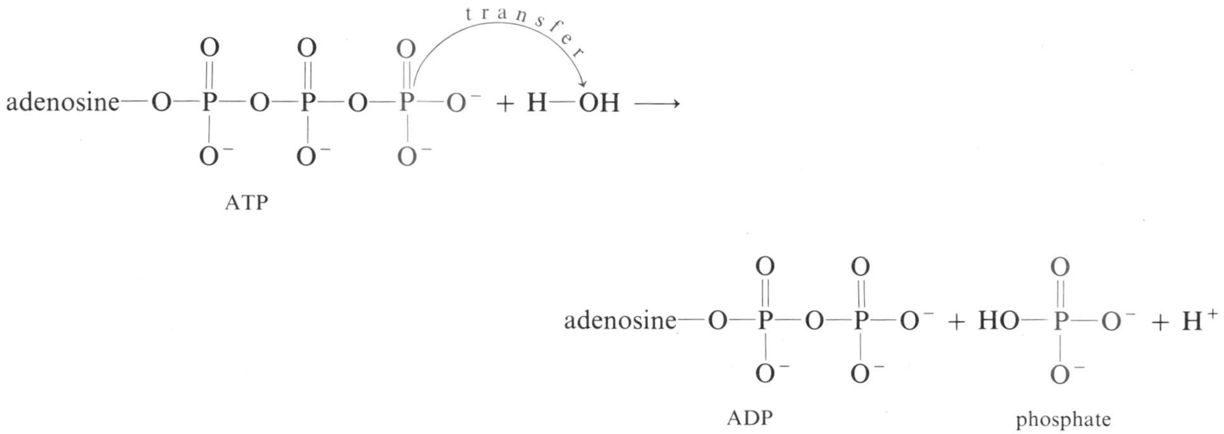

Transfer of a phosphoryl group from ATP to water is accompanied by a decrease in free energy, as shown in the following equation.

$$\text{ATP} + \text{H—OH} \longrightarrow \text{ADP} + \text{HPO}_4^{2-} \qquad \Delta G^0 = -7300 \text{ cal/mole}$$

ATP is but one of the many phosphate-containing compounds commonly found in biological systems. Several others, along with the free energy of hydrolysis of each, are listed in Table 23.1. Notice that the free energy of hydrolysis of ATP is larger than that of simple phosphate esters such as glucose 6-phosphate and glycerol 1-phosphate.

Table 23.1 Free energy of hydrolysis of some phosphate compounds present in biological systems.

Compound	Product	ΔG^0 (cal/mole)
phosphoenolpyruvate + H_2O $\longrightarrow$	pyruvate + phosphate	−14,800
1,3-diphosphoglycerate + H_2O $\longrightarrow$	3-phosphoglycerate + phosphate	−11,800
ATP + H_2O $\longrightarrow$	ADP + phosphate	−7,300
glucose 1-phosphate + H_2O $\longrightarrow$	glucose + phosphate	−5,000
fructose 6-phosphate + H_2O $\longrightarrow$	fructose + phosphate	−3,800
glucose 6-phosphate + H_2O $\longrightarrow$	glucose + phosphate	−3,300
glycerol 1-phosphate + H_2O $\longrightarrow$	glycerol + phosphate	−2,200

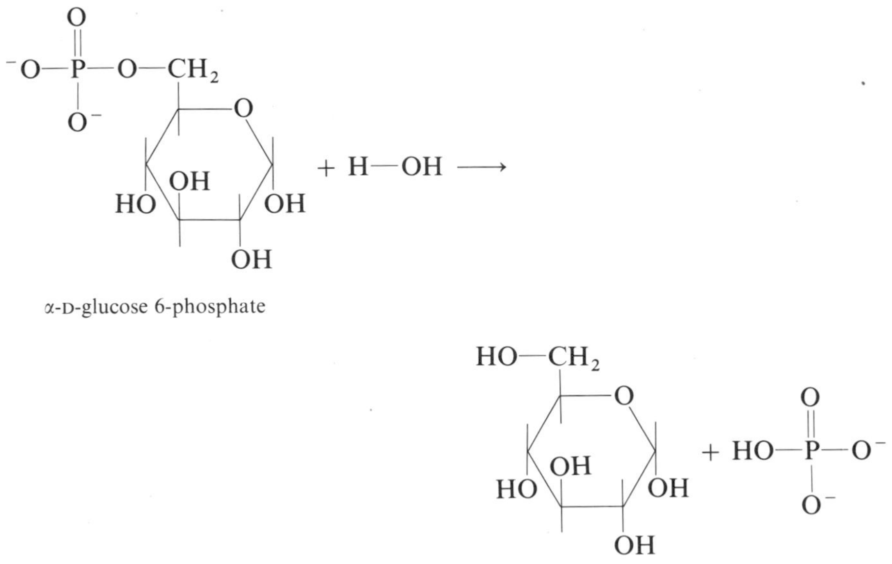

α-D-glucose 6-phosphate

$+ H-OH \longrightarrow$

α-D-glucose   phosphate $\Delta G^0 = -3,300$ cal/mole

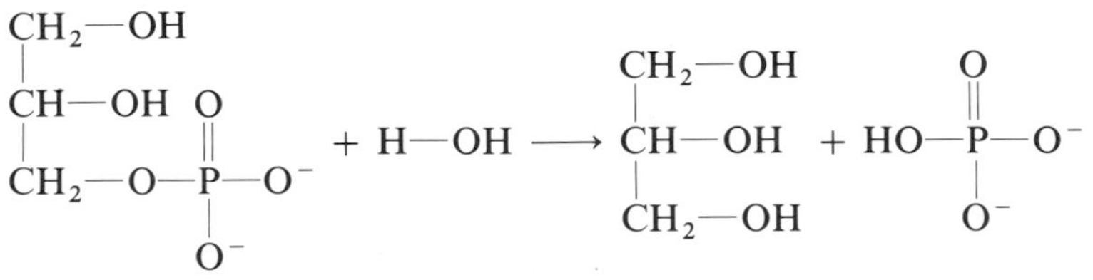

glycerol 1-phosphate   glycerol   phosphate $\Delta G^0 = -2,200$ cal/mole

Because of the size of its free energy of hydrolysis, ATP is called a **high-energy compound.** High-energy compounds have free energies of hydrolysis of 7000 cal/mole or greater; **low-energy compounds** have free energies of hydrolysis of less than 7000 cal/mole. As you can see from Table 23.1, there is no sharp line between high-energy and low-energy compounds.

Why is the standard free energy of hydrolysis of the phosphate anhydride bond of ATP so much larger than that of the hydrolysis of a phosphate ester bond, such as that in glucose 6-phosphate? The major reason lies in the structure of ATP itself. At the pH of a cell, the phosphate groups of ATP are fully ionized, giving the molecule a net charge of -4. These negative charges are very close to each other and create an electrostatic strain within the molecule. Hydrolysis of the terminal phosphate anhydride gives inorganic phosphate and ADP, a molecule with a net charge of -3. Thus, hydrolysis of ATP relieves some electrostatic strain. Because there is no such electrostatic strain in phosphate esters such as glucose 6-phosphate or glycerol 1-phosphate, their hydrolysis releases comparatively less energy.

C. Other High-Energy Compounds

Two other high-energy compounds are also listed in Table 23.1. The first of these is phosphoenolpyruvate. Hydrolysis of this phosphate ester gives the enol form of pyruvate and HPO_4^{2-}.

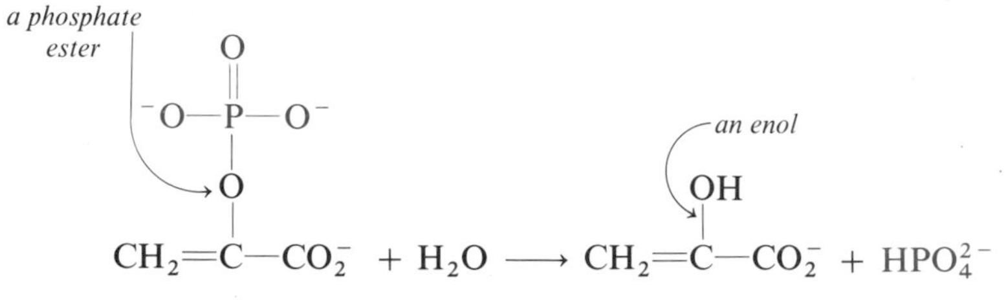

phosphoenolpyruvate

pyruvate
(*enol form*)

The enol form of pyruvate is in equilibrium with the keto form:

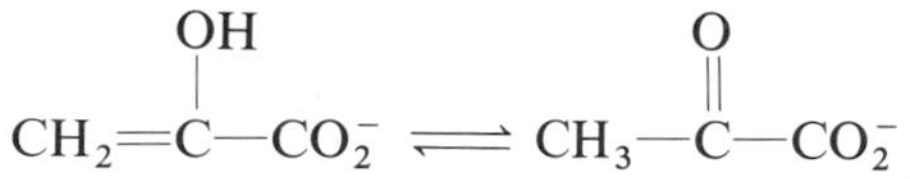

pyruvate
(*enol form*)

pyruvate
(*keto form*)

The equilibrium between this pair of keto and enol forms lies almost completely on the side of the keto form; accordingly, conversion to the keto form is spontaneous and accompanied by a large decrease in free energy.

The other high-energy compound listed in Table 23.1 is 1,3-diphosphoglycerate.

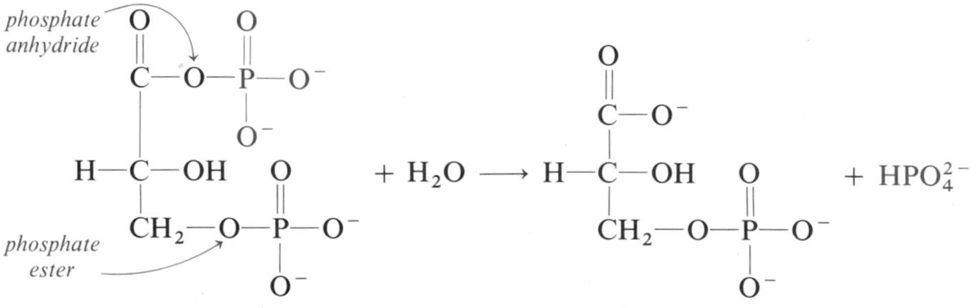

1,3-diphosphoglycerate

3-phosphoglycerate

1,3-Diphosphoglycerate contains a phosphate ester and a phosphate anhydride. Hydrolysis of the phosphate anhydride yields phosphate, 3-phosphoglycerate, and energy.

D. The Central Role of ATP in Cellular Energetics

In the preceding sections we discussed reactions involving transfer of a phosphoryl group to water. The same phosphate-containing compounds can, at least in principle, transfer a phosphoryl group to compounds of the type H—OR. Following is an equation for the transfer of a phosphoryl group from phosphoenolpyruvate to α-D-glucose to form pyruvate and α-D-glucose 6-phosphate.

phosphoenol-pyruvate	α-D-glucose	pyruvate	α-D-glucose 6-phosphate

The change in free energy for this reaction can be calculated by: (1) dividing the reaction into two separate equations for which changes in free energy are known; and (2) adding the separate equations and the free energy change for each.

$$\text{phosphoenolpyruvate} + \text{H—OH} \longrightarrow \text{pyruvate} + HPO_4^{2-} \qquad \Delta G^0 = -14{,}800 \text{ cal/mole}$$
$$\text{glucose} + HPO_4^{2-} \longrightarrow \text{glucose 6-phosphate} + \text{H—OH} \qquad \Delta G^0 = +\ 3{,}300 \text{ cal/mole}$$
$$\overline{\text{glucose} + \text{phosphoenol-pyruvate} \longrightarrow \text{pyruvate} + \text{glucose 6-phosphate}} \qquad \Delta G^0 = -11{,}500 \text{ cal/mole}$$

This calculation shows that transfer of a phosphoryl group from pyruvate to glucose results in a large decrease in free energy: it is a spontaneous reaction. Yet, although it is spontaneous, direct transfer of a phosphoryl group from phosphoenolpyruvate to glucose has not been observed in living systems. Rather, ATP is a common intermediate or "medium of exchange" that links this and other high-energy phosphate donors to phosphate acceptors.

$$\text{phosphoenolpyruvate} + \text{ADP} \longrightarrow \text{pyruvate} + \text{ATP}$$
$$\text{glucose} + \text{ATP} \longrightarrow \text{glucose 6-phosphate} + \text{ADP}$$

In order to appreciate the value to the cell of this means of phosphate transfer, consider the fact that virtually all reactions occurring in living systems are enzyme-catalyzed. In fact, a great deal of the work of cells is devoted to the

Figure 23.5 An illustration of the efficiency of using ATP as a common phosphate acceptor/donor. In (a), 15 different enzymes are required while in (b) only 7 different enzymes are required.

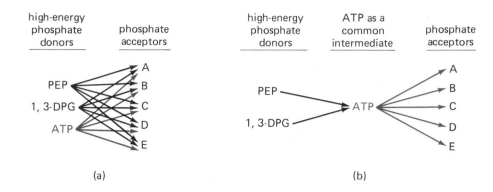

23.4 Stages in the Oxidation of Foodstuffs and the Generation of ATP

synthesis of biomolecules, including enzymes. Shown in Figure 23.5(a) is an illustration of the number of enzymes necessary to catalyze the transfer of phosphate from either phosphoenolpyruvate (PEP), 1,3-diphosphoglycerate (1,3-DPG), or ATP to five different phosphate acceptors. For the reactions in Figure 23.5(a), 15 different enzymes are required. By using ATP as a collector and common donor of phosphate groups to other low-energy acceptors, only 7 different enzymes are required.

As we have seen, ATP is the common intermediate or "medium of exchange" for energy in living systems. Next, let us see how cells extract energy from foodstuffs and use it for the synthesis of ATP. The basic strategy used by all cells is to oxidize foodstuffs and use a portion of the free energy released to convert ADP and HPO_4^{2-} to ATP. Oxidation of foodstuffs and the generation of ATP is accomplished in four stages.

A. Stage I: Digestion and Absorption of Fuel Molecules

Stage I is digestion of food (Figure 23.6). Digestion involves hydrolysis of carbohydrates to monosaccharides, proteins to amino acids, and fats and oils to fatty acids and glycerol.

Figure 23.6 Stage of the oxidation of foodstuffs and the generation of ATP.

polysaccharides $\xrightarrow{\text{hydrolysis}}$ monosaccharides

proteins $\xrightarrow{\text{hydrolysis}}$ amino acids

fats $\xrightarrow{\text{hydrolysis}}$ fatty acids + glycerol

As a result of digestion, the hundreds of thousands of different proteins, fats, oils, and carbohydrates ingested in the diet are converted to a relatively few, lower-molecular-weight compounds. The most common of these low-molecular-weight compounds are shown below.

from carbohydrates	*from fats and oils*	*from proteins*
D-glucose	palmitic acid	20 amino acids
D-fructose	stearic acid	
D-galactose	oleic acid	
	glycerol	

B. Stage II: Degradation of Fuel Molecules to Acetyl CoA

In Stage II (Figure 23.7), lower-molecular-weight compounds from stage I are converted to even fewer, smaller molecules. The carbon skeletons of glucose, fructose, and galactose along with those of fatty acids and glycerol are converted to acetate in the form of a thioester with coenzyme A. This ester is named **acetyl coenzyme A,** or more commonly, **acetyl CoA.** The carbon skeletons of several amino acids are also degraded to acetate in the form of acetyl CoA. The carbon skeletons of other amino acids are degraded to different small molecules, but eventually all molecules go through the reactions of stage III.

Figure 23.7 Stage II of the oxidation of foodstuffs and the generation of ATP.

Coenzyme A is a complex molecule derived from four subunits (Figure 23.8). On the left is a two-carbon unit derived from β-mercaptoethylamine. This unit is joined by an amide bond to β-alanine which is, in turn, joined by another amide bond to the carboxyl group of pantothenic acid. Pantothenic acid is a vitamin of the B group. Finally, the —OH group of pantothenic acid is joined by an ester bond to the terminal phosphate of ADP. A key feature in the structure of coenzyme A is the presence of a sulfhydryl group (—SH). Acetyl CoA is a thioester derived from the carboxyl group of acetic acid and the thiol group of coenzyme A.

C. Stage III: The Krebs or Tricarboxylic Acid Cycle

Stage III consists of a series of reactions known alternatively as the **tricarboxylic acid cycle,** the **citric acid cycle,** or the **Krebs cycle** (Figure 23.9). A major func-

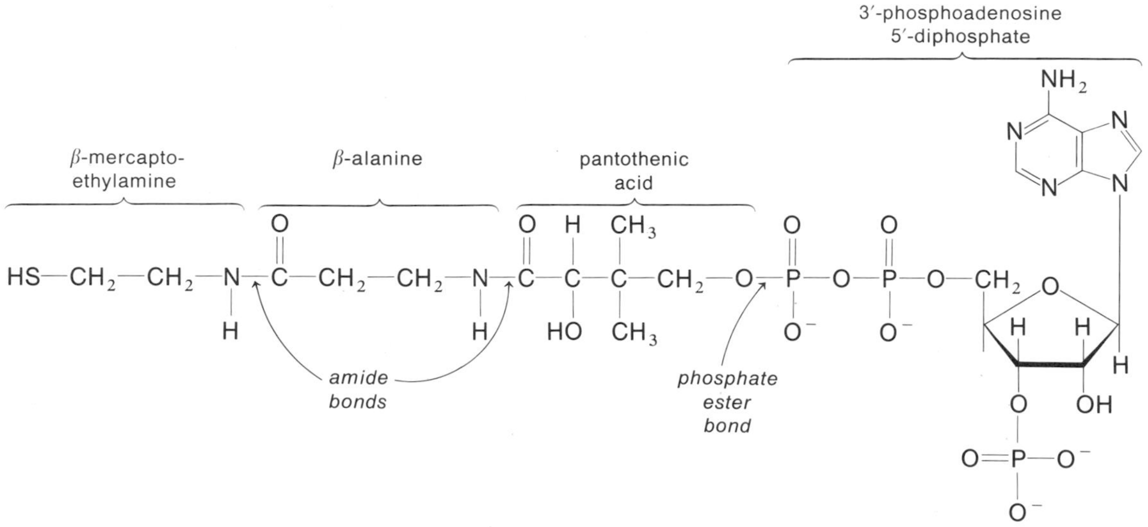

Figure 23.8 Coenzyme A. Pantothenic acid is one of the vitamins of the B group. The acetylated form of this coenzyme, designated acetyl coenzyme A, or acetyl CoA, is the thioester of acetic acid and the terminal sulfhydryl group.

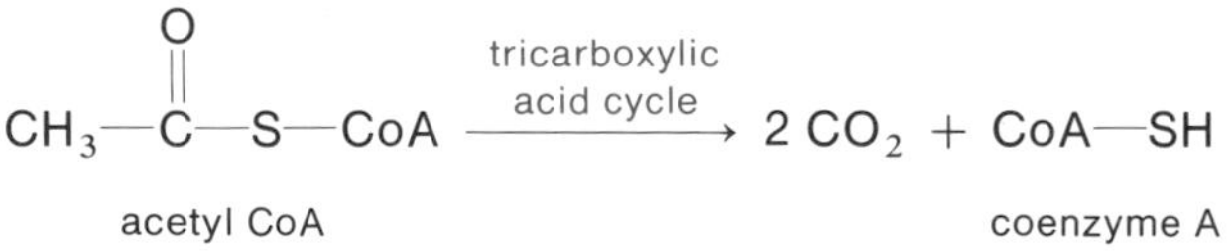

Figure 23.9 Stage III of the oxidation of foodstuffs and the generation of ATP. Small molecules derived from stages I and II are oxidized to carbon dioxide.

tion of the tricarboxylic acid cycle is oxidation of the two-carbon acetyl group of acetyl CoA to two molecules of carbon dioxide.

The biological oxidizing agents in stage III are nicotinamide adenine dinucleotide (NAD^+) and flavin adenine dinucleotide (FAD). The first of these, NAD^+ (Figure 23.10, next page), is the major acceptor of electrons in the oxidation of fuel molecules. The reactive group of NAD^+ is a pyridine ring, which accepts two electrons and one proton to form the reduced coenzyme NADH.

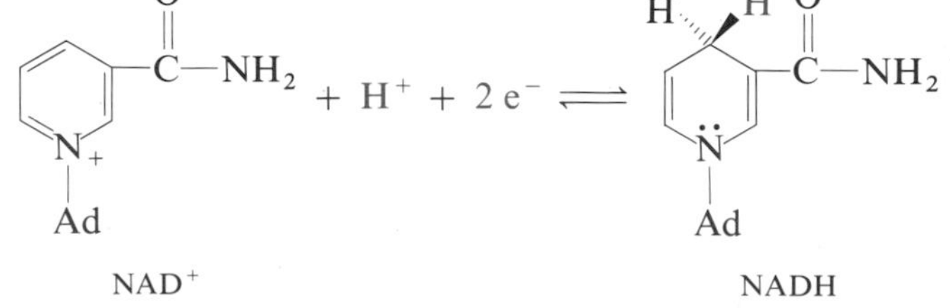

NAD$^+$ NADH

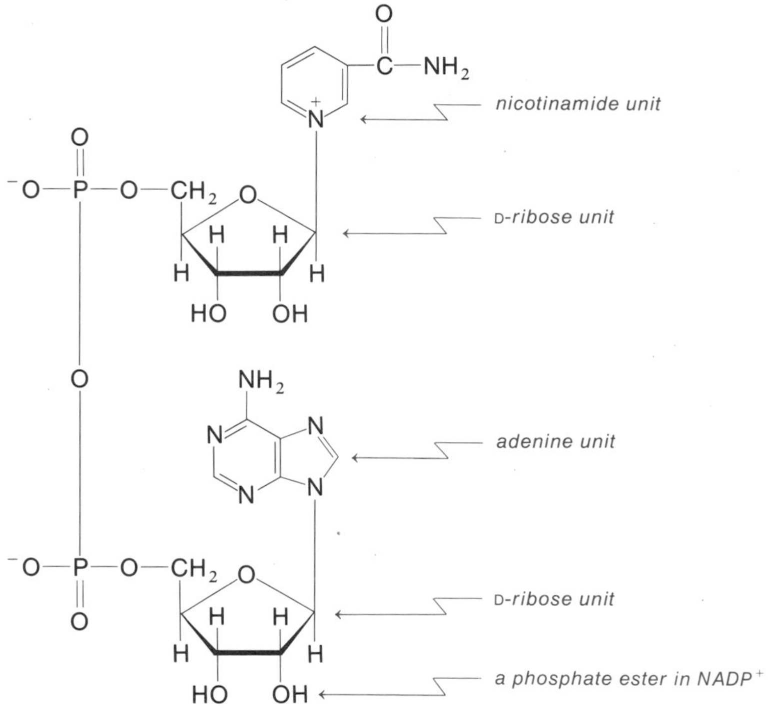

Figure 23.10 The structure of nicotinamide adenine dinucleotide (NAD$^+$). Nicotinamide is one of the water-soluble vitamins. In nicotinamide adenine dinucleotide phosphate, NADP$^+$, the 2'-hydroxyl of a D-ribose unit is esterified with phosphoric acid.

nicotinamide unit

D-ribose unit

adenine unit

D-ribose unit

a phosphate ester in NADP$^+$

The other electron acceptor in the oxidation of fuel molecules is FAD (Figure 23.11). The reactive group of this coenzyme is a flavin group, which accepts two electrons and two protons to form the reduced coenzyme FADH$_2$.

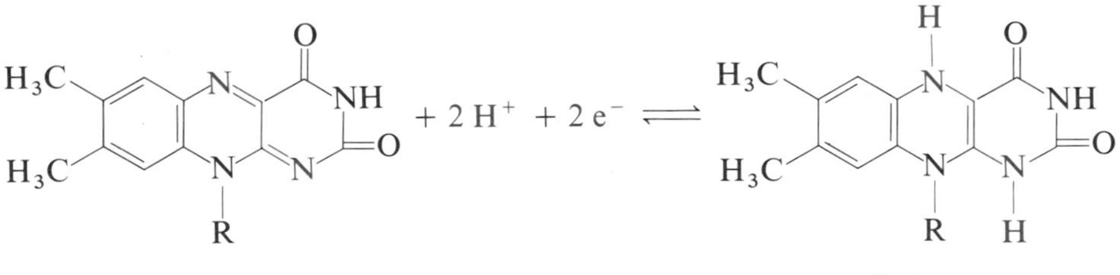

FAD   FADH$_2$

All reactions of the tricarboxylic acid cycle (stage III), as well as those of electron transport and oxidative phosphorylation (stage IV) take place within subcellular structures called **mitochondria** (singular: **mitochondrion**). To picture a mitochondrion, imagine two balloons, one larger than the other, and imagine that the larger balloon is extensively folded and stuffed inside the smaller balloon. A cross-section of such a double-walled structure is analogous to that of a mitochondrion (Figure 23.12). The outer membrane of a mitochondrion (which corresponds to the wall of the smaller balloon) is a phospholipid bilayer like that described in Chapter 18. The outer membrane is permeable to most small molecules. The inner membrane (corresponding to

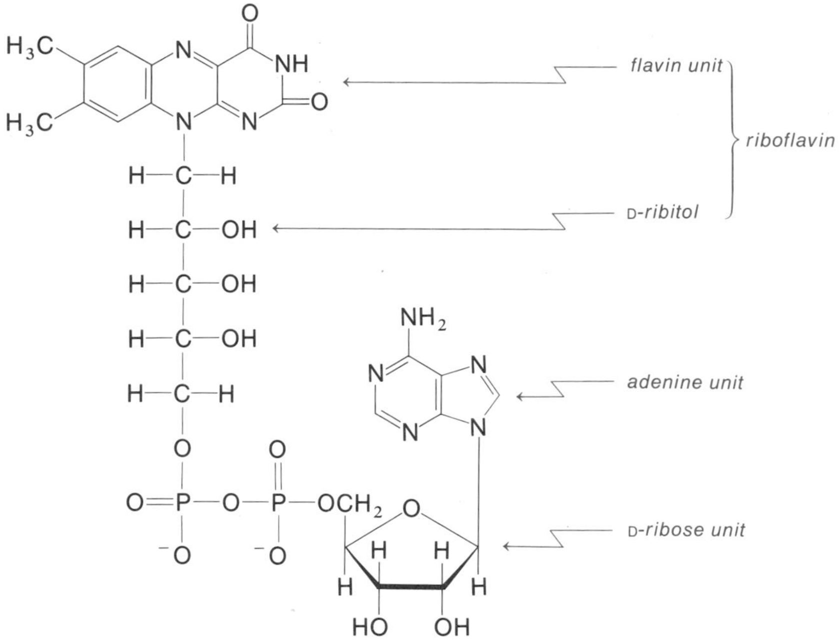

Figure 23.11 The structure of flavin adenine dinucleotide, FAD. Riboflavin is one of the B-group vitamins.

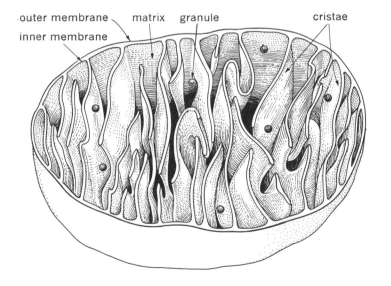

Figure 23.12 A mitochondrion. Both inner and outer membranes are phospholipid bilayers. The surface area of the inner membrane is highly folded and several thousand times larger than that of the outer membrane.

the wall of the larger balloon) is also a phospholipid bilayer. Because it is so extensively and irregularly folded, the surface area of the inner membrane is approximately 10,000 times that of the outer membrane. The folds of the inner membrane are called *cristea* and the space that surrounds them is called the *matrix*. All enzymes required for catalysis of the tricarboxylic acid cycle are located within the mitochondria, some in the matrix and others bound to the inner membrane.

D. Stage IV: Electron Transport and Oxidative Phosphorylation: A Central Pathway for the Oxidation of Reduced Coenzymes and the Generation of ATP

In stage IV, reduced coenzymes (NADH and $FADH_2$) accumulated from stages II and III are reoxidized by molecular oxygen and, in effect, this is the aerobic phase of metabolism. Reoxidation of NADH and $FADH_2$ is coupled with phosphorylation of ADP to ATP; for this reason, stage IV is called **oxidative phosphorylation.** The net reactions of stage IV are shown in Figure 23.13.

Figure 23.13 Stage IV of the oxidation of foodstuffs and the generation of ATP. The reoxidation of NADH and $FADH_2$ is coupled with the phosphorylation of ADP.

$$2\ NADH + O_2 + 2\ H^+ \longrightarrow 2\ NAD^+ + 2\ H_2O$$

$$2\ FADH_2 + O_2 \longrightarrow 2\ FAD + 2\ H_2O$$

$$ADP + HPO_4^{2-} \longrightarrow ATP + H_2O$$

Figure 23.14 A summary of the four stages in the oxidation of foodstuffs and the generation of ATP.

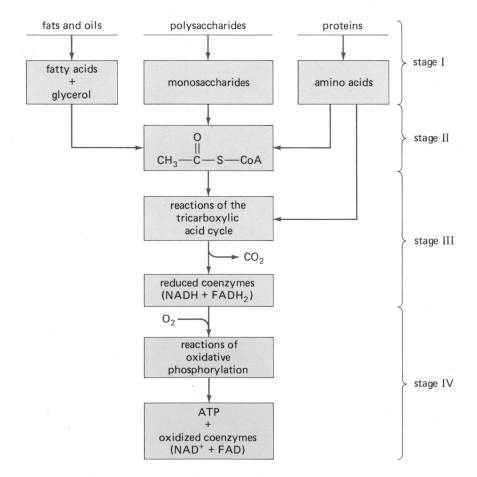

These four stages in the oxidation of foodstuffs and the generation of ATP are summarized in Figure 23.14.

23.5 Electron Transport and Oxidative Phosphorylation: A Closer Look

A. Electron Transport

The final stage in oxidation of foodstuffs and generation of ATP involves reoxidation of NADH and $FADH_2$ by molecular oxygen. As shown by the following equations, each of these oxidations is accompanied by a large decrease in free energy.

$$NADH + 1/2\,O_2 \longrightarrow NAD^+ + H_2O \qquad \Delta G^0 = -52,300 \text{ cal/mole}$$
$$FADH_2 + 1/2\,O_2 \longrightarrow FAD + H_2O \qquad \Delta G^0 = -43,400 \text{ cal/mole}$$

We have written these equations as single reactions. To better appreciate how cells bring about these oxidations, it is helpful to write balanced half-reactions for the oxidations of NADH and $FADH_2$.

$$NADH \longrightarrow NAD^+ + H^+ + 2\,e^-$$
$$FADH_2 \longrightarrow FAD + 2\,H^+ + 2\,e^-$$

The ultimate acceptor of electrons given up by these reduced coenzymes is molecular oxygen.

$$O_2 + 4\,H^+ + 4\,e^- \longrightarrow 2\,H_2O$$

Within mitochondria, the site of respiration, electrons are not passed directly from reduced coenzymes to molecular oxygen. Rather, they are passed from one electron acceptor to another and then to molecular oxygen. The series of reactions by which electrons are passed from NADH and $FADH_2$ to molecular oxygen is called the **electron transport chain,** or **respiratory chain.** All enzymes and cofactors required for the reactions of the respiratory chain are located on the inner membrane of mitochondria. As illustrated in Figure 23.15 (next page), there are six intermediate carriers of electrons between NADH and molecular oxygen.

In the first step of the respiratory chain, a pair of electrons is transferred from NADH to a flavoprotein, a molecule similar in structure to riboflavin and FAD. The flavoprotein can exist in both oxidized and reduced forms.

$$NADH + H^+ + \begin{array}{c}\text{oxidized}\\\text{flavoprotein}\end{array} \longrightarrow NAD^+ + \begin{array}{c}\text{reduced}\\\text{flavoprotein}\end{array} \qquad \Delta G^0 = -9,200 \text{ cal/mole}$$

There is a large decrease in free energy in this step.

Figure 23.15 Six carriers of electrons separate NADH (an electron donor) from molecular oxygen (an electron acceptor) in the respiratory chain.

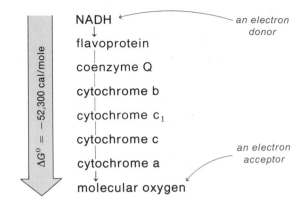

NADH ← an electron donor
↓
flavoprotein

coenzyme Q

cytochrome b

cytochrome c_1

cytochrome c

cytochrome a → an electron acceptor
↓
molecular oxygen

$\Delta G^0 = -52{,}300$ cal/mole

The second carrier of electrons in the respiratory chain is coenzyme Q (Figure 23.16). This molecule has a long hydrocarbon chain of between 6 and 10 isoprene units that serves to anchor it firmly in the nonpolar environment of the inner membrane of the mitochondrion. As you can see from the balanced half-reaction in Figure 23.16, the oxidized form of coenzyme Q is a two-electron oxidizing agent. In the second step of the respiratory chain, two electrons are transferred from the reduced form of flavoprotein to the oxidized form of coenzyme Q.

flavoprotein + coenzyme Q ⟶ flavoprotein + coenzyme Q

(*reduced form*)   (*oxidized form*)   (*oxidized form*)   (*reduced form*)

The remaining carriers of electrons in the respiratory chain are four structurally related proteins known as **cytochromes.** Cytochrome c, the most thoroughly studied of these electron carriers, is a globular protein of molecular weight 12,400 and consists of a single polypeptide chain of 104 amino acids folded around a single heme group (Figure 21.2). The iron atom of all four cytochromes can exist in either the Fe(II) or Fe(III) oxidation states. Thus, an atom of Fe(III) in a cytochrome molecule can accept an electron and be

Figure 23.16 Coenzyme Q. The nonpolar hydrocarbon chain of coenzyme Q consists of from six to ten ($n = 6$–10) isoprene units.

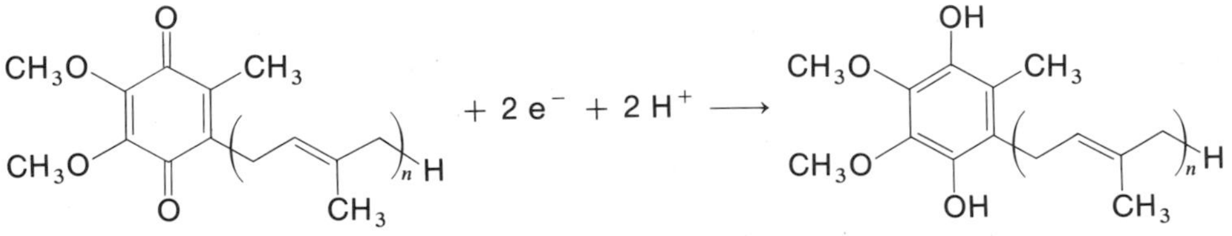

coenzyme Q
(*oxidized form*)

$+ 2e^- + 2H^+ \longrightarrow$

coenzyme Q
(*reduced form*)

reduced to Fe(II), which, in turn, gives up an electron to reduce the next cytochrome in the chain.

In the final step of the respiratory chain, electrons are transferred from cytochrome a to a molecule of oxygen. In the following equation, cytochrome a is abbreviated as cyt a.

$$2 \text{ cyt a} + 1/2 \, O_2 + 2 \, H^+ \longrightarrow 2 \text{ cyt a} + H_2O \qquad \Delta G^0 = -24{,}400 \text{ cal/mole}$$

(reduced
form)
(oxidized
form)

Of the seven steps in the respiratory chain, the transfer of electrons from cytochrome a to molecular oxygen occurs with the largest decrease in free energy.

The seven steps in the transfer of electrons from NADH to O_2 are summarized in Figure 23.17. Notice that the free energy decrease in four of these steps is larger than that required for the phosphorylation of ADP ($\Delta G^0 = +7{,}300$ cal/mole).

Electrons from $FADH_2$ are also transported via the intermediates of the respiratory chain to molecular oxygen. Electrons from $FADH_2$, however, enter the chain at coenzyme Q (Figure 23.17). There are only five intermediates in the transport of electrons from coenzyme Q to molecular oxygen, and only two steps with a decrease in free energy larger than that required for the phosphorylation of ADP.

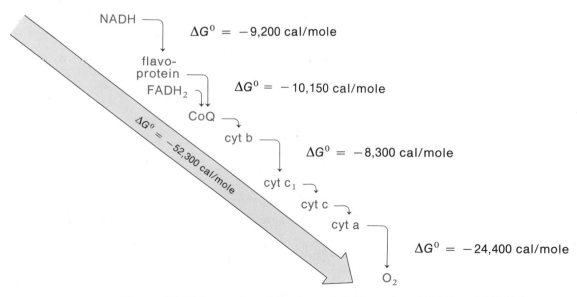

Figure 23.17 Energetics of the flow of electrons from NADH and $FADH_2$ to molecular oxygen in the respiratory chain.

B. Oxidative Phosphorylation

Cellular respiration occurs with a decrease in free energy; at four points in the flow of electrons from NADH to molecular oxygen, the decreases are large enough to drive the phosphorylation of ADP. Cells have evolved a mechanism by which a portion of the free energy decrease in three of these steps is conserved as ATP (Figure 23.18).

For each mole of NADH entering the respiratory chain, three moles of ATP are formed. The overall equation for oxidation of NADH and phosphorylation of ADP can be written as the sum of the exergonic oxidation of NADH and the endergonic phosphorylation of ADP:

$$
\begin{array}{lr}
\text{NADH} + \text{H}^+ + \tfrac{1}{2}\text{O}_2 \longrightarrow \text{NAD}^+ + \text{H}_2\text{O} & \Delta G^0 = -52{,}000 \text{ cal/mole} \\
3\,\text{ADP} + 3\,\text{HPO}_4^{2-} \longrightarrow 3\,\text{ATP} + 3\,\text{H}_2\text{O} & \Delta G^0 = +21{,}900 \text{ cal/mole} \\
\hline
\text{NADH} + \text{H}^+ + \tfrac{1}{2}\text{O}_2 + 3\,\text{ADP} + 3\,\text{HPO}_4^{2-} \longrightarrow \text{NAD}^+ + 3\,\text{ATP} + 4\,\text{H}_2\text{O} & \Delta G^0 = -30{,}100 \text{ cal/mole}
\end{array}
$$

Thus, coupling the oxidation and phosphorylation reactions conserves 22/52 or approximately 42% of the decrease in free energy during the reoxidation of NADH. Reoxidation of FADH$_2$ is coupled with phosphorylation of two

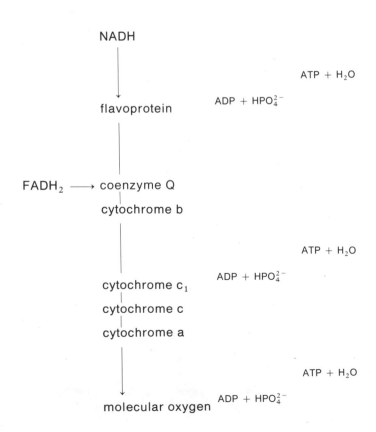

Figure 23.18 The three steps in the flow of electrons within the respiratory chain where a portion of the free energy released is conserved as ATP. For each molecule of NADH entering the electron transport chain, three molecules of ATP are formed. For each molecule of FADH$_2$ entering the respiratory chain, two molecules of ATP are formed.

moles of ADP and approximately 34% of the decrease in free energy is conserved as ATP.

The mechanism by which the cell couples the oxidation of reduced coenzymes with the phosphorylation of ADP is not understood at the present time and remains one of the great puzzles of modern biochemistry.

C. Inhibitors of Respiration

There are a number of chemicals that block specific steps in respiration. One of the best known of these, cyanide ion, is a powerful inhibitor of cytochrome oxidase, the enzyme that catalyzes the transfer of electrons from cytochrome c to cytochrome a. Cyanide ion also complexes with the iron atom of cytochrome a to form a complex that is unable to function as a carrier of electrons. The result of cyanide ion poisoning is a block in the flow of electrons from NADH and $FADH_2$ to molecular oxygen. These coenzymes, along with all carriers of electrons in the respiratory chain, accumulate in their reduced state and the generation of ATP ceases. Cyanide ion has the same effect on the cell as lack of oxygen; death is by asphyxiation.

Another inhibitor of respiration is rotenone, a compound extracted from the roots of certain tropical plants. Rotenone is a powerful inhibitor of NADH-dehydrogenase, the enzyme that catalyzes the transfer of electrons from NADH to a flavoprotein in the first step of respiration. Because it passes readily into the breathing tubes of insects and is intensely toxic to these organisms, rotenone is widely used as an insecticide. It is also toxic to fish because it passes readily into their gills. Rotenone is not readily absorbed through the skin and therefore has a relatively low toxicity for humans and other vertebrates.

23.6 Utilization of ATP for Cellular Work

As we have seen in Section 23.3, hydrolysis of the terminal phosphate anhydride bond of ATP results in a decrease in free energy. Cells are able to use a portion of this free energy to do three major types of work: chemical work, mechanical work, and osmotic work. Let us look in more detail at each type of work and see how it depends on ATP.

A. Chemical Work

As an example of how ATP is used for chemical work, consider the biosynthesis of proteins, nucleic acids, triglycerides, and polysaccharides, all polymers formed from smaller units. The formation of peptide, glycoside, and ester bonds requires energy. For example, formation of a glycoside bond between glucose and fructose to form sucrose requires 5500 cal for each mole of sucrose formed.

$$\text{glucose} + \text{fructose} \longrightarrow \text{sucrose} + H_2O \qquad \Delta G^0 = +5{,}500 \text{ cal/mole}$$

On the other hand, hydrolysis of ATP results in a decrease in energy.

$$\text{ATP} + \text{H}_2\text{O} \longrightarrow \text{ADP} + \text{HPO}_4^{2-} \qquad \Delta G^0 = -7,300 \text{ cal/mole}$$

Adding these reactions gives a net reaction that results in a decrease in free energy and is spontaneous in the direction written.

$$
\begin{array}{lr}
 & \Delta G^0 \\
 & \text{(cal/mole)} \\
\text{glucose} + \text{fructose} \longrightarrow \text{sucrose} + \text{H}_2\text{O} & +5,500 \\
\text{ATP} + \text{H}_2\text{O} \longrightarrow \text{ADP} + \text{HPO}_4^{2-} & -7,300 \\
\hline
\text{glucose} + \text{fructose} + \text{ATP} \longrightarrow \text{sucrose} + \text{ADP} + \text{HPO}_4^{2-} & -1,800
\end{array}
$$

If it were possible to capture a part of the free energy of the phosphate anhydride bond in ATP and channel it into glycoside bond formation, glucose and fructose could be converted into sucrose. Cells accomplish this by two sequential enzyme-catalyzed reactions involving a common intermediate.

$$
\begin{array}{c}
\text{glucose} + \text{ATP} \longrightarrow \text{glucose 1-phosphate} + \text{ADP} \\
\text{glucose 1-phosphate} + \text{fructose} \longrightarrow \text{sucrose} + \text{HPO}_4^{2-} \\
\hline
\text{glucose} + \text{fructose} + \text{ATP} \longrightarrow \text{sucrose} + \text{ADP} + \text{HPO}_4^{2-}
\end{array}
$$

Glucose 1-phosphate is the common intermediate. Together, these sequential reactions have a net free-energy change of $-1,800$ cal/mole. Thus, a portion of the energy stored in ATP is captured in the form of a common intermediate and is then used to form a glycoside bond.

B. Osmotic Work: Transport Across Membranes

The movement of molecules and ions across a membrane is called **transport** and is an essential process in all living organisms. There are two types of transport: passive and active. In **passive transport,** a molecule or ion moves

Figure 23.19 (a) In passive transport, molecules and ions flow with a concentration gradient. (b) In active transport, molecules and ions flow against a concentration gradient. Active transport is nonspontaneous and requires energy; passive transport is spontaneous.

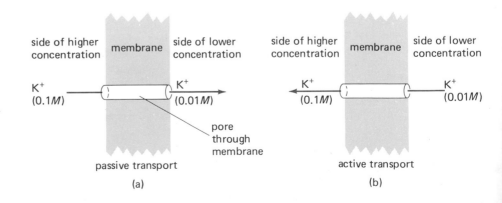

across a membrane from a side of higher concentration to a side of lower concentration. Passive transport is a spontaneous process and requires no energy. In **active transport,** a molecule or ion is moved across a membrane from a side of lower concentration to a side of higher concentration. Active transport is nonspontaneous and requires energy. Figure 23.19 illustrates transport with and against a concentration gradient.

For an example of the results of active transport, compare the relative concentrations of ions and molecules in the intracellular and extracellular fluids of skeletal muscle tissue (Figure 23.20). Note that the concentrations of

Figure 23.20 Relative concentrations of some molecules and ions in the intracellular and extracellular fluids of human skeletal muscle. (From David S. Page, *Principles of Biological Chemistry,* 2nd ed. Boston: Willard Grant Press, 1981.)

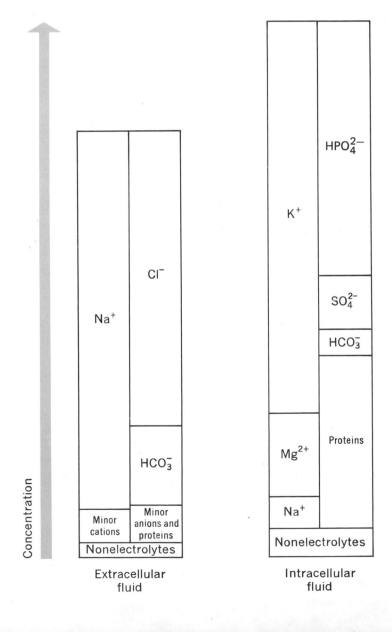

K^+, Mg^{2+}, HPO_4^{2-}, and SO_4^{2-} are all considerably higher inside the cells of skeletal muscle than they are in the surrounding fluid. It requires energy to concentrate these ions within skeletal muscle cells.

At the present time, little is known about the mechanism of active transport. However, it is known that active transport requires energy and that it is linked to the cleavage of high-energy phosphate bonds in ATP.

C. Mechanical Work: Muscle Contraction

Figure 23.21 shows a schematic diagram of a section of skeletal muscle fiber. A fiber consists of two types of protein-containing filaments. One type of filament, containing the protein **actin,** consists of thin rods connected to a protein plate or disc. Actin filaments connected to one plate do not make contact with those from an adjacent plate. A second type of filament, containing the protein **myosin,** consists of thicker rods that overlap actin filaments from adjacent plates.

Our best current model of muscle contraction is called the **sliding filament model.** According to this model, during contraction, actin filaments slide past myosin filaments and in the process, the free ends of actin filaments are pulled closer together. As actin filaments slide toward each other, they, in turn, pull the protein plates closer together and the entire muscle fiber contracts. Contraction of muscle fibers is coupled with the hydrolysis of ATP to ADP and phosphate, but how these two processes are coupled is almost totally unknown.

Although ATP is the immediate source of energy to power contraction of skeletal muscle, it is not the form in which energy for muscle contraction is stored. In resting muscle, energy is stored in the form of **creatine phosphate,** a high-energy compound containing a phosphate amide bond. Following is an

Figure 23.21 A section of skeletal muscle fiber.

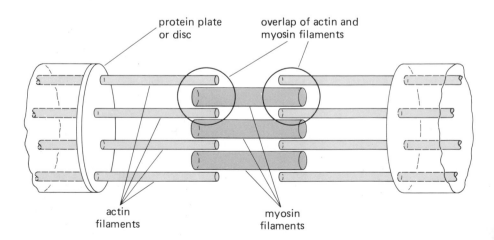

protein plate or disc

overlap of actin and myosin filaments

actin filaments

myosin filaments

equation for the hydrolysis of creatine phosphate. The free-energy change for this reaction is $-10,300$ cal/mole.

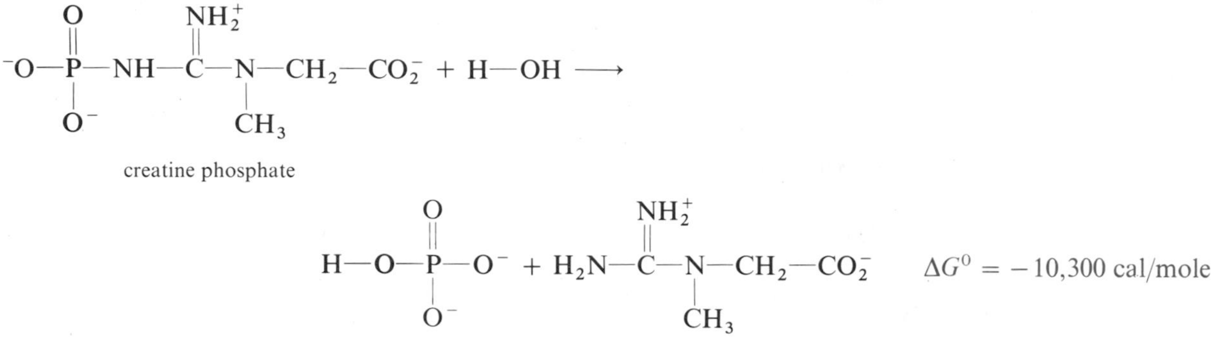

$$^-O-\overset{\overset{\displaystyle O}{\|}}{\underset{\underset{\displaystyle O^-}{|}}{P}}-NH-\overset{\overset{\displaystyle NH_2^+}{\|}}{\underset{\underset{\displaystyle CH_3}{|}}{C}}-N-CH_2-CO_2^- + H-OH \longrightarrow$$

creatine phosphate

$$H-O-\overset{\overset{\displaystyle O}{\|}}{\underset{\underset{\displaystyle O^-}{|}}{P}}-O^- + H_2N-\overset{\overset{\displaystyle NH_2^+}{\|}}{\underset{\underset{\displaystyle CH_3}{|}}{C}}-N-CH_2-CO_2^- \qquad \Delta G^0 = -10,300 \text{ cal/mole}$$

creatine

The immediate chemical change on muscle contraction is hydrolysis of ATP and an increase in the concentration of ADP. In response, a phosphoryl group is transferred from creatine phosphate to ADP and more ATP becomes available for muscle contraction. This reaction is catalyzed by the enzyme creatine kinase (CK).

$$\text{creatine phosphate} + \text{ADP} \xrightarrow{\text{CK}} \text{creatine} + \text{ATP} \qquad \Delta G^0 = -3,000 \text{ cal/mole}$$

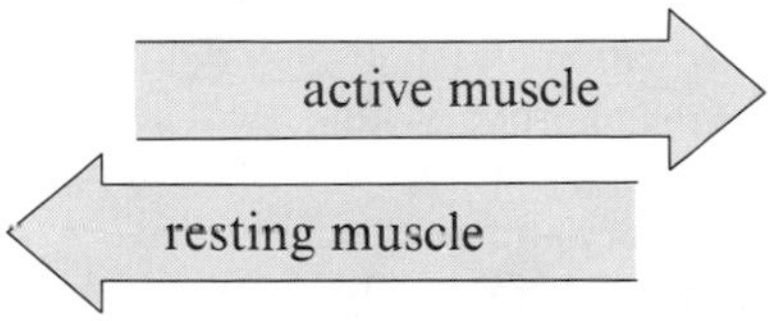

active muscle →

← resting muscle

During rest, the supply of ATP is regenerated and, in turn, used to regenerate the supply of creatine phosphate.

Key Terms and Concepts

acetyl coenzyme A (23.4B)

actin (23 6C)

active transport (23.6B)

adenosine triphosphate (23.3B)
 as a central carrier of biological
  energy (23.3D)

anabolism (23.1)

catabolism (23.1)

change in free energy under standard
 conditions, ΔG^0 (23.3A)

chemical work (23.6A)

coenzyme A (23.4B)

coenzyme Q (23.5C)

creatine phosphate (23.6C)

cytochromes (23.5C)

electron transport chain (23.5A)

free energy of hydrolysis (23.3B)

high-energy compound (23.3B)

inhibitors of respiration (23.5E)

low-energy compound (23.3B)

mechanical work (23.6C)

metabolic pathway (23.1)

metabolism (23.1)

mitochondrion (23.4C)

myosin (23.6C)

nicotinamide (23.4C)

osmotic work (23.6B)

oxidative phosphorylation (23.5B)

pantothenic acid (23.4B)

passive transport (23.6B)

photosynthesis (23.2)

respiration (23.2)

respiratory chain (23.5A)

riboflavin (23.4C)

sliding filament model (23.6C)

stages in oxidation of foodstuffs and
  generation of ATP (23.4)

tricarboxylic acid cycle (TCA) (23.4)

Problems

**The Concept of Free
Energy (Section
23.3)**

23.1 What is meant by the term *free energy*? By ΔG^0?

23.2 How is the sign of ΔG^0 related to the rate of a chemical reaction? To the position of equilibrium of a chemical reaction? To the spontaneity of a chemical reaction?

23.3 Draw structural formulas for the following. On each structural formula, label all phosphate ester bonds, phosphate anhydride bonds, and glycoside bonds.
  a. ATP   b. ADP   c. AMP

23.4 Define the term *high-energy compound* as it is used in biochemistry.

23.5 ATP is a phosphorylating agent. Explain what change in structural formula takes place when a molecule is phosphorylated.

23.6 Write equations for phosphorylation of the following compounds. Assume that ATP is the phosphorylating agent and that it is converted to ADP. Write structural formulas for the molecules named in parts (a), (b), and (c).
  a. phosphorylation of glucose to give α-D-glucose 6-phosphate
  b. phosphorylation of glycerol to give glycerol 1-phosphate
  c. phosphorylation of fructose 6-phosphate to give α-D-fructose 1,6-diphosphate

23.7 What does it mean to say that ATP is the central carrier of free energy in biological systems?

23.8 Calculate ΔG^0 for the following reactions. Which are spontaneous as written? Which are not spontaneous as written?
  a. phosphoenolpyruvate + ADP $\longrightarrow$ pyruvate + ATP
  b. 1,3-diphosphoglycerate + ADP $\longrightarrow$ 3-phosphoglycerate + ATP
  c. glucose + ATP $\longrightarrow$ glucose 6-phosphate + ADP
  d. glucose 1-phosphate + ADP $\longrightarrow$ glucose + ATP
  e. glucose 1-phosphate $\longrightarrow$ glucose 6-phosphate

23.9 The change in free energy for complete oxidation of glucose to carbon dioxide and water is $-686,000$ cal per mole of glucose.

$$C_6H_{12}O_6 + 6\,O_2 \longrightarrow 6\,CO_2 + 2\,H_2O \qquad \Delta G^0 = -686,000 \text{ cal/mole}$$

If all of this change in free energy could be channeled by the cell into conversion of ADP and HPO_4^{2-} into ATP, how many moles of ATP could be formed per mole of glucose oxidized?

Stages in the Oxidation of Foodstuffs and Generation of ATP (Section 23.4)

23.10 Outline the four stages by which cells extract energy from foodstuffs. Of these four stages, which are concerned primarily with:
a. degradation of fuel molecules
b. generation of NADH and FADH$_2$
c. generation of ATP
d. consumption of O$_2$

23.11 In stage 1 of the oxidation of foodstuffs, fuel molecules are hydrolyzed to smaller molecules, the bulk of which can be accounted for by 27 small molecules. List these 27 small molecules.

23.12 Name the separate units from which pantothenic acid is constructed.

23.13 a. Write an abbreviated structural formula for NAD$^+$ showing the portion of the molecule that functions as an oxidizing agent.
b. Write an abbreviated structural formula for NADH showing the portion of the molecule that functions as a reducing agent.
c. Which water-soluble vitamin is an essential part of NAD$^+$?
d. Complete and balance the following half-reaction:

$$NAD^+ + H^+ \longrightarrow$$

23.14 Write balanced equations for the oxidation of the following by NAD$^+$. Note that for each oxidation, the organic product is also given.

a. CH_3-CH_2-OH to $CH_3-\overset{\overset{\text{O}}{\|}}{C}-H$

b. $CH_3-\overset{\overset{\text{OH}}{|}}{CH}-CO_2^-$ to $CH_3-\overset{\overset{\text{O}}{\|}}{C}-CO_2^-$

c. $H-\overset{\overset{\text{OH}}{|}}{\underset{\underset{CH_2-CO_2^-}{|}}{\underset{\underset{|}{CH-CO_2^-}}{C}}}-CO_2^-$ to $\overset{\overset{\text{O}}{\|}}{\underset{\underset{CH_2-CO_2^-}{|}}{\underset{\underset{|}{CH-CO_2^-}}{C}}}-CO_2^-$

d. $CH_3-\overset{\overset{\text{O}}{\|}}{C}-H$ to $CH_3-\overset{\overset{\text{O}}{\|}}{C}-O^-$

23.15 Write the standard abbreviations for the oxidized and reduced forms of flavin adenine dinucleotide. Which water-soluble vitamin is an essential precursor for this molecule?

23.16 Write a balanced equation for the oxidation of succinate by FAD. The organic product is fumarate.

$$^-O_2C-CH_2-CH_2-CO_2^- \quad \text{to} \quad \underset{\text{H}}{\overset{^-O_2C}{>}}C=C\underset{CO_2^-}{\overset{H}{<}}$$

succinate   fumarate

Electron Transport and Oxidative Phosphorylation (Section 23.5)

23.17 What is the function of the respiratory chain?

23.18 The final stage in aerobic metabolism involves oxidative phosphorylation. What is oxidized? What is phosphorylated?

23.19 Name the functional groups in coenzyme Q.

23.20 Four of the carriers of electrons in the electron transport chain are structurally related proteins. Name the prosthetic group associated with each of these proteins. What metal is associated with each prosthetic group?

23.21 How many steps are there in the flow of electrons from NADH to molecular oxygen? In how many of these steps is the change in free energy $-7,300$ cal or larger?

23.22 How many steps are there in the flow of electrons from $FADH_2$ to molecular oxygen? In how many of these steps is the change in free energy $-7,300$ cal or larger?

23.23 Account for the fact that oxidation of NADH is coupled with formation of 3 ATP while oxidation of $FADH_2$ is coupled with the formation of only 2 ATP.

23.24 Explain why cyanide poisoning has the same effect on a cell as a lack of oxygen.

Utilization of ATP for Cellular Work (Section 23.6)

23.25 Name the three major types of cellular work that require ATP.

23.26 What is meant by the term *coupled reaction*? Give an example of the coupling of two biochemical reactions by a common intermediate.

23.27 What is the difference in terms of energy requirements between passive transport and active transport?

23.28 What is the role of creatine phosphate in skeletal muscle?

Clinical Enzymology – The Search for Specificity

A 35-year-old, muscular man complains of severe chest pains and is admitted to the hospital. He lifts heavy objects all day in the course of his work. Is his chest pain due to overexertion that particular day, a temporary muscle spasm, a heart attack, or some other cause? The attending physician must rely on several types of information in making a correct diagnosis: patient history; the clinical pattern of the chest pains; electrocardiogram findings; and the rise and fall in the concentration of certain blood serum enzymes. Enzyme assay procedures in use today, ones we will describe in this mini-essay, can be used to determine with virtually 100% certainty whether a patient has or has not had a heart attack as well as the extent of damage to cardiac muscle tissue. To understand the basis for these enzyme tests, we must describe the types of enzymes normally found in serum and then discuss isoenzymes.

Serum Enzymes

There are two types of enzymes normally found in serum, functional enzymes and nonfunctional enzymes. **Functional enzymes** are secreted into the circulatory system and there have clearly defined physiological roles. An example is thrombin, one of the enzymes involved in blood clotting. **Nonfunctional enzymes** have no apparent role in serum. They are largely confined to cells and appear in the surrounding extracellular fluids and blood only as a result of normal tissue breakdown. What makes the presence of nonfunctional enzymes significant is that their level in serum increases dramatically anytime there is cellular injury or increased breakdown due to:

1. Localized trauma, such as a blow, a surgical procedure, or even an intramuscular injection.

2. Inadequate flow of blood to a particular tissue or area.

3. Any condition, such as cancer, in which there is an increase in cell growth accompanied by a corresponding increase in cell destruction.

Isoenzymes

Many enzymes exist in multiple forms, and while all forms of a particular enzyme catalyze the same reaction or reactions, they often do so at different rates. The term **isoenzyme** is used to distinguish between multiple forms of the same enzyme. Two of the most extensively studied sets of isoenzymes are those of lactate dehydrogenase and creatine kinase.

Lactate dehydrogenase (LDH) catalyzes the following reversible reaction.

$$CH_3-\overset{\displaystyle OH}{\underset{\displaystyle |}{CH}}-CO_2^- + NAD^+$$

lactate

lactate
dehydrogenase

$$CH_3-\overset{\displaystyle O}{\overset{\displaystyle ||}{C}}-CO_2^- + NADH + H^+$$

pyruvate

In the forward reaction, lactate is oxidized by NAD^+, and in the reverse reaction, pyruvate is reduced by NADH. LDH is widely distributed

Figure 1 Absorbance versus wavelength for NADH and NAD⁺.

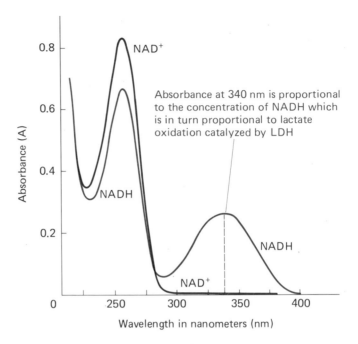

Absorbance at 340 nm is proportional to the concentration of NADH which is in turn proportional to lactate oxidation catalyzed by LDH

throughout the body. High levels are found in the liver, both skeletal and heart muscle, the kidneys, and erythrocytes. When extracts of these or other types of tissues are subjected to electrophoresis at pH 8.0, lactate dehydrogenase can be separated into five isoenzymes. By convention, the isoenzyme moving most rapidly toward the positive electrode is called LDH_1 and that moving most slowly is called LDH_5. LDH isoenzymes are not visible to the eye. To determine their locations, the electrophoresis strip is incubated with lactate and NAD⁺. At sites on the strip where LDH is present, NAD⁺ is reduced to NADH. Figure 1 shows a plot of absorbance versus wavelength for NAD⁺ and NADH. Notice that only

NADH absorbs radiation between 300–400 nm. In the clinical laboratory, absorbance at 340 nm is used to detect NADH.

Figure 2(a) shows a typical LDH isoenzyme pattern. In preparing this strip, the sample is spotted at the right and the direction of migration is toward the positive electrode at the left.

The five LDH isoenzymes are all tetramers composed of different combinations of two polypeptide chains. One chain is designated H because it is found in highest concentration in heart muscle LDH. The other, found in highest concentration in skeletal muscle, is designated M. The fastest moving LDH isoenzyme is a tetramer of four H-chains and the slowest is a tetramer of four

M-chains. The other three are hybrids of H- and M-chains.

Table 1 shows the distribution of LDH isoenzymes in several human tissues. Notice that each type of tissue has a distinct isoenzyme pattern. Liver and skeletal muscle, for example, contain particularly high percentages of LDH_5. Also notice that heart muscle and skeletal muscle have different ratios of LDH_1 / LDH_2.

Because of its wide distribution throughout the body, elevated serum levels of LDH are associated with a broad spectrum of diseases: anemias involving hemolysis of red blood cells, acute liver diseases, congestive heart failure, pulmonary embolism, and muscular diseases such as muscular dystrophy. This broad distribution makes an

Figure 2 An LDH isoenzyme assay. (a) The patterns of LDH isoenzymes in normal serum. (b) A densitometer plot showing the relative concentrations of each isoenzyme. The area under each peak is directly proportional to isoenzyme concentration.

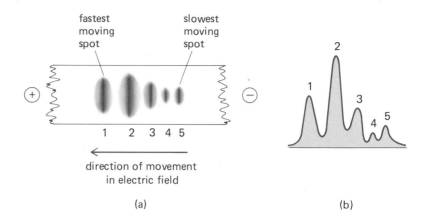

Table 1 Percentage distribution of LDH isoenzymes in several tissues.

tissue	LDH$_1$ (H$_4$)	LDH$_2$ (H$_3$M)	LDH$_3$ (H$_2$M$_2$)	LDH$_4$ (HM$_3$)	LDH$_5$ (M$_5$)
serum	25	35	20	10	10
heart	40	35	20	5	0
kidney	35	30	25	10	0
liver	0	5	10	15	70
brain	25	35	30	10	0
skeletal muscle	0	0	10	30	60

LDH assay a good initial test. If LDH activity is elevated, then an isoenzyme assay can be used to pinpoint the location and type of disease more accurately.

Creatine kinase (CK) catalyzes the transfer of a phosphate group from creatine phosphate to ADP (Section 23.6C) (see equation below). You should note that although the current name for this enzyme is creatine kinase, until recently the preferred name was creatine phosphokinase (CPK). It is likely that you will encounter both names and both abbreviations in your readings.

Creatine kinase consists of three isoenzymes, each of which is a dimer formed by combination of B and/or M polypeptide chains. The B subunit is so named because it was first isolated from brain tissue. The M subunit was first isolated from skeletal muscle tissue. Table 2 shows the percent distribution of CK isoenzymes in several tissues. Notice that heart muscle is the only tissue containing a high percentage of CK-MB isoenzyme.

Table 2 Percent distribution of creatine kinase isoenzymes in several human tissues.

tissue	CK_1 (B_2)	CK_2 (MB)	CK_3 (M_2)
serum	0	0	100
heart	0	40	60
lung	90	0	10
bladder	95	0	5
brain	90	0	10
skeletal muscle	0	0	100

Enzyme Profile of a Heart Attack

During a heart attack (myo-cardial infarction, or MI) a cor-onary artery is partially or completely blocked, reducing the flow of oxygen-rich blood to the heart muscle it serves. If the muscle is damaged because of oxygen starvation, cells die and release their contents into the sur-

Figure 3 LDH isoenzyme assay following a heart attack.

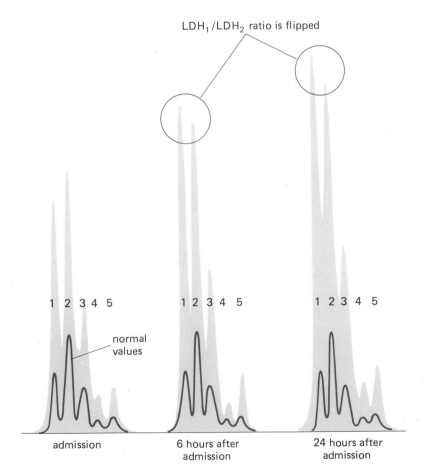

rounding extracellular fluid. Eventually, the cell contents find their way into the blood stream. LDH levels begin to rise approximately 6–24 hours following a heart attack and frequently reach two to three times normal serum levels. Peak LDH serum activity is usually reached within 2–3 days and may remain elevated for up to two weeks. It should be noted that many types of cell and tissue damage lead to elevation in serum LDH activity. What is unique about the pattern that follows heart damage is what happens to the ratio of LDH_1 to LDH_2. According to the data in Table 1, normal serum contains approximately 25% LDH_1 and 35% LDH_2. Thus, in normal serum, the LDH_1/LDH_2 ratio is

less than 1. In heart tissue, the LDH_1/LDH_2 ratio is greater than 1. Following a heart attack, there is an elevation in LDH_1 activity in the serum and the LDH_1/LDH_2 ratio "flips," that is, it changes from less than 1 to greater than 1. In 80% of patients with heart attacks, a flipped LDH_1/LDH_2 ratio appears within 24–48 hours following the attack. Although a flipped LDH_1/LDH_2 ratio is not always seen following a heart attack, there is almost invariably a significant increase in LDH_1 activity. It is for this reason that several companies have developed enzyme assay tests that are highly specific for LDH_1 only. Figure 3 shows an LDH isoenzyme pattern following a mild heart attack. Note that the LDH_1/LDH_2 ratio is

flipped at 12 and 24 hours following admission to the hospital.

Serum levels of creatine kinase also rise after many types of cell and tissue damage. As shown in Table 2, heart muscle is the only human tissue with a high percentage of CK-MB isoenzyme and for this reason, the serum CK isoenzyme pattern following heart damage is unique. Serum CK-MB begins to rise approximately 4–8 hours after myocardial infarction and reaches a peak at about 24 hours. Since CK isoenzymes are degraded quite rapidly, CK levels soon begin to drop and return to normal within a few days of the attack. Figure 4 shows a creatine kinase isoenzyme pattern following a

Figure 4 Creatine kinase (CK) isoenzyme patterns following a heart attack. The peak labeled ALB is serum albumin and is used as a reference and calibration point.

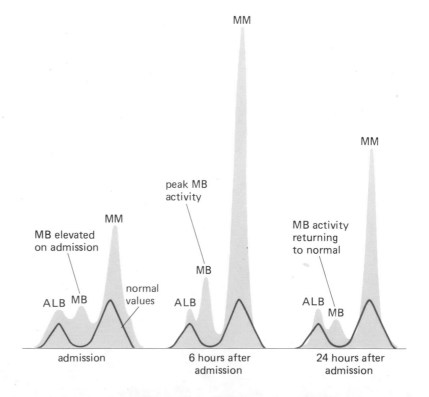

Figure 5 Typical plots of serum creatine kinase-MB activity following heart damage.

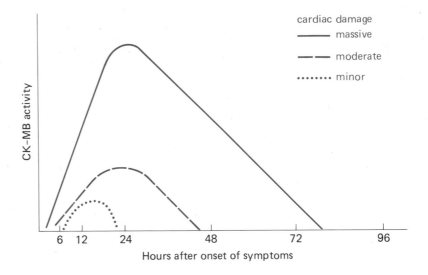

heart attack. Note that all samples, including the one on admission, are positive for CK-MB. The second sample shows greatest activity and the third shows CK-MB returning to normal.

It is often possible to determine the extent of heart damage from the CK-MB pattern. Shown in Figure 5 is a plot of CK-MB activity following minor, moderate, and massive heart damage. Notice that the more massive the heart damage, the sooner CK-MB appears in the serum, the higher its activity rises, and the slower it returns to normal.

What about our 35-year-old man admitted to the hospital with chest pains? Enzyme assays over the next 48 hours showed increased levels of LDH and CK enzymes. Isoenzyme assays, however, showed no elevation in CK-MB and no significant change in the LDH_1/LDH_2 ratio. Therefore, his pains were associated with trauma to skeletal muscle and there was no indication of heart damage.

The Metabolism of Carbohydrates

Glucose is the key food molecule for most organisms and virtually all organisms catabolize glucose by the same set of metabolic pathways. This fact suggests that glucose metabolism became a central feature at an early stage in the evolution of living systems. In this chapter, we will concentrate on the metabolic pathways by which cells extract energy from glucose.

24.1 Digestion and Absorption of Carbohydrates

The major function of dietary carbohydrate is as a source of energy. In a typical American diet, carbohydrates provide about 50–60% of daily energy needs. The remainder is supplied by fats and proteins. During digestion of carbohydrates, di- and polysaccharides are hydrolyzed to monosaccharides, chiefly glucose, fructose, and galactose.

$$\text{polysaccharides} \xrightarrow{\text{hydrolysis}} \text{glucose}$$

$$\text{sucrose} \xrightarrow{\text{hydrolysis}} \text{glucose} + \text{fructose}$$

$$\text{lactose} \xrightarrow{\text{hydrolysis}} \text{glucose} + \text{galactose}$$

In humans, hydrolysis begins in the mouth and is completed in the small intestine. Glucose, fructose, and galactose are then absorbed through the intestinal lining and transported via the blood stream to other parts of the body.

Under normal conditions, the concentration of glucose in blood is between 60 and 100 mg per 100 mL. This level usually rises following a meal and then falls to a fasting level, a point that usually is associated with the onset of hunger. If blood glucose falls below about 60 mg per 100 mL, the condition is known as **hypoglycemia.** In hypoglycemia, there is danger that cells of the central nervous system and other tissues that depend on glucose for nourishment may not receive adequate supplies of glucose. When blood glucose levels rise beyond about 160 mg per 100 mL, the condition is known as **hyperglycemia.**

The liver is the key organ for regulating the concentration of glucose in the blood. As glucose is absorbed after a meal, the liver counters this increase by removing glucose from the blood stream. Glucose removed from blood is used by the liver in two ways: (1) it can be converted to glycogen or triglycerides and stored in the liver; or (2) it can be catabolized to generate ATP and heat.

The concentration of glucose in the blood stream represents a balance between cellular intake, storage, and catabolism. Any defect in the regulation of blood glucose levels can be detected by a **glucose tolerance test,** which measures the ability of tissues to absorb glucose from the blood. One part of the test depends on the fact that the kidneys have only a limited ability to reabsorb glucose as they filter and purify the blood stream. When blood glucose levels are lower than approximately 160–180 mg per 100 mL, virtually all glucose is reabsorbed by the kidneys and returned to the blood stream. However, When blood glucose levels exceed 160–180 mg per 100 mL, the kidneys can no longer absorb the excess and it is passed into the urine. The condition in

Figure 24.1 A typical glucose tolerance curve for a normal individual and one with mild diabetes.

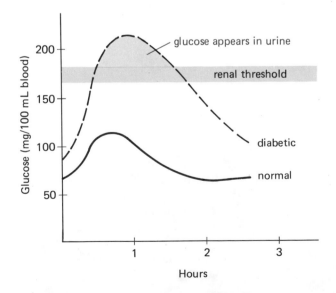

which glucose appears in the urine is called **glycosuria** and the blood glucose level at which this occurs is called the **renal threshold.**

A glucose tolerance test is done in the following way. After an overnight fast, the patient is given a single dose of glucose, typically between 50 and 100 grams in a fruit-flavored drink. Specimens of blood and urine are taken before administration of glucose and then at regular intervals for up to 3–4 hours after the test dose is taken. In normal individuals, the blood glucose level increases within the first hour from 80 mg per 100 mL to approximately 130 mg per 100 mL; at the end of 2–3 hours, it returns to normal levels. For persons with diabetes, blood glucose begins at an elevated level and rises considerably higher after ingestion of the glucose test solution. Furthermore, the return to pre-test levels is much slower than that observed in normal individuals. Figure 24.1 illustrates typical glucose tolerance curves for a normal individual and one with mild diabetes.

24.2 The Central Role of Glucose in Carbohydrate Metabolism

Because of the glucose requirements of cells, especially those of the central nervous system, the body has developed a set of interrelated metabolic pathways designed to use this compound efficiently and to ensure an adequate supply of it in the blood stream. Several of these pathways serve to oxidize glucose to carbon dioxide and water and to conserve a portion of the energy stored in glucose in the form of ATP and other high-energy compounds. Other pathways serve to "buffer" the concentration of glucose in the blood, that is, to maintain blood glucose levels within a rather narrow range. In this section, we will take an overview of the most important pathways of glucose metabolism. Then, in the following sections, we will discuss four of them (glycolysis, lactate fermentation, alcoholic fermentation, and the tricarboxylic acid cycle) in detail.

A. Glycolysis

Glycolysis is a series of ten consecutive reactions by which glucose is oxidized to two molecules of pyruvate. NAD^+ is the oxidizing agent. Furthermore, as we shall see in Section 24.3, two molecules of ATP are produced for each molecule of glucose oxidized to pyruvate. The net reaction for glycolysis is given in Figure 24.2.

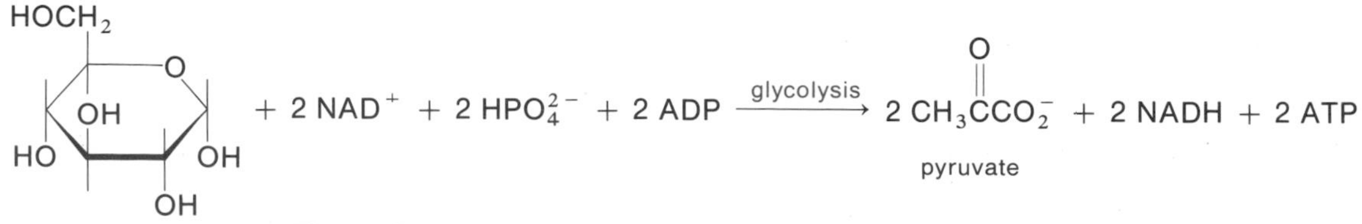

$$ + \; 2\,NAD^+ \; + \; 2\,HPO_4^{2-} \; + \; 2\,ADP \xrightarrow{\text{glycolysis}} 2\,CH_3\overset{O}{\overset{\|}{C}}CO_2^- \; + \; 2\,NADH \; + \; 2\,ATP $$

pyruvate

Figure 24.2 The net reaction for glycolysis.

B. Oxidation and Decarboxylation of Pyruvate

Following glycolysis, the carboxylate group of pyruvate is converted to carbon dioxide and the remaining two carbons are converted to an acetyl group in the form of a thioester with coenzyme A. Several coenzymes including NAD^+ and coenzyme A are required for this metabolic pathway. The net reaction for the oxidative decarboxylation of pyruvate is shown in Figure 24.3.

$$CH_3-\overset{\overset{\text{O}}{\|}}{C}-CO_2^- + NAD^+ + CoA-SH \longrightarrow CH_3-\overset{\overset{\text{O}}{\|}}{C}-SCoA + CO_2 + NADH$$

pyruvate   acetyl CoA

Figure 24.3 Net reaction for the oxidative decarboxylation of pyruvate.

C. The Tricarboxylic Acid Cycle

In the reactions of the tricarboxylic acid cycle (Section 24.5), the two-carbon acetyl group of acetyl CoA is oxidized to two molecules of carbon dioxide. The net reaction for the tricarboxylic acid cycle is shown in Figure 24.4.

   The combination of glycolysis, oxidation of pyruvate to acetyl CoA, and the tricarboxylic acid cycle brings about complete oxidation of glucose to carbon dioxide and water and generates 2 moles of $FADH_2$, 10 moles of NADH, and 4 moles of ATP for each mole of glucose oxidized.

$$CH_3-\overset{\overset{\text{O}}{\|}}{C}-SCoA + 3\,NAD^+ + FAD + HPO_4^{2-} + ADP \xrightarrow{\text{TCA cycle}}$$

$$2\,CO_2 + 3\,NADH + FADH_2 + ATP + CoA-SH$$

Figure 24.4 The net reaction of the tricarboxylic acid cycle.

D. Oxidative Phosphorylation

Glycolysis, oxidation of pyruvate to acetyl coenzyme A, and the tricarboxylic acid cycle are completely *anaerobic*, meaning that they do not involve molecular oxygen. Rather, there is a buildup of reduced coenzymes. Oxidation of NADH and $FADH_2$ is coupled with phosphorylation of ADP during electron transport and oxidative phosphorylation (Section 23.5). It is oxidative phosphorylation that generates the major share of the ATP produced during glucose catabolism. The net reaction for the oxidation of all reduced coenzymes formed during the catabolism of one mole of glucose is shown in Figure 24.5.

$$10\ NADH + 2\ FADH_2 + 6\ O_2 + 32\ ADP + 32\ HPO_4^{2-} + 10\ H^+$$

$$\xrightarrow[\text{phosphorylation}]{\text{oxidative}} 10\ NAD^+ + 2\ FAD + 32\ ATP + 44\ H_2O$$

Figure 24.5 The net reaction of oxidative phosphorylation following degradation of 1 mole of glucose to carbon dioxide and water.

Example 24.1

Glucose is oxidized to carbon dioxide and water by a combination of three metabolic pathways. How many molecules of CO_2 are produced in each pathway?

Solution

No CO_2 is produced during glycolysis. Two molecules of CO_2 are produced in the oxidation and decarboxylation of pyruvate to acetyl coenzyme A. The remaining four molecules of CO_2 are produced through the reactions of the tricarboxylic acid cycle.

Problem 24.1

a. During glycolysis, how many moles of NADH and $FADH_2$ are produced per mole of glucose converted to pyruvate?

b. During the conversion of pyruvate to acetyl CoA, how many moles of NADH are produced per mole of pyruvate?

c. During the tricarboxylic acid cycle, how many moles of NADH and $FADH_2$ are produced per mole of acetyl CoA entering the cycle?

E. Pentose Phosphate Pathway

The **pentose phosphate pathway** is an alternative pathway for the oxidation of glucose to carbon dioxide and water (Figure 24.6). At first glance, the pentose phosphate pathway appears to accomplish the same thing as a combination of glycolysis, oxidation of pyruvate to acetyl CoA, and the tricarboxylic acid cycle, namely, oxidation of glucose to carbon dioxide and water. While it is true that both sets of pathways bring about oxidation of glucose, there are important differences between them. The following reactions of the pentose phosphate pathway have been chosen to illustrate two of the most important differences. The first reaction of this pathway is oxidation of the aldehyde

Figure 24.6 The net reaction of the pentose phosphate pathway.

$$C_6H_{12}O_6 + 12\ NADP^+ + 6\ H_2O \xrightarrow[\text{pathway}]{\substack{\text{pentose}\\\text{phosphate}}} 6\ CO_2 + 12\ NADPH + 12\ H^+$$

group of glucose 6-phosphate to a carboxylate group. Oxidation requires $NADP^+$ (Figure 23.10), a phosphorylated form of NAD^+, and is catalyzed by glucose 6-phosphate dehydrogenase. Next, oxidation of the secondary alcohol on carbon-3 of 6-phosphogluconate by a second molecule of $NADP^+$ gives a β-ketoester, which undergoes decarboxylation (Section 16.5C) to form ribulose 5-phosphate. In one of several reactions that follow, ribulose 5-phosphate is isomerized to ribose 5-phosphate.

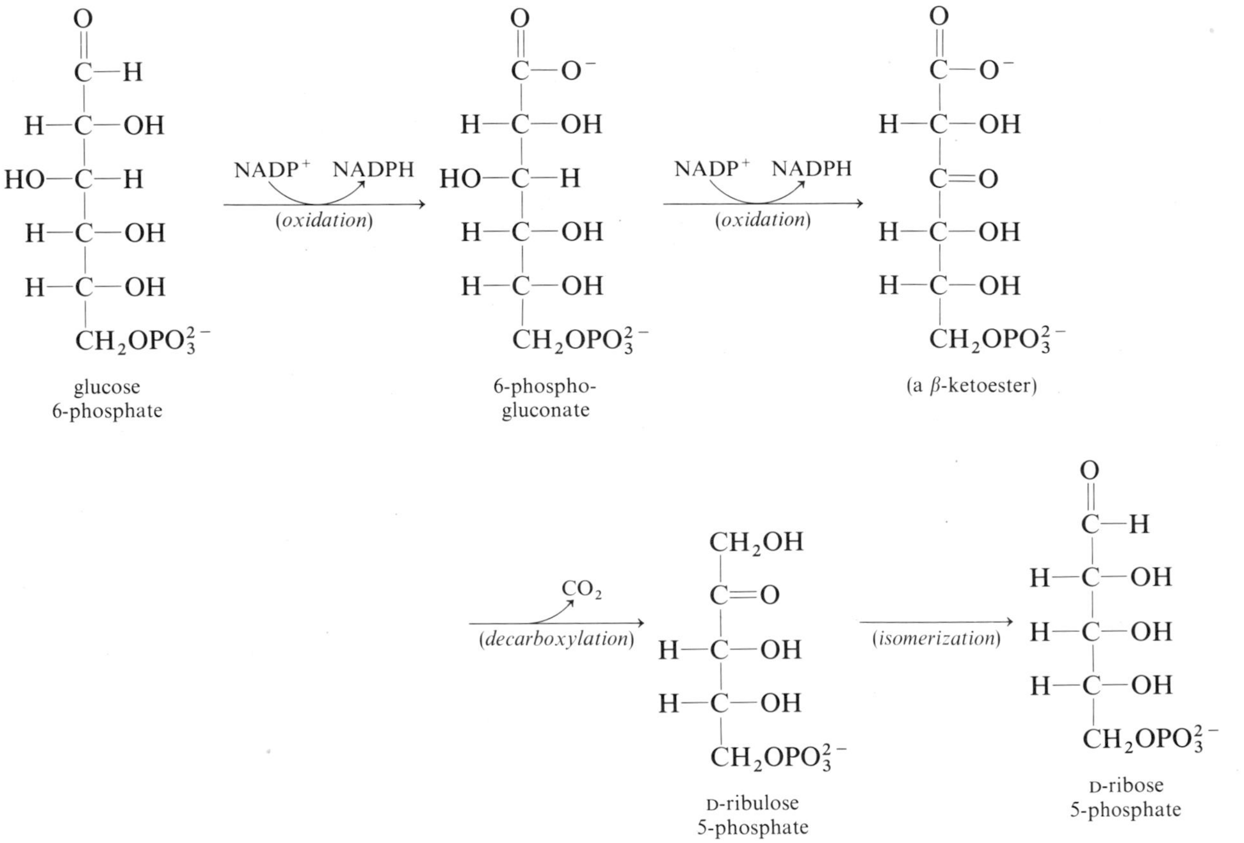

glucose 6-phosphate

6-phospho-gluconate

(a β-ketoester)

D-ribulose 5-phosphate

D-ribose 5-phosphate

These reactions illustrate two important characteristics of the pentose phosphate pathway. First, it provides a pool of pentoses for the synthesis of nucleic acids. Although not illustrated here, it also provides a pool of tetroses. Second, it uses $NADP^+$ as an oxidizing agent and generates the reduced coenzyme NADPH. The major function of NADPH is as a reducing agent in the biosynthesis of other molecules. For example, adipose tissue, which has a high demand for reducing power to support the synthesis of fatty acids (Chapter 25), is rich in $NADP^+/NADPH$. By comparison, glycolysis, oxidation of pyruvate, and the tricarboxylic acid cycle require NAD^+ and FAD as oxidizing agents, coenzymes that are reduced to NADH and $FADH_2$. The major uses of NADH and $FADH_2$ are for generation of ATP through electron transport and oxidative phosphorylation (Section 23.5).

The pentose phosphate pathway is especially important for the normal functioning of red blood cells. These cells depend on this pathway for a supply of NADPH as a reducing agent to maintain iron atoms of hemoglobin in the Fe^{2+} state.

The activity of the pentose phosphate pathway is controlled by regulation of the first enzyme in the pathway, glucose 6-phosphate dehydrogenase. There is a genetic disease, glucose 6-phosphate dehydrogenase deficiency, that affects over 100 million people, mostly in Mediterranean and tropical areas. In this disease, operation of the pentose phosphate pathway is decreased due to a defect in the regulatory enzyme. As a result, concentrations of NADPH are lower than normal, erythrocyte membranes are fragile and rupture more easily than normal, and the average lifespan of erythrocytes is reduced. The clinical condition that results from glucose 6-phosphate deficiency is called **hemolytic anemia.** Like sickle-cell anemia (see the mini-essay, "Abnormal Human Hemoglobins"), this disease appears to make individuals more resistant to certain forms of malaria parasites.

F. Glycogenesis and Glycogenolysis

Glycogenesis and **glycogenolysis** are probably the most important metabolic pathways contributing to a relatively constant blood glucose level. When dietary intake of glucose exceeds immediate needs, humans and other animals convert the excess to glycogen (Section 15.5B) which is stored in the liver and muscle tissue. In normal adults the liver can store about 110 grams of glycogen, and muscles can store about 245 grams. The pathway that converts glucose into glycogen is called **glycogenesis** (Figure 24.7).

$$\text{glycogen} \underset{\text{glycogenesis}}{\overset{\text{glycogenolysis}}{\rightleftharpoons}} \text{glucose}$$

Figure 24.7 Glycogenesis and glycogenolysis.

Liver and muscle glycogen are storage forms of glucose. When there is need for additional blood glucose, glycogen is hydrolyzed and glucose released into the blood stream. The pathway that hydrolyzes glycogen to glucose is called **glycogenolysis** (Figure 24.7). This process is stimulated by the pancreatic hormone glucagon (Problem 20.36). The counterbalancing actions of glucagon and insulin in regulating normal, resting levels of blood glucose are shown schematically in Figure 24.8 (next page).

Figure 24.8 Under normal circumstances, the rate of glucogen-stimulated hydrolysis of glycogen and release of glucose into the blood stream is balanced by insulin-stimulated uptake and metabolism of glucose by brain, muscle, adipose, and liver tissue.

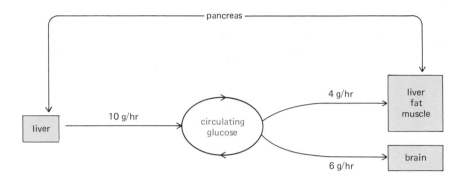

Figure 24.8 Under normal circumstances, the rate of glucogen-stimulated hydrolysis of glycogen and release of glucose into the blood stream is balanced by insulin-stimulated uptake and metabolism of glucose by brain, muscle, adipose, and liver tissue.

G. Synthesis of Fatty Acids and Triglycerides

When carbohydrate intake is greater than the body's immediate needs for energy and its capacity to store glycogen, the excess is converted into fatty acids, molecules that can be stored in almost unlimited quantities as triglycerides. To be stored as triglycerides, glucose is first catabolized to acetyl CoA, the acetyl group of which provides the carbon atoms for the synthesis of fatty acids.

$$C_6H_{12}O_6 \longrightarrow CH_3-\overset{\overset{\displaystyle O}{\|}}{\underset{\underset{\text{fatty acids}}{\Updownarrow}}{C}}-SCoA + CO_2$$

Fatty acids then combine with glycerol to form triglycerides. The synthesis of fatty acids from acetyl CoA represents a link between the metabolism of glucose and that of fatty acids. We shall discuss the biochemistry of fatty acid synthesis and degradation in Chapter 25.

H. Gluconeogenesis

The total supply of glucose in the form of liver and muscle glycogen and blood glucose can be depleted after about 12–18 hours of fasting. In fact, these stores of glucose often are not sufficient for the duration of an overnight fast between dinner and breakfast. Further, they also can be depleted in a short time during work or strenuous exercise. Without any way to provide additional supplies, nerve tissue, including the brain, would soon be deprived of glucose. Fortunately, the body has developed a metabolic pathway to overcome this problem.

Gluconeogenesis is the synthesis of glucose from noncarbohydrate molecules. During periods of low carbohydrate intake and when carbohydrate

Figure 24.9 Gluco-neogenesis, the synthesis of glucose from noncarbohydrate precursors.

$$\left\{ \text{lactate} + \genfrac{}{}{0pt}{}{\text{certain}}{\text{amino acids}} + \text{glycerol} \right\} \xrightarrow{\text{gluconeogenesis}} \text{glucose}$$

stores are being depleted rapidly, the carbon skeletons of lactate, glycerol (derived from the hydrolysis of fats), and certain amino acids are channeled into the synthesis of glucose (Figure 24.9).

The major pathways in the metabolism of glucose are summarized in Figure 24.10.

Figure 24.10 The flow of carbon atoms in the major metabolic pathways of glucose metabolism. The flow of energy (ATP generation and consumption) is not shown.

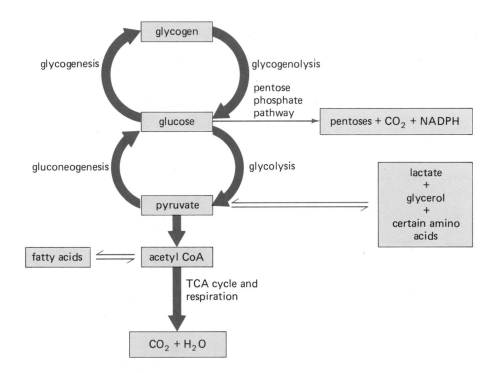

24.3 Glycolysis

A. The Reactions of Glycolysis

Although the net reaction of **glycolysis** is simple to write (Figure 24.2), it took several decades of patient, intensive research by scores of scientists to discover the separate steps by which glucose is degraded to pyruvate and to understand how this degradation is coupled with the production of ATP. But by 1940, the steps in glycolysis had been worked out. Glycolysis is frequently called the

Embden-Meyerhof pathway in honor of the two German biochemists, Gustav Embden and Otto Meyerhof, who contributed so greatly to our present knowledge of it.

The first step of glycolysis is phosphorylation of glucose by ATP to yield glucose 6-phosphate. Transfer of a phosphate group from ATP to an organic molecule is one of the basic reaction types in living systems and any enzyme that catalyzes this type of reaction is called a *kinase*. The enzyme that catalyzes the transfer of a phosphate group from ATP to glucose (a hexose) is called hexokinase.

$$
\begin{array}{c}
\text{CHO} \\
\text{H—C—OH} \\
\text{HO—C—H} \\
\text{H—C—OH} \\
\text{H—C—OH} \\
\text{CH}_2\text{OH}
\end{array}
\quad + \text{ ATP} \xrightarrow{\text{hexokinase}}
\begin{array}{c}
\text{CHO} \\
\text{H—C—OH} \\
\text{HO—C—H} \\
\text{H—C—OH} \\
\text{H—C—OH} \\
\text{CH}_2\text{OPO}_3^{2-}
\end{array}
\quad + \text{ ADP}
$$

glucose   glucose 6-phosphate

Formation of glucose 6-phosphate requires the expenditure of one mole of ATP for each mole of glucose entering glycolysis.

The second step of glycolysis is isomerization of glucose 6-phosphate to fructose 6-phosphate, catalyzed by the enzyme phosphoglucoisomerase. This isomerization involves formation of an enediol intermediate (see Problem 13.21) that then forms the carbonyl group of the ketone in fructose 6-phosphate.

$$
\begin{array}{c}
\text{H—C=O} \\
\text{H—C—OH} \\
\text{HO—C—H} \\
\text{H—C—OH} \\
\text{H—C—OH} \\
\text{CH}_2\text{OPO}_3^{2-}
\end{array}
\rightleftharpoons
\left[
\begin{array}{c}
\text{H—C—OH} \\
\text{C—OH} \\
\text{HO—C—H} \\
\text{H—C—OH} \\
\text{H—C—OH} \\
\text{CH}_2\text{OPO}_3^{2-}
\end{array}
\right]
\rightleftharpoons
\begin{array}{c}
\text{H} \\
\text{H—C—OH} \\
\text{C=O} \\
\text{HO—C—H} \\
\text{H—C—OH} \\
\text{H—C—OH} \\
\text{CH}_2\text{OPO}_3^{2-}
\end{array}
$$

glucose 6-phosphate   (*an enediol*)   fructose 6-phosphate
(*an aldohexose*)   (*a ketohexose*)

In the third step, a second mole of ATP is used to convert fructose 6-phosphate to fructose 1,6-diphosphate:

$$
\begin{array}{c}
\text{CH}_2\text{OH} \\
| \\
\text{C}=\text{O} \\
| \\
\text{HO}-\text{C}-\text{H} \\
| \\
\text{H}-\text{C}-\text{OH} \\
| \\
\text{H}-\text{C}-\text{OH} \\
| \\
\text{CH}_2\text{OPO}_3^{2-}
\end{array}
\quad + \text{ATP} \xrightarrow{\text{phosphofructokinase}}
\begin{array}{c}
\text{CH}_2\text{OPO}_3^{2-} \\
| \\
\text{C}=\text{O} \\
| \\
\text{HO}-\text{C}-\text{H} \\
| \\
\text{H}-\text{C}-\text{OH} \\
| \\
\text{H}-\text{C}-\text{OH} \\
| \\
\text{CH}_2\text{OPO}_3^{2-}
\end{array}
\quad + \text{ADP}
$$

fructose
6-phosphate

fructose
1,6-diphosphate

This phosphorylation is catalyzed by phosphofructokinase, a regulatory enzyme whose activity is a control point in the regulation of glycolysis. These first three steps convert glucose into a molecule that can be split into 2 three-carbon fragments.

In the fourth step of glycolysis, fructose 1,6-diphosphate is cleaved to dihydroxyacetone phosphate and glyceraldehyde 3-phosphate. Cleavage is catalyzed by the enzyme aldolase.

$$
\begin{array}{c}
\text{CH}_2\text{OPO}_3^{2-} \\
| \\
\text{C}=\text{O} \\
| \\
\text{HO}-\text{C}-\text{H} \\
| \\
\text{H}-\text{C}-\text{O}-\text{H} \\
| \\
\text{H}-\text{C}-\text{OH} \\
| \\
\text{CH}_2\text{OPO}_3^{2-}
\end{array}
\quad \overset{\text{aldolase}}{\rightleftharpoons}
\begin{array}{c}
\text{CH}_2\text{OPO}_3^{2-} \\
| \\
\text{C}=\text{O} \\
| \\
\text{CH}_2\text{OH} \\
+ \\
\text{H}-\text{C}=\text{O} \\
| \\
\text{H}-\text{C}-\text{OH} \\
| \\
\text{CH}_2\text{OPO}_3^{2-}
\end{array}
\qquad
\begin{array}{l}
\text{dihydroxyacetone} \\
\text{phosphate} \\
\\
\\
\\
\text{glyceraldehyde} \\
\text{3-phosphate}
\end{array}
$$

fructose 1,6-diphosphate

In the fifth step, dihydroxyacetone phosphate is converted into glyceraldehyde 3-phosphate by the same type of enzyme-catalyzed isomerization we have already seen in the case of the isomerization of glucose 6-phosphate to fructose 6-phosphate.

$$
\begin{array}{c}
\text{CH}_2\text{OH} \\
| \\
\text{C}=\text{O} \\
| \\
\text{CH}_2\text{OPO}_3^{2-}
\end{array}
\quad \overset{\substack{\text{triose} \\ \text{phosphate} \\ \text{isomerase}}}{\rightleftharpoons}
\left[
\begin{array}{c}
\text{OH} \\
| \\
\text{C}-\text{H} \\
|| \\
\text{C}-\text{OH} \\
| \\
\text{CH}_2\text{OPO}_3^{2-}
\end{array}
\right]
\rightleftharpoons
\begin{array}{c}
\text{O} \\
|| \\
\text{C}-\text{H} \\
| \\
\text{H}-\text{C}-\text{OH} \\
| \\
\text{CH}_2\text{OPO}_3^{2-}
\end{array}
$$

dihydroxyacetone
phosphate
(*a ketotriose*)

(*an enediol*)

glyceraldehyde
3-phosphate
(*an aldotriose*)

The sixth and seventh steps are two of the most important in glycolysis because they couple oxidation of the aldehyde group of glyceraldehyde 3-phosphate with phosphorylation of ADP. In the sixth step, glyceraldehyde 3-phosphate is oxidized by NAD$^+$. In this process, catalyzed by glyceraldehyde 3-phosphate dehydrogenase, one phosphate ion is required. The immediate product of the aldehyde oxidation is the mixed anhydride, 1,3-diphosphoglycerate.

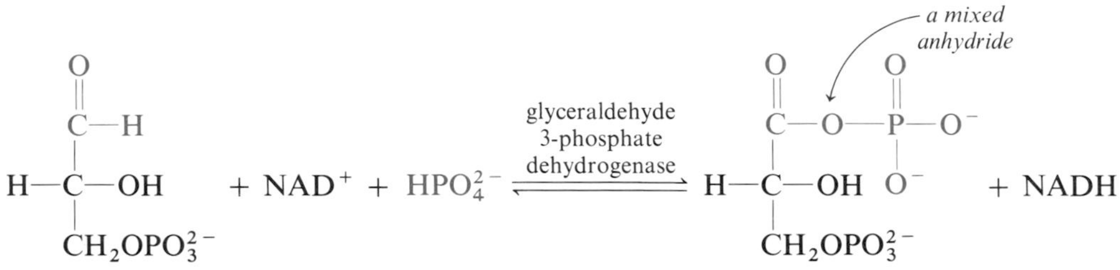

glyceraldehyde
3-phosphate

1,3-diphosphoglycerate

Transfer of a phosphate group from 1,3-diphosphoglycerate to ADP in the seventh step produces the first ATP generated in glycolysis.

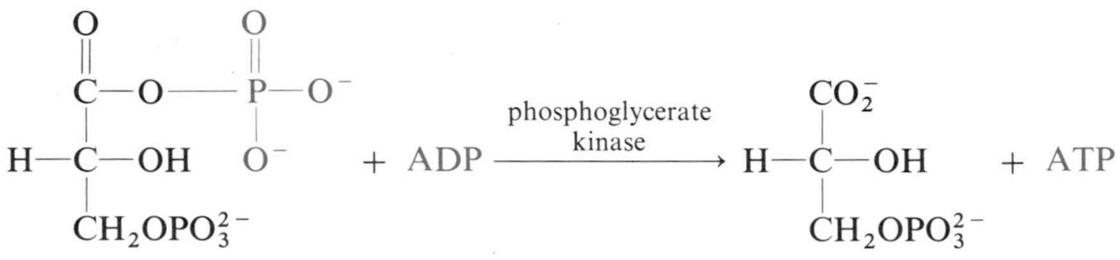

1,3-diphosphoglycerate

3-phosphoglycerate

Often, in writing biochemical reactions, there is a need to show reactants and products in a more compact manner; to do this, biochemists use the following system. A reactant may be shown at the tail of a curved arrow merging with the main arrow and a product may be shown at the head of an arrow branching off the main arrow. Using this shorthand, the overall equation for the oxidation of glyceraldehyde 3-phosphate and phosphorylation of ADP can be written in the following abbreviated form:

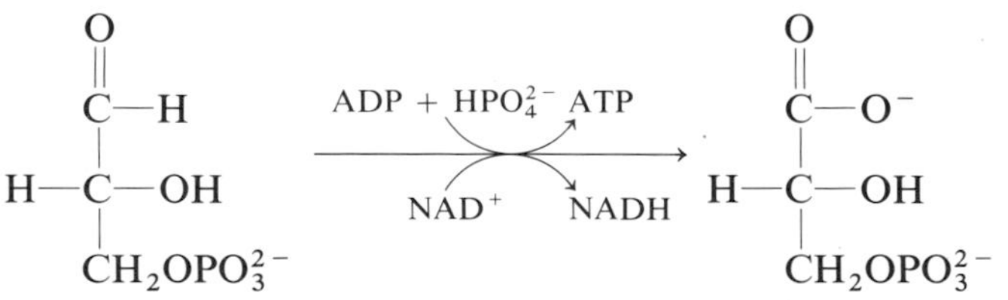

glyceraldehyde
3-phosphate

3-phosphoglycerate

Let us stop and look at the energy balance to this point. Two molecules of ATP were consumed in the conversion of glucose to fructose 1,6-diphosphate. Now, with the oxidation of two molecules of glyceraldehyde 3-phosphate to 3-phosphoglycerate (remember that the original glucose molecule has been split into 2 three-carbon fragments), two molecules of ATP have been generated. Thus, through the first seven steps of glycolysis, the energy expense and profit are balanced.

In steps 8 and 9, 3-phosphoglycerate is isomerized to 2-phosphoglycerate and then dehydrated to form phosphoenolpyruvate.

$$
\begin{array}{ccc}
\text{CO}_2^- & \text{CO}_2^- & \text{CO}_2^- \\
| & | & | \\
\text{H}-\text{C}-\text{OH} \underset{\text{phosphoglyceromutase}}{\rightleftharpoons} & \text{H}-\text{C}-\text{OPO}_3^{2-} \underset{\text{enolase}}{\rightleftharpoons} & \text{C}-\text{OPO}_3^{2-} + \text{H}_2\text{O} \\
| & | & || \\
\text{CH}_2\text{OPO}_3^{2-} & \text{CH}_2\text{OH} & \text{CH}_2 \\
\text{3-phosphoglycerate} & \text{2-phosphoglycerate} & \text{phosphoenolpyruvate}
\end{array}
$$

Phosphoenolpyruvate is a high-energy compound (Section 23.3D) and, in step 10, transfers a phosphate group to ADP to form ATP.

$$
\begin{array}{ccc}
\text{CO}_2^- & & \text{CO}_2^- \\
| & & | \\
\text{C}-\text{OPO}_3^{2-} + \text{ADP} \xrightarrow{\text{pyruvate kinase}} & & \text{C}=\text{O} + \text{ATP} \\
|| & & | \\
\text{CH}_2 & & \text{CH}_3 \\
\text{phosphoenolpyruvate} & & \text{pyruvate}
\end{array}
$$

The ten steps in the conversion of glucose to pyruvate, including those that consume NAD^+ and ATP as well as those that generate ATP, are summarized in Figure 24.11 (next page).

Example 24.2

List all reactions of glycolysis that involve isomerization.

Solution

An isomerization is one in which a reactant and product are isomers of each other. There are three isomerizations in glycolysis:

1. glucose 6-phosphate $\longrightarrow$ fructose 6-phosphate
2. dihydroxyacetone phosphate $\longrightarrow$ glyceraldehyde 3-phosphate
3. 3-phosphoglycerate $\longrightarrow$ 2-phosphoglycerate

Problem 24.2

List all reactions of glycolysis that involve:

a. oxidation

b. cleavage of carbon-carbon bonds

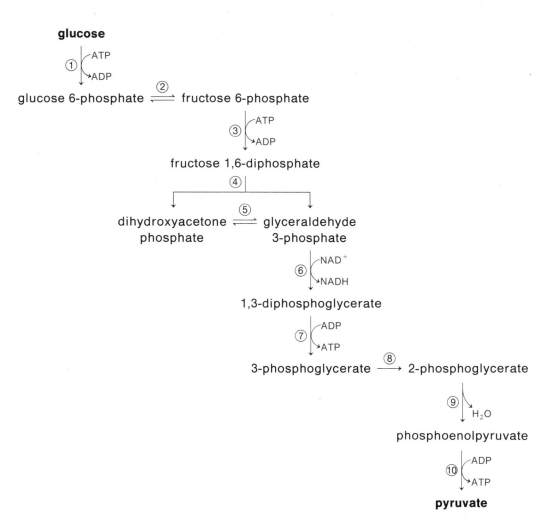

Figure 24.11 An overview of glycolysis.

B. Entry of Fructose and Galactose into Glycolysis

Transformation of fructose into glycolytic intermediates takes place in the liver and begins with phosphorylation of fructose, a reaction catalyzed by the enzyme fructokinase. Fructose 1-phosphate, the product, is then split into two trioses by the same type of reaction we have already seen for the conversion of fructose 1,6-diphosphate into trioses (Section 23.5C).

CH₂OH structures and reaction scheme:

$$\begin{array}{ccc}
\text{CH}_2\text{OH} & & \text{CH}_2\text{OPO}_3^{2-} \\
| & & | \\
\text{C}=\text{O} & & \text{C}=\text{O} \\
| & \xrightarrow[\text{fructokinase}]{\text{ATP ADP}} & | \\
\text{HO}-\text{C}-\text{H} & & \text{HO}-\text{C}-\text{H} \\
| & & | \\
\text{H}-\text{C}-\text{OH} & & \text{H}-\text{C}-\text{O}-\text{H} \xrightarrow[\text{aldolase}]{\substack{\text{fructose}\\\text{1-phosphate}}} \\
| & & | \\
\text{H}-\text{C}-\text{OH} & & \text{H}-\text{C}-\text{OH} \\
| & & | \\
\text{CH}_2\text{OH} & & \text{CH}_2\text{OH} \\
\text{fructose} & & \substack{\text{fructose}\\\text{1-phosphate}}
\end{array}$$

$$\begin{array}{ll}
\text{CH}_2\text{OPO}_3^{2-} & \\
| & \\
\text{C}=\text{O} & \text{dihydroxyacetone} \\
| & \text{phosphate} \\
\text{CH}_2\text{OH} & \\
+ & \\
\text{H}-\text{C}=\text{O} & \\
| & \\
\text{H}-\text{C}-\text{OH} & \text{glyceraldehyde} \\
| & \\
\text{CH}_2\text{OH} &
\end{array}$$

Dihydroxyacetone phosphate is a glycolytic intermediate and enters glycolysis directly. Glyceraldehyde is metabolized in several ways. It may be phosphorylated to glyceraldehyde 3-phosphate and then enter glycolysis. Alternatively, depending on the needs of the organism, it may be reduced to glycerol and then phosphorylated to glycerol phosphate, a metabolic intermediate required for the synthesis of phospholipids.

$$\begin{array}{ccccc}
\text{CHO} & & \text{CH}_2\text{OH} & & \text{CH}_2\text{OH} \\
| & \xrightarrow{\text{NADH NAD}^+} & | & \xrightarrow{\text{ATP ADP}} & | \\
\text{H}-\text{C}-\text{OH} & & \text{H}-\text{C}-\text{OH} & & \text{H}-\text{C}-\text{OH} \\
| & & | & & | \\
\text{CH}_2\text{OH} & & \text{CH}_2\text{OH} & & \text{CH}_2\text{OPO}_3^{2-} \\
\text{glyceraldehyde} & & \text{glycerol} & & \substack{\text{glycerol}\\\text{phosphate}}
\end{array}$$

As a result of genetic diseases, some humans cannot metabolize fructose in the normal way. Those lacking the enzyme fructokinase are unable to metabolize fructose and it is excreted in the urine. Those who lack the enzyme fructose 1-phosphate aldolase have a much more serious problem. Because it cannot be broken down to trioses, fructose 1-phosphate accumulates in the liver and interferes with the activity of several enzyme systems, including those of gluconeogenesis. Kidney function is also disturbed. This condition is known as **fructose intolerance.** Vomiting and loss of appetite are early symptoms of fructose intolerance.

Galactose enters glycolysis by way of a series of reactions that convert it to glucose 6-phosphate.

$$
\begin{array}{ccc}
\text{CHO} & & \text{CHO} \\
| & & | \\
\text{H}-\text{C}-\text{OH} & & \text{H}-\text{C}-\text{OH} \\
| & & | \\
\text{HO}-\text{C}-\text{H} & & \text{HO}-\text{C}-\text{H} \\
| & \xrightarrow[\text{steps)}]{\text{(several}} & | \\
\text{HO}-\text{C}-\text{H} & & \text{H}-\text{C}-\text{OH} \\
| & & | \\
\text{H}-\text{C}-\text{OH} & & \text{H}-\text{C}-\text{OH} \\
| & & | \\
\text{CH}_2\text{OH} & & \text{CH}_2\text{OPO}_3^{2-} \\
\text{D-galactose} & & \text{D-glucose} \\
& & \text{6-phosphate}
\end{array}
$$

The result of these steps is inversion of configuration at carbon-4 and phosphorylation of the hydroxyl group at carbon-6. Among humans, there is an inherited disease, **galactosemia,** that manifests itself by an inability to metabolize galactose. The genetic defect leading to galactosemia is the fact that the liver does not produce a key enzyme involved in the inversion of configuration at carbon-4 that converts galactose to glucose. In persons with this disease, galactose accumulates in the blood and various tissues, including those of the central nervous system, and causes damage to cells. Early symptoms are similar to fructose intolerance. Without treatment, an infant with galactosemia is likely to suffer irreversible brain damage and even death. If recognized in time, however, it is possible to treat galactosemia in a simple manner. Because it is the only source of galactose, milk is excluded from the infant's diet.

C. Regulation of Glycolysis

Within cells, the rate of glycolysis is controlled by the activity of two regulatory enzymes, hexokinase and phosphofructokinase. Hexokinase catalyzes the phosphorylation of glucose to glucose 6-phosphate.

$$\text{glucose} + \text{ATP} \xrightarrow[]{\text{hexokinase}} \text{glucose 6-phosphate} + \text{ADP}$$

inhibited by glucose 6-phosphate

Because it is only in the phosphorylated form that glucose is metabolically active, hexokinase represents a control point not only for glycolysis but for other pathways in the metabolism of glucose as well. Unless glucose is first phosphorylated, it cannot enter glycolysis, the pentose phosphate pathway, or glycogenesis. Hexokinase is inhibited by a high concentration of glucose

6-phosphate, the end product of the reaction it catalyzes. Thus, phosphorylation of glucose is under self-control by feedback inhibition.

The second and more important control point for glycolysis is phosphofructokinase. Once it is converted to fructose 1,6-diphosphate, the carbon skeleton of glucose is committed irreversibly to glycolysis. Because phosphorylation of fructose 6-phosphate represents a committed step, the enzyme that catalyzes it is ideally suited for the role of a regulatory enzyme. Phosphofructokinase is inhibited by high concentrations of ATP and citrate (an intermediate in the TCA) and it is activated by high concentrations of ADP and AMP.

$$\underset{\text{6-phosphate}}{\text{fructose}} + \text{ATP} \xrightarrow[\text{inhibited by ATP and citrate}]{\text{phosphofructokinase}} \underset{\text{1,6-diphosphate}}{\text{fructose}} + \text{ADP}$$

inhibited by ATP and citrate   *activated by AMP and ADP*

To understand the molecular logic of these means of enzyme regulation, remember that glycolysis provides fuel for the tricarboxylic acid cycle which, in turn, provides fuel for respiration and oxidative phosphorylation. When a cell or organism is in a state of low energy demand (supplies of ATP are adequate for immediate energy needs), there is no need to commit glucose to carbohydrate degradation. Hence, the inhibition of phosphofructokinase by ATP. On the other hand, when a cell or organism is using a great deal of energy and the concentration of ATP decreases and those of ADP and AMP increase, ADP and AMP serve as metabolic signals to speed up the degradation of carbohydrates so that more ATP is produced.

24.4 The Fates of Pyruvate

Glycolysis involves oxidation of glucose to pyruvate; in the process, NAD^+ is reduced to NADH:

$$C_6H_{12}O_6 + 2\,NAD^+ \longrightarrow 2\,CH_3-\overset{\overset{\displaystyle O}{\displaystyle \|}}{C}-CO_2^- + 2\,NADH$$

A key to understanding the fates of pyruvate is the fact that without a continuous supply of NAD^+, glycolysis would cease. Therefore, pyruvate is metabolized in ways that regenerate NAD^+.

A. Oxidation to Acetyl CoA

In most mammalian cells operating with a good supply of oxygen, O_2 is the terminal acceptor of electrons from NADH.

$$NADH + H^+ + \tfrac{1}{2}O_2 \longrightarrow NAD^+ + H_2O$$

During respiration, NADH is oxidized to NAD^+, oxygen is reduced to H_2O, and these processes are coupled with phosphorylation of ADP. Thus, under aerobic conditions, pyruvate is converted to acetyl coenzyme A and becomes a fuel for the tricarboxylic acid cycle.

$$CH_3-\overset{\overset{\displaystyle O}{\|}}{C}-CO_2^- + NAD^+ + CoA-SH \longrightarrow CH_3-\overset{\overset{\displaystyle O}{\|}}{C}-SCoA + NADH + CO_2$$

pyruvate   acetyl CoA

Coenzymes required for the conversion of pyruvate to acetyl CoA are NAD^+, coenzyme A, lipoic acid, FAD, and thiamine pyrophosphate.

B. Reduction of Lactate: Lactate Fermentation

During strenuous muscle activity or other conditions when the supply of oxygen is not adequate for reoxidation of NADH, cells turn to the reduction of pyruvate to lactate as a means of regenerating NAD^+.

$$CH_3-\overset{\overset{\displaystyle O}{\|}}{C}-CO_2^- + NADH + H^+ \longrightarrow CH_3-\overset{\overset{\displaystyle OH}{|}}{CH}-CO_2^- + NAD^+$$

pyruvate   lactate

Adding the reduction of lactate to the net reaction of glycolysis gives an overall reaction for a metabolic pathway called **lactate fermentation.**

$$C_6H_{12}O_6 + 2\,ADP + 2\,HPO_4^{2-} \xrightarrow[\text{fermentation}]{\text{lactate}} 2\,CH_3-\overset{\overset{\displaystyle OH}{|}}{CH}-CO_2^- + 2\,ATP$$

glucose   lactate

While lactate fermentation allows glycolysis to continue in the absence of oxygen and generates some ATP, it also results in an increase in the concentration of lactate in muscle tissue and in the blood stream. This buildup of lactate is associated with fatigue. When blood lactate reaches a concentration of about 0.4 mg per 100 mL, muscle tissue becomes almost completely exhausted.

A major portion of the lactate formed in active skeletal muscle is transported by the blood stream to the liver, where it is converted to glucose by the reactions of gluconeogenesis. This newly synthesized glucose is then returned to skeletal muscles for further anaerobic glycolysis and generation of ATP. In this way, a part of the metabolic burden of active skeletal muscle is shifted,

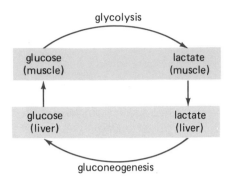

Figure 24.12 The Cori cycle.

at least temporarily, to the liver. Transport of lactate from muscle to the liver, resynthesis of glucose by gluconeogenesis, and return of glucose to muscle tissue is called the **Cori cycle** (Figure 24.12).

C. Reduction to Ethanol: Alcoholic Fermentation

Yeast and several other organisms have developed an alternative pathway for the regeneration of NAD^+ under anaerobic conditions. In the first step of this pathway, pyruvate is decarboxylated to acetaldehyde.

$$CH_3-\overset{\overset{\displaystyle O}{\|}}{C}-CO_2^- + H^+ \xrightarrow{\overset{\text{pyruvate}}{\text{decarboxylase}}} CH_3-\overset{\overset{\displaystyle O}{\|}}{C}-H + CO_2$$

pyruvate   acetaldehyde

The carbon dioxide produced in this reaction is responsible for the foam on beer and the carbonation of naturally fermented wines and champagnes. In a second step, acetaldehyde is reduced by NADH to ethanol.

$$CH_3-\overset{\overset{\displaystyle O}{\|}}{C}-H + NADH \xrightarrow{\overset{\text{alcohol}}{\text{dehydrogenase}}} CH_3CH_2OH + NAD^+$$

acetaldehyde   ethanol

Adding the reactions for the decarboxylation of pyruvate and the reduction of acetaldehyde to the net reaction of glycolysis gives the following overall reaction:

$$C_6H_{12}O_6 + 2\ ADP + 2\ HPO_4^{2-} \xrightarrow{\overset{\text{alcoholic}}{\text{fermentation}}} 2\ CH_3CH_2OH + 2\ CO_2 + 2\ ATP$$

glucose   ethanol

The metabolic pathway that converts glucose to ethanol and carbon dioxide is called **alcoholic fermentation.** Note that both alcoholic and lactate fermentation represent ways in which a cell can continue glycolysis under anaerobic conditions, that is, under conditions where NADH cannot be reoxidized by O_2.

Example 24.3

Under conditions of high oxygen concentration in muscle cells, which of the following compounds would you predict to be present in higher concentrations?

a. NAD^+ or NADH

b. acetyl CoA or coenzyme A

Solution

a. If there is an adequate supply of oxygen to cells, NADH can be oxidized to NAD^+ and therefore NAD^+ will be present in higher concentration than NADH.

b. If NAD^+ is present in high concentration, glycolysis can continue and pyruvate is produced. Pyruvate is, in turn, converted to acetyl CoA, and therefore the concentration of acetyl CoA increases and that of free coenzyme A decreases.

Problem 24.3

Under conditions of low oxygen concentration in muscle cells, which of the following compounds would you predict to be present in higher concentrations?

a. acetyl CoA or lactate

b. lactate or pyruvate?

24.5 The Tricarboxylic Acid Cycle

Under aerobic conditions, the central metabolic pathway for the oxidation of carbon atoms to carbon dioxide is the tricarboxylic acid cycle (TCA) or Krebs cycle (Figure 23.9). The latter name is in honor of Sir Adolph Krebs, the biochemist who first proposed the cyclic nature of this pathway in 1937. Through the reactions of this cycle, the carbon atoms of the acetyl group of acetyl CoA are oxidized to carbon dioxide. As you can see from the balanced half-reaction, this is an eight-electron oxidation.

$$CH_3-\overset{\overset{\displaystyle O}{\|}}{C}-SCoA + 3\,H_2O \longrightarrow 2\,CO_2 + CoA-SH + 8\,H^+ + 8\,e^-$$

This oxidation is brought about by three molecules of NAD^+ and one of FAD.

Following are balanced half-reactions for the reduction of NAD^+ to NADH and of FAD to $FADH_2$.

$$3\,NAD^+ + 3\,H^+ + 6\,e^- \longrightarrow 3\,NADH$$
$$FAD + 2\,H^+ + 2\,e^- \longrightarrow FADH_2$$
$$\overline{3\,NAD^+ + \quad FAD + 5\,H^+ + 8\,e^- \longrightarrow 3\,NADH + FADH_2}$$

Adding the balanced half-reactions for the oxidation of the two-carbon acetyl group and the reduction of three moles of NAD^+ and one of FAD gives the net reaction of the tricarboxylic acid cycle:

$$\overset{\overset{\displaystyle O}{\|}}{CH_3CSCoA} + 3\,NAD^+ + FAD + 3\,H_2O \longrightarrow$$
$$2\,CO_2 + CoA{-}SH + 3\,NADH + FADH_2 + 3\,H^+$$

As we study the individual reactions of the TCA cycle, we shall concentrate on the four reactions that involve oxidations and produce reduced coenzymes and the two that produce carbon dioxide.

A. Steps in the Tricarboxylic Acid Cycle

1. Formation of citrate. The two-carbon acetyl group of acetyl coenzyme A enters the TCA cycle by condensation between the α-carbon of acetyl CoA and the carbonyl group of oxaloacetate. The product of this reaction is citrate, the tricarboxylic acid from which the cycle derives one of its names. In this first reaction of the TCA cycle, carbonyl condensation is coupled with hydrolysis of the thioester to form free coenzyme A.

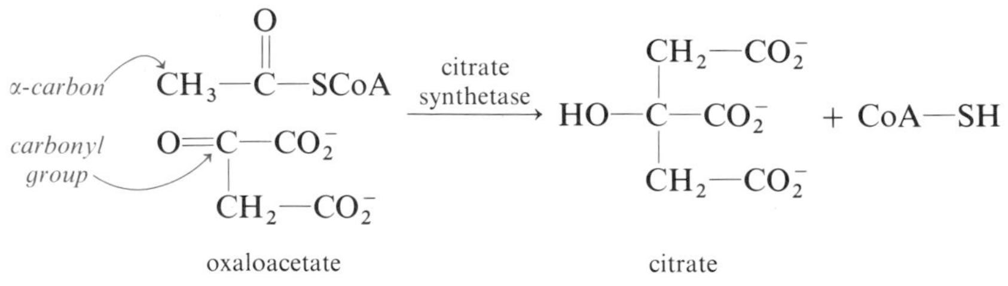

The entry of $CH_3CO{-}$ into the TCA cycle is catalyzed by the enzyme citrate synthetase.

2. Formation of isocitrate. In the next reactions of the cycle, citrate is converted into an isomer, isocitrate. This isomerization is accomplished in two steps. First, in a reaction analogous to acid-catalyzed dehydration of an alcohol (Section 12.5B), citrate undergoes enzyme-catalyzed dehydration to aconitate.

Then, in a reaction analogous to acid-catalyzed hydration (Section 11.5C), aconitate undergoes enzyme-catalyzed hydration to form isocitrate.

$$
\begin{array}{ccc}
\mathrm{CH_2-CO_2^-} & \mathrm{CH_2-CO_2^-} & \mathrm{CH_2-CO_2^-} \\
\mathrm{HO-C-CO_2^-} \xrightarrow{-\mathrm{H_2O}} & \mathrm{C-CO_2^-} \xrightarrow{+\mathrm{H_2O}} & \mathrm{H-C-CO_2^-} \\
\mathrm{H-CH-CO_2^-} & \mathrm{CH-CO_2^-} & \mathrm{HO-CH-CO_2^-} \\
\text{citrate} & \text{aconitate} & \text{isocitrate}
\end{array}
$$

3. Oxidation of isocitrate and generation of the first molecule of CO_2.

At this point, the secondary alcohol of isocitrate is oxidized to a ketone by NAD^+ in a reaction catalyzed by isocitrate dehydrogenase. The product, oxalosuccinate, is a β-ketoacid and undergoes decarboxylation (Section 16.5C) to produce α-ketoglutarate.

$$
\begin{array}{cc}
\mathrm{CH_2-CO_2^-} & \mathrm{CH_2-CO_2^-} \\
\mathrm{CH-CO_2^-} + NAD^+ \xrightarrow{\text{oxidation}} & \mathrm{CH-CO_2^-} + NADH \\
\mathrm{HO-CH-CO_2^-} & \mathrm{O{=}C-CO_2^-} \\
\text{isocitrate} & \text{oxalosuccinate} \\
& \text{(a } \beta\text{-ketoacid)}
\end{array}
$$

$$
\begin{array}{cc}
\mathrm{CH_2-CO_2^-} & \mathrm{CH_2-CO_2^-} \\
\mathrm{CH-CO_2^-} \xrightarrow{\text{decarboxylation}} & \mathrm{CH_2} + CO_2 \\
\mathrm{O{=}C-CO_2^-} & \mathrm{O{=}C-CO_2^-} \\
\text{oxalosuccinate} & \alpha\text{-ketoglutarate}
\end{array}
$$

Oxidation and decarboxylation of isocitrate can also be written in the following way.

$$
\begin{array}{cc}
\mathrm{CH_2-CO_2^-} & \mathrm{CH_2-CO_2^-} \\
\mathrm{CH-CO_2^-} \xrightarrow[]{NAD^+ \quad NADH} & \mathrm{CH_2} \\
\mathrm{HO-CH-CO_2^-} \qquad CO_2 & \mathrm{O{=}C-CO_2^-} \\
\text{isocitrate} & \alpha\text{-ketoglutarate}
\end{array}
$$

4. Oxidation of α-ketoglutarate and generation of the second molecule of carbon dioxide.

The second molecule of carbon dioxide is generated by the TCA cycle in the same type of oxidative decarboxylation we have already seen in Section 24.4A for the conversion of pyruvate to acetyl CoA and carbon dioxide. In oxidative decarboxylation of α-ketoglutarate, the carboxyl group is converted to carbon dioxide and the adjacent ketone is oxidized to a carboxyl group in the form of a thioester with coenzyme A.

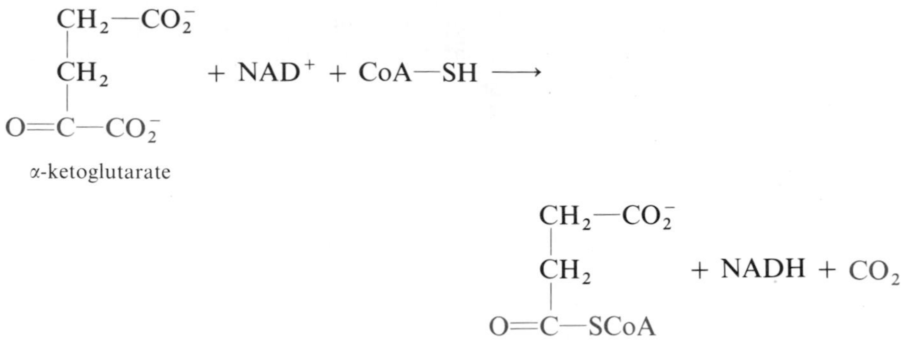

$$\begin{matrix} CH_2-CO_2^- \\ | \\ CH_2 \\ | \\ O{=}C-CO_2^- \end{matrix} \quad + NAD^+ + CoA-SH \longrightarrow$$

α-ketoglutarate

$$\begin{matrix} CH_2-CO_2^- \\ | \\ CH_2 \\ | \\ O{=}C-SCoA \end{matrix} \quad + NADH + CO_2$$

succinyl coenzyme A

In a series of coupled reactions, succinyl coenzyme A, HPO_4^{2-}, and guanosine diphosphate (GDP) react to form succinate, guanosine triphosphate (GTP), and coenzyme A.

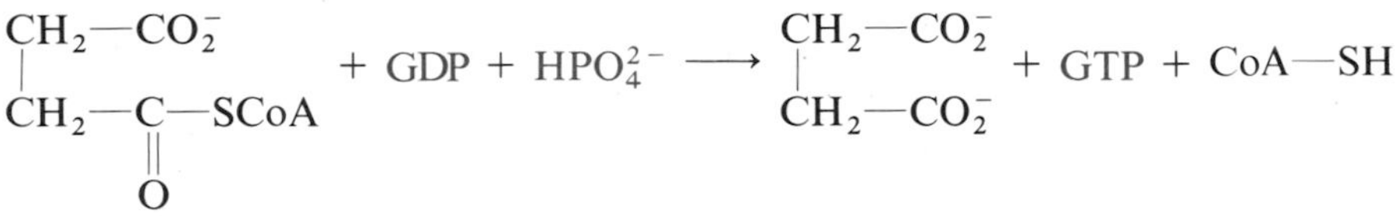

$$\begin{matrix} CH_2-CO_2^- \\ | \\ CH_2-\overset{\underset{\textstyle \|}{O}}{C}-SCoA \end{matrix} + GDP + HPO_4^{2-} \longrightarrow \begin{matrix} CH_2-CO_2^- \\ | \\ CH_2-CO_2^- \end{matrix} + GTP + CoA-SH$$

succinyl coenzyme A   succinate

The terminal phosphate group of GTP can be transferred to ADP according to the reaction:

$$GTP + ADP \rightleftharpoons GDP + ATP$$

Thus, one molecule of a high-energy compound (either GTP or ATP) is produced for each molecule of acetyl CoA entering the tricarboxylic acid cycle. This is the only reaction of the cycle that conserves energy as ATP.

5. *Oxidation of succinate.* In the third oxidation of the cycle, succinate is converted to fumarate. The oxidizing agent for this conversion is FAD. The enzyme catalyzing this reaction, succinate dehydrogenase, is inhibited by such compounds as malonate (Section 21.5E).

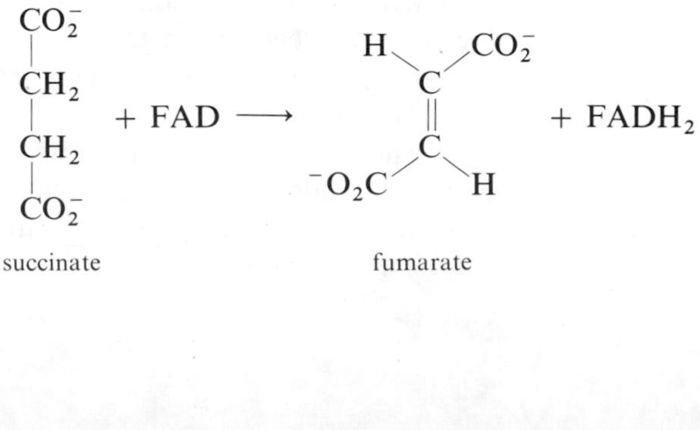

$$\begin{matrix} CO_2^- \\ | \\ CH_2 \\ | \\ CH_2 \\ | \\ CO_2^- \end{matrix} + FAD \longrightarrow \begin{matrix} H\diagdown \quad \diagup CO_2^- \\ C \\ \| \\ C \\ \diagup \quad \diagdown \\ {}^-O_2C \qquad H \end{matrix} + FADH_2$$

succinate   fumarate

6. *Hydration of fumarate*. In the second hydration of the tricarboxylic acid cycle, fumarate is converted to L-malate in a reaction catalyzed by the enzyme fumarase.

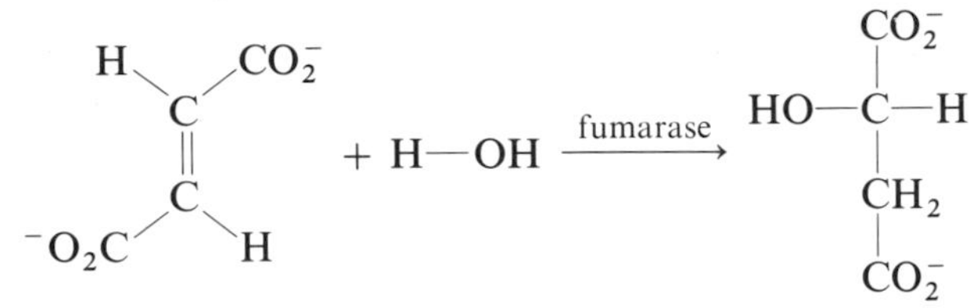

fumarate   L-malate

Fumarase shows a high degree of specificity: it recognizes only fumarate (a *trans* isomer) as a substrate and gives only L-malate (one member of a pair of enantiomers) as the product.

7. *Oxidation of L-malate*. In the fourth and final oxidation of the TCA cycle, L-malate is oxidized by NAD^+ to oxaloacetate.

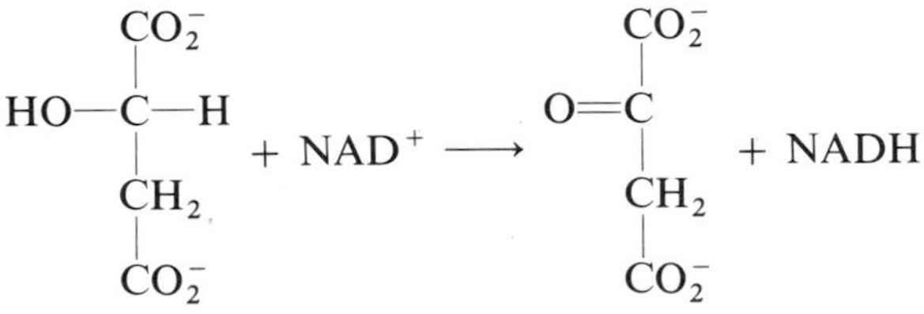

L-malate   oxaloacetate

With production of oxaloacetate, the reactions of the tricarboxylic acid cycle are complete. Continued operation of the cycle requires two things: (1) a supply of carbon atoms in the form of acetyl groups from acetyl CoA; and (2) a supply of oxidizing agents in the form of NAD^+ and FAD. In order to obtain acetyl CoA as a fuel, the TCA cycle is linked to glycolysis and the oxidation of pyruvate, and, as we shall see in the following chapters, it is also linked to the breakdown of fatty acids and amino acids. For a supply of NAD^+ and FAD, the cycle is dependent upon reactions of the electron transport system and oxidative phosphorylation. Recall from Section 23.5 that oxygen is the final acceptor of electrons in the electron transport system and for this reason, continued operation of the TCA cycle depends ultimately on an adequate supply of oxygen.

The reactions of the tricarboxylic acid cycle, including those that generate carbon dioxide, reduced coenzymes, and high-energy phosphates, are summarized in Figure 24.13. This figure also shows the central role of this cycle and its linkage to other metabolic pathways.

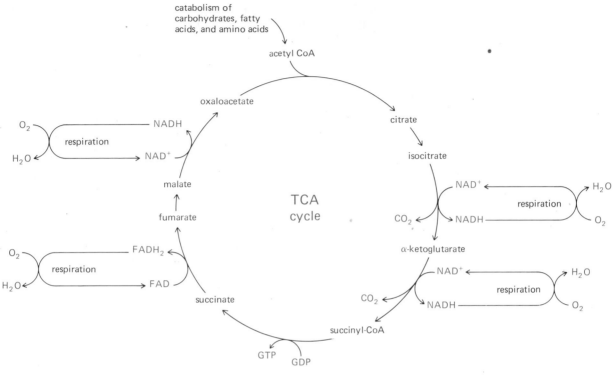

Figure 24.13 The tricarboxylic acid cycle. Fuel for the cycle is derived from catabolism of carbohydrates, fatty acids, and amino acids. For a continuing supply of NAD^+ and FAD, the TCA cycle is dependent on respiration and oxygen.

B. Control of the Tricarboxylic Acid Cycle

The major points of control of the TCA cycle are two regulatory enzymes:

enzyme name	*how regulated*
citrate synthetase	inhibited by ATP and NADH
isocitrate dehydrogenase	inhibited by ATP and NADH, activated by ADP

To appreciate the central role of these regulatory enzymes, remember that the major function of the TCA cycle is to provide fuel in the form of reduced coenzymes for stage IV in the oxidation of foodstuffs and the generation of ATP. Under conditions where supplies of ATP are adequate for the immediate needs of the cell, ATP interacts with citrate synthetase to reduce its affinity for acetyl coenzyme A. Thus, ATP acts as a negative modifier of citrate synthetase and inhibits entry of acetyl CoA into the TCA cycle. Similarly, NADH, a product of the TCA cycle, acts as a negative modifier of citrate synthetase. Isocitrate dehydrogenase, the primary control point in the cycle, is also inhibited by ATP and NADH. Because the activity of this enzyme is increased by interaction with ADP, this substance is said to be a positive modifier of isocitrate dehydrogenase.

24.6 Energy Balance for Glucose Catabolism

Now that we have examined the biochemical pathways by which the carbon atoms of glucose are oxidized to carbon dioxide, let us look at the energy changes in these transformations. Complete oxidation of glucose to carbon dioxide occurs with a large decrease in free energy.

$$C_6H_{12}O_6 + 6\,O_2 \longrightarrow 6\,CO_2 + 6\,H_2O \qquad \Delta G^0 = -686,000 \text{ cal/mole}$$

The number of moles of ATP derived from aerobic catabolism of glucose are summarized in Table 24.1. Note that glycolysis takes place in the cytoplasm

Table 24.1 Yield of ATP from the complete oxidation of glucose.

Reaction	Process	Yield of ATP (moles)
Glycolysis		
glucose $\longrightarrow$ glucose 6-phosphate	phosphorylation	-1
glucose 6-phosphate $\longrightarrow$ fructose 1,6-diphosphate	phosphorylation	-1
glyceraldehyde 3-phosphate $\longrightarrow$ 1,3-diphosphoglycerate	oxidation by NAD^+	$+4$
1,3-diphosphoglycerate $\longrightarrow$ 3-phosphoglycerate	phosphorylation	$+2$
phosphoenolpyruvate $\longrightarrow$ pyruvate	phosphorylation	$+2$
Oxidation of Pyruvate		
pyruvate $\longrightarrow$ acetyl CoA + CO_2	oxidation by NAD^+	$+6$
Tricarboxylic Acid Cycle		
isocitrate $\longrightarrow$ α-ketoglutarate + CO_2	oxidation by NAD^+	$+6$
α-ketoglutarate $\longrightarrow$ succinyl CoA + CO_2	oxidation by NAD^+	$+6$
succinyl CoA $\longrightarrow$ succinate	phosphorylation	$+2$
succinate $\longrightarrow$ fumarate	oxidation by FAD	$+6$
malate $\longrightarrow$ oxaloacetate	oxidation by NAD^+	$+6$
Net yield of ATP:		$+36$

and that NADH produced during oxidation of glyceraldehyde 3-phosphate must be transported into the mitochondria before it can be reoxidized by the reactions of the respiratory chain. Because transport requires hydrolysis of one ATP for each NADH transported across the membrane, each mole of NADH produced during glycolysis yields only 2 moles of ATP. Each mole of NADH generated within mitochondria during the oxidation of pyruvate and reactions of the tricarboxylic acid cycle yields 3 moles of ATP. Each mole of $FADH_2$ formed in the tricarboxylic acid cycle gives 2 moles of ATP. Further, 2 moles of ATP are produced during glycolysis and another 2 moles of ATP are produced during the tricarboxylic acid cycle.

We can write the net reaction and the associated energy changes for the complete oxidation of glucose as the sum of an exergonic oxidation of glucose to carbon dioxide and water and an endergonic phosphorylation of 36 moles of ADP.

Exergonic reaction:

$$C_6H_{12}O_6 + 6\,O_2 \longrightarrow 6\,CO_2 + 6\,H_2O \qquad \Delta G^0 = -686,000 \text{ cal/mole}$$

Endergonic reaction:

$$36\,ADP + 36\,HPO_4^{2-} \longrightarrow 36\,ATP + 36\,H_2O$$

$$\Delta G^0 = +263,000 \text{ cal/mole}$$

Overall reaction:

$$C_6H_{12}O_6 + 6\,O_2 + 36\,ADP + 36\,HPO_4^{2-} \longrightarrow$$

$$6\,CO_2 + 36\,ATP + 42\,H_2O \qquad \Delta G^0 = -423,000 \text{ cal/mole}$$

can be used for biochemical work   *liberated as heat*

The total energy conserved as a result of aerobic oxidation of one mole of glucose is 36 moles of ATP or 263,000 cal/mole. The efficiency of energy conservation during glucose metabolism is:

$$\frac{263,000}{686,000} \quad \text{or} \quad \text{approximately } 38\%$$

It is an impressive feat for living cells to trap this amount of energy as ATP!

The decrease in free energy for lactate fermentation is 47,000 cal/mole, a value considerably smaller than that for aerobic oxidation of glucose. We can write the net reaction and associated energy changes for lactate fermentation as the sum of the exergonic conversion of glucose to lactate and the endergonic phosphorylation of ADP.

Exergonic reaction:

$$glucose \longrightarrow 2\ lactate \qquad \Delta G^0 = -47,000\ cal/mole$$

Endergonic reaction:

$$2\ ADP + 2\ HPO_4^{2-} \longrightarrow 2\ ATP + 2\ H_2O \qquad \Delta G^0 = +14,600\ cal/mole$$

Overall reaction:

$$glucose + 2\ ADP + 2\ HPO_4^{2-} \longrightarrow 2\ lactate + 2\ ATP + 2\ H_2O$$

$$\Delta G^0 = -32,400\ cal/mole$$

can be used for   *liberated*
biochemical work   *as heat*

Approximately 30% of the decrease in free energy during lactate fermentation is conserved as ATP and is then available for the performance of biochemical work. The remainder of the decrease in free energy is liberated as heat. Lactate fermentation keeps glycolysis going, but at a cost. Only 2 moles of ATP are produced per mole of glucose by lactate fermentation, compared to 36 moles of ATP produced by complete oxidation of glucose to carbon dioxide and water. The same is true for alcoholic fermentation. Thus, in terms of use of fuel molecules for the production of heat and ATP, aerobic catabolism of glucose is 18 times more efficient than either lactate or alcoholic fermentation.

Key Terms and Concepts

alcoholic fermentation (24.4C)

Cori cycle (24.4B)

Embden-Meyerhof pathway (24.3)

fructose intolerance (24.3B)

galactosemia (24.3B)

gluconeogenesis (24.2H)

glucose tolerance test (24.1)

glycogenesis (24.2F)

glycogenolysis (24.2F)

glycolysis (24.2A and 24.3)

glycosuria (24.1)

hemolytic anemia (24.2E)

hyperglycemia (24.1)

hypoglycemia (24.1)

Krebs cycle (24.5)

lactate fermentation (24.4B)

oxidative decarboxylation (24.2B)

oxidative phosphorylation (24.2D)

pentose phosphate pathway (24.2E)

renal threshold (24.1)

tricarboxylic acid cycle (24.2C)

Key Reactions

1. Net reaction of the pentose phosphate pathway (Section 24.2E):

$$C_6H_{12}O_6 + 12\ NADP^+ + 6\ H_2O \longrightarrow 6\ CO_2 + 12\ NADPH + 12\ H^+$$

2. Glycogenesis (Section 24.2F):

$$glucose \longrightarrow glycogen$$

3. Glycogenolysis (Section 24.2F):

$$glycogen \longrightarrow glucose$$

4. Gluconeogenesis (Section 24.2H):

$$lactate \longrightarrow glucose$$
$$certain\ amino\ acids \longrightarrow glucose$$
$$glycerol \longrightarrow glucose$$

5. Net reaction of glycolysis (Section 24.3):

$$C_6H_{12}O_6 + 2\ NAD^+ + 2\ HPO_4^{2-} + 2\ ADP \longrightarrow$$

$$2\ CH_3\overset{\overset{\textstyle O}{\|}}{C}CO_2^- + 2\ NADH + 2\ ATP$$

6. Fates of pyruvate (Section 24.4):
   a. oxidation and decarboxylation:

$$CH_3\overset{\overset{\textstyle O}{\|}}{C}CO_2^- + NAD^+ + CoA-SH \longrightarrow CH_3\overset{\overset{\textstyle O}{\|}}{C}-SCoA + CO_2 + NADH$$

   b. lactate fermentation:

$$CH_3\overset{\overset{\textstyle O}{\|}}{C}CO_2^- + NADH + H^+ \longrightarrow CH_3\overset{\overset{\textstyle OH}{|}}{C}HCO_2^- + NAD^+$$

   c. alcoholic fermentation:

$$CH_3\overset{\overset{\textstyle O}{\|}}{C}CO_2^- + H^+ \longrightarrow CH_3\overset{\overset{\textstyle O}{\|}}{C}H + CO_2$$

7. Net reaction of lactate fermentation (Section 24.4B):

$$C_6H_{12}O_6 + 2\ ADP + 2\ HPO_4^{2-} \longrightarrow 2\ CH_3\overset{\overset{\textstyle OH}{|}}{C}HCO_2^- + 2\ ATP$$

8. Net reaction of alcoholic fermentation (Section 24.4C):

$$C_6H_{12}O_6 + 2\ ADP + 2\ HPO_4^{2-} \longrightarrow 2\ CH_3CH_2OH + 2\ CO_2 + 2\ ATP$$

9. Net reaction of the tricarboxylic acid cycle (Section 24.5):

$$CH_3\overset{\overset{\textstyle O}{\|}}{C}SCoA + 3\ NAD^+ + FAD + 3\ H_2O \longrightarrow$$

$$2\ CO_2 + CoA-SH + 3\ NADH + FADH_2 + 3\ H^+$$

Problems

The Central Role of Glucose in Carbohydrate Metabolism (Section 24.2)

24.4 Match the following names with the processes listed.
   a. glycolysis   b. gluconeogenesis
   c. glycogenolysis   d. glycogenesis
   _____ synthesis of glucose from a noncarbohydrate
   _____ breakdown of glucose to pyruvate
   _____ hydrolysis of glycogen to glucose
   _____ conversion of glucose to glycogen

24.5 How many moles of ATP are produced either directly or by oxidation of reduced coenzymes and phosphorylation of ADP when:
   a. two moles of glucose are oxidized to CO_2?
   b. two moles of glucose are oxidized to pyruvate?
   c. two moles of glucose 6-phosphate are oxidized to two moles of ribose 5-phosphate?
   d. two moles of acetyl CoA are oxidized to fumarate?

24.6 When liver stores of glycogen are very low (in the morning or during periods of vigorous exercise) and blood glucose levels are low, how can glucose be produced and energy supplied?

24.7 Name two important functions of the pentose phosphate pathway.

24.8 What is the difference in structural formula between NAD^+ and $NADP^+$?

24.9 The degradation of carbohydrates provides the cell with three things: energy; NADPH as reducing power for biosynthesis; and a pool of intermediates for the biosynthesis of other molecules. Which of these three are produced by the following pathways?
   a. the conversion of glucose to lactate   b. the tricarboxylic acid cycle
   c. the pentose phosphate pathway
   d. the conversion of glucose to ethanol

Glycolysis (Section 24.3)

24.10 Name one coenzyme required for glycolysis. From what vitamin is this coenzyme derived?

24.11 Name and draw a structural formula for the end product of glycolysis.

24.12 Number the carbon atoms of glucose 1 through 6 and show the fate of each atom in glycolysis.

24.13 Write equations for the two reactions of glycolysis that consume ATP.

24.14 Write equations for the two reactions of glycolysis that produce ATP.

24.15 Although glucose is the principal source of carbohydrates for glycolysis and other pathways, fructose and galactose are also metabolized for energy.
   a. What is the major dietary source of fructose? Of galactose?
   b. Explain how the carbon skeleton of fructose enters glycolysis.
   c. Explain how the carbon skeleton of galactose enters glycolysis.

24.16 Describe the genetic defect leading to:
   a. fructose intolerance   b. galactosemia

24.17 The feedback effects of ATP, ADP, and AMP are important in regulating both glycolysis and the tricarboxylic acid cycle. Explain the effect of ATP and ADP on:
   a. isocitrate dehydrogenase, a regulatory enzyme of the tricarboxylic acid cycle.
   b. phosphofructokinase, a regulatory enzyme of glycolysis.

The Fates of Pyruvate (Section 24.4)

24.18 Number the carbon atoms of glucose 1 through 6. Show the fate of each atom in:
a. alcoholic fermentation   b. lactate fermentation

24.19 In what ways are alcoholic fermentation and lactate fermentation similar? In what ways do they differ?

24.20 Write balanced half-reactions for the following conversions:
a. glucose $\longrightarrow$ pyruvate
b. glucose $\longrightarrow$ lactate
c. lactate $\longrightarrow$ pyruvate
d. pyruvate $\longrightarrow$ ethanol + carbon dioxide

24.21 What is the major function of the Cori cycle?

24.22 Based on your knowledge of glycolysis, the fates of pyruvate, and the tricarboxylic acid cycle, propose a series of steps for the following biochemical conversions.
a. glycerol $\longrightarrow$ lactate
b. phosphoglycerol $\longrightarrow$ ethanol + carbon dioxide
c. 3-phosphoglyceraldehyde $\longrightarrow$ glucose 6-phosphate
d. glycerol $\longrightarrow$ acetyl CoA
e. ethanol $\longrightarrow$ carbon dioxide

The Tricarboxylic Acid Cycle (Section 24.5)

24.23 What is the major function of the TCA cycle?

24.24 Write equations for the step in the citric acid cycle that results in the formation of a new carbon-carbon bond.

24.25 Write equations for the three steps of the TCA cycle that involve oxidation by NAD^+.

24.26 Write an equation for the step in the TCA cycle that involves oxidation by FAD.

24.27 Write an equation for the step in the TCA cycle that results in formation of a high-energy phosphate bond.

24.28 Why is GTP just as effective a high-energy compound as ATP?

24.29 What does it mean to say that the TCA cycle is catalytic? That it does not produce any new compounds?

24.30 The major control points of the TCA cycle are the regulatory enzymes, citrate synthetase and isocitrate dehydrogenase.
a. Write equations for the reactions catalyzed by each of these enzymes.
b. Each enzyme is inhibited by NADH and ATP. Explain the benefit to the cell of this means of regulation.

Energy Balance for Glucose Metabolism (Section 24.6)

24.31 A maximum of 36 moles of ATP can be formed as the result of complete metabolism of one mole of glucose to carbon dioxide and water. How many of the 36 moles are formed in:
a. glycolysis
b. the tricarboxylic acid cycle
c. the electron transport system

24.32 The total amount of energy that can be obtained from complete oxidation of glucose is 686,000 calories per mole. What fraction of this energy is conserved as ATP in alcoholic fermentation? (Note that although this fraction is small, it is sufficient for the survival of anaerobic cells.)

25

The Metabolism of Fatty Acids

In this chapter, we shall discuss the metabolic pathway for the degradation of fatty acids and how this degradation is coupled with the generation of ATP. In addition, we shall discuss the biosynthesis of fatty acids and then compare and contrast the steps by which cells degrade and synthesize these vital molecules. Finally, we will show some of the interrelationships between the metabolism of fatty acids and carbohydrates.

25.1 Fatty Acids as a Source of Energy

In terms of available energy, fatty acids have the highest caloric value of any food. Following are balanced equations for the complete oxidation of palmitic acid, one of the most abundant fatty acids, and glucose. As can be seen by comparing changes in free energy, complete oxidation of a gram of palmitic acid yields more than twice the energy obtained from a gram of glucose.

The larger yield of energy per gram stems from the fact that the hydrocarbon chain of a fatty acid is more highly reduced than the oxygenated carbon chain of a carbohydrate. This can be seen by comparing the number of molecules of oxygen consumed per carbon atom. One molecule of oxygen is consumed

778

	ΔG^0 (cal/mole)	ΔG^0 (cal/gram)

$$C_6H_{12}O_6 + 6\,O_2 \longrightarrow 6\,CO_2 + 6\,H_2O \qquad -686{,}000 \qquad -3{,}800$$

glucose

$$CH_3(CH_2)_{14}CO_2H + 23\,O_2 \longrightarrow 16\,CO_2 + 16\,H_2O \quad -2{,}340{,}000 \qquad -9{,}300$$

palmitic acid

per carbon atom of glucose, while 23/16 or 1.44 molecules of oxygen are consumed per carbon atom of palmitic acid.

Fatty acids constitute about 40% of the calories in a typical American diet. Further, because they can be stored in large quantities, fatty acids in the form of triglycerides are the major storage form of energy. Adipose tissue contains specialized cells, called adipocytes, whose sole function is to store fats.

25.2 Hydrolysis of Triglycerides

The first phase of catabolism of fatty acids involves their release from triglycerides by hydrolysis. Hydrolysis is catalyzed by a group of enzymes called lipases.

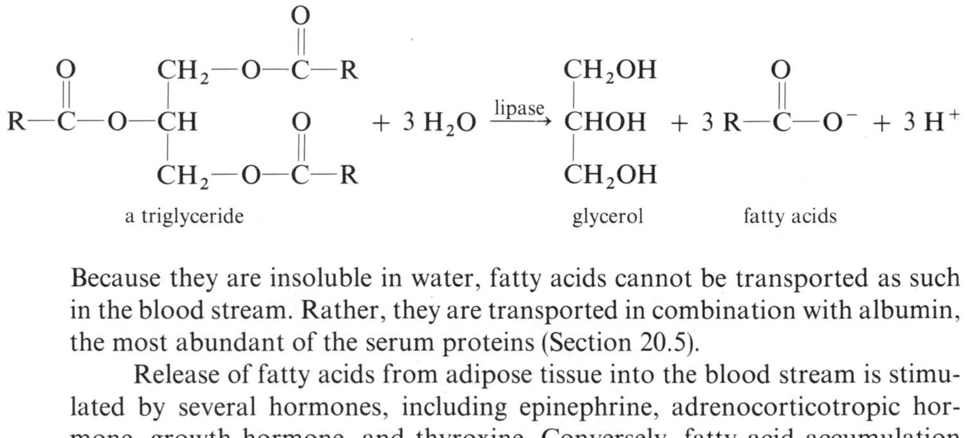

a triglyceride   glycerol   fatty acids

Because they are insoluble in water, fatty acids cannot be transported as such in the blood stream. Rather, they are transported in combination with albumin, the most abundant of the serum proteins (Section 20.5).

Release of fatty acids from adipose tissue into the blood stream is stimulated by several hormones, including epinephrine, adrenocorticotropic hormone, growth hormone, and thyroxine. Conversely, fatty acid accumulation in adipose tissue and storage as triglycerides is stimulated by high levels of glucose and insulin in the blood stream.

25.3 Oxidation of Fatty Acids

The three major stages in oxidation of fatty acids are: activation of free fatty acids in the cytoplasm by formation of a thioester with coenzyme A; transport across the inner mitochondrial membrane; and oxidation within mitochondria to carbon dioxide and water. Let us look at each of these stages separately.

A. Activation of Free Fatty Acids

To begin catabolism, a free fatty acid in the cytoplasm is converted to a thioester formed with coenzyme A. The product is called a fatty acyl CoA. Thioester formation is an endergonic reaction and the energy to drive it is derived by coupling thioester formation with hydrolysis of ATP. In this coupled hydrolysis, ATP is converted to AMP and pyrophosphate. Within the cytoplasm, pyrophosphate is further hydrolyzed to two HPO_4^{2-}.

Endergonic reaction:

$$\underset{\substack{\text{fatty} \\ \text{acid}}}{R-\overset{\overset{\textstyle O}{\|}}{C}-O^-} + HS-CoA \longrightarrow \underset{\substack{\text{fatty acyl} \\ \text{CoA}}}{R-\overset{\overset{\textstyle O}{\|}}{C}-S-CoA} + H_2O \qquad\qquad +7,400$$

ΔG^0 (cal/mole)

Exergonic reactions:

$$ATP + H_2O \longrightarrow AMP + {}^-O-\overset{\overset{\textstyle O}{\|}}{\underset{\underset{\textstyle O^-}{|}}{P}}-O-\overset{\overset{\textstyle O}{\|}}{\underset{\underset{\textstyle O^-}{|}}{P}}-O^- \qquad\qquad -7,600$$

pyrophosphate

$${}^-O-\overset{\overset{\textstyle O}{\|}}{\underset{\underset{\textstyle O^-}{|}}{P}}-O-\overset{\overset{\textstyle O}{\|}}{\underset{\underset{\textstyle O^-}{|}}{P}}-O^- + H_2O \longrightarrow 2\,H-O-\overset{\overset{\textstyle O}{\|}}{\underset{\underset{\textstyle O^-}{|}}{P}}-O^- \qquad\qquad -8,000$$

pyrophosphate · · · · · · · · · · · · · · · phosphate

Because activation of fatty acids is coupled with the hydrolysis of two high-energy phosphate anhydride bonds, the initial investment by a cell in fatty acid oxidation is equivalent to 2 moles of ATP for each mole of fatty acid oxidized.

B. Transport of Activated Fatty Acids into Mitochondria

Mitochondrial membranes do not contain systems for the transport of fatty acid thioesters of coenzyme A. They do, however, contain a system for transporting fatty acids in the form of esters with the molecule **carnitine.** Functional groups in carnitine are a carboxylate group, a secondary alcohol, and a quaternary ammonium ion. Fatty acyl CoA and carnitine undergo a reaction in which the fatty acyl group is transferred from the sulfur atom of coenzyme A to the oxygen atom of the secondary alcohol of carnitine. This reaction is an example of transesterification (Section 17.5D).

$$R-\overset{\overset{\displaystyle O}{\|}}{C}-S-CoA \ + \ HO-\underset{\underset{\underset{\underset{CH_3}{|}}{\overset{+}{N}-CH_3}}{\underset{|}{CH_2}}}{\overset{\overset{\overset{CO_2^-}{|}}{CH_2}}{\underset{|}{CH}}} \ \rightleftharpoons \ R-\overset{\overset{\displaystyle O}{\|}}{C}-O-\underset{\underset{\underset{\underset{CH_3}{|}}{\overset{+}{N}-CH_3}}{\underset{|}{CH_2}}}{\overset{\overset{\overset{CO_2^-}{|}}{CH_2}}{\underset{|}{CH}}} \ + \ HS-CoA$$

fatty acyl
CoA
          carnitine
          fatty acyl
carnitine

The fatty acyl carnitine is transported through the inner mitochondrial membrane and there the reaction is reversed: fatty acyl carnitine and a molecule of coenzyme A react to form a fatty acyl CoA and regenerate carnitine. The freed carnitine is then returned to the cytoplasm to repeat the cycle. The effect of these two transesterification reactions, one in the cytoplasm and the other in the mitochondria, is to transfer fatty acyl CoA from the cytoplasm into mitochondria.

C. The β-Oxidation Spiral

Once in mitochondria, the carbon chain of an activated fatty acid is degraded two carbons at a time. The metabolic pathway by which this is accomplished is called **β-oxidation** because in two separate steps, a β-carbon is oxidized.

1. Oxidation. In the first step of β-oxidation, the carbon chain is oxidized and a double bond is formed between the α- and β-carbons (carbons 2 and 3) of the hydrocarbon chain. FAD is the oxidizing agent. It is reduced to $FADH_2$ which is subsequently oxidized in the respiratory chain.

$$R-\overset{\beta}{C}H_2-\overset{\alpha}{C}H_2-\overset{\overset{\displaystyle O}{\|}}{C}-S-CoA \ + \ FAD \longrightarrow R-\overset{\beta}{C}H=\overset{\alpha}{C}H-\overset{\overset{\displaystyle O}{\|}}{C}-S-CoA \ + \ FADH_2$$

a fatty acyl CoA
          an α,β-unsaturated
thioester

2. Hydration. Next, in a reaction analogous to acid-catalyzed hydration of an alkene (Section 11.5C), water is added to the carbon-carbon double bond to form a β-hydroxythioester. The effect of the first two reactions of β-oxidation is conversion of a $-CH_2-$ group on carbon-3 of the hydrocarbon chain to a $-CHOH-$ group.

$$R-CH=CH-\overset{\overset{\displaystyle O}{\|}}{C}-S-CoA \ + \ H_2O \longrightarrow R-\overset{\overset{\displaystyle OH}{|}}{C}H-CH_2-\overset{\overset{\displaystyle O}{\|}}{C}-S-CoA$$

an α,β-unsaturated thioester
          a β-hydroxythioester

3. Oxidation. In the second oxidation of β-oxidation, the secondary alcohol is oxidized to a ketone. NAD^+ is the oxidizing agent and is reduced to NADH.

$$\underset{\text{a β-hydroxythioester}}{R-\overset{\overset{\displaystyle OH}{|}}{CH}-CH_2-\overset{\overset{\displaystyle O}{\|}}{C}-S-CoA} + NAD^+ \longrightarrow \underset{\text{a β-ketothioester}}{R-\overset{\overset{\displaystyle O}{\|}}{C}-CH_2-\overset{\overset{\displaystyle O}{\|}}{C}-S-CoA} + NADH$$

4. Cleavage of acetyl coenzyme A. In the final step of β-oxidation, reaction of the β-ketothioester with a molecule of coenzyme A results in cleavage of a carbon-carbon bond and gives a molecule of acetyl CoA and a fatty acyl CoA molecule now shortened by two carbon atoms.

$$\underset{\text{a β-ketothioester}}{R-\overset{\overset{\displaystyle O}{\|}}{C}-CH_2-\overset{\overset{\displaystyle O}{\|}}{C}-S-CoA} + CoA-SH \longrightarrow \underset{\substack{\text{a fatty acyl thioester}\\\text{shortened by}\\\text{two carbons}}}{R-\overset{\overset{\displaystyle O}{\|}}{C}-S-CoA} + \underset{\text{acetyl CoA}}{CH_3-\overset{\overset{\displaystyle O}{\|}}{C}-S-CoA}$$

The same series of steps is now repeated on the shortened fatty acyl chain and another molecule of acetyl CoA is cleaved. This series of steps is continued until the entire chain is degraded to acetyl CoA. The steps of β-oxidation are called a **spiral** because after each series of four reactions, the carbon chain is shortened by two carbon atoms. Figure 25.1 illustrates β-oxidation of the carbon chain of palmitic acid to 8 molecules of acetyl CoA.

D. Energetics of Fatty Acid Oxidation

Now that we have examined the steps of β-oxidation, let us calculate how much of the free energy available from complete oxidation of a fatty acid to carbon dioxide and water is conserved as ATP. Let us take as a specific example palmitic acid. Seven turns of β-oxidation converts one mole of palmitic acid to 8 moles of acetyl CoA and generates 7 moles of $FADH_2$ and 7 moles of NADH. As we have seen in Section 23.5, reoxidation of each $FADH_2$ is coupled with formation of 2 ATP and reoxidation of each NADH is coupled with formation of 3 ATP. Furthermore, oxidation of each acetyl CoA in the tricarboxylic acid cycle, followed by oxidation of all reduced coenzymes, generates another 12 ATP. Because two phosphate anhydride bonds (equivalent to 2 moles of ATP) are required to activate each mole of palmitic acid, 2 ATP must be subtracted. The ATP balance for the oxidation of one mole of palmitic acid is:

14 ATP	from oxidation of 7 $FADH_2$
21 ATP	from oxidation of 7 NADH
96 ATP	from oxidation of 8 acetyl CoA
−2 ATP	from activation of palmitic acid
129 ATP	

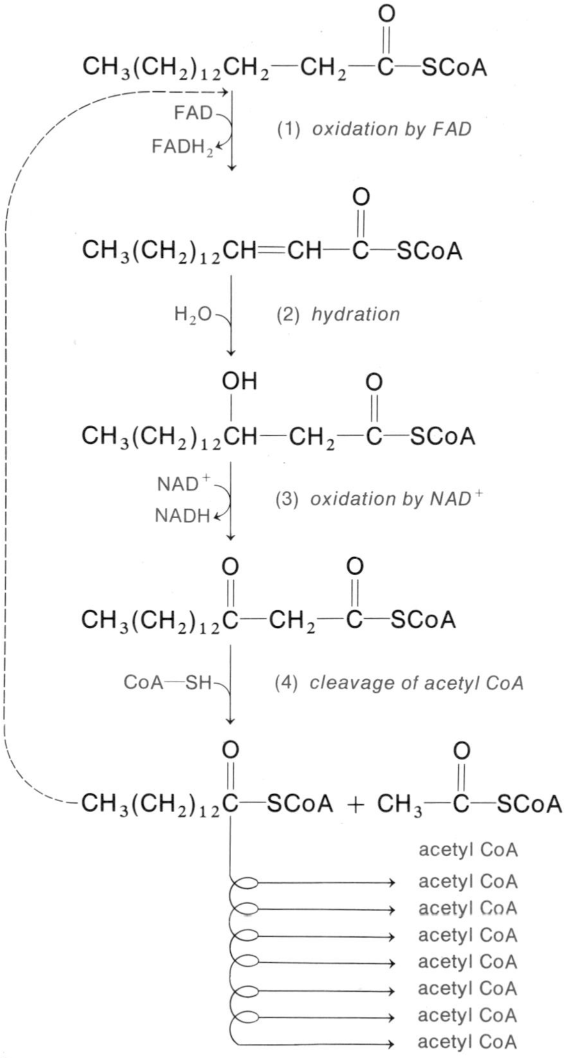

$$CH_3(CH_2)_{12}CH_2-CH_2-\overset{\overset{\displaystyle O}{\|}}{C}-SCoA$$

(1) *oxidation by FAD*

FAD → FADH₂

$$CH_3(CH_2)_{12}CH=CH-\overset{\overset{\displaystyle O}{\|}}{C}-SCoA$$

H_2O (2) *hydration*

$$CH_3(CH_2)_{12}\overset{\overset{\displaystyle OH}{|}}{CH}-CH_2-\overset{\overset{\displaystyle O}{\|}}{C}-SCoA$$

NAD^+ → NADH (3) *oxidation by* NAD^+

$$CH_3(CH_2)_{12}\overset{\overset{\displaystyle O}{\|}}{C}-CH_2-\overset{\overset{\displaystyle O}{\|}}{C}-SCoA$$

CoA—SH (4) *cleavage of acetyl CoA*

$$CH_3(CH_2)_{12}\overset{\overset{\displaystyle O}{\|}}{C}-SCoA + CH_3-\overset{\overset{\displaystyle O}{\|}}{C}-SCoA$$

acetyl CoA
acetyl CoA
acetyl CoA
acetyl CoA
acetyl CoA
acetyl CoA
acetyl CoA
acetyl CoA

Net reaction: $CH_3(CH_2)_{14}\overset{\overset{\displaystyle O}{\|}}{C}-SCoA + 7\ FAD + 7\ NAD^+ + 7\ CoA-SH \longrightarrow$

$8\ CH_3\overset{\overset{\displaystyle O}{\|}}{C}-SCoA + 7\ FADH_2 + 7\ NADH + 7\ H^+$

Figure 25.1 β-Oxidation of palmitic acid to 8 molecules of acetyl CoA.

Coupling the exergonic oxidation of palmitic acid with the endergonic phosphorylation of ADP gives:

Exergonic reaction: $CH_3(CH_2)_{14}CO_2H + 23\ O_2 \longrightarrow 16\ CO_2 + 16\ H_2O \qquad \Delta G^0 = -2,340\ \text{kcal/mole}$

Endergonic reaction:

$$129 \text{ ADP} + 129 \text{ HPO}_4^{2-} \longrightarrow 129 \text{ ATP} + 129 \text{ H}_2\text{O} \quad \Delta G^0 = +940 \text{ kcal/mole}$$

Net reaction:

$$CH_3(CH_2)_{14}CO_2H + 23 O_2 + 129 \text{ ADP} + 129 \text{ HPO}_4^{2-} \longrightarrow$$

$$16 CO_2 + 145 H_2O + \underset{\substack{\nearrow \\ available\ for\ work}}{129 \text{ ATP}} \quad \underset{\substack{\nearrow \\ liberated\ as\ heat}}{\Delta G^0 = -1,400 \text{ kcal/mole}}$$

Thus we see that some 940/2,340 or 40% of the standard free energy of oxidation of palmitate is conserved in the form of ATP and can be utilized by the cell for the performance of work. This fraction of energy conserved as ATP is comparable to that conserved in the complete oxidation of glucose to carbon dioxide and water (Section 24.6).

Example 25.1

β-Oxidation of stearic acid, $CH_3(CH_2)_{16}CO_2H$, produces 9 moles of acetyl CoA, 8 moles of NADH, and 8 moles of $FADH_2$. Calculate the number of moles of ATP produced:

a. by oxidative phosphorylation of 8 moles of NADH

b. by oxidative phosphorylation of 8 moles of $FADH_2$

c. by oxidation of 9 moles of acetyl CoA through the reactions of the tricarboxylic acid cycle

d. by oxidative phosphorylation of the NADH and $FADH_2$ produced during the oxidation of 9 moles of acetyl CoA in the tricarboxylic acid cycle

e. in parts (a), (b), (c), and (d), combined

Solution

a. Three moles of ATP are produced by oxidative phosphorylation of each mole of NADH. Therefore, oxidative phosphorylation of 8 moles of NADH gives 24 moles of ATP.

b. Two moles of ATP are produced per mole of $FADH_2$. Therefore, oxidative phosphorylation of 8 moles of $FADH_2$ gives 16 moles of ATP.

c. The reactions of the tricarboxylic acid cycle produce one mole of GTP (which can be converted to ATP) per mole of acetyl CoA entering the cycle. Therefore, oxidation of 9 moles of acetyl CoA gives 9 moles of ATP.

d. Each turn of the TCA cycle gives 3 moles of NADH and one mole of $FADH_2$ per mole of acetyl CoA entering the cycle. Therefore, 9 moles of acetyl CoA give 27 moles of NADH and 9 moles of $FADH_2$. Oxidative phosphorylation of these reduced coenzymes gives $81 + 18 = 99$ moles of ATP.

e. Complete oxidation of one mole of stearic acid and oxidative phosphorylation of the resulting NADH and $FADH_2$ give a total of $24 + 16 + 9 + 99 =$

148 moles of ATP. Since two high-energy bonds are required in the activation of a fatty acid for β-oxidation, the net yield per mole of stearic acid is 146 moles of ATP. Note that almost 65% of this ATP is produced by oxidative phosphorylation of the NADH and FADH$_2$ generated through the tricarboxylic acid cycle.

Problem 25.1

The structural formula of myristic acid is

$$CH_3(CH_2)_{12}\overset{\displaystyle O}{\overset{\displaystyle \|}{C}}OH$$

myristic acid

In the complete β-oxidation of myristic acid to acetyl CoA:

a. How many moles of ATP are required?

b. How many moles of NADH and FADH$_2$ are produced?

c. How many moles of acetyl CoA are produced?

E. Oxidation of Propanoate

In Section 25.3 we dealt with β-oxidation of palmitic acid, a fatty acid with an even number of carbon atoms. Fatty acids with even numbers of carbon atoms are degraded completely to acetyl CoA. While they are not nearly as common, fatty acids with odd numbers of carbon atoms do occur in nature. The final β-oxidation of an odd-numbered carbon chain gives the thioester of propanoic acid. The IUPAC name of this ester is propanoyl CoA (derived from propanoic acid). Its common name is propionyl CoA (derived from propionic acid). Propanoyl CoA is also produced in the catabolism of certain amino acids.

$$CH_3{-}CH_2{-}\overset{\displaystyle O}{\overset{\displaystyle \|}{C}}{-}S{-}CoA$$

propanoyl CoA
(propionyl CoA)

Propanoyl CoA is converted to succinyl CoA, an intermediate in the TCA cycle. In the first reaction of this conversion, propanoyl CoA is carboxylated to give methylmalonyl CoA.

$$O{=}C{=}O + \underset{\displaystyle CH_3}{\underset{\displaystyle |}{CH_2}}{-}\overset{\displaystyle O}{\overset{\displaystyle \|}{C}}{-}S{-}CoA \xrightarrow[\text{biotin}]{\text{ATP}} {}^{-}O{-}\overset{\displaystyle O}{\overset{\displaystyle \|}{C}}{-}\underset{\displaystyle CH_3}{\underset{\displaystyle |}{CH}}{-}\overset{\displaystyle O}{\overset{\displaystyle \|}{C}}{-}S{-}CoA$$

methylmalonyl CoA

This reaction requires ATP as a source of energy to form the new carbon-carbon bond. It also requires the coenzyme biotin. Next, methylmalonyl CoA is isomerized to succinyl CoA in an unusual reaction that requires vitamin B_{12} as a cofactor. In vitamin B_{12} deficiency, both propanoate and methylmalonate appear in the urine.

methylmalonyl CoA   succinyl CoA

Studies using radioisotopes have revealed that this isomerization involves migration of the entire thioester group to the methyl carbon and exchange of a hydrogen atom for it.

The importance of vitamin B_{12} was first realized when it was discovered that liver extracts could be used to treat patients with pernicious anemia. The active principle was isolated, purified, and crystallized in 1948, and its complete three-dimensional structure was worked out by Dorothy Hodgkin in 1956 (Figure 25.2). Central to the structure of vitamin B_{12} is an atom of cobalt embedded in a corrin ring, which, like the porphyrin rings of hemoglobin and myoglobin, has four pyrrole rings joined together. For her work in determining the structure of this complex molecule, Dorothy Hodgkin received the Nobel Prize in Chemistry in 1964.

Pernicious anemia is a disease caused by a deficiency of vitamin B_{12}. However, most cases of pernicious anemia are not due to lack of B_{12} in the diet, but rather to a deficiency of a substance called **intrinsic factor** that normally is present in gastric juice. Intrinsic factor is necessary for absorption of B_{12} through the walls of the gastrointestinal tract and into the blood stream. Since most B_{12}-deficiency diseases are due to a lack of intrinsic factor and reduced absorption of the vitamin, B_{12} taken orally often has little effect. The most common means of administration of B_{12} is by direct intramuscular injection.

25.4 Formation of Ketone Bodies

Acetoacetate, β-hydroxybutyrate, and acetone are classed as ketone bodies. These compounds are products of human metabolism and they are always present in blood plasma. However, under normal conditions, their concentration in plasma is low. In humans and most other animals, the liver is the only organ that produces any significant amounts of ketone bodies. Most other tissues, with the notable exception of the brain, have the capacity to use them as energy sources.

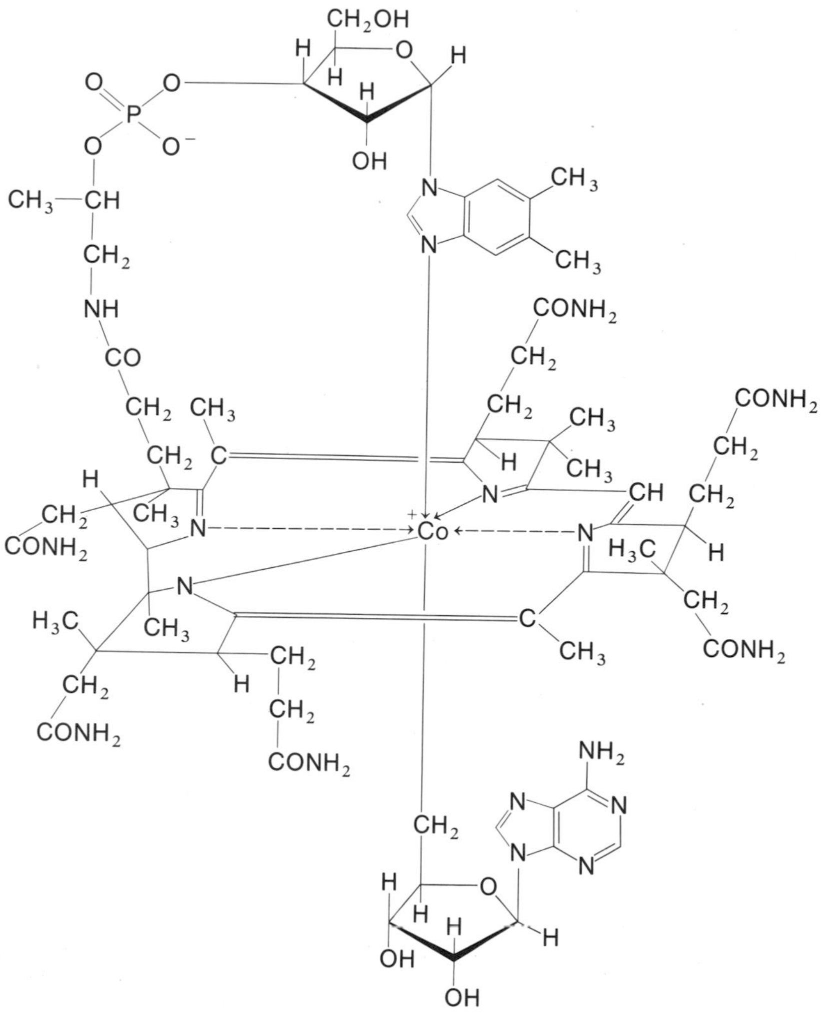

Figure 25.2 Structure of vitamin B_{12}. (From David S. Page, *Principles of Biological Chemistry*, Boston: Willard Grant Press, 1976.)

Ketone bodies are synthesized from acetyl coenzyme A. In a series of three reactions, acetyl CoA is converted to acetoacetate.

$$2 \; CH_3-\overset{\overset{\displaystyle O}{\|}}{C}-S-CoA + H_2O \longrightarrow \longrightarrow \longrightarrow$$

acetyl CoA

$$CH_3-\overset{\overset{\displaystyle O}{\|}}{C}-CH_2-\overset{\overset{\displaystyle O}{\|}}{C}-O^- + 2 \; CoA-SH$$

acetoacetate

In one subsequent reaction, the ketone group of acetoacetate is reduced to a secondary alcohol. The product is β-hydroxybutyrate.

$$CH_3 - \overset{\overset{\displaystyle O}{\|}}{C} - CH_2 - \overset{\overset{\displaystyle O}{\|}}{C} - O^- + NADH + H^+ \longrightarrow$$

acetoacetate

$$CH_3 - \overset{\overset{\displaystyle OH}{|}}{CH} - CH_2 - \overset{\overset{\displaystyle O}{\|}}{C} - O^- + NAD^+$$

β-hydroxybutyrate

In another reaction, acetoacetate loses carbon dioxide to give acetone. Recall that decarboxylation is a characteristic reaction of β-ketoacids (Section 16.5C).

$$CH_3 - \overset{\overset{\displaystyle O}{\|}}{C} - CH_2 - \overset{\overset{\displaystyle O}{\|}}{C} - O^- + H^+ \longrightarrow CH_3 - \overset{\overset{\displaystyle O}{\|}}{C} - CH_3 + CO_2$$

acetoacetate   acetone

When the production of acetoacetate, β-hydroxybutyrate, and acetone exceeds the capacity of the body to metabolize them, the condition is known as **ketosis.** Of the three ketone bodies, acetoacetic acid and β-hydroxybutyric acid are the most significant because they are acids and must be buffered in the blood and other body fluids to prevent their accumulation from disrupting normal acid-base balance. The acidosis that results from the accumulation of ketone bodies is called **ketoacidosis.** Under ketoacidosis, the effectiveness of hemoglobin to transport oxygen is reduced. In the extreme, a deficiency in the supply of oxygen to the brain can result in a fatal coma. The presence of ketone bodies in the urine indicates an advanced state of ketoacidosis and is a clear signal that immediate medical attention is essential.

Several abnormal conditions, including starvation, unusual diets, and diabetes mellitus, lead to increased production of ketone bodies, ketoacidosis, and spilling of ketone bodies into the urine.

After a short period of starvation, carbohydrate reserves become depleted and the cell or organism turns to β-oxidation of fatty acids as a source of energy. Fatty acid degradation in the liver is increased and, in turn, leads to an increase in the concentration of acetyl CoA. Under normal conditions, acetyl CoA in the liver can be channeled into three metabolic pathways: the tricarboxylic acid cycle for oxidation to carbon dioxide and water; the synthesis of fatty acids and triglycerides for storage in the liver; and, finally, the synthesis of ketone bodies. During starvation and under conditions where carbohydrate metabolism is drastically reduced, the synthesis of pyruvate and other intermediates of the tricarboxylic acid cycle are also reduced. Therefore, the increased supply of acetyl CoA generated during starvation is channeled into the production of ketone bodies (Figure 25.3).

Figure 25.3 As a result of starvation, an abnormal diet, or diabetes mellitus, β-oxidation of fatty acids and the synthesis of ketone bodies is increased.

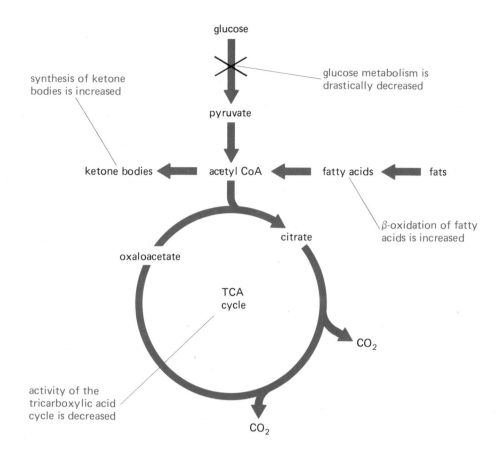

25.5 Synthesis of Fatty Acids

In Section 25.3, we saw how fatty acids are degraded to acetyl CoA by the reactions of β-oxidation. In this section, we will look at the synthesis of fatty acids. These two processes are quite separate, both in mechanism and in location within the cell. Among the major differences are the following:

1. Synthesis of fatty acids takes place in the cytoplasm, whereas degradation takes place in mitochondria.

2. Synthesis involves two reductions, one of a ketone to a secondary alcohol and the other of a carbon-carbon double bond to a carbon-carbon single bond. In a sense, these two reductions are the reverse of two steps in the oxidation of fatty acid hydrocarbon chains. Synthesis of fatty acids uses NADPH for both reductions; by comparison, fatty acid degradation uses FAD for one oxidation and NAD^+ for the other.

3. A key starting material for the synthesis of fatty acids is malonyl CoA.

$$\overset{\displaystyle O}{\overset{\displaystyle \|}{^-O-C}}-CH_2-\overset{\displaystyle O}{\overset{\displaystyle \|}{C}}-S-CoA$$

malonyl CoA

Thus, the starting material for synthesis is not the same molecule that is the end product of degradation.

A. Formation of Malonyl CoA

Malonyl CoA is formed by carboxylation of acetyl CoA in a reaction that requires biotin as a cofactor and acetyl CoA carboxylase as a catalyst. You might compare this reaction with the carboxylation of propanoyl CoA to give methylmalonyl CoA (Section 25.3E).

$$O{=}C{=}O + CH_3-\overset{\displaystyle O}{\overset{\displaystyle \|}{C}}-S-CoA \xrightarrow[\text{acetyl CoA}]{\text{biotin}} {^-O}-\overset{\displaystyle O}{\overset{\displaystyle \|}{C}}-CH_2-\overset{\displaystyle O}{\overset{\displaystyle \|}{C}}-S-CoA$$

acetyl CoA   carboxylase   malonyl CoA

Acetyl CoA carboxylase is a regulatory enzyme and the rate of fatty acid synthesis is controlled by modulation of its activity.

B. Synthesis of Fatty Acid Hydrocarbon Chains

Synthesis of fatty acid hydrocarbon chains is catalyzed by an enzyme complex called fatty acid synthetase. A key part of this enzyme complex is a low-molecular-weight protein called **acyl carrier protein (ACP).** Synthesis of a fatty acid chain begins with the transfer of an acetyl group from the sulfur atom of coenzyme A to a sulfur atom of ACP. This reaction is an example of trans-esterification (Section 17.5D). Next, the acetyl group is shifted to an adjacent sulfhydryl group (a second transesterification), and in a third transesterification a malonyl group is transferred from the sulfur atom of coenzyme A to the first sulfhydryl group of ACP. These three transesterifications are shown in Figure 25.4.

At this point, a pair of two-carbon fragments are activated, one as acetyl-ACP and the other as malonyl-ACP. They next undergo a series of four reactions that leads to elongation of the hydrocarbon chain by two atoms at a time. These four steps are condensation, reduction, dehydration, and reduction, each catalyzed by a separate enzyme component of the fatty acid synthetase enzyme complex. To visualize the operation of this complex, think of ACP in the center of a circle, surrounded by the enzymes that catalyze each step of chain elongation. Further, think of ACP as turning within this enzyme complex with the growing hydrocarbon chain as a flexible arm that gets longer and longer as

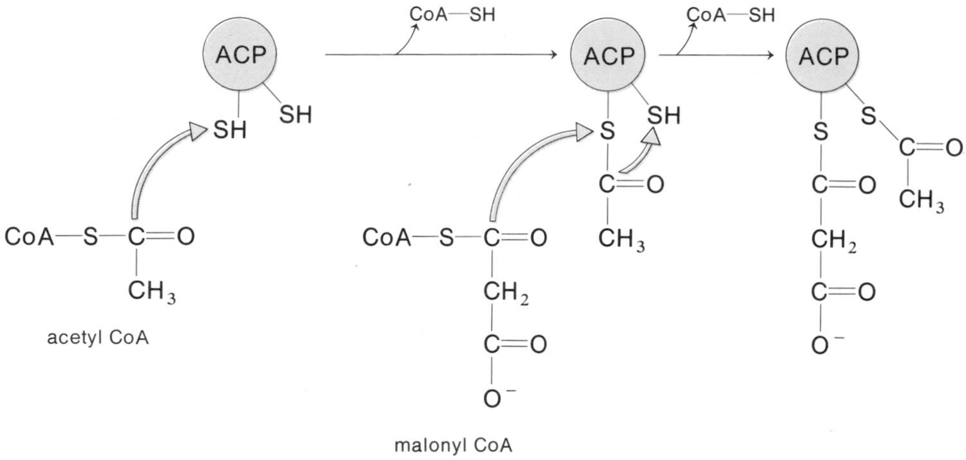

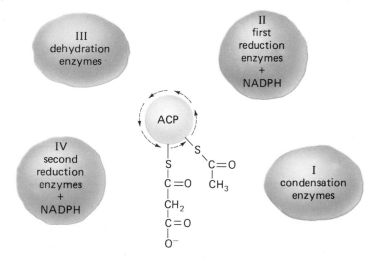

Figure 25.4 Binding of acetyl and malonyl groups to the acyl carrier protein (ACP).

Figure 25.5 The relationship between acyl carrier protein and the four enzyme systems responsible for chain elongation. In this arrangement, the growing chain is bound as a thioester to ACP and swings from one enzyme complex to the next.

the hydrocarbon chain is elongated. The relationship of acyl carrier protein to the four enzyme systems of chain elongation is illustrated in Figure 25.5.

1. Condensation. The first step in chain elongation is formation of a carbon-carbon bond between the carbonyl group of acetyl-ACP and the α-carbon of malonyl-ACP; this gives acetoacetyl-ACP.

$$CH_3-\overset{O}{\underset{\|}{C}}-S-ACP + \underset{\underset{CO_2^-}{|}}{CH_2}-\overset{O}{\underset{\|}{C}}-S-ACP \longrightarrow CH_3-\overset{O}{\underset{\|}{C}}-CH_2-\overset{O}{\underset{\|}{C}}-S-ACP + CO_2 + ACP-SH$$

acetyl-ACP   malonyl-ACP   acetoacetyl-ACP

This enzyme-catalyzed formation of a new carbon-carbon bond is analogous to a Claisen condensation (Section 17.11) and forms a β-ketothioester. Note that this condensation is coupled with the loss of carbon dioxide by decarboxylation. Thus, although carbon dioxide (actually bicarbonate) is required for fatty acid synthesis, it does not appear in the newly synthesized fatty acid.

2. Reduction of a ketone to a secondary alcohol. In step 2, the β-keto group of the growing fatty acid chain is reduced to a secondary alcohol.

$$CH_3-\underset{O}{\overset{O}{C}}-CH_2-\underset{O}{\overset{O}{C}}-S-ACP + NADPH \longrightarrow CH_3-\underset{\beta}{\overset{OH}{CH}}-\underset{\alpha}{CH_2}-\underset{O}{\overset{O}{C}}-S-ACP + NADP^+$$

acetoacetyl-ACP   β-hydroxybutyryl-ACP

The reducing agent is NADPH, consistent with the fact that the major use of this coenzyme is as a reducing agent in biosynthesis (Section 24.2E).

3. Dehydration. In step 3, the β-hydroxythioester is dehydrated to form an α,β-thioester in a reaction analogous to acid-catalyzed dehydration of an alcohol (Section 12.5B).

$$CH_3-\overset{OH}{CH}-CH_2-\overset{O}{C}-S-ACP \longrightarrow CH_3-CH=CH-\overset{O}{C}-S-ACP + H-OH$$

β-hydroxybutyryl-ACP   crotonyl-ACP

The configuration about the double bond formed in step 3 is *trans*. By comparison, the configuration in the unsaturated fatty acid components of triglycerides and phospholipids is *cis* (Section 16.6).

4. Reduction to a carbon-carbon single bond. In the final reaction of chain elongation, the carbon-carbon double bond of the growing chain is reduced to a carbon-carbon single bond. The reducing agent is another molecule of NADPH.

$$CH_3-CH=CH-\overset{O}{C}-S-ACP + NADPH \longrightarrow CH_3-CH_2-CH_2-\overset{O}{C}-S-ACP + NADP^+$$

crotonyl-ACP   butyryl-ACP

The thioester formed after steps 1–4 is shown in Figure 25.6. At this point, the acyl group of the four-carbon thioester is transferred to the adjacent —SH group and a second malonyl group is transferred to ACP.

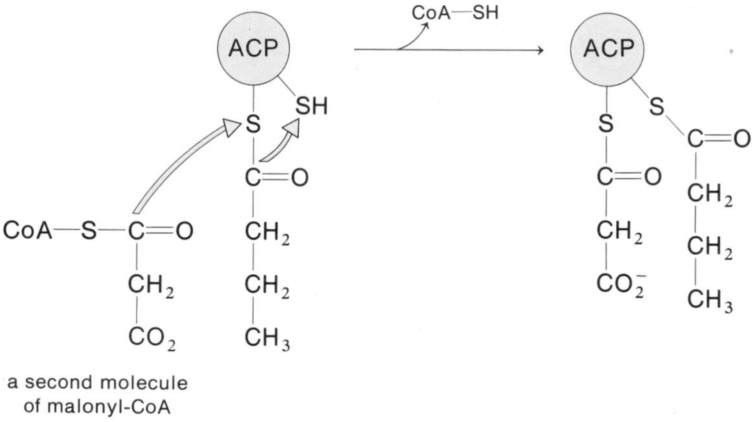

Figure 25.6 Prepara-
tion for a second
cycle of chain elonga-
tion reactions.

a second molecule
of malonyl-CoA

The second cycle of chain elongation begins with formation of a carbon-carbon bond between the carbonyl carbon of the four-carbon thioester and the α-carbon of malonyl-ACP (Figure 25.7).

The most common fatty acid synthesized by the series of reactions we have outlined is palmitic acid. After seven "turns" of the chain elongation process, the hydrocarbon chain of palmitic acid is complete, further elongation ceases, and palmitic acid is released from the acyl carrier protein by hydrolysis of the thioester bond.

$$CH_3(CH_2)_{14}\overset{O}{\overset{\|}{C}}-S-ACP + H_2O \longrightarrow CH_3(CH_2)_{14}\overset{O}{\overset{\|}{C}}OH + HS-ACP$$

palmityl-ACP   palmitic acid

Fatty acids of 18, 20, and larger numbers of carbon atoms are synthesized from palmitic acid in subsequent steps.

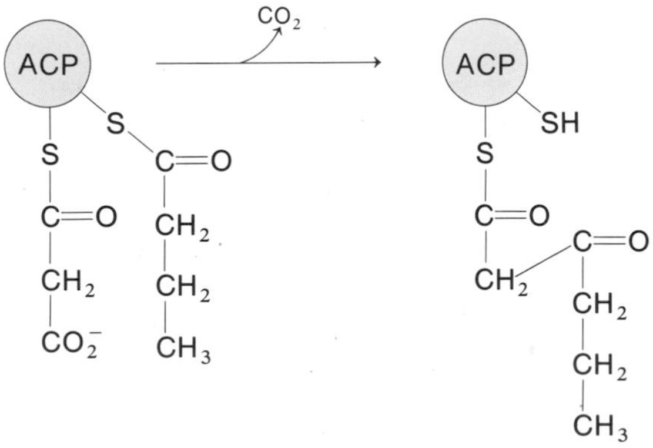

Figure 25.7 The be-
ginning of the second
cycle of chain elonga-
tion reactions.

C. Synthesis of Unsaturated Fatty Acids

Unsaturated fatty acids are synthesized from saturated fatty acids by oxidations that require O_2 and NADH. Insertion of unsaturation is catalyzed by an enzyme called fatty acid desaturase. Mammalian fatty acid desaturase has three important characteristics: (1) double bonds introduced in fatty acid chains have a *cis* configuration; (2) the enzyme is active on hydrocarbon chains of only 18 or fewer carbon atoms; and (3) double bonds cannot be introduced beyond carbons 9–10. By the action of fatty acid desaturase, palmitic acid is oxidized to palmitoleic acid and stearic acid is oxidized to oleic acid. (See Table 16.4 for structural formulas of these unsaturated fatty acids.)

$$CH_3(CH_2)_7CH_2-CH_2(CH_2)_7\overset{\overset{\displaystyle O}{\|}}{C}OH \xrightarrow{\text{oxidation}} CH_3(CH_2)_7 \overset{}{\underset{H}{}} C=C \underset{H}{} (CH_2)_7\overset{\overset{\displaystyle O}{\|}}{C}OH$$

stearic acid   oleic acid

Linoleic acid and linolenic acid are polyunsaturated fatty acids. Linoleic acid has double bonds between carbons 9–10 and 12–13. In linolenic acid, there are double bonds between carbons 9–10, 12–13, and 16–17. Because mammalian fatty acid desaturase cannot introduce double bonds beyond carbons 9–10, linoleic and linolenic acids cannot be synthesized in the body. They must be supplied in the diet and are called essential fatty acids (Section 16.6C).

D. Synthesis of Fatty Acids from the Carbon Atoms of Glucose

As we have pointed out several times (see, for example, Figure 24.10), there is a close relationship between the metabolism of glucose and that of fatty acids. Specifically, the body has only a limited capacity to store glucose. However, it has a very large capacity for the storage of fatty acids in the form of tri-glycerides. Through the metabolic pathways we have already studied, the carbon atoms of glucose can be used for the synthesis of fatty acids. The flow of carbon atoms from glucose to palmitic acid is traced in Figure 25.8.

Cleavage of glucose during glycolysis produces one molecule each of 3-phosphoglyceraldehyde and dihydroxyacetone phosphate. Dihydroxyacetone phosphate is isomerized to 3-phosphoglyceraldehyde and both continue in glycolysis to form pyruvate. Oxidation and decarboxylation of pyruvate (Section 24.4A) give acetyl CoA and carbon dioxide. The carbonyl carbons of acetyl CoA are derived from carbons 2 and 5 of glucose and the methyl carbons are derived from carbons 1 and 6 of glucose. Acetyl CoA in the form of malonyl CoA is the key building block for the synthesis of fatty acids, including palmitic acid. As shown in Figure 25.8, carbon atoms 1, 2, 5, and 6 of glucose can become incorporated into the carbon skeleton of a fatty acid.

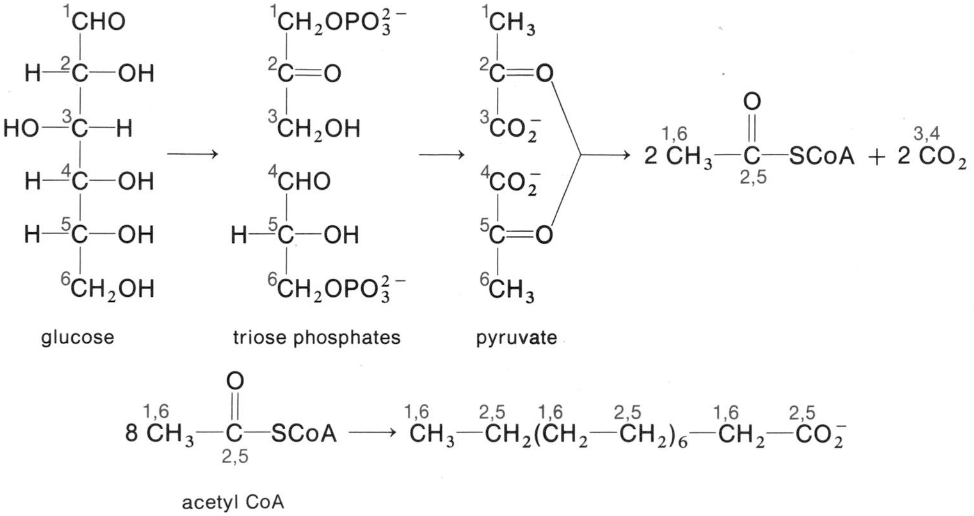

Figure 25.8 The synthesis of palmitate from glucose. The carbon atoms of glucose are numbered 1 through 6.

Key Terms and Concepts

acyl carrier protein (25.5B)

carnitine (25.3B)

hydrolysis of triglycerides (25.2)

intrinsic factor (25.3E)

ketoacidosis (25.4)

ketosis (25.4)

β-oxidation spiral (25.3C)

pernicious anemia (25.3E)

Key Reactions

1. Activation of fatty acids (Section 25.3A):

$$R-CH_2-CH_2-\overset{\overset{O}{\|}}{C}-OH + CoA-SH + ATP \longrightarrow$$

$$R-CH_2-CH_2-\overset{\overset{O}{\|}}{C}-SCoA + AMP + 2\ HPO_4^{2-}$$

2. Transport of fatty acids into the mitochondria (Section 25.3B):

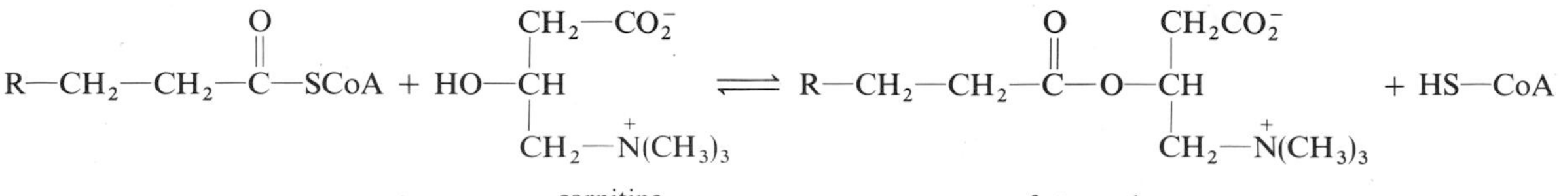

$$R-CH_2-CH_2-\overset{\overset{O}{\|}}{C}-SCoA + HO-\underset{\underset{CH_2-\overset{+}{N}(CH_3)_3}{|}}{\overset{\overset{CH_2-CO_2^-}{|}}{CH}} \rightleftharpoons R-CH_2-CH_2-\overset{\overset{O}{\|}}{C}-O-\underset{\underset{CH_2-\overset{+}{N}(CH_3)_3}{|}}{\overset{\overset{CH_2-CO_2^-}{|}}{CH}} + HS-CoA$$

carnitine

fatty acyl
carnitine

3. β-Oxidation of fatty acids (Section 25.3):

$$R-CH_2-CH_2-\overset{\overset{\displaystyle O}{\|}}{C}-SCoA + NAD^+ + FAD + CoA-SH \longrightarrow$$

$$R-\overset{\overset{\displaystyle O}{\|}}{C}-SCoA + CH_3-\overset{\overset{\displaystyle O}{\|}}{C}-SCoA + NADH + FADH_2$$

4. Carboxylation of propanoyl CoA (Section 25.3E):

$$CO_2 + CH_3-CH_2-\overset{\overset{\displaystyle O}{\|}}{C}-SCoA \xrightarrow[\text{biotin}]{\text{ATP}} {}^-O-\overset{\overset{\displaystyle O}{\|}}{C}-\underset{\underset{\displaystyle CH_3}{|}}{CH}-\overset{\overset{\displaystyle O}{\|}}{C}-S-CoA$$

propanoyl CoA   methylmalonyl CoA

5. Isomerization of methylmalonyl CoA to succinyl CoA (Section 25.3E):

$${}^-O-\overset{\overset{\displaystyle O}{\|}}{C}-\underset{\underset{\displaystyle CH_3}{|}}{CH}-\overset{\overset{\displaystyle O}{\|}}{C}-S-CoA \xrightarrow{\text{vitamin } B_{12}} {}^-O-\overset{\overset{\displaystyle O}{\|}}{C}-CH_2-CH_2-\overset{\overset{\displaystyle O}{\|}}{C}-S-CoA$$

methylmalonyl CoA   succinyl CoA

6. Formation of ketone bodies (Section 25.4):

$$2\,CH_3-\overset{\overset{\displaystyle O}{\|}}{C}-S-CoA + H_2O \longrightarrow CH_3-\overset{\overset{\displaystyle O}{\|}}{C}-CH_2-\overset{\overset{\displaystyle O}{\|}}{C}-O^- + 2\,CoA-SH$$

β-ketobutyrate

$$CH_3-\overset{\overset{\displaystyle O}{\|}}{C}-CH_2-\overset{\overset{\displaystyle O}{\|}}{C}-O^- + NADH + H^+ \longrightarrow CH_3-\underset{\underset{\displaystyle OH}{|}}{CH}-CH_2-\overset{\overset{\displaystyle O}{\|}}{C}-O^- + NAD^+$$

β-hydroxybutyrate

$$CH_3-\overset{\overset{\displaystyle O}{\|}}{C}-CH_2-\overset{\overset{\displaystyle O}{\|}}{C}-O^- \longrightarrow CH_3-\overset{\overset{\displaystyle O}{\|}}{C}-CH_3 + CO_2$$

acetone

7. Synthesis of fatty acids (Section 25.5):

$$CO_2 + CH_3-\overset{\overset{\displaystyle O}{\|}}{C}-S-CoA \xrightarrow[\text{biotin}]{\text{ATP}} {}^-O-\overset{\overset{\displaystyle O}{\|}}{C}-CH_2-\overset{\overset{\displaystyle O}{\|}}{C}-S-CoA$$

acetyl CoA   malonyl CoA

$$CH_3-\overset{\overset{\displaystyle O}{\|}}{C}-S-ACP + \underset{\underset{\displaystyle CO_2^-}{|}}{CH_2}-\overset{\overset{\displaystyle O}{\|}}{C}-S-ACP \longrightarrow CH_3-\overset{\overset{\displaystyle O}{\|}}{C}-CH_2-\overset{\overset{\displaystyle O}{\|}}{C}-S-ACP + CO_2 + ACP-SH$$

$$CH_3-\overset{\overset{\displaystyle O}{\|}}{C}-CH_2-\overset{\overset{\displaystyle O}{\|}}{C}-S-ACP + 2\,NADPH \longrightarrow CH_3-CH_2-CH_2-\overset{\overset{\displaystyle O}{\|}}{C}-S-ACP + 2\,NADP^+$$

Problems

25.2 Compare carbohydrates and fatty acids as energy sources. How do you account for the difference between the amount of energy released on complete oxidation of each?

25.3 Write structural formulas for palmitic, oleic, and stearic acids, the three most abundant fatty acids.

25.4 A fatty acid must be activated before it can be metabolized in the cell. Write a balanced equation for the reaction that activates palmitic acid.

25.5 Name three coenzymes necessary for the catabolism of fatty acids to acetyl CoA. What vitamin precursor is associated with each coenzyme?

25.6 Outline the four steps in the fatty acid oxidation spiral.

25.7 How much energy in the form of ATP is produced directly in the oxidation of palmitic acid to acetyl CoA?

25.8 Review the oxidation reactions in the catabolism of glucose and fatty acids. Prepare a list of (a) types of functional groups oxidized, and (b) the oxidizing agent used for each type. Compare the types of functional groups oxidized by FAD with those oxidized by NAD^+.

25.9 Calculate the number of moles of ATP produced when:
   a. palmitoyl CoA is oxidized to acetyl CoA and all of the reduced coenzymes produced in the process are reoxidized by molecular oxygen.
   b. palmitoyl CoA is oxidized to CO_2 and all of the reduced coenzymes produced in the process are reoxidized by molecular oxygen.

25.10 In patients with pernicious anemia, up to 50–90 mg of a dicarboxylic acid of molecular formula $C_4H_6O_4$ appear in the urine daily. Draw a structural formula for this dicarboxylic acid. How do you account for its formation?

25.11 The respiratory quotient (RQ) is used in studies of energy metabolism and exercise physiology. It is defined as the ratio of the volume of carbon dioxide produced to the volume of oxygen used:

$$RQ = \frac{\text{volume of } CO_2}{\text{volume of } O_2}$$

   a. Show that the RQ for glucose is 1.00. (*Hint:* Look at the balanced equation for the complete oxidation of glucose.)
   b. Calculate the RQ for triolein, a triglyceride of molecular formula $C_{57}H_{104}O_6$.
   c. Calculate an RQ on the assumption that triolein and glucose are oxidized in equal molar amounts.
   d. For an individual on a normal diet, the RQ is approximately 0.85. Would you expect this value to increase or decrease if ethanol supplies an appreciable portion of caloric needs?

25.12 What is the only organ in humans that produces any significant amounts of ketone bodies?

25.13 Explain why ketone body formation increases markedly when excessive amounts of fatty acids are being oxidized and carbohydrate availability is limited.

25.14 Explain why the accumulation of ketone bodies also produces acidosis.

Synthesis of Fatty Acids (Section 25.5)

25.15 Starting with butyryl-ACP, show all steps in the synthesis of hexanoic acid. Name the type of reaction involved in each step.

$$CH_3{-}CH_2{-}CH_2{-}\overset{\overset{\displaystyle O}{\|}}{C}{-}S{-}ACP$$

butyryl-ACP

25.16 For the synthesis of stearic acid from acetyl-ACP:
a. How many moles of NADPH are required?
b. How many moles of malonyl-ACP are required?

25.17 During fatty acid synthesis, NADPH is oxidized to $NADP^+$. Name the metabolic pathway that is primarily responsible for regeneration of NADPH.

25.18 If glucose were the only source of acetyl CoA, how many moles of glucose would be required for the synthesis of one mole of palmitate? How many grams of glucose are normally required for the synthesis of one mole of palmitate?

25.19 Explain why the vast majority of fatty acids have an even number of carbon atoms in an unbranched chain.

The Metabolism of Amino Acids

In the broadest sense, amino acids serve three vital functions in the human body: (1) building blocks for the synthesis of proteins; (2) sources of carbon and nitrogen atoms for the synthesis of other biomolecules; and (3) sources of energy. Compared with the metabolism of carbohydrates and fatty acids, the metabolism of amino acids is extremely complex. Unlike hexoses and fatty acids, which share common metabolic pathways (for example, glycolysis for hexoses and β-oxidation for fatty acids), each amino acid is degraded and synthesized by a separate pathway. In this chapter, we will not go into the details of the metabolic pathways for the degradation and synthesis of each amino acid. Rather, we will develop an overview of amino acid metabolism. Then, we will look specifically at how the carbon skeletons of amino acids can be used as a source of energy and how amino acid nitrogen atoms are collected, converted to urea, and excreted.

26.1 Amino Acid Metabolism — An Overview

A. Amino Acids Are Used for the Syntheis of Body Proteins

The most important function of amino acids, at least in terms of total amino acid utilization, is as building blocks for the synthesis of proteins. It is estimated that about 75% of amino acid metabolism in a normal, healthy adult is devoted to this purpose. The maintenance of body proteins is not a simple matter. Tissue proteins are being hydrolyzed constantly through normal "wear and tear." At the same time, we eat proteins that are hydrolyzed in the gut to amino acids. The amino acids from food combined with those from hydrolysis of tissue proteins form what is called the **amino acid pool,** from which new proteins are synthesized.

The use of radioisotopes has given us some idea of the extent of turnover of body proteins and the amino acid pool. For example, the **half-life** of liver proteins is about 10 days. This means that over a 10-day period, half of the proteins in the liver are hydrolyzed to amino acids and replaced by equivalent proteins. The half-life of plasma proteins is also about 10 days, hemoglobin about 120 days, and muscle protein about 180 days. The half-life of collagen is considerably longer. Some proteins, particularly enzymes and polypeptide hormones, have much shorter half-lives. That of insulin, once it is released from the pancreas, is estimated to be only 7–10 minutes.

Clearly, the stability of body protein is more apparent than real. It represents a dynamic balance between degradation and synthesis. In spite of this turnover, the total amount of body protein remains relatively constant in most adults. This means that the amount of amino acids required for the synthesis of body proteins and the amount obtained from hydrolysis of body proteins are roughly equivalent. Therefore, the amino acid pool of most adults contains a surplus approximately equal to the amino acids obtained in the diet. This surplus is used for the synthesis of other nonprotein compounds, as a source of fuel, or it is excreted.

B. Amino Acids are Sources of Carbon and Nitrogen Atoms for the Synthesis of Other Biomolecules

Tissues constantly draw on the amino acid pool for the synthesis of nonprotein biomolecules. These molecules include nucleic acids; porphyrin rings, such as those in the prosthetic groups of hemoglobin and myoglobin; choline and ethanolamine, which are building blocks of phospholipids; glucosamine and other amino sugars; and neurotransmitters such as acetylcholine, dopamine, norepinephrine, and serotonin. Like proteins, these compounds are also being broken down and replaced constantly.

Figure 26.1 An overview of the metabolism of amino acids. Average daily turnover of amino acids is approximately 400 grams.

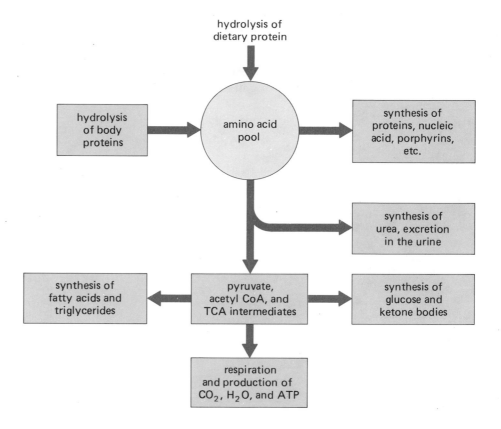

C. Amino Acids as Fuel

Unlike carbohydrates and fatty acids, amino acids in excess of immediate needs cannot be stored for later use. Their nitrogen atoms are converted to ammonium ions, urea, or uric acid, depending on the organism, and excreted. Their carbon skeletons are degraded to pyruvate, acetyl CoA, or one of the intermediates of the tricarboxylic acid cycle. The various metabolic pathways of amino acid metabolism are summarized in Figure 26.1.

26.2 Catabolism of Amino Acids

A. Transamination

Of all the reactions in the metabolism of amino groups of amino acids, **transamination** is one of the most important. It is generally the first reaction in the degradation of amino acids and the last reaction in their synthesis. In transamination, an amino group is transferred from a donor α-amino acid to an acceptor α-ketoacid. In the process, the acceptor α-ketoacid is transformed into a new α-amino acid. While several different α-ketoacids participate in transamination reactions, the most important are pyruvate and α-ketoglutarate. During the transamination phase of amino acid catabolism, all amino groups (with the possible exception of those of lysine and threonine) are channeled into either alanine or glutamate.

Alanine transaminase:

$$\underset{\substack{(a\ donor \\ \alpha\text{-amino acid})}}{R\overset{\overset{\displaystyle NH_3^+}{|}}{C}HCO_2^-} + \underset{pyruvate}{CH_3\overset{\overset{\displaystyle O}{\|}}{C}CO_2^-} \rightleftharpoons \underset{\substack{(a\ new\ \alpha\text{-} \\ ketoacid)}}{R\overset{\overset{\displaystyle O}{\|}}{C}CO_2^-} + \underset{L\text{-alanine}}{CH_3\overset{\overset{\displaystyle NH_3^+}{|}}{C}HCO_2^-}$$

Glutamate transaminase:

$$\underset{\substack{(a\ donor \\ \alpha\text{-amino acid})}}{R\overset{\overset{\displaystyle NH_3^+}{|}}{C}HCO_2^-} + \underset{\alpha\text{-ketoglutarate}}{^-O_2CCH_2CH_2\overset{\overset{\displaystyle O}{\|}}{C}CO_2^-} \rightleftharpoons \underset{\substack{(a\ new\ \alpha\text{-} \\ ketoacid)}}{R\overset{\overset{\displaystyle O}{\|}}{C}CO_2^-} + \underset{L\text{-glutamate}}{^-O_2CCH_2CH_2\overset{\overset{\displaystyle NH_3^+}{|}}{C}HCO_2^-}$$

Example 26.1

Draw structural formulas for the starting materials and products of the following transamination reactions.

a. tyrosine + α-ketoglutarate $\longrightarrow$ **b.** valine + pyruvate $\longrightarrow$

Solution

a. HO—⟨benzene ring⟩—$CH_2\overset{\overset{\displaystyle NH_3^+}{|}}{C}HCO_2^-$ + $^-O_2CCH_2CH_2\overset{\overset{\displaystyle O}{\|}}{C}CO_2^-$ $\longrightarrow$

tyrosine   α-ketoglutarate

HO—⟨benzene ring⟩—$CH_2\overset{\overset{\displaystyle O}{\|}}{C}CO_2^-$ + $^-O_2CCH_2CH_2\overset{\overset{\displaystyle NH_3^+}{|}}{C}HCO_2^-$

p-hydroxyphenyl-   L-glutamate
pyruvate

b. $CH_3\underset{\underset{\displaystyle CH_3}{|}}{\overset{\overset{\displaystyle NH_3^+}{|}}{C}}HCHCO_2^-$ + $CH_3\overset{\overset{\displaystyle O}{\|}}{C}CO_2^-$ $\longrightarrow$ $CH_3\underset{\underset{\displaystyle CH_3}{|}}{C}HCCO_2^-$ + $CH_3\overset{\overset{\displaystyle NH_3^+}{|}}{C}HCO_2^-$

valine   pyruvate   α-keto-   L-alanine
                                   isobutyrate

Problem 26.1

Draw structural formulas for the starting materials and products of the following transamination reactions.

a. phe + pyruvate $\longrightarrow$ **b.** his + α-ketoglutarate $\longrightarrow$

Transaminations are catalyzed by a specific group of enzymes called amino-transferases, or more commonly, transaminases. While transaminases are found in all cells, their concentrations are particularly high in heart and liver tissue. Damage to either of these organs leads to release of transaminases into the blood, and determination of serum levels of these enzymes can provide the clinician with valuable information about the extent of heart or liver damage. The two transaminases most commonly assayed for this purpose are serum glutamate oxaloacetate transaminase (SGOT) and serum glutamate pyruvate transaminase (SGPT).

For catalytic activity, all transaminases require pyridoxal phosphate, a coenzyme derived from pyridoxine, vitamin B_6 (Figure 26.2). In its role as a catalyst, this coenzyme undergoes reversible transformations between an aldehyde (pyridoxal phosphate) and a primary amine (pyridoxamine phosphate).

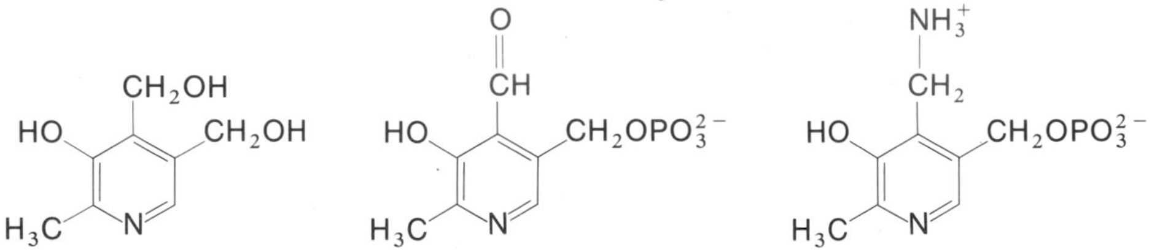

Figure 26.2 Pyridoxine, or vitamin B_6. Pyridoxal phosphate (PLP) and pyridoxamine phosphate (PMP) are coenzymes derived from pyridoxine.

Transaminations serve two vital functions. First, they provide a means to readjust the relative proportions of a number of amino acids to meet the particular needs of the organism; in most diets, the amino acid blend does not correspond precisely to what is needed. Second, transaminations serve to collect the nitrogen atoms of all amino acids as glutamate. Glutamate is the central source of nitrogen atoms for synthesis.

B. Oxidative Deamination

The major pathway by which amino groups are removed from amino acids is **oxidative deamination** of glutamate in the liver to form ammonium ion and α-ketoglutarate. The oxidation requires either NAD^+ or $NADP^+$ and is catalyzed by the enzyme glutamate dehydrogenase:

$$^-O_2C-CH_2-CH_2-\overset{\overset{\displaystyle NH_3^+}{|}}{CH}-CO_2^- + NAD^+ + H_2O \xrightleftharpoons{\text{glutamate dehydrogenase}}$$

glutamate

$$^-O_2C-CH_2-CH_2-\overset{\overset{\displaystyle O}{\|}}{C}-CO_2^- + NADH + NH_4^+ + H^+$$

α-ketoglutarate

In this way, amino groups collected from other amino acids are converted to ammonium ion.

Normal concentrations of ammonium ion in plasma are 0.025 to 0.04 mg/liter. Ammonium ion is extremely toxic and must be eliminated. The major pathway by which ammonium ion is detoxified and eliminated in humans is formation of urea, a neutral nontoxic compound, followed by excretion in the urine.

C. Synthesis of Urea

Urea synthesis in mammals occurs exclusively in the liver. The metabolic pathway that catalyzes the formation of urea is called the **urea cycle** or the Krebs-Henseleit cycle after Hans Krebs and Kurt Henseleit, who proposed it in 1932. This cycle accepts one carbon atom in the form of bicarbonate (or carbon dioxide) and two nitrogen atoms, one from ammonium ion and the other from aspartate, and, in a cycle process requiring five steps, generates urea and fumarate. The net reaction of the urea cycle is shown in Figure 26.3.

The five reactions of the urea cycle are shown in Figure 26.4. The first step in the formation of urea is synthesis of carbamoyl phosphate from bicarbonate, ammonium ion, and inorganic phosphate. This reaction is catalyzed by carbamoyl phosphate synthetase and is coupled with hydrolysis of two mole-

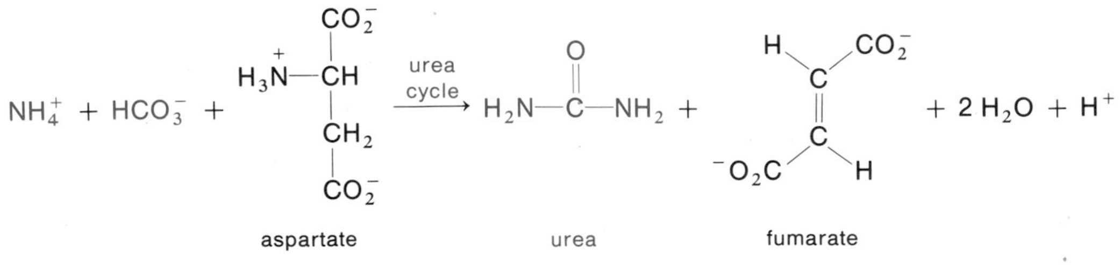

Figure 26.3 The net reaction of the urea cycle.

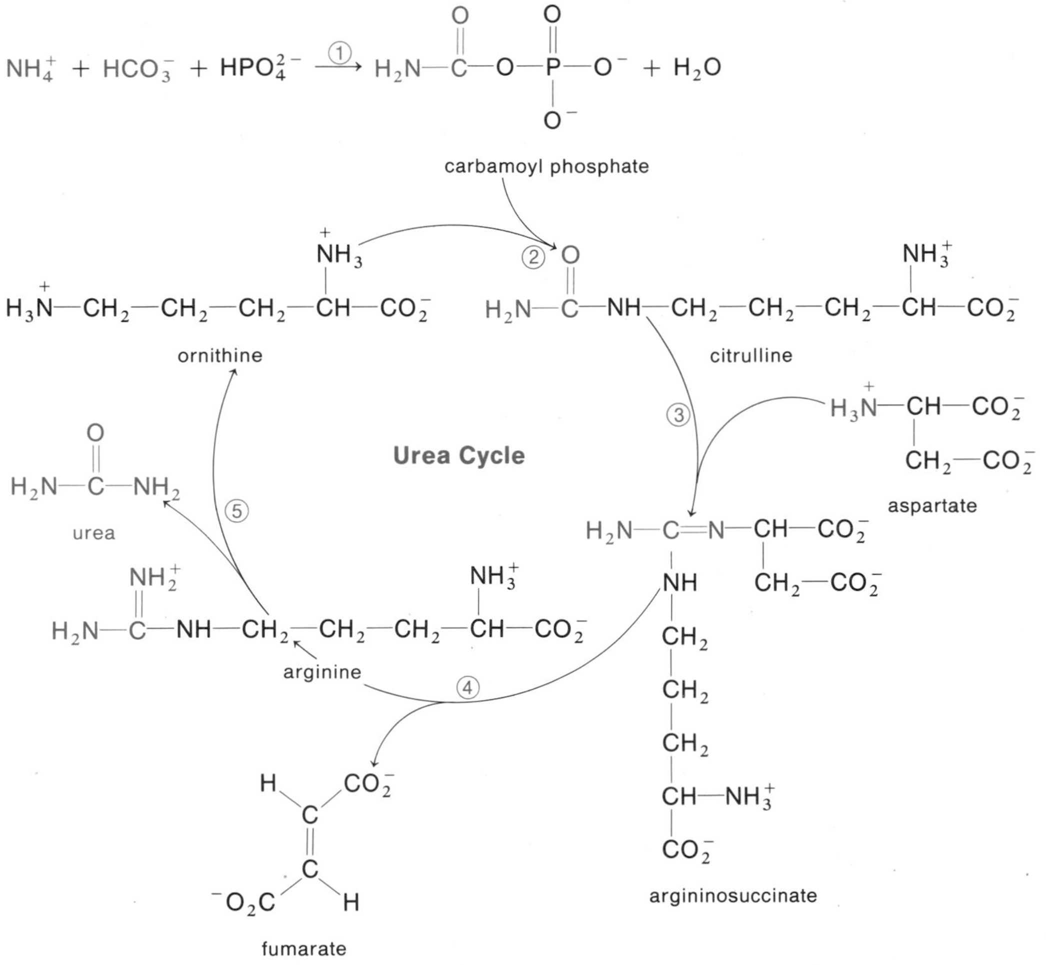

Figure 26.4 The urea cycle. The enzyme-catalyzed steps are numbered 1–5.

cules of ATP to ADP. In step 2, carbamoyl phosphate reacts with ornithine to form citrulline. The third step, condensation of citrulline with aspartate, is coupled with hydrolysis of a third molecule of ATP and incorporates a second nitrogen atom into the cycle. Cleavage of argininosuccinate produces arginine and fumarate (step 4). Finally, hydrolysis of arginine (step 5) yields one molecule of urea and regenerates ornithine.

Operation of the cycle requires a continuous supply of carbamoyl phosphate and aspartate. Carbamoyl phosphate is supplied by reaction of ammonium ion and bicarbonate. Aspartate is supplied from fumarate via reactions of the tricarboxylic acid cycle, followed by transamination. The conversion of

Figure 26.5 Aspartate, required for operation of the urea cycle, is synthesized from fumarate, a product of the urea cycle, via the tricarboxylic acid cycle, followed by transamination.

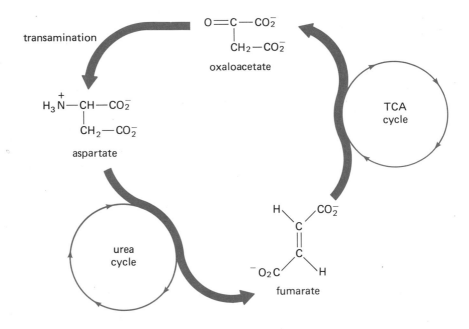

fumarate to aspartate via oxaloacetate demonstrates the interrelationship between the urea cycle and the tricarboxylic acid cycle (Figure 26.5).

Virtually all tissues produce NH_4^+ and, as we have seen in Section 26.2B, ammonium ion is highly toxic, especially to the nervous system. In humans, the synthesis of urea in the liver is the only major route for detoxification and elimination of NH_4^+. Failure of the urea-synthesizing pathway for any reason, including liver malfunction or inherited defects in any of the five enzymes of the urea cycle, results in an increase in blood, liver, and urinary levels of ammonium ion, a condition that produces **ammonia intoxication.** Symptoms of ammonia intoxication are protein-induced vomiting, blurred vision, tremors, slurred speech, and ultimately coma and death. In treating ammonia intoxication, it is essential to decrease the intake of dietary protein in order to decrease ammonium ion formation. Genetic defects in each of the five enzymes of the urea cycle do exist, but fortunately these inborn errors of metabolism are rare.

D. Oxidation of Carbon Skeletons

The central pathway for the oxidation of carbon skeletons of amino acids to carbon dioxide and water is the tricarboxylic acid cycle. There are five points of entry into the cycle for these skeletons: acetyl CoA, α-ketoglutarate, succinyl CoA, fumarate, and oxaloacetate. The specific entry points for carbon fragments from each of the 20 amino acids are summarized in Figure 26.6. As you study Figure 26.6, note the following:

1. Eleven amino acids (alanine, cysteine, glycine, isoleucine, leucine, lysine, phenylalanine, serine, threonine, tryptophan, and tyrosine) enter via acetyl CoA. Thus, acetyl CoA is the major point of entry.

2. Five (arginine, glutamate, glutamine, histidine, and proline) enter via α-ketoglutarate.

3. Three (isoleucine, methionine, and valine) enter via succinyl CoA.

4. Two (phenylalanine and tyrosine) enter via fumarate.

5. Two (asparagine and aspartate) enter via oxaloacetate.

6. The carbon skeletons of three amino acids are each degraded to two different fragments, each of which enters the cycle at a different point. Phenylalanine and tyrosine are degraded in part to acetoacetyl CoA, which enters the cycle via acetyl CoA, and in part to fumarate, which enters directly. Isoleucine is degraded to acetyl CoA and succinyl CoA, both of which enter the cycle directly.

Figure 26.6 Pathways by which carbon skeletons from amino acid degradation enter the tricarboxylic acid cycle.

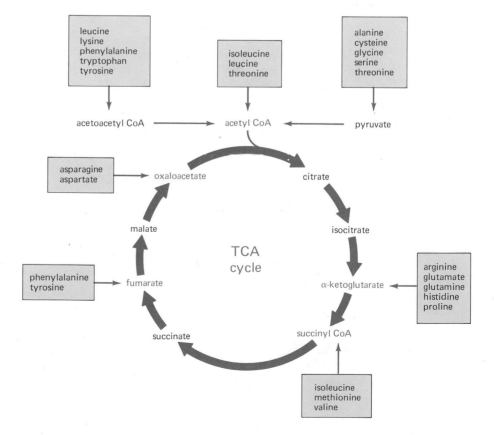

E. Conversion to Glucose and Ketone Bodies

Amino acids that can be degraded to pyruvate, α-ketoglutarate, succinyl CoA, and oxaloacetate can be converted to phosphoenolpyruvate and then into glucose. These amino acids are said to be **glycogenic** and the synthesis of glucose from them is called gluconeogenesis (Section 24.2D). Glycogenic amino acids, along with glycerol, provide alternative sources of glucose during periods of low carbohydrate intake or when stores are being rapidly depleted.

Amino acids degraded to acetyl CoA or acetoacetyl CoA cannot be converted into glucose because humans and other animals have no biochemical pathways for the synthesis of glucose from either of these intermediates. It is for this reason that fatty acids cannot serve as sources of carbon atoms for the synthesis of glucose. Acetyl CoA and acetoacetyl CoA derived from the degradation of amino acids can be transformed into ketone bodies, as we saw in Section 25.7. Amino acids that are degraded to either acetyl CoA or acetoacetyl CoA and then transformed into ketone bodies are said to be **ketogenic.**

The first experimental attempts to classify amino acids as glycogenic or ketogenic were carried out on test animals at a time when the metabolic pathways for the degradation of the individual amino acids were only poorly understood. In the earliest studies, laboratory dogs were made diabetic, either by selective chemical destruction of the insulin-producing capability of the pancreas, or by removal of the pancreas itself. Blood glucose levels were then controlled by injections of insulin. These diabetic dogs excreted glucose in the urine even when glycogen and fat stores had been depleted and when they were fed a diet containing protein as the sole source of metabolic fuel. They also excreted urea, the means by which amino acid-derived nitrogen atoms are detoxified and eliminated. The molar ratio of glucose to urea gives an

Table 26.1 Glycogenic and ketogenic amino acids.

Glycogenic		Glycogenic and Ketogenic	Ketogenic
alanine	glycine	isoleucine	leucine
arginine	histidine	lysine	
asparagine	methionine	phenylalanine	
aspartic acid	proline	threonine	
cysteine	serine	tryptophan	
glutamic acid	valine	tyrosine	
glutamine			

indication of the extent to which the carbon skeletons of amino acids can be used for the synthesis of glucose. These studies revealed that in diabetic dogs, a maximum of 58 grams of glucose can be derived from 100 grams of protein. In other words, 58% of protein is glycogenic.

To determine which of the 20 amino acids are glycogenic, diabetic test animals were fed pure amino acids, one at a time. If glucose was excreted in the urine following such a feeding, the amino acid was classified as glycogenic. If acetoacetate, β-hydroxybutyrate, or acetone was excreted, the amino acid was classified as ketogenic. Of the 20 protein-derived amino acids, only leucine is purely ketogenic. Six amino acids are both glycogenic and ketogenic at the same time, and the remaining thirteen amino acids are purely glycogenic. The glycogenic and ketogenic amino acids are given in Table 26.1.

Key Terms and Concepts

amino acid pool (26.1A)

ammonia intoxication (26.2C)

glycogenic amino acids (26.2E)

ketogenic amino acids (26.2E)

oxidative deamination (26.2B)

protein half-life (26.1A)

transamination (26.2A)

urea cycle (26.2C)

Key Reactions

1. Transamination (Section 26.2A):

$$\overset{\overset{\displaystyle NH_3^+}{|}}{^-O_2CCH_2CH_2CHCO_2^-} + \overset{\overset{\displaystyle O}{\|}}{CH_3CCO_2^-} \rightleftharpoons \overset{\overset{\displaystyle O}{\|}}{^-O_2CCH_2CH_2CCO_2^-} + \overset{\overset{\displaystyle NH_3^+}{|}}{CH_3CHCO_2^-}$$

$\qquad$ glutamate $\qquad\qquad$ pyruvate $\qquad\qquad$ α-ketoglutarate $\qquad$ alanine

2. Oxidative deamination (Section 26.2B):

$$\overset{\overset{\displaystyle NH_3^+}{|}}{^-O_2CCH_2CH_2CHCO_2^-} + NAD^+ + H_2O \longrightarrow$$

$\qquad$ glutamate

$$\overset{\overset{\displaystyle O}{\|}}{^-O_2CCH_2CH_2CCO_2^-} + NADH + NH_4^+ + H^+$$

$\qquad\qquad$ α-ketoglutarate

3. The urea cycle (Section 26.2C):

$$NH_4^+ + HCO_3^- + \underset{\underset{\displaystyle CH_2CO_2^-}{|}}{H_3\overset{+}{N}-CHCO_2^-} \longrightarrow \overset{\overset{\displaystyle O}{\|}}{H_2NCNH_2} + \begin{matrix} H \diagdown \quad \diagup CO_2^- \\ C \\ \| \\ C \\ \diagup \quad \diagdown \\ ^-O_2C \quad\quad H \end{matrix} + 2\,H_2O + H^+$$

$\qquad\qquad\qquad$ aspartate $\qquad\qquad\qquad\qquad$ urea $\qquad\qquad\qquad$ fumarate

Problems

26.2 List the three vital functions served by amino acids in the body.

26.3 Define the term *half-life* as it is applied in this chapter to tissue and plasma proteins.

26.4 Compare the degree to which carbohydrates, fats, and proteins can be stored in the body for later use.

26.5 What percentage of the total energy requirement of the average adult is supplied by carbohydrates? By fats? By proteins?

26.6 Complete the following reactions. Show structural formulas for products and reactants.

    a. leucine + oxaloacetate $\xrightarrow{\text{transaminase}}$

    b. alanine + α-ketoglutarate $\xrightarrow{\text{transaminase}}$

    c. glycine + pyruvate $\xrightarrow{\text{transaminase}}$

    d. glycine + pyridoxal phosphate $\xrightarrow{\text{transaminase}}$

    e. pyridoxamine phosphate + pyruvate $\xrightarrow{\text{transaminase}}$

26.7 a. Write an equation for the reaction catalyzed by the enzyme serum glutamate oxaloacetate transaminase (SGOT); for the reaction catalyzed by serum glutamate pyruvate transaminase (SGPT).

    b. Describe how an assay for the presence of these enzymes in blood can provide information about possible heart and liver damage.

26.8 In nutritional studies on rats, it has been found that certain α-ketoacids may substitute for essential amino acids. Shown here are structural formulas for three such α-ketoacids.

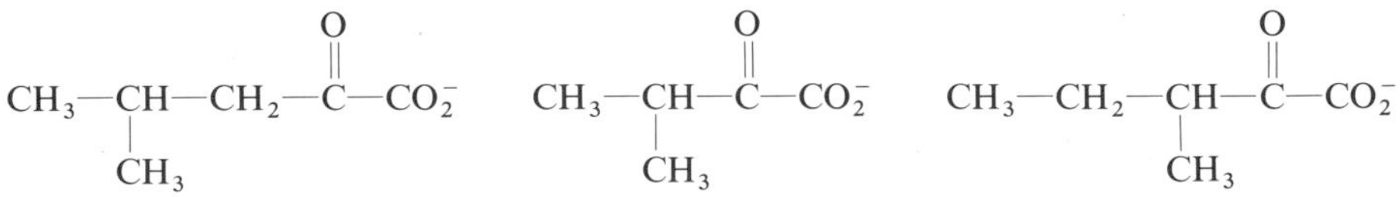

$$CH_3-CH-CH_2-\overset{\overset{\displaystyle O}{\|}}{C}-CO_2^- \qquad CH_3-CH-\overset{\overset{\displaystyle O}{\|}}{C}-CO_2^- \qquad CH_3-CH_2-CH-\overset{\overset{\displaystyle O}{\|}}{C}-CO_2^-$$
$$\qquad\quad \overset{|}{CH_3} \qquad\qquad\qquad\quad \overset{|}{CH_3} \qquad\qquad\qquad\qquad\quad \overset{|}{CH_3}$$

    (a)   (b)   (c)

    a. Account for the fact that in certain instances, these α-ketoacids may substitute for essential amino acids.

    b. For which essential amino acid might each substitute?

26.9 The following reaction is the first step in the degradation of ornithine. Propose a metabolic pathway to account for this transformation.

$$H_2N-CH_2-CH_2-CH_2-\overset{\overset{\displaystyle NH_3^+}{|}}{CH}-CO_2^- \longrightarrow H-\overset{\overset{\displaystyle O}{\|}}{C}-CH_2-CH_2-\overset{\overset{\displaystyle NH_3^+}{|}}{CH}-CO_2^-$$

    ornithine   glutamate semialdehyde

26.10 Write an equation for the oxidative deamination of glutamate.

26.11 What is the function of the urea cycle?

26.12 Using structural formulas, write an equation for the hydrolysis of arginine to ornithine and urea.

26.13 In what organ does the synthesis of urea take place?

26.14 Urea has two nitrogen atoms. What is the source of these nitrogen atoms? What is the source of the single carbon atom in urea?

26.15 Write a balanced equation for the net reaction of the urea cycle.

26.16 What is meant by the term *ammonia intoxication*?

26.17 List the five points at which carbon skeletons derived from amino acids enter the tricarboxylic acid cycle.

26.18 a. What does it mean to say that an amino acid is *glycogenic*?
       b. What does it mean to say that an amino acid is *ketogenic*?
       c. Is it possible for an amino acid to be both glycogenic and ketogenic? Explain.

26.19 Propose biochemical pathways to explain how the cell might carry out the following transformations. Name each type of reaction (for example, hydration, dehydration, oxidation, hydrolysis).
       a. phenylalanine $\longrightarrow$ phenylacetate
       b. 3-phosphoglycerate $\longrightarrow$ serine
       c. citrate $\longrightarrow$ glutamate
       d. ornithine $\longrightarrow$ glutamate
       e. methylmalonyl CoA $\longrightarrow$ oxaloacetate

26.20 Following is the metabolic pathway for the conversion of isoleucine to acetyl CoA and propionyl CoA. You have already studied each type of reaction shown here, though not necessarily in this chapter. For each step in this sequence, name the type of reaction and specify any coenzymes involved.

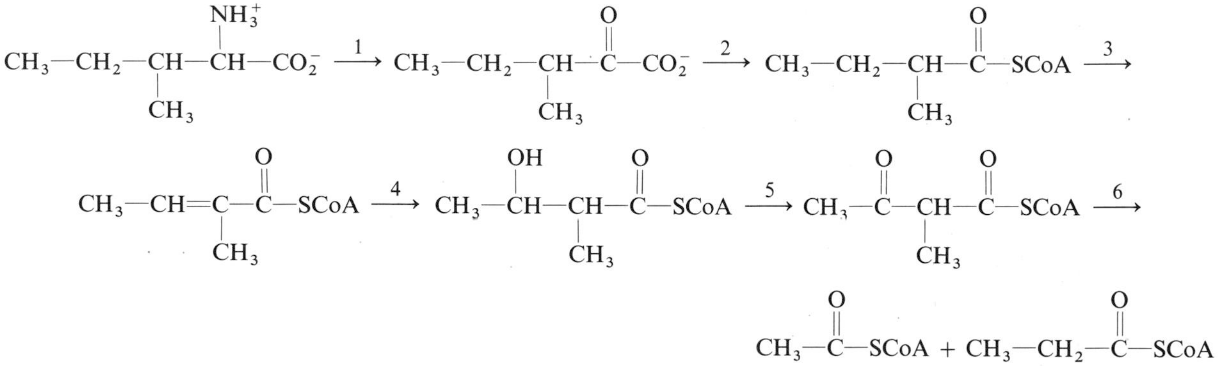

$$CH_3-CH_2-\underset{\underset{CH_3}{|}}{CH}-\underset{\underset{NH_3^+}{|}}{CH}-CO_2^- \xrightarrow{1} CH_3-CH_2-\underset{\underset{CH_3}{|}}{CH}-\overset{\overset{O}{\|}}{C}-CO_2^- \xrightarrow{2} CH_3-CH_2-\underset{\underset{CH_3}{|}}{CH}-\overset{\overset{O}{\|}}{C}-SCoA \xrightarrow{3}$$

$$CH_3-CH=\underset{\underset{CH_3}{|}}{C}-\overset{\overset{O}{\|}}{C}-SCoA \xrightarrow{4} CH_3-\underset{\underset{OH}{|}}{CH}-\underset{\underset{CH_3}{|}}{CH}-\overset{\overset{O}{\|}}{C}-SCoA \xrightarrow{5} CH_3-\overset{\overset{O}{\|}}{C}-\underset{\underset{CH_3}{|}}{CH}-\overset{\overset{O}{\|}}{C}-SCoA \xrightarrow{6}$$

$$CH_3-\overset{\overset{O}{\|}}{C}-SCoA + CH_3-CH_2-\overset{\overset{O}{\|}}{C}-SCoA$$

acetyl CoA   propionyl CoA

26.21 Following is the metabolic pathway for the conversion of proline to glutamate. Name the type of reaction involved in each step of this transformation and specify any coenzymes you think might be involved.

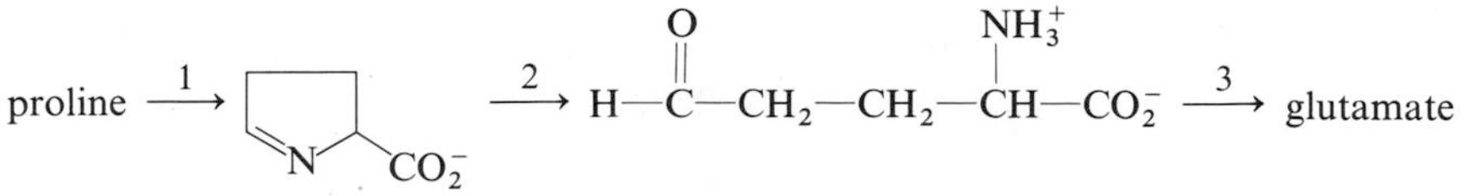

$$\text{proline} \xrightarrow{1} \underset{N}{\bigcirc}\text{-}CO_2^- \xrightarrow{2} H-\overset{\overset{O}{\|}}{C}-CH_2-CH_2-\underset{\underset{NH_3^+}{|}}{CH}-CO_2^- \xrightarrow{3} \text{glutamate}$$

26.22 The vast majority of substances in the biological world are built from just thirty or so smaller molecules. Review the biochemistry we have discussed in this text and make up your own list of these thirty or so fundamental building blocks of nature.

Answers to In-Chapter Problems

Chapter 1

1.1 a. 5.993×10^{24} g   b. 5.9082153×10^9 km

1.2 a. 1.67×10^{-24} g   b. 1.54×10^{-8} m

1.3 a. 1.10×10^5 b. 9.04×10^4 c. 6.10×10^{-7} d. 7.01×10^{-1}

1.4 15.5 cm^3 1.5 $2.1 \times 10^{-9}\%$ Zn   1.6 41 m   1.7 0.424 g   1.8 73.8 L

1.9 1.46 cm^3 1.10 a. 43°C   b. 316 K   c. 77°F   1.11 160 cal

Chapter 2

2.1 C_2H_6O 2.2 2 atoms carbon, 3 atoms hydrogen, 1 atom chlorine

2.3 15 protons, 16 neutrons, 15 electrons

2.4 uranium-238: 92 protons, 146 neutrons, 92 electrons
uranium-235: 92 protons, 143 neutrons, 92 electrons

2.5 19.5 g carbon   2.6 0.204 g   2.7 127 cm^3

2.8 27 protons, 27 electrons, 33 neutrons

2.9 $^{226}_{88}$Ra $\longrightarrow$ $^4_2$He + $^{222}_{86}$Rn; radon-222   2.10 8.6×10^{-10} g

Chapter 3

3.1 As: $1s^2 2s^2 2p^6 3s^2 3p^6 4s^2 3d^{10} 4p^3$

3.2 a. Sr: [Kr]$5s^2$ b. Br: [Ar]$4s^2 3d^{10} 4p^5$

3.3 a. gallium: representative element, inner core of 28 electrons, 3 valence electrons in 4th shell:

$$1s^2 2s^2 2p^6 3s^2 3p^6 4s^2 3d^{10} 4p^1$$

b. sulfur: representative element, inner core of 10 electrons, 6 valence electrons in 3rd shell:

$$1s^2 2s^2 2p^6 3s^2 3p^4$$

3.4 a. K·   b. ·Ȧs:

3.5 S: $1s^2 2s^2 2p^6 3s^2 3p^4$ S^{2-}: $1s^2 2s^2 2p^6 3s^2 3p^6$ Al: $1s^2 2s^2 2p^6 3s^2 3p^1$
Al^{3+}: $1s^2 2s^2 2p^6$

Chapter 4

4.1 a. $MgSO_4$ b. N_2O_3 4.2 6 oxygen atoms

4.3 glucose and formalin; CH_2O; glucose is $(CH_2O)_6$ 4.4 a. 180.2 b. 18.04

4.5 38.5 g 4.6 5.64×10^{20} atoms H 4.7 0.318 moles H_2SO_4 4.8 28.2% N

4.9 92.6% Hg 4.10 $FeCl_2$ 4.11 $C_5H_{10}O_5$ 4.12 a. 1.2 b. 0.8 c. 3.2

4.13 a. barium sulfate b. iron(II) nitrate c. chromium(III) oxide
     d. lithium fluoride

4.14 a. $PbCl_2$ b. NH_4NO_3 c. FeS d. $KHCO_3$ e. MgO f. $Co(NO_3)_3$

4.15 a.

$$H-\overset{\overset{\displaystyle H}{|}}{\underset{\underset{\displaystyle H}{|}}{C}}-\ddot{\underset{..}{C}}l:$$

b. $H-C\equiv N:$ c. $:\overset{..}{\underset{..}{S}}=C=\overset{..}{\underset{..}{S}}:$

4.16 a. $:\overset{..}{\underset{..}{O}}=C\overset{\displaystyle :\overset{..}{O}:^-}{\underset{\displaystyle :\overset{..}{\underset{..}{O}}:^-}{}}$ b. $^-:\overset{..}{\underset{..}{O}}-H$ c. $:\overset{..}{\underset{..}{O}}=\overset{\overset{\displaystyle :\overset{..}{O}:}{|}}{N}-\overset{..}{\underset{..}{O}}:^-$

4.17 $:\overset{..}{O}=\overset{..}{\underset{..}{O}}-\overset{..}{\underset{..}{O}}: \longleftrightarrow :\overset{..}{\underset{..}{O}}-\overset{..}{\underset{..}{O}}=\overset{..}{O}:$

4.18 a. in CH_3, 109.5°; in CHO, 120° b. 120° c. 109.5° d. 109.5°

4.19 polar covalent; C bears partial positive charge; Cl bears partial negative charge

4.20 a. polar b. nonpolar c. polar

4.21 $[H-\overset{..}{\underset{..}{S}}:]^- + H^+ \longrightarrow H-\overset{\overset{\displaystyle ..}{\underset{\underset{\displaystyle H}{|}}{S}}}{}:$

Chapter 5

5.1 $N_2 + O_2 \xrightarrow{\text{elect}} 2\,NO$ 5.2 $2\,C_8H_{18} + 25\,O_2 \longrightarrow 16\,CO_2 + 18\,H_2O$

5.3 $Cl_2 + 2\,Br^- \longrightarrow Br_2 + 2\,Cl^-$

5.4 a. $2\,C + O_2 \longrightarrow 2\,CO$ b. $BaO + H_2O \longrightarrow Ba(OH)_2$

5.5 a. $2\,Ag_2O \longrightarrow 4\,Ag + O_2$ b. $2\,AsH_3 \longrightarrow 2\,As + 3\,H_2$

5.6 a. $Zn + H_2SO_4 \longrightarrow H_2 + ZnSO_4$ b. $Hg + AgNO_3 \longrightarrow Ag + HgNO_3$

5.7 a. $Al(OH)_3 + 3\,HCl \longrightarrow AlCl_3 + 3\,H_2O$
    b. $CH_3CO_2^- + HCl \longrightarrow CH_3CO_2H + Cl^-$

5.8 a. $Ba(CH_3CO_2)_2$; soluble b. Ag_2S; insoluble c. $(NH_4)_3PO_4$; soluble
    d. $CaCO_3$; insoluble e. $Cr(NO_3)_3$; soluble f. Na_2SO_4; soluble
    g. $Al(OH)_3$; insoluble

5.9 a. $CrCl_3 + 3\,NaOH \longrightarrow Cr(OH)_3\downarrow + 3\,NaCl$

    chromium(III)   sodium
    hydroxide   chloride

    b. $H_2SO_4 + BaCl_2 \longrightarrow BaSO_4\downarrow + 2\,HCl$

    barium   hydrochloric
    sulfate   acid

5.10 a. Fe, +3; O, $U2$ b. Na, +1; Mn, +7; O, $N2$ c. N, +4; O, −2

5.11 a. not oxidation-reduction
   b. zinc is oxidized, is reducing agent; hydrogen is reduced, is oxidizing agent
5.12 copper is oxidized to Cu^{2+}, copper is reducing agent; Hg^{2+} is reduced to Hg, Hg^{2+} is oxidizing agent
5.13 $2 \text{ NaI} + Br_2 \longrightarrow 2 \text{ NaBr} + I_2$

$$2 \text{ I}^- \longrightarrow I_2 + 2 e^-$$
$$2 e^- + Br_2 \longrightarrow 2 \text{ Br}^-$$
$$\overline{2 \text{ I}^- + Br_2 \longrightarrow I_2 + 2 \text{ Br}^-}$$
$$2 \text{ NaI} + Br_2 \longrightarrow I_2 + 2 \text{ NaBr}$$

5.14 $2 \text{ Br}^- \longrightarrow Br_2 + 2 e^-$

$$4 H^+ + SO_4^{2-} + 2 e^- \longrightarrow SO_2 + 2 H_2O$$
$$\overline{4 H^+ + SO_4^{2-} + 2 \text{ Br}^- \longrightarrow SO_2 + Br_2 + 2 H_2O}$$

5.15 $13.8 \text{ g } O_2$ 5.16 $0.941 \text{ g } Cl_2$ 5.17 43% yield   5.18 82% pure
5.19 $2.33 \text{ g } C_2H_5Cl$
5.20 a. exothermic; reverse: $2 NH_3(g) \longrightarrow N_2(g) + 3 H_2(g)$ $\Delta H = +46.0 \text{ kJ}$
   b. exothermic; reverse: $4 HCl(g) + O_2(g) \longrightarrow 2 H_2O(g) + 2 Cl_2(g)$
      $\Delta H = +120 \text{ kJ}$
   c. exothermic; reverse: $3 H_2O(l) + 2 CO_2(g) \longrightarrow C_2H_5OH(l) + 3 O_2(g)$
      $\Delta H = +1.37 \times 10^3 \text{ kJ}$ 5.21 -725 kJ

Chapter 6

6.1 1.25×10^3 torr   6.2 127 cm^3 6.3 1.38 L   6.4 4.8 L   6.5 0.179 g/L
6.6 1.16 g oxygen   6.7 13 L oxygen
6.8 $19.6\% \, O_2, 0.04\% \, CO_2, 6.18\% \, H_2O, 74.2\% \, N_2$
6.9 1.49 for HCl to 1.00 for HBr   6.10 51.3 kcal
6.11 $1.2 \text{ kcal/mole } H_2O$ 6.12 6300 cal

Chapter 7

7.1 3.6 g NaCl in 396 g water
7.2 400 cm^3 ethyl alcohol in 600 cm^3 water   7.3 0.0563 mole glucose
7.4 11.0 g NaCl   7.5 12 cm^3 of 6.0M HCl   7.6 0.121M I^-
7.7 $0.031 \text{ L } CO_2$ 7.8 0.220M HCl

Chapter 8

8.1 $K_{eq} = \dfrac{[NO_2]^2}{[NO]^2[O_2]}$

8.2 a. $K_{eq} = \dfrac{[H_2]^2[O_2]}{[H_2O]^2}$ b. left   c. right   d. right

   e. b, c $\longrightarrow$ no change; d increases K_{eq}

8.3 a. $C_6H_8O_7 \rightleftharpoons H^+ + C_6H_7O_7^-$ b. $K_a = \dfrac{[H^+][C_6H_7O_7^-]}{[C_6H_8O_7]}$ 8.4 0.1M

8.5 $[H^+] = 4.2 \times 10^{-3}M$

8.6 a. pH = 3.3   b. $[H^+] = 6.9 \times 10^{-4}\ M$ 8.7 $1.0 \times 10^{-10}\ M$, pH = 10

8.8 $[H^+] = 1.8 \times 10^{-4} M$ 8.9 pH = 5.04

8.10 a. 4.04   b. 3.98   c. 4.11   8.11 7.26; final pH is pH 12

Chapter 9

9.1 Ether:

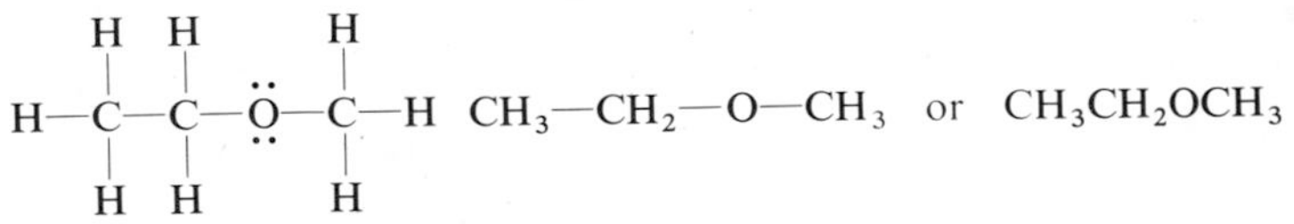

$$CH_3-CH_2-O-CH_3 \quad \text{or} \quad CH_3CH_2OCH_3$$

9.2 Ketones:

$$CH_3-\overset{O}{\overset{\|}{C}}-CH_2-CH_2-CH_3 \quad \text{or} \quad CH_3CCH_2CH_2CH_3$$

$$CH_3-\overset{O}{\overset{\|}{C}}-\overset{CH_3}{\underset{|}{CH}}-CH_3 \quad \text{or} \quad CH_3CCHCH_3 \ \ \underset{CH_3}{}$$

$$CH_3-CH_2-\overset{O}{\overset{\|}{C}}-CH_2-CH_3 \quad \text{or} \quad CH_3CH_2CCH_2CH_3$$

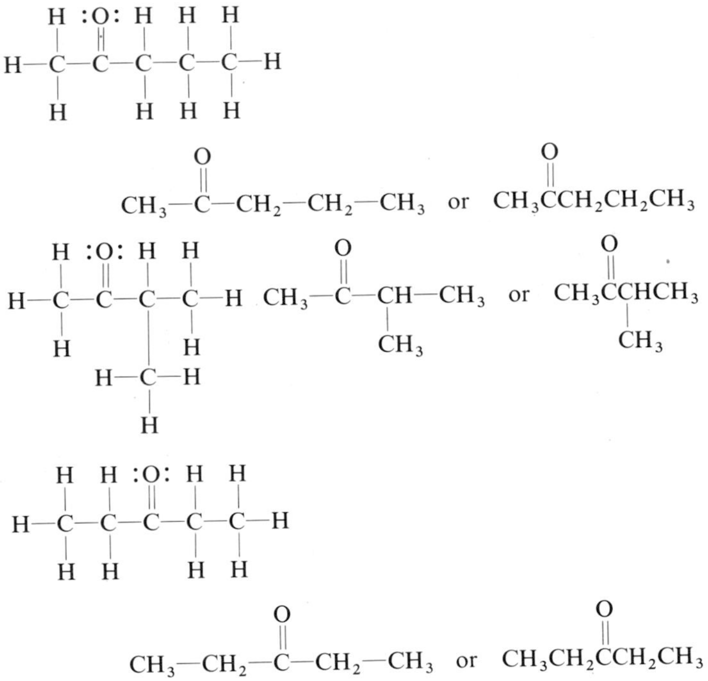

9.3 Carboxylic acids:

$$CH_3-CH_2-CH_2-\overset{O}{\overset{\|}{C}}-OH \quad \text{or} \quad CH_3CH_2CH_2CO_2H$$

$$CH_3-\underset{CH_3}{\underset{|}{CH}}-\overset{O}{\overset{\|}{C}}-OH \quad \text{or} \quad CH_3CHCO_2H \ \ \underset{CH_3}{}$$

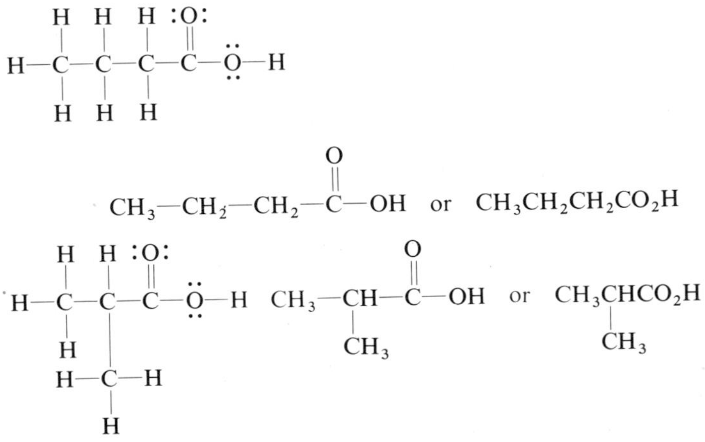

9.4 Following are molecular formulas for each.

a. C_4H_6O b. $C_7H_{12}O_2$ c. C_4H_6O

d. C_4H_6O e. $C_7H_{10}O_2$ f. $C_7H_{12}O_2$

Substances (a), (c), and (d) have the same molecular formula but different structural formulas and therefore are structural isomers. Substances (b) and (f) are also structural isomers. There are no structural isomers in this problem for (e).

9.5 a.

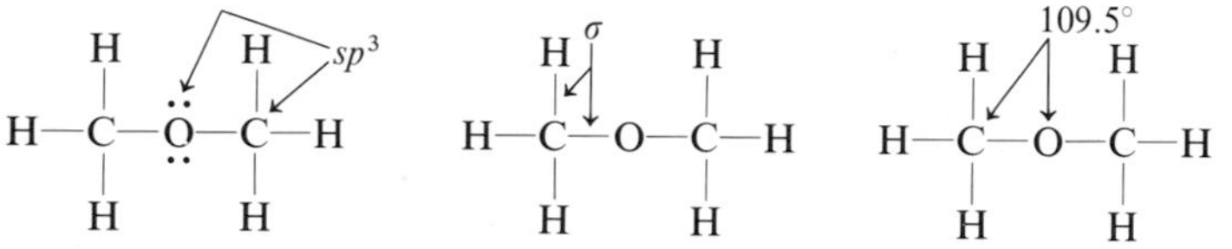

b.

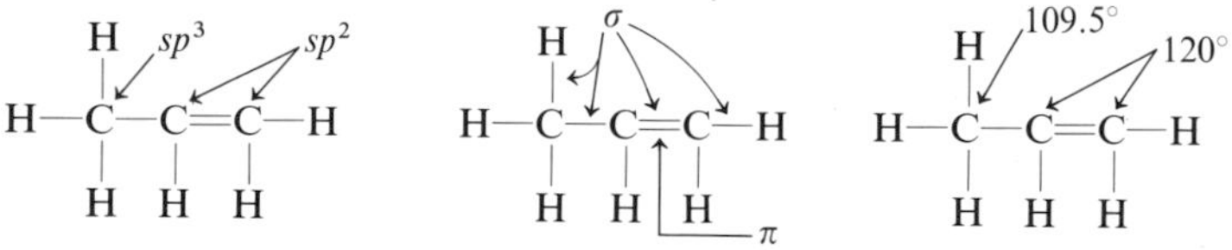

c.

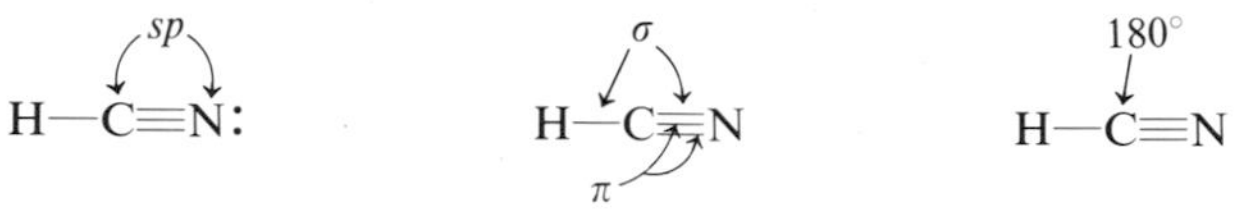

Chapter 10

10.1 a.

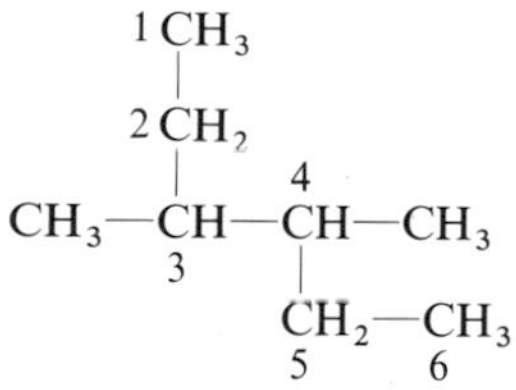

and

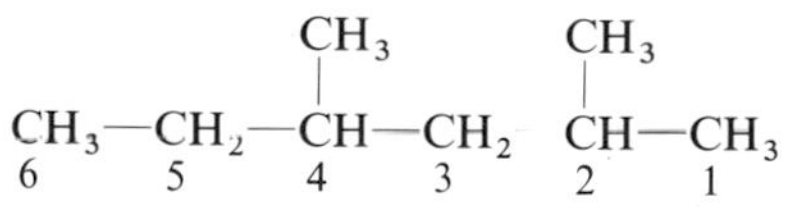

Each has the molecular formula C_8H_{18}, six carbons in the longest chain, and two $—CH_3$ groups on the chain. However, the $—CH_3$ groups are on carbons 3 and 4 of the left structure but on carbons 2 and 4 of the right structure. Therefore, these two structural formulas represent structural isomers.

b.

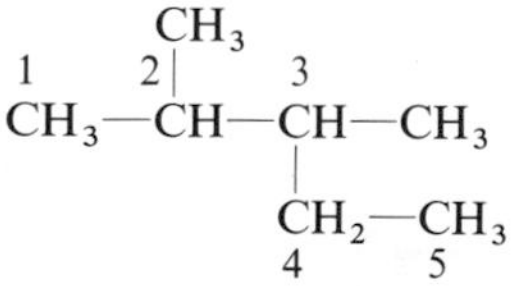

and

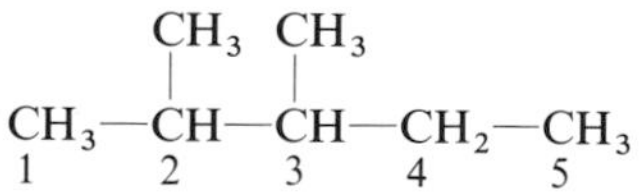

Each has the molecular formula C_7H_{16}, five carbons in the longest chain, and $—CH_3$ groups on carbons 2 and 3 of the chain. Therefore, the structural formulas are identical and represent the same alkane.

10.2 $CH_3—CH_2—CH_2—CH_2—CH_3$

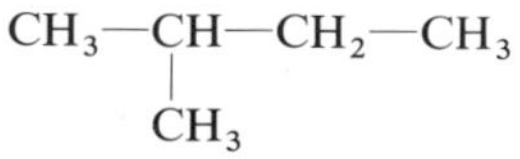

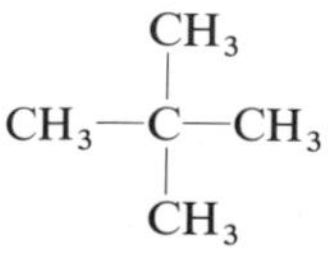

10.3   a. 5-isopropyl-2-methyloctane   b. 4-isopropyl-4-propylheptane

10.4   a. 1-ethyl-1-methylcyclopropane   b. 1,1,3-trimethylcyclohexane

10.5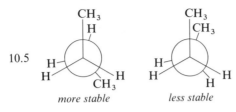

more stable   *less stable*

10.6   a.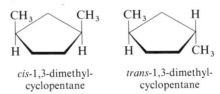

b.

c.

10.7   a. Methylcyclohexane does not show *cis-trans* isomerism because it has only one substituent on the ring.

b. 1,3-Dimethylcyclopentane shows *cis-trans* isomerism.

cis-1,3-dimethyl-   *trans*-1,3-dimethyl-
cyclopentane   cyclopentane

c. Ethylcyclopentane does not show *cis-trans* isomerism.

d. 1-Ethyl-2-methylcyclobutane shows *cis-trans* isomerism.

cis-1-ethyl-2-methyl-   *trans*-1-ethyl-2-methyl-
cyclobutane   cyclobutane

10.8   a. 2,2-dimethylpropane   2-methylbutane   pentane
          (bp 9.5°C)   (bp 29°C)   (bp 36°C)

b. 1,1-dimethylcyclohexane   ethylcyclohexane   cyclooctane
      (bp 120°C)   (bp 132°C)   (bp 149°C)

10.9 a.

$$CH_2-CH_2-\overset{\overset{\displaystyle CH_3}{|}}{\underset{\underset{\displaystyle CH_3}{|}}{C}}-CH_3$$
|
Br

4-bromo-2,2-di-
methylbutane

$$CH_3-CH-\overset{\overset{\displaystyle CH_3}{|}}{\underset{\underset{\displaystyle CH_3}{|}}{C}}-CH_3$$
|
Br

3-bromo-2,2-di-
methylbutane

$$CH_3-CH_2-\overset{\overset{\displaystyle CH_3}{|}}{\underset{\underset{\displaystyle CH_3}{|}}{C}}-CH_2-Br$$

1-bromo-2,2-di-
methylbutane

b.

chlorocyclo-
pentane

10.10 $C_6H_{14} + \dfrac{19}{2}O_2 \longrightarrow 6\,CO_2 + 7\,H_2O$ or

$2\,C_6H_{14} + 19\,O_2 \longrightarrow 12\,CO_2 + 14\,H_2O$

a. $\dfrac{6 \text{ moles C}}{\text{mole } C_6H_{14}} \times \dfrac{170 \text{ kcal}}{\text{mole C}} = \dfrac{1020 \text{ kcal}}{\text{mole } C_6H_{14}}$

b. $\dfrac{1020 \text{ kcal}}{\text{mole } C_6H_{14}} \times \dfrac{1 \text{ mole } C_6H_{14}}{86 \text{ g } C_6H_{14}} = \dfrac{11.9 \text{ kcal}}{\text{g } C_6H_{14}}$

Chapter 11   11.1 a. $CH_2{=}\overset{\overset{\displaystyle CH_3}{|}}{C}-CH_2-CH_2-CH_3 + CH_3-\overset{\overset{\displaystyle CH_3}{|}}{C}{=}CH-CH_2-CH_3$

$CH_3-\overset{\overset{\displaystyle CH_3}{|}}{CH}-CH{=}CH-CH_3 + CH_3-\overset{\overset{\displaystyle CH_3}{|}}{CH}-CH_2-CH{=}CH_2$

b. $CH_3-\overset{\overset{\displaystyle CH_3}{|}}{CH}-\overset{\overset{\displaystyle CH_3}{|}}{C}{=}CH_2 + CH_3-\overset{\overset{\displaystyle H_3C}{|}}{C}{=}\overset{\overset{\displaystyle CH_3}{|}}{C}-CH_3$

c. $CH_3-\overset{\overset{\displaystyle CH_3}{|}}{\underset{\underset{\displaystyle CH_3}{|}}{C}}-CH{=}CH_2$

11.2 $CH_3-CH_2-CH_2-CH_2-C{\equiv}CH$

$CH_3-\overset{\overset{\displaystyle }{}}{\underset{\underset{\displaystyle CH_3}{|}}{CH}}-CH_2-C{\equiv}CH$

$CH_3-\overset{\overset{\displaystyle CH_3}{|}}{\underset{\underset{\displaystyle CH_3}{|}}{C}}-C{\equiv}CH$

$CH_3-CH_2-CH_2-C{\equiv}C-CH_3$

$CH_3-\overset{\overset{\displaystyle }{}}{\underset{\underset{\displaystyle CH_3}{|}}{CH}}-C{\equiv}C-CH_3$

$CH_3-CH_2-C{\equiv}C-CH_2-CH_3$

$CH_3-CH_2-\overset{\overset{\displaystyle }{}}{\underset{\underset{\displaystyle CH_3}{|}}{CH}}-C{\equiv}CH$

11.3 a. 2-methyl-1-pentene; 2-methyl-2-pentene; 4-methyl-2-pentene; 4-methyl-1-pentene

b. 2,3-dimethyl-1-butene; 2,3-dimethyl-2-butene   c. 3,3-dimethyl-1-butene

11.4 a. $CH_3-CH_2-CH_2-CH_2-C\equiv CH$ b. $CH_3-CH_2-CH_2-C\equiv C-CH_3$

c. 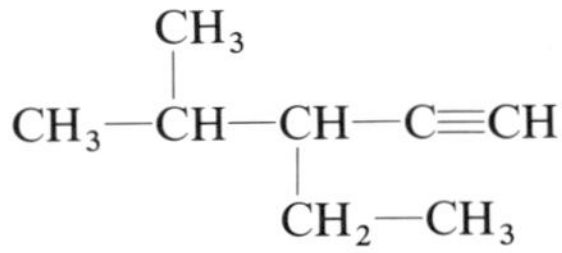 $CH_3-CH-CH-C\equiv CH$

with CH_3 on the first CH and CH_2-CH_3 below the second CH

11.5 Only part (c), 4-methyl-2-pentene, shows *cis-trans* isomerism.

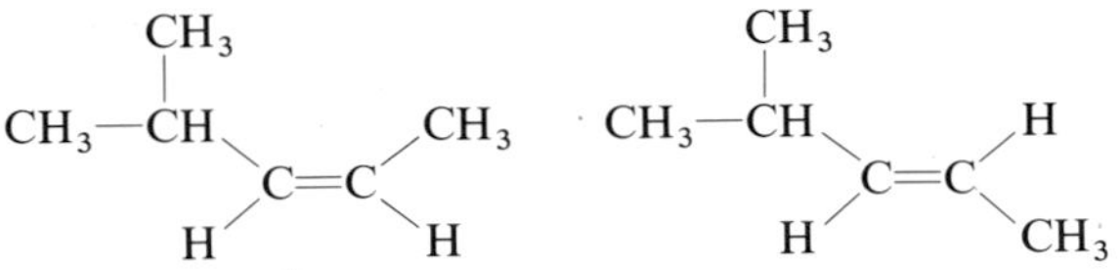

cis-4-methyl-2-pentene   *trans*-4-methyl-2-pentene

11.6 a. 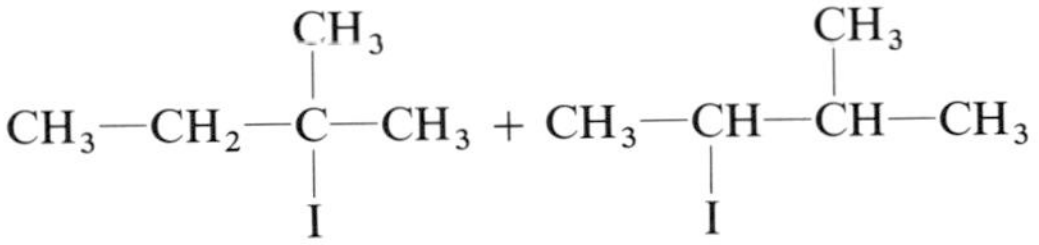 $CH_3-CH_2-\underset{I}{\overset{CH_3}{C}}-CH_3 + CH_3-\underset{I}{\overset{CH_3}{CH}}-CH-CH_3$

major product   *minor product*

b. 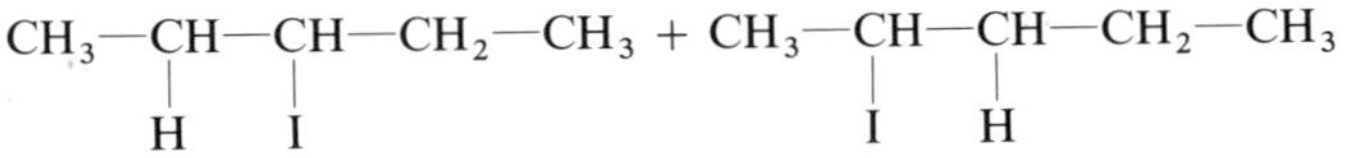 $CH_3-\underset{H}{CH}-\underset{I}{CH}-CH_2-CH_3 + CH_3-\underset{I}{CH}-\underset{H}{CH}-CH_2-CH_3$

Predict that 3-iodopentane and 2-iodopentane are formed in equal amounts.

11.7 a. 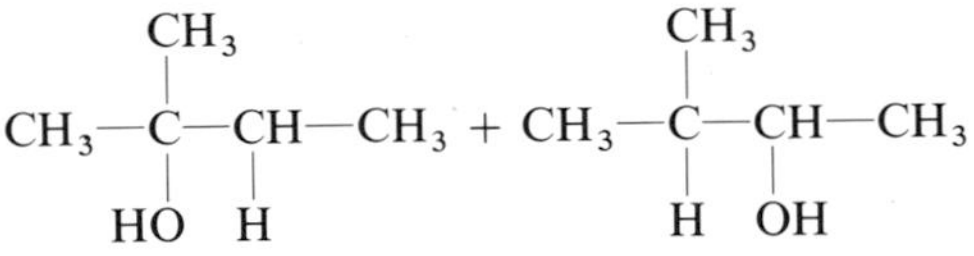 $CH_3-\overset{CH_3}{C}-\underset{HO\quad H}{CH}-CH_3 + CH_3-\overset{CH_3}{C}-\underset{H\quad OH}{CH}-CH_3$

major   *minor*

b. 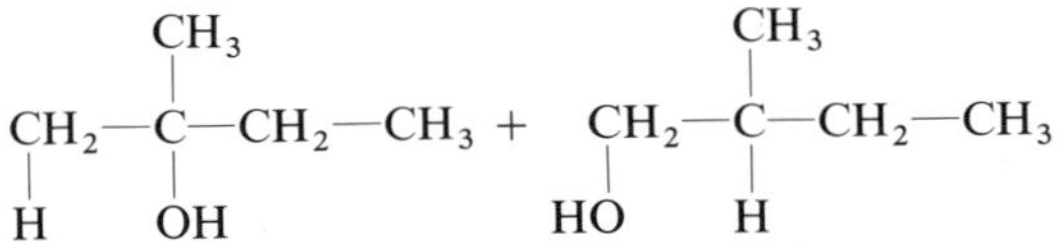 $CH_2-\underset{H\quad OH}{\overset{CH_3}{C}}-CH_2-CH_3 + CH_2-\underset{HO\quad H}{\overset{CH_3}{C}}-CH_2-CH_3$

major   *minor*

11.8 *Step 1:* $CH_3-\overset{CH_3}{C}=CH-CH_3 + \quad H^+ \quad \longrightarrow \quad CH_3-\overset{CH_3}{\underset{+}{C}}-CH_2-CH_3$

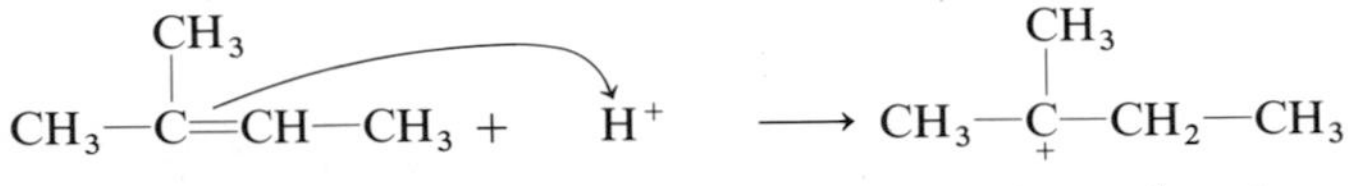

an electrophile   *a tertiary carbocation, also an electrophile*

Step 2: CH$_3$—$\overset{\overset{\displaystyle CH_3}{|}}{\underset{+}{C}}$—CH$_2$—CH$_3$ + H—$\ddot{\ddot{O}}$—H $\longrightarrow$ CH$_3$—$\overset{\overset{\displaystyle CH_3}{|}}{\underset{\underset{\displaystyle H^{\cdots}H}{\overset{+}{O}}}{C}}$—CH$_2$—CH$_3$

Step 3: CH$_3$—$\overset{\overset{\displaystyle CH_3}{|}}{\underset{\underset{\displaystyle H^{\cdots}H}{\overset{+}{O}}}{C}}$—CH$_2$—CH$_3$ $\longrightarrow$ CH$_3$—$\overset{\overset{\displaystyle CH_3}{|}}{\underset{\underset{\displaystyle :OH}{}}{C}}$—CH$_2$—CH$_3$ + H$^+$

11.9 a. CH$_3$—$\overset{\overset{\displaystyle CH_3}{|}}{CH}$—CH=CH$_2$ b. [cyclopentene with CH$_3$]

11.10 a. CH$_3$—$\overset{\overset{\displaystyle O}{||}}{C}$—OH + HO—$\overset{\overset{\displaystyle O}{||}}{C}$—$\overset{\overset{\displaystyle O}{||}}{C}$—OH + O=$\overset{\overset{\displaystyle CH_3}{|}}{C}$—CH$_3$

    carboxylic acids   a ketone

b. CH$_3$—$\overset{\overset{\displaystyle O}{||}}{C}$—CH$_2$—CH$_2$—$\overset{\overset{\displaystyle }{|}}{\underset{\underset{\displaystyle CH_3\quad CH_3}{CH}}{CH}}$—CH$_2$—$\overset{\overset{\displaystyle O}{||}}{C}$—OH

  a ketone   a carboxylic acid

Chapter 12 12.1 a. 2,2-dimethyl-1-propanol   b. 4-methyl-3-hexanol
     (neopentyl alcohol)

     c. 1-methylcyclopentanol

12.2 a. primary   b. secondary   c. primary   d. tertiary
12.3 a. 2-buten-1-ol   b. 2-cyclohexenol
12.4 ethyl isobutyl ether   b. cyclohexyl ethyl ether or ethoxycyclohexane

12.5 CH$_3$OCH$_2$CH$_2$OCH$_3$ CH$_3$OCH$_2$CH$_2$OH   HOCH$_2$CH$_2$OH
    bp 84°C   bp 125°C   bp 198°C

12.6 ClCH$_2$CH$_2$Cl   CH$_3$CH$_2$OCH$_2$CH$_3$ CH$_3$CH$_2$CH$_2$OH
   slightly   8 g/100 g H$_2$O   soluble in all
                 proportions

12.7 CH$_3$$\overset{}{\underset{\underset{\displaystyle OH}{|}}{CH}}CH_3$ CH$_3$(CH$_2$)$_6$$\overset{}{\underset{\underset{\displaystyle OH}{|}}{CH}}CH_3$ CH$_3$(CH$_2$)$_6$CH=CH$_2$

12.8 a. CH$_3$—$\overset{\overset{\displaystyle CH_3}{|}}{\underset{\underset{\displaystyle I}{|}}{C}}$—CH$_2$—CH$_3$ + H$_2$O   b. CH$_3$—$\overset{\overset{\displaystyle CH_3}{|}}{CH}$—$\overset{}{\underset{\underset{\displaystyle I}{|}}{CH}}$—CH$_3$ + H$_2$O

Tertiary alcohols react more rapidly with HI than secondary alcohols. Therefore, reaction (a) is more rapid than reaction (b).

12.9   a. $CH_3-CH_2-CH_2-CH_2-CH=CH_2$

b.
$$
\underset{\text{major}}{\overset{H_3C\quad CH_3}{CH_3-C=C-CH_3}} + \underset{\text{minor}}{\overset{CH_3\quad CH_3}{CH_3-CH-C=CH_2}}
$$

c.

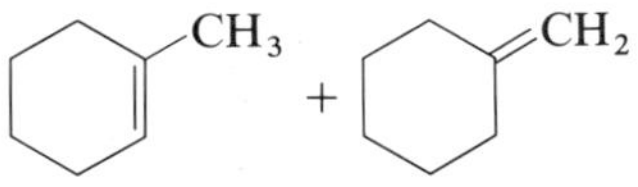

major   minor

12.10 Following are completed half-reactions. Those in which electrons appear on the right are oxidations; those in which electrons appear on the left are reductions. Those showing no electrons are neither oxidations nor reductions.

a.
$$
\underset{}{\overset{OH}{CH_3-CH-CH_2-\overset{O}{\overset{\|}{C}}-OH}} \longrightarrow
$$

$$
CH_3-\overset{O}{\overset{\|}{C}}-CH_2-\overset{O}{\overset{\|}{C}}-OH + 2\,H^+ + 2\,e^- \qquad \text{(oxidation)}
$$

b.
$$
\underset{}{\overset{OH}{CH_3-CH-CH_2-\overset{O}{\overset{\|}{C}}-OH}} \longrightarrow
$$

$$
CH_3-CH=CH-\overset{O}{\overset{\|}{C}}-OH + H_2O \qquad \text{(neither)}
$$

This reaction is balanced by the addition of H_2O to the right side of the equation and therefore is neither oxidation nor reduction. It is a dehydration.

c.
$$
CH_3-\overset{O}{\overset{\|}{C}}-CH_2-\overset{O}{\overset{\|}{C}}-OH \longrightarrow CH_3-\overset{O}{\overset{\|}{C}}-CH_3 + CO_2 \qquad \text{(neither)}
$$

Chapter 13

13.1 a. 3,3-dimethylbutanal   b. 3,3-dimethyl-2-butanone (*t*-butyl methyl ketone)
c. 2,5-dimethylcyclohexanone

13.2 a.

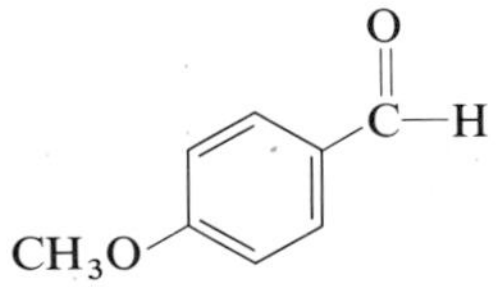

$+\ 2\ CH_3OH$

b.
$$
\underset{CH_3}{\overset{CH_3}{\diagdown}}C=O + HO-CH_2-CH_2-OH
$$

c.
$$
\underset{}{\overset{OH}{CH_3-CH-CH_2-CH_2-\overset{O}{\overset{\|}{C}}-H}} + CH_3OH
$$

13.3 a.

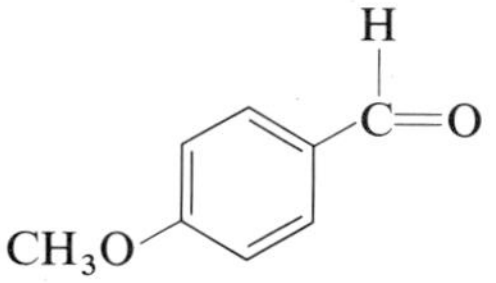

    + $H_2N-CH_2CH_3$

b.

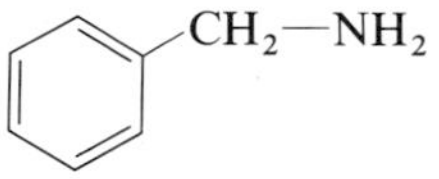

    + $CH_3-\overset{\overset{\displaystyle O}{\|}}{C}-CH_2-CH_3$

13.4 a.

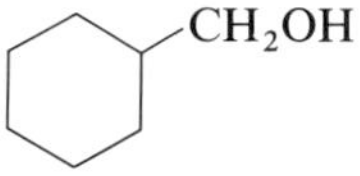

b.

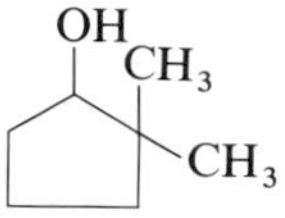

c.

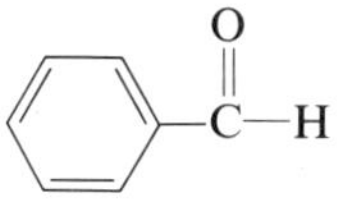

13.5 a. $CH_3-CH_2-CH_2-CH_2-\overset{\overset{\displaystyle O}{\|}}{C}-H$ b.

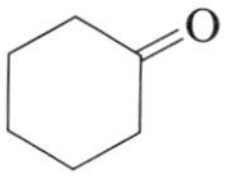

c.

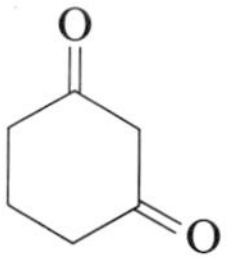

13.6 a.

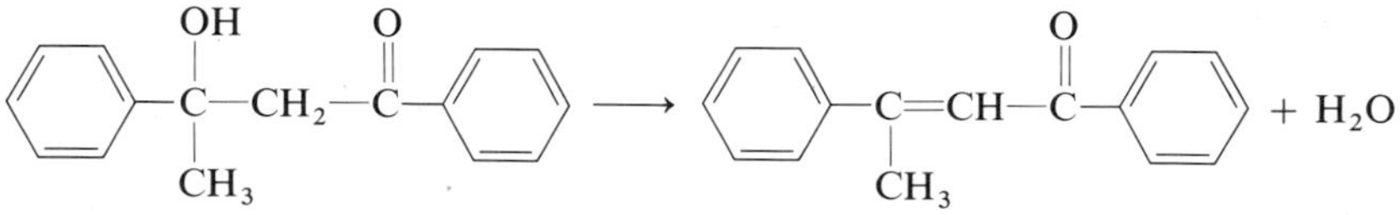

+ H_2O

b.

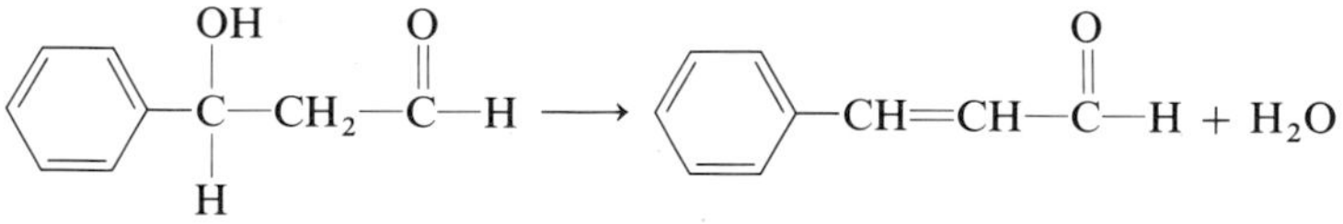

13.7 a. $2\ CH_3-CH_2-\overset{\overset{\displaystyle O}{\|}}{C}-H\ \xrightarrow[\text{(aldol)}]{\text{NaOH}}\ CH_3-CH_2-\overset{\overset{\displaystyle OH}{|}}{CH}-\overset{\underset{\displaystyle CH_3}{|}}{CH}-\overset{\overset{\displaystyle O}{\|}}{C}-H\ \xrightarrow[-H_2O]{H^+}$

$CH_3-CH_2-CH=\overset{\underset{\displaystyle CH_3}{|}}{C}-\overset{\overset{\displaystyle O}{\|}}{C}-H\ \xrightarrow{2\ H_2/Pt}\ CH_3-CH_2-CH_2-\overset{\underset{\displaystyle CH_3}{|}}{CH}-CH_2OH$

b. $CH_3-CH_2-CH_2-\overset{\overset{\displaystyle O}{\|}}{C}-H\ \xrightarrow[\text{(aldol)}]{\text{NaOH}}\ CH_3-CH_2-CH_2-\overset{\overset{\displaystyle OH}{|}}{CH}-\overset{\underset{\displaystyle CH_2-CH_3}{|}}{CH}-\overset{\overset{\displaystyle O}{\|}}{C}-H\ \xrightarrow[-H_2O]{H^+}$

$CH_3-CH_2-CH_2-CH=\overset{\underset{\displaystyle CH_2-CH_3}{|}}{C}-\overset{\overset{\displaystyle O}{\|}}{C}-H\ \xrightarrow{Ag(NH_3)_2^+}\ CH_3-CH_2-CH_2-CH=\overset{\underset{\displaystyle CH_2-CH_3}{|}}{C}-\overset{\overset{\displaystyle O}{\|}}{C}-OH$

Chapter 14

14.1 a.

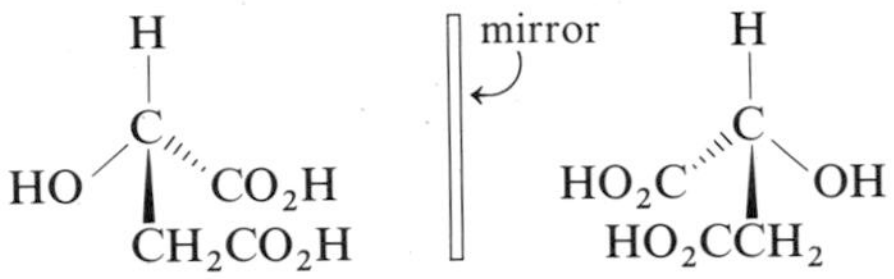

b.

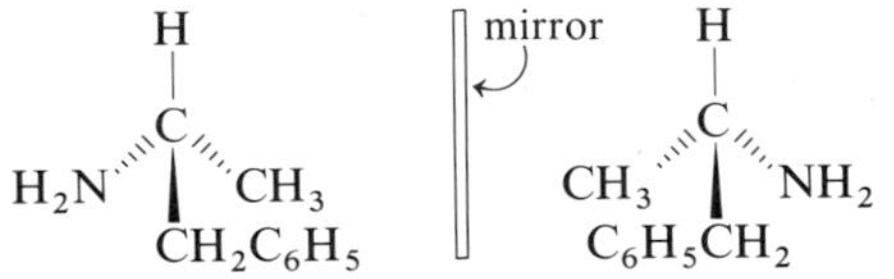

14.2 Compounds (a) and (b) are chiral and therefore show optical isomerism. Compound (c) is achiral.

a.

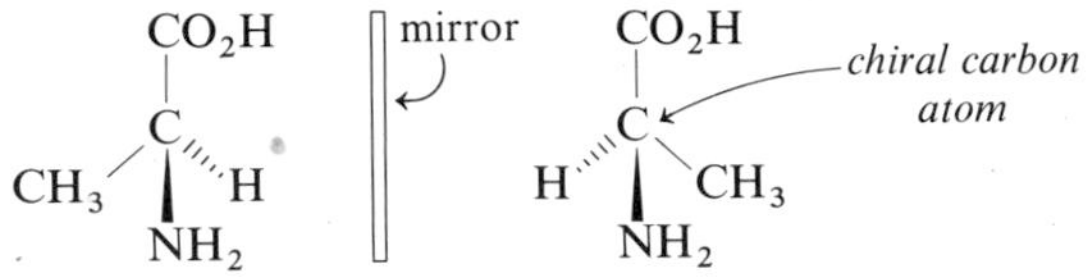

b.

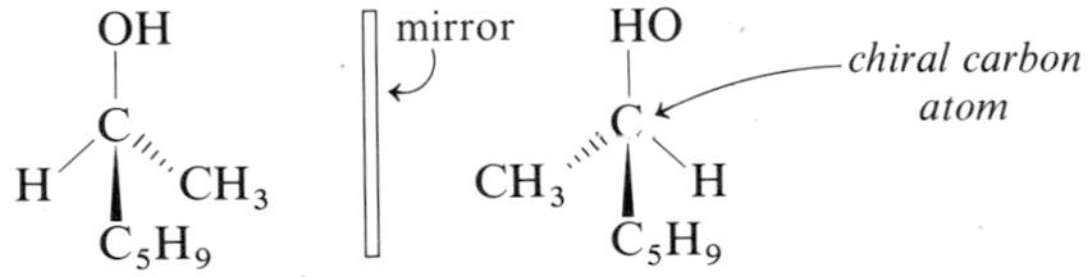

c.

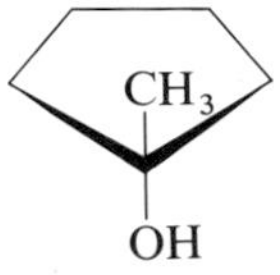

the plane of symmetry bisects the CH_3—C—OH *bonds*

14.3 a.

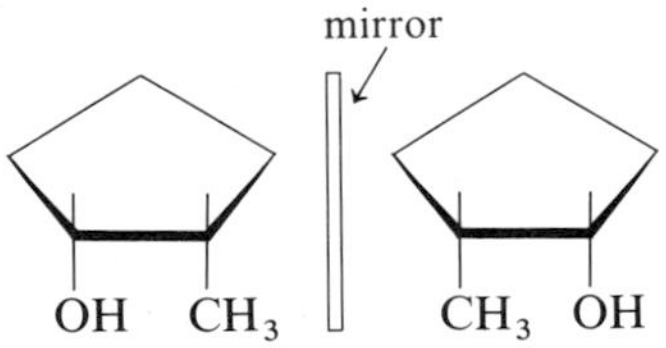

cis-2-methylcyclopentanol
(a pair of enantiomers)
(nonsuperimposable)

b.

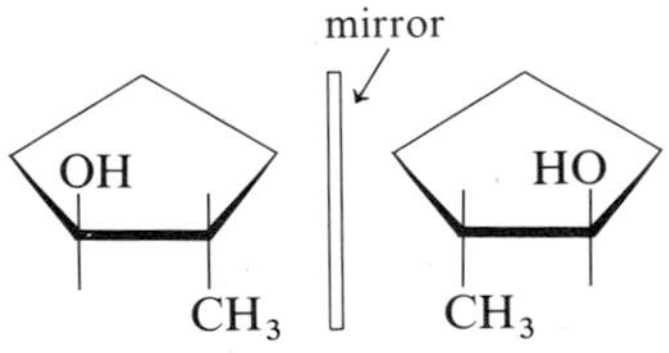

trans-2-methylcyclopentanol
(a pair of enantiomers)
(nonsuperimposable)

c. There are four stereoisomers of 2-methylcyclopentanol; two pairs of enantiomers.

Chapter 15

15.1

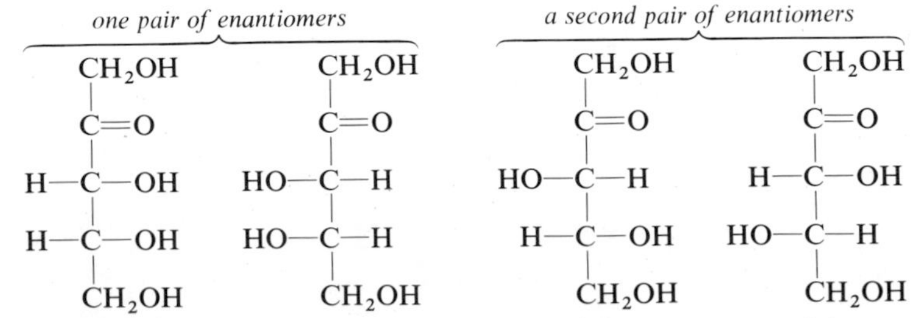

one pair of enantiomers		*a second pair of enantiomers*	

D-ribulose	L-ribulose	D-xylulose	L-xylulose
(a D-ketopentose)	(an L-ketopentose)	(a D-ketopentose)	(an L-ketopentose)

15.2 The configuration of D-mannose differs from that of D-glucose only at carbon-2. Therefore, the α- and β-forms of D-mannose differ from those of D-glucose only in the orientation of the —OH group on carbon-2.

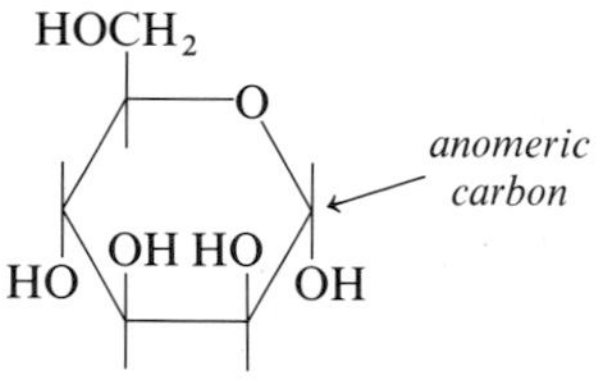

α-D-mannose

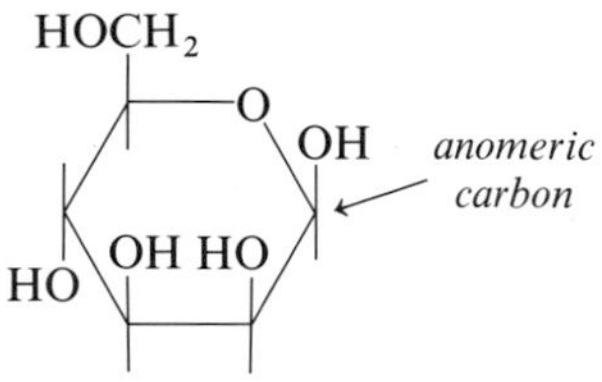

β-D-mannose

15.3 a.

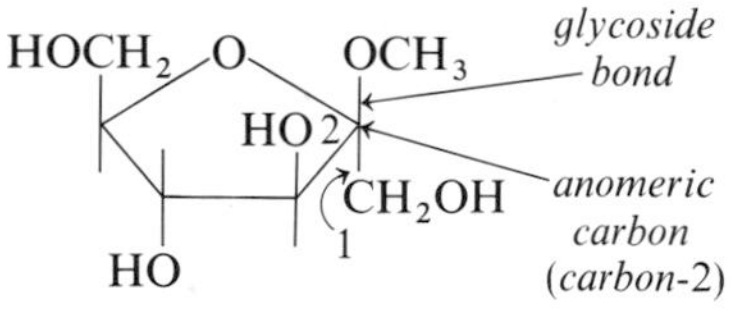

b.

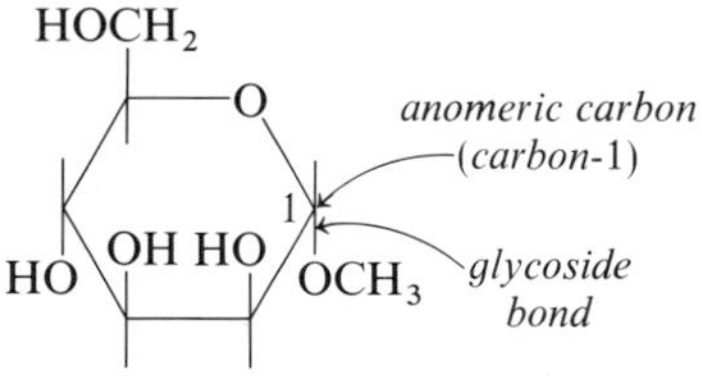

15.4 a.

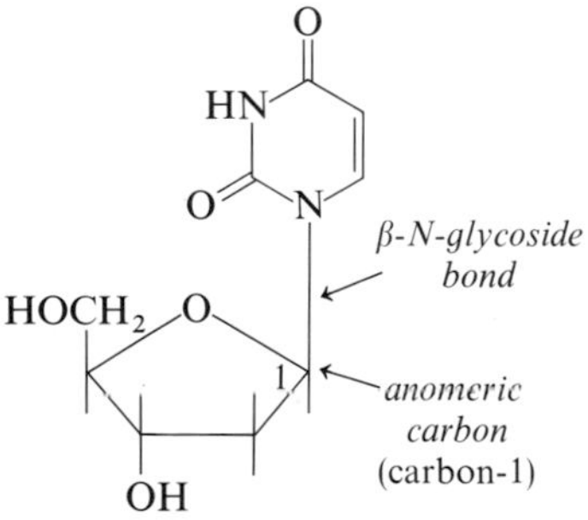

b.

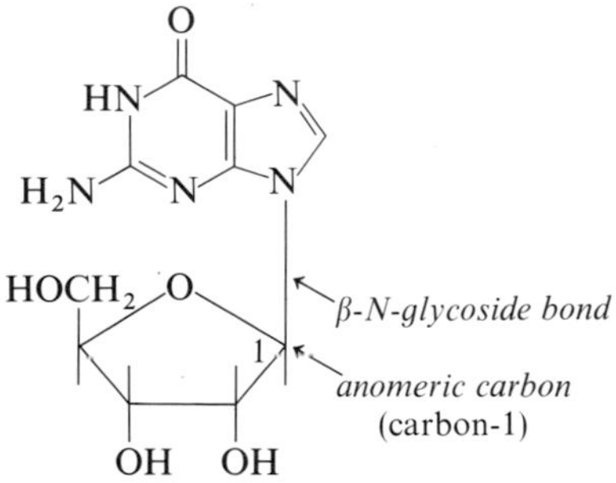

15.5

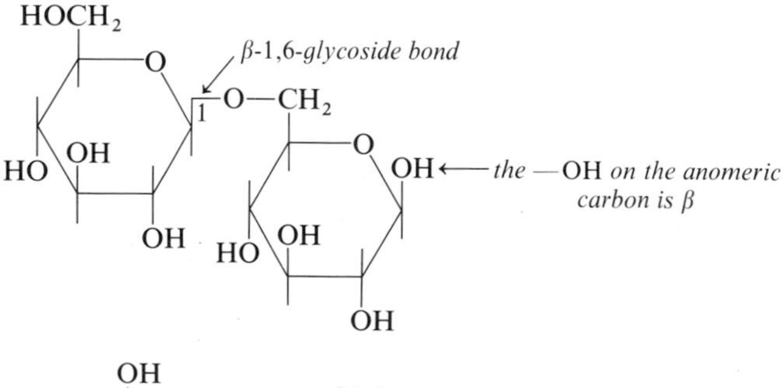

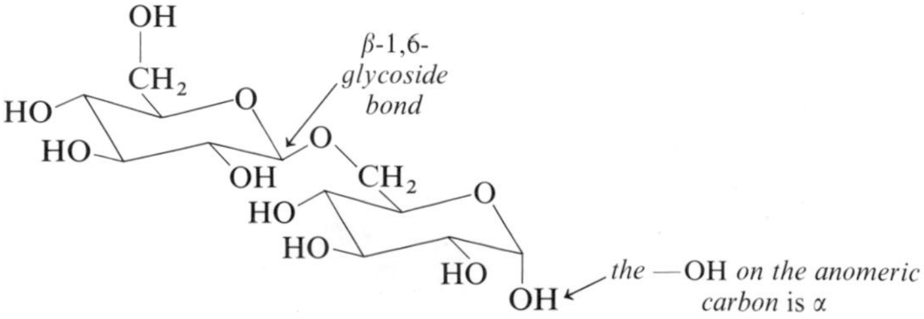

Chapter 16   16.1 (Common names given in parentheses, where appropriate.)
   a. *p*-methoxybenzoic acid   b. 2,2-dimethylpropanoic acid (trimethylacetic acid)
   c. 2-methylbutanedioic acid (α-methylsuccinic acid)
   d. 2,3-dihydroxybutanoic acid

16.2 a. potassium 2-methylpropanoate (potassium isobutyrate)
   b. sodium decanoate   c. ammonium acetate

16.3 $CH_3(CH_2)_8CO_2H$ $CH_3CH_2OCH_2CH_3$ $CH_3CH_2CH_2CO_2H$

   slightly   8 g/100 g H_2O soluble in all
                                                  proportions

16.4
$$\underset{a.}{} CH_3\overset{O}{\overset{\|}{C}}CH_3 + HO\overset{O}{\overset{\|}{C}}CH_2CH_2\overset{O}{\overset{\|}{C}}OH + CH_3\overset{O}{\overset{\|}{C}}CH_3$$

   b. $HO\overset{O}{\overset{\|}{C}}CH_2CH_2\overset{O}{\overset{\|}{C}}OH$

16.5 a. $CH_3CH_2CH{=}CHCO_2H + NaOH \longrightarrow CH_3CH_2CH{=}CHCO_2^-Na^+ + H_2O$

   b. $2\, C_6H_5CO_2H + Na_2CO_3 \longrightarrow 2\, C_6H_5CO_2^-Na^+ + CO_2 + H_2O$

   c.
$$HO\underset{\underset{CH_2-CO_2H}{|}}{\overset{\overset{CH_2-CO_2H}{|}}{C}}CO_2H \;+\; 3\,NaHCO_3 \longrightarrow HO\underset{\underset{CH_2CO_2^-Na^+}{|}}{\overset{\overset{CH_2CO_2^-Na^+}{|}}{C}}CO_2^-Na^+ \;+\; 3\,CO_2 + 3\,H_2O$$

Chapter 17   17.1 (Common names are given in parentheses where appropriate.)
   a. cyclohexyl acetate   b. diethyl propanedioate (diethyl malonate)
   c. *tert*-butyl pentanoate (*tert*-butyl valerate)

17.2 a. N-phenylacetamide   b. hexanamide
   c. *o*-hydroxybenzamide (salicylamide)

17.3 a.

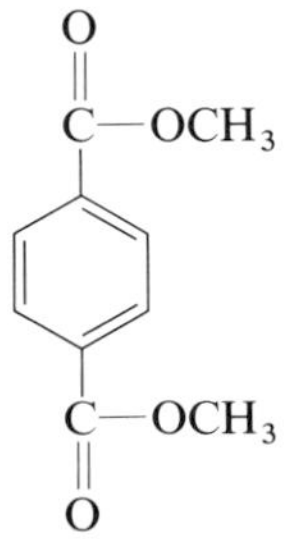

   dimethyl 1,4-benzenedicarboxylate
   (dimethyl terephthalate)

   b. $CH_3\overset{O}{\overset{\|}{C}}OCH_2CH_2CH_2CH_2CH_3$ c. $CH_3CH_2\overset{O}{\overset{\|}{C}}OCH_2CH_2\underset{\underset{CH_3}{|}}{CH}CH_3$

   pentyl acetate   3-methylbutyl propanoate
   (*n*-pentyl acetate)   (isopentyl propionate)

17.4 a.

$$\underset{\underset{\displaystyle OH}{|}}{\overset{\overset{\displaystyle OH}{|}}{H-C}}-O-\underset{\underset{\displaystyle CH_3}{|}}{CH}-CH_3$$

b.

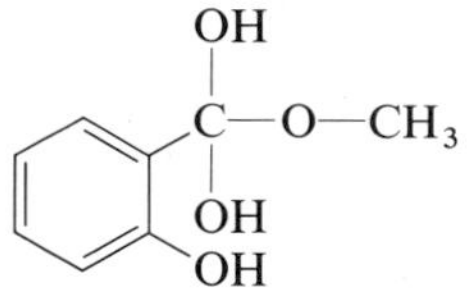

17.5 a.

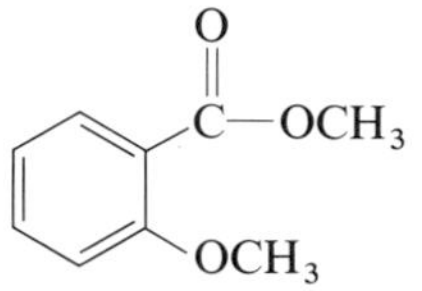

$+ H_2O \longrightarrow$ 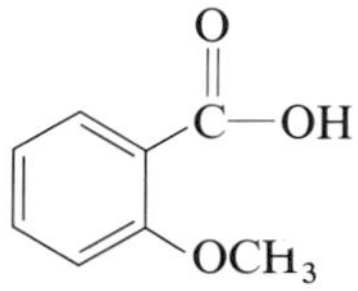 $+ CH_3OH$

b.

$$\begin{aligned}
&\overset{\displaystyle O}{\overset{\|}{CH_2-O-C-CH_3}}\\
&\overset{\displaystyle O}{\overset{\|}{CH-O-C-CH_3}} \quad + 3\,H_2O \longrightarrow\\
&\overset{\displaystyle O}{\overset{\|}{CH_2-O-C-CH_3}}
\end{aligned}
\qquad
\begin{aligned}
&CH_2-OH\\
&CH-OH\\
&CH_2-OH
\end{aligned}
\quad + 3\,\overset{\displaystyle O}{\overset{\|}{CH_3-C}}-OH$$

17.6 a.

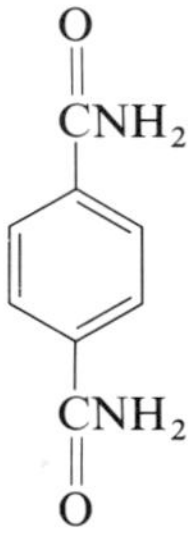

$+ 2\,CH_3OH$

b.

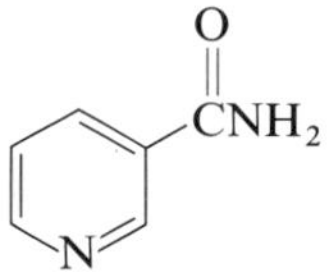

$+ CH_3CH_2OH$

c. $CH_3CH_2\overset{\displaystyle O}{\overset{\|}{C}}NH_2 + \underset{\underset{\displaystyle CH_3}{|}}{HOCHCH_3}$

17.7 a. $\underset{\underset{\displaystyle CH_2OH}{|}}{\overset{\overset{\displaystyle CH_2OH}{|}}{CHOH}} + 3\,CH_3(CH_2)_{14}\overset{\displaystyle O}{\overset{\|}{C}}-OCH_3$

b.

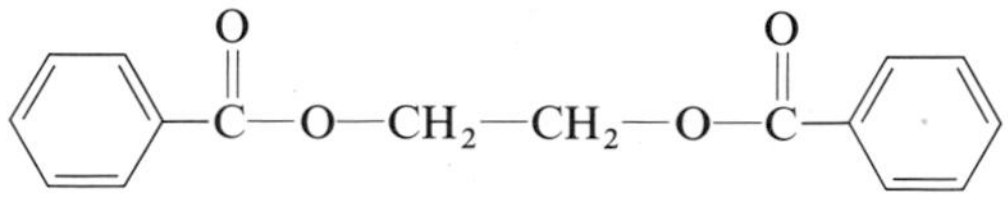

$+ 2\,CH_3OH$

17.8 a. $CH_2{=}CH-CH_2-O-\overset{\displaystyle O}{\overset{\|}{C}}-CH_3 + CH_3-\overset{\displaystyle O}{\overset{\|}{C}}-OH$

b. $CH_3-\overset{\displaystyle O}{\overset{\|}{C}}-O-CH_2-CH_2-NH-\overset{\displaystyle O}{\overset{\|}{C}}-CH_3 + 2\,CH_3-\overset{\displaystyle O}{\overset{\|}{C}}-OH$

17.9 a.

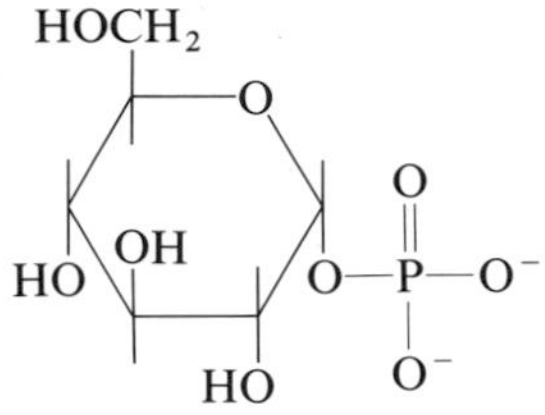

net charge is −2

b.

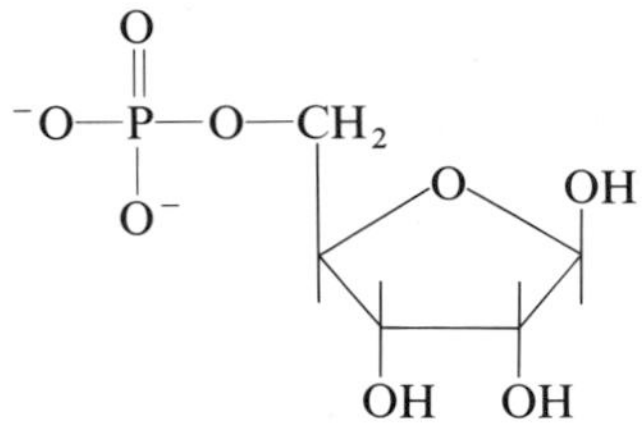

net charge is −2

17.10 a.

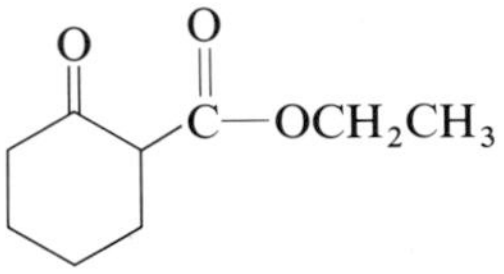

+ CH$_3$CH$_2$OH

b.

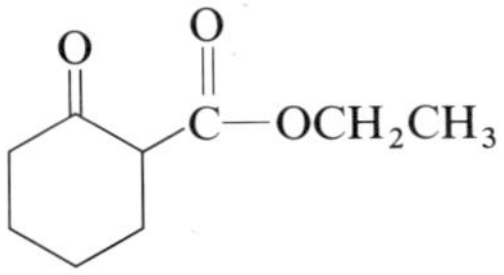

+ CH$_3$CH$_2$OH

Chapter 18

18.1 While the percentage of oleic acid is approximately the same for beef tallow and corn oil, the total percentages of unsaturated fatty acids are quite different: only 52.1% for beef tallow, but 85.1% for corn oil. The additional unsaturation in corn oil is due largely to linoleic acid.

Chapter 19

19.1 a. 1,6-diaminohexane (hexamethylenediamine)   b. cyclohexylmethylamine
      c. methyl p-aminobenzoate

19.2 a.

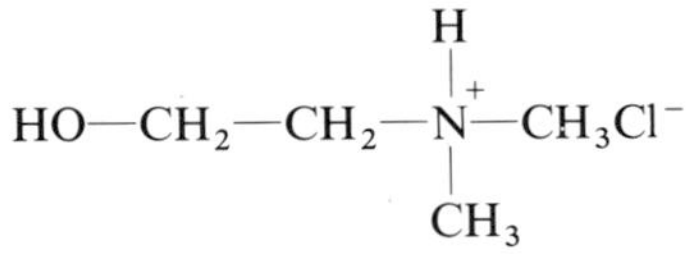

b.

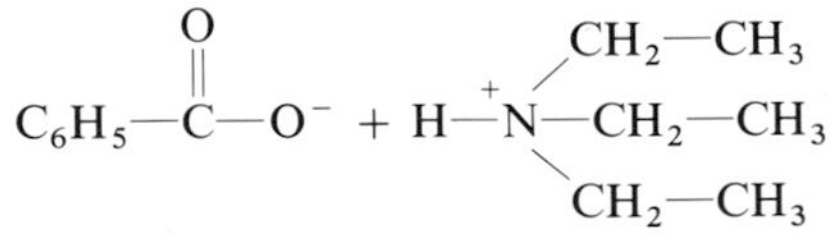

c.

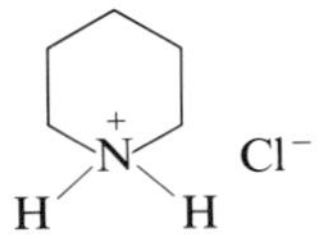

Chapter 20

20.1 a.

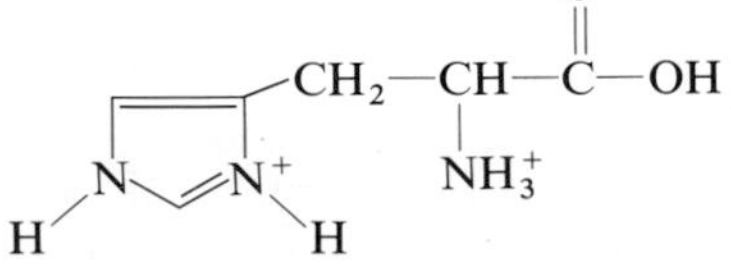

pH = 1.0
net charge = +1

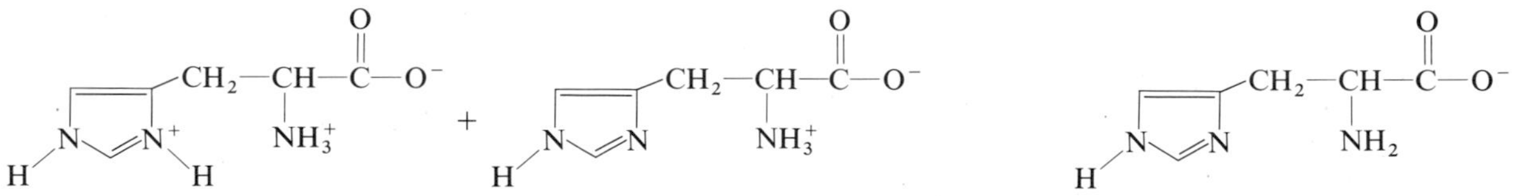

pH = 6.0
net charge approximately +0.5

pH = 12.0
net charge = −1

b. $H_3\overset{+}{N}(CH_2)_4\underset{\underset{NH_3^+}{|}}{CH}CO_2H$ $H_3\overset{+}{N}(CH_2)_4\underset{\underset{NH_3^+}{|}}{CH}CO_2^-$ $H_2N(CH_2)_4\underset{\underset{NH_2}{|}}{CH}CO_2^-$

pH = 1.0
net charge = +2

pH = 6.0
net charge = +1

pH = 12.0
net charge = −1

20.2 The isoelectric point of valine is 6.0. At this pH, valine has a net charge of zero and does not move from the origin. The isoelectric point of glutamic acid is 3.08 (Table 20.3); at pH 6.0, glutamic acid has a net negative charge and moves toward the positive electrode. The isoelectric point of arginine is 10.76 (Table 20.3); at pH 6.0, arginine has a net positive charge and moves toward the negative electrode.

20.3 Following is the structural formula of lys-phe-ala. The net charge on this tripeptide at pH 6.0 is +1.

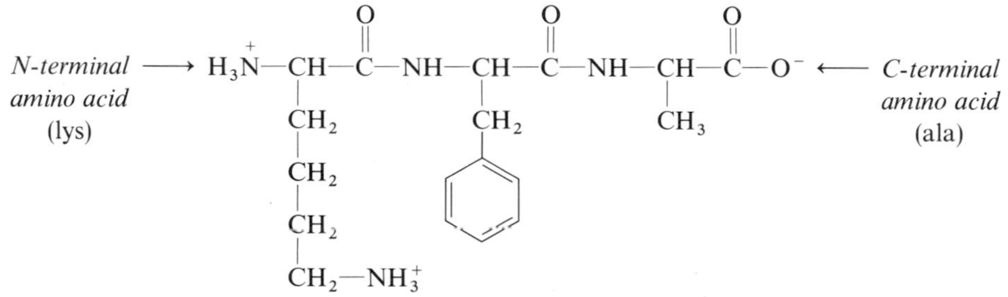

20.4 Sequences (a) and (b) are hydrolyzed by chymotrypsin. Sequence (c) is not hydrolyzed by either. None are hydrolyzed by trypsin.

20.5 Following is the hexapeptide amino acid sequence along with the points of cleavage by Edman degradation, trypsin, and chymotrypsin.

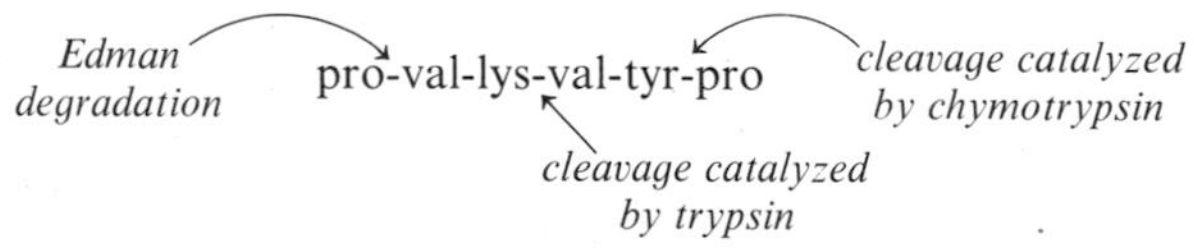

20.6 The side chain of lysine (—NH$_3^+$) can form salt linkages with the side chains of glutamic acid (—CO$_2^-$) and aspartic acid (—CO$_2^-$).

Chapter 21

21.1 a. $V_{max} = 38$ μmole/min $\frac{1}{2}V_{max} = 19$ μmole/min $K_m = 3.5 \times 10^{-3}M$
b. initial velocity = 26 μmole/min   c. new initial velocity = 52 μmole/min

21.2 pK_a of the imidazole ring of histidine is 6.10. At pH 4.0, the imidazole ring is completely in the protonated or acid form. Since the pH corresponds to the optimal pH, it must be that the enzyme requires the histidine side chain in the acid form.

Chapter 22   22.1 a.

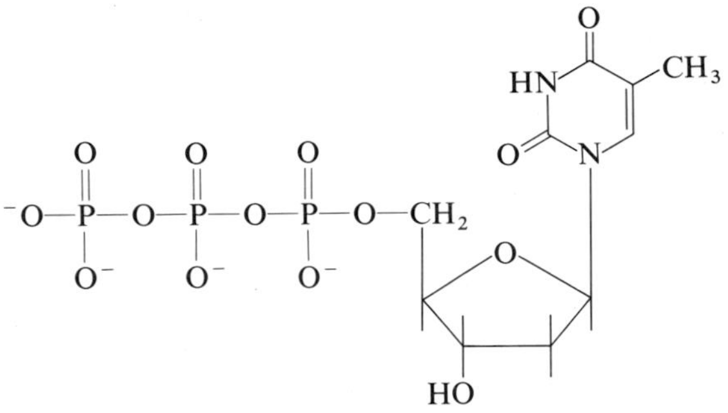

b.

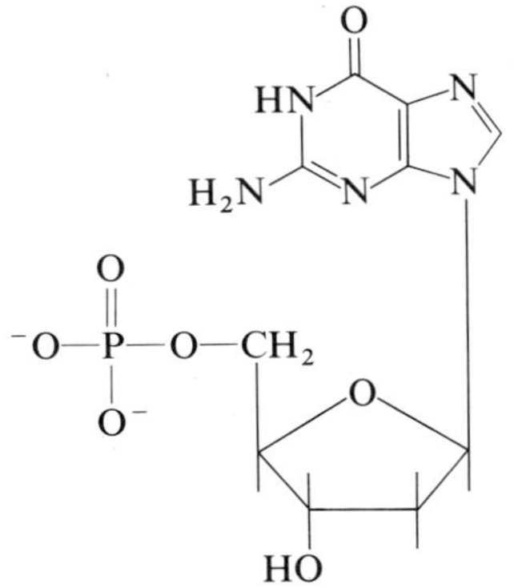

22.2

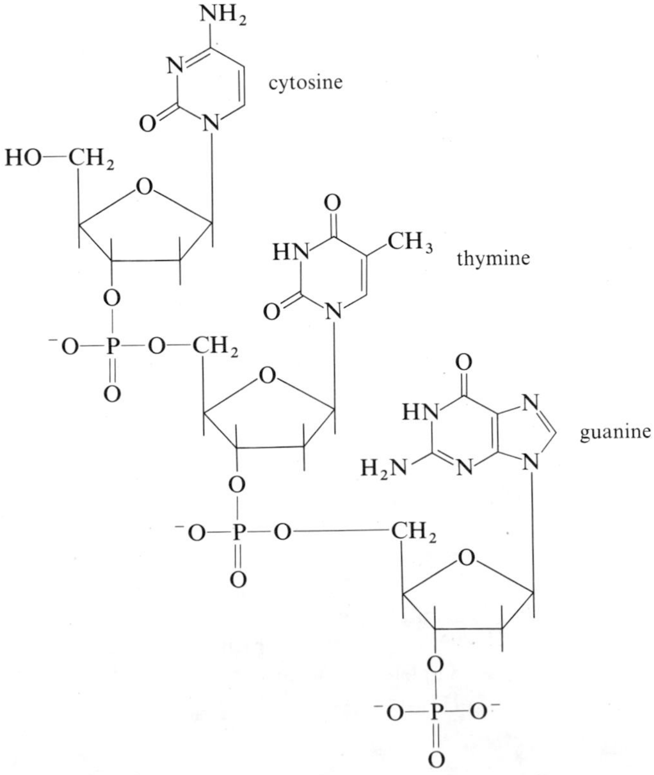

22.3 3'-G-G-C-A-T-G-C-T-5'
Alternatively, the complement can be written -TCGTACGG-.

22.4 3'-AGC-CAU-GUG-ACC-5'
Alternatively, the RNA strand can be written -CCA-GUG-UAC-CGA-.

22.5 a. 5'-UGC-UAU-AUU-CAA-AAU-UGC-CCU-CUU-GGU-UGA-3'
      b. cys-tyr-ile-gln-asn-cys-pro-leu-gly-stop

22.6 a. RNA complement: 5'-CUG-AGG-CUA-GCG-CUA-3'
          amino acid sequence: leu-arg-leu-ala-leu
      b. new mRNA: 5'-CUG-AAG-CUA-GCG-CUA-3'
          new amino acid sequence: leu-lys-leu-ala-leu

22.7 a. mRNA: 5'-CUA-CCC-UAC-AGA-3'   amino acid sequence: leu-pro-tyr-arg
      b. mRNA: 5'-CUC-CCU-ACA-GA-3'   amino acid sequence: leu-pro-thr

Chapter 23

No in-chapter problems.

Chapter 24

24.1 a. 2 moles NADH per mole of glucose. No FADH$_2$ is produced.
      b. 1 mole of NADH per mole of pyruvate.
      c. 3 moles of NADH and 1 mole of FADH$_2$.

24.2 a. There is only one oxidation, namely, oxidation of 3-phosphoglyceraldehyde to 1,3-diphosphoglycerate.
      b. Only one reaction involves cleavage of a carbon-carbon bond, namely the conversion of fructose 1,6-diphosphate to dihydroxyacetone phosphate and 3-phosphoglyceraldehyde.

24.3 a. lactate   b. lactate

Chapter 25

25.1 a. Only 1 mole of ATP is required, and that is in the first step. Note, however, that although only one mole of ATP is required, two high-energy bonds are hydrolyzed.
      b. 6 moles of NADH and 6 moles of FADH$_2$ c. 7 moles of acetyl CoA

Chapter 26

26.1 a.

phenylalanine   pyruvate   phenylpyruvate   alanine
                                        (an α-ketoacid)

b.

histidine   α-ketoglutarate

imidazolylpyruvate (an α-ketoacid)   glutamate

Index